College Algebra

Preliminary Edition

Michael Holtfrerich
Glendale Community College

Jack Haughn
Glendale Community College

THOMSON
BROOKS/COLE™

Australia • Canada • Mexico • Singapore • Spain • United Kingdom • United States

Printed in the United States of America
1 2 3 4 5 6 7 08 07 06 05 04

Printer: The P.A. Hutchison Company

0-534-39948-7

For more information about our products, contact us at:
Thomson Learning Academic Resource Center
1-800-423-0563

For permission to use material from this text or product, submit a request online at **http://www.thomsonrights.com.**
Any additional questions about permissions can be submitted by email to **thomsonrights@thomson.com.**

Thomson Brooks/Cole
10 Davis Drive
Belmont, CA 94002-3098
USA

Asia
Thomson Learning
5 Shenton Way #01-01
UIC Building
Singapore 068808

Australia/New Zealand
Thomson Learning
102 Dodds Street
Southbank, Victoria 3006
Australia

Canada
Nelson
1120 Birchmount Road
Toronto, Ontario M1K 5G4
Canada

Europe/Middle East/South Africa
Thomson Learning
High Holborn House
50/51 Bedford Row
London WC1R 4LR
United Kingdom

Latin America
Thomson Learning
Seneca, 53
Colonia Polanco
11560 Mexico D.F.
Mexico

Spain/Portugal
Paraninfo
Calle/Magallanes, 25
28015 Madrid, Spain

Table of Contents

To the Student

You are about to begin your journey through college algebra. The study of math in general is about learning to think in an orderly fashion using logic and common sense. It is a language and as such is not learned overnight, so be patient. College algebra typically covers a wide range of topics to address a variety of college majors. For this reason we include a lot of different applications in this text to help you see the usefulness of math in your area of interest.

Many concepts in math show up again and again in different applications and topics within algebra. You may recognize some them from prior math classes. This text has been constructed to help you absorb these reoccurring concepts in the following ways:

- We made this text easy to read and understand.
- We oriented the discussion portions of this text so that they read just as if an instructor was standing right in front of you, lecturing on the topics. We believe that if you take the time to read ahead before you go to class, you will find yourself understanding more than ever before in a math class, and more quickly too. If you happen to miss a lecture period, you will find the text has explained the topic well and it should be easy for you to get caught up.
- We designed each section to have four basic features: discussions, examples, question boxes, and homework problems. The discussion items are there to help you understand the concepts. They explain why math works the way it does. Examples are there to help you see how those concepts are used in concrete ways. Refer to them if you have questions when you are working on the homework exercises. The question boxes are there to help you measure whether or not you've gotten the concepts down, and will help you to think mathematically. The homework problems give you a chance to practice what you've learned. All four features will help you see connections between different mathematical concepts and how they relate to the world around you.
- We created plenty of review material for you. The chapter review is a list of the most important concepts and which sections they came from. The review homework gives you a chance to work even more problems from the chapter. And the chapter exam will serve as good practice for your own class exams.

We also created a student solution manual in case you have difficulty with the homework problems. The student solution manual has all of the odd exercises worked out in detail and will walk you through the process of solving those problems.

We believe that you are about to begin an enjoyable adventure learning the many facets of mathematics. Good luck on having a very successful semester!

Best wishes,
Michael Holtfrerich and Jack Haughn

To the Instructor

We know that college algebra can be a difficult course to teach for several reasons. Most students are taking it because they have to, not because they want to. Client disciplines only want us to teach the concepts they think their students need. As mathematicians, we would like our students to see the beauty of a discipline that is near and dear to our hearts. We want our students to experience the inspiration we find in mathematics and see its application in the world around them.

Because of these sometimes conflicting viewpoints, we have worked diligently to address each of them in our text. For the student, we have included examples from everyday life and explained the material in a way that should be interesting and easy to understand. Students should at the very least say, "Hey, this isn't so bad after all." For the client disciplines, we have included many applications from different areas of study. For the math instructors, we have focused on both understanding and skills, in that order. We have not sacrificed one for the other, but have a balanced mix between them.

We believe that to be successful in learning and applying mathematics, you need a book that focuses solely on *one aspect* of the world of mathematics at a time while incorporating material already learned. We arranged our chapters and material as follows with that belief in mind:

- Chapter 0 contains review material from previous algebra courses on which a student may be a little rusty. It also contains illustrated direction on how to use a TI-83 or TI-86 graphing calculator.
- Chapter 1 looks at the topic of functions. This topic is fundamental in the study of mathematics. It sets the foundation for understanding the rest of the course, so a complete discussion of functions, including inverse functions, is essential. Research shows that college algebra students have a hard time absorbing the concept of functions. By covering it right up front and continuing to use it throughout this text, we know that the students will get the best understanding possible.
- Chapter 2 takes time to make sure that students' manipulative skills are up to speed and second nature to them. Lack of these skills may hinder their ability to grasp concepts in future material.
- Chapter 3 covers the family of functions called polynomials. Students may have done a lot with these in their previous course work, but there is much more for them to learn. We have removed some of the traditional material like Descartes's rule of signs and the upper and lower bound theorem in the interest of spending more time giving students a complete understanding of the nature of polynomials.
- Chapter 4 covers the exponential and logarithmic family of functions. To help students really understand what these functions are all about, we use linear functions as a contrast. We also revisit inverses. We've found this to be a highly effective method of communicating the behavior of these functions while reinforcing prior knowledge. Because they show up so often in the real world, we offer a multitude of rich applications from many different disciplines.

- Chapter 5 involves data analysis. Again, we use all that your students have learned in the first several chapters to develop models for events in the world. It gives them a "big picture" sense of how all of the basic functions relate to each other.
- Chapter 6 covers matrices. We employ them to solve systems of equations with more than two variables.
- Chapter 7 covers the family of equations called conic sections. Though most of them aren't functions, their applications are many.
- Chapter 8 picks up a variety of interesting mathematical topics needed for finite math and calculus.

Features

Writing style This text was written by teachers, in a teacher's voice. We feel that it is extremely important to speak to the student, not the instructor. You already know the material and how to teach it.

Questions To add to the dynamic style of the writing, we interject questions throughout the reading (as you would do in the classroom). Answers to the questions are in the margin two pages ahead so that by flipping the page, the student can confirm or correct their own solution.

Homework We've given you an array of robust problems and applications to work with. Each odd problem is paired with an even companion. Within each exercise set are groups of exercises of a similar type. Each group progresses in difficulty from "easy" to more difficult. For example, in section 2.2, problems 1-53, the problems increase in difficulty as they progress from number 1 to number 53; as you begin a different type of homework problem, as in problems 57-67, they repeat the progression from easy to more difficult. This stays consistent with our philosophy that once students are working to learn a particular skill or concept, they should do it to completion before moving on to something different.

Activities We are working with Anne Dudley, Glendale Community College, to create activities that will accompany appropriate material. For the sake of this preliminary edition, the activities can be accessed on our text website found at www. math.brookscole.com. Each is a page in length and includes complete directions on how to implement it in your classroom.

We have taught from the manuscript of this text for several semesters. Other instructors have tested it in their classes as well. We have heard nothing but extreme praise from the students about the readability and ease of understanding this text. We are confident that you will too. Good luck in your coming semester!

Best wishes,
Michael Holtfrerich and Jack Haughn

Chapter 0 The Preliminaries

In the first half of this chapter you are going to review basic algebra topics studied in courses that are prerequisites to this course. In the second half of this chapter you will also see how to use the basic operations for the TI-83 and TI-86 graphing calculators. All of this material is essential to your success in college algebra.

Section 0.1 Exponents, Roots, and Scientific Notation

This section will give you a review of exponents, roots, scientific notation, and fractional exponents.

EXPONENTS

Let's begin with exponential notation. Exponential notation is a way to express repeated multiplication. This is similar to how multiplication expresses repeated addition.

The Meaning of Multiplication	The Meaning of Exponential Notation
$5 \cdot 3$ means 5 added to itself 3 times. $5 \cdot 3 = 5 + 5 + 5$.	5^3 means 5 times itself 3 times. $5^3 = 5 \cdot 5 \cdot 5$.

The exponential expression b^m, means b times itself m times. We call b the base and m the exponent. We read b^m as "b raised to the m power" or "b to the m."

Like multiplication, exponentiation has certain rules that must be followed. If b is a real number and m, n are integers, then the following rules are true (except for 0^0, and 0^n where n is negative, since that would imply dividing by 0):

The Product Rule	$b^m \cdot b^n = b^{m+n}$
The Quotient Rule	$\dfrac{b^m}{b^n} = b^{m-n}$
The Power Rule	$(b^m)^n = b^{mn}$
The Negative Rule	$b^{-m} = \dfrac{1}{b^m}$ or $\dfrac{1}{b^{-m}} = b^m$
The Zero Rule	$b^0 = 1$

Let's look at some examples of how we might apply these rules to simplify or rewrite an expression. Notice that b, the base, may not always be just an integer.

Example 1: Working with Exponential Expressions

Simplify the following expressions using the rules of exponents.

a) $b^4 \bullet b^5$ b) $\frac{x^{13}}{x^8}$ c) $(x^3 y^2)^5$ d) 7^{-3} e) $\frac{w^5}{x^{-2}w}$ f) $\frac{(xy^{-2})^3}{x^5 y^{-9} w^0}$

Solutions:

a) $b^4 \bullet b^5 = b^{4+5} = b^9$ (product rule)	b) $\frac{x^{13}}{x^8} = x^{13-8} = x^5$ (quotient rule)
c) $\left(\frac{x^3 y^2}{2}\right)^5 = \frac{x^{3\bullet 5} y^{2\bullet 5}}{2^{1\bullet 5}} = \frac{x^{15} y^{10}}{2^5}$ (power rule)	d) $7^{-3} = \frac{1}{7^3} = \frac{1}{343}$ (negative rule)
e) $\frac{w^5}{x^{-2}w} = w^{5-1} x^2 = w^4 x^2$ (quotient and negative rules)	f) $\frac{(xy^{-2})^3}{x^5 y^{-9} w^0} = \frac{x^3 y^{-6}}{x^5 y^{-9}(1)}$ $= \frac{x^{3-5} y^9}{y^6} = x^{-2} y^3 = \frac{y^3}{x^2}$ (every property used)

Typically, we don't want negative exponents in the final, simplified form. Notice the last step of example 1(f). There is one exception to this general rule when we are working with scientific notation. Scientists would rather have negative exponents in their answers than fractions. Remember that scientific notation is a way to rewrite very large or very small numbers in a concise manner.

SCIENTIFIC NOTATION

Scientific Notation: This is a way of expressing a number so that it is in the form $m \times 10^n$ where $1 \leq m < 10$ and n is an integer.

To change a number to its scientific notation form, we follow these steps:

1. First, decide where the decimal point needs to go to get a number between 1 and 10.
2. Count how many places the decimal has to be moved in order to put the decimal point where it needs to be. That will determine the exponent value, positive or negative.
3. If the original number is larger than 10, then the exponent will be positive in scientific notation form. It is positive because we would need to multiply the new number by powers of ten in order to get the original decimal form.
4. Likewise, if the original number is smaller than 1, then the exponent will be negative. It is negative because we would need to multiply the number by powers of one-tenth

($10^{-1} = \frac{1}{10}$) in order to get the original decimal form.

Discussion 1: Scientific Notation

How do we write the following numbers using scientific notation?

a) 40,000,000,000,000 km	Distance to a star in km
b) $5,700,000,000,000	Federal Debt in the year 2000 in dollars
c) 0.000000001 m	Diameter of an atom in meters
d) 0.0000000000000000000000000000091 g	Mass of an electron in grams

(a) We want the number 4 (a number between 1 and 10). To get 4 we need to move the decimal 13 places from where it is in the number.	40,000,000,000,000. 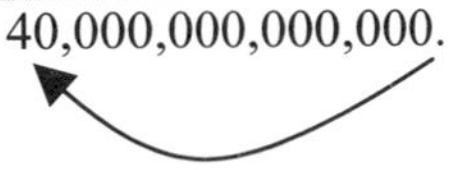
This will give us the answer of	4×10^{13}
The exponent, 13, is positive because our number was greater than 10 (in fact it was much greater than 10!). On a TI graphing calculator it would look like this:	4E13
(b) We want the number 5.7 (a number between 1 and 10), so we need to move the decimal 12 places from where it is in the number.	5,700,000,000,000. 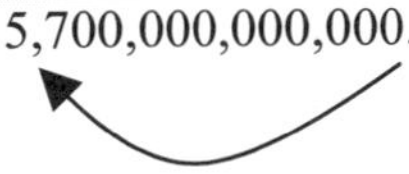
This will give us the answer of	5.7×10^{12}
The exponent, 12, is positive because our number was greater than 10 (again, much greater than 10). On a TI graphing calculator it would look like this:	5.7E12
(c) We want the number 1 (a number between 1 and 10 including one). To get 1 we need to move the decimal 9 places from where it is in the number.	0.000000001 
This will give us the answer of	1×10^{-9}
The exponent, –9, is negative because our number was smaller than 1. On a TI graphing calculator it would look like this:	1E –9

(d) We want the number 9.1 (a number between 1 and 10) so we need to move the decimal 28 places from where it is in the number.	0.000000000000000000000000000091
This will give us the answer of	9.1×10^{-28}
The exponent, –28, is negative because our number was smaller than 1 (much smaller). On a TI graphing calculator it would look like this:	9.1E –28

Question 1: Given the scientific number 3.45×10^{-6}, write it in the normal decimal form.

FRACTIONAL EXPONENTS

We've discussed integer exponents; it is now time to talk about fractional exponents. We will start with a few definitions.

> **Radical** – an expression of the form $\sqrt[n]{b}$, which means, "what raised to the n power equals b."
>
> **Radicand** – the b in the expression $\sqrt[n]{b}$.
>
> **Index** – n in the expression $\sqrt[n]{b}$.
>
> **Fractional Exponent** – $b^{\frac{p}{r}} = \left(\sqrt[r]{b}\right)^p = \sqrt[r]{(b)^p}$.

The simplest fractional exponent case is $b^{\frac{1}{2}}$ which equals $\sqrt[2]{b}$, commonly written as $\sqrt{b}$. This is called the principle square root, and refers only to the positive answer. For example, $49^{\frac{1}{2}} = \sqrt{49} = 7$ (the positive number that when squared equals 49). Of course we all know that you can also square –7 and get 49. This will be discussed in section 0.5 when we talk about absolute values.

Example 2: Simplifying Expressions with Fractional Exponents

Simplify the following fractional exponent expressions.

a) $64^{\frac{2}{3}}$ b) $\left(\frac{1}{25}\right)^{\frac{3}{2}}$ c) $16^{\frac{-3}{4}}$

Solutions:

(a) 2 is the power and 3 is the root, so	$64^{\frac{2}{3}} = \left(\sqrt[3]{64}\right)^2 = (4)^2 = 16$

(b) 3 is the power and 2 is the root, so	$\left(\frac{1}{25}\right)^{\frac{3}{2}} = \left(\sqrt{\frac{1}{25}}\right)^3 = \left(\frac{1}{5}\right)^3 = \frac{1}{125}$
(c), the "–" says to move the base, 3 is the power and 4 is the root, so	$16^{\frac{-3}{4}} = \frac{1}{16^{\frac{3}{4}}} = \frac{1}{\left(\sqrt[4]{16}\right)^3} = \frac{1}{2^3} = \frac{1}{8}$

Of course anything that has an exponent means we could use the rules of exponents to simplify it.

Example 3: Simplifying Expressions with Fractional Exponents

Simplify the following fractional exponent expressions using the rules of exponents.

a) $5^{\frac{1}{2}} \cdot 5^{\frac{3}{2}}$ b) $\frac{x^{\frac{2}{3}}}{x^{\frac{7}{3}}}$ c) $\left(y^{\frac{1}{4}}\right)^{\frac{2}{3}}$ d) $3^{\frac{1}{3}} \cdot 3^{\frac{2}{5}}$ e) $\frac{y^{\frac{7}{2}}}{y^{\frac{5}{7}}}$

Solutions:

(a) We are multiplying like bases so we add the exponents.	$5^{\frac{1}{2}} \cdot 5^{\frac{3}{2}} = 5^{\frac{1}{2}+\frac{3}{2}} = 5^{\frac{4}{2}} = 5^2$
(b) We are dividing like bases so we subtract the exponents.	$\frac{x^{\frac{2}{3}}}{x^{\frac{7}{3}}} = x^{\frac{2}{3}-\frac{7}{3}} = x^{\frac{-5}{3}} = \frac{1}{x^{\frac{5}{3}}}$
(c)We are raising to a power, so we multiply the exponents.	$\left(y^{\frac{1}{4}}\right)^{\frac{2}{3}} = y^{\frac{2}{12}} = y^{\frac{1}{6}}$
(d)We are multiplying like bases so we add the exponents. This time though, we need to get a common denominator first.	$3^{\frac{1}{3}} \cdot 3^{\frac{2}{5}} = 3^{\frac{5}{15}+\frac{6}{15}} = 3^{\frac{11}{15}}$
(e) We are dividing like bases so we subtract the exponents. This time though, we need to get a common denominator first.	$\frac{y^{\frac{7}{2}}}{y^{\frac{5}{7}}} = y^{\frac{49}{14}-\frac{10}{14}} = y^{\frac{39}{14}}$

If we have fractional exponents, it means we are dealing with a radical. With radicals, and thus with fractional exponent expressions, we have two significant rules.

Product rule for radicals	$\sqrt[n]{a} \cdot \sqrt[n]{b} = \sqrt[n]{ab}$
Quotient rule for radicals	$\frac{\sqrt[n]{a}}{\sqrt[n]{b}} = \sqrt[n]{\frac{a}{b}}$

It is understood in the world of mathematics that if we say, "simplify a radical" we are referring to three particular cases.

Case 1: Don't leave anything in the root symbol that has a power that is bigger than or equal to the root.

Case 2: Don't leave fractions in a radical.

Case 3: involves rationalizing the denominator, and we will get to that shortly.

Here is how these three cases come into play.

Discussion 2: Simplifying Expressions Involving Radicals

Answer Q1: The –6 exponent tells us that this is a small number (less than 1,) so we will move the decimal to the left 6 places to get 0.00000345.

How do we simplify the following radical expressions assuming that all variables are positive?

a) $x^{\frac{4}{3}}$ b) $\sqrt{50y^7}$ c) $\sqrt[3]{24w^5x^7}$ d) $\sqrt{\dfrac{k^3}{9x^4}}$

(a) 4 is the power and 3 is the root, so using our definition of $b^{\frac{p}{r}}$ and the properties of radicals we get (Case 1)	$x^{\frac{4}{3}} = \sqrt[3]{x^4} = \sqrt[3]{x^3} \bullet \sqrt[3]{x} = x\sqrt[3]{x}$
(b) We have the square root of 50 and y^7, neither of which is a perfect square, so we use our properties to simplify. (Case 1)	$\sqrt{50y^7} = \sqrt{25y^6} \bullet \sqrt{2y} = 5y^3\sqrt{2y}$
(c) We have the cube root of 24 w^5 and x^7, neither of which is a perfect cube, so we use our properties to simplify. (Case 1)	$\sqrt[3]{24w^5x^7} = \sqrt[3]{8w^3x^6} \bullet \sqrt[3]{3w^2x}$ $= 2wx^2\sqrt[3]{3w^2x}$
(d) We again use our properties to simplify. (Case 2)	$\sqrt{\dfrac{k^3}{9x^4}} = \dfrac{\sqrt{k^2} \bullet \sqrt{k}}{\sqrt{9x^4}} = \dfrac{k\sqrt{k}}{3x^2}$

In the examples we have just looked at we used the product property of radicals to separate the perfect square or cube factors from the non-perfect factors and then simplified the roots. In part (d) we used the quotient property to separate the numerator from the denominator so that we could approach them separately. Part (d) also demonstrates Case 2 (don't leave fractions in a radical).

OPERATIONS ON RADICALS

Next, let's look at adding, subtracting, multiplying and rationalizing radicals. A radical expression can in some sense be looked at like a variable expression. You can only add or subtract two expressions if they are like terms.

Note: A **term** is a constant or a variable (to whatever power) or a product of a constant and one or more variables (to whatever power). Let's define what it means for two terms to be like terms.

Two terms are considered **like terms** if every part of them is the same except for the coefficient (i.e. all the variables and roots etc. must be exactly the same to the same powers).

For example:

$3\sqrt{2}+4\sqrt{2}=7\sqrt{2}$ is just like	$3w+4w=7w$

Within both examples everything is the same except for the coefficients (the radicals are "like"). If the radicals aren't the same, then you can't add or subtract them. For example:

$3\sqrt{2}+4\sqrt{3}$ can't be simplified, just like	$3w+4x$ can't be simplified

You should notice the similarities between these ideas. Here is the big picture: Whenever you add or subtract terms in an expression or equation, you can't simplify them unless they have exactly the same variables in them to the same powers and, if there are any radicals in them, they must be identical.

Question 2: Simplify the expression $13\sqrt{5}-7\sqrt{5}+3\sqrt{3}$.

Example 4: Adding and Subtracting Radical Expressions

Simplify the following assuming that all variables are positive.

a) $\sqrt{27}-2\sqrt{3}$ b) $3\sqrt{50}+7\sqrt{2}$ c) $x\sqrt{28}-3\sqrt{7x^2}$ d) $\sqrt[3]{24}+x\sqrt[3]{3}$

Solutions:

(a) They don't appear at first to be likes, but if we simplify the first radical, they end up being likes.	$\sqrt{27}-2\sqrt{3}=\sqrt{9}\sqrt{3}-2\sqrt{3}$ $=\mathbf{3}\sqrt{3}-2\sqrt{3}=\sqrt{3}$
(b) After simplifying the first radical, we can add.	$3\sqrt{50}+7\sqrt{2}=3\sqrt{25}\sqrt{2}+7\sqrt{2}$ $=3\bullet\mathbf{5}\sqrt{2}+7\sqrt{2}$ $=15\sqrt{2}+7\sqrt{2}$ $=22\sqrt{2}$
(c) Here again is a situation where we can simplify the expressions and then subtract the two terms (like variables and like radicals).	$x\sqrt{28}-3\sqrt{7x^2}=x\sqrt{4}\sqrt{7}-3\sqrt{x^2}\sqrt{7}$ $=\mathbf{2}x\sqrt{7}-3\mathbf{x}\sqrt{7}$ $=-x\sqrt{7}$
(d) After we simplify the first term we find that these two terms aren't likes (one is	$\sqrt[3]{24}+x\sqrt[3]{3}=\sqrt[3]{8}\sqrt[3]{3}+x\sqrt[3]{3}$

missing an x), and thus we can't simplify this expression any further.	$= 2\sqrt[3]{3} + x\sqrt[3]{3}$

A good rule of thumb for dealing with radicals: think about what you would do if they were variables.

Question 3: Multiply $(\sqrt{2} + 3)(\sqrt{2} - 5)$ and simplify the result.

Example 5: Multiplying Radicals

Multiply the following expressions.

a) $\sqrt{3}\,(2\sqrt{5} + 5\sqrt{3})$ b) $(2\sqrt{3} + 7)(2\sqrt{3} - 7)$ c) $(4\sqrt{5} + \sqrt{3})^2$

Solutions:

(a) We distribute as we would with a monomial times a binomial. We can't simplify further since we don't have like terms.	$2\sqrt{3}\,\sqrt{5} + 5\sqrt{3}\,\sqrt{3}$ $= 2\sqrt{15} + 5\sqrt{9}$ $= 2\sqrt{15} + 5 \bullet 3$ $= 2\sqrt{15} + 15$
(b) We *could* FOIL as we would with two binomials times each other. We might notice instead, that this is the difference of two squares $[(x + y)(x - y) = x^2 - y^2]$.	$\left(2\sqrt{3}\right)^2 - (7)^2$ $= 4 \bullet 3 - 49$ $= 12 - 49$ $= -37$
In part (c), we use our formula for squaring a binomial $[(x + y)^2 = x^2 + 2xy + y^2]$.	$\left(4\sqrt{5}\right)^2 + 2 \bullet 4\sqrt{5} \bullet \sqrt{3} + \left(\sqrt{3}\right)^2$ $= 16 \bullet 5 + 8\sqrt{5} + 3$ $= 80 + 8\sqrt{5} + 3 = 83 + 8\sqrt{5}$

RATIONALIZING DENOMINATORS

We mentioned rationalizing the denominator earlier (Case 3). Rationalizing is where we multiply by some amount in order to eliminate the radical in the denominator of a fraction (it will always involve the radical itself). Here is Case 3: Don't leave radicals in the denominator. Let's take a moment here to recap the three cases where we need to simplify radicals.

- Case 1: Don't leave anything in the root symbol that has a power that is bigger than the root.
- Case 2: Don't leave fractions in a radical.
- Case 3: Don't leave roots in the denominator.

Example 6: Rationalizing the Denominator

Rationalize the denominators of the following expressions.

a) $\frac{2}{\sqrt{5}}$ b) $\frac{6}{\sqrt{12}}$ c) $\frac{1}{5+\sqrt{2}}$ d) $\frac{2-\sqrt{3}}{1-\sqrt{3}}$

Solutions:

(a) We need to multiply the numerator and the denominator by $\sqrt{5}$, since $\sqrt{5} \cdot \sqrt{5} =$ 5. We've eliminated the radical.	$\frac{2}{\sqrt{5}} \bullet \frac{\sqrt{5}}{\sqrt{5}} = \frac{2\sqrt{5}}{\sqrt{25}} = \frac{2\sqrt{5}}{5}$
(b) We should simplify the denominator first; then we'll multiply by $\sqrt{3}$ to eliminate the square root.	$\frac{6}{\sqrt{12}} = \frac{6}{\sqrt{4}\sqrt{3}} = \frac{6}{2\sqrt{3}} = \frac{3}{\sqrt{3}}$ $\frac{3}{\sqrt{3}} \bullet \frac{\sqrt{3}}{\sqrt{3}} = \frac{3\sqrt{3}}{3} = \sqrt{3}$

Answer Q2: The first two terms are like radicals since they both have a $\sqrt{5}$ in them. The third term isn't like the first two. So the answer is $6\sqrt{5} + 3\sqrt{3}$.

To do parts (c) and (d), we need to take a look again at example 5(b). Notice that when we multiply two expressions in parentheses that fit the formula for the difference of two squares, the square root symbol eventually disappears $(2\sqrt{3} + 7)(2\sqrt{3} - 7) = -37$. If we see a two-term denominator with one of the terms a square root, then we will need to multiply it by another two-term expression to create the difference of two squares.

Question 4: What do we call expressions that, when multiplied together, create the difference of two squares?

(c) We need to multiply the numerator and the denominator by the conjugate of $(5+\sqrt{2})$ to eliminate the square root symbol in the denominator.	$\frac{1}{5+\sqrt{2}} \bullet \frac{5-\sqrt{2}}{5-\sqrt{2}} = \frac{5-\sqrt{2}}{5^2-\left(\sqrt{2}\right)^2} = \frac{5-\sqrt{2}}{25-2}$ $= \frac{5-\sqrt{2}}{23}$
(d) We need to multiply the numerator and the denominator by the conjugate of $(1-\sqrt{3})$ to eliminate the square root symbol in the denominator.	$\frac{2-\sqrt{3}}{1-\sqrt{3}} \bullet \frac{1+\sqrt{3}}{1+\sqrt{3}} = \frac{2+2\sqrt{3}-\sqrt{3}-\sqrt{9}}{1^2-\left(\sqrt{3}\right)^2}$ $= \frac{2+\sqrt{3}-3}{1-3} = \frac{-1+\sqrt{3}}{-2}$ or $\frac{1-\sqrt{3}}{2}$

Section Summary

- b^n means b times itself n times (repeated multiplication).
- There are five major rules of exponents to remember.
  1. $b^m \bullet b^n = b^{m+n}$ Product Rule
  2. $\dfrac{b^m}{b^n} = b^{m-n}$ Quotient Rule
  3. $(b^m)^n = b^{mn}$ Power Rule
  4. $b^{-m} = \dfrac{1}{b^m}$ or $\dfrac{1}{b^{-m}} = b^m$ Negative Rule
  5. $b^0 = 1$ Zero Rule
- There are two major rules of radicals to remember.
  1. $\sqrt[n]{a} \bullet \sqrt[n]{b} = \sqrt[n]{ab}$ Product rule
  2. $\dfrac{\sqrt[n]{a}}{\sqrt[n]{b}} = \sqrt[n]{\dfrac{a}{b}}$ Quotient rule
- Scientific notation is a way of expressing a number, such that it is in the form $m \times 10^n$ where $1 \le m < 10$ and n is an integer.
- $b^{\frac{p}{r}} = \left(\sqrt[r]{b}\right)^p$ or $\sqrt[r]{(b)^p}$.
- When adding, subtracting, and multiplying, radical expressions behave similarly to variable expressions.
- Rationalizing means to eliminate a radical in a fractional expression by multiplying both the numerator and denominator by the appropriate expression.
- When asked to simplify a radical expression, remember the three cases:
  1) Don't leave anything in the radical that has a power larger than the root.
  2) Don't leave fractions in a radical.
  3) Don't leave radicals in the denominator.

Answer Q3: This is just like taking (x + 3)(x – 5) which is $x^2 - 2x - 15$. By using FOIL on this question, you get
$\left(\sqrt{2}\right)^2 - 5\sqrt{2} + 3\sqrt{2} - 15$
$= 2 - 2\sqrt{2} - 15$
$= -13 - 2\sqrt{2}$.
(FOIL, as you might remember, is an acronym (First Outer Inner Last) to help us to remember how to distribute when we have only two-term expressions (binomials).

SECTION 0.1 HOMEWORK

(1–8) Evaluate.

1. 3^4
2. 2^5
3. 3^{-5}
4. 7^{-3}
5. $25^{\frac{3}{2}}$
6. $8^{\frac{2}{3}}$
7. $16^{\frac{3}{4}}$
8. $27^{\frac{4}{3}}$

(9–12) Rewrite with fractional exponents.

9. $\sqrt[3]{x^2}$
10. $\sqrt[5]{y^3}$
11. $\sqrt[3]{x^2y^7}$
12. $\sqrt[4]{a^5b^3}$

(13–46) Simplify (write without negative exponents).

13. $x^5y^3x^3y$
14. $(3x^3y^5)(-2xy^4)$
15. $(-3a^2b^5)(-5a^3b)$
16. $(-15x^4y^3z^5)(12xy^4z^2)$
17. $(a^{-5}b^2)(a^{-3}b^{-1})$
18. $(x^5y^2)(x^{-5}y^{-7})$
19. $(x^{-3}y^3z)(x^{-5}y^{-1}z^{-4})$
20. $(a^{-5}b^4c)(a^{-8}b^{-4}c^5)$
21. $(-5xy^{-3}z^{-4})(3x^{-4}y^7)$
22. $(-4a^3b^{-3})(-7a^{-4}b^{-5}c^3)$
23. $(a^{\frac{3}{4}}b^{\frac{2}{3}})(a^{\frac{1}{8}}b^{\frac{1}{6}})$
24. $(x^{\frac{-2}{3}}b^{\frac{3}{5}})(x^{\frac{-2}{9}}b^3)$
25. $(x^{\frac{3}{4}}y^{\frac{2}{3}}z)(x^{\frac{-3}{8}}y^{\frac{-5}{12}}z^{\frac{-3}{4}})$
26. $(-3x^{\frac{-2}{3}}y^2)(5x^{\frac{1}{2}}y^{\frac{-4}{3}}z^{\frac{-1}{3}})$
27. $(x^3y^2)^3(2xy^3)^2$
28. $(-3xy^3)^2(-2x^2y^3)^3$
29. $(2x^2y^3)^3(3xy^4)^{-2}$
30. $(a^{-3}b^2c^3)^{-3}(ab^{-2}c^2)^4$
31. $(x^{\frac{2}{3}}y^{\frac{1}{4}})^{12}(x^{\frac{1}{5}}y^{\frac{3}{10}})^{10}$
32. $(a^{\frac{3}{4}}b^{\frac{1}{2}})^8(a^{\frac{1}{3}}b^{\frac{5}{6}})^{12}$
33. $(x^{\frac{2}{3}}y^{\frac{2}{5}})^{\frac{3}{4}}$
34. $(a^{\frac{3}{8}}b^{\frac{6}{7}})^{\frac{5}{6}}$
35. $(x^{\frac{2}{5}}y^{\frac{4}{3}})^{\frac{3}{4}}(x^{\frac{1}{4}}y^{\frac{2}{5}})^{\frac{3}{2}}$
36. $(a^{\frac{-2}{3}}b^{\frac{1}{6}})^{\frac{-1}{2}}(a^{\frac{3}{4}}b^{\frac{-1}{2}})^{\frac{2}{3}}$
37. $\dfrac{a^3b^4}{x^3y^2}\bullet\dfrac{x^5y}{ab^7}$
38. $\dfrac{5a^4b}{12x^4y^2}\bullet\dfrac{18xy^3}{25a^2b^4}$
39. $\dfrac{3x^{-4}y^2}{10a^{-3}b^{-2}}\bullet\dfrac{20a^{-2}b^4}{21x^{-3}y^4}$
40. $\dfrac{8ab^{-3}}{15x^3y^{-4}}\bullet\dfrac{25x^{-5}y^3}{12a^3b^{-5}}$
41. $\left(\dfrac{a^3b^2}{xy^3}\right)^3\left(\dfrac{x^2y}{ab^4}\right)^2$
42. $\left(\dfrac{x^2y^{-3}}{a^{-2}b^2}\right)^{-3}\left(\dfrac{a^{-3}b^{-4}}{x^3y^2}\right)^2$
43. $\dfrac{6xy^2}{5a^3b^2}\div\dfrac{18xy^4}{15a^4b}$
44. $\dfrac{-10a^3b^2}{7x^3y}\div\dfrac{15ab^4}{28x^4y^2}$
45. $\dfrac{4a^3b^2}{9xy^3}\div\dfrac{16x^2y}{27ab^3}$
46. $\dfrac{-5x^2y^{-3}}{7a^{-4}b^2}\div\dfrac{25a^{-2}b^{-3}}{49x^{-5}y^5}$

Answer Q4: Conjugate pairs. For example, $(2\sqrt{3}-7)$ is the conjugate of $(2\sqrt{3}+7)$.

(47–54) Rewrite with scientific notation.

47. 0.0000384
48. 0.00000358
49. 0.005948
50. 0.000000098341
51. 3,838,000,000
52. 9,830,000
53. 5,830,000,000,000
54. 830,000,000,000

(55–62) Rewrite without scientific notation.

55. 3.85×10^{-4} 56. 2.837×10^{-7} 57. 9.8×10^{-9}

58. 7.3564×10^{-10} 59. 2.17×10^{5} 60. 5.834×10^{7}

61. 8.3456×10^{9} 62. 6.93×10^{11}

(63–90) Simplify (assume all variables represent positive numbers).

63. $\sqrt{28}$ 64. $\sqrt{72}$ 65. $5\sqrt{75}$

66. $-\sqrt{243}$ 67. $\sqrt[3]{54}$ 68. $\sqrt[3]{256}$

69. $\sqrt{8x^3y^4}$ 70. $\sqrt{80x^6y^5}$ 71. $\sqrt[3]{x^5y^{13}z^{12}}$

72. $\sqrt[3]{a^6b^9c^{13}}$ 73. $\sqrt[4]{x^9y^{15}z^{12}}$ 74. $\sqrt[4]{a^{16}b^6c^{13}}$

75. $\sqrt{\frac{2}{3}}$ 76. $\sqrt{\frac{3}{5}}$ 77. $\sqrt{\frac{5}{8}}$

78. $\sqrt{\frac{7}{27}}$ 79. $\frac{3}{\sqrt{6}}$ 80. $\frac{5}{\sqrt{10}}$

81. $\frac{8}{\sqrt{12}}$ 82. $\frac{4\sqrt{5}}{\sqrt{48}}$ 83. $\frac{\sqrt{5}}{\sqrt{10}}$

84. $\frac{\sqrt{3}}{\sqrt{6}}$ 85. $\frac{2}{2+\sqrt{3}}$ 86. $\frac{-3}{3-\sqrt{2}}$

87. $\frac{3+\sqrt{5}}{2+\sqrt{5}}$ 88. $\frac{2+\sqrt{3}}{5-\sqrt{3}}$ 89. $\frac{2+\sqrt{3}}{5-\sqrt{2}}$

90. $\frac{3-\sqrt{2}}{2+\sqrt{5}}$

(91–98) Add or subtract.

91. $5\sqrt{3}+2\sqrt{5}-8\sqrt{3}+7\sqrt{5}$ 92. $9\sqrt{2}-8\sqrt{7}-9\sqrt{7}+5\sqrt{2}$

93. $3\sqrt{18}-4\sqrt{50}$ 94. $5\sqrt{12}+7\sqrt{75}$

95. $8\sqrt{3}-5\sqrt{2}-9\sqrt{27}+2\sqrt{98}$ 96. $3\sqrt{5}+5\sqrt{6}-9\sqrt{125}-3\sqrt{24}$

97. $5\sqrt{3}+7\sqrt{72}-8\sqrt{27}+2\sqrt{5}$ 98. $2\sqrt{5}+3\sqrt{3}+5\sqrt{20}+2\sqrt{32}$

(99–112) Multiply.

99. $3\sqrt{2}(5-7\sqrt{2})$ 100. $5\sqrt{3}(2\sqrt{2}+7\sqrt{3})$

101. $-2\sqrt{5}(3\sqrt{10}-5\sqrt{5})$ 102. $-3\sqrt{7}(5\sqrt{7}+2\sqrt{14})$

103. $(2+3\sqrt{3})(5-2\sqrt{3})$ 104. $(7-5\sqrt{2})(8-3\sqrt{2})$

105. $(3\sqrt{2}-5\sqrt{3})(2\sqrt{3}+4\sqrt{2})$ 106. $(4\sqrt{5}+3\sqrt{6})(3\sqrt{6}+2\sqrt{5})$

107. $(5-2\sqrt{3})(5+2\sqrt{3})$ 108. $(9+4\sqrt{5})(9-4\sqrt{5})$

109. $(2\sqrt{3}+5\sqrt{2})(2\sqrt{3}-5\sqrt{2})$ 110. $(4\sqrt{5}-3\sqrt{2})(4\sqrt{5}+3\sqrt{2})$

111. $(3\sqrt{3}+2\sqrt{5})(3+4\sqrt{6})$

112. $(5\sqrt{10}+3\sqrt{2})(3\sqrt{5}-4\sqrt{6})$

(113–116) Rationalize the numerator.

113. $\dfrac{5\sqrt{3}}{6}$

114. $\dfrac{3\sqrt{10}}{5}$

115. $\dfrac{6+\sqrt{3}}{11}$

116. $\dfrac{5-\sqrt{5}}{10}$

Section 0.2 Polynomial Operations

We have finished our review of exponent related topics. Let's turn now to polynomials.

> A **polynomial** in one variable is any expression of the form $a_nx^n + a_{n-1}x^{n-1} + a_{n-2}x^{n-2} + \ldots + a_2x^2 + a_1x + a_0$ where the a_i's are real numbers and n is a whole number.

In this section we are going to concentrate on adding, subtraction, multiplication and dividing polynomial expressions. Let's being with addition and subtraction. Here are a few examples.

Example 1: Adding and Subtracting Polynomials

Simplify the following expressions by combining like terms (terms with the same variables to the same powers).

a) $(-3x^2 + 8x) + (5x^2 - 13x + 5)$ b) $(12xy^2 - 5\sqrt{3}\,w) + (7xy^2 + 8\sqrt{3}\,w)$

b) $(k^3 - 4k^2 - 9) - (-5k^3 + 2k - 6)$ d) $(-2mn + 17m^2 - 14n^2) - (23m^2 - mn + 11)$

Solutions:

(a) The first terms in each parenthesis are likes as are the second terms, so we get	$(-\mathbf{3x^2} + 8x) + (\mathbf{5x^2} - 13x + 5)$ $= \mathbf{2x^2} - 5x + 5$
(b) The first terms in each parenthesis are likes as are the second terms, so we get	$(\mathbf{12xy^2} - 5\sqrt{3}\,w) + (\mathbf{7xy^2} + 8\sqrt{3}\,w)$ $= \mathbf{19xy^2} + 3\sqrt{3}\,w$
(c) The first terms in each parenthesis are likes as are the last terms, so we distribute the subtraction sign and then we get	$(\mathbf{k^3} - 4k^2 - 9) - (-\mathbf{5k^3} + 2k - 6)$ $= \mathbf{k^3} - 4k^2 - 9 + \mathbf{5k^3} - 2k + 6$ $= \mathbf{6k^3} - 4k^2 - 2k - 3$
(d) If we collect the like terms we get	$(-2mn + \mathbf{17m^2} - 14n^2) - (\mathbf{23m^2} - mn + 11)$ $= -2mn + \mathbf{17m^2} - 14n^2 - \mathbf{23m^2} + mn - 11$ $= -\mathbf{6m^2} - mn - 14n^2 - 11$

Question 1: Can we simplify the following expression? If not, why? $(2y + 7) + (y^2 - x)$

It is time to remind you of a few terms that are used when we talk about polynomial expressions.

> **Coefficient** – The constant in the front of the term (example: 3 is the coefficient in front of the term $3xy^2$).
>
> **Monomial** – A single-term polynomial (example: $3xy^2$).
>
> **Binomial** – A two-term polynomial (example: $5x^2 - 10$).
>
> **Trinomial** – A three-term polynomial (example: $-14y^2 + 35y + 1$).

> **Degree** – The sum of the exponents on the variables in a term (example: the degree of $3xy^2$ is $1 + 2 = \mathbf{3}$).
>
> **Degree of a polynomial** – The largest degree of the terms in the polynomial (example: the degree of $-14y^2 + 35y + 1$ is 2).
>
> **Descending order** – When a polynomial with respect to one variable is written in descending powers of that variable (example: $-6m^2 - mn - 14n^2 - 11$ is in descending order with respect to the variable m. With respect to the variable n it would look like this $-14n^2 - mn - 6m^2 - 11$).

Usually, we want everything simplified (like terms collected) as much as possible and then have those simplified expressions written in descending order.

Let's move on to multiplication. There are several different decisions we need to make when we are multiplying polynomials. We might need to distribute or FOIL or use some special formula. Some of the special formulas you need to remember are:

- $(x + y)(x - y) = x^2 - y^2$ **The difference of two squares**
- $(x + y)^2 = x^2 + 2xy + y^2$ } **Binomial squared**
- $(x - y)^2 = x^2 - 2xy + y^2$ }

Example 2: Multiplying Polynomials

Multiply the following polynomials together and simplify. (These three problems are similar to the ones in section 0.1, example 5.)

a) $2xy\,(-3x^2y + 5x - 8)$ b) $(4x + 9y)(4x - 9y)$ c) $(7x + wy)^2$

Solutions:

(a) We distribute, since we have a monomial times a trinomial. We can't simplify further since no two terms are alike.	$2xy\,(-3x^2y + 5x - 8)$ $-6x^3y^2 + 10x^2y - 16xy$
(b) We could FOIL. We might notice instead that this is the difference of two squares $[(x + y)(x - y) = x^2 - y^2]$.	$(4x + 9y)(4x - 9y)$ $(4x)^2 - (9y)^2$ $16x^2 - 81y^2$
(c) We use our formula for squaring a binomial $[(x + y)^2 = x^2 + 2xy + y^2]$.	$(7x + wy)^2$ $(7x)^2 + 2(7x)(wy) + (wy)^2$ $49x^2 + 14xwy + w^2y^2$

Let's do a little harder example.

Example 3: Multiplying a Binomial by a Trinomial

Multiply and simplify the binomial $(x - 2)$ times the trinomial $(3x^2 + 5x - 11)$.

Solution:

We need to take each term in the first expression (the binomial) times each term in the second expression (the trinomial).	$(x - 2)(3x^2 + 5x - 11)$ $3x^3 + 5x^2 - 11x$ $+ \quad - 6x^2 - 10x + 22$
Collect the like terms, write in descending order and we're done.	$3x^3 - x^2 - 21x + 22$

Dividing expressions is very much like dividing numbers.

Answer Q1: No, we can't simplify this expression. No two terms have the same variables to the same powers.

> Question 2: Simplify $\frac{38}{6}$ and leave as an improper fraction.

The following examples are similar to the problem in question 2.

Example 4: Simplifying Rational Polynomials

Simplify the following expressions.

a) $\frac{3xy^3}{xy}$ b) $\frac{-12y^4w}{4y^2w^2}$ c) $\frac{5x^2 - 15x}{5x}$ d) $\frac{36y^3 + 5y^2 - 9y + 2}{6y}$

Solutions:

(a) This is a monomial divided by a monomial, so we just cancel out like factor(s) as we did in question 2. We could also use our rules of exponents from section 0.1.	$\frac{3xy^3}{xy} = 3y^2$ $3x^{1-1}y^{3-1} = 3y^2$
(b) We cancel like factors as in part (a). Here there is a factor left in the denominator. Or we could use the rules of exponents from section 0.1 to get the same result.	$\frac{-12y^4w}{4y^2w^2} = \frac{-3y^2}{w}$ $-3y^{4-2}w^{1-2} = -3y^2w^{-1} = \frac{-3y^2}{w}$

(c) When you have a binomial or larger expression in the numerator, then you must separate the expression into individual fractions first, with monomials divided by monomials. Then cancel like factors.	$\frac{5x^2-15x}{5x}=\frac{5x^2}{5x}-\frac{15x}{5x}=x-3$
(d) We do the same as we did in part (c) but we have a few fractions remaining in our final simplified answer.	$\frac{36y^3+5y^2-9y+2}{6y}$ $=\frac{36y^3}{6y}+\frac{5y^2}{6y}-\frac{9y}{6y}+\frac{2}{6y}$ $=6y^2+\frac{5y}{6}-\frac{3}{2}+\frac{1}{3y}$

Any time we are dividing something by a monomial, we separate the expression into several fractions and simplify each one as much as possible. We can simplify them either by canceling out like factors or using the rules of exponents learned in 0.1. As you try to answer the next question, take time to reflect on what you are thinking about as you solve it.

Question 3: Divide 739 by 17.

As you were finding the answer to this question (divide 739 by 17), you probably began by thinking about what times 17 would get close to 73. You had a two-digit number, so you first tried to figure out how many times 17 would go into the first two digits of 739. Once you decided that 4 was the best choice, you multiplied the 17 by 4 and subtracted that answer from 73. Next, you thought what times 17 will get close to 59. That turned out to be 3 and you did the same as before but with the number 3 this time. Finally, you ended with the remainder of 8 and wrote out your final answer.

In a division problem, the number you are dividing by is called the **divisor** and the number you are dividing into is called the **dividend**. In question 3 the divisor is 17 and the dividend is 739 and the **quotient** (answer) is $43\frac{8}{17}$. This procedure is similar to what we do when we have an expression that is divided by a binomial or larger expression. Let's do an example of this type.

Example 5: Division by a Binomial

Divide $5x^2-6x-8$ by $(x-2)$ using long division.

Solution:

We think about what times x will be exactly $5x^2$ (the answer is $5x$). This is one place where dividing polynomials is a little different than dividing numbers. We then multiply the x and -2 by $5x$, change the signs and add.	$\mathbf{5x}$ $x-2\overline{)\,5x^2-6x-8}$ $+\quad \underline{\mathbf{-5x^2}+10x}$ $4x-8$
We now look for something times x that will be exactly $4x$ (the answer is 4). We again multiply the x and -2 by the 4, change the signs and add. We are now done since there aren't any more terms to bring down.	$5x\mathbf{+4}$ $x-2\overline{)\,5x^2-6x-8}$ $+\quad \underline{-5x^2+10x}$ $4x-8$ $+\quad \underline{\mathbf{-4x}-8}$ 0
Our final answer is	$5x+4$

Answer Q2: $\frac{19\cdot 2}{3\cdot 2}=\frac{19}{3}$. *The numerator and the denominator had a common factor that we cancelled out.*

As you can see, we followed essentially the same procedure as in question 3; we:

1) Figured out what times the first term in the divisor would equal the first term in the dividend (in this example it was what times x equals $5x^2$).
2) Multiplied the whole divisor by that amount ($5x$).
3) Changed all the signs on what we had just found and added it to the dividend and arrived at the expression $4x-8$.
4) Repeated the whole procedure by multiplying the divisor by 4.
5) Arrived at the final answer of $5x+4$.

Example 6: Dividing by a Binomial

Divide $6x^3-19x^2+23x-10$ by $(2x-3)$ using long division.

Solution:

We think about what times $2x$ will be exactly $6x^3$ (the answer is $3x^2$). We then multiply $2x$ and -3 by $3x^2$, change the signs and add.	$\mathbf{3x^2}$ $\mathbf{2x}-3\overline{)\,6x^3-19x^2+23x-10}$ $+\quad \underline{\mathbf{-6x^3}+9x^2}$ $-10x^2+23x$
We now look for something times $2x$ that will be exactly $-10x^2$ $(-5x)$.	$3x^2\ \mathbf{-5x}$ $\mathbf{2x}-3\overline{)\,6x^3-19x^2+23x-10}$ $+\quad \underline{-6x^3+9x^2}$

We multiply the $2x$ and -3 by the $-5x$, change the signs and add.	$\begin{array}{rr} & -10x^2 + 23x \\ + & \underline{10x^2 - 15x} \\ & 8x - 10 \end{array}$
We determine that 4 times $2x$ will equal the $8x$ in our remainder. Multiply the $2x$ and -3 by 4, change the signs and add. We have a remainder of 2.	$\begin{array}{rl} & 3x^2 - 5x + \mathbf{4} \\ \mathbf{2x} - 3 & \overline{)\, 6x^3 - 19x^2 + 23x - 10} \\ + & \underline{-6x^3 + 9x^2} \\ & -10x^2 + 23x \\ + & \underline{10x^2 - 15x} \\ & 8x - 10 \\ + & \underline{-\mathbf{8x} + 12} \\ & 2 \end{array}$
Our final answer is	$3x^2 - 5x + 4 + \dfrac{2}{2x-3}$

Section Summary

- When we add or subtract expressions, we put together only those terms that have the same variables to the same powers, only the coefficients may be different.
- When we multiply expressions, we must multiply each term in the first expression times every term in the second expression.
- When we divide expressions, we always compare the first term in the divisor with the first term in the dividend in order to figure out what to multiply by.

Answer Q3:

$$\begin{array}{rl} & 43 \\ 17 & \overline{)\,739} \\ & \underline{68} \\ & 59 \\ & \underline{51} \\ & 8 \end{array}$$

739 divided by 17 is $43\dfrac{8}{17}$.

SECTION 0.2 HOMEWORK

(1–6) Give the degree of each polynomial.

1. $5x^3 - 3x + 7$
2. $5y^4 + 2y^2 + 8$
3. $3a^2 + 2a^5 - 2$
4. $7b - 8b^3 + 9$
5. $5a^3b^3 + 7a^2b - 9ab^2$
6. $-3x^3y^4 + 7xy^3 - 9x^2y^2$

(7–10) Write each polynomial in descending order with respect to x.

7. $5x + 7x^3 - 8x^2 + 9$
8. $9 - 2x + 7x^2$
9. $3xy^3 - 5x^2y + 2x^3y^2$
10. $5x^2y + 2xy^3 - 5x^3y^2$

(11–22) Add or subtract each polynomial.

11. $(5x^3 - 7x + 8) + (2x^3 - 9x - 12)$
12. $(9y^2 + 5y - 9) + (7y^2 - 10y - 7)$
13. $(8x^3 - 9x^2 + 4) + (3x^2 - 5x + 7)$
14. $(7x^2 - 9x + 10) + (2x^3 - x^2 - 15)$
15. $(5a^3b + 7a^2b^2 - 10ab^3) + (7ab^3 + 3a^2b^2 - 2a^3b)$
16. $(12x^3y^2 + 9x^2y - 15xy^3) + (9x^3y^2 - 8x^2y^2 - 7xy^3)$
17. $(5x^2 - 10x + 7) - (8x^2 - 15x + 9)$
18. $(5a^3 - 7a^2 + 9a + 12) - (2a^3 + 8a^2 - 7a + 23)$
19. $(9y^3 - 12y + 7) - (8y^2 - 9y - 15)$
20. $(5b^2 - 8b + 10) - (7b^3 - 4b^2 + 9b + 15)$
21. $(8x^3y - 9x^2y^2 + 15xy^3) - (12x^3y - 3xy^3 + 8x^2y^2)$
22. $(5a^3b^3 + 7a^2b^2 - 9ab) - (10ab - 7a^3b^3 + 2a^2b^2)$

(23–60) Multiply and simplify each polynomial.

23. $3x(8x^2 - 5)$
24. $-5y^2(4y - 7)$
25. $6a^2b(9a^3b^2 + 5ab^3)$
26. $-7x^2y^3(3x^3y - 5xy^4)$
27. $3x^2(5x^2 - 8x - 9)$
28. $-5y^3(2y^2 + 3y - 9)$
29. $-5ab^2(2a^2b^2 - 7ab - 8)$
30. $5x^3y^2(7x^2 - 8xy + 6y^2)$
31. $(3x + 2)(5x + 7)$
32. $(3x - 5)(2x - 7)$
33. $(5x - 3)(7x + 2)$
34. $(7x - 2)(8x + 7)$
35. $(3x + 2y)(5x + 7y)$
36. $(5a - 4b)(7a - 3b)$
37. $(5a - 3b)(a + 9b)$
38. $(9x + 8y)(5x - 6y)$
39. $(5x - 3)(5x + 3)$
40. $(7y + 8)(7y - 8)$
41. $(5a + 3b)(5a - 3b)$
42. $(8x - 9y)(8x + 9y)$
43. $(2x - 3)(5x^2 + 7x - 4)$
44. $(5x + 4)(7x^2 - 3x + 2)$
45. $(5a + 2b)(7a^2 - 4ab + 2b^2)$
46. $(7x - 3y)(8x^2 - 3xy - 5y^2)$
47. $(3x - 2)(9x^2 + 6x + 4)$
48. $(5y + 3)(25y^2 - 15y + 9)$
49. $(2x + y)(4x^2 - 2xy + y^2)$
50. $(3a - b)(9a^2 + 3ab + b^2)$
51. $(3x + 5)^2$
52. $(5x + y)^2$
53. $(2y - 3)^2$
54. $(5a - 9)^2$

55. $(2x + 3y)^2$

56. $(9a + 2b)^2$

57. $(3a - 5b)^2$

58. $(6a - 11b)^2$

59. $(3x^2 - 5x + 2)(7x^2 + 3x - 8)$

60. $(5y^2 - 7y - 9)(4y^2 - 3y + 2)$

(61–86) Divide.

61. $\dfrac{15x^3y^2}{3x^2y}$

62. $\dfrac{-28a^5b^2}{4a^2b^2}$

63. $\dfrac{-21x^3y}{3xy^3}$

64. $\dfrac{5x^2y^4}{10x^3y}$

65. $\dfrac{10x^3 - 15x^2}{5x}$

66. $\dfrac{8y^3 - 16y^2 + 24y}{4y}$

67. $\dfrac{8a^4 - 6a^3 + 16a^2}{12a^2}$

68. $\dfrac{9x^4 + 12x^3 - 15x^2}{24x^2}$

69. $\dfrac{20a^3b^2 - 15a^2b^3}{5a^2b^2}$

70. $\dfrac{36x^4y^3 + 24x^3y^4}{6x^3y^3}$

71. $(2x^3 - 13x^2 + 17x - 10) \div (x - 5)$

72. $(3x^3 + 10x^2 - x - 12) \div (x + 3)$

73. $(10x^3 + 3x^2 - 7x - 3) \div (2x + 1)$

74. $(2x^3 - 7x^2 - 4x + 15) \div (2x - 3)$

75. $(12x^3 - 17x^2 + 4) \div (3x - 2)$

76. $(8x^3 - 6x - 18) \div (2x - 3)$

77. $(3x^3 + x^2 + 7x + 2) \div (x + 2)$

78. $(2x^3 + 5x^2 - 9x - 3) \div (x + 4)$

79. $(3x^3 + 7x^2 - 8x + 3) \div (x - 2)$

80. $(5x^3 - 8x^2 + 2x - 1) \div (x + 3)$

81. $(6x^3 + 25x^2 + 17x - 17) \div (2x + 5)$

82. $(15x^3 - 29x^2 + 18x - 3) \div (3x - 4)$

83. $(2x^4 - 5x^2 + 2) \div (x - 2)$

84. $(x^4 + 2x^3 - 5) \div (x + 3)$

85. $(3x^4 + 52x) \div (x + 2)$

86. $(4x^4 - 68) \div (x - 2)$

Section 0.3 Factoring Polynomials

In this section we will continue our discussion on topics related to polynomials. In the last section we focused on the basic operations involving polynomials; here we will focus on how to factor a polynomial. In your past math classes you probably learned how to:

1. Factor out a Greatest Common Factor (GCF)
2. Factor a trinomial
3. Factor special binomials
4. Factor by grouping

We are going to review each of these types of factoring. But before we begin, you need to remember that factoring is the "undoing" of multiplication. When you multiply, you are eliminating the parentheses from the expression $[5x \bullet (x-2) = 5x^2 - 10x]$, whereas, when you factor, you are putting parentheses back into the expression $[5x^2 - 10x = 5x(x-2)]$.

GREATEST COMMON FACTOR

The type of factoring that you should always look for first is the GCF. Factoring out the GCF is the undoing of the simple distributive property. Be sure to find every factor in common to all of the terms in the expression.

Example 1: Factoring by the Greatest Common Factor

Factor the following:

a) $35x^3y - 25xy^2$ b) $21wx^2 + 42w^2x - 7w$ c) $3x(x+2) - 5(x+2)$

Solutions:

(a) Each of the two terms is divisible by $5xy$, so that is what we will factor out. When you factor out the GCF, you divide each term by that amount in order to determine what is left in the parentheses. (Remember: To "undo" something, is to do the opposite. If distributing is multiplication then factoring out a GCF is dividing.)	$35x^3y - 25xy^2 = 5xy\left(\frac{35x^3y}{5xy} - \frac{25xy^2}{5xy}\right)$ $= 5xy\,(7x^2 - 5y)$
(b) Each of the three terms are divisible by $7w$, so we factor that out of the expression. Divide each term by $7w$.	$21wx^2 + 42w^2x - 7w = 7w(3x^2 + 6wx - 1)$
(c) Each of the two terms are divisible by the	

sum in parentheses $(x + 2)$, so that is what we factor out of this expression.	$3x(x + 2) - 5(x + 2) = (x + 2)\,(3x - 5)$

Notice that in every example, when we factored out the GCF, we went from having an expression with *many terms* to an expression with only *one term* because of the parentheses. There are times in mathematics when we want our expression to be written as a single term (monomial). Factoring will give us a way to change an expression into having just one term.

Question 1: Factor out the GCF from the expression $30x^3y^5z + 15x^4y^3z^3 - 20x^3y^2z^4$.

From question 1, you should have noticed that with each variable in the expression you were factoring out the smallest exponent. The smallest exponent on x was 3 and the smallest exponent on y was 2 and the smallest exponent on z was 1.

Example 2: Factoring by the Greatest Common Factor

Factor out the GCF on the following.

a) $-2x^{-2}y^3 - 8x^{-1}y^7 + 4x^{-3}y^4$ b) $x^{\frac{-2}{3}} + x^{\frac{1}{3}}$

Solutions:

(a)We can factor out a -2, x^{-3}, and y^3 because -3 is the smallest exponent on x and 3 is the smallest exponent on y.	$-2x^{-2}y^3 - 8x^{-1}y^7 + 4x^{-3}y^4$ $= -2x^{-3}y^3\left(\dfrac{-2x^{-2}y^3}{-2x^{-3}y^3} - \dfrac{8x^{-1}y^7}{-2x^{-3}y^3} + \dfrac{4x^{-3}y^4}{-2x^{-3}y^3}\right)$ $= -2x^{-3}y^3(x + 4x^2y^4 - 2y)$
(b) We can factor out an $x^{\frac{-2}{3}}$, since $\dfrac{-2}{3}$ is the smallest exponent on the variable x.	$x^{\frac{-2}{3}} + x^{\frac{1}{3}}$ $= x^{\frac{-2}{3}}\left(\dfrac{x^{\frac{-2}{3}}}{x^{\frac{-2}{3}}} + \dfrac{x^{\frac{1}{3}}}{x^{\frac{-2}{3}}}\right) = x^{\frac{-2}{3}}\left(1 + x^{\frac{1}{3}-\frac{-2}{3}}\right)$ $= x^{\frac{-2}{3}}(1 + x)$

THE FOIL METHOD

Let's tackle the next type of factoring, factoring a trinomial (polynomials of the form $ax^2 + bx + c$). This may very well be the hardest of the four types of factoring. Factoring a quadratic trinomial is the "undoing" of the FOIL method. The FOIL method is more involved than the simple distributive property, so it follows that factoring a trinomial will be more complex than factoring out a GCF. Let's start with a FOIL problem and then discuss how to undo it.

Discussion 1: "Undoing" the FOIL Method

Let's look at multiplying the two binomials $(x - 5)(x + 3)$ and then look at how we might go backwards.

We FOIL.	O F I L $(x-5)(x+3) = x^2 + 3x - 5x - 15$ First Outer Inner Last $= x^2 - 2x - 15$
Notice, how the outer and inner terms go together and have coefficients that are factors of the last term (–15). **Note:** This happens only when the coefficients on the first terms are one.	Outer $(\mathbf{3}x)$ + Inner $(\mathbf{-5}x) = -2x$ $\mathbf{3 \cdot (-5)} = -15$
If we were given the problem of factoring $x^2 - 2x - 15$, we would know that, if it factored, it must factor into two binomials.	$x^2 - 2x - 15 = (\quad)(\quad)$
Since the first term on the trinomial has a coefficient of 1, implying that the first terms of the binomials will have coefficients of 1, then the coefficients on the outer and inner terms are the factors of the last term. So, we look for factors of –15 that would add to –2.	$-1 \cdot 15$ $\quad -1 + 15 = 14$ $-3 \cdot 5$ $\quad -3 + 5 = 2$ $1 \cdot -15$ $\quad 1 + (-15) = -14$ $3 \cdot -5$ $\quad 3 + (-5) = -2$
3 and –5 are the only factors that equal –2 when added. So the binomials must be	$(x - 5)(x + 3)$

This is the easier of the two types of trinomials that we factor. If the coefficient on the first term of the trinomial, in descending order, is 1, then we know that adding factors of the coefficient of the last term must equal the coefficient on the middle term. This gives us what the Outer and Inner terms must have been in the FOIL process.

Question 2: Factor $x^2 + 7x + 12$.

If, in question 2, we were to make one little change (make $7x$ negative), we would get a similar answer. We would still be looking for factors that multiply to 12, but we'd be looking for them to add to −7 not 7. The factors of 12 in this case would be −3 and −4. Hence, the factors of $x^2 - 7x + 12$ are $(x - 3)$ and $(x - 4)$. Let's tackle the tougher kind of trinomial factoring, the one with a coefficient other than 1 in front of the first term.

Example 3: Multiplying Binomials with First Term Coefficients Greater than 1

Multiply the following two binomials $(2x + 3)(5x - 7)$.

Solution:

We FOIL.	$(2x+3)(5x-7) = 2\cdot 5x^2 - 2\cdot 7x + 3\cdot 5x - 3\cdot 7$ First Outer Inner Last $= 10x^2 - 14x + 15x - 21$ $= 10x^2 + x - 21$

Please note how the factors of both the 10 and the −21, in the final answer, are in the outer $(2\cdot -7)$ and inner $(3\cdot 5)$ terms, which added up to the middle term $(1x)$ in the answer. Whenever the first term in a trinomial of this type has a coefficient other than one, then you will need to look at factors of both the first and last terms in order to factor it.

Answer Q1: $5x^3y^2z(6y^3 + 3xyz^2 - 4z^3)$

Discussion 2: Factoring a Trinomial with a First Term Coefficient Greater Than 1

Factoring $10x^2 + x - 21$.

First we need to look at the factors of both the first and last terms.	First term's coefficient: $10 = 1\cdot 10$ or $2\cdot 5$ Second term's coefficient: $-21 = -1\cdot 21$ or $-3\cdot 7$ $1\cdot -21$ or $3\cdot -7$
We need to figure out how to multiply and add these factors, so that we get a middle coefficient of 1 (there are 16 possibilities). We will list a few here.	$1\cdot -1 + 10\cdot 21 = -1 + 210 = 209$ $1\cdot -3 + 10\cdot 7 = -3 + 70 = 67$ $2\cdot -1 + 5\cdot 21 = -2 + 105 = 103$ $2\cdot 21 + 5\cdot -1 = 42 + (-5) = 37$ $2\cdot 3 + 5\cdot -7 = 6 + -35 = 29$ $2\cdot -7 + 5\cdot 3 = -14 + 15 = \underline{1}$ (That's it!)
Well, we found what we already knew from example 3. We need inner and outer terms, such that, we get −14 and 15.	$2\cdot -7 = -14$ $5\cdot 3 = 15$ $-14x + 15x = 1x$ the middle term
We need to use 2 and 5 as the factors of 10 (the first term) so those	$(2x \quad)(5x \quad)$

numbers need to be in the first terms in each binomial as we factor. Also, we need to use 3 and –7 as the factors of the last term (–21) so, those numbers must come last in each binomial.	(3)(–7)
We therefore have only two choices.	$(2x + 3)(5x - 7)$ or $(2x - 7)(5x + 3)$
The correct factoring is	$10x^2 + x - 21 = (2x + 3)(5x - 7)$

Note that any time you are trying to factor a trinomial of the form $ax^2 + bx + c$, you will always get an answer like this $(Ax \pm B)(Cx \pm D)$ if it can be factored.

Example 4: Factoring a Trinomial with a First Term Coefficient Greater than 1

Factor $25x^2 + 40x + 16$.

Solution:

Answer Q2: We need factors that multiply to 12 and add to 7. That would be 3 and 4. Thus, $x^2 + 7x + 12$ factors into $(x + 3)(x + 4)$.

First, look for factors of 25 and 16.	$25 = 1 \bullet 25,\ 5 \bullet 5$ $16 = 1 \bullet 16,\ 4 \bullet 4,\ 2 \bullet 8$
Now, we need those factors to multiply and then add to 40. Here we show just a few of the 7 unique possibilities.	$1 \bullet 25 + 1 \bullet 16 = 41$ $25 \bullet 16 + 1 \bullet 1 = 401$ $5 \bullet 4 + 5 \bullet 4 = 40$
Our factors then appear to be	$(5x + 4)(5x + 4)$

Since $25x^2 + 40x + 16 = (5x + 4)(5x + 4) = (5x + 4)^2$, the original trinomial is what we call a perfect square trinomial. That is, it comes from a binomial being squared. In the future, we will be concerned about changing trinomials into perfect square trinomials.

Question 3: Factor $6x^2 - 19x - 20$.

We have covered two of our four types of factoring. Let's look at the special binomials next.

SPECIAL BINOMIALS

We have three special binomials to discuss and they are:

$x^2 - y^2$	The difference of two squares
$x^3 - y^3$	The difference of two cubes
$x^3 + y^3$	The sum of two cubes

These three special binomials factor easily if you memorize the following formulas:

$x^2 - y^2 = (x+y)(x-y)$	The difference of two squares
$x^3 - y^3 = (x-y)(x^2+xy+y^2)$	The difference of two cubes
$x^3 + y^3 = (x+y)(x^2-xy+y^2)$	The sum of two cubes

Example 5: Factoring Special Binomials

Factor the following binomials.

a) $16w^2 - 9p^2$ b) $h^3 - 8k^3$ c) $64x^6 + 27$ d) $x^6 - y^6$

Solutions:

(a) We see that we have the difference of two squares. We just need to determine what quantities squared will yield the two terms in the original binomial. Then we simply add and subtract those two quantities.	$16w^2 = (4w)^2$ $9p^2 = (3p)^2$ $(4w + 3p)\,(4w - 3p)$
(b) We see that we have the difference of two cubes. We just need to determine what quantities cubed will yield the two terms in the original binomial. Once we have done this we simply plug those two quantities into the general formula.	$h^3 = (h)^3$ $8k^3 = (2k)^3$ $(h - 2k)(h^2 + h(2k) + (2k)^2)$ $(h - 2k)(h^2 + 2hk + 4k^2)$
(c) We see that we have the sum of two cubes. We just need to determine what quantities cubed will yield the two terms in the original binomial. Once we have done this we simply plug those two quantities into the general formula.	$64x^6 = (4x^2)^3$ $27 = 3^3$ $(4x^2 + 3)((4x^2)^2 - 3(4x^2) + 3^2)$ $(4x^2 + 3)(16x^4 - 12x^2 + 9)$
(d) We see that we could view this two ways. Either look at it as the difference of two squares $(x^3)^2 - (y^3)^2$ or two cubes $(x^2)^3 - (y^2)^3$. Let's use the difference of two squares formula.	$x^6 - y^6 = (x^3)^2 - (y^3)^2 = \underline{(x^3 + y^3)}\,\underline{(x^3 - y^3)}$ But now these two factors factor as the sum and difference of two cubes. $= (x+y)(x^2 - xy + y^2)\,(x-y)(x^2 + xy + y^2)$

If we had chosen to view the original problem in part (d) as the difference of two cubes, we probably wouldn't have gotten this completely factored.

Question 4: Factor $x^6 - y^6$ by treating it like the difference of two cubes.

It is most unlikely that any of us would ever notice that the trinomial $(x^4 + x^2y^2 + y^4)$ factors into $(x^2 + xy + y^2)(x^2 - xy + y^2)$. Most of the time you won't see anything as tricky as this. But if you do, be sure to check both avenues before moving on.

This brings us to our last factoring type, by grouping. When you have more than three terms, this may very well be the only choice you will have.

FACTORING BY GROUPING

What you want to do in these situations is to group the terms into two groups and try to find a GCF for each group.

Example 6: Factoring by Grouping

Factor the following by grouping.

a) $3x^3 + x^2 + 6x + 2$ b) $10y^3 - 15y^2 - 4y + 6$ c) $x^2 + 10x + 25 - y^2$

Solutions:

Answer Q3: $6 = 1 \cdot 6$ or $2 \cdot 3$, $-20 = -1 \cdot 20$ or $1 \cdot -20$ or $-2 \cdot 10$ or $2 \cdot -10$ or $-4 \cdot 5$ or $4 \cdot -5$. $6 \cdot -4 + 1 \cdot 5 = -24 + 5 = -19$. Thus, $6x^2 - 19x - 20 = (6x + 5)(x - 4)$.

(a) We see that we have four terms, so let's try grouping the first two and the last two together. We now factor out a GCF in each grouping (**x^2 & 2**). Next, we look again for a GCF, but this time we look for one between the two groupings ($\underline{3x+1}$).	$3x^3 + x^2 + 6x + 2 = (3x^3 + x^2) + (6x + 2)$ $= \mathbf{x^2}(\underline{3x+1}) + \mathbf{2}\,(\underline{3x+1})$ $= (\underline{3x+1})\,(\mathbf{x^2 + 2})$
(b) We see that we have four terms again, so let's try grouping the first two and the last two together. Be sure to notice that we carried the sign of each term and then added the two groupings. We now factor out a GCF in each grouping (**$5y^2$ & -2**) and then again between the two groupings as before ($\underline{2y-3}$).	$10y^3 - 15y^2 - 4y + 6 = (10y^3 - 15y^2) + (-4y + 6)$ $= \mathbf{5y^2}(\underline{2y-3}) + \mathbf{(-2)}(\underline{2y-3})$ $= (\underline{2y-3})\,(5y^2 - 2)$
(c) We see that we have four terms, but this time it is better to group the first three together because they form a perfect square trinomial. After we factor the perfect square trinomial, we see that what is left forms the difference of two squares.	$x^2 + 10x + 25 - y^2 = (x^2 + 10x + 25) - y^2$ $= (x+5)^2 - y^2$ $= [(x+5) + y][(x+5) - y]$ $= (x + 5 + y)(x + 5 - y)$

Notice that in every factoring problem we have done, the final answer is a monomial. That is, the final answer is a single term (in these cases parentheses) with multiplication between them.

Question 5: Factor $18w^3 + 15w^2 + 12w + 10$.

Of course you know, from previous math classes, that these four different types of factoring can happen in combination. Let's do a few more examples.

Example 7: Using a Combination of Factoring Techniques

Factor completely the following polynomial expressions.

a) $4x^3 - 14x^2 + 12x$ b) $24w^3 + 3h^6$ c) $x^3 - 4x^2 - 9x + 36$

Solutions:

(a) We see that we can factor out a GCF of $2x$. Then, we can factor the remaining trinomial.	$4x^3 - 14x^2 + 12x = 2x\,(2x^2 - 7x + 6)$ $= 2x\,(2x - 3)(x - 2)$
(b) We see that we can factor out a GCF of 3. Then, we can factor the remaining sum of cubes.	$24w^3 + 3h^6 = 3\,(8w^3 + h^6)$ $= 3\,(2w + h^2)\,(4w^2 - 2wh^2 + h^4)$
(c) We see that we have four terms so we group the first two and the last two together. We now factor x^2 out of the first grouping and -9 from the second grouping. We notice that when we factor out the $(x - 4)$, we have the difference of two squares left which we then factor.	$x^3 - 4x^2 - 9x + 36 = (x^3 - 4x^2) + (-9x + 36)$ $= x^2(x - 4) + (-9)(x - 4)$ $= (x - 4)\,(x^2 - 9)$ $= (x - 4)\,(x + 3)\,(x - 3)$

Question 6: Factor $x^2 - 11x + 12$, if possible.

Section Summary

There are four basic factoring methods used with polynomial expressions:

- GCF. Always look to factor out the greatest common factor before doing any other factoring.

Answer Q4:
$x^6 - y^6$
$= (x^2)^3 - (y^2)^3$
$= (x^2 - y^2)(x^4 + x^2y^2 + y^4)$
$= (x - y)(x + y)(x^4 + x^2y^2 + y^4)$

- Trinomial. If you have a three-term polynomial expression, see if it factors into two binomials.
- Special Binomials. Does it fit the formulas for the difference of two squares ($x^2 - y^2$), or the sum and difference of two cubes ($x^3 + y^3$, $x^3 - y^3$)?
- Grouping. If you have four or more terms, try grouping the terms together into two groups (they don't need to have the same number of terms in each grouping) and look for a GCF in each grouping.

SECTION 0.3 HOMEWORK

Completely factor the following polynomials

COMMON FACTORS-

1. $5ab - 10a$
2. $3xy + 21x$
3. $24ax + 40ay$
4. $36ax - 63bx$
5. $10a^2b^2 + 25ab^3$
6. $27x^3y^3 - 15x^2y^4$
7. $35a^5b^2 - 28a^2b^4$
8. $12x^7y^4 + 8x^3y^7$
9. $15ax - 6ay + 9a$
10. $10ax + 6bx - 8x$
11. $10x^2y - 15xy^2 - 35xy$
12. $28a^3b + 35a^2b^2 - 42a^2b$
13. $30x^3y^2 - 12x^2y^3 - 24cx^2y^2$
14. $18a^2c + 15a^2bc^2 + 21abc^2$
15. $5a^{-3}b^{-2} + 20a^{-1}b^{-4}$
16. $6x^{-2}y^{-2} - 9x^{-3}y^2$
17. $10x^{\frac{-3}{2}} - 15x^{\frac{1}{2}}$
18. $24a^{\frac{5}{4}} + 30a^{\frac{-3}{4}}$

Answer Q5: $(18w^3 + 15w^2) + (12w + 10) = 3w^2(6w + 5) + 2(6w + 5) = (6w+5)(3w^2+2)$

TRINOMIALS-

19. $x^2 - 10x + 24$
20. $x^2 + 15x + 56$
21. $x^2 - 3x - 54$
22. $y^2 + 7y - 18$
23. $x^2 + 9xy + 20y^2$
24. $a^2 - 25ab + 150b^2$
25. $x^2 - 8ax - 48a^2$
26. $y^2 + 5by - 84b^2$
27. $5x^2 + 25x - 180$
28. $6y^2 - 66y + 180$
29. $3a^2x + 39abx + 108b^2x$
30. $7ax^2 + 35axy - 42ay^2$
31. $x^2 + 3x + 4$
32. $a^2 - 7ab - 6b^2$
33. $a^2x^2 + 23axy + 120y^2$
34. $a^2 - 19abx + 90b^2x^2$
35. $x^4 + 9x^2 + 14$
36. $a^4 + 3a^2 - 40$
37. $a^4 + 25a^2b + 24b^2$
38. $x^6 + 12abx^3 + 20a^2b^2$
39. $10x^2 + 19x + 6$
40. $6x^2 - x - 15$
41. $15x^2 - 29x + 12$
42. $21x^2 + 2x - 8$
43. $5x^2 - 16xy + 3y^2$
44. $7a^2 + 12ab + 5b^2$
45. $12x^2 - 7xy - 10y^2$
46. $24a^2 + 26ab - 5b^2$
47. $6x^2 + 24x + 15$
48. $10x^2 - 75x + 35$
49. $20ax^2 - 36ax - 8a$
50. $21by^2 + 12by - 33b$
51. $6a^4 + 13a^2b + 5b^2$
52. $14x^4 - 31x^2y + 15y^2$
53. $12a^2b^2 + abxy - 6x^2y^2$
54. $10x^2y^2 - 3a^2xy - 27a^4$

AnswerQ6: This doesn't factor. Many think that this factors into $(x - 12)(x + 1)$, but if you FOIL that possible answer out, you get $x^2 - 11x - 12$. This is not what we were asked to factor.

PERFECT SQUARES-

55. $x^2 + 8x + 16$
56. $y^2 - 10y + 25$
57. $a^2 - 12a + 36$
58. $b^2 + 20b + 100$
59. $4x^2 - 12x + 9$
60. $9x^2 + 30x + 25$
61. $9x^2 + 24xy + 16y^2$
62. $25a^2 - 60ab + 36b^2$

53. $3x^2 + 54x + 243$

64. $2y^2 - 32y + 128$
65. $x^2 - 18xy + 81y^2$
66. $x^2 + 24xy + 144y^2$
67. $4x^2 + 28xy + 49y^2$
68. $9a^2 - 30ab + 25b^2$
69. $x^4 - 6x^2y + 9y^2$
70. $a^4 + 10a^2b + 25b^2$

DIFFERENCE OF SQUARES-

71. $x^2 - 36$
72. $y^2 - 25$
73. $4a^2 - 25$
74. $9b^2 - 100$
75. $x^2 + 9$
76. $4y^2 + 25$
77. $5a^2 - 20$
78. $7ay^2 - 28a$
79. $25x^2 - 36y^2$
80. $81a^2 - 25b^2$
81. $y^4 - 64$
82. $a^4 - 36$

83. $x^4 - 16$

84. $b^4 - 81$

85. $x^4 - 13x^2 + 36$

86. $y^4 - 29y^2 + 100$

SUM AND DIFFRENCE OF CUBES-

87. $y^3 - 8$

88. $x^3 + 27$

89. $a^3 + 125$

90. $b^3 - 64$

91. $8x^3 - 27$

92. $125a^3 + 8b^3$

93. $64x^3 + 1$

94. $27x^3 - 8y^3$

95. $40x^3 - 5$

96. $3a^3 + 24$

97. $2ax^3 + 128a$

98. $3by^3 - 24b$

GROUPING-

99. $ab + 3a + bx + 3x$

100. $bx - 5b + 5x - 25$

101. $5x^2 + 10 + ax^2 + 2a$

102. $by^2 - 7b + 2y^2 - 14$

103. $ax - 3a - 5x + 15$

104. $by - 5b - 3y + 15$

105. $7x + 28 - ax - 4a$

106. $5ax + 100a - 7x - 14$

107. $ax^2 + 3x^2 - 9a - 27$

108. $by^2 + 5y^2 - 4b - 20$

109. $x^2 + 6x + 9 - y^2$

110. $a^2 + 4a + 4 - x^2$

111. $y^2 + 10y + 25 - 9a^2$

112. $b^2 - 8b + 16 - 25y^2$

113. $4x^2 + 20x + 25 - 4a^2$

114. $9y^2 - 24y + 16 - 25x^2$

Section 0.4 Rational Expressions

We have completed our review of some of the basic topics involving polynomials. We now turn our attention to the review of rational (fractional) expressions.

Rational expressions are considered to be any fraction made up of polynomials.

Of course, we are not allowed to divide by 0. When we are working with fractions that have variables in the denominator, those variables are not allowed to be any value that would make the denominator equal to 0. If there are values that make the denominator equal to 0, then we say that the expression is **undefined** for those values.

Let's begin our discussion by looking at how we reduce rational expressions. To reduce a rational expression first we need to factor both the numerator and denominator, and then cancel out any like factors between the numerator and the denominator.

Example 1: Reducing a Rational Expression to Simplest Terms

Reduce the fraction $\dfrac{x^2-16}{x^2+7x+12}$.

Solution:

Factor both polynomials.	$\dfrac{x^2-16}{x^2+7x+12}=\dfrac{(x-4)(x+4)}{(x+3)(x+4)}$
Now cancel the common factor $(x+4)$ from both the numerator and the denominator.	$\dfrac{(x-4)\cancel{(x+4)}}{(x+3)\cancel{(x+4)}}=\dfrac{(x-4)}{(x+3)}$, $x \neq -4$ (Because -4 would make the *original* denominator equal to 0.)

OPERATIONS ON RATIONAL EXPRESSIONS

We now turn our attention to the four basic operations (+, –, •, ÷) and how they apply to rational expressions.

Multiplication and Division of Rational Expressions

We start with multiplication and division. With multiplication, we follow the same two simple steps that we did when we were reducing fractions.

- First, we factor every polynomial in the fractions.
- Second, we cancel out like factors between the numerators and denominators.

We can only cancel like factors if the fractions involved are made up of only monomials. Remember from the last section, when we factor polynomials, we are changing the polynomial from a many-termed polynomial to a monomial. This allows us to then cancel the like factors and simplify the problem. Here are a few examples.

Example 2: Multiplying Rational Expressions

Multiply the following rational expressions.

a) $\left(\dfrac{3x^2y}{12x^5}\right)\bullet\left(\dfrac{2x^7}{5y^2}\right)$ b) $\left(\dfrac{x^2-16}{x^2+7x+12}\right)\bullet\left(\dfrac{x^2-4x-21}{x^2-4x}\right)$ c) $\left(\dfrac{x^3+64y^3}{x^3+4yx^2+5x+20y}\right)\bullet\left(\dfrac{x^2-5}{x^2-4xy+y^2}\right)$

Solutions:

(a) We cancel out the like factors. We do not need to factor anything, since this is already in the proper form, that is, a monomial divided by a monomial. When simplifying the variables, we could have used the rules of exponents that we discussed in section 0.1.	$\left(\dfrac{3x^2y}{12x^5}\right)\left(\dfrac{2x^7}{5y^2}\right)=\dfrac{x^4}{10y}$ (with x^2, 2, y written as cancellation results)
(b) We have binomials and trinomials. We look first for GCF's, and then we factor the trinomials and binomials if possible. After that, we cancel out the like factors. (Note: We have one GCF, one difference of squares, and two trinomials.)	$\left(\dfrac{x^2-16}{x^2+7x+12}\right)\left(\dfrac{x^2-4x-21}{x^2-4x}\right)$ $=\left(\dfrac{(x+4)(x-4)}{(x+4)(x+3)}\right)\left(\dfrac{(x+3)(x-7)}{x(x-4)}\right)$ $=\dfrac{(x-7)}{x}$
(c) We have two binomials, a trinomial and a four-term polynomial. We look first for GCF's, and then we factor the trinomial and binomials if possible. We also will need to factor the four-term polynomial by grouping. We then cancel out the like factors. (Note: $x^2 - 5$ is not the difference of two squares; 5 isn't a square. Also, $x^2 - 4xy + y^2$ doesn't factor.)	$\left(\dfrac{x^3+64y^3}{x^3+4yx^2+5x+20y}\right)\left(\dfrac{x^2-5}{x^2-4xy+y^2}\right)$ $=\left(\dfrac{(x+4y)(x^2-4xy+y^2)}{x^2(x+4y)+5(x+4y)}\right)\left(\dfrac{(x^2-5)}{(x^2-4xy+y^2)}\right)$ $=\left(\dfrac{(x+4y)(x^2-4xy+y^2)}{(x+4y)(x^2+5)}\right)\left(\dfrac{(x^2-5)}{(x^2-4xy+y^2)}\right)$ $=\dfrac{(x^2-5)}{(x^2+5)}$

Question 1: Multiply the two rational expressions $\left(\dfrac{x^2-x-6}{x^2+6x+9}\right)\left(\dfrac{x+3}{x^2-4}\right)$.

As you might remember, when you are dividing two fractions you simply invert (flip) the second fraction and change the problem to multiplication. Here is an example to refresh your memory.

Example 3: Dividing Rational Expressions

Divide the following rational expressions.

a) $\left(\dfrac{2x+3}{2x^2-50}\right)\div\left(\dfrac{6x+9}{2x^2+16x+30}\right)$ b) $\left(\dfrac{b-a}{x+1}\right)\div\left(\dfrac{a^2-b^2}{x^2-1}\right)$

Solutions:

(a) We first "flip" the second fraction and change this to a multiplication problem. We then factor all of the polynomials and cancel out any like factors.	$\left(\dfrac{2x+3}{2x^2-50}\right)\left(\dfrac{2x^2+16x+30}{6x+9}\right)$ $=\left(\dfrac{(2x+3)}{2(x^2-25)}\right)\left(\dfrac{2(x^2+8x+15)}{3(2x+3)}\right)$ $=\left(\dfrac{(2x+3)}{2(x-5)(x+5)}\right)\left(\dfrac{2(x+5)(x+3)}{3(2x+3)}\right)$ $=\dfrac{(x+3)}{3(x-5)}$
(b) We first "flip" the second fraction and change this to a multiplication problem. We now factor each polynomial. Notice that at first that the $(b-a)$ doesn't appear to cancel with the (a^2-b^2), but after factoring out a (-1) from $(b-a)$, they do have a common factor that can be canceled out.	$\left(\dfrac{b-a}{x+1}\right)\left(\dfrac{x^2-1}{a^2-b^2}\right)$ $=\left(\dfrac{(b-a)}{(x+1)}\right)\left(\dfrac{(x+1)(x-1)}{(a+b)(a-b)}\right)$ $=\left(\dfrac{(-1)(a-b)}{(x+1)}\right)\left(\dfrac{(x+1)(x-1)}{(a+b)(a-b)}\right)$ $=\dfrac{-(x-1)}{(a+b)}$

Question 2: Divide the two rational expressions $\left(\dfrac{x^2-4}{x-2}\right)\div\left(\dfrac{x+2}{4x-8}\right)$.

We have completed our review of multiplying and dividing rational expressions. They are probably the easiest operations to perform with rational expressions. Addition and subtraction, on the other hand, are a bit more involved.

Addition and Subtraction of Rational Expressions

Remember that when you add or subtract fractions, they must have the same denominators. The same is true when adding or subtracting rational expressions. The hardest part of adding or subtracting rational expressions is in changing the fractions so that they will have the same denominators. The steps for adding and subtracting fractions are as follows:

- Factor the denominators.
- Find a Least Common Denominator (LCD).
- *Within each fraction*, multiply the numerator and denominator by the same amount to make the denominator be the same as the LCD.
- Add or subtract the numerators as indicated.
- Try simplifying what you have just completed.

Let's do a couple of examples.

Example 4: Adding and Subtracting Rational Expressions

With the following, perform the indicated operation and then simplify the expression.

a) $\left(\frac{5y-7}{y+4}\right)-\left(\frac{4y-3}{y+4}\right)$ b) $\left(\frac{x}{x-4}\right)+\left(\frac{7x}{x+3}\right)$ c) $\left(\frac{3x}{x^2+3x-10}\right)-\left(\frac{2x}{x^2+x-6}\right)$

Solutions:

(a) We have two fractions that already have the same denominators so we simply combine the numerators. This problem won't simplify any further.	$\left(\frac{5y-7}{y+4}\right)-\left(\frac{4y-3}{y+4}\right)$ $=\frac{5y-7-(4y-3)}{y+4}=\frac{5y-7-4y+3}{y+4}=\frac{y-4}{y+4}$

(b) We need to find an LCD and then convert the two fractions so that they have the same denominator. The LCD is $(x-4)(x+3)$. We then multiply the first fraction by $(x+3)/(x+3)$ and the second one by $(x-4)/(x-4)$. Now, we simplify the numerators and add. This doesn't factor further.	$\left(\frac{x}{x-4}\right)+\left(\frac{7x}{x+3}\right)$ $=\left(\frac{x}{x-4}\right)\left(\frac{x+3}{x+3}\right)+\left(\frac{7x}{x+3}\right)\left(\frac{x-4}{x-4}\right)$ $=\left(\frac{x^2+3x}{(x-4)(x+3)}\right)+\left(\frac{7x^2-28}{(x+3)(x-4)}\right)$ $=\left(\frac{x^2+3x+7x^2-28}{(x-4)(x+3)}\right)$ $=\left(\frac{8x^2+3x-28}{(x-4)(x+3)}\right)$
(c) We have to factor each denominator first in order to determine the LCD. Find the LCD $[(x+5)(x-2)(x+3)]$ and multiply each fraction by the missing factor in its denominator. Next, subtract the numerators. The numerator factors, but not into any factors which would cancel with the denominator, so we are done.	$\left(\frac{3x}{x^2+3x-10}\right)-\left(\frac{2x}{x^2+x-6}\right)$ $=\left(\frac{3x}{(x+5)(x-2)}\right)-\left(\frac{2x}{(x-2)(x+3)}\right)$ $=\left(\frac{3x}{(x+5)(x-2)}\right)\bullet\frac{(x+3)}{(x+3)}-\left(\frac{2x}{(x-2)(x+3)}\right)\bullet\frac{(x+5)}{(x+5)}$ $=\left(\frac{3x^2+9x}{(x+5)(x-2)(x+3)}\right)-\left(\frac{2x^2+10x}{(x-2)(x+3)(x+5)}\right)$ $=\left(\frac{3x^2+9x-2x^2-10x}{(x+5)(x-2)(x+3)}\right)$ $=\left(\frac{x^2-x}{(x+5)(x-2)(x+3)}\right)$

Answer Q1:

$$\left(\frac{x^2-x-6}{x^2+6x+9}\right)\left(\frac{x+3}{x^2-4}\right)$$
$$=\left(\frac{(x-3)(x+2)}{(x+3)(x+3)}\right)\left(\frac{(x+3)}{(x+2)(x-2)}\right)$$
$$=\frac{(x-3)}{(x+3)(x-2)}$$

Answer Q2:

$$\left(\frac{x^2-4}{x-2}\right)\div\left(\frac{x+2}{4x-8}\right)$$
$$=\left(\frac{(x+2)(x-2)}{(x-2)}\right)\left(\frac{4x-8}{x+2}\right)$$
$$=\left(\frac{(x+2)(x-2)}{(x-2)}\right)\bullet\left(\frac{4(x-2)}{(x+2)}\right)$$
$$=4(x-2)$$

COMPLEX FRACTIONS

It is now time to look at what are called complex fractions.

A **complex fraction** is a fraction that contains fractions with in it.

Here are a few complex fractions: $\dfrac{\frac{2}{4}}{\frac{1}{3}}$, $\dfrac{1-\frac{1}{x}}{\frac{1}{x}+1}$, $\dfrac{5}{7+\frac{x-2}{x+3}}$.

We don't want complex fractions in a final answer. To simplify a complex fraction, we look for the LCD of all the denominators in the complex fraction and multiply the main fraction's numerator and denominator by that LCD.

Discussion 1: Simplifying Complex Fractions

Let's simplify the following complex fractions.

a) $\dfrac{\frac{2}{4}}{\frac{1}{3}}$ b) $\dfrac{1-\frac{1}{x}}{\frac{1}{x}+1}$ c) $\dfrac{1+\frac{1}{x-2}}{1-\frac{1}{x+3}}$ d) $\dfrac{b^{-1}-a^{-1}}{a-b}$

(a) We have one fraction divided by another. Let's try multiplying both the numerator and denominator by the LCD, which is 12.	$\dfrac{\frac{2}{4}}{\frac{1}{3}} = \dfrac{\frac{2}{4}\bullet\frac{12}{1}}{\frac{1}{3}\bullet\frac{12}{1}} = \dfrac{6}{4} = \dfrac{3}{2}$
(b) We need to multiply both the numerator and denominator by x.	$\dfrac{1-\frac{1}{x}}{\frac{1}{x}+1} = \dfrac{\left(1-\frac{1}{x}\right)\bullet\frac{x}{1}}{\left(\frac{1}{x}+1\right)\bullet\frac{x}{1}} = \dfrac{\left(x-\frac{x}{x}\right)}{\left(\frac{x}{x}+x\right)} = \dfrac{x-1}{1+x}$
c) We need to multiply the numerator and denominator by the LCD, $(x-2)(x+3)$. We now must simplify by multiplying and combining like terms. This final result factors, but none of the factors are common, so this doesn't simplify any further.	$\dfrac{1+\frac{1}{x-2}}{1-\frac{1}{x+3}} = \dfrac{\left(1+\frac{1}{x-2}\right)\bullet\frac{(x-2)(x+3)}{1}}{\left(1-\frac{1}{x+3}\right)\bullet\frac{(x-2)(x+3)}{1}}$ $= \dfrac{((x-2)(x+3)+(x+3))}{((x-2)(x+3)-(x-2))}$ $= \dfrac{x^2+x-6+x+3}{x^2+x-6-x+2} = \dfrac{x^2+2x-3}{x^2-4}$
(d) We need to change the numerator so that it doesn't have negative exponents in it. Multiply by the LCD, ab. This time what we have simplifies further. We can cancel out the $(a-b)$ factors.	$\dfrac{b^{-1}-a^{-1}}{a-b} = \dfrac{\frac{1}{b}-\frac{1}{a}}{a-b}$ $= \dfrac{\left(\frac{1}{b}-\frac{1}{a}\right)\bullet\frac{ab}{1}}{\left(\frac{a-b}{1}\right)\bullet\frac{ab}{1}}$ $= \dfrac{(a-b)}{ab(a-b)} = \dfrac{1}{ab}$

There are other ways to approach simplifying these fractions, but with this method you quickly get to the simplified form.

Question 3: Do you know a different way, as compared to example 5b, to simplify $\dfrac{\frac{2}{4}}{\frac{1}{3}}$?

Remember, any time you have a fraction divided by another fraction you can invert the second fraction and change the problem to a multiplication problem.

Question 4: Have you ever learned another way to simplify $\dfrac{x-\frac{1}{x}}{\frac{1}{y}-y}$?

Discussion 2: An Alternate Method for Simplifying a Complex Fraction

Here is another way to simplify the complex fraction in question 4.

LCD for the numerator is x, and the LCD for the denominator is y.	$\dfrac{x-\frac{1}{x}}{\frac{1}{y}-y} = \dfrac{\frac{x^2}{x}-\frac{1}{x}}{\frac{1}{y}-\frac{y^2}{y}} = \dfrac{\frac{x^2-1}{x}}{\frac{1-y^2}{y}}$
Now we invert the fraction from the bottom and multiply.	$= \dfrac{(x^2-1)}{x} \bullet \dfrac{y}{(1-y^2)} = \dfrac{y(x^2-1)}{x(1-y^2)}$

If we did this problem by multiplying both the numerator and denominator by the LCD of all of the denominators, then the simplification would look like this:

LCD for the numerator is x, and the LCD for the denominator is y.	$\dfrac{x-\frac{1}{x}}{\frac{1}{y}-y} = \dfrac{\left(x-\frac{1}{x}\right)\bullet\frac{xy}{1}}{\left(\frac{1}{y}-y\right)\bullet\frac{xy}{1}} = \dfrac{x^2y-y}{x-xy^2} = \dfrac{y(x^2-1)}{x(1-y^2)}$

We feel that the method of multiplying everything by the LCD of all of the fractions is a quicker method, and is needed when solving equations, but for now you can use whichever method makes the most sense to you.

Section Summary

- When simplifying, multiplying or dividing (flip and change to multiplication) rational expressions (fractions) you must:
    1. Factor all of the numerators and denominators in the expression.

2. Cancel all like factors in the numerators with the ones in the denominators.

- When adding or subtracting rational expressions (fractions) you must:
  1. Factor the denominators.
  2. Find a Least Common Denominator (LCD).
  3. *Within each fraction*, multiply the numerator and denominator by the same amount to make the denominator be the same as the LCD.
  4. Add or subtract the numerators as indicated.
  5. Then try simplifying what you have just completed.
- To simplify complex fractions, you multiply each fraction within the main fraction by the LCD of all of the little fractions and then simplify what remains.

SECTION 0.4 HOMEWORK

(1–26) Reduce each of the following rational polynomials:

1. $\dfrac{5x^2y^2}{-10xy^4}$
2. $\dfrac{-24ab^4}{15a^2b}$
3. $\dfrac{36x^3y^2}{24xy^2}$
4. $\dfrac{-30x^4y^2}{-27x^2y^5}$
5. $\dfrac{5x-10}{3x-6}$
6. $\dfrac{7x+21}{5x+15}$
7. $\dfrac{10x+30}{8x+24}$
8. $\dfrac{12x-24}{10x-20}$
9. $\dfrac{8x+16}{4}$
10. $\dfrac{12-36}{4}$
11. $\dfrac{6a-18}{9-3a}$
12. $\dfrac{20-5a}{10a-40}$
13. $\dfrac{5x-10}{x^2-4}$
14. $\dfrac{x^2-9}{4x-12}$
15. $\dfrac{9-y^2}{5y-15}$
16. $\dfrac{7x-14}{4-x^2}$
17. $\dfrac{a^2-25}{a^2-4a-5}$
18. $\dfrac{b^2+3b-18}{b^2-36}$
19. $\dfrac{x^2+3x-10}{x^2+x-6}$
20. $\dfrac{x^2-6x-27}{x^2+8x+15}$
21. $\dfrac{3x^2+8x-3}{3x^2-16x+5}$
22. $\dfrac{2y^2+9y+7}{2y^2-y-28}$
23. $\dfrac{x^3-27}{x^2+2x-15}$
24. $\dfrac{a^2+4a+14}{a^3+8}$
25. $\dfrac{x^2+9x+18}{x^2+2x-15}$
26. $\dfrac{x^2+7x-18}{x^2-3x-10}$

Answer Q3: Yes, we have one fraction divided by another, so we can flip the second fraction and multiply.

$$\frac{\frac{2}{4}}{\frac{1}{3}} =$$

$$\frac{2}{4}\bullet\frac{3}{1}=\frac{3}{2}$$

(27–54) Multiply or divide each of the following rational polynomials:

27. $\dfrac{5x^2y^2}{12x^4}\bullet\dfrac{18x^8y^2}{15y^5}$
28. $\dfrac{4a^3}{9ab^2}\bullet\dfrac{15b^5}{8a^4b}$
29. $\dfrac{25a^5b^2}{16a^2b}\div\dfrac{15a^3}{8b^4}$
30. $\dfrac{21x^3y}{20xy^2}\div\dfrac{7x^4y^2}{10x^2y^2}$
31. $\dfrac{x+2}{x^2-3x-10}\bullet\dfrac{x^2-5x+6}{4x+8}$
32. $\dfrac{a-5}{a^2-36}\bullet\dfrac{a^2-5a-6}{3a-15}$
33. $\dfrac{x^2-3x-10}{x^2+2x-8}\bullet\dfrac{x^2+5x-14}{x^2+2x-35}$
34. $\dfrac{x^2-25}{x^2-2x-15}\bullet\dfrac{x^2+7x+12}{x^2-4x-45}$
35. $\dfrac{2x^2+x-6}{3x^2+7x+2}\bullet\dfrac{3x^2-14x-5}{2x^2-9x+9}$
36. $\dfrac{6x^2-5x-6}{4x^2-11x-15}\bullet\dfrac{6x^2+17x+5}{6x^2+13x+6}$
37. $\dfrac{y^2+8y+15}{y^2-10y+21}\bullet\dfrac{y^2-4y-21}{y^2+3y-10}$
38. $\dfrac{a^2-15a-56}{a^2+14a+45}\bullet\dfrac{a^2+2a-63}{a^2-3a-40}$
39. $\dfrac{x^2-4}{x^2-3x-10}\bullet\dfrac{x-5}{x-2}$
40. $\dfrac{y^3-8}{y^2-4}\bullet\dfrac{y+2}{y^2+2y+4}$

Answer Q4: Yes, you find an LCD for the numerator and denominator separately and then flip the second fraction and multiply

41. $\frac{y^3+27}{y^2-9} \bullet \frac{y^2+2y-15}{y+5}$

42. $\frac{a^3-8}{a^2-4} \bullet \frac{a^2+11a+18}{a+9}$

43. $\frac{6x-12}{x+2} \bullet \frac{x+3}{3x+6}$

44. $\frac{x-5}{2x+4} \bullet \frac{4x-8}{x+1}$

45. $\frac{a+3}{a^2-3a-10} \div \frac{4a+12}{a^2-25}$

46. $\frac{y^2-9}{y+2} \div \frac{y^2+5y+6}{5y+10}$

47. $\frac{x^2-5x-14}{x^2-49} \div \frac{x^2+5x+6}{x^2+12x+35}$

48. $\frac{x^2-4}{x^2+15x+56} \div \frac{x^2-3x-10}{x^2+3x-40}$

49. $\frac{x^2-2x-15}{x^2+8x+7} \div \frac{x^2-4x-21}{x^2+6x+5}$

50. $\frac{a^2-19a+90}{a^2-3a-88} \div \frac{a^2-8a-9}{a^2+7a-8}$

51. $\frac{2x^2+11x-21}{3x^2+23x+14} \div \frac{2x^2+5x-12}{3x^2-13x-10}$

52. $\frac{6x^2+5x-6}{9x^2-3x-2} \div \frac{2x^2-15x-27}{3x^2+16x+5}$

53. $\frac{3x^2-16x-35}{2x^2-19x+35} \div \frac{3x^2+11x+10}{2x^2-x-10}$

54. $\frac{7x^2-8x+1}{5x^2-x-4} \div \frac{7x^2+13x-2}{5x^2+14x+8}$

(55–78) Add or subtract the following rational polynomials

55. $\frac{3y}{y+3} + \frac{9}{y+3}$

56. $\frac{2x}{x-5} - \frac{10}{x-5}$

57. $\frac{x^2}{x-3} - \frac{9}{x-3}$

58. $\frac{y}{y+5} - \frac{25}{y+5}$

59. $\frac{5}{x+2} + \frac{7}{x+3}$

60. $\frac{5}{x-4} - \frac{3}{x+2}$

61. $\frac{3x}{x-1} - \frac{2x}{x+5}$

62. $\frac{x}{x-2} + \frac{2x}{x+1}$

63. $\frac{5}{3-x} + \frac{2}{x-3}$

64. $\frac{8}{y-5} - \frac{2}{5-y}$

65. $\frac{3}{x^2-9} - \frac{2}{x^2+x-6}$

66. $\frac{1}{x^2-4x+3} + \frac{2}{x^2-1}$

67. $\frac{x}{x^2-9} + \frac{2}{5x-15}$

68. $\frac{3x}{x^2-4} - \frac{4}{3x+6}$

69. $\frac{5x}{x^2-x-2} - \frac{3x}{x^2+3x+2}$

70. $\frac{x}{x^2-3x} + \frac{2x}{x^2-2x-3}$

71. $\frac{3}{x^2-25} + \frac{x}{x+5}$

72. $\frac{y}{y-6} - \frac{5}{y^2-36}$

73. $\frac{a}{a^2-6x+9} - \frac{4}{a^2-5a+6}$

74. $\frac{5}{b^2-16} + \frac{3b}{b^2-8b+16}$

75. $\dfrac{5}{x^2-25}+\dfrac{3}{5-x}$

76. $\dfrac{3}{x^2-1}-\dfrac{5}{1-x}$

77. $\dfrac{5}{x^2-x-6}+\dfrac{2}{x^2-7x+12}+\dfrac{3}{x^2-2x-8}$

78. $\dfrac{2}{x^2+2x-3}-\dfrac{4}{x^2+5x+6}+\dfrac{3}{x^2+x-2}$

(79–92) Change each complex fraction to a simple fraction.

79. $\dfrac{\frac{1}{2}+\frac{2}{3}}{\frac{3}{4}+\frac{5}{6}}$

80. $\dfrac{\frac{2}{3}-\frac{1}{9}}{\frac{5}{9}+\frac{1}{3}}$

81. $\dfrac{\frac{3}{x}+2}{5+\frac{4}{x}}$

82. $\dfrac{5+\frac{3}{a}}{7-\frac{2}{a}}$

83. $\dfrac{\frac{2}{x-2}+3}{2-\frac{1}{x-2}}$

84. $\dfrac{2-\frac{3}{x-3}}{5+\frac{2}{x-3}}$

85. $\dfrac{a^{-1}+3}{b^{-1}+5}$

86. $\dfrac{5-x^{-1}}{y^{-1}+3}$

87. $\dfrac{\frac{3}{a^2}+\frac{2}{b}}{\frac{5}{a}-\frac{3}{b^2}}$

88. $\dfrac{\frac{5}{xy^2}-\frac{3}{x^2y}}{\frac{2}{xy}+\frac{3}{x^2}}$

89. $\dfrac{\frac{5}{3-x}+2}{5+\frac{3}{x-3}}$

90. $\dfrac{\frac{2}{b-3}-3}{\frac{5}{3-b}+2}$

91. $\dfrac{\frac{9}{x^2-9}}{\frac{3}{x^2-5x+6}}$

92. $\dfrac{\frac{y^3}{y^2-4}}{\frac{y^2}{y^2-3y+2}}$

Section 0.5 Solving Linear and Absolute Value Equations

So far in this chapter, we have reviewed exponents and radicals, operations on and factoring of polynomials, and how to simplify rational expressions. Now we will review how to solve linear equations and absolute value equations.

THE BIG PICTURE

In mathematics it is important to see the bigger picture as you learn new aspects of the subject. There is a list of steps for solving linear equations but first, let's talk about the bigger picture. Our goal in solving an equation is to find what values of the variable will make the equation true. In order to determine what those values might be, we need to isolate the variable (i.e. get the variable all alone by itself on one side of the equals sign). So, here is the first big picture concept.

Concept 1: Get the variable alone.

This first concept is true no matter what type of equation it is that you are trying to solve.

Concept 2: Whatever is done to one side of the equation must also be done to the other side.

A linear equation is like having a teeter-totter that is perfectly balanced [picture of several kids on a teeter totter that is horizontal], or if you prefer, a scale that is perfectly balanced [picture of scale with weights that are balanced]. Whenever you add to or take away from one side of the teeter-totter or scale, you must do the same thing to the other side in order to keep it in balance. Again, this is true no matter what kind of equation you are dealing with. Here is a partial list of examples of the kinds of operations you will need to do to both sides of equations:

- If you want to add 5 to one side, then you must add 5 to the other
- If you want to subtract 9 from one side, then you must subtract 9 from the other.
- If you want to multiply one side by -3, then you must multiply the other side by -3.
- If you want to square one side of the equation, then you must square the other side too.
- If you want to take the log of one side of the equation, then you must take the log of the other side also.

There are an infinite number of things you might want to do to both sides of an equation. The big question is, "How do I determine *what it is* that I what to do to both sides of the equation?" The answer to this question is really somewhat simple, but it is going to take a little discussion before we will get to it. You need to get the variable alone (concept #1) which means that you need to get all the variable terms together on one side of the equal sign, and everything else on the other side of the equal sign. Also, you should realize that ultimately you want just one variable to the first power to truly get the variable alone. Here is a question to illustrate our point.

Question 1: If I have some money in my pocket and you were to give me \$5, and then I tell you that I now have \$37 dollars total, could you figure out how much money was originally in my pocket?

You should have realized that if I now have \$37, and you had given me \$5, then I must have had \$32 in my pocket, since $32 + 5 = 37$. But notice how you arrived at the 32. You had to take 37, subtract 5, and get 32. Subtraction is the opposite operation of addition. If we set this up as a linear equation with the variable x being the amount of money in my pocket originally, we get the following:

Amount of money in my pocket	x
You give me 5 dollars.	$x + 5$
I tell you that I now have 37.	$x + 5 = 37$
You figure out how much was in my pocket by getting the variable alone. Most likely you did all of this calculating in your head.	$x + 5 = 37$ $\underline{\quad -5 \quad -5}$ $x \qquad = 32$

This brings us to the big picture concept #3.

Concept 3: Do the opposite to move or eliminate something from one side of the equation.

So we have three big concepts:

1. Get the variable alone.
2. Whatever is done to one side of the equation must also be done to the other side.
3. Do the opposite operation to move or eliminate something from one side of the equation.

Let's do several examples and see how these concepts work in practice.

Example 1: Solving Linear Equations Using the Big Concepts

Solve each of the following linear equations.

a) $x - 7 = 2$ b) $3x = 24$ c) $\dfrac{x}{5} = -1$ d) $2x + 11 = 5$ e) $4x + 7 = 9x - 13$

Solutions:

(a) We need to do the opposite of subtract 7 (Concept 3), which is to add 7, and we need to do it to both sides (concept #2).	$x - 7 = 2$ $\underline{\ +7 \quad +7}$ $x + 0 = 9$ $x = 9$
(b) We need to do the opposite of multiplying by 3 (Concept 3), which is to divide by 3, and we need to do it to both sides (Concept 2).	$\dfrac{3x}{3} = \dfrac{24}{3}$ $1x = 8$ $x = 8$

(c) We need to do the opposite of dividing by 5 (Concept 3), which is to multiply by 5, and we need to do it to both sides (Concept 2).	$5\left(\frac{x}{5}\right)=5(-1)$ $x \quad =-5$
(d) We need to do two things. We need to first subtract 11 (the opposite of adding 11). Then we need to divide by 2 (the opposite of multiplying by 2).	$2x+11 = 5$ $\underline{-11 \quad -11}$ $2x = -6$ $\frac{2x}{2} = \frac{-6}{2}$ $x = -3$
(e) We must first get the variables on the same side by subtracting $4x$ from both sides. Then we need to add 13 and then divide by 5.	$4x+7=9x-13$ $7=5x-13$ $20=5x$ $4=x$

We want to point out that in part (a), when we added 7 to both sides and solved the equation, we used the properties of real numbers called additive inverse and identity. 7 is the additive inverse of -7 and thus, they add to 0, the identity for addition. Then x plus 0, the identity, equals just x, so we end up with x equals 9 as our answer.

In part (b), we divided by 3 or better yet, we viewed the necessary step as multiplying by $\frac{1}{3}$, the multiplicative inverse of 3, so that we end up with 1 times x, which is just x by the identity rule for multiplication. But all you really need to remember is "do the opposite to move or eliminate something."

In part (e), notice that we could have started out by subtracting $9x$ from both sides instead of subtracting $4x$ from both sides of the equation. The reason we choose to subtract $4x$ instead of $9x$ from both sides of the equation is that it would have left us with a $-5x$ instead of a positive $5x$ in the equation. Consider this suggestion (which could save you some grief in the future when you are trying to get the variable terms together): always move the term with the smallest coefficient (4 is smaller than 9 in part (e)). This way the coefficient of the variable will always end up being positive. We believe that most people are less likely to make arithmetic mistakes if they can avoid negative numbers as much as possible.

STEPS FOR SOLVING LINEAR EQUATIONS

Here is a list of the steps that you can use to solve any linear equation. These steps are natural outcomes from the three big concepts we have just discussed. (Note: you may not need to use each step every time.)

1. Clear all parentheses (Distribute, FOIL, etc).
2. Eliminate any denominators or decimals (optional).
3. Simplify what remains by collecting like terms.
4. Add or subtract to get the variable terms on the same side of the equation.
5. Add or subtract to get the constant terms on the other side of the equation from the variable.
6. Multiply or divide to get one variable equal to the answer.

Answer Q1: Yes, $37 minus the $5 I gave you would be the amount originally in my pocket, $32.

Example 2: Solving Linear Equations

Solve the following linear equations.

a) $2(x-3)-5=4(x+5)$ b) $\frac{2}{3}x-5=\frac{1}{2}x-3$ c) $\frac{-5}{7}(p+11)+\frac{2}{5}(2p-5)=0$

Solutions:

In part (a), we need to follow steps 1, 3, 4, 5, and 6 for solving linear equations.	
Step 1: Clear the two parentheses.	$2x-6-5=4x+20$
Step 3: Collect the –6 and –5 (likes).	$2x-11=4x+20$
Step 4: Subtract the smaller coefficient variable term ($2x$).	$-11=2x+20$
Step 5: Subtract 20 to get numbers on the opposite side from the variable.	$-31=2x$
Step 6: Divide by 2 to get $+1x$ = answer.	$\frac{-31}{2}=x$
In part (b), we need to follow steps 2, 4, and 5 for solving linear equations.	
Step 2: Eliminate the denominators by multiplying by the LCD (6).	$\frac{6}{1}\left(\frac{2}{3}x\right)-\frac{6}{1}(5)=\frac{6}{1}\left(\frac{1}{2}x\right)-\frac{6}{1}(3)$ $4x-30=3x-18$
Step 4: Subtract the smaller coefficient variable term ($3x$).	$x-30=-18$
Step 5: Add 30 to get numbers on the opposite side from the variable.	$x=12$
In part (c), we need to follow steps 1, 2, 3, 5, and 6 for solving linear equations.	
Step 1: Clear the two parentheses.	$\frac{-5}{7}p+\frac{-55}{7}+\frac{4}{5}p-2=0$
Step 2: Eliminate the denominators by multiplying by the LCD (35).	$\frac{35}{1}\left(\frac{-5}{7}p\right)+\frac{35}{1}\left(\frac{-55}{7}\right)+\frac{35}{1}\left(\frac{4}{5}p\right)-\frac{35}{1}(2)=\frac{35}{1}(0)$

	$-25p - 275 + 28p - 70 = 0$
Step 3: Collect the like terms.	$3p - 345 = 0$
Step 5: Add 345 to get numbers on the opposite side from the variable.	$3p = 345$
Step 6: Divide by 3 to get $+1p$ = answer.	$p = 115$

In the process of following these 6 steps, we are also following our three big concepts. We need to clear parentheses in order to put the variable terms together so that we can eventually get the variable alone. We do the opposite operations in order to isolate the variable and we are doing the same thing to both sides of the equation in order to keep the equation balanced.

Question 2: Solve this linear equation $2 - (7x + 5) = 13 - 3x$.

The concept of moving terms around from one side of the equation to the other by doing the opposite operation is so very important in mathematics. You must move terms around in many different areas of study in mathematics. Make sure that while you do the homework in this section you think about our three big picture concepts.

There is one last piece of the puzzle: Not all equations have a solution or only one solution! Let's look at the following examples.

Example 3: Equations without a Solution or with Many Solutions

Solve the following equations.

a) $3(x + 2) - 5 = 4 + 3x$ b) $8x - 13 = 2(6 + 4x) - 25$

Solutions:

In part (a), as we go through the steps to solve this equation, we will find that there is a problem. When in the course of solving an equation you find yourself with a statement that is false, then the answer to the equation is that there is NO SOLUTION. We call this a <u>contradiction</u>.

Step 1: Clear the parentheses.	$3x + 6 - 5 = 4 + 3x$
Step 3: Collect the 6 and -5 (likes).	$3x + 1 = 4 + 3x$
Step 4: Get the variable terms together by subtracting $3x$ from both sides.	$1 = 4$
As you can see, we end up with $1 = 4$, a false statement. This means that there doesn't exist any possible value for x that could make the original equation true.	Final Answer is: No solution!

<table>
<tr><td colspan="2">In part (b), we see that a slightly different event will unfold. Here, when we go through the steps to solve the equation, we will find that we get a true statement without any variables left in it. This means that any value for x will make the equation true. We say that this type of equation is true for an infinite number of values. We call this an <u>identity</u>.</td></tr>
<tr><td>Step 1: Clear the parentheses.
Step 3: Collect the 12 and −25 (likes).
Step 4: Get the variable terms together by subtracting 8x from both sides.
As you can see, we end up with a true statement, regardless of the value of x. This means that there are an infinite number of values for x that could make the original equation true.</td><td>$8x - 13 = 12 + 8x - 25$
$8x - 13 = 8x - 13$

$-13 = -13$

We say that this has an infinite number of solutions (usually all real numbers).</td></tr>
</table>

SOLVING ABSOLUTE VALUE EQUATIONS

Let's turn our attention now to solving absolute value equations. We are going to concentrate on solving *linear* absolute value equations, but the approach used to solve these is the same for *any* absolute value equation. Here is the formal definition of absolute value.

> The **absolute value** of any real number *x*, is defined as
>
> $$|x| = \begin{cases} -x & x < 0 \\ x & x \geq 0 \end{cases}$$

Here are a few simple examples to illustrate the definition in use.

- $|6| = 6$, since $6 \geq 0$. $|x| = x$ when x is positive or zero.
- $|-3| = -(-3) = 3$, since $-3 \leq 0$. $|x| = -x$ when x is negative. A double negative makes a positive.
- $\left|\frac{-5}{7}\right| = -\left(\frac{-5}{7}\right) = \frac{5}{7}$, since $\left(\frac{-5}{7}\right) \leq 0$. $|x| = -x$ when x is negative.
- $-|13| = -(13) = -13$. The $|13| = 13$ but we are multiplying $|13|$ by a negative one, so the final answer turns out to be negative.

Question 3: Find $-|-2|$.

You should notice that, informally, when we take the absolute value of some number, the result will be positive. Of course, if we have other things going on around the absolute value signs, like multiplying by a negative, we can get a negative answer by the time we reach the end of the whole problem.

This is a good time to bring up a detail from section 0.1 involving square roots. We mentioned there that $\sqrt{49} = 7$ and not -7, even though we know that $(-7)^2$ also equals 49. We mentioned that the symbol $\sqrt{\ }$ means, "find the principle square root" (positive square root only). What if instead of just taking the square root of a number, we are taking the square root of a variable expression?

Answer Q2: $2 - 7x - 5 = 13 - 3x \Rightarrow$ $-7x - 3 = 13 - 3x \Rightarrow$ $-3 = 13 + 4x \Rightarrow$ $-16 = 4x \Rightarrow$ $-4 = x$

First of all, we wouldn't know whether it is positive or negative, and second, we wouldn't know whether it is a perfect square or not. The second problem is easy; if we don't know whether it is a perfect square or not, then we don't take the square root of it! As for the first problem, let's do an example.

If we had $\sqrt{x^2}$ we might be tempted to say that the answer is x, but that is incorrect. For example, what if the original value for x is -3? Then we would be saying that $\sqrt{(-3)^2} = \sqrt{9} = -3$, which is an impossibility since $\sqrt{\ }$ requires positive answers only. Thus it must be true that the $\sqrt{x^2} = |x|$. Now, no matter what the value for x is, the answer will be positive as required. We do this in order to avoid contradicting our definition that $\sqrt{\ }$ always means the principle square root. You might notice that if we had $\sqrt{x^4}$ we wouldn't need the absolute value symbol, since the answer x^2 is always positive anyway. It is time to return to solving linear absolute value equations.

Example 4: Solving a Linear Absolute Value Equation

Solve the equation $|y+7| = 5$.

Solution:

We are going to look at this problem two ways: by using the definition, and by applying a little logic.

First, we will use the definition which says $|x| = \begin{cases} -x & x<0 \\ x & x \geq 0 \end{cases}$, so either	$-(\underline{y+7}) = 5$ or $\underline{y+7} = 5$ We don't know if $(y + 7)$ is positive or negative, so we need to look at both possibilities.
We now solve both equations and arrive at two different values for x, which will make this equation true.	$-y - 7 = 5$ or $y + 7 = 5$ $-y = 12$ $\quad y = -2$ $y = -12$

Let's do this problem a second way where we apply a little logic. If the $|y+7|$ equals some number, then the $(y + 7)$ must be either that number or its negative. That is, if the answer to $|y+7|$ is supposed to be 5, then either (y + 7) is 5 or it is –5, since $|5| = 5 = |-5|$. There are two different numbers, for which the absolute value is equal to 5. Here is how we set up this problem if we were to apply some logic to the question.

Alternate Approach:

$\lvert y+7\rvert = 5$ so,	$y+7=5$ or $y+7=-5$
Now, solve each equation.	$y=-2$ or $y=-12$

This way is typically shorter than using the definition. We do want to caution you, though. There are instances where you need to know how to apply the definition of absolute value. We suggest you try solving problems both ways so as to gain the most understanding of this topic possible.

Answer Q3: $|-2| = 2$, *but we are multiplying that by –1, so the answer is –2* $(-|-2| = -2)$.

Example 5: Solving Linear Absolute Value Equations

Solve the following absolute value equations.

a) $|2x-5| = 11$ b) $\left|\frac{5x+3}{2}\right| = 6$ c) $|3x-7| = -2$

Solutions:

(a) Let's use the definition of absolute value here.	$-(2x-5)=11$ or $2x-5=11$ $-2x+5=11$ or $2x-5=11$ $-2x=6$ or $2x=16$ $x=-3$ or $x=8$
(b) Let's use the logical approach this time. The $\frac{5x+3}{2}$ must either equal 6 or –6, since it is inside of absolute value signs.	$\frac{5x+3}{2}=6$ or $\frac{5x+3}{2}=-6$ $5x+3=12$ or $5x+3=-12$ $5x=9$ or $5x=-15$ $x=\frac{9}{5}$ or $x=-3$
(c) We should notice that we have an absolute value equal to a negative number. That isn't possible. Thus the solution to this problem must be that we have **No Solution**. Let's work through it anyway and see what happens. Notice that we got two possible solutions, but when we go to check them, we get an answer of 2, not –2.	$3x-7=-2$ or $3x-7=2$ $3x=5$ or $3x=9$ $x=\frac{5}{3}$ or $x=3$ Check answers: $\lvert 3(3)-7\rvert = \lvert 9-7\rvert = \lvert 2\rvert = 2.$ $\left\lvert 3\left(\frac{5}{3}\right)-7\right\rvert = \lvert 5-7\rvert = \lvert -2\rvert = 2.$

Part (c) is a good example to remind us that we need to check and see if a problem is even possible before we begin to try to solve it.

There is one more vital piece to solving these types of equations. Our logical approach, as well as the definition approach, requires that the *absolute value part of the equation be isolated first* before we can begin to solve.

Example 6: Solving an Absolute Value Equation

Solve the equation $|5x+9|-27=-8$.

Solution:

Your first thought might be that this isn't possible because of the –8. But once you isolate the absolute value, you will see that it is possible because you have an absolute value equal to a positive number.

First, we need to add 27 to both sides of the equation in order to isolate the absolute value.	$\|5x+9\| = 19$
Now, we can separate this into its two equations and solve them.	$5x+9=19$ or $5x+9=-19$ $5x=10$ or $5x=-28$ $x=2$ or $x=\frac{-28}{5}$

Section Summary

- We have three big concepts we need to know when we are solving linear equations:
    1. Get the variable alone.
    2. Whatever is done to one side of the equation must also be done to the other side.
    3. Do the opposite operation to move or eliminate something from one side of the equation.
- The three big concepts lead us to these 6 universal steps used to solve linear equations.
    1. Clear all parentheses (Distribute, FOIL, etc).
    2. Eliminate any denominators or decimals (optional).
    3. Simplify what remains by collecting like terms.
    4. Add or subtract to get the variable terms on the same side of the equation.
    5. Add or subtract to get the constant terms on the other side of the equation from the variable.
    6. Multiply or divide to get one variable equals the answer.
- When we solve linear equations we can get results ranging from no solution, to one solution, to an infinite number of solutions.

- The definition of absolute value is $|x| = \begin{cases} -x & x < 0 \\ x & x \geq 0 \end{cases}$.
- Remember to isolate the absolute value before you apply either the definition or logic in order to solve the equation.

SECTION 0.5 HOMEWORK

(1–38) Solve each of the following linear equations.

1. $2x - 3 = 23$
2. $3x + 7 = 25$
3. $5x + 8 = 2x - 7$
4. $3x - 21 = 7x + 15$
5. $3(5x + 3) - 4 = -25$
6. $5(2 - 3x) + 8 = 3$
7. $3x + 2 = 5x - 3$
8. $7x - 3 = 3x + 6$
9. $7x + 8 = 5x + 9 + 2x$
10. $3x - 4 + x = 5x + 7 - x$
11. $7x + 9 - 3x = 4x + 9$
12. $3 - 2x + 7x = 5x - 4 + 7$
13. $7(3 - 2x) + 8 = 4(3x + 5) - 10$
14. $-4(3x - 7) + 2 = 5(x + 4) - 9$
15. $3(2 - 3x) + 5x = 8 - 2x$
16. $-2(3x + 4) - 2x = 5 - 2(3x - 1)$
17. $2(3x + 2) - 8 = 6x + 5$
18. $2(3 - 2x) + 5 = 2(x + 5) - 6x$
19. $5(x + 4) + 3x = 4(2x + 3) + 8$
20. $3(5 - 2x) + 3 = 3(x + 6) - 9x$
21. $\frac{3}{4}x - 5 = \frac{2}{3}x + \frac{1}{4}$
22. $\frac{3}{5}x + \frac{7}{10} = x - \frac{4}{5}$
23. $\frac{1}{2}x + 8 = 3x - \frac{3}{4}$
24. $3x - \frac{5}{6} = \frac{1}{4}x - 5$
25. $\frac{3x-4}{8} = \frac{2x+3}{12}$
26. $\frac{x+5}{4} = \frac{3x-5}{6}$
27. $\frac{3x+2}{5} - 3 = \frac{2x+1}{10}$
28. $\frac{5x-1}{9} + 5 = \frac{2x+3}{3}$
29. $\frac{2x+3}{2} - \frac{3x}{4} = \frac{x+5}{4}$
30. $\frac{2x+5}{2} + 4 = \frac{3x+2}{3}$
31. $\frac{x+3}{5} = \frac{2x-4}{10} + 1$
32. $\frac{4x-3}{4} = \frac{5x-3}{5} - \frac{3}{20}$
33. $\frac{2}{3}(3x-4) + 2 = 5$
34. $\frac{-3}{5}(5-2x) - 1 = 7$
35. $\frac{-5}{6}(2x+3) = \frac{3}{4}(3x-2)$
36. $\frac{1}{5}(3-2x) = \frac{3}{10}(3x+4)$
37. $\frac{1}{3}(3x+1) + 2 = \frac{2}{5}(1-2x) - 4$
38. $\frac{3}{4}(x-5) - 3 = \frac{1}{6}(3x+1) + 1$

(39–64) Solve each of the following absolute value equations.

39. $|3 - x| = 11$
40. $|y + 8| = 2$
41. $|9 - x| = -4$
42. $|y + 9| = -3$
43. $|2x - 1| = 7$
44. $|3 - 2x| = 9$
45. $|3 - 7x| = 11$
46. $|5x + 7| = 8$
47. $|3x + 9| + 2 = 8$
48. $|3 - 5x| + 3 = 5$
49. $|9 - 2x| + 7 = 3$
50. $|7x + 4| + 5 = 1$
51. $2|3x - 1| - 3 = 5$
52. $3|5 - 2x| + 1 = 10$
53. $|7 - 3x| + 5 = 5$
54. $|3x + 11| + 4 = 4$
55. $\left|\frac{2}{3}x + 5\right| = 7$
56. $\left|\frac{3}{4}x - 2\right| = 10$
57. $\left|\frac{3x-1}{5}\right| = 4$
58. $\left|\frac{5-2x}{7}\right| = 3$
59. $\left|\frac{4}{5}x - \frac{3}{10}\right| + 3 = 7$
60. $\left|\frac{5}{4} - \frac{3}{8}x\right| = 4$
61. $\left|\frac{5}{16} - \frac{3}{8}x\right| + 7 = 3$
62. $\left|\frac{5}{9}x + \frac{7}{18}\right| + 4 = 1$

63. $-2\left|\frac{3}{5}x+2\right|+5=1$

64. $-3\left|\frac{1}{3}x-3\right|+8=1$

Section 0.6 Graphing Calculator Basics

In the next three sections we are going to discuss many of the graphing calculator techniques you will need to know as we move through this text. Our directions will be specifically geared towards the TI-83 and TI-86. Though many of the techniques we will show you apply to other graphing calculators as well. We believe that by reading your graphing calculator manual along with the following sections, you will be able to effectively use your graphing calculator with this text, no matter which calculator you are using. Let's begin our discussion on how to use this tool, the graphing calculator, to explore College Algebra.

Topic	TI-83	TI-86
Calculator Keys Notice that the ON key is on the lower left corner of the keypad. To turn the calculator **off**, push the 2nd key (upper left corner) followed by the ON key.	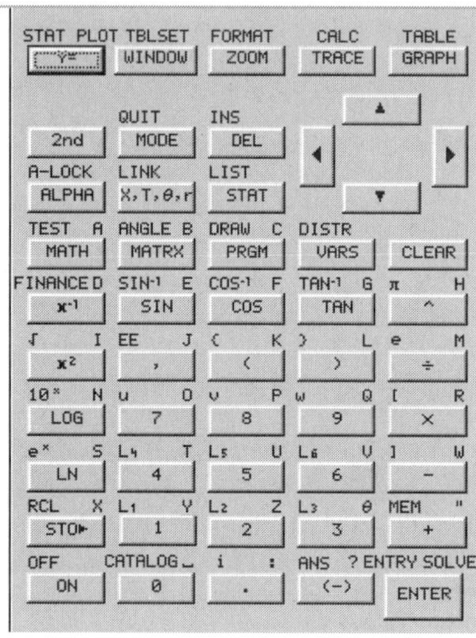	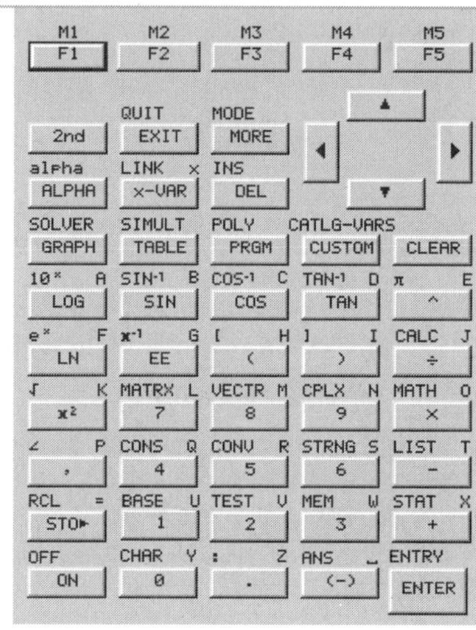
Brightness of screen (contrast) MODE key is on the top left part of the keypad along with the 2nd key.	You will need something on the screen in order to see how bright or dim the screen is, so we suggest you push the MODE key first. Now, to adjust how bright or dim your view screen is, you simply push the 2nd key, then hold down either the up arrow to make it brighter or the down arrow to make it dimmer.	The mode command on the TI-86 requires that you push the 2nd key, then the MORE key, which is located on the top center part of the keypad. Adjusting the contrast is the same as with the TI-83.

<table>
<tr><td colspan="3">While you are adjusting the contrast (brightness) of your screen, you will notice numbers that appear in the upper right corner of the screen. If those numbers, at the point where you think that the screen is at the proper brightness, are high, like 8 or 9, then this means that your batteries are very weak and that you will need to replace them soon.</td></tr>
<tr><td>**MODE**

Read the manual or ask your instructor if you want to know more about each of the items in the MODE menu.</td><td>While you are in the MODE menu, make sure that all of the left side items are highlighted. If something isn't highlighted, then arrow down to it and push the ENTER key on the lower right corner of the keypad.
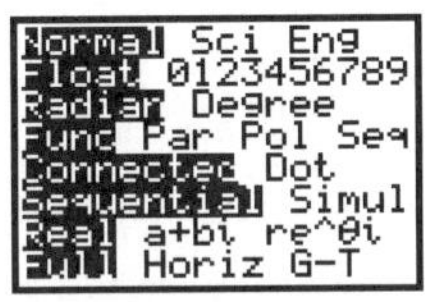</td><td>While you are in the MODE menu, make sure that all of the left side items are highlighted. If something isn't highlighted, then arrow down to it and push the ENTER key on the lower right corner of the keypad.
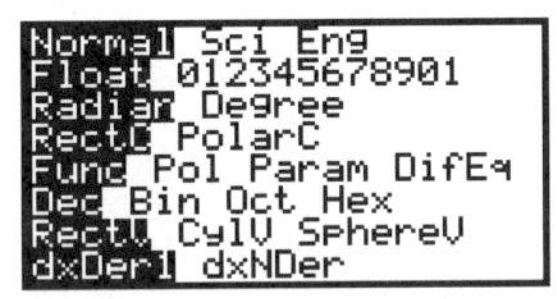</td></tr>
<tr><td>**Home** Screen

The QUIT command is on the top left side of the keypad above the MODE key on the TI-83 and the EXIT key on the TI-86.</td><td>This is the screen we need to be on in order to do operations with numbers, etc. To get to the home screen, you simply need to push the 2nd key followed by the MODE key (Quit). Pushing the 2nd key first makes the calculator execute the yellow operations above the black keys.</td><td>On the TI-86 you need to push the EXIT key, possibly several times, to get to the home screen. The EXIT key is on the upper left side of the keypad. You can also get to the home screen by using the QUIT command, which you get to by pushing the 2nd key followed by the EXIT key (quit).</td></tr>
<tr><td>Performing **operations** Numbers are in the center; operators are on the right side in blue.</td><td>To perform basic operations like add, subtract, multiply and divide, you simply type in your problem exactly the way you would right it on a piece of paper.</td><td>You do the same as with the TI-83</td></tr>
</table>

<table>
<tr>
<td rowspan="2">To get the $\sqrt{\ }$ you must push the [2nd] key followed by the [x^2] key. The [x^2] key is on the center left side of the keypad</td>
<td>EX: 6(5 – 7)/4 + (8 – 3)5
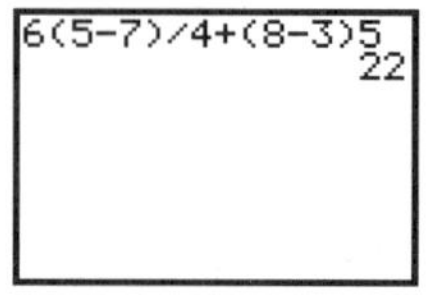

Notice that what you type in starts on the left side of the screen and what you get for answers starts on the right side of the screen.</td>
<td>EX: 6(5 – 7)/4 + (8 – 3)5
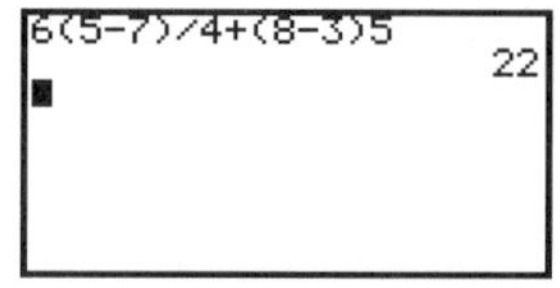

Much of what is done on the TI-83 is the same on the TI-86.</td>
</tr>
<tr>
<td>EX: 500* $\sqrt{200}$
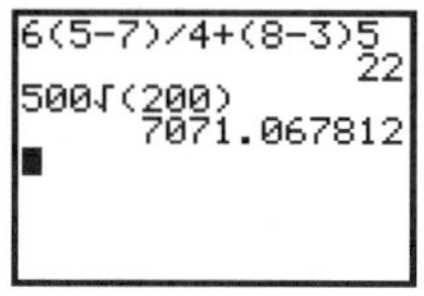

Notice the parentheses; they are very important since the calculator can't read your mind about what it should do when.</td>
<td>EX: 500* $\sqrt{200}$
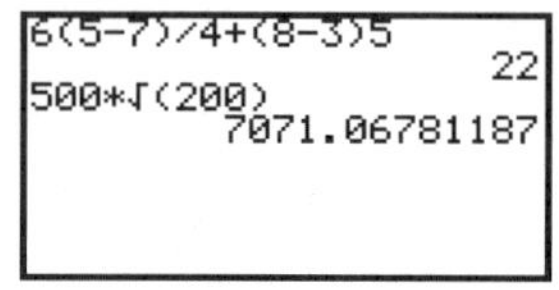

Notice that we put parentheses in the square root. This is a good habit to get into. Also notice that we have the multiplication sign between the two numbers. We did this just to show that it can be done both ways, with and without the sign.</td>
</tr>
<tr>
<td colspan="3">Let's take a moment here to remind you of the order of operations.
First perform operations within the innermost parentheses and work your way outwards. If the expression has division in it, treat the numerator and the denominators as if they had parentheses around them. Within the parentheses and after all parentheses have been eliminated, you perform operations in the following order:
• Simplify exponent expressions.
• Next, perform multiplication and division working from the left to the right.
• Last, perform addition and subtraction working from the left to the right.</td>
</tr>
<tr>
<td>This is an example of a major mistake that many make with the graphing calculators.</td>
<td>EX: $\frac{5+7}{6-3}$
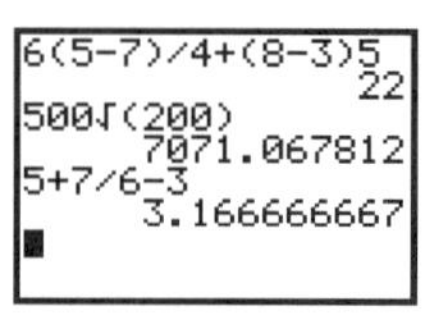
</td>
<td>EX: $\frac{5+7}{6-3}$
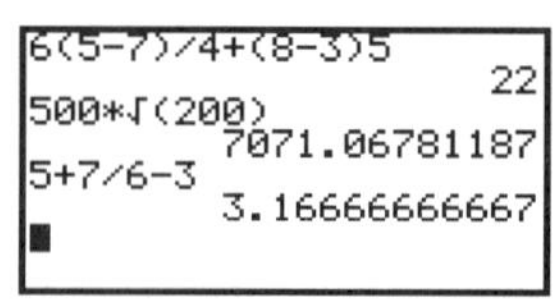
</td>
</tr>
</table>

Question 1: Did both calculators give us the correct answer?

<table>
<tr>
<td>It is very important for you to remember this rule about graphing calculators!</td>
<td>When you type in a fraction, be sure to put parentheses around the numerators and denominators if they have more than one term in them.

$\frac{(5+7)}{(6-3)}$

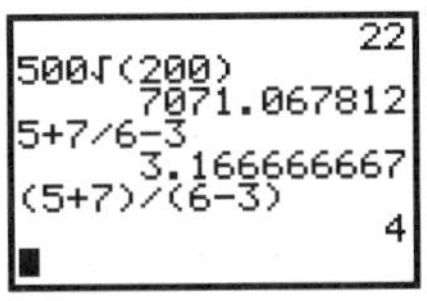
</td>
<td>When you type in a fraction, be sure to put parentheses around the numerators and denominators if they have more than one term in them.

$\frac{(5+7)}{(6-3)}$

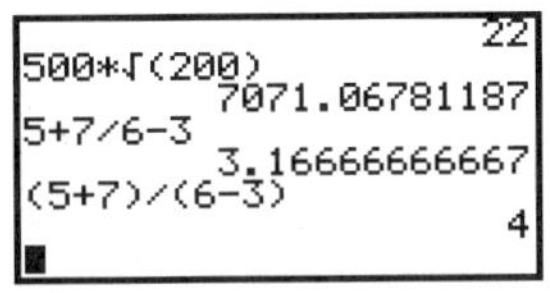
</td>
</tr>
</table>

<table>
<tr>
<td>Another mistake you might make is to confuse the minus sign key [−] with the negative sign key [(−)].</td>
<td>On the TI-83 the minus sign is on a blue key and the negative sign is on a gray key.
And the negative sign has parentheses around it.

Minus sign [−]

Negative sign [(−)]</td>
<td>On the TI-86 the minus sign is on a black key and the negative sign is on a gray key.
And the negative sign has parentheses around it.

Minus sign [−]

Negative sign [(−)]</td>
</tr>
<tr>
<td>Examples of how these two signs get misused.</td>
<td>EX: 4 − 5
Incorrectly using the negative sign instead of the minus sign will produce this:

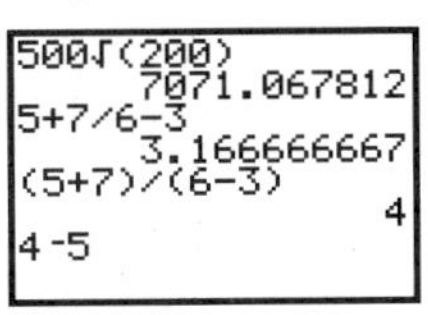

</td>
<td>EX: 4 − 5
Incorrectly using the negative sign instead of the minus sign will produce this:

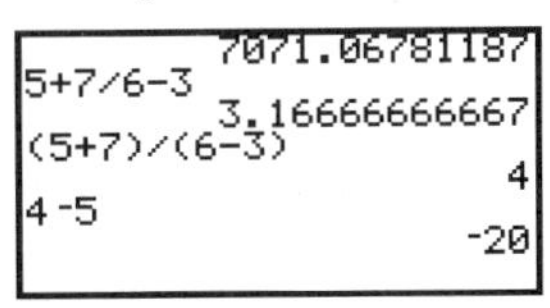

On the TI-86 when it doesn't see an operation between numbers, it assumes that you wanted to multiply so you get the answer of −20.</td>
</tr>
</table>

<table>
<tr>
<td></td>
<td>The calculator doesn't understand what you want because you don't have an operator between 4 and –5.
The correct use of the minus sign yields
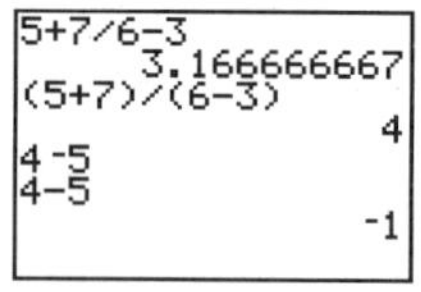
</td>
<td>The correct use of the minus sign yields
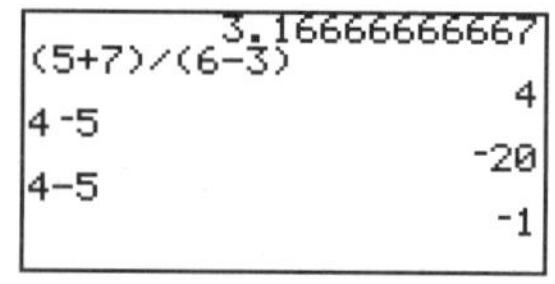
</td>
</tr>
<tr>
<td></td>
<td>EX: –3 + 7
The incorrect way to enter this problem into your calculators
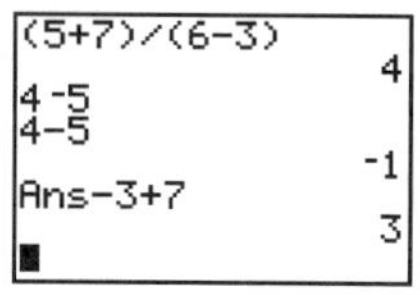

The TI-83 did something. Can you figure out what that was? It assumed that since you pushed the minus sign, you wanted to subtract 3 from the previous answer in the calculator of –1 and then add 7.
The correct way to input this is
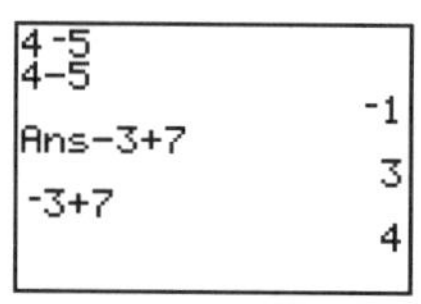
</td>
<td>EX: –3 + 7
The incorrect way to enter this problem into your calculators
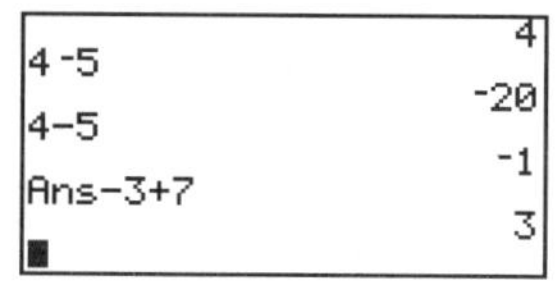

The TI-86 did the exact same thing that the TI-83 did.
The correct way to input this is
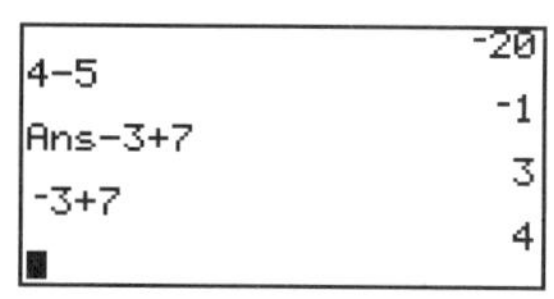
</td>
</tr>
</table>

<table>
<tr>
<td>The importance of the ^ key

"The ^ key, raise to the power key, is located on the upper half of the</td>
<td>EX: 5^4
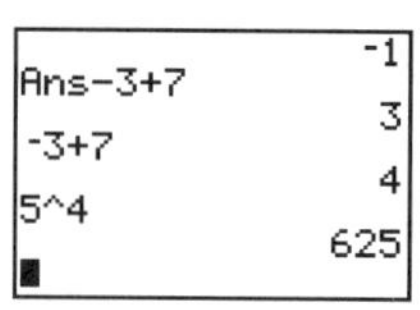

EX: $125^{\frac{2}{3}}$
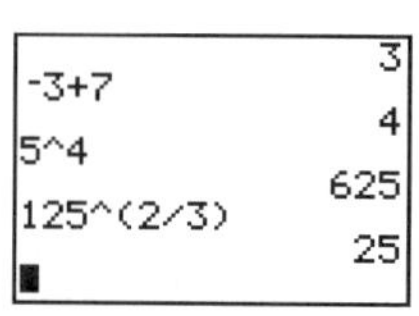

Notice the parentheses on the</td>
<td>EX: 5^4
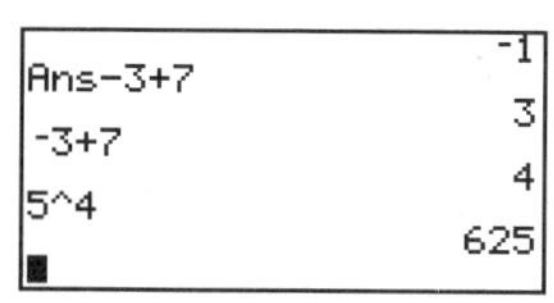

EX: $125^{\frac{2}{3}}$
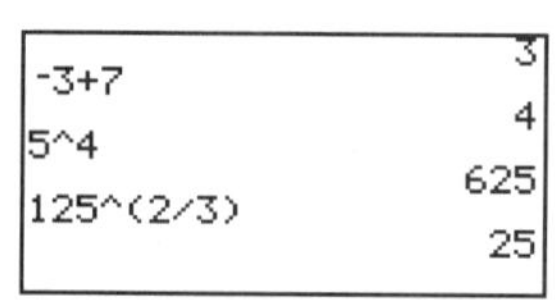

Notice the parentheses on the</td>
</tr>
</table>

<table>
<tr><td>right side of the calculator"</td><td>exponent. Here is what happens if you forget them.
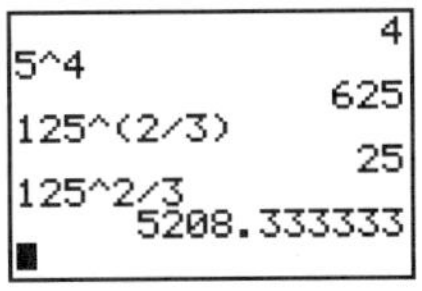
Did you see what the calculator did? It squared 125 and then divided by 3. Once again, this is not what we wanted. You should be getting the picture that the use of parentheses is vital with a graphing calculator!</td><td>exponent. Here is what happens if you forget them on a TI-86.
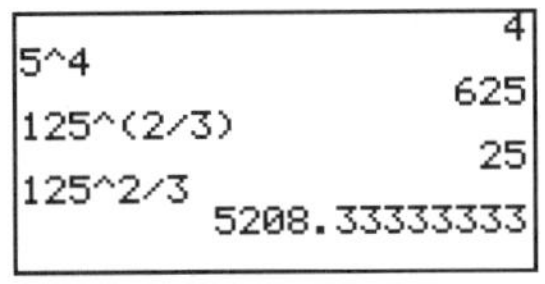
Notice that it acts just like the TI-83. The use of parentheses is vital for ALL graphing calculators.</td></tr>
<tr><td colspan="3">There are special keys and menu choices, in addition to using the ^ key, that can assist you in raising to a power or evaluating roots.
• If you want to just square something, you only need to use the x^2 key on the left side of the calculator.
• On the TI-83 there are other choices under the MATH menu; use the button on the left side of the calculator labeled MATH. There you can cube things and take the cubed root them, and n^{th} root them.
• On the TI-86, if you need to take a root of something you go to the MATH menu by pushing the 2nd key followed by the × (times) key and then press the F5 key on the upper right corner above the arrows. Then push the MORE key and then the F4 key will give you the root symbol.
We will show a few examples here.</td></tr>
<tr><td>EX: 7^3</td><td>On the TI-83 type 7, then push the MATH key, then the number 3 to select cubing, and then push ENTER.
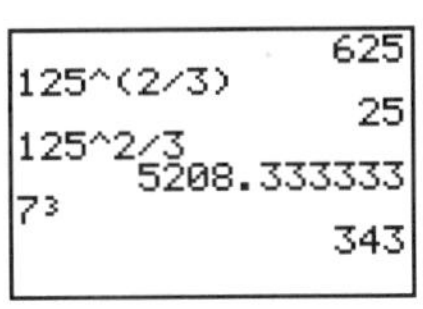</td><td>On the TI-86 type 7, then push the ^ key, then the number 3 because you want to cube 7, and then push ENTER. (No special cube button)
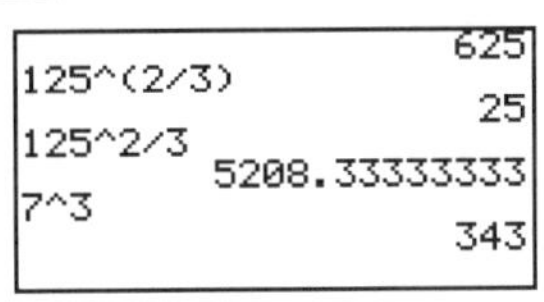</td></tr>
<tr><td>EX: $\sqrt[4]{81}$</td><td>On the TI-83 type 4 first to tell the calculator that you want a fourth root, then push the MATH key, then the number 5 to select n^{th} rooting,</td><td>On the TI-86 type 4 first to tell the calculator that you want a fourth root, then push the 2nd key followed by the × key to get the MATH</td></tr>
</table>

Answer Q1: No. They should have answered 12/3 = 4. Notice that both the TI-83 and the TI-86 have the same problem; they can't read your mind about which operation to do first. The calculators follow the order of operations, but to the calculator this looks like 7/6 first; then take 5 plus that answer followed by subtracting 3. This is not what we wanted to do.

<table>
<tr>
<td></td>
<td>then push [8] [1], and then push [ENTER].

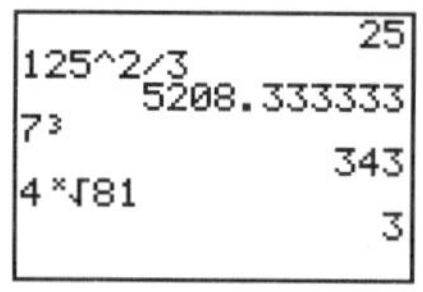

</td>
<td>menu, then push [F5] for the MISC menu, then the [MORE] key to see more options, then the [F4] key to select n^{th} rooting, then push [8] [1], and then push [ENTER].

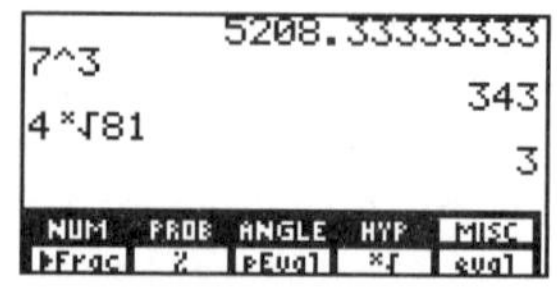

</td>
</tr>
<tr>
<td colspan="3">Our suggestion to you would be to use the [^] key most of the time. That way you only have to remember one way to do any kind of powers. Of course, to do this you need to remember that the $\sqrt[n]{a}$ is equivalent to $a^{\frac{1}{n}}$, which was discussed in section 0.1.

EX: 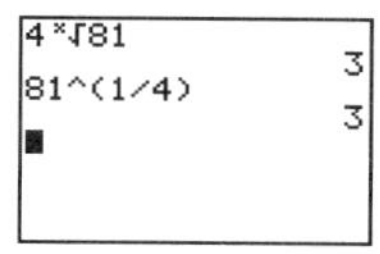

</td>
</tr>
</table>

Question 2: Find the answer to $\dfrac{-13+7}{45-(-21)}$ by using the graphing calculator.

<table>
<tr>
<td>FRAC key

The FRAC command is used to turn decimals into fractions, if the number is rational. If the number isn't rational, then it will just remain a decimal.</td>
<td>Let's "FRAC" the answer to question 2. The FRAC command is located under the MATH menu. Push the [MATH] key on the upper left side of the calculator, then the number [1] to select FRAC. Now push the [ENTER] key to see your answer in fraction form.

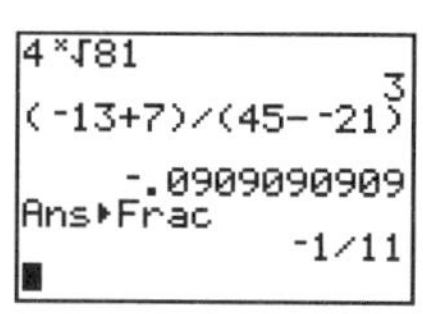

</td>
<td>The FRAC operation is located under the MATH menu. We push the [2nd] key followed by the [×] key, then [F5] for MISC, then the [MORE] key, then [F1] for FRAC. Now push the [ENTER] key to see your answer in fraction form.

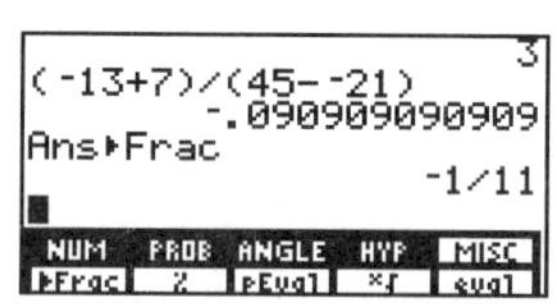

</td>
</tr>
<tr>
<td>Absolute Value</td>
<td>EX: $|-5|$

On the TI-83, push the [MATH] key, then arrow to the right to NUM, and</td>
<td>EX: $|-5|$

On the TI-86, push the [2nd] key, then the [×] key (MATH), then [F1] for</td>
</tr>
</table>

<table>
<tr>
<td></td>
<td>then arrow to the right to NUM, and then push the number 1 to select the absolute value symbol (abs). Now, enter in what you want to take the absolute value of ((−) 5)and then push ENTER. Of course don't forget your parentheses.
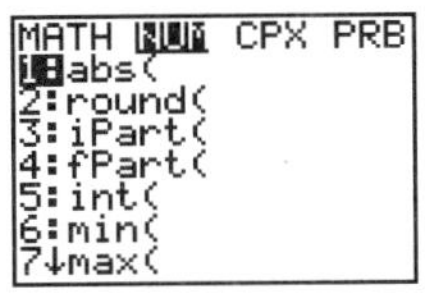

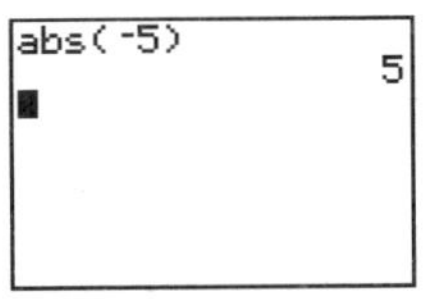
</td>
<td>the x key (MATH), then F1 for NUM, and then F5 for abs. Now, enter in what you want to take the absolute value of ((−) 5) and then push ENTER. Of course don't forget your parentheses.
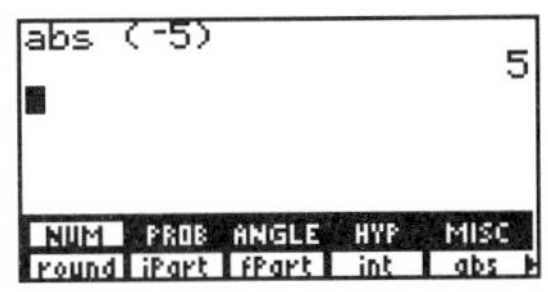
</td>
</tr>
</table>

Section Summary

- Know how to turn the calculator on (ON) and off (2nd ON).
- Know how to adjust the brightness (contrast) of the calculator screen (2nd, arrows).
- Make sure that your MODE menu has everything on the left highlighted, for this text.
- Know how to get to the HOME screen (2nd MODE on the TI-83; 2nd EXIT on the TI-86).
- Remember that numerators and denominators need parentheses, or the calculator could misinterpret what you want it to do. The same is true for roots.
- − is the *minus* key and (−) is the *negative* key. Don't use one in place of the other!
- The ^ key can be used to raise something to a power or take a root of something. Or instead, you can use special keys under the MATH menu.
- The FRAC command, located under the MATH menu, can be used to turn decimal rational numbers into their fraction form.

SECTION 0.6 HOMEWORK

(1–24) Each of the following problems is written the way it would appear on the calculator.

a.) Evaluate with the calculator.

b.) Explain the order in which the calculator does the operations.

1. 2+10/5–3
2. 9–6/3+3
3. (2+10)/5–3
4. (9–6)/3+3
5. (2+10)/(5–3)
6. (9–6)/(3+3)
7. 3^2+5
8. 4^3–1
9. 3^(2+5)
10. 4^(3–1)
11. 4^3/2
12. 8^2/3
13. 4^(3/2)
14. 8^(2/3)

Answer Q2:

TI-83

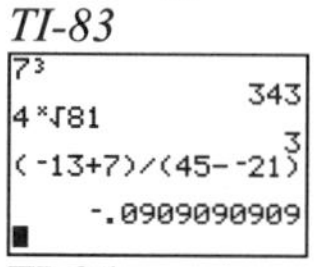

TI-86

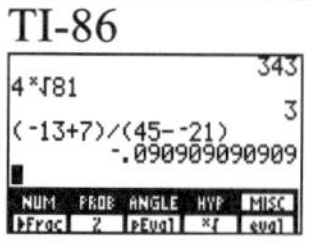

15. 5+3*6+9
16. 9–2*6–4
17. (5+3)*6+9
18. (9–2)*6–4
19. (5+3)*(6+9)
20. (9–2)*(6–4)
21. 500(1+0.08/12)^(12*30)
22. 1000(1+0.07/52)^(52*30)
23. 20000(0.08/12)/(1–(1+0.08/-12)^(-12*5))
24. 150000(0.072/12)/(1–(1+0.072/12)^(-12*30))

(25–44) a.) Write each of the following problems as you would input them in the calculator to evaluate using the appropriated symbols and parentheses.

b.) Compute each problem using your calculator.

25. 5^2
26. -3^2
27. 2^{-5}
28. 5^{-4}
29. $\dfrac{3+5}{6+4}$
30. $\dfrac{8+12}{7+1}$
31. $9^{\frac{3}{2}}$
32. $81^{\frac{3}{4}}$
33. $\dfrac{3\cdot 2+8}{7}$
34. $\dfrac{24}{2\cdot 4-4}$
35. $\dfrac{5\cdot 3^2-7}{4\cdot 2+11}$
36. $\dfrac{8\cdot 2^3+4}{5\cdot 3-7}$
37. $\dfrac{5}{8}+\dfrac{3}{4}$
38. $\dfrac{7}{40}-\dfrac{11}{16}$
39. $\sqrt[3]{13824}$
40. $\sqrt[5]{16807}$
41. $5000(1+\dfrac{0.09}{12})^{12\cdot 15}$
42. $10000(1+\dfrac{0.075}{2})^{2\cdot 15}$

43. $\dfrac{24000(\frac{0.085}{12})}{1-(1+\frac{0.085}{12})^{-12\cdot 5}}$

44. $\dfrac{180000(\frac{0.0675}{12})}{1-(1+\frac{0.0675}{12})^{-12\cdot 30}}$

(45–50) Compute each of the following problems using your calculator and convert the answer to a fraction.

45. $\dfrac{3\bullet 7+11}{5-2\bullet 6}$

46. $\dfrac{5\bullet 8-71}{3\bullet 9-83}$

47. $\dfrac{3^2\bullet 4}{3\bullet 2+5}$

48. $\dfrac{5\bullet 7-9}{2^3+7}$

49. $\dfrac{4}{9}+\dfrac{7}{8}$

50. $\dfrac{9}{13}-\dfrac{17}{21}$

Section 0.7 Graphing with Your Calculator

We are now ready to discuss how to use the graphing part of the calculator, which can do more for you than just draw graphs. Here we will show you how to graph plus a little bit more, and then, as you go through the text, we will show you other uses as well. The graphing calculator only graphs what are called functions, which will be studied in chapter 1. For now though, we can learn how to enter and manipulate functions in the calculator. Let's begin our discussion.

<u>Topic</u>	<u>TI-83</u>	<u>TI-86</u>
	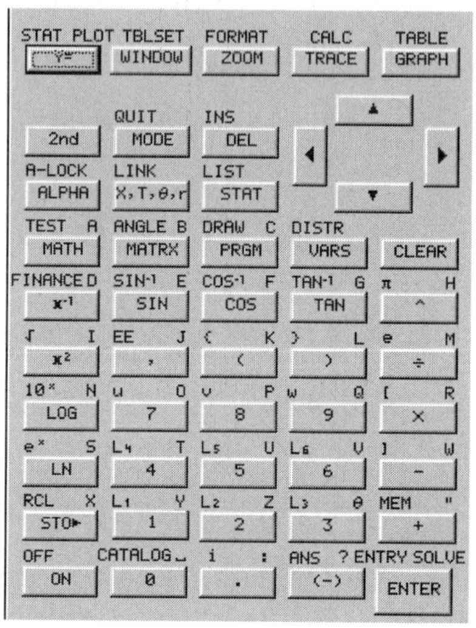	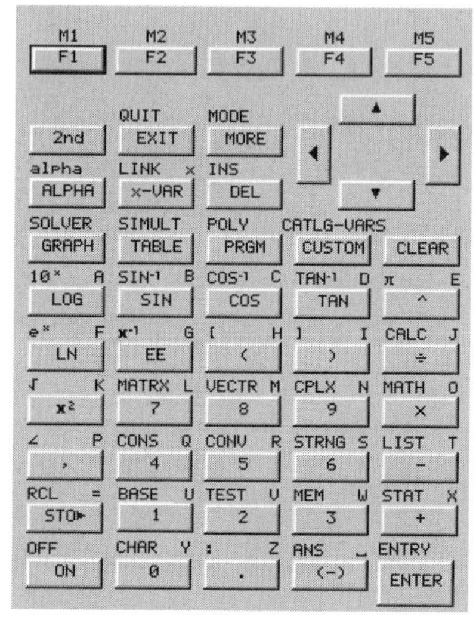
Getting to the **Y**= screen on the calculator. This is where you type in functions to be graphed.	To get to the point where you can input a function into the calculator, you need to push the [Y=] key on the top left corner of the calculator. 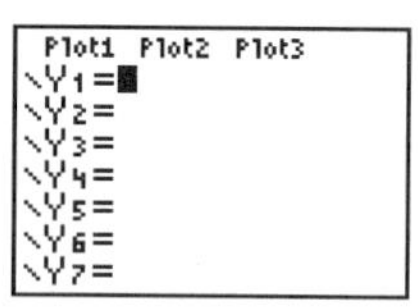	To get to the point where you can input a function into the calculator, you need to push the [GRAPH] key on the upper left part of the calculator, then push the [F1] key to go to the Y= screen. 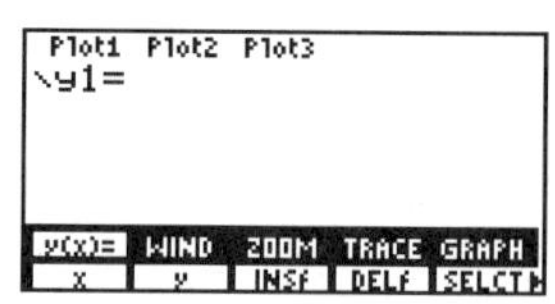
Making sure that you are **ready** to input your new function.	If there are already functions typed into the calculator, then simply arrow down to them and push the [Clear] button to erase them. Also make sure that "Plot1," "Plot2" and "Plot3" at the top of the screen are not highlighted. If they are, then arrow up and over to the ones that are	The exact procedure for getting ready to input a function into the calculator on a TI-83 works for TI-86's.

<table>
<tr><td></td><td>highlighted and push the ENTER key to un-highlight them. If there is nothing typed in and the Plots are not highlighted, then you are ready to proceed to inputting your new function.</td><td></td></tr>
<tr><td>To type in a function, you just type it in as it looks, except you can only use x as the variable and you must use the designated x keys.</td><td>To get the x on a TI-83 you push the XTθn key on the top center part of the calculator.
EX: $y = x^3 - 2x$
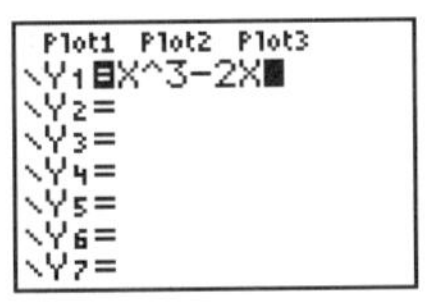
</td><td>To get the x on a TI-86, you push the X-VAR key on the top center part of the calculator. This isn't the only way to get x on the TI-86 but it's good enough for now.

Notice that on the screen on the menu at the bottom is another place to get the variable x.</td></tr>
<tr><td colspan="3">Before we can continue, we need to talk about how the calculator graphs. The calculator is just a machine that is really fast at computing! If you look real closely at your screen, you will see that it is made up of a lot of little dots (pixels). Each dot will be a point on a graph. In order to graph a function, the calculator first determines what values to plug in for the variable x, from the size of the window you have defined, and then plugs those numbers into the function to determine the y-values. Next, it plots these points on the screen and connects the dots if you are in "connect" mode. This can present us with a few problems since most functions usually are made up of an infinite number of points and our calculator screen isn't.</td></tr>
<tr><td colspan="3">• The calculator only plots points. That means there are holes between those points that the calculator knows nothing about and thus it can make mistakes and not give you a complete, accurate graph.
• What size window do you need? This is how you tell the calculator what values to use for the variable x. This also determines what y-values you will be able to see. This is one of the hardest parts about finding the graph. We will be spending a lot of time on this topic.</td></tr>
<tr><td>ZOOM Standard is a good place to begin to look for</td><td>Continuing with our example, push the ZOOM key, then the 6 key to see the graph of $x^3 - 2x$ in a window that goes 10 units in all direction out</td><td>On the TI-86 we recommend that you always push the GRAPH key first to get the menu choices, then push F3 for ZOOM, then F4 for zoom</td></tr>
</table>

<table>
<tr>
<td>a “complete” graph of the function.

If this graph weren’t near the origin we might not have even seen it.</td>
<td>that goes 10 units in all direction out from the origin.

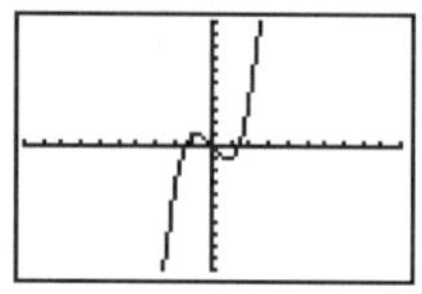</td>
<td>for ZOOM, then F4 for zoom standard (ZSTD). Following these steps will help eliminate mistakes. This isn’t the only way to get ZOOM standard.

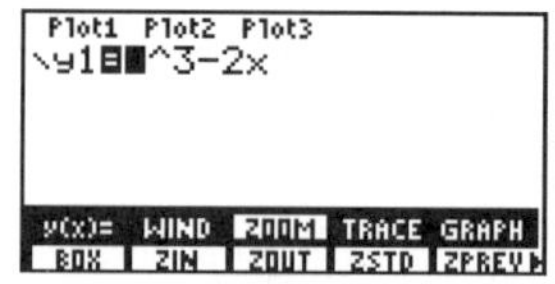

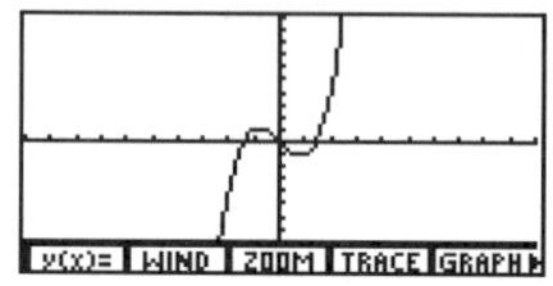
</td>
</tr>
<tr>
<td>The **TRACE** command is how we move from point to point on the graph of the function.</td>
<td>Push the TRACE key at the very top of the calculator. Then push the left and right arrow keys and notice how the calculator displays the coordinates of the points at the bottom of the screen. It also shows you where you are on the graph and in the top left corner of the screen it shows you what function it is that you are graphing.

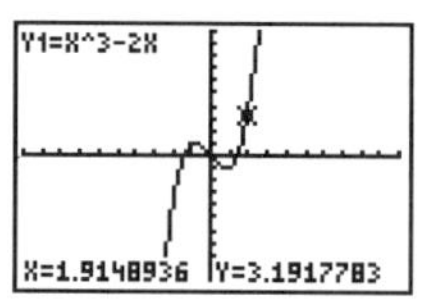
</td>
<td>On the TI-86, push the GRAPH key to get the menu choices, then F4 for TRACE, then the arrows to move along the graph. Like the TI-83 it shows you the coordinates of the points at the bottom of the screen. But unlike the TI-83, it only tells you, in the upper right corner of the screen, which function number you are graphing, not the function formula itself. Once again you could push keys in a different order, to trace along a graph, if you are already looking at a graph (see manual).

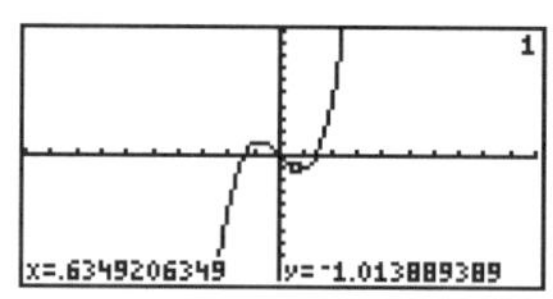
</td>
</tr>
</table>

Perhaps you noticed that the points the calculator graphed are awful (*many* decimal places!). The problem is that the TI-83 calculator has 94 dots across the screen but we asked it to use numbers between –10 and 10. That means the calculator takes 20 (10 – (–10) = 20), divides that by 94 (dots) and you get 0.2127659574… units between each dot or x-value. If we were to use an interval like –4.7 to 4.7 instead, then everything might work out a lot better (4.7 – (–4.7) = 9.4 then divide by 94 you get 0.1). Let's use something the calculator calls ZOOM decimal. The TI-86 has a similar problem but it has 126 dots across the screen.

The **ZOOM decimal** command creates a more "viewer friendly" window size.	To get a ZOOM decimal window push the ZOOM key, then the 4 key.  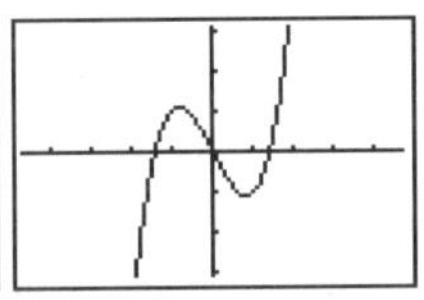	On the TI-86 push the GRAPH key again to get the menu choices, then push the F3 key for ZOOM, then the MORE key, and then the F4 key for zoom decimal (ZDECM). 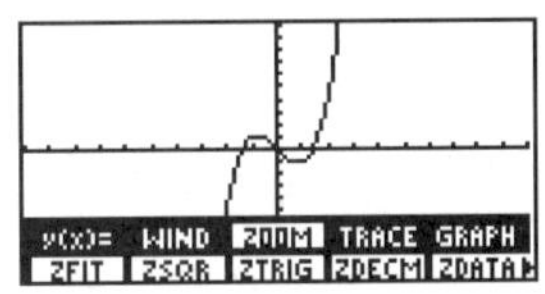
Use the **TRACE** command again. Isn't that a lot better?	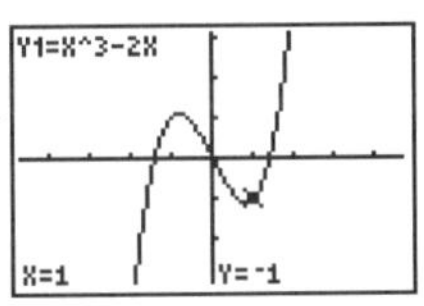	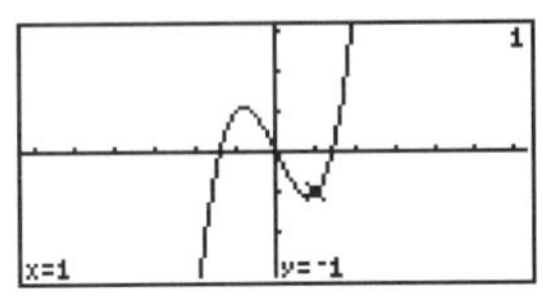
The **WINDOW** menu is where we can manually adjust the size of the window. Ymin and Xmin mean the smallest y and x value that the calculator will use in the window. Ymax	Let's say that we would like a ZOOM decimal type of window but bigger. One way to do this is to go to the WINDOW menu and change it ourselves. (Push WINDOW.)At the WINDOW menu let's double the size of all the X and Y mins and maxs by using our arrows to move around the screen and then multiply the values by 2 (push ENTER after each multiplication).	On the TI-86 push the GRAPH key again to get the menu choices, then F2 for window (WIND), then double all of the X and Y max and min.'s. 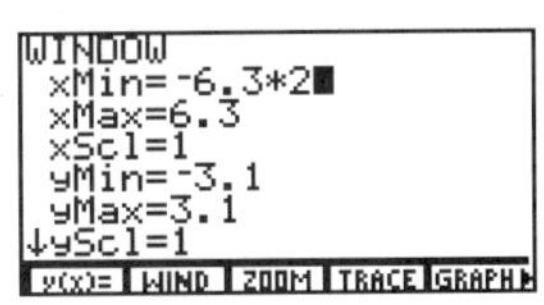 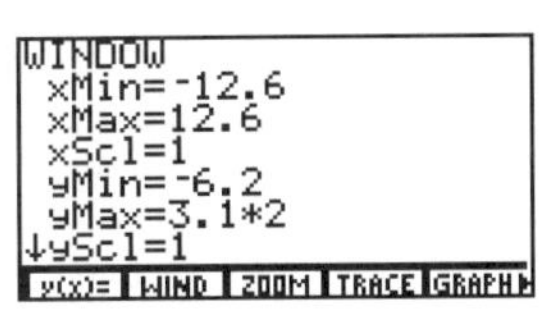

and Xmax mean the largest values used. Yscl and Xscl mean how often the calculator will put a mark on the axes to show the scaling of them.	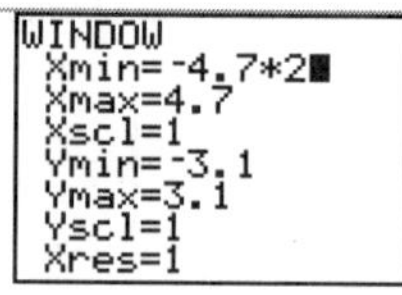 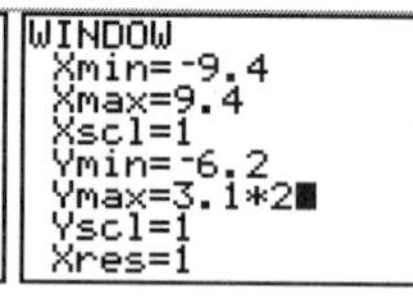 Now push GRAPH 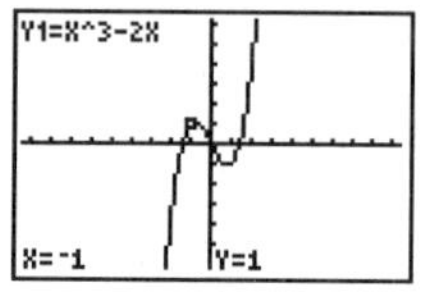 Zoom decimal has been preserved, but the window edges are now twice as far away from the origin.	Notice that ZOOM decimal is different on a TI-86. It has a different number of dots on the screen. Now push F5 for GRAPH. 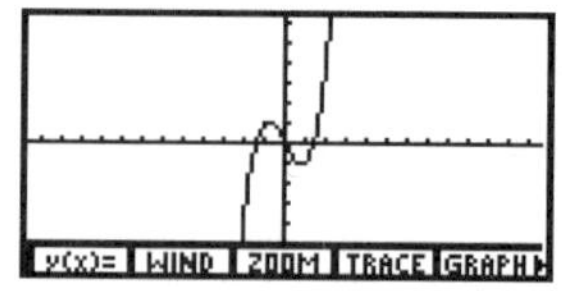
Finding a **specific point** on the graph is easy as long as the *x*-value is on the screen. Let's find the point on this graph when *x* is 5. The value of *y* when *x* is 5 is 115.	To plug in a value for *x* we need to go to the CALC menu by pushing the 2nd key, then the TRACE key. We now push 1 for value, then type in 5 (the number we want for *x*), and then push ENTER. 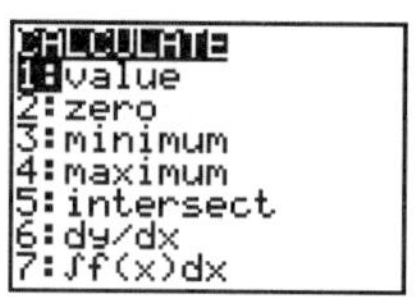 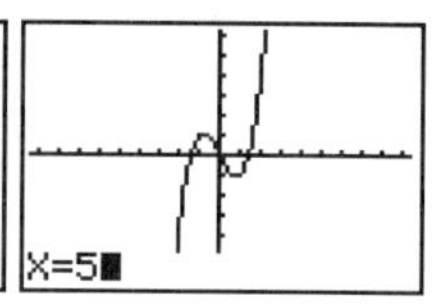 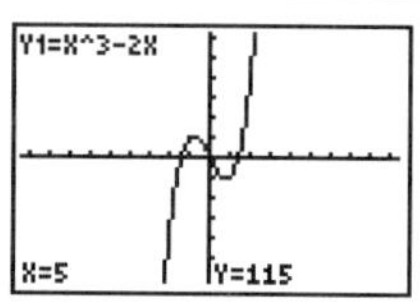	On the TI-86 you must push the GRAPH key, then the MORE key *twice*, then F1 for evaluate (EVAL), then push 5 (the number we want for *x*), and then push ENTER. 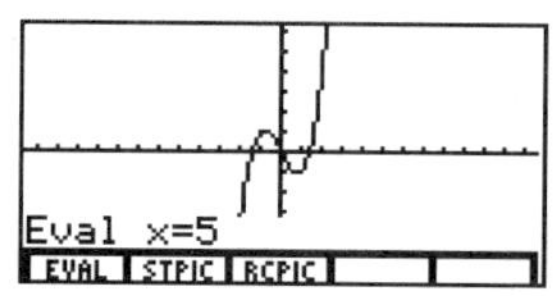 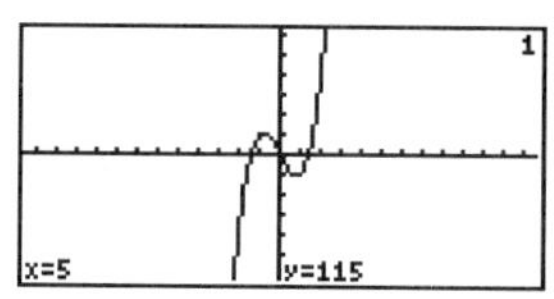

Question 1: We can't see the point on the screen. Why not?

If we want to see the point, we go to the window and change the Ymax to something	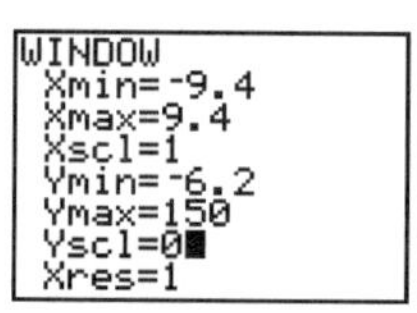 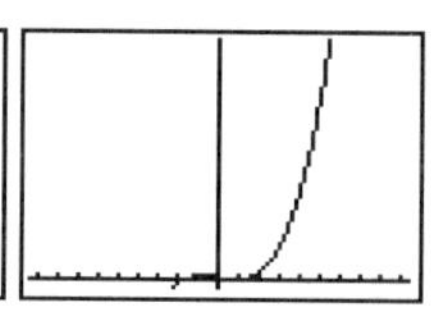 This is the graph. Now, to see the point we wanted, we need to plug the	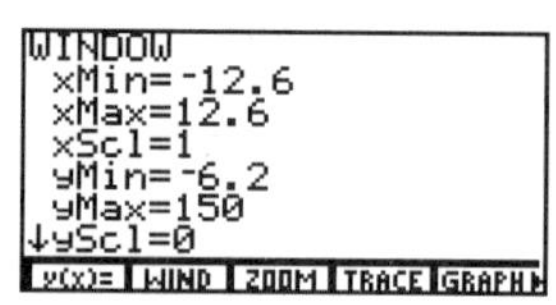

<table>
<tr>
<td>larger than 115 (say 150) and we should change the Yscl to zero (no marks on the axes). We will explain more on this later.</td>
<td>x-value in again. Repeat previous procedure.
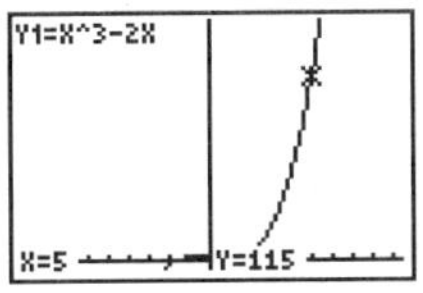</td>
<td>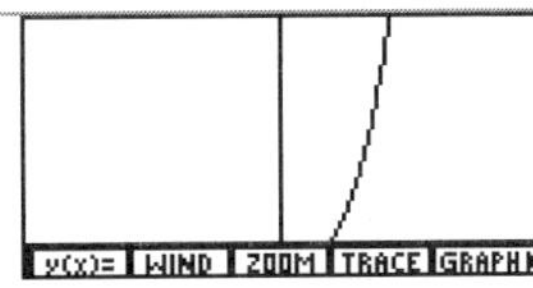
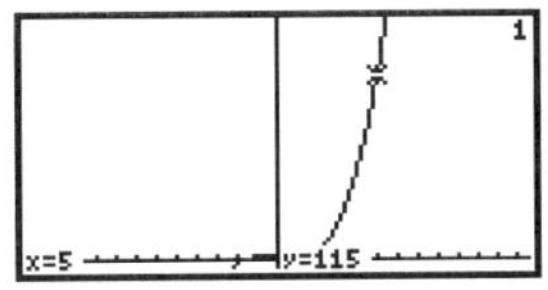</td>
</tr>
<tr>
<td>Let's try finding the point when x is −12 from the beginning.</td>
<td>First go back to ZOOM standard.
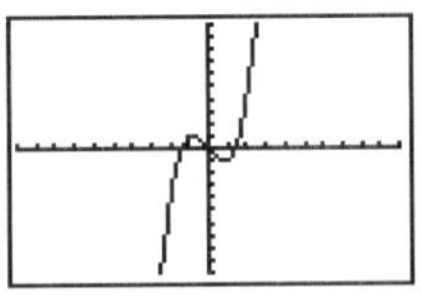
Then, go to the CALC menu choose value, then input the number −12 and ENTER.
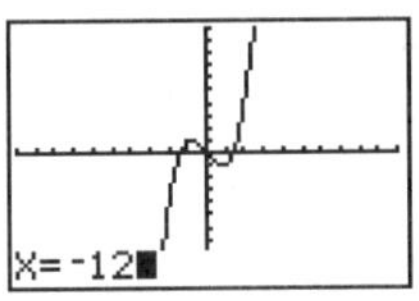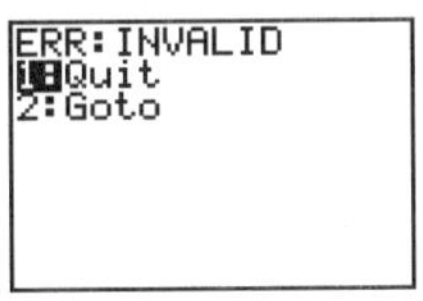</td>
<td>First go back to ZOOM standard.
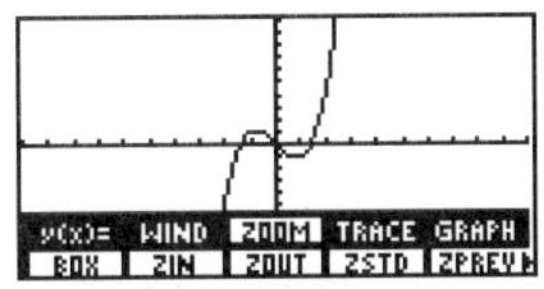
Now you must push the GRAPH key, then the MORE key twice, then F1 for evaluate (EVAL), input the number −12, and push ENTER.
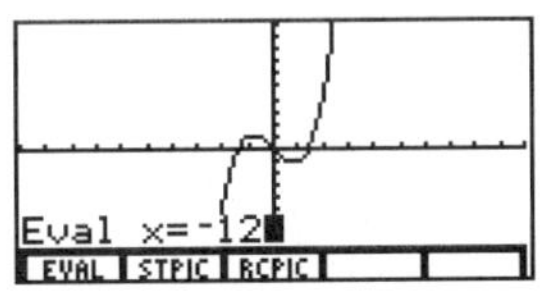
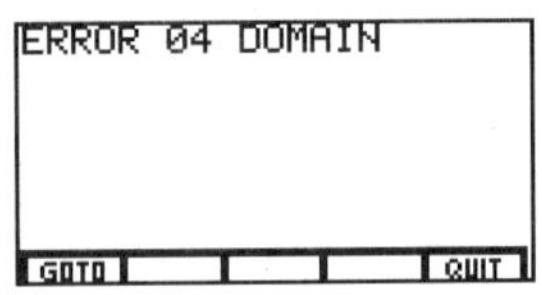</td>
</tr>
<tr>
<td colspan="3">Can you guess what went wrong? The number −12 is outside of our window, so the calculator doesn't know that it exists. If you want to find this point, then you will need to adjust the Xmin in the WINDOW menu. Let's make the Xmin = −20 and try again. Push ENTER to QUIT, then go to the WINDOW menu, make Xmin = −12, then GRAPH, then go to value/eval etc.</td>
</tr>
<tr>
<td>The value of y when x is −12 is −1704. Notice that the point is too low for us to see on the graph.</td>
<td>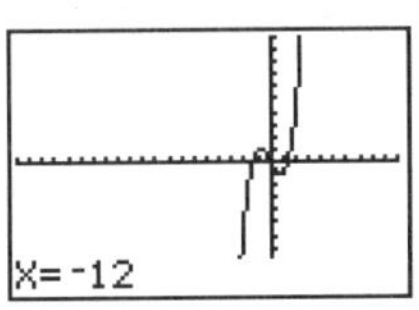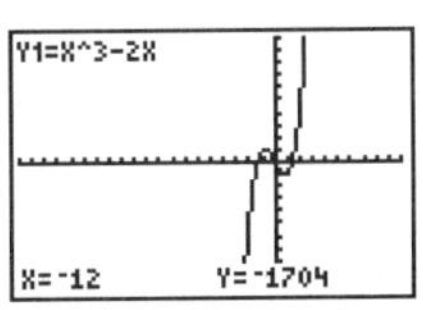</td>
<td>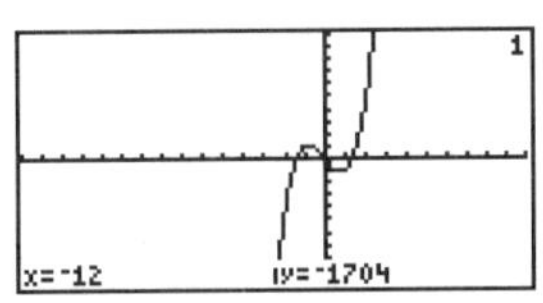</td>
</tr>
</table>

Not every graph is going to be near the origin. Let's do another example where it isn't so easy to find a complete graph.	EX: $y = -x^3 + 39x^2 - 505x + 2188$ Zoom Standard 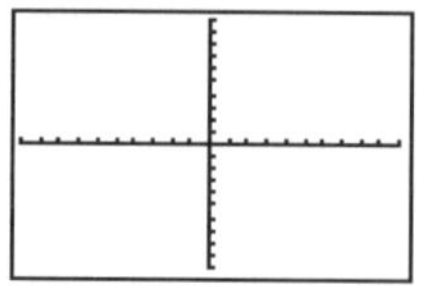 We can't see it because it isn't near the origin.	Zoom Standard 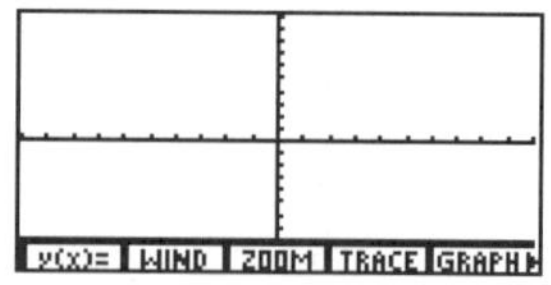 We can't see it because it isn't near the origin.
One method for finding a graph is to push the TRACE key, then the ENTER key.	TRACE ENTER 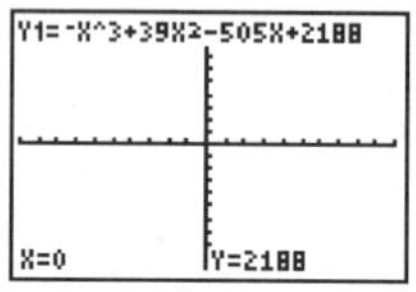 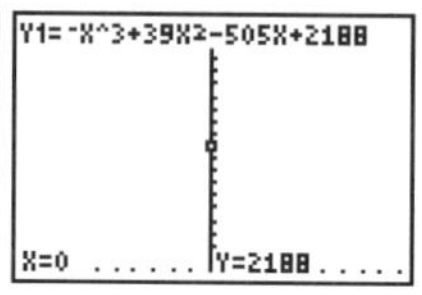 This adjusts the screen so that we are centered on the y-intercept $(0, y)$. As you can see, it isn't all that helpful on this function.	TRACE 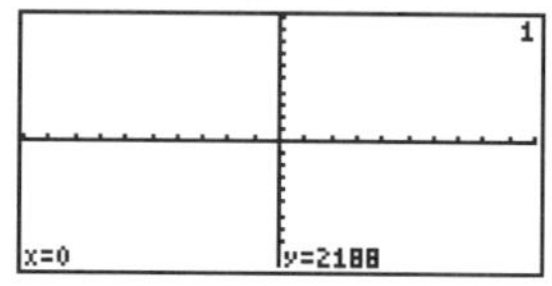 ENTER 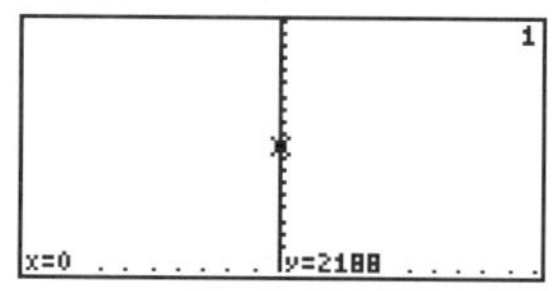
Another method for finding a graph is to push the TRACE key in order to get an idea of how high or low the graph goes.	ZOOM Standard, TRACE 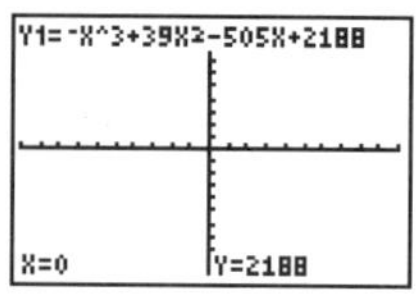 We now go to the WINDOW screen and adjust the y-values. We will try changing the Ymin to –4000 and the Ymax to 4000 (since we see a y-value of 2188) and Yscl to 0. Old Window New Window 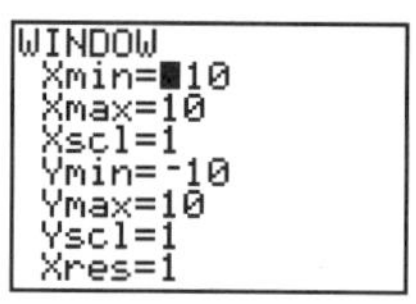 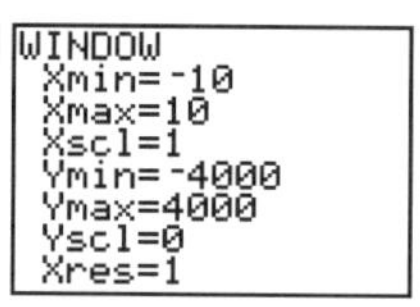 Now push GRAPH	ZOOM Standard, TRACE 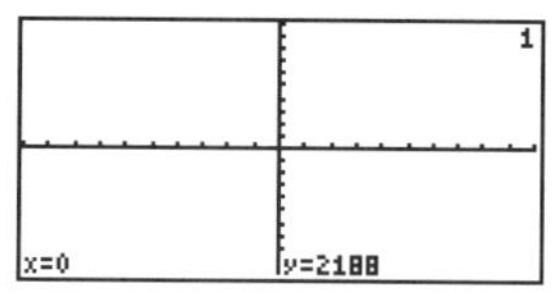 Push the GRAPH key, then F2 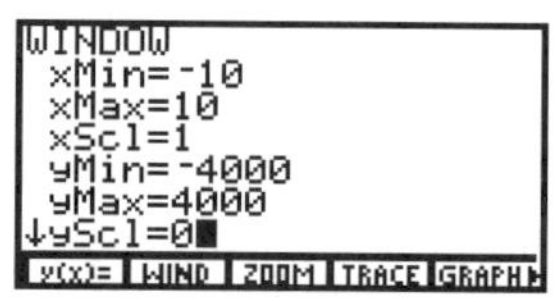 Now push F5 for GRAPH 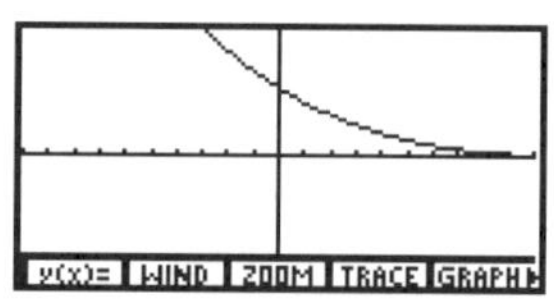 We see a graph now, but it is only a

Answer Q1: The y-value is 115 and our window size stops at around 6 for the y-values.

<table>
<tr>
<td></td>
<td>
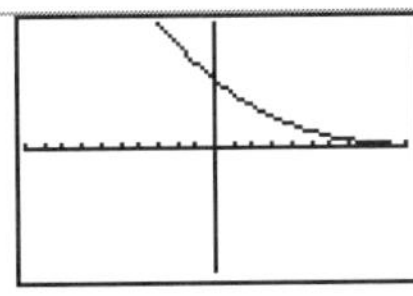

We see a graph now but it is only a portion of the whole graph. We can see that the graph goes to the right, so let's increase the Xmax to 40 and see how much more we can see. As you learn more algebra it will become easier for you to determine if you have a complete graph or not.

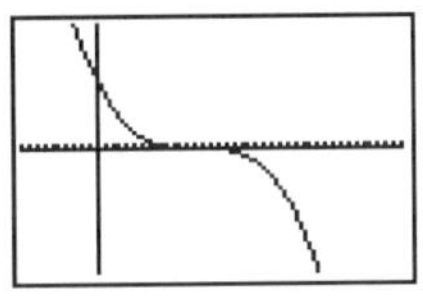

This *looks* like it's complete, but it's not. We chose this example because it is a hard one. Instead of continuing this way, let's try yet another method.
</td>
<td>
portion of the whole graph. We can see that the graph goes to the right so let's increase the Xmax to 40 and see how much more we can see.

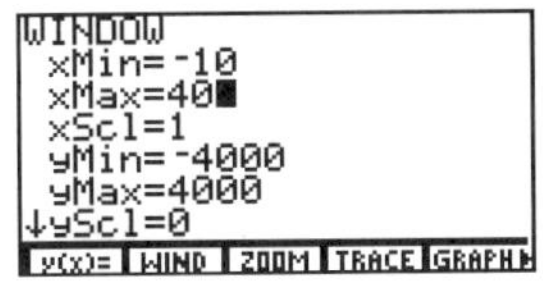

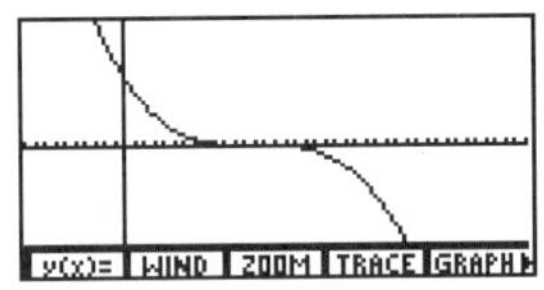

This *looks* like it's complete, but it's not. We chose this example because it is a hard one. Instead of continuing this way, let's try yet another method.
</td>
</tr>
<tr>
<td>Another method for finding a complete graph is to start at ZOOM standard or decimal, and then ZOOM OUT. We might need to do this one or more times until we see something or move on to another approach.</td>
<td>
Zoom Standard

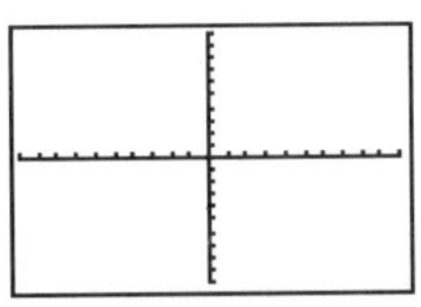

To ZOOM OUT push the [ZOOM] key, then [3] (ZOOM OUT), then push the [ENTER] key.

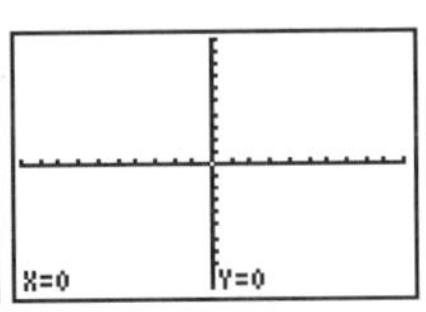

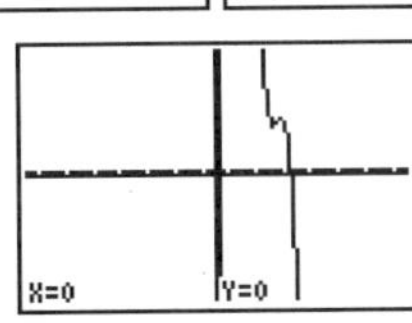

This time we are close to a complete graph. The graph is a little too skinny though and consequently hard to see.
</td>
<td>
Zoom Standard

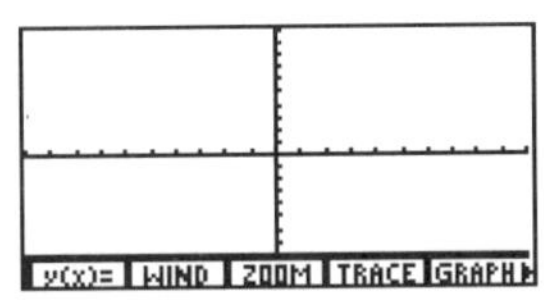

To ZOOM OUT push the [GRAPH] key, then [F3] for ZOOM, and then [F3] key for ZOUT.

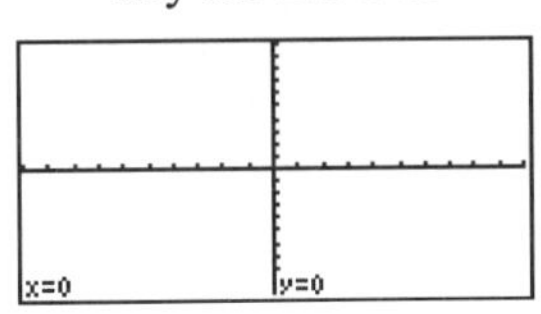

Then push the [ENTER] key.

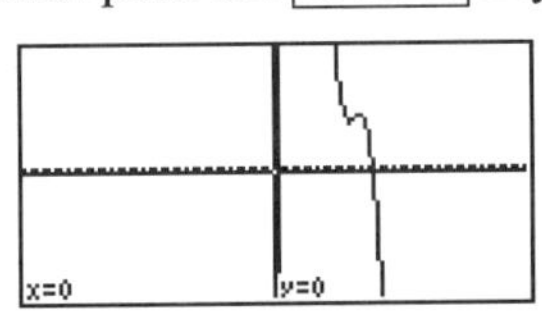

Since the graph is very narrow we will reduce the Xmax and Xmin in order to expand the graph in the
</td>
</tr>
</table>

<table>
<tr>
<td>There isn't just one "best" way to hunt for the graph.

Xmin = −10
Xmax = 20</td>
<td>though and consequently hard to see. We really would prefer that the graph fill more of the screen with all of its interesting parts (complete). Since the graph is very narrow we will reduce the Xmax and Xmin in order to expand the graph in the horizontal direction (fewer x-values).
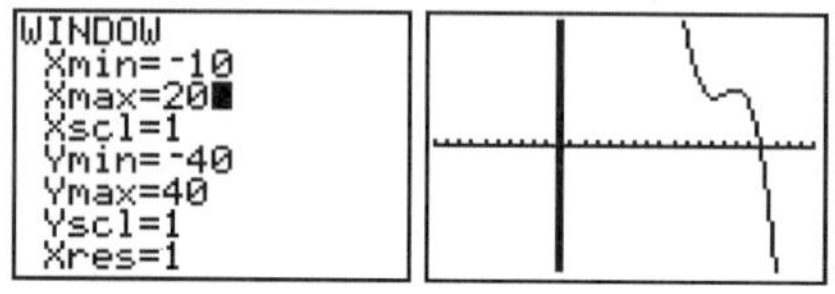
</td>
<td>horizontal direction or, in other words, widen the graph.
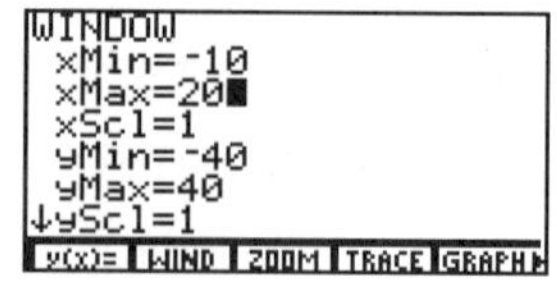

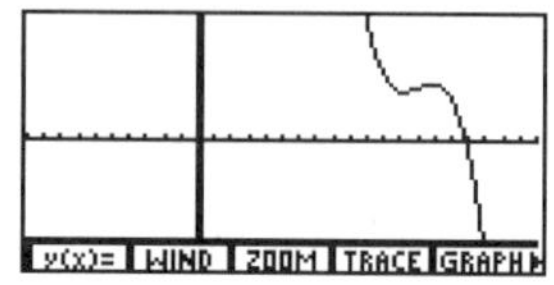
</td>
</tr>
<tr>
<td colspan="3">There <u>isn't a right way</u> to approach finding a complete graph of a function. You can try TRACE ENTER, or just TRACE and then adjust the WINDOW manually until you are satisfied, or you might just ZOOM OUT (you may need to ZOOM IN instead). And of course, there are other ways in addition to these. You will just need to experiment. Let's do another example.</td>
</tr>
<tr>
<td>Yscl and Xscl are the ways to tell the calculator how often to place a mark on the axes.</td>
<td>EX: $y = x^4 + 5x^3 - 209x^2 - 465x + 6300$
Adjust window to this
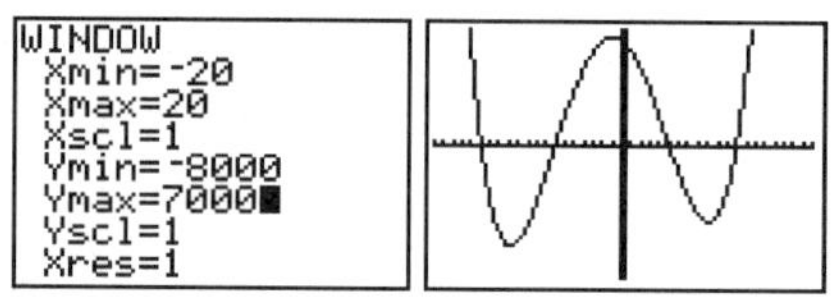

Notice that there are marks on the x-axis, representing one-unit distances, and a wide line for the y-axis. The y-axis is wide because we have asked the calculator to put a mark every one unit on a window size that runs from −8000 to 7000. The calculator doesn't have enough dots on its screen for that. Let's change the Yscl to 1000 and see what happens.
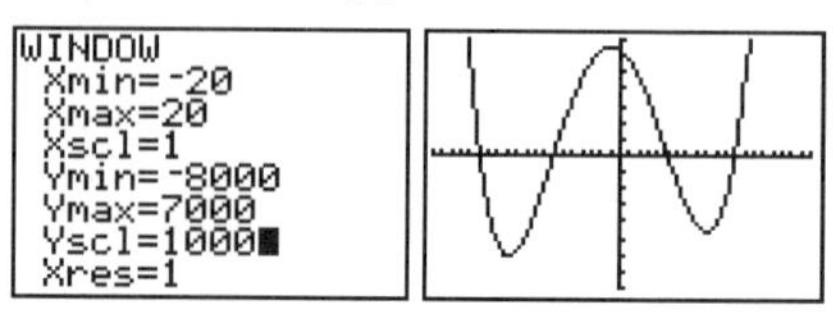

Now on the y-axis there is a mark</td>
<td>Adjust window to this
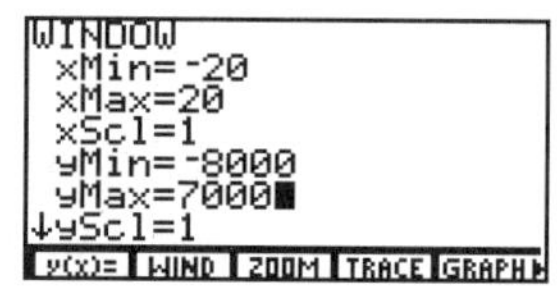

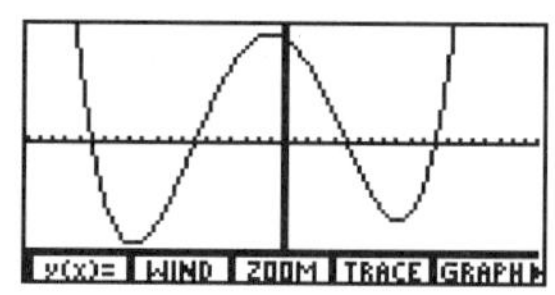

The TI-86 is just like the TI-83. Let's change the Yscl to 1000 and see if it acts similarly to the TI-83.
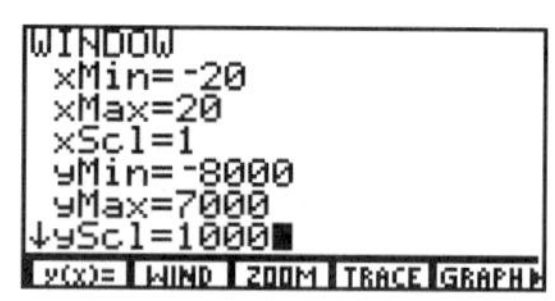

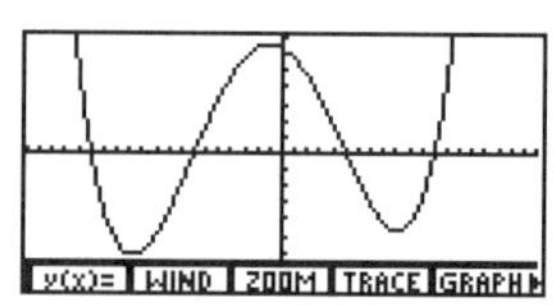

Now on the y-axis there is a mark</td>
</tr>
</table>

	every 1000 units. This scaling helps to give us perspective.	every 1000 units. This scaling helps to give us perspective.
Finding a complete graph.	EX: $y = \sqrt{x-25} - 50$ If we try ZOOM standard we see nothing, and if we then push TRACE we again see nothing. 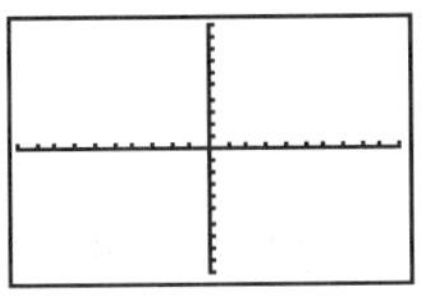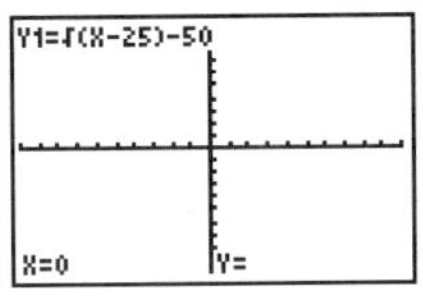 Notice, X = 0 but Y = . This means that when x is zero there isn't a y answer (value). So this graph doesn't cross the y-axis. In this case we will need to look at the formula and guess about where to find the graph. Since the numbers in the formula are large, let's adjust the WINDOW by hand. 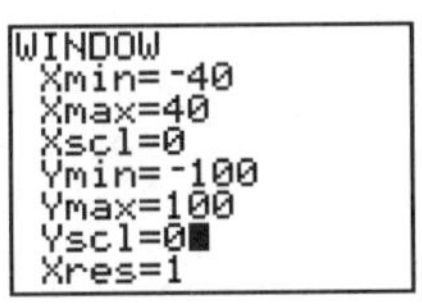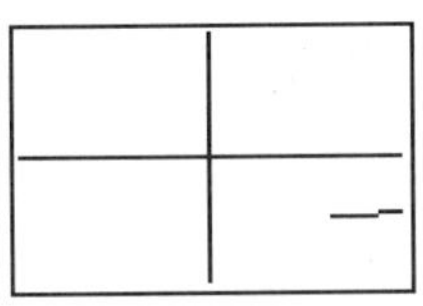 We finally see something. Let's trace along the graph and then readjust the WINDOW. 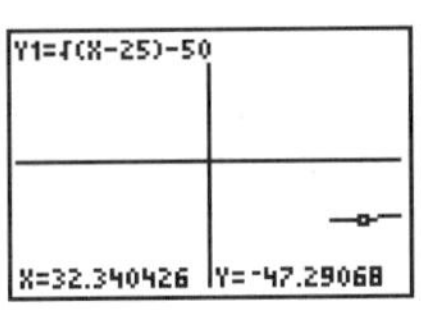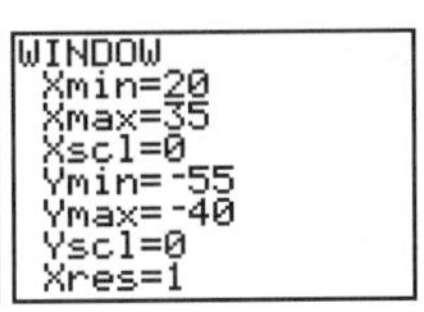 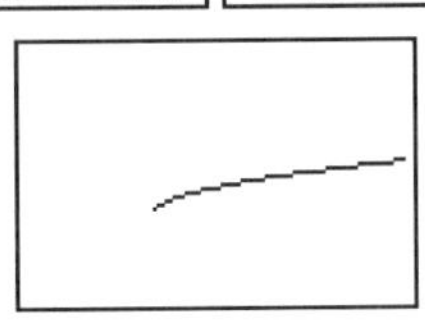 It is usually nice to see the axes, so let's make one more adjustment.	If we try ZOOM standard we see nothing, and if we then push TRACE we again see nothing. 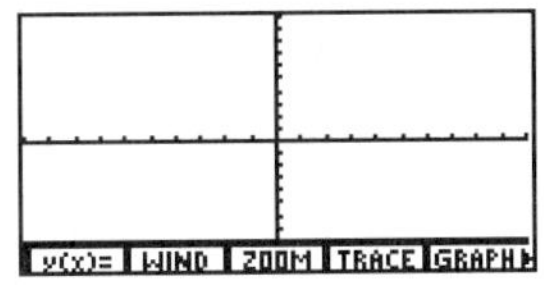 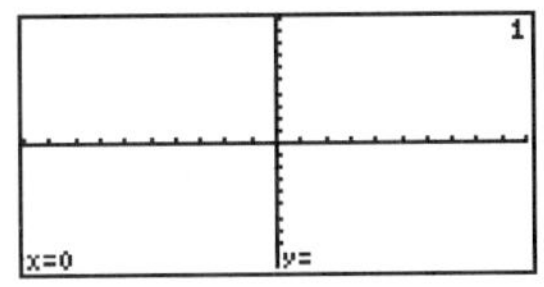 Notice, X = 0 but Y = . This means that when x is zero there isn't a y answer (value). So this graph doesn't cross the y-axis. In this case we will need to look at the formula and guess about where to find the graph. Since the numbers in the formula are large, let's adjust the WINDOW by hand. 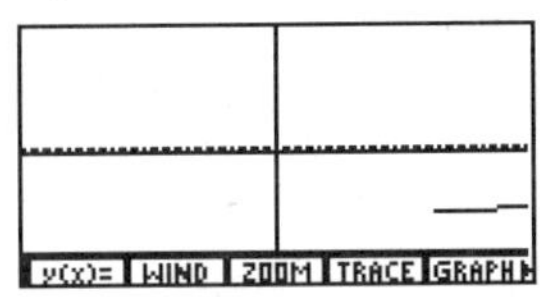 We finally see something. Let's trace along the graph and then readjust the WINDOW. 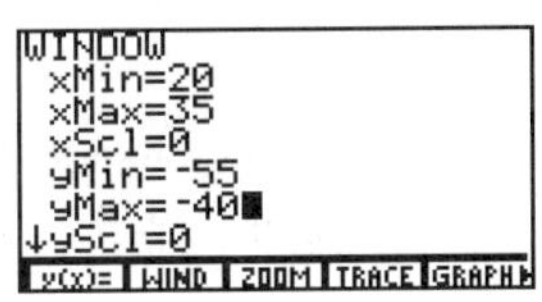 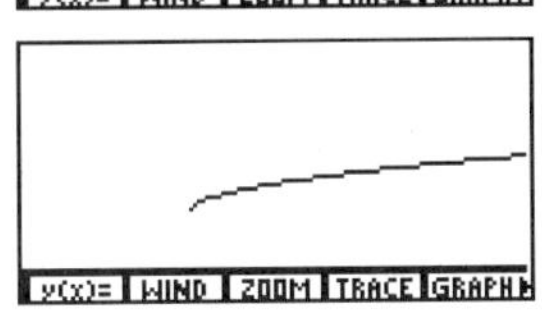 It is usually nice to see the axes, so let's make one more adjustment.

<table>
<tr>
<td></td>
<td>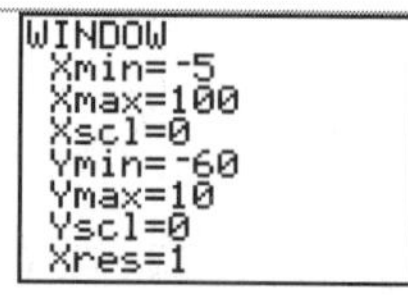

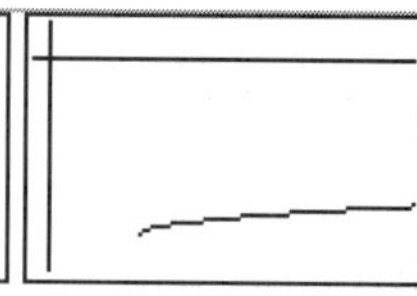</td>
<td>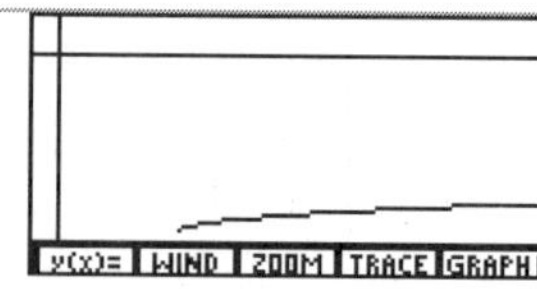
</td>
</tr>
<tr>
<td colspan="3">As you have seen, finding a complete graph of a function isn't easy. It will take practice and a good knowledge of the different types of functions in order to be efficient at obtaining their complete graphs on the calculator.</td>
</tr>
<tr>
<td>Yet another example:</td>
<td>EX: $y = x^3 - 7x^2 - 60x$
ZOOM standard, then use TRACE
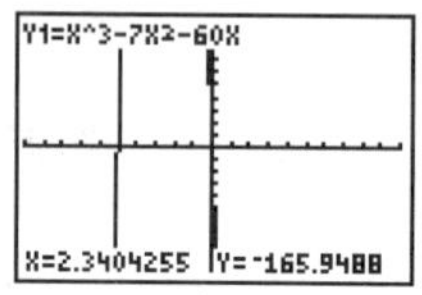

It looks as though this graph goes very high and low, so let's adjust the Ymax, Ymin and Yscl
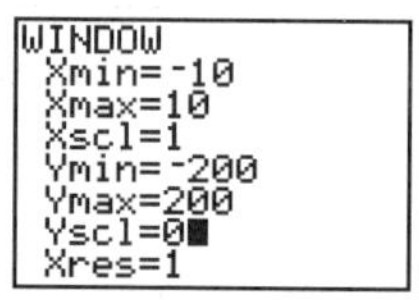

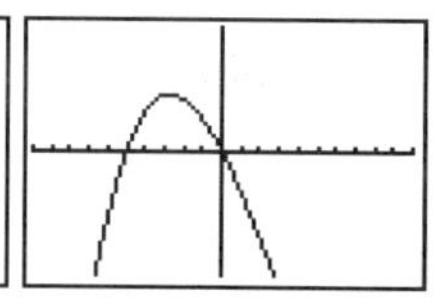
You should notice that this function has an x^3 as its highest power as did some of the other examples so far. We would expect it to look similar. We need to keep looking for a complete graph. The graph is still going down, so this time let's change the Ymin only.
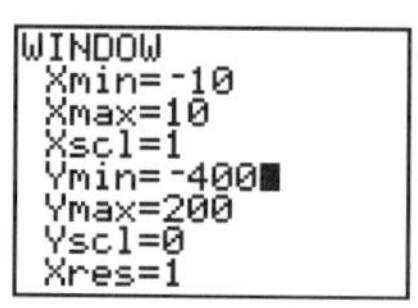

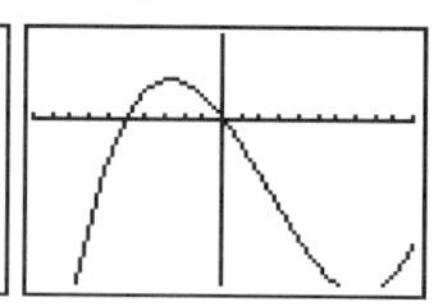
We are almost there. The graph goes slightly lower than –400 and to the right, so let's change the Ymin and the Xmax, and then we will probably be done.</td>
<td>ZOOM standard, then use TRACE
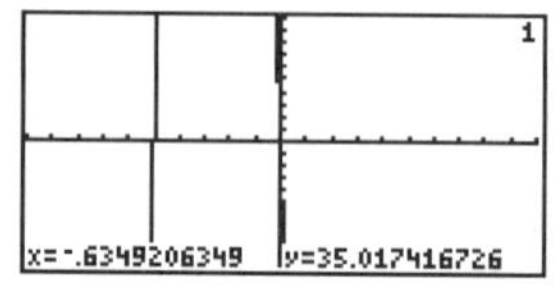

It looks as though this graph goes very high and low, so let's adjust the Ymax, Ymin and Yscl
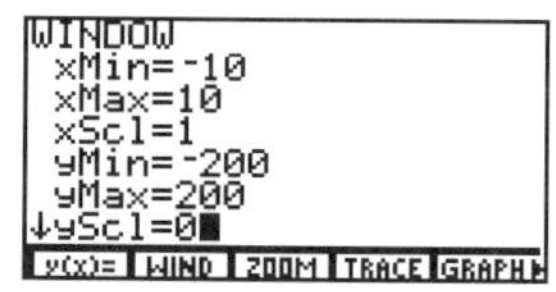

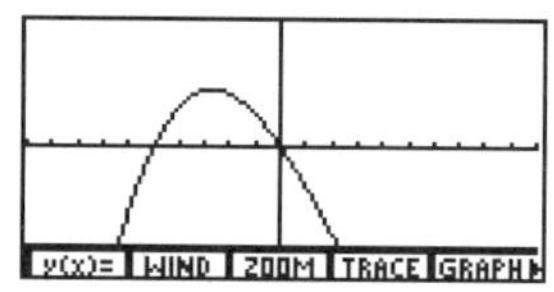

Let's continue as we have with the TI-83 to the left.
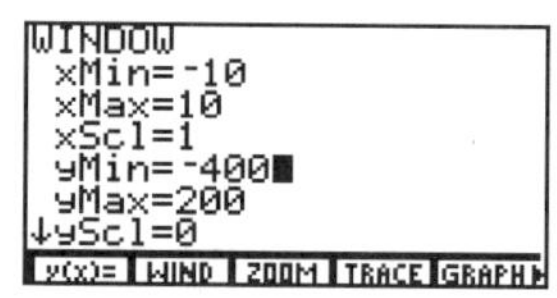

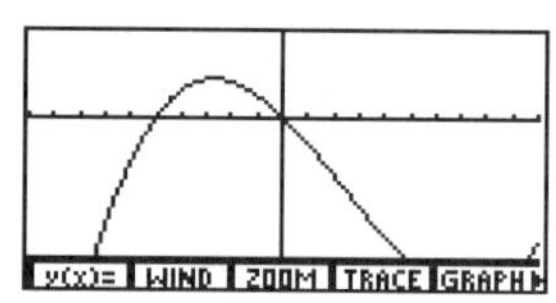

Continuing we have,
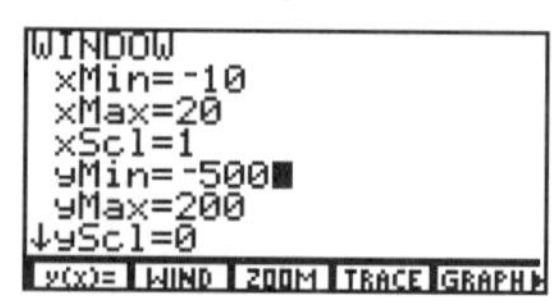
</td>
</tr>
</table>

	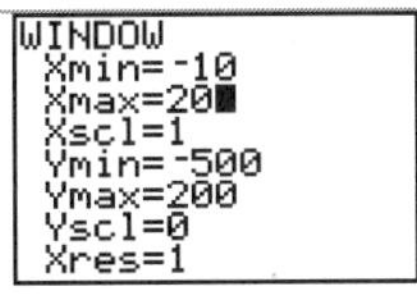 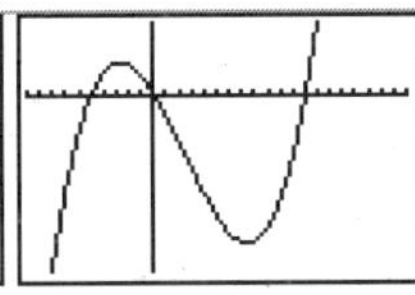	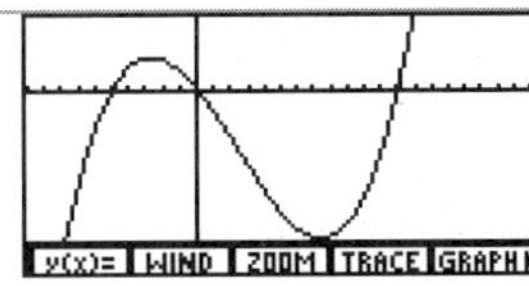
Here is one example where the number of dots on the calculator screen causes a problem with the graph.	EX: $y = \sqrt{7 - x^2}$ A half circle. ZOOM decimal 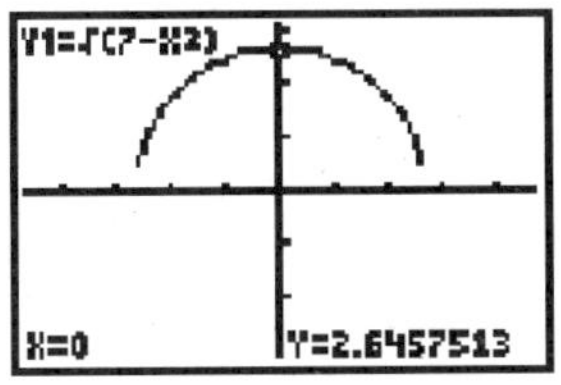 The graph should touch the *x*-axis but it doesn't because the last points (end points) that the calculator knows exists on the graph of the function are (2.6, 0.489…), (-2.6, 0.489…) which isn't the end points. The end points happen when *y* equals zero at $x = \pm\sqrt{7} \approx \pm 2.6457...$ 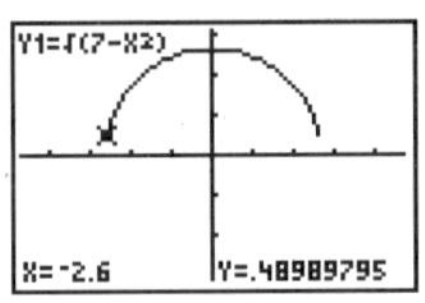 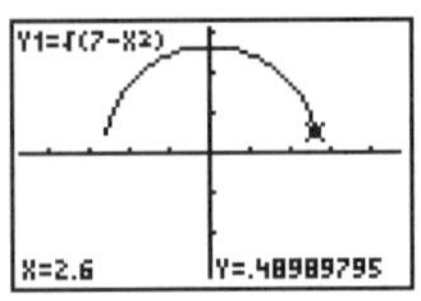	To get to ZOOM decimal, push GRAPH, then F3 for ZOOM, then MORE, and then F4 for ZDECM. ZOOM decimal 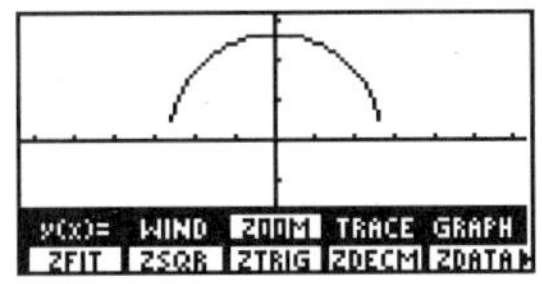 TRACE 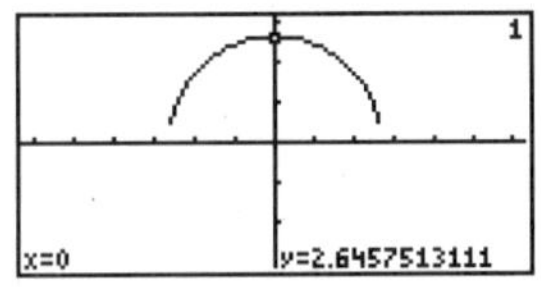 The same event happens as with the TI-83. 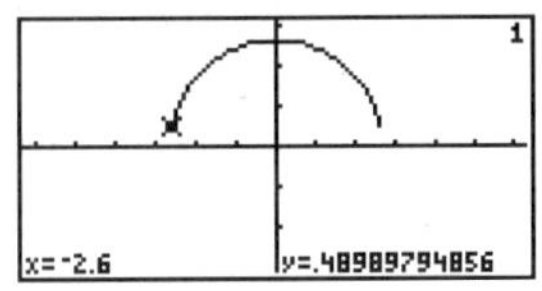 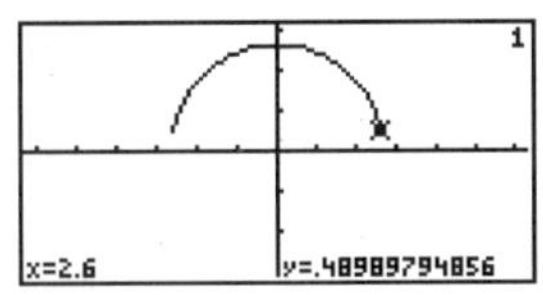
A few trivialities.	If, while the calculator is graphing, you want to pause the calculator, just push the ENTER key. If, while the calculator is graphing, you want to stop the calculator, just push the ON key.	

Section Summary

- Make sure that the Plots are not highlighted and to clear out any old functions.
- Do a ZOOM standard or ZOOM decimal as a first window size.
- If you cannot see the graph, then push the TRACE key and then adjust the window size so that you can at least see part of your function.

- Keep adjusting the window until you are certain that you have a complete graph (you are seeing all of the interesting parts of the graph).

SECTION 0.7 HOMEWORK

(1–14) Graph each function in the zoom standard window first, then graph in the given windows and scales.

1. $y = x^2 - 3x + 2$ a. $-30 \le x \le 30$ Scale 3 $-30 \le y \le 30$ Scale 3 b. $-3 \le x \le 6$ Scale 1 $-5 \le y \le 15$ Scale 1 c. Which window do you think best fits the function?	2. $y = x^2 - 2x + 5$ a. $-30 \le x \le 30$ Scale 3 $-30 \le y \le 30$ Scale 3 b. $-5 \le x \le 7$ Scale 1 $-30 \le y \le 30$ Scale 3 c. Which window do you think best fits the function?
3. $y = x^2 + 3x + 15$ a. $-7 \le x \le 5$ Scale 1 $-40 \le y \le 40$ Scale 4 b. $-5 \le x \le 5$ Scale 1 $-5 \le y \le 5$ Scale 1 c. Which window do you think best fits the function?	4. $y = -2x^2 + 4x - 20$ a. $-10 \le x \le 10$ Scale 1 $-20 \le y \le 20$ Scale 2 b. $-6 \le x \le 8$ Scale 1 $-100 \le y \le 100$ Scale 10 c. Which window do you think best fits the function?
5. $y = x^4 - 16x^2 + 2$ a. $-8 \le x \le 8$ Scale 1 $-20 \le y \le 20$ Scale 2 b. $-8 \le x \le 8$ Scale 1 $-80 \le y \le 80 \le$ Scale 8 c. Which window do you think best fits the function?	6. $y = -2x^4 + 50x + 15$ a. $-10 \le x \le 10$ Scale 1 $-100 \le y \le 100$ Scale 10 b. $-5 \le x \le 5$ Scale 0.5 $-330 \le y \le 330$ Scale 30 c. Which window do you think best fits the function?
7. $y = 2x^3 - 3x^2 + 2x - 1$ a. $-5 \le x \le 5$ Scale 0.5 $-10 \le y \le 10$ Scale 1 b. $-4 \le x \le 4$ Scale 0.4 $-100 \le y \le 100$ Scale 10 c. Which window do you think best fits the function?	8. $y = -3x^3 + x^2 - 5x + 3$ a. $-6 \le x \le 6$ Scale 0.6 $-30 \le y \le 30$ Scale 3 b. $-3 \le x \le 3$ Scale 0.3 $-40 \le y \le 40$ Scale 4 c. Which window do you think best fits the function?
9. $y = \sqrt{x^2 - 100}$ a. $-6 \le x \le 6$ Scale 0.6 $-4 \le y \le 4$ Scale 0.4 b. $-30 \le x \le 30$ Scale 3	10. $y = \sqrt{4 - x^2}$ a. $-20 \le x \le 20$ Scale 2 $-20 \le y \le 20$ Scale 2 b. $-3 \le x \le 3$ Scale 0.3

$-30 \le y \le 30$ Scale 3 c. Which window do you think best fits the function?	$-3 \le y \le 3$ Scale 0.3 c. Which window do you think best fits the function?
11. $y=(9-x^2)^{\frac{3}{4}}$ a. $-15 \le x \le 15$ Scale 1.5 $-12 \le y \le 12$ Scale 1.2 b. $-4 \le x \le 4$ Scale 0.4 $-6 \le y \le 6$ Scale 0.6 c. Which window do you think best fits the function?	12. $y=(x^2-4)^{\frac{3}{4}}$ a. $-30 \le x \le 30$ Scale 3 $-10 \le y \le 10$ Scale 1 b. $-8 \le x \le 8$ Scale 0.8 $-20 \le y \le 20$ Scale 2 c. Which window do you think best fits the function?
13. $y=5,000\left(1+\frac{0.08}{12}\right)^{12x}$ a. $0 \le x \le 30$ Scale 3 $0 \le y \le 10,000$ Scale 1,000 b. $0 \le x \le 30$ Scale3 $0 \le y \le 50,000$ Scale 5,000 c. Which window do you think best fits the function?	14. $y = 10,000(1 + 0.08x)$ a. $0 \le x \le 40$ Scale 4 $0 \le y \le 80,000$ Scale 8,000 b. $0 \le x \le 40$ Scale 4 $0 \le y \le 40,000$ Scale 4,000 c. Which window do you think best fits the function?

(15–30) For each of the following functions find a good window for graphing. (Answers may vary.)

15. $y = 9 - 4x - 2x^2$
16. $y = 3x^2 + 6x - 12$
17. $y = -2x^4 - 6x^3 + x^2 - 12x$
18. $y = x^4 + 10x^2 + 24$
19. $y = x^4 + 5x^3 + 2x^2 + 3$
20. $y = x^4 - 3x^3 - 7x^2$
21. $y = x^4 - 8x^2 - 25$
22. $y = -2x^4 + 12x^2 + 13$
23. $y = |2x-3| + 12$
24. $y = 18 - |3x+2|$
25. $y = \sqrt{144 - x^2}$
26. $y = \sqrt{x^2 - 144}$
27. $y = 500(1 + 0.08)^x$ (Set $0 \le x \le 30$ Scale of 3 and $0 \le y \le ?$ Scale of ?)
28. $y = 10,000\left(1+\frac{0.09}{12}\right)^{12x}$ (Set $0 \le x \le 30$ Scale of 3 and $0 \le y \le ?$ Scale of ?)
29. $y = 100(x-1)^{\frac{3}{2}}$
30. $y = 50(x+1)^{\frac{3}{2}} + 100$

(31–38) Graph each of the following functions and find the y–value for the given x–value using the graphing calculator. (Use value under the Calc menu and make sure your window contains the x–value your are looking for.)

31. $y = 3x^2 - 5x + 3;\ x = 15$
32. $y = -5x^2 + 3x - 10;\ x = -20$
33. $y = x^3 - 3x^2 - 5x + 3;\ x = 10$
34. $y = 2x^3 - 5x^2 + 3x - 2;\ x = -10$
35. $y = -2x^4 + 5x^2 + 3x - 2;\ x = -12$
36. $y = 3x^4 - 10x^2 - 5x + 10;\ x = 9$
37. $y = 5,000(1 + 0.08)^x;\ x = 20$ (Approximate y to two decimal places.)

38. $y = 5{,}000\left(1+\dfrac{0.09}{12}\right)^{12x}$; $x = 20$ (Approximate y to two decimal places.)

39. For problem number 31, graph with a standard window.

a. Use the calculator to try to find the value of y when $x = 15$. What happens?

b. What is the problem?

Section 0.8 Finding Roots and Graphing Piecewise Functions

Let's continue our discussion on how to use the graphing calculator. In this section we will talk about how to find zeros, also called roots, and maximums and minimums, and show you how to graph what are called piecewise functions.

Topic	TI-83	TI-86
How to find **zeros/roots** on a graphing calculator. Basically, zeros/roots are just the x-values of the points where the graph crosses the x-axis, in other words, they occur at x-intercepts. They are simply the x-values that make the function equal zero, thus the name "zeros." More about this will be discussed in chapter 3.	EX: $y = x^3 - 7x^2 - 60x$ This is an example we used in the last section. We found the "best" window size to be the following. 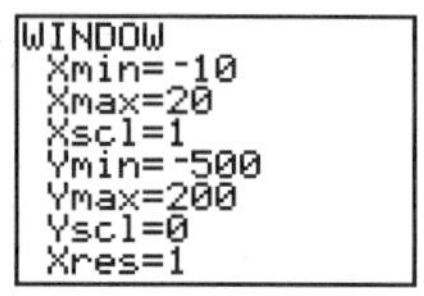 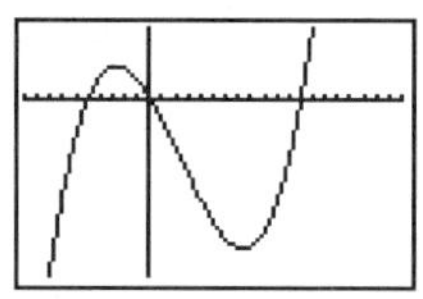 To find the zeros (x-intercept, x-values) push the 2nd key, then the TRACE key, and then the 2 key for zero. We are now ready to begin to find the first zero. Notice that the calculator asks, "Left Bound?" which means it wants you to arrow over to a point where you are to the left of the zero you are looking for. Let's arrow over to the far left side of the screen to find the zero on the left. Now, push ENTER. It now asks for "Right Bound?" which means arrow over to the right of the zero you want, but not too far, then push ENTER. Now it asks, "Guess?" which means arrow close to the zero and push ENTER to finally find it.	Window size from last section: 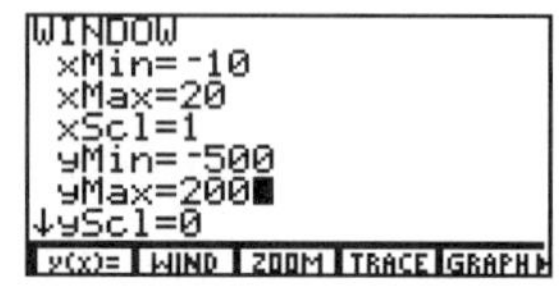 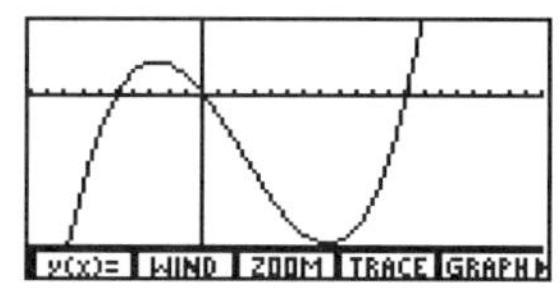 To find the roots (x-intercept, x-values) push the GRAPH key, then the MORE key, then the F1 key for MATH, and then the F1 key for ROOT. We are now ready to begin to find the first root. The procedure, from this point on, is the same for a TI-86 as it is for a TI-83.

<table>
<tr>
<td>We must go through all of these steps in order to tell the calculator where to look for the zero.

For this particular function the leftmost zero is at the point (–5, 0). Thus the zero or root is $x = -5$. This is where the y-value is zero.</td>
<td>

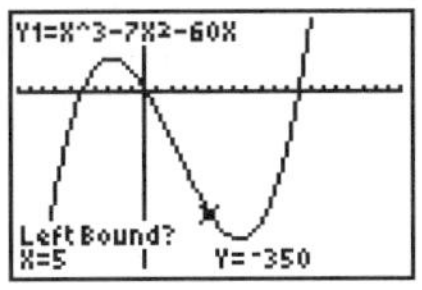

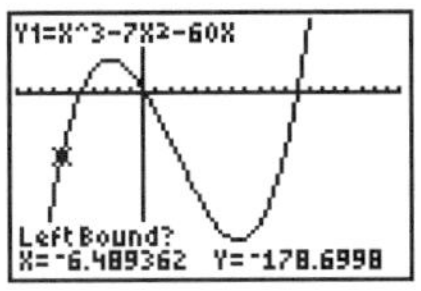

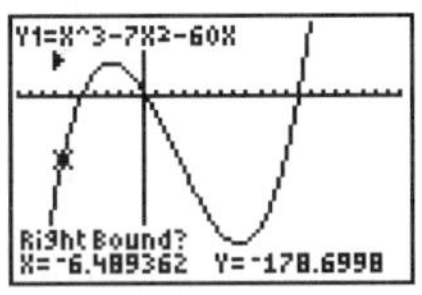

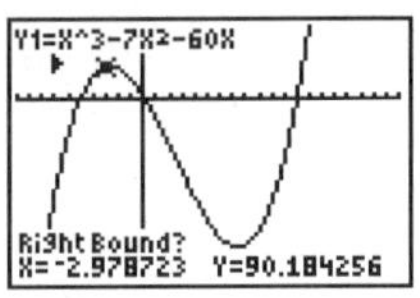

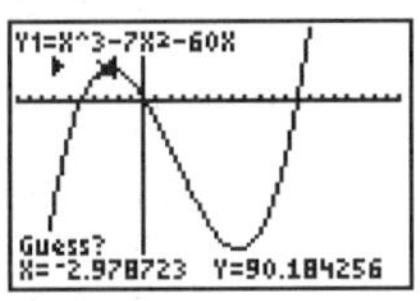

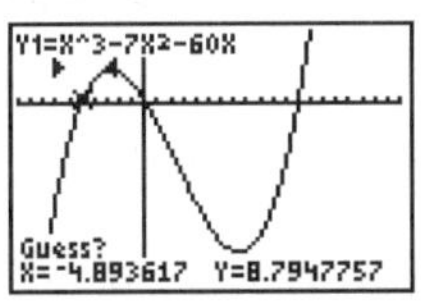

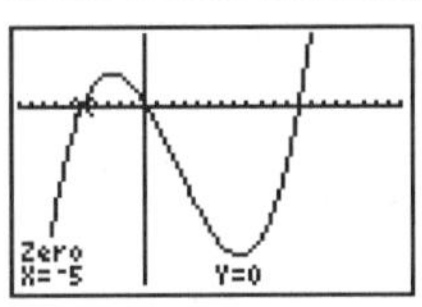

You must go through this same procedure for each zero you want to find.</td>
<td>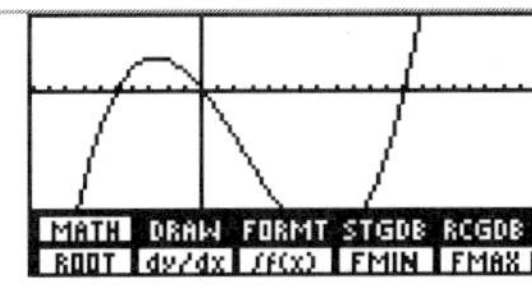

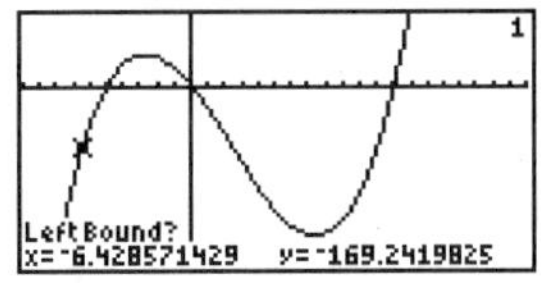

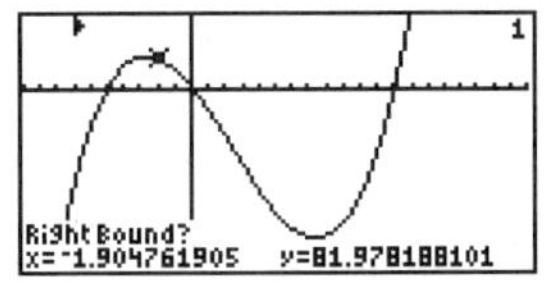

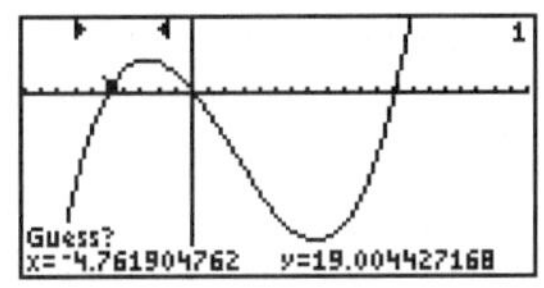

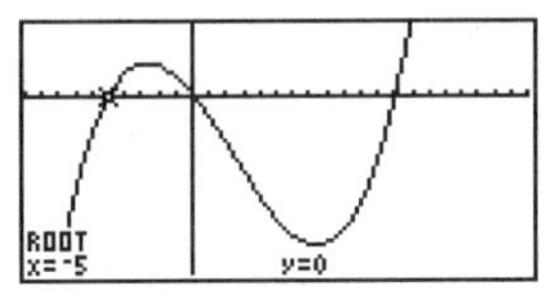

You must go through this same procedure for each zero you want to find.</td>
</tr>
</table>

Question 1: Find the zero that is farthest to the right.

How to find relative **maximums** and relative **minimums** of	The graph of our function has one hill and one valley, so it will have one relative maximum and one relative	The graph of our function has one hill and one valley, so it will have one relative maximum and

<table>
<tr>
<td>a function. Simply stated, relative max. and min.'s happen at the tops of humps and the bottoms of valleys on a graph. In other words, a max occurs at the point where the *y*-value of that point is greater than any *y*-values on either side of the max. Likewise with the min, the *y*-value is less than any *y*-values near by. Let's continue to work with the last example.</td>
<td>minimum (our calculators just call them maximums and minimums). The procedure for finding a maximum or a minimum is similar to finding zeros. First, push the [2nd] key, then the [TRACE] key, and then the [4] key for maximum. We are now ready to begin searching for the maximum. Notice that the calculator asks "Left Bound?" which means the same as before. Let's arrow over to the left side of the hump, and then push [ENTER]. It now asks for "Right Bound?" which means we arrow over to the right side of the hump, and then push [ENTER]. Now it asks "Guess?" which means we arrow back close to the maximum itself and push [ENTER].
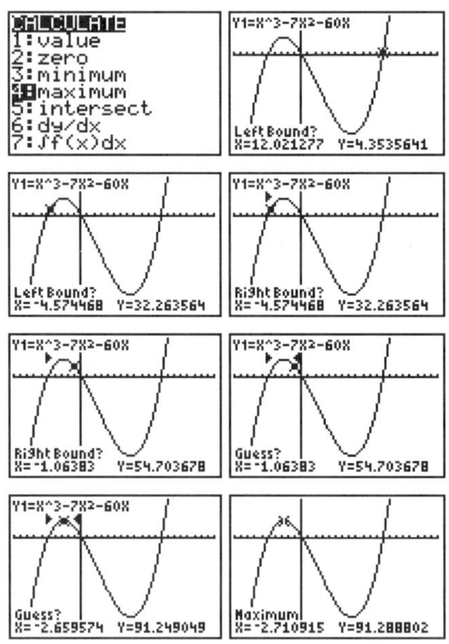

The relative maximum, the highest point on graph of any point near it, is the *y*-value, which in this case is approximately 91.288.</td>
<td>one relative minimum (our calculators just call them maximums and minimums). The procedure for finding a maximum or a minimum is similar to finding zeros. First, push the [GRAPH] key, then the [MORE] key, then the [F1] key for MATH, and then the [F5] key for FMAX (maximum). Note: for minimums push the [F4] key (FMIN). From here on out the procedure is just like it was for finding roots.
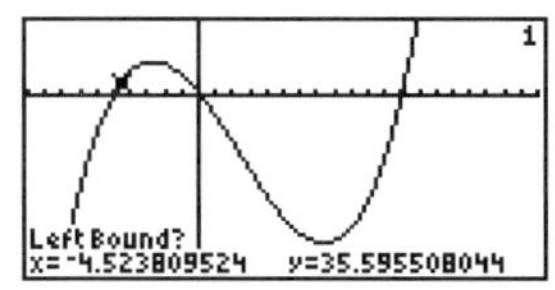

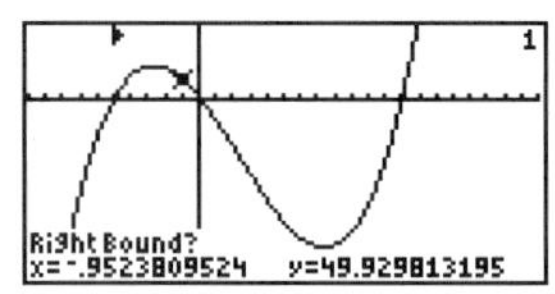

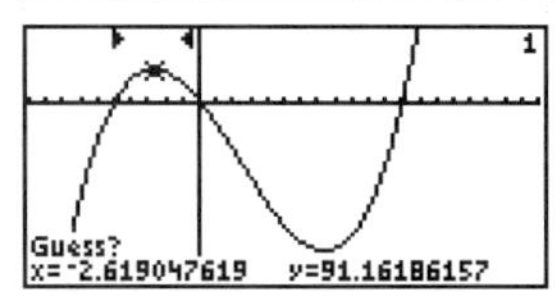

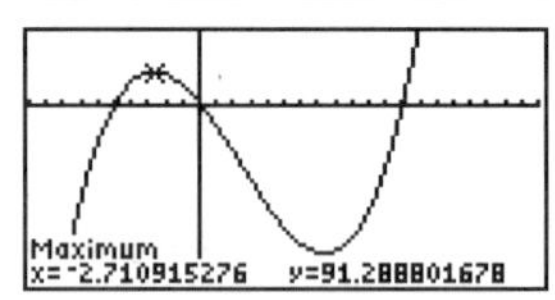

The relative maximum, the highest point on graph of any point near it, is the *y*-value, which in this case is approximately 91.288.</td>
</tr>
</table>

Question 2: Find the relative minimum.

Here is another example of finding maximums or minimums (using ZOOM standard to graph).

EX: $y = \dfrac{5}{x^2+1}$

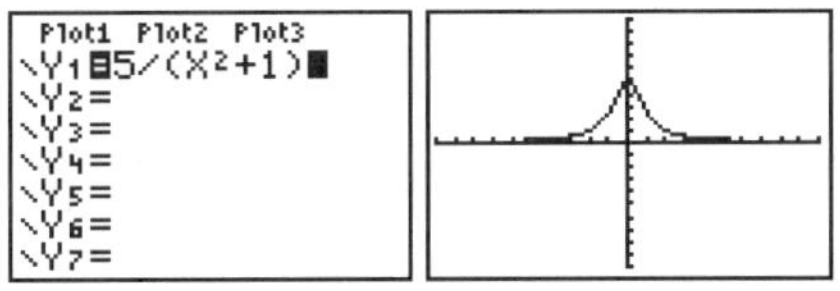

It looks as though this example only has a maximum (we only see a hill, no valleys). We need to go through the whole procedure again (2nd, TRACE, maximum, left bound, right bound, guess, then done).

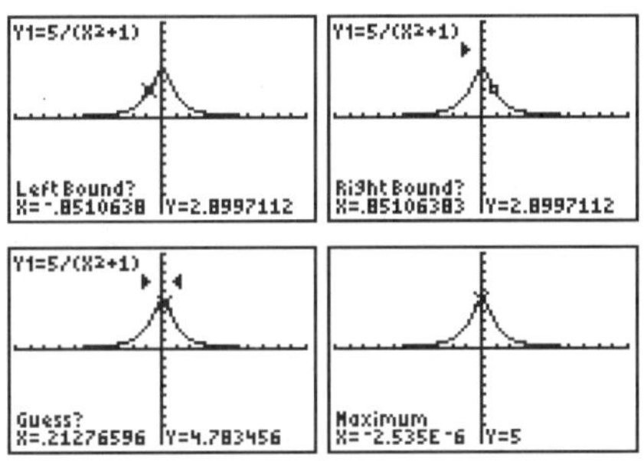

The maximum is 5. Notice that the x-value is –2.535E–6 (number in scientific notation). This means the number –0.000002535, which in essence is the number zero. Many times the calculator will give answers of this type for the number zero. Any time you see numbers like this on the graphing screen, you can safely assume that it is really zero.

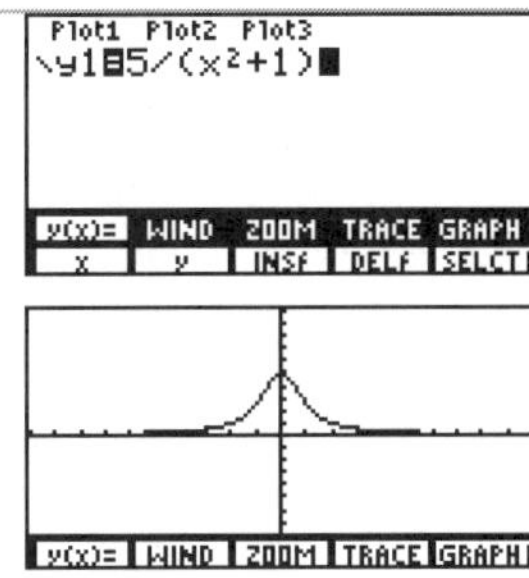

It looks as though this example only has a maximum (we only see a hill, no valleys). We need to go through the whole procedure again (GRAPH, MORE, math, fmax, left bound, right bound, guess, then done).

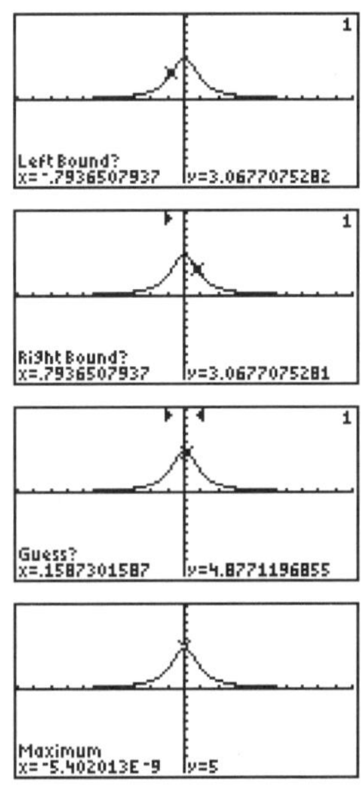

The maximum is 5. Notice that the x-value is –5.402013E–9 (number in scientific notation). This means the number –0.000000005402013, which in essence is the number zero.

Answer Q1:
x = 12

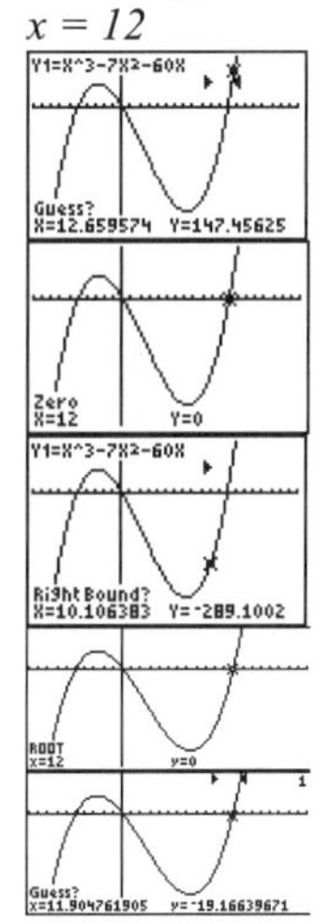

Graphing **piecewise** functions is graphing functions made up of two or more other functions that have their own specific range of allowable x-values.

The calculator checks each x-value it plugs into the pieces of the function against the restriction for each piece, in this example that would be $x < 2$ for the first piece of the piecewise function. If it is true that the x-value is less than 2, then the calculator will put a 1 into the denominator and plot a point since anything divided by 1 is itself. But if it is false, that is, the x-value is greater than or equal to 2, then the calculator will put a 0 in the denominator and this will make the expression undefined. Thus, the calculator does not plot a point there. Therefore, the calculator only plots points where it should.

EX: $y = \begin{cases} 2x+1 & x<2 \\ x^2-8x+17 & x \geq 2 \end{cases}$

We have two functions. The first is $2x + 1$, but we only use this function when our x-value is less than 2. The second is $x^2 - 8x + 17$, but we only use this function when our x-value is greater than or equal to 2. To input these into the calculator and get an accurate graph, we will need to type in both the functions and there x restrictions. In Y1 we type this:

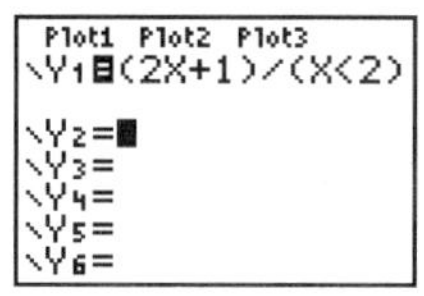

We get the < from the test menu. In order to get the <, push the [2nd] key, then the [MATH] key, and then [5]. The parentheses are very important here because we are dividing. Here is what we now type in for the second function under Y2.

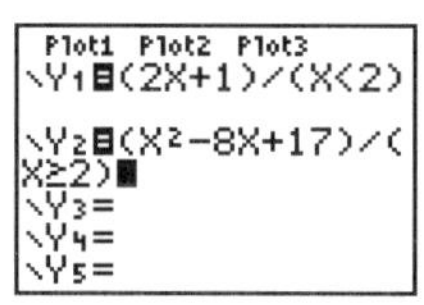

This time you had to push the number [4] key to get the ≥ symbol. Now all we need to do is go to the ZOOM standard window and see our graph.

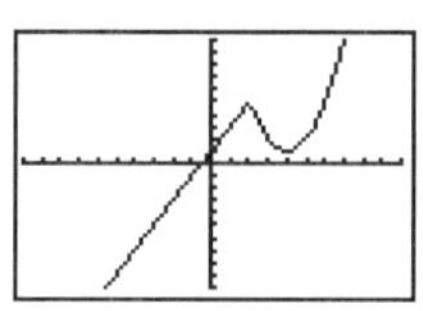

We have two functions. The first is $2x + 1$, but we only use this function when our x-value is less than 2. The second is $x^2 - 8x + 17$, but we only use this function when our x-value is greater than or equal to 2. To input these into the calculator and get an accurate graph, we will need to type in both the functions and there x restrictions. In Y1 we type in this:

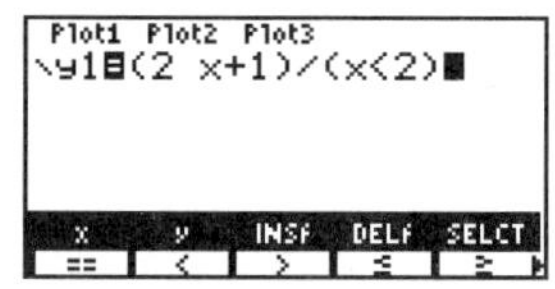

We get the < from the test menu. In order to get the <, push the [2nd] key, then the [2] key for TEST, and then [F2]. The parentheses are very important here because we are dividing. Here is what we now type in for the second function under Y2.

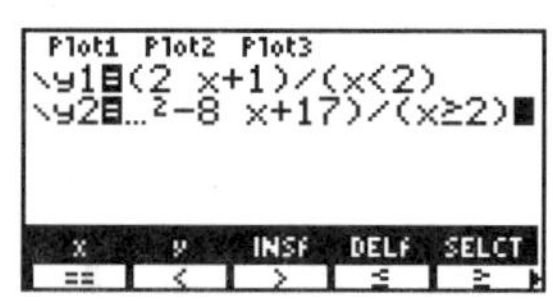

This time you had to push the [F5] key to get the ≥ symbol. Now all we need to do is go to the ZOOM standard window and see our graph.

This is a trick we can use to get the calculator to graph the correct graph."	If you watch closely, you can see it graph the line and then the curve. In this case the two separate graphs meet, but piecewise functions do not have to meet.	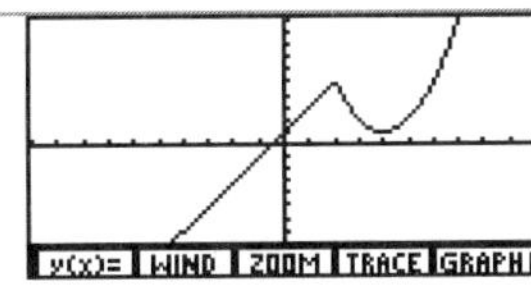 If you watch closely, you can see it graph the line and then the curve. In this case the two separate graphs meet, but piecewise functions do not have to meet.

Here is another example of a piecewise graph.	EX: $y = \begin{cases} x^3 + 9x^2 + 23x + 15 & x \le -1 \\ -2\sqrt{x+1} + 5 & x > -1 \end{cases}$ Once again as we type these in as Y1 and Y2 on our calculators, we must put parentheses around the functions (numerators) and the restrictions on the x-variables (denominators). 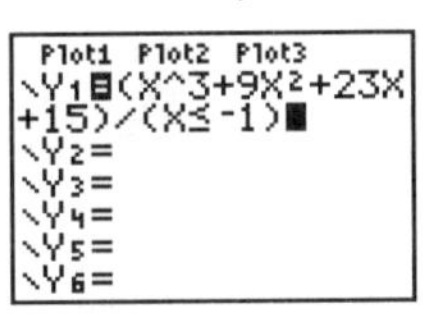 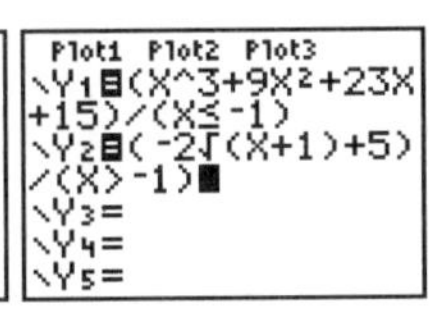 Now graph the function (Zstandard). 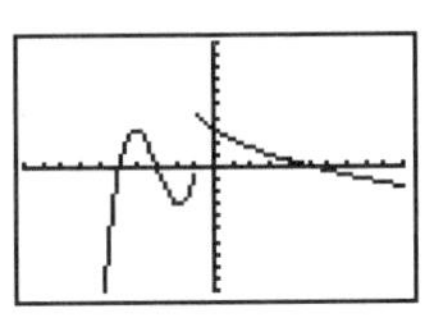 This time you can easily see the two distinct graphs that make up this piecewise function. Let's see how tracing works for piecewise functions. Remember that if you see Y= , it means that the function is undefined for	Once again as we type these in as Y1 and Y2 on our calculators, we must put parentheses around the functions (numerators) and the restrictions on the x-variables (denominators). Don't forget to put parentheses around the $x + 1$ in the square root. 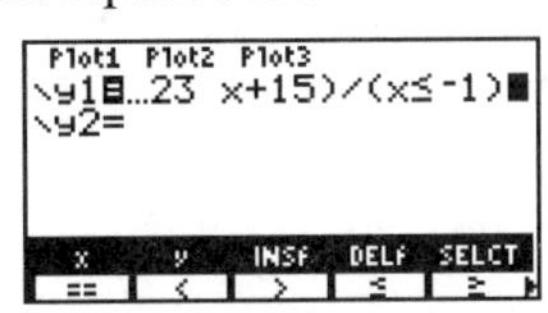 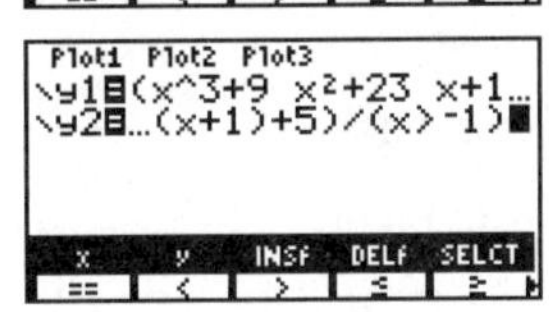 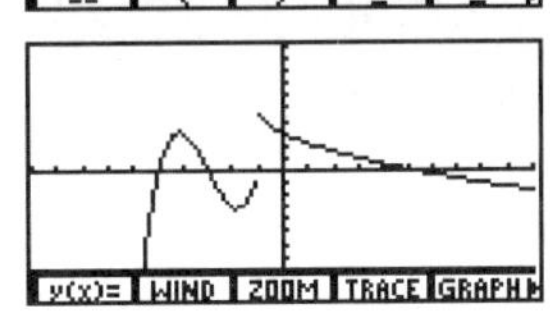 This time you can easily see the two distinct graphs that make up this piecewise function. Let's trace see how tracing works for piecewise functions. Remember

Answer Q2: Minimum = –422.10...

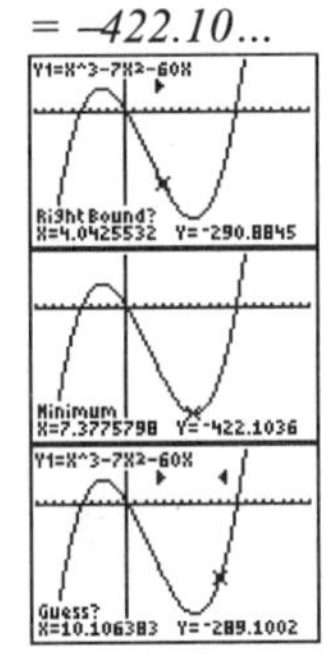

<table>
<tr>
<td></td>
<td>that x-value. If you trace to the left, you will find that we are on the first piece of the piecewise function.

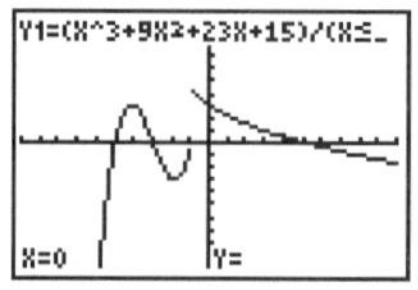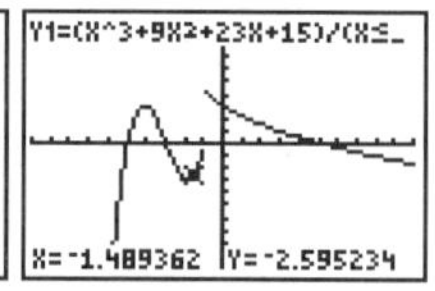

To trace the other piece of the function, push the down arrow, and then trace over to the right to get to a place where it is defined.

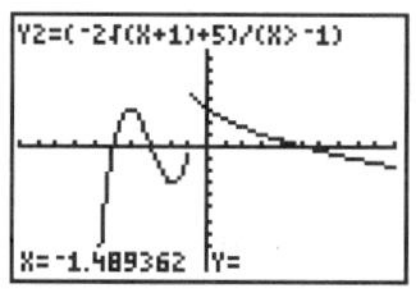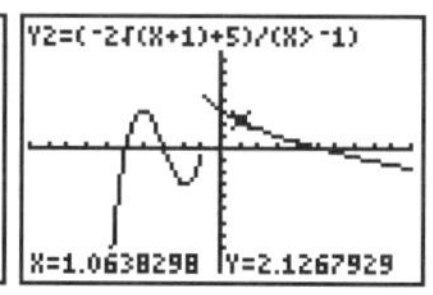

You can find max and min.'s on piecewise function the same way as you do on non-piecewise functions.</td>
<td>that if you see Y= , it means that the function is undefined for that x-value. If you trace to the left, you will find that we are on the first piece of the piecewise function.
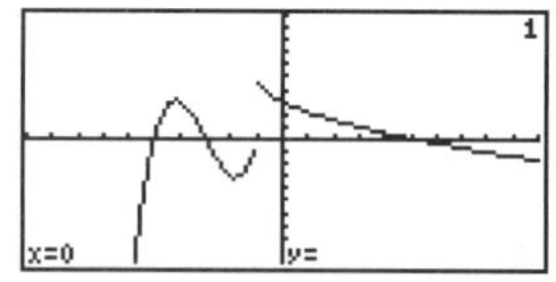
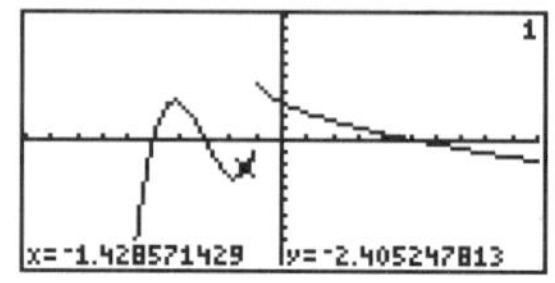
To trace the other piece of the function, push the down arrow, and then trace over to the right to get to a place where it is defined.
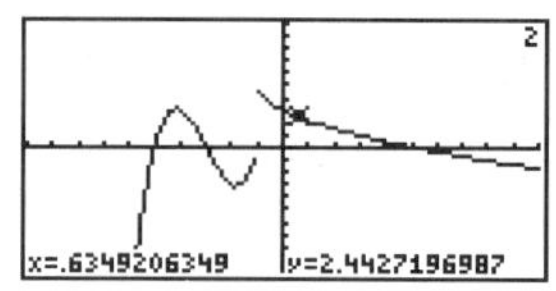</td>
</tr>
<tr>
<td>What to do if you have one part of a piecewise function that has a restriction on both ends. That is, it has an interval where there is a distinct starting and stopping point. As in this example $(-x+1)$ must be between</td>
<td>EX: $y=\begin{cases} -3 & x<-2 \\ -x+1 & -2\le x<3 \\ \frac{5}{x} & x\ge 3 \end{cases}$

We first must type in the function.
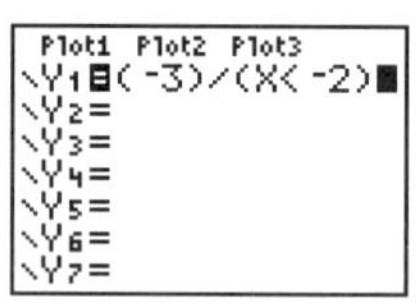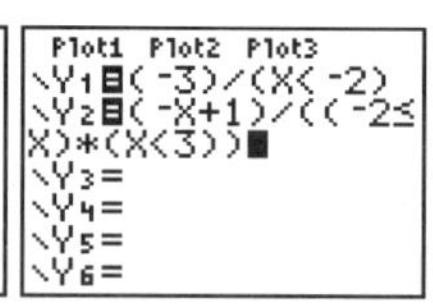
Take a good look at how we typed in the second part of this function. We had to input the restriction on the x-variable in two parts $((-2\le x)(x<3))$.</td>
<td>We first must type in the function.

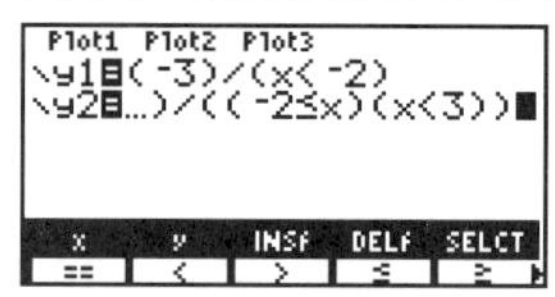

Take a good look at how we typed in the second part of this function. We had to input the restriction on the x-variable in two parts $((-2\le x)(x<3))$.</td>
</tr>
</table>

<table>
<tr>
<td>−2 and 3. The x-value must be greater than or equal to −2 and less than 3. We might say that this piece has a double restriction.</td>
<td>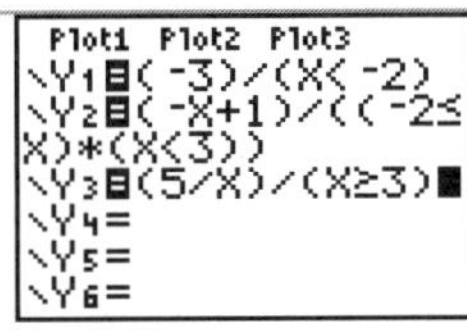

Now we go to ZOOM standard to take a look at our function.
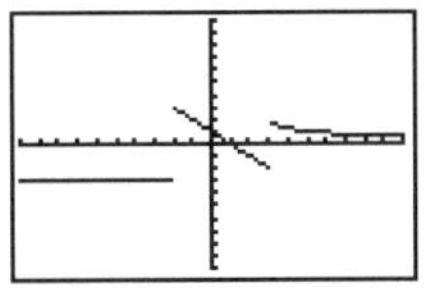</td>
<td>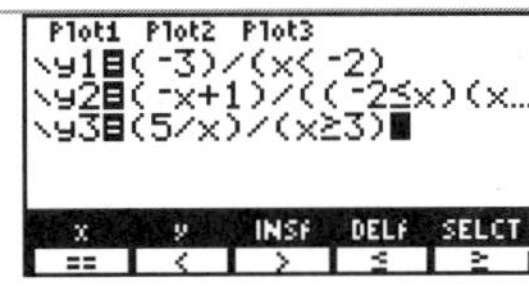

Now we go to ZOOM standard to take a look at our function.
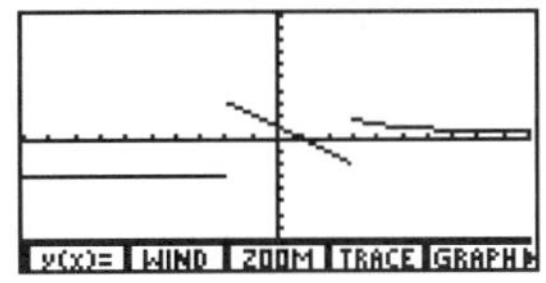
</td>
</tr>
</table>

Section Summary

- To find zeros/roots of a function, type in the function, go to the zero or root key on the calculator, identify for the calculator where to look for it, and then you are done.
- To find max or mins, you follow the same procedure as with finding zeros/roots.
- To graph a piecewise function, type in each piece as a separate function in parentheses and divide each piece by the restriction on x for that piece.

SECTION 0.8 HOMEWORK

(1–14) Use the zero/root finder on your calculator to find all the zeros/roots of each of the following functions. (If necessary, approximate to 4 decimal places.)

1. $y = 8x^2 - 2x - 15$
2. $y = 10x^2 + 41x - 77$
3. $y = 20x^3 - 53x^2 - 297x + 126$
4. $y = 10x^3 - 17x^2 - 172x + 35$
5. $y = 4x^4 - 11x^3 - 55x^2 + 47x + 15$
6. $y = 40x^4 - 62x^3 - 385x^2 - 61x + 132$
7. $y = x^3 + 7x^2 + 17x + 35$
8. $y = x^3 + 2x - 33$
9. $y = x^4 - 3x^3 + 3x^2 - 12x - 4$
10. $y = x^4 - 5x^3 + x^2 - 15x - 6$
11. $y = x^2 - 3x - 7$
12. $y = 3x^2 + 5x - 1$
13. $y = x^3 + 2x^2 - 6x - 9$
14. $y = -2x^3 + 9x^2 - 2x - 16$

(15–28) Use the maximum and minimum capability of your calculator to find all the different maximum and minimum values of y for each of the following functions. (If necessary, approximate to 4 decimal places.)

15. $y = 3x^2 + 12x - 1$
16. $y = x^2 - 3x + 2$
17. $y = -2x^2 + 16x - 3$
18. $y = -4x^2 + 9x + 5$
19. $y = 7x^2 - 13x + 1$
20. $y = -11x^2 + 5x - 9$
21. $y = 2x^3 - 3x^2 - 36x + 3$
22. $y = 2x^3 - 3x^2 - 72x - 5$
23. $y = x^4 - 4x^3 - 26x^2 + 60x - 9$
24. $y = x^4 + 4x^3 - 26x^2 - 60x + 15$
25. $y = 3x^4 + 22x^3 + 12x^2 - 30x + 7$
26. $y = 9x^4 - 4x^3 - 228x^2 - 288x + 100$
27. $y = 2x^3 - 9x^2 - 30x + 5$
28. $y = -2x^3 + 5x^2 + 8x - 10$

(29–36) Graph each of the following piecewise functions:

29. $y = \begin{cases} 2x+3 & x \le -3 \\ x^2 & x > -3 \end{cases}$
30. $y = \begin{cases} 3x+2 & x \le 1 \\ x^2 - 5 & x > 1 \end{cases}$
31. $y = \begin{cases} x^3 + 3 & x \le -1 \\ -x^2 + 5 & x > -1 \end{cases}$
32. $y = \begin{cases} x^2 - 12 & x \le -2 \\ 3x+1 & x > -2 \end{cases}$
33. $y = \begin{cases} 3x+2 & -3 \le x \le 1 \\ 5x-12 & 1 < x \le 4 \end{cases}$
34. $y = \begin{cases} x^2 - 3 & -3 \le x < 3 \\ 3x-5 & 3 \le x \le 5 \end{cases}$
35. $y = \begin{cases} 3x+1 & -2 \ge x \\ 2x+1 & -2 < x < 4` \\ x-9 & x \ge 4 \end{cases}$
36. $y = \begin{cases} x^2 - 9 & -3 \ge x \\ 5x+9 & -3 < x < -1 \\ 3x-4 & x \ge -1 \end{cases}$

Chapter 0 Review

Topic	Section	
Exponents	0.1	b^n means b times itself n times (repeated multiplication).
Properties of exponents	0.1	There are five major properties of exponents to remember. $b^m \bullet b^n = b^{m+n}$ Product Rule $\dfrac{b^m}{b^n} = b^{m-n}$ Quotient Rule $(b^m)^n = b^{mn}$ Power Rule $b^{-m} = \dfrac{1}{b^m}$ and $\dfrac{1}{b^{-m}} = b^m$ Negative Rule $b^0 = 1$ Zero Rule
Properties of radicals	0.1	There are two major properties of radicals to remember. $\sqrt[n]{a} \bullet \sqrt[n]{b} = \sqrt[n]{ab}$ Product rule $\dfrac{\sqrt[n]{a}}{\sqrt[n]{b}} = \sqrt[n]{\dfrac{a}{b}}$ Quotient rule
Scientific Notation	0.1	A way of expressing a number, such that it is in the form $m \times 10^n$ where $1 \leq m < 10$ and n is an integer.
Fractional exponents	0.1	$b^{\frac{p}{r}} = \left(\sqrt[r]{b}\right)^p$ or $\sqrt[r]{(b)^p}$
Guidelines for when to simplify radicals	0.1	When asked to simplify a radical expression, remember the three cases: 1) Don't leave anything in the radical that has a power larger than the root. 2) Don't leave fractions in a radical. 3) Don't leave roots in the denominator.
Adding and subtracting expressions	0.2	When we add or subtract expressions, we put together only those terms that have the same variables to the same powers.
Multiplying expressions	0.2	When we multiply expressions, we must multiply each term in the first expression times every term in the second expression.
Dividing expressions	0.2	When we divide expressions, we always compare the first term in the divisor with the first term in the dividend in order to figure out what to multiply by.
GCF factoring	0.3	Always look to factor out the greatest common factor before doing any other factoring.
Trinomial	0.3	If you have a three-term polynomial expression, try looking to see if it

factoring		factors into two binomials.
Special Binomial factoring	0.3	If it fits the formulas for the difference of two squares $(x^2 - y^2)$, or the sum and difference of two cubes $(x^3 + y^3, x^3 - y^3)$
Factoring by Grouping	0.3	If you have four or more terms, try grouping the terms together into two groups and look for a GCF in each grouping.
Simplifying, multiplying or dividing rational expressions	0.4	1. Factor all of the numerators and denominators in the expression. 2. Cancel all like factors in the numerators with the ones in the denominators.
When adding or subtracting rational expressions you must	0.4	1. Factor the denominators. 2. Find a Least Common Denominator (LCD). 3. Within each fraction, multiply the numerator and denominator by the same amount so as to make the denominator be the same as the LCD. 4. Add or subtract the numerators as indicated. 5. Then try simplifying what you have just completed.
Complex fractions	0.4	To simplify complex fractions, you multiply each fraction within the main fraction by the LCD of all of the fractions within it and then simplify what remains.
The big three concepts for solving linear equations	0.5	1. Get the variable alone. 2. Whatever is done to one side of the equation must also be done to the other side. 3. Do the opposite operation to move or eliminate something from one side of the equation.
Six universal steps used to solve linear equations.	0.5	1. Clear all parentheses (Distribute, FOIL, etc.) 2. Eliminate any denominators or decimals (optional). 3. Simplify what remains by collecting like terms. 4. Add or subtract to get the variable terms on the same side of the equation. 5. Add or subtract to get the constant terms on the other side of the equation from the variable. 6. Multiply or divide to get one variable equals the answer.
Results possible from	0.5	1. No solution 2. One solution

solving linear equations		3. An infinite number of solutions.
Definition of Absolute Value	0.5	$\lvert x\rvert = \begin{cases} -x & x < 0 \\ x & x \geq 0 \end{cases}$ Remember to isolate the absolute value before you apply either the definition or logic in order to solve an equation.
Be sure to know how to do the following on the graphing calculator	0.6-0.8	• Turn the calculator on (ON)and off (2nd ON). • Adjust the brightness (contrast) of the calculator screen (2nd, arrows). • Get to the HOME screen (2nd MODE on the TI-83, 2nd EXIT on the TI-86). • Use − for minus and this (-) for negative. • Use the ^ key to raise something to a power or root. • Use the FRAC command, located under the MATH menu, to turn decimal rational numbers into their fraction form. • Use parentheses with numerators, denominators and roots • Graph new functions by first making sure that Plots are not highlighted and that you have cleared out any old functions. • Use ZOOM standard or ZOOM decimal as a first window size. • Use the TRACE key and then by adjusting the window size find a complete graph of the function. • Find zeros/roots of a function. • Find max and/or min.'s. • Graph a piecewise function.

CHAPTER 0 REVIEW HOMEWORK

Section 0.1

(1–4) Evaluate.

1. 3^6 2. 5^{-3} 3. $9^{\frac{3}{2}}$ 4. $81^{\frac{-3}{4}}$

(5–14) Simplify. (Write without negative exponents)

5. $a^8b^7ab^6$ 6. $(4x^3y^2)(-3x^4y^3)$ 7. $(-2x^{-4}y^5)(-3x^{-3}y^{-3})$

8. $\left(a^{\frac{3}{4}}b^{\frac{2}{3}}\right)\left(a^{\frac{2}{5}}b^{\frac{1}{2}}\right)$ 9. $(2x^3y^{-2})^{-3}(3xy^2)^3$ 10. $\left(x^{\frac{3}{4}}y^{\frac{1}{2}}\right)^8\left(x^{\frac{-2}{3}}y^{\frac{3}{4}}\right)^{12}$

11. $\dfrac{9a^5b^{-3}}{20xy^3}\bullet\dfrac{25x^3y^5}{12ab^5}$ 12. $\left(\dfrac{a^3b^5}{x^2y}\right)^2\left(\dfrac{xy^4}{ab^3}\right)^4$ 13. $\dfrac{10x^3y}{21a^4b}\div\dfrac{5xy^4}{7a^3b^5}$

14. $\dfrac{7a^3b^{-2}}{15x^{-3}y^2}\div\dfrac{21ab^3}{5xy^4}$

(15–16) Rewrite each of the following with scientific notation.

15. 0.00000832 16. 5,980,000,000

(17–18) Rewrite each of the following without scientific notation.

17. $2.93 \text{ x } 10^{-8}$ 18. $9.34 \text{ x } 10^9$

(19–28) Simplify (assume all variables represent positive numbers).

19. $\sqrt{80}$ 20. $5\sqrt{63}$ 21. $\sqrt[3]{135}$

22. $\sqrt{50a^3b^6}$ 23. $\sqrt[4]{x^{15}y^9z^8}$ 24. $\sqrt{\dfrac{5}{3}}$

25. $\dfrac{6}{\sqrt{3}}$ 26. $\dfrac{9}{\sqrt{27}}$ 27. $\dfrac{5}{2+\sqrt{3}}$

28. $\dfrac{3+\sqrt{2}}{7-\sqrt{2}}$

(29–30) Add or subtract.

29. $3\sqrt{5}+8\sqrt{7}-9\sqrt{5}+4\sqrt{7}$ 30. $5\sqrt{3}+7\sqrt{50}-13\sqrt{2}-9\sqrt{27}$

(31–34) Multiply.

31. $3\sqrt{7}\left(4+2\sqrt{7}\right)$ 32. $\left(5-2\sqrt{3}\right)\left(2+3\sqrt{3}\right)$ 33. $\left(8-\sqrt{3}\right)\left(8+\sqrt{3}\right)$

34. $\left(2\sqrt{3}+5\sqrt{2}\right)\left(2\sqrt{3}-5\sqrt{2}\right)$

Section 0.2

(35–38) Add or subtract the following polynomials:

35. $(5x^3-7x^2+9)+(8x^3-12x^2-15)$ 36. $(7a^3b-9a^2b^2+12ab^3)+(9a^3b-13a^2b^2+9ab^3)$

37. $(8y^3+9y-10)-(5y^3-7y+12)$ 38. $(5x^2y-9xy-7xy^2)-(7x^2y+8xy-15xy^2)$

(39–48) Multiply and simplify each of the following polynomials:

39. $5a(3a^2 - 8a + 2)$
40. $-5x^2y(3x^3y - 4xy - 9xy^2)$
41. $(5a - 4)(3a - 2)$
42. $(5x - 2y)(7x + 3y)$
43. $(2x - 3)(5x^2 + 2x + 3)$
44. $(2x - 3y)(4x^2 + 6xy + 9y^2)$
45. $(3x - 5)(3x + 5)$
46. $(5a^2b + 9x)(5a^2b - 9x)$
47. $(5b + 3)^2$
48. $(3x - 7y)^2$

(49–52) Divide each of the following polynomials:

49. $\dfrac{24a^3b}{16a^2b^4}$
50. $\dfrac{9x^5 - 15x^4 + 24x^3}{12x^2}$
51. $(3x^3 + 7x^2 - 3x + 2) \div (x + 2)$
52. $(3x^4 - 19x^2 + 4) \div (x - 3)$

Section 0.3

(53–80) Factor completely each of the polynomials. (If not factorable, write "prime.")

53. $9ax - 27a$
54. $20b^3y^3 + 15b^5y^2$
55. $x^2 - 2x - 63$
56. $y^2 + 7y - 120$
57. $a^2 + 23ab + 120b^2$
58. $a^2x^2 - 19abx + 190b^2$
59. $3x^2 + 14x + 15$
60. $6a^2 - 23a + 21$
61. $8y^2 - 14by - 15b^2$
62. $15a^2 + 13ax - 20x^2$
63. $x^4 + 12x^2 + 35$
64. $9x^2 + 30x + 25$
65. $16a^2 - 56ab + 49b^2$
66. $x^2 + 5x - 4$
67. $y^2 - 121$
68. $4x^2 - 25y^2$
69. $9a^2 + 16b^2$
70. $a^3 - 64$
71. $8x^3 + 27y^3$
72. $ax + 3a + bx + 3b$
73. $by - 3b + 3y - 9$
74. $x^2 - 6x + 9 - y^2$
75. $ax^2 - 5x^2 - 9x + 45$
76. $3ax^2 - 9ax - 162a$
77. $12by^2 - 2by - 30b$
78. $x^4 - 81$
79. $x^4 - 13x^2 + 36$
80. $5y^3 + 625$

Section 0.4

(81–86) Reduce each of the rational polynomials.

81. $\dfrac{28x^3y^5}{24x^6y^2}$
82. $\dfrac{7x - 14}{5x - 10}$
83. $\dfrac{3x - 12}{8 - 2x}$
84. $\dfrac{5x + 20}{5}$
85. $\dfrac{y^2 - 16}{y^2 - 4y - 32}$
86. $\dfrac{6y^2 + 5y - 6}{9y^2 - 3y - 2}$

(87–94) Multiply or divide each of the following rational polynomials:

87. $\dfrac{9a^3b^3}{20ab^2} \bullet \dfrac{25a^3b}{27a^2b^4}$
88. $\dfrac{36x^3y}{7xy^4} \div \dfrac{18xy^2}{35x^3y^5}$
89. $\dfrac{x + 3}{x^2 - 25} \bullet \dfrac{x^2 + 6x + 5}{5x + 15}$
90. $\dfrac{y^2 - 36}{y^2 - y - 30} \bullet \dfrac{y^2 + 8y + 15}{y^2 + 9y + 18}$
91. $\dfrac{6x^2 - 7x - 3}{15x^2 + 4x - 4} \bullet \dfrac{6x^2 + 13x + 6}{4x^2 - 9}$
92. $\dfrac{10 - 5x}{x^2 + 11x + 24} \div \dfrac{x^2 - 4}{x^2 - 9}$

93. $\dfrac{x^2+12x+27}{x^2+4x-45} \div \dfrac{x^2-9}{x^2-25}$

94. $\dfrac{a^3-27}{a^2-10a+21} \div \dfrac{a^2+3a+9}{a^2+2a-63}$

(95–101) Add or subtract the following rational polynomials:

95. $\dfrac{5x}{x+2}+\dfrac{10}{x+2}$

96. $\dfrac{y^2}{y-5}-\dfrac{25}{y-5}$

97. $\dfrac{2x}{x+3}+\dfrac{x}{x-2}$

98. $\dfrac{5}{2-y}-\dfrac{3}{y-2}$

99. $\dfrac{3a}{a^2-9}+\dfrac{5}{2a-6}$

100. $\dfrac{7}{b^2-25}+\dfrac{3}{b^2+10b+25}$

101. $\dfrac{5}{x^2-9}+\dfrac{2}{x^2+2x-15}-\dfrac{7}{x^2+8x+15}$

(102–107) Change each of the following complex fractions to simple fractions:

102. $\dfrac{\frac{3}{8}+\frac{1}{6}}{\frac{3}{4}-\frac{1}{2}}$

103. $\dfrac{\frac{5}{b}+3}{7-\frac{8}{b}}$

104. $\dfrac{5-\frac{3}{x-3}}{\frac{2}{x-3}+3}$

105. $\dfrac{x^{-1}+2}{3-x^{-2}}$

106. $\dfrac{\frac{2}{a^2}-\frac{3}{b^3}}{\frac{5}{a^3}+\frac{2}{b}}$

107. $\dfrac{\frac{y^4}{y^2-25}}{\frac{y^2}{y^2+10y+25}}$

Section 0.5

(108–118) Solve each of the following linear equations:

108. $5x+7=27$

109. $7x-5=3x+7$

110. $7(3-2x)+4=-3$

111. $3(2x-1)+4=5-2(5-2x)$

112. $2(3-2x)-13=5-4(3+x)$

113. $3(3x+5)-5x=4(x-3)+1$

114. $\dfrac{3}{4}x-\dfrac{5}{12}=x+\dfrac{3}{8}$

115. $\dfrac{2x+5}{2}=\dfrac{3x-2}{3}$

116. $\dfrac{3x+1}{3}+\dfrac{2x}{4}=\dfrac{2x-3}{6}$

117. $\dfrac{2}{3}(2x+1)+3=\dfrac{5}{6}(x-3)+2$

118. $5-\dfrac{3}{4}(3x-1)=\dfrac{5}{8}(3-2x)$

(119–126) Solve each of the following absolute value equations:

119. $|3x+2|=7$

120. $|2x-3|+4=9$

121. $|5x-2|+7=2$

122. $|3x+5|+8=8$

123. $2|5x+3|-2=12$

124. $\left|\dfrac{3}{4}x-\dfrac{2}{3}\right|=4$

125. $-3|2x+1|+7=1$

126. $-2|3x+5|+2=8$

Section 0.6

(127–136) Evaluate each of the following problems with your calculator:

127. 5^7

128. $\sqrt[4]{194481}$

129. $\dfrac{5\bullet 3+7}{2\bullet 8-5}$

130. $\dfrac{9-3^4}{4^3-48}$

131. $125^{\frac{2}{3}}$

132. 5^{-2}

133. $\dfrac{4-2\bullet 3^2}{3\bullet 2^3-2\bullet 11}$

134. $\left(\dfrac{1}{81}\right)^{\frac{-3}{4}}$

135. $20000\left(1+\dfrac{0.08}{6}\right)^{6\bullet 20}$

136. $\dfrac{30000\left(\dfrac{0.075}{12}\right)}{1-\left(1+\dfrac{0.075}{12}\right)^{-12\bullet 5}}$

(137–142) Compute each of the following problems with your calculator and convert the answers to fractions:

137. $\dfrac{5}{23}+\dfrac{7}{15}$

138. $\dfrac{9}{52}-\dfrac{7}{30}$

139. $\dfrac{3\bullet 8+7}{7-2\bullet 9}$

140. $\dfrac{5\bullet 13+11}{3\bullet 19-4}$

141. $\dfrac{5+3^2}{8-2^{-3}}$

142. $\dfrac{7+81^{\frac{-1}{2}}}{5+3^{-2}}$

Section 0.7

(143–146) For each of the following functions, graph first in the zoom standard window and then graph in the given windows and scales:

143. $y = x^2 - 8x - 10$

a. $-6 \le x \le 10$ Scale 1
   $-30 \le y \le 30$ Scale 3

b. $-6 \le x \le 10$ Scale 1
   $-20 \le y \le 20$ Scale 2

c. Which window do you think best fits the function?

144. $y = -x^2 + 2x + 20$

a. $-8 \le x \le 10$ Scale 1
   $-20 \le y \le 20$ Scale 2

b. $-8 \le x \le 10$ Scale 1
   $-30 \le y \le 30$ Scale 3

c. Which window do you think best fits the function?

145. $y = x^4 - 3x^2 + 12$

a. $-4 \le x \le 4$ Scale 0.4
   $-30 \le y \le 30$ Scale 3

b. $-4 \le x \le 4$ Scale 0.4

$-5 \le y \le 5$ Scale 0.5

c. Which window do you think best fits the function?

146. $y = 20{,}000\left(1 + \frac{0.09}{12}\right)^{12x}$

a. $0 \le x \le 20$ Scale 2

$0 \le y \le 120{,}000$ Scale 10,000

b. $0 \le x \le 20$

$0 \le y \le 80{,}000$ Scale 5,000

c. Which window do you think best fits the function?

(147–150) For each of the following functions find a good window for graphing. (Answers may vary)

147. $y = -x^2 + 10x - 3$

148. $y = x^2 - 6x - 15$

149. $y = -x^4 + 4x^2 + 12$

150. $y = 15{,}000(1 + 0.08)^x$ (Set $0 \le x \le 20$, Scale 2 and $0 \le y \le ?$, Scale ?)

(151–154) Graph each of the following functions, and find the y–value for the given x–value using the graphing calculator.

151. $y = x^3 - 5x^2 + 2x + 3;\ x = 8$

152. $y = -x^4 + x^3 + x^2 - 2;\ x = -6$

153. $y = 5{,}000(1 + 0.08x);\ x = 15$

154. $y = 5{,}000(1 + 0.08)^x;\ x = 15$

Section 0.8

(155–158) Use the zero/root finder on your calculator to find all the zeros/roots of each of the following functions. (If necessary, approximate to 4 decimal places.)

155. $y = 5x^2 + 12x - 32$

156. $y = 8x^3 - 14x^2 - 53x + 14$

157. $y = x^2 - 5x - 2$

158. $y = x^3 - 14x + 15$

(159–162) Use the maximum and minimum capability of your calculator to find all the different maximum and minimum values of y for each of the following functions. (If necessary, approximate to 4 decimal places.)

159. $y = x^2 - 6x + 4$

160. $y = -2x^2 - 10x + 3$

161. $y = x^3 + 3x^2 - 45x + 4$

162. $y = 3x^4 + 4x^3 - 60x^2 - 144x + 9$

(163–165) Graph each of the following piecewise functions.

163. $y = \begin{cases} 3x + 2 & x \le -1 \\ x^2 - 3 & x > 1 \end{cases}$

164. $y = \begin{cases} x^2 - 4 & -3 \le x < 3 \\ 3x - 7 & x \ge 3 \end{cases}$

165. $y = \begin{cases} 2x + 5 & x \le -2 \\ x + 3 & -2 \le x < 5 \\ x - 9 & x \ge 5 \end{cases}$

CHAPTER 0 EXAM

(1–2) Evaluate.

1. 4^{-3}
2. $16^{\frac{3}{4}}$

(3–5) Simplify (write without negative exponents).

3. $(-5x^3y^2z)(4x^4y^{-3}z^{-5})$
4. $(-2a^3b^2)^2(3ab^3)^{-3}$
5. $\left(x^{\frac{2}{3}}y^{\frac{1}{2}}\right)\left(x^{\frac{3}{4}}y^{\frac{5}{8}}\right)$

(6–7) Rewrite with scientific notation.

6. 583,000,000,000
7. 0.00002857

(8–13) Simplify (assume variables represent positive numbers).

8. $5\sqrt{98}$
9. $\sqrt[3]{875}$
10. $\sqrt{108x^5y^6}$
11. $\sqrt{\frac{5}{7}}$
12. $\frac{8}{\sqrt{6}}$
13. $\frac{3-\sqrt{2}}{5+\sqrt{2}}$

(14–15) Add or subtract.

14. $7\sqrt{5}-8\sqrt{3}+9\sqrt{5}-2\sqrt{3}$
15. $3\sqrt{12}+4\sqrt{50}-9\sqrt{48}+7\sqrt{8}$

(16–17) Multiply.

16. $3\sqrt{3}\left(8-7\sqrt{3}\right)$
17. $\left(5+3\sqrt{2}\right)\left(8-2\sqrt{2}\right)$

(18–19) Add or subtract the following polynomials.

18. $(3y^2-7y+9)+(8y^2-9y-15)$
19. $(7x^2+8xy-12y^2)-(9x^2-5xy+3y^2)$

(20–23) Multiply the following polynomials.

20. $3x^2(5x^2-2x+3)$
21. $(3x+2y)(5x-7y)$
22. $(3a-2)(5a^2-4a+7)$
23. $(7a-3b)^2$

(24–25) Divide the following polynomials

24. $\frac{8x^3-4x^2}{2x}$
25. $\left(3x^3-8x^2+3\right)\div(x-3)$

(26–33) Factor each of the following polynomials.

26. $10a^3b^2-15a^4b^3+20a^3b^4$
27. $y^2-7y-60$
28. $a^2-14ab+33b^2$
29. $3x^2+26x+16$
30. $4x^2+20ax+25a^2$
31. $25x^2-49$
32. $ax+3a+5x+15$
33. y^3-64

(34–35) Reduce each of the following rational polynomials.

34. $\frac{15-5x}{x^2-9}$
35. $\frac{x^2-2x-15}{2x^2-7x-15}$

(36–39) Multiply or divide each of the following rational polynomials.

36. $\dfrac{10a^5b^2}{27x^3y} \bullet \dfrac{21xy^3}{25a^2b^2}$

37. $\dfrac{33x^5y^2}{19ab^5} \div \dfrac{55x^2y^3}{38a^3b}$

38. $\dfrac{x^2-2x-15}{x^2+10x+21} \bullet \dfrac{2x^2+11x-21}{2x^2+7x-15}$

39. $\dfrac{2x^2-7x-15}{x^2-25} \div \dfrac{6x^2+7x-3}{3x^2+14x-5}$

(40–41) Add or subtract the following rational polynomials:

40. $\dfrac{5}{x^2-9} + \dfrac{-3}{6-2x}$

41. $\dfrac{5}{x^2-4} - \dfrac{3}{x^2-5x+6}$

(42–43) Change each of the following complex fractions to simple fractions:

42. $\dfrac{\frac{3}{x}+2}{5-\frac{2}{x}}$

43. $\dfrac{3a^{-2}+2}{5b^{-1}-3}$

(44–45) Solve each of the following linear equations:

44. $3(2-4x)+8x=6-4x$

45. $\dfrac{3}{5}x+4=\dfrac{7}{10}x-2$

(46–49) Solve each of the following absolute value equations:

46. $|2x-1|=5$

47. $|3x+2|+3=7$

48. $|5x-1|+7=2$

49. $3|2x-7|-4=8$

(50–54) Evaluate the following with your calculator. (Write all answers as fractions.)

50. $\dfrac{5 \bullet 7-3}{17-2 \bullet 5}$

51. $\left(\dfrac{2}{5}\right)^{-3}$

52. $\sqrt[5]{6436343}$

53. $1296^{\frac{3}{4}}$

54. $\dfrac{5-2^5}{3^4-7}$

(55–56) For each of the following functions, find a good window for graphing. (Answers may vary.)

55. $y=2x^2-12x+3$

56. $y=15{,}000(1+0.09)^x$ (Set $0 \le x \le 30$, scale 3 and $0 \le y \le ?$, scale?)

(57–58) Graph each of the following functions and find the y–value for the given x–value using the graphing calculator:

57. $y=2x^3-5x^2-3x+4;\ x=8$

58. $y=2{,}000(1+0.095)^x$

(59–60) Use the zero/root finder on your calculator to find all the zeros/roots of each of the following functions. (If necessary, approximate to four decimal places.)

59. $y=4x^2+9x-9$

60. $y=3x^2-x-5$

(61–62) Use the maximum and minimum commands in your calculator to find all the different maximum and minimum values of the following functions. (If necessary, approximate to four decimal places.)

61. $y = -3x^2 + 12x + 2$

62. $y = 4x^3 + 3x^2 - 36x + 4$

(63–64) Graph each of the following piecewise functions.

63. $y = \begin{cases} x^2 - 5 & x < -1 \\ 2x + 3 & x \geq -1 \end{cases}$

64. $y = \begin{cases} 2x + 5 & x \leq -2 \\ 3x + 1 & -2 < x \leq 2 \\ x - 4 & x > 2 \end{cases}$

Chapter 1 Functions

In this chapter you are going to learn about a topic called functions. Functions are rules that relate one idea to another in a special way. They show up in many aspects of life. You will be looking at this topic from several different perspectives involving graphs, formulas and tables.

Section 1.1 Relations and Functions – An Informal Introduction

We begin our discussion with the idea of **relations**. A relation is any rule with inputs that cause outputs. Let's look at a few discussion items.

Discussion 1: Soda Pop Machine

Consider a soda pop machine that charges a dollar for a can of soda pop. Let's pretend you have fed a dollar into the machine, pushed a button and out came your favorite soda pop. In math we would call the dollar you put into the machine an "**input**" and the soda pop that came out of the machine an "**output**."

Question 1: Does only one type of soda pop come out of a typical pop machine or can you get several different kinds of soda pops from your input of a dollar?

You can look at this discussion item from another viewpoint as well. If you walk up to a soda pop machine and a person is standing next to it with a Pepsi in his or her hand (just purchased from this machine), do you know how much it cost the person to purchase that soda? The answer is yes, because a soda pop purchased from this machine costs one dollar. This viewpoint treats the Pepsi as the input and the dollar as the output.

Question 2: Could you answer that same question if the person had a Sprite in his or her hand instead, or maybe a Mountain Dew?

So, in the first viewpoint, an input of one dollar can yield *many outputs* (kinds of soda pop) and in the second viewpoint one soda pop in the hand (input) yields *only one possible answer or output*, one dollar. In this chapter, you are going to learn how important this difference is.

Discussion 2: Sales Tax

Think about sales tax. When you purchase something at a store, you usually have to pay some sales tax. For example, if you buy a shirt at a clothing store for $20 and the sales tax rate is 7% then the entire purchase would cost you $20 + .07 • $20 = $21.40. Again you may look at this situation from two viewpoints. You may consider the price of the shirt ($20) as the input and the after tax, or final, cost of the item ($21.40) as the output.

Question 3: Can there be more than one possible final cost (output) if you consider the cost of the item as the input?

Now consider the after tax cost, or final cost, of an item as the input and the price of an item itself as the output. If you know that your friend purchased something for a final cost of $35.24, would you know what he or she bought? Well, of course you wouldn't.

Let's compare the two viewpoints again. From the first viewpoint, if you have an input of the price of a $20 shirt, then there is only one output possible, a final cost of $21.40. But, if you treat the after tax, or final, cost as the input, say $35.24, there may be hundreds of things (outputs) that might have been purchased for that final cost. Once again, notice the difference in how many outputs come from a single input in the two viewpoints.

Discussion 3: Pumping Gasoline

Think about the last time you put gas into your car. First you pumped the gas into your car; then you paid for it. From one viewpoint you can consider the number of gallons pumped into your tank as the input and the cost of the gas as the output. If gas costs $1.39 per gallon, an input of 13.24 gallons would cause you to pay $18.40 (the output).

You also can look at this situation from another viewpoint. You may treat the amount paid as the input and the number of gallons pumped as the output. Assuming that a person at the same gas station that same day told you he or she paid $22.24 for gas, could you know exactly how much gas they pumped into the car? Yes, because if you divide $22.24 by $1.39 you would get 16 gallons. So this is a discussion item where, in both viewpoints, a single input yields a single output. This is slightly different than the first two discussion items.

Question 4: Look back at the three discussion items and the two different points of view for each discussion item. Are there any similarities between any of the six viewpoints?

Let us introduce some vocabulary for any events like the ones above. We've discovered two different types of relations. In the first type at least some of the inputs (if not all the inputs) cause *many outputs*. This is simply called a relation.

We will call anything with inputs that cause one or more outputs a **relation**.

The second type, where each input causes *only one output*, is called a function.

Functions are simply special relations that have only one output for each input.

Let's go through a few additional discussion items and see if you can decide whether each is just a relation or more specifically a function.

Answer Q1: You can get several different soda pops (outputs) for a dollar (input).

Discussion 4: The Dow Jones Industrial Average

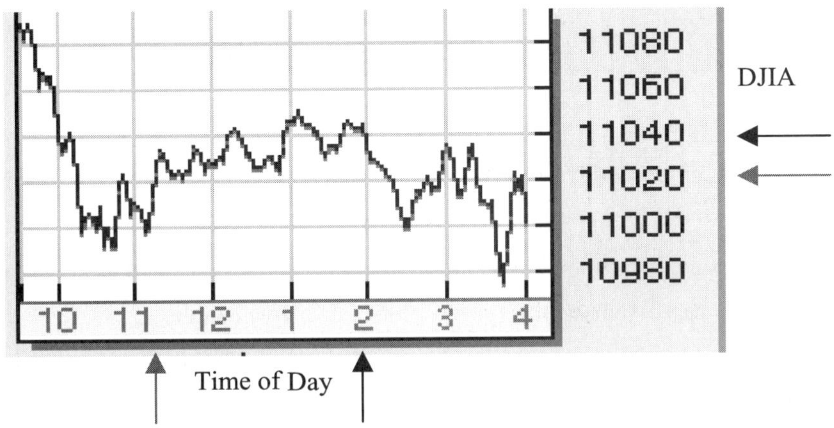

This graph gives information about the Dow Jones Industrial Average (DJIA) during a particular day in 1999. The numbers along the bottom of the graph represent the time of day and the numbers along the right hand side of the graph represent the value in dollars of the DJIA. We can consider the time of day as the input, and the value of the DJIA as the output. Looking at the graph from this viewpoint, when given an input of 2:00 p.m. you can see that the output, the DJIA, is approximately \$11,040 (See black arrows). Given an input of 11:15 a.m. you can see from the graph that the output is approximately \$11,020 (See red arrows).

Answer Q2: Yes all soda pops in this machine cost one dollar.

Question 5: Is this viewpoint, which treats the time of day as inputs and the DJIA as outputs, a relation or more specifically a function? Why?

Now consider the viewpoint where the value of the DJIA is the input and the time when that value occurs is the output. Given an input of a DJIA of \$11,040, the outputs, the times of day that the Dow was at this value, are 2:00 p.m., 1:40 p.m., 1:20 p.m., 12:50 p.m., 12:20 p.m., 12:10 p.m., 10:10 a.m., and 10:00 a.m. Similarly, given an input of a DJIA of \$10,980, the outputs, the times of day that the Dow was at this value, are 3:40 p.m., and 3:45 p.m. Notice that for some inputs in the last point of view you get several outputs.

Question 6: Is this viewpoint of the DJIA inputs and time outputs a relation or more specifically a function? Why?

Discussion 5: Mortgage Rates

Answer Q3: No, you pay only one final price at the checkout counter.

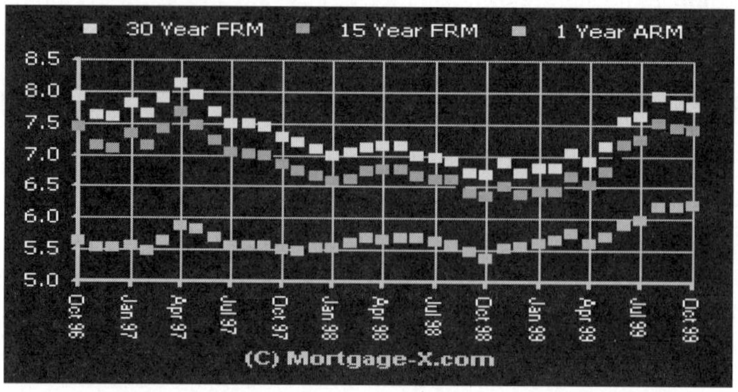

This graph contains average interest rates for 15-year and 30-year fixed rate mortgages and 1-year adjustable rate mortgages. It specifically shows the average new home interest rate for each quarter from October 1996 through October 1999. Let us investigate the two different ways, or viewpoints, in which we can look at the inputs and outputs of this graph. We will concentrate on the yellow graph, which shows 30-year fixed rate mortgages. Notice that given an input of July '97 the output is a 7.5% interest rate. Also, if given an input of July '98, the output is a 7.0% interest rate.

Question 7: Is this a relation or more specifically a function? Why?

If you now take a look at the other viewpoint, that is, given an input of an 8.0% 30-year interest rate, you should see that the output (date) from this graph is May 97 and August 99.

Question 8: Is this a relation or more specifically a function? Why?

Answer Q4: There are two different groups: one where, given an input, there are many outputs and another where, given an input, there is only one output.

Discussion 6: Temperature in Degrees Fahrenheit and Celsius

The formula $F=\frac{9}{5}C+32$ changes Celsius degrees (inputs) to Fahrenheit degrees (outputs). If you input a Celsius temperature of 25° your output is a temperature of 77° Fahrenheit ($\frac{9}{5}(25^\circ)+32=9(5)+32=45+32=77^\circ$). So traveling from the United States of America to just about any other country in the world, you could use this formula to change Celsius temperatures (inputs) to Fahrenheit temperatures (outputs). Doing this would give you a better idea of how to dress for the weather. Likewise, someone from Europe could use the formula $C=\frac{5}{9}(F-32)$ when visiting the U.S.A to convert a Fahrenheit temperature to a Celsius temperature. If a person was visiting Buffalo, New York in January and the temperature was -13° Fahrenheit outside, he or she could input the -13° into the formula and get an output of -25° Celsius. That person would then realize that their visit should have been in June, not January.

Question 9: Are these two formulas relations, or more specifically, functions - or one of each? Why?

Discussion 7: Student Enrollment in Two Year Colleges

# of Students in Two Year Colleges (in Thousands)	Year
3616	1998
3272	1997
3329	1996
3121	1995
3318	1994
3352	1993
3442	1992
3496	1991
3190	1990

Answer Q5: Function, because at any given time, input, there is only one output, DJIA.

This table shows the number of students enrolled in two-year colleges in thousands for the years 1990 - 1998. For example, there were 3,272,000 students in two-year colleges in 1997. When using a table it is customary to treat the first column as inputs and the second column as outputs. So for this table an input of 3121 (that is, 3,121,000 students) yields an output of 1995.

Question 10: Is this a relation or more specifically a function? Why?

It is possible, with more data gathered from the real world, that a different table of the same data, might turn out not to represent a function. If, for example, we added data for 1999, and it happened that in 1999 the number of students was 3,318,000 then that table wouldn't represent a function anymore, since an input of 3318 in that table would yield two possible outputs, 1994 and 1999.

Answer Q6: A relation, because for some of the inputs (the DJIA), it is possible to get more than one output, the times of day.

Discussion 8: Average Temperatures in Phoenix, Arizona

Month	Jan.	Feb.	Mar.	Apr.	May	June	July	Aug.	Sept.	Oct.	Nov.	Dec.
Average Temp.	54	58	62	70	79	88	94	92	86	75	62	54

This is a table of the average temperatures in Phoenix, Arizona by month in degrees Fahrenheit rounded off to the nearest degree. Notice that an input of May yields an output of 79°. But unlike the last Discussion item, if you switch viewpoints, then an input of 62° would yield two outputs, March and November.

Question 11: Are both viewpoints relations, or more specifically functions, or do you get one of each type?

Answer Q7
This is a function because one input, month, yields only one output, interest rate.

We can identify functions visually by using something called the **vertical line test**. We can apply this test to a graph of a relation. Imagine drawing vertical lines up and down everywhere on the graph of the relation. If these imaginary lines don't cross the graph more than once anywhere, then the relation is a function. If an imaginary vertical line does cross the graph of the relation more than once, that would mean that one input yields two or more outputs, and we would just call it a relation. Here are two examples:

Not a Function

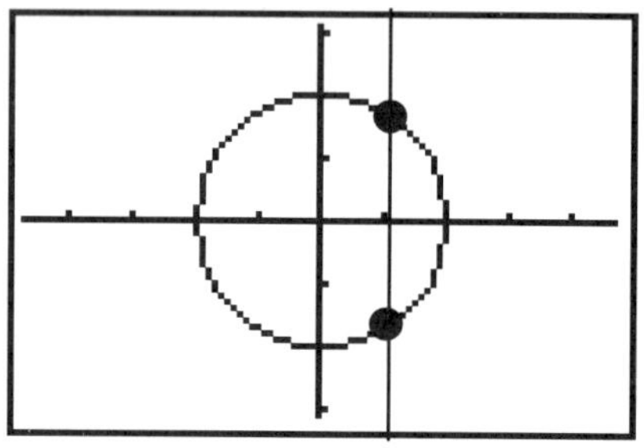

Function

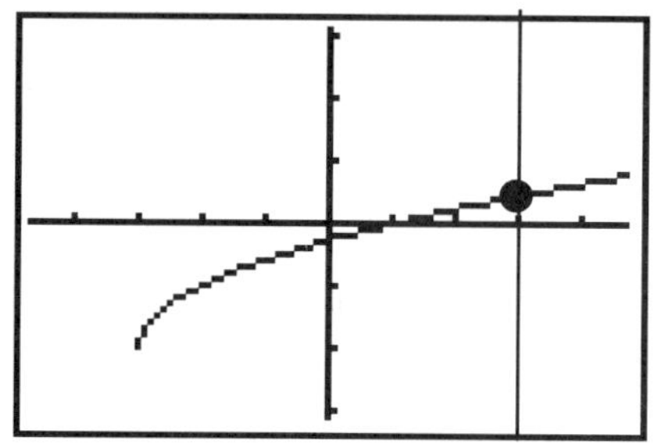

Answer Q8:
This is a relation because some of the inputs yield several outputs.

Let's turn our attention to formulas. Consider the formula $x = y^2$. This formula represents a relationship between x and y because given a value for x you can calculate a value for y. For example, if you plug in 9 for x $(9 = y^2)$ you get 3 and -3 for y. (Notice that $9 = (3)^2$ and $9 = (-3)^2$).

Question 12: Considering x as the input and y as the output, is $x = y^2$ a function?

You might now ask the question, "How can you determine if a formula is a function without having to plug in every possible input to see if any of them have more than one output?"

One guideline to follow is to remember that most elementary formulas are functions unless certain terms are present. The terms $|y|$ and y^2 or y to any even power will cause a formula to have more than one output for some of the inputs. So formulas that contain terms such as y^2 or $-7y^{12}$ or $|y+3|$ will not be functions.

For example, if the formula is $x = |y|$ where x is the input and y the output, then an input of 2 yields outputs 2 and -2 since $2 = |2|$ and $2 = |-2|$. This formula describes a relation, not a function. Most formulas that we will see in this book will represent functions.

Question 13: Which of the following are functions of x inputs and y outputs?

a.) $y = 3x - 5$ b.) $3x = 7 - y^2$ c.) $y = 4x^2 - 5x + 7$

d.) $x^2 - y^4 = 1$ e.) $y = -4$ f.) $y = |2x - 5|$

FUNCTION NOTATION

You have looked at many relations and functions in the form of graphs, formulas, and tables. It's time to introduce notation that will be used throughout the rest of this book. Look at the graph from Discussion 4 (DJIA). We'll use the notation $V(t)$ (read "V of t") to represent the real-life problem symbolically. The t represents the input value and stands for time in hours. $V(t)$, the whole symbol, represents the output value of the DJIA at the time t. The V itself is just the name of the function. We have chosen this letter because it's the first letter of the words "Value of the DJIA". To recap, V is the function's name, t is the input and $V(t)$, the whole symbol, is the output.

Example 1: Function Notation

Let's revisit Discussion 4 and the DJIA using the new notation. Find V(2:00), which means, find the DJIA at 2:00 p.m..

Solution:

Your answer is V(2:00) = \$11,040. Notice that this notation tells you the input (2:00 p.m.) and the output (\$11,040), all in one nice, neat concise form.

Answer Q9: These are both functions because one input yields only one output in both cases.

Question 14: What does V(11:15) equal?

Answer Q10: This is a function because one input yields only one output in every case.

Recall from Discussion 4 that every value of t yields only one value for $V(t)$; thus it is a function. Mathematicians say, "**V is a function of t**" which means that the value of the DJIA is a function of the time of day. Both t and $V(t)$ are called **variables** because their values can vary. Consider the following observation: we can choose as input any value we want for t from the range of values on the graph between 9:30 a.m. and 4:20 p.m. Since we are free to choose any of those values, t is considered an **independent variable**. The values of $V(t)$ are always dependent on what we choose for t, therefore we call **$V(t)$** a **dependent variable.** Now let's reverse the notation and talk about another relation, $t(V)$ (read "t of V"), which may or may not turn out to be a function.

Question 15: Which symbol *in this opposite viewpoint* represents the input value, and hence is called the independent variable and which symbol represents the output value and hence is called the dependent variable?

Example 2: The Dow Jones Industrial Average Revisited

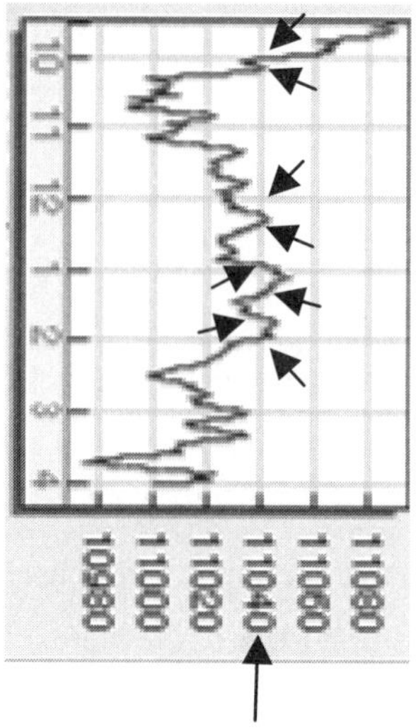

Answer Q11: One of each type. The first is a function and the second is a relation.

Answer Q12: No, because with an input of 9 you get two outputs 3 and –3.

Find t($11,040).

Solution:

An input value for V of $11,040 would cause t to have several outputs. Those outputs are 2:00 p.m., 1:40 p.m., 1:20 p.m., 12:50 p.m., 12:20 p.m., 12:10 p.m., 10:10 a.m., and 10:00 a.m. Note that t($11,040) = 2:00, 1:40, 1:20, 12:50, 12:20, 12:10, 10:10, and 10:00. Since for a single input value of V there can be more than one output value for $t(V)$, t is not a function of V.

Question 16: What is t($10,980)?

Example 3: Function Notation

Rewrite the equation from discussion 6, $F=\frac{9}{5}C+32$, with our new notation.

Solution:

$F(C)=\frac{9}{5}C+32$, where C is the input and represents the Celsius temperature and $F(C)$ is the output and represents the Fahrenheit temperature. For example, $F(20^\circ)=\frac{9}{5}(20^\circ)+32=68^\circ$ means that 20° Celsius converts to 68° Fahrenheit. Since the formula doesn't include terms like C^2 or $|C|$, it seems fairly obvious that for every value of C there is only one value for $F(C)$. To verify this, we can sketch the graph of $F(C)$ and apply the vertical line test.

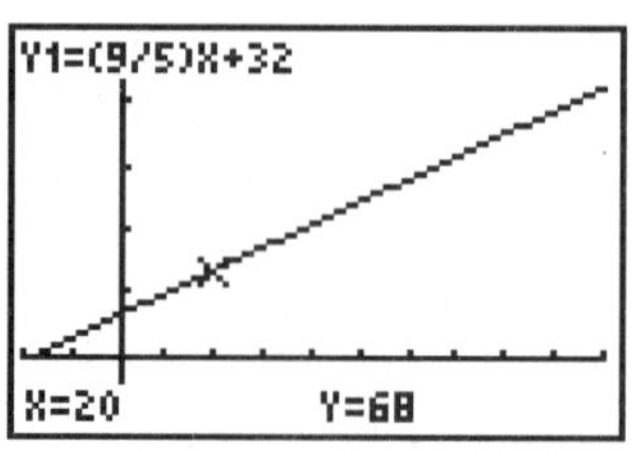

No imaginary vertical line would cross the graph more than once. Thus, this is a function, and we can say ***F***, Fahrenheit**, is a function of *C*,** Celsius. *C* is the **independent variable** and ***F*(*C*)** is the **dependent variable.**

Answer Q13: Each of the examples a,c,e,f are functions, but examples b and d are not. You should be able to tell this since y is raised to an even power in both cases. Notice that by letting x = 2 in formula (b), for example, we get 3(2) = 7 – y^2. Then solving this for y we get y^2=1 and thus y = 1 or y = -1. So a single value for x yields two values for y.

Question 17: What is the value of *F*(60°)?

Question 18: For what temperature *C* does *F*(*C*) = 32°?

Discussion 9: Temperature

Look back at discussion 8 and the average temperature of Phoenix by month. Try using the notation *T*(*M*), where *M* will be the input (month) and *T*(*M*) will be the output, the average temperature for that month.

We discovered earlier that this is a function, so you would say, "***T* is a function of *M*.**" As you saw in discussion 8: *T*(May) = 79° and *T*(November) = 62°.

Question 19: Find *T*(August).

Let's reverse the notation and look at *M*(*T*), where *T* is the input value and *M*(*T*) is the output value. You observed earlier that this is not a function, because for some values of *T* there were more than one value for *M*(*T*). For example, *M*(54°) = December and January. So, from this viewpoint, *T* is the independent variable and *M*(*T*) is the dependent variable, but *M* is *not* a function of *T*.

Answer Q14: V(11:15) = $11,020

Notice that in the first viewpoint the inputs were months, not numbers. In the second viewpoint the outputs were not numbers. They were the months of the year. **The idea of functions does not limit us to the use of numbers only**. Anything, numbers or concepts like dates and places, can be inputs or outputs. Inputs and outputs must simply make sense in the context of a relationship.

Check your understanding of the concepts presented in this section by trying to answer the following questions.

Answer Q15: V is the input or independent variable and t(V) is the output value and we would call t(V) the dependent variable.

Question 20: Let *A* be the label for a relationship between area of a square and the length of a side of a square. Let *s* represent the length of a side of the square, and *A*(*s*) be the output, area of the square.

a) Which is the input variable and which is the output variable?

b) Which variable is independent and which is dependent?

c) What is required for *A* to be a function of *s*?

d) From your past experiences do you believe this to be a function or not? Explain.

Example 4: Function Notation

Let's now do a few examples to see if you're comfortable with your algebra skills. Use the formulas to find the following values: $f(x) = -2x^2 + 3x$, $g(x) = 7 - 5x$, $h(x) = \sqrt{x+5}$, $k(x) = \dfrac{x+3}{7-x}$

Solutions:

a.) $\mathbf{f(-2)} = -2(\underline{-2})^2 + 3(\underline{-2}) = -2(4) + (-6) = -8 + (-6) = -14$	Substituted -2 in for the x in the $f(x)$ formula.
b.) $\mathbf{h(11)} = \sqrt{\underline{11}+5} = \sqrt{16} = 4$	Substituted 11 in for the x in the $h(x)$ formula.
c.) $\mathbf{g(-3)} = 7 - 5(\underline{-3}) = 7 + 15 = 22$	Substituted -3 in for the x in the $g(x)$ formula.
d.) $\mathbf{k(2)} = \dfrac{\underline{2}+3}{7-\underline{2}} = \dfrac{5}{5} = 1$	Substituted 2 in for the x in the $k(x)$ formula.
e.) $\mathbf{f(k)} = -2(\underline{k})^2 + 3(\underline{k}) = -2k^2 + 3k$	Substituted k in for the x in the $f(x)$ formula.
f.) $\mathbf{g(a+h)} = 7 - 5(\underline{a+h}) = 7 - 5a - 5h$	Substituted $(a+h)$ in for the x in the $g(x)$ formula.
g.) Find x so that $g(x) = -3$. Means "implies" (arrow pointing to $\Rightarrow$) $7 - 5x = -3 \Rightarrow -5x = -10 \Rightarrow x = 2$	Set the $g(x)$ formula equal to -3 and then solve for x.
h.) Find x so that $k(x) = -2$. $\dfrac{x+3}{7-x} = -2 \Rightarrow x+3 = -2(7-x) \Rightarrow x+3 = -14+2x \Rightarrow$ $3 = -14 + x \Rightarrow 17 = x$	Set the $k(x)$ formula equal to -2 and then solve for x.

Answer Q16: t($10,980) = 3:40 pm, 3:45pm

Section Summary

- Input – The value (numeric or other) plugged into a rule (relationship) that produces an output.
- Output – The value (numeric or other) that is the result of an input.
- Relation (informal) – Any rule with inputs that cause outputs.
- Function (informal) – Any relation where for each input there is only one output.
- Vertical line test – Imagine drawing vertical lines up and down everywhere on the graph of a relation, and if no vertical line crosses the graph more than once then the relation is a function.
- *M* as a function of *t* – *M*(*t*), the whole symbol, is the output and the *t* is the input.
- Independent Variable – The variable that represents the inputs which is freely chosen.
- Dependent Variable – The variable that represents the outputs which depend on the inputs.

SECTION 1.1 HOMEWORK

1. The following graph represents the growth of an investment. The numbers along the x-axis represent years while the numbers along the y-axis represent dollars. You need to notice that the investment is worth approximately \$500 at 0 years.

Answer Q17: F(60°) = 140°

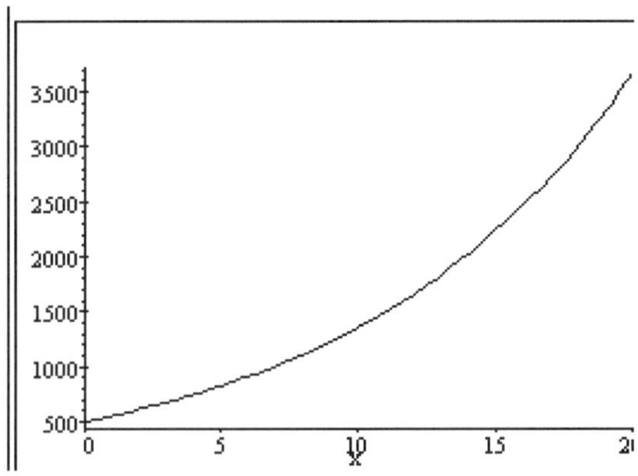

Answer Q18: C = 0

Answer Q19: T(August) =92°

a. Approximately how much money is the investment worth after 5 years?

b. Approximately how much money is the investment worth after 10 years?

c. Does it appear that the value of the investment is a function of the number of years? Why or why not?

2. The following is a graph representing the growth of bacteria per hour. The values along the x-axis represent the number of hours and the numbers along the y-axis represent the number of bacteria, in 1,000's.

Answer Q20: a). The input variable is "s", and the output variable is A(s). b). The "s" is the independent variable and A(s) is the dependent variable. c). A will be a function of "s" if for each input "s" there is only one possible output values for A(s). d). Yes, this is a function, because for each length of a side of a square you will get only one area.

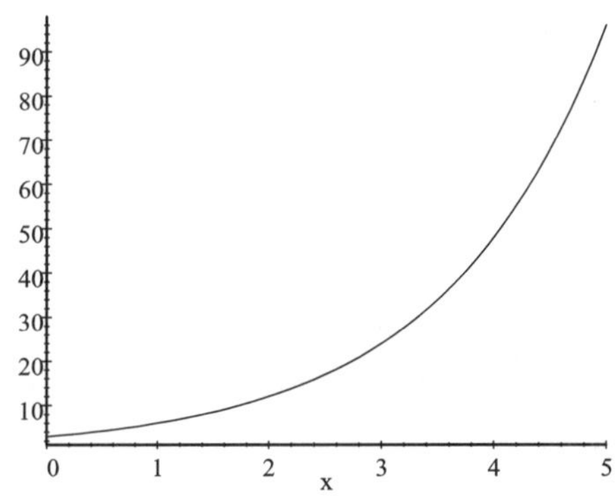

a. Approximately how many bacteria are there after 1 hour?

b. Approximately how many bacteria are there after 3 hours?

c. Does it appear that the number of bacteria is a function of time? Why or why not?

3. The following is a graph of immigration into the U.S. from 1989 to 1996.

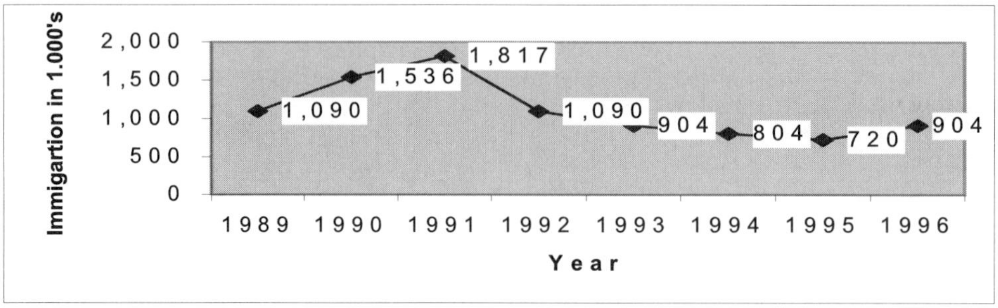

a. How many immigrants (in 1000's) came to the U.S. in 1991?

b. How many immigrants (in 1000's) came to the U.S. in 1993?

c. How many immigrants (in 1000's) came to the U.S. in 1996?

d. Does it appear the number of immigrants to the U.S. is a function of the year? Why or why not?

4. The following is a graph of Robberies in 1,000's for the years 1987-1997:

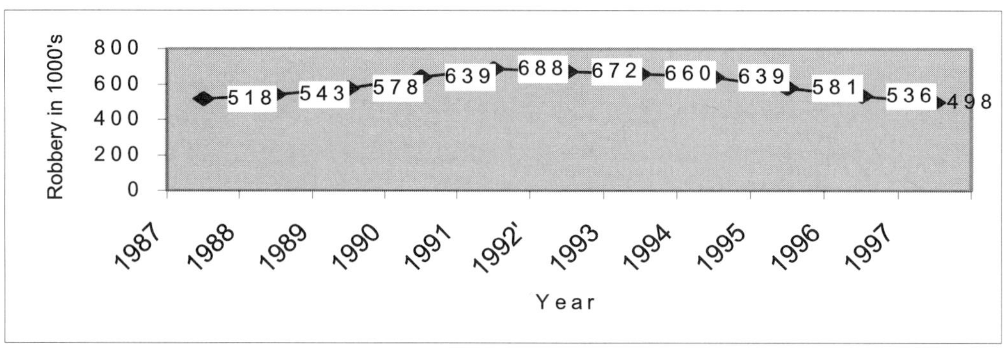

a. How many robberies occurred in the year 1988?

b. How many robberies occurred in the year 1995?

c. How many robberies occurred in the year 1997?

d. Does it appear the number of robberies is a function of time? Why or why not?

5. Here is a graph of the number of state prisoners per 100,000 population from 1970 to 1997:

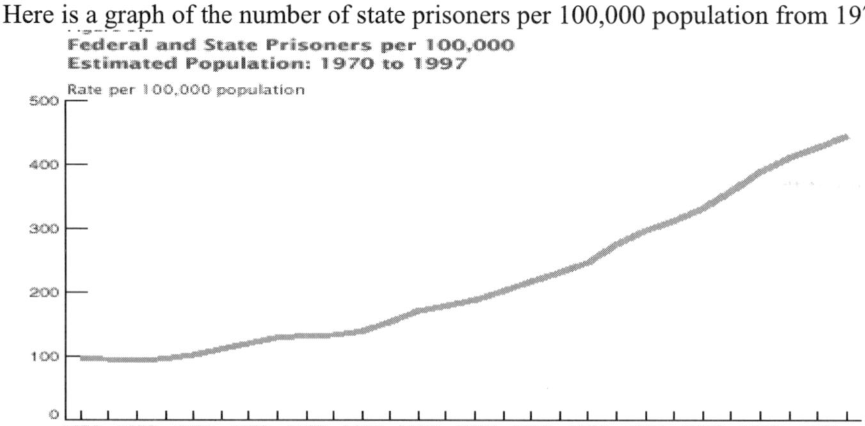

a. Approximately how many state prisoners per 100,000 were there in 1980?

b. Approximately how many state prisoners per 100,000 were there in 1996?

c. Does it appear that the number of state prisoners is a function of the year? Why or why not?

d. Approximately what year had 200 state prisoners per 100,000?

e. Approximately what year had 300 state prisoners per 100,000?

f. Is the year a function of the number of state prisoners? Why or why not?

6. National debt in billions of dollars for the years 1980-1999

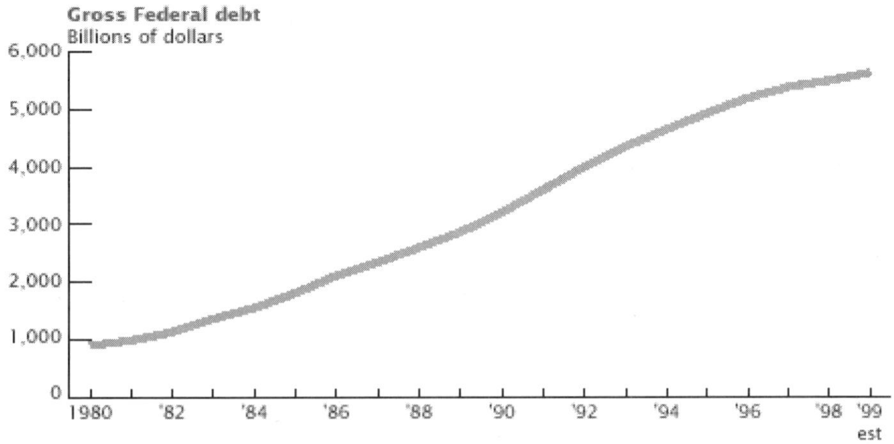

Source: Chart prepared by U.S. Census Bureau. For data, see Table 542.

a. Approximately what is the national debt in billions for the year 1986?

b. Approximately what is the national debt in billions for the year 1990?

c. Does is appear the amount of national debt is a function of the year? Why or why not?

d. Approximately what year is the national debt 4,000 in billions?

e. Approximately what year is the national debt 1,000 in billions?

f. Is the year a function of the national debt? Why or why not?

7. The following chart represents the value in dollars of an investment after a number of years:

Year	5	10	15	20	25	30	35
Dollars	\$822.65	\$1353.52	\$2226.96	\$3664.04	\$6028.47	\$9918.70	\$16319.33

a. What is the value of the investment after 20 years?

b. What is the value of the investment after 35 years?

c. Is the value of the investment a function of the number of years? Why or why not?

d. How many years before you have \$6028.47 in the investment?

e. How many years before you have \$1353.52 in the investment?

f. Is the number of years a function of the amount of money in the investment? Why or why not?

8. The following chart represents the number of bacteria in a culture after a number of days:

Days	1	2	3	4	5	6	7	8
Bacteria	6	12	24	48	96	192	384	768

a. How many bacteria are in the culture after 3 days?

b. How many bacteria are in the culture after 6 days?

c. Is the number of bacteria in the culture a function of the number of days? Why or why not?

d. How many days before there are 96 bacteria in the culture?

e. How many days before there are 384 bacteria in the culture?

f. Is the number of days a function of the number of bacteria in the culture? Why or why not?

9. Table for school population from 1965-1996 in thousands:

Year	Total
1965	54,394
1970	59,838
1975	61,004
1980	58,305
1981	57,916
1982	57,591
1983	57,432
1984	57,150
1985	57,226
1986	57,709
1987	58,254
1988	58,485
1989	59,436
1990	60,267
1991	61,605
1992	62,686
1993	63,241
1994	63,986
1995	64,803
1996	65.674

a. What is the school population in thousands for 1983?

b. What is the school population in thousands for 1990?

c. Is the school population a function of the year? Why or why not?

d. What year is the school population 64,803 in thousands?

e. What year is the school population 54,394 in thousands?

f. Is the year a function of the school population in thousands? Why or why not?

g. Is it possible that this may turn out not to be a function for future values? Explain your answer.

10. Expenditures for Secondary and Elementary schools in the U.S. by millions of dollars from 1980-1995.

Year	Total
1960	131,342
1970	293,123
1975	339,348
1980	345,070
1985	378,726
1986	400,554
1987	424,554
1988	437,542
1989	462,947
1990	485,990
1991	498,390
1992	506,716
1993	517,563
1994	527,909
1995	541,898

a. What was the expenditure in millions for 1980?

b. What was the expenditure in millions for 1990?

c. Are the dollars spent a function of the year? Why or why not?

d. In what year were 517,583 in millions spent?

e. In what year were 293,123 in millions spent?

f. Is the year a function of the millions spent? Why or why not?

g. Is it possible that sometime in the future this would not be a function? Explain why.

11. Presidential election year retirements from House of Representatives:

Year	Retire-ments [1]
PRESIDENTIAL-YEAR ELECTIONS	
1964	33
1968	23
1972	40
1976	47
1980	34
1984	22
1988	23
1992	65
1996	50

a. How many members of the House of Representatives retired in the 1968 election?

b. How many members of the House of Representatives retired in the 1988 elections?

c. If the number of House of Representatives retirements a function of the election year? Why or why not?

d. What election year did 40 members of the House of Representatives retire?

e. What election year did 65 members of the House of Representatives retire?

f. Is the election year a function of the number of members of the House of Representatives that retire? Why or why not?

12. Privately-owned housing units, in 1000's, on which construction began in 1996:

State	1996
MO	28.5
MT	2.8
NE	10.0
NV	38.5
NH	4.8
NJ	23.6
NM	10.8
NY	31.2
NC	68.6
ND	2.5
OH	47.8
OK	13.5
OR	28.8
PA	37.4
RI	2.4
SC	30.0
SD	3.9
TN	42.2
TX	125.0
UT	24.3
VT	2.2
VA	47.0
WA	40.6
WV	4.9
WI	32.5
WY	2.5

a. How many houses in 1000's began construction in North Dakota in 1996?

b. How many houses in 1000's began construction in Wyoming in 1996?

c. Is the number of beginning constructions a function of the state?

d. Explain your answer for (c).

e. What state had beginning constructions of 10.8, in 1000's?

f. What state had beginning constructions of 125.0, in 1000's?

g. Is the state a function of the number of beginning constructions in 1000's? Why or why not?

13. $I = 500(.08)\,t$ is an equation that computes the amount of simple interest earned on a \$500 investment at 8% after t years. (Where I = amount of interest earned in dollars; t = time in years.)

a. How much simple interest is earned after 5 years?

b. How much simple interest is earned after 10 years?

c. Is the amount of simple interest earned a function of the number of years? Why or why not?

14. In a certain housing development the cost of a house in dollars, C, according to the square footage of that house, S, is given by the equation $C = 78S$.

a. If the house measures 1500 square feet, how much does it cost?

b. If the house measures 2000 square feet, how much does it cost?

c. Is the cost of the house a function of the square feet? Why or why not?

15. The volume of a sphere is given by the following equation:

$V = \frac{4}{3}\pi r^3$, (V = volume and r = radius of the sphere)

a. If the sphere's radius is 6 inches what is the volume?

b. If the radius is 4 feet, what is the volume?

c. Is the volume a function of the radius? Why or why not?

16. The volume of a cube is given by the following equation:

$V = S^3$, where V = volume and S = length of a side

a. What is the volume of a cube with a side length of 2 meters?

b. What is the volume of a cube with a side length 23 centimeters?

c. Is the volume of a cube a function of the length of a side? Why or why not?

17. Refer to problem #3: Let $I(t)$ be the notation where t represents the date and $I(t)$ represents the number of immigrants:

a. What is the independent variable?

b. What is the dependent variable?

c. What is $I(1991)$?

d. What is $I(1994)$?

e. What are the input values?

f. What are the output values?

18. Refer to problem #3: Let $t(I)$ be the notation where I represents the number of immigrants and $t(I)$ represents the date:

a. What is the independent variable?

b. What is the dependent variable?

c. What is $t(904)$?

d. What is $t(804)$?

e. What are the input values?

f. What are the output values?

19. Refer to problem #9: Let $P(t)$ be the notation where t represents the date and $P(t)$ represents the student population:

a. What is the independent variable?

b. What is the dependent variable?

c. What is $P(1982)$?

d. What is $P(1995)$?

e. What are the input values?

f. What are the output values?

20. Refer to problem #11: Let $Y(R)$ be the notation where R represents the number of retired representatives and $Y(R)$ represents the election year:

a. What is the independent variable?

b. What is the dependent value (output)?

c. What is $Y(65)$?

d. What is $Y(23)$?

e. What are the input values?

f. What are the output values?

21. Refer to problem #13: Let $I(t) = 500(.08)t$ be the notation where t represents the number of years and $I(t)$ represents interest earned:

a. What is the independent variable?

b. What is the dependent variable?

c. What is $I(25)$?

d. What is $I(9)$?

22. Let $t(I)=\frac{I}{40}$ be the notation where I represents interest earned and $t(I)$ represents the number of years:

a. What is the independent variable?

b. What is the dependent variable?

c. What is $t(200)$?

d. What is $t(3000)$?

23. Refer to problem #15: Let $V(r) = \frac{4}{3}\pi r^3$ be the notation where r represents radius and $V(r)$ represents volume of a sphere:

a. What is the independent variable?

b. What is the dependent variable?

c. What is V(9 inches)?

d. What is V(3 feet)?

24. Refer to problem #16: Let $V(S) = S^3$ be the notation where S represents length of a side of a cube and $V(S)$ represents volume of the cube:

a. What is the independent variable?

b. What is the dependent variable?

c. What is V(10 meters)?

d. What is V(15 centimeters)?

25. If you live in a city with sales tax:

a. Would sales tax be a function of purchase price?

b. Explain your answer to (a).

c. Would purchase price be a function of sales tax?

d. Explain your answer to (c).

26. If you are asked to measure the temperature outside every hour for 24 hours:

a. Would temperature be a function of time of the day?

b. Explain your answer to (a).

c. Would the time of day be a function of temperature?

d. Explain your answer to (c).

(27 – 42) Use the following equations to find all real number answers, if they exist.

$f(x)=3x^2-2x+4$ $\quad g(x)=|5-2x|$ $\quad r(x)=\sqrt{x-3}$ $\quad t(x)=\dfrac{x-2}{x+5}$

27. $f(2)=$ 28. $f(5)=$ 29. $f(-3)=$ 30. $f(-4)=$
31. $g(-3)=$ 32. $g(-5)=$ 33. $g(4)=$ 34. $g(7)=$
35. $r(19)=$ 36. $r(39)=$ 37. $t(4)=$ 38. $t(8)=$
39. $t(-3)=$ 40. $t(-12)=$ 41. $t(-5)=$ 42. $r(-6)=$

(43-48) Use the equations from 27–42 to find:

43. $f(a)$ 44. $r(a)$ 45. $f(a+h)$
46. $g(a+h)$ 47. $t(b+3)$ 48. $f(b+2)$

49. a. For each of the problems (27 – 48) what was the independent value (the input)?
b. For each of the problems (27 – 48) what was the dependent value (the output)?

(50-63) For each of the equations determine if y is a function of x.

50. $y=x^2-3x$ 51. $y=3x^3-5x+8$ 52. $y^2=3x^2+2x-3$

53. $y^2=2x^3-5x^2+8$ 54. $|y|=3x-4$ 55. $|y|=x^2-4x+8$

56. $y=|x|+9$ 57. $y=|3x^2-9|$ 58. $3x^2+4y^2=25$

59. $5x^2-5y^2=36$ 60. $3x^3+5y^3=4$ 61. $7y^3-8x^3=25$

62. $2x^2+3y^2-8x-24y-6=0$ 63. $3x^2+y^2-18x+10y+3=0$

64. $2x^3+4y^4=24$ 65. $4x^3-3x^2+6y^4-5y^3=36$

66. $4xy^2+5x^2y=8$

Answer questions (67-78) using the following functions.

$f(x)=3x^2-2x+3$ $\quad g(x)=9-3x$ $\quad h(x)=\dfrac{x+3}{x-1}$ $\quad k(x)=\sqrt{x+1}$

67. $f(-3)=$ 68. $f(3)=$ 69. $g(-4)=$
70. $g(6)=$ 71. $h(5)=$ 72. $h(-3)=$
73. $k(8)=$ 74. $k(24)=$ 75. $g(a+b)=$
76. $f(a+h)=$ 77. Find x so that $g(x)=-12$
78. Find x so that $f(x)=8$

Section 1.2 Relations and Functions

We have just finished looking at relations and functions from an informal perspective. It is now time to give you some formal mathematical definitions. First up is the term "relation."

> A **relation** is any set of ordered pairs (x, y).

You should remember from previous math classes that an ordered pair is of the form (x, y) where x and y are related to each other in some way. Putting this definition together with what you learned in the last section tells us that the first coordinate in an ordered pair is considered to be the value for the input or independent variable. The second coordinate in the ordered pair is the output or dependent variable. We have special names for the set of all possible inputs and outputs.

> We use the word **domain** to refer to the set of all possible inputs, or independent values, and the word **range** to refer to the set of all possible outputs, or dependent values, that result from evaluating the relation at each input.

All of the examples that have been presented to you so far in this chapter have been relations. But some of the relations in the preceding section were also functions.

> A **function** is a set of ordered pairs where no two ordered pairs have the same first coordinate but a different second coordinate.

In the last section you learned an informal definition of a function that stated, "For each input you get only one output." The formal definition states that the input, the first coordinate, of the ordered pair must not have two different corresponding second coordinates; otherwise, it is not a function.

Discussion 1: Ordered Pairs

The ordered pairs $\{(1, 2), (3, 2)\}$ form a function because although the two points have the same second coordinate, they have different first coordinates (that is, there is only one output for each input). On the other hand, the ordered pairs $\{(1, 2), (1, 3)\}$ don't form a function because the two points have the same first coordinate but a different second coordinate (two outputs for the same input).

Let us look at some of the examples from the last section to better understand these formal definitions.

Discussion 2: Ordered Pairs and the DJIA

Look at the discussion item about DJIA (discussion 4 of the previous section).

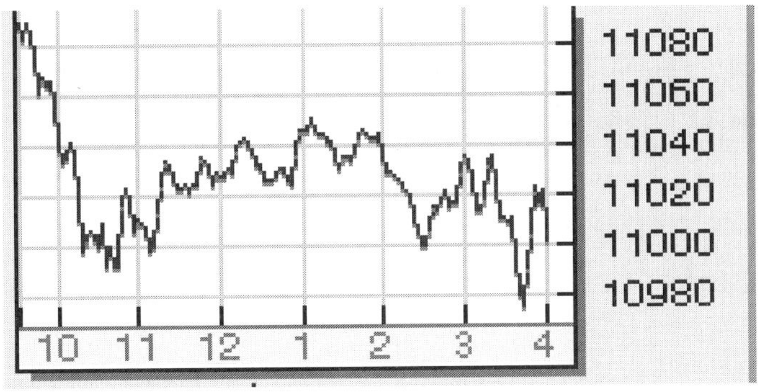

We will consider the values along the bottom of the graph to be the x-coordinates of the ordered pairs. Thus, each ordered pair would be of the form (time of day, Value of the DJIA). Any "time value" is a member of the domain (set of inputs) and any DJIA value is a member of the range (set of outputs). Some examples of ordered pairs might be as follows: (2:00, \$11,040), (11:15, \$11,020), and (3:40, \$10,980). If we were to list all of the ordered pairs that exist in this example, we would find that no two would have the same first coordinate but a different second coordinate. In fact, no two pairs would even have the same first coordinate. This is easy to recognize as a function because there isn't any way for it to violate the formal definition. Also if you look at this graph it passes the vertical line test, discussed in the last section.

Discussion 3: Ordered Pairs and Temperature

Consider the example from section 1.1 that related Celsius temperature to Fahrenheit temperature (discussion 6). One ordered pair in that example was (25 C°, 77 F°). You can use the formula $F(C) = \frac{9}{5}C + 32$ to produce any number of ordered pairs. (Try it!) If you produced all possible ordered pairs, you wouldn't find any two pairs with the same first coordinate but a different second coordinate. Therefore, by the formal definition, you can say that F is a function of C. The domain of the function is every Celsius temperature and the range is every possible Fahrenheit output.

Discussion 4: Ordered Pairs and the Average Temperature in Phoenix

Look at discussion 8 from the previous section, but consider the point of view where average temperature is the input and months are the outputs.

Average Temp.	54	58	62	70	79	88	94	92	86	75	62	54
Month	Jan.	Feb.	Mar.	Apr.	May	June	July	Aug.	Sept.	Oct.	Nov.	Dec.

From the table we can list all of the ordered pairs as follows: (54, January), (58, February), (62, March), (70, April), (79, May), (88, June), (94, July), (92, August), (86, September), (75, October),

(62, November), (54, December). Notice that the domain is {54, 58, 62, 70, 75, 79, 86, 88, 92, 94}, and the range is the months of the year.

Question 1: Are there any ordered pairs that have the same first coordinate but a different second coordinate? If so, what would this mean?

FINDING DOMAIN AND RANGE

Look at the following table from the Mortgage discussion item number 5 in section 1.1 (remember, we are only looking at the yellow dots):

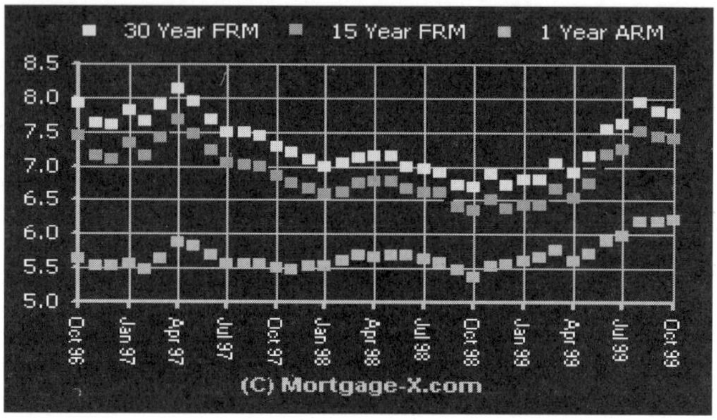

You should consider the values on the horizontal axis of every graph to be the x-coordinates unless you are specifically told otherwise. This being the case, the domain for this graph is any month from October 1996 to October 1999. You can see this from the "x-axis" and the fact that there is a yellow dot for each month as you go across the graph. The range is the collection of the heights of each of the yellow dots. Looking at the graph we can make a good estimate as to what the range values are: {8.1, 8.0, 7.9, 7.8, 7.7, 7.6, 7.5, 7.4, 7.3, 7.2, 7.1, 7.0, 6.9, 6.8, 6.7}.

As you have seen, it's relatively easy to determine the domain and range of a function when you are looking at its graph or its table of values. We will see that it's not as easy when we have a formula. So, let's talk about how to find the domain and range of a function given its formula. For now, we will only look at problems where the inputs and outputs are real numbers. But remember, they are not limited to just numbers.

Domain

First, let us tackle domain, the easier of the two to find. Simply stated, the domain will consist of all real value inputs that **produce real value outputs**. This will usually mean that the domain consists of all real numbers, but not always. For instance, if a formula includes a denominator, we have to exclude from the domain any value that would the make the denominator 0. As an example, given $f(x) = \frac{3}{x-2}$, the domain is all real numbers excluding $x = 2$.

We should also consider the effect that square roots may have on domain. Formulas with square roots might have inputs that cause negative values to be under the square root. These would yield complex number answers, not real ones. Thus we have to exclude from the domain any values that cause negative numbers to be under the square root. An example is $f(x) = \sqrt{x+1}$; the domain is all real numbers greater than, or equal to, -1.

Example 1: Domain of All Real Numbers

Find the domain of $y = 3x - 5$.

Solution:

This is an example of a function (where y is a function of x) that has all real numbers as its domain. We know that the domain is all real numbers because there isn't a fraction or a square root in the problem. Specifically any real number that is multiplied by 3 and has 5 subtracted from it will be another real number.

Example 2: Restricted Domain

Find the domain of $y = \sqrt{x+3}$.

Solution:

This is an example of a function with a restricted domain. The input is x and the domain would be all real numbers greater than or equal to –3. Any real number smaller than –3 would cause the number under the square root to become negative.

Example 3: Domain of All Real Numbers

Find the domain of $y = \sqrt{x^2+1}$.

Solution:

This is an example of a function with all real numbers as its domain. There is a square root in this problem so the domain might not be all real numbers. But with a closer look you will notice that any real number that is squared and then has 1 added to it will always be positive and thus we can take the square root of it to yield a real number answer.

These previous three examples did not use function notation. So, let us do a few examples with function notation:

Example 4: Function with Domain of All Real Numbers

Find the domain of $f(x) = x^2$.

Solution:

This is an example of a function that has all real numbers as its domain. You should notice that the problem does not have a fraction or a square root. This almost always tells you that the domain is all real numbers. Specifically, you know this because any real number that you square will give you a real answer.

Answer Q1: Yes there are ordered pairs where the first coordinates are the same but the second coordinates are different (54, January), (54, December), or (62, March), (62, November). This means that this relation is not a function.

Example 5: Function with a Restricted Domain

Find the domain of $f(x) = \dfrac{5}{x-2}$.

Solution:

This is an example of a function with a restricted domain. Here the domain would be all real numbers except $x = 2$. Once again, to find the domain of a problem with a fraction, ignore the numerator and find the values of x that make the denominator equal to zero. Solving $x - 2 = 0$ we get $x = 2$.

Example 6: Function with Domain of All Real Numbers

Find the domain of $g(x) = \dfrac{7}{x^2+2}$.

Solution:

This is an example of a function with all real numbers as its domain. The reason is that there are no values for x that make the denominator equal to zero. With a closer look you will see that any real number that is squared and then has 2 added to it would be positive, and thus could never be zero.

Example 7: Function with Domain of All Real Numbers

Find the domain of $k(x) = |x-5|$.

Solution:

This is an example of a function with all real numbers as its domain. The absolute value of any real number minus 5 is a real number.

There is another thing to think about when you are trying to determine the domain of a function. It occurs when you are looking at a formula that is describing a real-world phenomenon. In that case there might be some further restrictions on the domain. For example, consider the formula for the area of a square ($A(s) = s^2$). The domain consists of only positive numbers. It makes no sense in the real world to have a side of a square with a length that is negative or zero.

Range

Now let's address the problem of finding the range of a function. Finding the range is more difficult than finding the domain, particularly when all you have is a formula. To do this by hand you would need to think about three things: what is the smallest answer you could get, what is the largest answer you could get, and are there any values in between those that you can't get? We are not going to emphasize finding the range by hand in this book. Instead, let's make use of our graphing calculators. We will be able to find the range from looking at a graph of the function as well as being able to verify the domain. Let's revisit example 1 but this time we will begin by finding the range.

Example 8: Linear Function

Find the range of the function $y = 3x - 5$. Here is the graph of $y = 3x - 5$ on a TI-83.

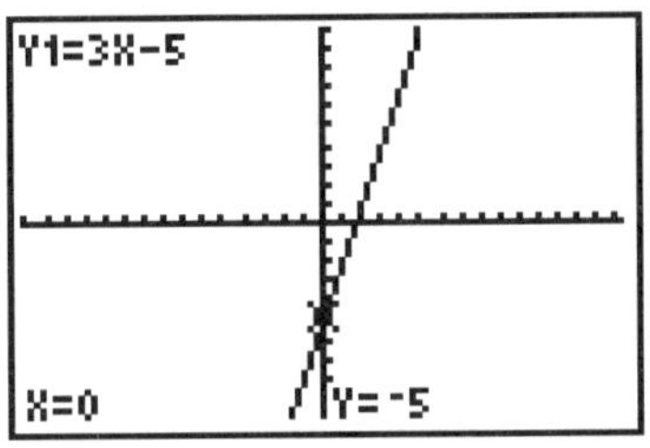

Solution:

To determine the range of a function from a graph we scan the graph from bottom to top. The range consists of all values of y that you can draw a horizontal line through and cross the graph. In this particular function, every horizontal line that you can draw will cross the graph. Therefore the range is all real numbers. The domain, which we discussed in example 1, will consist of all values of x that you can draw a vertical line through and cross the graph. Notice that it seems as though every vertical line you can draw will cross the graph somewhere. Thus we can be pretty sure that the domain is all real numbers also. Try typing the function into your own calculator and see it for yourself. Then try zooming out and notice that there are no vertical or horizontal lines that will miss the graph.

Example 9: Function with a Square Root

Find the domain and range of the function $y = \sqrt{x+3}$. Here is the graph of it on a TI-83.

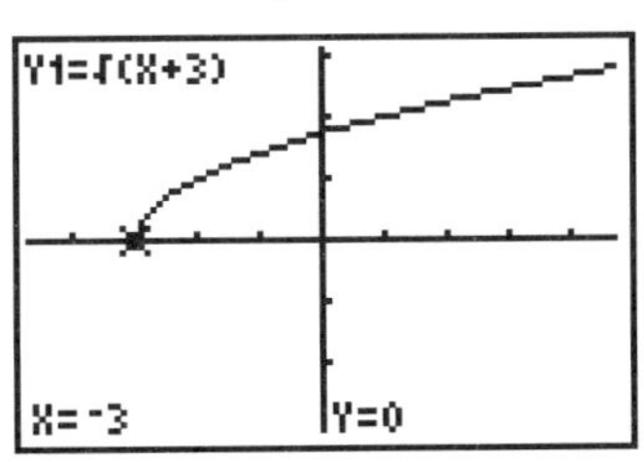

Solution:

If you use "zoom decimal" to graph this function, the graph appears as above. By tracing to the left you find that the graph starts at the point (–3,0) and goes upwards and to the right. Let's find the range. Notice that no horizontal line drawn below the x-axis will touch the graph. But every horizontal line that you draw above the x-axis, including $y = 0$ (the x-axis), will touch the graph somewhere. Therefore the range is $y \geq 0$. To verify the domain, notice that no vertical line you can draw to the left of $x = -3$ will touch the graph. Every vertical line that you draw to the right of $x = -3$ (including $x = -3$) will touch the graph. Thus the domain is $x \geq -3$.

Example 10: Function with an Absolute Value

Find the domain and range of $k(x) = |x - 5|$.

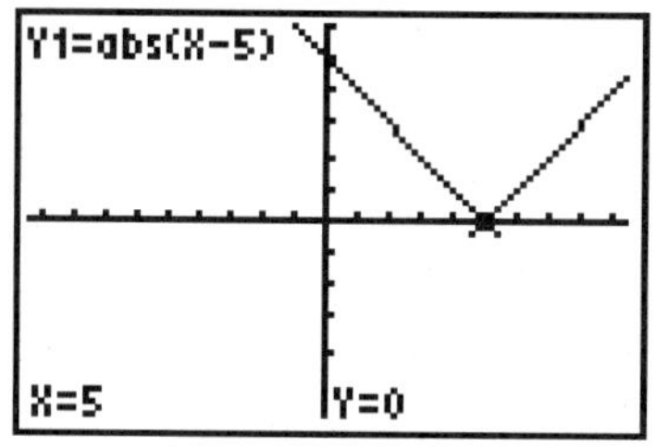

Solution:

The graph is starting from $y = 0$, the <u>lowest</u> point, and it goes up forever. Any horizontal line would touch the graph when $y \geq 0$. So the range is $y \geq 0$. From this graph you can also see that any vertical line you might draw will touch the graph, so the domain is all real numbers.

Example 11: Function in Fractional Form

Find the domain and range of $f(x) = \dfrac{5}{x-2}$.

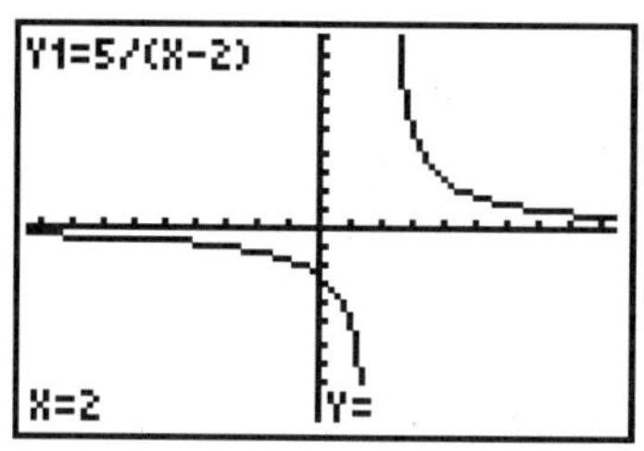

Solution:

This one is a little harder. For the range, notice that if you were to draw horizontal lines they would touch the graph at every y-value except at $y = 0$ (the x-axis). So, the range is all real numbers except $y = 0$. We cannot find an input for x that would produce an output of 0. You can't get a function like this to equal 0 if you can't make the numerator equal 0. For the domain, you might notice that if you were to draw vertical lines they would touch the graph at every x-value except $x = 2$. At $x = 2$ the denominator equals 0. Thus, the domain is all real numbers except for $x = 2$. Also notice that when $x = 2$ on the graphing calculator, $y =$ "blank". This is because the function is undefined when $x = 2$.

INTERVAL NOTATION

It's time to discuss interval notation. Interval notation is something you have probably heard about before. It is used in many areas of mathematics for expressing a range of values. It is also the way we express, in a concise manner, the domain and range of a function. Here are some examples of how mathematicians express a range of values:

	Number line	Inequality	Interval Notation
a.)		$-2 \le x < 5$	$[-2, 5)$
b.)	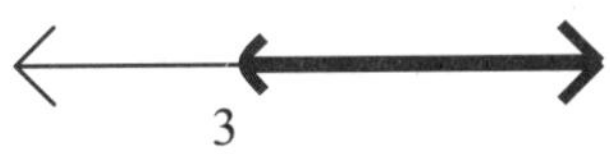	$x > 3$	$(3, \infty)$
c.)		$x \le -4$	$(-\infty, -4]$
d.)		$x < -1$ or $x \ge 7$	$(-\infty, -1) \cup [7, \infty)$

A collection of values written in interval notation form (right column) always begin with the smallest possible value of the variable in any one section of the domain. The second number in the notation is the largest possible value of the variable in that same section of the domain. So in example (a) above, the x's begin at -2 and go up to (but don't include) 5.

Notice that we used both a parenthesis and a bracket. We use a parenthesis when the variable can't equal the first or last number and a bracket when it can. In example (d) we have two intervals. This is because the possible values of x were split into two sections, as seen on the number line.

Lastly, we use the infinity symbol (∞) when the set of possible values goes on forever to the right and negative infinity ($-\infty$) when they go on forever to the left.

Question 2: If we know that $x < -4$ or $x \ge 9$ then what would this look like on a number line, and what would the interval notation be?

PIECEWISE DEFINED FUNCTIONS

Sometimes in order to graph a real-world situation you need to put together several different functions into one. The way this works is to break up the domain into intervals and use a different function for each interval. In other words, each piece of the piecewise function is only good with certain inputs (certain domain values).

Discussion 5: Piecewise Defined Function

a) Traced equation 1 over to $x = 1$

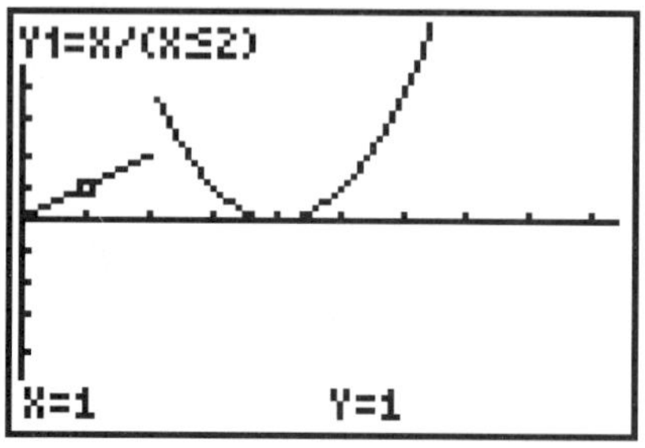

b) Traced equation 1 over to $x = 3$

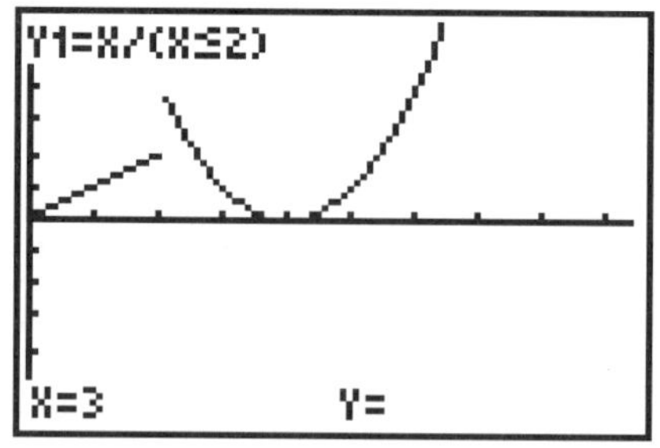

c) Traced equation 2 over to $x = 1$

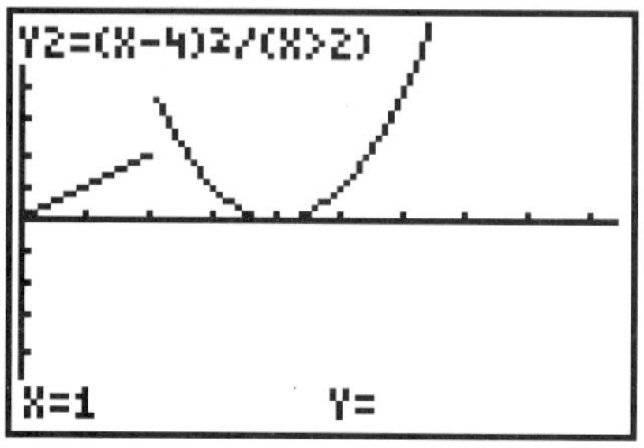

d) Traced equation 2 over to $x = 3$

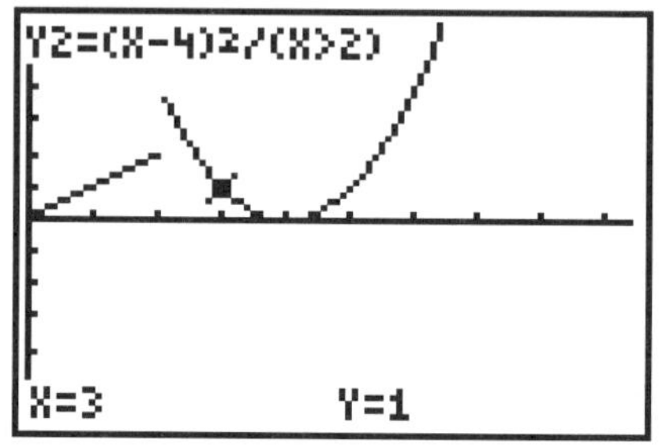

The function graphed above is $f(x) = \begin{cases} x & x \le 2 \\ (x-4)^2 & x > 2 \end{cases}$. Notice that in graph (a) when $x = 1$, $y = 1$. You can see this from the trace symbol on the graph. But notice that when you trace over to $x = 3$, $y =$ "blank" (graph b). This happens because in order to use the first part of the piecewise function, $f(x) = x$, the input (x) must be less than or equal to 2. Three is too big an x-value and falls outside of the domain for that part of the piecewise function. To trace the whole function you must switch to the other piece by using the down arrow on your calculator.

Now if you look at graphs c) and d), notice that when you trace over to $x = 3$, $y = 1$ (when tracing the second piece). But when you trace back to $x = 1$, $y =$ "blank". This happens because in order to use the second part of the piecewise function, $f(x) = (x - 4)^2$, x must be greater than 2. Hence, $x = 3$ yields an answer, because it's in the domain for the second piece, but $x = 1$ does not, because it's outside of the domain for the second part of the piecewise function.

When working with piecewise functions on your graphing calculators be very careful about which piece you are viewing and what it is you are trying to find.

Discussion 6: Federal Income Tax Formulas

Another example of a piecewise function is the Federal Income Tax Formulas. We will use the formula for a single person in the year 1998. The input is the taxable income (in dollars), and the output is the tax owed to the federal government (in dollars). The formula is as follows:

$$T(i) = \begin{cases} .15i & 0 < i \le 25350 \\ 3802.50 + .28(i - 25350) & 25350 < i \le 61400 \\ 13896.50 + .31(i - 61400) & 61400 < i \le 128100 \\ 34573.50 + .36(i - 128100) & 128100 < i \le 278450 \\ 88699.50 + .396(i - 278450) & 278450 < i \end{cases}$$

(We bet you never knew that taxes were this awful.)

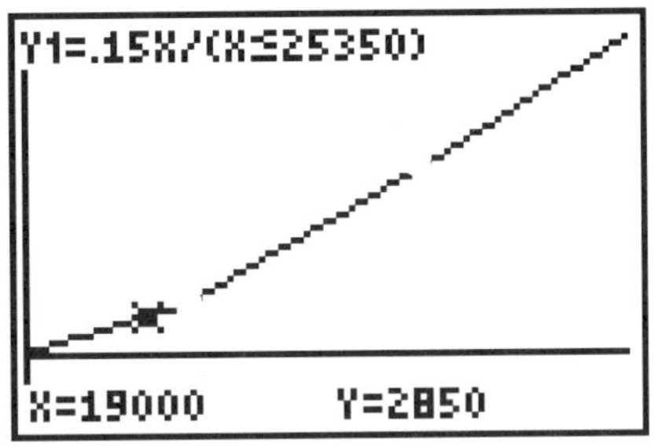

We have only given you part of the graph above. What you see is the first three parts of the graph of the tax formula, and we have created two holes in the graph to help you to better see where the different pieces of this graph start and stop. Notice how the graph gets steeper. The more you earn (income), the larger the percentage you pay (taxes).

Answer Q2: On the number line we would have a picture like this

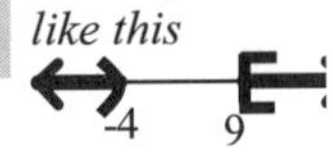

and the interval notation would be $(-\infty, -4) \cup [9, \infty)$

Question 3: Find $T(35460)$. Hint: use your calculator to evaluate the appropriate part of the function.

Section Summary

So far we have formalized the terms relation, function, domain and range. Once again, a **function** is a relation where no two ordered pairs have the same first coordinate but a different second coordinate. **Domain** is the set of all possible inputs (x's) and **range** is the set of all resulting outputs (y's) from those inputs. We use **interval notation** to express the values of the domain and range. And here are some final examples of how to find domain and range by hand.

1.) $f(x) = 6 - 5x$ 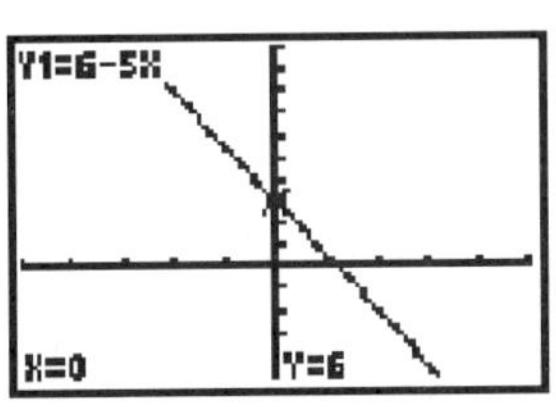	Domain is all real numbers since we don't have a variable in a denominator or a square root. The range is all real numbers because you can get answers as large or small as you want. (There isn't a "largest" or "smallest" answer.) The interval notation is $(-\infty, \infty)$ for both the domain and range.

<table>
<tr>
<td>

2.) $g(x)=\sqrt{3x-12}$

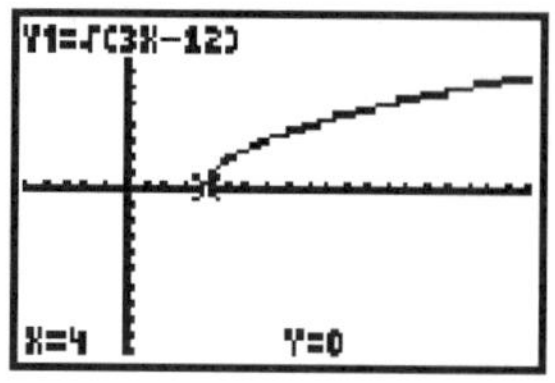

</td>
<td>

Domain is $x \geq 4$, because $3x - 12$ must be ≥ 0, so $3x-12\geq 0 \;\rightarrow\; 3x\geq 12 \;\rightarrow\; x\geq 4$. In interval notation that is [4, ∞). The range is $y \geq 0$ since the smallest output possible is 0 and there isn't a limit to the largest possible output value. The interval notation is [0, ∞).

</td>
</tr>
<tr>
<td>

3.) $h(x)=\dfrac{5}{2x-5}$

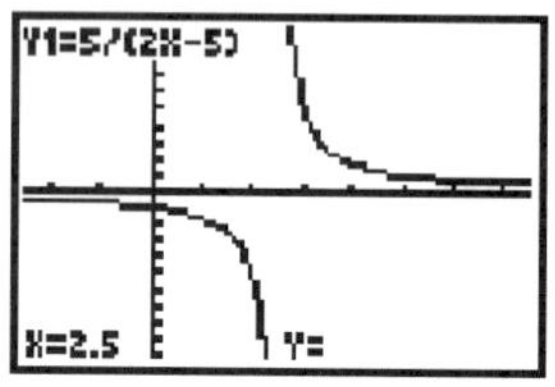

</td>
<td>

Domain is all real values excluding $x=\frac{5}{2}$, since the only input not allowed is one that causes the denominator to equal zero. So the algebra is:

$2x-5=0 \;\Rightarrow\; 2x=5 \;\Rightarrow\; x=\frac{5}{2}$. The interval notation is (–∞, 5/2)∪(5/2, ∞). The range includes all real values excluding $y=0$ because you can't get the fraction to equal 0 with a numerator of 5. Also you can get answers as large or small as you want. There isn't a largest or smallest answer. The interval notation is (–∞, 0)∪(0, ∞).

</td>
</tr>
<tr>
<td>

4.) $k(x)=\lvert 7-2x\rvert$

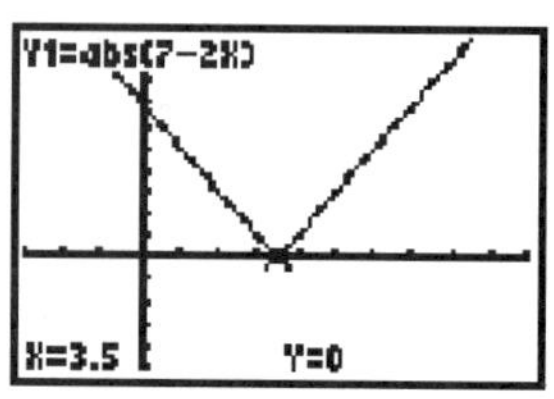

</td>
<td>

Domain is all real numbers since we don't have a variable in a denominator or a square root. The interval notation is (–∞, ∞). The range is $y\geq 0$ because the smallest possible value for $k(x)$ is 0 and there isn't a largest value. The interval notation is [0, ∞).

</td>
</tr>
<tr>
<td>

5.) $J(x)=\dfrac{x-2}{\sqrt{x^2-9}}$

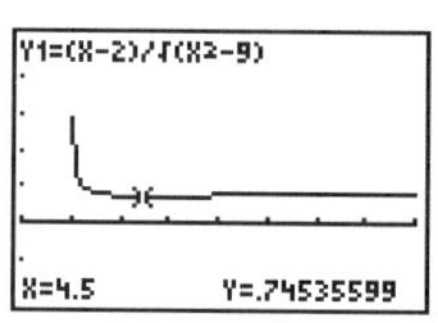

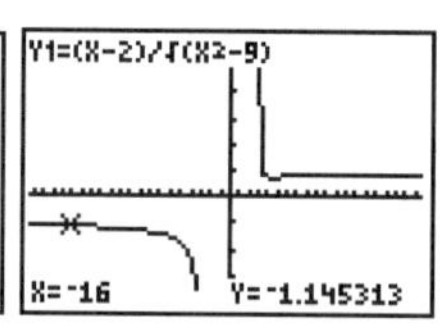

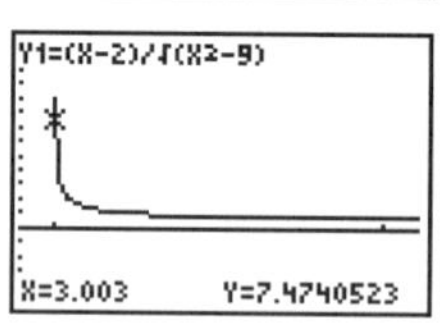

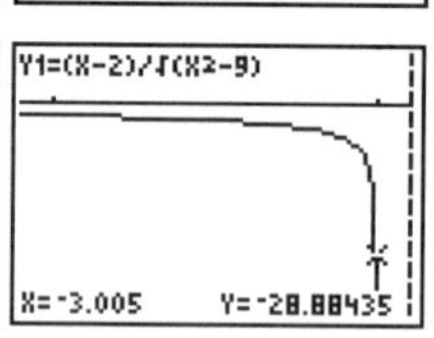

</td>
<td>

Domain is $x<-3$ or $x>3$ since we can't have zero in the denominator or a negative under a square root. The algebra looks like this:

$x^2-9=0 \rightarrow (x-3)(x+3)=0 \rightarrow x=3, x=-3$ and

$x^2-9\geq 0 \rightarrow (x-3)(x+3)\geq 0 \rightarrow x\leq -3, x\geq 3$ for this product to be positive. The interval notation is (–∞, –3)∪(3, ∞). Using the graphing calculator and its trace feature we can estimate the range which is: $y<-1$ and $.74535599\ldots<y<$ infinity. The interval notation is (–∞, -1)∪(.74535599…, ∞).

</td>
</tr>
</table>

SECTION 1.2 HOMEWORK

(1-4) In the first four exercises you will see scatter plots (graphs of points not connected). The horizontal axis represents the independent values and the vertical axis represents the dependent values.

1.

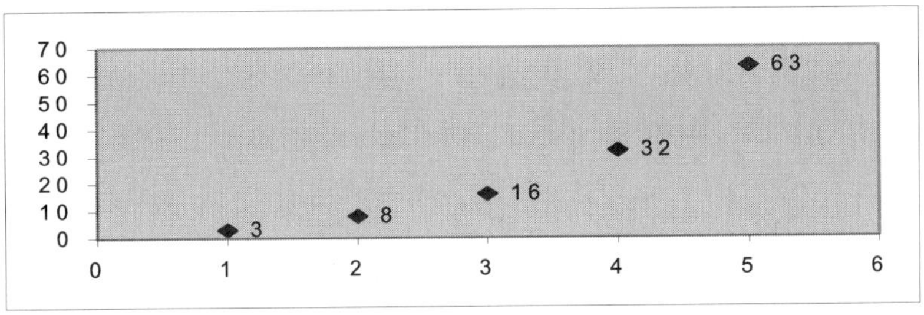

a. List all the ordered pairs for the relation created by the scatter plot.

b. What is the domain of the relation?

c. What is the range of the relation?

d. Does the scatter plot represent a function? Why or why not?

2.

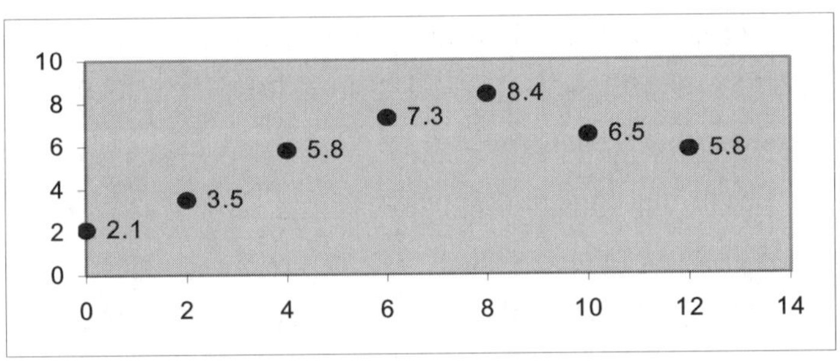

a. List all the ordered pairs for the relation created by the scatter plot.

b. What is the domain of the relation?

c. What is the range of the relation?

d. Does the scatter plot represent a function? Why or why not?

3.

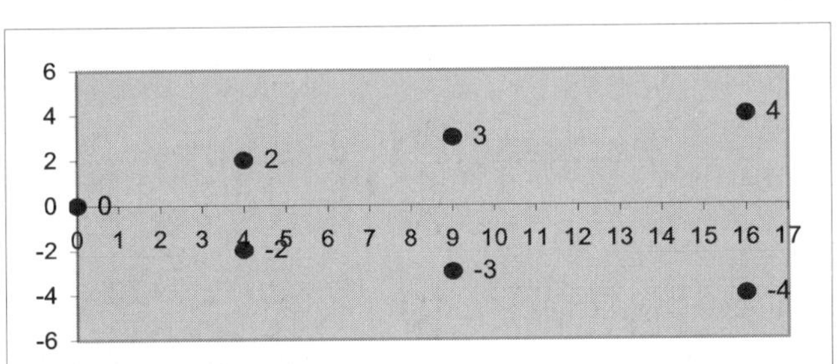

a. List all the ordered pairs for the relation created by the scatter plot.

b. What is the domain of the relation?

Answer Q3: T(35460) = $6633.30. Notice that the domain value is 35460, which is between 25350 and 61400, so you use this part of the formula (3802.50 + .28(i – 25350)) to calculate the taxes.

c. What is the range of the relation?

d. Does the scatter plot represent a function? Why or why not?

4.

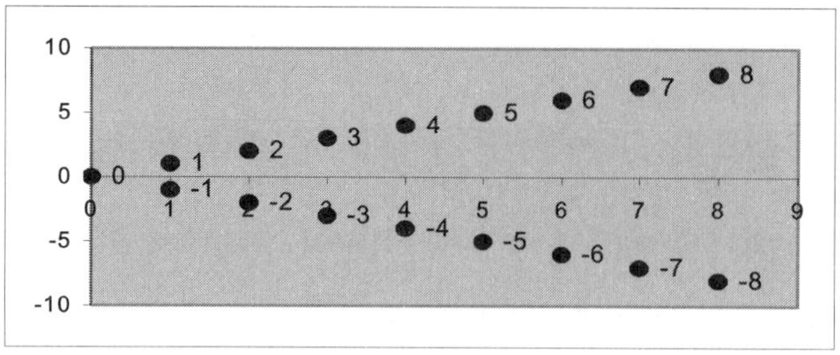

a. List all the ordered pairs for the relation created by the scatter plot.

b. What is the domain of the relation?

c. What is the range of the relation?

d. Does the scatter plot represent a function? Why or why not?

(5-8) The top line of the table is the independent values and the bottom line is the dependent values.

5. Below is a table representing the vote by percentage for the Democratic Presidential Nominee in every presidential election since 1960.

Year	1960	1964	1968	1972	1976	1980	1984	1988	1992	1996
%	49.7	61.1	42.7	37.5	50.1	41.0	40.1	45.6	43.0	49.2

a. List all the ordered pairs for the relation created by the table.

b. What is the domain of the relation? What is the range of the relation?

c. Does the table represent a function? Why or why not?

6. Below is a table representing the vote by percentage for the Republican Presidential Nominee in every presidential election since 1960.

Year	1960	1964	1968	1972	1976	1980	1984	1988	1992	1996
%	49.5	38.5	43.4	60.7	48.0	50.7	58.8	53.4	37.4	40.7

a. List all the ordered pairs for the relation created by the table.

b. What is the domain of the relation?

c. What is the range of the relation?

d. Does the table represent a function? Why or why not?

7. Using the table below, answer the questions:

1	1	16	16	64	64	125	125	216	216	343	343
1	–1	2	–2	4	–4	5	–5	6	–6	7	–7

a. List all the ordered pairs for the relation created by the table.

b. What is the domain of the relation?

c. What is the range of the relation?

d. Does the table represent a function? Why or why not?

8. Using the table below, answer the questions:

2	2	4	4	6	6	8	8	10	10	12	12
2	–2	4	–4	6	–6	8	–8	10	–10	12	–12

a. List all the ordered pairs for the relation created by the table.

b. What is the domain of the relation?

c. What is the range of the relation?

d. Does the table represent a function? Why or why not?

(9-30) State the domain of the function: (Write the answers in interval notation.)

9. $f(x) = x^2 + 3$

10. $f(x) = x^2 - 1$

11. $y = \dfrac{2x}{x-3}$

12. $y = \dfrac{2x+1}{x+5}$

13. $g(x) = |x+3|$

14. $g(x) = |x-5|$

15. $y = \dfrac{x+2}{x^2-9}$

16. $y = \dfrac{x-2}{x^2-1}$

17. $f(x) = \sqrt{x-2}$

18. $f(x) = \sqrt{3x+5}$

19. $g(x) = \dfrac{x+3}{x^2+9}$

20. $g(x) = \dfrac{x-5}{x^2+1}$

21. $y = \sqrt{2-x}$

22. $y = \sqrt{8-3x}$

23. $f(s) = \dfrac{s+2}{|s-3|}$

24. $f(r) = \dfrac{r-3}{|r+5|}$

25. $y = \sqrt{x^2+4}$

26. $y = \sqrt{2x^2+9}$

27. $g(t) = 3t^3 - 5t + 7$

28. $g(t) = t^4 - 3t^3 - 9t + 5$

29. $f(w) = \dfrac{w+1}{\sqrt{w-5}}$

30. $f(w) = \dfrac{w-3}{\sqrt{2w+3}}$

(31- 42) Find the range for each function: (Write the answers in interval notation.)

31. $f(x) = |x-4|$

32. $f(x) = |3x+16|$

33. $y = x^2 + 3$

34. $y = x^2 - 9$

35. $f(t) = \sqrt{t+3}$

36. $f(t) = \sqrt{t-9}$

37. $y = |x-5| + 2$

38. $y = |x+3| - 8$

39. $y = 3 - x^2$

40. $y = 5 - \sqrt{x+3}$

41. $y = \sqrt{x^2+16}$

42. $y = \sqrt{4x^2+25}$

(43–60) Use your graphing calculator to graph the following functions. Determine the domain and range by inspecting the graph. (Write the answers in interval notation.)

43. $y = x^2 + 2$

44. $y = x^2 - 3$

45. $f(x) = x^2 - 6x + 11$

46. $f(x) = x^2 + 4x - 1$

47. $g(x) = |2x-3|$

48. $g(x) = |3x+2|$

49. $y = |4x+5| + 3$

50. $y = |3x-4| - 2$

51. $f(x) = x^3$

52. $f(x) = x^3 + 2$

53. $y = \sqrt{x-4} + 2$

54. $y = \sqrt{x+3} - 3$

55. $g(x) = \sqrt{x^2 - 9}$

56. $g(x) = \sqrt{4x^2 - 9}$

57. $y = \dfrac{5}{x+4}$

58. $y = \dfrac{2}{x-3}$

59. $f(x) = 3x + 5$

60. $f(x) = 3 - 2x$

61. $y = 2x^3 - 3x^2 - 6x$

62. $y = 2x^3 + 3x^2 - 36x$

Problems 63-72 are examples of piecewise functions, for each function:

(Reminder: Section 0.8 shows how to input these types of functions into your calculators)

a) Graph the function with your graphing calculator

b) Give the domain of the function in interval notation

c) Give the range of the function in interval notation

63. $f(x) = \begin{cases} x+1 & x < 3 \\ x^2 + 1 & x \geq 3 \end{cases}$

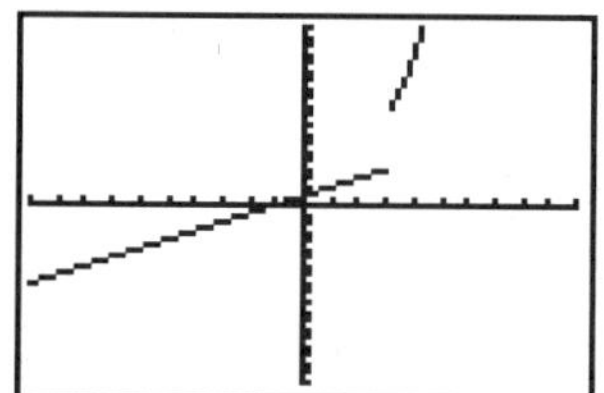

64. $f(x) = \begin{cases} x^2 - 3 & x \leq 2 \\ x - 3 & x > 2 \end{cases}$

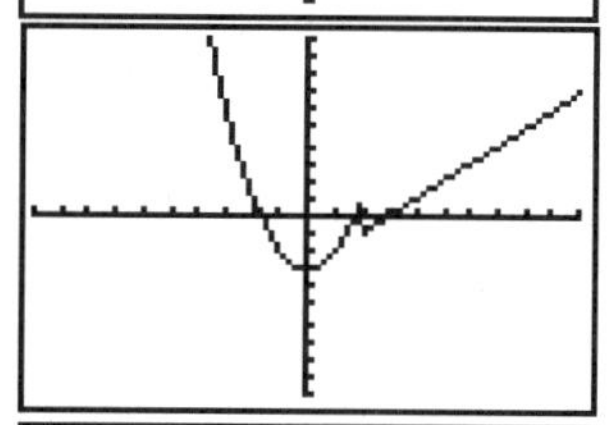

65. $g(x) = \begin{cases} x+1 & x < 3 \\ x+3 & x \geq 3 \end{cases}$

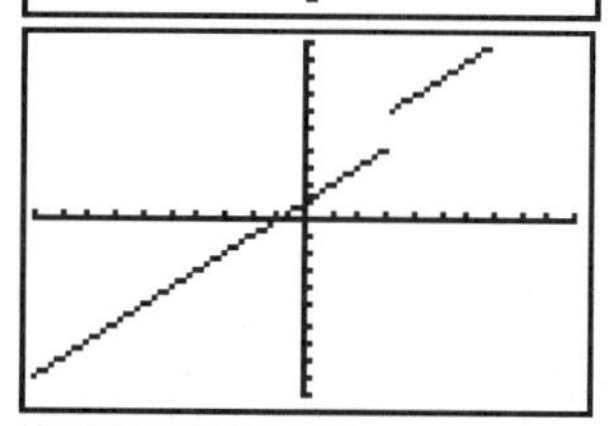

66. $g(x) = \begin{cases} x-3 & x \leq 1 \\ x+2 & x > 1 \end{cases}$

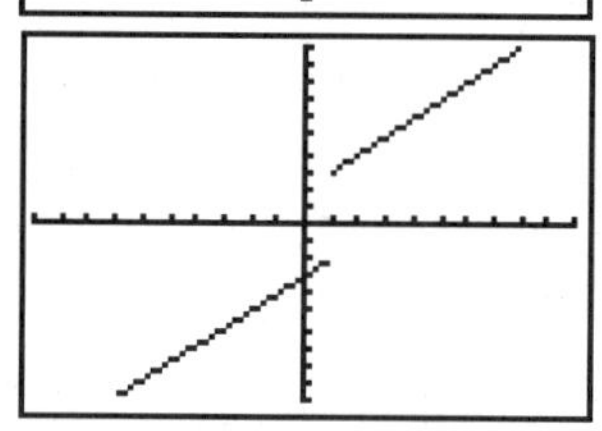

67. $t(x) = \begin{cases} 2x-1 & -3 \leq x < 1 \\ 2x & 1 \leq x \leq 5 \end{cases}$

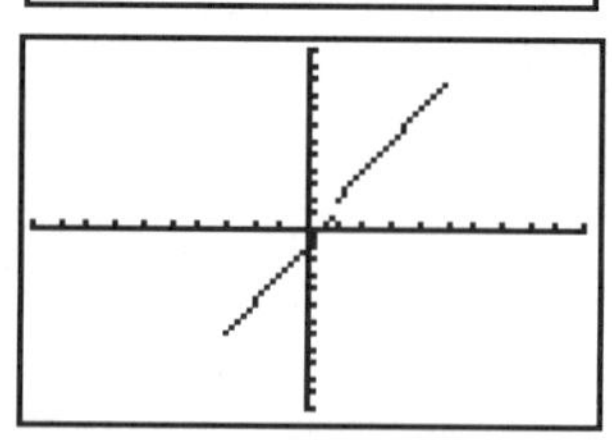

68. $t(x) = \begin{cases} 3x+1 & -4 \leq x < 0 \\ 2x-3 & 0 \leq x < 6 \end{cases}$

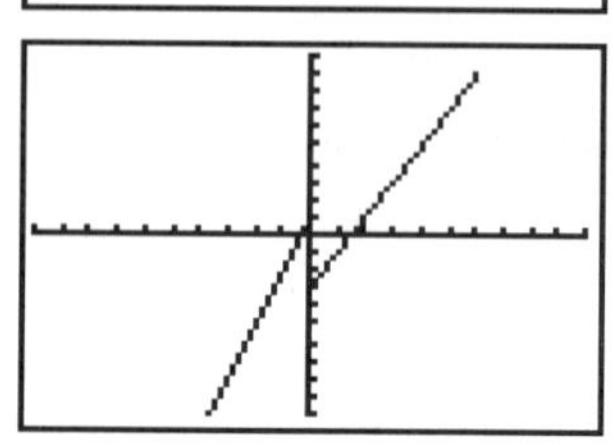

69. $f(x)=\begin{cases}|x+2| & x<-3\\ |2x-1| & x\geq -3\end{cases}$

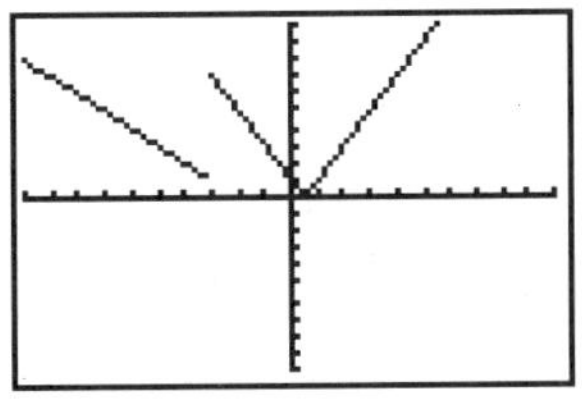

70. $f(x)=\begin{cases}|x-3| & x<-1\\ |3x+1| & x\geq -1\end{cases}$

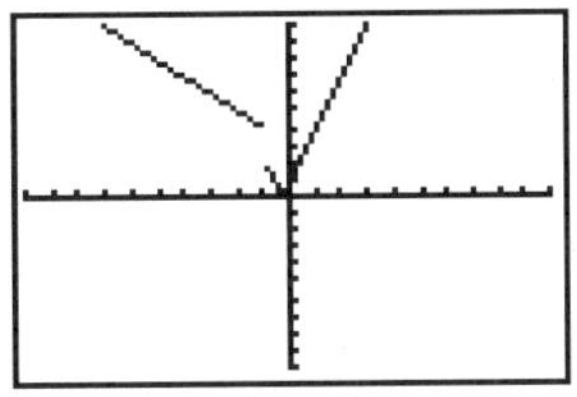

71. $g(x)=\begin{cases}x+2 & x<-3\\ 2x+1 & -3\leq x<1\\ x^2 & x\geq 1\end{cases}$

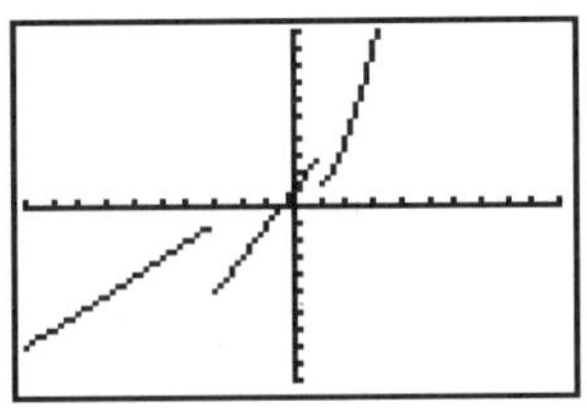

72. $g(x)=\begin{cases}2x+3 & x<-3\\ x^2+1 & -3\leq x\leq 2\\ 3x+1 & x>2\end{cases}$

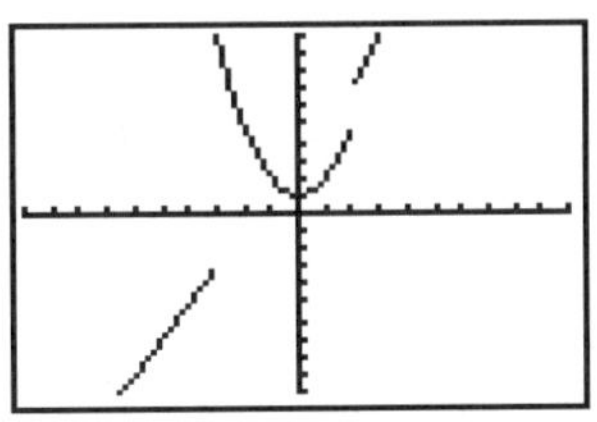

73. A box is created from a 10 inch by 10 inch square piece of cardboard by cutting equally sized squares from each corner then bending up the resulting flaps. We can express the volume of the box as a function of the length (x) of the sides of the cut squares. The volume formula for the box thus created will be $V(x)=4x^3-40x^2+100x$.

   a. What is the domain for the problem?

   b. Approximately what is the range for the problem?

74. A box is created from a 10 by 12 inch rectangular piece of cardboard by cutting equally sized squares from each corner then bending up the resulting flaps. We can express the volume of the box as a function of the length (x) of the sides of the cut squares. The volume formula for the box thus created will be $V(x)=4x^3-44x^2+120x$.

   a. What is the domain of the problem?

   b. Approximately what is the range for the problem?

Section 1.3 Maximums and Minimums

We are going to expand our look at functions to include the concepts of increasing and decreasing functions and local maximums and minimums. These concepts have many applications in the real world because we try to increase what's good and decrease what's bad, and often wish to get the least or the most of something.

Let us begin by taking a look at what it means for a function to be increasing or decreasing.

> A function f is said to be **increasing** when x-values getting larger cause $f(x)$-values to get *larger.*

> A function f is said to be **decreasing** when x-values getting larger cause $f(x)$-values to do the opposite and get *smaller.*

One way to determine if a function is increasing or decreasing is to graph it with a graphing calculator. If, as you look from left to right the graph goes uphill, it is increasing. If, as you look from left to right it goes downhill, it is decreasing. A function may do a little of both. It may increase in some places and decrease in others. You may see these terms defined more formally in this manner:

> If $x_1 < x_2$ implies that $f(x_1) < f(x_2)$ for all x's in an interval (a, b), then we say that f is increasing over that interval.

> If $x_1 < x_2$ implies that $f(x_1) > f(x_2)$ for all x's in an interval (a, b), then we say that f is decreasing over that interval.

Discussion 1: Increasing Functions

Here are a couple of functions that are always **increasing**:

a.) Exponential with b>1 (in chapter 4)

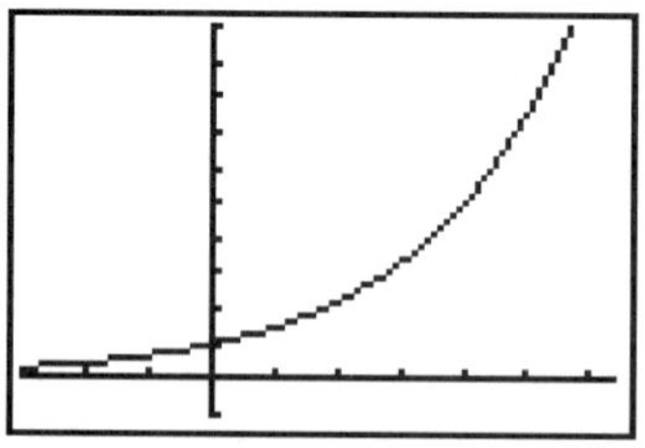

b.) Logarithmic with b>1 (in chapter 4)

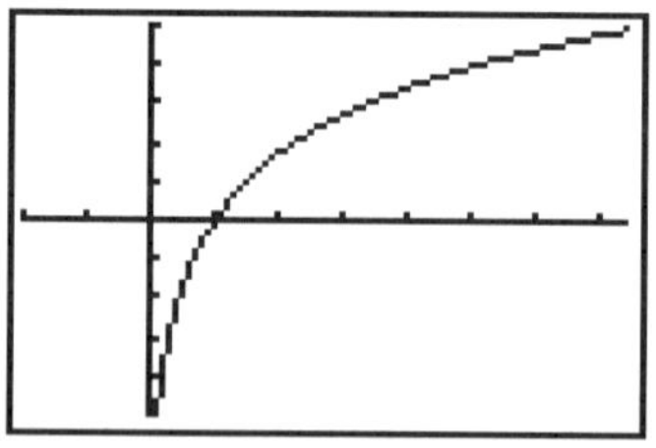

Notice in the above two graphs as you move from left to right, both the x's and the y's are increasing in value.

Discussion 2: Decreasing Functions

Here are a couple of functions that are always **decreasing**:

c.) Exponential with $0<b<1$ (in chapter 4)

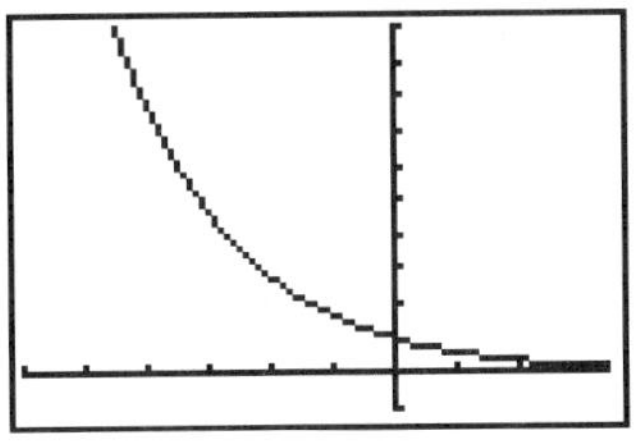

d.) Logarithmic with $(-x)$, $b>1$ (in chapter 4)

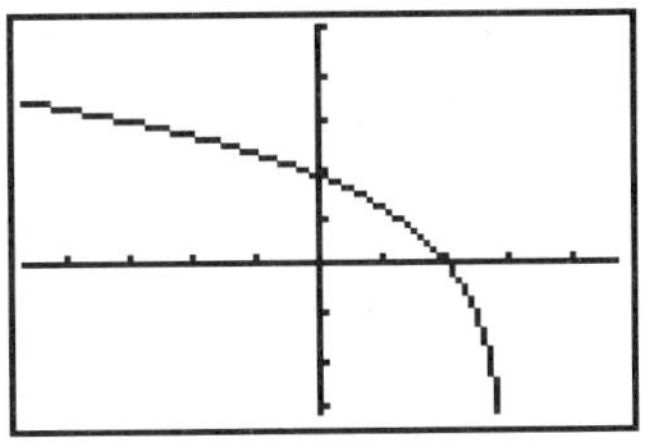

Notice in the above two graphs as you move from left to right the x's are increasing in value, but the y's are decreasing in value.

Look at graphs (a) and (c); they look like half of a "cup". We say that they are **concave up**. But graph (a) is increasing and graph (c) is decreasing. Likewise, graphs (b) and (d) look like half of an "n," so we say that they are **concave down**. Yet graph (b) is increasing and graph (d) is decreasing. This idea of concave up and down is covered more thoroughly in a calculus class. But do notice how it has no bearing on whether the function is increasing or decreasing.

Discussion 3: Increasing and Decreasing Tables

Here is a table which is an example of an increasing function:

a.)

X	Y1
1	1
2	4
3	9
4	16
5	25
6	36
7	49

X=1

Here is a table which is an example of a decreasing function:

b.)

X	Y1
0	50
1	49
2	46
3	41
4	34
5	25
6	14

X=0

Notice in table (a) that as the x-values increase the y-values increase. This is an increasing function. In table (b) as the x-values increase the y-values decrease. This is a decreasing function.

Question 1: Is the following graph an increasing or decreasing function?

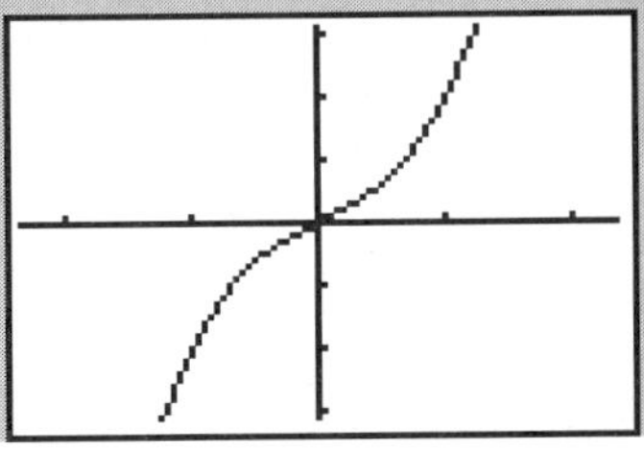

LOCAL MAXIMUMS AND MINIMUMS

We need to talk about how to find local (or relative) maximums and minimums. They help us to find the best solutions to real world problems.

> A **local maximum** is an output value (or range value) that is larger than any of the output values near it within a chosen x-interval (called a neighborhood). These cannot occur at the endpoints of your chosen interval.

On a graph of a function or relation, the local maximums will be the tops of the peaks on the graph. In a table of values, a local maximum will be the largest output value, $f(x)$, within a chosen x-interval.

> A **local minimum** is an output value (or range value) that is smaller than any of the output values near it within a chosen x-interval (called a neighborhood). These cannot occur at the endpoints of your chosen interval.

On a graph of a function or relation, the local minimums will be the bottoms of the valleys on the graph. In a table of values, a local minimum will be the smallest output value, $f(x)$, within a chosen x-interval. More formally stated:

> If there exists an "x_i" in an open interval (a,b) such that for all x's in that same interval you get $f(x_i) \geq f(x)$, then we say that $f(x_i)$ is a **local maximum**.

> If there exists an "x_i" in an open interval (a,b) such that for all x's in that same interval you get $f(x_i) \leq f(x)$, then we say that $f(x_i)$ is a **local minimum**.

Discussion 4: Local Maximums

Here are two examples of a local maximum, one in a graph the other in a table:

a)

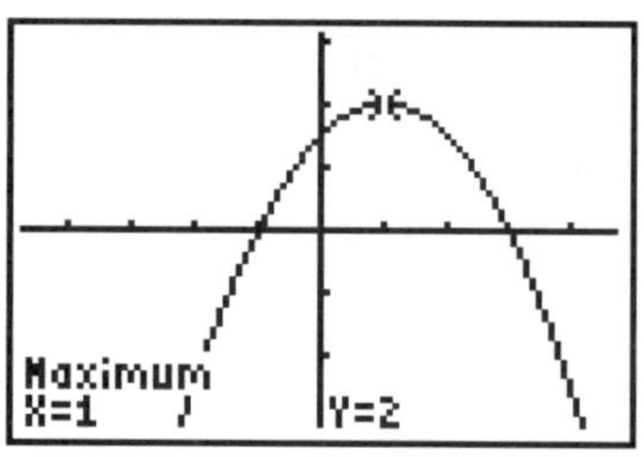

b)

X	Y1
.25	1.7188
.5	1.875
.75	1.9688
1	2
1.25	1.9688
1.5	1.875
1.75	1.7188

X=1

Notice that with an input of 1, in both the graph and the table, we get a local maximum output value of 2. No other input value near 1 yields an output value *greater than* 2.

Discussion 5: Local Minimums

Here are two examples of a local minimum, one in a graph the other in a table:

a)

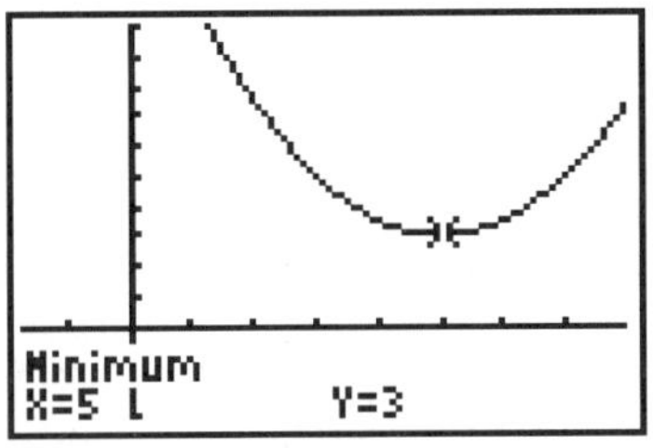

b)

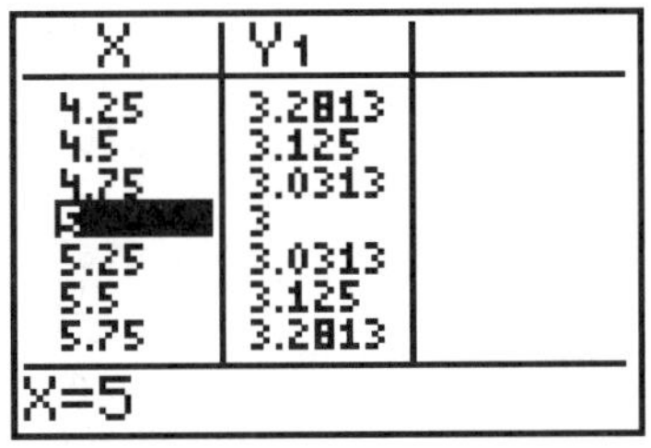

In both figures above, notice that with an input of 5 we get a local minimum output value of 3. No other input value near 5 yields an output value smaller than 3.

Example 1: Local Maximum and Minimum Temperatures (in Phoenix)

Consider the function defined in the table below. Find all the local maximum(s) or minimum(s) and determine the intervals where the function increases and/or decreases.

Answer Q1: Increasing

Month	Jan.	Feb.	Mar.	Apr.	May	June	July	Aug.	Sept.	Oct.	Nov.	Dec.
Average Temp.	54	58	62	70	79	88	94	92	86	75	62	54

Solution:

Here again is discussion 8 from section 1.1. It is the average temperature in Phoenix, Arizona, by month over the course of a year. If you remember, we called this function T, where M was our input (months) and $T(M)$ was our output (average temperature). Notice that with an input of July we have an output of 94 degrees. You can see from this table that 94 is the largest value of all the values around it (79, 88, 94, 92, 86), so we call 94 our local maximum. It also happens to be the largest of all the values (absolute maximum), but that is another topic. Absolute maximums are usually covered more thoroughly in a Calculus course.

You might also see, as one would expect, that as you move through the year from January to July the average temperatures are increasing, so we say that this function is increasing from January to July. Likewise, the temperatures from July to December are decreasing, so we say that this function is decreasing from July to December. This table doesn't have a local minimum because January and December are endpoints of the year. Recall that to be a local maximum or minimum, the input values must be within an interval, not at the endpoints.

Example 2: Finding Local Maximums and Minimums on a Graph

Find the local maximum(s) or minimum(s) and determine the intervals where the function increases and/or decreases.

Solution:

Below are three graphs of the same function. The last two graphs each have a point labeled. These points should help in determining the local maximums and minimums.

a) 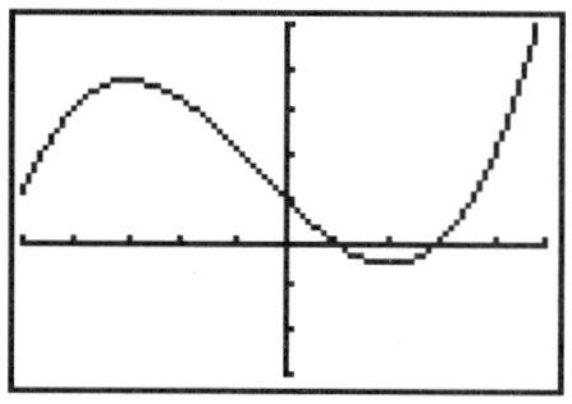

b) 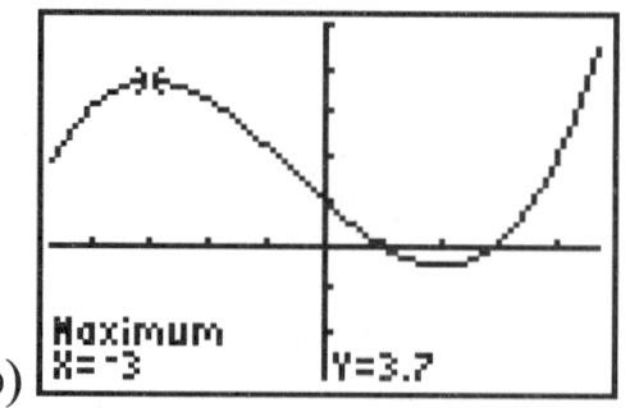

c) 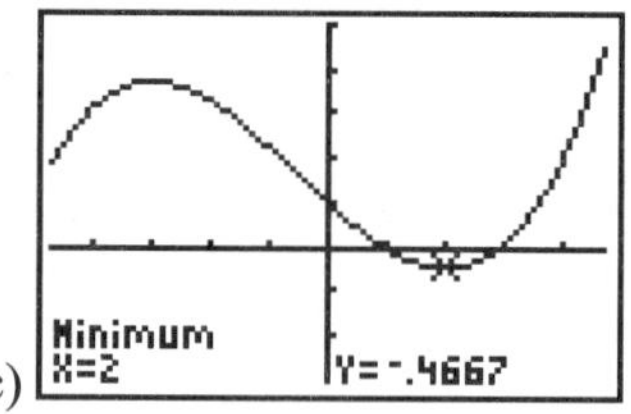

You can see from graph (c) that the minimum is $y = -0.4667$ and from graph (b) that the maximum is $y = 3.7$. Those are the low (valley) and high (peak) points on the graph. The minimum happens when $x = 2$ and the maximum happens when $x = -3$. When we describe *where* a function is increasing and decreasing, we focus our attention on the x-values, *not* the y-values. So the answer to the question, "*Where* is a function increasing or decreasing?" is the x-intervals (inputs) that are causing the function (outputs) to either increase or decrease. In other words it's the x-values that are causing the graph to go either uphill or downhill.

In following chapters we'll learn how to predict the behavior of a graph as it goes to infinity in either direction. For now, let's assume that we see the graph in its entirety. Look at the far left-hand portion of the graph and start tracing along the function. Start at $x = -5$ and trace along the curve to the right. Notice that the graph tops out and starts decreasing when you get to the point (–3, 3.7), where the local maximum is located. So we will say that the graph is increasing on the x-interval (–5, –3) (notice how the y-value of 3.7 does not get shown in our answer).

Now continue tracing along the graph. Notice that the graph decreases from the point (–3, 3.7) our local maximum to the point (2, –0.4667) our local minimum. We will say that the graph is decreasing on the x-interval (–3, 2) (once again the y-value of –0.4667 does not get shown in our answer). Continue tracing along the graph, from the location of the local minimum (2, –0.4667) to the end of the graph. The function is increasing over this final section of the graph. Since the graph stops when $x = 4$ we will say that the graph is increasing on the interval (2, 4). To recap, see the figure:

```
-5                          -3                             2                               4
|---------------------------|------------------------------|-------------------------------| x
       Increasing           |         Decreasing           |         Increasing
                      Local Max at                   Local Min at
                       (–3, 3.7)                     (2, –0.4667)
```

Question 2: Approximately, what are the local maximum(s) or minimum(s) for this graph?

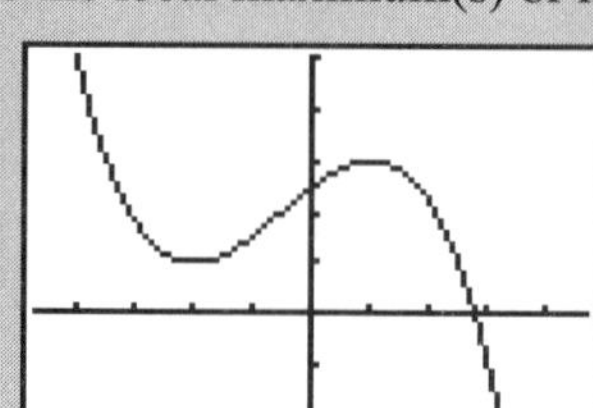

Question 3: Using the graph in Question 2, and assuming that the graph goes up forever on the left and down forever on the right, over what intervals is this function increasing and decreasing?

Example 3: Investigating a Function Graphically

Find the local maximums and local minimums of $f(x) = 0.2x^4 + 0.4x^3 - 2.4x^2 - 2.6x + 2$, if they exist, and determine where the function is increasing and decreasing. Assume that the graph keeps going in the same direction as it goes off the screen (to infinity).

Solution:

Using a graphing calculator we can find good approximations for the maximums and minimums. First we graph $f(x)$, and then we zoom in on the minimum on the left side of the graph.

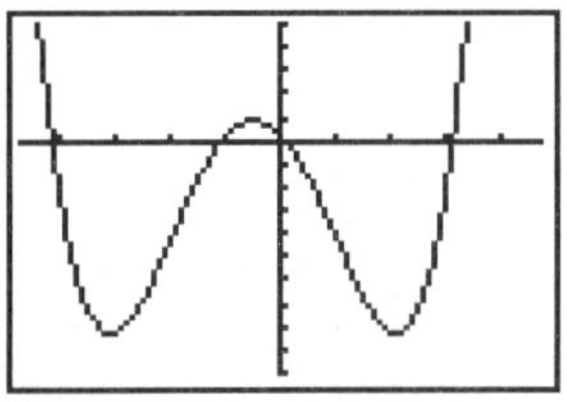

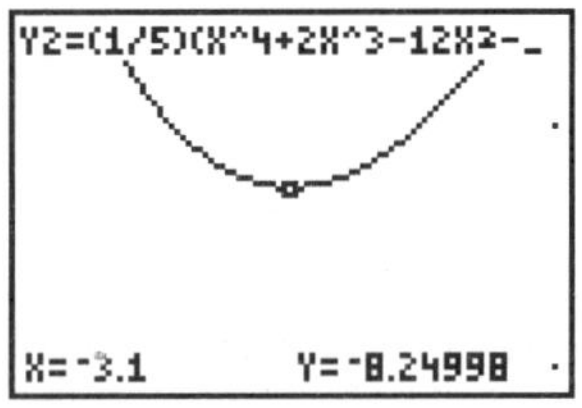

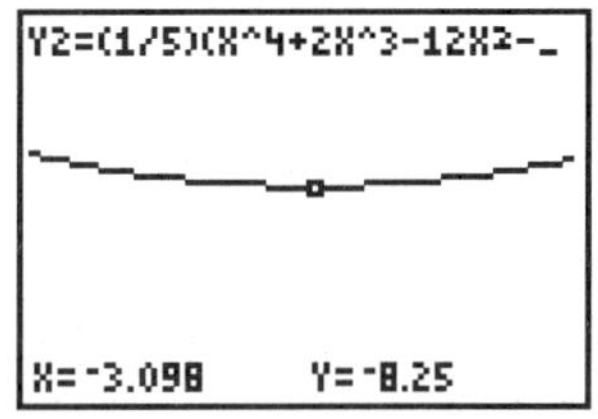

It appears that a minimum is $y = -8.25$ and is located at $x = -3.098$. We will do the same as above to find the other minimum and the one maximum that this function has.

The other min. 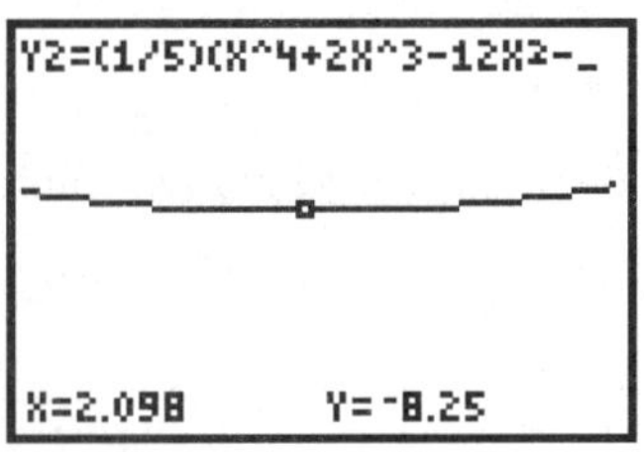

the max. 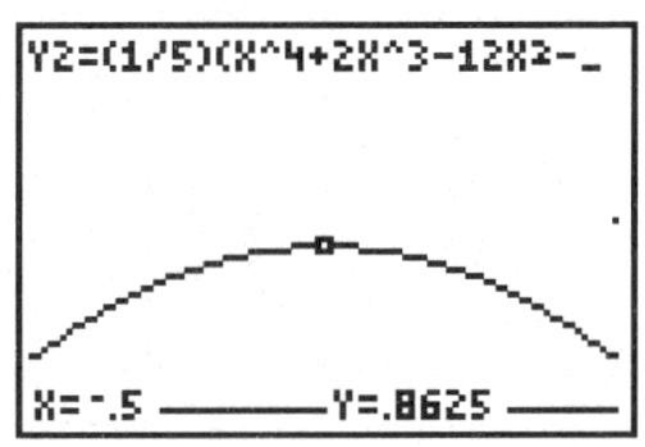

We have found two minimums that, it turns out, have the same value ($y = -8.25$), and one maximum ($y = .8625$). Now from the original graph we can see that the function increases on the x-intervals $(-3.098, -0.5)$ and $(2.098, \infty)$ and decreases on the intervals $(-\infty, -3.098)$ and $(-0.5, 2.098)$.

Remember that the intervals over which a function increases or decreases are expressed as x-intervals (domain inputs). The intervals begin and/or end at the local maximums and minimums.

Example 4: Investigating a Function Using a Graphing Calculator

Find the maximums and minimums if they exist and also find where the function is increasing and decreasing for $f(x) = \left|\sqrt[3]{x-1}\right| + 2$.

Solution:

First let us look at a graph of this function with our usual assumption that the graph continues on.

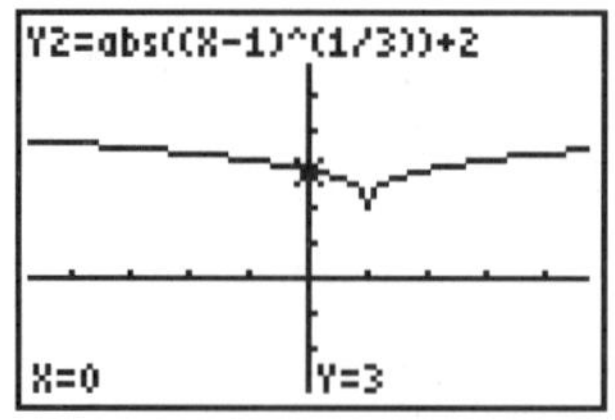

This time we will use the calculate menu on our calculators to find the minimum that we can see on this graph. To get to this menu (on a TI-83) push the $\boxed{2^{nd}}$ key followed by the $\boxed{\text{TRACE}}$ key. Choose minimum (3) and you will see the graph below on your screen. Next, move your cursor to the left of the minimum and push $\boxed{\text{ENTER}}$; then move the cursor to the right of the minimum and push $\boxed{\text{ENTER}}$; then move the cursor close to the minimum and push $\boxed{\text{ENTER}}$. You should now see your answer.

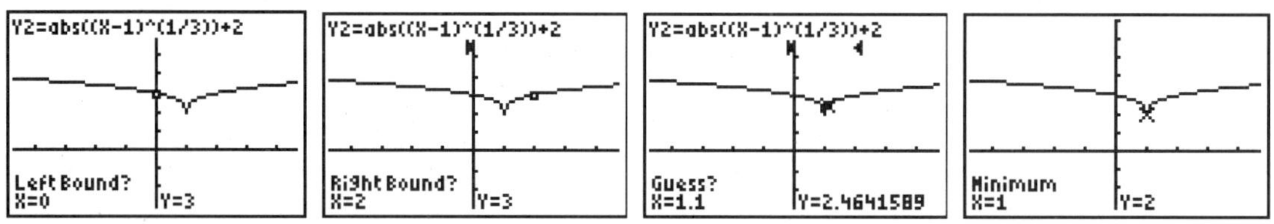

So the local minimum happens at $x = 1$ and the minimum value is $y = 2$. The x-interval where the function is decreasing is $(-\infty, 1)$ and the x-interval where the function is increasing is $(1, \infty)$.

Answer Q2:

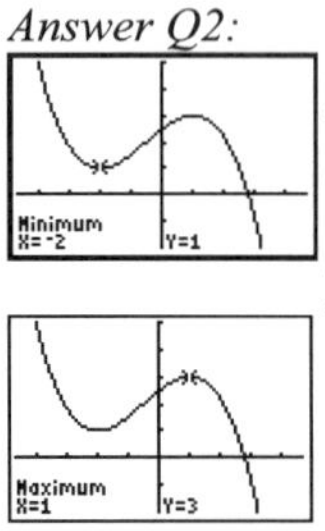

At this point you might be wondering if every function has a local maximum or minimum or both. The answer is, "No." Below is the graph from question 1 of this section. As you can see, it has no peaks or valleys. So, it has no local maximums or minimums.

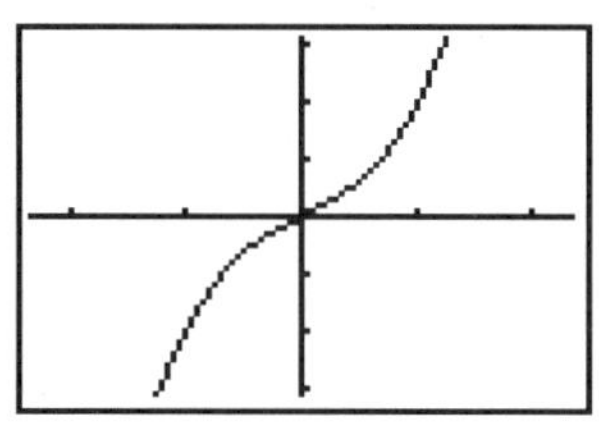

Example 5: Finding a Minimum Car Payment

Let's consider the formula that will give us the monthly car payment on a \$20,000 loan at 7% interest compounded monthly. The variable x represents the number of years over which we'll repay the loan. Of course, we'd like to make the *minimum* payment. The formula and graph are below.

$$MonthlyPayment = \frac{20000\left(\frac{.07}{12}\right)}{\left(1-\left(1+\frac{.07}{12}\right)^{-12x}\right)}$$

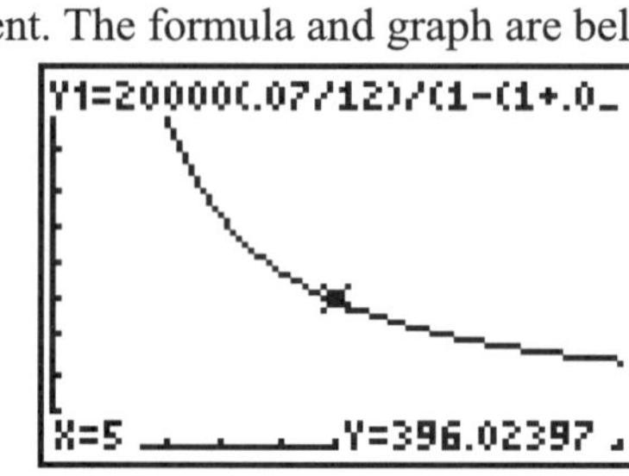

What are the local maximum(s) or minimum(s), if they exist, and over what intervals is the function increasing and decreasing?

Solution:

You might first notice that there doesn't appear to be a maximum or minimum. This is correct. The graph doesn't have any peaks or valleys. From the graph we can see that this function is always decreasing. So what is the interval where this function is decreasing? Well, we have to consider what the domain is for this problem. Since this is a "real-life" problem, and most lenders wouldn't let you have a car loan for more than 5 years, an interval of [0,5] is a reasonable answer. Notice on the graph that we have used the trace key to show you that for a five year loan ($x = 5$) the monthly payment is \$396.02 ($y = 396.02$). This is the smallest car payment you could have on a loan set up as we've described.

Answer Q3: Increasing on the interval $(-2, 1)$ and decreasing on the intervals $(-\infty, -2)$ and $(1, \infty)$.

Example 6: Maximizing Profit

Suppose that a certain toy maker produces a cute little stuffed animal to sell. And suppose that the profit equation for the cute little toy is $P(x) = -0.1x^2 + 5x - 30$, where x is the number of stuffed animals sold (in millions) and $P(x)$ is the total profit earned by the company (in millions of dollars). Find the number of stuffed animal toys that should be produced and sold to maximize profit. In other words, find the maximum. Also, find the intervals where the profit is increasing and where the profit is decreasing.

Solution:

First, let's graph this function. Of course, the x-values need to be positive since it's not possible to produce a negative number of toys. Here is the graph.

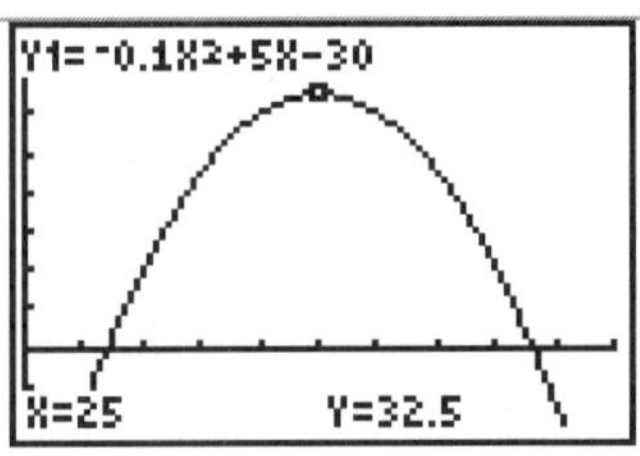

The window for this graph is Xmin = 0, Xmax = 50, Xscl = 5, Ymin = -5, Ymax = 35 and Yscl = 5.

As you can see, the maximum happens when $x = 25$, and the maximum is $y = 32.5$. Also the profit function is increasing on the interval $(0, 25)$ and decreasing on the interval $(25, \infty)$. Thus the company will make a maximum profit of \$32.5 million when it produces and sells 25 million toys. By this kind of analysis a company can decide what their production rate should be to make the most money.

Example 7: Decibel Measure

Let's look at the formula that will give us the decibel measure for sounds that we hear everyday. The formula and the graph are below. What is the local maximum or minimum and when is the function increasing and decreasing?

$dB = 10\log\left(\frac{X}{10^{-12}}\right)$	Here X is the acoustic power of the sound we are hearing and dB is the decibel number associated with that acoustic power.	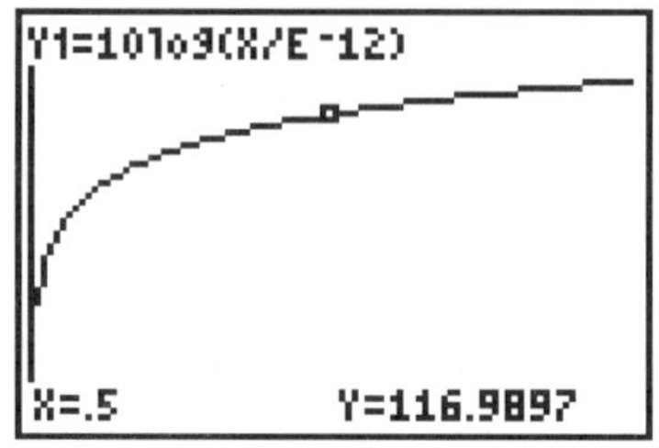

Solution:

The 10^{-12}, in the formula, is the acoustic power of the smallest sound audible to the human ear. The window for the graph above is Xmin = 0, Xmax = 1, Xscl = 0, Ymin = 90, Ymax = 125 and Yscl = 0. Here, similar to the last example, there isn't a maximum or minimum; but this time the function is always increasing. This should make sense, because as a sound gets stronger the dB level gets higher. So the function is increasing on the x-interval $(0, \infty)$.

Question 4: What are the local maximum(s) or minimum(s) for the function $f(x) = x^3 + 6x^2 - 15x - 46$ and over what intervals is the function increasing and decreasing? Graph this function on your calculator, be sure that you have a complete picture and then use one of the techniques discussed.

Section Summary

- Local (relative) maximums or minimums are found at the peaks or valleys on the graph of the function.
- A function will change from increasing to decreasing or vice versa anywhere that you find a local (relative) maximum or minimum.
- The x-values between local maximums and minimums create the intervals on which a function increases and decreases. We express the local maximum or minimum in terms of its y-value.
- For increasing functions, as the x-values get larger the $f(x)$-values also get larger.
- For decreasing functions, as the x-values get larger the $f(x)$-values do the opposite and get smaller.

SECTION 1.3 HOMEWORK

1. The following table is the average temperatures monthly for Apalachicola Florida. Let *T*(*M*) be a function for this table, where *M* = the month and *T* = the temperature.

Jan.	Feb.	Mar	Apr	May	Jun	Jul	Aug	Sep	Oct	Nov	Dec
52.2	54.5	60.7	67.3	74.0	79.7	81.4	81.3	78.6	70.2	61.9	55.3

a. Where is the function increasing?

b. Where is the function decreasing?

c. Is there any relative (local) minimums and if so where?

d. Is there any relative (local) maximums and if so where?

2. The following table is the average temperatures monthly for Kotzebue, Alaska. Let *T*(*M*) be a function for this table, where *M* = the month and *T* = the temperature.

Jan.	Feb.	Mar	Apr	May	Jun	Jul	Aug	Sep	Oct	Nov	Dec
–1.0	–4.8	0.4	11.3	31.2	43.8	53.8	52.1	42.0	22.9	7.8	–.09

a. Where is the function increasing?

b. Where is the function decreasing?

c. Is there any relative minimums and if so what are they?

d. Is there any relative maximums and if so what are they?

(3 - 4) For each of the following graphs:

a. Approximate the intervals where the graph seems to be increasing.

b. Approximate the intervals where the graph seems to be decreasing.

c. Approximate any relative maximum value(s).

d. Approximate an relative minimum value(s).

3. The following graph is the stock market results for EXXON MOBILE CORP for one day. The horizontal values are hourly starting at 9:30 a.m. to 4:30 p.m. (From Market Watch Interactive)

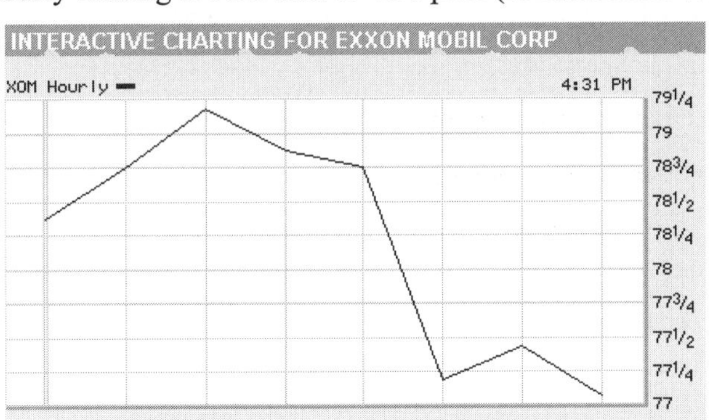

4. The following graph shows the price of WAL-MART stock, by quarter for a two-year period. The horizontal values are quarterly from the 1st quarter to the 8th quarter. (From Market Watch Interactive).

Answer Q4: There is a minimum at the point (1, –54) and a maximum at the point (–5, 54). So the maximum is f(x) = 54 and the minimum is f(x) = –54. This function is decreasing on the x interval (–5, 1) and increasing on the x-intervals (–∞, –5) and (1, ∞).

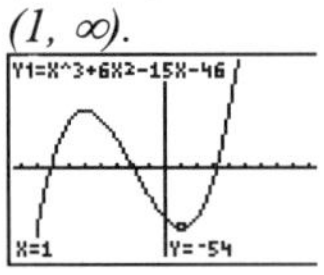

(5-32) For each of the following function use a graphing calculator to approximate:

a. Any local maximum(s)

b. Any local minimum(s)

c. The interval(s) where the function in increasing

d. The interval(s) where the function is decreasing

5. $f(x) = x^2 + x - 2$
6. $f(x) = x^2 + 8x - 3$
7. $f(x) = 2x^2 + 5x - 3$
8. $f(x) = 3x^2 - 2x + 5$
9. $f(x) = -x^2 + 8x - 8$
10. $f(x) = -x^2 - 12x - 6$
11. $f(x) = -2x^2 - 5x + 3$
12. $f(x) = -3x^2 + 11x + 1$
13. $f(x) = .2x^3 + .3x^2 - 3.6x + 1$
14. $f(x) = .08x^3 - .3x^2 - .36x + 2$
15. $f(x) = 2x^3 - 3x^2 - 12x + 4$
16. $f(x) = 2x^3 + 3x^2 - 12x - 5$
17. $f(x) = x^4 - 5x^2 + 4$
18. $f(x) = .6x^4 - 1.9x^3 - 3.7x^2 + 6.2x + 2$
19. $f(x) = x^4 + x^3 - 2x + 2$
20. $f(x) = 2x^4 + 3x^3 - 20x + 3$
21. $f(x) = -2x^4 - 3x^3 + 8x - 3$
22. $f(x) = -3x^4 - x^3 + 10x - 5$
23. $f(x) = |2x - 5| + 3$
24. $f(x) = |3x + 4| - 5$
25. $f(x) = -3|2x + 3| + 5$
26. $f(x) = -2|3 - 2x| - 4$
27. $f(x) = 0.1x^3 - 0.02x^2 + 2x - 3$
28. $f(x) = -0.03x^3 + 0.01x^2 - 0.5x + 1$
29. $f(x) = 3^{0.3x+1}$
30. $f(x) = 2e^{-0.5x} - 1$
31. $f(x) = \ln(x)$
32. $f(x) = \ln(-x)$
33. A toy rocket is shot up vertically from ground level with an initial velocity of 240 ft. per second. The height of the rocket $s(t)$, in feet, after time t, in seconds, is given by the function $s(t) = -16t^2 + 240t$ (Use your graphing calculator to approximate answers.):

   a. At what time does the rocket reach its maximum height?

   b. What is the maximum height reached by the rocket?

   c. Over what time interval is the rocket increasing in height?

   d. Over what time interval is the rocket decreasing in height?

   e. What is the domain of this function?

f. What is the range of this function?

34. A ball is thrown vertically upward with an initial velocity of 48 feet per second. If the ball started at a height of 8 feet, then its height $s(t)$ (in feet) after time t (in seconds) is given by the function $s(t) = -16t^2 + 48t + 8$.

    Use your graphing calculator to approximate the answers for the following questions:

    a. What is the maximum height reached by the ball?

    b. At what time does the ball reach the maximum height?

    c. During what time interval is the ball increasing in height?

    d. During what time interval is the ball decreasing in height?

    e. What is the domain of this function?

    f. What is the range of this function?

35. A gutter is to be formed by bending up equal pieces on either side of a rectangular piece of sheet metal that is 15 inches by 18 inches along the side that is 15 inches long. The volume of this piece of gutter is given by the function $V(x) = 18(15 - 2x)(x)$, where x is the width of the bent up pieces.

    a. What is the logical domain for this function?

    (Use your graphing calculator to approximate the answers for the following.):

    b. What value of x gives the maximum volume of the guttering?

    c. What is the maximum volume the guttering will hold?

    d. Over what interval for x is the volume increasing?

    e. Over what interval for x is the volume decreasing?

    f. What is the range of this function?

36. Using a piece of cardboard that is a 12 inch by 12 inch square you are to create a box by cutting equal squares from each corner and bending up the resulting flaps. The volume of the resulting box is given by the function $V(x) = (12 - 2x)(12 - 2x)x$, where x is the length of the sides of the squares.

    a. What is the logical domain for this function?

    (Using your graphing calculator to approximate the following answers.):

    b. What value of x will create the largest volume box?

    c. What is the volume of the largest box created?

    d. Over what interval of x is the volume increasing?

    e. Over what interval of x is the volume decreasing?

    f. What is the range for this function?

37. The profit for a certain company is given by the function $P(x) = 230 + 20x - 0.5x^2$, where x is the amount (in hundreds of dollars) spent on advertising and $P(x)$ is the profit (in hundreds of dollars).

(Use your graphing calculator to approximate the following answers.):

a. How much should the company spend on advertising to maximize its profit?

b. What will that maximum profit be?

c. Over what interval of advertising dollars is its profit increasing?

d. Over what interval of advertising dollars is its profit decreasing?

e. What is a logical domain for this problem?

f. What is the range for this problem?

38. The following function represents the profit of a certain company $P(x) = -x^3 + 2700x - 5392$, where $P(x)$ represents the profit and x the number of items sold.

(Use your graphing calculator to approximate the following answers.):

a. How many items should the company produce to maximize profit?

b. What will be the maximum profit?

c. Over what interval of x is the profit increasing?

d. Over what interval of x is the profit decreasing?

39. A manufacturer of lighting fixtures has a daily production cost given by the function $C(x) = 800 - 75x + 0.25x^3$, where x is the number of fixtures produced and $C(x)$ is the cost of production.

(Use your graphing calculator to approximate the following answers.):

a. How many fixtures should the company produce to minimize cost for the day?

b. What will be the minimum cost for the day?

c. Over what interval of x is cost increasing?

d. Over what interval of x is cost decreasing?

40. A company finds the average cost per unit for producing x units is given by the following function: $C(x) = \dfrac{800}{x} + 20 + 0.05x^2$, where $C(x)$ represents the average cost per unit for producing x units. (Use your graphing calculator to approximate the following answers. Hint: use the zoom fit feature of your calculator.):

a. How many units should be produced to minimize average cost?

b. What will be the smallest average cost?

c. Over what intervals of x is average cost decreasing?

c. Over what intervals of x is average cost increasing?

41. The number of hamburgers produced every 10 minutes by x workers is given by the function $B(x) = 50\left(\dfrac{1}{0.1\sqrt{2\pi}}\right)e^{-(x-10)^2(0.02)} - 27$, where $B(x)$ is the number of hamburgers produced.

(Use your graphing calculator to approximate the following answers)

a. How many workers will it take to produce the largest number of burgers every 10 minutes?

b. What is the largest number of burgers that can be produced every 10 minutes?

c. Over what interval is the number of hamburgers produced increasing?

d. Over what interval is the number of hamburgers produced decreasing?

42. Every day the number of tires produced by x workers is given by the function

$T(x) = 300\left(\frac{1}{0.05\sqrt{2\pi}}\right)e^{-(x-20)^2(2(0.05))^2} - 324$, where $T(x)$ represents the number of tires produced.

(Use your graphing calculator to approximate the following answers)

a. How many workers will it take to maximize the output of tires?

b. What is the maximum amount of tires that can be produced?

c. Over what interval is the output of tires increasing?

d. Over what interval is the output of tires decreasing?

Section 1.4 Combining Functions

In this section we'll continue to look at functions in the forms of graphs, tables and formulas. But we'll see how to combine these different forms with each other to answer questions or come up with new functions. We are going to focus on how to add, subtract, multiple, divide and compose functions. Note: When we do these operations using functions the resulting answers will always be functions.

In past algebra courses, you probably did some of these operations with polynomials. In effect, you were doing those operations with functions. For example, look at the following comparison between adding two polynomials and adding two functions. Notice that the formulas are equivalent in both forms; only the notation is different.

Adding polynomials:	Adding functions:
Add $(2x^2 - 5x + 7)$ to $(-3x^2 + 8x + 4)$	Find $(f + g)(x)$ given that
	$f(x) = 2x^2 - 5x + 7$ and $g(x) = -3x^2 + 8x + 4$
Answer: $(-x^2 + 3x + 11)$	Answer: $(f + g)(x) = -x^2 + 3x + 11$

So, you probably already know how to do most of what is in this section. It's only the way the problems are stated that are different than what you might be used to.

Rules of Operations on Functions	
Addition	$(f+g)(x) = f(x) + g(x)$.
Subtraction	$(f-g)(x) = f(x) - g(x)$
Multiplication	$(fg)(x) = f(x)g(x)$
Division	$(f/g)(x) = f(x)/g(x)$

Example 1: Operations on Functions

Use these functions to answer the following questions:

$$f(x) = -2x^2 + 3x, \quad g(x) = 7 - 5x, \quad h(x) = \sqrt{x+2}, \quad k(x) = \frac{2x}{x-3}$$

a.) Find $(f - g)(x)$.

Solution:

$$\begin{aligned}(f - g)(x) &= f(x) - g(x)\\ &= (-2x^2 + 3x) - (7 - 5x)\\ &= -2x^2 + 3x - 7 + 5x\\ &= -2x^2 + 8x - 7\end{aligned}$$

b.) Find $(g \cdot k)(x)$.

Solution:

$(g \bullet k)(x) = g(x) \bullet k(x)$

$$= (7 - 5x)\left(\frac{2x}{x-3}\right)$$

$$= \frac{-10x^2 + 14x}{x-3}$$

c.) Find $(h + f)(x)$.

Solution:

$(h + f)(x) = h(x) + f(x)$

$$= (\sqrt{x+2}\,) + (-2x^2 + 3x)$$

$$= -2x^2 + 3x + \sqrt{x+2}\,.$$

This expression can't be simplified because there are no like terms.

Question 1: Find $(g + f)(x)$.

We *could* ask you to find $(g + f)(2)$ instead of $(g + f)(x)$ in question 1. The function operations are no different than before; just replace all of the x's with the number "2."

Example 2: Operations on Functions with Real-Valued Inputs

Use these functions to answer the following questions:

$$f(x) = -2x^2 + 3x, \quad g(x) = 7 - 5x, \quad h(x) = \sqrt{x+2}, \quad k(x) = \frac{2x}{x-3}$$

a.) Find $(h/g)(7)$.

Solution:

$$(h/g)(7) = \frac{h(7)}{g(7)}$$

$$= \frac{\sqrt{(7)+2}}{7-5(7)}$$

$$= \frac{\sqrt{9}}{7-35}$$

$$= \frac{3}{-28}$$

b.) Find $(k \bullet h)(-1)$.

Solution:

$$(k \cdot h)(-1) = \left(\frac{2(-1)}{(-1)-3}\right) \cdot \sqrt{(-2)+3}$$

$$= \frac{-2}{-4} \cdot \sqrt{1}$$

$$= \frac{1}{2}$$

c.) Find $(f \cdot g)(a)$.

Solution:

$$(f \cdot g)(a) = (-2a^2 + 3a) \cdot (7 - 5a)$$

$$= -14a^2 + 10a^3 + 21a - 15a^2$$

$$= 10a^3 - 29a^2 + 21a$$

Question 2: Find $(k - f)(9)$.

Let's not forget about the idea of domain and range. What happens to the domain and range of functions when they are combined by the operations (+, –, *, ÷)?

DOMAIN OF COMBINED FUNCTIONS

Generally, when you add, subtract, or multiply functions, the domain of the new function will be the *intersection of the domains* of the two or more original functions.

Example 3: Finding the Domain of the Difference of Two Functions

Given $f(x) = x - 5$ and $g(x) = \sqrt{x}$, find $(f-g)(x)$ and its domain.

Solution: $f(x) = x - 5$ {domain $(-\infty, \infty)$}

$g(x) = \sqrt{x}$ {domain $[0, \infty)$}

$(f-g)(x) = x - \sqrt{x} - 5$ {domain $[0, \infty)$}

Note that the new domain is the intersection of the two original domains as shown below.

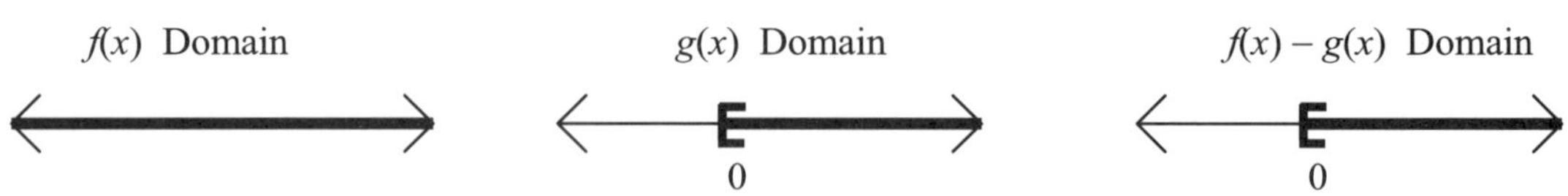

The domain for a division problem is often just the intersection of the functions domains minus any x-values that would make the denominator of the new function equal to zero. For example, if we were asked to find $(g/f)(x)$, using the two functions just mentioned, $f(x) = x - 5$ and $g(x) = \sqrt{x}$, then

$(g/f)(x) = \dfrac{\sqrt{x}}{x-5}$. The intersections of the two domains is still $[0, \infty)$; however, the new function is not defined at $x = 5$ so we must omit 5 from the intersection of the domains. Here is how we show that 5 is not part of the domain using interval notation: $[0, 5) \cup (5, \infty)$. Shown here, are two graphs of the function $(g/f)(x)$:

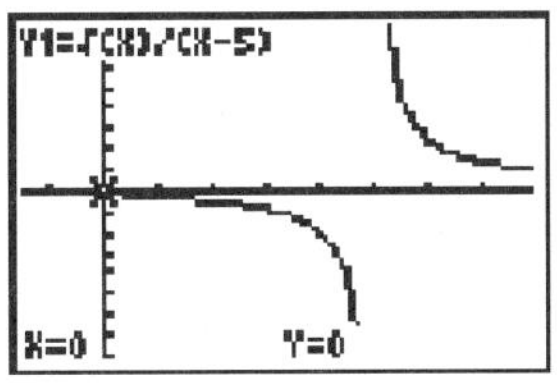

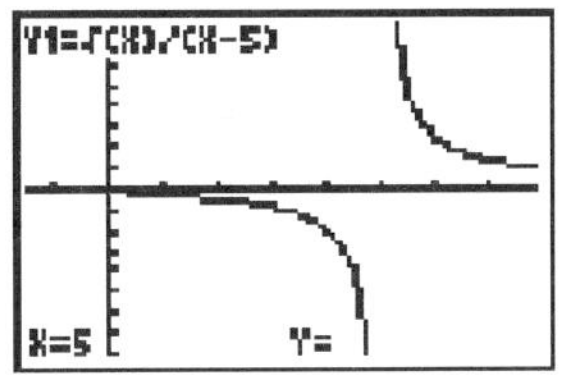

Question 3: What is the domain for $(f/g)(x)$ given $f(x) = \sqrt{x+4}$ and $g(x) = x - 1$?

RANGE OF COMBINED FUNCTIONS

To determine the range, the easiest thing to do is to look at the graph of the function. Remember, we can find the range of a function by identifying all values on the y-axis that have the property that a horizontal line drawn at that value will pass through the graph. In question 3 above, where $f(x) = \sqrt{x+4}$ and $g(x) = x - 1$, the range of $(f/g)(x)$ would be $(-\infty, \infty)$ since every horizontal line that can be drawn will touch the graph.

Answer Q1:
$(g + f)(x)$
$= g(x) + f(x)$
$= (7 - 5x) +$
$(-2x^2 + 3x)$
$= -2x^2 - 2x + 7$

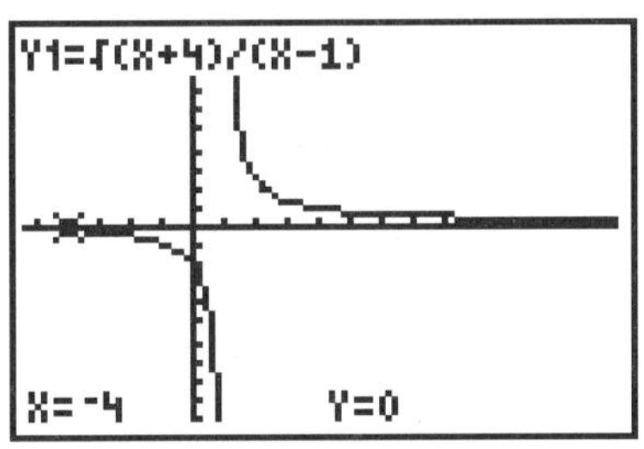

Example 4: Domain and Range of a Combined Function

Given $h(x) = \sqrt{x+2}$ *and* $k(x) = \dfrac{2x}{x-3}$ find $(h + k)(x)$ and its domain and range.

Solution:

$h(x) = \sqrt{x+2}$ {domain $[-2, \infty)$, since $x + 2 \geq 0$ is a must}

$k(x) = \dfrac{2x}{x-3}$ {domain is $(-\infty, 3) \cup (3, \infty)$, since $x \neq 3$}

When $h(x)$ and $k(x)$ are added together the result is,

$$(h + k)(x) = \sqrt{x+2} + \frac{2x}{x-3}.$$

This new function's domain, as we mentioned earlier, will be the intersection of the two domains of $h(x)$ and $k(x)$, which is $[-2, 3) \cup (3, \infty)$.

We can verify this answer by looking at the graph of $(h + k)(x)$.

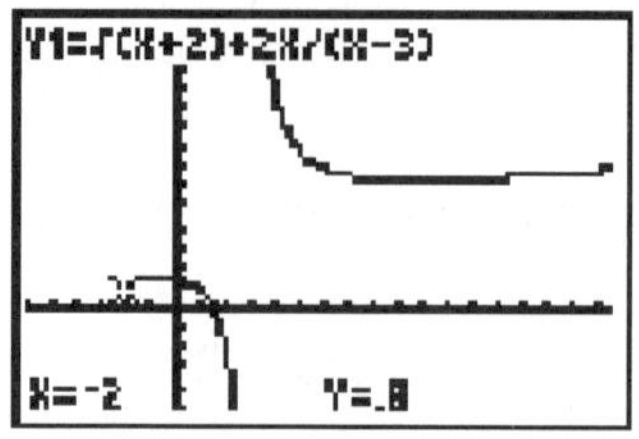

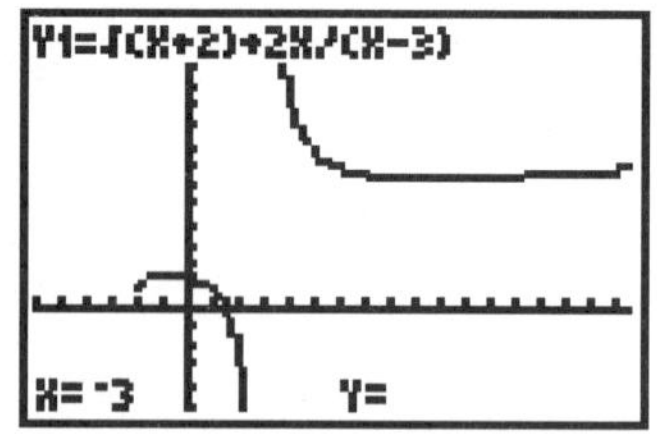

Let's think about the range. Notice that there is a gap between the upper and lower parts of the graph. We need to omit these values from the range. By finding the maximum and minimum values of each part, we see that any horizontal line drawn between $y = 1.5185$ to $y = 6.3138561$ would miss the graph.

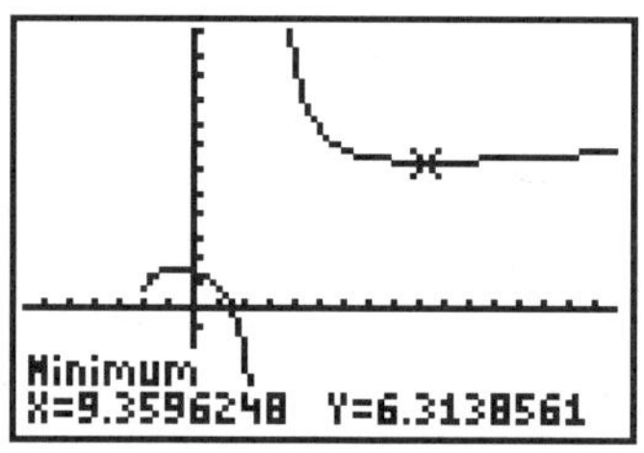

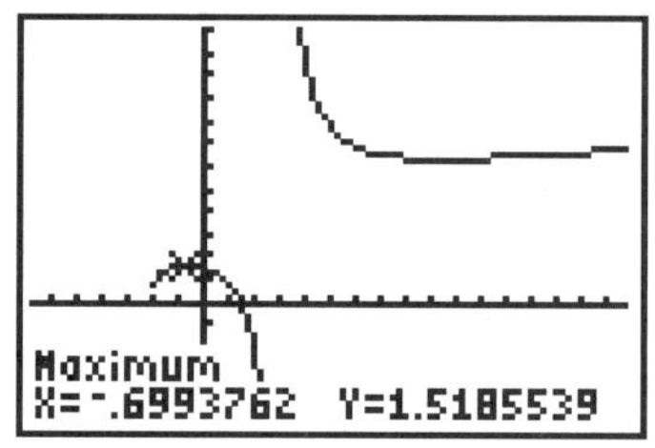

Answer Q2:
$k(9) - f(9)$
$= \frac{2\cdot 9}{9-3} - (2\cdot 9^2 + 3\cdot 9)$
$= \frac{18}{6} - (162+27)$
$= 3 - 189$
$= -186$

Therefore the range is: $(-\infty, 1.5185) \cup (6.3138, \infty)$.

COMBINING FUNCTIONS OF DIFFERENT FORM

Do you remember how to find values from a table? If you were asked to find $f(-2)$ from the tables below, you would look for a -2 in the left column, under the x, and next to it, in the right column, under $f(x)$, you would see your answer, which in this case is 7. So we would write $f(-2) = 7$.

x	$f(x)$
–4	11
–2	7
0	3
2	1
4	5
6	9

x	$g(x)$
–4	3
–2	2
0	1
2	0
4	–1
6	–2

Example 5: Using Tables to Combine Functions

Using the tables above:

a) Find $(f - g)(0)$; b) Find $(g/f)(-2)$; c) Find $(g \cdot f)(6)$

Solutions:

a) $(f-g)(0) = f(0) - g(0)$
$= 3 - 1 = 2$

b) $(g/f)(-2) = \dfrac{g(-2)}{f(-2)}$
$= \dfrac{2}{7}$

c) $(g \cdot f)(6) = g(6) \cdot f(6)$
$= (-2) \cdot (9) = -18$

Notice that we are doing the same things we did with formulas, but we're getting our outputs (y-values) directly from the table instead of having to plug numbers into formulas.

Question 4: Find $(g+f)(4)$.

Answer Q3: The domain of f(x) is [-4,∞), the domain of g(x) is (−∞,∞) and (f/g)(x) = $\dfrac{\sqrt{x+4}}{x-1}$*. The intersection of the domains is [-4, ∞) and since the fraction is undefined at x = 1, we must omit x=1 from the domain. So, the domain of (f/g)(x) is [-4,1) ∪ (1, ∞).*

Let's try doing more of the same using graphs. Look at the graphs of $f(x)$ and $g(x)$ below.

$f(x)$

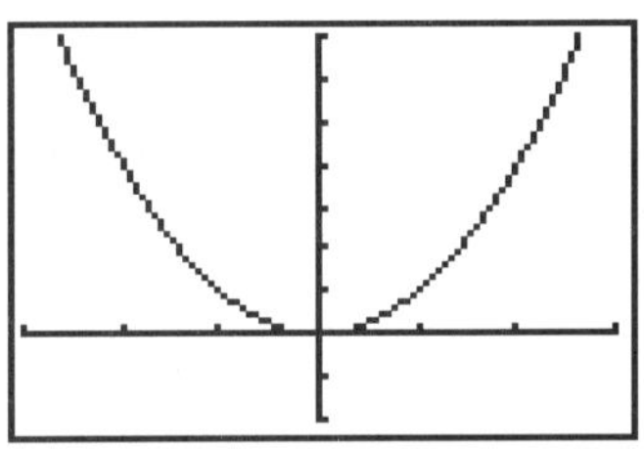

$g(x)$

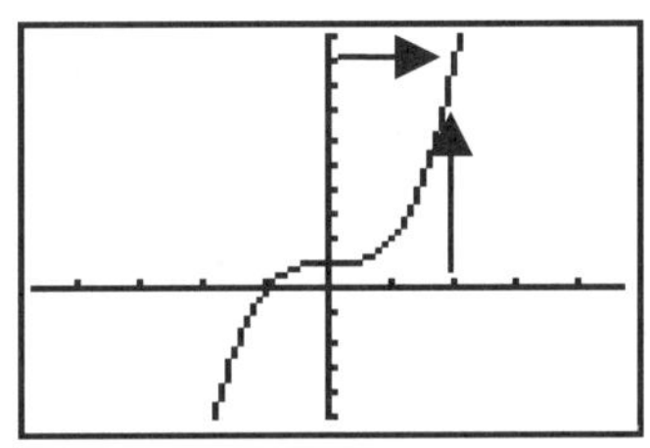

You can tell that $f(0) = 0$, because the point (0, 0) is on the graph of $f(x)$. But, can you find $g(2)$? (Here we need to do some estimating from the graphs.) The value of $g(2)$ seems to be somewhere between 7 and 10. We choose to say that $g(2) = 9$. Your estimate may be different than ours, but as long as you are using the same procedure to get your answers as we are, yours is correct, too.

Example 6: Using Graphs to Combine Functions

Use the graphs of $f(x)$ and $g(x)$ above to find the following values, $(g+f)(-2)$, $(f \cdot g)(2)$, $(f/g)(-1)$.

Solutions:

a) $(g+f)(-2) = g(-2) + f(-2)$	b) $(f \cdot g)(2) = f(2) \cdot g(2)$	c) $(f/g)(-1) = \dfrac{f(-1)}{g(-1)}$
$= -5 + 4$	$= 4 \cdot 9$	
$= -1$	$= 32$	$= \dfrac{1}{0} =$ undefined

Since, in part c) the answer was undefined, we would say that $x = -1$ is not in the domain of (f/g).

Question 5: Find $(g-f)(0)$.

Example 7: Combining Functions Using a Table, Formula, and Graph

Use the table, formula and graph to find the following function values, $(f \cdot h)(0)$, $(g-f)(3)$, $(h/g)(3)$.

x	$f(x)$
0	2
1	0
2	−2
3	0
4	2

$g(x) = \sqrt{x+1}$

$h(x)$

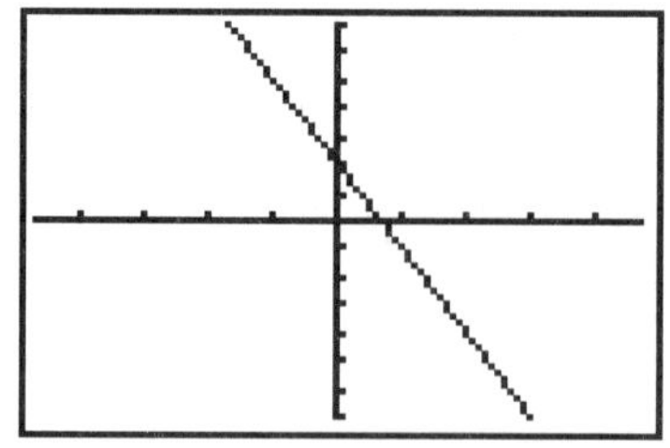

Solutions:

a.) $(f \cdot h)(0) = f(0) \cdot h(0)$	b.) $(g - f)(3) = g(3) - f(3)$	c.) $(h/g)(3) = \dfrac{h(3)}{g(3)}$
$= 2 \cdot 2$	$= 2 - 0$	
$= 4$	$= 2$	$= \dfrac{-7}{2}$

Question 6: Find $(h + g)(-1)$?

If we were to look for the domain and range of $f(x) + g(x)$ from example 7, we would do exactly the same things as earlier. The domain of $f(x)$ is $\{0, 1, 2, 3, 4\}$ the domain of $g(x) = (-1, \infty)$. The intersection of the two domains is $\{0, 1, 2, 3, 4\}$. The range includes the five answers you get from evaluating the new function at those five input values from the domain. The range is $\{3, \sqrt{2}, -2+\sqrt{3}, 2, 2+\sqrt{5}\}$.

The next example is one where you must subtract functions to express the height of a rectangle between two function's graphs.

Discussion 1: Subtraction of Functions

In future math classes it will be necessary to express the h (height of the little rectangle) in the picture to the right as a function of x. The h is equal to the difference in the heights of the two graphs. This means we need to subtract the y-values to obtain h. Hence $h(x) = y_1 - y_2 = \dfrac{1}{x+1} - x^2$. One use of this idea is in a Calculus class where we use the rectangles created at each input value to find the area between the two functions.

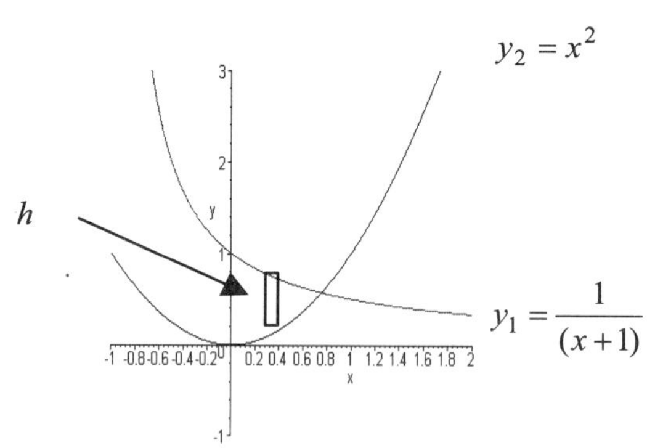

COMPOSITION OF FUNCTIONS

[picture of composition function machine] Our next topic is the composition of functions where, instead of substituting in a number for a variable, we'll substitute an entire function in for the variable. We'll use a new symbol "$\circ$" that means "**composed of**." The expression $(f \circ g)(x)$ (which is read "f composed of g of x") is equivalent to $f(g(x))$.

Example 8: Composing Two Functions

Suppose $f(x) = 2x + 3$ and $g(x) = \sqrt{x-5}$. Find $(f \circ g)(x)$.

Solution:

This means that we are to replace each x in the $f(x)$ function with the $g(x)$ function. So,

$$\begin{aligned}(f \circ g)(x) &= f(g(x)) \\ &= f(\sqrt{x-5}) \\ &= 2\sqrt{x-5} + 3.\end{aligned}$$

Answer Q4:
$g(4) + f(4)$
$= -1 + 5$
$= 4$

Example 9: Composite Functions

Let $f(x) = -2x^2 + 3x$, $g(x) = 7 - 5x$, $h(x) = \sqrt{x+2}$, $k(x) = \dfrac{2x}{x-3}$

Answer Q5:
$g(0) - f(0) = 1$
$- 0 = 1$

Use these functions to find the following compositions, $(g \circ f)(x)$, $(k \circ g)(x)$, $(h \circ k)(x)$.

Solutions:

a) $(g \circ f)(x) = g(f(x))$	b) $(k \circ g)(x) = k(g(x))$	c) $(h \circ k)(x) = h(k(x))$
$= 7 - 5\underline{(-2x^2 + 3x)}$ $= 7 + 10x^2 - 15x$ $= 10x^2 - 15x + 7$	$= \dfrac{2(7-5x)}{(7-5x)-3}$ $= \dfrac{14-10x}{4-5x}$	$= \sqrt{\dfrac{2x}{x-3} + 2}$ $= \sqrt{\dfrac{2x}{x-3} + \dfrac{2x-6}{x-3}}$ $= \sqrt{\dfrac{4x-6}{x-3}}$

Question 7: Using functions from example 9, find $(f \circ g)(x)$.

Notice that example 10(a) asked for $(g \circ f)(x) = 10x^2 - 15x + 7$ and question 7 asked for $(f \circ g)(x) = -50x^2 + 125x - 77$. The two aren't equal. In general $(f \circ g)(x)$ and $(g \circ f)(x)$ will not be equal to each other, but in section 1.6 we will talk about a special situation in which two such compositions will be equal.

Example 10: Composing Functions and Finding Real-Valued Solutions

$f(x) = -2x^2 + 3x, \quad g(x) = 7 - 5x, \quad h(x) = \sqrt{x+2}, \quad k(x) = \dfrac{2x}{x-3}$

Use these functions, from example 9 to evaluate the following compositions, $f(k(2))$, $(h \circ g)(1)$.

Solutions:

a)	b)
$f(k(2)) = f\left(\dfrac{2 \cdot (2)}{(2)-3}\right)$ $= f(-4)$ $= -2(-4)^2 + 3(-4)$ $= -2(16) - 12$ $= -44$	$(h \circ g)(1) = h(g(1))$ $= h(7 - 5(1))$ $= h(2)$ $= \sqrt{(2)+2}$ $= \sqrt{4} = 2$

Question 8: Using functions from example 10, find $(f \circ h)(-1)$.

DOMAIN AND RANGE OF COMPOSED FUNCTIONS

We want to talk about the domain of $(f \circ g)(x)$ to further cement your understanding of how composition works. You should notice that in order for some x to be in the domain of $(f \circ g)(x)$, the value of x must first be in the domain of g so that you can get $g(x)$. Then that answer $g(x)$ must be in the domain of f so that you can get a value for $f(g(x))$.

Discussion 2: Finding the Domain of a Composite Function

Let $f(x) = \sqrt{x-3}$, and $g(x) = \sqrt{x-1}$. We are going to look at how to find the domain and range of $(f \circ g)(x)$.

Answer Q6: $h(-1) + g(-1) = 5 + 0 = 5$

To do this we will follow these steps:

1) Begin by finding the domain of $g(x)$ which is $[1, \infty)$ ($\underline{x - 1}$ must be ≥ 0).

2) Next find the range of $g(x)$ which is $[0, \infty)$ (when you take the square root of something, the answer is always positive).

3) Intersect the range of $g(x)$ with the domain of $f(x)$ which is $[3, \infty)$ ($\underline{x - 3}$ must be ≥ 0). The intersection is $[3, \infty)$.

4) Then decide what values from the domain of g will cause the outputs of g to be in the interval you have just found $[3,\infty)$. In this case that would be the interval $[10,\infty)$ because $g(x)$ must be greater than or equal to 3. $g(x) \geq 3 \Rightarrow \sqrt{x-1} \geq 3 \Rightarrow x - 1 \geq 9 \Rightarrow x \geq 10$.

The graphing calculator can make this task much easier as long as we're careful. One way to find the domain is to graph the composition without doing any simplification. If you do simplify, you might not get the correct domain. Here are the graphs that go with this example.

g(*x*) graph

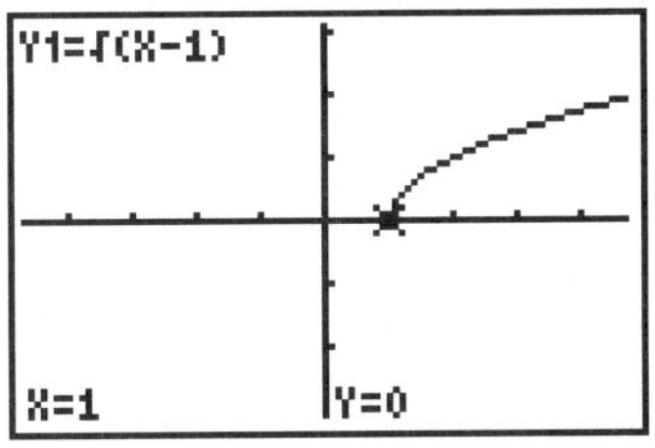

f(*x*) graph

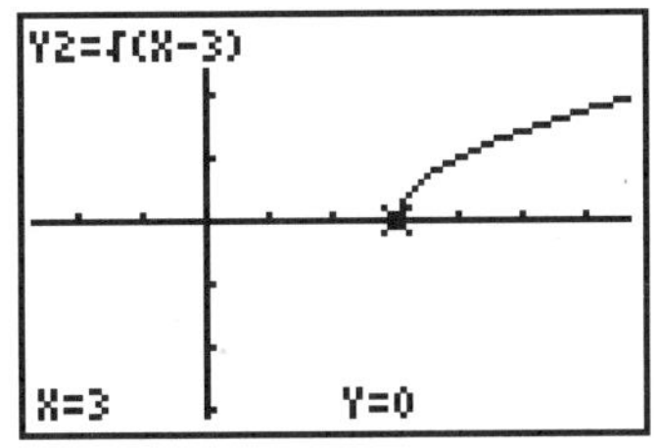

The graph of $(f \circ g)(x)$ is as follows.

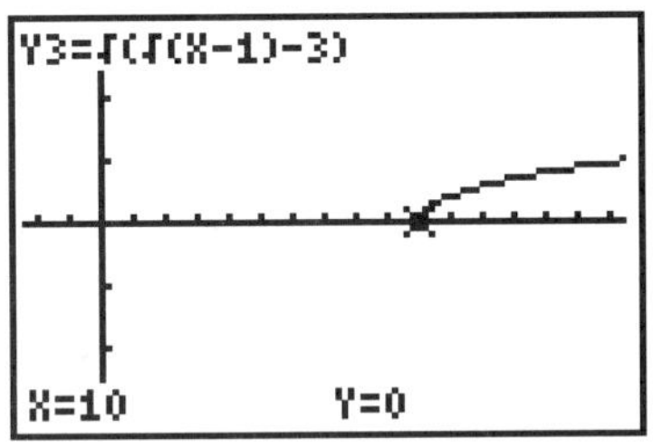

Notice, from the graph, we can see that the domain of the composite function is $[10, \infty)$ and the range is $[0, \infty)$.

Let's try another example to show that if you simplify the composition before you graph, you can arrive at the wrong answer for the domain and range of the composition function.

Answer Q7:
f(g(x))
$= -2(7-5x)^2 + 3(7-5x)$
$= -2(49-70x + 25x^2) + 21 - 15x$
$= -50x^2+125x - 77$

Example 11: Finding the Domain and Range of a Composite Function

Find the domain of $(f \circ g)(x)$ when $f(x) = x^2 - 6$ and $g(x) = \sqrt{x+2}$.

Solution:

$(f \circ g)(x) = \left(\sqrt{x+2}\right)^2 - 6$; its graph is:

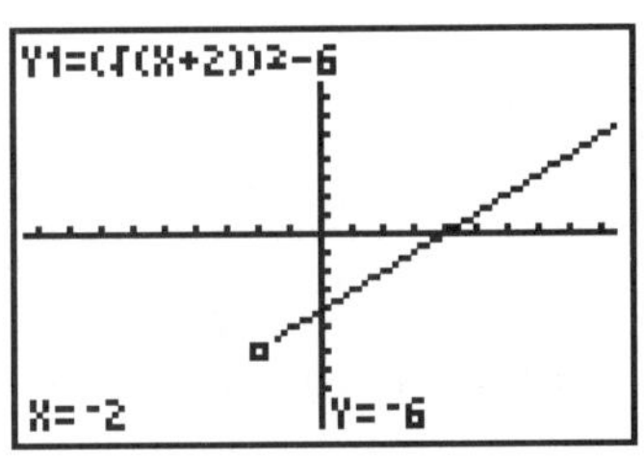

Note that the domain is $[-2, \infty)$ and the range is $[-6, \infty)$.

If we had simplified the composite function, we'd have $x + 2 - 6$, which simplifies to $x - 4$, the graph of which is at right.

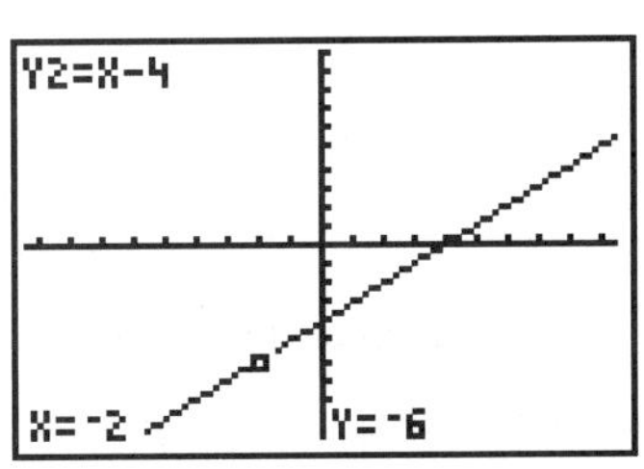

The domain of the "simplified" version of the composite function is (-∞, ∞) and the range is (-∞, ∞). So you can see that by simplifying, we would have lost vital information about the restrictions on the domain and range of the composite function.

If we had analyzed this example without a graphing calculator we would have seen the following. The domain of $g(x)$ is [2, ∞) ($\underline{x+2}$ must be ≥ 0). The range of $g(x)$ is [0, ∞) (when you take the square root of something the answer is positive). The domain of $f(x)$ is (−∞, ∞) so the intersection between the domain of $f(x)$ and the range of $g(x)$ is the range of $g(x)$ [0, ∞). Thus the domain of $(f \circ g)(x)$ is the whole domain of $g(x)$. Hence, you can see that the simplified version above is incorrect because it shows a domain of all real numbers.

Answer Q8:
$f(h(-1)) = f(1) = -2 + 3 = 1$

Example 12: Composing Functions in Table Form

Use these tables to find, $(f \circ g)(3)$, $(g \circ f)(0)$, $(g \circ g)(4)$.

x	$f(x)$
−3	3
−2	2
−1	1
0	0
1	−1
2	−2
3	−3

x	$g(x)$
−2	3
−1	2
0	1
1	0
2	1
3	2
4	3

Solutions:

a) $(f \circ g)(3) = f(g(3)) = f(2) = -2$	b) $(g \circ f)(0) = g(f(0)) = g(0) = 1$	c) $(g \circ g)(4) = g(g(4)) = g(3) = 2$

Question 9: Find $(f \circ g)(-2)$.

Example 13: Composing Functions in Graph Form

Use the graphs to find $(g \circ f)(2)$, and $(f \circ g)(-6)$ (here we will need to do some estimating from the graphs to answer the questions).

$f(x)$

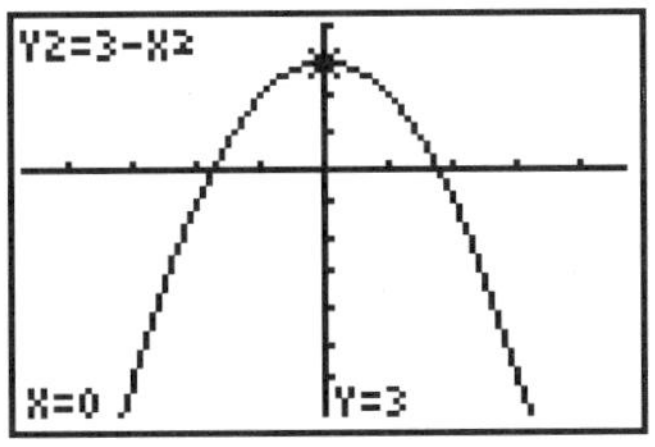

$g(x)$

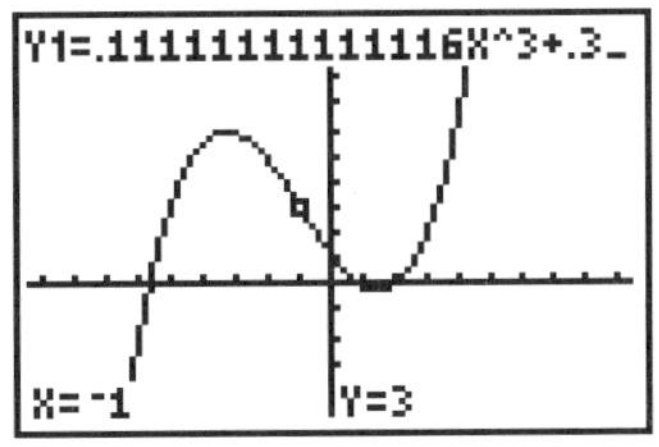

Solutions:

a) $(g \circ f)(2) = g(f(2))$	b) $(f \circ g)(-6) = f(g(-6))$
$= g(-1)$	$= f(-2)$
$= 3$	$= -1$

Question 10: Find $(g \circ f)(1)$.

Example 14: Composing Functions from a Table, Formula, and Graph

Use the table, formula, and graph below to find, $(g \circ h)(0)$, $(f \circ g)(0)$, and $(f \circ f)(2)$.

x	$f(x)$
0	2
1	0
2	–2
3	0
4	2

$g(x) = \sqrt{x+1}$

$h(x)$

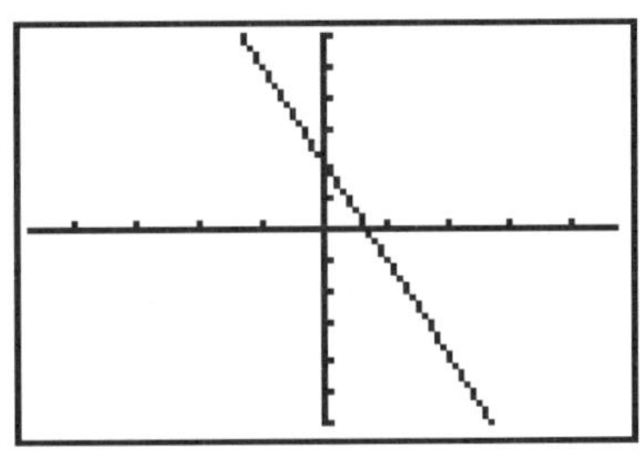

Solutions:

a) $(g \circ h)(0) = g(h(0))$	b) $(f \circ g)(0) = f(g(0))$	c) $(f \circ f)(2) = f(f(2))$
$= g(3)$	$= f(1)$	$= f(-2)$
$= 2$	$= 0$	= undefined

Notice, in example 14 (c) that the answer is undefined because -2 is not in the domain of $f(x)$.

Question 11: Find $(h \circ f)(4)$.

Section Summary

- When you are asked to add, subtract, multiply or divide two or more functions, you perform the operation and simplify.

- The domain of the resulting new function will be the intersection of the old functions' domains, less any numbers that would make the denominator of the new function equal to zero.

Answer Q9:

$f(g(-2)) = f(3) = -3$

- To compose functions you simply replace the x's in one function with the entire other function.
- We can find, or at least estimate, the domain and range of a composite function by using a graphing calculator.

SECTION 1.4 HOMEWORK

(1–12) For each pair of functions defined below, find $(f+g)(x)$, $(f-g)(x)$, $(f\bullet g)(x)$, $(f/g)(x)$, $(f\circ g)(x)$, $(g\circ f)(x)$. Give the domain of each resulting function.

1. $f(x) = 3x + 2,\ \ g(x) = 5x - 3$
2. $f(x) = 8 - 7x,\ \ g(x) = 3x - 12$
3. $f(x) = 3x^2 - 2,\ \ g(x) = x^2 + 4x$
4. $f(x) = 5x - 3x^2,\ \ g(x) = 2x^2 - 7x + 4$
5. $f(x) = x + 2,\ \ g(x) = x^2 - 16$
6. $f(x) = 3 - 2x,\ \ g(x) = x^2 - 5x + 6$
7. $f(x) = 3x + 2,\ \ g(x) = \dfrac{x-2}{3}$
8. $f(x) = x^2 - 4,\ \ g(x) = \sqrt{x+4}$
9. $f(x) = 3x - 1,\ \ g(x) = \sqrt{x+2}$
10. $f(x) = 2x + 3,\ \ g(x) = \sqrt{x-5}$
11. $f(x) = \sqrt{3x-1},\ \ g(x) = \dfrac{x^2+1}{3}$
12. $f(x) = \dfrac{x^2-3}{4},\ \ g(x) = \sqrt{4x+3}$

Answer Q10:

$(g\circ f)(1)$
$= g(f(1))$
$= g(2)$
$= 0$

(13–42) Use these functions to find the following:

Let $f(x) = 2x - 3,\ \ g(x) = x^2 + 2x + 1,\ \ h(x) = \dfrac{2}{x+1},\ \ r(x) = \sqrt{x}$.

13. $(f+g)(2)$
14. $(f+r)(4)$
15. $(g+h)(1)$
16. $(g+h)(9)$
17. $(f-g)(3)$
18. $(g-f)(3)$
19. $(g-h)(-2)$
20. $(r-f)(25)$
21. $(f\bullet g)(2)$
22. $(f\bullet r)(3)$
23. $(g\bullet h)(3)$
24. $(g\bullet r)(2)$
25. $(f/g)(3)$
26. $(f/h)(3)$
27. $(h/r)(16)$
28. $(g/r)(2)$
29. $(f/h)(-3)$
30. $(r/h)(5)$
31. $(f+g)(a)$
32. $(g+f)(b)$
33. $(f\bullet h)(2c)$
34. $(f\bullet g)(-3a)$
35. $(g\bullet r)(a^2)$
36. $(r\bullet f)(b^2)$
37. $(f\circ g)(-2)$
38. $(g\circ f)(-2)$
39. $(h\circ r)(4)$
40. $(r\circ h)(4)$
41. $(f\circ f)(-2)$
42. $(g\circ g)(-3)$

Answer Q11:

$(h\circ f)(4)$
$= h(f(4))$
$= h(2)$
$= -4$

(43–68) Use these tables to answer the following questions:

x	$f(x)$
–3	5
–2	0
–1	–3
0	–4
1	–3
2	0
3	5
4	12

x	$g(x)$
–3	–8
–2	–5
–1	–2
0	1
1	4
2	7
3	10
4	13

x	$h(x)$
–6	–8
–4	–5
–2	–2
0	1
2	4
4	7
6	10
8	13

43. $(f+g)(-3) =$
44. $(f+h)(-2) =$
45. $(g+h)(-2) =$
46. $(f+h)(4) =$
47. $(g-f)(-3) =$
48. $(g-h)(2) =$

49. $(f \cdot g)(0) =$

50. $(f \cdot g)(4) =$

51. $(f \cdot h)(-4) =$

52. $(g \cdot h)(6) =$

53. $\left(\frac{f}{g}\right)(3) =$

54. $\left(\frac{h}{g}\right)(-2) =$

55. $\left(\frac{g}{f}\right)(2) =$

56. $\left(\frac{h}{f}\right)(2) =$

57. $(g \circ f)(-1)=$

58. $(g \circ h)(-2) =$

59. $(f \circ g)(-3) =$

60. $(h \circ f)(4)=$

61. $(f + f)(3) =$

62. $(h - h)(-6) =$

63. $(g \cdot g)(-3) =$

64. $(h \cdot h)(8) =$

65. For addition, subtraction and multiplication operations between $f(x)$ and $g(x)$ what is the domain?

66. What is the domain of the function $\left(\frac{g}{f}\right)(x)$?

67. For addition, subtraction and multiplication operations between $g(x)$ and $h(x)$ what is the domain?

68. What is the domain of the function $\left(\frac{h}{f}\right)(x)$?

(69–80) Use these graphs to approximate the values. The algebraic equations for the graphs are $f(x) = 2x - 3$ and $g(x) = |2x|$. Use this information to check your approximations.

$f(x)$

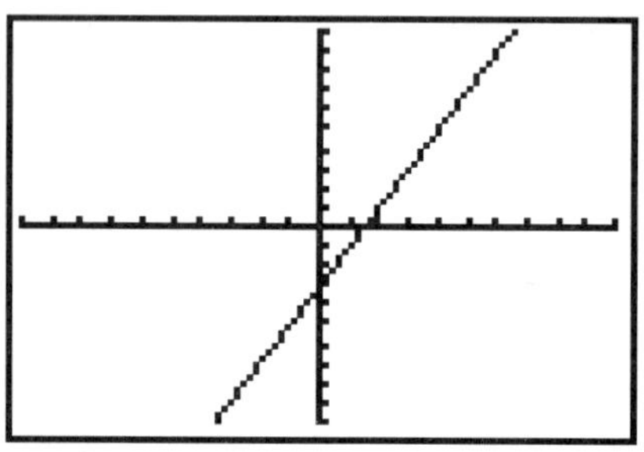

$g(x)$

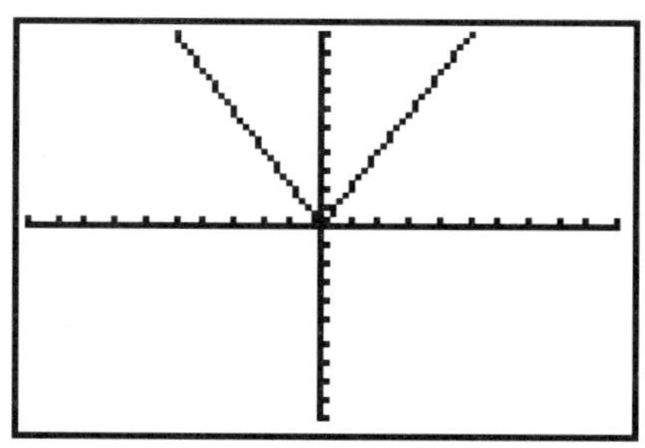

69. $(f + g)(2) =$

70. $(f + g)(-2) =$

71. $(f - g)(-1) =$

72. $(f - g)(1) =$

73. $(f \cdot g)(3) =$

74. $(f \cdot g)(-3) =$

75. $\left(\frac{f}{g}\right)(-3) =$

76. $\left(\frac{g}{f}\right)(2) =$

77. $\left(\frac{g}{f}\right)(0) =$

78. $\left(\frac{f}{g}\right)(0) =$

79. $(f \circ g)(2)=$

80. $(g \circ f)(-1)=$

(81–94) Use these graphs to approximate the following:

$f(x)$

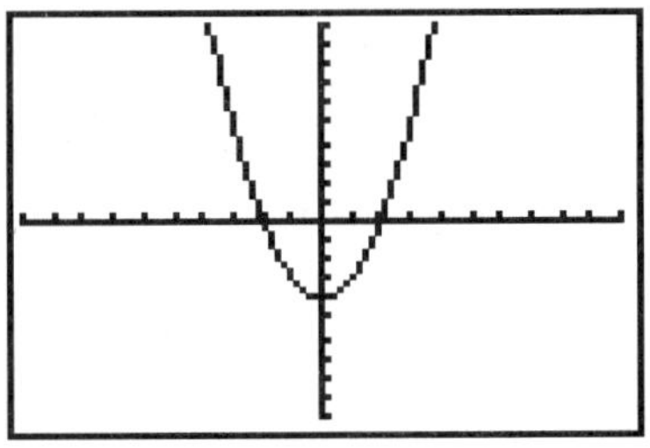

$g(x)$

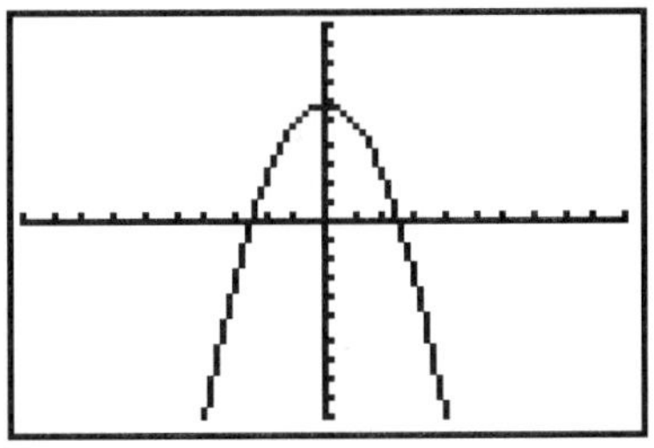

81. $(f+g)(1)=$ 82. $(f+g)(-3)=$ 83. $(f-g)(0)=$

84. $(g-f)(-1)=$ 85. $(f\cdot g)(3)=$ 86. $(f\cdot g)(-1)=$

87. $\left(\dfrac{f}{g}\right)(3)=$ 88. $\left(\dfrac{g}{f}\right)(-3)=$ 89. $(f+g)(2)=$

90. $(g+f)(-2)=$ 91. $\left(\dfrac{f}{g}\right)(2)=$ 92. $\left(\dfrac{g}{f}\right)(-2)=$

93. $(f\circ g)(-1)=$ 94. $(g\circ f)(-2)=$

The algebraic equations for the above graphs is $f(x)=x^2-4$ and $g(x)=-x^2+6$. Use this information to check your approximations made in the problems above.

(95–108) Use the graph, table and formula to approximate the following:

$f(x)$

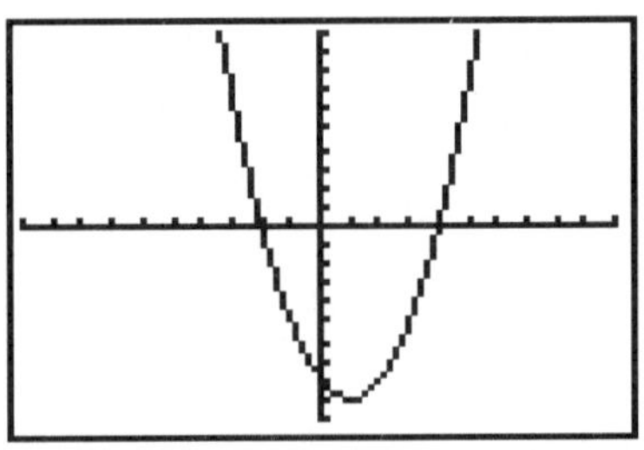

x	$g(x)$
–4	0
–3	4
–2	12
–1	15
0	8
1	–3
2	–8
3	–15
4	–20

$h(x) = 3x - 1$

95. $(f+g)(2)=$ 96. $(f-h)(2)=$ 97. $(g+h)(3)=$

98. $(g+f)(3)=$ 99. $(hf)(2)=$ 100. $(fg)(2)=$

101. $(g\circ f)(4)=$ 102. $(f\circ h)(1)=$ 103. $(h\circ g)(-4)=$

104. $(g\circ f)(-2)=$ 105. $(f+g+h)(2)=$ 106. $(f+g-h)(3)=$

107. $(fgh)(3)=$ 108. $(fgh)(4)=$

($f(x)= x^2 - 2x - 8$ for problems (95–108), you can use this to check yourself.)

(109- 112) Find the expression for the height h in terms of x.

109. Express h as a function of x:

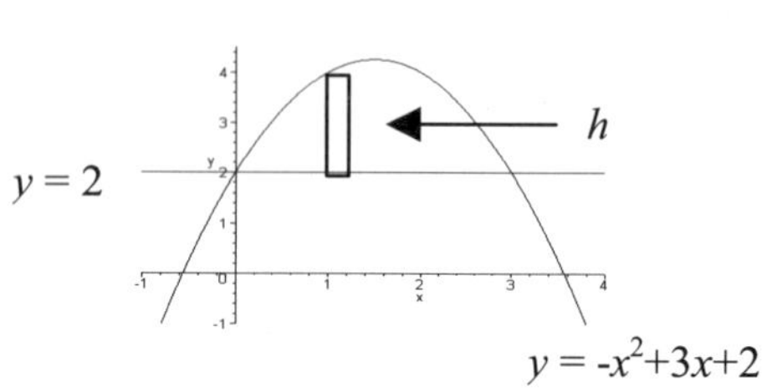

110. Express h as a function of x:

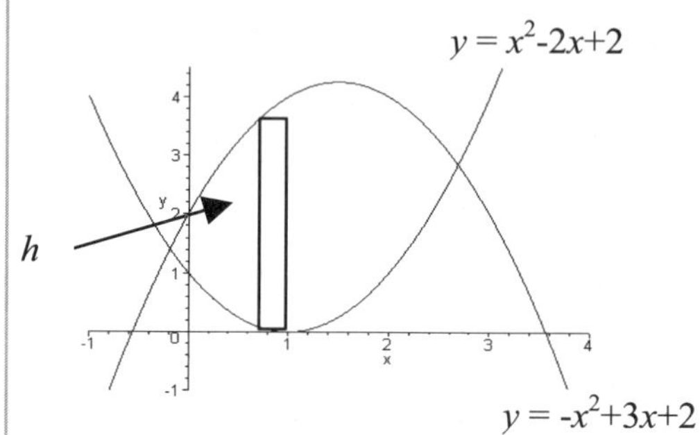

111. Express h as a function of x:

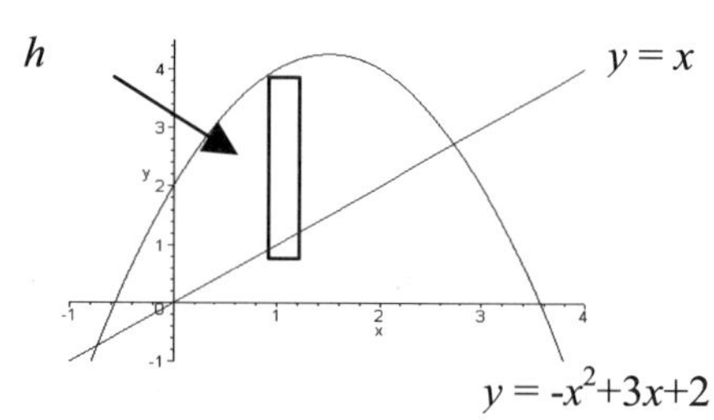

112. Express h as a function of x:

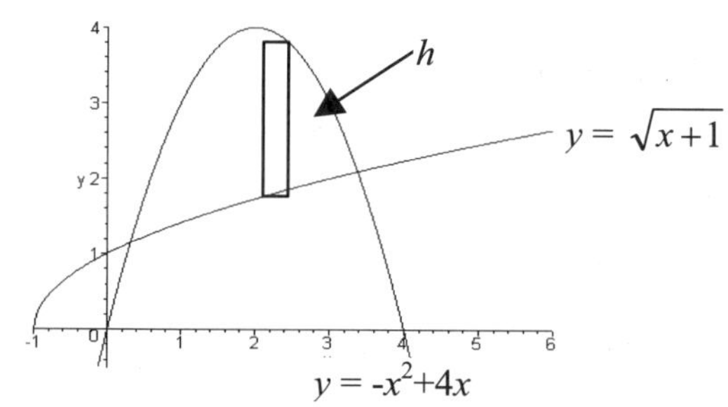

Section 1.5 The Application of Composition

People in many occupations run into problems that need to be quantified, so they need to create functions to describe the phenomena they are studying. Manipulating existing formulas is one way to go about obtaining a function to describe the phenomena.

In this section we will cover the ideas of solving for a variable in a single function and rearranging several functions in order to get a new function. We will also learn about concepts called the difference quotient and the average rate of change.

SOLVING A FUNCTION FOR THE DEPENDENT VARIABLE

Discussion 1: Manipulating the Fahrenheit-Celsius Conversion Formula

Suppose you have the following formula: $F = \frac{9}{5}C + 32$, where F (Fahrenheit) is a function of C (Celsius), but you are in a situation where degrees are given in Fahrenheit not Celsius. If you are looking for Celsius measurements then you need a different formula where C is a function of F. How would you go about doing this? All you need to do is solve the formula for C.

First, multiply by 5.	$5F = 9C + 160$
Second, subtract 160.	$5F - 160 = 9C$
Last, divide by 9.	$\frac{5}{9}F - \frac{160}{9} = C$
We might factor out a 5/9 to get	$\frac{5}{9}(F - 32) = C$

This new equation expresses the Celsius temperature as a function of Fahrenheit temperature. Using function notation, the answer is $C(F) = \frac{5}{9}(F - 32)$.

Discussion 2: The Area of a Square

Recall that the area of a square is $A(s) = s^2$. With this formula, if you were given the length of a side of a square, then you could calculate the area by plugging in your input value and finding your output value area ($A(s)$). What if instead you would like to calculate the length of the side of a square (output) given its area as the input? Here are the necessary steps required to find this new formula.

Write without function notation.	$A = s^2$
Now take the square root of both sides.	$\sqrt{A} = s$. Notice, no $\pm$ because s must be positive (it is a length).
Write in function notation.	$s(A) = \sqrt{A}$

Note: If s were not required to be positive, then our answer $s(A)=\sqrt{A}$ would have been $s(A)=\pm\sqrt{A}$ instead and would not be a function. The input for A would then yield two answers for s, one positive the other negative. But, that is not the case since you can't have negative distances. The moral of the story is that when you solve a function for the other variable, your result may or may not be a function. Also, if you are working on a "real-life" application you must be concerned about what is and isn't possible. More will be said about this when we get to section 6 in this chapter. Right now we are only going to concentrate on the mechanics of solving a function for a variable.

Example 1: Solving a Function for the Dependent Variable

Suppose that you have the function $s(t)=\sqrt{2t-13}+5$ and you want to have the function $t(s)$. Solve the function $s(t)$ for t.

Solution:

First, write without function notation.	$s=\sqrt{2t-13}+5$
Second, subtract 5.	$s-5=\sqrt{2t-13}$
Third, square both sides.	$(s-5)^2=2t-13$
Fourth, add 13.	$(s-5)^2+13=2t$
Last, divide by 2.	$\frac{1}{2}(s-5)^2+\frac{13}{2}=t$
Now, write in function notation.	$t(s)=\frac{1}{2}(s-5)^2+\frac{13}{2}$

As you can see, by working logically, going step-by-step, even a rather complicated formula can be manipulated.

Example 2: Solving a Function for the Dependent Variable

Let $y(x)=\frac{2x-1}{3x+2}$, where y is a function of x. Let's write x as a function of y. That is, solve for x.

Solution:

Write without function notation.	$y=\frac{2x-1}{3x+2}$
Multiply both sides of the equation by the denominator.	$y(3x+2)=2x-1$
Distribute the y.	$3xy+2y=2x-1$
Get all the x terms on the same side of the equation.	$3xy-2x=-2y-1$

Factor out an x.	$x(3y-2)=-2y-1$
Divide by the expression in the parenthesis.	$x=\dfrac{-2y-1}{3y-2}$
Write in function notation.	$x(y)=\dfrac{-2y-1}{3y-2}$

We now have x written as a function of y. We will see many problems like these in section 6 of this chapter. In fact, we will call these pairs of functions inverses of each other.

Question 1: Given the function $A(r) = 100\ (1+r)$ find A when $r = 0.05$.

The function in question 1 allows you to calculate how much a \$100 investment would be worth after one year at a given interest rate r. Specifically, you found how much your \$100 would grow into at a 5% interest rate. It would grow by \$5 to yield a total of \$105.

Question 2: Given the function $A(r) = 100\ (1+r)$ find the function $r(A)$.

Question 3: Using the answer to question 2, find the interest rate required for a \$100 investment to grow to \$113 in one year.

We are now going to look at some situations where we might have two or three functions that need to be used to create a new function.

APPLICATIONS OF COMPOSITION

Example 3: Area as a Function of Perimeter

Given the functions for the area and perimeter of a square, $A(s)=s^2$ and $P(s) = 4s$, find the function $A(P)$. Remember that s (length) must be positive.

Solution:

This new function will allow you to calculate the area of a square given its perimeter. Here are the steps needed to find our answer:

Write without function notation.	$A=s^2, P=4s$
We need to eliminate the s in the Area formula. We do this by solving for s in the perimeter formula first.	$\dfrac{P}{4}=s$ This is the function $s(P)=\dfrac{P}{4}$

Now substitute $\frac{P}{4}$ in for the s in the $A = s^2$ formula (This is $(A \circ s)(P)$).	$A = \left(\frac{P}{4}\right)^2 \rightarrow A = \frac{P^2}{16}$
Write in function notation.	$A(P) = \frac{P^2}{16}$

If you were told that the perimeter of a square was 20 meters, you could plug 20 into our new function to find the area of the square:

$$A(20) = \frac{20^2}{16} = \frac{400}{16} = 25\text{ m}^2.$$

Question 4: Use the above functions and find $P(A)$. This new function will allow you to calculate the perimeter of a square given its area.

Question 5: What would the perimeter of a square be with an area of 81 square meters (m^2)?

Example 4: Carbon Dioxide and Global Warming

Let's look at a real-life problem involving the concentration of carbon dioxide in the atmosphere, which is believed to be a cause of global warming. One possible mathematical model that can be used to predict how the CO_2 concentration changes over time is $C(t) = 1.32t + 280$ where $t = 0$ corresponds to the year 1939. We used C as the variable for the carbon dioxide concentration measured in parts per million, and we used t as the variable for time measured in years after 1939.

A possible model for the global temperature of the earth is $GT(C) = .016C - 4.48$, where we use GT as the variable for global temperature, and C as the variable for the carbon dioxide concentration measured in parts per million. What if we, as scientists, wanted a formula for global temperature in terms of time?

Solution:

We can compose the two equations, $C(t) = 1.32t + 280$ and $GT(C) = .016C - 4.48$, to come up with a function for $GT(t)$.

Write without function notation to get	$GT = .016\mathbf{C} - 4.5$, $C = \mathbf{1.32t + 280}$
Now replace C in the GT formula with the formula for C (composition again).	$GT = .016(\mathbf{1.32t + 280}) - 4.48$
Then simplify.	$GT = 0.02112t$
Now rewrite this in function notation.	$GT(t) = 0.02112t$

Now you have a useful function, GT(t), that relates global temperature to the years since 1939.

Question 6: Using our answer above for global temperature in terms of time, find $GT(65)$.

Question 7: A possible model for determining the change in the oceans level since 1939 in terms of global temperature is the function $OL(GT) = .3GT$. Many scientists think that warmer global temperatures will cause higher ocean levels because of the melting of the polar ice caps. Find $OL(t)$. $OL(t)$ is measured in feet.

Answer Q1:
$A(0.05) = 100(1 + 0.05) = 100(1.05) = \$105$

This new function should allow us to predict the ocean levels for any year. Remember $t = 0$ represents 1939.

Question 8: Find the change in the ocean level in the year 2004.

Example 5: Change in Ocean Levels

Using your answer from Question 7, find $OL(C)$, the change in the ocean level in terms of carbon dioxide concentration.

Answer Q2:

$$r(A) = \frac{A}{100} - 1$$

Solution:

Write without function notation to get	$OL = 0.006336t$, $C = 1.32t + 280$
Now solve for t in the $C = 1.32t + 280$.	$t = \dfrac{C - 280}{1.32}$
Substitute into $OL = 0.006336t$.	$OL = 0.006336(\dfrac{C - 280}{1.32})$
Simplify.	$OL = 0.0048C - 1.344$
Write in function notation.	$OL(C) = 0.0048C - 1.344$

Answer Q3:
$r(113) = 113/100 - 1 = .13$ or 13%.

This function would allow you to calculate the ocean level given the concentration of carbon dioxide in the atmosphere. A scientist armed with this information could make predictions about how storms might affect the coastline.

Discussion 3: Exchange Rates

Here we will use the following table of money exchange rates (from 3/13/00) to create functions that model each relation in the table. Also, we can use these to create new functions.

One U.S. Dollar = 1.04 Euros
One Euro = 0.6 British Pounds
One U.S. Dollar = 9.3 Mexican Pesos
One Canadian Dollar = 0.7 U.S. Dollars

First we will create functions for each of the different exchange rates given above.

Let D stand for U.S. Dollars and E for Euros.	$E(D) = 1.04\ D$
Let P stand for British Pounds	$P(E) = 0.6\ E$
Let Pe stand for Mexican Pesos	$Pe(D) = 9.3\ D$
Let C stand for Canadian Dollars	$D(C) = 0.7\ C$

Now let us use the functions we have just created to answer a couple of questions. First let's find a function that will change U.S. dollars into British pounds. We will use the functions involving U.S. dollars and Euros along with the function involving Euros and British pounds. The procedure is as follows:

Answer Q4: $P(A) = 4\sqrt{A}$. ($S = \sqrt{A}$, *substituted* $\sqrt{A}$ *in for S in* $P = 4S$)

Write the two equations without function notation.	$E = 1.04\ D$, $P = 0.6\ E$
Replace the E in the second equation with what E equals in the first equation.	$P = .6\ (1.04\ D) = 0.624\ D$
Write with function notation.	$P(D) = 0.624\ D$

This new function can tell us what the U.S. dollar is worth in British pounds. Now we're ready for that field trip to England.

Next, let's find an equation that would change Euros into Mexican pesos. We already know a function that changes U.S. dollars into pesos ($Pe(D) = 9.3D$), and a function that changes U.S. dollars into Euros ($E(D) = 1.04\ D$). But we need one that changes Euros into pesos. We will do this in steps. First we will find an equation to convert Euros to dollars. We will then substitute this function into the equation that converts dollars into pesos. Our first problem is that we only have a function that changes U.S. dollars into Euros ($E(D) = 1.04\ D$). We will need to solve our $E(D) = 1.04D$ function for D. Here is how we will go about doing this:

Answer Q5: $P(81) = 4\sqrt{81} = 36$ *meters*

Write equations without function notation.	$Pe = 9.3\ D$, $E = 1.04\ D$
Solve the second equation for D. (This is the equation that will allow us to convert from Euros to dollars.)	$D = \dfrac{E}{1.04}$
Substitute into the first equation. (This is the equation that will allow us to convert the Euros to pesos.)	$Pe = 9.3\left(\dfrac{E}{1.04}\right) = \left(\dfrac{9.3}{1.04}\right)E \approx$ $8.94\ E$
Now write in function notation.	$Pe(E) = 8.94\ E$

Using this new function we can tell what the Euro is worth in Mexico. Notice that we used the U.S. dollar as the intermediate variable that linked pesos to Euros.

To summarize, you need to find formulas with common intermediate variables so that you can, by a series of substitutions, get a formula in terms of the two variables that you desire. The homework in this section should supply enough practice for you to become proficient with this skill.

Answer Q6: GT(65) = 0.02112(65) = 1.3728 F°. This says that the global temperature will have increased by 1.3728 F° by the year 2004.

Question 9: Find an equation that will change Canadian dollars into Mexican pesos.

Question 10: How many pesos would you receive for 380 Canadian dollars?

THE DIFFERENCE QUOTIENT AND AVERAGE RATE OF CHANGE

Now let's turn our attention to the difference quotient and the average rate of change. These ideas require the combining of functions. Both of these have formulas that are quite similar. Each of the formulas is related to the formula used to calculate the slope of a line $\left(\dfrac{y_2 - y_1}{x_2 - x_1}\right)$. The formula for the difference quotient is $\left(\dfrac{f(x+h) - f(x)}{h}\right)$. The formula for the average rate of change is $\left(\dfrac{f(x_2) - f(x_1)}{x_2 - x_1}\right)$. Notice that in all three formulas above, that each formula's numerator is simply a difference in y-values and each denominator is a difference of x-values. In the difference quotient the h in the denominator comes from $[(x + h) - x]$, still the difference between x-values. The following are three graphs to illustrate these three formulas:

Answer Q7: OL(t) = .3(0.02112t) which simplifies to OL(t) = 0.006336t.

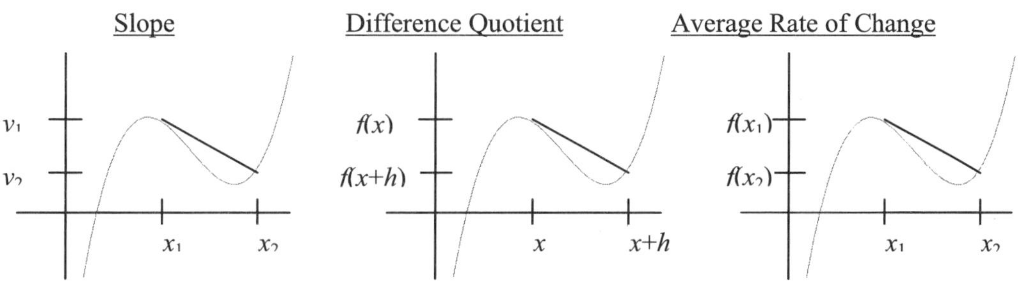

Answer Q8: OL(65) = 0.006336(65) = .41184 feet. According to this model the oceans will be approximately 0.4 ft higher than they were in 1939.

As you go through the following pages, you will learn how the slope between two points can be interpreted as a division of two differences, a difference quotient, and as an average rate of change. Let's start with an example that uses the slope formula.

Example 6: Using the Slope Formula

Given two points (–3, 5) and (4,–1), find the slope between them.

Solution:

The point (x_1, y_1) stands for the first point given to you and (x_2, y_2) stands for the second point given to you. So in this example, $x_1 = -3$, $x_2 = 4$ and $y_1 = 5$ and $y_2 = -1$. To find the slope, we plug these numbers into the slope formula and get m $= \left(\dfrac{y_2 - y_1}{x_2 - x_1}\right) = \dfrac{(-1) - 5}{4 - (-3)} = \dfrac{-6}{7}$.

Discussion 4: Time vs. Distance - The Slope Formula

Think about taking a trip from Chicago to Kansas City. Let's say that you left Chicago at 9:00 am and arrived in Kansas City around 7:00 p.m., and the distance you traveled for this trip was 530 miles. We can describe this trip using coordinates. The point (0, 0) will represent your starting point in Chicago. You've neither spent any time on the road nor traveled any distance yet. We'll use the point (10, 530) to describe your stopping point in Kansas City. You've traveled for 10 hours and covered 530 miles. The slope of the line between these two points is $\dfrac{y_2 - y_1}{x_2 - x_1} = \dfrac{530 - 0}{10 - 0} = 53$.

Discussion 5: Time vs. Distance Traveled - The Difference Quotient

Now if you use the difference quotient for this problem, the $f(x)$ in the formula refers to the number of miles traveled in zero hours. The $f(x + h)$ refers to the number of miles traveled after 10 hours. The h refers to the number of hours spent traveling. Because the difference quotient is made up of the difference of the miles traveled divided by the difference in the time spent traveling, it looks like this

$$\frac{f(x+h) - f(x)}{(x+h) - x} = \frac{f(10) - f(0)}{h} = \frac{530 - 0}{h} = \frac{530}{10} = 53.$$

Discussion 6: Time vs. Distance Traveled - Average Rate of Change

The average rate of change, or in this case the average speed traveled, is the difference in the miles traveled (530 – 0) divided by the difference in the time traveled (9:00 am – 7:00 pm), which is $\dfrac{f(x_2) - f(x_1)}{x_2 - x_1} = \dfrac{530miles - 0miles}{10hours - 0hours} = 53mph$. Notice that all three calculations yielded the same result. The key is to see that these two new formulas, difference quotient and average rate of change, are found by basically calculating the slope between two points.

The difference quotient is used in Calculus classes, which are required by many science and business degrees. We want to be sure that you can use these formulas without any difficulties. In a Calculus class you would learn how the difference quotient leads to something called the derivative. Using the difference quotient will also give you a chance to practice your function notation again. Let's do a few examples.

Example 7: Finding the Difference Quotient

Given $f(x) = 7 - 3x$, find the difference quotient.

Solution:

Formula for difference quotient is	$\frac{f(x+h)-f(x)}{h}$
Find $f(x+h)$ and simplify.	$7 - 3\,(x+h)$ $\mathbf{7 - 3x - 3h}$
Find $\boldsymbol{f(x+h)} - f(x)$, and simplify.	$(\mathbf{7 - 3x - 3h}) - (7 - 3x)$ $7 - 3x\ - 3h - 7 + 3x$ $-3h$
Now simplify $\frac{f(x+h)-f(x)}{h}$.	$\frac{-3h}{h} = -3$

Answer Q9: Write equations Pe = 9.3 D, D = .7 C. Substitute the second equation into the first Pe = 9.3(.7 C) = 6.51 C. Function notation: Pe(C) = 6.51C.

So the difference quotient is $\frac{f(x+h)-f(x)}{h} = -3$ for the function $f(x) = 7 - 3x$. It is also the slope of the line and the average rate of change of the line.

Example 8: Finding the Difference Quotient

Given $f(x) = 2x^2 - 3x + 5$, find the difference quotient.

Solution:

Difference quotient	$\frac{f(x+h)-f(x)}{h}$
Find $f(x + h)$ and simplify.	$2(x+h)^2 - 3(x+h) + 5$ $2(x^2 + 2xh + h^2) - 3x - 3h + 5$ $2x^2 + 4xh + 2h^2 - 3x - 3h + 5$
Find $f(x + h) - f(x)$.	$(2x^2 + 4xh + 2h^2 - 3x - 3h + 5) - (2x^2 - 3x + 5)$ $2x^2 + 4xh + 2h^2 - 3x - 3h + 5 - 2x^2 + 3x - 5$ $4xh + 2h^2 - 3h$
Now simplify $\frac{f(x+h)-f(x)}{h}$.	$\frac{4xh+2h^2-3h}{h} = \frac{h(4x+2h-3)}{h} = 4x+2h-3$

Answer Q10: Pe(380) = 6.51(380) = 2473.8

So the difference quotient $\frac{f(x+h)-f(x)}{h} = 4x + 2h - 3$ for the function $f(x) = 2x^2 - 3x + 5$. Notice that the answer for the difference quotient is itself a function. We could use that function to find the slope or average rate of change for any two points on the graph. Let's do some examples involving the average rate of change.

Example 9: Finding Average Speed - Average Rate of Change

If you leave New York City at 10:00 am and arrive in Washington DC at 1:30 p.m. and the odometer read 10,435 miles at the beginning of the trip and 10,665 miles at the end of the trip, then what was your average speed?

Solution:

Time	Odometer
10:00 a.m.	10435
1:30 p.m.	10665

This question is asking for the average rate of change. The formula is $\left(\dfrac{f(x_2)-f(x_1)}{x_2-x_1}\right)$. The $f(x_2)$ stands for the second y-value, which is the final odometer number (10,665). The $f(x_1)$ stands for the first y-value, which is the starting odometer number (10,435). The x_2 is the time at the end of the trip and x_1 is the time at the beginning of the trip. Thus,

$$Average = \frac{f(x_2)-f(x_1)}{x_2-x_1} = \frac{10665-10435}{1:30pm-10:00am} = \frac{230}{3.5} = 65.7 \text{ mph}$$

Example 10: Average Rate of Change of Temperature

Given that the low in Green Bay, Wisconsin, was 39°F at 4:00am and the high was 62°F at 3:00pm, what was the average rate of change in the temperature between 4:00am and 3:00pm?

Solution:

The average rate of change was $\dfrac{62-39}{3:00pm-4:00am} = \dfrac{23}{11} = 2.1$ °/hr. This tells us that, on average, the temperature in Green Bay increased by roughly 2.1° each hour from 4:00am to 3:00pm.

Section Summary

- By solving a formula for the independent variable you can create a new relation, possibly a function, in terms of the old dependent variable.
- To create a composite function from multiple functions, you may need to find formulas with common intermediate variables so that you can, by a series of substitutions, get a formula in terms of the two variables that you desire.

- Average rate of change is determined by the slope between two points and can be used to describe a real-life event.
- The difference quotient is just what its name says, the division of the difference between y's and x's.
- Both average rate of change and difference quotient formulas have a similar formula to the slope formula.

SECTION 1.5 HOMEWORK

1. $C = \pi D$ is an equation that gives circumference of a circle as a function of the diameter of the circle, where C = circumference and D = diameter. Rewrite the equation so that D is written as a function of C, that is, solved for D.
2. $A = P + .08P$ is an equation that gives the value after one year of an investment A as a function of the original investment P at 8%. Rewrite the equation with the beginning investment P as a function of the ending value A.
3. $A = \pi r^2$ is an equation that gives the area of a circle as a function of radius, where A is the area of the circle and r is the radius of the circle. Rewrite the equation so that radius is a function of area, that is, solved for r.
4. $d = 16t^2$ is an equation that measure the distance an object has fallen in feet as a function of number of seconds the object has fallen, where d is the distance the object has fallen and t is the time it has fallen. Rewrite the equation so that time is a function of distance.
5. $C = \dfrac{100}{R}$ is an equation that measures the current in a circuit as a function of resistance, where C is the current in amperes and R is the resistance in ohms. Rewrite the equation so that resistance is a function of current.
6. $I = \dfrac{3600}{d^2}$ is an equation that measures the intensity of illumination on a surface as a function of the distance the object is from the light source, where I is the measure of intensity of illumination and d is the measure of the distance the surface is from the light source. Rewrite the equation so that distance is a function of intensity of illumination.
7. $V = s^3$ is an equation that measures volume of a cube as a function of the length of a side of that cube, where V is the volume and s is the length of a side. Rewrite the equation so that the length of a side is a function of volume.
8. $V = \dfrac{4}{3}\pi r^3$ is an equation that measures the volume of a spheres as a function of the radius of the sphere, where V is the volume and r is the radius. Rewrite the equation so that the radius of the sphere is a function of volume.
9. $y(x) = 3x–5$. Write x as a function of y.
10. $r(s) = 5s–3$. Write s as a function of r.
11. $w(t) = 3 + \sqrt{3t+2}$. Write t as a function of w.
12. $y(x) = 5–\sqrt{2x+7}$. Write x as a function of y.
13. $m(s) = \dfrac{3}{4}s - \dfrac{5}{8}$. Write s as a function of m.

14. $t(m) = \dfrac{3m+2}{2} - 8$. Write m as a function of t.

15. $y(x) = \dfrac{3x+1}{2x-3}$. Write x as a function of y.

16. $p(a) = \dfrac{2a-3}{3a-5}$. Write a as a function of p.

17. $r(w) = \dfrac{3w-1}{w+2} - 3$. Write w as a function of r.

18. $r(t) = \dfrac{5t+3}{t-3} + 2$. Write t as a function of r.

19. $r(t) = 3t–5$ and $t(s) = 2s + 4$. Write r as a function of s.

20. $y(p) = 2p+4$ and $p(x) = 5x–7$. Write y as a function of x.

21. $y(t) = 2t^2 + 3$ and $t(x) = 3x–1$. Write y as a function of x.

22. $r(m) = \sqrt{3m+2}$ and $m(w) = 2w–3$. Write r as a function of w.

23. $y(r) = r^2–3$ and $r(x) = \dfrac{3}{x}$. Write y as a function of x.

24. $m(p) = p–4$ and $p(r) = \dfrac{3}{r+2}$. Write m as a function of r.

25. $m(n) = \sqrt{n+3}$ and $n(w) = w^2–3$. Write m as a function of w.

26. $r(t) = \sqrt[3]{t-5}$ and $t(s) = s^3 + 5$. Write r as a function of s.

27. $R = px$ is the revenue equation used where R is the revenue, p is the price per unit, and x is the number of units. The demand function which gives the value of price according to the number of units is $p(x) = 80–x$.
    a. Write revenue as a function of the number of units.
    b. Write revenue as a function of price.

28. Using problem 27 suppose the demand function is $p(x) = 6000–0.4x^2$.
    a. Write revenue as a function of the number of units.
    b. Write revenue as a function of price.

29. $P = R – C$ is the equation for computing profit, where P is profit, R is revenue and C is cost. As in problem 27 $R = px$ is the revenue equation, $C = 300–30x + 0.01x^2$ is the cost equation and $p(x) = 40 - \dfrac{x}{50}$ is the demand function. Write the profit as a function of the number of units.

30. Using problem 29 but with $C= 0.7x + 400$, the cost function and $p(x) = \dfrac{40}{\sqrt{x}}$, the demand function, write the profit as a function of the number of items.

31. The volume of a cylinder is given by the equation $V = \pi r^2 h$, where r is the radius of the cylinder, h is the height of the cylinder and V is the volume of the cylinder. If the height is always three times the radius write the following functions:

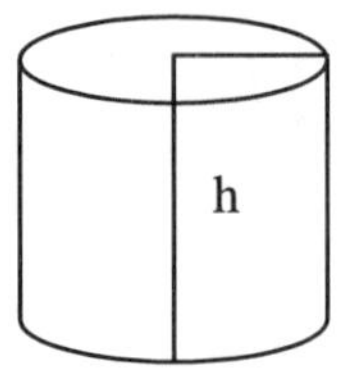

a. Volume as a function of radius
b. Volume as a function of height

32. The volume of water in a water tank in the shape of a cone is given by the equation $V = \frac{1}{3}\pi r^2 h$, where V is the volume of the water in the cone, r is the radius and h the height of the cone formed by the filled part of the tank. If the tank is shaped such that the height is always three times the radius, write the following functions:
a. Volume as a function of radius.
b. Volume as a function of height.

33. The volume of a rectangular box that is 6 inches high is given by the equation $V = 6LW$, where L is the length of the base and W is the width of the base. If the length is always three more than twice the width, write the following function:

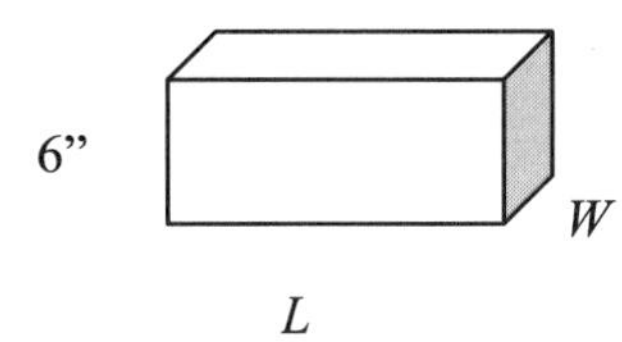

a. Volume as a function of width
b. Volume as a function of length

34. Volume of a box with a square base is given by the formula $V = s^2 h$, where s is the length of a side of the square base; h is the height of the box and V is the volume. If the height is twice as long as the length of the square base, write the following function:

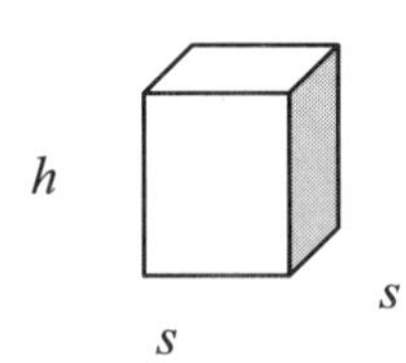

a. Volume as a function of the length of the side of the square base.
b. Volume as a function of the height.

35. You want to enclose a rectangular plot of land with 10,000 feet of fencing.

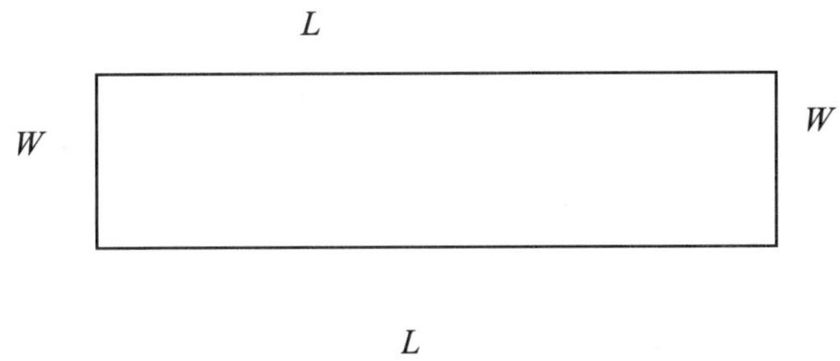

$2L + 2W = 10{,}000$ is the formula for the relationship between the amount of fencing, L (the length of the rectangular plot), and W (the width of the rectangular plot).

$A = LW$ is the formula for the area of the rectangular plot, where A is the area.

a. Write area as a function of width.

b. Write area as a function of length.

36. You have 10,000 feet of fencing to fence in three adjacent rectangular corrals.

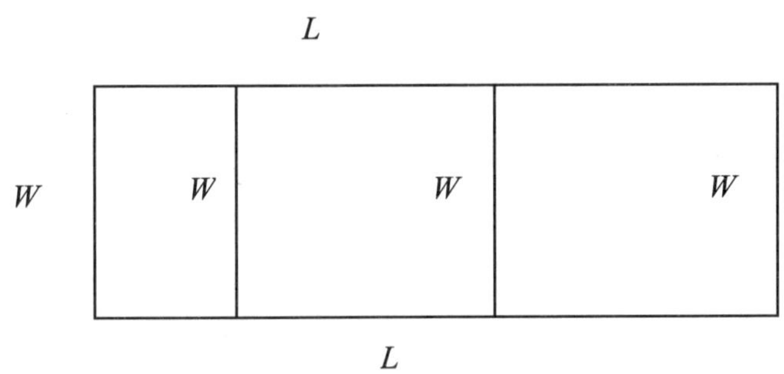

$4W + 2L = 10{,}000$ is the formula for the relationship between the amount of fencing, L (the length of the rectangular plot), and W (the width of the rectangular plot).

$A = LW$ is the formula for the area of the rectangular plot, where A is the area.

a. Write area as a function of width.

b. Write area as a function of length.

37. You want to fence in a rectangular lot of 40,000 square feet. $LW = 40{,}000$ is the area formula for the lot, where L is the length of the lot and W is the width of the lot. $F = 2L + 2W$ is the formula for the amount of fencing needed for the lot, where F is the amount of fencing needed.

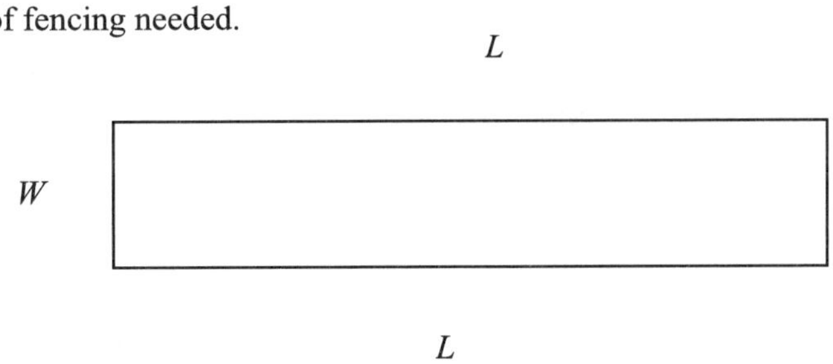

a. Write the amount of fencing as a function of width.

b. Write the amount of fencing as a function of length.

38. You want to enclose a rectangular lot with fencing. One side of this lot runs along the highway. The fencing along the highway costs \$8 per foot and the rest of the fencing costs \$6 a foot; the total money available for fencing is \$5,000.

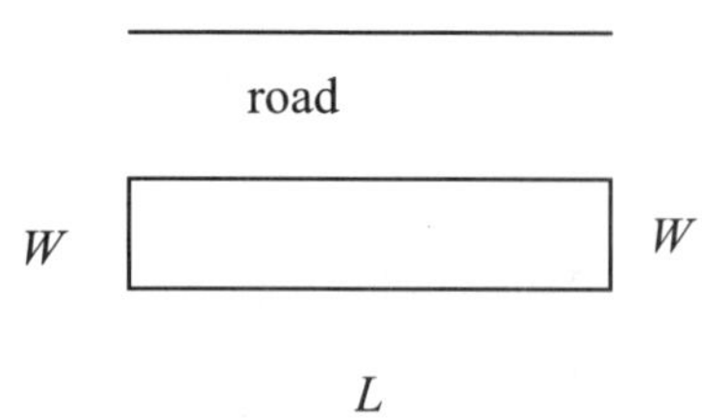

$12W + 14L = 5{,}000$ is the cost equation for the fencing, where L is the length of the lot and W is the width of the lot. $A = LW$ is the area formula for the lot, where A is the area.

a. Write the area of the lot as a function of length

b. Write the area of the lot as a function of width.

39. By the Pythagorean Theorem, $a^2 + b^2 = 4^2$ is the relationship in a right triangle between the lengths of the two legs, a and b, and the hypotenuse of length 4.

$A = \frac{1}{2}ab$ is the area of the above right triangle where A is the area.

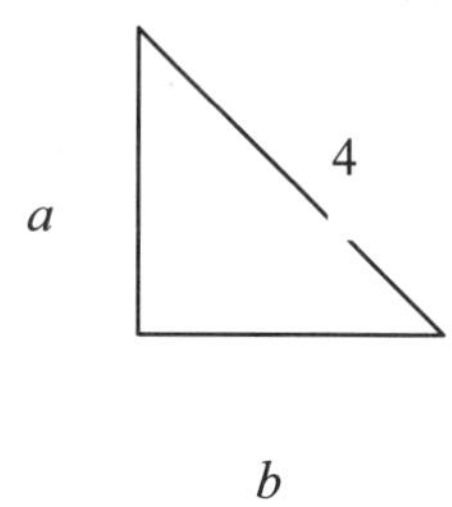

a. Write the area as a function of leg a.

b. Write the area as a function of leg b.

40. By the Pythagorean Theorem, $a^2 + 5^2 = c^2$ is the relationship in a right triangle between the lengths of the two legs, a and 5, and the length of the hypotenuse c.

$A = \frac{1}{2}5a$ is the area formula for the right triangle.

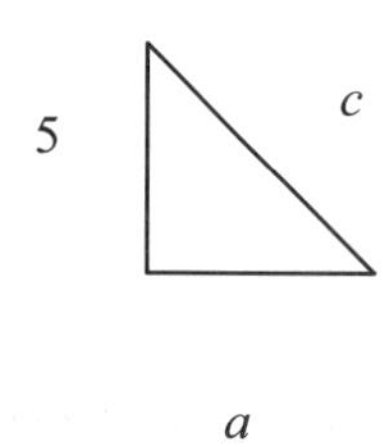

Write the area as a function of the hypotenuse.

41. A conical shaped water tank is filled with water. $V = \frac{1}{3}\pi r^2 h$ is a formula for the volume of the filled part, where V is the volume; r is the radius and h is the height of the filled part. The tank itself has a radius of 20 inches and a height of 40 inches. By similar right triangles the relationship between the height and radius of the filled part is $\frac{r}{20} = \frac{h}{40}$.

   a. Write the volume of the filled part as a function of the radius.
   b. Write the volume of the filled part as a function of the height.

42. A conical shaped water tank is filled with water. $V = \frac{1}{3}\pi r^2 h$ is a formula for the volume of the filled part, where V is the volume; r is the radius and h is the height of the filled part. The tank itself has a radius of 15 meters and a height of 45 meters. By similar right triangles the relationship between the height and radius of the filled part is $\frac{15}{r} = \frac{45}{h}$.

   a. Write the volume of the filled part as a function of the radius.
   b. Write the volume of the filled part as a function of the height.

43. A rectangular poster is created with 900 square inches of printed material and a border all the way around of 6 inches. The formula for the area of the printed material is $LW = 900$, where L is the length of the printed material and W is the width. The formula for area of the entire poster is $A = (L + 12)(W + 12)$.

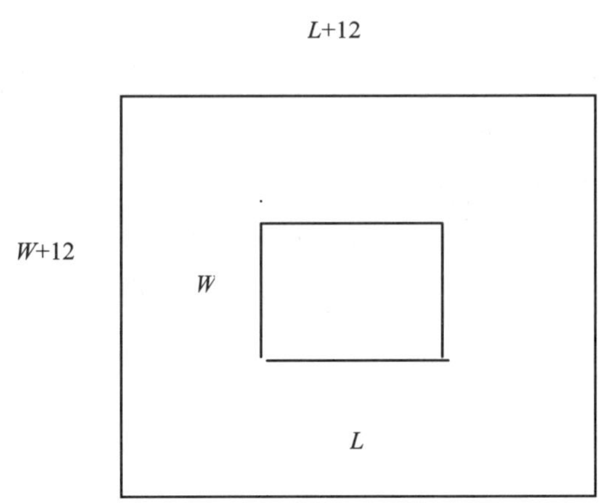

   a. Write the area of the entire poster as a function of length of the printed material.
   b. Write the area of the entire poster as a function of width of the printed material.

44. A rectangular flower bed has an area of 1200 square feet. There is a brick sidewalk all the away around the flower bed that is 6 feet wide on the ends and 5 feet wide on the sides. The formula for the area of the flower bed is $LW = 1200$, where L is the length of the flower bed and W is the width of the flower bed. The area of the rectangle formed by the flower bed and the brick sidewalk is $A = (L + 10)(W + 12)$.

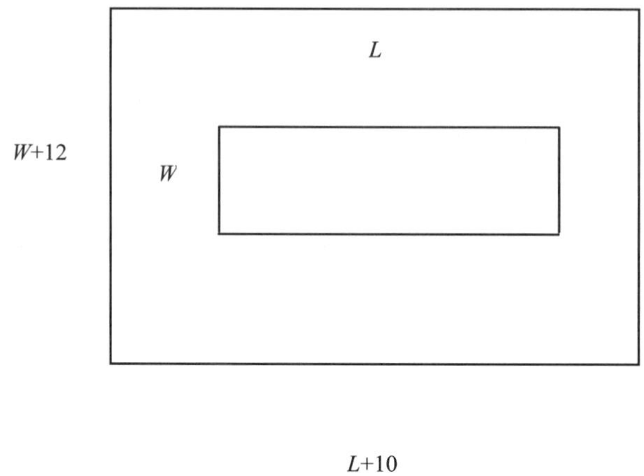

a. Write the area of the rectangle formed by the flower bed and sidewalk as a function of the length of the flower bed.

b. Write the area of the rectangle formed by the flower bed and sidewalk as a function of the length of the flower bed.

(45–50) Find the difference quotient for each of the following:

45. $f(x) = 3x - 5$ 46. $f(x) = 5x + 3$ 47. $f(x) = 3x^2 - 2x + 3$

48. $f(x) = 2x^2 + 5x - 8$ 49. $f(x) = 2x^3 - 3x^2 + 3$ 50. $f(x) = 3x^3 + 2x^2 - x - 5$

51. A projectile is shot in the air from ground level. The equation for the height, $S(t)$, of that projectile after time t is $S(t) = -16t^2 + 256t$ (where $S(t)$ is in feet and t is in seconds). What is the average rate of change of height to time (speed) for the object on the time interval $t = 1$ second to $t = 3$ seconds?

52. A balloon is ascending vertically from a point 3000 ft. from an observation station. The distance from the balloon to the observation station is given by the formula $D(t) = \sqrt{90000 + 40000t^2}$. Where $D(t)$ is the distance in feet and t is the time in seconds. What is the average rate of change of height to time for the balloon on the time interval $t = 0$ to $t = 2$?

53. You are renting apartments and the equation that gives the revenue in terms of the number of apartments rented is $R(x) = -8x^2 + 1400x$. Where $R(x)$ is the revenue in dollars and x is the

number of rented apartments. What is the average rate of change of dollars received to apartments rented on the interval $x = 60$ units rented to $x = 80$ units rented?

54. Using problem 53 above the profit in terms of apartments rented is $P(x) = -8x^2+1300x$, where $P(x)$ is the profit in dollars and x is the number of rented apartments. What is the average rate of change of dollars received to apartments rented on the interval $x = 60$ units to $x = 80$ units rented?

55. The volume of a cube is changing with time by the equation $V(t) = 8t^3$. Where $V(t)$ is the volume in cubic feet and t is the time in seconds, what is the average change in volume with respect to time on the interval $t = 1$ second to $t = 4$ seconds?

56. The area of a circle is changing with time by the equation $A(t) = 9\pi t^2$. Where $A(t)$ is the area of the circle in square inches and t is the time in seconds, what is the average change in area with respect to time on the interval $t = 2$ seconds to $t = 4$ seconds?

Section 1.6 Inverse Functions

When you get up in the morning to attend school, you have to do several things in a particular order so that you can get to class on time. One example of the activities that you might do to attend class on any given day is as follows:

1. You have to exit your house
2. Get in your car
3. Drive to school
4. Get out of your car
5. Walk to class
6. Enter the classroom

The inverse of traveling from home to class is traveling from class to home. You have to *undo* what you did in order to get from class to home. To get home you have to do the *opposite activities in the opposite order*. For example, to go back home you would do the following:

1. Exit the classroom (opposite of enter the classroom, step 6 above)
2. Walk to your car (opposite of walking from your car to the class, step 5 above)
3. Get into your car (opposite of getting out of your car, step 4 above)
4. Drive home (opposite of driving to school, step 3 above)
5. Get out of your car at home (opposite of getting into your car at home, step 2 above)
6. Enter your house (opposite of exiting your house, step 1 above)

In order to get home you had to do the opposite activities in the opposite order. This is precisely the concept of an inverse, which is the last topic of discussion in Chapter 1. This example clearly shows how the inverse is just "undoing!" Other examples that exhibit this idea of inverse would be basic arithmetic operations. In the world of mathematics, addition and subtraction, multiplication and division, squaring and taking the square root, are inverses of each other. They all undo each other. For example, if we add 3 to 7 and then subtract 3, we will be right back where we started, at 7.

Discussion 1: Arithmetic Inverses

Additive Inverse: You might remember that the additive *identity* is 0. An additive *inverse* is a number that, when added to the original number, gives you 0. What is the additive inverse of 5? The answer is (-5). Why? Because $5 + (-5) = 0$, so 5 and (-5) are additive inverses of each other.

Multiplicative Inverse: The multiplicative *identity* is 1. The multiplicative *inverse* is the number that when multiplied by the original number gives you 1. What is the multiplicative inverse of 7? The answer is $\frac{1}{7}$. Why? Because $7 \bullet \frac{1}{7} = 1$, so 7 and $\frac{1}{7}$ are multiplicative inverses of each other.

INVERSE FUNCTIONS

The way in which we define the inverse of a function is similar to how we define inverses of numbers. From above, "0" is the additive identity and "1" the multiplicative identity. When working with functions, x is the composition identity. If two **one-to-one** functions (discussed in just a moment) are composed with one another ($(f \circ g)(x)$ and $(g \circ f)(x)$), and both compositions equal x then they are inverses of each other. A formal definition of the inverse of a function follows:

> If $f(x)$ is a one-to-one function and $g(x)$, another function, is such that $(f \circ g)(x) = (g \circ f)(x) = x$, then we say that $g(x)$ is the inverse function of $f(x)$. We replace the symbol $g(x)$ with $f^{-1}(x)$ to signify that it is the inverse of $f(x)$. Note: the range (outputs) of one function is the domain (inputs) of the other and vice versa.

Before continuing our discussion of inverse functions, let's investigate the idea of "one-to-one." The informal definition of a function said that "for each *input* there can be only one *output*". The informal definition of a one-to-one function extends that rule to include "for each *output* there can be only one *input*". Thus, for a function to be one-to-one, it must satisfy *both* of these definitions, that is, for each input there is only one output, and for each output there is only one input. It should not be surprising that this concept of one-to-one should show up in a discussion about inverses, since it is the inverse of the definition of a function. Below are examples of functions that are and aren't one-to-one:

One-to-one	Not one-to-one
$f(x) = 2x - 7$	$h(x) = x^2 + 1$
$g(x) = \sqrt{x-2}$	$k(x) = x^3 - 13x - 1$
$j(x) = x^3 - 3x^2 + 3x - 1$	$L(x) = \sqrt{25 - x^2}$

The easiest way to tell if a function is one-to-one or not is to graph it and see if it passes the **horizontal line test**. (Remember that the *vertical line test* determines if a graph represents a function or not). Imagine drawing horizontal lines through the graph of the function in question. If no horizontal line would touch the graph in more than one place, we say that the function is one-to-one. Here are the graphs of the functions from the above table:

One-to-one **Not one-to-one**

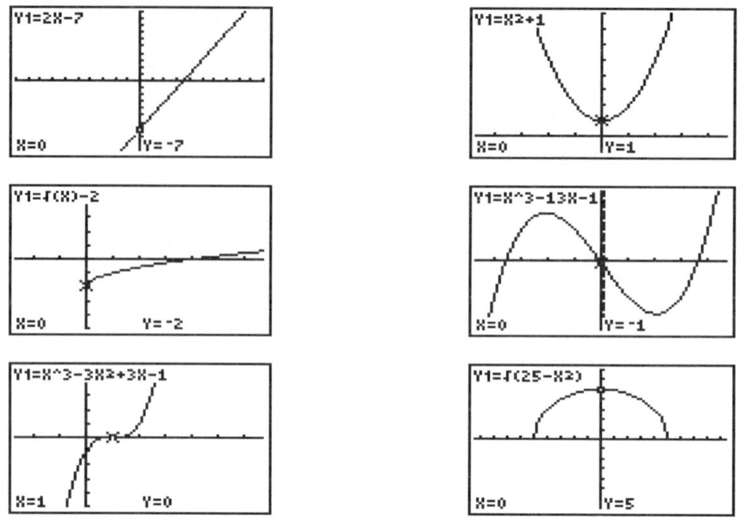

Question 1: Is the following a graph of a one-to-one function?

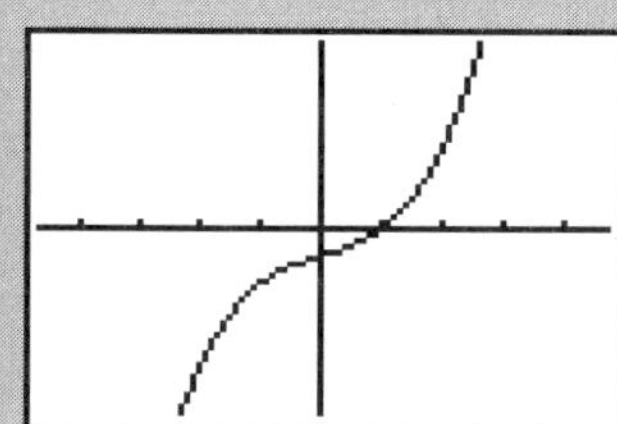

Let's do a few examples and see how inverses relate to each other.

Discussion 2: Observations of Inverse Functions

Let $f(x)$ consist of the ordered pairs $\{(-3, -8), (-1, -5), (1, -2), (3, 1), (5, 4), (7, 7), (9, 10)\}$ and $g(x)$ consist of the ordered pairs $\{(-8, -3), (-5, -1), (-2, 1), (1, 3), (4, 5), (7, 7), (10, 9)\}$. If we put these functions into tables we can more easily make some observations.

x	$f(x)$
−3	−8
−1	−5
1	−2
3	1
5	4
7	7
9	10

x	$g(x)$
−8	−3
−5	−1
−2	1
1	3
4	5
7	7
10	9

Notice that the function g is the "reverse" of f. That is, if you switch the x and y coordinates of f, you get g. In addition, the domain of f is the range of g, and vice versa. (Note: $f(x)$ and $g(x)$ are both functions and one-to-one, that is, for each input there is only one output and vise versa.)

a) If we evaluate $f(1)$ we get $f(1) = -2$. We can tell that this is the answer from the first table above. With an input of 1 in the f function we get an output of -2. Does the function g undo this? Yes it does. With an input of -2 in the g function you get an output of 1. Now if we were to apply the definition above, we would want to see $(f \circ g)(x) = (g \circ f)(x) = x$. In this example that means $(f \circ g)(-2) = (g \circ f)(-2) = -2$.

b) If we evaluate $f(5)$ we get $f(5) = 4$.

Question 2: Does the function g undo $f(5) = 4$?

c) If we evaluate $g(-8)$ we get $g(-8) = -3$.

Question 3: Does the function f undo $g(-8) = -3$?

What might we conclude from this evidence? That $f(x)$ and $g(x)$ are inverses of each other because they undo each other. The formal definition stated that $(f \circ g)(x) = (g \circ f)(x) = x$ when f and g are inverses. Let's show you how to prove that two functions are inverses of each other.

Example 1: Proving Two Functions Are Inverses

Find $(f \circ g)(x)$ and $(g \circ f)(x)$ given $f(x) = \dfrac{3x-7}{2}$ and $g(x) = \dfrac{2x+7}{3}$. By the way, these are the formulas that were used to produce the values in the tables in Discussion 2.

Solution:

$$(f \circ g)(x) = f(g(x)) = \frac{3\left(\frac{2x+7}{3}\right)-7}{2} = \frac{(2x+7)-7}{2} = \frac{2x}{2} = x \text{ and}$$

$$(g \circ f)(x) = g(f(x)) = \frac{2\left(\frac{3x-7}{2}\right)+7}{3} = \frac{(3x-7)+7}{3} = \frac{3x}{3} = x\,.$$

Since the answer to $(f \circ g)(x)$ and $(g \circ f)(x)$ is x, and because f is a one-to-one function, we may conclude that f and g are inverses of each other.

Questions 4: Show that $f(x) = 5x + 2$ and $g(x) = \dfrac{x-2}{5}$ are inverse functions of each other? Hint: look at graphs first to verify that $f(x)$ and $g(x)$ are both functions (vertical line test) and one-to-one (horizontal line test).

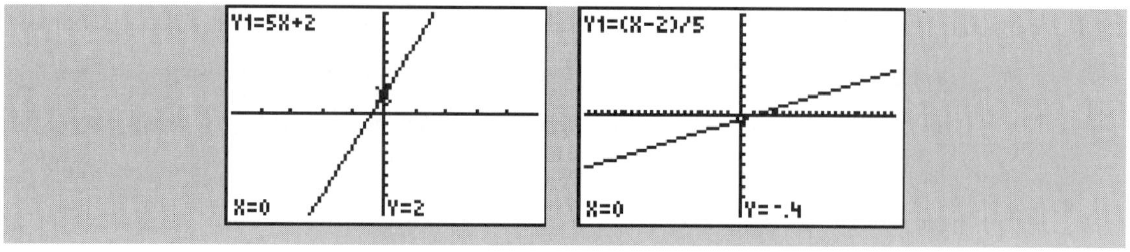

Example 2: Inverse Relations

Let $f(x) = x^2 + 1$ and $g(x) = \sqrt{x-1}$. Are f and g inverse functions of each other?

Solution:

First graph both functions. Notice that $f(x)$ fails the horizontal line test and thus isn't one-to-one; therefore, $f(x)$ does not have an inverse function. So $g(x)$ can't be its inverse function.

Graph of $f(x)$

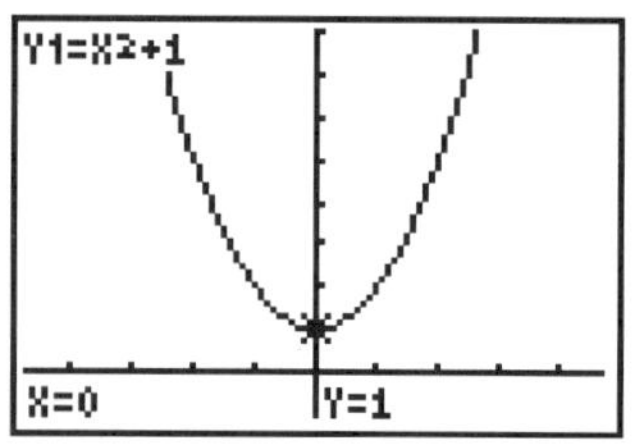

Graph of $g(x)$

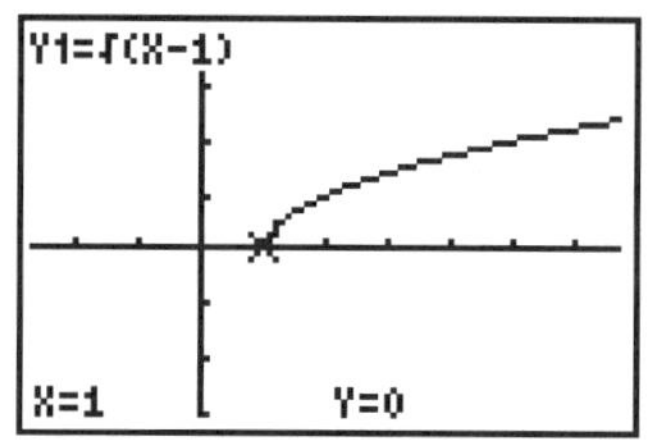

But if we were to look at $f(x) = x^2 + 1$ and $y = \pm\sqrt{x-1}$, then y would turn out to be the **inverse relation** of $f(x)$. Notice that y isn't a function; it's just a relation (one input yields more than one output). It fails the vertical line test. Notice that we didn't use function notation for y because it isn't a function.

Answer Q1: Yes.

Graph of y

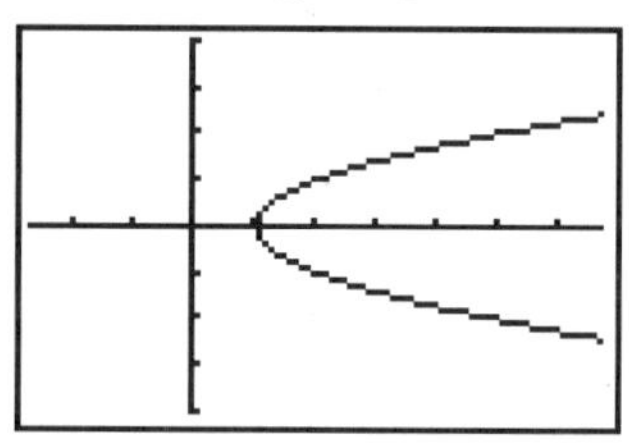

In this book when we talk about finding an inverse of some function, we will always be referring to finding the inverse function, never the inverse relation.

FINDING INVERSE FUNCTIONS

One way of finding an inverse function is as follows.

Discussion 3: Finding the Inverse Logically

Assume that you have the function $f(x) = \dfrac{3x-7}{2}$. We could deduce what its inverse must be just by thinking about doing the opposite operations in the opposite order (like the home-to-school story). If we were to input a value in for the x in the $f(x)$ function, then to find the output we would need to first multiply by 3; second, subtract 7; and third, divide by 2. To undo that we would need to first multiply by 2 (opposite of divide by 2); second, add 7 (opposite of subtract 7); and third, divide by 3 (opposite of multiply by 3). So the function that would multiply by 2 then add 7 and then divide by 3 would be the inverse of $f(x)$ and look like $g(x) = \dfrac{2x+7}{3}$.

Answer Q2: Yes. With an input of 4 in the g function you will get 5 for your output. g(4) = 5. So, $(f \circ g)(4) = (g \circ f)(4) = 4$

Question 5: Using this technique for finding inverses, what would be the inverse of $f(x) = 9x + 5$?

Given a more complicated function, it can be tricky to get the inverse of a function using this technique. Here is a more general procedure that we can use to find an inverse of any one-to-one function.

1.) Rewrite your function without using function notation.
2.) Solve your function for the dependent variable.
3.) Finally, rewrite your answer in inverse function notation form.

Answer Q3: Yes. With an input of –3 in the f function you will get –8 for your output. f(–3) = –8. So, $(g \circ f)(-3) = (f \circ g)(-3) = -3$

Example 3: Finding the Inverse of a One-to-One Function

Find the inverse of the one-to-one function $h(x) = \dfrac{5-4x}{7}$.

Solution:

First, rewrite as	$y = \dfrac{5-4x}{7}$
Second, multiply by 7.	$7y = 5 - 4x$
Third, subtract 5.	$7y - 5 = -4x$
Fourth, divide by –4.	$\dfrac{5-7y}{4} = x$
Last, write in inverse function notation.	$h^{-1}(y) = \dfrac{5-7y}{4}$
It is customary to write your function using x as the independent variable so we'll change $h^{-1}(y)$ to…	$h^{-1}(x) = \dfrac{5-7x}{4}$

Answer Q4: $(f \circ g)(x) = 5\left(\dfrac{x-2}{5}\right) + 2 = (x-2) + 2 = x$

$(g \circ f)(x) = \dfrac{(5x+2)-2}{5} = \dfrac{5x}{5} = x$

We can check ourselves by calculating $(h \circ h^{-1})(x)$ and $(h^{-1} \circ h)(x)$.

$$(h \circ h^{-1})(x) = h(h^{-1}(x)) = h\left(\frac{5-7x}{4}\right) =$$

$$\frac{5-4\left(\frac{5-7x}{4}\right)}{7} = \frac{5-(5-7x)}{7} = \frac{5-5+7x}{7} = \frac{7x}{7} = x$$

$$(h^{-1} \circ h)(x) = h^{-1}(h(x)) =$$

$$h^{-1}\left(\frac{5-4x}{7}\right) = \frac{5-7\left(\frac{5-4x}{7}\right)}{4} = \frac{5-(5-4x)}{4} = \frac{5-5+4x}{4} = \frac{4x}{4} = x$$

This proves that these are inverses.

As you have seen, inverse functions have the property that the range (outputs) of one function is the domain (inputs) of the other and vice versa. If you take a moment to look back at the first eight "real life" discussion items in section 1 of this chapter, you will notice that the two different viewpoints used in each discussion item were relations and their inverse relations. To be more specific, some of the relations in section 1 (discussions 2, 4, 5, 8) were functions, but they were not one-to-one and so the other viewpoints were not inverse functions but only inverse relations. Other examples from section 1 (discussions 3, 6, 7) were one-to-one functions and thus, the other viewpoints were inverse functions.

Discussion 4: Trying to Find Inverses of "Non" One-to-One Functions

If we were to try finding the inverse of $g(x) = x^2 + 3$, we would run into a problem. By inspection, notice that if we were to use –2 and 2 as inputs, the output would be the same: 7, thus this isn't a one-to-one function. There are two inputs for an output. If we were to graph this function, we would see that it doesn't pass the horizontal line test.

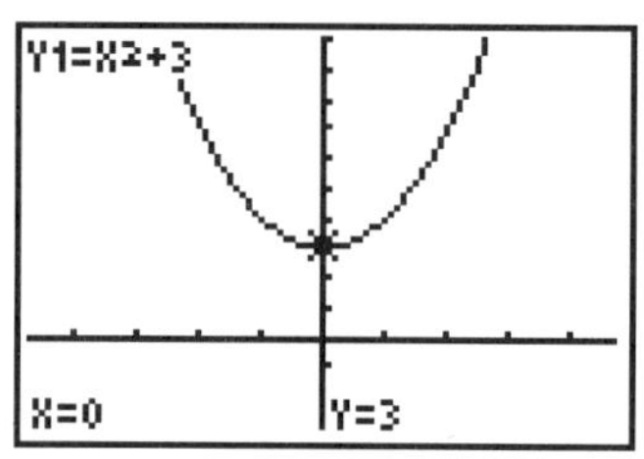

Also, as we know from our earlier discussions, the domain and range must switch in order for a function f to have an inverse function. But if there are two inputs for one output in a function f, then that implies that by switching the inputs (domain) and the outputs (range) we would get a relation that would have two outputs for an input, and this means it couldn't be a function. Hence, it would fail the definition for being a function (for each input only one output). So, let's look at a question that we can do.

Example 4: Finding the Inverse of a Function with a Restricted Domain

Find the inverse of $g(x) = x^2 + 3$ (from discussion 4) where the domain is restricted to only the x-interval $[0, \infty)$.

Solution:

Look below at the graph of this function. It is now a one-to-one function because of the restriction on the domain. We only get half of the graph.

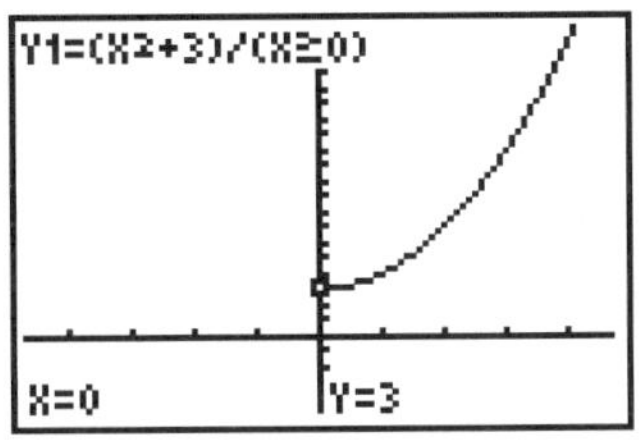

Now we can find the inverse of this one-to-one function.

First, rewrite without function notation.	$y = x^2 + 3$, x must be ≥ 0 Domain $[0, \infty)$; Range $(3,\infty)$
Second, subtract 3.	$y - 3 = x^2$, x must be ≥ 0
Third, take the square root of both sides.	$\pm\sqrt{y-3} = x$, x must be ≥ 0
Fourth, drop the negative sign, since $x \geq 0$.	$\sqrt{y-3} = x$
So, in function form we get	$g^{-1}(y) = \sqrt{y-3}$
Rewriting with a variable of "x"	$g^{-1}(x) = \sqrt{x-3}$ Domain $(3,\infty)$; Range $[0, \infty)$

Answer Q5: $f^{-1}(x) = \dfrac{x-5}{9}$.

The opposite of multiply by 9 add 5 is to subtract 5 divide by 9.

Remember our informal discussion about how inverses relate to each other. That is, the inverse of a function must do the opposite operations in the opposite order. Is this happening here? Yes. In the original function, you square first, and then add 3 to get the output. With the inverse function, you subtract 3 first, then square root. Those are the opposite operations in the opposite order. Note that the domain and range of the first becomes the range and domain of the other.

Question 6: If $k(x)$ has a Domain interval of $(-3, 10)$ and a Range interval of $(0, 5)$, then what would be the Domain and Range of $k^{-1}(x)$?

Question 7: If the point (5, –7) is on the graph of $g(x)$, then what would be one point on the graph of $g^{-1}(x)$ that we would know?

Example 5: Evaluating $g^{-1}(a)$ without Finding $g^{-1}(x)$

Given the function $g(x)$ find a) $g^{-1}(-2)$ and b) $g^{-1}(8)$.

x	$g(x)$
–1	13
0	8
1	3
2	–2
3	–7

Solution:

a) Remember that the inputs of g are the outputs of g^{-1} and vice verse. So, being asked to find $g^{-1}(-2)$ is the same as being asked to find $g(x) = -2$ (domain and range are switched).	$g^{-1}(-2) = y$ means, solve for y, but that means the same as $g(x) = -2$ where you solve for x. From the table we see that $x = 2$ for $g(x)$ to equal –2. Thus, $g^{-1}(-2) = 2$.
b) We apply the same understanding of how inverses relate to each other.	$g^{-1}(8)$ means find the output of g^{-1} when the input is 8. $g(x) = 8$ means find the input of g when the output is 8, which from the table is $x = 0$. So, $g^{-1}(8) = 0$.

Example 6: Finding $f^{-1}(a)$ from the Graph of $f(x)$

Given the function $f(x)$ find $f^{-1}(1)$.

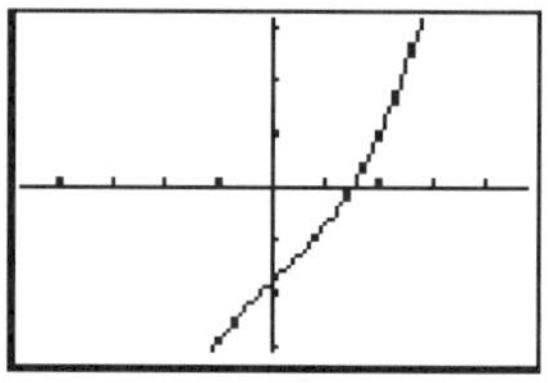

Solution:

Once again, asking for $f^{-1}(1)$ is asking for the same thing as $f(x) = 1$. If you can answer one of them you've answered them both.	$f(x) = 1$ when $x = 2$ from what we can see in the graph of $f(x)$. So, $f^{-1}(1) = 2$.

Example 7: Finding the Inverse of a Function

Find the inverse of $F(C) = \frac{9}{5}C + 32$ (you've seen this function before!). This is a one-to-one function. Check it by graphing it.

Solution:

Write without function notation.	$F = \frac{9}{5}C + 32$
Multiply by 5.	$5F = 9C + 160$
Subtract 160.	$5F - 160 = 9C$
Divide by 9.	$\frac{5F - 160}{9} = C$
Factor out $\frac{5}{9}$.	$\frac{5}{9}(F - 32) = C$
Write in function notation.	$C(F) = \frac{5}{9}(F - 32)$

You should recognize both of these as the formulas for Celsius and Fahrenheit. Rather than changing our answer to inverse notation form ($F^{-1}(F) = \frac{5}{9}(F - 32)$), it's better to keep it in its more meaningful, real-world form, where C indicates degrees Celsius and F indicates degrees Fahrenheit.

Finally, let's compare the graphs of inverse functions.

Answer Q6: The domain is (0, 5) and the range is (–3, 10) since inverses have their domains and ranges switched.

Discussion 5: Graphs of Inverse Functions

Look at the graphs of the Fahrenheit and Celsius formulas.

Celsius to Fahrenheit

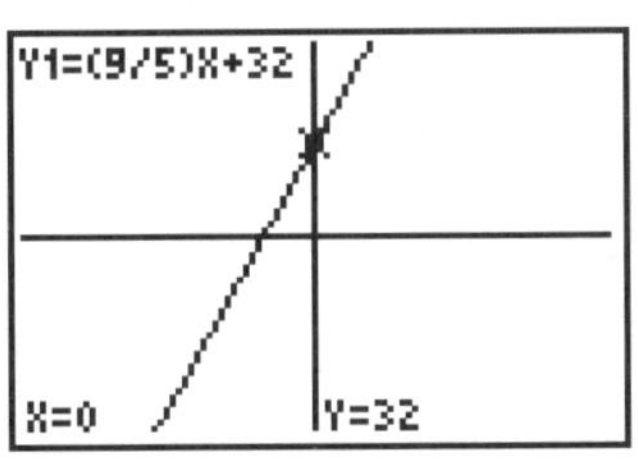

Fahrenheit to Celsius

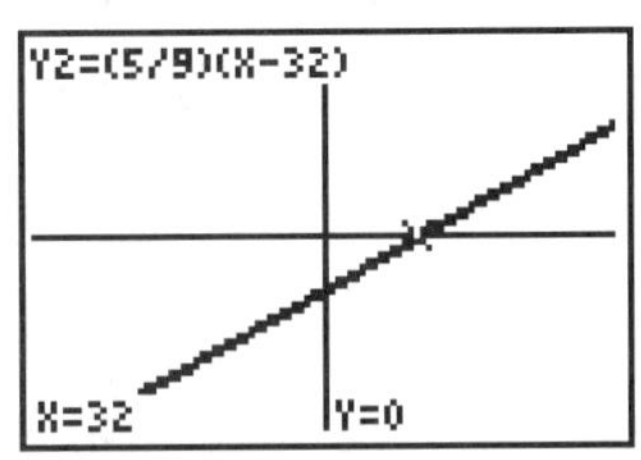

If we put both graphs together on the same graph, they will form a picture which is symmetric to the line $y = x$, in other words, you get a mirror reflection around the line $y = x$. Here are the previous two graphs on the same axis (notice how they form this mirror reflection):

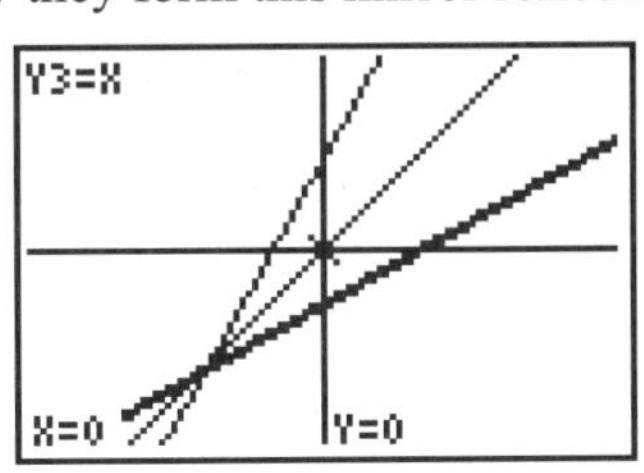

Question 8: Below are the graphs of two functions. From the third graph, with both pictures on the same axis, can you say that they *might* be inverses?

Question 9: Can you tell, by the graphs, if these two functions might be inverses of each other?

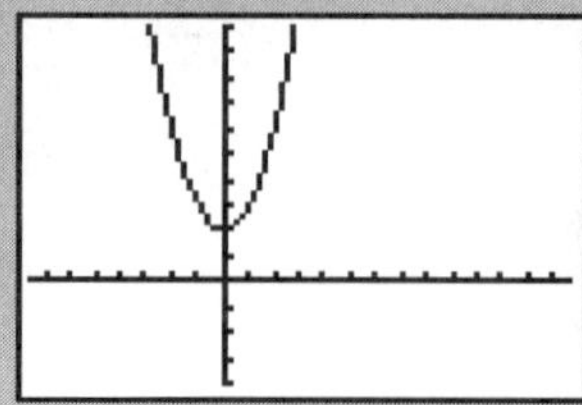

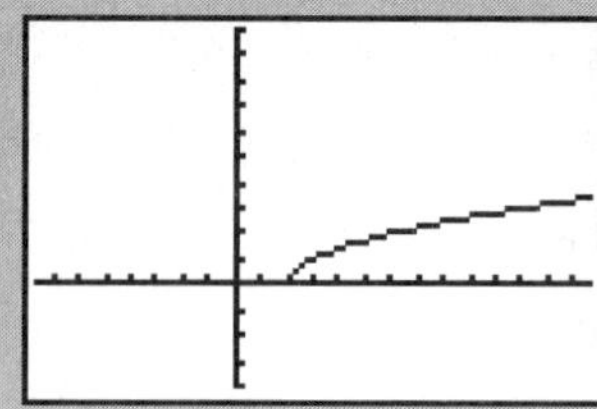

Example 8: Difficult Inverses

Can you find the inverse of $f(x) = x^3 + x$?

Solution:

If you look at the graph of this function, you will see that it does have an inverse since it passes the horizontal line test.

Answer Q7: The point (−7, 5), since domain and range switch.

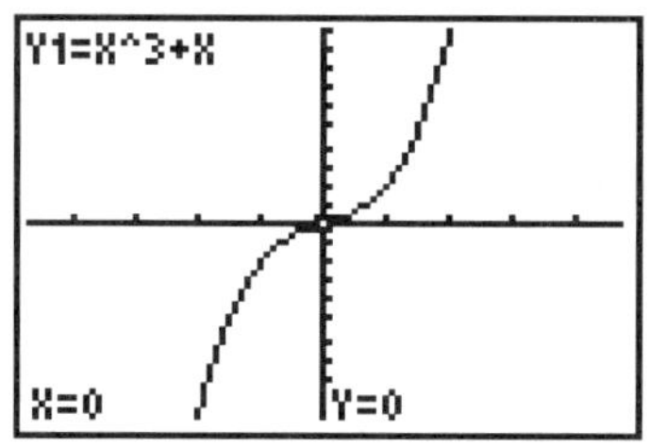

But if we try to find the inverse, we find that we can't solve for x in order to get the inverse. We don't have any easy way to solve a cubic equation. So this example shows us that even though we might know that an inverse exists, finding it could be rather difficult.

Section Summary

- The inverse of something is a process that undoes it.
- The first function must be one-to-one in order to have an inverse function.
- The domain and range of inverse functions are switched.
- To find the inverse of a one-to-one function, we solve the formula for the dependent variable.

SECTION 1.6 HOMEWORK

Answer Q8: Yes, they might be, but a graph isn't enough to prove it.

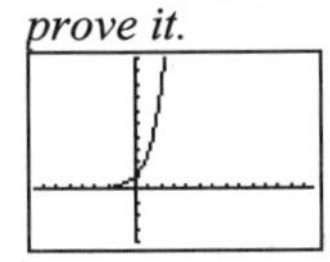

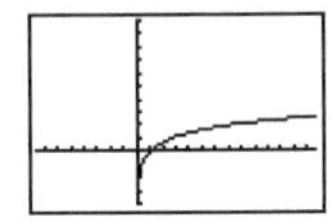

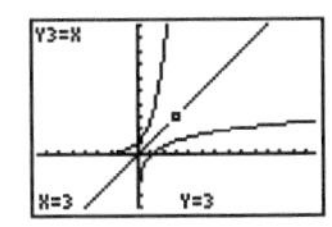

(1-2) State the inverse of the following:

1. To solve an equation you did the following steps:
   a. Multiplied both sides of the equation by $x-5$.
   b. Subtracted $2x$ from both sides of the equation.
   c. Added 1 to both sides of the equation.
   d. Divided both sides of the equation by 3.

   The answer is $x = -3$, now do the inverse of the above steps and find the original equation.

2. Give the inverse steps to get back home:
   a. Leave your house
   b. Get in your car
   c. Start your car
   d. Back out of your driveway
   e. Drive North 3 blocks
   f. Turn East and drive East 2 blocks
   g. Turn North and drive ½ block
   h. Stop your car at the house address 7755 Clairmont
   i. Get out of the car and walk to the house

(3–10) Find $f^{-1}(a)$ given $f(x)$.

3. Find $f^{-1}(3)$ and $f^{-1}(-1)$.

x	$f(x)$
−3	5
−2	3
−1	1
0	−1
1	−3
2	−5
3	−7

4. Find $f^{-1}(10)$ and $f^{-1}(-2)$.

x	$f(x)$
−5	−10
−3	−6
−1	−2
0	2
1	6
3	10
5	14

5. Find $f^{-1}(1)$ and $f^{-1}(4)$.

x	$f(x)$
49	7
36	6
25	5
16	4
9	3
4	2
1	1

6. Find $f^{-1}(2)$ and $f^{-1}(4)$.

x	$f(x)$
−3	14
−2	7
−1	2
0	1
1	4
2	11
3	22

7. Find $f^{-1}(-3)$ and $f^{-1}(-1)$.

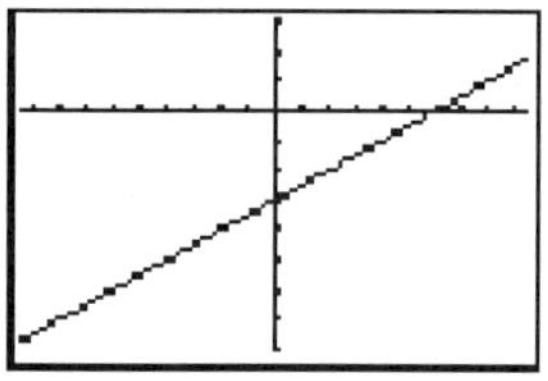

8. Find $f^{-1}(1)$ and $f^{-1}(-2)$.

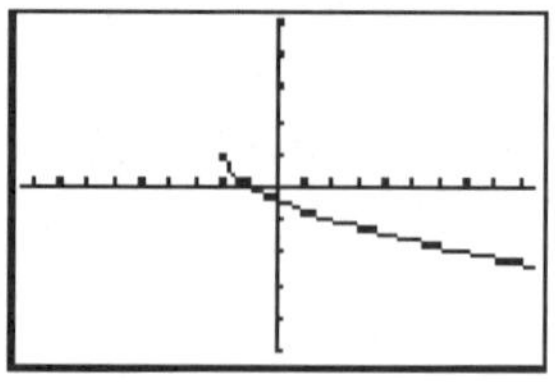

(9–14) Find the inverse relation for problems 3–8 and state whether it is a function or not. If $f(x)$ is in the form of a table then express the inverse in the form of a table. Do likewise if $f(x)$ is in the form of a graph.

Answer Q9: Yes, you can tell that they are clearly ***not*** *inverses of each other.*

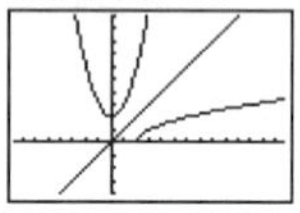

(15–26) Graph each of the following functions and check for one-to-one by the horizontal line test.

15. $f(x) = 3x - 5$
16. $f(x) = 4 - 5x$
17. $f(x) = 2x^2 - 3x + 1$
18. $f(x) = 3x^2 + 5x - 1$
19. $f(x) = x^3 - 3x - 5$
20. $f(x) = 2x^3 - 3x - 5$
21. $f(x) = x^3 + 2x + 1$
22. $f(x) = x^3 - x^2 + x$
23. $f(x) = |3x + 1|$
24. $f(x) = |3 - 2x|$
25. $f(x) = \sqrt{3x+2}$
26. $f(x) = \sqrt{2-x}$

(27–38) For each of the following pairs of functions:

a.) Check their graphs compared with the graph $y = x$ and guess if they are inverses.

b.) Prove they are or are not inverses by checking the values $f(g(x))$ and $g(f(x))$.

27. $f(x) = 3x - 1$ $\quad g(x) = \dfrac{x+1}{3}$
28. $f(x) = 2x + 5$ $\quad g(x) = \dfrac{x-5}{2}$
29. $f(x) = x^3 - 3$ $\quad g(x) = \sqrt[3]{x+3}$
30. $f(x) = 3x^3 + 4$ $\quad g(x) = \sqrt[3]{\dfrac{x-4}{3}}$
31. $f(x) = \dfrac{3x+1}{2}$ $\quad g(x) = \dfrac{2x-1}{3}$
32. $f(x) = \dfrac{2x-3}{5}$ $\quad g(x) = \dfrac{5x+3}{2}$

33. $f(x) = \frac{2x+3}{5}$ $\quad g(x) = \frac{2x-5}{3}$

34. $f(x) = \frac{3x-4}{7}$ $\quad g(x) = \frac{4x-7}{3}$

35. $f(x) = x^3 + 4$ $\quad g(x) = \sqrt[3]{8x-32}$

36. $f(x) = x^3 - 10$ $\quad g(x) = \sqrt[3]{2x+20}$

37. $f(x) = \frac{x-1}{x+2}$ $\quad g(x) = \frac{2x+1}{1-x}$

38. $f(x) = \frac{2x+3}{x-2}$ $\quad g(x) = \frac{2x+3}{x-2}$

(39–54) Find the inverse of each of the following functions. (If the inverse is not a function, give the inverse relation and state the inverse is not a function)

39. $f(x) = 3x - 5$

40. $f(x) = 2x + 3$

41. $f(x) = \frac{2x-3}{5}$

42. $f(x) = \frac{3x+4}{2}$

43. $f(x) = \sqrt{2x-1}$

44. $f(x) = \sqrt{3x+5}$

45. $f(x) = 3x^2 + 1$

46. $f(x) = 2x^2 - 3$

47. $f(x) = \sqrt[3]{2x+3}$

48. $f(x) = \sqrt[3]{3x-5}$

49. $f(x) = 2x^3 - 1$

50. $f(x) = 5x^3 + 3$

51. $f(x) = \frac{x-1}{x+3}$

52. $f(x) = \frac{x+5}{x-2}$

53. $f(x) = \frac{2x+3}{x-2}$

54. $f(x) = \frac{3x+1}{x-3}$

55. You invest \$5000 at 8% simple interest. The equation that gives the amount in the account, $A(t)$, after a number of years, t, is $A(t) = 5000 + 400t$.

    a. Find the inverse of the equation.

    b. What does the inverse equation allow you to now do?

56. A rock is tossed into a pool. The radius of the first circular ripple increases at the rate of 5 ft./sec. The equation for the circumference of the circle in terms of time is $C(t) = 5\pi t$, where $C(t)$ is the circumference and t is the time.

    a. Find the inverse of the equation.

    b. What does the inverse equation allow you to do?

57. A spherically shaped balloon is being inflated so that the radius, r, is changing at the constant rate of 3 inches per second. Assume $r = 0$ at time $t = 0$, the equation that determines volume, $V(t)$, at time t is $V(t) = 36\pi t^3$.

    a. Find the inverse of the equation

b. What does the inverse equation allow you to do?

58. A cubed figure is expanding so that the length of a side of the cube, s, is changing at a constant rate of 2 inches per second. Assume $s = 0$ at $t = 0$, the equation that determines volume, $V(t)$, at time t is $V(t) = (2t)^3$.

a. Find the inverse of the equation.

b. What does the inverse allow you to do?

59. A function has a domain of $[-3, 5]$ and a range of $[1, 9]$, what is the domain and range of the inverse of the function?

60. A function has a domain of $[5, \infty)$ and a range of $(-\infty, -4]$, what is the domain and range of the inverse of the function?

61. A function has a solution of $(-2, 3)$, what is a solution for the inverse of the function?

62. A function has a solution of $(-3, 1)$, what is a solution for the inverse of the function?

Chapter 1 Review

Topic	Section	
Input	1.1	The object plugged into a rule (relationship) that produces an output. The first coordinate of an ordered pair.
Output	1.1	The object that is the result of an input. The second coordinate of an ordered pair.
Relation (informal)	1.1	Any rule with inputs that cause outputs. One input may have any number of outputs.
Relation (formal)	1.2	Any set of ordered pairs.
Function (informal)	1.1	Any rule where for each input there is only one output.
Function (formal)	1.2	Any set of ordered pairs where no two pairs have the same first coordinate and a different second coordinate.
Vertical Line Test	1.1	Imagine drawing vertical lines up and down everywhere on the graph of a relation, and if no vertical line crosses the graph more than once then the relation is a function.
M as a function of t	1.1	$M(t)$. $M(t)$, the whole symbol, is the output and the t is the input.
Independent Variable	1.1	The input. Freely chosen.
Dependent Variable	1.1	The output. Depends on the input.
Domain	1.2	Set of all possible inputs.
Range	1.2	Set of all possible outputs that are produced from the domain.
Interval Notation	1.2	A way to express answers for the Domain and Range.
Increasing Function	1.3	As the x-values get larger the $f(x)$-values also get larger.
Decreasing Function	1.3	As the x-values get larger the $f(x)$-values do the opposite and get smaller.
Local (relative) Maximum	1.3	The largest y-value of any y-values near it. The top of a peak in the graph.
Local (relative) Minimum	1.3	The smallest y-value of any y-values near it. The bottom of a valley in the graph.
Combining Functions	1.4	We can add, subtract, multiply, divide, and compose functions. Composing functions is simply plugging in one function for every x in the other.

Creating New Functions	1.5	Here is where you solved one function for a particular variable and then substituted into another function. Remember, you need an intermediate variable in-order to get a new function in terms of the variables you want.
Difference Quotient	1.5	$\frac{f(x+h)-f(x)}{h}$ Just the same as the formula for slope.
Average Rate of Change	1.5	$\frac{f(x_2)-f(x_1)}{x_2-x_1}$ Also, just the same as the formula for slope.
Inverse Functions	1.6	One function that *undoes* another, both of which must be one-to-one functions. A function that does the opposite operations in the opposite order from another function. If f and g are inverses then $(f \circ g)(x) = (g \circ f)(x) = x$ and their domains and ranges "switch."
One-to-one Function	1.6	A function where for each output there is only one input. The reverse idea of what a function is.
Horizontal Line Test	1.6	Imagine drawing horizontal lines everywhere across your graph of the relation, if no horizontal lines cross the graph more than once then the relation is one-to-one.
Finding An Inverse	1.6	To find an inverse of a one-to-one function you just solve for the other variable, or in the case of a table just switch your columns.
Symmetry to the line $y = x$	1.6	Inverse functions form a mirror image around the line $y = x$.

CHAPTER 1 REVIEW HOMEWORK

Section 1.1

1. The following formula can be used to estimate the average attendance for each home game the Arizona Diamondbacks play based on the number of wins (w) they achieve in a season. $A(w) = 2.5w^2 - 47.86w + 19943$
   a. Estimate the average per game attendance in a year the Diamondbacks win 90 games.
   b. Estimate the average per game attendance in a year the Diamondbacks win 60 games.
   c. Is the average per game attendance for the Diamondbacks a function of the number of wins?
   d. What is the independent variable?
   e. What is the dependent variable?
2. The number of calories burned on a treadmill in 30 minutes of exercise is related to the level of difficulty (d) chosen. The following function can be used to model this relation. $C(d) = 50d$
   a. How many calories would be burned in 30 minutes at a difficulty level of 6?
   b. How many calories would be burned in 30 minutes at a difficulty level of 10?
   c. Is the number of calories burned in 30 minutes a function of the difficulty level?
   d. What is the independent variable?
   e. What is the dependent variable?
3. The sales tax rate on non-food items in Phoenix Arizona is 7.01% of the price. A formula that will allow one to calculate the price of an item including tax is:
   $A(p) = 1.0701p$, where p is the before tax price.
   a. Calculate the after tax price of a $30 pair of jeans.
   b. Is the after tax price a function of the before tax price?
   c. What is the independent variable?
   d. What is the dependent variable?
4. The following table shows the average speed of the winning car in the Indianapolis 500 race.

Year	1920	1930	1940	1950	1960	1970	1980	1990	1999
Average speed (mph)	88	100	114	124	139	156	143	186	153

   a. What was the average speed for the winning car in 1980?
   b. What year was the average speed the greatest?
   c. Is the average speed a function of the year?
5. The following table shows the number of earthquakes in the world with a magnitude of 7 or greater on the Richter scale for selected years.

Number of earthquakes	13	36	8	13	23	39	22	29
Year	1900	1910	1920	1930	1940	1950	1960	1970

   a. How many earthquakes of a magnitude greater than 7 were there in 1960?
   b. What years had 13 earthquakes of a magnitude of 7 of more?

c. Is the year a function of the number of earthquakes?

6. The following chart shows the estimated amount of Carbon Dioxide emitted each year in the United States from 1990 – 1997 in Millions of Metric tons.

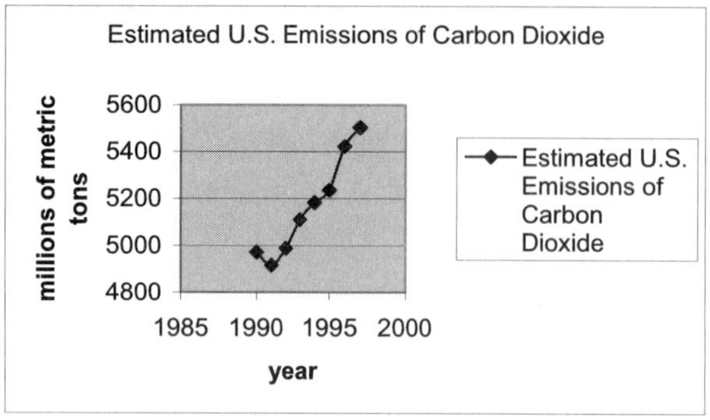

a. How many metric tons of Carbon Dioxide were emitted in 1995?

b. Is the number of metric tons of carbon dioxide emitted a function of the year?

7. The following table provides a method for translating dollar values from the past into 1998 dollars. Notice that the graph passes through the point (1970, 4.20). This means that \$4.20 in 1998 would have the same purchasing power as \$1.00 in 1970.

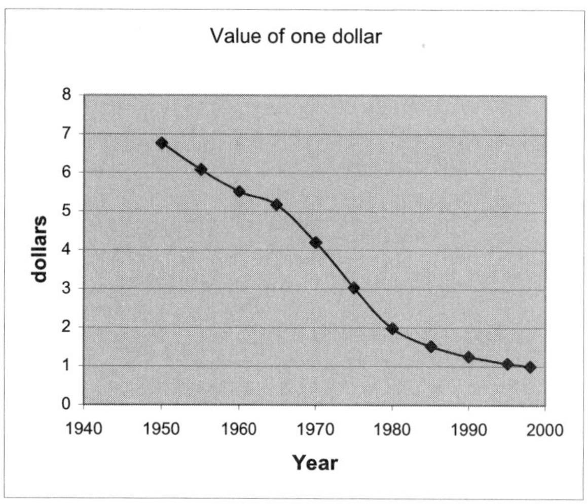

a. About how much money in 1998 has the same purchasing power as a dollar in 1960?

b. About how much money in 1998 has the same purchasing power as a dollar in 1950?

c. Does it appear that 1998 purchasing power is a function of time?

(8–12) Use the following functions to answer questions:

$$f(x) = x^2,\ g(x) = \frac{x+2}{x-3},\ h(x) = \sqrt{2x-6}$$

8. $h(5) =$ 9. $g(2) =$ 10. $f(-3) =$

11. $f(a) =$ 12. $h(a+2) =$

(13–15) Determine if y is expressed as a function of x:

13. $y = 3x - 4$ 14. $|y+2| = 3x$ 15. $x = y^2 - 6$

Section 1.2

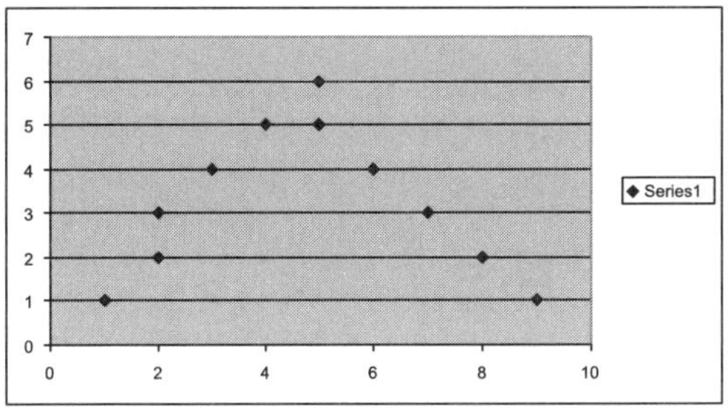

16. Consider the following scatter plot:
    a. List all the ordered pairs for the relation created by the scatter plot.
    b. What is the domain of the relation?
    c. What is the range of the relation?
    d. Does the scatter plot represent a function?
17. Below is a table showing the percent of 18 – 24 year olds males living at home

Years	1960	1970	1980	1985	1990	1995	1998
Percent of 18-24 year old males at home	52	54	54	60	58	58	59

(U.S. Statistical Abstract)

a. List all the ordered pairs created by the table.
b. What is the domain of the relation?
c. What is the range of the relation?
d. Does the table represent a function?

(18–22) State the domain of the following functions:

18. $f(x) = x^3$ 19. $y = \dfrac{x+2}{x-3}$ 20. $f(s) = \sqrt{2s-6}$

21. $y = \dfrac{\sqrt{x-3}}{x^2-16}$ 22. $y = \sqrt{9-x^2}$

(23–26) State the range of the following functions:

23. $f(x) = |x-4|$ 24. $f(w) = 2 + \sqrt{w-4}$ 25. $y = x^2 - 4$

26. $y = 3$

(27–30) State the domain and range of the following functions:

27. $g(x) = \begin{cases} 2x & x < 0 \\ x+1 & x \geq 0 \end{cases}$ 28. $f(x) = \begin{cases} 3x-6 & x < 2 \\ x^2-4 & x \geq 2 \end{cases}$

29. $h(x) = \sqrt{x-5}$ 30. $k(m) = 5$

Section 1.3

31. The following table shows the number of births per 1,000 people in the United States.

Year	1975	1980	1985	1990	1994	1998
Births per 1000 people	14.0	16.2	15.7	15.5	15.2	14.4

a. Where is the function increasing?

b. Where is the function decreasing?

c. Are there any relative maximums?

d. Are there any relative minimums?

32. The following table shows the number of births per 1000 teens for various years.

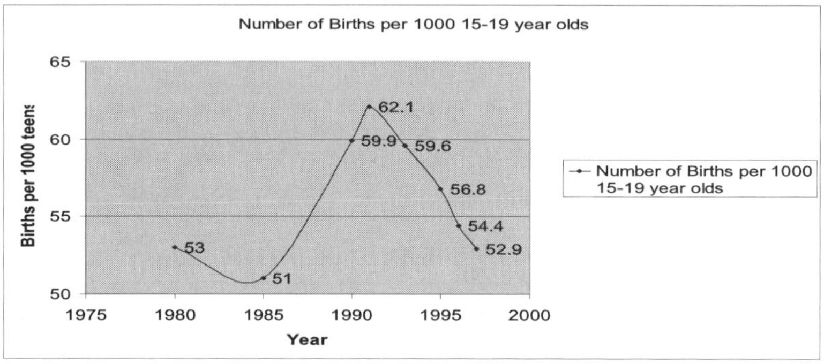

a. Over what interval is the function is increasing.

b. Over what interval is the function is decreasing.

c. Find all relative maximums.

d. Find all relative minimums.

33. A toy rocket is shot up vertically from the top of a building with an initial velocity of 240 feet per second. The height of the rocket $s(t)$ (in feet) after t seconds is given by the function: $s(t) = -16t^2 + 240t + 80$

a. At what time does the rocket reach its maximum height?

b. What is the rocket's maximum height?

c. Over what time interval is the rockets height increasing?

d. Over what time interval is the rockets height decreasing?

e. What is the domain for this function?

f. What is the range for this function?

34. A small company manufactures and sells bicycles. The company manager has determined that the profit for selling x bicycles can be modeled by the following equation:

$$P(x) = \frac{-1}{60}x^3 - x^2 + 105x - 10$$

a. How many bikes should be sold to maximize profit?

b. What will the maximum profit be?

c. Over what interval will the profit be increasing?

d. Over what interval will the profit be decreasing?

e. What is the logical domain for this problem?

f. What is the range?

(35–42) Use a graphing calculator to estimate:

a. Over what interval will the function be increasing?

b. Over what interval will the function be decreasing?

c. Any local (relative) maximums?

d. Any local (relative) minimums?

35. $y = x^2 - 3x + 12$
36. $y = -3x^2 + 5x + 6$
37. $y = x^3 - 4x$
38. $y = -x^3 + 8$
39. $y = x^4 - 13x^2 + 36$
40. $f(x) = x^4 + 5x^3 + 8x^2 + 5x + 1$
41. $f(x) = |2x - 5| + 7$
42. $y = -4|x + 2| - 5$

43. A group of researchers found that people prefer training films of moderate length: shorter films contain too little information, while longer films are boring. For a training film on the care of exotic birds, the researchers determined that the ratings people gave for the film could be approximated by $R(t) = \dfrac{20t}{t^2 + 100}$ where t is the length of the film in minutes.

(Assume the maximum length of a training film is 30 minutes.)

a. Over what time interval is the film ratings increasing.

b. Over what time interval is the film ratings decreasing.

c. What is the optimal film length?

Section 1.4

$f(x) = 2x - 3$ $\quad g(x) = x^2 + 4$ $\quad r(x) = \sqrt{x-4}$ $\quad t(x) = \dfrac{x+3}{2}$

x	$m(x)$
–3	10
–2	7
–1	4
0	1
1	–2
2	–5
3	–8

$n(x)$

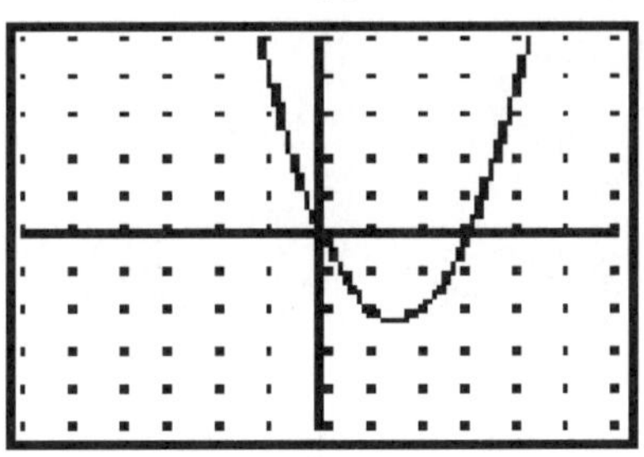

(44–58) Use the above functions to answer the following:

44. $(f + t)(3)$
45. $(g - r)(5)$
46. $(fm)(-2)$
47. $(g/n)(0)$
48. $(f \circ n)(4)$
49. $(g \circ m)(2)$
50. $(f/r)(4)$
51. $(f + t)(x)$
52. $(fg)(x)$
53. $(f - g)(a)$
54. $(g/t)(2b+1)$
55. $(r \circ g)(x)$
56 $(f \circ t)(x)$
57. What is the domain of $(f + r)(x)$?
58. What is the domain of $(g - m)(x)$?

Section 1.5

59. $y(x) = 3x - 1$. Write x as a function of y.

60. $y(x) = \frac{2}{3}x - \frac{5}{6}$. Write x as a function of y.

61. $r(t) = 5 + \sqrt{t+3}$. Write t as a function of r.

62. $m(w) = \frac{3w-2}{2w+1}$. Write w as a function of m.

63. $r(t) = 3t + 4$ and $t(s) = 2s - 5$ write r as a function of s.

64. $y(r) = r^2 - 2r + 3$ and $r(x) = 2x + 1$. Write y as a function of x.

65. $r(w) = \sqrt{w-4}$ and $w(t) = t^2 + 4$. Write r as a function of t.

66. Find the difference quotient for $f(x) = 5x + 3$.

67. Find the difference quotient for $2x^2 - 5x + 3$.

68. $C = 2\pi r$ is an equation for the circumference of a circle as a function of the radius of the circle, where C = circumference and r = radius. Rewrite the equation with radius a function of circumference.

69. $W = \frac{3000}{l}$ is the equation that gives the weight, W, that can be supported by a 2 inch by 4 inch piece of pine lumber as a function of l, the length of the board. Rewrite the equation with length as a function of weight.

70. $V = \frac{1}{3}\pi r^2 h$ is the equation of a cone; where V = volume of the cone, r = radius of the cone, and h = height of the cone. If the radius is always twice the height write the equation with volume a function of height.

71. $R = px$ is the revenue equation used when R is the revenue, p is the price per unit, and x is the number of units. The demand function is $p(x) = 100 - x$. Write R as a function of x.

72. A baseball is hit in the air. The equation for the height, $S(t)$, of the ball after time t is given by the formula $S(t) = -16t^2 + 128t + 2$ (where $S(t)$ is in feet and t is in seconds). What is the average rate of change of height to time for the baseball on the time interval $t = 1$ second to $t = 3$ seconds?

73. You are selling widgets and the equation that gives the revenue in terms of the number of widgets sold is $R(x) = 900x - 0.1x^2$, where $R(x)$ is the revenue in dollars and x is the number of widgets sold. What is the average rate of change in dollars received to widgets sold on the interval $x = 10$ to $x = 20$?

Section 1.6

Table for 74 and75:

x	$f(x)$
–2	4
–1	1
0	0
1	1
2	4
3	9
4	16

74. a) Find $f^{-1}(9)$.

    b) Using the above table create the inverse relation.

75. Is the inverse relation a function?

Table for 76 and 77

x	$f(x)$
–3	–27
–2	–8
–1	–1
0	0
1	1
2	8
3	27

76. a) Find $f^{-1}(8)$.

    b) Using the above table create the inverse relation.

77. Is the inverse relation a function?

78. $f(x) = 2x + 5$ Check to see if this function is one–to–one by the horizontal line test.

79. $f(x) = x^3 - 2x - 3$ Check to see if this function is one–to–one by the horizontal line test.

80. Are $f(x) = 3x - 2$ and $g(x) = \dfrac{x+2}{3}$ inverse functions? Prove your answer.

81. Graph $f(x) = x^3 + 3$, $g(x) = \sqrt[3]{x-3}$ and $y = x$ and from these graphs determine if you think $f(x)$ and $g(x)$ are inverses of each other.

82. Are $f(x) = x^3 + 1$ and $g(x) = \sqrt[3]{x-1}$ inverse functions? Prove your answer.

83. Are $f(x) = x^2$ and $g(x) = \sqrt{x}$ inverse functions? Prove your answer.

84. Find the inverse of $f(x) = 3x - 5$. Is the inverse a function?

85. Find the inverse of $f(x) = \dfrac{x+1}{x-3}$. Is the inverse a function?

86. Find the inverse of $f(x) = x^2 - 2$. Is the inverse a function?

87. The cost for manufacturing x items is $C(x) = 500x + 3000$, where $C(x)$ is the cost of production in dollars.

    a. Find the inverse of the function.

    b. How many items can be produced for $12,000?

88. A rock is thrown in a pool of water and the first circular ripple expands is such a way that the radius is changing at the rate of 3 inches per second. $A(t) = 9\pi t^2$ is the function that determines the area of the circle, $A(t)$ in square inches per second, at time t.

    a. Find the inverse function.

    b. At what time is the area 144π square inches?

CHAPTER 1 EXAM

1. The following chart shows the percent of 18-24 year old females living at home in various years.

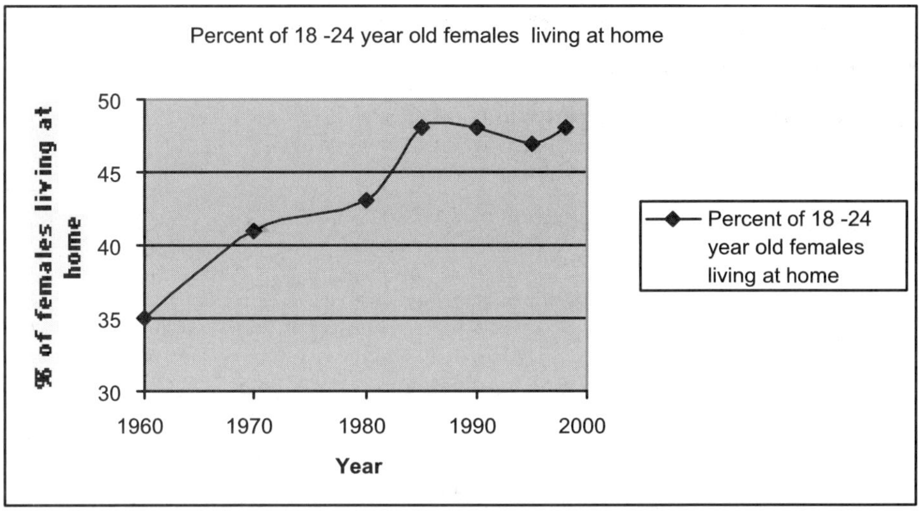

a. About what percent of females between 18 – 24 years old were living at home in 1990?

b. Is the number of females living at home a function of the year?

2. The following table show the percent of students initially enrolled in a College Algebra class that passes with a grade of C or better.

Year	1995	1996	1997	1998	1999	2000
% passing College Algebra	62	64	66	61	59	70

a. During what years did at least 60% of the college algebra students pass with a grade of C or better?

b. Is the number of students that pass college algebra a function of the year?

3. Let $f(x)=\sqrt{x-4}$. Find $f(9)$.

4. Let $g(x)=\dfrac{2x+3}{2x-3}$. Find $g(a+2)$

5. Determine if y is expressed as a function of x: $x^2+y^2=9$

6. Consider the following scatter plot:

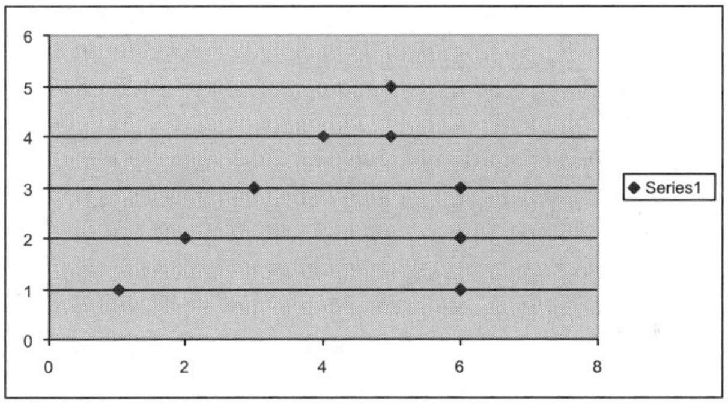

a. List all the ordered pairs for the relation created by the scatter plot.

b. What is the domain of the relation?

c. What is the range of the relation?

d. Does the scatter plot represent a function?

7. State the domain of the following function: $y = \sqrt{x+2}$

8. State the range of the following function: $f(x) = x^2 - 9$

9. State the domain and range of the following function: $\begin{cases} 2x-3 & x \geq 1 \\ x^2 & x < 1 \end{cases}$

10. State the domain and range of the following function: $y = |x+2| + 5$

11. A toy rocket is shot up vertically from the ground with an initial velocity of 120 feet per second. The height of the rocket $s(t)$, (in feet), after t seconds is given by the function

    $s(t) = -16t^2 + 120t$.

    a. At what time does the rocket reach its maximum height?

    b. What is the rockets maximum height?

    c. Over what time interval is the rocket's height increasing?

    d. Over what time interval is the rocket's height decreasing?

    e. What is the domain for this function?

    f. What is the range of this function?

(12 – 15) Find the intervals where the function is increasing and decreasing. Also identify any relative maximums or minimums.

12. $y = -x^2 + 2x + 4$

13. $f(x) = x^2 + 64$

14. $y = -|3x+5| - 7$

15. $f(x) = x^4 - 16$

(16–21) Use the functions below to answer the questions:

$f(x) = 2x - 3$ $\qquad$ $g(x) = x^2 - 2x + 1$

x	$r(x)$
–3	5
–2	1
–1	–5
0	–3
1	4
2	0

16. $(f+g)(x)=$ $\qquad$ 17. $(fg)(x)=$ $\qquad$ 18. $(f \circ g)(x)$

19. $(f+r)(-3)=$ $\qquad$ 20. $(gr)(2)=$ $\qquad$ 21. $(f-g)(a+3)=$

22. $y(x) = \sqrt{3x-4}$. Write x as a function of y.

23. $s(t) = \dfrac{2t-3}{t+3}$. Write t as a function of s.

24. $f(x) = 2x^2 - 3$ and $x(t) = 2t - 3$. Write f as a function of t.

25. $A(r) = \pi r^2$ gives area of a circle as a function of the radius of the circle. $r(t) = 3t$ gives the radius of the circle as a function of time. Write the area of a circle as a function of time.

26. Find the difference quotient for $f(x) = x^2 - 5x$.

27. Find $f^{-1}(x)$

x	$f(x)$
−2	−9
−1	−4
0	−1
1	0
2	7

28. $f(x) = 3x - 2$ Find $f^{-1}(x)$.

29. $f(x) = \dfrac{x+1}{2x-3}$ Find $f^{-1}(x)$.

30. Check with the horizontal line test if $f(x) = x^3 - 2x^2 + 1$ is $f(x)$ one-to-one?

31. $A(t) = 10000 + 1000t$ is a simple interest function. $A(t)$ is the amount of dollars in the account and t is the number of years the money has been in the account.
    a. Find the inverse of the function.
    b. How many years will it take for the account to be worth \$23,300?

Chapter 2 Algebra Skills

In this chapter you are going to practice and learn how to solve all kinds of equations and inequalities. The skills you learn here are necessary for your success in this and future courses.

Section 2.1 Complex Numbers

First, we need to spend some time studying the topic of complex numbers. As you might already know, you were unable to get an answer to problems like $\sqrt{-4}$ before learning about complex numbers. $\sqrt{-4}$ does not yield a real number. But with the following definition

$$\boxed{i = \sqrt{-1}}$$

you can handle problems like the one just mentioned: ($\sqrt{-4} = \sqrt{4} \bullet \sqrt{-1} = 2i$).

Question 1: What is equivalent to i^2?

Let's do a few examples using the definition of "i." We will use this definition to convert square roots into their proper imaginary form. Numbers of the form " bi " are called **imaginary numbers** (where b is any real number). We will also show what happens when you get an i^2.

Example 1: Simplifying Square Roots That Yield Imaginary Numbers

Simplify the following problems

Solutions:

a.) $\sqrt{-49} = \sqrt{49} \bullet \sqrt{-1} = 7i$

b.) $\sqrt{-144} = \sqrt{144} \bullet \sqrt{-1} = 12i$

c.) $\sqrt{-50} = \sqrt{50} \bullet \sqrt{-1} = \sqrt{25} \bullet \sqrt{2} \bullet i = 5i\sqrt{2}$

d.) $\sqrt{-12} = \sqrt{4} \bullet \sqrt{3} \bullet \sqrt{-1} = 2i\sqrt{3}$

e.) $-\sqrt{-36} = -36 \bullet \sqrt{-1} = -6i$

f.) $4i^2 = 4\,(-1) = -4$

g.) $-13i^2 = -13(-1) = 13$

The definition of a **complex number** is,

> **A complex number** is any number of the form $(a + bi)$, where a and b are real numbers and i equals $\sqrt{-1}$.

Notice that complex numbers are made up of a real number added to an imaginary number.

In mathematics, things sometimes happen to be **cyclic**. What we mean by this is that they have a cycle, that is, they repeat after a while. For example, the following illustrates cyclic behavior:

$i = \sqrt{-1}$	$i^5 = i\,(i^4) = i\,(1) = i = \sqrt{-1}$
$i^2 = -1$	$i^6 = (i^2)^3 = (-1)^3 = -1$
$i^3 = i\,(i^2) = i\,(-1) = -i$	$i^7 = i\,(i^6) = i\,(-1) = -i$
$i^4 = i^2\, i^2 = (-1)(-1) = 1$	$i^8 = (i^4)^2 = (1)^2 = 1$

See how every fourth answer, after the initial $i = \sqrt{-1}$, bring us right back to where we started ($\sqrt{-1}$). We have a cycle. Every fourth step is equal to the first. So for example, $i^4 = i^8 = i^{12} \ldots$

Question 2: What would i^9 and i^{13} equal?

OPERATIONS WITH COMPLEX NUMBERS

Let's look at how you add and subtract complex numbers. To add or subtract complex numbers you simply add the like terms. In essence, you treat the *i* like you would the variable *x*. Here are some examples.

Example 2: Simple Operations with Complex Numbers

Add and subtract the following complex numbers.

a.) $(5 - 2i) + (-7 + 5i)$

b.) $(-7 + 6i) - (-9 + 11i)$

c.) $(17) + (-13 + 7i)$

Solutions:

a.) $(5 - 2i) + (-7 + 5i) = (5 + (-7)) + (-2i + 5i) = -2 + 3i$

b.) $(-7 + 6i) - (-9 + 11i) = (-7 - (-9)) + (6i - 11i) = 2 + (-5i) = 2 - 5i$

c.) $(17) + (-13 + 7i) = 4 + 7i$

You may very well have been told, when you first learned algebra, that the way to add $3x$ and $5x$, for example, was to use the distributive property:

Factor out x when adding $3x$ and $5x$	$(3 + 5)x = 8x$

Or you might have been taught to think that $3x$ says that you have three x's, and $5x$ says that you have five x's, so together they give you a total of eight x's ($8x$). Both ways are correct. The same way of thinking works with complex numbers.

Question 3: What is the answer to $(2 - 5i) - (7 - 8i)$?

Multiplication

Discussion 1: Multiplying Complex Numbers

Let's look at a problem like $\sqrt{-4} \bullet \sqrt{-25}$. Here, we must simplify the square roots before we perform the multiplication. For this example we get $\sqrt{-4} \bullet \sqrt{-25} = 2i \bullet 5i = 10i^2 = -10$. Notice that if you were to multiply the square roots first, you would get $\sqrt{-4} \bullet \sqrt{-25} = \sqrt{100} = 10$, an ***incorrect*** *answer* (the rule that says $\sqrt{a}\sqrt{b} = \sqrt{ab}$, requires that $\sqrt{a}$ and $\sqrt{b}$ both be real numbers).

Since the two original square roots are not real numbers, the rule about multiplying roots doesn't apply. We are not allowed to just multiply the two square roots. It is only after we convert these square roots into complex numbers that we can then multiply them together. Also notice again how we combined like terms (multiplied likes together) and treated the i like we would have an x.

Answer Q1: $i^2 = \left(\sqrt{-1}\right)^2 = -1$. *Remember, squaring and taking the square root are inverses of each other, so they undo each other or, in other words, cancel each other out.*

Example 3: Simplifying Complex Numbers

Simplify and perform the indicated operations.

a.) $\sqrt{-121} \bullet \sqrt{-9}$

b.) $\sqrt{-27} \bullet \sqrt{-12}$

c.) $-\sqrt{48} \bullet \sqrt{-18}$

d.) $7i\,(5 - 2i)$

e.) $(-3 + 4i)(2 - i)$

Solutions:

a.) $\sqrt{-121} \bullet \sqrt{-9} = 11i \bullet 3i = 33i^2 = -33$

b.) $\sqrt{-27} \bullet \sqrt{-12} = \sqrt{9}\sqrt{3}\sqrt{-1} \bullet \sqrt{4}\sqrt{3}\sqrt{-1} = 3i\sqrt{3} \bullet 2i\sqrt{3} = 6i^2\sqrt{9} = -6 \bullet 3 = -18$

c.) $-\sqrt{48} \bullet \sqrt{-18} = -\sqrt{16}\sqrt{3} \bullet \sqrt{9}\sqrt{2}\sqrt{-1} = -4\sqrt{3} \bullet 3i\sqrt{2} = -12i\sqrt{6}$

d.) $7i\,(5 - 2i) = 35i - 14i^2 = 35i - (-14) = 14 + 35i$

e.) $(-3 + 4i)(2 - i) = -6 + 3i + 8i - 4i^2 = -6 + 11i - (-4) = -2 + 11i$

As you can see, when we have one complex number times another, as in example 3(e), we "foil," just like we would if there had been x's in the problem instead of i's. Similarly with example 3(d), we distribute just like we would have if the i's were a variable. Also notice that in examples 3(b) and 3(c) the square roots didn't work out conveniently. Remember that when you are trying to simplify square roots that are not perfect squares, you have to factor the numbers into perfect square factors times non-perfect square factors. For example, $\sqrt{28}$ equals $\sqrt{4}$ a perfect square, times $\sqrt{7}$, not a perfect square, which simplifies to $2\sqrt{7}$ ($\sqrt{28} = \sqrt{4} \bullet \sqrt{7} = 2\sqrt{7}$).

Question 4: What is the product of $(-1 + i)(4 - 2i)$?

Example 4: Multiplying Complex Numbers

What is the product of $(5-3i)(5+3i)$?

Solution:

$(5-3i)(5+3i) = 25 + 15i - 15i - 9i^2 = 25 - 9(-1) = 25 + 9 = \underline{34}$

Notice how the product of these two complex numbers yields a real answer. There isn't an i in the final answer. This happened because the two numbers are what we call **conjugates**. The only thing different about the two numbers is the sign; one has a subtraction sign, the other a plus. The product of conjugates is always the difference of two squares $[(a+b)(a-b) = a^2 - b^2]$. This example is similar to problems like:

Answer Q2: We know that $i^9 = i^{13}$ (they are four steps apart). $i^{13} = i^9 = i(i^2)^4 = i(-1)^4 = i$.

$(2x+7)(2x-7) = 4x^2 - 49$
or
$(4-\sqrt{3})(4+\sqrt{3}) = 16 - 3 = 13$

Division

With division, we need to rationalize the denominator. If we are dividing one number by a complex number, then we have a square root in the denominator (since $i = \sqrt{-1}$). You have learned in previous algebra courses that when there is a square root in the denominator you must rationalize it. With complex numbers we want the final form to be $(a + bi)$ where a or b, or both, might be fractions. When we have a complex number we *don't* want it to be a single fraction .

Example 5: Dividing Complex Numbers

Simplify the division of these two complex numbers $\frac{1+2i}{3-4i}$.

Solution:

Answer Q3: $(2-5i)-(7-8i) = (2-7) + (-5i-(-8i)) = -5 + 3i$.

First, multiply both the numerator and the denominator by the conjugate of 3 – 4i which is 3 + 4i.	$\frac{1+2i}{3-4i} \bullet \frac{3+4i}{3+4i}$
The denominator turns out to be the difference of two squares and so the i's vanish.	$\frac{(1+2i)(3+4i)}{9+12i-12i-16i^2} = \frac{(1+2i)(3+4i)}{9-16i^2}$
We foil the numerator and notice how the i^2 causes a sign change.	$\frac{3+4i+6i+8i^2}{9+16} = \frac{3+10i-8}{25} = \frac{-5+10i}{25}$
Now simplify what is left and write it in complex number form $(a + bi)$.	$\frac{5(-1+2i)}{25} = \frac{-1+2i}{5} = \frac{-1}{5} + \frac{2}{5}i$

Here are a few more examples to help you understand the required algebraic steps.

Example 6: Dividing Complex Numbers

Divide these two complex numbers $\frac{-2+3i}{-5+7i}$.

Solution:

First, multiply both the numerator and the denominator by the conjugate of $-5+7i$ which is $-5-7i$.	$\frac{-2+3i}{-5+7i} \bullet \frac{-5-7i}{-5-7i}$
The denominator turns out to be the difference of two squares and so the *i*'s vanish.	$\frac{(-2+3i)(-5-7i)}{25+35i-35i-49i^2} = \frac{(-2+3i)(-5-7i)}{25-49i^2}$
We foil the numerator and notice how the i^2 causes a sign change.	$\frac{10+14i-15i-21i^2}{25+49} = \frac{10-i+21}{74} = \frac{31-i}{74}$
Now simplify what is left and write it in complex number form ($a+bi$).	$\frac{31-i}{74} = \frac{31}{74} - \frac{1}{74}i$

Answer Q4:
$(-1+i)(4-2i)$
$= -4+2i+4i-2i^2$
$= -4+6i-2(-1)$
$= -2+6i$.

Example 7: More Division of Complex Numbers

Divide these two complex numbers $\frac{7-2i}{i}$.

Solution:

First, multiply both the numerator and the denominator by the conjugate of i which is –i.	$\frac{7-2i}{i} \bullet \frac{-i}{-i}$
The denominator turns out to be just (i) times ($-i$).	$\frac{(7-2i)(i)}{-i^2} = \frac{(7-2i)(i)}{1}$
Now, by distributing the numerator and simplifying, we're done.	$\frac{(7-2i)(i)}{1} = \frac{7i-2i^2}{1} = \frac{7i+2}{1} = 2+7i$

Question 5: What is the answer to $\frac{-2+i}{3+2i}$?

FRACTALS

If you have ever heard of **fractals**, then you have already heard of a place where complex numbers are used. Some fractal pictures are created by using a function $f(z)$, where the domain contains numbers of the form $a+bi$. These numbers come from a **defined area** (a box drawn in the complex plane) and are plugged into the function $f(z)$.

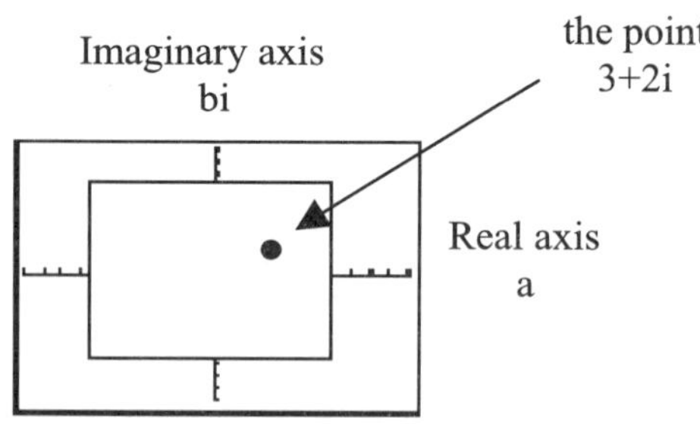

Discussion 2: Fractal Calculations

Let's take the function $f(z) = z^2 + c$ (Mandelbrot Set), where c is a point in the complex plane of the form (a, b), from $c = a + bi$. We will start with the point $c = (1.1, 0.6)$, which represents a complex number, in this case $1.1 + 0.6i$. Every point (a, b) in our defined area is of the form $a + bi$.

Then we use that number $(1.1 + 0.6i)$ as our first value plugged into the function $f(z)$ to get a new complex number: $(1.1 + 0.6i)^2 + (1.1 + 0.6i) = 1.95 + 1.92i$. We then plug this new number back into the original function to get another complex number: $(1.95 + 1.92i)^2 + (1.1 + 0.6i) = 1.2161 + 8.088i$. We keep plugging our answer back into the function again and again and again, until our answer ends up outside of the defined area. The number of times it takes to end up outside of the defined area will determine the color given to the original point you started with. Now do this with every point in the defined area and you will get the following pictures (we used a program call fractint, found on the internet, to create these pictures). Each different function will yield a different picture.

Mandelbrot Set $f(z) = z^2 + c$ c is the number you start with	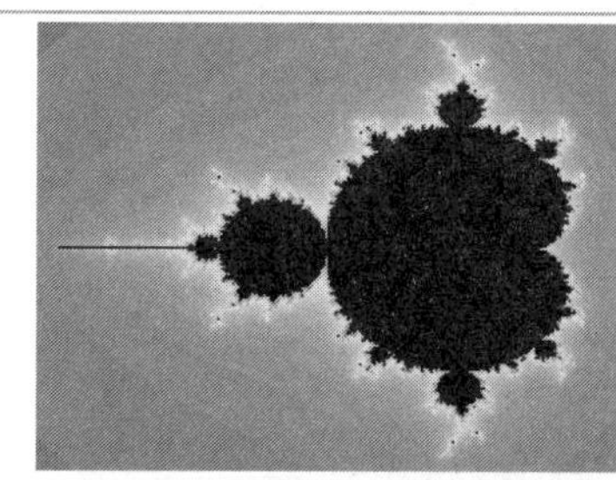
Julia Set $f(z) = z^2 + (-0.20 + 0.75i)$	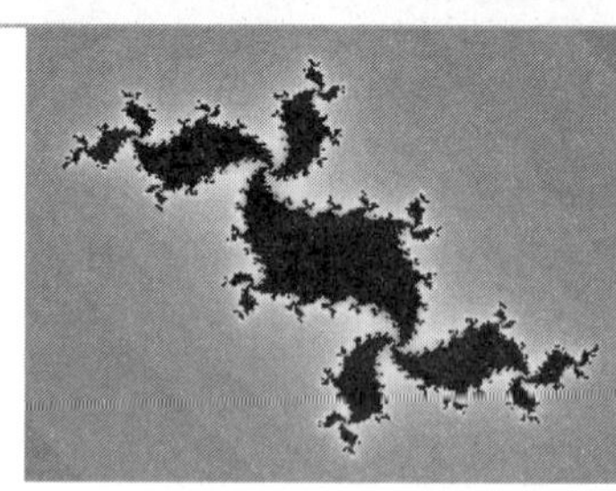

Julia Set $f(z) = 2.97\cos(z)$	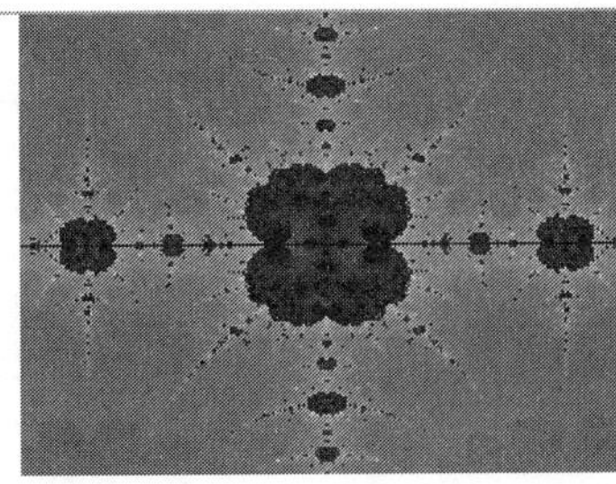

Below are other kinds of fractals created in a different way. The following pictures are just a sample of what you can get by using different methods.

Fern Iterated Function System Deterministic	
Sierpinski's triangle Iterated Function System Random	

Answer Q5:

$$\frac{-2+i}{3+2i} = \frac{-2+i}{3+2i}\cdot\frac{3-2i}{3-2i} = \frac{(-2+i)(3-2i)}{3^2-(2i)^2} = \frac{-6+4i-3i-2i^2}{9-4i^2} = \frac{-4+i}{9+4} = \frac{-4}{13}+\frac{1}{13}i$$

A defining property of fractals is that they are usually **self-similar**. That is, if you zoom in on a part of them, you get a picture that looks like the first one. Keep zooming in and you just keep on getting the same looking picture. They also are usually infinitely complex and detailed. A formal definition of fractals would involve something called **fractional dimension**.

Fractals have a dimension that is not a whole value. For example, a line is one-dimensional; a plane is two-dimensional and a cube is three-dimensional. But fractals can be 1.23487 dimensional, hence the name fractals (a fraction of). Fractals are being used everywhere. They are used in biology, medicine, and mapping, just to name a few areas. Let's work within this setting of fractals to practice our ability to work with complex numbers.

Example 8: Calculating Iterations

How many times must we keep recomputing an output, starting with the number $(.5 + i)$, until the output goes outside the defined area, or window, (x range $[-2, 2]$ and y range $[-2, 2]$) for the Julia set $f(z) = z^2 - 1$?

Solution:

Here is a picture of the fractal created by this function:

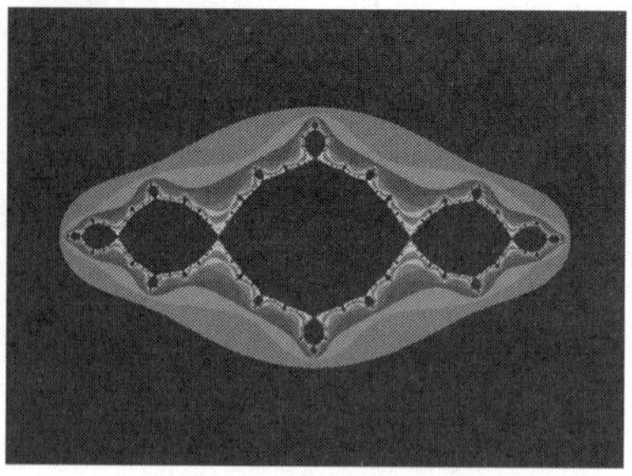

First Iteration (that means the first time the number is plugged into the function)	$(0.5 + i)^2 - 1 = -1.75 + i$, point (−1.75, 1)
Second Iteration	$(-1.75 + i)^2 - 1 = 1.0625 - 3.5i$, point (1.0625, −3.5) → outside the window

So it took two iterations for the point $(0.5 + i)$ to escape the defined area.

Question 6: How many times must we keep re-computing an output starting with the number $(.5 - 0.5i)$ until the output goes outside the defined area, or window, (x range [−2, 2] and y range [−2, 2]) for the Julia set $f(z) = z^2 - 1$?

Section Summary

- We must convert any square root with negative numbers into its imaginary form before we can perform mathematical operations such as multiplication.
- Also, the adding, subtracting, and multiplying of complex numbers is just like operations with functions; you simply combine like terms as always.
- With division involving a complex number in the denominator, we must multiply by a conjugate in order to eliminate the square root in the denominator.

SECTION 2.1 HOMEWORK

(1–8) Write each of the following as a complex number in standard form $a + bi$.

1. $\sqrt{-25}$ 2. $\sqrt{-36}$ 3. $\sqrt{-40}$ 4. $\sqrt{-32}$

5. $5 + \sqrt{-49}$ 6. $7 - \sqrt{-4}$ 7. $\dfrac{8 + \sqrt{-12}}{2}$ 8. $\dfrac{9 + \sqrt{-18}}{3}$

(9–12) Understanding Complex Numbers:

9. How would you write 8 as a complex number in standard form?

10. How would you write $-3i$ as a complex number in standard form?

11. When is $a + bi$ a real number?

12. When is $a + bi$ an imaginary number?

(13 –48) Simplify and write the answers in standard form $a + bi$.

13. $(5 + 9i) + (8 - 4i)$ 14. $(6 - 11i) + (8 + 3i)$ 15. $(4 - 7i) - (9 - 10i)$

16. $(9 + 6i) - (12 - 4i)$ 17. $\left(3 + \sqrt{-9}\right) + \left(5 - \sqrt{-25}\right)$ 18. $\left(2 - \sqrt{-36}\right) - \left(7 - \sqrt{-49}\right)$

19. $5(3 - 2i)$ 20. $8(2 + 4i)$ 21. $3i(6 + 3i)$

22. $2i(5 - 9i)$ 23. $\left(\sqrt{-5}\right)\left(\sqrt{-20}\right)$ 24. $\left(\sqrt{-12}\right)\left(\sqrt{-20}\right)$

25. $\left(3\sqrt{-8}\right)\left(2\sqrt{-2}\right)$ 26. $\left(5\sqrt{-10}\right)\left(-3\sqrt{-40}\right)$ 27. $(5 - 3i)(2 + 5i)$

28. $(3 + 2i)(2 + 9i)$ 29. $(3 + 2i)(2+3i)$ 30. $(5 - 2i)(2 + 5i)$

31. $\left(3 + \sqrt{-4}\right)\left(2 - \sqrt{-16}\right)$ 32. $\left(5 - \sqrt{-9}\right)\left(7 - \sqrt{-1}\right)$ 33. $(3 - 2i)(3 + 2i)$

34. $(5 - 4i)(5 + 4i)$ 35. $(6 + 8i)(6 - 8i)$ 36. $(5 + 12i)(5 - 12i)$

(37–44) Simplify:

37. i^{12} 38. i^{27} 39. i^{33} 40. i^{22}

41. i^{34} 42. i^{16} 43. i^{19} 44. i^{14}

(45–50) Answer questions (a) and (b) for the following:

a.) Give the complex conjugate of each of the following.

b.) Multiply together the conjugates and give the answer in standard form $a + bi$.

45. $3 - 2i$ 46. $5 - 3i$ 47. $2 + 3i$

48. $3 + 7i$ 49. $3i$ 50. $-2i$

(51–62) Simplify and write each of the following in standard form $a + bi$.

51. $\dfrac{3}{5 + 2i}$ 52. $\dfrac{-2}{4 - 3i}$ 53. $\dfrac{3 + 2i}{2 - 3i}$

54. $\dfrac{5 - i}{3 + 2i}$ 55. $\dfrac{3i}{4 - 7i}$ 56. $\dfrac{2i}{3 + 8i}$

57. $\dfrac{2 + 3i}{5i}$ 58. $\dfrac{5 - 2i}{3i}$ 59. $\dfrac{3 + \sqrt{-64}}{2 + \sqrt{-9}}$

60. $\dfrac{2-\sqrt{-4}}{5+\sqrt{-1}}$ 61. $\dfrac{3+\sqrt{-8}}{5+\sqrt{-2}}$ 62. $\dfrac{3-\sqrt{-20}}{2-\sqrt{-5}}$

(63–66) Applications:

An equation used in electronics is $V = IR$; this is called ohm's law. V represents Voltage (in volts); I represents current (in amperes); and R represents resistance (in ohms).

63. Find V, when $I = 5 - 3i$ amperes and $R = 5 + 7i$ ohms
64. Find V, when $I = 7 - 2i$ amperes and $R = 4 + 9i$ ohms
65. Find R when $I = 5 - 2i$ amperes and $V = 31 + 5i$ volts
66. Find I when $R = 3 + 5i$ ohms and $V = 30 + 16i$ volts

(67-68) Fractals

67. Given the Julia Set $f(z) = z^2 - 2$ with x range $[-4, 4]$ and y range $[-4, 4]$ find the number of iterations it takes to get outside this given window for the given starting value.
    a. Starting value of $0.3 + i$.
    b. Starting value of $0.4 - 0.5i$
68. Given the Mandelbrot Set $f(z) = z^2 + 1$ with x range $[-8, 8]$ and y range $[-8, 8]$ find the number of iterations it takes to get outside this given window for the given starting value.
    a. Starting value of $1 - i$.
    b. Starting value of $0.5 - 0.5i$.

Answer Q6: Four iterations are required. $(0.5-0.5i)^2-1 = -1-0.5i$, $P_1(-1, -0.5)$ $(-1-0.5i)^2-1 = -0.25+1.9375i$, $P_2(-0.25, 1.9375)$ $(-0.25+1.9375i)^2-1 = -1.9375-0.5i$, $P_3(-1.9375, 0.5)$ $(-1.9375-0.5i)^2-1 = 2.5039+1.9375i$, $P_4(2.5039, 1.9375)$ outside the defined area

Section 2.2 Solving Quadratic Equations

We are now going to solve equations where some of the solutions might be complex numbers. Polynomials, as you might remember, are any combination (+,–, ·) of variables to whole number powers with real coefficients (reviewed in chapter 0). The most basic polynomial functions would be:

Just a number (constant)	$f(x) = c$	(no variable)
Linear (first power)	$f(x) = bx + c$	
Quadratic (second power)	$f(x) = ax^2 + bx + c$	

In this section we are going to concentrate on **quadratic** polynomial functions. Specifically, we are going to concentrate on how to solve quadratic equations (of the form $ax^2 + bx + c = 0$). You have most likely seen these before, so this will just serve as a refresher for you. Here in this section we are going to describe three different methods that can be used to solve quadratic equations.

SOLVE BY FACTORING

We will refer to the first method as the **factoring method**. In the factoring method you need to get all the terms on the same side of the equal sign. This will leave a zero on the other side. From there you factor the quadratic by using the different methods discussed in chapter 0.3. These different factoring techniques are:

1.) Greatest Common Factor	Factor out anything in common.
2.) Trinomial Factoring	Factor into two binomials.
3.) Special Binomials	Factor difference of square and sum or difference of cubes.
4.) Factor by Grouping	When you have four or more terms, group into two groups and try factoring each group.

Then set each factor equal to zero and solve the two linear equations to find the solutions.

We'll do a few examples that can be solved by the factoring method. Next to each example is the graph of the quadratic function. Make a note of how the real solutions we find by algebraic methods show up as x-intercepts on the graph. This happens because we are solving $y = 0$ where y is our quadratic function. This is exactly how you find x-intercepts (let $y = 0$). The graphs will give us a quick way to confirm that our algebraic solutions are correct.

Example 1: Solving by Factoring

Solve the following equations: a) $x^2 - 2x - 15 = 0$; b) $12y^2 - 41y = -24$; c) $x(3x - 23) = 8$.

Solutions:

a.) $x^2 - 2x - 15 = 0$

First, factor the left side of the equation.	$(x-5)(x+3)=0$	Notice that when you graph a function on the graphing calculator the formula is displayed on the screen.
Second, set each factor equal to 0.	$x-5=0$ or $x+3=0$	
Last, solve each equation.	$x=5$ or $x=-3$	

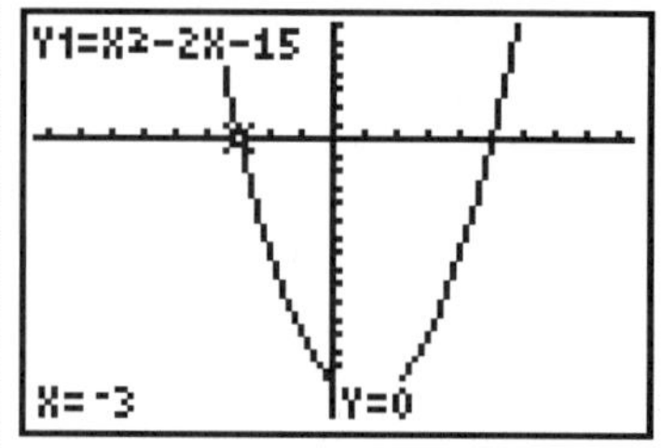

b.) $12y^2 - 41y = -24$

First, add 24 to get all terms on the same side of the equation and to ensure that the y^2 term is positive.	$12y^2 - 41y + 24 = 0$	
Second, factor the left side of the equation.	$(3y-8)(4y-3)=0$	
Third, set each factor equal to 0.	$3y-8=0$ or $4y-3=0$	
Last, solve each equation.	$x=\frac{8}{3}$ or $x=\frac{3}{4}$	

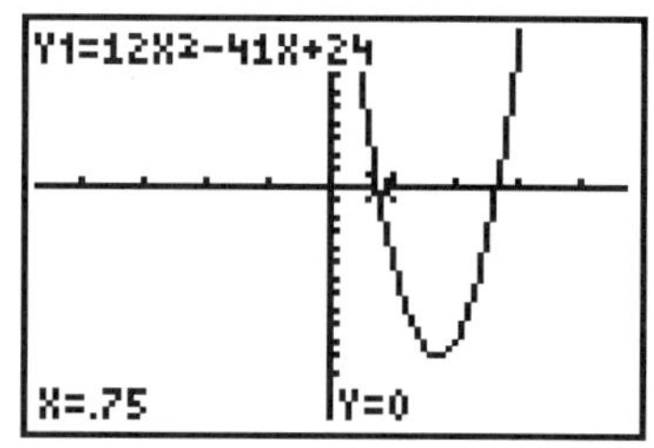

c.) $x(3x-23)=8$

First, distribute x through the parentheses.	$3x^2 - 23x = 8$	
Second, add 8 to get all terms on the same side of the equation and to ensure that the x^2 term is positive.	$3x^2 - 23x - 8 = 0$	
Third, factor the left side of the equation.	$(3x+1)(x-8)=0$	
Fourth, set each factor equal to 0.	$3x+1=0$ or $x-8=0$	
Last, solve each equation.	$x=\frac{-1}{3}$ or $x=8$	

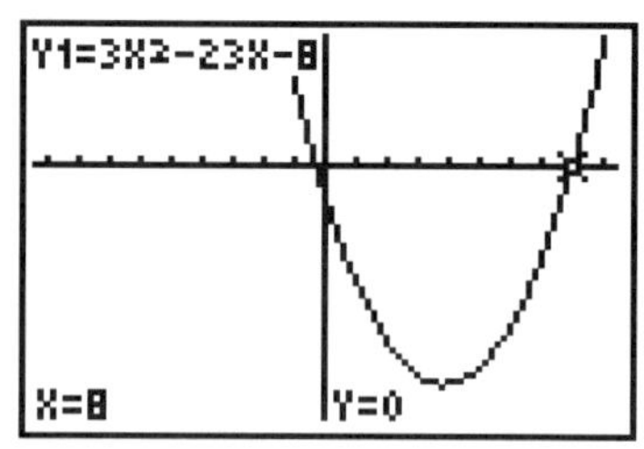

Question 1: Solve $10z^2 - 13z - 3 = 0$?

The main problem with the factoring method is that it can be hard to factor some quadratics and, in fact, some are virtually impossible to factor. So let's look at another method.

SOLVE BY TAKING THE SQUARE ROOT

We'll call the next method the **Square Root Method**. In order for this method to work you must have your quadratic equation in the form of $(bx + c)^2 = d$ (a perfect binomial square equal to a number). This form can be simple, like $x^2 = 4$, to something more complicated, like $(3x-7)^2 = 12$ (a linear expression squared equal to a number). If your equation doesn't fit this mold, then this method won't work. So, just like the factor method, this will not always work. To use this method, you must get the squared piece alone on one side of the equal sign, then take the square root of both sides. Finally solve the resulting two equations. (Don't forget the plus or minus symbol ($\pm$), because $\sqrt{x^2} = |x|$).

Let's look at a few examples that can be solved by this square root method. Next to each example is the graph of the quadratic equation set equal to 0. You will notice that if the answers are irrational, it is hard to know what they are exactly. The calculator can only estimate them for you. Complex solutions on the other hand don't even show up (i.e. the calculator can't show them to you on the graph).

Example 2: Solve by Taking the Square Root

Solve the following equations: a) $4x^2 - 20 = 0$; b) $3x^2 = -12$; c) $(2x - 5)^2 - 49 = 0$; d) $(x - 4)^2 = 0$.

Solutions:

a.) $4x^2 - 20 = 0$

First, isolate the x^2 term by adding 20 to both sides of the equation.	$4x^2 = 20$	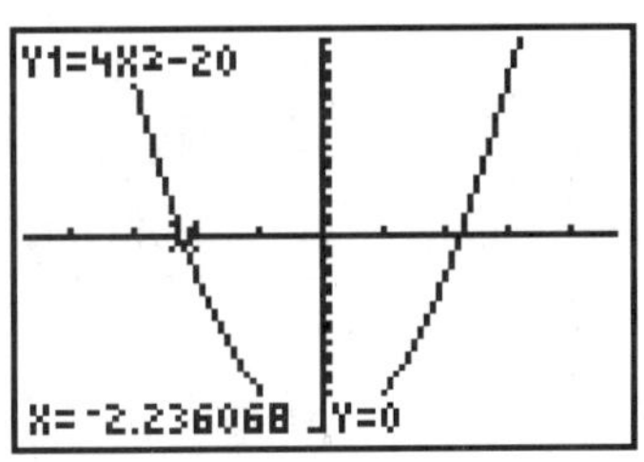
Second, divide by 4 to get just x^2	$x^2 = 5$	
Third, take the square root of both sides.	$x = \pm\sqrt{5}$	

b.) $3x^2 = -12$

First, divide by 3 to get just x^2.	$x^2 = -4$	

Second, take the square root of both sides.	$x = \pm\sqrt{-4}$	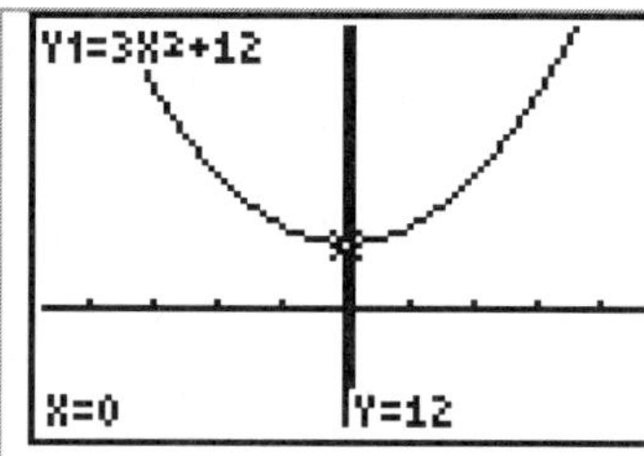
Last, remember that the square root of a negative number gives us an i.	$x = \pm 2i$	

c.) $(2x-5)^2 - 49 = 0$

First, isolate the squared term by adding 49 to both sides of the equation.	$(2x-5)^2 = 49$	
Second, take the square root of both sides.	$(2x-5) = \pm 7$	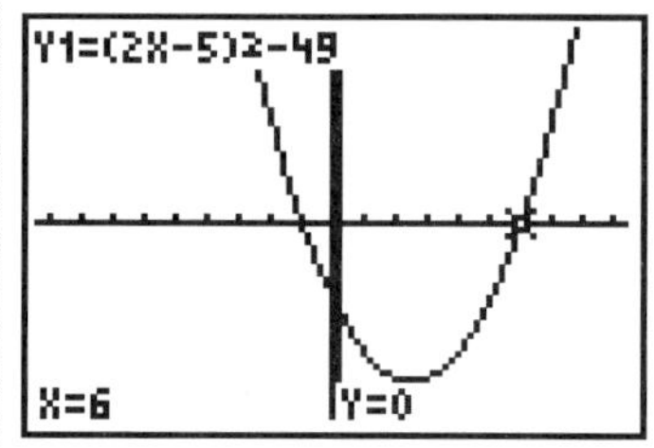
Third, solve the two resulting equations.	$2x-5=7$ or $2x-5=-7$	
Answers:	$x = \frac{12}{2} = 6$ or $x = \frac{-2}{2} = -1$	

d.) $(x-4)^2 = 0$

First, the equation is already equal to 0 so we take the square root of both sides.	$\sqrt{(x-4)^2} = \pm\sqrt{0} \Rightarrow x-4=0$	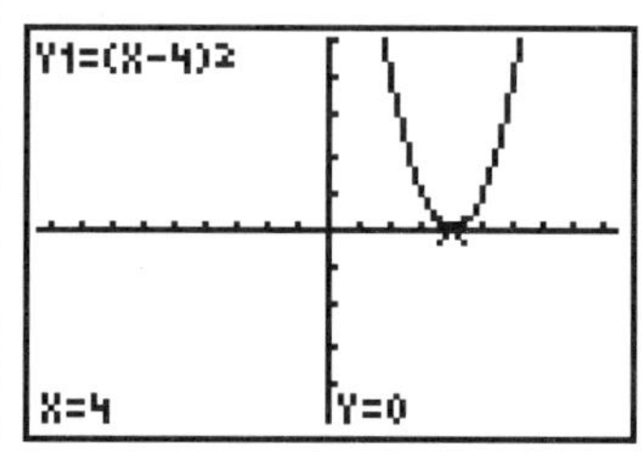
Second, add 4 to both sides.	$x = 4$	

Notice that we got only one answer this time. (We might refer to this as a double root or as having **multiplicity** 2, but we will talk more about this in section 3.2.)

Question 2: Solve $(x-3)^2 = 25$?

COMPLETING THE SQUARE

We need to take a moment here to talk about how to "complete the square" before we can present the third method used for solving quadratic equations. Completing the square is used to change an equation from the form $ax^2 + bx + c = 0$ into the form $(x+m)^2 + k = 0$ (a perfect binomial square plus a number) which can then be solved by the square root method.

If you want an equation in the form of a binomial square, you first need to create a perfect square trinomial by completing the square on it and then factoring. Let's do a few examples to show you what we mean.

Example 3: Completing the Square

Complete the square on the following problems in order to change them into binomial squares.

a) $x^2 - 6x$ b) $y^2 + 8y$ c) $-2x^2 + 4x$ d) $\frac{1}{3}y^2 + 5y$

Solutions:

Now, in each part of this problem we are looking to add some amount to what we already have in order to create a perfect square trinomial, which then can be factored into a binomial square. This process is called completing the square. In part (a) we need to take $x^2 - 6x$ and add some number to it to get $x^2 - 6x + c$, which will then factor into $(x - h)^2$. Since we know that $(x - h)^2$ is equivalent to $x^2 - 2hx + h^2$ by multiplying it out, we will do the following.

Answer Q1: $10z^2 - 13z - 3 = (5z+1)(2z-3) = 0$ so, $5z + 1 = 0$ or $2z - 3 = 0$ thus, $z = \frac{-1}{5}$ or $\frac{3}{2}$

a) $x^2 - 6x$

What we need to get	What we have
$x^2 - 2hx + h^2$	$x^2 - 6x$
First we look at the fact that we need $-2h$	to be equal to -6
Clearly h must be 3 (notice that h will always be half of the number, in this case 6, because of the 2 in the formula $-2h = -6 \rightarrow 2h = 6 \rightarrow h = 3$).	
Therefore, $h^2 = 9$.	So we need to add 9 ($c = 9$).
By adding 9, what we had becomes	$x^2 - 6x + \mathbf{9} = (x - 3)^2$

We've completed the square. We have added 9 to $x^2 - 6x$ and it then factored into a binomial square $(x - 3)^2$ which we can solve easily.

Question 3: Try completing the square on part (b) above.

You should notice that the binomial square has the same sign as the linear term in the original expression (– in part (a) from the $-6x$, and + in part (b) from the $+8y$). Also, the h value in the final binomial square is always half of the coefficient on the linear term (-3 from the $-6x$, and $+4$ from the $+8y$).

In part (c) we need to factor out the coefficient on the squared term first, before we can begin to complete the square. Remember, we need $x^2 - 2hx + h^2$ so that it will factor into $(x - h)^2$. As you can see, this will not allow for any coefficient other than a one to be in front of the squared term.

c) $-2x^2 + 4x$

What we need to get	What we have
$x^2 - 2hx + h^2$	$-2x^2 + 4x$
First, factor out a factor of –2. This gives us a coefficient of 1; thus we can now complete the square on the part inside the parentheses.	$-2(x^2 - 2x)$
$-2h$	needs to be equal to -2
Clearly h must be 1. Therefore, h^2 must be 1. So, we need to add 1 ($c = 1$) in the parentheses	
By adding 1 inside the parentheses, what we had becomes	$-2(x^2 - 2x + \mathbf{1}) = -2(x-1)^2$

We have now completed the square.

In part (d) we need to factor out the coefficient on the squared term first, before we can begin to complete the square. We need $y^2 + 2ky + k^2$ so that it will factor into $(y+k)^2$. (We have used the letter k instead of h because of what you will see in section 7.4)

d) $\frac{1}{3}y^2 + 5y$

Answer Q2: Square root both sides and get $x - 3 = \pm 5$. Now that gives us two equation $[(x - 3 = 5), (x - 3 = -5)]$ where we need to add 3 to solve for the x in both equations giving us $x = 8$ and $x = -2$

What we need to get	What we have
$y^2 + 2ky + k^2$	$\frac{1}{3}y^2 + 5y$
First, factor out a factor of $\frac{1}{3}$. This gives us a one coefficient; thus we can now complete the square on the part inside the parentheses.	$\frac{1}{3}(y^2 + 15y)$
$2k$	needs to be equal to 15
So k must be $\frac{15}{2}$. Therefore, k^2 must be $\frac{225}{4}$. So we need to add $\frac{225}{4}$ ($c = \frac{225}{4}$) in the parentheses	
By adding $\frac{225}{4}$ inside the parentheses, what we had becomes	$\frac{1}{3}(y^2 + 15y + \frac{225}{4}) = \frac{1}{3}(y + \frac{15}{2})^2$

To summarize, in order to complete the square you will need to do the following steps:

- Concentrate on just the squared term and the linear term.
- Factor out the coefficient on the squared term in the expression, if there is one.

- Take half of the coefficient on the linear term and square it and add that amount to your expression.
- Now factor the perfect square trinomial, which will always be the variable plus the half amount from the previous step.

Question 4: Complete the square on the expression $5x^2 + 15x$.

SOLVING USING THE QUADRATIC FORMULA

The last method described in this section uses the **Quadratic Formula** to solve quadratic equations. First let's show you how the formula is derived.

First, start with the standard form of a quadratic equation.	$ax^2 + bx + c = 0$
Second, divide both sides by a, the coefficient in front of the squared variable.	$x^2 + \frac{b}{a}x + \frac{c}{a} = 0$
Third, subtract $\frac{c}{a}$ from both sides.	$x^2 + \frac{b}{a}x = -\frac{c}{a}$
Fourth, add half of $\frac{b}{a}$ squared, to both sides $\left(\frac{b}{2a}\right)^2$. This comes from completing the square, which was just discussed.	$x^2 + \frac{b}{a}x + \left(\frac{b}{2a}\right)^2 = -\frac{c}{a} + \left(\frac{b}{2a}\right)^2$
Fifth, the left side now factors as a perfect square trinomial and the right side can be simplified.	$(x + \frac{b}{2a})^2 = \frac{-4ac}{4a^2} + \frac{b^2}{4a^2}$ $(x + \frac{b}{2a})^2 = \frac{b^2 - 4ac}{4a^2}$
Sixth, square root both sides since it now fits the form for the square root method already discussed.	$x + \frac{b}{2a} = \frac{\pm\sqrt{b^2 - 4ac}}{2a}$
Lastly, subtract $\frac{b}{2a}$ from both sides to solve for x.	$x = \frac{-b \pm \sqrt{b^2 - 4ac}}{2a}$

Answer Q3:
Need
$y^2 + 2ky + k^2$
Have
$y^2 + 8y$
"k" must be 4

Therefore,
k^2 must be 16
So we need to add 16 ($c = 16$) and get
*$y^2 + 8y$ **+ 16***
$= (y + 4)^2$

We used the square root method in conjunction with completing the square to solve the general form of a quadratic equation.

Quadratic Formula – any quadratic equation of the form ($ax^2 + bx + c = 0$) with real coefficients (a, b, c) can now be solved with this formula

$$x = \frac{-b \pm \sqrt{b^2 - 4ac}}{2a}.$$

Notice that this means ***any*** quadratic equation can be solved with this formula. The formula does look a little complicated, but it ALWAYS WORKS. Let's do a few examples to show you how the Quadratic Formula is used.

Example 4: Solve by Quadratic Formula

Solve the following equations: a) $x^2 + 2x = 15$; b) $2x^2 - 10x - 3 = 0$; c) $4x^2 - 3x + 9 = 0$

Solutions:

a.) $\mathbf{x^2 + 2x = 15}$

First, get equal to 0 by subtracting 15.	$x^2 + 2x - 15 = 0$	
Second, determine what *a*, *b*, and *c* are	$a = 1, b = 2, c = -15$	
Third, plug these numbers into the formula.	$x = \dfrac{-(2) \pm \sqrt{(2)^2 - 4(1)(-15)}}{2(1)}$	
Next, simplify the formula.	$x = \dfrac{-2 \pm \sqrt{4 + 60}}{2} = \dfrac{-2 \pm 8}{2}$	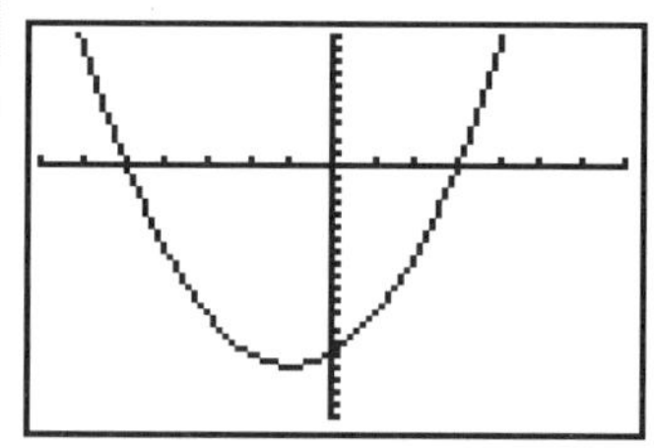
You can see the answers on the graph; they are where the graph crosses the *x*-axis.	$x = \dfrac{-2+8}{2} = 3$ or $x = \dfrac{-2-8}{2} = -5$	

You should realize that since the answers are rational numbers $\{3, -5\}$, the original problem could have been factored ($x^2 + 2x - 15 = (x - 3)(x + 5) = 0$). Whenever you are not instructed on how to do a particular problem, you should first try to do it the easiest way possible.

b.) $\mathbf{2x^2 - 10x - 3 = 0}$

First, determine what *a*, *b*, and *c* are since it is already equal to 0.	$a = 2, b = -10, c = -3$	
Second, plug these numbers into the formula.	$x = \dfrac{-(-10) \pm \sqrt{(-10)^2 - 4(2)(-3)}}{2(2)}$	
Third, simplify the formula.	$x = \dfrac{10 \pm \sqrt{100 + 24}}{4} = \dfrac{10 \pm \sqrt{124}}{4}$	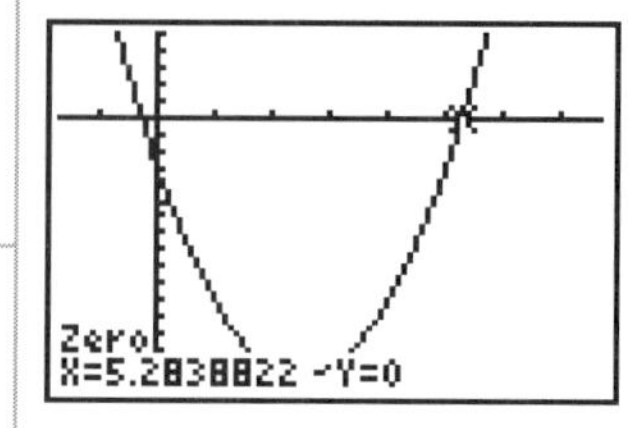
Next, simplify the square root.	$x = \dfrac{10 \pm \sqrt{4 \cdot 31}}{4} = \dfrac{10 \pm 2\sqrt{31}}{4}$	

Answers. You can see the answers on the graph; they are where the graph crosses the x-axis.	$x = \frac{10+2\sqrt{31}}{4} = \frac{5+\sqrt{31}}{2}$ or $x = \frac{10-2\sqrt{31}}{4} = \frac{5-\sqrt{31}}{2}$	$\frac{5+\sqrt{31}}{2} \approx 5.28388\ldots$

The answers are not rational so you *couldn't* have done this problem by factoring. Also the graphing calculator can only give you an *approximate* answer by using the zero/root command.

c.) $\mathbf{4x^2 - 3x + 9 = 0}$

First, determine what a, b, and c are since it is already equal to 0.	$a = 4,\ b = -3,\ c = 9$	
Second, plug these numbers into the formula.	$x = \frac{-(-3) \pm \sqrt{(-3)^2 - 4(4)(9)}}{2(4)}$	
Third, simplify the formula.	$x = \frac{3 \pm \sqrt{9-144}}{8} = \frac{3 \pm \sqrt{-135}}{8}$	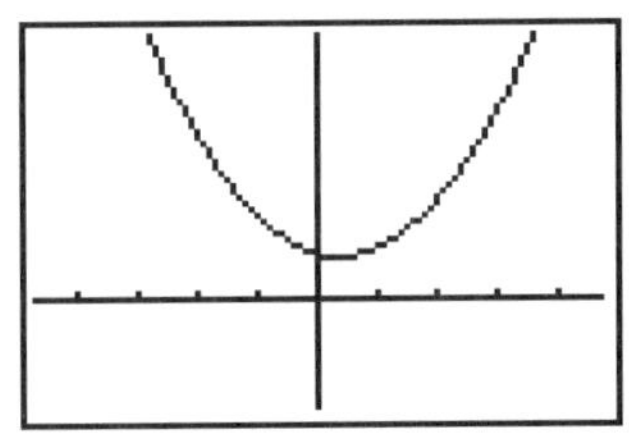
Next, simplify the square root.	$x = \frac{3 \pm \sqrt{-9 \bullet 15}}{8} = \frac{3 \pm 3i\sqrt{15}}{8}$	
Answers. Notice the graph doesn't cross the x-axis.	$x = \frac{3+3i\sqrt{15}}{8}$ or $\frac{3-3i\sqrt{15}}{8}$	

Answer Q4: Factor out 5 to get $5(x^2 + 3x)$. Now, complete the square in the parentheses by taking half of 3 and squaring it and adding the result $5(x^2 + 3x + \mathbf{9/4}) = 5(x + 3/2)^2$

The answers are not rational so you *couldn't* have done this problem by factoring. In fact, the answers were complex, so you don't even see the graph cross the x-axis. No real number (input) can be plugged into the x's and cause an output equal to zero ($y = 0$). This means that the function has **no real roots**.

As you have just seen, some of the work needed to solve quadratic equations using the quadratic formula is quite involved. The moral of the story is to first see if you can factor or use the square root method. If these don't easily work, then use the quadratic formula.

Question 5: Solve $x(3x + 8) = -2$ using the quadratic formula.

There are going to be several places throughout this book where you will need to solve a quadratic equation. So be sure that you are comfortable with these three methods, especially with the quadratic formula since it will always get an answer for you.

Section Summary

- When you see an equation where the highest power of a variable is 2 (a second degree equation), then you know that it is a quadratic equation.
- In order to complete the square you will need to do the following steps:
  1. Concentrate on the squared and linear terms (those involving a variable).
  2. Factor out the coefficient on the squared term in the expression, if there is one.
  3. Take half of the coefficient on the linear term, square it and add that amount to your expression.
  4. Now factor the perfect square trinomial, which will always be the variable plus the half amount from the previous step.
- When solving quadratic equations
  1. You want to first see if you can use the square root method.
  2. If the problem is not of that type, then get it equal to zero and try to factor it.
  3. If after 30 to 60 seconds you haven't figured out how to factor it, or it can't be factored, then use the quadratic formula.

SECTION 2.2 HOMEWORK

(1–52) Solve by any method that is appropriate: **Factor, Square Root Method, Quadratic Formula.** (Use your graphing calculator to confirm your answer.)

1. $x^2 - 6x - 7 = 0$
2. $x^2 + 3x - 10 = 0$
3. $3x^2 + 14x + 15 = 0$
4. $2x^2 - 17x + 21 = 0$
5. $9x^2 - 12x + 4 = 0$
6. $25x^2 + 40x + 16 = 0$
7. $3x^2 = 24$
8. $2x^2 = 14$
9. $5x^2 - 1 = 79$
10. $3x^2 + 7 = 34$
11. $4x^2 + 7 = 3$
12. $2x^2 + 9 = 1$
13. $(3x-5)^2 = 9$
14. $(2x+3)^2 = 36$
15. $(3x+1)^2 = 5$
16. $(5x-1)^2 = 6$
17. $(2x-1)^2 + 10 = 1$
18. $(4x+3)^2 + 17 = 1$
19. $x^2 - 3x - 5 = 0$
20. $x^2 + 4x - 3 = 0$
21. $2x^2 + 5x - 1 = 0$
22. $3x^2 + 5x - 7 = 0$
23. $x^2 - 6x + 10 = 0$
24. $x^2 + 10x + 26 = 0$
25. $4x^2 - 16x + 17 = 0$
26. $25x^2 - 20x + 5 = 0$
27. $3x^2 - 11x + 12 = 0$
28. $5x^2 + 9x + 7 = 0$
29. $5x^2 - 35x - 40 = 0$
30. $7x^2 - 21x - 70 = 0$
31. $24x^2 - 16x + 40 = 0$
32. $33x^2 - 55x + 99 = 0$
33. $30x^2 - 36x - 12 = 0$
34. $8x^2 + 44x + 12 = 0$
35. $\frac{3}{4}x^2 - \frac{1}{2}x - \frac{1}{4} = 0$
36. $\frac{5}{8}x^2 - \frac{3}{4}x - 1 = 0$
37. $\frac{3}{5}x^2 + \frac{2}{3}x - 1 = 0$
38. $\frac{3}{4}x^2 + 5x - \frac{3}{8} = 0$
39. $\frac{5}{3}x^2 + \frac{3}{2}x + \frac{5}{6} = 0$
40. $\frac{3}{2}x^2 - \frac{1}{4}x + \frac{1}{2} = 0$
41. $x^2 + 7x = 2$
42. $5x^2 - 3x = 4$
43. $2x^2 = 3x - 5$
44. $3x^2 + 3 = 2x$
45. $\frac{3}{4}x^2 = \frac{7}{8}x + \frac{1}{2}$
46. $\frac{2}{3}x^2 - \frac{1}{3} = \frac{5}{9}x$
47. $\sqrt{2}x^2 + 3x + \sqrt{2} = 0$
48. $\sqrt{3}x^2 + 4x + \sqrt{3} = 0$
49. $2x^2 + \sqrt{5}x - 3 = 0$
50. $3x^2 + \sqrt{2}x - 1 = 0$
51. $3x^2 + 5ix - 2 = 0$
52. $x^2 + 6ix - 5 = 0$

Answer Q5:
$3x^2 + 8x + 2 = 0$
then x =

$$\frac{-8 \pm \sqrt{64 - 4(3)(2)}}{6} = \frac{-8 \pm \sqrt{40}}{6} = \frac{-8 \pm 2\sqrt{10}}{6} = \frac{-4 \pm \sqrt{10}}{3}$$

(53–56) Use your calculator to solve each quadratic equation to three decimal places:

53. $2.58x^2 - 3.75x - 2.83 = 0$
54. $3.73x^2 + 9.74x + 2.34 = 0$
55. $5.3x^2 + 7.08x + 1.02 = 0$
56. $3.04x^2 + 4.35x + 1.234 = 0$

(57–66) Complete the square for each of the following quadratics.

57. $x^2 + 8x +$ ____
58. $x^2 - 12x +$ ____
59. $x^2 - 5x +$ ____
60. $x^2 + 11x +$ ____
61. $2x^2 + 12x +$ ____
62. $3x^2 - 12x +$ ____
63. $-x^2 + 16x +$ ____
64. $-x^2 - 24x +$ ____
65. $2x^2 - 9x +$ ____
66. $-3x^2 - 11x +$ ____

(67-74) Using the Quadratic equations to solve mathematical modeling problems.

67. A rectangular plot of land has the short side one-third the length of the long side. If the area of the plot of land is 120,000 ft^2, what are the dimensions of the lot? **(Area of a rectangle.)**

68. A rectangular plot of land, where the long side is twice the short side, is purchased for \$5 per square foot. What are the dimensions of the plot if the cost is \$64,000?

69. A rectangular flower garden is 6 ft. longer then it is wide. If the area is 520 square feet what are the dimensions of the flower garden?

70. It costs \$0.50 per square foot to fertilize a flower garden. The flower garden to be fertilized is rectangular shaped with the long side 4 ft longer than the short side. If the cost to fertilize the flower garden is \$240, what are the dimensions of the flower bed?

71. A box with a square base has a height of 6 inches. What are the dimensions of the square base if the volume is?
    a) 600 in^3
    b) 1350 in^3

72. A box with a rectangular base, where the length if 2 inches more than the width, has a height of 4 inches. What are the dimensions of the base if the volume is?
    a) 192 in^3
    b) 1440 in^3

73. Two cars are approaching an intersection at a right angle to each other. When the distance between them is 50 feet, one of the cars is 10 feet farther away from the intersection then the other. How far is each car from the intersection? **(Pythagorean Theorem)**

74. Two airplanes are flying towards each other at the same altitude, one heading east and the other north. At the time one airplanes is 20 miles from the point where the airplanes could meet and the other airplane's distance from that point is 50 miles less than the twice the distance between the planes.
    a) What is the distance between the planes? (Approximate to 2 decimal places.)
    b) How far is the second plane from the intersection point? (Approximate to two decimal places.)

(75–76) An object thrown or fired straight upward will reach a height of h feet after t seconds, where h and t are related by $h(t) = -16t^2 + v_0t + s_0$. v_0 = the initial velocity of the object s_0 = the initial height of the object, use this information to answer the following questions.

75. A ball is thrown straight up; the release point is 6 feet above the ground and the initial speed is 128 ft/sec.
    a) At what time will the ball be 198 feet above the ground?
    b) At what time will the ball be 246 feet above the ground?
    c) At what time will the ball be 262 feet above the ground?
    d) At what time will the ball reach its maximum height and what is that height? (Maximizing, section 1.3)
    e) When will the ball hit the ground? (Approximate to two decimal points.)

76. A rocket is shot off at ground level after 2 seconds the rocket propellant burns out. The rocket is 540 feet above ground and traveling 320 feet per second when the propellant burns out.
    a) At what time will the rocket be 1356 feet above the ground?
    b) At what time will the rocket be 1740 feet above the ground?
    c) Using the information above and looking at the graph of the equation with the graphing calculator, at what time will the rocket reach its maximum height and what is that height?

(77-80) Using the Quadratic equations to solve mathematical modeling problems.

77. A certain company produces widgets; the weekly profit equation for the company is $P = -x^2 + 200x - 100$. P = profit in dollars and x = the number of widgets produced.
    a) When the profit is \$5,000 how many widgets were produced?
    b) When the profit is \$6,300 how many widgets were produced?
    c) When the profit is \$9,900 how many widgets were produced?
    d) Using the information above and looking at the graph of the profit equation with the graphing calculator, how many widgets should be produce to maximize profit and what is the profit?

78. An apartment owner has 60 apartment units for rent. The equation that relates the number of apartments, x, rented to the rent charged, p, is $p(x) = -30x + 2100$. The revenue, R, the apartment owner receives, is given by the equation $R = px$. Write R as a function of x and then answer the following:
    a) For a revenue of \$34,320 how many apartments were rented and what rent was charged?
    b) For a revenue of \$36,000 how many apartments were rented and what rent was charged?
    c) For a revenue of \$36,750 how many apartments were rented and what rent was charged?
    d) Using the information above and looking at the graph of the revenue equation with the graphing calculator, how many apartments should be rented to maximize revenue and what is the maximum revenue?

79. Squares with a dimension of 2 inches are cut out of a square piece of cardboard and the resulting flaps are folded up to form a pizza box. The resulting box has a volume of 200 in^3.
    a. What is the size of the square piece of cardboard?
    b. Using your graphing calculator, approximately what size squares should be cut out to maximize volume?

80. Squares with a dimension of 4 inches are cut out of a rectangular piece of cardboard and the resulting flaps are folded up to form a storage box. If the rectangular cardboard has the length 4 inches more than the width and the volume of the resulting box is 128 in^3.
    a. What are the dimensions of the piece of cardboard?
    b. Using your graphing calculator approximately what size squares should be cut out to maximize volume?

Section 2.3 Solving Square Roots and Fractional Exponent Equations

Now that you have honed your skills at solving quadratic equations and have practiced using complex numbers, let's learn how to solve other kinds of equations.

SOLVING SQUARE ROOT EQUATIONS

In this section you will learn how to solve equations with square roots and equations that have fractional exponents. First we will look at equations that have square roots.

Discussion 1: Simple Square Root Equation

If you remember what the square root symbol means, then you could figure out the answer to $\sqrt{x} = 8$ in your head. The square root symbol means, "What squared equals the expression under the radical?" In this case it's x under the radical. So in this problem we already know the "what," it's 8. Thus, x must be 64.

The difficulty is that when a problem is more complicated, it isn't as easy to get the answer in your head. So let's talk about the procedure that we will use to solve anything like this.

- First, you will want to isolate the square root in the problem (get the square root alone on one side of the equal sign, just like you must do when solving absolute value equations (section 0.5)).
- Then, square both sides of the equation to eliminate the square root symbol. Remember, squaring and taking the square root are inverses of each other.
- Next, solve for the variable and check your answers. So in discussion 1, if we square both sides, we get $x = 64$, which checks in the original problem.

Example 1: Square Root Equation

Solve $\sqrt{7-2x} = 3$.

Solution:

First, square both sides. Notice the square root is already isolated (alone).	$7 - 2x = 9$	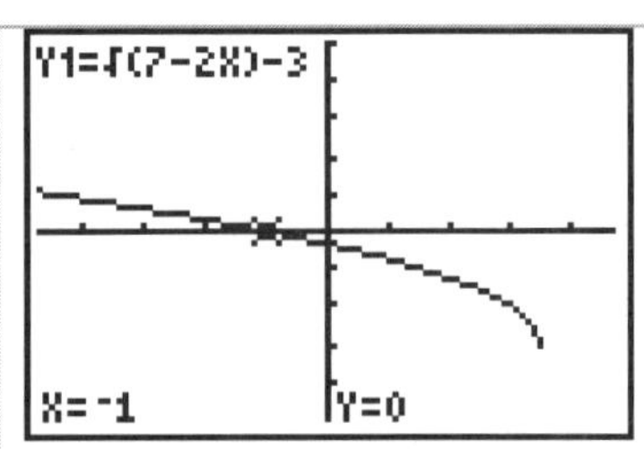
Next, subtract the 7 to isolate the x term.	$-2x = 2$	
Now divide by -2.	$x = -1$	

Check to see that the answer works.	$\sqrt{7-2(-1)}=3 \rightarrow \sqrt{9}=3$ True!	The graph confirms our answer.

Let's do some slightly more complicated examples now.

Example 2: Square Root Equation

Solve $\sqrt{3x-5}-6=1$.

Solution:

First, add 6 to both sides to isolate the square root.	$\sqrt{3x-5}=7$	
Square both sides of the equation to eliminate the square root symbol.	$3x-5=49$	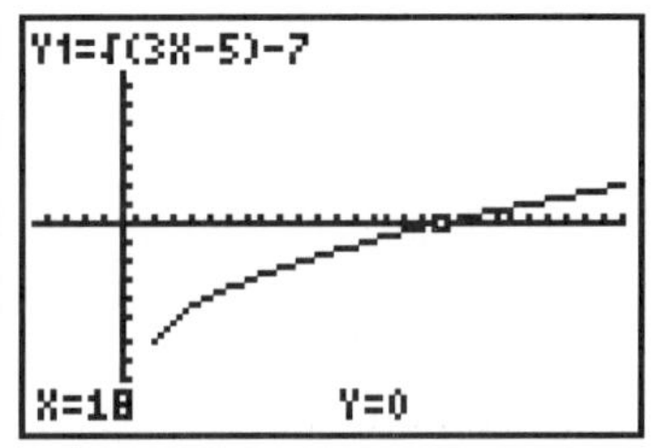
Add 5 to both sides.	$3x=54$	
Divide by 3.	$x=18$	
Check.	$\sqrt{3(18)-5}-6=\sqrt{49}-6=7-6$ True!	The graph confirms our answer.

Example 3: Square Root Equation with No Solution

Solve $\sqrt{x+3}+4=2$.

Solution:

First, subtract 4 from both sides to isolate the square root.	$\sqrt{x+3}=-2$	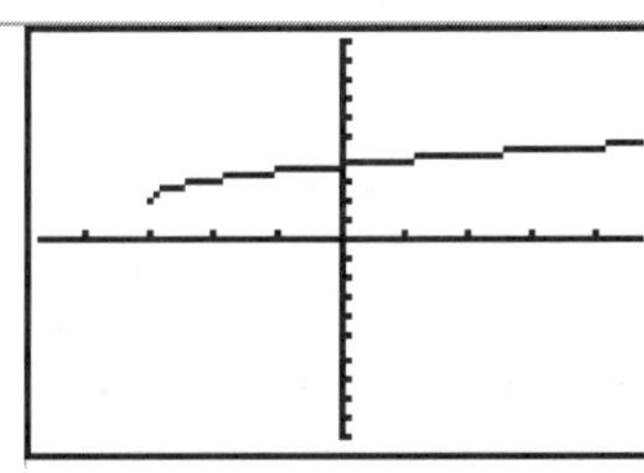
Next, square both sides.	$x+3=4$	
Now, subtract 3 from both sides.	$x=1$	
Check.	$\sqrt{1+3}+4=\sqrt{4}+4$ $=2+4=6\neq 2$ False!	The graph confirms our answer; it doesn't cross the x-axis anywhere.

Our final answer is that this equation has <u>No Solution</u>. (Good thing we checked our answer!)

Now, here is an example where the answer we found didn't check. Why? Because at our first step, $\sqrt{x+3}=-2$, we have an equation that is impossible to solve. You can't take the square root of a *real number* and get a *negative* answer. Notice that when you square both sides of an equation to eliminate the square root sign you will always eliminate any negative signs. That can cause us to get answers that don't work in the original equation that had a square root in it. We call these answers that don't check **extraneous solutions**.

Question 1: What is the solution of $\sqrt{10-x}=4$?

Let's now solve the most difficult of these kinds of equations.

Example 4: Two Square Roots in an Equation

Solve $\sqrt{x+4}-\sqrt{x-5}=1$.

Solution:

First, add $\sqrt{x-5}$ to both sides to isolate one of the square roots.	$\sqrt{x+4}=\sqrt{x-5}+1$	**Note:** $(\sqrt{x-5}+1)(\sqrt{x-5}+1)$ $=(\sqrt{x-5}+1)^2$ and $(a+b)^2=a^2+2ab+b^2$
Next, square both sides. Notice, the right side is squaring a binomial.	$x+4=(x-5)+2\sqrt{x-5}+1$	
Now, isolate the square root that is still in the equation by subtracting x and adding 4 to each side.	$x+4=x-4+2\sqrt{x-5}$ $-x+4 \quad -x+4$ $8 = 2\sqrt{x-5}$	
Divide by 2 since that will simplify the equation.	$4=\sqrt{x-5}$	
Now we can square both sides to eliminate the last square root.	$16=x-5$ so $x=21$	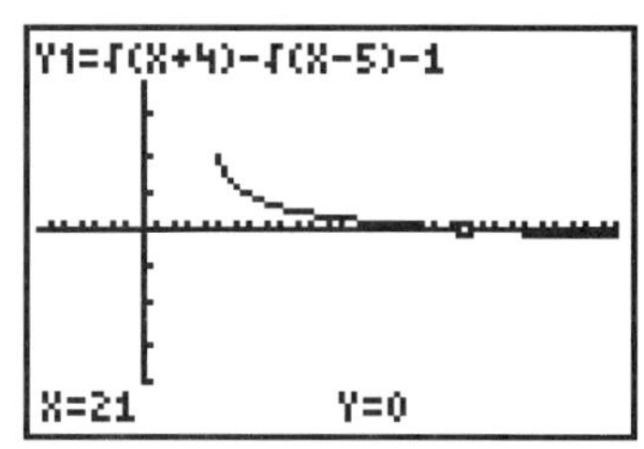
Check.	$\sqrt{(21)+4}-\sqrt{(21)-5}$ $=\sqrt{25}-\sqrt{16}$ $=5-4=1$ True!	The graph confirms our answer.

Example 5: Two Square Roots in an Equation

Solve $\sqrt{x+2}+\sqrt{3x+4}=2$.

Solution:

First, subtract $\sqrt{3x+4}$ from both sides to isolate one of the square roots.	$\sqrt{x+2} = 2 - \sqrt{3x+4}$	
Next, square both sides. Notice, the right side is squaring a binomial.	$x + 2 =$ $4 - 4\sqrt{3x+4} + (3x+4)$	
Now, collect like terms and isolate the square root that is still in the equation by subtracting 3x and subtracting 8 from each side.	$x + 2 = 3x + 8 - 4\sqrt{3x+4}$ $\underline{-3x - 8 \quad -3x - 8}$ $-2x - 6 = \quad -4\sqrt{3x+4}$	
Divide by –2 since that will simplify the equation.	$x + 3 = 2\sqrt{3x+4}$	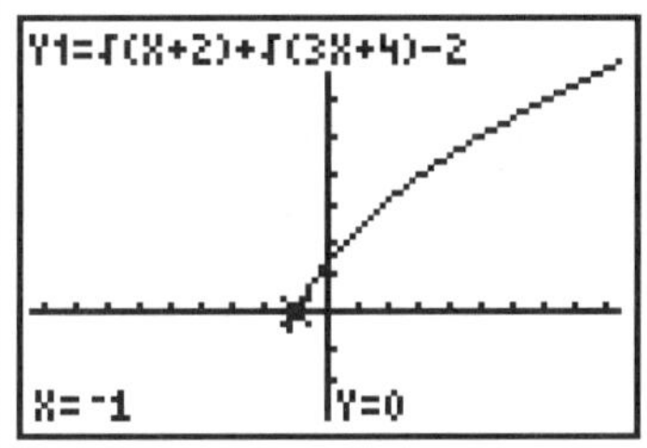
Now, square both sides to eliminate the last square root.	$x^2 + 6x + 9 = 4(3x + 4)$	
We now have a quadratic to solve so we must get the equation equal to 0.	$x^2 + 6x + 9 = 12x + 16$ which becomes $x^2 - 6x - 7 = 0$	The graph confirms our answer.
It does factor so we will.	$(x - 7)(x + 1) = 0$	
So our potential answers are:	$x = 7$ or $x = -1$	
7 doesn't check but –1 does.	$\sqrt{(7)+2} + \sqrt{3(7)+4} = \sqrt{9} + \sqrt{25} = 8 \neq 2$ $\sqrt{(-1)+2} + \sqrt{3(-1)+4} = \sqrt{1} + \sqrt{1} = 2$	

Let's now look at an example that isn't a square root. It is a fourth root and we solve it similarly.

Example 6: Solving a 4th Root Equation

Solve $\sqrt[4]{2x-8} + 5 = 7$.

Solution:

First subtract 5 from both sides to isolate the fourth root.	$\sqrt[4]{2x-8} = 2$	
Next, raise both sides to the fourth power.	$(\sqrt[4]{2x-8})^4 = 2^4 \Rightarrow$ $2x - 8 = 16$	

Now, add 8 to both sides then divide by 2.	$2x = 24$ then $x = 12$	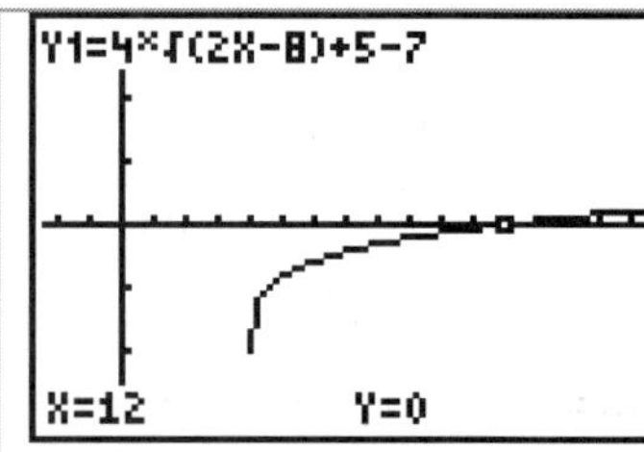 The graph confirms our answer
Check.	$\sqrt[4]{2(12)-8}+5=\sqrt[4]{16}+5=7$ True!	

Answer Q1: 10 − x = 16, then −x = 6: thus x = −6 and it checks $\sqrt{10-(-6)}=\sqrt{16}=4$

As you can see, we can solve any kind of root we might have.

EQUATIONS WITH FRACTIONAL EXPONENTS

Now let's take some time to look at equations with fractional exponents. We will look at some of the basic ones in this section and then do some harder ones in the next section. To start with, let's refresh your memory about the definition of a fractional exponent.

$b^{\frac{p}{r}}=\left(\sqrt[r]{b}\right)^p$ or $b^{\frac{p}{r}}=\left(\sqrt[r]{b^p}\right)$	b is the base being raised to a power; in the exponent, p is the power and r is the root. It can be viewed in either of the two forms shown.

As you can see fractional exponents have roots as part of them. So they fit very well here with our discussion of how to solve roots. Let's look at an example.

Example 7: Solving a Fractional Exponent Equation

Solve $x^{3/2} = 8$.

Solution:

There are two ways to approach this problem and we are going to show you both.

Approach #1	
Use the definition of fractional exponent.	$\left(\sqrt{x}\right)^3=8$
Take the cubed root of both sides.	$\sqrt{x}=2$
Square both sides.	$x=4$
Check.	$4^{3/2}=2^3=8$ True!

Approach #2	
Use the properties of exponents to eliminate the exponent. Raise both sides to the 2/3 power.	$\left(x^{\frac{3}{2}}\right)^{\frac{2}{3}}=8^{\frac{2}{3}}$
Simplify.	$x=\left(\sqrt[3]{8}\right)^2$
Final answer:	$x=2^2=4$
Check.	$4^{3/2}=2^3=8$

In the first approach we used the definition of fractional exponents and then followed that with inverse operations until we got the x alone. In the second approach we used the properties of exponents and then the definition of fractional exponents. **With the second method you must be careful – if you exponentiate each side with an exponent that has an even number in the denominator, you will need a $\pm$ symbol.** (Remember, if you take the square root or 4th root (etc.) to solve a problem, you get two possible answers ($\pm$).)

Depending on the complexity of the problem one approach could turn out to be easier or take fewer steps than the other. Regardless of what you choose to do, you should be comfortable using either approach.

Example 8: Solving a Fractional Exponent Equation

Solve $x^{2/3} - 5 = 11$.

Solution:

Approach #1	
Isolate the variable term by adding 5.	$x^{2/3} = 16$
Use the definition of fractional exponent.	$\left(\sqrt[3]{x}\right)^2 = 16$
Take the square root of both sides.	$\sqrt[3]{x} = \pm 4$
Cube both sides.	$x = \pm 64$
Check.	$64^{2/3} - 5 = 4^2 - 5$ $= 16 - 5 = 11$ True!
	$(-64)^{2/3} - 5 = (-4)^2$ $- 5 = 16 - 5 = 11$ True!

Approach #2	
Isolate the variable term by adding 5.	$x^{2/3} = 16$
Use the properties of exponents to eliminate the exponent. Raise both sides to the 3/**2** power. Notice the **square root.**	$\left(x^{\frac{2}{3}}\right)^{\frac{3}{2}} = 16^{\frac{3}{2}}$
Simplify.	$x = \pm\left(\sqrt{16}\right)^3$
Final answer:	$x = \pm 4^3 = \pm 64$
Check.	$64^{2/3} - 5 = 4^2 - 5$ $= 16 - 5 = 11$
	$(-64)^{2/3} - 5$ $= (-4)^2 - 5$ $= 11$

Question 2: What are the solutions of $x^{4/3} = 16$?

Let's look at a slightly harder example.

Example 9: Solving a Fractional Exponent Equation

Solve $x^{-4/3} = 81$.

Solution:

This time we will just show one approach.

First, raise both sides of the equation to the reciprocal power which does have an even root (4).	$\left(x^{\frac{-4}{3}}\right)^{\frac{-3}{4}} = \pm 81^{\frac{-3}{4}}$	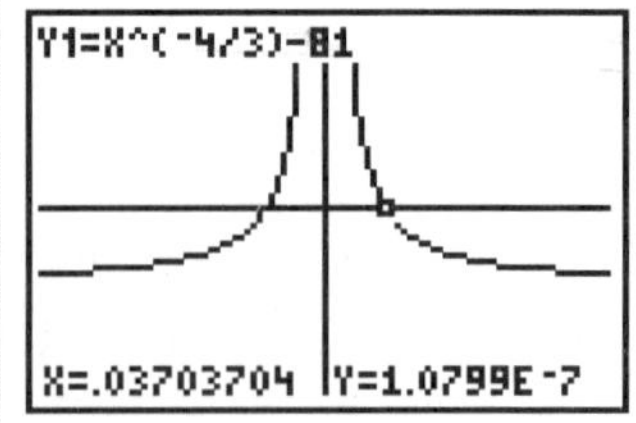
Simplify by first eliminating the negative exponent.	$x = \pm\dfrac{1}{81^{\frac{3}{4}}}$	
Now use the definition of fractional exponents.	$x = \pm\dfrac{1}{\left(\sqrt[4]{81}\right)^3}$	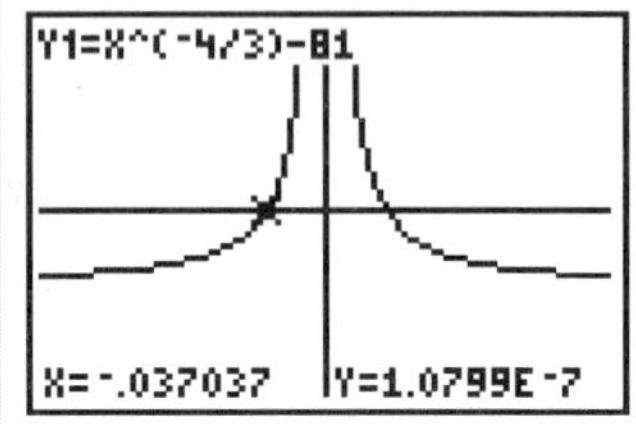
Now simplify to get our answer.	$x = \pm\dfrac{1}{3^3} = \pm\dfrac{1}{27}$	
Check.	$\left(\pm\dfrac{1}{27}\right)^{\frac{-4}{3}} = (\pm 27)^{\frac{4}{3}} = (\pm 3)^4$ $= 81$ True!	1/27 = .037037… The graph confirms our answers

In this last example notice how the negative exponent caused the base to invert (i.e. flip upside-down). Also notice that we got two answers; seeing these exact answers on the graphing calculator is hard, but the graph serves as a good double-check of our answers.

FACTORING TO SOLVE EXPONENT EQUATIONS

Next, we are going to show you an example where you need to factor using a greatest common factor. You should remember that when you factor out a greatest common factor (GCF) you look for the *smallest exponent on the common variables*. Below is a list of expressions and how they factor. Pay special attention to the examples below with negative and fractional exponents – they can be tricky.

Expression	Factored Form
$3x^3 + 7x^2$	$x^2(3x + 7)$
$-2x^{-3} + 5x^{-5}$	$x^{-5}(2x^{-3-(-5)}+5x^{-5-(-5)})$ $=x^{-5}(2x^{-3+5}+5x^{-5+5})$ $=x^{-5}(-2x^2 + 5)$
$9x^{2/3} - 4x^{5/3}$	$x^{2/3}(9 - 4x)$
$5x^{-5/2} - 15x^{-3/2}$	$5x^{-5/2}(1 - 3x)$

There are several things you should take note of from the table above.

- When you factor out a common variable, you subtract the exponents; this is how you determine what is left in the parentheses. You subtract the exponents because you are dividing the original terms by the common factor. You are dividing, because factoring out a common factor is the opposite, the inverse, the undoing of the distributive property, which involves the operation multiplication (multiplication and division are inverse operations).
- In the last example –5/2 is smaller than –3/2 and so we factor out the $x^{-5/2}$.
- Also in the last example when we factor out $x^{-5/2}$, the second term becomes $3x^{2/2}$ which simplifies to just $3x^1$.

Answer Q2:

$\left(x^{\frac{4}{3}}\right)^{\frac{3}{4}} = \pm 16^{\frac{3}{4}}$

so

$x = \pm\left(\sqrt[4]{16}\right)^3$

$= \pm 2^3$

$= \pm 8$

Question 3: What is the factored form of $x^{3/5} - 4x^{1/5}$?

Example 10: Solving a Fractional Exponent Equation

Solve $x^{3/5} - 4x^{1/5} = 0$.

Solution:

First, factor since we are already equal to 0.		$x^{1/5}(x^{2/5} - 4) = 0$	
Next, set each factor equal to 0 like we did when we solved quadratics by factoring in the last section.		$x^{1/5} = 0$ and $x^{2/5} - 4 = 0$	
First Equation		Second Equation	
Raise both sides to the 5th power.	$\left(x^{\frac{1}{5}}\right)^5 = 0^5$	Get the x term alone by adding 4 to both sides.	$x^{2/5} = 4$
Answer:	$x = 0$	Raise both sides to the $\frac{5}{2}$ power. Even root.	$\left(x^{\frac{2}{5}}\right)^{\frac{5}{2}} = \pm 4^{\frac{5}{2}}$
Check.	$(0)^{3/5} - 4(0)^{1/5} = 0$	Answer.	$x = \pm 2^5 = \pm 32$
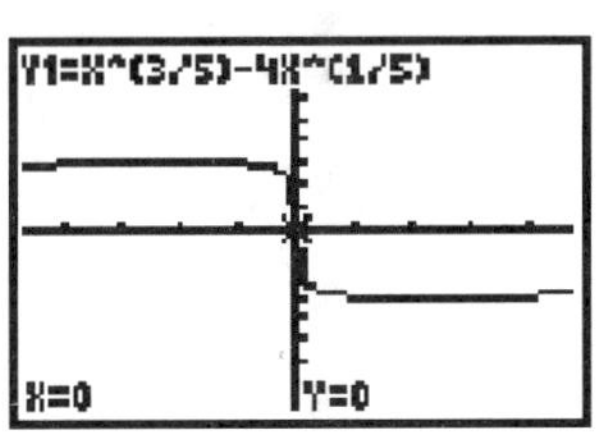		Check.	$(32)^{3/5} - 4(32)^{1/5}$ $= \left(\sqrt[5]{32}\right)^3 - 4\sqrt[5]{32}$ $= 2^3 - 8 = 0$ $(-32)^{3/5} - 4(-32)^{1/5}$ $= \left(\sqrt[5]{-32}\right)^3 - 4\sqrt[5]{-32}$ $= (-2)^3 + 8 = 0$

Section Summary

When you solve equations with roots or equations with fractional exponents you need to

- Isolate the variable part first.
- Eliminate the root or fractional exponent second.
- Finally check to see that all of your answers make the original equation true and that none of them are extraneous solutions.

SECTION 2.3 HOMEWORK

(1–34) Solve each radical equation for real number values of x. (Use your graphing calculator to confirm your answers.):

1. $\sqrt{2x-1}=5$
2. $\sqrt{3x+2}=12$
3. $\sqrt{2x+3}+5=9$
4. $\sqrt{5x-1}-11=7$
5. $\sqrt{3x+2}+8=2$
6. $\sqrt{2x+3}+9=4$
7. $\sqrt{2x+6}=x+3$
8. $\sqrt{3x+1}=x+1$
9. $\sqrt{3x+7}=x+1$
10. $\sqrt{2x+3}=3x+2$
11. $\sqrt{3x+5}=x+2$
12. $\sqrt{3x+2}=2x-1$
13. $\sqrt{2x+10}-x=1$
14. $\sqrt{3x-2}+x=4$
15. $\sqrt{x+1}-x=3$
16. $\sqrt{2x-1}-3x=2$
17. $\sqrt[3]{3x-1}=2$
18. $\sqrt[4]{2x+13}=3$
19. $\sqrt[3]{x^3+26}=x+2$
20. $\sqrt[3]{x^3-63}=x-3$
21. $\sqrt{2-x}=x-2$
22. $\sqrt{18-2x}=x-5$
23. $\sqrt{x-3}+\sqrt{x+5}=4$
24. $\sqrt{x+7}-\sqrt{x-5}=2$
25. $\sqrt{2x+3}-\sqrt{x+1}=1$
26. $\sqrt{3x+1}-\sqrt{x-1}=2$
27. $\sqrt{x+2}-\sqrt{2x-10}=1$
28. $\sqrt{2x+3}+\sqrt{x+1}=5$
29. $\sqrt{3x-1}-\sqrt{2x+3}=1$
30. $\sqrt{2x-1}+\sqrt{3x+2}=2$
31. $\sqrt{x+5}=x+1$
32. $\sqrt{2x-1}=2x-5$
33. $\sqrt{2x+7}-\sqrt{x+3}=\sqrt{2x-1}$
34. $\sqrt{2x-1}+\sqrt{3x+1}=\sqrt{8x+9}$

Answer Q3: $x^{1/5}(x^{2/5}-4)$

(35-64) Solve each fractional exponent equation for real number values of x:

35. $x^{\frac{3}{2}}=8$
36. $x^{\frac{3}{4}}=27$
37. $x^{\frac{2}{3}}=16$
38. $x^{\frac{4}{3}}=81$
39. $x^{\frac{3}{4}}+5=69$
40. $x^{\frac{5}{2}}-3=240$
41. $x^{\frac{2}{3}}-4=12$
42. $x^{\frac{4}{3}}+3=4$
43. $(2x-1)^{\frac{3}{2}}=1$
44. $(3x+1)^{\frac{5}{2}}=32$
45. $(3x+1)^{\frac{2}{5}}=4$
46. $(5x-1)^{\frac{4}{3}}=256$
47. $x^{\frac{-3}{5}}=8$
48. $x^{\frac{-2}{3}}=64$
49. $x^{\frac{-2}{3}}=25$
50. $x^{\frac{-4}{3}}=16$
51. $(3x-1)^{\frac{-1}{2}}=4$
52. $(2x+1)^{\frac{-3}{2}}=1$
53. $x^{\frac{3}{4}}-x^{\frac{1}{4}}=0$
54. $x^{\frac{5}{6}}+x^{\frac{1}{6}}=0$
55. $x^{\frac{4}{5}}-8x^{\frac{1}{5}}=0$
56. $x^{\frac{5}{7}}+8x^{\frac{2}{7}}=0$
57. $x^{\frac{4}{5}}-9x^{\frac{2}{5}}=0$
58. $x^{\frac{5}{7}}-4x^{\frac{3}{7}}=0$
59. $x^{\frac{3}{4}}=4$
60. $x^{\frac{3}{5}}=9$
61. $x^{\frac{2}{3}}=8$

62. $x^{\frac{4}{5}} = 12$

63. $x^{\frac{2}{3}} - 4x^{\frac{1}{6}} = 0$

64. $x^{\frac{3}{4}} - x^{\frac{3}{8}} = 0$

(65-66) Use the distance formula to answer the following.

65. On a coordinate system, the distance from point (6, 2) to two different points on the line $y = 2x$ is 5. Find the two points.

(Distance formula from point (x_1, y_1) to point (x_2, y_2): $d = \sqrt{(x_2 - x_1)^2 + (y_2 - y_1)^2}$.)

66. On a coordinate system, the distance from point (6, –5) to two different points on the line $y = 3x + 4$ is 13. Find the two points.

(67-71) Mathematical models that use radical equations.

67. You own a company that manufacturers furniture. A certain type of chair has a cost formula $C(x) = 100(20 + 10\sqrt{x})$, where x is the number of chairs produced and $C(x)$ is the cost in dollars. How many chairs are manufactured when the cost is $17,000?

68. The demand function (the number of items desired at a certain price) is given by the following equation: $p(x) = 50(80 - \sqrt{x})$, where x is the number of items sold and $p(x)$ is the price per item when that number of items is sold. How many items are desired at a price of $2500?

69. There is an oilrig 3 miles off shore. You want to lay a pipeline from the oilrig to the docks that are 10 miles down the shoreline from the oilrig. The cost of laying the pipe under water is $5000 per mile and the cost of laying the pipe over ground is $3000 per mile. To what point on the shore was the pipeline laid if the total cost was $43,000? (Two answers.)

70. Using your graphing calculator approximate at what point along the shoreline the pipeline should be laid to minimize the cost? (Minimizing, section 1.3)

71. The position s (in feet) of an accelerating car in terms of t (time in seconds) is $s(t) = 10t^{\frac{3}{2}}$. How long has the car traveled if the position is 640 feet?

Section 2.4 Quadratic-Like and Fractional Equations

In this section we will look at equations that have fractions in them (called "rational expressions") and equations that are quadratic-like.

RATIONAL EQUATIONS

Let's take a look at equations that have fractions in them first. We call these kinds of equations rational equations. The key to success when solving rational equations, is to eliminate the fractions from the equation as soon as possible. You can do this is by multiplying both sides of the equation by the lowest common denominator (LCD). Also, since the denominator may have a variable in it, we must check our answers to see if we obtained solutions that were not in the *domain of the original equation*. Remember, we discussed in Chapter 1 that division by 0 is not allowed. If there is a value (or values) for the variable that would cause the denominator of the original problem to equal 0, then that number (or numbers) is not part of the domain and can't be an answer to the equation we are trying to solve.

Example 1: Constant Denominators

Solve $\frac{4x}{2}-\frac{5}{3}=\frac{13}{3}$.

Solution:

First, the LCD is 6 so we'll multiply both sides by 6.	$12x - 10 = 26$	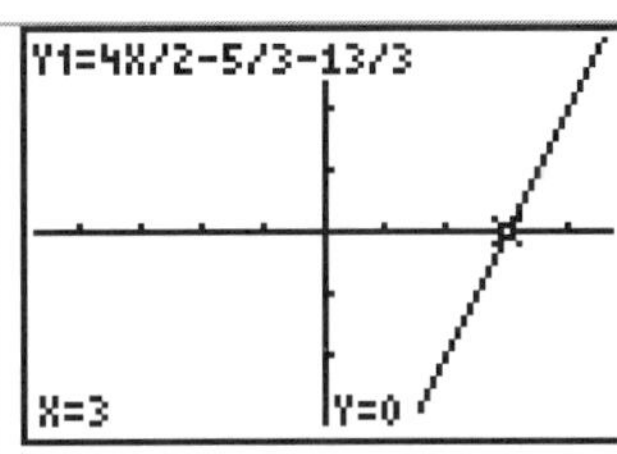
Now add 10 and divide by 3.	$12x = 36 \rightarrow x = 3$	

Usually you need to check your answer any time you solve a fraction equation. But this time there weren't any variables in the denominators, so we know that the answer is okay. There isn't any way to get division by 0. Of course the graphing calculator can easily confirm our answer for us.

Example 2: Single Variable Term in the Denominator

Solve $\frac{1}{y}+1=\frac{3}{y}-\frac{1}{2y}$.

Solution:

First, the LCD is $2y$ so we'll multiply both sides by $2y$.	$2y\bullet\frac{1}{y}+2y\bullet 1=2y\bullet\frac{3}{y}-2y\bullet\frac{1}{2y}$ $2+2y=6-1$

Now subtract 2 from both sides.	$2y = 3$
Divide by 2 to get y alone.	$y = \frac{3}{2}$
Check. Notice that the domain for this equation is $y \neq 0$, so 3/2 will work.	$\frac{1}{\left(\frac{3}{2}\right)} + 1 = \frac{3}{\left(\frac{3}{2}\right)} - \frac{1}{2\left(\frac{3}{2}\right)} \Rightarrow \frac{2}{3} + 1 = 2 - \frac{1}{3} \Rightarrow \frac{5}{3} = \frac{5}{3}$

Example 3: Variables in the Denominators

Solve $\frac{25}{x(x+5)} - \frac{x}{x+5} = -3$.

Solution:

First, the LCD is $x(x + 5)$ so we'll multiply both sides by $x(x + 5)$.	$\frac{25}{x(x+5)} \bullet \frac{x(x+5)}{1} - \frac{x}{x+5} \bullet \frac{x(x+5)}{1} = -3 \bullet \frac{x(x+5)}{1}$	
Simplifying we get	$25 - x^2 = -3x^2 - 15x$	
Notice that it is now a quadratic equation so we need to get it equal to 0 by adding $3x^2$ and $15x$ to both sides.	$2x^2 + 15x + 25 = 0$	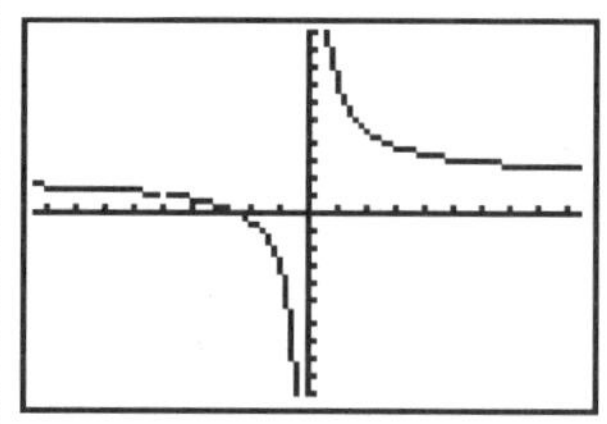
This can be solved by the factoring method.	$(x + 5)(2x + 5) = 0$	
Set each factor equal to 0.	$(x + 5) = 0$ or $2x + 5 = 0$	
Potential answers:	$x = -5$ or $x = \frac{-5}{2}$	The graph confirms our answer

The domain for this equation is $x \neq -5$ and $x \neq 0$. These numbers would make the denominators equal 0. In interval notation form, the domain is $(-\infty, -5) \cup (-5, 0) \cup (0, \infty)$. Since one of our potential answers is $x = -5$, not in the domain, we only get one answer: $x = \frac{-5}{2}$.

The potential answer $x = -5$ is an extraneous solution. Once again, the graphing calculator easily verifies this. You might also notice a hole in the graph at $x = -5$. This will be discussed in more detail in section 3.5.

Question 1: Solve $\frac{5}{x+2} - \frac{2x-1}{5} = 0$.

QUADRATIC-LIKE EQUATIONS

Let's turn our attention now to solving "quadratic-like" equations. Equations that are "quadratic-like" look similar to the ones we have just solved in section 2.2. The form we are looking for is ($a[Y]^2 + b[Y] + c = 0$) where Y is some, possibly complicated, expression. Some examples of equations that are quadratic-like are as follows:

$5x^4 - 7x^2 + 4 = 0$	See how the first term variable (x^4) is the middle term variable (x^2) squared. $Y = x^2$
$-3z^6 + 2z^3 = 7$ Could be made to look like $-3z^6 + 2z^3 - 7 = 0$	You can move the 7 to the other side and then you would see that the first term variable (z^6) is the middle term variable (z^3) squared. $Y = z^3$
$x^{2/3} - 3x^{1/3} + 8 = 0$	Once again notice how the first term variable ($x^{2/3}$) is the middle term variable ($x^{1/3}$) squared. $Y = x^{1/3}$
$\left(\frac{2x-5}{3}\right)^2 + 4\left(\frac{2x-5}{3}\right) - 12 = 0$	Again notice how the first term variable $\left(\frac{2x-5}{3}\right)^2$ is the middle term variable $\left(\frac{2x-5}{3}\right)$ squared. $Y = \left(\frac{2x-5}{3}\right)$
$5x + \sqrt{x} - 7 = 0$	Here the first term variable, x, is the middle term variable $\sqrt{x}$ squared. $Y = \sqrt{x}$

In the above examples, notice how each one has three terms in the equation. Also, when written in descending order, the first term variable is the second term variable squared or, in other words, the first exponent is the second one multiplied by 2. Let's solve an equation like the ones above.

Example 4: Solving a Quadratic-Like Equation

Solve $4t^4 - 5t^2 = -1$.

Solution:

Get equal to 0 by adding 1.	$4t^4 - 5t^2 + 1 = 0$
First term variable (t^4) is the second (t^2) squared, so we will solve it like a quadratic. Let $Y = t^2$. (Like the form $a[Y]^2 + b[Y] + c = 0$ mentioned above.) This method of solving is called **substitution**.	$4(Y)^2 - 5Y + 1 = 0$
Factor.	$(4Y - 1)(Y - 1) = 0$

Set each factor equal to 0.	$4Y-1=0$ or $Y-1=0$
Solve each equation and get	$Y=\frac{1}{4}$ or $Y=1$
Now these look like answers but they're answers for Y not t. Substitute $Y=t^2$ back into our answers.	$t^2=\frac{1}{4}$ and $t^2=1$
Now take the square root of both sides	$t=\pm\frac{1}{2}$ and $t=\pm 1$

The domain for this problem is all real numbers and we didn't have any fractions or even roots in the original equation, so we have four real-numbered solutions (no complex numbers). Remember, we only get extraneous solutions when we have square roots (or more specifically even roots) or fractions in our original equation. You should still check your answers in this case to see if you might have made any mistakes in your arithmetic. Don't forget that your calculator can be a big help in checking your answers.

Question 2: What is the power of the first term in the last example and how many solutions did we get?

Example 5: Quadratic-Like Equation

Solve $x^6+27=-28x^3$.

Solution:

Get equal to 0 by adding $28x^3$.	$x^6+28x^3+27=0$
First term variable (x^6) is the second (x^3) squared, so we will solve it like a quadratic. Use substitution where we let $Y=x^3$.	$(Y)^2+28Y+27=0$
Factor.	$(Y+27)(Y+1)=0$
Set each factor equal to 0.	$Y+27=0$ or $Y+1=0$
Solve each equation and get	$Y=-27$ or $Y=-1$
Now these look like answers but they're answers for Y not x. Substitute $Y=x^3$ back into our answers.	$x^3=-27$ and $x^3=-1$
Now take the cubed root of both sides.	$x=-3$ and $x=-1$

Answer Q1: Multiply by $5(x+2)$ to get $25-(2x-1)(x+2)=0$. Simplifies to $0=2x^2+3x-27$. Factors to $0=(2x+9)(x-3)$. Solutions are $x=\frac{-9}{2}$ or $x=3$.

In this example we got just two real answers. The truth of the matter is that there are six total answers. The other four are complex. To find the other four requires learning several topics we will cover later in this book in chapters 2 and 3.

Note: It is possible to work the last two examples *without* substituting.

Discussion 1: Alternative Method for Solving Quadratic-Like Equations

Let's look at example 4 again, $4t^4 - 5t^2 = -1$.

Get equal to 0 by adding 1.	$4t^4 - 5t^2 + 1 = 0$
First term variable (t^4) is the second (t^2) squared, so we will solve it like a quadratic. But instead of substituting we will leave the t^2 in the equation and just factor as normally treating the t^2 like a single variable.	$(4t^2 - 1)(t^2 - 1) = 0$
Set each factor equal to 0.	$4t^2 - 1 = 0$ or $t^2 - 1 = 0$
Solve each equation for t^2.	$t^2 = \frac{1}{4}$ or $t^2 = 1$
Now take the square root of both sides of each equation.	$t = \pm\frac{1}{2}$ and $t = \pm 1$

As you can see, this problem works out well either way we choose to solve it. This shortcut method works well when the variable isn't too complicated.

Example 6: Solving a Quadratic-Like by Two Approaches

Solve $y^{-2} - y^{-1} - 2 = 0$.

Solution:

	Substitution Method	Direct Approach
Let $Y = y^{-1}$ and substitute Y for y^{-1} and Y^2 for y^{-2}. $[(y^{-1})^2] = y^{-2}$. Notice that if you double the second variable exponent it equals the first variable exponent. (It's quadratic-like.)	$Y^2 - Y - 2 = 0$	$y^{-2} - y^{-1} - 2 = 0$
Factor.	$(Y - 2)(Y + 1) = 0$	$(y^{-1} - 2)(y^{-1} + 1) = 0$
Set each factor equal to 0.	$Y - 2 = 0$ or $Y + 1 = 0$	$y^{-1} - 2 = 0$ or $y^{-1} + 1 = 0$
Solve each equation.	$Y = 2$ and $Y = -1$	$y^{-1} = 2$ or $y^{-1} = -1$
Now that's great but we need the answer for y not Y. Substitute $Y = y^{-1}$ back into our answers.	$y^{-1} = 2$ and $y^{-1} = -1$	

Eliminate the negative exponent, which results in a fraction.	$\frac{1}{y}=2$ and $\frac{1}{y}=-1$	$\frac{1}{y}=2$ and $\frac{1}{y}=-1$
Eliminate the fractions as we discussed earlier in this section.	$1=2y$ and $1=-y$	$1=2y$ and $1=-y$
Get y alone.	$\frac{1}{2}=y$ and $-1=y$	$\frac{1}{2}=y$ and $-1=y$

Now we do need to think about checking our answers because y has a negative exponent, which means the equation really has fractions in it. The domain for this equation is all real numbers except for $y = 0$, since that is the only number that could make y^{-1} ($\frac{1}{y}$) and y^{-2} ($\frac{1}{y^2}$) undefined. Thus ½ and −1 are the solutions.

Example 7: Solving a Quadratic-Like by Two Approaches

Solve $x^{4/3} + 3x^{2/3} = 28$.

Solution:

Answer Q2: 4 in both cases. There is a connection here that will be discussed in 3.4.

	Substitution Method	Direct Approach
Get equal to 0 by subtracting 28.	$x^{4/3} + 3x^{2/3} - 28 = 0$	$x^{4/3} + 3x^{2/3} - 28 = 0$
First term variable ($x^{4/3}$) is the second ($x^{2/3}$) squared, so we will solve it like a quadratic. Use substitution where we let $Y = x^{2/3}$.	$(Y)^2 + 3Y - 28 = 0$	
Factor.	$(Y+7)(Y-4)=0$	$(x^{2/3}+7)(x^{2/3}-4)=0$
Set each factor equal to 0.	$Y+7=0$ or $Y-4=0$	$x^{2/3}+7=0$ or $x^{2/3}-4=0$
Solve each equation and get	$Y=-7$ or $Y=4$	$x^{2/3}=-7$ or $x^{2/3}=4$
Now this looks like answers but they're answers for Y not x. Substitute $Y = x^{2/3}$ back into our answers.	$x^{2/3}=-7$ and $x^{2/3}=4$	
Now raise both sides to the 3/2 power. Even root.	$x = \pm(-7)^{3/2}$ and $x = \pm 4^{3/2}$	$x = \pm(-7)^{3/2}$ and $x = \pm 4^{3/2}$
Use the rules of fractional exponents	$x = \pm\sqrt{(-7)^3} = \pm 7i\sqrt{7}$ $x = \pm(\sqrt{4})^3 = \pm 2^3 = \pm 8$	$x = \pm\sqrt{(-7)^3} = \pm 7i\sqrt{7}$ $x = \pm(\sqrt{4})^3 = \pm 2^3 = \pm 8$

We have found 4 possible solutions ($7i\sqrt{7}$, $-7i\sqrt{7}$, 8, –8). Since this equation has a domain of all real numbers and there aren't any fractions or even roots in the original equation, our answers will all work. But you might want to check your arithmetic!

Example 8: Solving a Quadratic-Like Using the Quadratic Formula

Solve $x^4 - 6x^2 + 2 = 0$.

Solution:

Notice, that this quadratic-like equation doesn't factor. So we will need to use the quadratic formula.	$x^4 - 6x^2 + 2 = 0$
Plug into the formula $x = \dfrac{-b \pm \sqrt{b^2 - 4ac}}{2a}$. Notice we used x^2 instead of x because this is a quadratic-like equation and it's the middle variable that is used.	$x^2 = \dfrac{6 \pm \sqrt{36 - 4(1)(2)}}{2}$
Now simplify.	$x^2 = \dfrac{6 \pm \sqrt{24}}{2} = \dfrac{6 \pm 2\sqrt{6}}{2} = 3 \pm \sqrt{6}$
We now have two quadratic equations to solve.	$x^2 = 3 + \sqrt{6}$ and $x^2 = 3 - \sqrt{6}$
These both can be done by the square root method, so we square root both sides of the equations.	$x = \pm\sqrt{3+\sqrt{6}}$ and $x = \pm\sqrt{3-\sqrt{6}}$

As you can see, WE CAN SOLVE JUST ABOUT ANY EQUATION but they can get quite complicated.

Section Summary

Let's take a look at the steps one might take in order to solve the kinds of equation we have looked at so far through this book.

<u>Steps to solve equations</u>

1. Eliminate parentheses, brackets, etc.
2. Eliminate fractions, decimals, square roots.
3. Simplify what remains.
4. Ask yourself the question "Is what I now have linear or quadratic-like?"
5. If it is linear, isolate the variable.
6. If it is quadratic-like, then look to see if you can use any of the three methods we have discussed so far in this chapter.
7. Confirm your solution with your graphing calculator.

SECTION 2.4 HOMEWORK

(1–40) For each fractional equation give the domain and solve for x: (Use your graphing calculator to confirm your answers)

1. $\frac{3x}{2}-\frac{5}{3}=\frac{1}{6}$
2. $\frac{2x}{5}-\frac{3}{4}=\frac{3}{10}$
3. $\frac{5x}{3}+4=\frac{1}{5}$
4. $\frac{3x}{4}-\frac{2}{9}=3$
5. $\frac{3}{x-2}+\frac{5}{3}=2$
6. $\frac{2}{x+3}+4=\frac{2}{5}$
7. $\frac{3}{x+2}+\frac{2}{x}=1$
8. $\frac{2}{x-2}+\frac{1}{2x}=\frac{13}{6}$
9. $\frac{3}{x}-\frac{2}{x+1}=\frac{1}{2}$
10. $\frac{3}{x-2}-\frac{5}{x}=\frac{1}{4}$
11. $\frac{3}{x-1}+\frac{2}{3}=\frac{9}{3x-3}$
12. $\frac{5}{x+2}+\frac{3}{4}=\frac{20}{4x+8}$
13. $\frac{2}{x-1}+\frac{3}{x+1}=\frac{9}{x^2-1}$
14. $\frac{5}{x+4}-\frac{2}{x-4}=\frac{2}{x^2-16}$
15. $\frac{2}{x-2}+\frac{3}{x+2}=\frac{8}{x^2-4}$
16. $\frac{3}{x-3}+\frac{2}{x+3}=\frac{-12}{x^2-9}$
17. $\frac{3}{x-3}+\frac{1}{x+2}=\frac{11}{x^2-x-6}$
18. $\frac{5}{x+3}-\frac{2}{x-1}=\frac{4}{x^2+2x-3}$
19. $\frac{3}{x-4}+\frac{2}{x+5}=\frac{-18}{x^2+x-20}$
20. $\frac{21}{x^2-5x-6}-\frac{1}{x+1}=\frac{3}{x-6}$
21. $\frac{x}{x-1}+\frac{2}{x+1}=\frac{8}{3}$
22. $\frac{x}{x+2}-\frac{3}{x-1}=\frac{-9}{10}$
23. $\frac{2x}{x-1}+\frac{5}{x+1}=\frac{17}{x^2-1}$
24. $\frac{3x}{x+2}-\frac{2x}{x-2}=\frac{-25}{x^2-4}$
25. $\frac{x}{x-3}+\frac{2}{x+3}=\frac{18}{x^2-9}$
26. $\frac{2}{x-2}+\frac{3x}{x+2}=\frac{24}{x^2-4}$
27. $\frac{x}{x+5}+\frac{3}{x+2}=\frac{11}{x^2+7x+10}$
28. $\frac{x}{x+3}-\frac{2}{x-2}=\frac{-6}{x^2+x-6}$
29. $\frac{2x}{x-3}-\frac{x}{x+2}=\frac{30}{x^2-x-6}$
30. $\frac{x}{x-1}+\frac{x}{x+3}=\frac{4}{x^2+2x-3}$
31. $\frac{3}{x-1}+\frac{x}{x+3}=\frac{12}{x^2+2x-3}$
32. $\frac{x}{x+2}+\frac{2}{x-1}=\frac{6}{x^2+x-2}$
33. $\frac{2}{x+3}-\frac{3}{x+5}=2$
34. $\frac{3}{x+5}+\frac{4}{x-2}=3$
35. $\frac{x}{2x-1}+\frac{2x}{x+3}=4$
36. $\frac{3x}{x+2}-\frac{x}{3x+1}=2$
37. $\frac{4}{x-3}+\frac{2}{x+1}=\frac{3}{x+2}$
38. $\frac{1}{x-2}+\frac{2}{x+3}=\frac{-4}{3x-3}$
39. $\frac{2}{x+2}-\frac{3}{x-3}=\frac{4}{x+1}$
40. $\frac{3}{x+1}+\frac{2}{x-2}=\frac{2}{x-3}$

(41– 84) Find all real and complex roots for each of the following equations. (Use your graphing calculator to confirm your answers.)

41. $x^4-5x^2+4=0$
42. $x^4-13x^2+36=0$
43. $x^4-3x^2=28$
44. $x^4-21x^2=100$
45. $x^4+20x^2+64=0$
46. $x^4+17x^2+16=0$
47. $3x^4=30x^2-72$
48. $2x^4=28x^2-90$
49. $x^2=20-x^4$
50. $x^4=11x^2+80$
51. $4x^4-29x^2+25=0$
52. $9x^4-85x^2+36=0$

53. $25x^4 = 21x^2 + 4$
54. $16x^4 = 32x^2 + 9$
55. $81x^4 + 85x^2 + 4 = 0$
56. $49x^4 + 58x^2 + 9 = 0$
57. $25 = 53x^2 - 6x^4$
58. $9 = 14x^2 - 5x^4$
59. $x^{\frac{1}{2}} - 5x^{\frac{1}{4}} + 6 = 0$
60. $x^{\frac{2}{3}} - 7x^{\frac{1}{3}} + 12 = 0$
61. $3x^{\frac{1}{2}} - 11x^{\frac{1}{4}} + 10 = 0$
62. $6x^{\frac{2}{3}} - 19x^{\frac{1}{3}} + 15 = 0$
63. $5x = 11x^{\frac{1}{2}} + 12$
64. $8x = 18 - 10x^{\frac{1}{2}}$
65. $2x^{\frac{4}{3}} + 72 = 26x^{\frac{2}{3}}$
66. $x^{\frac{4}{5}} + 100 = 29x^{\frac{2}{5}}$
67. $5x^{\frac{3}{2}} = 35x^{\frac{3}{4}} + 40$
68. $2x^{\frac{4}{5}} = 54 - 52x^{\frac{2}{5}}$
69. $x^{-2} - 8x^{-1} + 15 = 0$
70. $x^{-2} - 12x^{-1} + 32 = 0$
71. $\frac{1}{4}x^{-2} + \frac{5}{2}x^{-1} + 6 = 0$
72. $\frac{1}{2}x^{-2} + \frac{9}{2}x^{-1} + 9 = 0$
73. $x^{-4} + 36 = 13x^{-2}$
74. $x^{-4} - 65x^{-2} + 64 = 0$
75. $x^{\frac{-2}{3}} + 5x^{\frac{-1}{3}} - 24 = 0$
76. $x^{\frac{-3}{4}} + 7x^{\frac{-3}{2}} - 8 = 0$
77. $x^{-1} - 10x^{\frac{-1}{2}} + 25 = 0$
78. $x^{\frac{-1}{2}} - 12x^{\frac{-1}{4}} + 36 = 0$
79. $4x^{\frac{-2}{3}} + 4x^{\frac{-1}{3}} + 1 = 0$
80. $4x^{\frac{-2}{3}} + 12x^{\frac{-1}{3}} + 9 = 0$
81. $x^4 - 3x^2 - 2 = 0$
82. $x^4 = 3 - 5x^2$
83. $2x^4 = 2x^2 + 3$
84. $3x^4 + 5x^2 + 1 = 0$

(85–90) For each of the following equations find only real roots: (There are complex roots but you will not be asked to find these)

85. $x^6 - 9x^3 + 8 = 0$
86. $x^6 - 28x^3 + 27 = 0$
87. $x^6 + 65x^3 + 64 = 0$
88. $x^6 + 126x^3 + 125 = 0$
89. $\frac{1}{4}x^6 + \frac{3}{4}x^3 = 10$
90. $\frac{1}{2}x^6 = \frac{25}{2}x^3 + 27$

(91-94) Mathematical modeling involving the concept of work.

91. You have a business that takes surveys over the phone. Each worker has a quota of surveys to do each day. One worker can do his or her quota in 8 hours. The worker needs to leave early so he or she brings in a helper and gets to go home in 6 hours. How long would it have taken the inexperienced helper to meet the quota working alone?

92. You own a cheese plant. An experienced person can clean one vat in 20 minutes. If you put an inexperienced helper with the experienced person, they can clean the same vat in 15 minutes. How long would it take the inexperienced person working alone to clean the vat?

93. You own a printing company. A customer wants a print job done within 10 hours. One machine can do the print job in x hours and other will take 15 fewer hours. How fast does each machine do the job working alone if you can promise the job done in 10 hours with both machines working together?

94. Your business is processing loans. An experienced loan processor can process the loans you need done 3 hours faster than a processor in training. If the two working together can process the loans in two hours, how quickly can each process the loans working by themselves?

(95-96) Mathematical modeling of business applications

95. The revenue generated by producing x items is given by the formula: $R(x) = -x + 100x^{\frac{1}{2}}$. How many items are manufactured to produce a revenue of $900?

96. The profit generated by producing x items is given by the formula: $P(x) = -x^{\frac{1}{2}} + 50x^{\frac{1}{4}}$, where $P(x)$ is the profit in 1000's of dollars. How many items are manufactured to produce a profit of $400,000?

Section 2.5 Inequalities

We have spent three sections talking about solving equations. Let's take this section and discuss how to solve inequalities.

LINEAR INEQUALITIES

First, let's refresh our memories about how to solve linear inequalities. Linear inequalities are very similar to linear equations that you solved in section 0.5. The only difference between solving a linear equation and a linear inequality is that when you multiply or divide both sides by a negative, you must turn the inequality around. Here are some concrete examples to show you why this is.

Discussion 1: When to Reverse an Inequality

Given $[-3 < 2]$, when you multiply both sides by **−2** you get	$-2(-3)\ ?\ (-2)2 \Rightarrow 6\ ?\ -4$ Which is only true if we use the **<u>opposite</u>** inequality sign >. Hence we get $6 > -4$. The inequality sign must turned around inorder to get a true statement
Given $[-6 > -12]$, when you divide both sides by **−3** you get	$\frac{-6}{-3}\ ?\ \frac{-12}{-3} \Rightarrow 2\ ?\ 4$ Which is only true if we use the **<u>opposite</u>** inequality sign <. Hence we get $2 < 4$. The inequality sign must turned around in order to get a true statement
Given $[5 < 7]$, when you multiply both sides by **4** you get	$4(5)\ ?\ 7(4) \Rightarrow 20\ ?\ 28$ Which is only true if we use the **<u>same</u>** inequality sign <. So we get 20<28. The inequality sign must stay the same to get a true statement. Multiplying or dividing by a positive number doesn't change the order of the numbers.

Another way to think about what is happening here is that if you are changing the sign on something, by multiplying or dividing by a negative, you are changing big numbers into smaller ones (like above $-2(\underline{2}) = -4$) and small numbers into bigger ones (like above $-2(\underline{-3}) = 6$). Remember from your past that the negative sign means "opposite of". Let's now do our first example.

Example 1: Linear Inequality

Find the solution of $2(x+3) - 7 \geq 6 - 3(4x - 1)$, then graph the solution and write the interval notation form of the solution.

Solution:

First, just as with equations, clear parentheses.	$2x+6-7 \geq 6-12x+3$
Collect like terms.	$2x-1 \geq -12x+9$
Put x's on same side.	$14x-1 \geq 9$
Get numbers on the other side.	$14x \geq 10$
Now divide by a 14 (positive number so don't change the inequality).	$x \geq \frac{10}{14} = \frac{5}{7}$
The graph looks like this: (This was discussed in 1.2.)	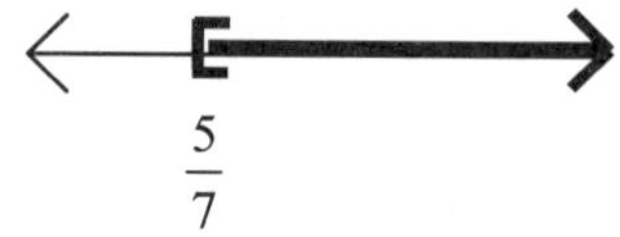
The interval notation is: (This was discussed in 1.2.)	$\left[\frac{5}{7}, \infty\right)$

Question 1: Solve the inequality $-3x+5>17$; then graph the solution and write the interval notation form of the solution. (Reminder: When you multiply or divide an inequality by a negative number you must turn the inequality symbol around.)

Example 2: Linear Double Inequality

Solve the inequality $-5<-4x-9 \leq 15$; then graph the solution and write the interval notation form of the solution.

Solution:

This is a double inequality. We can treat it just like any other linear inequality. Whatever you want to try to get the x alone, in the middle of the problem, be sure to do it to every different part of the problem (left, middle, right).	$-5<-4x-9 \leq 15$
First, add 9 to all parts.	$-5\underline{+9}<-4x-9\underline{+9} \leq 15\underline{+9}$ $4 < -4x \leq 24$
Now divide by –4. Notice this will cause the inequality symbols to turn around.	$-1 > x \geq -6$
The graph looks like this:	
The interval notation is:	$[-6, -1)$

ABSOLUTE VALUE INEQUALITIES

Let's turn our attention to solving absolute value inequalities. In section 0.5 we talked about how to solve absolute value equations. When you have an absolute value equation, you must solve two equations because of the definition of absolute value. Let's do one example just to remind ourselves how it works.

Example 2: Absolute Value Equation

Solve $|2x-5|=7$.

Solution:

Change into two equations (the absolute value is already isolated) because $\|7\|$ and $\|-7\|$ both equal 7. There are two ways to get 7 using absolute value.	$2x-5=7$	$2x-5=-7$
Solve each linear equation by dividing by 2. So our two solutions are:	$2x=12$ $x=6$	$2x=-2$ $x=-1$

Now things start to get interesting. There are many different types of inequalities. We have seen some linear ones already. We will talk about absolute value inequalities here. In a few more pages we will talk about the polynomial and rational kinds. Later in the book we will even see two-variable inequalities. What is interesting is that there are different ways in which to approach solving all these different types of inequalities. In our book though, we want to show you *one way* to do them *all*.

The **Test Point Method** can be used to solve all kinds of inequalities. The reason we like this method is that it turns every inequality into an equation, which you already know how to solve. Only one additional step is required: we'll test points to determine the correct solution intervals or shaded regions. Let's do example 1 again by this new method.

Discussion 2: Test Point Method

Change the problem to an equality.	$2x+6-7=6-12x+3$
Collect like terms.	$2x-1=-12x+9$
Move x's to the same side.	$14x-1=9$
Move numbers to the other side.	$14x=10$
Now divide by a 14. (Notice, we don't have an inequality symbol so you don't need to worry about whether to change the inequality symbol.)	$x=\frac{10}{14}=\frac{5}{7}$

5/7 makes us equal. In an inequality problem we want to be either < or > so 5/7 must be on the border between the shaded region we want and the region we don't want.	$2x + 6 - 7 > 6 - 12x + 3$
So plug a point into the inequality that is on one side or the other of our border to decide where to shade. Let's try $x = 0$ (easy to plug in).	$2(0) + 6 - 7 > 6 - 12(0) + 3$ $0 + 6 - 7 > 6 - 0 + 3$ $-1 > 3$
This is not true so 0 must be on the side we don't want to shade, so shade the other side of 5/7. It might be a good idea to try a number on that side to be certain that it is the correct side.	
$2(1) + 6 - 7 > 6 - 12(1) + 3$ $2 + 6 - 7 > 6 - 12 + 3$ $1 > -3$ (x = 1 checks out to be true) So the graph looks like this:	
The interval notation is:	$\left[\frac{5}{7}, \infty\right)$ We have a bracket because the original problem had a ≥ in it.

We arrived at the same answer as before, just by a different method. Now for linear inequalities, it is probably easier to just do them how you were taught (like in example 1 above). But this new way does eliminate the problem of forgetting to change the inequality symbol when you multiply or divide by a negative. We will leave it up to you to choose which way to solve linear inequalities.

Answer Q1:
$-3x > 12 \Rightarrow$
$x < -4$,
$(-\infty, -4)$

However, with all of the other types of inequalities it might be easier to learn one method of doing them rather than learning different methods for each one. By the way, you might be thinking, "Can I use my calculator?" The answer is "yes" and we will get to that soon.

In our next example, solving an absolute value inequality, we'll look at how you were most likely shown to solve it versus how we are suggesting you should solve it.

Example 3: Absolute Value Inequality

Solve $|2x-5| > 7$; then graph the solution. (Note: this is just like example 2 above, but this is an inequality.)

Solution:

The method you were most likely taught		Test point method	
You had to memorize that if the absolute value is > some number, it's a union problem (like this example), but if the absolute value is < some number, it's an intersection problem.	$2x-5>7 \cup 2x-5<-7$ $2x>12 \quad \cup \quad 2x<-2$ $x>6 \quad \cup \quad x<-1$ 	$\|2x-5\|=7$ $2x-5=7 \quad 2x-5=-7$ $2x=12 \quad 2x=-2$ $x=6 \quad x=-1$ 	1. Change to an equals 2. Solve two equations 3. Graph borders then check. We used 0. $\|2(0)-5\|>7 \Rightarrow \|-5\|>7$ $5>7$ failed! By checking –2 and 7 you would see that they work, so shade left and right.

Both methods work well. They get you to the correct answer. In this book we are going to stress the Test Point Method for the reasons stated earlier. Note, you must test every interval to be certain where you should shade and where you shouldn't.

Question 2: Solve the inequality $|3-2x|\le 1$; graph the solution and write the interval notation form of the solution.

Example 4: Solve with Graphing Calculator

Find the solution to question 2 by using your calculator.

Solution:

There are different ways to find the answers to inequalities on your calculator. We are going to show you a way that we think is easy and that stresses the concepts we want you to learn.

Move terms around to get the inequality compared to <u>0</u>. In this example, subtract 1 from both sides.	$\|3-2x\|\le 1$ $\|3-2x\|-1\le 0$
Graph $y=\|3-2x\|-1$. Notice once again how, when you graph something, its "Y =" (which in this case is, <u>abs(3 – 2*x*) – 1)</u> and that represents the height of the graph (outputs). Now we want <u>abs(3 – 2*x*) – 1</u> to be ≤ 0. This means we want y to be less than 0. On the graph that means locate where the graph is below the x-axis. In this case that	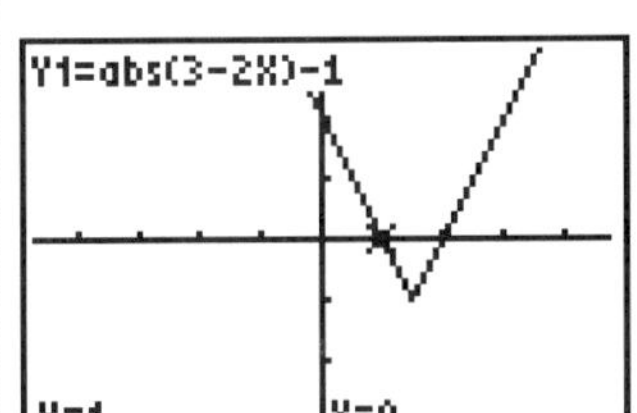

happens when the x's (inputs) are between 1 and 2.	
As far as the endpoints are concerned, if the inequality is $\leq$ or $\geq$, then you would include the ends. In this case our final answer is	[1, 2]

Example 5: Absolute Value Inequality

Find the solution to the inequality $|5x+6|>21$ by using the calculator.

Solution:

Graph $|5x+6|-21=0$. It is difficult to see where y is equal to 0. But as long as the important x-values are rational numbers you can determine the exact answer from you calculator. If you use the root or zero key that we talked about in section 0.8, then you can find where the output values are zero. In this example that happens at $x=\frac{-27}{5}$ and $x=3$. Since in this example $y>0$, the inputs that cause the outputs to be positive are $x<\frac{-27}{5}$ or $x>3$ (this is where the graph is above the x-axis). So the solution in interval notation form is $\left(-\infty,\frac{-27}{5}\right)\cup(3,\infty)$.	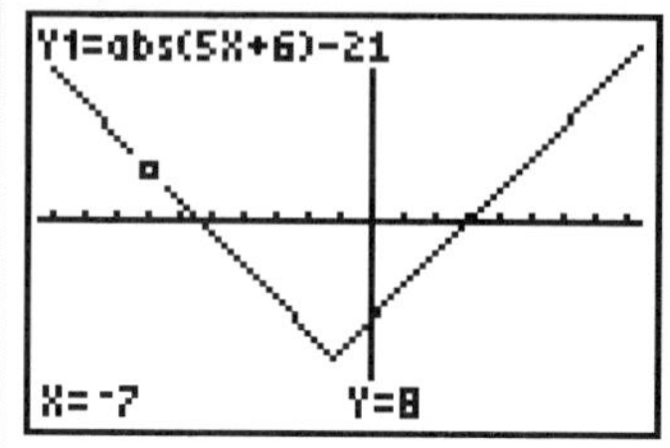 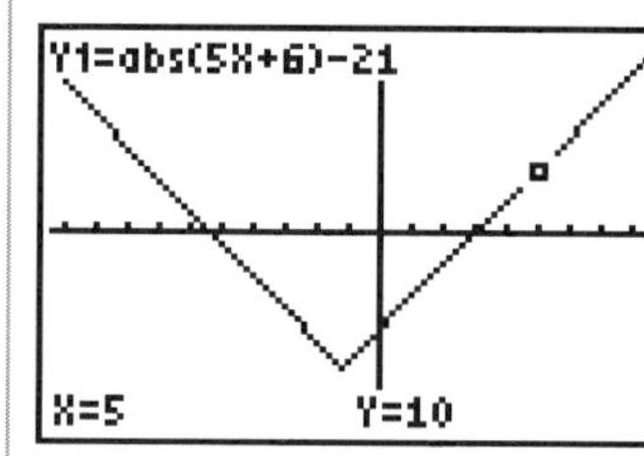

Question 3: Find the solution to the inequality $|1-4x|+5<0$ using your calculator.

Example 6: Absolute Value Inequality

Find the solution to the inequality $9+|3-x|\geq 0$.

Solution:

Since the graph is always above the x-axis and we are looking for positive y-values, our solution must be all real numbers or in interval form $(-\infty,\infty)$.

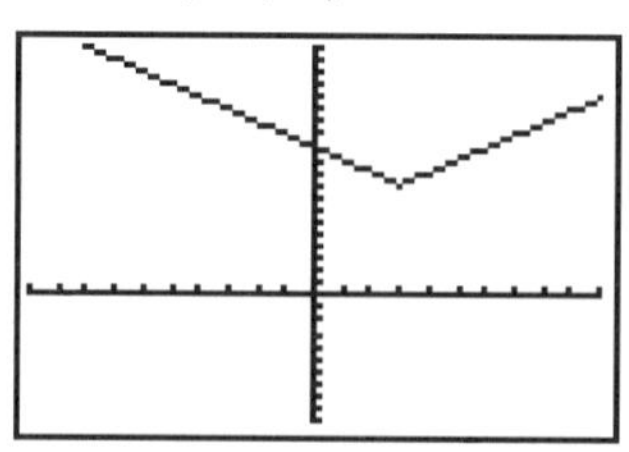

POLYNOMIAL INEQUALITIES

It is time for us to see examples of how we use the test point method to solve quadratic and larger polynomial inequalities and rational inequalities. Let's begin by looking at how we solve quadratic inequalities.

Example 7: Solving a Quadratic Inequality

Solve the inequality $x^2 + x \geq 2$; then write the solution in interval notation form.

Solution:

The first thing we do when using the Test Point Method is change the inequality to an equation after everything has been moved to one side.	$x^2 + x \geq 2$ $x^2 + x - 2 \geq 0$ $x^2 + x - 2 = 0$
Now solve this quadratic equation by the methods discussed in this chapter (in this case factoring works great).	$(x + 2)(x - 1) = 0$ $x + 2 = 0 \qquad x - 1 = 0$ $x = -2 \qquad x = 1$
Check where to shade (we'll use $x = 0$ as a test point in the original inequality). Notice "0" doesn't work. If you try –3 and 2 they do work.	$0^2 + 0 - 2 \geq 0$ $-2 \geq 0$ False! 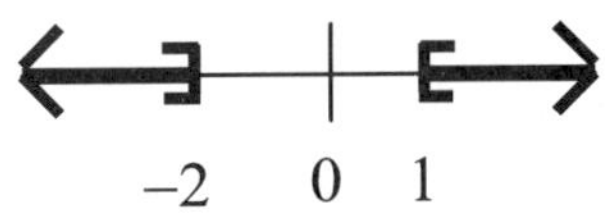
The interval notation is:	$(-\infty, -2] \cup [1, \infty)$
The calculator verifies our answer. Notice that as before, if we put$\text{Y1} = \text{X}^2 + \text{X} - 2$ into our calculators, then what we are looking for on the graph is where the Y's are ≥ 0 (positive).	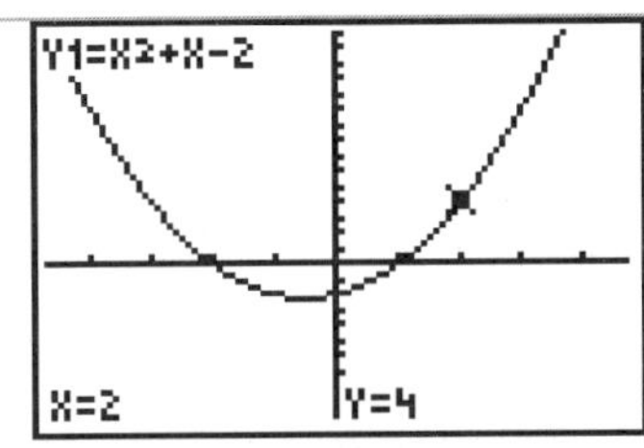

Answer Q2:
$|3 - 2x| = 1$
$3 - 2x = 1$
$-2x = -2$
$x = 1$
or
$3 - 2x = -1$
$-2x = -4$
$x = 2$
[1,2]

Remember, in order to use the Test Point Method, as described in this book, you need to get the inequality compared to 0 first, then solve the equation and finally test where to shade.

Question 4: Solve the inequality $x^2 + 6x < -8$; then write the solution in interval notation form.

Example 8: Solving a Polynomial Inequality

Solve the inequality $x(x - 5)(x + 3) \leq 0$; then write the solution in interval notation form.

Solution:

Notice this is a cubic. But we can solve all inequality problems by this method.	$x(x - 5)(x + 3) = 0$

Answer Q3: No solution. There aren't any x-values which could cause $|1-4x|+5$ *to be negative.*

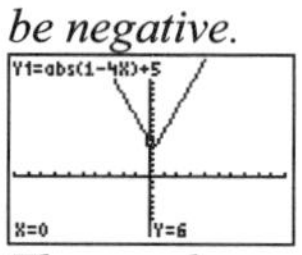

The graph never goes below the x-axis.

Now solve this equation by setting each factor equal to 0.	$x(x-5)(x+3)=0$ $x=0 \quad x-5=0 \quad x+3=0$ $x=0 \quad x=5 \quad x=-3$
Check where to shade (we will use $x = 1$ as a test point in the original inequality). Notice that 1 does work and so does –4. If you try –1 or 7 they don't work.	$1(1-5)(1+3) \le 0$ $-16 \le 0$ True! 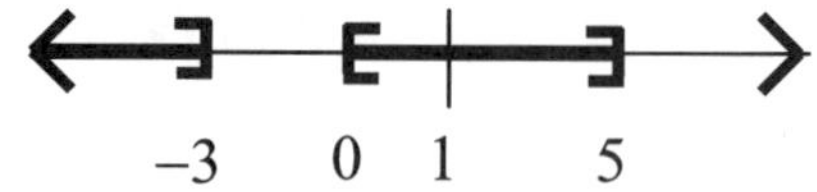
The interval notation is:	$(-\infty, -3] \cup [0, 5]$
The calculator verifies our answer. What we are looking for on the graph is where the Y's are ≤ 0 (negative).	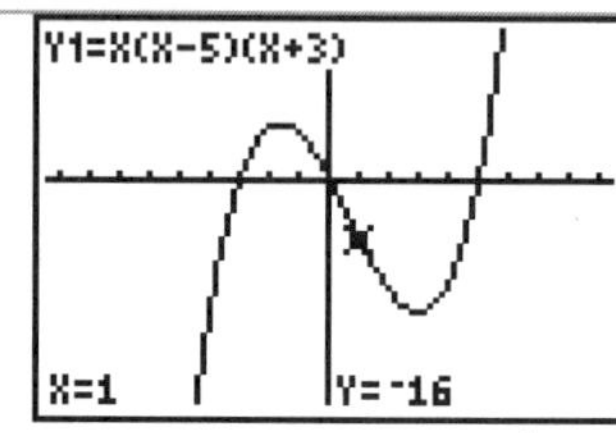

If you have a cubic or larger polynomial inequality and it's not factored, then in order to solve the equation needed to find the border points, you would need to use methods that we are going to discuss in the next chapter.

Rational Inequalities

Let's do some examples of how to use the Test Point Method on rational inequalities.

Example 9: Rational Inequality

Solve the inequality $\frac{x+2}{x-1} \le 0$.

Solution:

The first thing we do when using the Test Point Method is change the inequality to an equation.	$\frac{x+2}{x-1}=0$
Now solve this rational equation by first multiplying by the LCD, then subtract 2. We have found a border point.	$(x-1)\frac{x+2}{x-1}=0(x-1)$ $x+2=0$ $x=-2$
Rational equations have one major difference from polynomials: they can't have values for x that aren't in the domain. (Polynomials always have a domain of all real numbers.)	$x-1=0$ $x=1$

Because of this we must also find values for x that aren't in the domain.	This is the x-value that isn't in the domain (it makes the denominator = 0). But it *will be* another *border point* for us to consider
Check where to shade (in this case we will let $x = 0$ in the original inequality). Notice "0" does work. If you try 2 or −3, they don't work.	$\frac{0+2}{0-1} = -2 \leq 0$ $-2 \leq 0$ True! 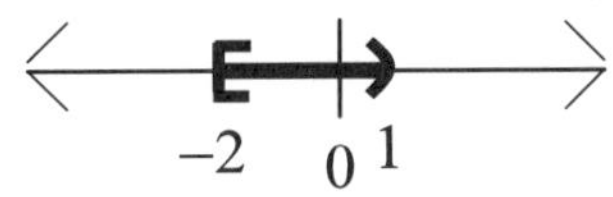
The interval notation has a bracket at −2 because the inequality was $\leq$ (−2 makes it equal). It has a parenthesis at 1 because we can't use $x = 1$; it's not in the domain.	[−2, 1)
The calculator verifies our answer (Y's are ≤ 0, negative).	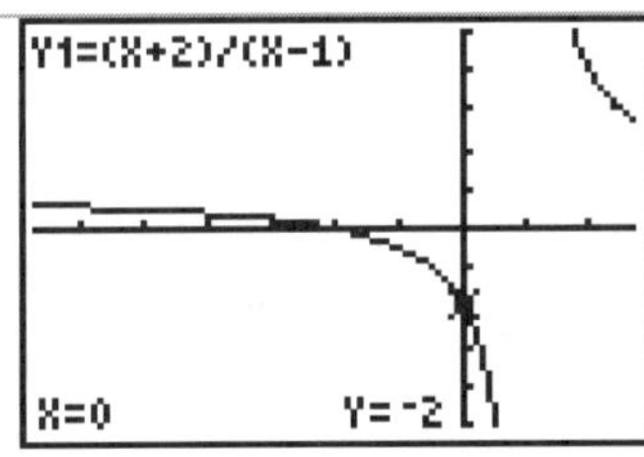

Answer Q4:
$x^2 + 6x + 8 = 0$
$(x+2)(x+4)=0$
$x = -2$ and
$x = -4$
$0^2+0<-8$, False

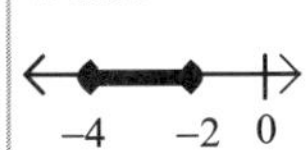

(−4, −2)

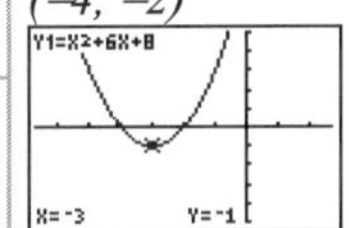

Question 5: What makes this rational expression $\frac{11x+7}{x^2-4x+3}$ undefined? (That is, what makes the denominator equal 0?)

Example 10: Rational Inequality

Solve the inequality $\frac{11x+7}{x^2-4x+3} \geq -1$; then write the interval notation form of the solution.

Solution:

The first thing we do is get everything on one side and change the inequality to an equation (notice that the rational expression part of the problem is the same as question 5).	$\frac{11x+7}{x^2-4x+3} + 1 = 0$
Now solve this rational equation by first multiplying by the LCD { $(x-3)(x-1)$ } then simplifying. Collect the like terms.	$11x + 7 + (x-3)(x-1) = 0$ $11x + 7 + (x^2 - 4x + 3) = 0$ $11x + 7 + x^2 - 4x + 3 = 0$ $x^2 + 7x + 10 = 0$

Factor. We have found the border points.	$x^2 + 7x + 10 = 0$ $(x + 2)(x + 5) = 0$ $x = -2 \quad x = -5$
From question 5 above we know that $x = 1$ and 3 are not in the domain so those two numbers are border points also.	$x = 1 \quad x = 3$
Check where to shade (in this case we will let $x = 0$ in the original inequality). Notice "0" does work. If you try –3 or 2 they don't work. –6 and 4 also work. $\dfrac{11(2)+7}{(2)^2-4(2)+3} = \dfrac{29}{-1} = -29 \geq -1?$ False.	$\dfrac{11(0)+7}{(0)^2-4(0)+3} = \dfrac{7}{3} \geq -1$ True 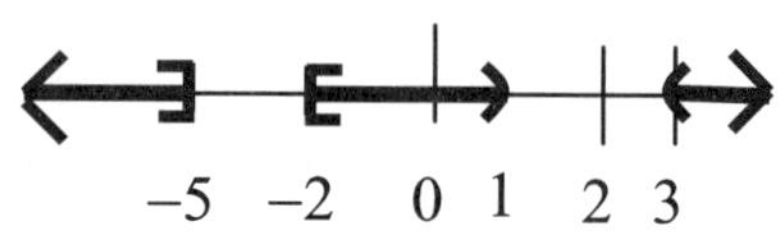
The interval notation has a bracket at –5 and –2 because the inequality was ≥ (–5 and –2 make it equal). It has parentheses at 1 and 3 because we can't use $x = 1$ or 3; they're not in the domain. The calculator verifies our answer (Y's are ≥ 0, positive).	$(-\infty, -5] \cup [-2, 1) \cup (3, \infty)$
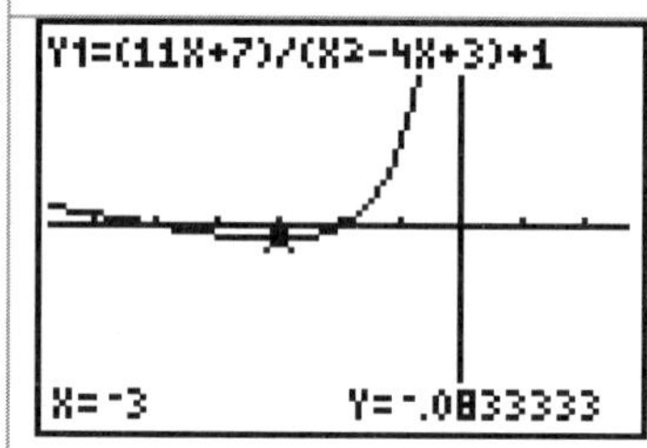	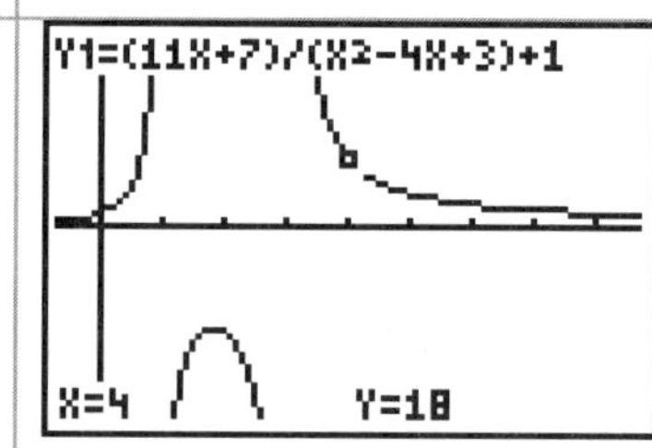

Section Summary

When you see an inequality problem think "test point method" and do the following steps:

- Change the inequality to an equation
- Solve the resulting equation (watch for x-values not in the domain, that is, values for x that would cause the denominator to equal zero)
- Test points, in between your border point, in the original inequality to find the places where to shade
- Write your final answer in interval notation form
- Once again your calculator can be a great help either in confirming your solution or assisting you in finding the solution.

SECTION 2.5 HOMEWORK

(1–8) Solve each linear inequality. (Express your answer with a number line and interval notation.) (Use your graphing calculator to confirm your answer.)

1. $4 - 3x > 5x + 12$
2. $-3(2x - 5) + 7 \geq 2(2x + 3) - 4$
3. $\frac{2}{3}x + \frac{3}{4} < \frac{5}{6}x - 3$
4. $\frac{5-3x}{2} + \frac{3}{4} \leq \frac{2x+3}{8} - 2$
5. $.35x - .13 \geq .05x + .73$
6. $.09x + .23 > .17x - .83$
7. $.3x + 9 \leq .8x - 3.5$
8. $2.3x - 3 < 1.5x + .4$

(17–30) Solve each absolute value equation. **(Use your graphing calculator to confirm your answer.)**

9. $|3x-5| = 15$
10. $|2x+3| = 7$
11. $2|3x-1| - 8 = 6$
12. $-3|2x-1| + 9 = 3$
13. $|5x-3| + 9 = 4$
14. $|2x+8| + 6 = 2$
15. $|3x-9| + 10 = 10$
16. $|2x-8| + 7 = 7$
17. $|2x-3| = |x+5|$
18. $|3x-7| = |7x+1|$

(19–32) Solve each absolute value inequality. Express your answer with a number line and interval notation. Use your graphing calculator to confirm your answer.

19. $|3x-2| > 17$
20. $|2x+7| - 9 > 23$
21. $|2x+3| < 8$
22. $3|3x-1| - 4 < 5$
23. $|2x+3| - 5 \geq 12$
24. $2|3x-1| + 4 \geq 20$
25. $|3-2x| \leq 7$
26. $3|3x+2| - 5 \leq 10$
27. $|2-3x| + 5 \leq 1$
28. $-2|2x+3| - 3 \geq 5$
29. $|3x+5| > -3$
30. $-3|2x-9| < 15$
31. $|3x-8| + 10 \geq 10$
32. $3|2x+5| + 12 \leq 12$

Answer Q5:
$x^2 - 4x + 3 = 0$
$(x-1)(x-3) = 0$
$x = 1$ or 3

(33–48) Solve each quadratic inequality. Express your answer with a number line and interval notation. Use your graphing calculator to confirm your answer.

33. $x^2 - 3x \geq 28$
34. $x^2 + 7x + 10 \geq 0$
35. $x^2 > 36 - 5x$
36. $3x^2 + 10 > 11x$
37. $x^2 \leq 25$
38. $5x^2 \leq 9x + 18$
39. $\frac{1}{2}x^2 + 3x < \frac{27}{2}$
40. $\frac{2}{3}x^2 + 2x > \frac{9}{2}$
41. $x^2 + 25 > 0$
42. $x^2 + 3x + 7 \geq 0$
43. $x^2 + 5x + 9 \leq 0$
44. $2x^2 + 3x + 5 < 0$
45. $x^2 + 4x + 2 \geq 0$
46. $2x^2 > 3x + 4$
47. $x^2 + 3x < 5$
48. $3x^2 \leq 5x + 1$

(49–70) Solve each rational polynomial inequality. Express your answer with a number line and interval notation. Use your graphing calculator to confirm your answer.

49. $\frac{x-3}{x+2} \geq 0$
50. $\frac{2x+3}{3x-2} > 0$
51. $\frac{x-5}{x+1} < 0$

52. $\frac{2x-3}{3x+1} \le 0$

53. $\frac{x^2-9}{x^2-3x-10} > 0$

54. $\frac{x^2-16}{x^2-5x+6} \ge 0$

55. $\frac{x^2+7x+12}{x^2+3x-18} \le 0$

56. $\frac{2x^2-11x+15}{3x^2-12} < 0$

57. $\frac{x^2+3x+1}{x^2-3} \ge 0$

58. $\frac{x^2-3x-5}{x^2-5} < 0$

59. $\frac{x^2+9}{x^2-16} \le 0$

60. $\frac{x^2-25}{x^2+4} > 0$

61. $\frac{x+2}{x-3} \ge 2$

62. $\frac{x-3}{x+2} < 3$

63. $3 - \frac{x-4}{x-3} > 0$

64. $5 + \frac{x+2}{x-4} \le 0$

65. $\frac{3}{x-2} + \frac{2}{x+3} \le 0$

66. $\frac{5}{x+1} - \frac{3}{x-4} > 0$

67. $\frac{2x}{x-2} + \frac{1}{x+1} \ge 0$

68. $\frac{x}{x+1} - \frac{2}{x-1} < 0$

69. $\frac{(x-1)(x+2)(x-3)}{x(x-2)(x+4)} \le 0$

70. $\frac{(x+2)(2x-1)(x-3)}{(3x+1)(x+3)(x-2)} \ge 0$

(71-74) Solve each higher degree polynomial inequality. Express your answer with a number line and interval notation. Use your graphing calculator to confirm your answer.

71. $(x-1)(x+2)(x-5) \ge 0$

72. $(x+2)(3x-1)(x-4) > 0$

73. $(x-1)(x+2)(2x-3)(3x+5) \le 0$

74. $x^3 - 5x^2 - 6x < 0$

(75-80) Mathematical modeling using quadratic inequality for solving.

75. The fish population of a lake in 1,000's over a number of years is give by the equation,

$P(t) = \frac{600t}{t^2+35} + 20$, where t is the time in years and $P(t)$ is the fish population. Find the time interval where the population is greater than 70,000.

76. The population in 100's of a certain town in the Midwest is given by the equation,

$P(t) = \frac{12000t}{4t^2+16} + 400$ where t is the time in years from 1982. From what year to what year is the population greater than 100,000?

77. A rocket is shot in the air from ground level with an initial velocity of 320 ft./sec. Over what time interval is the rocket greater than 816 feet above the ground?
(Hint: To set up the equation look at problem #75 on page 238.)

78. Using the problem above, on what time interval is the rocket 1,344 feet above the ground?

79. You own a small business that sells go-carts. The profit you make from selling x go-carts is given by the following equation: $P(x) = -0.1x^2 + 60x - 100$. What is the range of go-carts you must sell if you want your profit to exceed $7900?

80. You own 50 rental properties. If you charge $600 rent you can rent all 50, but if you raise the rent to $800 you can rent only 40 of the properties. The revenue you receive from the rentals is given by the following equation: $R(x) = -20x^2 + 1600x$ where x is the number of properties

rented. If you want your revenue to be more than \$31,500 what is the range of the number of properties you must rent?

Chapter 2 Review

Topic	Section	
Square roots of negative numbers	2.1	We must convert any square root with negative numbers into its imaginary form before we can perform mathematical operations such as multiplication.
Simplifying complex numbers (+, –, •)	2.1	Adding, subtracting, and multiplying of complex numbers are just like operations with functions; you simply combine like terms as always.
Division with complex numbers	2.1	With division, we must multiply by a conjugate in order to eliminate the square root in the denominator.
Definition of a quadratic equation	2.2	Any equation where the highest power on the variable is 2.
Solving quadratic equations	2.2	1. You want to first see if you can use the square root method. 2. If the problem is not of that type, then get it equal to zero and try to factor it. 3. If after 30 to 60 seconds you haven't figured out how to factor it, or it can't be factored, then use the quadratic formula.
Solving equations with roots or equations with fractional exponents	2.3	1. Isolate the variable part first. 2. Eliminate the root or fractional exponent second. Watch for the need of a $\pm$ symbol. 3. Finally check to see that all of your answers make the original equation true and that none of them are extraneous solutions.
Steps to solve equations	2.4	1. Eliminate parentheses, brackets, etc. 2. Eliminate fractions, decimals, square roots. 3. Simplify what remains. 4. Ask yourself the question "Is what I now have linear or quadratic like?" 5. If it is linear, isolate the variable. 6. If it is quadratic-like, then look to see if you can use any of the three methods we have discussed so far in this chapter. 7. Confirm your solution with your graphing calculator.
Solving inequality problems using the test point method	2.5	1. Change the inequality to an equation. 2. Solve the resulting equation (watch for x-values not in the domain, that is, values for x that would cause the denominator to equal zero).

		3. Test points, in between your border point, in the original inequality to find the places where to shade. 4. Write your final answer in interval notation form. 5. Once again your calculator can be a great help either in confirming your solution or assisting you in finding the solution.

CHAPTER 2 REVIEW HOMEWORK

Section 2.1

(1–4) Write each of the following complex numbers in standard form.

1. $\sqrt{-49}$ 2. $5+\sqrt{-16}$ 3. $7-\sqrt{-50}$ 4. $\dfrac{5+\sqrt{-45}}{3}$

(5–14) Simplify and write the answers in the form $a + bi$.

5. $(7 + 16i) + (12 - 8i)$ 6. $(7 - 9i) - (9 + 6i)$ 7. $(8 + \sqrt{-16}) + (7 + \sqrt{-25})$

8. $(3 + 2i)(5 - 3i)$ 9. $(8 - 9i)(8 + 9i)$ 10. $(\sqrt{-7})(\sqrt{-38})$

11. $(5 - \sqrt{-4})(3 - \sqrt{-49})$ 12. $\dfrac{2}{3-5i}$ 13. $\dfrac{2+3i}{5+2i}$ 14. $\dfrac{3+\sqrt{-18}}{2+\sqrt{-5}}$

(15–16) Simplify

15. i^{13} 16. i^{27}

Section 2.2

(17–28) Using an appropriate method, give real and complex solutions.

17. $x^2 - 3x = 10$ 18. $5x^2 + 24 = 26x$ 19. $3x^2 - 10 = 0$

20. $2x^2 + 9 = 0$ 21. $(5x - 3)^2 = 7$ 22. $3x^2 = 2x + 3$

23. $5x^2 + 9x + 3 = 0$ 24. $4x^2 + 20x + 25 = 0$ 25. $3x^2 + 2x + 4 = 0$

26. $\dfrac{2}{3}x^2 + \dfrac{3}{4}x = \dfrac{5}{12}$ 27. $\sqrt{5}x^2 + 7x + \sqrt{5} = 0$ 28. $8x^2 + 48x - 16 = 0$

(29–30) Use the quadratic formula and your calculator to solve each quadratic equation to 3 approximate decimal places.

29. $3.81x^2 = 2.73x + 1.85$ 30. $6.2x^2 + 4.23x = 2.83$

31. You are manufacturing guttering for homes. The piece of tin you use to make the gutter from is a rectangular piece where the length is 20 inches more than the width. You are going to bend along the length equal four inch pieces on each side to form the gutter. If the gutter's volume is 1152 in^3 what are the dimensions of the piece of tin you used?

32. Using the length and width of the piece of tin you found in problem 31, use the calculator to find how large the bend should be to maximize the volume of the gutter and give that maximum volume?
(Maximizing, section 1.3.)

33. Two ships are on a collision course approaching each other at a right angle. When the distance between them is 15 km, one ship is 3 km closer to the point of intersection than the other. How far from the point of intersection is each ship?

34. An arrow is shot straight up in the air. The arrow is 4 feet above the ground when it leaves the bow and is traveling at 124 feet per second. (Use the formula from section 2.2, problem 65.).
    a. How many seconds have passed when the arrow is 188 feet above the ground?

b. Approximately, to two decimal places, when will the arrow hit the ground?

c. Using the graphing calculator, approximate when the arrow will reach its maximum height.

(Maximizing, section 1.3.)

35. You own 20 homes that you rent. If you charge $850 you can rent all 20. The relation of number of homes rented to the rent you charge is given by the formula: $p = -10x + 1050$, where p is the rent charged and x is the number of home rented. The formula for revenue, R, is $R = px$.

a. What is the revenue formula in terms of x? $R(x) =$

b. The revenue is $14,240. How many homes did you rent and what rent did you charge?

c. Using $R(x)$ and the graphing calculator approximate how many homes you need to rent to maximize revenue and find the maximum revenue .

(Maximizing, section 1.3.)

Section 2.3

(36–47) Solve each radical equation for real number answers only. (Use your graphing calculator to confirm your answers)

36. $\sqrt{3x-2} = 4$ 37. $\sqrt{7x+3} - 2 = 8$ 38. $\sqrt{3x+2} + 8 = 5$

39. $\sqrt{3x+13} = x+3$ 40. $\sqrt{2x-5} + x = 4$ 41. $\sqrt{x+3} = x+3$

42. $\sqrt[3]{2x+1} = 3$ 43. $\sqrt{3x+1} - \sqrt{2x-1} = 1$ 44. $\sqrt{x+5} - \sqrt{3x+4} = 7$

45. $\sqrt{2x+3} + \sqrt{3x+1} = 2$ 46. $\sqrt{3x+3} - \sqrt{x+2} = \sqrt{3x-5}$ 47. $\sqrt{2x+3} - \sqrt{x-2} = \sqrt{x+1}$

(48–55) Solve each fractional exponential equation for real number only.

48. $x^{\frac{2}{3}} = 16$ 49. $x^{\frac{3}{5}} = 27$ 50. $x^{\frac{4}{3}} - 5 = 11$

51. $(3x-1)^{\frac{3}{2}} = 64$ 52. $(2x+3)^{\frac{2}{5}} = 4$ 53. $x^{\frac{-2}{3}} = 16$

54. $x^{\frac{-3}{4}} = 8$ 55. $(2x-1)^{\frac{-3}{2}} = 27$

56. The distance from the point (–2, 5) on a coordinate system to two different points on the line $y = -2x + 3$ is 5. What are the two points?

57. A shoe manufacturer of a very expensive shoe has found the cost of producing x shoes is given by the formula $C(x) = 200(10 + 5\sqrt{x})$. If the cost for a week was $12000, how many shoes did the manufacturer produce that week?

Section 2.4

(58–67) For each fractional equation give the domain and solve for x. (Use your graphing calculator to confirm your answers.)

58. $\frac{2}{5}x - \frac{3}{4} = \frac{7}{20}$ 59. $\frac{3}{x-3} + \frac{2}{x+3} = \frac{18}{x^2-9}$ 60. $\frac{2}{x-2} + \frac{3}{x} = 1$

61. $\frac{3}{x+2}-\frac{2}{x+3}=\frac{3}{2}$

62. $\frac{x}{x-2}-\frac{12}{x^2-4}=\frac{3}{x+2}$

63. $\frac{3}{x-3}+\frac{2}{x^2-5x+6}=\frac{x}{x-2}$

64. $\frac{2x}{x-4}+\frac{3}{x+4}=\frac{9}{x^2-16}$

65. $\frac{2}{x+2}-\frac{3}{x+3}=3$

66. $\frac{x}{x-1}-\frac{2}{x+3}=\frac{2}{3}$

67. $\frac{2}{x-1}+\frac{1}{x+2}=\frac{3}{x-3}$

(68–77) Find the real and complex roots for each of the following equations. (Use your graphing calculator to confirm your answers)

68. $x^4-8x^2+15=0$

69. $x^4+100=29x^2$

70. $x^4+10x^2+9=0$

71. $x^4+9x^2+20=0$

72. $x^4=36-5x^2$

73. $16x^4=24x^2+27$

74. $x^{\frac{1}{2}}-7x^{\frac{1}{4}}+10=0$

75. $x^{\frac{2}{3}}-6x^{\frac{1}{3}}+8=0$

76. $x^{-4}+8x^{-2}=9$

77. $x^{-2}+10x^{-1}+21=0$

78. It takes the hot water faucet one minute longer to fill a tub then the cold-water faucet. If both faucets can fill the tub in 5 minutes, how long does it take the cold-water faucet to fill the tub? (Approximate to one decimal place.)

79. One pump takes 3 hours longer to drain a swimming pool then a second pump. Working together they can drain the pool in 24 hours. How long will it take the second pump working by itself to drain the pool? (Approximate to one decimal place.)

80. Your company is hired to clean houses. You can ordinarily clean this particular person's house in x hours. To put the house back in the same shape it was before you started cleaning it takes the children of the house two more hours than the time you took to clean it. If the couple left the children home while you were cleaning the house, it took 4 hours for your company to clean the house. How long does it normally take your company to clean the house?

Section 2.5

(81–100) Solve each inequality. (Express your answer with a number line and interval notation.) Confirm your answers with your graphing calculator for problems (83–96).

81. $5(3-2x)+3\le -2(2-3x)+4$

82. $\frac{2}{3}(2x-3)+2>\frac{3}{4}(6x+1)-3$

83. $|2x-3|+8>15$

84. $3|2x-1|+4\le 25$

85. $|3x-1|+9<6$

86. $|2x-1|+8>5$

87. $x^2-2x\le 15$

88. $x^2+3>5x$

89. $x^2+3x+7<0$

90. $(x-2)(x+3)(x+1)\ge 0$

91. $x^3<3x^2+40x$

92. $\frac{x+3}{x-4}<0$

93. $\frac{3x-1}{2x+5}\ge 0$

94. $\frac{x-2}{x^2-9}\le 0$

95. $\frac{x-3}{x+2}>3$

96. $\frac{x+2}{x-3}+\frac{x+2}{x-2}>0$

97. $\frac{(x-3)(x-1)(x+3)}{x(x-2)(x+1)}\ge 0$

98. $\dfrac{x^2+4}{x^2-16}<0$

99. $\dfrac{x^2-3}{x^2+4x+1}\le 0$

100. $\dfrac{4^2x-9}{x^2-3x+5}>0$

CHAPTER 2 EXAM

1. Simplify

   a. $(4 - 2i) + (6 - 5i)$ b. $(3 + 5i) - (8 - 6i)$ c. $(3 - 5i)(4 + 3i)$ d. $\dfrac{1+3i}{3-2i}$

   e. $\left(2+\sqrt{-4}\right)-\left(5+\sqrt{-25}\right)$

2. Solve by an appropriate method:

   a. $3x^2 - 4x = 15$ b. $9x^2 = 1 - 12x$ c. $4x^2 + 13 = 8x$ d. $\dfrac{4}{9}x^2 = \dfrac{4}{3}x + 4$

3. Use your calculator to solve $3.12x^2 - 2.14x - 1.2 = 0$ to 3 decimal places.

(4 – 24) Solve.

4. $\sqrt{2x-1} = x-2$
5. $\sqrt{3x+2} = x+1$
6. $\sqrt{2x+3} - \sqrt{x+3} = 1$
7. $x^{\frac{3}{2}} + 1 = 9$
8. $x^{\frac{2}{3}} = 81$
9. $(2x-3)^{\frac{-2}{3}} = 9$
10. $\dfrac{2}{x-4} + \dfrac{6}{x+1} = 1$
11. $\dfrac{3}{x+2} - \dfrac{2}{x+3} = 5$
12. $\dfrac{2}{x+3} - \dfrac{5}{x+1} = 1$
13. $x^4 - 3x^2 = 54$
14. $x^{\frac{2}{3}} - 9x^{\frac{1}{3}} + 20 = 0$
15. $x^{-2} - x^{-1} = 20$
16. $2|3x-1| - 3 \geq 7$
17. $3|2x+5| + 4 < 13$
18. $x^2 - 3x \leq 28$
19. $x^2 > 5x + 36$
20. $(x - 2)(x + 3)(2x + 1) \geq 0$
21. $3x^2 + 5x + 4 < 0$
22. $\dfrac{x-3}{x+2} \geq 0$
23. $\dfrac{x+2}{x-3} < 2$
24. $\dfrac{x^2-4}{x^2-3x-4} \leq 0$

25. A certain company produces umbrellas. The weekly profit the company receives for selling x umbrellas is given by $P(x) = -0.01x^2 + 20x - 3000$; $P(x)$ is the profit.

    a. If the profit for the week is \$6600, how many umbrellas were sold?

    b. Using the graphing calculator, determine how many umbrellas need to be sold to maximize profit?

26. Two different points of the line $y = 3x - 2$ are a distance 5 from the point $(-1, 8)$. What are the two points?

27. A car manufacture needs to paint 30 cars. The manufacture has two paint machines. One paint machine can paint the 30 cars by itself 2 hours faster then the other paint machine. The two paint machines working together can do the 30 cars in 4 hours. Approximately how long does it take each paint machine working alone to paint the 30 cars? (Give your answer to two decimal places.)

28. The population of a small town in California in 100's is given by the formula

    $P(t) = \dfrac{100t}{t^2+4} + 300$, where $t = 0$ in 1998.

a. In what year will the population be 32,000?

b. Using the graphing calculator, find what year the population will be a maximum and what is the maximum population?

Chapter 3 Polynomial Functions

In this chapter we are going to look at a family of functions called polynomials. The members of the family are made up of terms that have variables to whole number powers. These are functions that you have undoubtedly spent a lot of time doing already in your math lives.

Section 3.1 Introduction to Polynomials

Let's begin this chapter by spending some time looking informally at the first three types of polynomial functions, which are:

Type	General Form	Degree	Example
1.) Constant	$f(x) = c$	degree 0	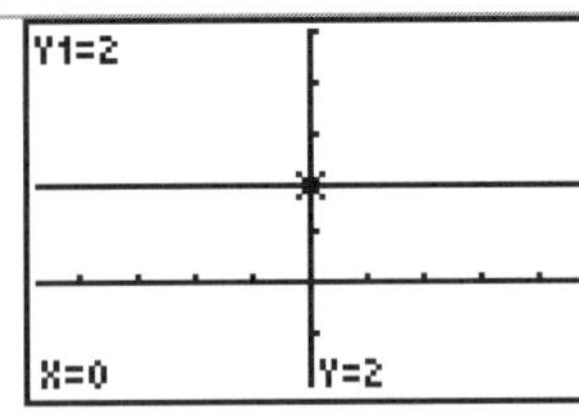
2.) Linear	$f(x) = mx + b$	degree 1	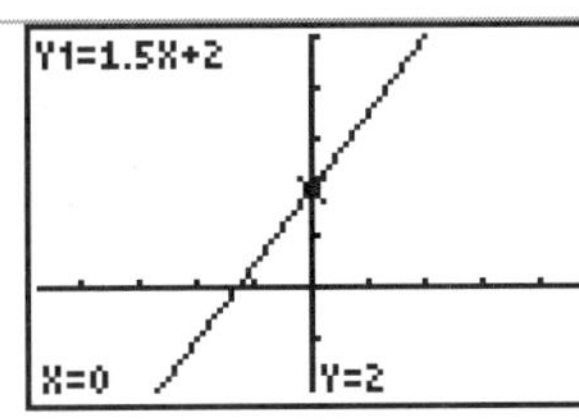
3.) Quadratic	$f(x) = ax^2 + bx + c$	degree 2	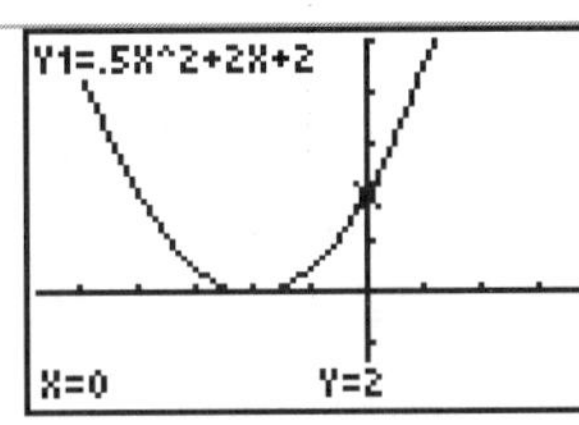

NOTE: Remember that the degree of a polynomial, when there is only one variable in it, is basically the highest power of the variable in the polynomial function.

CONSTANT POLYNOMIALS (x^0)

A constant polynomial is of the form $f(x) = c$.

Discussion 1: Distance from Home to School

The word **constant** probably gives you the impression that "it," whatever "it" is, isn't changing. Or it might give you a sense that the rate of change of the thing is zero. A good example of something that is a constant function is the distance between your house and the school you are currently attending. Your house isn't moving away or towards your school and your school isn't moving away or towards your house. The distance between them, we would say, is remaining constant. The math formula for our example might look like this: $d(t) = c$, where d is the name of the function (distance) and $d(t)$ the output variable, t (time) is the input variable and c is the exact distance between your house and school. Notice, if we asked you, "What is $d(3)$?" your answer would be, "$d(3) = c$." The output is always the same regardless of the input.

Question 1: Find the following function values $d(0)$, $d(6)$, $d(9)$ from discussion 1.

It doesn't matter what value you plug in for t; you always get c for the answer. Your house and school are always the same distance apart. This function is constant; it always yields the same output regardless of the input.

One way to investigate how things change is to look at something called **common differences**. To calculate common differences you *subtract output values over equal time intervals*.

Example 1: Common Differences

Find the common differences given the function $d(t) = c$, the aforementioned formula.

Solution:

t	d	common differences
0	c	
3	c	$c - c = 0$
6	c	$c - c = 0$
9	c	$c - c = 0$

So the common differences in this case are all the same; they equal 0. We get 0 as the common difference because the outputs aren't changing. When you subtract equal amounts, you get 0.

Question 2: Does the following appear to be an example of a constant function? Why or why not? (Evaluate the common differences to help you justify your answer.)

t	B
1	38
3	43
5	48
7	53

LINEAR POLYNOMIALS (x^1)

A linear polynomial is of the form $f(x) = mx + b$.

You may have noticed in the last question that the common differences were the same *but didn't equal 0*. This is an example of a linear function. In the real world very little remains constant. So, the constant function, or what we would call the constant model, isn't one you will see very often.

Let's move on to a type of function that is much more prevalent in the real world, the linear function (model). A linear function, as mentioned earlier, is of the form $f(x) = mx + b$. From previous math courses you know that m represents the slope of the line and b is the y-intercept.

Discussion 2: Finding Slope

Looking again at the table in question 2, let's find the equation of the linear function that fits the table of values.

First find the slope using the slope formula $m = \frac{y_2 - y_1}{x_2 - x_1}$. Notice how it won't matter which points we use – we'll get the same answer every time.	$m = \frac{43-38}{3-1} = \frac{5}{2} = 2.5$ $m = \frac{48-43}{5-3} = \frac{5}{2} = 2.5$ etc…(Discussed more in section 5.1)
Next, after you find the slope you must use the point slope formula for a line $y - y_1 = m(x - x_1)$. (Discussed more in section 5.1)	$B - 38 = 2.5\,(t - 1)$ $B - 38 = 2.5t - 2.5$ $B = 2.5t + 35.5$
So, in function notation we write our final answer in this form	$B(t) = 2.5t + 35.5$

Notice how the common difference of the output values, in this case 5, divided by the unit of time intervals, in this case 2, equals the slope of the linear function $\left(m = \frac{5}{2}\right)$. This will always happen if you are looking at something that is linear in nature.

If the common differences are the same between every point, with equal time intervals, then what you are seeing is something that is *changing at a constant rate*. Or more simply, it means that you are adding the same amount each time. It has the same rate of change at all times. That is quite different than a constant function, which *never changes*. Constant functions have a zero rate of change (they could be written $y = 0x + b$, slope = 0) while a linear function has a constant rate of change. So what about quadratic functions? Well, let's discuss them for a while.

QUADRATIC POLYNOMIALS (x^2)

A quadratic polynomial is of the form $f(x) = ax^2 + bx + c$.

Example 2: Baseball

Find the common differences for the table below. The function $f(t)$ in this table gives the height of a baseball, in feet, t seconds after being hit. This is only one of many different functions that could describe the height of a baseball since the ball can be hit in so many different ways.

T (sec.)	$f(t)$ (feet)
0	3
1	83
2	131
3	147
4	131
5	83
6	3

Answer Q1:
d(0) = c,
d(6) = c,
d(9) = c

Solution:

t (sec.)	$f(t)$ (feet)	Common Difference
0	3	
1	83	83 – 3 = 80
2	131	131 – 83 = 48
3	147	147 – 131 = 16
4	131	131 – 147 = –16
5	83	83 – 131 = –48
6	3	3 – 83 = –80

The common differences were not the same and they were not all 0, so this can't be linear or constant. What if we tried doing the common differences of the common differences?

Answer Q2:
No, the common differences don't equal 0.

B	*Common Difference*
38	
43	43–38 = 5
48	48–43 = 5
53	53–48 = 5

Question 3: Find the common differences of the common differences.

The common difference of the common differences is –32. This is exactly what happens with quadratic polynomial functions. In this example the function just happens to be $f(t) = -16t^2 + 96t + 3$.

Now from what we looked at earlier in this section, the common differences divided by the time interval gave us the rate of change. Maybe second common differences give us rate of change of the rate of change? As it turns out that is exactly what it means and in Physics you would call that **acceleration**. The first common differences, divided by the equal time intervals, give you rate of change (**velocity**) and the second common differences, divided by the equal time intervals twice, give you acceleration.

To recap:

- A *zero* rate of change means we have a *constant* function.
- A *constant* rate of change means we have a *linear* function (add the same amount each time).
- A changing rate of change that is constant (where the second common differences are equal, called a *linear rate of change*) means we have a *quadratic* function.

Now we could continue on with this discussion and talk about how the third common differences, if they are the same, tell you that you have a cubic function (quadratic rate of change), and fourth common differences, if they are the same, tell you that you have a fourth-degree polynomial (cubic rate of change) and on and on and on. But let's stop our discussion for now at quadratic functions.

Example 3: Looking at Common Differences

Determine whether the following table is an example of a constant, linear or quadratic polynomial function.

X	$C(x)$	Common difference
10	13.9	
11	15.29	15.29 – 13.9 = 1.39
12	16.68	16.68 – 15.29 = 1.39
13	18.07	18.07 – 16.68 = 1.39
14	19.46	19.46 – 18.07 = 1.39

Solution:

As you can see from the common differences, this is an example of a linear function. This is true because the first common differences are all the same. You might recognize this problem from chapter 1, section 1. This was the third example about how gas price and total amount paid for the gas you put into your car is a function. As you can see now, that was an example of a real-life problem that is a linear polynomial function.

Question 4: Determine whether the table at right is an example of a constant, linear or quadratic polynomial function. Use common differences to help you decide.

X	$P(x)$
10	10
15	22.5
20	30
25	32.5
30	30
35	22.5
40	10

This last question, again, was a look back at section 1.3, example 16. Notice that since the second common differences were the same (–5), this must be a quadratic polynomial function.

Question 5: Determine, to the best of your ability, whether the following graph is an example of a constant, linear or quadratic polynomial function. Explain your answer.

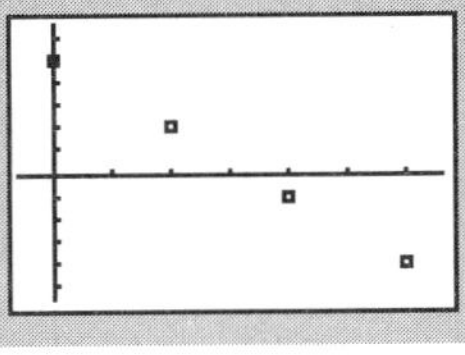

What we want you to see at this point is how similar and yet how different the polynomial functions are. In one respect they are similar because of how the degree (power) of the polynomial function is the same as the number of successive common differences you must find before the answers are the same.

Answer Q3:

cd	*cd of cd*
80	
48	-32
16	-32
−16	-32
−48	-32
−80	-32

Also,

- No common differences are needed for a constant polynomial. You can see that the outputs are all the same.
- One common difference is needed for a linear polynomial. After finding the first common differences you see that the answers are all the same.
- Two common differences are needed for a quadratic polynomial. After finding the second common differences you see that the answers are all the same, and so on.

They are also similar in that a zero rate of change is a constant. A constant rate of change is linear. A linear rate of change is quadratic and so on. Each type of polynomial function is one step above the previous.

In another respect they are different in that with each increase in degree (power) the polynomial functions behave much more erratically. That is, the outputs change from one to the other in a manner that is more and more difficult to figure out. The polynomial function, depending on its power, increases, then decreases, then increases again and on and on, such that it becomes more and more difficult to discern the pattern. Later on, in this book, we will see functions that are different than polynomials.

Section Summary

- You subtract successive pairs of output to find the common differences.
- If all of the common differences are the same then the function is linear.
- If the common differences of the common differences are the same, the function is quadratic.
- The common difference divided by the time interval equals the slope of the line.

SECTION 3.1 HOMEWORK

(1–12) Using the following table of values:

a. Create a stat plot using your graphing calculator and use the graph to see if you think the points form a constant, linear, quadratic or none of these types of function.

b. Using the common difference idea on the table, verify your guess.

1.

1	9
2	−4
3	−17
4	−30
5	−43

2.

2	−5
4	1
6	7
8	13
10	19

3.

1	−2
2	7
3	22
4	43
5	70

4.

2	11
4	−1
6	−21
8	−49
10	−85

5.

3	−7
4	−7
5	−7
6	−7
7	−7

6.

−3	5
0	5
3	5
6	5
9	5

Answer Q4: quadratic

cd	*cd of cd*
12.5	
7.5	−5
2.5	−5
−2.5	−5
−7.5	−5
−12	−5

7.

5	17
0	2
−5	−13
−10	−28
−15	−43

8.

9	−29
6	−17
3	−5
0	7
−3	19

9.

0	1
1	3
2	9
3	27
4	81

10.

1	2
2	4
3	8
4	16
5	32

11.

0	−3
2	1
4	13
6	33
8	61

12.

0	2
−1	1
−2	−2
−3	−7
−4	−14

Answer Q5: Linear, the common difference is −3.

(13–22) For each function, create a table of values using $x = \{0, 1, 2, 3, 4, 5\}$and check the differences to verify what type of function each is:

13. $f(x) = 3x - 2$

14. $f(x) = -5x + 3$

15. $f(x) = -2x^2 + 3$

16. $f(x) = 3x^2 - 5$

17. $f(x) = -8$

18. $f(x) = 3$

19. $f(x) = 3x^2 - 5x + 1$

20. $f(x) = -2x^2 + 5x + 3$

21. $f(x) = \frac{2}{3}x - 4$

22. $f(x) = \frac{-3}{4}x + 5$

(23–34) Using the information given in each of the problems:

a. Create a stat plot using your graphing calculator for each problem and make a guess at whether the problem is a constant function, a linear function, a quadratic function or none of these types of function.

b. Create a table of values and with the common difference verify your guess.

23. Using a spring, you run an experiment and you find the following information:

    When there is no weight on the spring the length of the spring is 15 inches.

    With a 1-pound weight on the spring the length of the spring is 13 inches.

    With a 2-pound weight on the spring the length of the spring is 11 inches.

    With a 3-pound weight on the spring the length of the spring is 9 inches.

24. You have two thermometers to measure temperature; one measures in Fahrenheit and the other in Celsius. On different days you measure the temperature with both thermometers and find the following:

    On the first day, when the Fahrenheit measure was 77 degrees, the Celsius measure was 25 degrees.

    On the second day, when the Fahrenheit measure was 68 degrees, the Celsius measure was 20 degrees.

    On the third day when the Fahrenheit measure was 59 degrees, the Celsius measure was 15 degrees.

    On the fourth day when the Fahrenheit measure was 50 degrees, the Celsius measure was 10 degrees.

25. A rocket is shot in the air and the height of the rocket above ground, in feet, is measured with respect to time in the air, in seconds. The following are the results of those measurements:

    At 0 seconds the rocket is on the ground (0 feet in height).

    At 2 seconds the rocket is 448 feet above the ground.

    At 4 seconds the rocket is 768 feet above the ground.

    At 6 seconds the rocket is 960 feet above the ground.

    At 8 seconds the rocket is 1024 feet above the ground.

26. A ball is thrown in the air and the height of the ball above ground, in feet, is measured after a given number of seconds. The following is the results of those measurements:

    At 0 seconds the height of the ball above the ground is 8 feet.

    At 0.5 seconds the height of the ball above the ground is 28 feet.

    At 1.0 seconds the height of the ball above the ground is 40 feet.

    At 1.5 seconds the height of the ball above the ground is 44 feet.

    At 2.0 seconds the height of the ball above the ground is 40 feet.

At 2.5 seconds the height of the ball above the ground is 28 feet.

27. The following information is the amount it costs a manufacturer to produce a number of sofas of the same type.

    To manufacture 10 sofas costs $7,000.

    To manufacture 20 sofas costs $12,000.

    To manufacture 30 sofas costs $17,000.

    To manufacture 40 sofas costs $22,000.

    To manufacture 50 sofas costs $27,000.

28. Money is invested in an IRA for one year; the following is the interest earned on different amounts of investments:

    An investment of $500 earns $25 interest.

    An investment of $1000 earns $50 interest.

    An investment of $1500 earns $75 interest.

    An investment of $2000 earns $100 interest.

    An investment of $2500 earns $125 interest.

29. Guttering is to be created from a rectangular piece of sheet metal with dimensions of 48 inches by 12 inches by turning up equal pieces along the side that is 48 inches long. The volume of the guttering according to the length of the pieces turned up is given by the following:

    If 1 inch is turned up on each side, the volume will be 480 cubic inches.

    If 2 inches are turned up on each side, the volume will be 768 cubic inches.

    If 3 inches are turned up on each side, the volume will be 864 cubic inches.

    If 4 inches are turned up on each side, the volume will be 768 cubic inches.

    If 5 inches are turned up on each side, the volume will be 480 cubic inches.

30. A garden is to be created in the shape of a rectangle where the length is always three more than the width. The cost of the garden according to the length of a side is given by the following:

    If the length is 5 meters, then the cost is $30.

    If the length is 10 meters, then the cost is $210.

    If the length is 15 meters, then the cost is $540.

    If the length is 20 meters, then the cost is $1020.

    If the length is 25 meters, then the cost is $1650.

31. The following is the amount of the paycheck a salaried worker receives every two weeks:

    The first 2-week period the worker received a check for $1,240.28.

    The second 2-week period the worker received a check for $1,240.28.

    The third 2-week period the worker received a check for $1,240.28.

    The fourth 2-week period the worker received a check for $1,240.28.

    The fifth 2-week period the worker received a check for $1,240.28.

32. With a certain HMO insurance plan the following represents how much a patient paid to a doctor for each visit:

    The first visit cost you $15.

    The second visit cost you $15.

    The third visit cost you $15

    The fourth visit cost you $15.

    The fifth visit cost you $15.

33. A life insurance company did a survey on life expectancy. The following are the results of that survey:

    At age 10, there were 100,000 people alive for the survey.

    At age 30, 85,441 of those 100,000 were still alive.

    At age 50, 69,804 of those 100,000 were still alive.

    At age 70, 38,569 of those 100,000 were still alive.

    At age 90, 847 of those 100,000 were still alive.

34. You have agreed to work for an employer using the following pay scale for one 40hour week period. You have agreed to start with $1.00 for the first hour if your wages are doubled for every hour after that.

    For the first hour you are paid $1.00.

    For the second hour you are paid $2.00.

    For the third hour you are paid $4.00.

    For the fourth hour you are paid $8.00.

    For the fifth hour you are paid $16.00.

Section 3.2 Polynomial Characteristics

Now let's look at some other characteristics of polynomial functions. Polynomial functions are functions of the form $f(x) = a_n x^n + a_{n-1}x^{n-1} + a_{n-2}x^{n-2} + \ldots + a_2x^2 + a_1x + a_0$ where the a_i's are real numbers and n is a whole number. We will leave polynomial functions with complex coefficients for another book.

Discussion 1: A Representative Polynomial

An example of what a typical graph of a polynomial function might look like is as follows:

$f(x) = 2x^6 + x^5 - 10x^4 - 5x^3 + 8x^2 + 4x + 1$

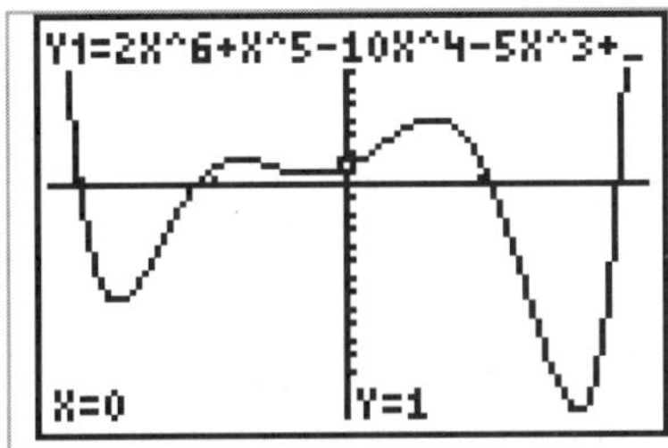	Window: xmin = –2.35 xmax = 2.35 ymin = –20 ymax = 10

As you can see, this graph doesn't have any sharp corners. We call this characteristic **smooth** (not having a corner). This is true for all polynomials. But not true for all relations. $Y = |x|$, not a polynomial, is an example of a function that does have a corner.

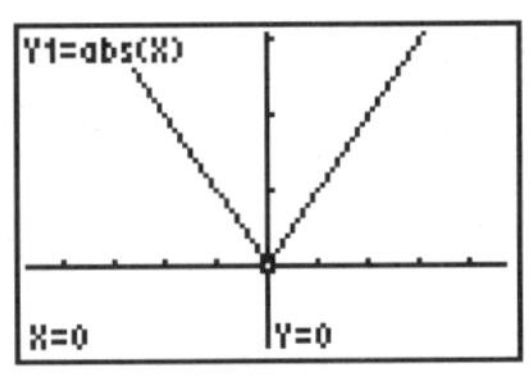

Also, polynomial functions are **continuous** everywhere, which means you can draw them without having to lift your pencil (no breaks in the graph). $Y = \dfrac{2x+1}{x-1}$, not a polynomial, is a good example of a function that isn't continuous everywhere (has a break in it).

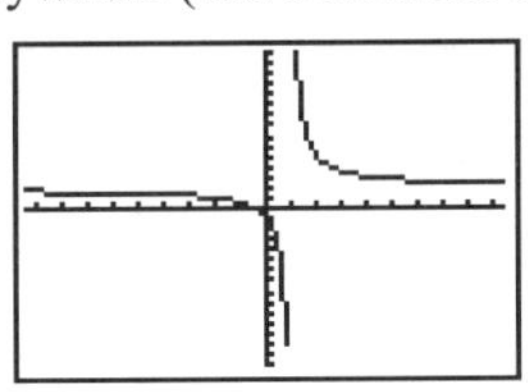

Discussion 2: Examples of Polynomials

Using your graphing calculator, see if you create graphs similar to ours for each of the following polynomial functions.

$f(x) = 2x - 3$

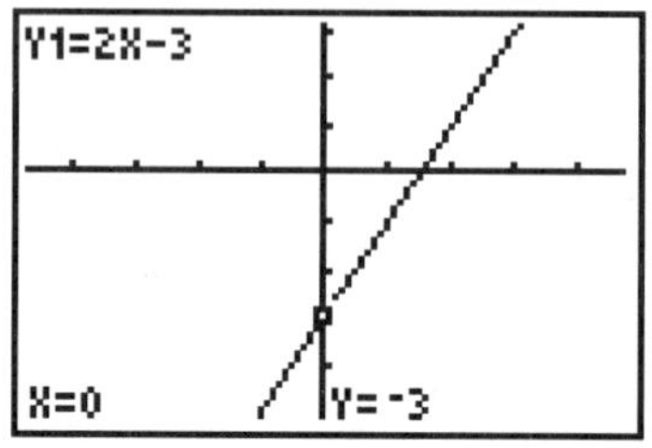

$g(x) = -x^3 + 2x^2 + 5x - 7$

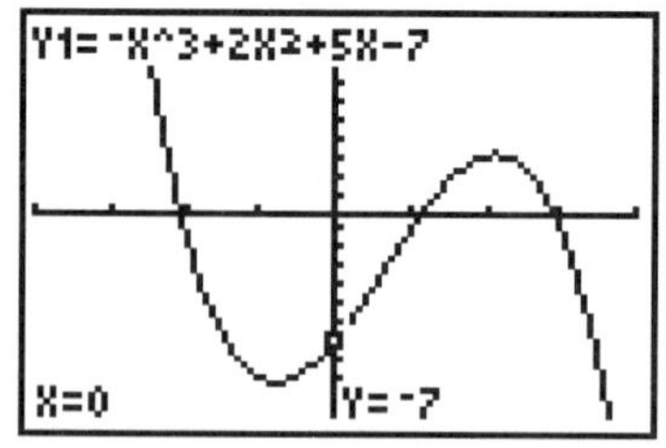

$h(x) = x^6 + 5x^3 - 2x^2 + x$

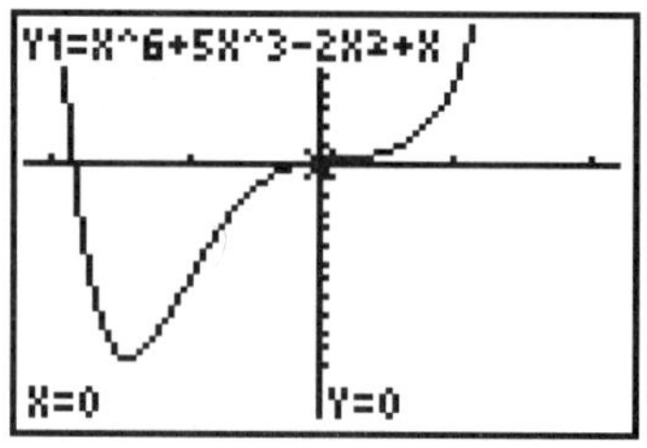

$k(x) = 0.1x^{10} - 2x^7 + x^2 - 1$

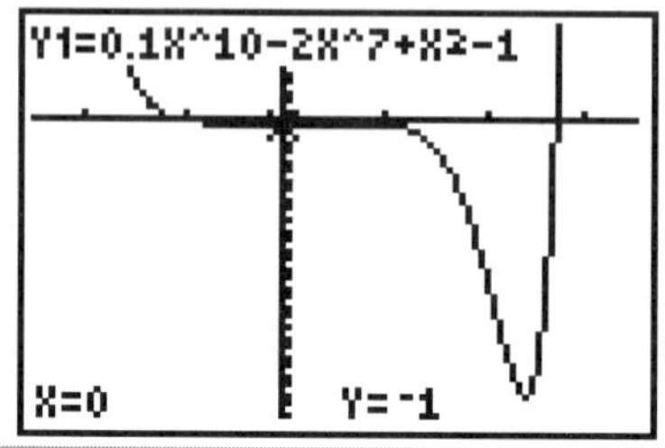

Notice, in each example above, that the graphs are smooth (no sharp corners) and continuous (no breaks). If you were to make up any polynomial function and graph it on your calculators, you would not be able to create an example where it had a break or a sharp corner in it. In mathematical terms, the family of functions called polynomials is a really "well-behaved" family. So, by learning a few basic characteristics of their behavior, we'll be able to use polynomials in a wide variety of applications.

We are going to concentrate on three characteristics of polynomial functions in this section. The three characteristics are the shape of the graph, how they behave at the ends (as the magnitude of the x-values get large), and how they behave near an x-intercept. Magnitude means $|x|$ (the value of x without the sign).

SHAPE OF GRAPHS OF POLYNOMIALS

Let's begin with a sampling of possible graphs for the first few polynomial function subfamilies.

Constant Family

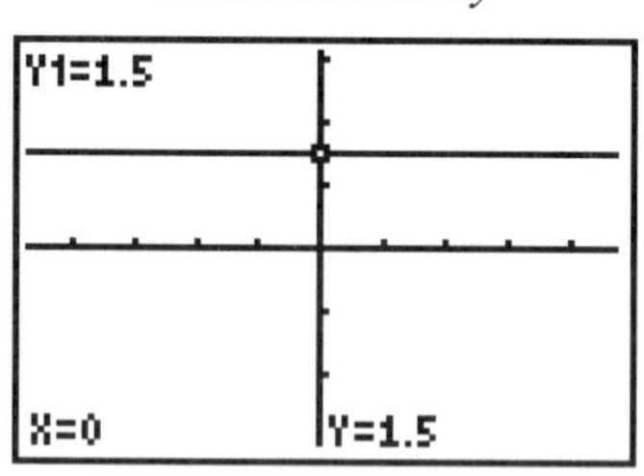

Linear Family

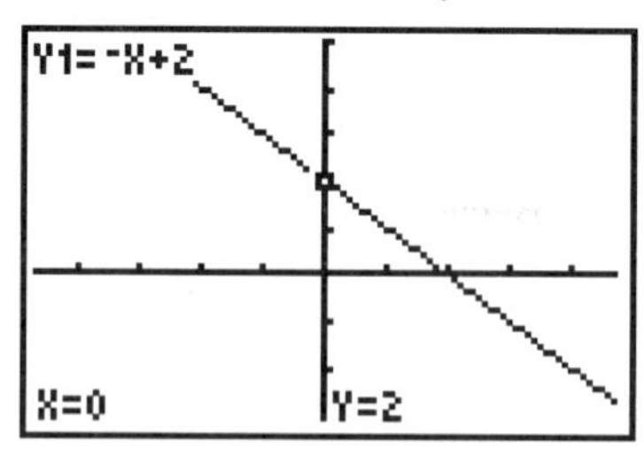

Quadratic Family

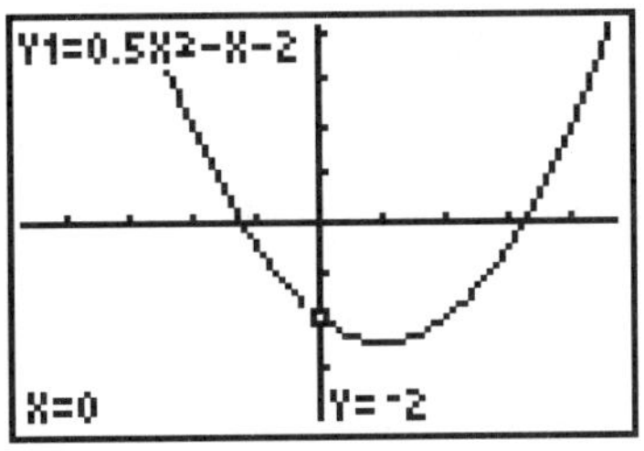

Cubic Family

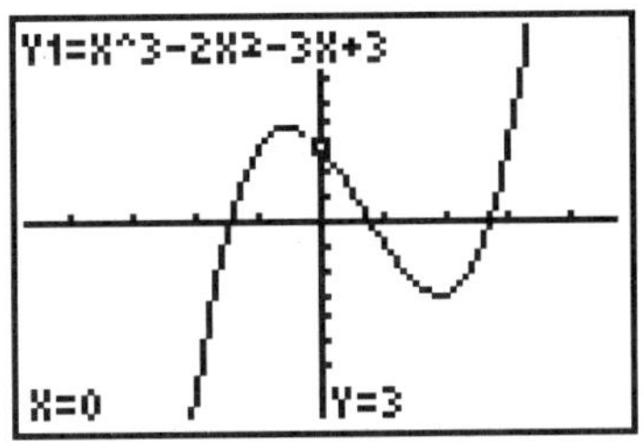

Quartic Family

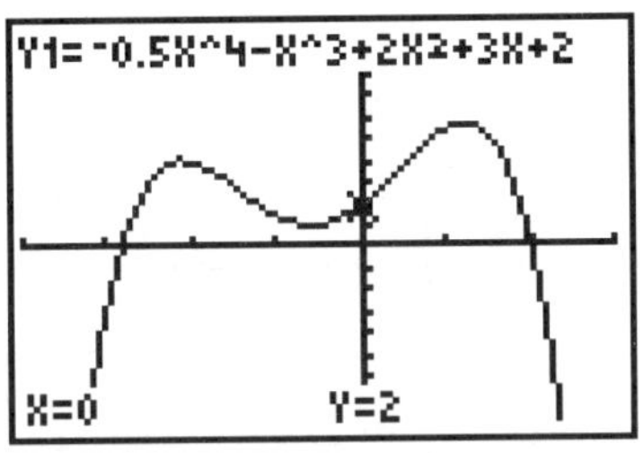

Fifth Power Family

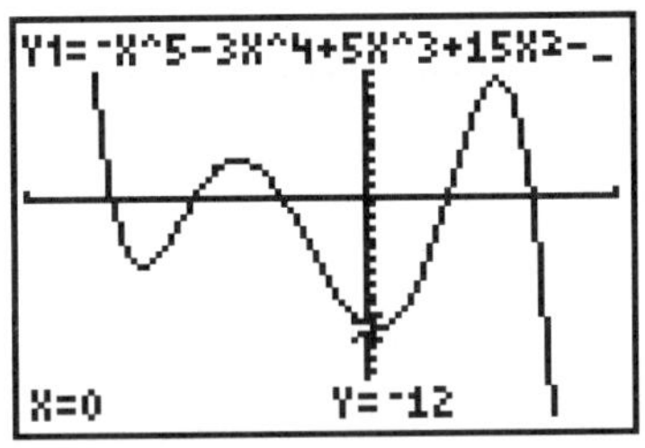

When we talk about graphs, we view them from left to right just like we read a book. So if we talk about where the graph is entering the graphing calculator window, we are talking about where the graph of the function enters the screen on the left. Likewise, with exiting, we are talking about where it leaves the screen on the right. Here are some things you should notice in the graphs above:

1.) The left column of graphs all have a largest degree (power) that is even (0, 2, 4).

2.) The right column of graphs all have a largest degree (power) that is odd (1, 3, 5).

3.) The even-degree graphs (left column) in every case, enter and exit from the same part of our window. That is, they either enter and exit from the bottom of the screen or enter and exit from the top of the screen. They never enter from the top and exit from the bottom or vise versa.

4.) On the other hand, the odd-degree graphs (right column) in every case, enter and exit from opposite places. That is they either enter from the top and exit on the bottom or they enter from the bottom and exit out the top.

5.) Because of observations 3 and 4 above, the odd-degree polynomials must always cross the x-axis; the even polynomials don't necessarily have to cross the x-axis.

6.) The number of hills and valleys (humps and troughs) in any particular graph is *never more than 1 less than the degree* (power) of the polynomial. Notice that the *fourth* powered polynomial above has two hills and one valley for a total of *three hills and valleys* $(4 - 1 = 3)$.

> Question 1: Does the following polynomial have an even or odd degree (i.e. does it have a largest degree that is an even power or an odd power)? What is the total number of hills and valleys in the given polynomial? Does it demonstrate the characteristics mentioned above?
>
> $$F(x) = 2x^4 + 7x^3 - 26x^2 - 49x + 30.$$

In observation #6 above, notice we said "…never more than 1 less than the degree of the polynomial." It is possible, though, to have fewer hills and valleys than this. Here are some examples:

No Hills or Valleys

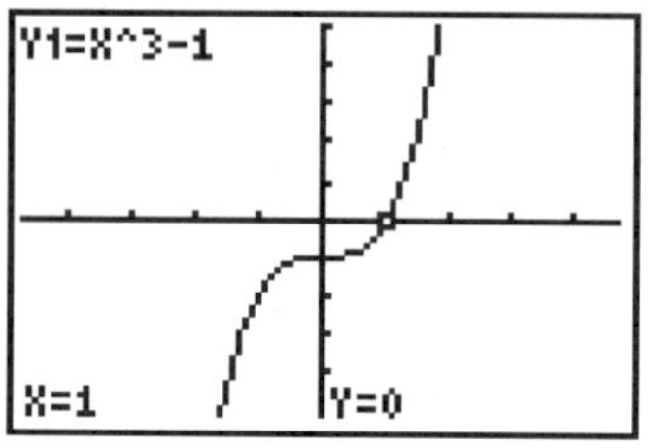

Three Hills and Valleys Total

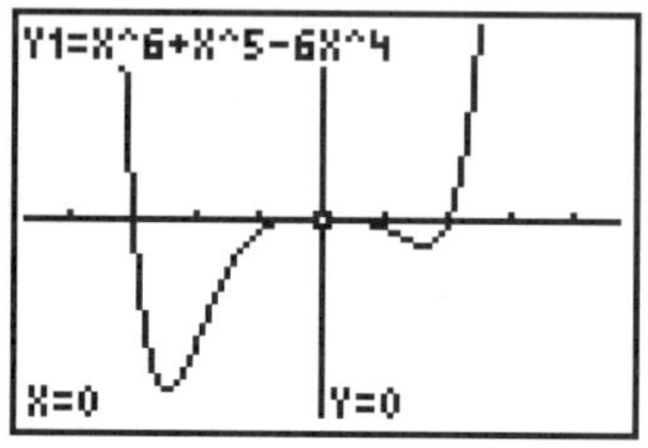

One Valley Only

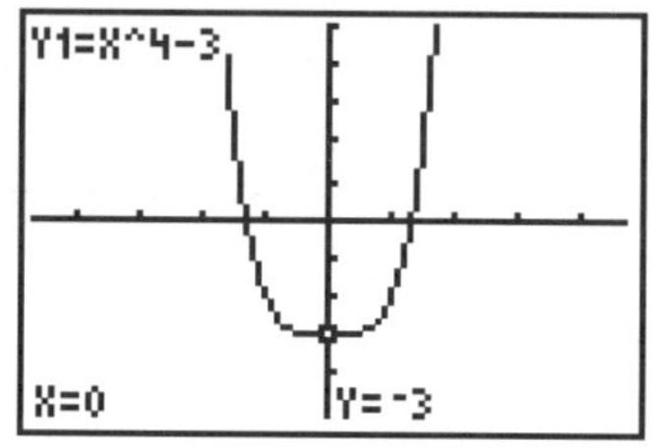

From *all* of the examples so far in this section, you might have noticed how one fourth-degree polynomial can have three hills and valleys total ($2x^4 + 7x^3 - 26x^2 - 49x + 30$), while another may have only one hill or valley total ($x^4 - 3$). Similarly, one third-degree polynomial can have two hills or valleys ($x^3 - 2x^2 - 3x + 3$), but another may have none ($x^3 - 1$).

The total maximum number of hills and valleys *possible* will always be 1 less than the degree of the polynomial or any multiple of 2 less than that, down to 0 (…, 4, 2, 0) or 1 (…, 5, 3, 1). Why this happens is discussed more completely in section 3.4.

END BEHAVIOR

Now let's turn your attention to the ends of the graphs. Look at the graphs of the linear, quadratic, cubic, and quartic polynomials above. Here are the equations that produced those graphs in the tables above.

- Linear: $y = -x + 2$
- Quadratic: $y = 0.5x^2 - x - 2$
- Cubic: $y = x^3 - 2x^2 - 3x + 3$
- Quartic: $y = -0.5x^4 - x^3 + 2x^2 + 3x + 2$

Notice how in every case where the highest-degree term is positive, the graph exited upward to the right. Likewise, when the highest-degreed term is negative, the graph exited downward to the right. This happens because as you raise larger and larger numbers (x-values) to a power, the larger powered term dominates the final answer; that is, it contributes the most to the final result. Here is an example:

x-values	x^2	x^3
10	100	1,000
100	10,000	1,000,000
1000	1,000,000	1,000,000,000

Notice, the x^3 column has much larger values than the x^2 column as the x-values get larger.

Furthermore, when positive large values are plugged into the highest-degree term, the final answer is positive and for a negative-degree term the answer is negative. Also, as mentioned earlier, even-degree polynomials have graphs that both enter and exit on top of the screen, or both enter and exit on the bottom of the screen, because even powers always yield the same answer whether the x-value is positive or negative.

x-values	x^2	x^4
10	100	10,000
−10	100	10,000
20	400	160,000
−20	400	160,000

Odd-degree polynomials, on the other hand, do not yield the same answers for positive and negative inputs of the same magnitude, because we know that negatives raised to odd powers are negative and positives raised to odd powers are positive.

x-values	x	x^3
10	10	1,000
−10	−10	−1,000
20	20	8,000
−20	−20	−8,000

So, the graph of an odd degreed polynomial must enter from the bottom and exit out the top or vice versa.

Example 1: An Odd-Power Polynomial

What happens at the ends of the graph of the function $g(x) = -x^3 + 2x^2 + 5x - 7$? (This was graphed earlier in this section.)

Solution:

It is going up on the left side of the graph because the term of largest degree ($-x^3$) is negative, so the output for negative input (x-values) is positive. It turns downward on the right side of the graph because again, the term of largest degree ($-x^3$) is negative, which causes the output to be negative for positive inputs.

Example 2: An Even-Power Polynomial

What happens at the ends of the graph of the function $f(x) = 2x^4 + 7x^3 - 26x^2 - 49x + 30$? (This was graphed in the answer to question 1.)

Solution:

It is going up on the right and left sides of the graph because the term of highest degree ($2x^4$), yields positive output for any x-values used.

x-INTERCEPTS

Let's now focus on how the factors of a polynomial affect what the x–intercepts look like. Remember that an x–intercept is the point on the graph where the graph *crosses* or *touches* the x–axis. You should remember that you find an x–intercept by letting $y = 0$.

Discussion 3: A Graph's Appearance Near an x-Intercept

$f(x) = (x-2)^2(x+1)$ is a polynomial in factored form. Here is its graph:

Answer Q1: Even degree (observation 3 above). Three. Yes.

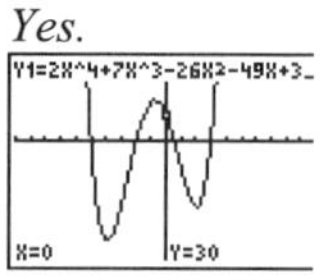

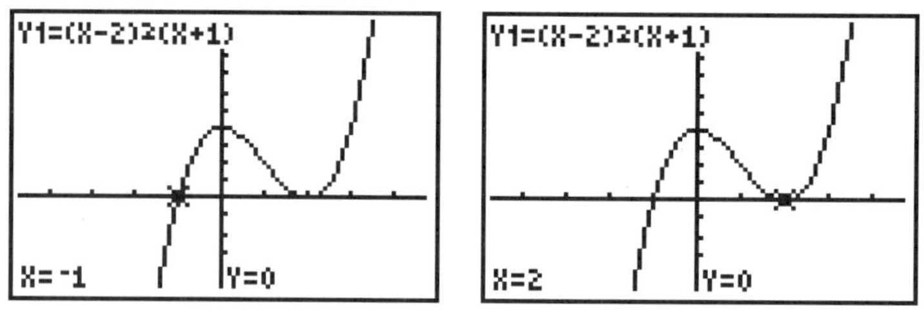

Do you see that we have an x-intercept ($y = 0$) at $x = -1$ and $x = 2$? Also, notice that if we zoom in on the $x = 2$ intercept, the graph has a shape similar to the graph of $y = x^2$.

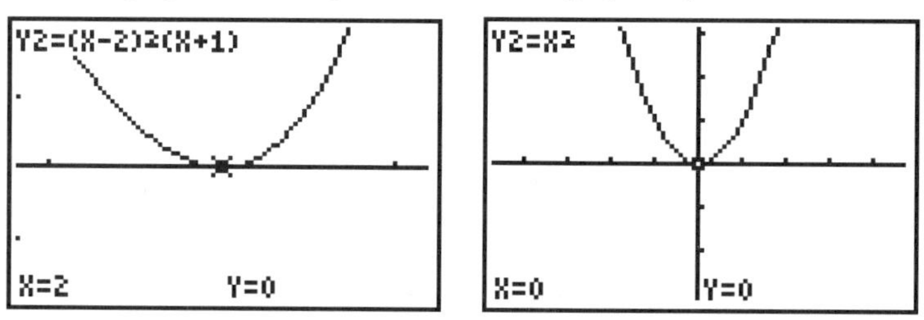

These two graphs look quite similar because when you plug in x–values near 2, the *squared* factor $(x-2)$ dominates the function $f(x)$, which means it contributes the most or has the most effect on the answers to the polynomial near $x = 2$. This is similar to what we saw earlier in this section with regard to how the highest-degree terms dominate for really large x–values. Also, both graphs above touch but don't cross the x-axis. This happens because when you square something you can only get positive answers. Likewise, the graph of $f(x) = (x-2)^2(x+1)$ near $x = -1$ will look a lot like the graph of $y = x$, a line that crosses the x–axis.

It turns out that all odd-power factors will cause the polynomial to cross the x–axis at the x–value that would make that factor equal 0. We might refer to these as *odd x-intercepts*. Any even-power factors will cause the polynomial to just touch the x–axis at the x–value that would make that factor equal 0. We might refer to these as *even x-intercepts*. Here is another example.

Discussion 4: A Graph's Appearance at an x-Intercept

Let $g(x) = (x - 1)^3(x + 2)^2$. Here are two graphs of this function:

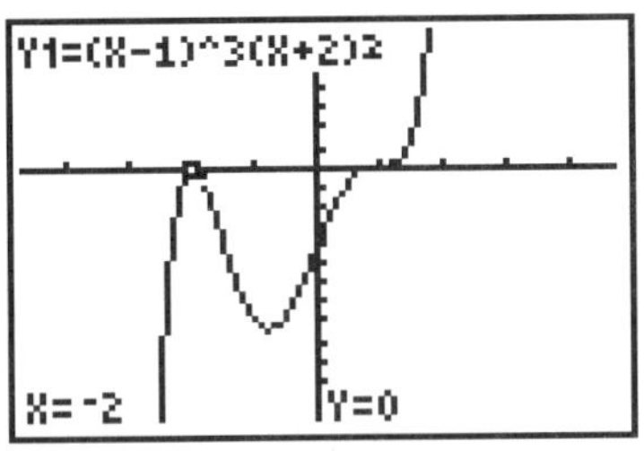

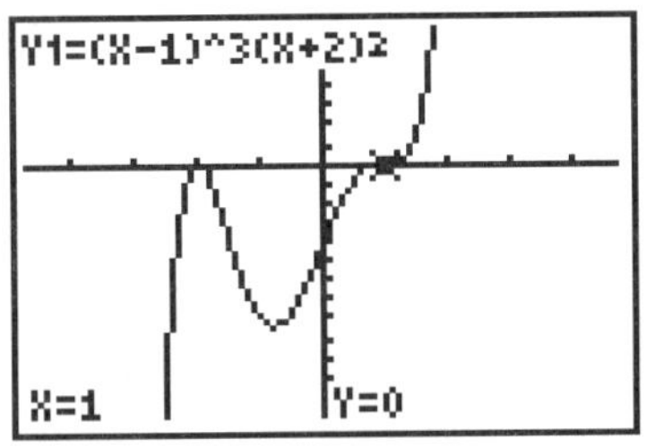

Notice how at $x = -2$ the graph touches the x–axis and at $x = 1$ it crosses. Also at $x = -2$ it looks like $y = x^2$ upside down and at $x = 1$ it looks like $y = x^3$. Below are the graphs of $g(x)$ and the two functions it looks like when you zoom in on the x–intercepts:

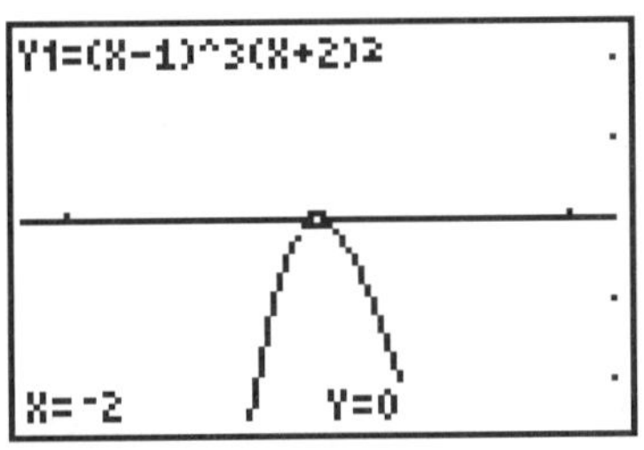

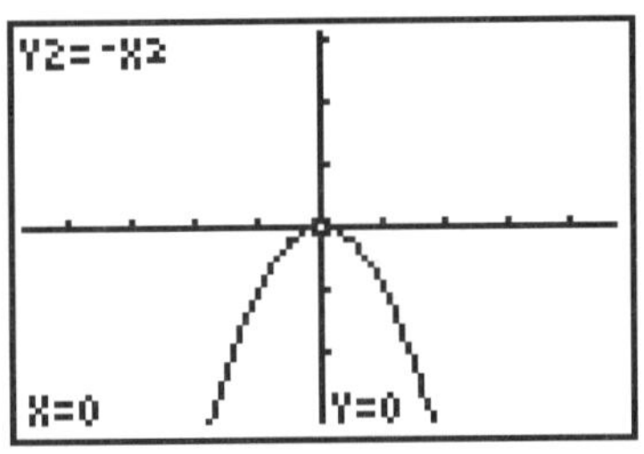

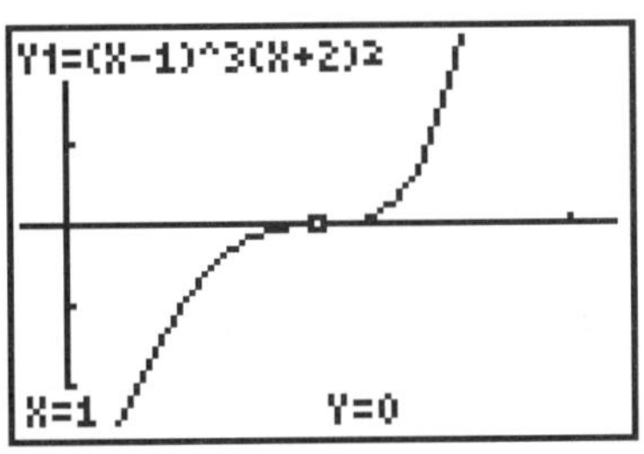

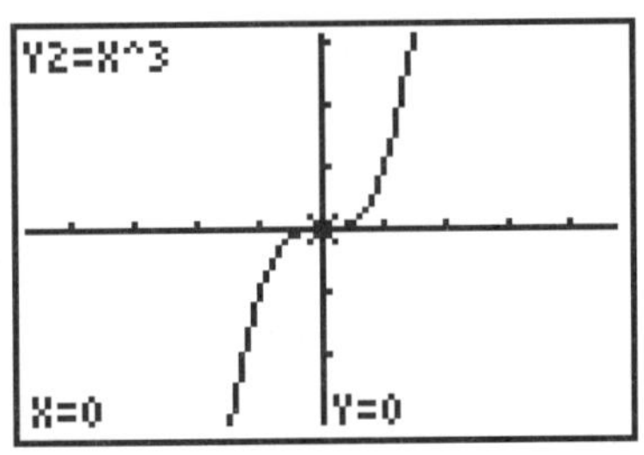

Question 2: Given $f(x) = (x + 4)(x - 3)^3(x + 1)^2$ what will the graph look like near the x–values of -4, 3, and -1? (Hint: will they look like $y = x$, $y = x^2$, or $y = x^3$?) For the moment, don't worry about whether they are right-side-up or upside-down. Check yourself with your graphing calculator.

At this point you might be saying to yourself, "But at $x = -1$ it was upside down compared to $y = x^2$." Yes, it was and let's talk about that. Even though the factor $(x + 1)$ near $x = -1$ dominates how the graph looks near $x = -1$, the other factors that are part of the function still have an effect. The other factors affect whether we are right-side-up or upside-down at an x–intercept because they cause a negative or positive coefficient to be in front of the factor. So, this makes the graph in question 2 upside down near $x = -1$; $(x + 4)$ will equal a positive number near $x = -1$, and $(x - 3)^3$ will equal a negative number near $x = -1$, so the product of the two factors will be negative.

Let's do another example and try to pull together all the parts we have talked about in this section.

Example 3: Sketching a Polynomial by Hand

Given the function $f(x) = (x + 5)(x - 3)(x + 1)^2$, sketch a rough graph of this function **without** using your graphing calculator.

Solution:

First, think about what the highest power would be if you multiplied all the factors of the polynomial together. You would have an x times an x times an x^2, which would give you an x^4 leading term. So we know that this function will enter and exit the window on the top, since the x^4 has a positive coefficient and is raised to an even power.

Second, at $x = -5$, $x = 3$ and $x = -1$ we have x–intercepts. So at $x = -5$, since that factor $(x + 5)$ is raised to the first power, it will look like a line. At $x = 3$, since that factor $(x - 3)$ is raised to the first power, it will also look like a line. At $x = -1$, since that factor $(x + 1)$ is raised to the second power, it will look like $y = x^2$.

So our chore now is to get our sketch to cross at $x = -5$ and 3, touch at $x = -1$, and begin and end at the top of the graph. As we sketch the graph, let's go backwards, that is, start up at the top of the graph paper on the right and trace the graph backwards, touching and crossing the x-axis at the appropriate places.

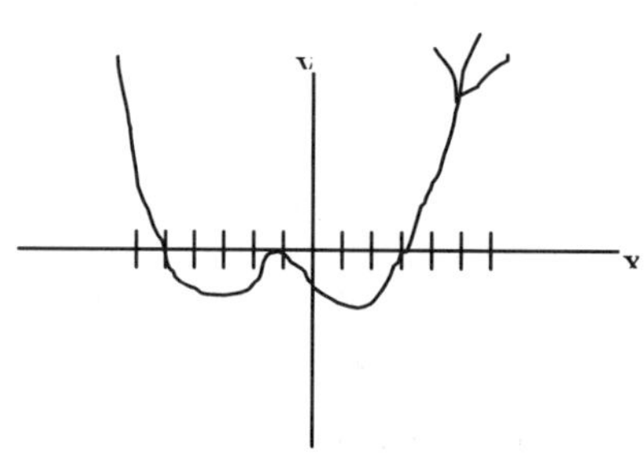

Question 3: Sketch $f(x) = -(x - 4)^3(x + 2)$. Try to do this without the aid of your calculator.

All of this is important to know and understand since it will help us make decisions about what type of functions might best model a real-world phenomenon. The more we understand, the better our decisions. When we reach chapter 5, we will try to model a lot of real-world problems. This can also serve as a good way to mentally check our graphing calculator graphs to make doubly sure that when we typed in what we wanted, we did it correctly.

MULTIPLICITY

Here is a good place to bring up a topic called **multiplicity**. Let's look again at the polynomials from discussion 3, $f(x) = (x - 2)^2(x + 3)$, and 4, $g(x) = (x - 1)^3(x + 2)^2$. We would say that the function $f(x)$ has an x-intercept at $x = 2$ with a multiplicity of 2 and an x-intercept at $x = -3$ with a multiplicity of 1. Likewise with the function $g(x)$, we would say that the x-intercept at $x = 1$ has a multiplicity of 3 and the x-intercept at $x = 2$ has a multiplicity of 2. Multiplicity basically says how many times the same factor occurs (the exponent on the factor).

Question 4: What are the x-intercepts of $k(x) = (x + 7)^4(x - 4)^3(x + 1)$ and what are their multiplicities?

SYMMETRY

One last thing we will talk about is **symmetry**. We say that a graph has symmetry if there is a point or line about which you can rotate the graph and still get exactly the same graph again. Two special kinds of symmetry that are often mentioned in math books are **even** and **odd** functions.

An even function is one that is symmetric about the y-axis. Some good examples are:

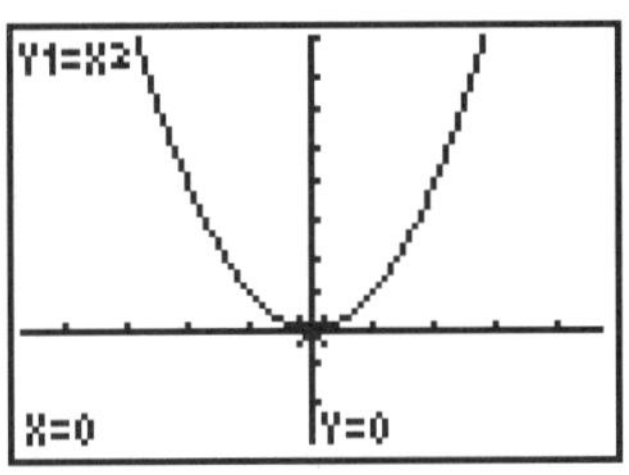

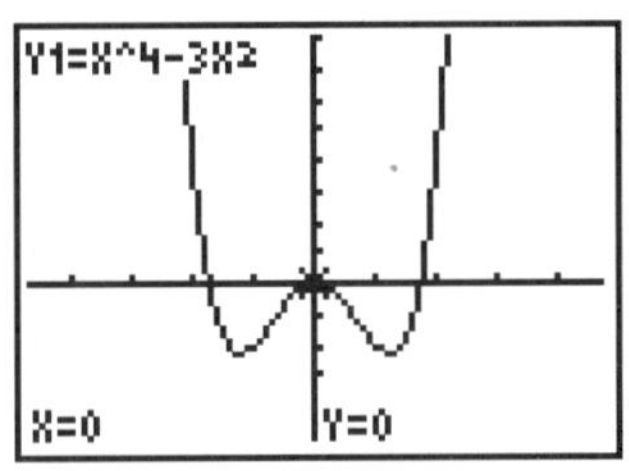

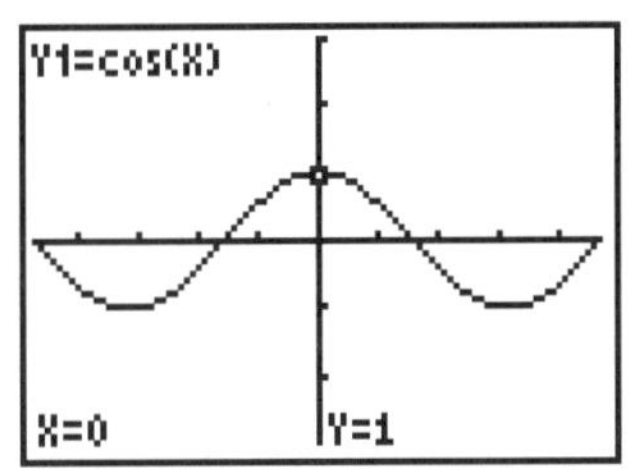

Notice that if you revolve the three graphs around the y-axis (turn them 180 degrees), you get exactly the same graphs again.

Answer Q2: At $x = -4$ it will look like $y = x$; at $x = 3$ it will look like $y = x^3$; at $x = -1$ it will look like $y = x^2$.

In formal math terms, if you can replace the x's in the function by $-x$ and, once simplified, your answer looks exactly the same as before, then the original function is an even function $[f(-x) = f(x)]$ (the algebraic way of proving something to be even).

An odd function is one that is symmetric to the origin. Some good examples are:

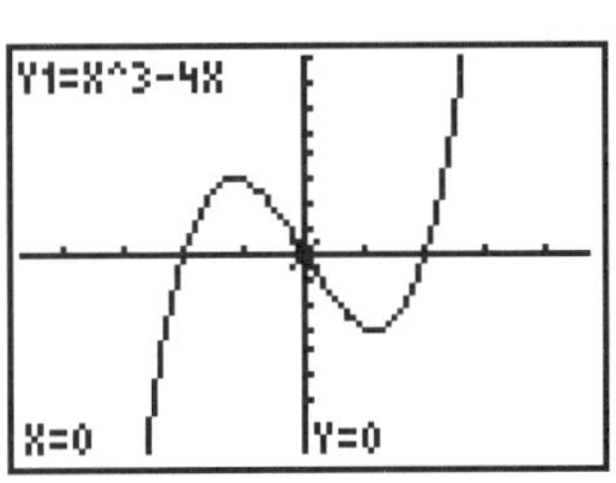

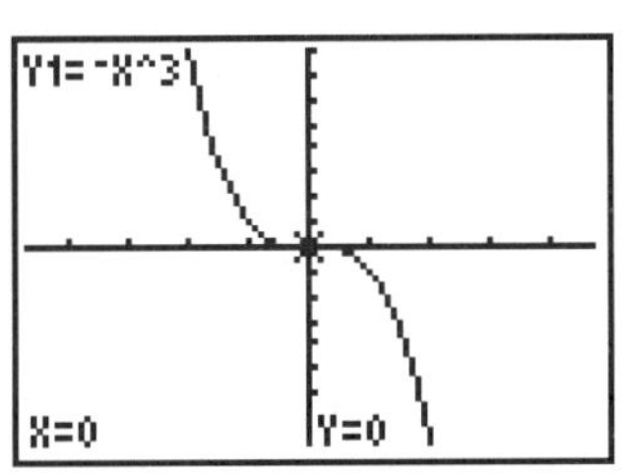

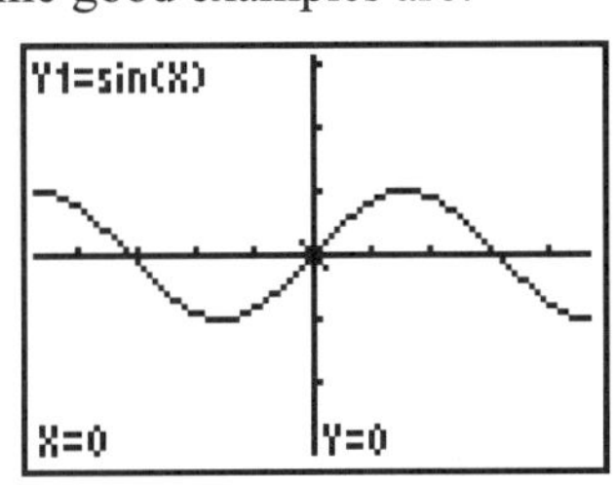

Notice here that if you revolve these around the origin (turn them 180 degrees), you get exactly the same graphs again.

In formal math terms, if you can replace the x's in the function by $-x$ and, once simplified, your answer looks exactly the same as before but every term has an opposite sign, then the original function is an odd function $[f(-x) = -f(x)]$ (the algebraic way of proving something to be odd).

Question 5: Graph $f(x) = x^3 + 2$, $g(x) = -x^4 + 2x^2 + 5$, and $h(x) = \frac{1}{5}x^5 - \frac{3}{5}x^3 + \frac{7}{5}x$. Look at each graph and decide whether they are even, odd, or neither.

Answer Q3: The graph should look like $y = x^3$ near $x = 4$ and like $y = x$ near -2. Also the biggest power would be $-x^4$ so it must begin and end down

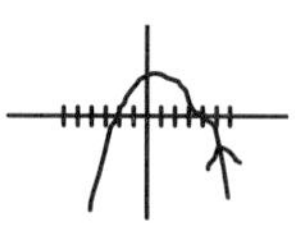

Section Summary

- Polynomials are smooth (no sharp corners) continuous (no breaks), "nicely behaved" functions.
- The highest-degree terms determine how the graph looks near the ends of the window (if you have found a good complete graph).
- Also, if the polynomial is factored, then the graph *near each x–intercept* will look like the graph of the factor as if it were the only factor in the function.

The following is a table summarizing some of the information about polynomials and their graphs:

Highest- degree term in the polynomial	Enters the graphing window from	Exits the graphing window from	Number of possible hills and valleys	Number of possible x-intercepts
x^n (n odd)	below	above	$n-1,\ldots,4,2,0$	$n,\ldots,5,3,1$
$-x^n$ (n odd)	above	below	$n-1,\ldots,4,2,0$	$n,\ldots,5,3,1$
x^m (m even)	above	above	$m-1,\ldots,5,3,1$	$m,\ldots,4,2,0$
$-x^m$ (m even)	below	below	$m-1,\ldots,5,3,1$	$m,\ldots,4,2,0$

SECTION 3.2 HOMEWORK

(1–16) For each of the following graphs:

a. State if the graph could represent a polynomial or not

b. If the graph could represent a polynomial, would the power of the polynomial be even or odd?

Answer Q4: $x = -7$ multiplicity 4 and $x = 4$ multiplicity 3 and $x = -1$ multiplicity 1.

1. 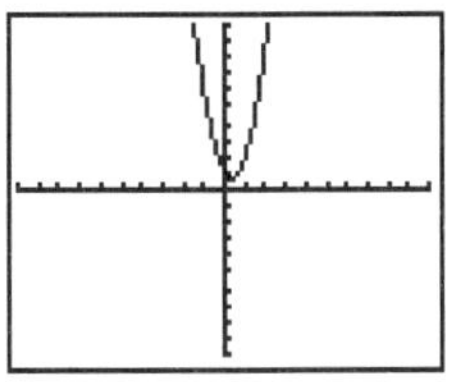

2. 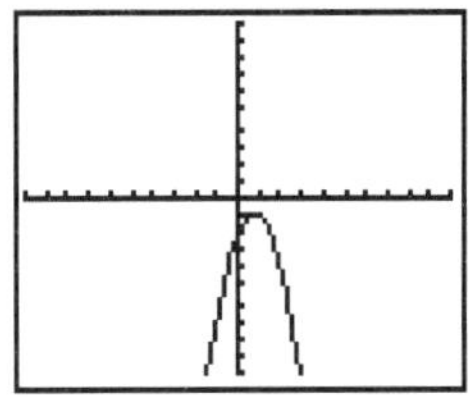

3. 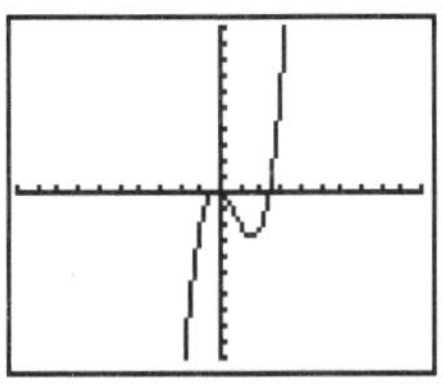

4. 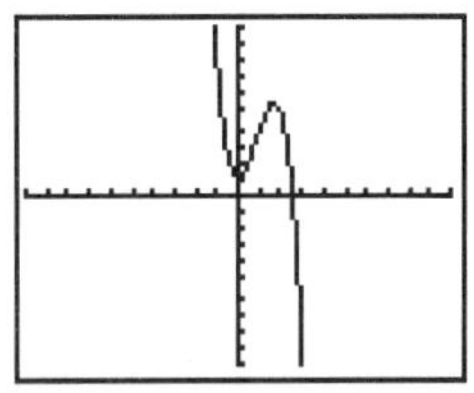

5. 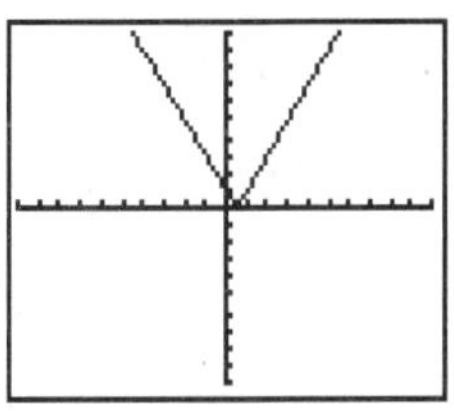

6. 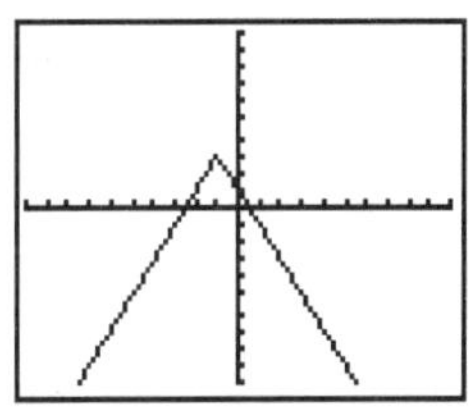

7. 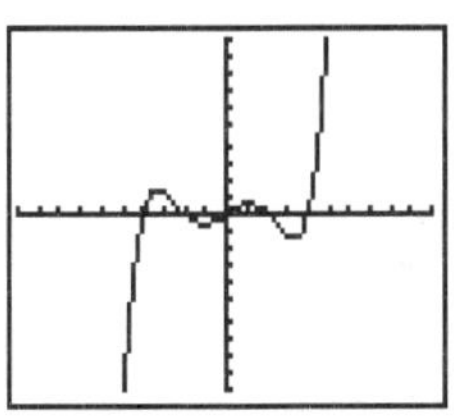

8. 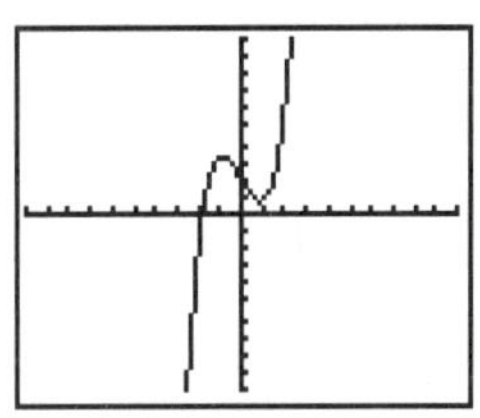

9. 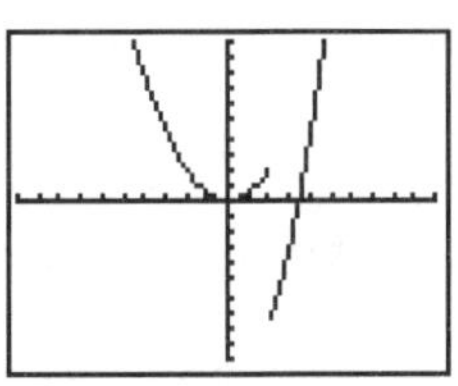

10. 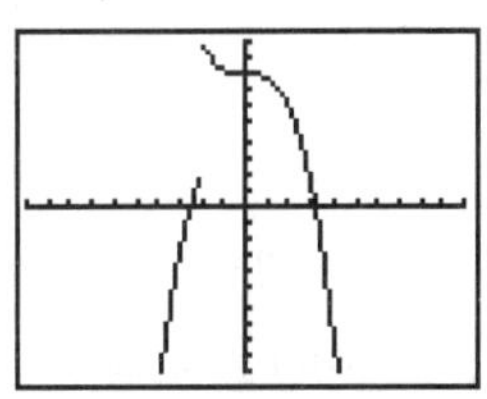

11. 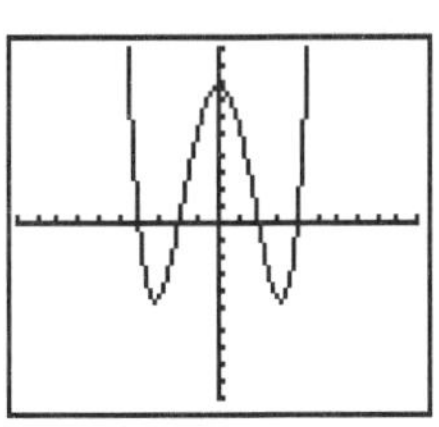

12. 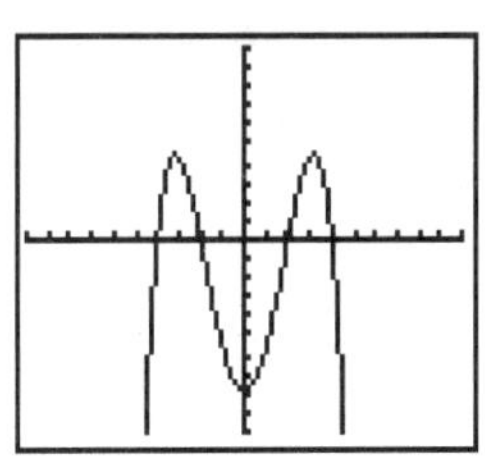

13. 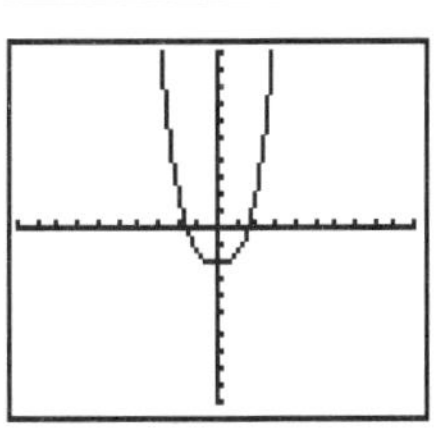

14. 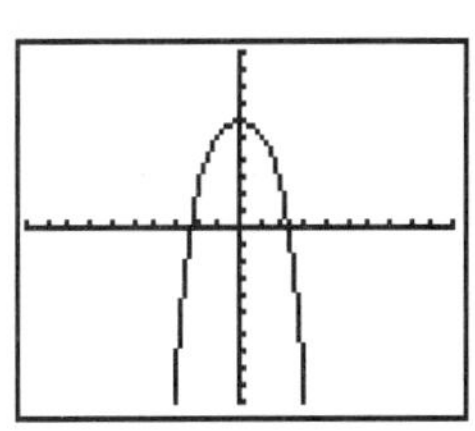

15. 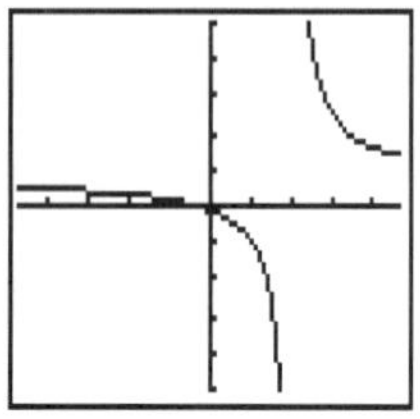

16. 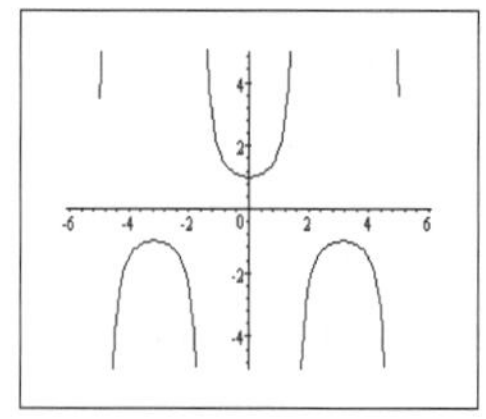

(17–30) Without using the graphing calculator for (a) thru (d) answer the following questions about the graph of each of the polynomial functions:

a. How many possible humps (minimum number and maximum number)?

b. What are the maximum number of places the graph could cross the x–axis?

c. Where will it enter the window on your graphing calculator?

d. Where will it exit the window on your graphing calculator?

e. Confirm your answers with your graphing calculator.

f. Using what you learned in section 1.3, use the graphing calculator to approximate the answers to the questions below to three decimal places.

  i. Over what intervals is the function increasing?

  ii. Over what intervals is the function decreasing?

  iii. What are the relative maximums, if any?

  iv. What are the relative minimums, if any?

Answer Q5: Is Neither An Odd or Even

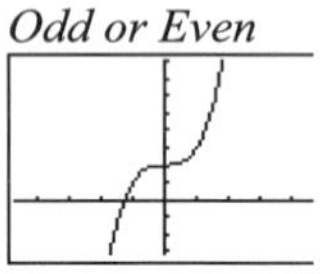

An Even Function

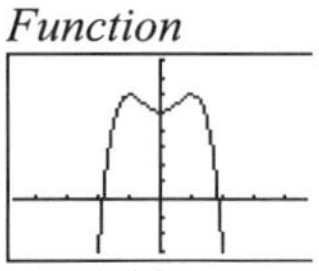

An Odd Function

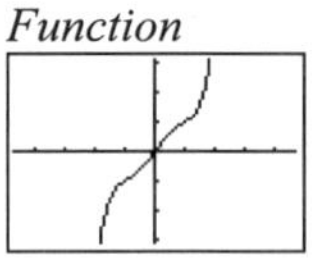

17. $f(x) = 3x^2 - 5x + 1$
18. $t(x) = -2x^2 + 5x - 4$
19. $w(x) = x^3 - 3x + 2$
20. $g(x) = -2x^3 + 4x^2 + 1$
21. $m(x) = -x^3 + 2x^2 - 3x$
22. $r(x) = 2x^3 - 3x^2 + 5x + 1$
23. $f(x) = -x^4 + x^3 - 4$
24. $g(x) = x^4 - 3x + 1$
25. $f(x) = x^4 + 3x^3 - 3x - 1$
26. $r(x) = -2x^4 + 5x^2 - 1$
27. $t(x) = x^5 - 5x^3 + 4x$
28. $s(x) = -x^5 + 5x^3 - 3x$
29. $f(x) = x^5 + 3x^4 + 4x^3 + 3x^2 - 5x - 6$
30. $g(x) = -x^5 - x^4 - x^3 - x^2 - x - 1$

(31–38) Sketch the graph of each of the following functions without using the graphing calculator. Then confirm your sketch with your graphing calculator.

31. $f(x) = (x - 1)^2 (x + 2)$
32. $f(x) = -(x + 3)^2 (x - 3)$
33. $g(x) = (x - 1)^3 (x + 1)$
34. $g(x) = -(x + 1)^3 (x - 2)$
35. $t(x) = -x (x + 1)^2 (x - 2)^3$
36. $t(x) = x (x + 1)^3 (x - 2)^2$
37. $r(x) = -x^3 (x - 1)^2 (x + 2)^2$
38. $r(x) = -x^2 (x + 1)^3 (x - 2)^2$

(39–44) Create an equation using the following information and then check the equation with your graphing calculator.

39. x–intercept at 2, and at the intercept the graph looks like an x^2 graph.
    X–intercept at –3, and at the intercept the graph looks like a line graph.
    Enters at the top and exits at the bottom.

40. x–intercept at –2, and at the intercept the graph looks like a line graph.

    X–intercept at 1, and at the intercept the graph looks like an x^2 graph.

    Enters at the bottom and exits at the top.

41. x–intercept at 3, and at the intercept the graph looks like an x^3 graph.

    X–intercept at 1, and at the intercept the graph looks like an x^2 graph.

    Enters at the bottom and exits at the top.

42. x–intercept at –3, and at the intercept the graph looks like an x^2 graph.

    X–intercept at 1, and at the intercept the graph looks like an x^3 graph.

    Enters at the top and exits at the bottom.

43. x–intercept at 0, and at the intercept the graph looks like a line graph.

    X–intercept at –3, and at the intercept the graph looks like an x^3 graph.

    X–intercept at 2, and at the intercept the graph looks like a line graph.

    Enters at the top and exits at the bottom.

44. x–intercept at 2, and at the intercept the graph looks like an x^2 graph.

    X–intercept at –2, and at the intercept the graph looks like an x^2 graph.

    X–intercept at 0, and at the intercept the graph looks like an x^2 graph.

    Enters at the bottom and exits at the bottom.

(45–56) Are the following polynomial functions even, odd or neither? Graph each with your graphing calculator and, using the idea of symmetry for even and odd, decide if you believe that they might be even or odd (you need not prove your answer algebraically).

45. $f(x) = 2x^3$
46. $f(x) = -3x^2$
47. $g(x) = 3x^4 - 2x^2$
48. $g(x) = 3x^3 - 4x$
49. $f(x) = x^3 - 2x + 3$
50. $f(x) = -5x^4 + 2x^2 - 3x$
51. $r(x) = 8$
52. $r(x) = -2x$
53. $g(x) = (x + 2)^2$
54. $g(x) = (x - 3)^3$
55. $t(x) = x^5 - 3x^3 - 2x$
56. $t(x) = x^4 - 3x^2 - 2$

Section 3.3 Synthetic Division and Factors

We have finished talking about some of the characteristics of the graphs of polynomial functions. Now, in the next two sections, we are going to discuss the relationships that exist between long division, function values, zeros, x-intercepts, factors and roots. Let's begin by reviewing how to do long division.

Discussion 1: Long Division of Polynomials

Use long division to find the answer to $\dfrac{4x^2+7x-10}{x+3}$.

First, you need to figure out what times x (the first term in the divisor) will equal $4x^2$ (the first term in the dividend). For this example the answer to that question is $4x$.	$\begin{array}{r} 4x \\ x+3\overline{)4x^2+7x-10} \end{array}$
Now, multiply the $4x$ times the $x+3$.	$\begin{array}{r} 4x \\ x+3\overline{)4x^2+7x-10} \\ 4x^2+12x \end{array}$
Next, change the signs and add and then bring down the next term.	$\begin{array}{r} 4x \\ x+3\overline{)4x^2+7x-10} \\ \underline{-4x^2-12x} \\ -5x-10 \end{array}$
Now ask yourself the question again "What times x equals $-5x$"? Then multiply the answer to that question -5 times the $x+3$.	$\begin{array}{r} 4x-5 \\ x+3\overline{)4x^2+7x-10} \\ \underline{-4x^2-12x} \\ -5x-10 \\ -5x-15 \end{array}$
Change the signs and add.	$\begin{array}{r} 4x-5 \\ x+3\overline{)4x^2+7x-10} \\ \underline{-4x^2-12x} \\ -5x-10 \\ \underline{5x+15} \\ 5 \end{array}$
So our answer is	$4x-5+\dfrac{5}{x+3}$

Long division is a skill that is needed from time to time, but when the polynomial that you are *dividing by* (the divisor) is "simple," then we can use something called **synthetic division** instead of the whole, long division process. What we mean by "simple" is that the polynomial is of the form $(1x \pm c)$. We can use a little finesse to get polynomials of the form $(ax \pm c)$ to work in synthetic division but that isn't our concern at the moment.

Synthetic division is a shortcut way of doing long division. Because long division has so many stipulations like descending order, every term represented, etc., it is well ordered, and so much writing *could* be left out while doing the process. In the following illustration we will try to show how and why synthetic division works. Let's look again at discussion 1 and see how it could be shortened.

Here is what the final work looked like when we used long division to solve the problem:	$\begin{array}{r} 4x-5 \\ x+3\overline{)4x^2+7x-10} \\ \underline{-4x^2-12x} \\ -5x-10 \\ \underline{5x+15} \\ 5 \end{array}$
Notice that since long division requires the polynomials to be in descending order, and that every term is represented, we really don't need to write the x's (_). We know where to stick the numbers without the x's. **Long division is well ordered.**	$\begin{array}{r} 4\_-5 \\ 1\_+3\overline{)4\_+7\_-10} \\ \underline{-4\_-12\_} \\ -5\_-10 \\ \underline{5\_+15} \\ 5 \end{array}$
Notice also how the first terms always cancel each other out. So why should we bother doing the multiplication of the first term?	$\begin{array}{r} 4\_-5 \\ +3\overline{)4\_+7\_-10} \\ \underline{-12\_} \\ -5\_-10 \\ \underline{+15} \\ 5 \end{array}$
You should also notice how the numbers in our answer at the top of the long division are the same as the first number we are trying to get each time in the long division process.	$\begin{array}{r} 4\_-5 \\ +3\overline{)4\_+7\_-10} \\ \underline{-12\_} \\ -5\_-10 \\ \underline{+15} \\ 5 \end{array}$

Last, in long division we have to change the signs and then add. So why don't we just change the sign first on what we are going to multiply and save a step?	$$\begin{array}{r} 4\ \ -5 \\ -3\overline{)4\ \ +7\ \ -10} \\ -12 \\ \hline -5\ \ -10 \\ +15 \\ \hline 5 \end{array}$$
So here is what this problem would look like if it had been done using synthetic division. Notice that this new format is just the old long division format compressed, with the answer on the bottom instead of on top.	$$\begin{array}{rrrr} -3) & 4 & 7 & -10 \\ & & -12 & 15 \\ \hline & 4 & -5 & 5 \end{array}$$
Now, understanding how organized long division is, we can look at this answer and know that the last number is the remainder; then next, going from right to left, is a number and the next is the x term.	$4x - 5 + \dfrac{5}{x+3}$

Let's do another example.

Discussion 2: Synthetic Division

Using synthetic division to find the answer to the long division problem, $\dfrac{2x^3 - 3x^2 + 5x + 1}{x-1}$.

We set up the synthetic division problem in this manner. The first number is the opposite sign of the –1 in the polynomial $(x - 1)$. Next we write down all of the coefficients of the top polynomial (the dividend). Last we bring down the first number.	$$\begin{array}{rrrrr} 1) & 2 & -3 & 5 & 1 \\ & & & & \\ \hline & 2 & & & \end{array}$$
Now, take the 1 times the 2 and write the answer under the –3 and add.	$$\begin{array}{rrrrr} 1) & 2 & -3 & 5 & 1 \\ & & 2 & & \\ \hline & 2 & -1 & & \end{array}$$
Next, take the 1 times the –1 and write the answer under the 5 and add.	$$\begin{array}{rrrrr} 1) & 2 & -3 & 5 & 1 \\ & & 2 & -1 & \\ \hline & 2 & -1 & 4 & \end{array}$$

Now, take 1 times the 4 and write the answer under the 1 and add.	$\begin{array}{r\|rrrr} 1 & 2 & -3 & 5 & 1 \\ & & 2 & -1 & 4 \\ \hline & 2 & -1 & 4 & 5 \end{array}$
Our final answer is	$2x^2 - x + 4 + \dfrac{5}{x-1}$

As you can see, we can do long division problems in $\frac{1}{3}$ the time when using synthetic division.

Notice it only took three multiplication operations and three addition operations to get the complete answer to the long division question.

Example 1: Synthetic Division

Find the long division answer to $\dfrac{3x^3 - 15x + 7}{x+2}$.

Solution:

In setting up the synthetic division this time, we need to add a 0 after the 3 because long division problems need to be in descending order and every term must be represented. We are missing an x^2 term.	$\begin{array}{r\|rrrr} -2 & 3 & 0 & -15 & 7 \\ & & & & \\ \hline & 3 & & & \end{array}$
Now, working through the 3 multiplications and 3 additions we get	$\begin{array}{r\|rrrr} -2 & 3 & 0 & -15 & 7 \\ & & -6 & 12 & 6 \\ \hline & 3 & -6 & -3 & 13 \end{array}$
So, our final answer is	$3x^2 - 6x - 3 + \dfrac{13}{x+2}$

So, $\dfrac{3x^3 - 15x + 7}{x+2} = 3x^2 - 6x - 3 + \dfrac{13}{x+2}$

Discussion 3: Finding Function Values

Given $f(x) = 3x^3 - 15x + 7$ we will find $f(-2)$.

If you remember from our discussions about function notation in chapter 1, to find the answer to $f(-2)$, you simply plug in -2 everywhere there is an x in the $f(x)$ function.

$$F(-2) = 3(-2)^3 - 15(-2) + 7 = 3(-8) + 30 + 7 = -24 + 30 + 7 = 13$$

But, notice that $f(x) = 3x^3 - 15x + 7$ is the same as the dividend in example 1, and –2 is the same number we used in the synthetic division. Also, our answer $f(-2) =$ **13** is the same as the remainder in example 1. Could there be a connection? Yes.

We know that, from long division	$\frac{3x^3-15x+7}{x+2} = (3x^2 - 6x - 3) + \frac{13}{x+2}$
This means that if we multiply both sides of the equation by $(x + 2)$ we would get	$3x^3 - 15x + 7 = (x + 2)(3x^2 - 6x - 3) + 13$
Now, discussion 3 had $f(x) = 3x^3 - 15x + 7$ which also means that	$f(x) = (x + 2)(3x^2 - 6x - 3) + 13$
So the term with the factor $(x + 2)$ will equal 0, when –2 is plugged in for x.	$f(-2) = (-2 + 2)[3(-2)^2 - 6(-2) - 3] + 13 = 0[21] + 13 = 13$

Question 1: Using synthetic division, find the answer to the long division problem $\frac{3x^3-15x+7}{x-3}$ and then the value of $f(3)$, given that $f(x) = 3x^3 - 15x + 7$.

What you have just seen, in action, is called the **Remainder Theorem**. So in summary,

> If $f(x)$ is a polynomial and you divide it by $(x - c)$, using synthetic division, then the answer to f is R (the remainder from the synthetic division). **Notice that it's $-c$ in the factor, but just c in the function notation (a sign change).**

Example 2: Using the Remainder Theorem

Find $f(-1)$ using synthetic division on the polynomial $f(x) = x^3 + 2x^2 + 3x - 5$.

Solution:

Do synthetic division.	$\begin{array}{r|rrrr} -1 & 1 & 2 & 3 & -5 \\ & & -1 & -1 & -2 \\ \hline & 1 & 1 & 2 & -7 \end{array}$
Now we know that $f(-1) =$ remainder from the remainder theorem.	$f(-1) = -7$

A by-product of doing synthetic division to find f is that if we wanted to divide $f(x)$ by $(x + 1)$ we would have the answer already. The answer would be $x^2 + x + 2 - \frac{7}{x+1}$.

Question 2: Find $f(4)$ using synthetic division given that $f(x) = x^3 - 4x^2 + 4x - 16$.

Let's take a moment to do a problem from your past. If we divide by hand, the number 221 by 13, we would get

$$\begin{array}{r} 17 \\ 13\overline{)221} \\ \underline{13} \\ 91 \\ \underline{91} \\ 0 \end{array}$$

When we get 0 as a remainder, what does it tell us? It says that 13 divides evenly into 221 or, in other words, that 13 is a factor of 221. We know that the answer 17 is also a factor of 221 since there wasn't a remainder.

With this idea in mind, what does it tell us when the remainder in question 2 is 0? It tells us that if we were doing the long division problem $\dfrac{x^3 - 4x^2 + 4x - 16}{x-4}$ the answer would be $x^2 + 4$ with no remainder. So, $(x - 4)$ and $(x^2 + 4)$ must be factors of $f(x) = x^3 - 4x^2 + 4x - 16$.

Here is what we call the **Factor Theorem**.

> If the remainder equals 0 after synthetic division with the number c, then you know that $(x - c)$ and the quotient (the answer to the division problem) are both factors of the polynomial $f(x)$.

Example 3: Determining If Something Is a Factor

Is $(x + 2)$ a factor of $x^3 - x^2 - 10x - 8$? If it is, then see if the resulting quotient can be factored further.

Solution:

Do synthetic division.	$\begin{array}{r\|rrrr} -2 & 1 & -1 & -10 & -8 \\ & & -2 & 6 & 8 \\ \hline & 1 & -3 & -4 & 0 \end{array}$
Now we know that $(x + 2)$ and $(x^2 - 3x - 4)$ are factors of the polynomial since the remainder in the synthetic division equals 0.	$X^3 - x^2 - 10x - 8 = (x + 2)(x^2 - 3x - 4)$
Now factor $(x^2 - 3x - 4)$.	$(x - 4)(x + 1)$

So, $(x + 2)$, $(x - 4)$, and $(x + 1)$ are all factors of the original polynomial $x^3 - x^2 - 10x - 8$. We have now discovered a way to factor polynomials that have a degree greater than 2. Notice also how the number of factors equaled the degree of the polynomial, 3. It will turn out, as we go through the next section, that this will always be the case if you take multiplicity into account.

Question 3: Is $(x - 5)$ a factor of $2x^3 - 10x^2 - 18x + 90$? If it is, then see if the resulting quotient can be factored further.

Example 4: Determining If Something Is a Factor

Is $(x - 4)$ a factor of $x^3 - 6x^2 + 10x - 8$? If it is, then see if the resulting quotient can be factored further.

Solution:

Answer Q1:

$$\begin{array}{r|rrrr} 3 & 3 & 0 & -15 & 7 \\ & & 9 & 27 & 36 \\ \hline & 3 & 9 & 12 & 43 \end{array}$$

$3x^2 + 9x + 12 + \frac{43}{x-3}$, $f(3) = 43$

Do synthetic division	$\begin{array}{r\|rrrr} 4 & 1 & -6 & 10 & -8 \\ & & 4 & -8 & 8 \\ \hline & 1 & -2 & 2 & 0 \end{array}$
Now we know that $(x - 4)$ and $(x^2 - 2x + 2)$ are factors of the polynomial since the remainder in the synthetic division equals 0.	$X^3 - 6x^2 + 10x - 8 = (x - 4)(x^2 - 2x + 2)$
Now the factor $(x^2 - 2x + 2)$ isn't factorable into factors with only real coefficients so we are done, for now☺	$(x - 4)(x^2 - 2x + 2)$

In the next section, we will see how to factor the polynomial $x^2 - 2x + 2$ if we are allowed to use complex numbers.

Section Summary

- Synthetic division is a way to shorten the long division process.
- Synthetic division is a way to find factors.
- If we use a number c in synthetic division, then the remainder will equal *f*l, the same value you'd get if you were to plug the number c into the function $f(x)$.
- If the remainder equals 0, we know that what we have just divided by (the divisor) and the answer (the quotient) are factors of the function we have divided into (the dividend).

SECTION 3.3 HOMEWORK

(1–16) Use synthetic division to divide.

1. $(x^3 - 2x^2 + 3x - 2) \div (x + 2)$
2. $(x^4 + 3x^3 - 2x^2 + 7x - 2) \div (x - 2)$
3. $(x^4 - 3x^3 + 2x^2 - 5x - 1) \div (x - 3)$
4. $(x^3 - 7x^2 - 8x + 3) \div (x + 3)$
5. $(2x^3 + 7x^2 - 3x + 8) \div (x + 5)$
6. $(4x^3 - 13x^2 - 33x - 4) \div (x - 5)$
7. $(x^4 - 3x^2 + 3) \div (x - 3)$
8. $(x^5 - 2x^2 + 7) \div (x - 4)$
9. $(x^4 - 19) \div (x - 1)$
10. $(x^5 + 1) \div (x + 1)$
11. $(x^8 + 4x^7 - x^4 + 5) \div (x + 4)$
12. $(x^7 - 5x^6 - 37x^4 + 35x^2 - 2) \div (x - 6)$
13. $(3x^4 - 7x^3 - 5x^2 + 8) \div \left(x - \frac{4}{3}\right)$
14. $(8x^4 - 6x^3 - x^2 + 2x + 4) \div (x + \frac{3}{4})$
15. $(4x^3 + 3x^2 + 5) \div (x + \frac{3}{2})$
16. $(3x^3 - 4x^2 - 8) \div (x - \frac{2}{3})$

Answer Q2:

4)	1	−4	4	−16
		4	0	16
	1	0	4	0

f(4) = 0

(17 – 28) Use synthetic division and the remainder theorem to find *f*I.

17. $f(x) = 3x^2 - 11x + 3,\ f(2)=$
18. $f(x) = 2x^2 + 12x - 4,\ f(-5)=$
19. $f(x) = 3x^3 + 8x^2 + 2,\ f(-4) =$
20. $f(x) = 7x^3 - 9x^2 - 3x + 15,\ f(3) =$
21. $f(x) = x^4 - x^3 + 2x^2 - 3,\ f(-2)=$
22. $f(x) = 2x^4 - 13x^3 + 5x^2 - 5,\ f(6)=$
23. $f(x) = x^7 - 3x^6 + 3x^5 - 9x^4 + 9,\ f(3)=$
24. $f(x) = x^8 + 2x^7 - 3x^6 + x^3 + 3,\ f(-3)=$
25. $f(x) = x^3 - 13x^2 + 11x + 9,\ f(12)=$
26. $f(x) = x^4 - 17x^3 + 30x^2 + 9,\ f(15) =$
27. $f(x) = 2x^5 - 33x^3 - 53,\ f(-4)=$
28. $f(x) = x^5 - 125x^2 - 15,\ f(5)=$

(29–48) Use synthetic division and the factor theorem to see if the given binomial is a factor of $f(x)$. If it is a factor, write the polynomial in factored form.

29. $f(x) = 3x^3 + 10x^2 - 22x + 15,\ x + 5$
30. $f(x) = 5x^3 - 32x^2 - 42x + 147,\ x - 7$
31. $f(x) = x^4 - 6x^3 + 10x^2 - 5x + 6,\ x - 3$
32. $f(x) = 2x^4 + 13x^3 + 17x^2 - 8x + 16,\ x + 4$
33. $f(x) = 3x^3 + 5x^2 - 8x + 3,\ x + 3$
34. $f(x) = 2x^3 - 4x^2 - 2x - 5,\ x - 3$
35. $f(x) = x^5 - 6x^4 + 7x^3 - 11x^2 + 8x - 15,\ x - 5$
36. $f(x) = 2x^5 + 11x^4 + 8x^3 - 13x^2 + 7x - 20,\ x + 4$
37. $f(x) = x^4 - 34x^2 + 225,\ x + 5$
38. $f(x) = x^4 - 61x^2 + 900,\ x - 6$
39. $f(x) = x^5 - 1,\ x + 1$
40. $f(x) = x^4 - 18,\ x - 2$
41. $f(x) = 6x^3 - 13x^2 + 11x - 3,\ x - \frac{1}{2}$
42. $f(x) = 12x^3 + 13x^2 - 3x - 2,\ x + \frac{1}{3}$
43. $f(x) = 2x^4 - 3x^3 - 4x^2 + 12x - 9,\ x - \frac{3}{2}$
44. $f(x) = 3x^4 + 2x^3 - 9x^2 + 4,\ x + \frac{2}{3}$
45. $f(x) = x^4 - x^3 + 2x^2 - 4x - 8,\ (x - 2i)$
46. $f(x) = 2x^4 + 3x^3 + 16x^2 + 27x - 18,\ (x + 3i)$
47. $f(x) = x^4 - x^3 + 5x^2 + x + 10,\ (x - 1 + 2i)$
48. $f(x) = x^4 - x^3 - 2x^2 + 51x - 39,\ (x - 2 - 3i)$

(Hint: For 49–50 you may not want to use synthetic division)

49. Find the remainder of $19x^{90} - 23x^{45} + 8x^{13} + 2$ divided by $x - 1$.

50. Find the remainder of $33x^{93} + 53x^{80} - 35x^{19} + 58$ divided by $x + 1$.

51. Show that $x^n + 1$, when n is a positive odd integer, has a factor of $x + 1$.

52. Show that $x^n - 1$, when n is a positive even integer, has a factor of $x - 1$.

53. Explain why $5x^6 + 9x^4 + 10$ has no factor of the form of $x - c$, where c is a real number.

54. Explain why $-8x^4 - 12x^2 - 24$ has no factor of the form of $x - c$, where c is a real number.

Answer Q3:

$$\begin{array}{r|rrrr} 5 & 2 & -10 & -18 & 90 \\ & & 10 & 0 & -90 \\ \hline & 2 & 0 & -18 & 0 \end{array}$$

$2x^3 - 10x^2 - 18x + 90$
$= (x - 5)(2x^2 - 18)$
$= 2(x - 5)(x + 3)(x - 3)$

Section 3.4 Zeros, Roots, and x-Intercepts

We have talked about how to do synthetic division, use the remainder theorem, and find the factors of a polynomial by using synthetic division. Now we are going to look at zeros, roots and x-intercepts. From the previous section we have learned that when the remainder is equal to 0 we have found a factor of the polynomial. Let's again take a look at example 3 from the previous section.

Example 3 (Section 3.3):

Is $(x + 2)$ a factor of $x^3 - x^2 - 10x - 8$? If it is, then see if it can then be factored further.

Solution:

Do synthetic division.	$\begin{array}{r\|rrrr} -2 & 1 & -1 & -10 & -8 \\ & & -2 & 6 & 8 \\ \hline & 1 & -3 & -4 & 0 \end{array}$

Since the remainder was 0, $(x + 2)$ is a factor of $f(x) = x^3 - x^2 - 10x - 8$. By the remainder theorem, this also means that $f(-2) = 0$. This means that when you plug in -2 for x in the function $f(x)$ you get 0 for an answer. You should remember from chapter 1 that x is the input. In this case, the input is -2, and $f(x)$ is the output; which is 0. This means that the point $(-2, 0)$ is on the graph of $f(x) = x^3 - x^2 - 10x - 8$. This point is an x-intercept.

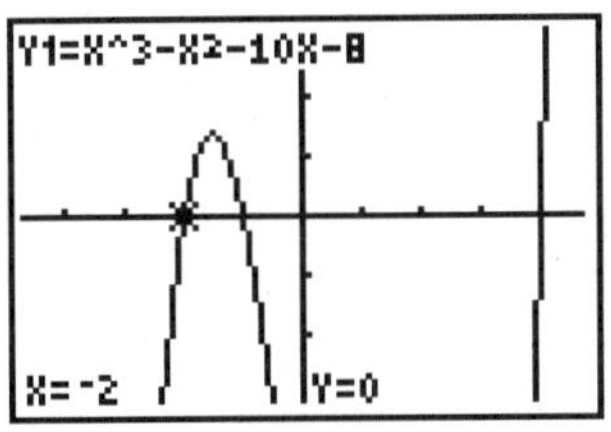

So, the ***x*-Intercept Theorem** states,

> If the remainder in the synthetic division is equal to 0, then the number used in the synthetic division will also be an x-intercept for the graph of $f(x)$ as long as that number is a real number.

One way to find the x-intercepts is to find all the real numbers that, when used in the synthetic division, cause a remainder of 0. You might have noticed that in the graph above there were three x-intercepts. We have shown that $x = -2$ is one of them. We know that the other two are $x = 4$ and $x = -1$ because in example 3 of the previous section, we found that $(x + 2)$, $(x - 4)$ and $(x + 1)$ were all factors of $x^3 - x^2 - 10x - 8$.

This brings together several points: if you know that the remainder of the synthetic division is equal to 0 ($fI = 0$, remainder theorem), then you know that x minus that number $(x - c)$ is a factor; therefore you know that x equal to that number $(x = c)$ is an x-intercept (as long as c is a real number).

We need to define two more terms.

Zero of a Polynomial	**Root** of a Polynomial
Any input I for x that causes $f(x) = 0$.	Any input I for x that makes an equation true.

Let's take a look at the definition of the zero of a polynomial. Notice that the definition talks about an x value that causes the function $f(x)$ to equal 0. This is just the same as the remainder being equal to 0 (from the remainder theorem). So this isn't any different than our discussion about x-intercepts. If we find a number that causes the remainder from the synthetic division to equal *zero*, then we also know that that number is called a zero of the polynomial.

We'll illustrate the second definition, the root of a polynomial, by doing an example.

Discussion 1: Roots of a Polynomial

Let's look at how we find all of the roots for the equation $x^3 - 5x^2 = 9x - 45$.

Now a root is an x value, which would make this equation true. To solve this problem we will change it to an equivalent form $x^3 - 5x^2 - 9x + 45 = 0$ (by subtracting $9x$ and adding 45 to both sides of the equation). To solve this we would like to find all of the x-values that would make the function on the left equal to 0. But notice if the equation is now *equal to 0*, then what we really want to find are all of the zeros of this polynomial. So it's time to use synthetic division again.

Do synthetic division.	$\begin{array}{r\|rrrr} 3 & 1 & -5 & -9 & 45 \\ & & 3 & -6 & -45 \\ \hline & 1 & -2 & -15 & 0 \end{array}$
Now we know that $(x - 3)$ and $(x^2 - 2x - 15)$ are factors of the polynomial since the remainder in the synthetic division equals 0.	$X^3 - 5x^2 - 9x + 45 = (x - 3)(x^2 - 2x - 15)$
Now factor $(x^2 - 2x - 15)$.	$(x - 5)(x + 3)$ $x = 5,\ x = -3$

So from the two factors and the synthetic division above, we know that $x = 3, 5, -3$ are zeros and x-intercepts of the equation $x^3 - 5x^2 - 9x + 45 = 0$, and thus roots of the original equation.

We can now put everything together.

If we use synthetic division with the number I on a polynomial and we get a remainder of 0, then we know **all** of the following:

- fI = 0
- $(x - c)$ is a factor of the polynomial.
- $x = c$ is an x-intercept (**if c is a real number**).
- $x = c$ is called a zero of the polynomial.
- $x = c$ is a root of the equation $f(x) = 0$.

Now, you might be wondering why we chose to use the number "3" in the synthetic division of the last example. That is exactly the million-dollar question: "How do you decide what numbers to try in synthetic division in order to find zeros, roots, x-intercepts and factors of a polynomial?" Before there were graphing calculators, the "Final Answer" to this question was "It's a long hard process!" In our book it will be fairly easy. We are going to use something called the Rational Zeros Theorem, along with the graphing calculator to answer the above million-dollar question.

The **Rational Zeros Theorem** states:

> If the polynomial function $f(x) = a_n x^n + a_{n-1}x^{n-1} + a_{n-2}x^{n-2} + \ldots + a_2x^2 + a_1x + a_0$ has integer coefficients, then every rational zero of $f(x)$ is of the form $\frac{p}{q}$ where p is a factor of a_0 and q is a factor of a_n .

Discussion 2: Finding Zeros

Here is how we will find the zeros for the function $f(x) = x^3 + x^2 - 5x - 5$.

First, understand that from the rational zeros theorem the possible rational zeros are factors of 5 divided by factors of 1, plus or minus.	$\pm\frac{1, 5}{1}$
Second, graph the function and see if it looks like any of the possible rational zeros could be an x-intercept.	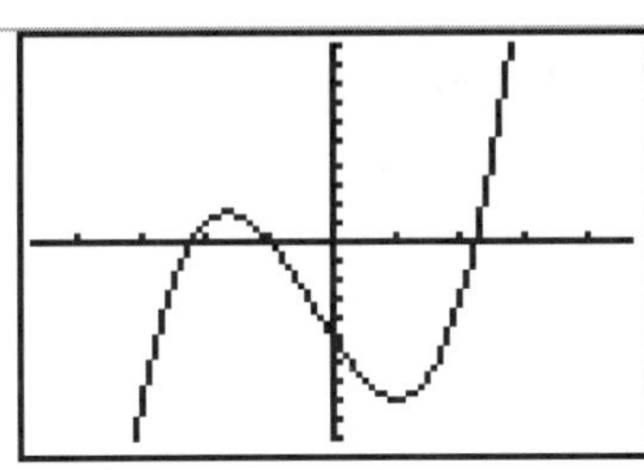
From the graph it looks as though $x = -1$ could be an x-intercept, so we now do synthetic division using -1.	$\begin{array}{r\|rrrr} -1 & 1 & 1 & -5 & -5 \\ & & -1 & 0 & 5 \\ \hline & 1 & 0 & -5 & 0 \end{array}$
Now we know that $(x + 1)$ and $(x^2 - 5)$ are	

factors of the polynomial since the remainder in the synthetic division equals 0.	$X^3 + x^2 - 5x - 5 = (x + 1)(x^2 - 5)$
Since $(x^2 - 5)$ doesn't factor nicely, we need to solve the equation $\underline{x^2 - 5 = 0}$ to find the other zeros.	$X^2 - 5 = 0$ $x^2 = 5$ $x = \pm\sqrt{5}$
We have found three zeros. No surprise; we had a third-degree equation to begin with. We also have all of the factors.	Zeros are $x = -1,\ \sqrt{5}, -\sqrt{5}$ $(x + 1)(x - \sqrt{5})(x + \sqrt{5}) = x^3 + x^2 - 5x - 5$

Notice that with the three zeros we get three factors of the form $(x - c)$ which, when multiplied together, equal the original polynomial.

Question 1: Given zeros of $x = -1$ and $x = 3$ what are the factors, and what is a possible polynomial that would have these zeros?

As you have seen in this last example, after synthetic division you might have something left that is either hard to factor or even impossible to factor. In this case, you need to solve the remaining factor by other means as in the above example, where we used the square root method to solve the resulting quadratic factor. It might also be that after you find one number that causes the remainder to equal 0 in the synthetic division, what remains isn't quadratic.

> NOTE: There do exist cubic and quartic (fourth-degree) formulas for solving cubic and quartic equations but they are very long and involved, so no one uses them very often. It has also been proven that <u>no</u> formula(s) can exist for solving equations whose degree is greater than 4.

So, in these cases, you must do synthetic division again. Let's look at an example of this type.

Example 1: Finding Zeros of a Fourth-Degree Polynomial

Find the zeros for the function $g(x) = x^4 - x^3 - 16x^2 - 4x - 80$.

Solution:

First, from the rational zeros theorem, the possible rational zeros are factors of 80 divided by factors of $\pm$ 1.	$\pm\dfrac{1, 2, 4, 5, 8, 10, 16, 20, 40, 80}{1}$

Second, graph the function and see if it looks like any of the possible rational zeros could be an x-intercept.	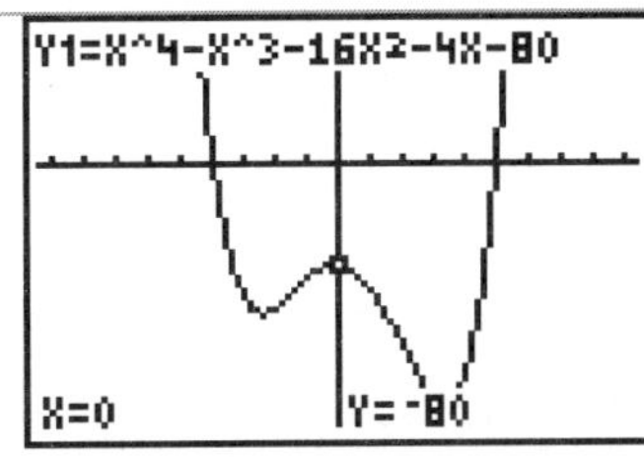
From the graph, it looks as though $x = -4$ or 5 could be x-intercepts, so we now do synthetic division using one of these numbers.	-4)$\;1 \quad -1 \quad -16 \quad -4 \quad -80$ $-4 \quad 20 \quad -16 \quad 80$ $1 \quad -5 \quad 4 \quad -20 \quad 0$
Now, we know that $(x + 4)$ and $(x^3 - 5x^2 + 4x - 20)$ are factors of the polynomial since the remainder in the synthetic division equals 0.	$X^4 - x^3 - 16x^2 - 4x - 80$ $= (x + 4)(x^3 - 5x^2 + 4x - 20)$
Since $(x^3 - 5x^2 + 4x - 20)$ isn't a quadratic, we need to do synthetic division in order to factor it further. We will use the other number, 5, which looked like it might be an x-intercept in the old function $g(x)$.	5)$\;1 \quad -5 \quad 4 \quad -20$ $5 \quad 0 \quad 20$ $1 \quad 0 \quad 4 \quad 0$
Now, we know that $(x - 5)$ and $(x^2 + 4)$ are factors of the polynomial $(x^3 - 5x^2 + 4x - 20)$.	$X^4 - x^3 - 16x^2 - 4x - 80$ $= (x + 4)(x^3 - 5x^2 + 4x - 20)$ $= (x + 4)(x - 5)(x^2 + 4)$
Since $(x^2 + 4)$ doesn't factor nicely but is quadratic, we need to solve $\underline{x^2 + 4 = 0}$ to find the other zeros, which in this case turn out to be complex.	$X^2 + 4 = 0$ $x^2 = -4$ $x = \pm\sqrt{-4} = \pm 2i$
We have found four zeros. This is no surprise; we had a fourth-degree equation to begin with. We also have all of the factors of $x^4 - x^3 - 16x^2 - 4x - 80$.	Zeros are $x = -4, 5, 2i, -2i$ $(x + 4)(x - 5)(x - 2i)(x + 2i)$

Look back at the graph of $g(x)$. Notice how it only had two x-intercepts. Now look at our four answers. Notice how two of them are real numbers and two of them are complex. Only real zeros give you x-intercepts. This happens because when we graph, we are using two number lines, one for the x values (horizontal axis) and one for the y values (vertical axis), which are made up of real numbers, not complex ones. So, the complex zeros don't show up on the graph.

Also notice how the two complex zeros were conjugates of each other (section 2.1). That will always happen if the original function's coefficients are all real numbers.

> **Complex Conjugate Zeros Theorem**
> If the polynomial function $f(x)$ has real coefficients and there is one complex zero, then its conjugate must also be a zero of $f(x)$.

Question 2: If $(3 + 5i)$ is a zero of $f(x)$, with real coefficients, then what other zero must it have?

If you look back at example 2, you will notice how the irrational zeros were conjugates of each other, too. That will always happen if the original function's coefficients are all rational numbers (we could call this the irrational conjugate zeros theorem). Let's do an example.

Example 2: Finding Roots of an Equation

Find the roots of the equation $6x^4 - 5x^3 + 43x^2 + 76x - 30 = 0$.

Solution:

Answer Q1: $(x + 1)(x - 3) = x^2 - 2x - 3$ is a possible answer, as is $5(x + 1)(x - 3) = 5x^2 - 10x - 15$. There are many possibilities.

First, finding the roots of this equation is just like finding the zeros of the function $f(x) = 6x^4 - 5x^3 + 43x^2 + 76x - 30$. Here are the possible rational zeros.	$\pm\dfrac{1,2,3,5,6,10,15,30}{1,2,3,6}$
Second, graph the function and see if it looks like any of the possible rational zeros could be an x-intercept.	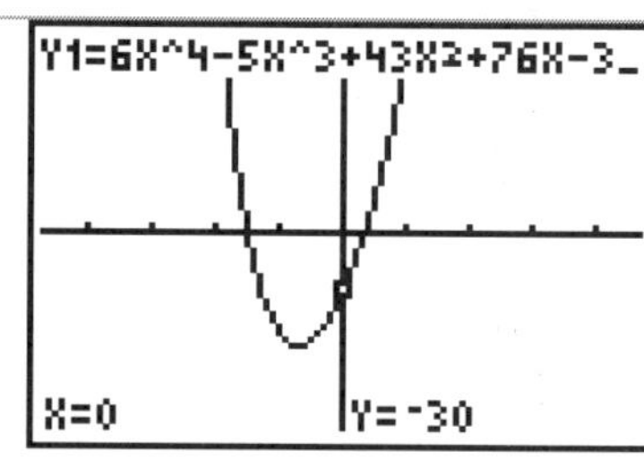
From the graph, it looks as though the x-intercepts must be some number between -2 and -1, and some number between 0 and 1. From the possible zeros, let's try $-\dfrac{3}{2}$.	$\begin{array}{r\|rrrrr} -\frac{3}{2} & 6 & -5 & 43 & 76 & -30 \\ & & -9 & 21 & -96 & 30 \\ \hline & 6 & -14 & 64 & -20 & 0 \end{array}$
Now we know that $(x + \frac{3}{2})$ and $(6x^3 - 14x^2 + 64x - 20)$ are factors of the polynomial, since the remainder in the synthetic division equals 0.	$6x^4 - 5x^3 + 43x^2 + 76x - 30$ $= (x + \frac{3}{2})(6x^3 - 14x^2 + 64x - 20)$

Since $(6x^3 - 14x^2 + 64x - 20)$ isn't a quadratic we need to do synthetic division to it in order to factor it further. We will try using the other possible number, $\frac{1}{3}$, which looked like it might be an x-intercept in the old function $f(x)$.	$\frac{1}{3}$) $6 \quad -14 \quad 64 \quad -20$ $\quad 2 \quad -4 \quad 20$ $6 \quad -12 \quad 60 \quad 0$
Now we know that $(x - \frac{1}{3})$ and $(6x^2 - 12x + 60)$ are factors of the polynomial $6x^3 - 14x^2 + 64x - 20$.	$6x^4 - 5x^3 + 43x^2 + 76x - 30$ $= (x + \frac{3}{2})(6x^3 - 14x^2 + 64x - 20)$ $= (x + \frac{3}{2})(x - \frac{1}{3})(6x^2 - 12x + 60)$
Since $(6x^2 - 12x + 60)$ doesn't factor but is quadratic, we need to solve the equation $6x^2 - 12x + 60 = 0$ to find the other zeros, which in this case turn out to be complex.	$x = \frac{12 \pm \sqrt{12^2 - 4(6)(60)}}{12} = 1 \pm 3i$
We have found four zeros of this fourth-degree polynomial which are the four roots of the original equation. Notice the two complex roots were conjugates of each other. We also have all four of the linear factors.	Zeros are $x = -\frac{3}{2}, \frac{1}{3}, 1 \pm 3i$ $(x + \frac{3}{2})(x - \frac{1}{3})(x - (1 + 3i))(x - (1-3i))$

Now maybe you have noticed that in every example so far the number of zeros equals the degree of the polynomial. This is what we will call the **Zeros Theorem**:

> Every polynomial of degree $n \geq 1$ has $\underline{n \text{ zeros}}$ if any zeros of multiplicity k are counted k times.

Question 3: From a graph of $f(x)= x^5 + x^4 - 5x^3 - x^2 + 8x - 4$ how many real zeros does it *appear* to have? How many complex ones? Why?

You may *think* that this has only two real zeros but since complex zeros come in conjugate pairs (Complex Conjugate Zeros Theorem), this can't be right. From the discussion on multiplicity and the shapes of polynomials in section 3.2, and with our discussions in this section, there must be five real zeros and no complex ones. The correct answer to question 3 follows:

Since the graph near $x = -2$ looks like x^2 and near $x = 1$ it looks like x^3, we must have just two factors, one with multiplicity 2 the other with multiplicity 3.

Question 4: Write the factors of a fourth-degree polynomial with a zero of $(1 + I)$ and a zero of 3 (with multiplicity 2).

Let's take a moment to show you how to multiply out the answer to question 4.

Clear the inner most parentheses first.	$(x-3)^2(x-1-I)(x-1+I)$
Regroup with the I's separated.	$(x-3)^2((x-1)-I)((x-1)+I)$
Multiply the second two factors (difference of two squares) then simplify.	$(x-3)^2((x-1)^2-I^2)$ $=(x-3)^2((x^2-2x+1)+1)$ $=(x-3)^2(x^2-2x+2)$
Finish clearing parentheses.	$(x^2-6x+9)(x^2-2x+2)$ $=(x^4-2x^3+2x^2-6x^3+12x^2-12x+9x^2-18x+18)$ $=x^4-8x^3+23x^2-30x+18$
So a function that could have 3 (multiplicity 2) and (1 + I) as zeros is	$f(x)=x^4-8x^3+23x^2-30x+18$

Answer Q2:
3 – 5i

Example 3: Finding Zeros of a Fifth-Degree Polynomial

Find all the zeros of the polynomial in question 3, $f(x)=x^5+x^4-5x^3-x^2+8x-4$.

Solution:

First, from the rational zeros theorem the possible rational zeros are factors of 4 divided by factors of ± 1.	$\pm\frac{1,2,4}{1}$
From the graph it looks as though $x = -2$ or 1 could be x-intercepts, so we now do synthetic division using one of these numbers; we'll use 1.	1) 1 1 −5 −1 8 −4 1 2 −3 −4 4 1 2 −3 −4 4 0
We know from the discussion above that $x = 1$ appears to have multiplicity 3, so let's do synthetic division with the number 1 two more times.	1) 1 2 −3 −4 4 1 3 0 −4 1 3 0 −4 0
	1) 1 3 0 −4 1 4 4 1 4 4 0

Now we know that $(x-1)^3$ and (x^2+4x+4) are factors of the polynomial so we factor (x^2+4x+4).	$X^2+4x+4=(x+2)(x+2)$ So $x=-2$ is a zero.
We have found five zeros. We also have all of the factors.	Zeros are $x=1$, multiplicity 3, and $x=-2$, multiplicity 2, for a total of $(3+2=5)$ five zeros. $(x-1)^3(x+2)^2$

We are going to look at one more example where you can't find the answers by hand but must approximate them with the calculator.

Answer Q3: f(x) ***appears*** *to have 2 real zeros. Since the polynomial is of degree 5, this would imply that the rest are complex (3).*

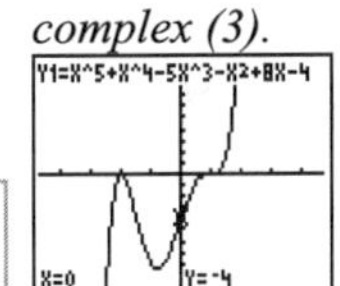

Example 4: Approximating Zeros Using Your Calculator

Find the zeros of $f(x)=2x^5-13x^4+18x^3+6x^2-4x-1$.

Solution:

By the Rational Zeros Theorem, only $1, -1, \frac{1}{2}, -\frac{1}{2}$ could be rational zeros.

<table>
<tr>
<td>From the graph it looks as though $x=\pm\frac{1}{2}$ could be rational x-intercepts.
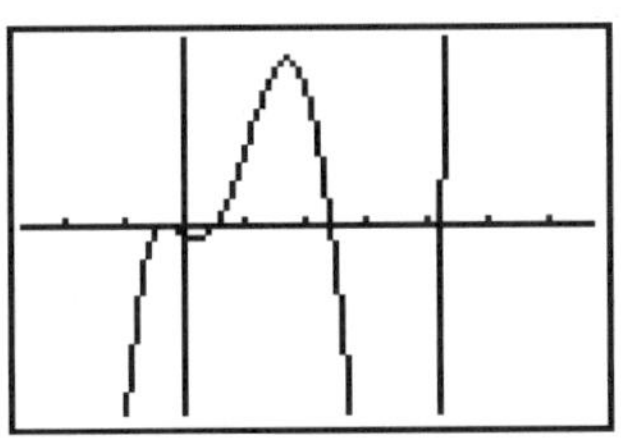</td>
<td>$$\begin{array}{r|rrrrrr} -\frac{1}{2} & 2 & -13 & 18 & 6 & -4 & -1 \\ & & -1 & 7 & -12.5 & 3.25 & .375 \\ \hline & 2 & -14 & 25 & -6.5 & -.75 & -.625 \end{array}$$
$$\begin{array}{r|rrrrrr} \frac{1}{2} & 2 & -13 & 18 & 6 & -4 & -1 \\ & & 1 & -6 & 6 & 6 & 1 \\ \hline & 2 & -12 & 12 & 12 & 2 & 0 \end{array}$$</td>
</tr>
<tr>
<td>However, only $\frac{1}{2}$ worked. Notice that something $(x=-\frac{1}{2})$ can look like an x-intercept on the graph but turn out not to be one. This is why we use synthetic division to prove our suspicions. So $(x-\frac{1}{2})$ is a factor and the rest of the x-intercepts must be irrational, since we see five of them (when you zoom in) and this is a fifth-degree polynomial</td>
<td>$(x-\frac{1}{2})(2x^4-12x^3+12x^2+12x+2)$
$=2x^5-13x^4+18x^3+6x^2-4x-1$</td>
</tr>
</table>

We can approximate the remaining x-intercepts and thus the zeros by using the root or zero key on our graphing calculators. The other four zeros rounded to four decimals are -0.4142, -0.2361, 2.4142, 4.2361.	 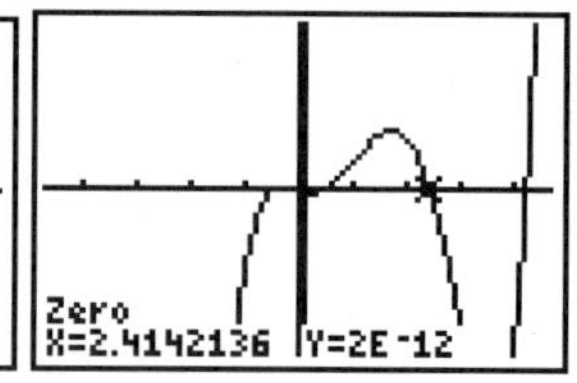 (Only two of the four graphs displayed here.)

Section Summary

Answer Q4:
$f(x) = (x-3)^2$
$(x-(1+i))$
$(x-(1-i))$

- If we use synthetic division with the number c on a polynomial and we get a remainder of 0, then we know **all** of the following:
  1. $f(c) = 0$
  2. $(x - c)$ is a factor of the polynomial.
  3. $x = c$ is an x-intercept (**if c is a real number**).
  4. $x = c$ is called a zero of the polynomial.
  5. $x = c$ is a root of the equation $f(x) = 0$
- Complex zeros must come in conjugate pairs if all the coefficients in the polynomial are real.
- The irrational zeros must also come in conjugate pairs if all the coefficients in the polynomial are rational.
- In examples 4 and 6 we see that, finding zeros can be hard, if not impossible, by hand. Fortunately, as we will see in chapter 5, most models of real-life phenomena don't usually require the use of higher-degree polynomials.

SECTION 3.4 HOMEWORK

(1–6) For each polynomial function find all the possible rational roots, using the rational root theorem (don't try to find the roots).

1. $f(x) = x^3 - 5x^2 + 2x - 10$
2. $f(x) = x^4 + 3x^2 - 8x + 12$
3. $f(x) = 5x^5 - 3x^2 + 15$
4. $f(x) = 3x^5 - 8x^4 + 5x^2 - 24$
5. $f(x) = 6x^4 - 8x^2 + 7x - 16$
6. $f(x) = 12x^5 - 8x^3 + 18$

(7–40) For each polynomial function find all the roots and then write the function in its factored form.

7. $f(x) = x^3 - 2x^2 - 5x + 6$
8. $f(x) = x^3 + x^2 - 9x - 9$
9. $f(x) = x^3 - x^2 - 18x - 10$
10. $f(x) = x^3 + x^2 - 5x - 2$
11. $f(x) = x^3 + 6x^2 + 13x + 20$
12. $f(x) = x^3 - 2x - 21$
13. $f(x) = x^4 - 4x^3 - 14x^2 + 36x + 45$
14. $f(x) = x^4 - 9x^3 + 21x^2 + x - 30$
15. $f(x) = x^4 + 8x^3 + 20x^2 + 16x + 3$
16. $f(x) = x^4 - 6x^3 + x^2 + 28x - 12$
17. $f(x) = x^4 + x^3 - 3x^2 - 17x - 30$
18. $f(x) = x^4 - 3x^3 - 5x^2 + 19x - 60$
19. $f(x) = 12x^3 - 8x^2 - 3x + 2$
20. $f(x) = 36x^3 - 27x^2 - 4x + 3$
21. $f(x) = 3x^3 - 7x^2 - 12x - 4$
22. $f(x) = 4x^3 + 17x^2 - 7x - 6$
23. $f(x) = 8x^3 + 10x^2 + 13x - 4$
24. $f(x) = 15x^3 - 7x^2 + 13x + 3$
25. $f(x) = x^3 - 4x^2 - 3x + 18$
26. $f(x) = x^3 + x^2 - 8x - 12$
27. $f(x) = x^4 + x^3 - 3x^2 - 5x - 2$
28. $f(x) = x^4 - 3x^3 - 6x^2 + 28x - 24$
29. $f(x) = x^4 + 2x^3 - 11x^2 - 12x + 36$
30. $f(x) = x^4 + 10x^3 + 33x^2 + 40x + 16$
31. $f(x) = x^4 + 3x^3 - 8x^2 - 21x + 9$
32. $f(x) = x^4 - 3x^3 - 21x^2 + 56x + 48$
33. $f(x) = x^4 + 8x^2 - 64x + 80$
34. $f(x) = x^4 + 4x^3 + 2x^2 + 12x + 45$
35. $f(x) = x^5 - 6x^4 - x^3 + 63x^2 - 108x + 27$
36. $f(x) = x^5 + 4x^4 - 4x^3 - 40x^2 - 64x - 32$
37. $f(x) = 144x^4 + 120x^3 - 47x^2 - 30x + 9$
38. $f(x) = 36x^4 + 12x^3 - 179x^2 - 30x + 225$
39. $f(x) = 8x^3 + \frac{34}{3}x^2 - \frac{19}{3}x - 5$
40. $f(x) = 9x^3 + \frac{15}{2}x^2 - \frac{103}{2}x - 35$

(41–60) Give a polynomial function with integer coefficients that has the following roots. (There is not just one correct answer.)

41. –2, 3, 5
42. –3, –2, 3
43. –1, –3, 2, 4
44. 3, –1, 2, 0
45. 2 (multiplicity of 2), –1
46. –3 (multiplicity of 2), 2
47. 4 (multiplicity of 3), –4
48. –5 (multiplicity of 3), 5
49. –2 (multiplicity of 2), 2 (multiplicity of 2)
50. –4 (multiplicity of 2), 0 (multiplicity of 2)

51. 2, $3-\sqrt{5}$, $3+\sqrt{5}$

52. −3, $2+\sqrt{3}$, $2-\sqrt{3}$

53. −4, $3+i$

54. 3, $-2+i$,

55. $2+3i$, $5-2i$

56. $-3+2i$, $2+5i$

57. −2, $3+\sqrt{2}\,i$

58. 3, $2-\sqrt{3}\,i$

59. $-\frac{1}{3}, \frac{3}{4}, 1$

60. $-\frac{1}{2}, \frac{3}{5}, -\frac{1}{4}$

(61–70) For each of the following polynomial functions find the real roots; where necessary use the graphing calculator to approximate the real roots to three decimal places.

61. $f(x) = x^3 + 3x^2 - 3x + 3$

62. $f(x) = x^3 - x^2 + 5x + 1$

63. $f(x) = x^4 - x^3 - 2x^2 - 5$

64. $f(x) = x^4 - x^3 - 5x^2 - 4$

65. $f(x) = x^5 - 11x^4 + 40x^3 - 48x^2 - x - 3$

66. $f(x) = x^5 - x^4 + 2x^3 - 1$

67. $f(x) = x^5 + 3x^4 - 15x^3 + 4$

68. $f(x) = x^5 + 2x^4 - 12x^3 + x^2 + 3$

69. $f(x) = x^4 - 6x^3 + 18x^2 - 30x + 25$

70. $f(x) = x^4 - 12x^3 + 59x^2 - 138x + 130$

71. Refer to problem 69 above:
   a. How many real solutions did you find?
   b. What can you conclude about the roots of this polynomial function?
   c. One of the roots is $2 + i$. Find the other roots.

72. Refer to problem 70 above:
   a. How many real roots did you find?
   b. What can you conclude about the roots of this polynomial function?
   c. One of the roots is $3 + 2i$. Find the other roots.

(73–80) Solve each of the following inequalities. Express your answer with a number line and interval notation. (Hint: Read section 2.5.)

73. $x^3 - 3x^2 - 10x + 24 < 0$

74. $x^3 + 2x^2 - 11x - 12 > 0$

75. $x^4 - 2x^3 - 16x^2 + 2x + 15 \geq 0$

76. $x^4 + x^3 - 11x^2 - 9x + 18 \leq 0$

77. $x^3 + x^2 - 5x + 3 < 0$

78. $x^3 + 4x^2 - 3x - 18 > 0$

79. $x^4 - 4x^3 - 2x^2 + 12x + 9 \geq 0$

80. $x^4 + 2x^3 - 11x^2 - 12x + 36 \leq 0$

81. A company manufactures air conditioners; the cost of manufacturing x air conditioners is given by $C(x) = \dfrac{x^3}{20} - x^2 + 800x + 2400$. How many air conditioners are produced for a cost of \$36,000? (Hint: use zero or root finder on the graphing calculator.)

82. A company is manufacturing computers; the cost of manufacturing x computers is given by: $C(x) = \dfrac{x^3}{240} - 2x^2 + 600x + 2420$. How many computers will be produced for a cost of \$32,120? (Hint: use zero or root finder on the graphing calculator.)

(83–84) Use the following relationships to answer the questions:

- **Profit = Revenue – Cost**
- **Revenue = px, where p is the price each item is sold for and x is the number sold.**

83. From problem 81, suppose each air conditioner sells for $1800.
    a. What is the revenue formula?
    b. How many air conditioners must be sold to make a profit of $58,400, assuming you sell all you produce?
    c. Use the graphing calculator to find out how many air conditioners must be produced to make a maximum profit. (See section 1.3.)

84. From problem 82, suppose you computers sell for $1200 each.
    a. What is the revenue formula?
    b. How many computers must be sold to make a profit of $39880, assuming you sell all you produce?
    c. Use the graphing calculator to find out how many air conditioners must be produced to make a maximum profit. (See section 1.3.)

85. An open box with a rectangular base is constructed for storage. If the short side of the base is 1 foot shorter then the long side and the height of the box is 3 feet less than the long side, what are the dimensions of a box that has a volume of 90 ft^3?
    (Hint: The volume is the base area times height, and the base is a rectangle with area = length times width.)

86. An open box is constructed to store bottles of chemicals. The box is constructed from a 16 inch by 30 inch piece of tin by cutting out squares of equal size from the four corners and bending up the sides.
    What should the size of the cut squares be if the volume is 720 in^3?
    (Hint: The volume is the base area times height, and the base is a rectangle with area = length x width.)

Section 3.5 Rational Functions

We have discussed much about polynomial functions. In this section we are going to discuss rational functions. The division of one polynomial by another creates a **rational function**. Here are some examples:

$f(x) = \dfrac{x+3}{x^2-4}$	$g(x) = \dfrac{7-3x}{2x+1}$	$h(x) = \dfrac{x^2+3}{x-3}$
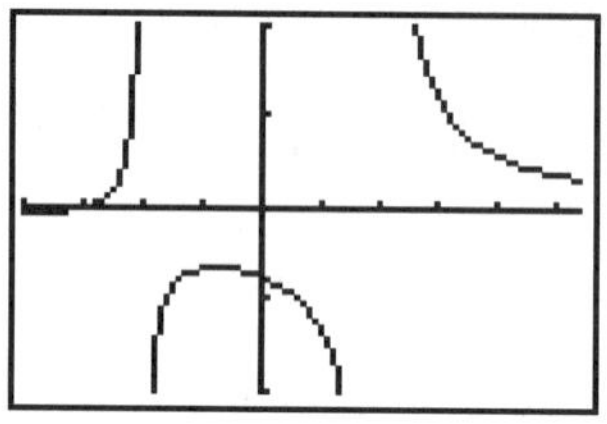	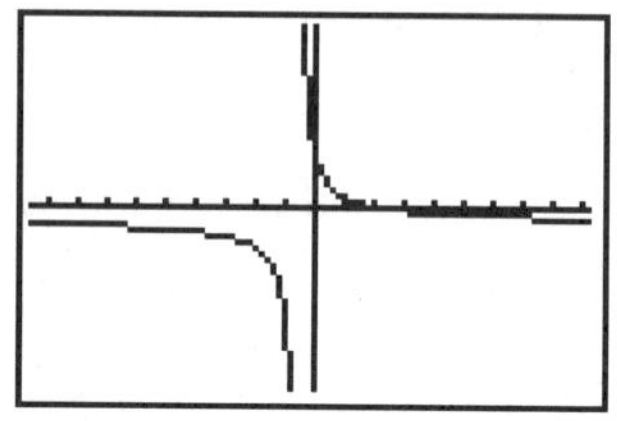	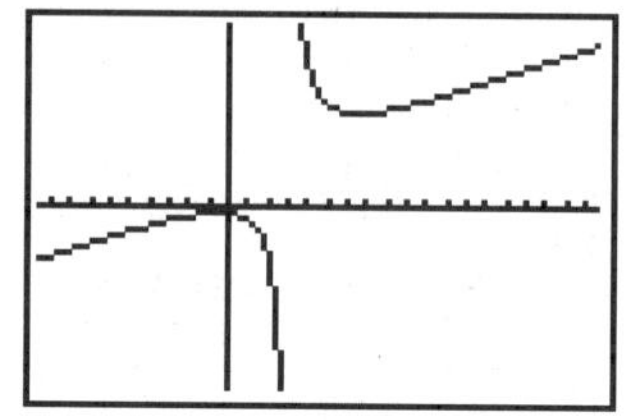

The major topic to discuss with regards to rational functions is the topic of **asymptotes**.

> **Asymptotes** are themselves polynomial functions or relations that the graphs of rational functions approach (get close to) either as x approaches some value a or as the magnitude of x gets very large.

There are three types of asymptotes. They are vertical, horizontal, and oblique. The first type we will discuss is vertical.

VERTICAL ASYMPTOTES

> A **vertical asymptote** is a vertical line at $x = a$ (not a function) where, as the x values get closer and closer to a, the magnitude of the $f(x)$ outputs get very large.

To find vertical asymptotes we look for values of x that cause the function to be undefined. To be more specific, we want to find values that cause the denominator to equal 0 when the rational function is in its *most simplified form*. When your x-values are close to a number that would make the denominator equal 0, you'll find that the output values, $f(x)$, are large. Let's look at the three examples from above.

Discussion 1: Finding Vertical Asymptotes

Let's find the vertical asymptotes for the functions shown above.

$f(x) = \dfrac{x+3}{x^2-4}$. Notice in the table to the right that as the x values get closer to 2, the $f(x)$ outputs get larger and larger in magnitude. This is a good indication that we have a vertical asymptote at $x = 2$. If we were to look at a table of values as x gets closer to -2, we would see the same situation.

	A	B
1	x	f(x)
2	2.1	12.43902439
3	2.050000000	24.93827160
4	2.025000000	49.93788820
5	2.012500000	99.93769470
6	2.006250000	199.9376015
7	2.003125000	399.9375368

To actually find the vertical asymptotes, or you might say to prove what they are, we do the following:

Set the denominator equal to 0 and solve.

$x^2 - 4 = (x+2)(x-2) = 0$

So, the vertical lines $x = 2$ and $x = -2$ are called vertical asymptotes.

So if $x = 2$ or -2, the denominator would equal 0.

$g(x) = \dfrac{7-3x}{2x+1}$. Notice in the table to the right that as the x values get closer to $-\dfrac{1}{2}$, the $g(x)$ outputs get larger and larger in magnitude. This is a good indication that we have a vertical asymptote at $x = -\dfrac{1}{2}$. To actually find the vertical asymptotes, set the denominator equal to 0 and solve as follows:

$$2x + 1 = 0 \Rightarrow 2x = -1 \Rightarrow x = -\frac{1}{2}$$

Hence the vertical line $x = -\dfrac{1}{2}$ is called a vertical asymptote.

	A	B
1	x	g(x)
2	-.8	-15.66666667
3	-.6500000000	-29.83333333
4	-.5750000000	-58.16666667
5	-.5375000000	-114.8333333
6	-.5187500000	-228.1666667
7	-.5093750000	-454.8333333
8	-.5046875000	-908.1666667

$h(x) = \dfrac{x^2+3}{x-3}$. Notice in the table to the right that as the x values get closer to 3, the $h(x)$ outputs get larger and larger in magnitude. This is a good indication that we have a vertical asymptote at $x = 3$. To actually find the vertical asymptotes set the denominator equal to 0 and solve as follows:

$$x - 3 = 0 \Rightarrow x = 3$$

So the vertical line $x = 3$ is a vertical asymptote.

	A	B
1	x	h(x)
2	3.3	46.30000000
3	3.150000000	86.15000000
4	3.075000000	166.0750000
5	3.037500000	326.0375000
6	3.018750000	646.0187499
7	3.009375000	1286.009375
8	3.004687500	2566.004687

Question 1: What are the vertical asymptotes for the following rational functions?

$$\frac{-5}{x+7} \qquad \frac{x^3-4x^2+5x-9}{x^2-5x+4}$$

HORIZONTAL ASYMPTOTES

A **horizontal asymptote** is a horizontal line at $y = b$ where, as the magnitude of the x-values get larger, the graph of the rational function levels off near a height of b.

This is more easily discussed using a topic from calculus called **limits**. But here we are going to come up with the three cases that exist which we will need to know in order to find these in a college algebra class. The three cases are as follows:

Case 1	If the degree of the denominator is greater than the degree of the numerator, then the horizontal asymptote will be $y = 0$.
Case 2	If the degree of the denominator is equal to the degree of the numerator, then the horizontal asymptote will be $y = \dfrac{\text{coefficient of largest degree term in the numerator}}{\text{coefficient of largest degree term in the denominator}}$.
Case 3	If the degree of the denominator is less than the degree of the numerator, then there isn't a horizontal asymptote; we get an oblique one instead (discussed next).

Why do we have three cases? Well, from our discussions about **dominating terms** in section 3.2, what will be important in rational functions is where the highest-degree term is, because that will dominate our answers (outputs) as our x-values get large. So, this can happen in three ways: Case 1) denominator degree > numerator degree, Case 2) denominator degree = numerator degree or Case 3) denominator degree < numerator degree.

Question 2: What happens when you divide a smaller number by a larger number? For example,

$$\frac{35}{7249}.$$

In Case 1, if the dominating term is in the denominator, then the denominator will be much larger than the numerator and so our answers will get closer and closer to 0 as the x's get larger and larger in magnitude (see the graph of $f(x) = \frac{x+3}{x^2-4}$ above).

In Case 2, if the denominator and numerator have an equally dominating term, then they tend to offset each other so the result is that you level out at some height $y = b$. The b will equal the division of the coefficients of the two largest terms, the one in the numerator with the one in the denominator (see the graph of $g(x) = \frac{7-3x}{2x+1}$ above).

Question 3: What happens when you divide a larger number by a smaller number? For example,

$$\frac{9268473150}{385}.$$

Discussion 2: Finding Horizontal Asymptotes

Let's find the horizontal asymptotes of the three rational functions from the beginning of this section ($f(x) = \frac{x+3}{x^2-4}$, $g(x) = \frac{7-3x}{2x+1}$, $h(x) = \frac{x^2+3}{x-3}$).

$f(x) = \frac{x+3}{x^2-4}$ has a horizontal asymptote at $y = 0$, since the degree of the denominator ($x^2 - 4$) is 2, which is larger than the degree of the numerator ($x + 3$), which is 1. Look at

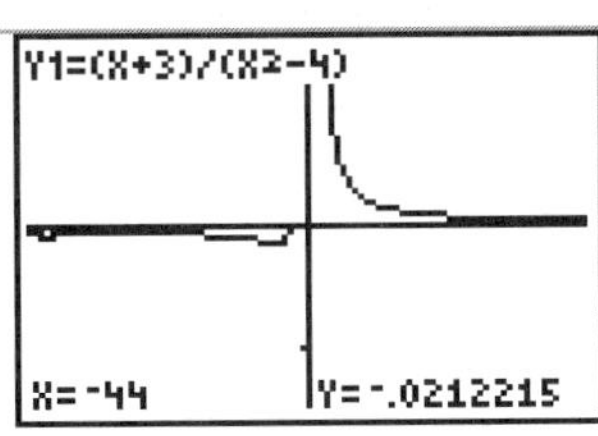

Zoomed out

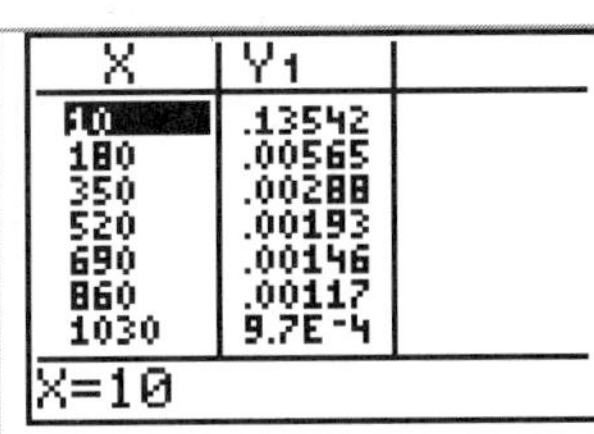

<table>
<tr><td>the graph and tables to the right. See how the graph of the rational function is leveling out near a height of 0 ($y = 0$) and how the table values are getting closer and closer to 0.</td><td></td><td>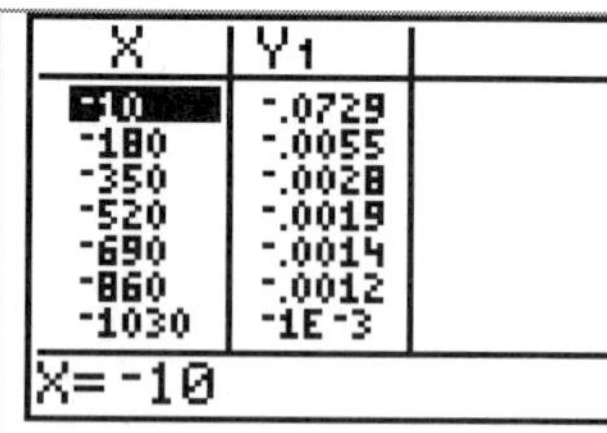
</td></tr>
<tr><td>$g(x) = \frac{7-3x}{2x+1}$ has a horizontal asymptote at $y = -\frac{3}{2}$. Case 2 is demonstrated here since the denominator and numerator have the same degree. We must divide the –3 coefficient of the largest term in the numerator by the coefficient of 2 of the largest term in the denominator. Look at the graph and tables to the right. See how the graph of the rational function is leveling out near a height of –1.5 ($y = -\frac{3}{2}$) and the table values are getting closer and closer to –1.5.</td><td>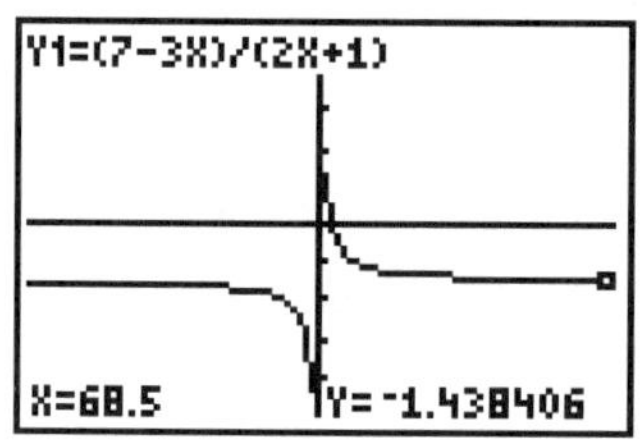

Zoomed out</td><td>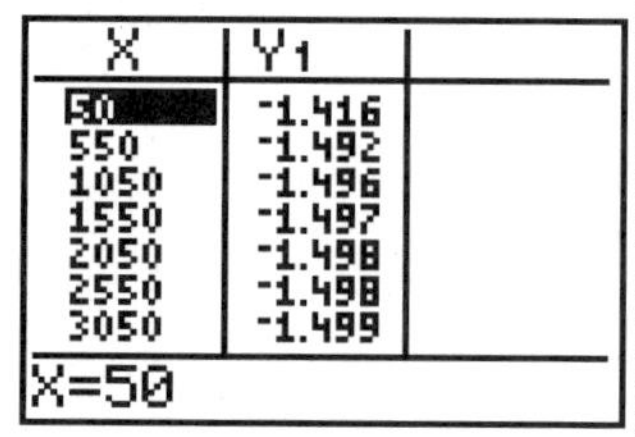

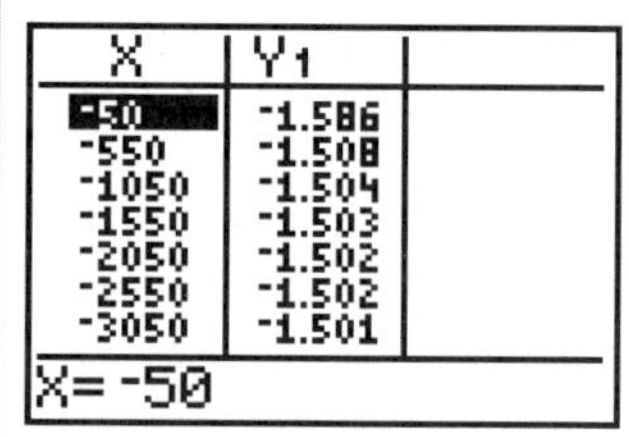
</td></tr>
<tr><td>$h(x) = \frac{x^2+3}{x-3}$ doesn't have a horizontal asymptote because the degree of the denominator ($x - 3$) is 1, which is smaller than the degree of the numerator ($x^2 + 3$), which is 2. This is an example of</td><td></td><td></td></tr>
</table>

Answer Q1: denominator $(x + 7)$ so $x = -7$ and denominator $x^2 - 5x + 4 = (x - 1)(x - 4)$ so, $x = 1, x = 4$

Case 3 where you get an oblique asymptote. Look at the graph and tables to the right. Notice that as the magnitude of x gets larger, the graph seems to level out along a diagonal line when we zoom out. Also, the table values just keep getting larger and larger without end.

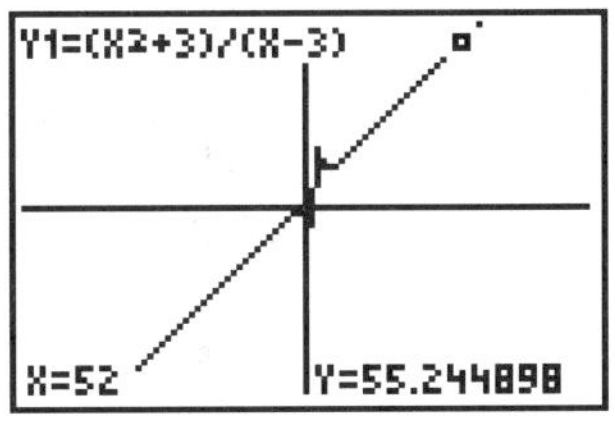

Zoomed out

X	Y1
45	48.286
10045	10048
20045	20048
30045	30048
40045	40048
50045	50048
60045	60048

X=45

X	Y1
-45	-42.25
-10045	-10042
-20045	-20042
-30045	-30042
-40045	-40042
-50045	-50042
-60045	-60042

X= -45

Question 4: What are the vertical and horizontal asymptotes for the rational function $f(x) = \dfrac{6x-5}{3x+9}$?

In Case 3, if the dominating term is in the numerator, then the numerator will be much larger than the denominator, so the output will get very large in magnitude as the x's get large in magnitude. The graph will continue upwards or downwards forever; it won't level out (see the graph of $h(x) = \dfrac{x^2+3}{x-3}$ above). What you get is a graph that gets closer and closer to the graph of some other polynomial. In this case, that polynomial is called an **oblique asymptote**.

Answer Q2: You get a small answer close to zero. In the example we get 0.004828....

OBLIQUE ASYMPTOTES

An **oblique asymptote** is the polynomial that the graph of the rational function gets closer to as the magnitude of x gets very large.

Answer Q3: You get a large answer. In this example we get 24,073,956.2

Discussion 3: Finding Oblique Asymptotes

Let's find the oblique asymptote for the rational function $h(x) = \dfrac{x^2+3}{x-3}$ from above.

First, make sure that the function is simplified before you do the long division. Here, since we are dividing by $(x - 3)$ we can do the long division problem by using synthetic division.	$\begin{array}{r\|rrr} 3 & 1 & 0 & 3 \\ & & 3 & 9 \\ \hline & 1 & 3 & 12 \end{array}$
So the answer to the long division is	$x+3+\dfrac{12}{x-3}$

The fraction $\frac{12}{x-3}$ will get very small as x gets very large. So, it must be that $(x + 3)$ is the function that $\frac{x^2+3}{x-3}$ gets close to as the x's get large. We have graphed both of the functions $\frac{x^2+3}{x-3}$ and $x + 3$ on the same axis at right. Notice that you can't tell the difference between the two graphs as x gets far away from the origin. In the table, notice how the outputs of each function get closer and closer as the x-values get larger.

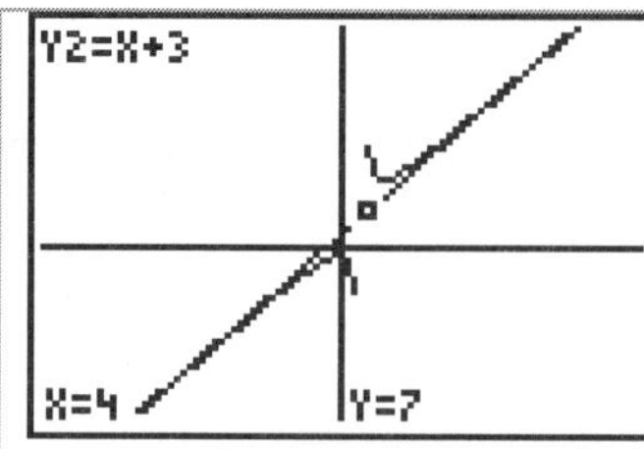

X	Y1	Y2
4	19	7
154	157.08	157
304	307.04	307
454	457.03	457
604	607.02	607
754	757.02	757
904	907.01	907

X=4

Since in this case the oblique asymptote is a linear function we will call it a **linear asymptote**. Many books will call this a slant asymptote.

Question 5: Find all the asymptotes for the following functions:

a.) $f(x)=\frac{2x-5}{(x-1)(x+4)}$ b.) $g(x)=\frac{x^2-2}{x+1}$

We want to take a moment here to discuss the shortcomings of your graphing calculator. The graphing calculator is a wonderful and maybe enjoyable tool, but it has some technical barriers that are hard to overcome. For example, how many numbers are there on the x-axis between 0 and 1? Answer: an infinite number. The problem is that your calculator screen is like your TV. It is made up of a bunch of dots (called pixels), but not an infinite number of them. So, sometimes when you graph a function on the calculator, you get a picture that isn't quite right.

Discussion 4: Rational Functions on the Graphing Calculator

Look at the following graphs of $f(x)=\frac{x}{x-2}$.

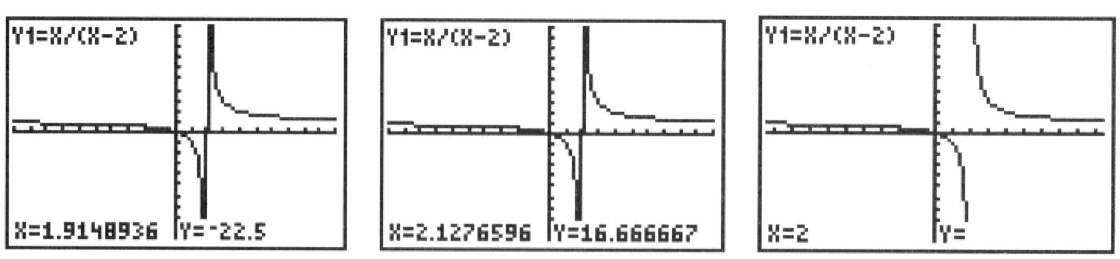

Notice how in the first two graphs of $f(x)$, using zoom standard, we see a vertical line that's not in the last graph. This is happening because when the calculator is connecting dots to make the graph, it connects the point (1.9148936…,-22.5) with the point (2.1276596…,16.66666…) the next pixel over. But, in the third graph, when you use zoom decimal, zoomed out by a factor of two (-9.4 to 9.4), you don't get a vertical line in the picture because the pixels work out just right and $x = 2$, the vertical asymptote, falls exactly on one of the pixels. So when the calculator is trying to connect the points it sees a blank pixel between a couple of points and so it does not try to connect across the blank pixel.

Now look at the graph of the function $g(x) = \dfrac{x+1}{3x+2}$; zoom decimal doesn't help because the vertical asymptote is $-\dfrac{2}{3}$.

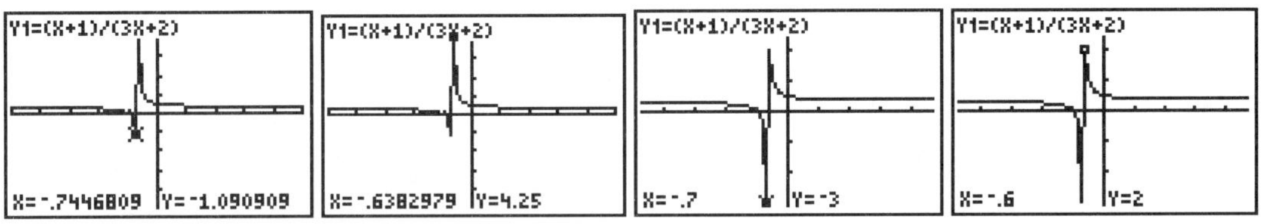

In fact, when you try to adjust the zoom decimal type window to land on $-\dfrac{2}{3}$ it still won't work (see the following graph).

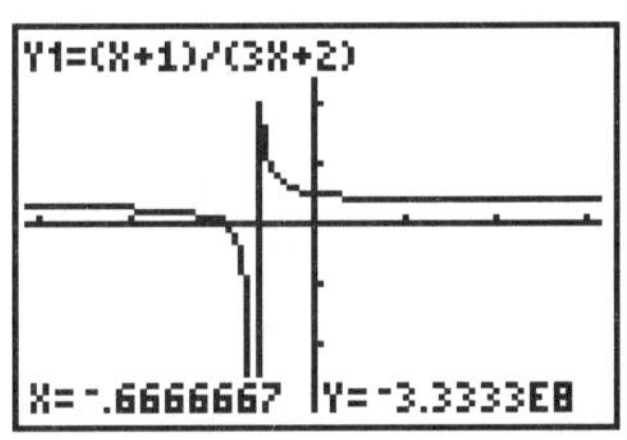

Answer Q4: The vertical asymptote is $x = -3$ since $3x+9=0$ when $x=-3$. The horizontal asymptote is $y = 2$, since $6/3 = 2$.

It is really important for you to understand asymptotes so that you can intelligently use your technology.

Example 1: Finding Asymptotes

Find all the asymptotes for the following functions; then sketch a graph with the help of your graphing calculator:

a.) $f(x) = \dfrac{x+4}{x^2-16}$ b.) $g(x) = \dfrac{4-3x^2}{4x^2+x+1}$ c.) $h(x) = \dfrac{3x^3-2x^2+x+5}{x+3}$

Solutions:

a) Since $x^2 - 16 = 0$ at $x = \pm 4$ you might think that this function should have two vertical asymptotes. But remember, when we started this section we mentioned that the rational function must be simplified as much as possible before you look for vertical asymptotes. We have only one vertical asymptote at $x = 4$ since the factor $(x + 4)$ cancels out with the $(x + 4)$ in the numerator. Notice that when this happens we get a **hole** in the graph because we still can't use $x = -4$ in the function since it's not in the domain.

Also, since the degree of the denominator is greater than the degree of the numerator, the horizontal asymptote is $y = 0$.

When you sketch a graph, you should draw in dashed lines where the asymptotes are.

What is on the calculator

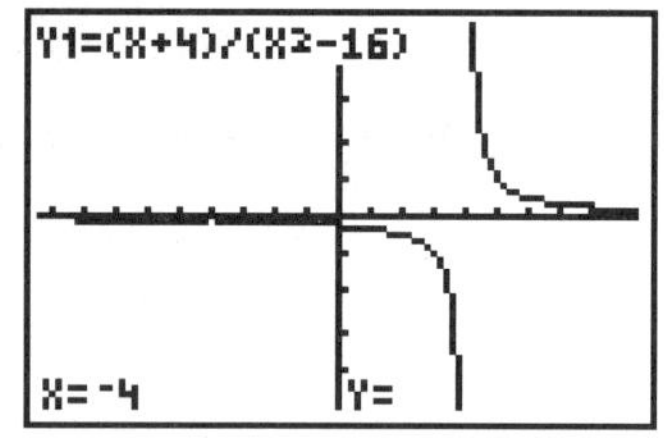

Sketch you might make

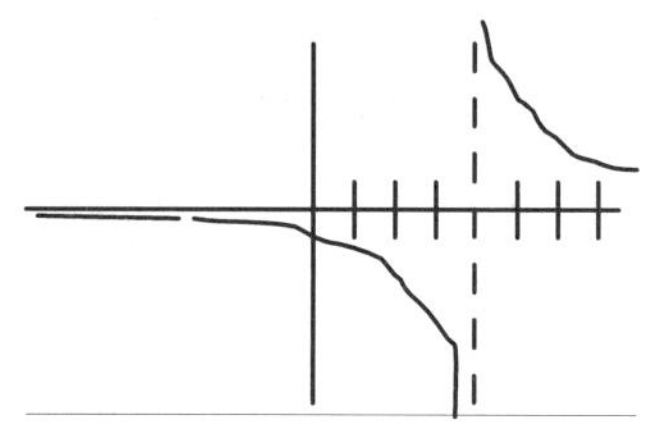

Answer Q5:
a) $x = 1$ and $x = -4$ are the vertical asymptotes (denominator= 0),degree numerator < degree denominator, so $y = 0$ is the horizontal asymptote.
b) $x = -1$ is the vertical asymptote, and we have an oblique asymptote since the degree numerator > degree denominator and is $y = x - 1$ since

$$\begin{array}{r|rrr} -1 & 1 & 0 & -2 \\ & & -1 & 1 \\ \hline & 1 & -1 & -1 \end{array}$$

b) In this example there isn't a vertical asymptote because there is no real number that will make $4x^2 + x + 1 = 0$. The horizontal asymptote is $y = -\frac{3}{4}$, since the two highest-degree terms are $-3x^2$ in the numerator and $4x^2$ in the denominator.

What is on the calculator

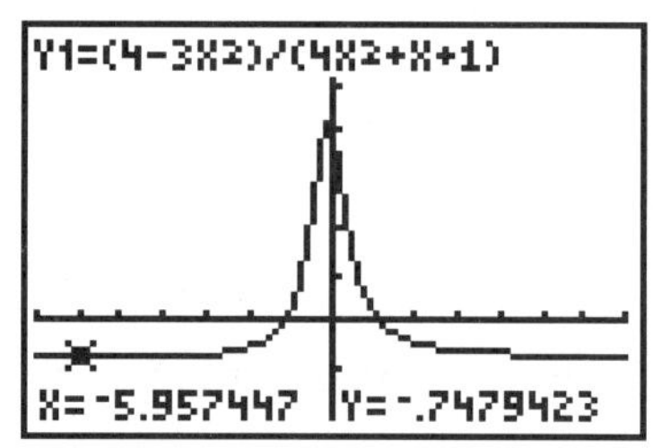

Sketch you might make

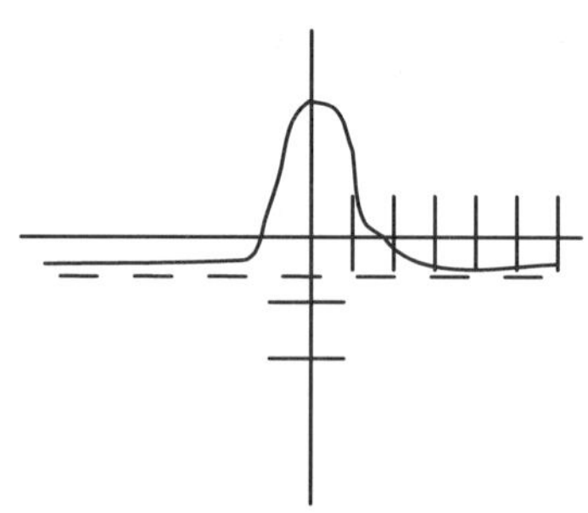

c) In this last example there is a vertical asymptote at $x = -3$ since there is a $(x + 3)$ factor in the denominator. But this time there isn't a horizontal asymptote because the degree of the numerator is greater than the degree of the denominator. So we have an oblique asymptote.

$$\begin{array}{r|rrrr} -3 & 3 & -2 & 1 & 5 \\ & & -9 & 33 & -102 \\ \hline & 3 & -11 & 34 & -97 \end{array}$$

Also notice that the asymptote is $3x^2 - 11x + 34$. It is a **quadratic asymptote**, not a linear one like we saw earlier in this section. Oblique asymptotes can turn out to be any kind of polynomial.

What is on the calculator

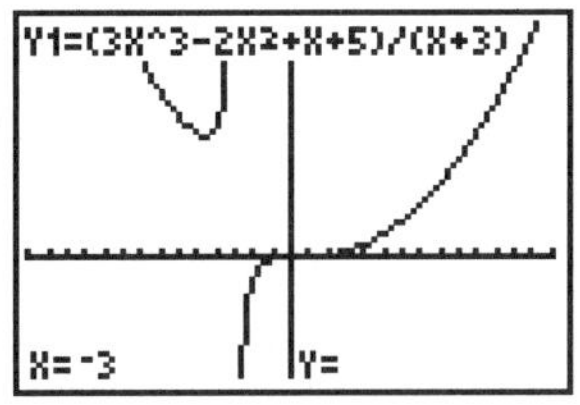

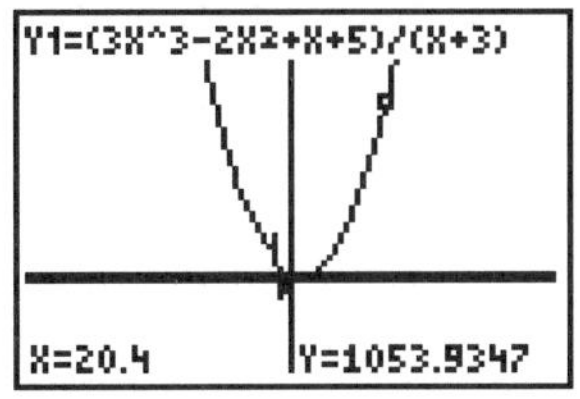

Sketch you might make

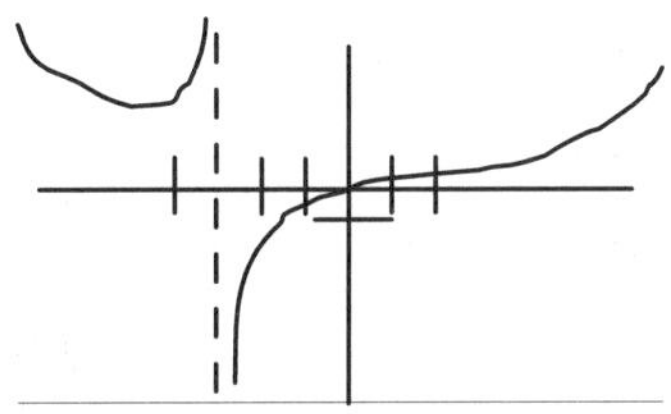

Example 2: x–Intercepts for Rational Functions

Find the x-intercepts for the three rational functions in the beginning of this section:

$$f(x) = \frac{x+3}{x^2-4}, \quad g(x) = \frac{7-3x}{2x+1}, \quad h(x) = \frac{x^2+3}{x-3}.$$

Solution:

Remember that an x-intercept happens whenever the y-value (the output value) is equal to 0. For fractions to equal 0 requires that the numerator be equal to 0. For example, the only way that $\frac{x-1}{5} = 0$ is for x to be 1. Let's look at each function.

For $f(x)$ to equal 0, $x + 3$ would have to equal 0; so $f(x) = 0$ when $x = -3$.	For $g(x)$ to equal 0, $7 - 3x$ would have to equal 0; so $f(x) = 0$ when $x = \frac{7}{3}$.	For $h(x)$ to equal 0, $x^2 + 3$ would have to equal 0; so, $f(x)$ will never be 0 since $x^2 + 3$ will always be positive.

Section Summary

Asymptote	Type of formula	Definition	How to find
Vertical	Vertical line ($x = a$)	As x's get close to a, y's get very large in magnitude.	After function is simplified, set denominator equal to 0.
Horizontal	Horizontal line ($y = b$)	As x's get very large in magnitude, y's level out (graph becomes flat).	Three cases: Case 1) Degree denominator > degree numerator, $y = 0$ Case 2) Degree denominator = degree numerator, $y = b = \dfrac{\text{coef of largest degree term}}{\text{coef of largest degree term}}$ Case 3) Degree denominator < degree numerator: is an oblique asymptote, not horizontal.
Oblique	Function that could be a linear, quadratic or higher-degree polynomial	As x's get very large in magnitude, the graph of the rational function gets closer to the graph of the asymptote.	Use synthetic division; the quotient minus the remainder is the oblique asymptote.

When you sketch the graph of a rational function, remember to draw in a dashed line where the asymptotes are and label them.

SECTION 3.5 HOMEWORK

(1–18) For each of the following rational functions:

a. Find any horizontal asymptotes.

b. Find any vertical asymptotes.

c. Find any x-intercepts.

d. Find the domain.

e. Use your graphing calculator to confirm your answers.

1. $f(x) = \dfrac{2}{x-4}$ 2. $f(x) = \dfrac{-7}{x+5}$ 3. $f(x) = \dfrac{-3x}{5x+6}$

4. $f(x) = \dfrac{5x}{3x-5}$ 5. $f(x) = \dfrac{9x-2}{3x}$ 6. $f(x) = \dfrac{9-8x}{2x}$

7. $f(x) = \dfrac{x-3}{x+2}$ 8. $f(x) = \dfrac{x+2}{x-1}$ 9. $f(x) = \dfrac{2x-3}{x^2-4}$

10. $f(x) = \dfrac{3x+1}{x^2-9}$ 11. $f(x) = \dfrac{4x^2-16}{2x^2-3x-5}$ 12. $f(x) = \dfrac{9x^2-16}{3x^2+4x-7}$

13. $f(x) = \dfrac{3}{x^2-3x-10}$ 14. $f(x) = \dfrac{-2}{x^2+5x-36}$ 15. $f(x) = \dfrac{2x^2-1}{5x^2+3}$

16. $f(x) = \dfrac{3x^2-5}{2x^2+5}$ 17. $f(x) = \dfrac{3x^3-27}{x^3+64}$ 18. $f(x) = \dfrac{27x^3+1}{8x^3-1}$

(Hint: for 17 and 18 above you will need to remember how to factor difference and sum of cubes.)

(19-28) From the following information create a rational function that would fit the information. (There can be more than one correct answer.)

Use your graphing calculator to confirm your answers.

19. a. Vertical asymptote: $x = 2$ b. Horizontal asymptote: $y = 2$	20. a. Vertical asymptote: $x = -3$ b. Horizontal asymptote: $y = -3$
21. a. Vertical asymptotes: $x = 3$ and $x = -1$ b. Horizontal asymptote: $y = 0$	22. a. Vertical asymptotes: $x = 2$ and $x = -3$ b. Horizontal asymptote: $y = 0$
23. a. Vertical asymptotes: $x = -3$ and $x = 1$ b. Horizontal asymptote: $y = 2$	24. a. Vertical asymptotes: $x = -2$ and $x = 2$ b. Horizontal asymptote: $y = -2$
25. a. No vertical asymptote. b. Horizontal asymptote: $y = 3$	26. a. No vertical asymptote. b. Horizontal asymptote: $y = -2$
27. a. Vertical asymptotes: $x = -2$ and $x = 2$ b. Horizontal asymptote: $y = 0$	28. a. Vertical asymptotes: $x = -1$ and $x = 3$ b. Horizontal asymptote: $y = 2$

(29–34) Give a rational polynomial function that fits each graph. Use your graphing calculator to confirm your answers.***Vertical and horizontal lines should be dotted.***

29.

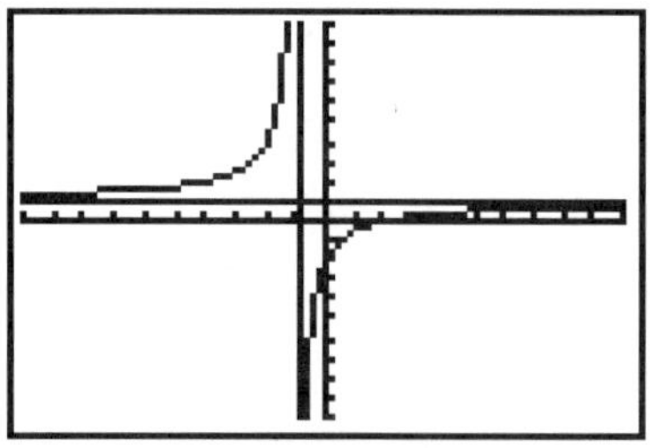

30.

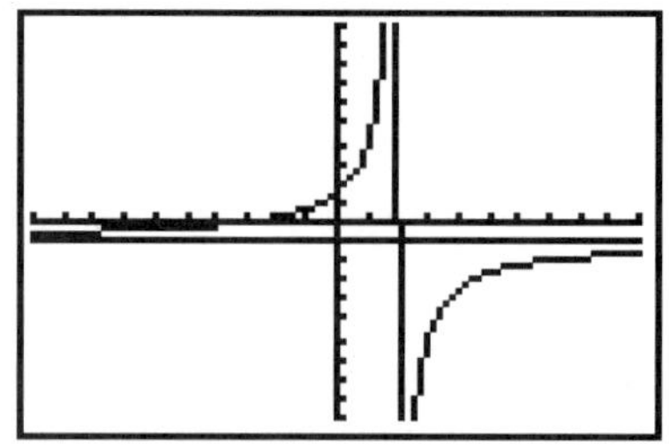

31.

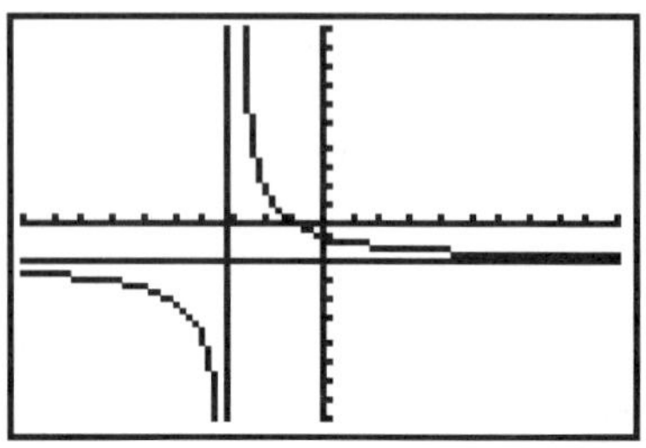

32.

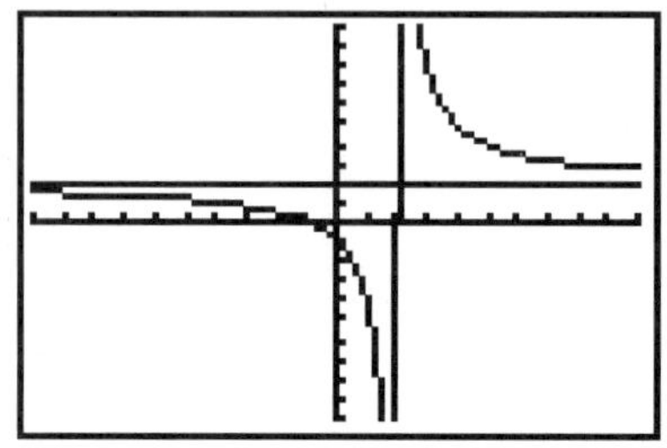

33.

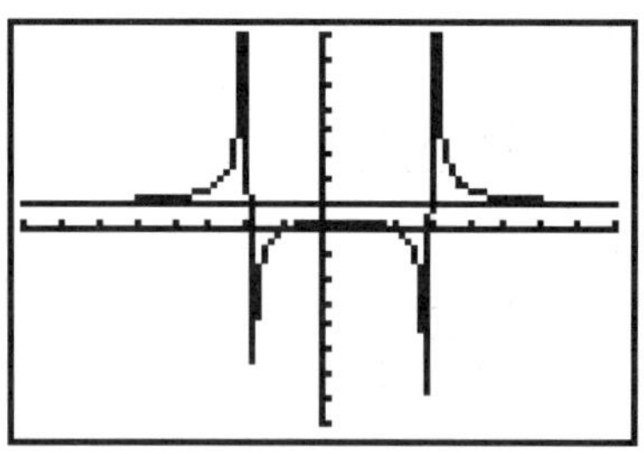

34.

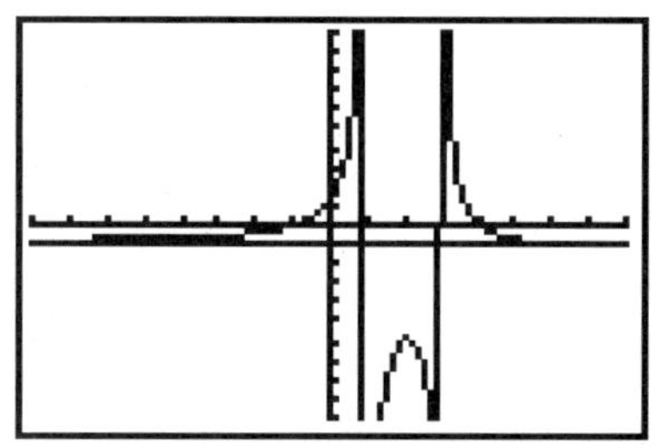

(35–40) Find the oblique asymptote. (Use your graphing calculator to confirm your answers.)

35. $f(x) = \dfrac{x^2 - 3x + 1}{x - 2}$

36. $f(x) = \dfrac{x^2 + 2x - 2}{x + 3}$

37. $f(x) = \dfrac{3x^2 - 5x + 2}{x + 2}$

38. $f(x) = \dfrac{-4x^2 + 12x + 3}{x - 3}$

39. $f(x) = \dfrac{x^3 - 3x^2 - 5x - 2}{x - 5}$

40. $f(x) = \dfrac{2x^3 + 7x^2 - 2x - 5}{x + 4}$

(41–44) Find the hole in the graph.

41. $f(x) = \dfrac{x^2 - 4}{x + 2}$

42. $f(x) = \dfrac{x^2 - 9}{x - 3}$

43. $f(x) = \dfrac{x + 5}{x^2 + 4x - 5}$

44. $f(x) = \dfrac{x - 2}{x^2 + 5x - 14}$

(45–62) For each of the following rational functions:

a. Find any horizontal asymptotes.

b. Find any vertical asymptotes.

c. Find any oblique asymptotes.

d. Find any holes in the graph.

e. Sketch a rough graph.

f. Use your graphing calculator to confirm your answers.

45. $f(x) = \dfrac{5}{2x-5}$

46. $f(x) = \dfrac{3}{3x+7}$

47. $f(x) = \dfrac{2x+1}{3x-2}$

48. $f(x) = \dfrac{3x-5}{5x+1}$

49. $f(x) = \dfrac{2x^2-3x+2}{3x^2+x-2}$

50. $f(x) = \dfrac{5x^2+2x+1}{4x^2-x-3}$

51. $f(x) = \dfrac{x^2-4}{3x-6}$

52. $f(x) = \dfrac{x^2-x-6}{x+2}$

53. $f(x) = \dfrac{-2x^2+2x-1}{x+3}$

54. $f(x) = \dfrac{2x^2-3x-2}{x-4}$

55. $f(x) = \dfrac{x^2-9}{x^2-4x+3}$

56. $f(x) = \dfrac{3x-1}{6x^2+13x-5}$

57. $f(x) = \dfrac{(x-3)(x+2)(x-1)}{(x+2)(x+1)(x-2)}$

58. $f(x) = \dfrac{(2x-1)(x-3)(x+3)}{(3x+1)(x-3)(x-2)}$

59. $f(x) = \dfrac{x(x+2)(x+3)}{(x+3)(x-2)}$

60. $f(x) = \dfrac{x(x-2)(x+2)}{(x+2)(x+1)}$

61. $f(x) = \dfrac{3x(x-1)}{(x+2)(x-1)(x-2)}$

62. $f(x) = \dfrac{5x(x+2)}{(x+3)(x+2)(x-1)}$

63. How can a rational polynomial function not have a vertical asymptote?
64. How can a rational polynomial function not have any x-intercepts?
65. What is required for a rational polynomial function to have an oblique asymptote?
66. What is required for a rational polynomial function to have a horizontal asymptote of $y = 0$?
67. What is required for a rational polynomial function to have a horizontal asymptote of $y = 3$?
68. What causes a hole in the graph of a rational polynomial function?
69. If a rational polynomial function has three different values that make the numerator 0, explain how it could have only two x-intercepts.
70. If a rational polynomial function has two different values that make the denominator 0, explain how it could have only one vertical asymptote.
71. If a rational polynomial function has a quadratic numerator that factors and a quadratic denominator that has no zeros, how can the function have only one x-intercept?
72. If a rational polynomial function has a quadratic denominator that factors and a quadratic numerator that has no zeros, how can the function have one vertical asymptote?

Chapter 3 Review

<table>
<tr><th>Topic</th><th>Section</th><th></th></tr>
<tr><td>Common difference</td><td>3.1</td><td><ul><li>You subtract the previous output from the next output to find the common difference.</li><li>If all of the common differences are the same then the function is linear.</li><li>The common difference divided by the time interval equals the slope of the line.</li></ul></td></tr>
<tr><td>Polynomial characteristics</td><td>3.2</td><td><ul><li>Polynomials are smooth (no sharp corners), continuous (no breaks), "nicely-behaved" functions.</li><li>The largest-degree terms determine how the graph looks near the ends of the window (if you have found a good complete graph).</li><li>Also, if the polynomial is factored, then the graph near each x–intercept must look like the graph of the factor as if it were the only factor in the function.</li></ul></td></tr>
<tr><td>Synthetic division</td><td>3.3</td><td><ul><li>Synthetic division is a way to shorten the long division process.</li><li>Synthetic division is a way to find factors.</li><li>If we use a number c in synthetic division, then the remainder will equal $f(c)$. The same as if you were to plug the number c into the function $f(x)$.</li><li>If the remainder equals 0, we know that what we have just divided by (the divisor) and the answer (the quotient) are factors of the function we have divided into (the dividend).</li></ul></td></tr>
<tr><td>More on synthetic division</td><td>3.4</td><td>If we use synthetic division with the number c on a polynomial and we get a 0 for the remainder, then we know all of the following:<ul><li>$f(c) = 0$</li><li>$(x - c)$ is a factor of the polynomial.</li><li>$x = c$ is an x-intercept (if c is a real number).</li><li>$x = c$ is called a zero of the polynomial.</li><li>$x = c$ is a root of the equation $f(x) = 0$.</li></ul>Complex zeros must come in conjugate pairs if all the coefficients in the polynomial are real.
The irrational zeros must also come in conjugate pairs if all the coefficients in the polynomial are rational.</td></tr>
</table>

The following is a table summarizing some of the information about polynomials discussed in section 3.2:

Highest-degree term in the polynomial	Enter the graphing window from	Exiting the graphing window	Number of possible hills and valleys	Number of possible x-intercepts
x^n (n odd)	below	above	$n-1,\ldots,4,2,0$	$n,\ldots,5,3,1$
$-x^n$ (n odd)	above	below	$n-1,\ldots,4,2,0$	$n,\ldots,5,3,1$
x^m (m even)	above	above	$m-1,\ldots,5,3,1$	$m,\ldots,4,2,0$
$-x^m$ (m even)	below	below	$m-1,\ldots,5,3,1$	$m,\ldots,4,2,0$

Section 3.5

Asymptote	Type of formula	Definition	How to find
Vertical	Vertical line ($x = a$)	As x's get close to a, y's get very large in magnitude.	After function is simplified, set denominator equal to 0.
Horizontal	Horizontal line ($y = b$)	As x's get very large in magnitude, y's level out (graph becomes flat).	Three cases: Case 1) Degree denominator > degree numerator, $y = 0$ Case 2) Degree denominator = degree numerator, $y = b = \dfrac{\text{coef of largest degree term}}{\text{coef of largest degree term}}$ Case 3) Degree denominator < degree numerator: is an oblique asymptote, not horizontal.
Oblique	Function that could be a linear, quadratic or higher-degree polynomial	As x's get very large in magnitude, the graph of the rational function gets closer to the graph of the asymptote	Use synthetic division; the quotient minus the remainder is the oblique asymptote.

When you sketch the graph of a rational function, remember to draw in a dashed line where the asymptotes are and label them.

CHAPTER 3 REVIEW HOMEWORK

Section 3.1

(1–4) Using the following table of values, verify whether the values form a constant, linear, quadratic or none of these types of function

1.

2	5
4	–1
6	–7
8	–13
10	–19

2.

1	–1
3	1
5	11
7	29
29	55

3.

–5	–3
–1	–3
3	–3
7	–3
10	–3

4.

2	–3
3	6
4	–12
5	24
6	–48

(5–8) Using the information given in each of the problems below:

a.) Create a stat plot using your graphing calculator for each problem and determine if you think the problem is a constant function, linear function, quadratic function or none of those types of functions.

b.) Create a table of values for each problem and with the common difference, verify your guess.

5. You have found the following information about how much revenue you receive for producing a certain number of items:

   10 items produce a revenue of \$495.

   20 items produce a revenue of \$980.

   30 items produce a revenue of \$1455.

   40 items produce a revenue of \$1920.

   50 items produce a revenue of \$2375.

6. You apply a certain amount of pressure to stretch a 12-inch spring. By careful measurement you come up with the following information:

   A pressure of 5 pounds stretches the spring to 13.5 inches.

   A pressure of 10 pounds stretches the spring to 15 inches.

   A pressure of 15 pounds stretches the spring to 16.5 inches.

   A pressure of 20 pounds stretches the spring to 18 inches.

   A pressure of 25 pounds stretches the spring to 19.5 inches.

7. Each day you take the subway to go to work.

   Monday a subway token costs you \$1.00

   Tuesday a subway token costs you \$1.00

   Wednesday a subway token costs you \$1.00

   Thursday a subway token costs you \$1.00

   Friday a subway token costs you \$1.00

8. You invest \$10,000 at 9% compounded yearly. The following is the worth of that investment after a number of years.

   After 10 years the investment is worth \$23,674.

   After 15 years the investment is worth \$36,425.

   After 20 years the investment is worth \$56,044.

   After 25 years the investment is worth \$86,230.

   After 30 years the investment is worth \$132,676.

Section 3.2

(9–13) For each of the following graphs, could the graph represent a polynomial (yes or no) and, if yes, would the power of the polynomial be even or odd?

9.

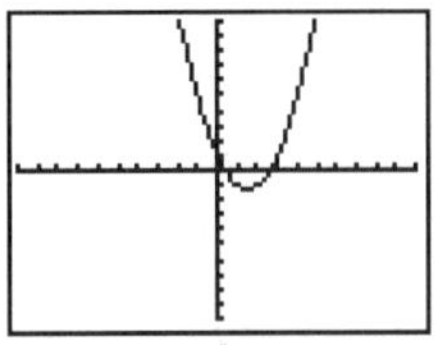

10.

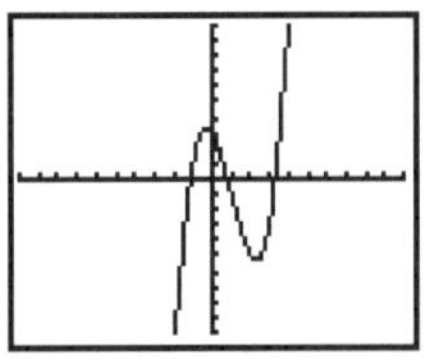

11.

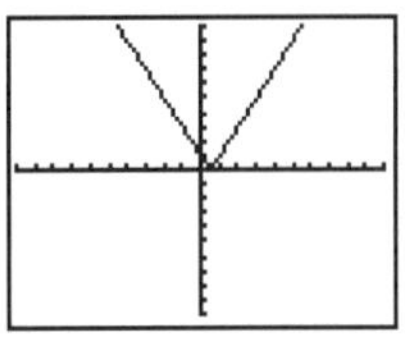

12.

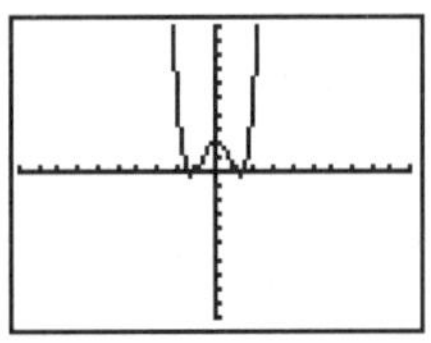

13.

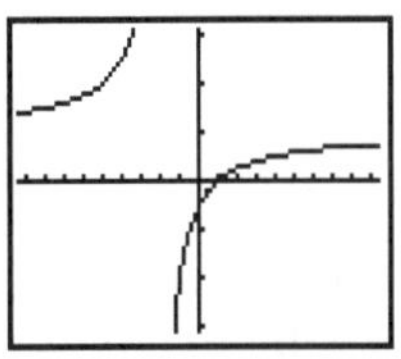

(14–17) Without the graphing calculator answer the questions (a) through (d) about the graph of the function given. Use your graphing calculator to confirm your answers.

a. How many possible hills are there (minimum number and maximum number)?

b. What are the maximum number of places the graph could cross the x–axis?

c. Where will the graph enter the window of the graphing calculator?

d. Where will the graph exit the window of the graphing calculator?

14. $f(x) = 2x^2 + 3x - 1$

15. $g(x) = -3x^3 + 7x - 1$

16. $r(x) = -x^4 + 3x^2 + 2$

17. $t(x) = x^5 - 4x^3 + 5$

(18–20) Sketch the graph of each function without the graphing calculator and then confirm your sketch with the graphing calculator.

18. $f(x) = (x - 2)(x + 1)^2$

19. $w(x) = -(x - 2)(x + 1)^3$

20. $m(x) = x(x - 1)^3(x + 2)^2$

(21–23) Create a function from the following information and then use your graphing calculator to confirm your answers.

21. x–intercept at 3; at this intercept the graph looks like a line.
 x–intercept at –2; at this intercept the graph looks like x^2.
 Enters at the bottom and exits at the top.
22. x–intercept at 1; at this intercept the graph looks like x^2.
 x–intercept at –2; at this intercept the graph looks like x^2.
 Enters at the top and exits at the top.
23. x–intercept at 0; at this intercept the graph looks like x^3.
 x–intercept at 2; at this intercept the graph looks like x^2.
 x–intercept at –2; at this intercept the graph looks like a line.
 Enters at the bottom and exits at the bottom.

(24–26) Are the following polynomials even, odd, or neither?

24. $f(x) = 3x^3 - 2x$ 25. $g(x) = x^4 - x^2 + 2$ 26. $r(x) = x^3 - 3x + 1$

Section 3.3

(27–30) Use synthetic division to divide.

27. $(2x^3 - 5x^2 + 2x - 3) \div (x - 2)$ 28. $(-3x^3 - 3x + 1) \div (x + 3)$

29. $(x^4 - 8) \div (x + 2)$ 30. $(9x^3 + 6x^2 - 12x + 1) \div (x - \frac{2}{3})$

(31–34) Use synthetic division and the remainder theorem to find $f(c)$.

31. $f(x) = -4x^3 - 10x^2 - 8x + 1$; find $f(-2)$.
32. $f(x) = 3x^3 - 11x^2 + 9$, find $f(3)$.
33. $f(x) = x^4 - 18x^3 + 32x^2 - 8$, find $f(16)$.
34. $f(x) = x^4 + 8x^2 + 3$, find $f(-8)$.

(35–38) Use synthetic division and the factor theorem to see if the binomial is a factor of the polynomial. If it is a factor, write the factored form of the polynomial using the binomial and the quotient.

35. Is $(x - 5)$ a factor of $x^3 - 3x^2 - 13x + 15$?
36. Is $(x + 3)$ a factor of $x^4 - 2x^3 - 13x^2 + 2x - 12$?
37. Is $(x - 3)$ a factor of $3x^3 - 2x^2 - 23x + 9$?
38. Is $(x - 2)$ a factor of $x^4 - 2x^3 - 3x + 6$?

Section 3.4

(39–42) For each polynomial function give all the possible rational roots, using the rational root theorem. (Don't try to find the roots.)

39. $f(x) = 3x^3 - 7x^2 + 4$ 40. $g(x) = 6x^4 - 8x^3 + 5$
41. $r(x) = 6x^3 + 8x^2 - 9x - 15$ 42. $m(x) = 24x^3 + 7x^2 + 9$

(43–53) For each polynomial function find all the roots and then write the function in factored form.

43. $f(x) = x^3 - 2x^2 - 9x + 18$

44. $g(x) = x^4 - 4x^3 - 14x^2 + 36x + 45$

45. $t(x) = x^3 - 6x^2 + 9x - 2$

46. $w(x) = 9x^3 + 15x^2 - 44x - 24$

47. $f(x) = x^3 + x^2 - 5x + 3$

48. $m(x) = x^5 + 2x^4 - 8x^3 - 16x^2 + 16x + 32$

49. $g(x) = 3x^4 - 7x^3 + 3x^2 + 3x - 2$

50. $r(x) = 12x^3 - 28x^2 - 7x + 5$

51. $f(x) = x^3 - 6x^2 + 13x - 10$

52. $g(x) = x^3 + x^2 + 4x + 30$

53. $s(x) = x^4 - 6x^3 + 26x^2 - 46x + 65$ (Hint: One root of this polynomial is $1 - 2i$.)

(54–57) Give a polynomial function with integer coefficients that have the following roots:

54. $-2,\ 3,\ 5$

55. $1, -2, -4, 3$

56. $3, 2+\sqrt{5}, 2-\sqrt{5}$

57. $2, 3 + i$

(58–59) Use the graphing calculator to approximate all real roots of the polynomial functions to three decimal places.

58. $f(x) = x^3 - 5x^2 + 2x + 3$

59. $f(x) = x^3 - 3x^2 - 2$

60. Sending a package by U. S. Mail costs \$.025 per cubic inch. If you send a Christmas package that has a rectangular base where the length is twice the width and the height is 4 inches, less than the width, the cost of the package is \$128.00. What are the dimensions of the package?

61. Your business sells a certain type of tool kit wholesale to other businesses; you sell more than 50 tool kits each week. The profit you make each week for selling x of these tool kits is given by the formula: $P(x) = -.01x^3 + x^2 + 20x + 20$.

a. How many of the tool kits do you need to sell in a week to make a profit of \$2020?

b. Use the graphing calculator to find the approximate number of kits you need to sell to maximize profit. (Maximizing, section 1.3.)

c. Using the answer from (b), what is the maximum profit you make in a week?

Section 3.5

(62–66) For each of the rational polynomial functions:

a. Find any vertical asymptote(s).

b. Find any horizontal asymptote(s).

c. Find any oblique asymptote(s).

d. Find any hole(s) in the graph.

e. Find any x–intercepts.

f. Write the domain.

g. Use your graphing calculator to confirm your answers.

h. Sketch the graph.

62. $f(x) = \dfrac{4x^2 - 25}{x^2 - 1}$

63. $g(x) = \dfrac{3x - 1}{x^2 - 5x - 6}$

64. $h(x) = \dfrac{3x^2 - 5x - 12}{x + 2}$

65. $r(x) = \dfrac{x^2 - 9}{x^2 - 5x + 6}$

66. $s(x) = \dfrac{(x-2)(2x+1)(x-3)}{(x-1)(x-3)(3x+2)}$

(67–70) Find a rational polynomial function that fits the following information. Use your graphing calculator to confirm your answers.

67. Vertical asymptote: $x = -3$

    Horizontal asymptote: $y = -1$

    x–intercept: (2, 0)

68. Vertical asymptotes: $x = -2$ and $x = 3$.

    Horizontal asymptote: $y = 3$

    x–intercepts: (–3, 0) and (1, 0)

69. Vertical asymptotes: $x = -2$ and $x = 2$

    No horizontal asymptote.

    x–intercept: (0, 0)

70. Vertical asymptote: $x = -2$

    Horizontal asymptote: $y = 3$

    x–intercept: (3, 0)

    Hole in the graph at (1, –2).

(71–74) Each of the following is a graph of a rational polynomial function; inspect each graph and:

a. Find any vertical asymptote(s).

b. Find any horizontal asymptote(s).

c. Find any x–intercept(s).

71.

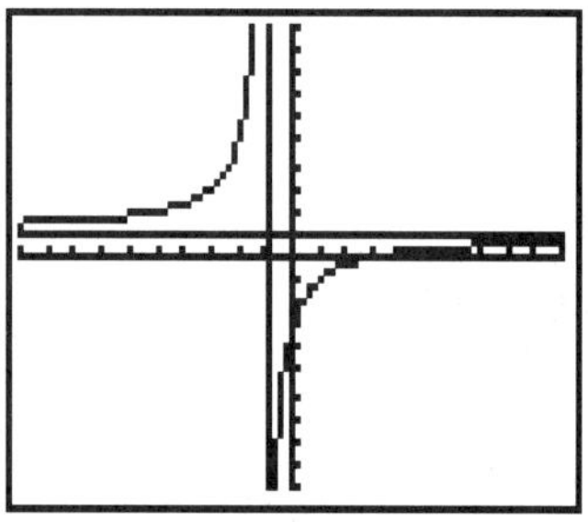

72.

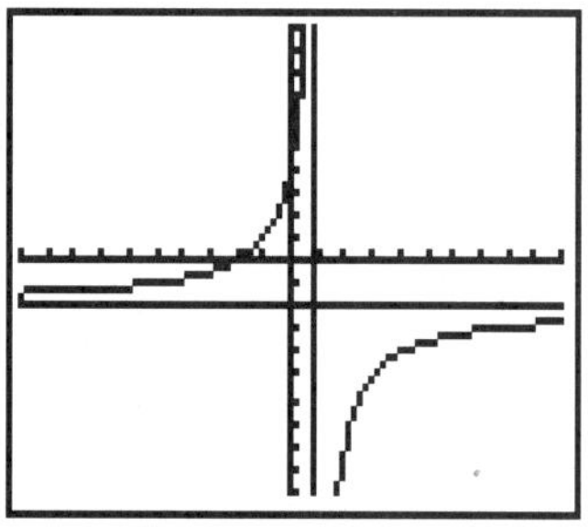

73.

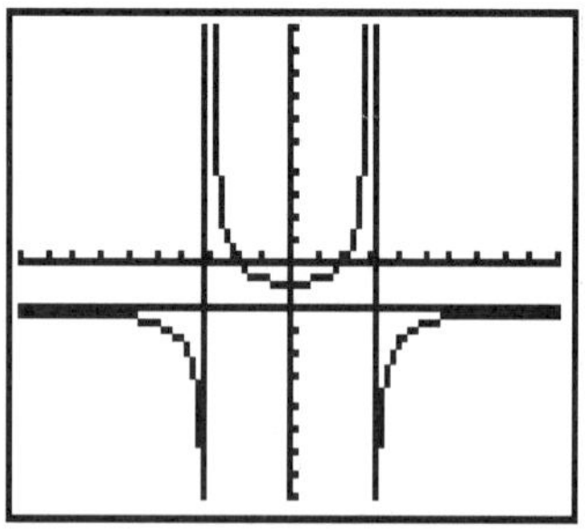

74.

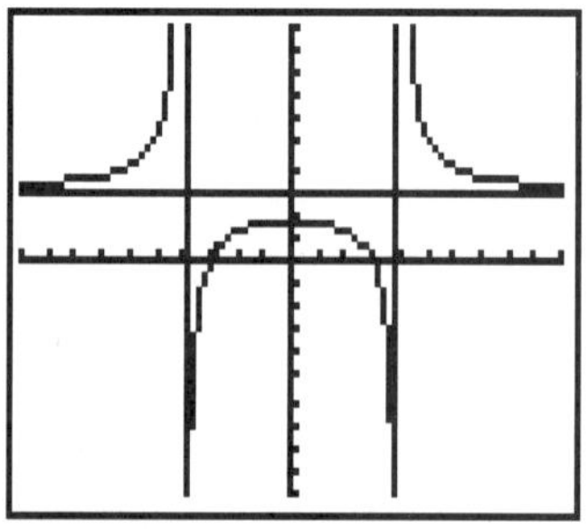

(*note vertical and horizontal lines should be dotted*)

CHAPTER 3 EXAM

1. What type of a function does the following information indicate (constant, linear, quadratic, or none)?
   a. The manufacture of 30 tables costs \$6,000.
      The manufacture of 40 tables costs \$ 8,000.
      The manufacture of 50 tables costs \$10,000.
      The manufacture of 60 tables costs \$12,000.
   b. The construction of a rectangular garden with the longest side 20 feet costs \$400.
      The construction of a rectangular garden with the longest side 30 feet costs \$900.
      The construction of a rectangular garden with the longest side 40 feet costs \$1,600.
      The construction of a rectangular garden with the longest side 50 feet costs \$2,500.
   c. An accountant who is paid a salary receives a check for \$583.50 the first week.
      The accountant's check the second week is \$583.50.
      The accountant's check the third week is \$583.50
      The accountant's check the fourth week is \$583.50
   d. A construction worker is paid hourly; his check the first week is \$483.75.
      The construction worker's check the second week is \$523.58.
      The construction worker's check the third week is \$385.78.
      The construction worker's check the fourth week is \$502.35.
2. $f(x) = ax^4 + bx^3 + cx^2 + dx + e$
   a. How many possible hills exist (relative minimum and maximum numbers)?
   b. What are the maximum number of places the graph can cross the x–axis?
   c. If $a > 0$ where will the graph enter the window of the graphing calculator?
   d. If $a < 0$ where will the graph exit the window of the graphing calculator?
3. $f(x) = ax^3 + bx^2 + cx + d$
   a. How many possible hills exist (relative minimum and maximum numbers)?
   b. What are the maximum number of places the graph can cross the x–axis?
   c. If $a < 0$ where will the graph enter the window of the graphing calculator?
   d. If $a > 0$ where will the graph exit the window of the graphing calculator?
4. Create a function from the following information:
   x–intercept at –2; at this intercept the graph looks like a line.
   x–intercept at 3; at this intercept the graph looks like x^3.
   Enters at the bottom and exits at the bottom.
5. Create a function from the following information:
   x–intercept at –3; at this intercept the graph looks like x^2.
   x–intercept at 2; at this intercept the graph looks like a line.

x–intercept at 0; at this intercept the graph looks like x^2.

Enters at the top and exits at the bottom

6. For each of the graphs below:

   a. State if the graph could be a polynomial or not.

   b. If the graph could represent a polynomial, would the power be even or odd?

A.

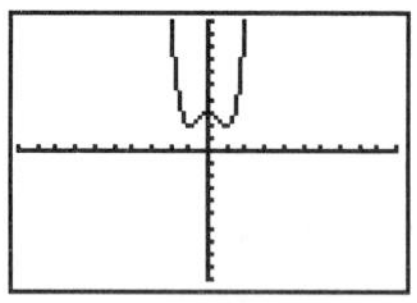

B.

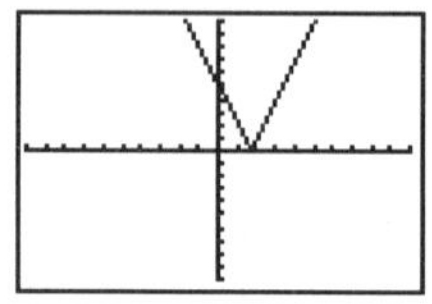

C.

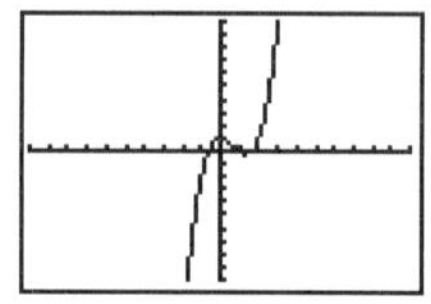

D.

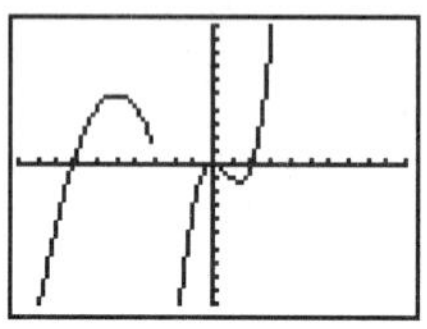

E.

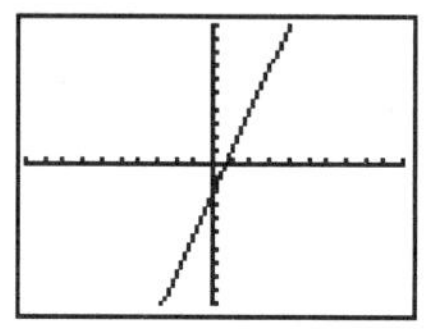

7. Use synthetic division and the remainder theorem to find $f(c)$. (Show work)

   a. $f(x) = 3x^3 - 5x^2 + 12x - 15$, $f(3) =$

   b. $g(x) = 2x^4 + 9x + 23$, $g(-3) =$

8. Using the rational root theorem give all possible rational roots of $f(x) = 6x^5 - 4x^2 - 15$. (Don't try to find the roots.)

9. For each of the polynomial functions find the roots and then write the function in factored form.

   a. $g(x) = x^3 + 2x^2 - 13x - 6$ b. $r(x) = 3x^3 - 8x^2 + 19x - 10$ c. $f(x) = x^5 + x^4 - 5x^3 - x^2 + 8x - 4$

10. For the following give **a** polynomial function with integer coefficients that has the given roots:

    a.. $-3, 1, 2$ b. $-2, 2, 2i$

11. For each of the following rational polynomial functions:

    a. $f(x) = \dfrac{2x-1}{x+3}$ b. $g(x) = \dfrac{x-2}{x^2-4}$ c. $r(x) = \dfrac{x^2-3x-4}{x+2}$

    i. Find any horizontal asymptotes.

    ii. Find any vertical asymptotes.

    iii. Find any oblique asymptotes.

    iv. Find any holes in the graph.

    v. Find any x–intercepts.

    vi. Write the domain.

12. From the following information create a rational function that would fit the information.

    a. Vertical asymptote $x = -3$ and $x = 2$

       Horizontal asymptote $y = 0$

       x–intercepts $(-1,0)$

    b. Vertical asymptote $x = 1$ and $x = -2$

       Horizontal asymptote $y = 2$

       x–intercepts $(-1, 0)$ and $(3, 0)$

13. Give a rational function that matches the graph below:

Vertical and horizontal lines should be dotted.

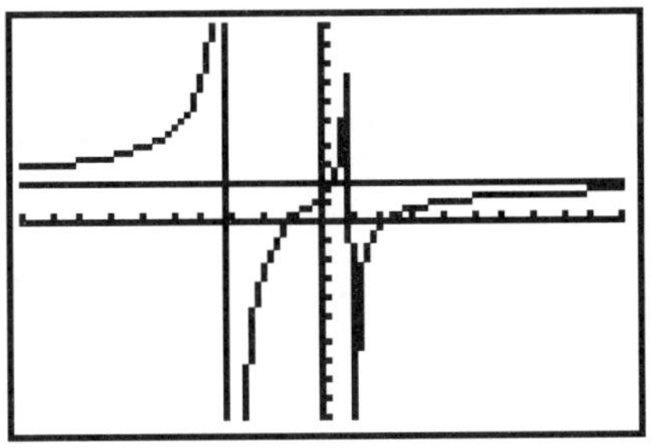

14. A company produces up to 90 mattresses each week. $P(x) = -0.1x^3 + 10x^2 + 10x - 2000$ gives the amount of profit the company makes for producing x mattresses.

    a. The profit for the week is \$4,600. How many mattresses were produced?

    b. Using the graphing calculator, find how many mattresses must be produced to maximize profit.

Chapter 4 Exponential and Logarithmic Functions

It is time to look at functions that are vastly different from those we have been learning about. This chapter will discuss a new family of functions that have a lot of applications in the world around us.

Section 4.1 Exponential Functions

Let's begin by reviewing a past example from section 3.1. In that section we looked at something called common differences, which gave us a way to determine if data represented some kind of a polynomial.

(Example 2, section 3.1): Find the common differences for the table below. The function $f(t)$ in this table gives the height of a baseball, in feet, t seconds after being hit. This is only one of many different functions that could describe the height of a baseball since the ball can be hit in so many different ways.

t (sec.)	$f(t)$ (feet)	Common Difference	Common Difference of the Common Differences
0	3		
1	83	80	
2	131	48	$48 - 80 = -32$
3	147	16	$16 - 48 = -32$
4	131	−16	$(-16) - 16 = -32$
5	83	−48	$(-48) - (-16) = -32$
6	3	−80	$(-80) - (-48) = -32$

Remember that if we have equal time (input) intervals and we find the common differences and they turn out to be equal, then we know that the data represents a linear function. If we must go to the common differences of the common differences, as in this example, then we know that we are looking at data that represents a quadratic function.

In section 3.1 we also mentioned that "…the 3rd common differences, if they are the same, tell you that you have a cubic function, and 4th common differences, if they are the same, tell you that you have a 4th powered polynomial and on and on and on." But we will see that with the new functions presented in this chapter, the common differences would never turn out to be the same, no matter how many times we calculated them.

Discussion 1: Common Differences

Look at the following table of values and several common differences.

x	$f(x)$	Common Difference	Common Difference of the Common Differences	3rd Common Differences
0	1			
1	3	$3-1=2$		
2	9	$9-3=6$	$6-2=4$	
3	27	$27-9=18$	$18-6=12$	$12-4=8$
4	81	$81-27=54$	$54-18=36$	$36-12=24$
5	243	$243-81=162$	$162-54=108$	$108-36=72$
6	729	$729-243=486$	$486-162=324$	$324-108=216$
7	2187	$2187-729=1458$	$1458-486=972$	$927-324=648$

Notice that the first common differences are not all equal. For that matter, none of the common differences calculated so far have values that are all the same or even close to each other. Now, we could keep on trying to determine if this table represents a polynomial by finding more common differences, but it looks like that might be futile. Instead of calculating the common differences, to see if we *add* the same amount each time to get to the next output value, maybe we should try a different approach.

We could try finding **common ratios (*r*)** instead. By calculating common ratios we would be looking to see if there is a number that we could *multiply by* each time to get to the next output value. So instead of subtracting the table values, let's try dividing them and see what happens.

x	$f(x)$	Common Ratio
0	1	
1	3	$r=\frac{3}{1}=3$
2	9	$r=\frac{9}{3}=3$
3	27	$r=\frac{27}{9}=3$
4	81	$r=\frac{81}{27}=3$
5	243	$r=\frac{243}{81}=3$
6	729	$r=\frac{729}{243}=3$
7	2187	$r=\frac{2187}{729}=3$

The way we calculated the common ratios was similar to the way we calculated the common differences. When we calculated the common ratios we took the next number (output) in the table and divided it by the previous number (output). Also just like before, we needed the x-values (inputs) to increase by equal amounts. The common ratios in this example were all the same value ($r = 3$).

This is what happens when you are looking at something that is an example of an exponential function or what we would call an exponential model. Let us try to demonstrate to you how incredibly different this kind of a function is compared to a linear function.

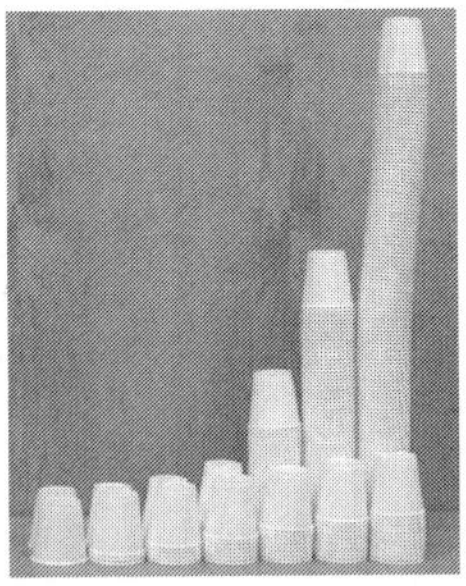

Notice that we are adding two cups each time to the linear example and multiplying by two each time in the exponential example. Also notice that when we multiply by two each time in essence we are adding 1 then 2 then 4 then 8… In both procedures the amount added keeps getting larger. However, you can see in the picture above that when you are adding the same amount each time to get to the next output value (linear model, common differences equal), the stack of cups doesn't grow nearly as much as when you multiply by the same amount each time to get the next output value (exponential model, common ratios equal).

Question 1: What calculations can you make to help you determine whether a table of values might be an example of a linear model? Exponential model?

Example 1: Calculating Common Differences and Common Ratios

Calculate the common differences and common ratios for both of the tables below. Then decide from these calculations whether the tables are closer to examples of linear models or exponential models, or one of each.

Population of the World	
t (years)	$P(t)$ billions)
1950	2.556
1960	3.039
1970	3.707
1980	4.454
1990	5.279
2000	6.083

2000, 2001 The Learning Network Inc.

Tuition & Fees 4-yr Colleges	
t (years)	$C(t)$
1994	\$2681
1995	\$2811
1996	\$2975
1997	\$3111
1998	\$3247
1999	\$3356

The World Almanac 2001

Solution:

First we have to calculate the common differences and common ratios.

Population of the World		Common difference	Common ratio (r)
t	$P(t)$		
1950	2.556		
1960	3.039	0.483	1.189
1970	3.707	0.668	1.220
1980	4.454	0.747	1.202
1990	5.279	0.825	1.185
2000	6.083	0.804	1.152

Tuition & Fees 4-yr Colleges		Common difference	Common ratio (r)
t	$C(t)$		
1994	$2681		
1995	$2811	130	1.048
1996	$2975	164	1.058
1997	$3111	136	1.046
1998	$3247	136	1.044
1999	$3356	109	1.034

From the table it appears that an exponential model might best describe the population of the world because the common ratios are roughly the same ($r = 1.19$ on average). Also the common differences are steadily increasing; they are not fairly close to the same value. Remember the stacks of cups (exponential example multiplying by two) we discussed earlier. We saw that the same thing happened there, too (common differences were 1 then 2 then 4 then 8 etc.).

The average tuition and fees for 4-year colleges, on the other hand, might be linear since the common differences are all fairly close (135 on average). Also notice that the common ratios are basically decreasing. This is true for linear models because we are adding the same amount each time to larger and larger amounts, so the ratios become less and less. The percent change keeps getting less each time. For example, adding 5 to 10 makes a bigger impact than adding 5 to 15. One is a 50% increase, whereas the other is only a 33.3% increase, but in both cases, we were adding the same amount.

Question 2: Would a linear or exponential model best represent the data in this table?

Students per Computer in Public Schools	
t (years)	$SC(t)$
1988	25
1990	20
1992	16
1994	10.5
1996	7.8
1998	5.7

The World Almanac 2001

Look at question 2, since the values in the table were getting smaller instead of larger, the opposite occurs as compared to the world population example (where common differences increased). Also notice that the values (outputs) of the world population example were increasing and its common ratios were about 1.19; we call that an example of a **growth** function.

Compare that to the table of values (outputs) in question 2 where the values were decreasing and the common ratios were about 0.744 on average. We call that an example of a **decay** function. Notice that in one of the examples the common ratio is greater than 1 and the values were increasing, while in the other the common ration is less than 1 and the values were decreasing. This happens because, as discussed earlier, the common ratio is the number we are multiplying by each time. If we are multiplying by a number larger than 1, then our next answer will be larger. On the other hand, if we are multiplying by numbers less than 1, then our next answer will be smaller.

Answer Q1: Calculations that can be helpful are common differences for linear, and common ratios for exponential.

Example 2: Exponential Growth

Let's suppose that your rich uncle decides to give you some money. What he is going to do is give you \$1 today and then, each day for the next 19 days, he'll give you double the amount he gave to you the day before. How much money will he give you on the 19th day from now?

Solution:

Day	\$	Day	\$	Day	\$	Day	\$
0	1	5	32	10	1024	15	32768
1	2	6	64	11	2048	16	65536
2	4	7	128	12	4096	17	131072
3	8	8	256	13	8192	18	262144
4	16	9	512	14	16384	19	524288

From the table, we can see that on the 19th day from today, he'll give you \$524,288.

Let's look at how we would create a formula for this problem.

One day after today you will receive double the original amount (\$1).	$\$1 \cdot \underline{2} = \$2$
Two days after today you will receive double yesterday's amount, which was double the original, so you have now had the original doubled twice.	$(\$1 \cdot \underline{2}) \cdot \underline{2} = \$2 \cdot 2 = \$4$
Three days after today you will receive the original doubled for the third time.	$(\$1 \cdot \underline{2} \cdot \underline{2}) \cdot \underline{2} = \$4 \cdot 2 = \$8$
Four days after today you will receive the original doubled for the fourth time.	$(\$1 \cdot \underline{2} \cdot \underline{2} \cdot \underline{2}) \cdot \underline{2} = \$8 \cdot 2 = \$16$

Notice that the number of times you double the original amount is equal to the number of days since you received the $1. Thus, we can calculate the amount we will receive on any give day by the formula to the right.	($1 • 2 • 2 • 2 …) • 2 = $1(2)^t$ t = the days since we received the $1

So, for example, if you wanted to know what would happen on the 23rd day, you'd simple plug in 23 for t and use your calculator ($\$1(2)^{23} = \$8{,}388{,}608$). Our formula in its final form might look like $p(t)=(2)^t$ where t stands for the days since today and $p(t)$ stands for the amount you would receive on that day.

Question 3: If my current paycheck is $1 and my employer offers to make each successive paycheck triple the previous paycheck, then how much would I receive in the 14th paycheck from now and what would the formula be to describe this example?

Discussion 2: Exponential Decay

Answer Q2: Since common ratios are close and common differences were decreasing, the best model might be exponential.

CD (d)	CR (r)
–5	0.80
–5	0.75
–4.5	0.70
–2.7	0.74
–2.1	0.73

Let's assume that 65,536 people are sick with the flu. What if after each day only 1/4 of those who were sick remain sick? Make a table and a graph to show how many are sick each day and come up with a formula to describe this event.

Day	$	Day	$
0	65536	5	64
1	16384	6	16
2	4096	7	4
3	1024	8	1
4	256	9	0

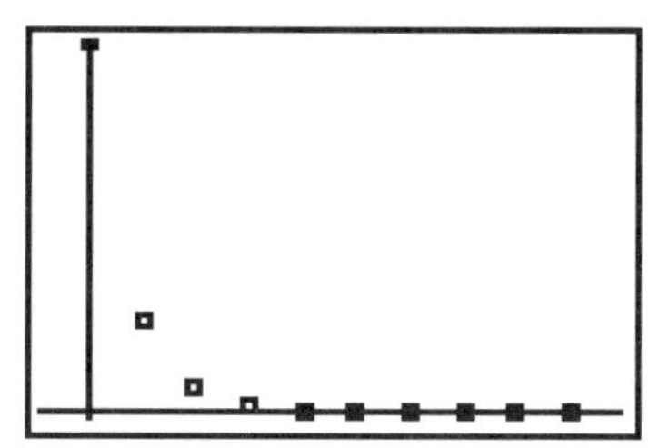

When $t = 0$ the number of sick people is 65536.	$S(0) = 65536$
One day later ¼ as many are sick.	$S(1) = 65536 \cdot ¼ = 16384$
Two days later ¼ as many are sick as the day before.	$S(2) = (65536 \cdot ¼) \cdot ¼ = 16384 \cdot ¼ = 4096$
You can see the same pattern as before, so our formula will be	$S(t) = (65536 \cdot ¼ \cdot ¼…) = 65536\,(¼)^t$ t = the days since 65536 people were sick.

You should have noticed, from our last two examples, that the first number in the formula is what you start with and the multiplier (r) is the number being raised to the "t" power. So, whether you are looking at an example where the values in the table are increasing or decreasing, the formulas have the same form: $f(x)$ = **start amount (multiplier)^{t}**. The formula would need some adjustment if we didn't begin with day 0, but we'll discuss that later.

Question 4: If you know that there are 100 grams of a particular substance and that you lose half of what remains every hour, how would you write a formula to describe this event?

Now, what if our time frame isn't an hour at a time or a day at a time or a paycheck at a time? For instance, in example 1 earlier in this section, the world population was recorded every 10 years, or in question 2 earlier, students per computer in public schools were recorded every two years. The time intervals between each value in the table were the same, which we needed in order to calculate the common differences and common ratios, but to create a formula when the time intervals aren't one unit at a time is a little more complex.

If we would like our formula to have a variable that represents one-unit intervals, we must adjust the exponent on the multiplier in order to convert the formula to handle one-unit intervals. The way we do this is by dividing the variable in the exponent position by the time interval itself. Let's revisit example 1.

Example 3: World Population Formula

Find a formula to represent the world population in terms of years.

Solution:

We know the starting amount from our table in example 1: it's 2.556. Since the first value in the table is from 1950, we will call that our starting point and label 1950 as time zero ($t = 0$). Our time interval for this example was 10 years and our multiplier was roughly 1.19. Our first thought might be to write the formula $P(t) = 2.556\ (1.19)^{t}$ but we wanted a formula that went a year at a time, not 10 at a time. Since our multiplier doesn't take complete effect until 10 years have passed, maybe we could think of it as being 1/10th effective each one of the 10 years.

Hence, our exponent should be $t/10$. So in 1951, the exponent would be 1/10. In 1952 it would be 2/10 and so on until 1960 when the exponent would be 10/10 = 1, which implies that the multiplier has now happened *one complete* time. So our formula would now be $P(t) = 2.556\ (1.19)^{t/10}$. The following table show the values obtained from this formula. The value for 1960 is very close to the real value found in example 2.

	1950	1951	1952	1953	1954	1955	1956	1957	1958	1959	1960
t	**0**	**1**	**2**	**3**	**4**	**5**	**6**	**7**	**8**	**9**	**10**
$P(t)$	2.556	2.6	2.646	2.693	2.74	2.788	2.837	2.887	2.937	2.99	3.039

Section Summary

- Exponential functions are very different from polynomials. The stacks of cups gave us a good picture of that fact.
- We have also seen that exponential functions come from a multiplying process and not an addition process like the polynomials do.
- We have seen that by calculating common ratios we can make a good guess as to whether a table of values describes an exponential function or not.
- Lastly, we have seen that one possible general formula for exponential functions has the form $f(t) = a(r)^{t/n}$ where, a = start amount at $t = 0$; r = common ratio; t = variable, and n = time interval.

Answer Q3:
My 14th paycheck would be $4,782,969 and the formula is $p(n) = (3)^n$ where n is the number of paychecks since now and p(n) is the paycheck amount.

Pay check	$	Pay check	$	Pay check	$
0	1	5	243	10	59049
1	3	6	729	11	177147
2	9	7	2187	12	531441
3	27	8	6561	13	1594323
4	81	9	19683	14	4782969

SECTION 4.1 HOMEWORK

(1–12) For the following tables use common difference or ratio to state if the tables represent linear functions, exponential functions or neither.

Answer Q4: f(t) = start amount (multiplier)^t = 100 (1/2)^t.

1.

0	1
1	2
2	4
3	8
4	16

2.

0	1/25
1	1/5
2	1
3	5
4	25

3.

0	1
1	0.5
2	0.25
3	0.125
4	0.0625

4.

0	1
1	0.3
2	0.09
3	0.027
4	0.0081

5.

0	–5
1	–2
2	1
3	4
4	7

6.

0	2
1	–3
2	–8
3	–13
4	–18

7.

0	–3
1	1
2	–2
3	–1
4	–3

8.

0	3
1	6
2	18
3	72
4	360

9.

9.

0	2
1	6
2	18
3	54
4	162

10.

0	3/16
1	3/4
2	3
3	12
4	48

11.

0	–3
1	–7/3
2	–5/3
3	–1
4	–1/3

12.

0	4
1	13/4
2	5/2
3	7/4
4	1

(13–20) For the following tables use common difference or ratio to choose the best fit function for the data (exponential or linear).

13.

0	1.35
2	2.47
4	3.58
6	4.68
8	5.79

14.

0	3.85
2	5.04
4	6.65
6	8.75
8	11.43

15.

0	283.4
5	694.33
10	1694.17
15	4154.10
20	10169.24

16.

0	32.85
5	43.56
10	54.25
15	64.95
20	75.67

17.

0	2,835.47
100	2,511.69
200	2,187.88
300	1,864.13
400	1,540.34

18.

0	3,854.25
100	3,588.31
200	3,347.89
300	3,113.20
400	2,897.46

19.

0	58,434.27
500	57,674.62
1,000	56,976.76
1,500	56,228.66
2,000	55,514.55

20.

0	33,285.79
500	32,032.01
1000	30,778.20
1,500	29,524.44
2,000	28,270.65

(21–28) Mathematical modeling for exponential growth and decay problems.

21. The following table shows how much money will be in an account after t years.

t = years	0	1	2	3
A = amount of money	\$5,000	\$5,400	\$5,832	\$6,298.56

a. Use the common ratio to show this is an exponential function.

b. What is the common ratio?

c. Give a function $A(t)$ that gives the amount of money in the account after t years.

d. How much money will be in the account after 20 years?

e. Is the function a growth or decay function?

22. A colony of bacteria is in a jar and the colony is doubling every minute. You started with 4 bacteria in the jar.

a. Create a table of values for the number of bacteria from 0 to 6 minutes.

b. Use this table to calculate a common ration to show the growth is exponential.

c. What is the common ratio?

d. Give a function $B(t)$ for the number of bacteria after t minutes.

e. How many bacteria will be in the jar after 10 minutes?

f. Is the function a growth or decay function?

23. The following table represents the number of fish, F, in a lake after t years.

t = years	0	1	2	3
F = The number of fish	100,000	110,000	121,000	133,100

a. Use the common ratio to show this is an exponential function.

b. What is the common ratio?

c. Give a function $F(t)$ that gives the number of fish after t years.

d. Approximately how many fish will be in the lake after 5 years?

e. Is the function a growth or decay function?

24. The following table represents the population of a city after time t. The time is in 5 years increments from 1980 to 1995 ($t = 0$ is 1980, $t = 1$ is 1985, etc.).

t = years in 5 year increments	0	1	2	3
P = population of the city	500,000	510,000	520,200	530,604

a. Use the common ratio to show this is an exponential function.

b. What is the common ratio?

c. Give a function $P(t)$ that gives the population as a function of time.

d. Approximately what will the population be in 2020?

e. Is the function a growth or decay function?

25. The following table represents the population of a certain city after time t. Time is in 5-year increments from 1980 to 1995.

t = years in 5 year increments	0	1	2	3
P = population of the city.	1,000,000	960,000	921,600	884,736

a. Use the common ratio to show this is an exponential function.

b. What is the common ratio?

c. Give a function $P(t)$ that gives the population of the city after time t.

d. Approximately what is the population in the year 2020?

e. Is the function a growth or decay function?

26. The following table represents the number of trees in a forest after a time t.

t = years	0	1	2	3
F = number of trees in the forest	200,000	190,000	180,500	171,475

a. Use the common ratio to show this is an exponential function.

b. What is the common ratio?

c. Give a function $F(t)$ that gives the number of trees as a function of time.

d. Approximately how many trees will be in the forest after 6 years?

e. Is the function a growth or decay function?

27. The following table gives the number of sperm whales, in thousands, in the ocean after t years (t is measured in five year increments where $t = 0$ in 1980).

t = years in 5 year increments	0	1	2	3
S = the number of sperm whales	150	120	96	76.8

a. Use the common ratio to show this is an exponential function.

b. What is the common ration?

c. Give a function $S(t)$ that gives the number of sperm whale as a function of time.

d. Approximately how many sperm whales will there be in the ocean in 2030?

e. Is the function a growth or decay function?

28. The following table shows how much radioactive substance, in grams, there is after t days.

t = the number of days	0	1	2	3
R = the amount of radioactive substance	20	16.4	13.448	11.02736

a. Use the common ratio to show this is an exponential function

b. What is the common ratio?

c. Give a function $R(t)$ that gives the amount of radioactive substance after time t.

d. Approximate to 4 decimal places how much radioactive substance there is after 10 days.

e. Is the function a growth or decay function?

(29–34) Problems 1, 2, 3, 4, 9, and 10 are examples of exponential functions.

29. a. Give an equation for problem 1.

    b. Is the function a growth or decay function?

30. a. Give an equation for problem 2.

    b. Is the function a growth or decay function?

31. a. Give an equation for problem 3.

    b. Is the function a growth or decay function?

32. a. Give an equation for problem 4.

    b. Is the function a growth or decay function?

33. a. Give an equation for problem 9.

    b. Is the function a growth or decay function?

34. a. Give an equation for problem 10.

    b. Is the function a growth or decay function?

Section 4.2 Characteristics of Exponential Functions

We saw in the last section that exponential functions are much different than linear ones. In this section we are going to discuss some of the specific characteristics of exponential functions.

Discussion 1: Exploring 2^x

Let's look at the function $f(x) = 2^x$ which is the same function that we saw in example 2 in the last section. We will start by looking at a table and a graph of this function.

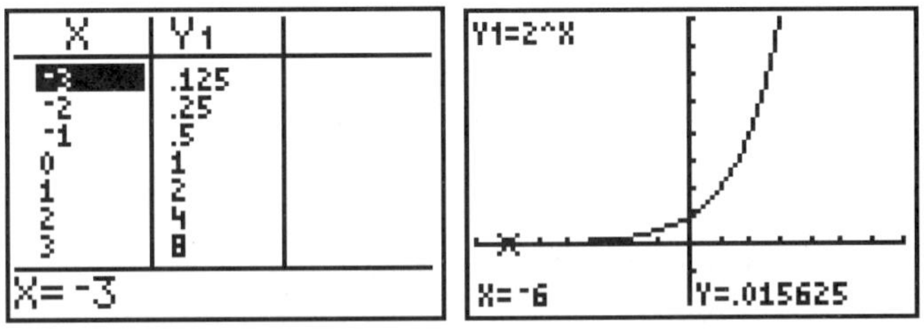

Let's work out a couple of the values that we see in the table.

If $x = 3$, then $f(3)$ will equal	$f(3) = 2^3 = 8$
If $x = 0$, then $f(0)$ will equal (any number raised to the zero power will always equal one)	$f(0) = 2^0 = 1$
If $x = -2$, then $f(-2)$ will equal (a negative exponent causes a base to flip upside down or, in other words, invert)	$f(-2) = 2^{-2} = \dfrac{1}{2^2} = \dfrac{1}{4} = 0.25$

If you look at the graph, you will see the same shape that we saw with the stack of cups in the last section. When you see a graph that starts out flat and then really shoots upwards near the end that's a good sign that what you are looking at is exponential growth. We discussed in the last section how, if the base is larger than one, then you have growth. Let's examine the graph more closely.

Question 1: What do the domain and range appear to be?

If we take some time to think about this function, the domain and range make total sense. First, if we are raising a positive number, in this case 2, to a power, there isn't anyway to get a negative answer. Positive numbers times themselves could never give a negative, or for that matter, even zero as an answer. So the range has to be all positive numbers $(0, \infty)$. Second, since the range is $(0, \infty)$ and we see the graph flattening out to the left, there must be a horizontal asymptote at $y = 0$. Third, we can raise a positive number to any power so the domain is obviously all real numbers $(-\infty, \infty)$.

One last thing: since we can raise numbers to any power, for example ½ power, we must only allow ourselves to use positive numbers as bases. If we did allow ourselves to use numbers like –2 for the base, then we would get answers like $(-2)^{1/2} = \sqrt{-2}$, which don't yield real solutions. As mentioned earlier in this book, we are only graphing real numbers at this point, so we can't allow there to be imaginary solutions. In summary, we have found out the following about exponential functions with bases larger than one (growth models):

1.) The base of an exponential function must be positive.
2.) The domain is all real numbers $(-\infty, \infty)$.
3.) The range is only positive numbers $(0, \infty)$.
4.) There is a horizontal asymptote at $y = 0$.

Now it is possible for the four characteristics just stated to be slightly different if we are not talking about the *basic* exponential functions. Here is an example.

Discussion 2: Expanding on the Basic Function 2^x

Using the graphing calculator to graph $y = 2^{x-3} + 1$ we get,

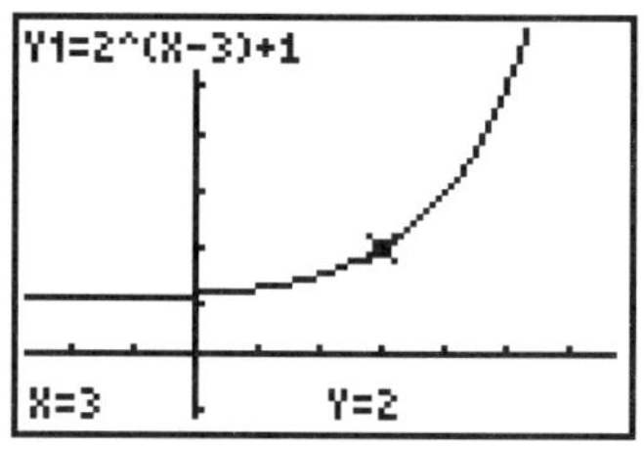

Notice that there now appears to be a horizontal asymptote at $y = 1$ instead of at $y = 0$. Also the range is now $(1, \infty)$. What caused this to happen is that the basic function 2^{x-3} is shifted up one unit because of the + 1 in the function (this was discussed in chapter 3).

We saw in the last section that not all exponential functions look like the one we have just seen. If the base is less than 1, then the graph will decrease, not increase, and we called that an example of decay.

Discussion 3: Decay

Let's investigate the function $g(x) = \left(\frac{1}{3}\right)^x$. Here is the table and the graph of this function.

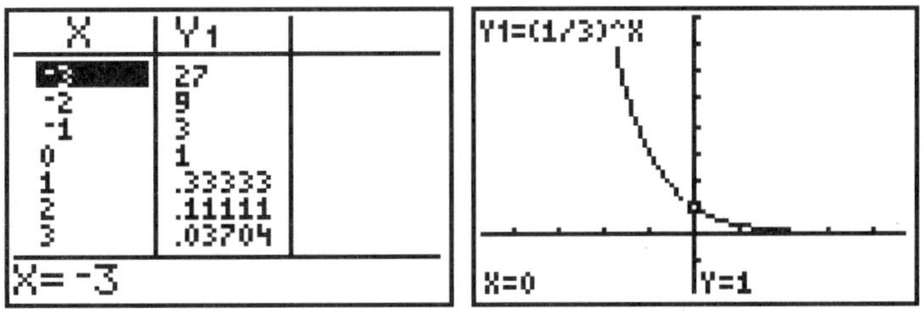

Notice that the base is $\frac{1}{3}$ for this function, a number less than 1, so this example is one that demonstrates decay. You can see from both the table and the graph that decay is the opposite of growth. It decreases rapidly at the beginning and then flattens out. But you can also tell that it has many of the same characteristics as growth does. The domain is $(-\infty,\infty)$ the range is $(0, \infty)$ and it has $y = 0$ as a horizontal asymptote.

So let's now give you the formal definition of an exponential function.

> An **exponential function** is of the form $f(x) = a(b)^x$, where a is a coefficient, b is a base greater than 0 but not equal to 1 (or what we called a multiplier in the last section), and x is the input variable.

Any time we see a function with a positive number as a base and a variable in the exponent position, then we will know that what we are looking at is an exponential function. Notice that this is different from the polynomials in chapter 3, where the variable x was in the base position and the positive numbers were in the exponent position.

Answer Q1: Domain is $(-\infty, \infty)$ and the Range is $(0, \infty)$.

Question 2: In discussion 1 and 3, what were the x and y intercepts for the graphs?

The y-intercept shouldn't have been a surprise since the way we find a y-intercept is by letting the x variable equal zero and any number raised to the zero power will be 1. Likewise, since we find x-intercepts by letting y equal zero, we can't get an x-intercept in these two examples because the output variables $f(x)$ and $g(x)$ can't ever equal 0.

Discussion 4: Comparing Exponential Functions

How do different exponential functions compare with each other? Graph the following exponential functions and look for similarities and differences: $f(x) = 2^x$, $g(x) = 3^x$, $h(x) = 5^x$.

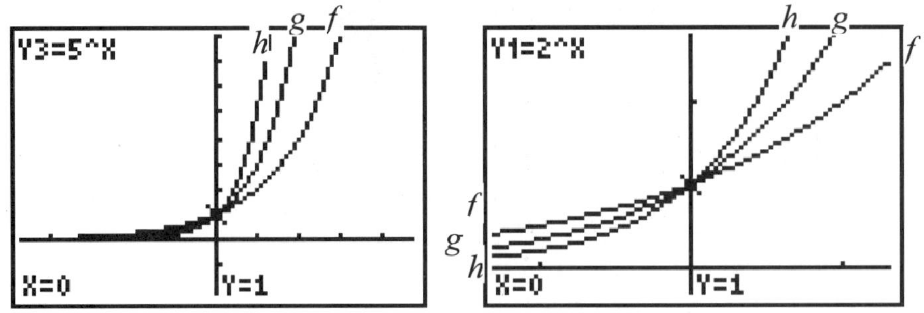

First, see how they all have the same growth shape ($b>1$). Next, notice how, as the base gets larger in value, the graph gets steeper on the right side of the y-axis and lower on the left side of the y-axis. This relationship expresses itself even better in a table (a negative exponent inverts the base so larger values become smaller and vice verse).

X	$g(x)$	$h(x)$
-3	.03704	.008
-2	.11111	.04
-1	.33333	.2
0	1	1
1	3	5
2	9	25
3	27	125

$Y_3 = h(3)$

Notice how $g(x) < h(x)$ when $x>0$ and $g(x) > h(x)$ when $x<0$. Likewise, with the next 3 graphs:

$$j(x) = \left(\frac{1}{2}\right)^x, \quad k(x) = \left(\frac{1}{3}\right)^x, \quad p(x) = \left(\frac{1}{5}\right)^x$$

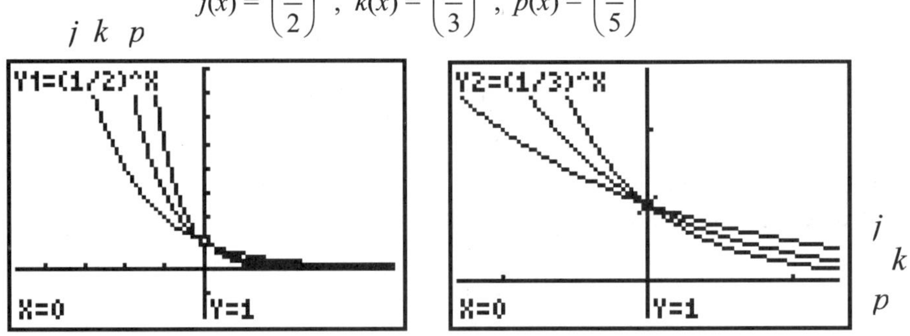

First, see how they all have the same decay shape ($0<b<1$). Next, notice how as the base gets smaller in value, the graph gets steeper on the left side of the y-axis and lower on the right side of the y-axis. This relationship expresses itself even better in a table (once again a negative exponent inverts the base so smaller values become larger and vice verse).

X	$j(x)$	$k(x)$
-3	8	27
-2	4	9
-1	2	3
0	1	1
1	.5	.33333
2	.25	.11111
3	.125	.03704
Y2=k(-3)		

Now remember that everything we have talked about so far has to do with exponential functions that are basic. That is, they have only an x in the exponent position and their form is as stated earlier $f(x) = a(b)^x$. They can be a bit more complicated than this (like in example 2), but for now we'll let our calculators take care of the more complicated examples.

Question 3: What would $f(x) = 1^x$ look like?

This function is the one exactly between growth and decay. The base is the number one, which is larger than decay bases and smaller than growth bases. Also the graph of 1^x is a horizontal line, which as we saw in section 3.1, was a graph of a constant function (one that doesn't ever change).

Answer Q2: The y-intercepts in both examples were (0, 1) and there weren't any x-intercepts.

THE NUMBER *e*

We should take a moment to discussing a special number in mathematics called the number "*e*." This number is defined as follows:

> *e* is equal to the number $(1+x)^{1/x}$, as the values for x get very close to zero, which is 2.71828… (this is from a calculus concept called Limit).

Here is a table and a graph that we will use to illustrate what *e* is equal to in decimal form.

x	f(x)
.1000000000	2.5937424600
.0100000000	2.7048138290
.0010000000	2.7169239320
.0001000000	2.7181459270
.0000100000	2.7182682370
.1000000000 10^{-5}	2.7182804690
.1000000000 10^{-6}	2.7182816920
.1000000000 10^{-7}	2.7182818150
.1000000000 10^{-8}	2.7182818270

x	f(x)
-.1000000000	2.8679719910
-.0100000000	2.7319990260
-.0010000000	2.7196422160
-.0001000000	2.7184177550
-.0000100000	2.7182954200
-.1000000000 10^{-5}	2.7182831880
-.1000000000 10^{-6}	2.7182819640
-.1000000000 10^{-7}	2.7182818420
-.1000000000 10^{-8}	2.7182818300

This number is an important one in mathematics and we will be learning more and more about it as we go through this chapter. This number e shows up a lot in real life applications like finance.

THE INVERSE OF THE EXPONENTIAL FUNCTION

We need to do one more thing in this section. We need to remind ourselves about the concept of inverse and then find the inverse of the exponential function. Back in section 1.6 we learned about inverses. We learned how inverses "undo" each other. If you remember, we talked about going from home to school and then back again. We saw in that illustration that in order to "undo" something, we have to do the opposite operations in the opposite order. In mathematics, we like to be able to undo everything we can, so now we would like to find an inverse for exponential functions.

Question 4: Find the inverse of $y = 3x + 5$.

You should have remembered that in order to be able to find an inverse of a function, the original function must be a one-to-one (for each output there is only one input) function. One-to-one is the reverse of what it means to be a function (for each input there is only one output). An exponential function is one-to-one (passes the horizontal line test), so all we need to do is switch the x (input variable) and the y (output variable) and then solve for the y.

Our exponential function:	$y = b^x$
Switch the x and y to find the inverse.	$x = b^y$

Now what we have is the inverse; all we need to do is solve for the y. But we don't know any algebraic procedures to solve for the y. Up until now we haven't had to solve for a variable that was in the exponent position; our variables have always been in the base position. In mathematics, if we have a dilemma like this, we make a definition. This is the definition of a **logarithm**.

Informal: $f(x) = \log_b x$, means b raised to what power equals x.

Formal: $f(x) = \log_b x$, where $y = \log_b x$ if and only if $x = b^y$, for $b>0$

A logarithm is the *exponent* that makes $x = b^y$ true. Notice that by making this definition we have solved for the y variable (the exponent of the inverse) and thus have found the expression for the inverse of the exponential function. By the way that we have defined a logarithm we have guaranteed that it will be the inverse of the exponential function. Let's investigate our newly found friend☺.

Question 5: Thinking in terms of the informal definition, what would be the answers to,

a) $\log_{10} 10 =$ b) $\log_{10} 100 =$ c) $\log_{10} 1000 =$

Notice that the answers to question 5 were the exponents that you needed to raise 10 to, in order to get the number you were trying to log.

Example 1: Estimating a Log

Answer Q3: A horizontal line at y = 1.

Estimate the $\log_{10} 30$.

Solution:

First, there isn't a whole number that will answer this problem. 10 raised to a whole number power can't equal 30. But let us look at it like this:

Look at an exponent of 1 and 2.	$10^1 = 10$ and $10^2 = 100$
$\log_{10} 30$, means $10^? = 30$ and that means that "?" must be between 1 and 2.	$10^1 < 10^? < 10^2$ $10 < 30 < 100$
So, let's guess that ? = 1.3.	$10^{1.3} = 19.95$, oops, way off
Well, let's try again ? = 1.5.	$10^{1.5} = 31.62$, closer
How about trying 1.47 now?	$10^{1.47} = 29.51$, getting closer 10^1.3 19.95262315 10^1.5 31.6227766 10^1.47 29.51209227

If we were to keep on trying different numbers, eventually we would find that the answer is approximately 1.477121255!

Question 6: Estimate the following to two decimal places.

a) $\log_{10} 85$ b) $\log_{10} 50$ c) $\log_{10} 15$

Example 2: Evaluating Logs

What are the answers to the following logs?

a.) $\log_{10} 1$ b.) $\log_{10}(1/10)$

Solution:

For part (a) we are looking for an exponent, which will cause 10, when raised to an exponent, to equal 1. This means that the exponent must be zero, since any number raised to the zero power equals 1. For part (b) we are also looking for an exponent that would yield 1/10 as an answer.

So we are looking to solve this:	$10^? = 1/10$
Which means that the following is also true since $1/10 = 10^{-1}$.	$10^? = 10^{-1}$
It looks like from this that ? = –1.	$10^{-1} = 10^{-1}$

Question 7: Estimate the following to two decimal places.

a.) $\log_2 \frac{1}{2}$ b.) $\log_2 4$ c.) $\log_2 8$ d.) $\log_2 16$ e.) $\log_2 9$ f.) $\log_2 15$

Section Summary

Answer Q4: Switch x and y then solve for y and you will get $y = \dfrac{x-5}{3}$.

In the next section we are going to look more closely at this new function $f(x) = \log_b x$.

For now, just remember,

- That the function $\log_b x$ is the inverse of the function b^x.
- That the answer to a logarithm is an exponent that satisfies the expression $x = b^y$.
- Also, from this section remember that all of the following are true for an exponential function:

Answer Q5: a.) 1 since $10^1=10$ b.) 2 since $10^2=100$ c.) 3 since $10^3=1000$. The informal definition says that you are looking for the exponent that you raise the 10 to, in order to get the number you are logging (10,100,1000).

The base of an exponential function must be positive ($b>0$).	The domain of an exponential function is $(-\infty, \infty)$.
The range of an exponential function is $(0, \infty)$.	The horizontal asymptote is $y = 0$.
An exponential function like this one is an example of growth.	An exponential function like this one is an example of decay.
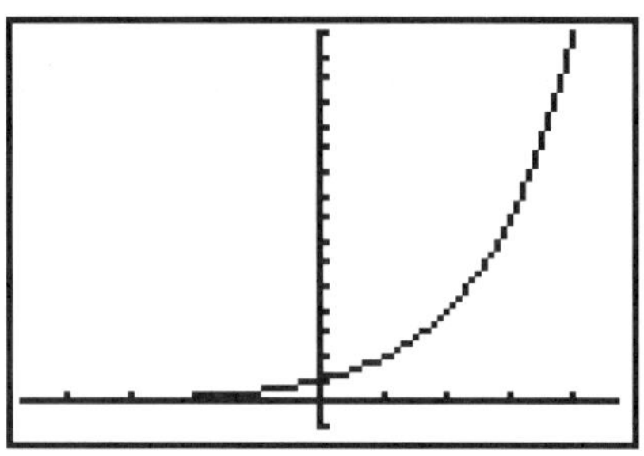	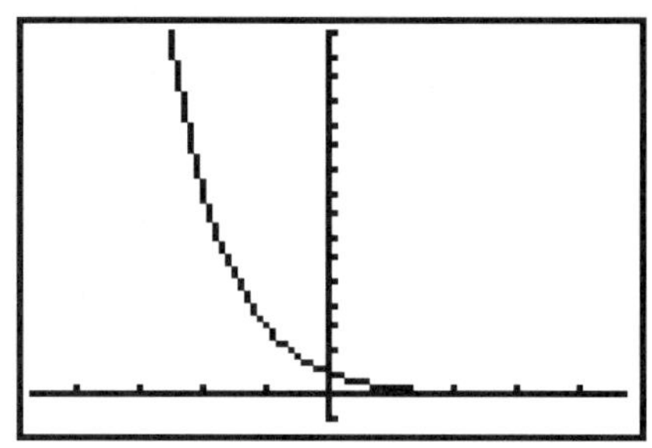

- One last thing: the number e (2.71828…) is special and we will be using it quite a bit in the future.

SECTION 4.2 HOMEWORK

(1–14) For each of the exponential functions create a table of values using $x = -1, 0, 1, 2, 3$. Graph each function using the table of values and check the graph with your graphing calculator.

1. $f(x) = 2^x$
2. $f(x) = 3^x$
3. $f(x) = (1/4)^x$
4. $f(x) = (0.5)^x$
5. $f(x) = 4^{-x}$
6. $f(x) = 2^{-x}$
7. $f(x) = 2^x + 3$
8. $f(x) = 3^x - 2$
9. $f(x) = (1/3)^x - 1$
10. $f(x) = (0.5)^x + 2$
11. $f(x) = 3(2^x)$
12. $f(x) = -2(2^x)$
13. $f(x) = 2^{x+1}$
14. $f(x) = (1/2)^{x-1}$

(15–28) Graph each of the functions with a graphing calculator and answer the following questions:

a. Does the function represent a growth or decay exponential function?
b. What is the domain of the function?
c. What is the range of the function?
d. What is the y-intercept?

15. $y = 20(0.8)^x$
16. $y = 15(2/3)^x$
17. $y = 50(3)^x$
18. $y = 10(1.5)^x$
19. $A(t) = 5000(1.08)^t$
20. $A(t) = 10000(1.10)^t$
21. $R(t) = e^{-.03t}$
22. $R(t) = e^{-.0134t}$
23. $f(x) = (0.7)^x + 5$
24. $f(x) = (3/4)^x - 2$
25. $f(x) = (5/3)^x - 3$
26. $f(x) = (3.4)^x + 2$
27. $f(x) = 10(2)^x + 10$
28. $f(x) = 20(0.3)^x - 10$

(29–50) Evaluate the following (i.e. find the exponent, that the base would be raised to, so that it would equal what we are taking the log of):

EXAMPLE: $\log_7(49) = \log_7 7^2$ which means $7^2 = 49$.

29. $\log_3 81$
30. $\log_5 (25)$
31. $\log_2 (64)$
32. $\log_4(1024)$
33. $\log_2(1/4)$
34. $\log_4 (1/64)$
35. $\log_3 (1/81)$
36. $\log_5 (1/625)$
37. $\log_5 1$
38. $\log_3 1$
39. $\log_5 5$
40. $\log_3 3$
41. $\log_5 (-25)$
42. $\log_4 (-64)$

Answer Q6:
a.) 1.93 since $10^{1.93} = 85.1$
b.) 1.70 since $10^{1.70} = 50.1$
c.) 1.18 since $10^{1.18} = 15.1$

43. a. $\log_4 (1/4)$ b. $\log_4 (1)$ c. $\log_4 (4)$ d. $\log_4 (16)$ e. $\log_4 (64)$
44. a. $\log_2 (1/2)$ b. $\log_2 (1)$ c. $\log_2 (2)$ d. $\log_2 (4)$ e. $\log_2 (8)$
45. a. $\log_5 (1/5)$ b. $\log_5 (1)$ c. $\log_5 (5)$ d. $\log_5 (25)$ e. $\log_5 (125)$
46. a. $\log_3 (1/3)$ b. $\log_3 (1)$ c. $\log_3 (3)$ d. $\log_3 (9)$ e. $\log_3 (27)$
47. a. $\log_{\frac{1}{2}} (2)$ b. $\log_{\frac{1}{2}} (1)$ c. $\log_{\frac{1}{2}} (1/2)$ d. $\log_{\frac{1}{2}} (1/4)$ e. $\log_{\frac{1}{2}} (1/8)$
48. a. $\log_{\frac{1}{3}} (3)$ b. $\log_{\frac{1}{3}} (1)$ c. $\log_{\frac{1}{3}} (1/3)$ d. $\log_{\frac{1}{3}} (1/9)$ e. $\log_{\frac{1}{3}} (1/27)$
49. a. $\log_{\frac{1}{4}} (4)$ b. $\log_{\frac{1}{4}} (1)$ c. $\log_{\frac{1}{4}} (1/4)$ d. $\log_{\frac{1}{4}} (1/16)$ e. $\log_{\frac{1}{4}} (1/64)$
50. $\log_{\frac{1}{5}} (5)$ b. $\log_{\frac{1}{5}} (1)$ c. $\log_{\frac{1}{5}} (1/5)$ d. $\log_{\frac{1}{5}} (1/25)$ e. $\log_{\frac{1}{5}} (1/125)$

(51–58) Using the values found in problems (43–50), set up a table of values and plot the points to sketch a graph of the functions.

51. $y = \log_4(x)$
52. $y = \log_2(x)$
53. $y = \log_5(x)$
54. $y = \log_3(x)$
55. $y = \log_{\frac{1}{2}}(x)$
56. $y = \log_{\frac{1}{3}}(x)$
57. $y = \log_{\frac{1}{4}}(x)$
58. $y = \log_{\frac{1}{5}}(x)$

Answer Q7:
a.) $2^? = 1/2 = 2^{-1}$ means $? = -1$
b.) $2^? = 4 = 2^2$ means $? = 2$
c.) $2^? = 8 = 2^3$ means $? = 3$
d.) $2^? = 16 = 2^4$ means $? = 4$
e.) $2^? = 9$ means that $3 < ? < 4$ since $8 < 9 < 16$ so $? \approx 3.18$
f.) $2^? = 15$ means that $3 < ? < 4$ since $8 < 15 < 16$ so $? \approx 3.91$

(59–64) Approximate the value of the following logs to two decimal places:

59. $\log_3(12)$
60. $\log_4(23)$
61. $\log_2(100)$
62. $\log_5(500)$
63. $\log_{\frac{1}{3}}(1/29)$
64. $\log_{\frac{1}{5}}(1/34)$

(65-73) Understanding concepts of translations involving exponential functions.

65. Using the graphing calculator, compare the graphs of $y = 3^x$ and $y = 10(3^x)$.
    How do the y–intercepts of the graphs compare?
66. Using the graphing calculator, compare the graphs of $y = 2^x$ and $y = 5(2^x)$.
    How do the y–intercepts of the graphs compare?
67. Using the graphing calculator, compare the graphs of $y = (1/2)^x$ and $y = 8(1/2)^x$.
    How do the y–intercepts compare?
68. Using what you have observed from problems 65 through 67, how will the y–intercepts compare for the functions $y = 5^x$ and $y = 3(5^x)$? (Check your guess with the graphing calculator.)
69. Using the graphing calculator, compare the graphs of $y = 2^x$ and $y = 2^x + 5$.
    a. How do the y–intercepts compare?
    b. How do the ranges compare?
70. Using the graphing calculator, compare the graphs of $y = 3^x$ and $y = 3^x - 4$.
    a. How do the y–intercepts compare?
    b. How do the ranges compare?
71. Using the graphing calculator, compare the graphs of $y = (1/2)^x$ and $y = (1/2)^x + 3$.
    a. How do the y–intercepts compare?
    b. How do the ranges compare?
72. Using what you have observed from problems 69 through 71, how will the y–intercepts and ranges compare for the functions $y = 4^x$ and $y = 4^x + 2$? (Check with your graphing calculator.)
73. How do you think the y–intercepts and the range will compare for the functions $y = 2^x$ and $y = 3(2^x) + 2$? (Check with your graphing calculator.)

Section 4.3 Logarithmic Functions

Let's continue looking at our new function $f(x) = \log_b x$. In the previous section we stated the definition of a logarithm, which was $\log_b x = y$ if and only if $x = b^y$, for $b>0$. We will use this definition to help us do the first example.

Discussion 1: Graphing and Creating Tables of Log Functions

We saw in the last section one way to go about estimating the answers to logs. But typically it will be easier to work with logs if we use the definition and convert them back to an exponential form, which is more familiar to us. Here we will have the function $f(x) = \log_3 x$ which we will change into $3^y = x$ to work with it. Here is a table and a graph of this function.

x	f(x)
27.0000	3.0000
9.0000	2.0000
3.0000	1.0000
1.0000	0
.3333	-1.0000
.1111	-2.0000
.0370	-3.0000
.0123	-4.0000
.0041	-5.0000
.0013	-6.0000
.0004	-7.0000

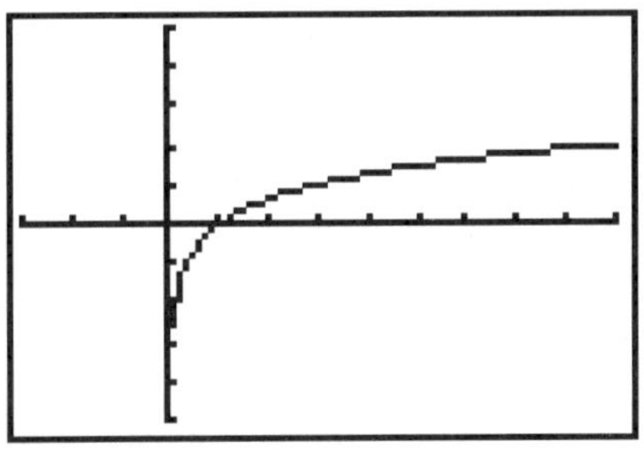

Let's work out a couple of the values that we see in the table. Notice though, in order to do this, we must plug in y-values and find the x-values, since the y variable is in the exponent position.

If $y = 3$, then x will equal	$x = 3^3 = 27$
If $y = 0$, then x will equal (any number raised to the zero power will always equal one)	$x = 3^0 = 1$
If $y = -2$, then x will equal (a negative exponent causes a base to flip upside down or in other words invert)	$x = 3^{-2} = \frac{1}{3^2} = \frac{1}{9} = 0.11111...$

In the last section we saw that $y = \log_b x$ is the inverse of $y = b^x$ because of the way it was defined. From our discussions so far in this chapter, along with section 1.6, we now know several important facts about how a logarithm relates to the exponential when they have the *same base*.

- First, because they are inverses their domains and ranges must be switched.
- Second, their graphs are mirror images to each other around the line $y = x$.

- Third, since the graphs are mirror images, you should get a vertical asymptote since the exponential function has a horizontal asymptote.
- Lastly, if $f(x) = \log_b x$ and $f^{-1}(x) = b^x$ then $(f \circ f^{-1})(x) = \log_b b^x = x$ and $(f^{-1} \circ f)(x) = b^{\log_b x} = x$, hence we can see that $(f \circ f^{-1})(x) = (f^{-1} \circ f)(x) = x$.

Let's look further at the functions $y = \log_3 x$ and what we are talking about.

The exponential function $y = 3^x$ has domain $(-\infty, \infty)$, range $(0, \infty)$.	The graph of $y = \log_3 x$ shows that it has domain $(0, \infty)$, range $(-\infty, \infty)$.
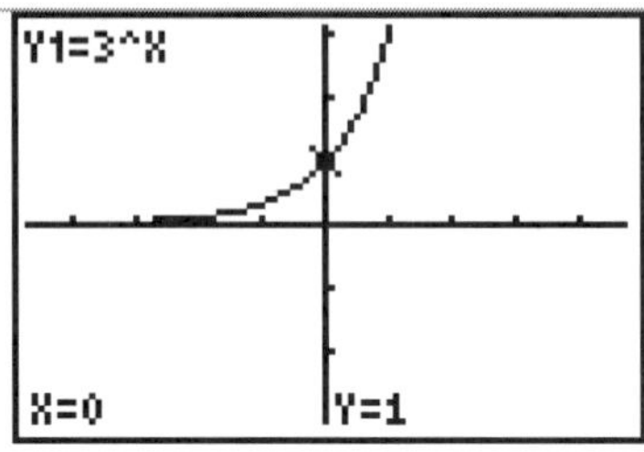	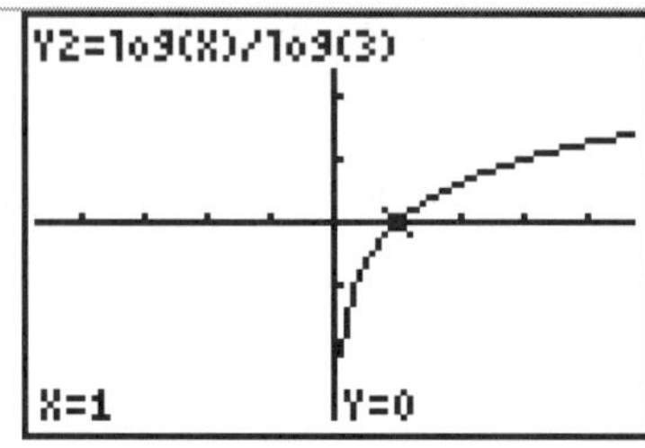
$3^{\log_3 x} = \log_3 3^x = x$	

Notice that the domain and ranges of these two inverse functions are just switched with each other. The two graphs are mirror images of each other around the line $y = x$.

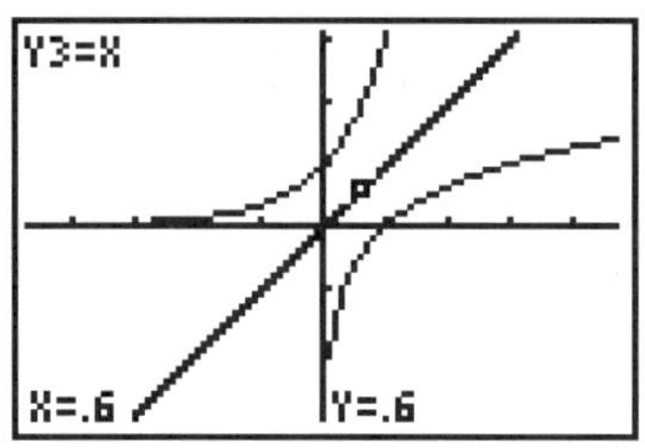

The exponential graph has a horizontal asymptote ($y = 0$) and so the logarithmic graph has a vertical asymptote ($x = 0$). Lastly, let's test a few values in this ($3^{\log_3 x} = \log_3 3^x = x$).

$x = 3$	$x = 1$	$x = 1/3$	$x = 1/9$
$3^{\log_3 3} = 3^1 = 3$	$3^{\log_3 1} = 3^0 = 1$	$3^{\log_3 1/3} = 3^{-1} = 1/3$	$3^{\log_3 1/9} = 3^{-2} = 1/9$
$\log_3 3^3 = \log_3 27 = 3$	$\log_3 3^1 = \log_3 3 = 1$	$\log_3 3^{1/3} = 1/3$	$\log_3 3^{1/9} = 1/9$

Let's summarize what we know about the basic logarithms.

- Informally, a logarithm is an exponent.
- Formally, $f(x) = \log_b x$, where $y = \log_b x$ if and only if $b^y = x$ for $b > 0$.
- The base must be positive.

- The domain is all positive numbers $(0, \infty)$, the same as the range of $y = b^x$ the exponential function.
- The range is all real numbers $(-\infty, \infty)$, the same as the domain of $y = b^x$ the exponential function.
- There is a vertical asymptote at $x = 0$, the opposite of the horizontal asymptote $y = 0$ of the exponential function.

Question 1: Find a few values for $y = \log_5 x$ and sketch a graph. Remember to use the definition and try plugging in numbers for y and getting answers for x.

Example 1: Graphing a Log Function

Graph the function $g(x) = \log_{0.5} x$.

Solution:

We will change this to its exponential form using the definition of a logarithm.

$$y = \log_{0.5} x \Leftrightarrow x = 0.5^y$$

x	y
.1250	3.0000
.2500	2.0000
.5000	1.0000
1.0000	-0.0000
2.0000	-1.0000
4.0000	-2.0000
8.0000	-3.0000

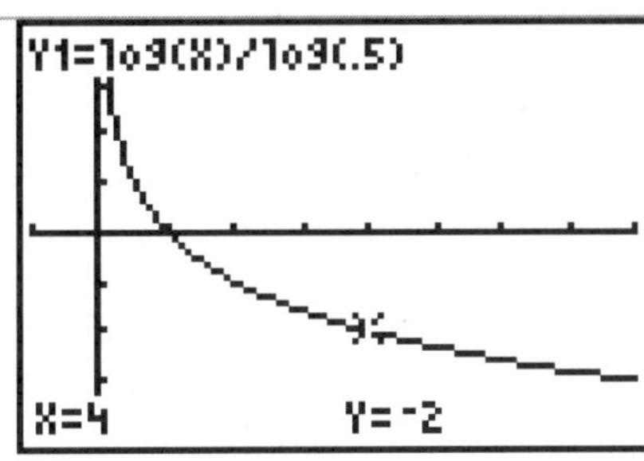

So here is a graph of a logarithm with a base less than 1. Let's look at how the base being greater or less than 1 affects the graphs of both the exponential and logarithmic functions and how similar the effect is in each of them.

$$f(x) = 2^x, \quad g(x) = \left(\frac{1}{2}\right)^x$$

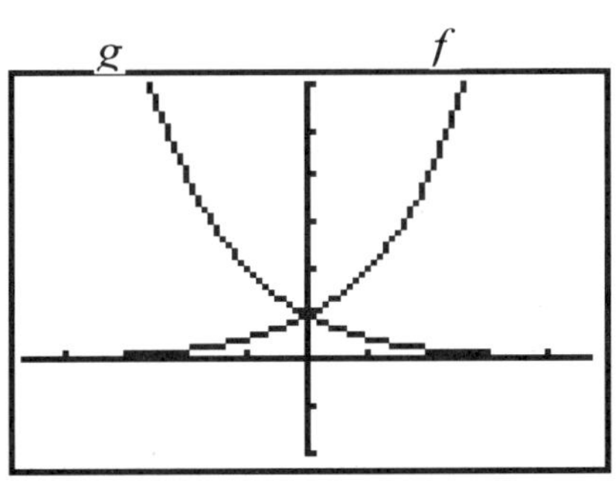

$$f^{-1}(x) = \log_2 x, \quad g^{-1}(x) = \log_{1/2} x$$

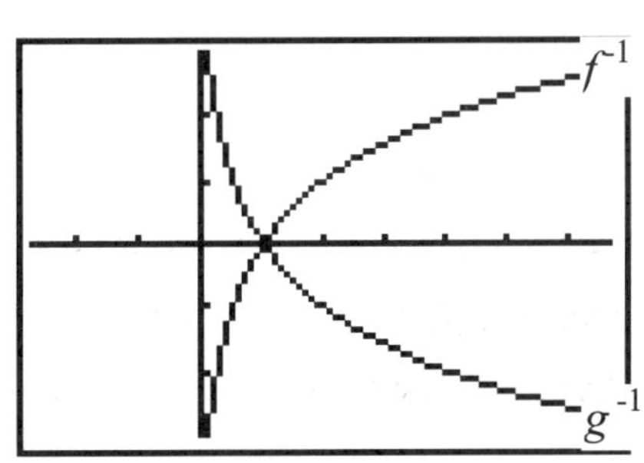

You should recognize $f(x)$, growth, and $g(x)$, decay, from the last section and how together they form a picture that is symmetric to the y-axis when they have reciprocal bases. Likewise notice how the inverses ($f^{-1}(x)$, $g^{-1}(x)$) form a picture that is symmetric to the x-axis. This should be of no surprise given all that we have talked about with regards to inverses. Remembering these two pictures will help you to remember the basic graphs of exponential and logarithmic functions and how the base affects the graphs.

So far in this section you have seen many graphs of logarithms that were done on the graphing calculator. Let's take a moment to talk about what your graphing calculator can do and how to go about graphing logarithms on it. First, let's define two special logarithms.

> The **common** logarithm is a logarithm whose base is 10 and would normally be written as $\mathbf{\log_{10} x}$ but we write it as $\mathbf{\log x}$ instead.

If you look at your graphing calculator, you will see a button that has **LOG** on it. This is the log base 10 key. Here is what your calculator screen would look like if you evaluated the log (85).

```
log(85)
         1.929418926
■
```

> The **natural** logarithm is a logarithm whose base is e and would normally be written as $\mathbf{\log_e x}$ but we write it as $\mathbf{\ln x}$ instead.

If you look at your graphing calculator, you will see a button that has **LN** on it. This is the log base e key. Remember e is a special number that we talked about in the last section ($e = 2.71828\ldots$). Here is what your calculator screen would look like if you evaluated the ln (3).

```
ln(3)
         1.098612289
■
```

So your calculator will graph any logarithm or find the value for a logarithm as long as it is base 10 or base e. The big question then is, "How do we get it to graph and evaluate other bases?" The answer is that we have the following formula called the **Change of Base** formula:

> $\log_b x = \dfrac{\log_a x}{\log_a b}$, where $a, b > 0$. Usually we will choose a to be 10 or e.

So the end result of this formula is that when we have a log that isn't base10 or *e*, then we will change it to 10 or *e* whenever we want to use our calculators.

$$\log_b x = \frac{\log x}{\log b} \quad \text{or} \quad \log_b x = \frac{\ln x}{\ln b}$$

Look at the answer to question 1 in this section. You will notice that in the graph you have Y1=log(*x*)/log(5). Notice the similar Y1 on the calculator graph in example 1 above.

Answer Q1:

x	y
.0080	-3.0000
.0400	-2.0000
.2000	-1.0000
1.0000	0.0000
5.0000	1.0000
25.0000	2.0000
125.0000	3.0000

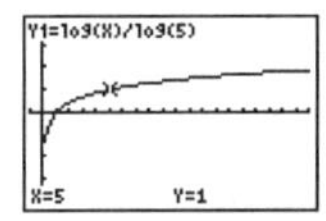

Discussion 2: Change of Base

Here are some numerical examples to help you to see that this really does work.

We will evaluate $\left(\frac{\log 17}{\log 5}\right)$, and $\left(\frac{\ln 17}{\ln 5}\right)$; then we'll find $5^{1.760374428}$. On your calculator $\left(\frac{\log 17}{\log 5}\right) =$ 1.760374428, $\left(\frac{\ln 17}{\ln 5}\right) = 1.760374428$ and $5^{1.760374428} = 17.00000001$. This last computation, $5^{1.760374428} = 17.00000001$, just uses the definition of a logarithm ($\log_5 17 = x \Rightarrow 5^x = 17$). Notice how the exponent is the same as the answers to the two log expressions.

```
log(17)/log(5)
         1.760374428
5^1.760374428
         17.00000001
■
```

Question 2: Try this yourself. Find $\log_7 53$ by first finding $\left(\frac{\log 53}{\log 7}\right)$ then $\left(\frac{\ln 53}{\ln 7}\right)$ and then check to see that your answers really work by using the definition of a log like we did in discussion 2 above.

Question 3: Try graphing the following functions on your graphing calculators.
$f(x) = \log_4 x$ $\quad g(x) = \log_7 x$ $\quad h(x) = \log_{1/5} x$ $\quad k(x) = \log_e x$

PROPERTIES OF LOGARITHMS

We now need to spend some time discussing properties of logarithms. We have seen two of these properties already in this discussion.

Property 1: $\log_b b^x = x$	Property 2: $b^{\log_b x} = x$

These two properties are called inverse properties, because they are the composition of inverse functions that "undo" each other. Next, we have four more properties, which are useful for combining and expanding logarithms or for eliminating exponents or coefficients.

Property 3: $\log_b MN = \log_b M + \log_b N$	Product property
Property 4: $\log_b \dfrac{M}{N} = \log_b M - \log_b N$	Quotient property
Property 5: $\log_b M^r = r \log_b M$	Power property
Property 6: $\log_b 1 = 0$	Log of 1 property

Now, logarithms and exponents are related, so it shouldn't be too surprising that these properties are similar to the properties of exponents. Let's take a moment to look at the properties of exponents that you learned in a previous algebra class.

1.) $b^M \bullet b^N = b^{M+N}$

2.) $\dfrac{b^M}{b^N} = b^{M-N}$

3.) $\left(b^M\right)^N = b^{MN}$

Question 4: State the similarities you see between the logarithm and exponent properties.

Let's take a moment here and prove the third property ($\log_b MN = \log_b M + \log_b N$).

Let	$A = \log_b M$ and $B = \log_b N$
Rewrite each logarithm in its exponential form using the definition of a log.	$b^A = M$ and $b^B = N$
Now, multiply the two exponentials together.	$M \bullet N = b^A \bullet b^B = b^{A+B}$ (property 1 of exponents)
Now, convert $M \bullet N = b^{A+B}$ into its logarithm form.	$\log_b M \bullet N = A + B$ (Definition of logs)
Now, substitute back into this statement that A and B are equal to from above.	$\log_b M \bullet N = \log_b M + \log_b N$

The other two properties are proved similarly and the fourth is easy to prove using the definition of a logarithm. Here are some numerical examples that might help you to see why these do work.

$\log_3 8 - \log_3 4 = \log_3 2$	$\log_3 5^4 = 4\log_3 5$	$\log_3 1 = 0$
Checked on our calculator.	Checked on our calculator.	Checked on our calculator.
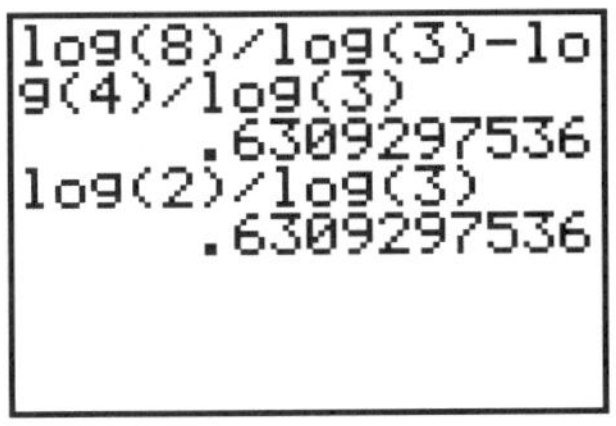	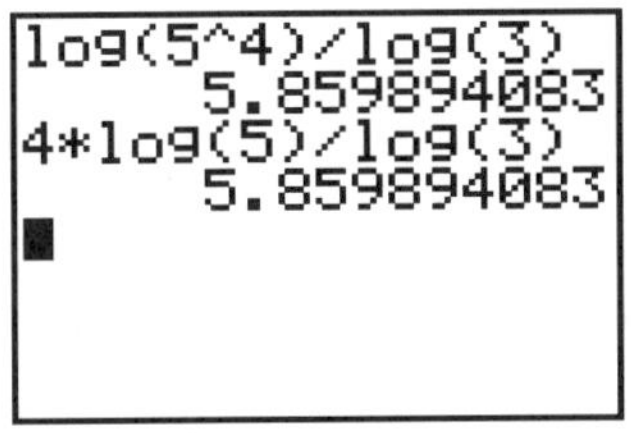	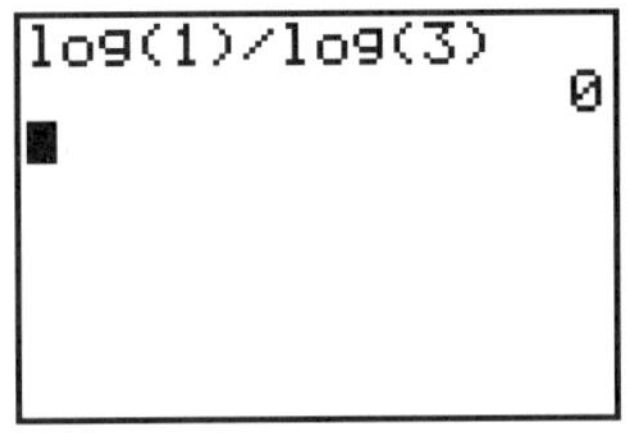

Answer Q2:

2.040326433...

Let's look at some examples of how we will be using these properties.

Answers Q3

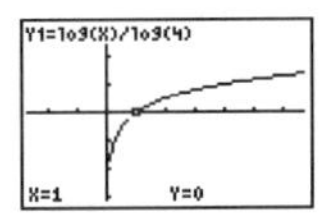

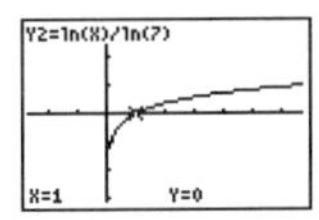

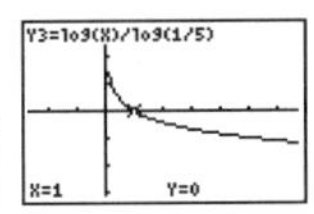

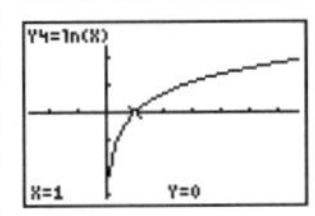

Example 2: Using Properties of Logarithms

Use the properties of logarithms to combine the following logs into one log.

a.) $\log_2 u + \log_2 p$

b.) $\log_8 w^2 - \log_8 n$

c.) $\log 5 + \log x - \log z$

d.) $\frac{1}{2}\log_5 g - 3\log_5 h$

Solution:

a) is the sum of two logs so (use: $\log_b M + \log_b N = \log_b MN$)	$\log_2 u + \log_2 p = \log_2 (up)$
b) is the difference of two logs so (use: $\log_b M - \log_b N = \log_b \frac{M}{N}$)	$\log_8 w^2 - \log_8 n = \log_8 \frac{w^2}{n}$
c) is a combination of properties so (use product then quotient properties)	$\log 5 + \log x - \log z = \log (5x) - \log z$ $= \log \frac{5x}{z}$
d) is a combination of properties so (use power then subtraction properties)	$\frac{1}{2}\log_5 g - 3\log_5 h = \log_5 g^{1/2} - \log_5 h^3$ $= \log_5 \frac{g^{1/2}}{h^3}$

You may have noticed in example 2(c) that the net result of combining the logs into one log was that any log that was positive to begin with had what it was logging end up in the numerator of the answer. Any log that was negative to begin with had what it was logging end up in the denominator. This little observation will help you to combine and expand logs more quickly.

Example 3: Combining Logs

Combine the following logs into one log.

$\ln x^2 - \ln y^{1/3} - \ln w + \ln z^4$

Solution:

Since the first and last natural logs are preceded with a positive coefficient, the x^2 and z^4 will be in the numerator of the final natural log. Also since the middle two natural logs are preceded with a negative, the $y^{1/3}$ and w will be in the denominator. So the final answer is,

$$\ln \frac{x^2 z^4}{y^{1/3} w}.$$

Answer Q4:
1) Multiplying like bases means you add exponents which is similar to adding two logs which are exponents.
2) Dividing like bases means you subtract exponents which is similar to subtracting two logs which are exponents.
3) Raising to a power means you multiply the exponents which is similar to multiplying an exponent times a log which is an exponent.

Question 5: Combine the following logs into one log.
$\log h^3 + \log q^{1/2} - \log t - \log p^5$

Question 6: Combine log 3 + 2log 5 – 3log 7 into a single log and then find the numerical answer to both on your graphing calculator to see that they do equal each other.

Now, let's do the opposite and take one log and expand it into a sum, difference, or multiple of logs.

Example 4: Expanding Logarithmic Expressions

Expand the following logs into a sum, difference, or multiple of logs.

a) $\log_8 \frac{x^{3/2} w}{y^4}$ b) $\log_2 \frac{\sqrt{p}\, f^7}{g^2 k}$

Solution:

We now apply the reverse logic as compared to the last couple of examples.

a) We see that we have two variables in the numerator and one in the denominator, so we will get one log with a negative coefficient and two with positive coefficients.	$\log_8 \frac{x^{3/2} w}{y^4} = \log_8 x^{3/2} + \log_8 w - \log_8 y^4$
Next, we can move the exponents to the front of the logs and they will become coefficients.	$= \frac{3}{2} \log_8 x + \log_8 w - 4\log_8 y$
b) We see that we have two variables in both the numerator and the denominator, so we will get two logs	

with a negative coefficient and two with positive coefficients.	$\log_2 \dfrac{\sqrt{p}\, f^7}{g^2 k} = \log_2 \sqrt{p} + \log_2 f^7 - \log_2 g^2 - \log_2 k$
Next, we can move the exponents to the front of the logs and they will become coefficients. (Note, that the square root is the same as an exponent of ½ .)	$= \log_2 p^{1/2} + 7\log_2 f - 2\log_2 g - \log_2 k$ $= \dfrac{1}{2}\log_2 p + 7\log_2 f - 2\log_2 g - \log_2 k$

Section Summary

- Informally a logarithm is an exponent
- Formally $f(x) = \log_b x$, where $y = \log_b x$ if and only if $b^y = x$ for $b > 0$
- The base must be positive
- The domain is all positive numbers $(0, \infty)$, the same as the range of $y = b^x$ the exponential function
- The range is all real numbers $(-\infty, \infty)$, the same as the domain of $y = b^x$ the exponential function
- There is a vertical asymptote at $x = 0$, the opposite of the horizontal asymptote $y = 0$ of the exponential function
- The basic two types of graphs are

Base > 1 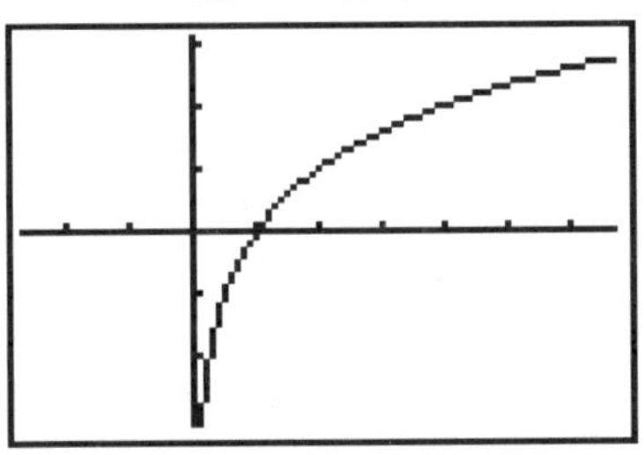

Base < 1 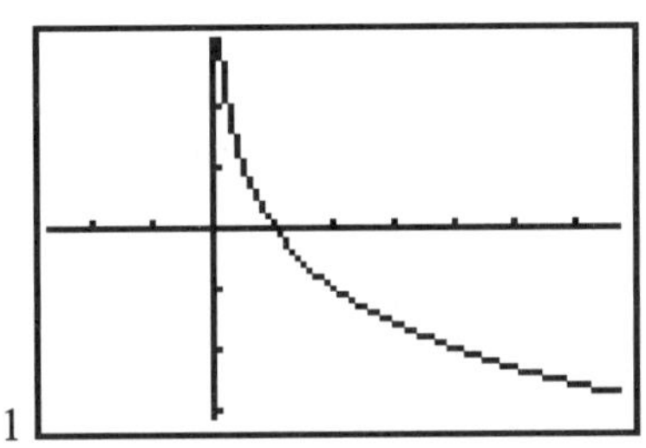

- And lastly all the properties we will need to know to manipulate logarithms:

1.) $\log_b b^x = x$	Inverse property
2.) $b^{\log_b x} = x$	Inverse property
3.) $\log_b M + \log_b N = \log_b MN$	Addition property
4.) $\log_b M - \log_b N = \log_b \dfrac{M}{N}$	Difference property
5.) $r \log_b M = \log_b M^r$	Product property
6.) $\log_b 1 = 0$	
7.) $\log_b x = \dfrac{\log x}{\log b} \; or \; \dfrac{\ln x}{\ln b}$	Change of base formula

SECTION 4.3 HOMEWORK

(1–6) Change each of the logarithmic statements to exponential statements:

1. $\log_3 x = y$ 2. $\log_a w = 3$ 3. $\log_5 2 = x$
4. $\log_3 x = -2$ 5. $\log_2 (x + y) = a$ 6. $\log_a 3 = (x + 2)$

(7–12) Change each of the exponential statements to logarithmic statements:

7. $3^x = m$ 8. $y^3 = r$ 9. $a^{3x-1} = 5$
10. $3^{2x+3} = 6$ 11. $a^5 = (x + y)$ 12. $3^b = (2x - 3)$

(13–26) Expand each of the following:

13. $\log_3 (3x)$ 14. $\log_3 (ar)$ 15. $\log_2 (x/3)$
16. $\log_2 (5/b)$ 17. $\log\left(\frac{xy}{5z}\right)$ 18. $\log\left(\frac{3a}{2b}\right)$

Answer Q5: $\log \frac{h^3 q^{1/2}}{tp^5}$

19. $\log_5 (x^2 y)$ 20. $\log_5 \left(x^3 \sqrt{y}\right)$ 21. $\ln\left(\frac{a^3}{x^2 y}\right)$
22. $\ln\left(\frac{a^3 b^2}{x}\right)$ 23. $\log_4\left(\frac{3y\sqrt[3]{x}}{ab^2}\right)$ 24. $\log_4\left(\frac{x^2 y^3}{3\sqrt[4]{a}}\right)$

Answer Q6: $\log 3 + 2\log 5 - 3\log 7 = \log \frac{3 \cdot 5^2}{7^3} \approx -0.66023$

25. $\ln\left(\sqrt[3]{\frac{x^2 y}{a}}\right)$ 26. $\log\left(\sqrt[4]{\frac{3a^2}{bc^2}}\right)$

(27–42) Write each of the following as a single logarithm:

27. $\log_2 3 + \log_2 x$ 28. $\log_2 a + \log_2 5$ 29. $\log_5 r - \log_5 4$
30. $\log_5 7 - \log_5 x$ 31. $\log x + \log (2x - 1)$ 32. $\log(2x - 1) + \log(3x + 2)$
33. $\log x - \log(3x + 2)$ 34. $\log (3x + 1) - \log(2x + 3)$ 35. $\ln 3 + \ln (2x - 1) - \ln (x + 3)$
36. $\ln 2 + \ln (3x + 1) - \ln (2x - 3)$ 37. $\log_7 3 + 2\log_7 x - 3\log_7 y$
38. $3\log_9 a + \frac{1}{2}\log_9 b - \frac{1}{3}\log_9 c$ 39. $2\log x + \log (2x-1) - 3\log y$
40. $2\log a - 2\log b - \log (b+3)$ 41. $2\ln x + 3\ln x - 5\ln x$
42. $8\log a - 2\log a - 6\log a$

(43–46) Approximate parts (a), (b), (c), and (d) with your calculator and determine if they are all equal. State which log properties were used to come up with (b), (c), and (d). (Approximate to 4 decimal places):

43. a.) log 36 b.) log 4 + log 9 c.) log 2 + log 18 d.) log 3 + log 12
44. a.) ln 100 b.) ln 2 + ln 50 c.) ln 5 + ln 20 d.) ln 4 + ln 25
45. a.) ln 5 b.) ln 10 – ln 2 c.) ln 20 – ln 4 d.) ln 35 – ln 7
46. a.) log 8 b.) log 16 – log 2 c.) log 40 – log 5 d.) log 64 – log 8

(47–50) Approximate parts (a) and (b) with your calculator, and determine if they are all equal. State which log properties were used to come up with (9). (Approximate to 4 decimal places.)

47. a.) $\log 125$ b.) $3\log 5$

48. a.) $\ln 243$ b.) $5\ln 3$

49. a.) $\ln \frac{1}{25}$ b.) $-2\ln 5$

50. a.) $\log \frac{1}{8}$ b.) $-3\log 2$

(51-56) Understanding concepts of logarithmic properties.

51. Rewrite ln 15 as a sum of logarithms and check your answer with your calculator by approximating both logarithm statements.

52. Rewrite log 33 as a sum of logarithms and check your answer with your calculator by approximating both logarithm statements.

53. Rewrite log 3 as a difference of logarithms and check your answer with your calculator by approximating both logarithm statements.

54. Rewrite ln 7 as a difference of logarithms and check your answer with your calculator by approximating both logarithm statements.

55. Expand $\log\left(\frac{5^2 \cdot 8^2}{10^2}\right)$ and check your answer with the calculator by approximating both logarithm statements.

56. Expand $\log\left(\frac{2^3 \cdot 3^2}{6^3}\right)$ and check your answer with the calculator by approximating both logarithm statements.

(57–68) Use the change of base property and the graphing calculator to approximate each of the following logarithms to 4 decimal places. (Check your answer with an equivalent exponential statement.)

Example: $\log_3 23 \approx 2.8540$ check: $3^{2.8540} \approx 22.9987$

57. $\log_5 59$ 58. $\log_4 123$ 59. $\log_7 159$ 60. $\log_6 233$

61. $\log_{\frac{1}{2}} \frac{1}{23}$ 62. $\log_{\frac{1}{5}} 15$ 63. $\log_3 \frac{1}{12}$ 64. $\log_5 \frac{3}{19}$

65. $\log_7 (-8)$ 66. $\log_4 (-33)$ 67. $\log_{(-2)} 8$ 68. $\log_{(-3)} 25$

(69–76) Use your graphing calculator to graph each of the functions.

a.) What is the domain? b.) What is the range?

69. $y = \log_3 x$ 70. $y = \log_5 x$ 71. $y = \log_{\frac{1}{2}} x$ 72. $y = \log_{\frac{1}{3}} x$

73. $y = 2 + \log_2 x$ 74. $y = 3 + \log_5 x$ 75. $y = 2\log_3 x$ 76. $y = -3\log_4 x$

(77–82) Find the domain of each of the following logarithm functions:

77. $f(x) = \log_4 (-3x)$ 78. $g(x) = \log_8(5x)$ 79. $y = \log (3x + 1)$ 80. $y = \ln (5x - 8)$

81. $t(x) = \ln (x^2 - 16)$ 82. $r(x) = \log (4x^2 - 25)$

Section 4.4 Solving Exponential and Logarithmic Equations

In the last several sections we have discussed the characteristics of exponential and logarithmic functions. Now that we have an understanding of the basics, let's discuss how to solve equations involving exponential or logarithmic functions.

There are four types of equations that we might run into when we are solving equations involving exponential or logarithmic functions. The four basic types are as follows:

1. $b^x = b^y$ — Bases are the same in the equation
2. $b^x = a$ — Bases are not the same in the equation
3. $\log_b x = a$ — Some logs and some numbers in the equation
4. $\log_b x = \log_b y$ — All logs in the equation

Let's take these one at a time and look at examples of how we solve each type.

EXAMPLES OF THE TYPE $b^x = b^y$

Example 1: Equation of the Type $b^x = b^y$

Solve the equation $3^{x+1} = 3^{-2}$.

Solution:

First, notice that the bases are the same (type 1). If the bases are the same on both sides of the equal sign, then surely their exponents will have to be equal also in order for both sides of the equation to turn out to be equal. This means that:

Exponents must be equal.	$x + 1 = -2$
Subtract one from both sides in order to solve for x.	$x = -3$
Check.	$3^{-3+1} = 3^{-2}$

Sometimes in order to solve an equation you will have to use some of the rules and properties of exponents in order to get the equation you have into one of the forms above that you can solve.

Example 2: Equation of the Type $b^x = b^y$

Solve the equation $2^x = \dfrac{1}{8}$.

Solution:

Change $\frac{1}{8}$ so that it has a 2 in it ($8 = 2^3$).	$\frac{1}{8} = \frac{1}{2^3}$

Next, we don't want a fraction (negative exponent rule).	$\frac{1}{2^3} = 2^{-3}$
Now, rewrite the equation.	$2^x = 2^{-3}$
So, this equation now looks like what it should in order for us to solve it by our first method ($b^x = b^y$), hence we know that	$x = -3$

Example 3: Equation of the Type $b^x = b^y$

Solve the equation $25^x = 125^4$.

Solution:

Change both the 25 and the 125 so that they will both have a 5 in them.	$25 = 5^2$ and $125 = 5^3$
Next, rewrite the equation.	$(5^2)^x = (5^3)^4$
Simplify the equation using the power rule of exponents.	$5^{2x} = 5^{12}$
We now know that	$2x = 12$
Hence,	$x = 6$

Question 1: Solve the equation $7^{3-2x} = 49$.

EXAMPLES OF THE TYPE $b^x = a$

Next, what about when the bases are not the same (type 2)? When we see this, we will need to log both sides of the equation. Let's do some examples.

Example 4: Equation of the Type $b^x = a$

Solve the equation $2^x = 15$.

Solution:

Now this doesn't fit the first type we looked at. The bases are not the same, and we don't know any way by which to solve for a variable in the exponent position. Remember the last time this happened to us (section 4.2)? We made a definition of a logarithm. Maybe logs could be of some use to us here. The property of logs that says $\log_b x^r = r \log_b x$ could be just the idea we need. This property allows us to move something from an exponent position to a coefficient position.

First, lets log both sides of the equation to get logs into this problem since we need to use a property of logs to help us solve this equation. (Note: We are using a base 10 log. We could have used a base e log just as well.	$\log 2^x = \log 15$

Either way we want to use something our calculator likes.)	
Now, use the property of logs to move the exponent x.	$x \log 2 = \log 15$
Now, we have x times log 2 so we just need to divide in order to solve for the x.	$x = \dfrac{\log 15}{\log 2}$
Using our calculators we can approximate the solution.	$x \approx 3.9069$

By the way, the problem we have just solved was the same as Question 7(f) in section 4.2. There we approximated the answer with the help of our calculators, guessed an exponent to use and then saw how close we got to the answer, but now we have seen a method for getting the answer exactly! Let's do some more.

Example 5: Equation of the Type $b^x = a$

Solve the equation $5^{3-x} = 7$.

Solution:

Take the natural log of both sides since the bases aren't the same. (This time we'll use **ln** just so you can see both types of logs in use.)		$\ln 5^{3-x} = \ln 7$	
Move the exponent.		$(3 - x) \ln 5 = \ln 7$	
There are two ways to solve at this point. We'll show both here.			
Divide both sides by ln 5.	$3 - x = \dfrac{\ln 7}{\ln 5}$	Distribute the ln 5	$3\ln 5 - x\ln 5 = \ln 7$
Subtract 3.	$-x = \dfrac{\ln 7}{\ln 5} - 3$	Subtract 3ln 5	$-x\ln 5 = \ln 7 - 3\ln 5$
Multiply by –1.	$x = -\dfrac{\ln 7}{\ln 5} + 3$	Divide by –ln 5	$x = -\dfrac{\ln 7}{\ln 5} + 3$

So if we use our calculators, we find that the approximate solution is $x \approx 1.790938045$. This might be a good time to take a moment to remind you that you can use the graphing ability of your calculators to solve any equation you might encounter. For example, with this last problem $5^{3-x} = 7$, we could have graphed it and used the zero key to approximate the solution.

First, get the equation equal to 0.	$5^{3-x} - 7 = 0$

Now, graph it and look for the x-intercepts. Remember that the zero or root key is 2nd trace then 5.	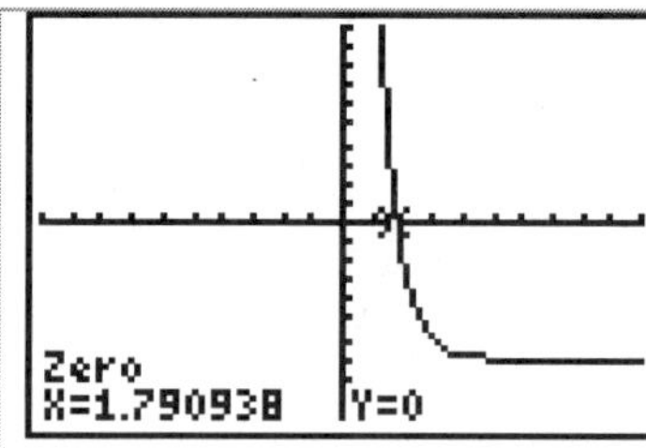
Our solution to this equation appears to be	$x \approx 1.790938$

Example 6: Equation of the Type $b^x = a$

Solve the equation $(1 + .005)^{12x} = 2$.

Solution:

Bases aren't the same, so take the natural log both of sides.	$\ln(1 + .005)^{12x} = \ln 2$
Move the exponent.	$12x\ln(1.005) = \ln 2$
Divide by 12 and ln (1.005).	$x = \dfrac{\ln 2}{12\ln(1.005)} \approx 11.58$ years

So we have solved the equation and have found that $x \approx 11.58$. In the next section, if we were asked, "How long would it take for money to double if invested at 6% compounded monthly?" this would be the equation we would need in order to solve the problem. Our answer, $x \approx 11.58$, would be in years. The meaning of our answer, and its ramifications for banking, will be discussed in the next section.

Question 2: Solve the equation $7^x = -2$.

Answer Q1: $7^{3-2x} = 7^2 \Rightarrow 3 - 2x = 2$ *since the bases are the same* $\Rightarrow -2x = -1 \Rightarrow x = ½$.

EXAMPLES OF THE TYPE $\log_b x = a$

Example 7: Equation of the Type $\log_b x = a$

Solve the equation $\log_3 x = 2$.

Solution:

Here is an equation where we have a log equal to a number. What we want to do in this case is use the definition of a logarithm to change this to its exponent form.

Use the definition of a logarithm.	$3^2 = x$
So by simplifying we get	$9 = x$

Question 3: What if the above problem was $\log_3 x - \log_3 5 = 2$? How would we get it to look like a type 3 problem ($\log_b x = a$, one log only in the problem)?

Example 8: Equation of the Type $\log_b x = a$

Solve the equation $\log_6 x - \log_6 (x - 3) = 1$.

Solution:

First, this does look like $\log_b x = a$ because we can combine the two logs into one log by our properties of logs (difference property).	$\log_6 \dfrac{x}{x-3} = 1$
Now, change to exponent form.	$6^1 = \dfrac{x}{x-3}$
Guess what? Everything we have learned or reviewed so far this semester can help us in this chapter. Here we see that we need to remember how to solve fraction equations from back in section 2.4 (multiply by LCD).	$6(x-3) = x$ $6x - 18 = x$
Get x's on the same side (linear equation now).	$-18 = -5x$
Divide by -5.	$\dfrac{18}{5} = x$

The last thing you will need to do when solving log equations is to check to make sure that your answers don't cause any of the values being logged to be negative. In this last example our answer works well.

Example 9: Equation of the Type $\log_b x = a$

Solve the equation $\log_3 (x - 5) + \log_3 (x + 3) = 2$.

Solution:

Here is an example where some terms of the equation are logs and some are just numbers.

First, combine the two logs into one log by our properties of logs (addition property).	$\log_3 [(x-5)(x+3)] = 2$
Now, change to exponent form.	$3^2 = (x-5)(x+3)$ $9 = x^2 - 2x - 15$
Here we see that we need to remember how to solve quadratic equations from back in section 2.2 (factor or square root or quadratic formula).	$0 = x^2 - 2x - 24$ $0 = (x-6)(x+4)$
This time factoring worked so our *potential* answers are	$x = 6$ or -4

Question 4: Do both of the potential answers in example 9 work?

So this example only has one correct solution, which is $x = 6$. How would this show itself on a graphing calculator? Well, let's see.

First, get the equation equal to 0.	$\log_3(x-5) + \log_3(x+3) - 2 = 0$
Now, graph it and look for the x-intercepts. Remember that you will need to use the change of base formula since our calculators only do base 10 or e. (Y1 = log(x–5)/log(3) + log(x+3)/log(3) –2)	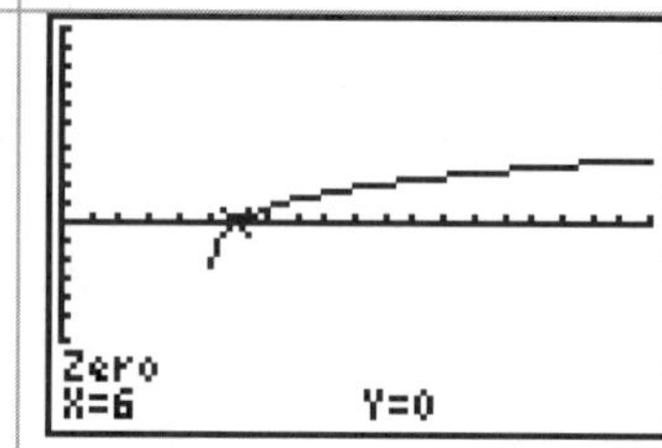
Our solution to this equation appears to be	$x = 6$

When we use the calculator, the extraneous solution doesn't show up.

EXAMPLES OF THE TYPE $\log_b x = \log_b y$

Example 10: Equation of the Type $\log_b x = \log_b y$

Solve the equation $\log_7 x + \log_7(x+1) = \log_7(x+9)$.

Solution:

First, combine the two logs on the left side of the equation into one log by our properties of logs (addition property).	$\log_7[x(x+1)] = \log_7(x+9)$
Since the two logs are equal, what they are logging must also be equal if the two sides of the equation are going to be equal.	$x(x+1) = (x+9)$ $x^2 + x = x + 9$
Here we see that we need to remember how to solve quadratic equations from back in section 2.2 (factor or square root or quadratic formula).	$x^2 - 9 = 0$ $x^2 = 9;\ x = \pm 3$
This time the square root method worked, so our *potential* answers are	$x = 3$ or -3
But if we check them, we see that –3 doesn't work.	$\log_7(\underline{-3})$ doesn't exist
So we get only one solution.	$x = 3$

Answer Q2 No Solution. Remember a basic exponential can't equal a negative numbers. So if you try solving this, here's what happens: when you log both sides you get $\log 7^x = \log(-2)$ Trouble! You can't log a negative number. So the answer must be No Solution!

Example 11: Equation of the Type $\log_b x = \log_b y$

Solve the equation $\ln x - \ln(x-3) = \ln(x+2) + \ln 2$.

Solution:

First, combine the two logs on the left and right sides of the equation into one log each by our properties of logs.	$\ln\left[\frac{x}{x-3}\right] = \ln\left[(x+2)(2)\right]$
So,	$\ln\left[\frac{x}{x-3}\right] = \ln\left[2x+4\right]$
Since the two logs are equal, what they are logging must also be equal if the two sides of the equation are going to be equal.	$\frac{x}{x-3} = 2x+4$
Multiply by the LCD.	$x = (2x+4)(x-3)$ $x = 2x^2 - 2x - 12$
Now, this is quadratic, so we get it equal to 0.	$0 = 2x^2 - 3x - 12$
To solve this we need the quadratic formula.	$x = \frac{3 \pm \sqrt{9-4(2)(-12)}}{4}$ $x = \frac{3 \pm \sqrt{105}}{4}$
So our two potential solutions approximated are	$x \approx 3.3117$ or -1.8117
But x ≈ -1.8117 can't work because	ln (-1.8117) doesn't exist
So we get one solution.	$x \approx 3.3117$

Answer Q3: Put logs together by using the difference property: $\log_b x - \log_b y = \log_b \frac{x}{y}$, *so we would then have* $\log_3 \frac{x}{5} = 2$.

Section Summary

We have now learned how to solve all kinds of equations in this book. Let's see if we can't summarize everything up to this point for you in the following flow charts on how to solve equations that have exponentials, logarithmic or polynomial terms in them. It might be a good idea to copy these last two pages and have them handy as you do your homework. Over time see if you can get yourself to think along the same lines as the flow charts, so that you will automatically go through the thought process necessary to be successful at solving equations.

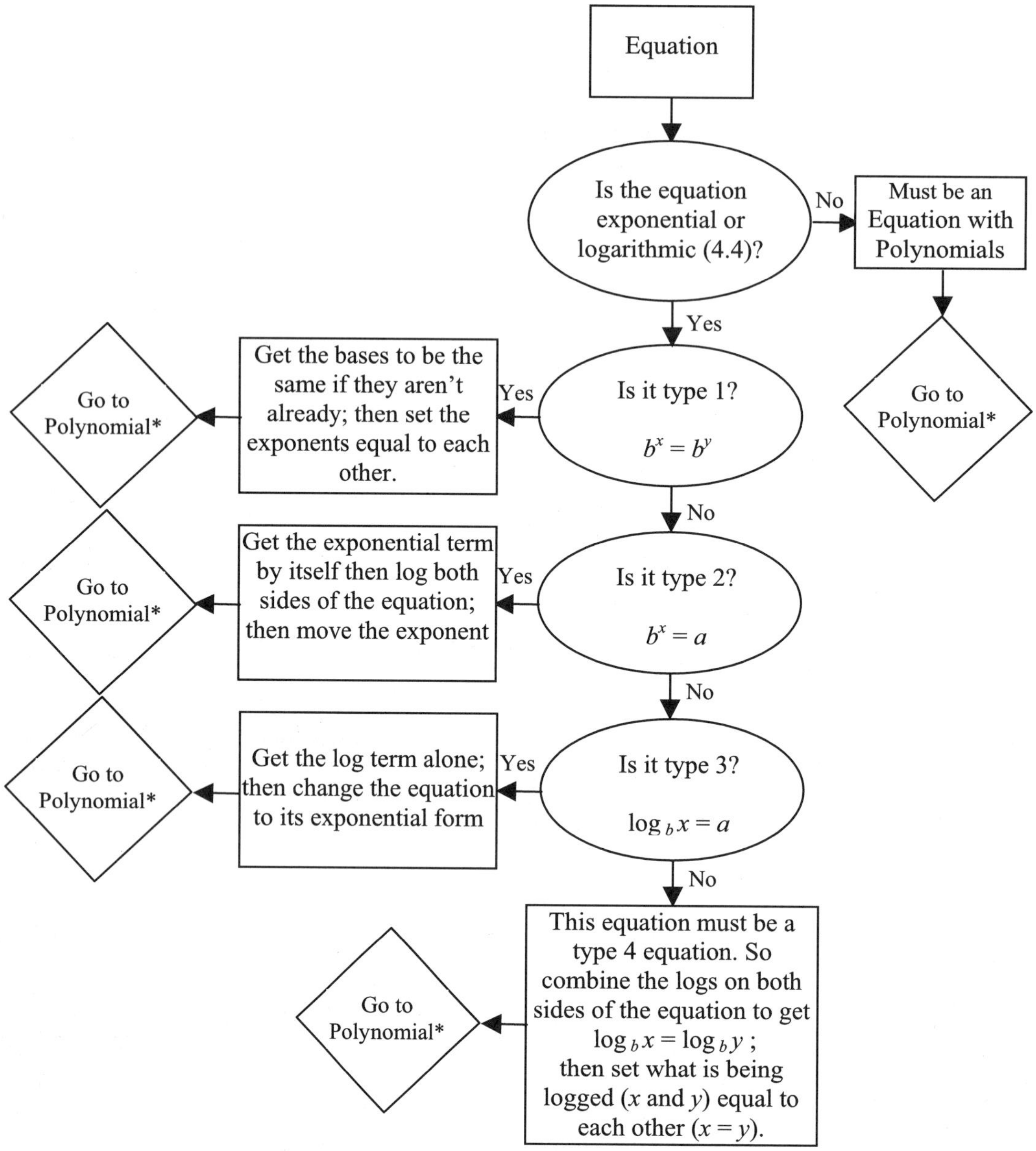

Answer Q4: *No. When you put –4 into the equation, you get* $\log_3(-4-5) = \log_3(-9)$.

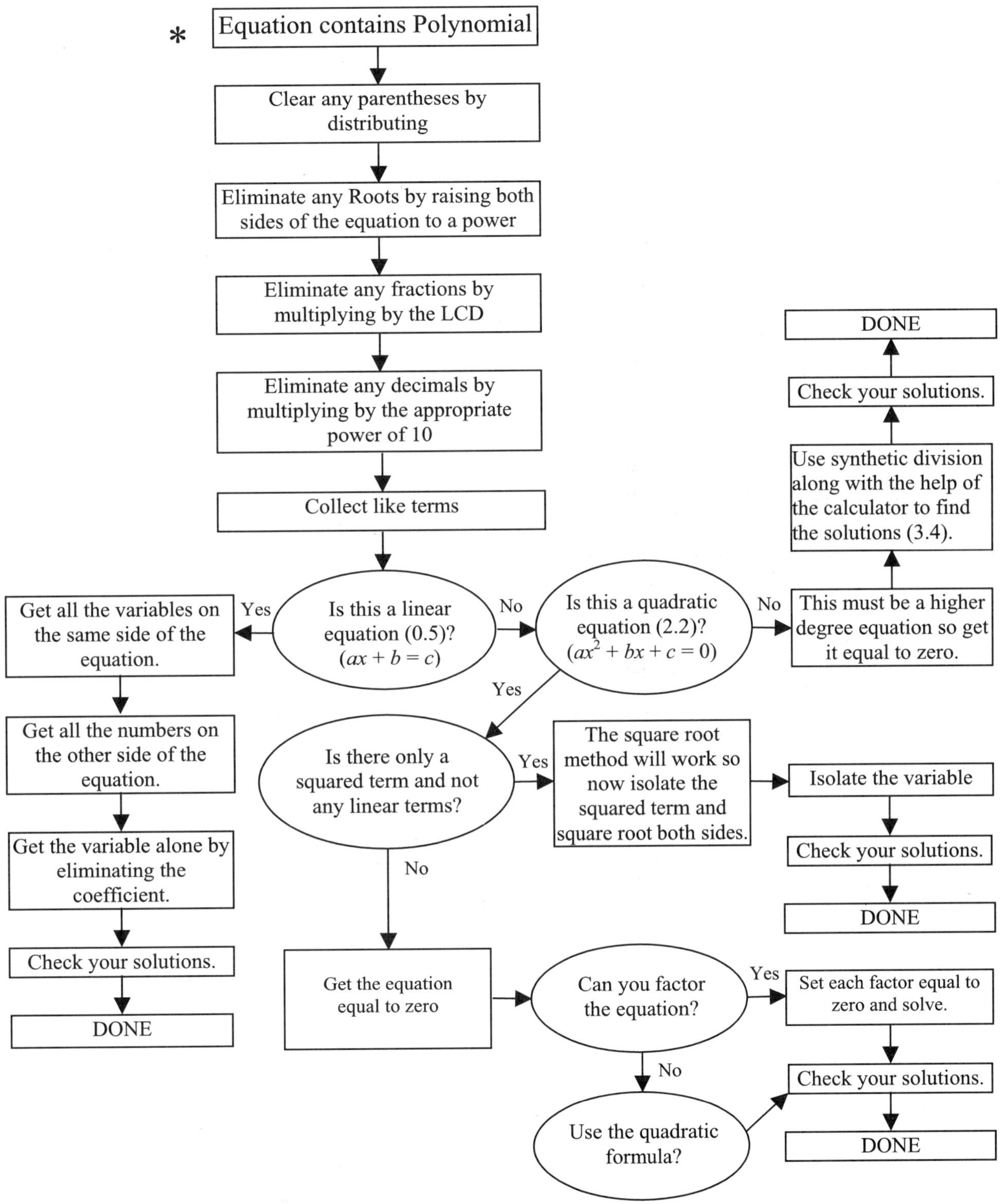
*
Equation contains Polynomial
Clear any parentheses by distributing
Eliminate any Roots by raising both sides of the equation to a power
Eliminate any fractions by multiplying by the LCD
Eliminate any decimals by multiplying by the appropriate power of 10
Collect like terms
Is this a linear equation (0.5)? ($ax + b = c$)
Yes
Get all the variables on the same side of the equation.
Get all the numbers on the other side of the equation.
Get the variable alone by eliminating the coefficient.
Check your solutions.
DONE
No
Is this a quadratic equation (2.2)? ($ax^2 + bx + c = 0$)
No
This must be a higher degree equation so get it equal to zero.
Use synthetic division along with the help of the calculator to find the solutions (3.4).
Check your solutions.
DONE
Yes
Is there only a squared term and not any linear terms?
Yes
The square root method will work so now isolate the squared term and square root both sides.
Isolate the variable
Check your solutions.
DONE
No
Get the equation equal to zero
Can you factor the equation?
Yes
Set each factor equal to zero and solve.
Check your solutions.
DONE
No
Use the quadratic formula?

SECTION 4.4 HOMEWORK

(1–26) Solve each exponential equation (For any irrational answers approximate to 3 decimal places.):

1. $3^{2x} = 27$
2. $2^{5x} = 32$
3. $5^{x+3} = \frac{1}{25}$
4. $4^{x-2} = \frac{1}{64}$
5. $3^{2x-3} = 81$
6. $2^{3x+2} = 64$
7. $4^{2x-1} = 8^{3x-2}$
8. $9^{2x+3} = 27^{3x+1}$
9. $5^{3x} = 23$
10. $7^{2x} = 81$
11. $3^{x+2} = 19$
12. $4^{x-3} = 93$
13. $2^{2x-1} = 51$
14. $6^{3x+2} = 29$
15. $e^{3x+2} = 15$
16. $e^{2x-5} = 59$
17. $3 \cdot 5^{2x+1} = 27$
18. $5 \cdot 3^{3x-1} = 50$
19. $15 \cdot e^{2x-3} = 45$
20. $9 \cdot e^{3x+2} = 54$
21. $5000(1+.09)^x = 10000$
22. $4000(1+.08)^x = 100000$
23. $3000e^{.08x} = 9000$
24. $4000e^{.10x} = 40000$
25. $3000\left(1+\frac{.08}{12}\right)^{12x} = 9000$
26. $4000\left(1+\frac{.10}{4}\right)^{4x} = 40000$

(27–60) Find all real number solutions for the following logarithmic equations:

27. $\log_3(2x - 1) = \log_3 9$
28. $\log_5(5x - 2) = \log_5 21$
29. $\log(2x + 15) - \log(x + 2) = \log 3$
30. $\ln(3x + 2) - \ln(x - 3) = \ln 5$
31. $\ln(3x - 4) - \ln(x + 2) = \ln 4$
32. $\log(2x - 3) - \log(3x + 2) = \log 9$
33. $\log_5(2x - 1) + \log_5(x) = \log_5 6$
34. $\log_4(3x - 2) + \log_4(x) = \log_4 5$
35. $\log(x + 2) - \log(2x - 3) = \log(x + 3)$
36. $\ln(2x - 1) - \ln(3x - 2) = \ln(x + 2)$
37. $\ln(3x - 1) + \ln(2x + 3) = \ln(x + 2)$
38. $\log(x + 3) + \log(5x - 1) = \log(x + 2)$
39. $\ln(2x - 1) - \ln(3x + 2) = \ln(x + 3)$
40. $\log(3x + 1) - \log(2x + 3) = \log(x + 5)$
41. $\log(3x - 1) + \log(2x + 3) = \log(x - 5)$
42. $\ln(2x + 1) + \ln(5x + 3) = \ln(x - 1)$
43. $2\log_2(x) - \log_2(4) = \log_2(2x + 5)$
44. $2\log(x) - \log 3 = \log(3x + 12)$
45. $\log_3(2x - 1) = 3$
46. $\log_2(3x - 1) = 3$
47. $\log_5(2x + 9) = 1$
48. $\log_4(2x + 3) = 2$
49. $\log(5x - 3) = 1$
50. $\log(12x + 3) = 2$
51. $\log_3(x + 2) - \log_3(2x - 3) = 1$
52. $\log_4(x + 3) - \log_4(x - 2) = 2$
53. $\log_2(x - 2) + \log_2(x) = 3$
54. $\log_4(x + 12) + \log_4(x) = 3$
55. $\log_3(2x - 1) + \log_3(x + 2) = 2$
56. $\log_2(x + 3) + \log_2(x + 4) = 4$
57. $\log_2(3x - 1) - \log_2(2x + 3) = 1$
58. $\log_3(2x + 1) - \log_3(x + 2) = 2$
59. $\log_5(2 - x) + \log_5(x) = 1$
60. $\log_3(5 - x) + \log_3(x) = 2$

(61-70) Mathematical modeling that uses logarithmic equations.

61. In chemistry the pH of a substance is defined by the equation $pH = -\log[H^+]$, where H^+ is the hydrogen ion concentration in moles per liter.

  a. If the hydrogen ion concentrate (H^+) of tap water is 10^{-7}, what is the pH?

  b. If the hydrogen ion concentrate (H^+) of orange juice is 2.8×10^{-4}, what is the pH (approximately)?

  c. If the pH of milk is 6.4, what is the hydrogen ion concentrate (approximately)?

  d. If the pH of orange juice is 3.2, what is the hydrogen ion concentrate (approximately)?

62. The magnitude R, measured on the Richter scale, of an earthquake of intensity I is defined by the equation $R = \log\frac{I}{I_0}$, where I_0 is a minimum intensity used for comparison.

  a. What is the magnitude (R) of an earthquake with intensity of $10^{4.5} \cdot I_0$?

  b. What is the magnitude (R) of an earthquake with intensity of $10^{6.5} \cdot I_0$?

  c. What is the intensity $I \cdot I_0$ of an earthquake that has a magnitude of 3?

  d. What is the intensity $I \cdot I_0$ of an earthquake that has a magnitude of 5?

  e. How many times stronger is an earthquake of magnitude 5 than an earthquake of magnitude 3?

63. $t = k\ln\left(\frac{T_0 - C}{T - C}\right)$ is the formula for Newton's law of cooling. t equals the time it will take for an object to cool to a temperature T; T_0 is the initial temperature; C is the temperature of the surrounding air, and k is a constant.

  a. How long will it take for a hot cup of tea with an initial temperature of 200° to cool to 90° if the room temperature is 70° and $k = 30$ (time is measured in minutes)?

  b. The coroner determines the time of death using the above formula. The temperature of a body is 83.5°and the room temperature is 78°. The normal temperature of a body is 98.6° and $k = 12$ (for time measure in hours). How long has the victim been dead?

  c. In part (b) above, if the body was found 6 hours after death, what would the approximate temperature of the body be?

64. The population of fish in a lake is given by $P(t) = 10000+2000\ln(t+1)$, where t is the time in months after the first stocking and $P(t)$ is the number of fish in the lake.

  a. Approximately how many fish will be in the lake after 24 months?

  b. Approximately how many months will it take for there to be 18,000 fish?

65. Four African snails were introduced to the Florida Everglades. The number of snails present in the Florida Everglades t months after the introduction of the African snails is given by the function $S(t) = 4(2)^{.8t}$.

  a. Approximately how many snails will be in the Florida Everglades after 8 months?

b. Approximately how many months before there would be 20,000 snails?

66. Ten bacteria are placed in a jar and allowed to grow. The number of bacteria in the jar after t minutes is given by the function $B(t) = 10e^{.13t}$.

a. Approximately how many bacteria in the jar after 10 minutes?

b. Approximately how many minutes must elapse before there are 120 bacteria in the jar?

67. \$10,000 is invested at 8% compounded continuously. $A(t) = 10000e^{.08t}$ is the function that gives the amount of money, $A(t)$, that will be in the account after t years.

a. Approximately how much money will be in the account after 20 years?

b. Approximately how many years will it be before the account is worth \$50,000?

68. \$20,000 is invested at 9% compounded monthly. $A(t) = 20000(1+\frac{.09}{12})^{12t}$ is the function that gives the amount of money, $A(t)$, that will be in the account after t years.

a. Approximately how much money will be in the account after 20 years?

b. Approximately how many years will it be before the account is worth \$100,000?

69. Na–24 has a half–life of 15 hours. You start out with 10 grams of Na–24; the following function gives the amount of Na–24 remaining after time t in hours. $R(t) = 10e^{-.04621t}$.

a. Approximately how much is left after 10 hours?

b. Approximately how long will it be before there are 2 grams left?

70. CS–137 has a half–life of 30 years. You start out with 100 grams of CS–137; the following function gives the amount of CS–137 remaining after time t in years. $R(t)=100e^{-.0231t}$.

a. Approximately how much is left after 50 years?

b. Approximately how long will it be before there are 10 grams left?

Section 4.5 Applications of Compound Interest

We have now covered all the skills needed to solve almost any equation we might encounter. Consequently, for the next several sections and the whole next chapter, we are going to look at how algebra is used in many different parts of life. In this section we are going to start a discussion about how algebra is an integral part of the financial world.

Discussion 1: Investing

You have \$1,000 in the bank that makes 6% interest. Let's investigate much money will be in the bank at the end of one year. Two years? Three years? Four years?

First, we need to remember how to compute interest. Interest equals principle (the amount you start with) times the interest rate (in decimal form). So, for this example at the end of one year, you would have the \$1,000 you started with, plus the interest you made (\$1000(0.06)=\$60) which would be \$1,060. Now the \$1,060 would be the principle for the second year (the amount you start the second year with) and so on for years 3 and 4. This chart illustrates what's happening.

Year 1	$\$1000 + \$1000(0.06) = \$1000(1 + 0.06) = \underline{\$1000(1.06)} = \underline{\$1060}$ (factored out \$1000)
Year 2 But \$1060 = \$1000(1.06) so,	$\$1060 + \$1060(0.06) = \$1060(1 + 0.06) = \$1060(1.06) = 1123.60$ $\underline{\$1123.60} = \$1060(1.06) = \$1000(1.06)(1.06) = \underline{\$1000(1.06)^2}$
Year 3 But as in year two we can rewrite to get	$\$1123.60(1.06) = \$1191.016$ $\underline{\$1191.016} = \$1123.60(1.06) = \$1000(1.06)^2(1.06) = \underline{\$1000(1.06)^3}$
Year 4 Again we can rewrite this with the original \$1,000 in the formula.	$\$1191.016(1.06) = \$1262.48$, rounded to the nearest penny $\underline{\$1262.48} = \$1191.016(1.06) = \$1000(1.06)^3(1.06) = \underline{\$1000(1.06)^4}$

So to summarize, we see the following:

Year 1	$\$1060 = \$1000(1.06)^1$
Year 2	$\$1123.60 = \$1000(1.06)^2$
Year 3	$\$1191.016 = \$1000(1.06)^3$
Year 4	$\$1262.48 = \$1000(1.06)^4$

Question 1: From the pattern that is visible in the table above, write the equation for finding the money we would have t years from the time we put \$1,000 in the bank at 6% interest.

What we have just discovered is called the **Annual Compound Interest Formula**.

The annual compound interest formula is $\boldsymbol{A = P(1 + r)^t}$ where A = final amount, P = principle (start amount), r = interest rate (decimal form), t = years invested

From here we can get more general by asking, "What would the formula be if the interest is not compounded annually"? For example, what if the interest is compounded every month? To figure out how much money we would have at the end of each month, let's make a few changes to example 1 and see what would happen:

Month 1	$\$1000(1 + \frac{0.06}{12}) = \underline{\$1000(1.005)} = \underline{\$1005}$ (the interest is only 1/12th of the whole 6% since only one month of the year has gone by. The interest rate is an annual rate.)
Month 2	$\$1005(1.005) = \$1010.025$ so, $\underline{\$1010.025} = \$1000(1.005)(1.005) = \underline{\$1000(1.005)^2}$
Month 3	$\$1010.025(1.005) = \underline{1015.075} = \underline{\$1000(1.005)^3}$
Month 12 (So by the time we get to the twelfth month, we have:)	$\$1000(1.005)^{12} = 1061.68$ rounded to the nearest penny

From this example we can see that if we are compounding more often than annually, we have more money at the end of one year (\$1060 verses \$1061.68). Also we notice that our formula will have to be adjusted. First, we must divide the interest rate by the number of times we compound per year. Second, we must multiply the years by the number of times we compound each year. In this example we were compounding monthly, so we had to divide the 6% by 12 and multiply the one-year by 12.

So in general, our formula is as follows:

The general **Compound Interest Formula** is $A = P\left(1+\frac{r}{n}\right)^{nt}$ where A = final amount, P = principle (start amount), r = interest rate (decimal form), t = years invested, n = number of compound periods per year.

Example 1: Compounded Weekly

Find the final amount of money you would have if you invested \$3000 at 5.6% annual interest for 6 years compounded weekly.

Solution:

We are looking for A the final amount. We know that $P = \$3000$, the amount invested, $r = 0.056$, the interest rate, $t = 6$, the years, and $n = 52$, since there are 52 weeks in a year. So we plug in what we know into our compound interest formula and get:

$$A = \$3000\left(1+\frac{0.056}{52}\right)^{52(6)} = \$4197.26$$

Question 2: How much money will you have in 10 years if you put \$5,000 in a Certificate of Deposit (CD) that is compounded monthly and has a 7.75% annual interest rate?

At this point you may be wondering, " Just how often can I compound - daily, hourly, by the second?" The answer is "continuously."

Answer Q1: *$A(t) = \$1000(1.06)^t$*

Discussion 2: Compound Continuously

Let's put a table together, with some concrete numbers and see what happens. We will let $t = 1$ and $r = 0.1$ and $P = 100$.

Yearly	$A = 100(1 + \frac{0.1}{1})^1$	=	110
Monthly	$A = 100(1 + \frac{0.1}{12})^{12}$	=	110.4713
Daily	$A = 100(1 + \frac{0.1}{365})^{365}$	=	110.515578
Hourly	$A = 100(1 + \frac{0.1}{8760})^{8760}$	=	110.5170287
Now try this formula.	$A = 100e^{0.1}$	=	110.5170918

Notice how, as we go down the table, the numbers get closer and closer to the one at the bottom. Let's walk through why this works out the way it does and why e shows up in the formula.

We begin with the formula for compounding, then we raise the parentheses part by both a r and $1/r$.	$A = P\left(1+\frac{r}{n}\right)^{nt} = P\left(\left(1+\frac{r}{n}\right)^{nt}\right)^{r\frac{1}{r}}$
This simplifies to the following by rearranging the exponents.	$A = P\left(1+\frac{r}{n}\right)^{\frac{nrt}{r}} = P\left(\left(1+\frac{r}{n}\right)^{\frac{n}{r}}\right)^{rt}$

Now if we choose to let $m = \frac{r}{n}$, we get	$A = P\left((1+m)^{\frac{1}{m}}\right)^{rt}$
If n is getting larger then m must be getting smaller (close to zero).	$(1+m)^{\frac{1}{m}}$ as m gets close to zero is exactly the definition of e stated in section 4.2
So, our formula now looks like	$A = Pe^{rt}$

We have proven now that the **Continuous Compound Formula** is

$A = Pe^{rt}$ where A = final amount, P = principle (start amount), r = interest rate (decimal form), and t = years invested.

Question 3: Given \$1000 invested at 6% compounded continuously, find the amount of money you will have accumulated by the end of one year.

Notice how we would have more money at the end of the year than either compounding annually (example 1, \$1060) or compounding monthly (\$1061.68). The more often you compound, the faster your money grows. Of course the biggest change occurred when we went from annually to monthly (\$1.68); after that we only gained a few cents. Let's look at a graph of example 1.

<table>
<tr>
<td>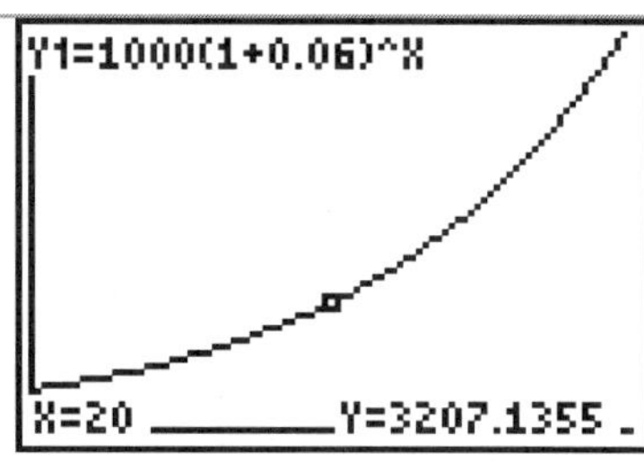
</td>
<td>Xmin = 0
Xmax = 40
Xscl = 0
Ymin = 0
Ymax = 10000
Yscl = 0</td>
</tr>
</table>

Notice that the variable is time, which is in the exponent position, so we get a graph that has the shape of a growth exponential function. We could compare this to compounding monthly which has this for a graph.

<table>
<tr>
<td>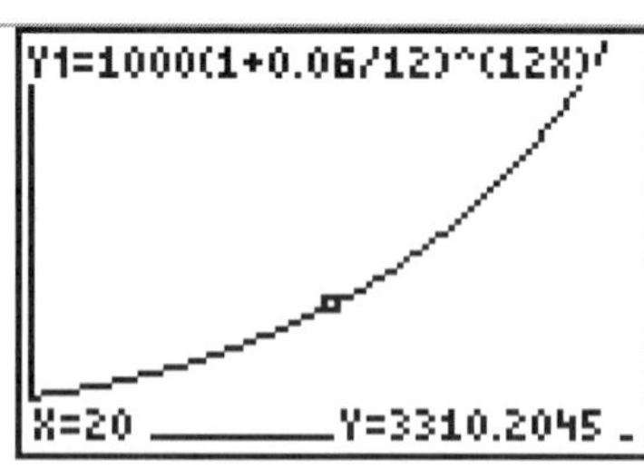
</td>
<td>Xmin = 0
Xmax = 40
Xscl = 0
Ymin = 0
Ymax = 10000
Yscl = 0</td>
</tr>
</table>

From the trace key on the calculator we can see that after 20 years annually, we would have \$3,207.14, whereas monthly we would have \$3,310.20, a difference of \$103.06.

Example 2: Difference Between Interest Rates

How much of a difference will there be between 5.5% interest and 6.25% interest if we are investing \$2000 compounded monthly for 5 years?

Solution:

First, let's type into our calculators the compound interest formula with r as the variable. Our formula is $A = \$2000(1 + r/12)^{12(5)}$. But of course our calculators only use X and Y as variables, so we'll call A the variable Y and r the variable X in the calculator.

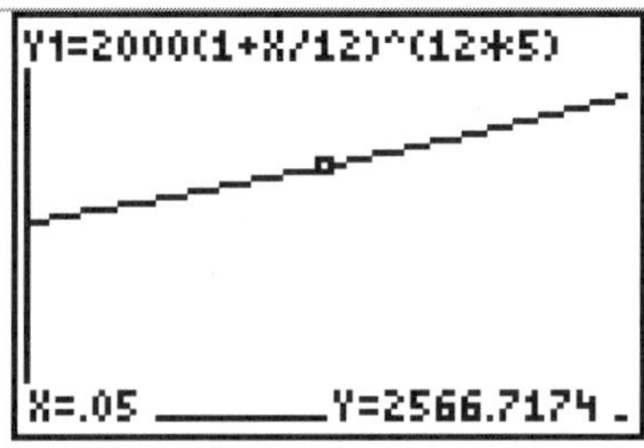

Xmin = 0
Xmax = 0.1
Xscl = 0
Ymin = 0
Ymax = 4000
Yscl = 0

Answer Q2:
$A = 5000 (1+\frac{.0775}{12})^{12 \cdot 10} = \$10,825.95$

Notice that we used Xmax as 0.1 since we weren't going past an interest rate of 10%. Also we adjusted the Ymax in order to get a good graph of the function. You should also notice that this graph looks a little curved.

Question 4: Is this graph a graph of an exponential or polynomial function?

So, to answer the question, "What is the difference between 5.5% and 6.25% interest?" we will use the value key (remember: push 2nd trace then 1 on the TI83, or GRAPH, MORE, then MORE again, then F1 on the TI86) on our calculators and get the following results.

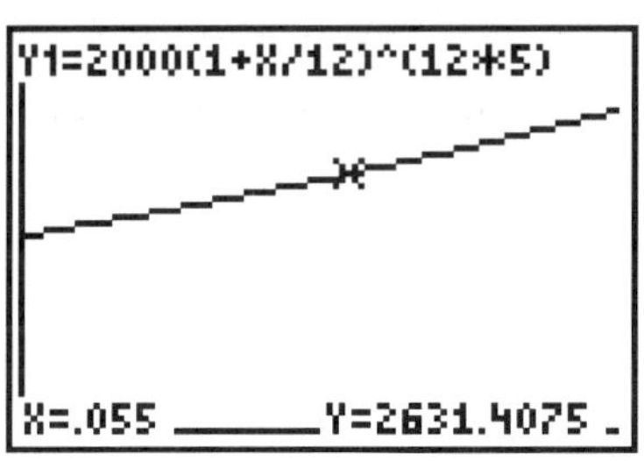

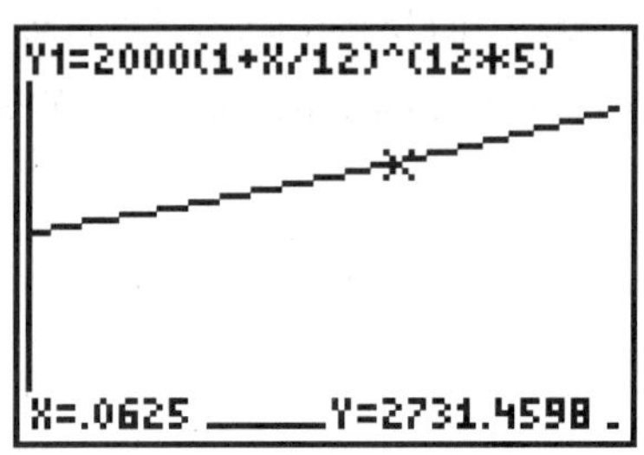

The difference then is \$2731.46 – \$2631.41 = \$100.05.

Discussion 3: Principle Question

What if we wanted to know how much we needed to invest now, if we could invest it at 8% interest compounded annually, so that in 10 years we would have $20,000?

From reading the problem, we can see that our question is about the principle (the amount to start with). Our formula would be $A = P(1 + 0.8)^{10}$ and we could look at a graph on our calculators by letting Y be the A, which we want to be $20,000 and X the P. The graphs would be as follows:

Answer Q3: $A = 1000e^{0.06(1)}$ $= \$1061.84$

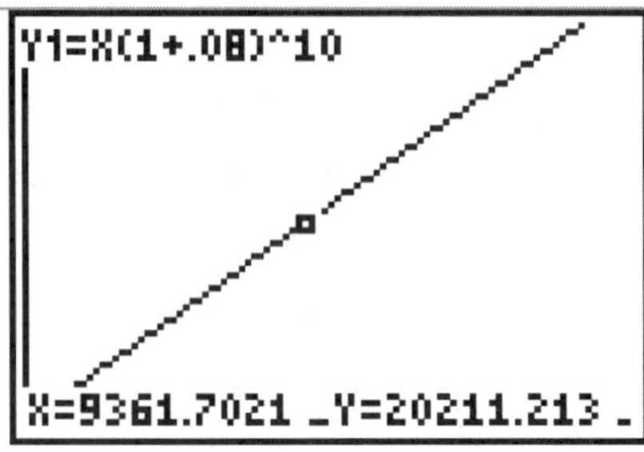	Xmin = 0 Xmax = 20000 Xscl = 0 Ymin = 0 Ymax = 40000 Yscl = 0 Notice this looks like a line. In fact it is because we see X raised to the 1 power! X^1 times $(1+.08)^{10}$ is just a value.
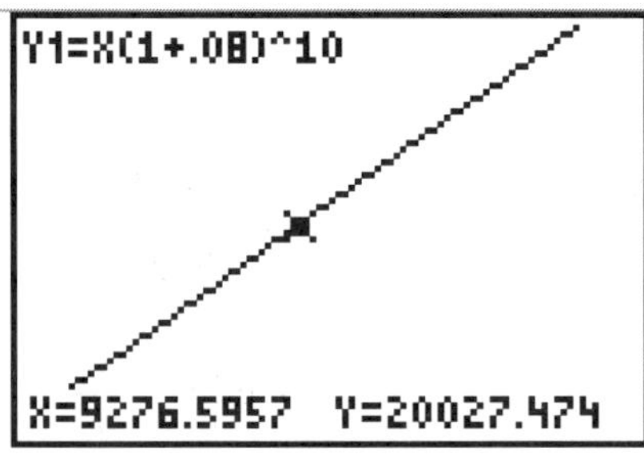	We have zoomed in by a factor of 10 and then traced closer to Y = 20,000 since that is our aim after 10 years of saving. Let's zoom in one more time.
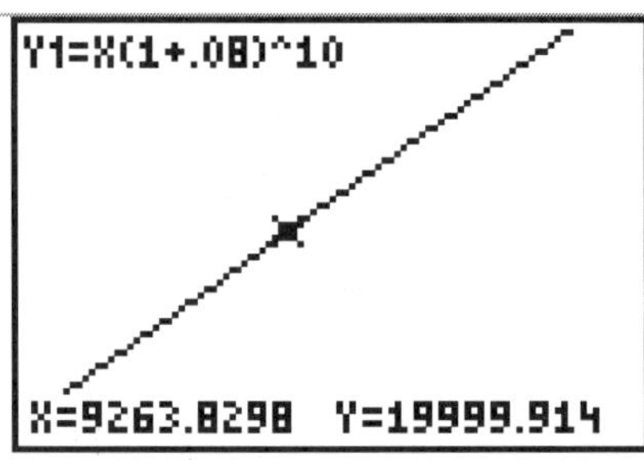	It looks like we've gotten pretty close now. It appears that if we invest $9,263.83 today, in 10 years at 8% compounded annually, we will have about $20,000.

Now of course, we could have solved this problem by hand. Let's do that and see if we come up with the same answer.

Our initial formula	$20000 = P(1 + 0.08)^{10}$
Divide both sides by the parentheses.	$\dfrac{20000}{(1+0.08)^{10}} = P$
So our answer is	$P = \$9263.87$

As you can see, our two answers aren't quite the same. Doing things by hand will be more accurate; of course depending on the question, being exactly right may not be necessary in which case the graphing calculator does a good job of approximating solutions for us.

Example 3: Doubling

How long will it take an investment of $3500 to double at 11% interest compounded daily?

Solution:

First, if we are looking for when $3500 is doubled, then we are looking for when we will have $7000. So we now know that A = $7000; P = $3500; $r = .11$ and $n = 365$. Let's work this out by hand first.

Our formula to start with is	$7000 = 3500(1 + .11/365)^{365t}$
This is an exponential equation since the variable t is in the exponent position. Next, we have to decide if this is a No Log, Some Log or All Log type of problem. In this case it's a No Log type so we need to use some algebra manipulations to get this into the form $a = b^x$. So, we first divide by 3500.	$\frac{7000}{3500} = \left(1+\frac{.11}{365}\right)^{365t}$ $2 = \left(1+\frac{.11}{365}\right)^{365t}$
Now that we have the proper form, let's take the log of both sides of the equation.	$\log 2 = \log\left(1+\frac{.11}{365}\right)^{365t}$
By using the properties of logs we get	$\log 2 = 365\,t\log\left(1+\frac{.11}{365}\right)$
Finally, we just need to divide by everything that is multiplied by t and we're done.	$\frac{\log 2}{365\log\left(1+\frac{.11}{365}\right)} = t$
If we use our calculator and round off, we get	$t \approx 6.3$ years

Answer Q4: Polynomial function since we have a variable (x) raised to a number power(12 • 5).

To do this problem on a graphing calculator, we would use the formula $A = 3500(1 + \frac{.11}{365})^{365t}$, then trace and zoom near $7000. Here is what we would see.

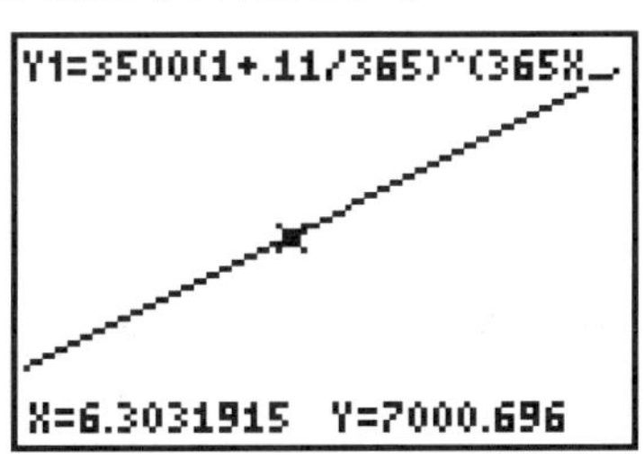

The calculator shows that the number of years (X) needed for our money to double would be about 6.3.

Section Summary

We have two important formulas from this section. They are:

1. The compound interest formula $A = P\left(1+\dfrac{r}{n}\right)^{nt}$
2. The continuous compound interest formula $A = Pe^{rt}$

In both of these formulas A is the final amount; P is the starting amount called Principle; r is the interest rate in decimal form; n is the number of times you compound in a year and t is the number of years the money will be invested. If our variables are A and t, then these do fit our definition of an exponential function; that is $y = ab^x$. So we would say that A is a function of t and the formulas written in function notation would be $A(t) = P\left(1+\dfrac{r}{n}\right)^{nt}$ and $A(t) = Pe^{rt}$. Here are two examples similar to ones found in this section.

In question 2 of this section, if instead of 10 years, we didn't know how long the time would be, then the formula would look like this in function notation form.	In question 3 of this section, if instead of one year, we didn't know how long the time would be, then the formula would look like this in function notation form.
$A(t) = 5000(1+\dfrac{.0775}{12})^{12*t}$	$A(t) = 1000e^{\mathbf{0.06}(t)}$ ↓
$A(t) = 5000(1.00646)^{12t}$	$A(t) = 1000(\mathbf{1.06184})^{t}$
↖ ↑ ↑ ↗ $y = a\,(b)^{x}$	↖ ↑ ↑ ↗ $y = a\,(b)^{x}$

SECTION 4.5 HOMEWORK

1. Fill in the missing table values:

Initial Investment	Compounded	Annual Percentage Rate	Doubling time	Amount after 10 years	Amount after 20 years	Amount after 40 years
$10,000	Yearly	8%				
$30,000	Monthly	9.5%				
	Quarterly	10%		$30,000		
	Weekly	7.5%			$50,000	
$12,000	Yearly		8.5 years			
$6,000	Monthly		5.8 years			
$10,000	Yearly			$27,200		
$5,000	Monthly			$15,000		

Last row is a challenge problem

2. Fill in the missing table values for continuous compound interest:

Initial Investment	Annual Percentage Rate	Doubling time	Amount after 10 years	Amount after 20 years	Amount after 40 years
$40,000	9%				
$25,000	7.5%				
	8%		$11,227.70		
	9.5%		$25,857.10		
$80,000		9.9 years			
$10,000		6.6 years			
$20,000				$60,000	
$5,000				$25,000	
			$10,000	$24,596	

Last row is a challenge problem

(3-26) Mathematical modeling that use compound formulas.

3. At their births you invest $5,000 for your grandchildren in an account that pays 9% interest compounded monthly. How much will be in the account for their college education at age 18?
4. At their births you invest $10,000 for your grandchildren in an account that pays 8.5% interest compounded weekly. How much will be in the account for their retirement at age 55?
5. At age 30 you invest $15,000 for your retirement in an account that pays 8.5% interest compounded continuously. If the fund is for retirement at age 60, how much will be in the retirement fund when you are ready to retire?

6. You invest \$5000 at 9.8% compounded continuously. How much will you have as a down payment for a new car 6 years later?
7. You bought a new house and the lender has agreed to a carry-back of \$5,000 due in 5 years. How much money would you have to invest at 9.5% compounded continuously to pay off the \$5,000 in 5 years?
8. Your brother has agreed to give you an interest free loan of \$10,000 to buy a business, if you will pay him back in 10 years. How much would you need to invest now at 8.9% compounded monthly to have the \$10,000 in 10 years to pay back your brother?
9. How long will it take for a \$10,000 investment to become \$1,000,000?
   a. Invested at 8% compounded monthly
   b. Invested at 10% compounded monthly
   c. Invested at 8% compounded continuously
   d. Invested at 10% compounded continuously
10. How long will it take for a \$25,000 investment to become \$100,000?
    a. Invested at 9.5% compounded weekly
    b. Invested at 12.25% compounded weekly
    c. Invested at 9.5% compounded continuously
    d. Invested at 12.25% compounded continuously
11. How long will it take an investment that is compounded yearly to double if the interest rate is
    a. 8%
    b. 9%
    c. 10%
    d. 12%
12. How long will it take an investment that is compounded monthly to double if the interest rate is
    a. 8.5%
    b. 9.5%
    c. 10.5%
    d. 12.5%
13. How long will it take an investment that is compounded continuously to triple if the interest rate is
    a. 7%
    b. 9%
    c. 11%

14. How long will it take an investment that is compounded continuously to double if the interest rate is
    a. 6.5%
    b. 8.5%
    c. 10.5%
15. The inflation rate since 1960 has averaged 4.8% a year. In 1960 the wage of a high school teacher averaged $4,860. In order to keep up with inflation what would the teacher's wages have to be in the year 2010? (Hint: This would be compound interest compounded yearly.)
16. **Using the information given in problem 15**: The average cost of a home in the U.S. in 1960 was $12,500. Due to inflation, what would we expect the average cost of a home in 2005 would be?

(17–26) **Approximate each solution with your calculator:**

17. If $5,000 is compounded semi-annually at 9.5%, how long will it take for it to become $9,000?
18. If $5,000 is compounded continuously at 9.5%, how long will it take for it to become $9,000?
19. Compounded continuously at 10.3%, how long will it take to double your money?
20. Compounded daily at 10.3%, how long will it take to triple your money?
21. How much money needs to be invested at 8.9% compounded monthly to become $15,000 after 15 years?
22. How much money needs to be invested at 7.5% compounded continuously to become $38,000 after 10 years?
23. What is the value of the yearly interest rate for $5,000 to become $8,000 when the investment is compounded continuously for 4 years?
24. What is the value of the yearly interest rate for $8,000 to become $20,000 when the investment is compounded quarterly for 10 years?
25. How much difference will there be between 8.3% interest and 10.5% interest if you are investing $3,000 compounded continuously for 8 years?
26. How much difference will there be between 9.25% interest and 10.25% interest if you are investing $5,000 compounded monthly for 10 years?

Section 4.6 Applications of Annuities and Amortization

In this section we'll continue to look at how exponential functions are used in the world of finance. We have discussed the compound interest formulas; now let's discuss two formulas called the Annuity formula and the Amortization formula.

ANNUITIES

First we will take a look at the annuity formula. An annuity is an account into which you make regular payments. Here is an example of an annuity.

Discussion 1: Making Regular Payments

You have a job and you have decided to save some money for when you will retire. You have decided to put \$50 out of every paycheck into a mutual fund (buying stocks in the stock market). Let's look at how your money might accumulate over a few paychecks, assuming you get paid every two weeks and that you are receiving a steady rate of return on your investments (10%).

Paycheck #1	$A = \$50$ deposited (A is the total money accumulated)
Paycheck #2 (\$50 deposited plus the original \$50 plus the interest on the original \$50) (PC – paycheck)	$A = \underline{\$50} + \underline{50(1 + .1/26)^1}$ ↑ PC#2 ↑ PC#1 ($P = \$50$, $r = .1$, $n = 26$ $nt = 26 \cdot \frac{1}{26} = 1$)
Paycheck #3 (another \$50 plus all the earlier amounts plus all the interest they have made)	$A = \underline{\$50} + \underline{\$50(1 + .1/26)^1} + \underline{\$50(1 + .1/26)^2}$ ↑ PC#3 ↑ PC#2 ↑ PC#1 compounded twice now
Paycheck #4	$A = \underline{\$50} + \underline{\$50(1 + .1/26)^1} + \underline{\$50(1 + .1/26)^2} + \underline{\$50(1 + .1/26)^3}$ ↑ PC#4 ↑ PC#3 ↑ PC#2 ↑ PC#1
Let's stop here and do some algebra. Multiply the equation by $(1 + .1/26)$; then subtract the old equation from this new one.	
$(1 + .1/26)A = 50(1 + .1/26)^1 + 50(1 + .1/26)^2 + 50(1 + .1/26)^3 + \mathbf{50(1 + .1/26)^4}$ $- \quad A = \mathbf{50} + 50(1 + .1/26)^1 + 50(1 + .1/26)^2 + 50(1 + .1/26)^3$ $(1 + .1/26)A - A = -50 + 50(1 + .1/26)^4$ (All the other terms canceled each other out)	
Factor each side of the equation.	$A[(1 + .1/26) - 1] = 50[-1 + (1 + .1/26)^4]$
Simplify the left side, rearrange the right.	$A(.1/26) = 50[(1 + .1/26)^4 - 1]$

Solve for A by dividing by (.1/26).	$A = \dfrac{50\left[\left(1+\dfrac{.1}{26}\right)^4 - 1\right]}{\dfrac{.1}{26}}$

Now we have created a formula for this example that will figure out how much money we have in our mutual fund after our 4th paycheck.

Question 1: What does A equal in the formula we have just found (use your calculator)?

Without having to go through all the algebra again to prove the general case, we think you can see the pattern from our little example above. So, here is the general form of the **Annuity Formula.**

$$A = R\left[\frac{\left(1+\frac{r}{n}\right)^{nt} - 1}{\frac{r}{n}}\right]$$

where A = the final amount accumulated, R = the amount of the regular payment, r = annual interest rate, n = compound periods per year and t = number of years you continue to make a regular payment.

In our last example the A, of course, was \$201.16; the R was \$50; the r was .1; the n was 26 and t was 4/26, since we had only made 4 of the 26 payments in the year.

Continuing discussion 1, we will look at how much money you might accumulate after 30 years of working.

Plug in our information into the annuity formula.	$A = 50\dfrac{\left[\left(1+\dfrac{.1}{26}\right)^{26*30} - 1\right]}{\dfrac{.1}{26}}$
Use your calculator to simplify.	50((1+.1/26)^(26*30)-1)/(.1/26) = \$246,613.73

You would have almost a quarter of a million dollars after 30 years, assuming that the stock market would average a rate of return on your money of 10%.

Question 2: How much of the \$246,613.73 did you actually deposit into your account?

Example 1: Annuity Formula

Graph the function $A(t) = 50\dfrac{\left[\left(1+\dfrac{.1}{26}\right)^{26t}-1\right]}{\dfrac{.1}{26}}$ then find $A(10)$, $A(20)$, $A(40)$.

Solution:

<table>
<tr><td>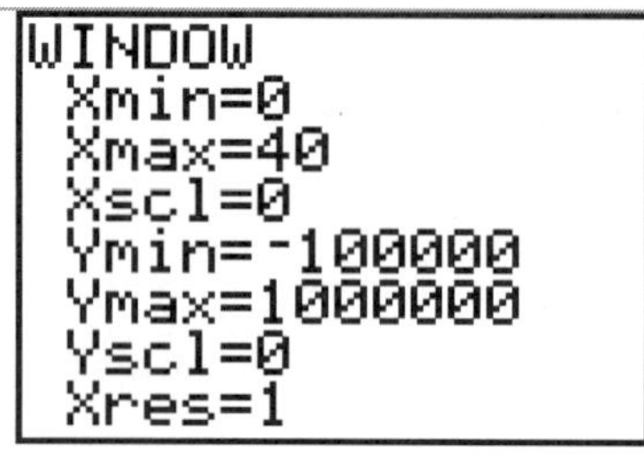
</td><td>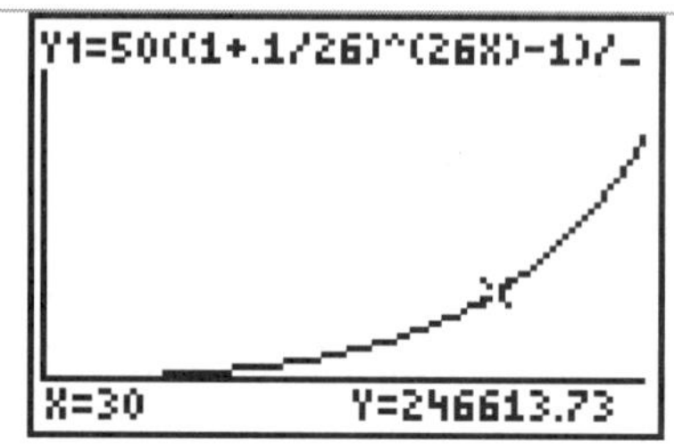
</td></tr>
<tr><td>Now use the evaluate function on the calculator to find $A(10)$, $A(20)$, $A(40)$.</td><td>$A(10) = \$\ 22{,}269.95$
$A(20) = \$\ 82{,}689.93$
$A(40) = \$691{,}350.93$</td></tr>
</table>

You should notice that the variable in this question was t and that it is in the exponent position. Hence we see the typical graph for a growth exponential function, and as mentioned in other sections, the further you go to the right the steeper the graph becomes. The greatest growth takes place at the right end of the graph.

So, in this real-life example, if we can get an early start saving money we can, in the end, accumulate great sums of money. What this means to you is that if you start saving for retirement at age 45, in 20 years you retire with only \$82,689.93, whereas if you start at age 25, in 40 years you retire with \$691,350.93.

Example 2: Late Savings

So what happens if you have waited too long already to start saving? How much would your regular payment have to be now at age 45, so that in 20 years you will have the same amount of money that you would have had if you had started at age 25, when paying \$50 a paycheck?

Solution:

We will look at two ways to answer this question - one by hand, the other with the calculator.

Set up the equation.	$\$691{,}350.93 = R\dfrac{\left[\left(1+\dfrac{.1}{26}\right)^{26*20}-1\right]}{\dfrac{.1}{26}}$		Type this equation into your calculator.	$\mathrm{Y1} = x\dfrac{\left[\left(1+\dfrac{.1}{26}\right)^{26*20}-1\right]}{\dfrac{.1}{26}}$

Simplify the fraction.	$691,350.93 = R(1653.7985)$	Trace and zoom until Y is the number we want.	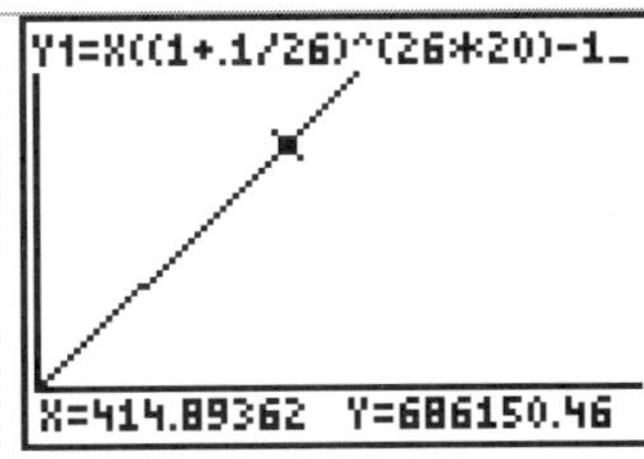
Divide by 1653.7985.	$418.04 = R	The answer is	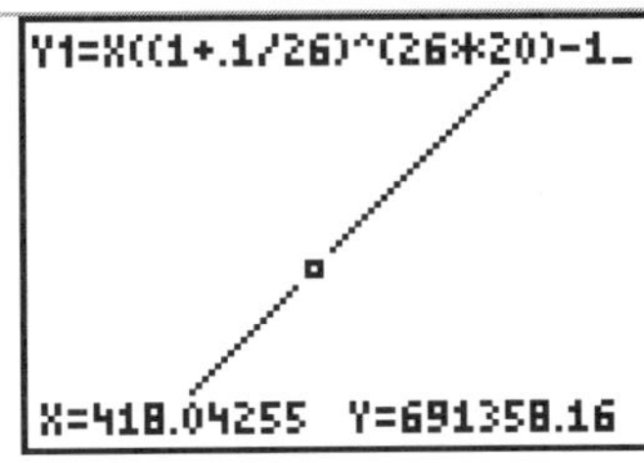

So it looks like if you wait until age 45 to start saving, you will have to take out of your paychecks over 8 times as much money ($\frac{418}{50} = 8.36$) as compared with starting at age 25. Also, let's look at the total amount deposited. If you started at age 25, the total deposits would be $52,000 (50*26*40) but if you start at 45, the total deposits would be $217,360 (418*26*20).

Answer Q1:
$A \approx \$201.16$

Question 3: How much would you need to deposit each paycheck to have $1,000,000 in 30 years if the interest rate is 10%?

Example 3: Reaching a Goal

If you are able to deposit $250 per paycheck at 10% interest and you would like to have at least $800,000 when you retire, then how long will you need to work?

Answer Q2:
$50 <u>times</u> 26 payments per year <u>times</u> 30 years = $39,000. The rest is interest!

Solution:

Let's do this one on the calculator.

Type in this equation for Y1.	Y1 = 250((1+.1/26)^(26X)-1)/(.1/26)
Graph with a window of: Xmin = 0 Xmax = 50 Xscl = 0 Ymin = -200000 Ymax = 1500000 Yscl = 0	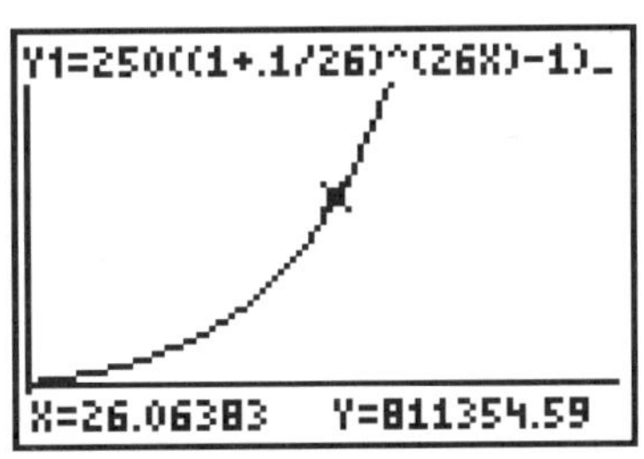

Zoom in.	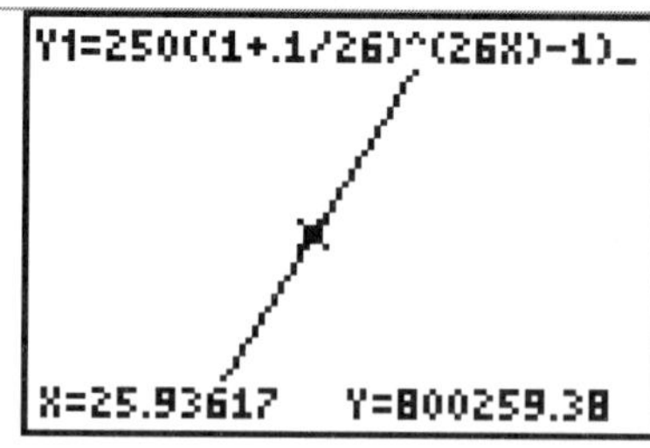

You could retire in roughly 26 years.

With the graphing calculator, you can let the R, or the, r or t be variables and answer all kinds of "What if?" questions using this annuity formula.

AMORTIZATION

Let's move to the next formula, the amortization formula. This formula is involved with loans. Let's do an example.

Discussion 2: Amortization Formula

Let's see how we would calculate our monthly car payment if we borrowed $20,000 at 7% interest for 5 years in order to buy a car.

First, we will introduce some new terms. From the compound interest formula $A = P\left(1+\frac{r}{n}\right)^{nt}$, we might call A the **future value (FV)** since it refers to the amount of money we will have in the future. We might refer to P as the **present value (PV)** because it's the amount of money we have right now. So one way to write the same formula is $FV = PV\left(1+\frac{r}{n}\right)^{nt}$. Likewise we could rewrite our annuity formula as $FV = R\left[\frac{\left(1+\frac{r}{n}\right)^{nt}-1}{\frac{r}{n}}\right]$. Let's think about what a loan is really all about. We borrow some money now (PV) and we make regular payments (annuity), then in the end, the bank has received a final sum of money from us (FV). So we simply need to make a substitution and then solve for R to figure out what our regular payments will need to be to buy a car.

Formulas involved PV = loan amount n = 12 (monthly) $t = 5$ FV and R are unknown for the moment.	$FV = \$20000\left(1+\frac{0.07}{12}\right)^{12(5)}$, $FV = R\left[\frac{\left(1+\frac{0.07}{12}\right)^{12(5)}-1}{\frac{0.07}{12}}\right]$
We don't know the FV so lets do a substitution. (The FV is being paid off by an annuity.)	$R\left[\frac{\left(1+\frac{0.07}{12}\right)^{12(5)}-1}{\frac{0.07}{12}}\right] = \$20000\left(1+\frac{0.07}{12}\right)^{12(5)}$
Solve for R since it's the only variable left. Also, it is what our question was asking about and probably one of our greatest concerns when buying a car.	$R = \frac{\$20000\left(1+\frac{0.07}{12}\right)^{60}}{\left[\frac{\left(1+\frac{0.07}{12}\right)^{60}-1}{\frac{0.07}{12}}\right]}$
To simplify first multiply the numerator and denominator by $\frac{0.07}{12}$.	$R = \frac{\$20000\left(1+\frac{0.07}{12}\right)^{60}}{\left[\frac{\left(1+\frac{0.07}{12}\right)^{60}-1}{\frac{0.07}{12}}\right]} \bullet \frac{\frac{0.07}{12}}{\frac{0.07}{12}} =$ $\frac{\$20000\left(\frac{0.07}{12}\right)\left(1+\frac{0.07}{12}\right)^{60}}{\left(1+\frac{0.07}{12}\right)^{60}-1}$
Now divide the numerator and denominator by $\left(1+\frac{0.07}{12}\right)^{12(5)}$.	$R = \frac{\frac{\$20000\left(\frac{0.07}{12}\right)\left(1+\frac{0.07}{12}\right)^{60}}{\left(1+\frac{0.07}{12}\right)^{60}}}{\frac{\left(1+\frac{0.07}{12}\right)^{60}-1}{\left(1+\frac{0.07}{12}\right)^{60}}} = \frac{\$20000\left(\frac{0.07}{12}\right)}{1-\left(1+\frac{0.07}{12}\right)^{-60}}$
Now use your calculator to find R.	R = 20000(.07/12)/(1 – (1 + .07/12)^(-60)) ≈ \$396.02

Answer Q3: $\$1000000 = R\left[\frac{\left(1+\frac{.1}{26}\right)^{26*30}-1}{\frac{.1}{26}}\right]$

$\Rightarrow$ *\$1000000 = R(4932.27)*

$\Rightarrow$

R = \$202.74

So, it appears that the monthly payment for our car is \$396.02. Hence, our car will end up costing us roughly \$23,761.20 (\$396.02 • 12 months • 5 years) in principle and interest. In the final equation that we used to find our car payment, we see that \$20,000 was the loan amount, 0.07 was the interest, 12 was the number of payments per year, and 60 was the number of payments total (n • t). From this example we will write the general **Amortization** formula as follows:

$$R = \frac{L\left(\frac{r}{n}\right)}{1-\left(1+\frac{r}{n}\right)^{-nt}}$$ where L = the loan amount, R = the amount of the regular payment, r = annual interest rate, n = compound periods per year and t = number of years you continue to make a regular payment.

Once again we have a practical formula that we could use in conjunction with our graphing calculators to answer just about any "What if?" questions we might have with regards to loans.

Question 4: What would a monthly house payment (principle and interest) be if our loan were \$120,000 at 6.5% interest for 30 years?

Example 4: Buying a House

If we know that we can afford a house payment of \$950, for the principle and interest amount, and the current interest rate for 30-year loan is 7.5%, then how big can our loan be?

Solution:

We will look at two ways to answer this question, one by hand the other with the calculator.

Set up the equation.	$\$950 = \frac{L\left(\frac{0.075}{12}\right)}{1-\left(1+\frac{0.075}{12}\right)^{-12(30)}}$	Type this equation into your calculator.	$Y1 = \frac{x\left(\frac{0.075}{12}\right)}{1-\left(1+\frac{0.075}{12}\right)^{-12(30)}}$
Simplify the fraction.	$\$950 = L(0.006992)$	Trace and zoom until Y is the number we want.	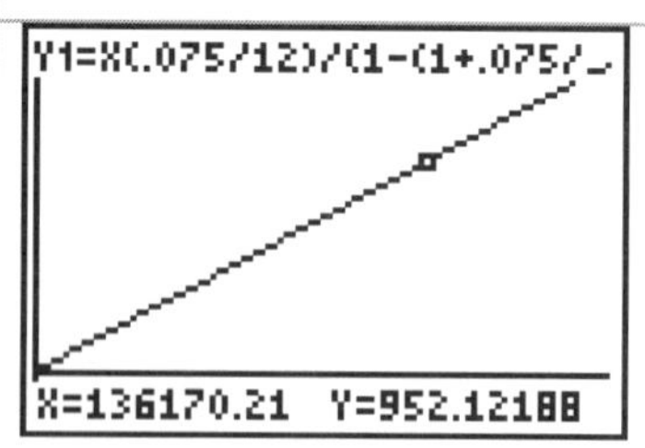

Divide by 0.006992.	$135,870 ≈ L$	The answer is	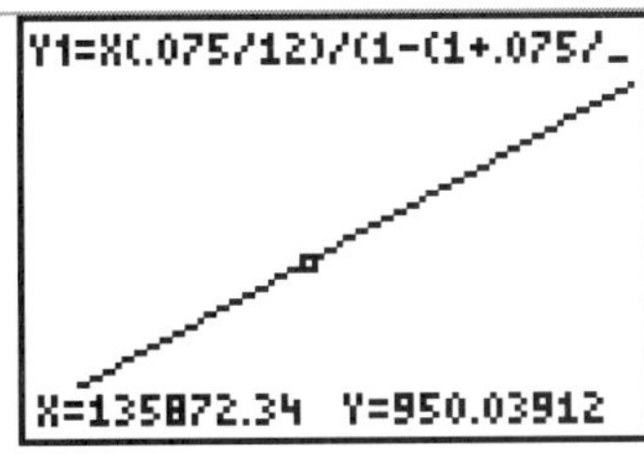

So it appears that we can afford to buy a house that would need a loan of up to a total of $135,870.

Example 5: Thirty-Year Verses a Fifteen-Year Loan

If we wanted to buy a house that required a loan of $150,000 at 7.25% interest with monthly payments, what would be the differences between using a 30-year loan and a 15-year loan?

Solution:

Let's graph the amortization formula and do some investigating.

Y1 = 150000(.725/12)/(1−(1+.725/12)^(-12X)) 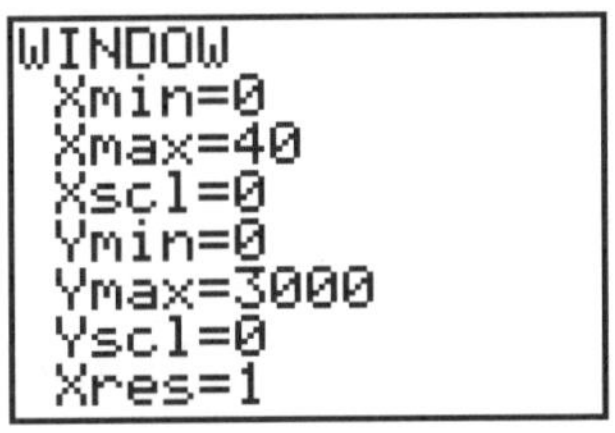	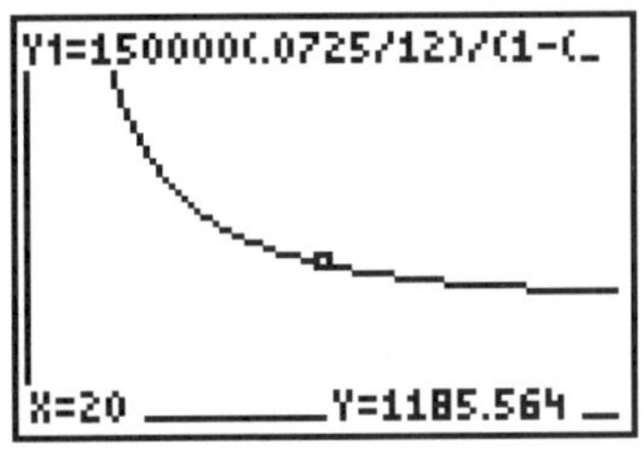
Use the evaluate function on your calculator to find the monthly payments for 15 and 30 year loans.	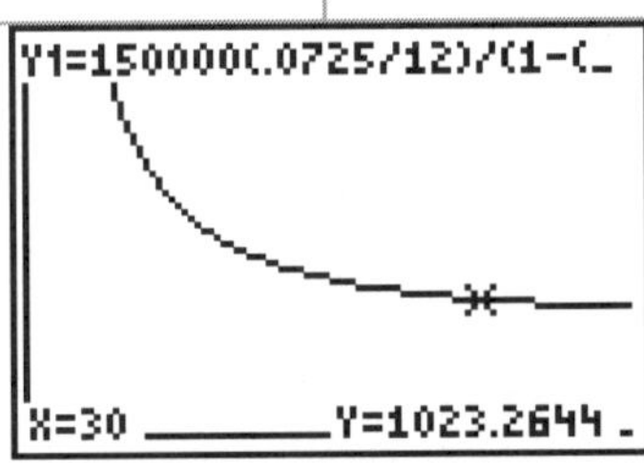 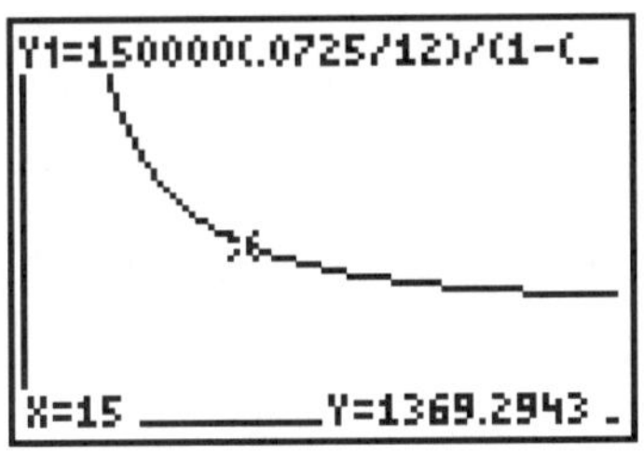

Now we can find out how much this house will end up costing us.

For the 30 year loan we have a monthly payment of $1023.26, so the total we would pay is	$\$1023.26 \bullet 12 \bullet 30 = \$368,373.60$ More than double the loan of $150,000 which means we would pay $218,373.60 in interest.
For the 15 year loan we have a monthly	$\$1369.29 \bullet 12 \bullet 15 = \$246,472.20$

payment of \$1369.29, so the total we would pay is	Less than double the loan of \$150,000 which means we would pay only \$96,472.20 in interest. That is \$121,901.40 less interest than we would have to pay for the 30-year loan. Quite a savings!

Remember, when you borrow money it is like the bank is depositing money with you and so you must pay the interest to the bank. This means that the longer the bank has its money with you the more interest it will receive from you. Hence a 30-year loan is much more costly then a 15-year loan. But, what if we can't quite make a monthly payment of \$1369.29? Let's say that we could afford a monthly payment of \$1200. Then how long would it take to pay off the loan?

We will look at two ways to answer this question - one by hand, the other with the calculator.

Answer Q4: $R = \dfrac{\$120000\left(\frac{0.065}{12}\right)}{1-\left(1+\frac{0.065}{12}\right)^{-12(30}}$ *= \$758.48 for principle and interest.*

Set up the equation.	$1200=\dfrac{150000\left(\frac{0.0725}{12}\right)}{1-\left(1+\frac{0.0725}{12}\right)^{-12(t)}}$	Type this equation into your calculator.	$\dfrac{150000\left(\frac{0.0725}{12}\right)}{1-\left(1+\frac{0.0725}{12}\right)^{-12(x)}}$
Eliminate the fraction; then divide by 1200.	$\left(1-\left(1+\frac{0.0725}{12}\right)^{-12t}\right)1200=150000\left(\frac{0.0725}{12}\right)$ $1-\left(1+\frac{0.0725}{12}\right)^{-12t}=125\left(\frac{0.0725}{12}\right)$	Trace and zoom until Y is the number we want.	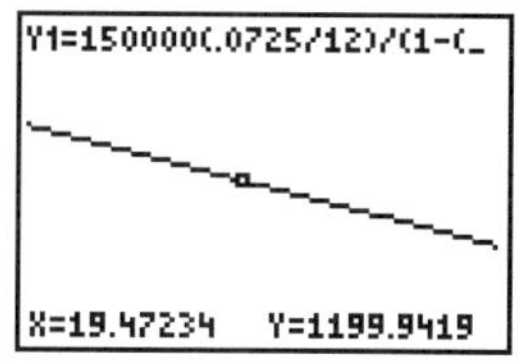
Now subtract 1 multiply by −1; then log both sides.	$-\left(1+\frac{0.0725}{12}\right)^{-12t}=125\left(\frac{0.0725}{12}\right)-1$ $\left(1+\frac{0.0725}{12}\right)^{-12t}=-125\left(\frac{0.0725}{12}\right)+1$ $\log\left(1+\frac{0.0725}{12}\right)^{-12t}=\log\left(1-125\left(\frac{0.0725}{12}\right)\right)$	The answer is about 19.5 years.	

Simplify the log on the right and move the exponent on the left; then get the t alone.	$-12\,t\log\left(1+\dfrac{0.0725}{12}\right)\approx -0.6112$ $t\approx\dfrac{-0.6112}{-12\log\left(1+\dfrac{0.0725}{12}\right)}\approx 19.47$	

Question 5: How much will our house cost us now?

Section Summary

- These formulas are very important in your life and you can save yourself a lot of money if you learn how to play "What if?" games with them when it comes time for you to start saving for retirement or buying big ticket items on time.
- Annuity Formula, $A = R\left[\dfrac{\left(1+\dfrac{r}{n}\right)^{nt}-1}{\dfrac{r}{n}}\right]$
- Amortization Formula, $R=\dfrac{L\left(\dfrac{r}{n}\right)}{1-\left(1+\dfrac{r}{n}\right)^{-nt}}$

SECTION 4.6 HOMEWORK

(1-18) Mathematical modeling that use the annuity formula.

1. When your child is born, you take out an annuity to provide for her college education when she reaches the age of 18. The annuity is \$50 a week. How much money will be available for her college education if the interest rate is 8%?
2. From problem (1) how much money would be available if the interest rate is 9%?
3. At age of 23 you decide to prepare for retirement by taking out an annuity of \$150 a month at 9.5%. How much money would you have available for retirement at age 64?
4. From problem (3) suppose you took out the annuity at age 35. How much money will be available for retirement at age 64?
5. You take out an annuity for \$75 a month at 8.3% for 30 years.
   a. How much money will you have in the annuity?
   b. How much of that money is interest?
6. You take out an annuity for \$125 every two weeks at 9.75% for 30 years.
   a. How much money will you have in the annuity?
   b. How much of that money is interest?
7. At age 64 you want to retire, but you will still owe \$58,000 on your house. If your present age is 44,
   a. How much money will you have to put in an annuity at 7.5% every month so you will be able to pay off your home?
   b. How much of that money will be your money and how much interest?
8. From problem (7), if the interest rate is 8.9%,
   a. How much money each month will you have to put in the annuity?
   b. How much of that money will be your money and how much interest?
9. You want to buy a new car, which you compute will cost \$24,000, in 6 years. How much money will you need to put in an annuity at 9.2% every week to pay cash for the car? When you pay for the car, how much of the payment will be your money and how much of the payment will be interest earned on the annuity?
10. From problem (9), if you decide to wait for 8 years, how much money will you need to put in the annuity every week?
11. At age 30 you decide that in order to live a life style you are comfortable with, you will need \$250,000 to supplement your other retirement benefits. You plan on retiring at age 60. What do you need to put away every 2 weeks at 7.9% to have the \$250,000?
12. From problem (11), if the interest rate is 9.2%, how much will you need to put in the annuity every two weeks?
13. From problem (11), suppose you wait until you are 40 and the interest rate is still 7.9%, how much money will you need to put in the annuity every two weeks?

14. From problem (11) suppose you want to put the money in the annuity every month instead of every two weeks.
    a. How much would you have to put in the annuity every month?
    b. Compare the total amount of money you put in the annuity in both cases.
15. How long will it take for a monthly annuity of \$150 at 7.5% to be worth \$131,589?
    a. Use the graphing calculator to approximate the time.
    b. Find the time algebraically.
16. How long will it take a weekly annuity of \$75 at 8.4% to be worth \$202,349?
    a. Use the graphing calculator to approximate the time.
    b. Find the time algebraically.
17. What will the interest rate be for a yearly annuity of \$1,000 to be worth \$15,192.93 after 10 years?
    a. Use the graphing calculator to approximate the interest rate.
    b. Find the interest rate algebraically.
18. What will the interest rate be for a bi–weekly annuity of \$80 to be worth \$12,933.45 after 5 years?
    a. Use the graphing calculator to approximate the interest rate.
    b. Find the interest rate algebraically.

Answer Q5: \$1200 a month *<u>times</u> 12 months a year <u>times</u> 19.5 years = \$280,800.*

(19-32) Mathematical modeling that use the amortization formula.

19. You are going to buy a \$165,000 home with a down payment of \$15,000. (Notice how interest rate and number of years affects the payment and total interest paid.)
    a. What are your monthly payments if the interest rate is 8% for 30 years? (How much interest do you pay?)
    b. What are your monthly payments if the interest rate is 7.5% for 30 years? (How much interest do you pay?)
    c. What are your monthly payments if the interest rate is 7% for 15 years? (How much interest do you pay?)
    d. What are your monthly payments if the interest rate is 6.5% for 15 years? (How much interest do you pay?)
20. You want to purchase an automobile for \$31,000 with a down payment of \$4,000.
    a. What are your monthly payments if the interest rate is 5.9% for 36 months? (How much interest do you pay?)
    b. What are your monthly payments if the interest rate is 6.5% for 48 months? (How much interest do you pay?)
    c. What are your monthly payments if the interest rate is 8% for 60 months? (How much interest do you pay?)
    d. What are your monthly payments if the interest rate is 8.5% for 72 months?

(How much interest do you pay?)

21. A car dealer is running a promotion of the car you want to buy. You can purchase the car for $21,000 with 60 monthly payments at 8% and receive a discount of $3,000 or you can purchase the car for $21,000 and receive a special interest rate of 1.9% for the same 60 monthly payments. Assuming you will use the $3,000 for a down payment, what is the best deal for you financially?
22. Referring to the problem (21) above, suppose the discount is $2,000 and the special interest rate is 2.5%. Which is the best deal for you financially?
23. A loan company has determined your debt ratio is such that you can afford house payments of $1,080 per month.
    a. If the interest rate is 8% for 30 years, what price home can you purchase?
    b. If the interest rate is 7.25% for 30 years, what price home can you purchase?
24. You have determined you can afford car payments of $425 a month.
    a. For 60 monthly payments at 7.5%, what is the most expensive car you can afford?
    b. For 72 monthly payments of 8.3%, what is the most expensive car you can afford?
25. You feel you have $150 extra a month you can pay for the purchase of furniture. How much furniture can you buy?
    a. If you are willing to pay for 36 months at 9.5% interest?
    b. If you are willing to pay for 48 months at 10.2% interest?
26. A loan company has determined your debt ratio is such that you can afford house payments of $1350 per month.
    a. If the interest rate is 7.2% for 15 years, what price home can you afford?
    b. If the interest rate is 7.9% for 15 years, what price home can you afford?
27. If the house payments are $872 a month for 30 years and the original value of the home was $120,000, use your calculator to compute the interest rate?
28. The payments for a big screen television that originally cost $1,483.00 are $48.20 a month for 30 months, use your calculator to compute the interest rate?
29. The payments for furniture that cost $7,500 are $192 a month at an interest rate of 10.5%, use your calculator to compute how many months you financed the furniture?
30. The payments for a new watch that costs $330 are $20 a month at an interest rate of 11.5%, use your calculator to compute how many months you financed the watch?
31. A husband and wife have got themselves deep in dept with credit cards. The balance on one of the cards was $10,000.00 and the interest is 18%. The credit card company has agreed to let them make payments of $220.00 a month until the card is paid off.
    a. How long will it take to pay off this credit card debt?
    b. How much money total will be paid, to pay off this credit card debt?
    c. How much interest will have been paid by the time this credit card debt has been paid off?

32. An individual has purchased \$8,400.00 of furniture on a revolving credit card. If the card has an interest rate of 20% per year and the monthly payments on the \$8,400 furniture is \$180.00,
    a. How long will it take for the individual to pay off the furniture?
    b. How much money total will be paid for the furniture?
    c. How much interest will be paid on the credit card for the furniture?

(33– 38) For each of the following:

a.) Create a function that will generate values for the unknowns.

b.) Graph the functions and determine if you think the function is exponential or linear.

33. A monthly annuity over a 30 year period has an interest rate of 8%. (Let R be the independent variable and A the dependent variable.)

    (Consider a domain of \$0 to \$100)

34. A monthly annuity of \$120 has an interest rate of 8.5%. (Let t be the independent variable and A be the dependent variable.)

    (Consider a domain of 0 to 20 years)

35. A \$25 annuity has an interest rate of 7.9% over a 20 year time period. (Let n be the independent variable and A the dependent variable.)

    (Consider a domain of 1 through 26)

36. A monthly annuity has an interest rate of 9.5% and a value of \$50,000. (Let t be the independent variable and R the dependent variable.)

    (Consider a domain of 1 through 20)

37. A loan of \$20,000 is to be paid back in monthly payments at 8%. (Let t be the independent variable and R the dependent variable.)

    (Consider a domain of 1 through 15)

    Use the graph and your calculator to give the monthly payment for:
    a. A four-year loan.
    b. A five-year loan.
    c. A ten-year loan.

38. A loan is to be paid back in 10 years with monthly payments at 9.5% (Let L be the independent variable and R the dependent variable.)

    (Consider a domain of 0 through 10,000)

    Use the graph and your calculator to give the monthly payment for:
    a. A loan of \$7,000
    b. A loan of \$8,000
    c. A loan of \$4,000

Section 4.7 Applications of Growth and Decay

We have spent the last couple of sections on finance; now let's turn our attention to some other topics of concern in the world. One such topic concerns being able to determine how the number of people or animals or things change over time. This idea is referred to as the study of populations, and it happens that exponential growth and decay functions seem to be at the core. You see, quite often populations are either growing or decaying exponentially. We'll use exponential formulas to predict what will happen in the future regarding the populations we might be studying.

Example 1: The Tale of Two Cities!

Find functions to describe the populations of Flint, Michigan, and Orlando, Florida, over the last 18 years using the following population numbers:

Year	Flint, Michigan	Orlando, Florida
1980	160,000	128,000
1998	132,000	181,000

2000 Statistical Abstract of the United States

Solution:

As you can see, it looks like the people from up north (Flint), where it's colder, seem to be moving south (Orlando) to where it's warmer. Flint is an example of a city with a decaying population and Orlando is an example of a city with a growing population. Assuming that the decay of Flint is exponential (we will talk more about how we might determine that in the next chapter), we will use the formula $\boldsymbol{y = a(b)^x}$ to come up with a function we can use to describe how the population of Flint is changing. This formula should seem very familiar to you from what we discussed in section 4.1. In that section we talked about how we could come up with a formula for things that behaved exponentially if we knew the *starting amount*, the *common ratio*, and the *time intervals* between data points ($y = a(r)^{t/n}$). So, let's use that knowledge to answer this question.

The starting amount for Flint is	160,000 in 1980
The common ratio is (next number divided by the previous)	$\frac{132{,}000}{160{,}000} = 0.825$
The time interval is	18 years (1998 – 1980)
Hence our formula is	$y = 160000(0.825)^{t/18}$
Which we can simplify by manipulating the exponent term.	$y = 160000\left[(0.825)^{\frac{1}{18}}\right]^t = 160000(0.9894)^t$
In function notation we'd have	$F(t) = 160000(0.9894)^t$, F = Flint pop., t = time

Question 1: How many people according to our formula will still live in Flint come 2010 (Hint: find $F(30)$)?

All of what we have done is fine and dandy, but there is another formula that is a little more convenient: $y = ae^{kt}$. In this formula, a is still the starting amount, but e is the special number we talked about in section 4.2, and k is the rate of change (like in the interest rate problems we discussed in sections 4.5 and 4.6). Let's use the algebra that we have learned in this chapter to put our example into this new form.

The starting amount, as before, is	160,000 in 1980
Now, we must find k by plugging in the values for y and t from the last year we have, which is $y = 132{,}000$ and $t = 18$ (1998-1980).	$132{,}000 = 160{,}000(e)^{k(18)}$
This looks like a No Log type of problem, so we need to isolate the exponential by dividing by 160,000.	$0.825 = e^{18k}$
Now we need to take the log of both sides; as you might remember, since we see an e, let's use ln.	$\ln(0.825) = \ln(e^{18k})$ [remember $\ln e^x = x$] $\ln(0.825) = 18k$
Divide by 18, then we've got our answer.	$k = \dfrac{\ln 0.825}{18} \approx -0.0107$
Our formula now reads	$y = 160000e^{-0.0107t}$
In function notation we'd have	$F(t) = 160000e^{-0.0107t}$

The k, which we found to be -0.0107, tells us that Flint is decaying at the rate of 1.07% per year. The negative tells us it's decaying and the number is the rate of change in decimal form. We have found two different formulas for the change in the population of Flint and they are virtually the same except for the rounding off we did. Here are the graphs of both.

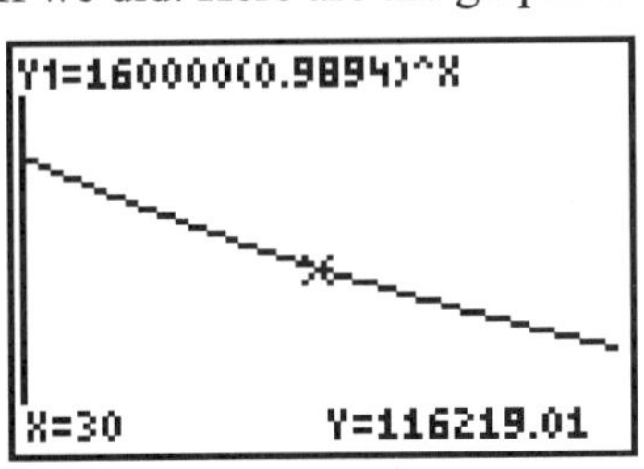

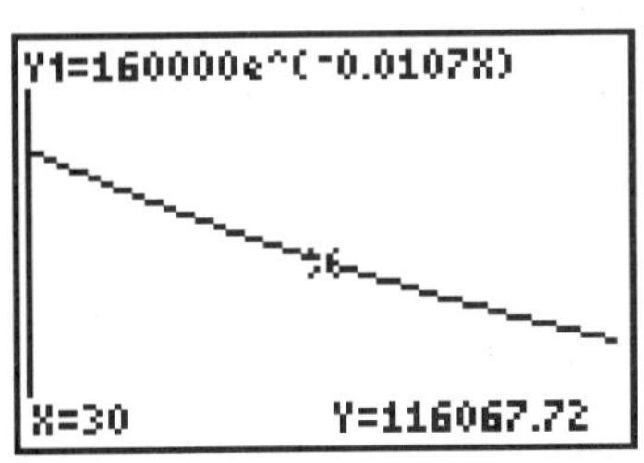

You might notice that on the graphs is the answer to question 1 above. They aren't exactly the same, but remember, we had to round off in both formulas. If we could enter the exact values in both cases the graphs would be exactly the same. You might also notice that if we calculate e^k it will be the same as b. In the discussion about Flint, b is 0.9894 and $e^{-0.0107} = 0.9894$ which is the same as b.

Question 2: Now find both formulas for Orlando.

Notice that $b = e^k$ ($1.01944 = e^{0.01925}$). With the formula $O(t) = 128000e^{0.01925t}$ we can now make a prediction about the population of Orlando in the year 2010 ($O(30) = 128000e^{0.01925*30} = 228{,}042$). Also, notice that Orlando has a growth rate of 1.925%.

Example 2: Ozone

Assuming that ozone is decaying exponentially, find a formula that we could use to make predictions about the concentration of ozone in the atmosphere over Miami, Florida.

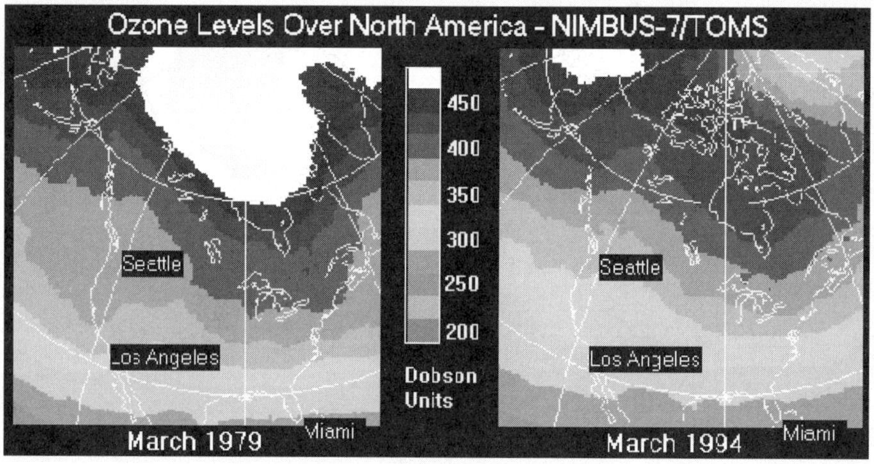

Source: NASA Goddard Space Fight Center

Solution:

Let's take a look at the ozone concentrations over the city of Miami. In 1979 the measurement of ozone over Miami was 303 dobson units (DU), but in 1994 the concentration had fallen to 296 DU. We will use this formula ($y = a\,e^{kt}$) for exponential growth and decay.

First, let's make 1979 be our starting year, so we know that	$t = 0$ (in 1979) $a = 303$ (the start amount)
Plug in 296 for y and 15 for t since those are our values in 1994, our second point.	$296 = 303\,e^{k*15}$
Divide by 303 and take the natural log of both sides.	$\ln\left(\frac{296}{303}\right) = k \cdot 15$

Divide by 15.	$\dfrac{\ln\left(\dfrac{296}{303}\right)}{15} = k \approx -0.001558$
Our final formula is	$y = 303\ e^{-0.001558\ t}$
In function notation we'd have	$MO(t) = 303\ e^{-0.001558\ t}$ (MO = Miami's Ozone, t = time)

So with this formula we can see that Miami has a rate of decay in ozone of 0.1558% per year and we can now make some predictions about the future if our assumptions are true.

Question 3: Approximately what will the ozone concentration be over Miami in the year 2009?

Example 3: Half-life of Barium

Half-life means the time it takes for half of the original amount to change (decay) into something else. Given that the half-life of Barium–131 is 12.0 days, and assuming we started with 4 grams, how much barium will be left after 40 days?

Answer Q1: $160000(0.9894)^{30}$ = 116,219. (the 30 comes from 2010 – 1980 the start date)

Solution:

Our general exponential equation $y = ae^{kt}$ becomes $y = 4\ e^{kt}$, because we know that a (the starting amount) is 4 grams. We also know the half-life: after 12 days (time, t), the 4 grams we started with will have decayed into 2 grams (amount of barium, y). This gives us the point we need to plug into the formula to find k, (12, 2).

The formula we have so far:	$y = 4\ e^{kt}$
Plug in the point (12, 2).	$2 = 4\ e^{k(12)}$
Divide both sides by 4.	$\frac{1}{2} = e^{k(12)}$
Take the natural log of both sides to eliminate the e.	$\ln(1/2) = 12k \quad (\ln e^{12k} = 12k)$
So, we find k to be	$k = \dfrac{\ln\left(\dfrac{1}{2}\right)}{12} \approx -0.05776$
Our, final function is	$y = 4\ e^{-0.05776t}$
In function notation we have	$B_{131}(t) = 4\ e^{-0.05776t}$

Now we can answer the question of how much Barium–131 is left after 40 days. We simply need to plug in 40 for t and evaluate. We get $B_{131}(40) = 0.397$ grams. This answer seems reasonable if we consider that 36 days would be 12 three times. 36 days would mean that we halve the 4 amount 3 times. So, 4 halved once is 2, which halved again is 1, which halved again is 0.5. This 0.5 is close to

0.397, which was what was left after 40 days. Since 40 days is longer than 36 days, it seems reasonable that out answer must be correct.

Example 4: Half-life of Cobalt

Let's assume that we started with 12 grams of Cobalt–60 and after 1.3836 years we find that there are only 10 grams left. What is the half-life of Cobalt–60?

Solution:

First, we know the starting amount, *a*, which is 12 grams. From the other information given to us we can find *k*.

The formula we have so far	$y = 12\,e^{kt}$
Plug in the point (1.3836, 10).	$10 = 12\,e^{k(1.3836)}$
Divide both sides by 12	$\frac{10}{12} = e^{k(1.3836)}$
Take the natural log of both sides to eliminate the *e*.	$\ln\left(\frac{10}{12}\right) = 1.3836k$
So, we find *k* to be	$k = \frac{\ln\left(\frac{10}{12}\right)}{1.3836} \approx -0.13177$
Our, final function is	$y = 12\,e^{-0.13177t}$
In function notation we have	$Cb_{60}(t) = 12\,e^{-0.13177t}$

Answer Q2:

Section 4.1 Method
128,000 in 1980
$\frac{181{,}000}{128{,}000} \approx 1.4141$
18 years (1998 – 1980)
$y = 128000(1.4141)^{t/18}$
$y = 128000\left[(1.4141)^{\frac{1}{18}}\right]^t$
$y = 128000(1.01944)^t$
In function notation we'd have (O = Orlando, t = time)
New Method
128,000 in 1980
$181000 = 128000(e)^{k(18)}$
$1.4141 \approx e^{18k}$
$\ln(1.4141) = \ln(e^{18k}) = 18k$
$\frac{\ln 1.4141}{18} \approx 0.01925$
$y = 128000e^{0.01925t}$
$O(t) = 128000e^{0.01925t}$

Now, if we plug in a 6 (half of the original amount of 12) for $Cb_{60}(t)$ and then solve for *t* we'll have the half-life of Cobalt–60.

Equation to solve	$6 = 12\,e^{-0.13177t}$
Divide by 12	$0.5 = e^{-0.13177t}$
Take the natural log of both sides to eliminate the *e*	$\ln(0.5) = -0.13177t$
Divide by –0.13177 to solve for *t*	$t = \frac{\ln(0.5)}{-0.13177} \approx 5.26$ years

The half-life of Cobalt–60 must be 5.26 years.

Example 5: Potassium-Argon Dating

Potassium-argon dating is frequently used to date the age of volcanic material. It is believed that when hot lava cools no argon remains in the material, so this method of dating is considered fairly reliable. Of course our assumption must be true for this to work. What happens is that Potassium-40, an isotope of regular potassium, decays into argon. If we know the half-life of Potassium-40, then we can determine the age of the rock by the ratio of argon to Potassium-40. The half-life of Potassium-40 has been determined to be 1.3 billion years; that is, it takes 1.3 billion years for half of the Potassium-40 to change into argon. Using the formula ($y = ae^{kt}$), find the k (rate of decay) of Potassium-40.

Solution:

First, let a be our starting amount at year zero (the year the lava cooled).	$t = 0$ (year lava cooled) a (the start amount)
So our formula is	$y = ae^{k*t}$
Half-life means half is left after 1.3 billion years so,	$\frac{a}{2} = a\,e^{k*1,300,000,000}$
Divide by a and take the natural log of both sides.	$\ln\left(\frac{a/2}{a}\right) = k \cdot 1,300,000,000$
Divide by 1,300,000,000.	$\frac{\ln\left(\frac{1}{2}\right)}{1,300,000,000} = k \approx -5.3319 \text{ X } 10^{-10}$ $k \approx -0.0000000053319$
Our final formula is	$y = ae^{-0.00000000053319\,t}$
In function notation we'd have (A = Argon, t = time)	$A(t) = ae^{-0.00000000053319\,t}$

Answer Q3: 2009 – 1979 = 30 = t, so MO(30) = $303e^{-0.001558*30}$ ≈ 289.2 DU

It should be no surprise that the rate of decay turned out to be such a small number; the half-life was 1.3 billion years. Let's now assume we are scientists who have discovered that a sample of rock has a ratio of potassium to argon of 1 to 2. This would mean that of the 3 total parts of material (potassium and argon), only one of the three is still potassium. So, there is 1/3 as much potassium as there must have been in the beginning. Hence, we plug in ***a*/3** for the y and solve for t.

Substitute (a = amount present many years ago).	$\frac{a}{3} = ae^{-0.00000000053319t}$
Divide by a.	$\frac{1}{3} = e^{-0.00000000053319t}$
Take the natural log of both sides.	$\ln(1/3) = -0.0000000053319t$
Divide to solve for t.	$\frac{\ln(1/3)}{-0.00000000053319} = t$
The rock is about this old:	t = 2,060,451,788 years or roughly 2 billion years

Example 6: Rumors

Rumors, believe it or not, grow exponentially. You tell two people something; then they tell a few others and so on. An equation used to describe the spread of a rumor is $R(t) = P(1 - e^{kt})$ where $R(t)$ is the number of people who have heard the rumor; P is the total number of people who could hear the rumor; k is the rate of decay of those who don't know the rumor, and t is the number of days since the rumor started. Assuming that there are 18,000 people who could potentially hear a rumor, and after three days 1440 have heard the rumor, calculate the rate of decay of those who haven't heard the rumor yet.

Solution:

Substitute.	$1440 = 18000(1 - e^{k*3})$
Divide by 18000.	$\frac{1440}{18000} = (1 - e^{k*3})$ $0.08 = 1 - e^{k*3}$
Subtract 1 and multiply by –1.	$0.92 = e^{k*3}$
Take the natural log of both sides and divide by 3.	$\frac{\ln(0.92)}{3} = k \approx -0.0278$
For this example the equation is	$R(t) = P(1 - e^{-0.0278\,t})$

Question 4: How many days will have pasted when 15000 people have finally heard the rumor?

Discussion 1: Advertising

Let's take a look at total money spent on advertising. Here is a table of values of total money spent on advertising over the years 1992-1999.

Year	Total Amount Spent on Advertising (in billions)
1992	132.7
1993	139.5
1994	151.7
1995	162.9
1996	175.2
1997	187.5
1998	201.6
1999	215.2

Source: Statistical Abstract of the United States

For simplicity, let's just take the two values from the years 1992 and 1999 and use them to find an exponential function to describe how money is spent on advertising. In the next chapter we will learn how to use *all* the values in the table to decide if this is exponential or not.

First, call 1992 our starting time ($t = 0$) (132.7 is the start amount) so our equation is	$y = 132.7\, e^{k\,t}$
Now, we need to find k so we will plug in the values $y = 215.2$ and $t = 7$ (1999 – 1992 = 7).	$215.2 = 132.7\, e^{k*7}$
Now, it is time to solve for k like we have done many times already.	$\dfrac{215.2}{132.7} = e^{k*7} \Rightarrow 1.6217 \approx e^{k*7}$ $\ln(1.6217) \approx 7k$ $k \approx \dfrac{\ln(1.6217)}{7} \approx 0.0691$
Now, that we have found k our formula is	$y = 132.7\, e^{0.0691\,t}$
In function notation we'd have	$AD(t) = 132.7\, e^{0.0691\,t}$ (AD = advertisement, t = time)

Question 5: Using this formula how much would you predict would be spent on advertising in the year 2012? (Hint: find $AD(20)$.)

Section Summary

- The basic formula for exponential growth and decay is $y = ae^{k\,t}$, where a is the starting amount and k is the rate of growth (if k is positive) or the rate of decay (if k is negative).
- To find a formula to use to predict how things change you first need to know the starting amount a; then you plug in another point into our basic formula and solve for the k.
- To say this another way, you need to perform two steps and they are:
  1) Find a, the starting amount, at time $t = 0$.
  2) Plug in the other point you know and solve for k.

SECTION 4.7 HOMEWORK

Answer Q4:
$15000 = 18000(1 - e^{-0.0278\,t})$
$\Rightarrow$
$5/6 = 1 - e^{-0.0278\,t}$
$\Rightarrow$
$e^{-0.0278\,t} = 1/6$
$\Rightarrow$
$-0.0278t = \ln(1/6)$
$\Rightarrow$
$t = \ln(1/6)/-0.0278$
≈ 64.5 days

(1-16) Radioactive decay problems.

1. Barium–131 has a half–life of 12 days. If you start with 10 grams of Barium–131,
   a. Give the exponential equation that determines how much Barium–131 is left after time *t*, in days.
   b. How much Barium–131 is left after 33 days?
2. Bismuth–210 has a half–life of 5 days. If you start with 20 grams of Bismuth–210,
   a. Give the exponential equation that determines how much Bismuth–210 is left after time *t*, in days.
   b. How much Bismuth–210 is left after 28 days?
3. Radium–226 has a half-life of 1,620 years. If you start with 15 grams of Radium–226,
   a. Give the exponential equation that determines how much Radium–226 is left after time *t*, in years.
   b. How much Radium–226 is left after 900 years?
4. Potassium–40 has a half-life of 1.28×10^9 years. If you start with 30 grams of Potassium–40,
   a. Give the exponential equation that determines how much Potassium–40 if left after time *t*, in years.
   b. How much Potassium–40 is left after 100,000 years?
5. A radioactive substance starts with 10 grams and after 20 years there are 6.3 grams left,
   a. Give the exponential equation that determines how much radioactive substance is left after time *t*, in years.
   b. What is the half–life of the radioactive substance?
   c. How much radioactive substance is left after 50 years?
   d. From the chart of radioactive substances below, determine which radioactive substance you think you are working with.
6. A radioactive substance starts with 40 grams and after 60 years there are 4.7 grams left,
   a. Give the exponential equation that determines how much radioactive substance is left after time *t*, in years.
   b. What is the half–life of the radioactive substance?
   c. How much radioactive substance is left after 30 years?
   d. From the chart of radioactive substances below, determine which radioactive substance you think you are working with.
7. A radioactive substance starts with 50 grams and after 7 days there are 15.75 grams left,
   a. Give the exponential equation that determines how much radioactive substance is left after time t, in days.
   b. What is the half–life of the radioactive substance?
   c. How much radioactive substance is left after 30 days?

d. From the chart of radioactive substances below, determine which radioactive substance you think you are working with.

8. A radioactive substance starts with 100 grams and after 150 days there are 30.3 grams left,
   a. Give the exponential equation that determines how much radioactive substance is left after time t, in days.
   b. What is the half–life of the radioactive substance?
   c. How much radioactive substance is left after 60 days?
   d. From the chart of radioactive substances below, determine which radioactive substance you think you are working with.

(For problems 5–8 and 30–33 refer to the following table)

Selected Radioisotopes and their Half–Life

Barium–131 12.0 days	Bismuth–210 5.0 days	Calcium–54 165 days
Cesium–137 30 years	Cobalt–60 5.26 years	Hydrogen–3 12.3 years
Iodine–131 8.14 days	Iron–59 46.3 days	Lead–210 19.4 years
Phosphorus–32 14.3 days	Sulfur–35 97.1 days	Thallium–206 4.20 days

Answer Q5: $y=132.7e^{0.0691(20)}$ = 528.5 billion dollars.

9. Using Example 3 and Potassium–argon dating, find how old a rock is that has a ratio of potassium to argon of 2 to 7.
10. Using Example 3 and Potassium–argon dating, find how old a rock is that has a ratio of potassium to argon of 3 to 5.
11. Another example of dating living material occurs by the use of carbon dating. Carbon–14 is a radioactive isotope that is found in living material, and the half–life of carbon–14 is 5,730 years. Approximate the age of a fossilized leaf that has 40% of its normal amount of carbon–14.
12. Using Carbon–14 dating approximate the age of a fossilized bone that has 10% of its normal amount of Carbon–14.
13. Nuclear plants require that spent fuel, radioactive material from power plants, need to be stored safely for a time equal to 10 times the half–life of the material in order not to be considered hazardous. Strontium–90 is one of those spent fuels.
    a. If 10 grams of Strontium–90 are stored and it decays to 2.9 grams after 50 years find the approximate half–life of Strontium–90.
    b. How long will Strontium–90 need to be stored before it is considered non–hazardous?
14. Another type of spent fuel from a power plant is Plutonium–239.
    a. If 100 grams of Plutonium–239 are stored and it decays to 97.2 grams after 1000 years, find the approximate half–life of Plutonium–239.
    b. How long will Plutonium–239 need to be stored before it is considered non– hazardous?
15. Radioactive isotopes are also used in medicine. Technetium–99 is used in the treatment of prostate cancer.

a. If 10 milligrams are used in the treatment and after 12 hours they decay to 2.5 milligrams, find the approximate half–life of Technetium–99.

b. How long will Technetium–99 need to be stored before it is considered non–hazardous?

16. Iodine–131 is used to treat hyperthyroidism.

 a. If 5 milligrams are used in the treatment and after 1 day they decay to 4.585 milligrams, find the approximate half–life of Iodine–131.

 b. How long would Iodine–131 need to be stored before it is considered non–hazardous?

(17-29) Mathematical modeling that use exponential growth and decay.

17. Consider an ant population that changes exponentially. If there is no predators or disease 20 ants become 100 in just 5 days.

 a. Give an equation $A = A_0e^{kt}$ that will determine the number of ants after t days.

 b. Is this a growth or decay function?

 c. Approximately how many ants would be in the colony after 30 days?

18. Consider a rabbit population changes exponentially. If there are no predators or disease, two female and one male rabbit become 60 rabbits in 4 months.

 a. Give an equation $R = R_0e^{kt}$ that will determine the number of rabbits after t months.

 b. Is this a growth or decay function?

 c. Approximately how many rabbits will be present after one year?

19. A small southwest town has a population of 12,500 in the year 1990. The same town has a population of 18,000 in the year 2000. Assume the population is changing exponentially

 a. Give an equation $P=P_0e^{kt}$ that will approximate the population for any year ($t = 0$ for 1990).

 b. Is this a growth or decay function?

 c. Approximately what will the population be for the year 2020?

 d. How long will it take for the population to double?

20. A large eastern city has lost some of its manufacturing jobs and hence its workforce. The population in the year 1995 was 280,000 and in the year 2000 the population was 220,000. Assume the population is changing exponentially

 a. Give an equation $P = P_0e^{kt}$ that will approximate the population for any year ($t = 0$ for 1995 and approximate k to four decimal places).

 b. Is this a growth or decay function?

 c. In what year will the population be approximately 172,862 people?

21. A prairie dog colony had a population of 5,000 but fleas carrying bubonic plague infected the colony. After 3 days there were only 4,000 prairie dogs left. Assume their death rate was exponential

 a. Give an equation $P=P_0e^{kt}$ that will approximate the prairie dog population after t days. (Approximate k to five decimal places)

 b. Is this a growth or decay function?

c. How many days will pass before the population is approximately 537 prairie dogs?

22. If left alone, the number of fleas changes exponentially. There were 10 fleas to start with and 10 days later there were 200 fleas.
    a. Give an equation $F = F_0e^{kt}$ that will approximate the flea population after t days (approximate k to five decimal places).
    b. Is this a growth or decay function?
    c. How many days will pass before the population is approximately 894 fleas?
23. Newton's law of cooling is given by the equation $u = T+(u_0-T)e^{kt}$ where u is the temperature of a heated object, u_0 is the initial temperature, T is the constant temperature of the surroundings, and k is a constant. An object is heated to 110° Celsius and then allowed to cool in a room whose air temperature is 55° Celsius.
    a. If the temperature of the object is 100° Celsius after 2 minutes, approximately when will the temperature be 90° Celsius?
    b. Approximately when will the temperature be 70° Celsius?
24. A thermometer outside on a hot day has a reading of 103° F. It is brought inside and placed in a refrigerator that has a constant temperature of 36°. After 3 minutes the thermometer reads 76°.
    a. What will the thermometer read after 6 minutes?
    b. When will the thermometer read 48°?
25. A roast is cooked in an oven at 350°. The roast's original temperature was 50° when it was placed in the oven and 70 minutes later, the temperature was 100°. However, your meat thermometer suddenly broke and now you need Newton's law to finish the roast.
    a. What will the temperature of the roast be after 90 minutes?
    b. If you want the roast to be rare the temperature should be 135°. How long do you need to cook the roast to make it rare?
    c. If you want the roast to be medium, the temperature should be 150°. How long do you need to cook the roast to make it medium?
26. A frozen chicken breast has a temperature of 26°. It is placed in a room with a constant temperature of 72°. After 10 minutes, the temperature of the chicken breast has risen to 30°.
    a. What will the temperature of the chicken breast be after 20 minutes?
    b. How long will it take for the temperature of the chicken breast to be 50°?

The number of votes cast in the elections held in 1996 and 1998 for Arizona and Ohio are given in the following table (Information from 2000 Statistical Abstract of the U.S.):

YEAR	VOTES IN ARIZONA	VOTES IN OHIO
1996	3,321,000	8,334,000
1998	3,362,000	8,263,000

(Use the above table for problems 27 and 28)

27. Assuming the information for the Arizona voting is exponential, look at example 1 in the reading and answer the following:
    a. Write an equation $y = a(b)^t$ that will find the number of votes cast in any year.
    b. Write an equation $y = ae^{kt}$ that will find the number of votes cast any year.
    c. Use both equations to predict the number of votes that will be cast in 2004 and compare the answers.
28. Assuming the information for the Ohio voting is exponential, look at Example 1 in the reading and answer the following:
    a. Write an equation $y = a(b)^t$ that will find the number of votes cast in any year.
    b. Write an equation $y = ae^{kt}$ that will find the number of votes cast in any year.
    c. Use both equations to predict the number of votes that will be cast in 2004 and compare the answers.
29. For problems 27 and 28 what type of exponential functions did you find, increasing or decreasing?

(30-33) Challenging radioactive decay problems.

30. After 10 years there are 11.38 grams of a radioactive substance left and after 20 years there are 6.48 grams of the same radioactive substance left.
    a. Give the value of k for the exponential function that determines how much radioactive substance is left after time t, in years.
    b. Find the value of the original amount of radioactive substance, R_0.
    c. Give the exponential equation that determines how much radioactive substance is left after time t, in years.
    d. What is the half-life of the radioactive substance?
    e. From the chart of radioactive substances used for (5–8) determine which radioactive substance you think you are working with.
31. After 5 days there are 37.4581 grams of a radioactive substance left and after 25 days there are 11.799 grams of the same radioactive substance left
    a. Give the value of k for the exponential function that determines how much radioactive substance is left after time t, in days.
    b. Find the value of the original amount of radioactive substance, R_0.
    c. Give the exponential equation that determines how much radioactive substance is left after time t, in days.
    d. What is the half-life of the radioactive substance?
    e. From the chart of radioactive substances used for (5–8) determine which radioactive substance you think you are working with.
32. After 10 days there are 6.1588 grams of a radioactive substance left and after 20 days there are 3.7931 grams of the same radioactive substance left

a. Give the value of k for the exponential function that determines how much radioactive substance is left after t, in days.

b. Find the value of the original amount of radioactive substance, R_0.

c. Give the exponential equation that determines how much radioactive substance is left after time t, in days.

d. What is the half-life of the radioactive substance?

e. From the chart of radioactive substances used for (5–8) determine which radioactive substance you think you are working with.

33. After 12 years there are 6.171 grams of a radioactive substance left and after 20 years there are 6.98 grams of the same radioactive substance left

a. Give the value of k for the exponential function that determines how much radioactive substance is left after time t, in years.

b. Find the value of the original amount of radioactive substance, R_0.

c. Give the exponential equation that determines how much radioactive substance is left after time t, in years.

d. What is the half-life of the radioactive substance?

e. From the chart of radioactive substances used for (5–8) determine which radioactive substance you think you are working with.

Section 4.8 Applications of Logarithms

In the last section we saw a lot of examples of how exponential functions are useful in our world. In this section we will be looking at examples of how we can use logarithmic functions to make predictions, and understand our world better.

Discussion 1: Lemon Juice

What makes a lemon taste sour is the fact that it has a lot of hydrogen ions in it compared to pure water, which is considered neutral. In other words, lemons are very acidic. These hydrogen ions mean that the taste of things can vary enormously - from Milk of Magnesia, which is basic and tastes bitter and has a hydrogen ion concentration of roughly 2.5 X 10^{-11} moles per liter, to sauerkraut, which tastes sour and has a concentration of roughly 2.5 X 10^{-4} moles per liter. Chemists measure the amount of acidity or basicity by what they call pH. The formula for pH is

$pH = -\log_{10}[H^+]$, where H^+ represents the concentration of hydrogen ions in moles per liter

Chemists know that lemon juice has a pH of 2.2, so we should be able to figure out how concentrated the H^+ is in lemon juice. Let's solve the pH formula to find H^+.

Plug in our information on pH.	$2.2 = -\log[H^+]$
Multiply by −1.	$-2.2 = \log[H^+]$
To solve a log equal to a number type of equation we have to change to exponential form (section 4.4).	$[H^+] = 10^{-2.2} = 0.00631$ moles per liter $= 6.31 \text{ X } 10^{-3}$

We now know what the hydroxide ion concentration is for lemon juice. Logarithms are often used when the numbers we are trying to measure or quantify vary by a large amount. For example, when we want to talk about how acidic a substance is the numbers can really vary as demonstrated above (2.5 X 10^{-11} Milk of Magnesia verses 6.31 X 10^{-3} lemon juice). But when we use logs to write a number to describe the acidity of a substance, we get numbers like 10.6 for Milk of Magnesia or 2.2 for lemon juice, which are much more convenient numbers to work with.

Numbers of measurement can also vary by a great deal when we compare the intensity of earthquakes, or sound. When we do this comparison, we use what is called the intensity model, which is $S = k \log \frac{I}{I_0}$. When we use this type of model we are coming up with numbers that describe how one thing compares to a standard. For example, in earthquakes we compare the shaking of the ground (I in the formula) to a value of 1 (I_0 in the formula). With regard to sound, we compare every sound (I) to the smallest sound audible to the human ear (I_0). Let's look at some examples of these types.

Discussion 2: Earthquakes

There are several different ways in which to look at earthquakes. One way, which was developed in 1935 and has been used for many years, is called the Richter scale. The Richter scale is a way to quantify the magnitude of an earthquake. The magnitude of earthquakes, or you might say the strength of earthquakes, can vary considerably so we use logs to make the numbers that represent the magnitudes of earthquakes more manageable. Here is the formula used to come up with Richter scale numbers. Notice, it is an intensity formula.

$R = \log \frac{I}{1}$, where I is the measured intensity of the earthquake on a seismograph and 1 is the standard which is 1 micron of motion or $1 \text{ X } 10^{-4}$ cm of motion.

Let's look at a few Richter scale numbers from the past. On July 8, 2001 near Australia, there was an earthquake of 6.2 on the Richter scale. Let's see how much more intense that is compared to the standard of 1.

Plug in our information of $R = 6.2$.	$6.2 = \log I$
To solve change to exponential form (section 4.4).	$I = 10^{6.2} = 1{,}584{,}893$ times stronger than the normal shaking of the earth. Wow!

Question 1: On the same day there was an earthquake very near Salt Lake City, Utah, with a Richter scale value of 3.4. How much more intense was that compared to the standard of 1?

Now you see how different earthquakes can be in intensity and how we can make the numbers more manageable by using logs.

Discussion 3: Sound Intensity

Sound is another instance in which we compare things. We compare the loudness of something to the smallest sound audible to the human ear, which is $I_o = 1 \text{ X } 10^{-12}$ watts per square meter. We call the measurement of sound intensity **decibels** (dB). The formula is once again an intensity formula and it looks like this:

$\text{dB} = 10 \log \frac{I}{1\text{X}10^{-12}}$, where I is the measured intensity of the sound

The sound intensity, or what is called acoustic power, of the typical library is $1.28 \text{ X } 10^{-9}$. Let's figure out the dB value.

Plug in our information, $I = 1.28 \text{ X } 10^{-9}$.	$\text{dB} = 10 \log \dfrac{1.28\text{X}10^{-9}}{1\text{X}10^{-12}} = 31$

Question 2: Compare our answer for a library with that of a firecracker that has an intensity of $I = 40$.

Once again, notice how far apart the two sounds are in intensity; a firecracker is more than 31 billion times stronger in acoustic power than a library, but the dB's are 136 and 36 respectively. The dB formula makes the values more reasonable to work with.

Logarithms, besides making a large range in values easier to manage, also make good models for growths or decays that tend to level out over time.

Discussion 4: Books Sold in the U.S.A.

The number of books sold in the U.S.A. over the past 20 plus years has followed a log model. A possible formula is $B(t) = 366\ln(t) + 1188$, where $t = 0$ in the year 1980. Remember our discussions in chapter 1; we like to use letters in our function notation that will help us remember what the equation is about (B = Books, t = years or any **time** unit). Here is a table of values and a graph on the number of books sold.

Years *(t)* *(t=0, 1980)*	1985 5	1990 10	1995 15	1998 18
Number of books sold in millions	1,788	2,005	2,186	2,254

Source: 2000 Statistical Abstract of the United States

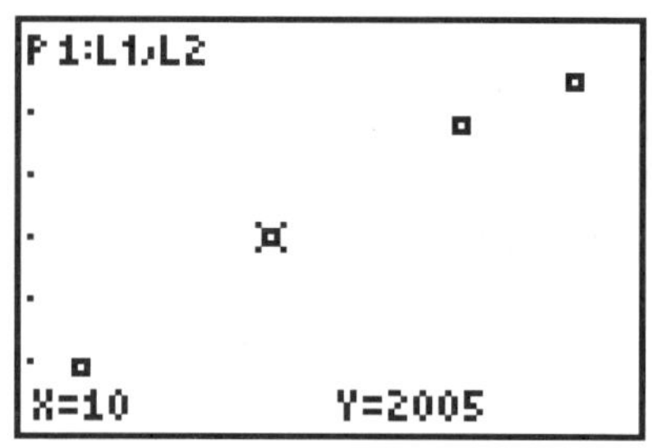

To estimate when we could expect 2,500 million books to be sold, we plug in that number for $B(t)$ and find the year that this might happen.

Plug in our information of $B(t) = 2{,}500$.	$2{,}500 = 366 \ln(t) + 1188$
To solve, first we must isolate the $\ln(t)$.	$1312 = 366 \ln(t)$ $3.585 = \ln(t)$
To solve change to exponential form.	$t = e^{3.585} = 36$ or in the year 2016.

Example 1: Consumption of Skim Milk

How has the consumption of skim milk changed over the recent years?

Solution:

Years *(t)*	1985	1990	1995	1996	1997	1998
(t=0, 1980)	5	10	15	16	17	18
Amount of skim milk used per person per year (in gallons)	3.2	4.9	6.2	6.4	6.6	6.6

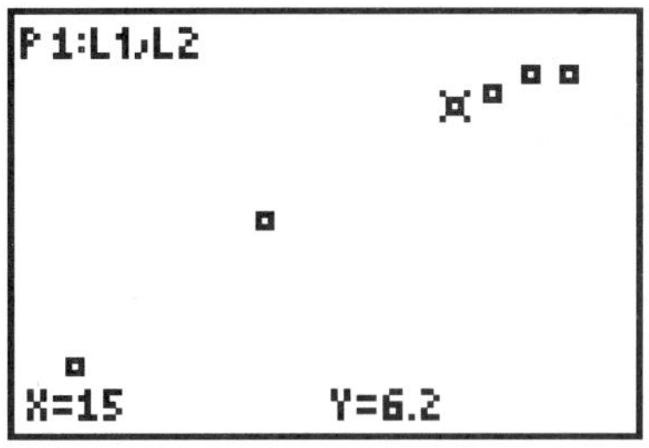

Source: 2000 Statistical Abstract of the United States

As you can see from the chart and the graph, logs might provide a good model for predicting the near future of skim milk consumption since it has been increasing and leveling out. (In chapter 5 we will talk about how to get an equation to model this example by using our graphing calculators, but for now we will just give you one.) Here is one equation that fits the data fairly well [$SM(t) = 2.786\ln(t) - 1.35$ when $t = 0$ in 1980]. Once again we like to use letters in our function notation that will help us remember what the equation is about (SM = skim milk, t = years).

Question 3: When might we be using 8 gallons per person per year?

So, it might be that in the year 2008 (1980 + 28.7) we, as a country, will be consuming 8 gallons of skim milk per person per year.

Answer Q1: $3.4 = \log I$ $\Rightarrow$ $I = 10^{3.4}$ = *2,511 times stronger than normal.*

Section Summary

- Log functions are good to use when we have values that really vary greatly. They make the numbers more manageable to work with and easier to compare. Also, they can be helpful when trying to model events that are leveling out.

SECTION 4.8 HOMEWORK

(1–8) **The intensity of sound:** Decibel rating, L, is given by $L = 10\log\frac{I}{I_0}$ where I is the intensity of sound and $I_0 = 1\text{x}10^{-12}$:

Answer Q2: dB = $10\log\frac{40}{1\text{X}10^{-12}}$ *= 136*

1. Find the value of L in Decibels for a whisper with an intensity of $115I_0$.
2. Find the value of L in Decibels for a library with an intensity of $2510I_0$.
3. Find the value of L in Decibels for rock music at a concert with an intensity of $895{,}000{,}000I_0$.
4. Find the value of L Decibels for a busy street with an intensity of $9{,}500{,}000I_0$.
5. Express the intensity level for conversational speech as a multiple of I_0 when the Decibel rating is 60.
6. Express the intensity level for a dishwasher as a multiple of I_0 when the Decibel rating is 63.98.
7. Express the intensity level for a jetliner at takeoff as a multiple of I_0 when the Decibel rating is 140.37.
8. Express the intensity level for a heavy truck 60 feet away as a multiple of I_0 when the Decibel rating is 90.79.

(9–24) **Magnitude of an earthquake on the Richter scale:** $R(I) = \log\frac{I}{I_0}$, where I is the amplitude registered on a seismograph located 100km from the epicenter of the earthquake and I_0=1.

9. Find the measure on the Richter scale when the amplitude is equal to $10{,}000I_0$.
10. Find the measure on the Richter scale when the amplitude is equal to $100{,}000I_0$.
11. Find the measure on the Richter scale when the amplitude is equal to $1{,}000{,}000I_0$.
12. Find the measure on the Richter scale when the amplitude is equal to $10{,}000{,}000I_0$.
13. Find the measure on the Richter scale for the earthquake in Southern Alaska on July 28, 2001 that had an amplitude measure of $3{,}162{,}277.66I_0$.
14. Find the measure on the Richter scale for the earthquake in Arkansas on August 4, 2001, that had an amplitude measure of $1584.9I_0$.
15. Find the measure on the Richter scale for the earthquake in Utah on July 19, 2001, that had an amplitude measure of $19{,}952.62I_0$.
16. Find the measure on the Richter scale for the earthquake in the Aegean Sea on July 26, 2001, that had an amplitude measure of $1{,}995{,}262.32I_0$.
17. Express amplitude as a multiple of I_0 when the Richter scale measurement is 4.0.
18. Express amplitude as a multiple of I_0 when the Richter scale measurement is 5.0.
19. Express amplitude as a multiple of I_0 when the Richter scale measurement is 3.7.
20. Express amplitude as a multiple of I_0 when the Richter scale measurement is 5.7.
21. Express the amplitude as a multiple of I_0 for an earthquake in Nevada July 17, 2001, that measured 4.3 on the Richter scale.

22. Express the amplitude as a multiple of I_0 for an earthquake in the Gulf of California August 8, 2001 that measured 4.5 on the Richter scale.
23. Express the amplitude as a multiple of I_0 for an earthquake East of the north islands in New Zealand August 21, 2001, that measured 6.9 on the Richter scale.
24. Express the amplitude as a multiple of I_0 for an earthquake in the Central Pacific East Rise July 30, 2001, that measured 5.2 on the Richter scale.

(25–32) **The measure of Acidity Solution with pH:** pH = $-\log[H^+]$, where H^+ measures the concentration of hydrogen ions in the solution.

25. The concentration of hydrogen ions in pineapple juice is .00016. What is the pH of pineapple juice?
26. The concentration of hydrogen ions in mouthwash is .00000063. What is the pH of mouthwash?
27. The concentration of hydrogen ions in eggs is .000000016. What is the pH of eggs?
28. The concentration of hydrogen ions in hair rinse is .0013. What is the pH of hair rinse?

Answer Q3: $8 = 2.786\ln(t) - 1.35 \Rightarrow 9.35 = 2.786\ln(t) \Rightarrow 3.356 = \ln(t) \Rightarrow t = e^{3.356} \Rightarrow t = 28.7$

29. The pH of pure water is 7. What is the concentration of hydrogen ions in pure water?
30. The pH of acid rain is about 4. What is the concentration of hydrogen ions in acid rain?
31. The pH of laundry soap is about 11. What is the concentration of hydrogen ions in laundry soap?
32. The pH of black coffee is about 5. What is the concentration of hydrogen ions in black coffee?

(33–36) **The height of male children:** $h(A) = 29 + 48.8\log(A+1)$, where $h(A)$ is the percent of the adult height of a child of age A.

33. a. If the child is 10 years old, what percent of his adult height will he have reached?
    b. If the child above is 59 inches tall, how tall will he be as an adult?
34. a. If the child is 8 years old, what percent of his adult height will he have reached?
    b. If the child above is 57 inches tall, how tall will he be as an adult?
35. a. If the child has reached approximately 86.4 percent of his adult height, how old is the child to the nearest year?
    b. If the child above is 68 inches tall how tall will he be as an adult?
36. a. If the child has reached approximately 63.1 percent of his adult height, how old is the child to the nearest year?
    b. If the child above is 45 inches tall, how tall will he be as an adult?

(37–42) **Doubling time for compound interest problems:** $t = \dfrac{\ln 2}{n\ln(1+\frac{r}{n})}$ where t is the time in years; r is the yearly interest rate and n is the number of time interest is compounded per year.

37. How long will it take an investment of 8% compounded monthly to double?
38. How long will it take an investment of 6% compounded weekly to double?
39. How long will it take an investment of 9.5% compounded quarterly to double?
40. How long will it take an investment of 7.8% compounded yearly to double?

41. If an investment compounded semi-annually takes approximately 14 years to double, approximately what is the interest rate?
42. If an investment compounded monthly takes approximately 7.7 years to double, approximately what is the interest rate?

(43– 47) **Index of diversity for a species in an ecological community:**

$H=\frac{-1}{\ln n}\left[P_1\ln P_1+P_2\ln P_2+P_3\ln P_3+...+P_n\ln P_n\right]$ where H is the index of diversity and $P_1...P_n$ are the percentage population of each species in the ecological community and n = the number of different species. i.e. If P is the total population $P_1=\frac{Total\ population\ of\ P_1}{P}$ to find the percentage. Perfect diversity exists when $H=1$.

43. The ecological community has two species with a population of 80 and 20. What is the index of diversity for this ecological community?
44. The ecological community consists of four species with populations of 100, 500, 400 and 300. What is the index of diversity for this ecological community?
45. The ecological community consists of two types of tree with populations of 10,500 and 10,500. What is the index of diversity for this ecological community?
46. The ecological community consists of four types of species with populations of 500, 500, 500 and 500. What is the index of diversity for this ecological community?
47. If there are two species in an ecological community and one species has a population of 100 and the index of diversity is approximately .8113, approximately how many are there of the other species? Don't try to solve algebraically; use your graphing calculator to approximate.

Chapter 4 Review

Topic	Section	
Basics about exponentials	4.1	• Exponential functions come from a multiplying process and not an addition process like the polynomials do. • By calculating common ratios we can make a good guess as to whether a table of values describes an exponential function or not. • We have seen that one possible general formula for exponential functions has the form $f(t) = a(r)^{t/n}$ where a = starting amount at $t = 0$; r = common ratio; t = variable, and n = time interval.
Exponential functions	4.2	• The base of an exponential function must be positive (b>0). • The domain of an exponential function is $(-\infty, \infty)$. • The range of an exponential function is $(0, \infty)$. • The horizontal asymptote is $y = 0$. Growth 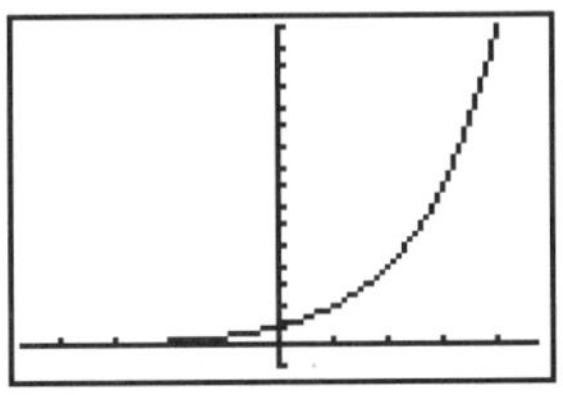 Decay 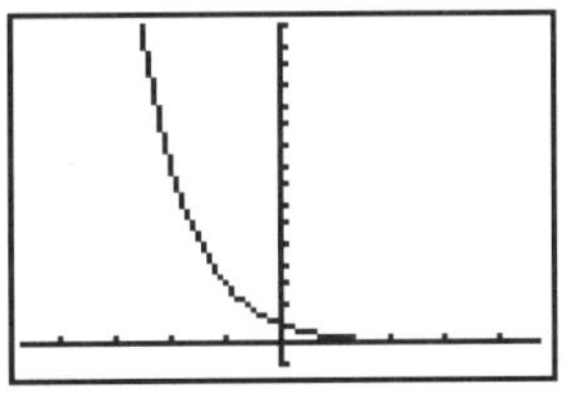
Facts about logarithms	4.2	• The function $\log_b x$ is the inverse of the function b^x. • The answer to a logarithm is an exponent that satisfies the expression $x = b^y$.
Logarithmic functions	4.3	• Informally a logarithm is an exponent. • Formally $f(x) = \log_b x$, where $y = \log_b x$ if and only if $b^y = x$ for $b > 0$. • The base must be positive. • The domain is all positive numbers $(0, \infty)$, the same as the range of $y = b^x$. • The range is all real numbers $(-\infty, \infty)$, the same as the domain of $y = b^x$. • There is a vertical asymptote at $x = 0$, the opposite of a horizontal asymptote. Growth 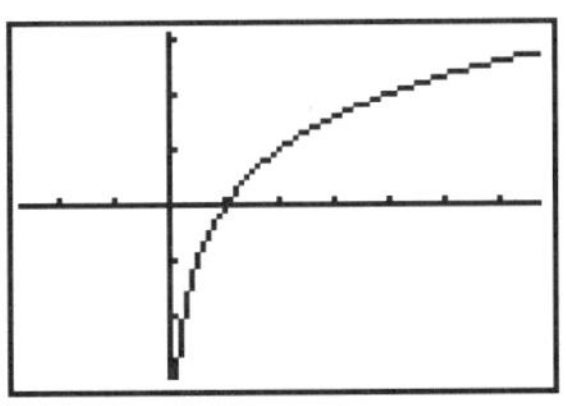 Decay 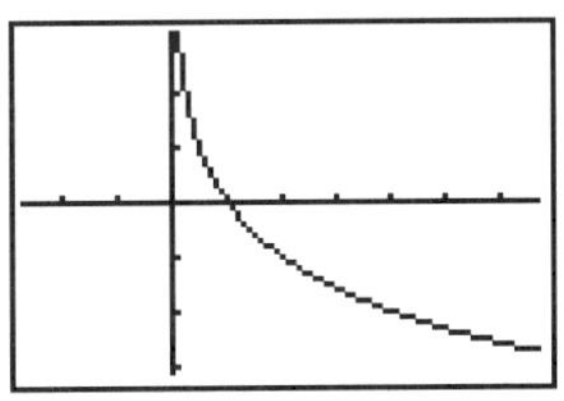

Properties of logarithms	4.3	1.) $\log_b b^x = x$ Inverse property 2.) $b^{\log_b x} = x$ Inverse property 3.) $\log_b M + \log_b N = \log_b MN$ Addition property 4.) $\log_b M - \log_b N = \log_b \dfrac{M}{N}$ Difference property 5.) $r \log_b M = \log_b M^r$ Product property 6.) $\log_b 1 = 0$
Solving equations	4.4	See flow charts in section 4.4
Compound interest formula	4.5	$A = P\left(1+\dfrac{r}{n}\right)^{nt}$
Continuous compound interest formula	4.5	$A = Pe^{rt}$
Annuity Formula	4.6	$\mathrm{A} = R\left[\dfrac{\left(1+\dfrac{r}{n}\right)^{nt} - 1}{\dfrac{r}{n}}\right]$
Amortization Formula	4.6	$R = \dfrac{L\left(\dfrac{r}{n}\right)}{1-\left(1+\dfrac{r}{n}\right)^{-nt}}$
Using $y = ae^{kt}$	4.7	• Find a, the starting amount, at time $t = 0$. • Plug in the other point you know and solve for k.
About k	4.7	k is the rate of growth (if k is positive) or the rate of decay (if k is negative).
Common uses of logs	4.8	$\mathrm{pH} = -\log_{10}[\mathrm{H}^+]$ Measure acidity $R = \log \dfrac{I}{1}$ Measure earthquakes $\mathrm{dB} = 10 \log \dfrac{I}{1\mathrm{X}10^{-12}}$ Measure sound $y = a\log x + b$ Basic log equation

CHAPTER 4 REVIEW HOMEWORK

Section 4.1

(1–4) For the following tables use common difference or ratio to state if the tables represent linear functions or exponential functions.

1.

0	5
1	–2
2	–9
3	–16
4	–23

2.

0	4
1	12
2	36
3	108
4	324

3.

0	16
1	36
2	56
3	76
4	96

4.

0	9
1	3
2	1
3	1/3
4	1/9

5. The following table shows how much money will be in an account after t years.

t = years	0	1	2	3
A = amount of money	\$20,000	\$21,600	\$23,328	\$25,194.24

a. Use the common ratio to show this is an exponential function.

b. What is the common ratio?

c. Give a function $A(t)$ that gives the amount of money in the account after t years.

d. How much money will be in the account after 20 years?

e. Is the function a growth or decay function?

6. The following table represents the population of a state after time t. The time is in 10-year increments from 1970 to 2000. ($t = 0$ is 1970, $t = 1$ is 1980)

t = years	0	1	2	3
P = population of the state	8,000,000	7,800,000	7,605,000	7,414,875

a. Use the common ratio to show this is an exponential function.

b. What is the common ratio?

c. Give a function $P(t)$ that gives the population as a function of time.

d. Approximately what will the population be in 2030?

e. Is the function a growth or decay function?

Section 4.2

(7–10) Graph each of the following with a graphing calculator:

a. Does the function represent a growth or decay exponential function?

b. What is the domain of the function?

c. What is the range of the function?

d. What is the y–intercept of the function?

7. $y = 30(0.6)^x$ 8. $y = 10(3.5)^x$ 9. $f(x) = (2.5)^x + 10$ 10. $f(x) = (2/3)^x + 9$

(11–18) Evaluate the following:

11. $\log_4 (1/16)$ 12. $\log_3 (243)$ 13. $\log_8 (8)$ 14. $\log_2(1/64)$

15. $\log_7 (1)$ 16. $\log_{12} (1/144)$ 17. $\log_{\frac{1}{3}} (27)$ 18. $\log_{\frac{1}{5}}\left(\frac{1}{125}\right)$

(19–22) Approximate the value of the following logs to two decimal places:

19. $\log_5 (23)$ 20. $\log_3 (85)$ 21. $\log_{\frac{1}{4}}\left(\frac{1}{74}\right)$ 22. $\log_{\frac{1}{2}}\left(\frac{1}{84}\right)$

Section 4.3

(23–24) Change each of the logarithmic statements to an exponential statement:

23. $\log_a 5 = x$ 24. $\log_3 (y + 3) = \text{b}$

(25–26) Change each of the exponential statements to a logarithmic statement:

25. $5^x = y$ 26. $a^{x+3} = 9$

(27–30) Expand each of the following:

27. $\log_3(3x^3 y^2)$ 28. $\log\left(\frac{\sqrt{x}}{c^2}\right)$

29. $\log_5\left(\frac{3x^4}{a^3}\right)$ 30. $\ln\left(\frac{5x^5 \sqrt[3]{y}}{ab^4}\right)$

(31–34) Write each of the following as a single logarithm:

31. $\log_3 a + 5\log_3 \text{b}$ 32. $2\log_7 x - \frac{1}{2}\log_7 y$

33. $\log 3 + 5\log x - \frac{1}{3}\log a - 5 \log b$ 34. $\frac{1}{5}\ln(x + 3) + 2 \ln y - (-\ln 8) - 4 \ln \text{b}$

(35–36) Expand each of the following and check your answer by approximating both logarithm statements:

35. $\log \frac{3^2\sqrt{4}}{5^3}$ 36. $\ln \frac{5^3 \cdot 2^4}{10^2}$

(37–40) Use the change of base property and the graphing calculator to approximate each of the following to four decimal places:

37. $\log_6 (59)$ 38. $\log_{12} (387)$ 39. $\log_{\frac{1}{3}} (54)$ 40. $\log_{\frac{2}{3}} \frac{7}{123}$

(41–44) Find the domain of each of the following logarithm functions:

41. $f(x) = \log (2x)$ 42. $g(x) = \log (-5x)$

43. $y = 2\ln (2x - 3)$ 44. $y = 3\ln (x^2 - 4)$

Section 4.4

(45–52) Solve each exponential equation (For any irrational answers approximate to three decimal places):

45. $5^{2x-1} = 125$ 46. $3^{3x+1} = \frac{1}{81}$ 47. $4^{5x} = 27$ 48. $8^{-4x} = 62$

49. $6^{2x+3} = 54$ 50. $e^{3x-5} = 75$ 51. $5 \cdot 3^{5x-1} = 120$ 52. $500 \cdot e^{.09x} = 3000$

(53–60) Find all real number solutions for the following logarithmic equations:

53. $\log_5 (5x - 3) = \log_5 (2x + 12)$ 54. $\log_2(2x - 7) = \log_2(7x + 3)$

55. $\log (2x - 1) - \log (x - 3) = \log 5$ 56. $\ln (x - 2) + \ln (x + 3) = \ln 14$

57. $\log_4 (3x + 1) = 2$ 58. $\log_{25} (4x - 7) = \frac{1}{2}$

59. $\log_3 (2x + 1) - \log_3 (x - 2) = 3$ 60. $\log (x + 3) + \log (x) = 1$

61. $R(t) = 10e^{-.1386t}$ is a function that determines how many grams of a radioactive substance Bismuth is left after t days.
    a. How much Bismuth is left after 30 days?
    b. How much Bismuth did you start with?
    c. Approximately how long will it be before there are 2 grams left?

62. The amount of a radioactive substance remaining after time t, is given by the formula $R(t) = R_0e^{kt}$. The relationship between the half-life and k is given by $k = \frac{\ln(0.5)}{t}$, where t is the amount of time it takes the radioactive substance to decay by half.
    a. If the half-life is 1500 years, what is the value of k?
    b. If the value of k is –0.000356 and time is in days, what is the half-life of the isotope?

Section 4.5

63. You invest $25,000 in an account for retirement. If the interest you receive is 8.5% compounded monthly, how much will be in the account after 30 years?

64. How much would you need to invest in an account at 10% compounded semi–annually to have $265,329.77 in the account after 10 years?

65. How long will it take a $50,000 invested at 9.5% compounded weekly to double?

66. You invest $50,000 in an account. If the interest you receive is 7.8% compounded continuously, how much will be in the account after 25 years?

67. You are trying to save $23,000 to pay cash for a new car. If you want to purchase the new car in 3 years and want to invest your money in an account paying 6.5% compounded continuously, how much will you need to invest?

68. How long will it take $50,000 invested at 9.5% compounded continuously to double?

Section 4.6

69. You need to borrow \$185,000 to purchase a new home. If the interest rate for a 30–year mortgage is 7.3% a year, what will your monthly payments be for principle and interest?

70. You can afford to pay \$430.00 a month for a new car. If you finance the car for 60 months with an interest rate of 7.5% a year, how much can you pay for the new car?

71. For retirement you take out an annuity of \$50 a week with an interest rate of 8.5% a year. How much will you have in your retirement fund after 35 years?

72. You want to provide a \$40,000 college fund for your new–born child. If you think it will be 18 years before he or she goes to college, how much each month will you need to put away in an annuity that has an interest rate of 10.5% a year?

Section 4.7

73. A radioactive substance has a half–life of 8 days and you start out with 5 grams.
    a. Give the exponential equation that determines how much is left after t days.
    b. How long will it take for the radioactive substance to decay to 2.3 grams?

74. A radioactive substance starts with 10 grams and after 50 years there are 8 grams left.
    a. Give the exponential equation that determines how much is left after t days.
    b. How long will it take for the radioactive substance to decay to 4 grams?

75. A small city has a population of 45,000 in 1990 and 46,500 in 1995. Assuming the population change is exponential:
    a. Give an equation $P = P_0e^{kt}$ that will approximate the population for any year. ($t = 0$ for 1990)
    b. Is this a growth or decay exponential function?
    c. In what year will the population reach 52,000?

76. Before a spoon was sprayed with a disinfectant, there were 1,000,000 germs on the spoon. If two minutes later there were 635,000 germs and the number of germs is changing exponentially:
    a. Give an equation $G=G_0e^{kt}$ that will approximate the number of germs after t minutes.
    b. Is this a growth or decay exponential function?
    c. Approximately how many minutes will pass before there will be only 355,000 germs?

Section 4.8

77. The decibel rating, L, is given by $L = 10\log\frac{I}{I_0}$ where I is the intensity of sound:
    a. Find the value of L in Decibels for an intensity of sound of $150000I_0$.
    b. Find the intensity of sound as a multiple of I_0 when the Decibel rating is 100.

78. The percent of adult height, $h(A)$, of a male child's age A is given by the function $h(A)=29 + 48.8 \log(A+1)$
    a. Find the percent of the adult height of a male child 14 years old.

b. If a male child 12 years old is 64 inches tall, how tall will he be when he is an adult?

CHAPTER 4 EXAM

1. A survey finds the population of a city every five years starting in 1990. The following table represents the results of that survey:

Year of survey	1990	1995	2000
Population	123,000	125,829	128,723

   a. Show by using common ratio why or how this seems to represent an exponential function.

   b. Let $t = 0$ represent 1990. Find the equation $P = P_0e^{kt}$ that predicts the population any year.

   c. Predict the population in the year 2003.

   d. According to the equation, during what year will the population reach 135,000?

2. $f(x) = 20(3)^x + 5$

   a. Does the function represent a growth or decay function?

   b. What is the domain?

   c. What is the y–intercept?

3. Evaluate $\log_2 (1/32)$

4. Use your calculator and the change of base theorem to approximate $\log_{12} (856)$ to three decimal places.

5. Expand $\log \dfrac{3x^3}{a^2(\sqrt[3]{b})}$.

6. Write as a single logarithm $\log 5 + \frac{1}{4}\log x - 3\log y - \log z$.

7. Find the domain of $y = \log (3x + 2)$.

(8–14) Solve for x. (If necessary approximate the answer to 4 decimal places.)

8. $3^{3x+2} = 243$ 9. $5^{3x-2} = 71$ 10. $20e^{x+2} = 44$ 11. $\log_5 (2x–3) = 2$

12. $\log_3(5x–7) = \log_3(2x + 8)$ 13. $\log(7x - 2) - \log(x - 3) = 1$ 14. $\ln(x - 3) + \ln(x + 4) = \ln(8)$

15. You want to invest enough money to pay off your home 15 years from now when you plan to retire. If the computed balance left on your home mortgage in 15 years will be \$97,000, how much do you need to invest in an account paying 7.9% compounded monthly to pay off the home at retirement?

16. What interest rate is necessary for \$50,000, compounded continuously to become \$95,777 after 10 years?

17. You have borrowed \$10,000 from your brother and have agreed to pay him back in 3 years. How much money a month do you need to put in an annuity at 8.5% to have enough to pay your brother back?

18. Your mortgage for a new home is \$185,000.

   a. What are your monthly payments for a 30-year mortgage at 7.2%?

   b. What are your monthly payments for a 15-year mortgage at 6.8%?

c. How much do you save in interest with the 15-year instead of 30-year mortgage?

19. Radium–226 has a half–life of 1,620 years. If you start with 20 grams of Radium–226;

    a. How much will be left after 900 years?

    b. How long will it be before it has decayed to 8 grams?

20. The population of a Midwest town is 24,000 in 1970 and 26,850 in 1980. Assume the population is changing exponentially.

    a. Give an exponential function for finding the population in any given year. ($t = 0$ in the year 1970)

    b. Predict the population for the year 2010.

    c. Approximately when will the population be 33,000?

(21–22) Use the formula $\text{pH} = -\log[\text{H}^+]$, where H^+ measures the concentration of hydrogen ions in the solution.

21. Most bacterial growth is inhibited in foods that have high acidity. All foods below a pH of 7 are considered to be acidic. Any foods below a pH of 4.6 seriously inhibit the growth of bacteria and would be quite safe when not refrigerated. Find the pH of the following and state which is the safest when left out.

    a. The concentration of hydrogen ions in oranges is .0005.

    b. The concentration of hydrogen ions in squash is .000005

22. The following foods have the given pH: What is the concentration of hydrogen ions in each?

    a. Lemons with a pH of 2.2.

    b. Squash with a pH of 5.3.

    c. Oysters with a pH of 6.4.

Chapter 5 Data Analysis

We have made it to a point where we can take all the functions we have learned and use them to model real-life events. In the real world we create models to help us make decisions about the future or come up with ideas about why things happened in the past. We will take the first couple of sections in this chapter to talk about how to do this by hand, and then we will talk about how to do this on the graphing calculators.

Section 5.1 Linear Models

Let's begin with the easiest of the functions we have studied, linear functions. In previous math classes they taught you how to create linear models when they taught you how to find an equation of a line. But let's take some time here to review this topic. As you might remember there were several formulas that were presented to you with regard to linear functions. Here is a list of them.

- $m = \dfrac{y_2 - y_1}{x_2 - x_1}$ — The slope formula, we talked about this in section 1.5
- $y = mx + b$ — The slope-intercept form of a line
- $y - y_1 = m(x - x_1)$ — The point-slope form of a line
- $Ax + By = C$ — The standard form of a line
- $y = D$ — The form of a horizontal line
- $x = D$ — The form of a vertical line

Let's talk about each of these.

The slope formula, $m = \dfrac{rise}{run} = \dfrac{y_2 - y_1}{x_2 - x_1}$ (or in function notation $\dfrac{f(x_2) - f(x_1)}{x_2 - x_1}$).

We use this formula to find the slope of a line between two points. As mentioned in section 1.5, the numerator is the difference of the output values (dependent variables, the rise) and the denominator is the difference of the input values (independent variables, the run).

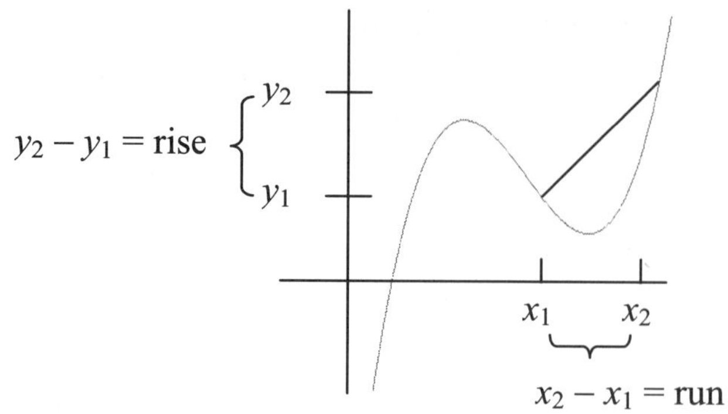

$$m = \frac{rise}{run} = \frac{y_2 - y_1}{x_2 - x_1}$$

The slope of a line is by far the most important aspect of a line. It tells you the rate at which something is growing (positive slope – the line is going up hill as you move left to right) or decaying (negative slope – the line is going down hill) as we mentioned in section 3.1. Of course, in the discussion following question 2 in section 3.1, we saw that the slope (m) of a line is the same as the common difference (d) divided by the time interval. So, $m = \frac{d}{time\ interval}$, where a positive common difference means growth and a negative means decay.

Example 1: Slope Between Two Points

Find the slope of the line between the points (-1, 5) and (2, 3).

Solution:

Plug the points into the slope formula.	$m = \frac{3-5}{2-(-1)} = \frac{-2}{3}$
Remember that you can treat either point as the second one.	$m = \frac{5-3}{(-1)-2} = \frac{2}{-3} = \frac{-2}{3}$

Question 1: Is the line passing through the two points in example 1 going up hill or down? How do you know?

The slope-intercept form of a line, $y = mx + b$.

We use this form of a line to graph lines and to find equations when we know the slope and y-intercept. If you solve a linear equation for y (the output variable) then the coefficient in front of x (the input variable) will always be the slope of the line and the number (b) will be the y-value of the y-intercept.

Example 2: Finding the Slope Intercept Form of a Line

Given that the slope of a line is –1 and the y-intercept is the point (0, 3) find the equation and graph the line.

Solution:

By plugging $m = -1$ and $b = 3$ (the y-value of the y-intercept) into our formula we get	$y = -1x + 3$
We use our graphing calculator to get this graph:	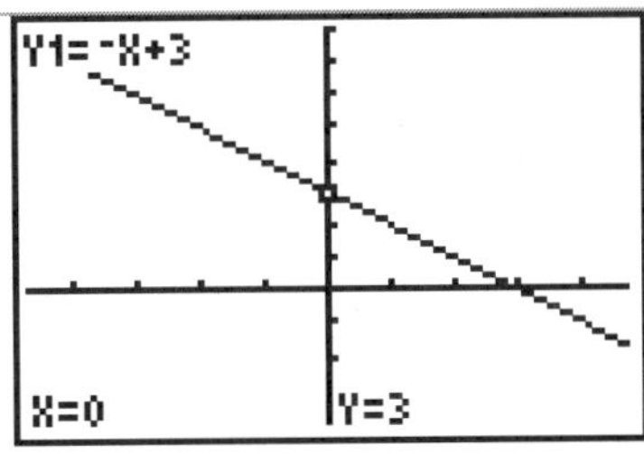

Question 2: Given a slope of 4/5 and a y-intercept of (0, –2) find the equation of this line and graph it.

The point-slope form of a line, $y - y_1 = m(x - x_1)$.

We use this form of a line to find equations when we know the slope and at least one point. Once again, m stands for the slope of the line, and x_1 and y_1 are the values of the known point (x_1, y_1).

Example 3: Finding the Equation of a Line Given Slope and a Point

Find the equation of the line given a slope of –0.5 and the point (–3, –1).

Solution:

Plug our information into the formula $y - y_1 = m\,(x - x_1)$: $m = -0.5$, $x_1 = -3$, $y_1 = -1$.	$y - (-1) = -0.5\,(x - (-3))$
Simplify.	$y + 1 = -0.5\,(x + 3)$

Usually we like to solve this for y (put it into the slope-intercept form) since that makes it easier to graph.

Solve for y (clear parenthesis and subtract the 1).	$y = -0.5x - 1.5 - 1$ $= -0.5x - 2.5$

Let's take a moment to remind you how to graph this by hand.

First, find the y-intercept on the graph.	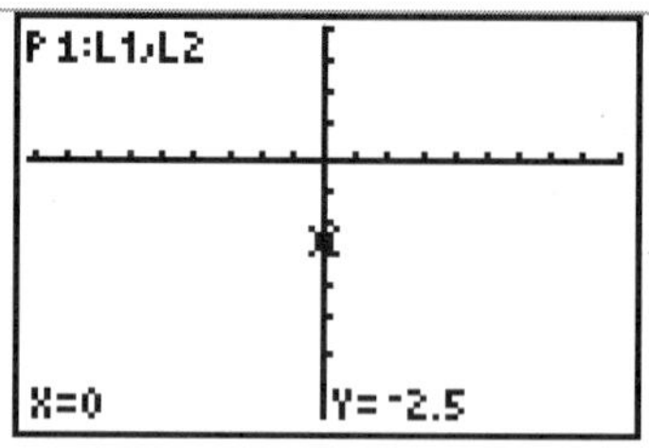
Now, write the slope in fraction form, using the idea of $\frac{rise}{run} = m$, and move up or down and across the appropriate amount. For this example the fraction is $\frac{-1}{2}$ so move to the right 2 and down 1 (negative rise) from the y-intercept, then connect the dots.	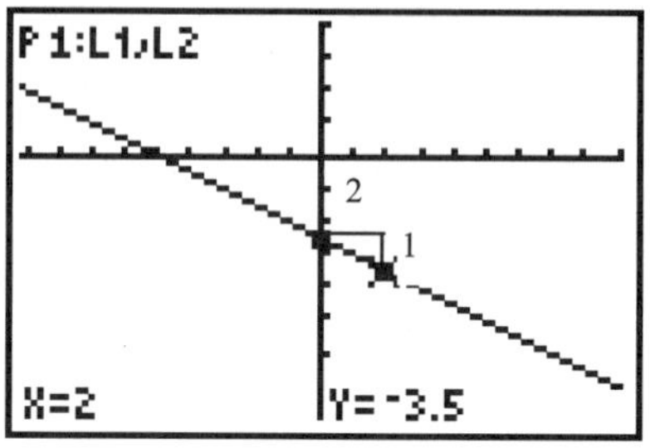

The standard form of a line, $Ax + By = C$.

The A, B, and C are non-fractional numbers. This form is most useful when working with system of equations or matrices that are covered later in this chapter and in chapter 6.

Question 3: Take our last equation $y = -0.5x - 2.5$ and put it into standard form.

Answer Q1: The line is going down hill because the slope is negative.

The form of a horizontal line, $y = D$, where D is a real number.

The absence of an x in this kind of linear equation tells you that the input value has no effect on the output. This means that y is a constant, and the graph is always at a constant height. This also tells us that the slope of this line must be zero since it is flat. The change in the y's is equal to zero. If you know that the slope is zero then you know that the equation of the line will be $y = D$.

The form of a vertical line, $x = D$, where D is a real number.

Here, the exact opposite is happening as compared to the horizontal line. There isn't a y in the equation so we can have any value we want for the output. Since x is now constant, we have only one input (so this is not a function). The graph of this equation is a vertical line that infinitely rises above, and falls below, one particular x-value. Also notice that the change in the x's is now zero, which causes the slope $\frac{y_2 - y_1}{x_2 - x_1}$ to be undefined (division by 0).

Example 4: Finding the Equation of a Line Given Slope and a Point

Find the equations of the following lines.

a.) $m = 0$ and goes through the point $(-2, 5)$ b.) $m =$ undefined and goes through the point $(3,-4)$

Solutions:

a) Since the slope is zero we know that this is a horizontal line hence the equation must be $y = 5$ (the y-value of our point).	b) Since the slope is undefined we know that this is a vertical line hence the equation must be $x = 3$ (the x-value of our point).

Question 4: Find the equations of the following lines:

a.) $m = -2$, goes through the point $(5, -1)$.

b.) $m = 0$, goes through the point $(-4, -7)$.

Example 5: Finding the Equation of a Line Given Two Points

Find the equation of the line going through the points $(1, 3)$ and $(-2, 6)$.

Solution:

First, we need to find the slope. The slope is the most important piece of information to know.	$m = \frac{6-3}{-2-1} = \frac{3}{-3} = -1$	
Plug the slope and one of the points into the point-slope form of a line, then put the answer into slope-intercept form. (We have shown that no matter which point you use you get the same answer.)	$y - \underline{3} = \underline{-1}(x - \underline{1})$ $y - 3 = -x + 1$ $y = -x + 4$	$y - \underline{6} = \underline{-1}(x - \underline{(-2)})$ $y - 6 = -x - 2$ $y = -x + 4$

Discussion 1: Modeling Linear Data

Given the table of data below, let's find the linear function that could be used to model this data. (quad. Btu stands for quadrillion Btu's a measurement of energy.)

Year	Energy Consumption by Residential and Commercial users in (quad. Btu)
1979 $t = 0$	26.6
1989 $t = 10$	30.4
1999 $t = 20$	34.2

Source: 2000 Statistical Abstract of the United States

Answer Q2:

$y = \frac{4}{5}x - 2$

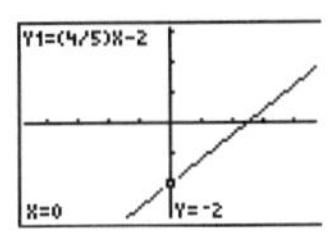

Question 5: Find the common differences (section 3.1) to confirm that this data is really linear.

To find a linear function to model this real-life situation we will need to first find the slope since we are given points but no slope. Then we can use the point-slope formula to get the function.

Find the slope from two of the three points (10, 30.4) and (20, 34.2).	$m = \dfrac{34.2 - 30.4}{20 - 10} = \dfrac{3.8}{10} = 0.38$ The same as the common difference divided by the time interval
Plug the slope and one of the points into the point-slope form of a line, then put the answer into the slope-intercept form.	$y - \underline{34.2} = \underline{0.38}\,(x - \underline{20})$ $y = 0.38x + 26.6$
In function notation it is written as	$E(t) = 0.38t + 26.6$ ($t = 0$ in 1979)

Notice, common difference divided by time interval and starting amount, makes up an equation in a similar manner to what we saw with exponential equations.

With exponentials, the basic function is $f(t) = a(r)^{t/n}$. For linear models it will be

$f(t) = \left(\dfrac{d}{time}\right)t + b$, where $\left(\dfrac{d}{time}\right)$ is the common difference divided by the time interval (slope)

and b is the starting amount at $t = 0$.

Question 6: How much energy will we be using in the year 2009 if we let $t = 0$ in 1979?

Answer Q3:

$y = \dfrac{-1}{2}x - \dfrac{5}{2}$

$2y = -x - 5$

$x + 2y = -5$

The above example is a real-world problem so the slope has a real meaning. The meaning of slope comes directly from the units given for the numerator and the denominator. In discussion 1 the numerator was in quad Btu's and the denominator was in years.

Question 7: What does the slope 0.38 mean in discussion 1?

Note: real-world data usually won't turn out to be exactly linear like this example, so in later sections of this chapter we will talk about how to evaluate a more common data distribution, that is, one that doesn't work out exactly right.

PARALLEL AND PERPENDICULAR LINES

We should discuss one more thing, which is the idea of parallel and perpendicular lines.

Parallel lines are those that have the same slope $(m_1 = m_2)$

Example 6: Finding a Parallel Line

Find the equation of a line parallel to $3x - 5y = 10$.

Solution:

Begin by finding the slope of $3x - 5y = 10$ by writing the equation in slope-intercept form. In this case we get $-5y = -3x + 10$ which becomes $y = \frac{3}{5}x - 2$. So, the slope must be $\frac{3}{5}$.

Now, any line parallel to $y = \frac{3}{5}x - 2$ will have the same slope, but a different y-intercept. So, for instance, the line $y = \frac{3}{5}x + 1$ (same slope; different y-intercept) is parallel.

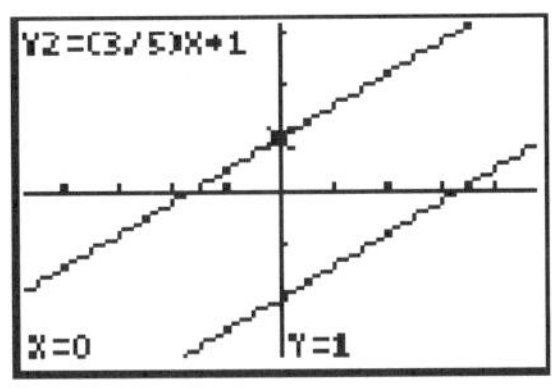

Answers Q4:

a) y+1=–2(x–5)

b) y = –7

In general, *any* line of the form $y = \frac{3}{5}x + b$, where b is any real number, is parallel to our original line. Test this by plotting the original line and several others using various values of b on the same graph.

Question 8: Find the slope of the two lines and determine if they are parallel or not? Why?

a.) $5x + y = -3$ b.) $3y = -15x - 27$

Perpendicular lines are those that have negative reciprocal slopes $\left(m_1 = \frac{-1}{m_2}\right)$.

Answer Q5: The common differences are all the same d = 3.8

Example 7: Finding the Slope of a Perpendicular Line

If we have the line $4x - 7y = 1$ then what would be the slope of a line perpendicular to it?

Solution:

We need to solve the given equation for y so that we can find its slope.	$-7y = 1 - 4x$ $y = \frac{-1}{7} + \frac{4x}{7} = \frac{4}{7}x - \frac{-1}{7}$
The slope of the line is $m_1 = \frac{4}{7}$ so the slope of the line perpendicular is	$m_2 = \frac{-7}{4}$, the negative reciprocal.

Example 8: Finding Parallel and Perpendicular Lines

Find the equation of the lines parallel and perpendicular to the line $5x + 2y = 12$ through the point $(-2, 3)$.

Solution:

Answer Q6:
E(30)
=0.38(30)+26.6
= 38 quad Btu

First we need to find the slope of the line given to us.	$2y = 12 - 5x$ $y = 6 - \frac{5x}{2} = \frac{-5}{2}x + 6$
So the slope for this line is	$m = \frac{-5}{2}$
Parallel lines have the same slope. The slopes of perpendicular lines are negative reciprocals.	A line parallel would have $m_{//} = \frac{-5}{2}$. A line perpendicular would have $m_{\perp} = \frac{2}{5}$.

Answer Q7:
A 0.38 slope means that the United States uses 0.38 more quad Btu per year in Energy for residential and commercial users.

To find the parallel line use the point slope form of a line with slope = $\frac{-5}{2}$ and the point (–2, 3). $y - 3 = \frac{-5}{2}(x - (-2))$ $y - 3 = \frac{-5}{2}(x + 2)$ $y - 3 = \frac{-5}{2}x - 5$ $y = \frac{-5}{2}x - 2$	To find the perpendicular line use the point slope form of a line with slope = $\frac{2}{5}$ and the point (–2, 3). $y - 3 = \frac{2}{5}(x - (-2))$ $y - 3 = \frac{2}{5}(x + 2)$ $y - 3 = \frac{2}{5}x + \frac{4}{5}$ $y = \frac{2}{5}x - \frac{4}{5} + 3$ $= \frac{2}{5}x + \frac{19}{5}$

Here is a graph of all three lines.

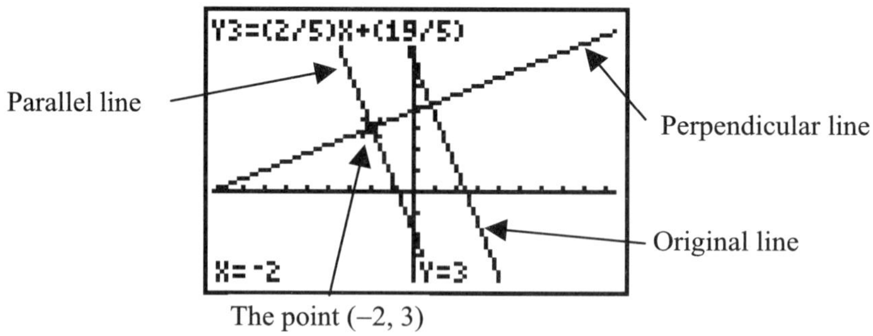

Discussion 2: Modeling Daily Car Rental

You can still rent a car at some places where you pay a daily fee for renting the car, and a cost per mile driven. If we know that it will cost us $25 a day to rent a car and $0.30 per mile driven, then let's find an equation to model this problem.

First, we should notice that even if we don't drive a single mile it will cost us $25. That means when we have driven "0" miles it costs us $25, that sounds like a y-intercept (like example 6).	$b = 25$
The cost per mile must be a slope, since it is giving to us as a rate of change of cost to miles driven (cost on top is the dependent or output variable and miles on the bottom is the independent or input variable). The words used in the expression of slope tells you which variable is the output and which is the input.	$m = 0.3$
By using the slope-intercept form of a line we get	$y = 0.3x + 25$
In function notation we have (Notice that we use the variable C for cost and m for miles so $C(m)$ represents the output cost.)	$C(m) = 0.3m + 25$

Answer Q8:

a.) $y = -5x - 3$ so by the slope–intercept formula $m_1 = -5$,
b.) $y = -5x - 9$ so by the slope–intercept formula $m_2 = -5$. They are parallel since they have the same slope.

Question 9: If we were to drive 128 miles in one day, how much would we pay to rent the car?

Example 9: Modeling Demand for a Product

Lines show up in the business world when developing demand models (formulas used to determine how much people will buy depending on the price charged). Companies hire people to figure out how much they can sell at a particular price. For example, if you discovered that by charging $10 for a particular t-shirt you could sell 100,000 t-shirts and if you charged $15 you could only sell 20,000 t-shirts, then what is the linear equation that would model such a situation?

Solution:

First, we see two points ($10, 100000) and ($15, 20000). With these two points we can compute the slope. We know that the 100,000 and 20,000 belong in the numerator because of the words "how much people will buy *depending on* the price charged." This tells us that demand for shirts **depends on** price charged.	$m = \dfrac{20000 - 100000}{15 - 10}$ $= -1600\ \dfrac{t-shirts}{price}$
Now use one of the points along with the slope to find the equation.	$y - 100000 = -1600(x - 10)$ $y - 100000 = -1600x + 16000$ $y = -1600x + 116000$
In function notation (p = price, x = number of items) it is	$x(p) = -1600p + 116{,}000$

What we have just found is the demand function for this particular type of t-shirt. We can use this function to help us determine how many items we could expect to sell at different prices.

Question 10: In example 10, what would 12 represent in $x(12)$, and what would $x(12)$ mean?

Section Summary

- $m = \dfrac{y_2 - y_1}{x_2 - x_1}$ The slope formula (need **two points** to compute).
- $y = mx + b$ The slope-intercept form of a line (need the **slope** and **y-intercept**).
- $y - y_1 = m(x - x_1)$ The point-slope form of a line (need a **point** and a **slope**).
- $Ax + By = C$ The standard form of a line (needed for systems of equations, A, B, C not fractions).
- $y = D$ The form of a horizontal line ($m = 0$).
- $x = D$ The form of a vertical line ($m =$ undefined).
- $m_1 = m_2$ The lines are parallel.
- $m_1 = \dfrac{-1}{m_2}$ The lines are perpendicular.

Also remember that slope is the most important aspect of a line, so look for it first. Then, when you are trying to find an equation, the name of the formula will tell you the information you need in order to use it.

SECTION 5.1 HOMEWORK

(1–6) Using the graphs below give the slope of each line:

1.

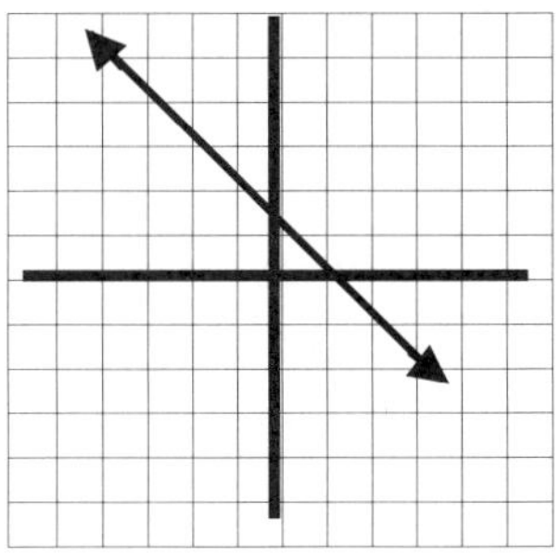

2.

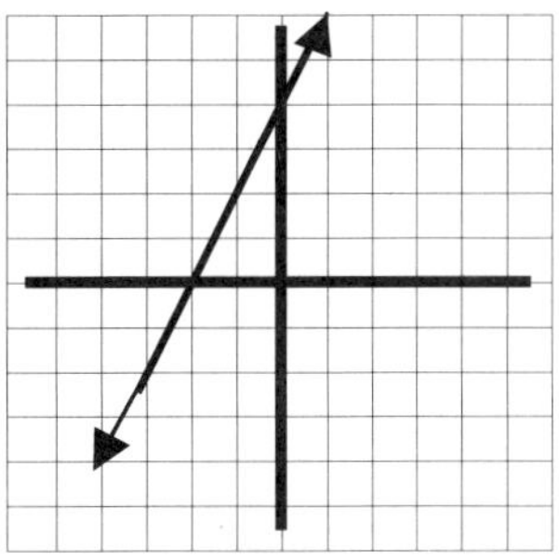

3.

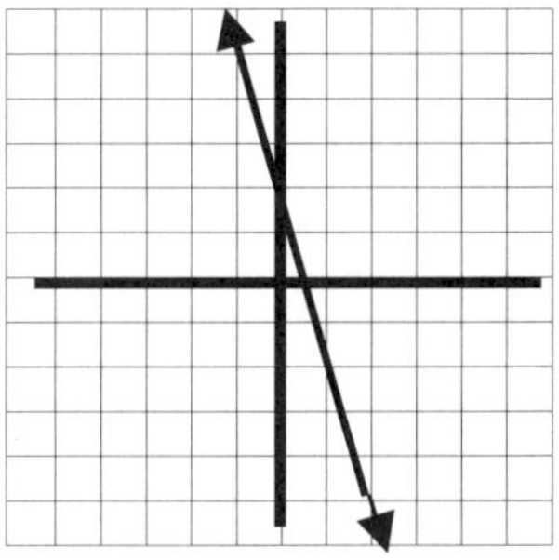

4.

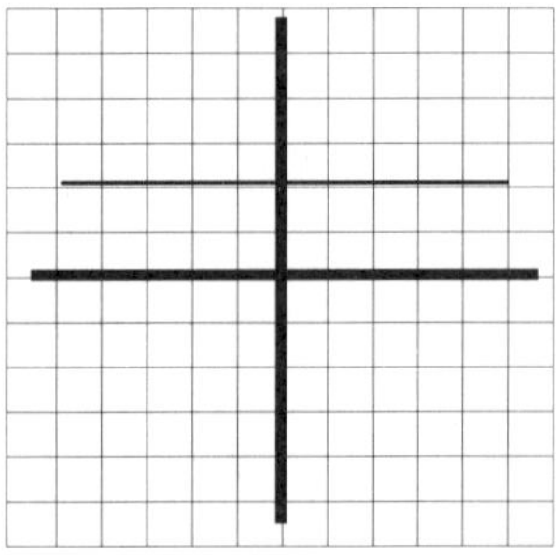

Answer Q9:
$y=0.3(128)+25$
$= \$63.4$.

5.

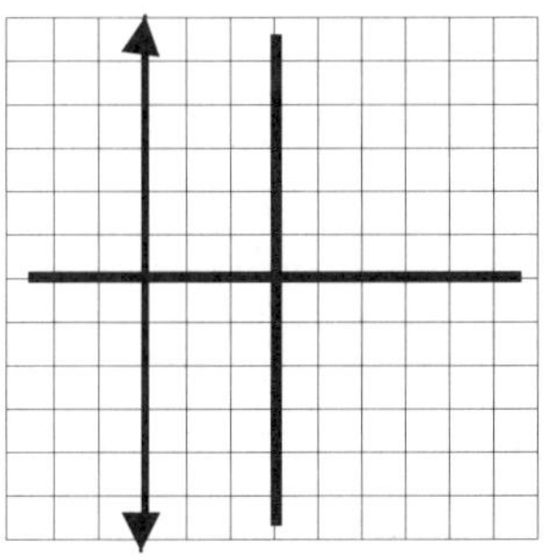

6.

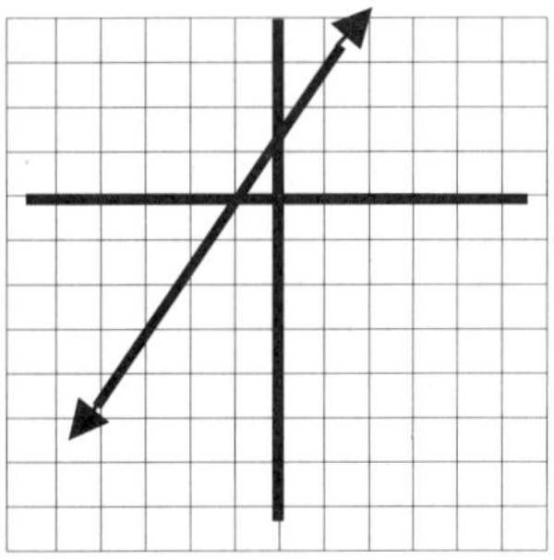

(7–14) Find the slope of the line through the given points:

7. (–2, 3) and (5, –4)

8. (–3, –1) and (5, 3)

9. (2, –3) and (–3, –11)

10. (1, 7) and (10, –3)

11. $(\frac{3}{4}, \frac{-4}{3})$ and $(\frac{-2}{8}, \frac{5}{6})$

12. (–2.34, 5.31) and (–3.54, 2.41)

13. (2, –8) and (–12, –8)

14. (5.2, –7) and (5.2, 12)

(15–26) For each of the following equations find the slope and the y–intercept:

15. $y = \frac{2}{3}x + 7$

16. $y = \frac{-3}{5}x - 9$

17. $y = 7 - \frac{8}{3}x$

18. $y = 12 + \frac{3}{11}x$

19. $y = -4$

20. $x = \frac{3}{4}$

21. $3x - 2y = 8$

22. $2x + 5y = 9$

23. $5x = 8 - 2y$

24. $3y + 4 = 7x$

25. $\frac{3}{4}x + \frac{2}{3}y = 3$

26. $\frac{5}{6}x - \frac{3}{4}y = \frac{7}{12}$

(27–46) Write the slope-intercept form equation of each line that fits the given information. (If you write the original equation in point slope form change it to slope-intercept form. If the line is a vertical or horizontal line write in the appropriate form.)

Answer Q10: The 12 would represent a \$12 shirt and x(12) would mean how many shirts you could expect to sell if you charged \$12 a shirt.

27. Slope = $\frac{-2}{3}$, y-intercept is (0, –4)
28. Slope = $\frac{3}{7}$, y-intercept is (0, 9)
29. Slope = 4, y-intercept is (0, $\frac{3}{8}$)
30. Slope = –9, y-intercept is (0, –3.84)
31. Slope = 0, y-intercept is (0, –5)
32. Slope is undefined, y-intercept is (0, 8)
33. Slope = 3, passes through (–4, 5)
34. Slope = –5, passes through (3, –4)
35. Slope = $\frac{-3}{5}$, passes through (–2, –8)
36. Slope = $\frac{2}{7}$, passes through (3, 7)
37. Slope is undefined, passes through (–10, 3)
38. Slope = 0, passes through (12, –11)
39. Passes through (–4, 5) and (1, –15)
40. Passes through (–3, 8) and (–5, –4)
41. Passes through (2, 8) and (–3, –1)
42. Passes through (–3, –5) and (–7, –2)
43. Passes through (–1.2, 3.2) and (0.3, –5.8)
44. Passes through (2.3, –3.4) and (5.2, –1.3)
45. Passes through (2.3, –5.4) and (–3.8, –5.4)
46. Passes through (3.8, 2.7) and (3.8, –5.5)

(47–52) Determine whether the pair of lines is parallel, perpendicular or neither.

47. $y = \frac{2}{3}x + 3$

 $2x - 3y = 5$

48. $2x + 5y = 11$

 $5x - 2y = 8$

49. $3x + 2y = 7$

 $2x + 3y = 11$

50. $y = \frac{3}{5}x + 4$

 $3x + 5y = 10$

51. $\frac{3}{4}x - \frac{3}{5}y = 4$

 $\frac{9}{5}x + \frac{9}{4}y = 3$

52. $\frac{3}{5}y = \frac{1}{10}x + \frac{3}{10}$

 $\frac{2}{3}y = \frac{7}{12} + \frac{1}{9}x$

(53–66) Find the equation of a line through the given point that is parallel to the given line. Then write the equation of a line through the given point that is perpendicular to the given line.

53. $(-2, 5)$, $y = \frac{2}{3}x + 3$

54. $(3, -3)$, $y = 5 - \frac{3}{5}x$

55. $(-3, -7)$, $y = 8 - 0.7x$

56. $(2, 5)$, $y = 1.3x + 4$

57. $(3,9)$, $f(x) = 3x + \frac{3}{5}$

58. $(-4, -8)$, $g(x) = \frac{2}{7} - 5x$

59. $(2, -4)$, $3x + 2y = 8$

60. $(-4, 9)$, $5x - 4y = 3$

61. $(-2, -5)$, $-10x + 5y = 7$

62. $(4, 3)$, $12x - 3y = 11$

63. $(-5, 7)$, $\frac{3}{8}x - \frac{5}{4}y = 6$

64. $(3, -8)$, $\frac{2}{5}x = 2 - \frac{3}{10}y$

65. $(3, -4)$, $x = 5$

66. $(-2, -5)$, $y = 3$

67. A market survey has found that in a week 50 pairs of shoes are sold when the price per pair of shoes is \$85.00. A market survey done in a similar market finds that in a week 60 pairs of shoes are sold when the price per pair of shoes is \$80.
 a. Assuming the relationship between pairs of shoes sold and price is linear, give a linear function $P(x)$, where x is the cost for a pair of shoes and P is the number of shoes sold.
 b. How many pairs of shoes would be sold at a price of \$95.00 per pair?
 c. What is the price per pair of shoes if you sell 48 pairs of shoes?
 d. What does the slope tell you about the relationship between price and pairs of shoes sold?

68. A company selling wigs is willing to produce 120 of their premium wigs in one week if the price they receive for a premium wig is \$250.00. If the price they receive for a premium wig goes up to \$310.00 they will produce 300 wigs. (This is a supply question.)

a. Assuming the relationship between the price the company receives for a premium wig and the number they are willing to produce is linear, give a linear function $W(x)$, where x is the price received for a premium wig and W is the number that will be produced.

b. How many premium wigs will the company produce for \$220.00 per wig?

c. If the company is willing to produce 60 premium wigs, how much are they receiving per wig?

d. What does the slope tell you about the relationship between price and number of premium wigs produced?

69. There are 50 apartments in a complex. If the rent is \$560.00 per apartment all 50 units can be rented. If the rent is raised to \$600.00 per apartment only 45 units are rented.

a. Assuming the relationship between rent and the number of apartments rented is linear, find the linear function $A(x)$, where x is the rental price and A is the number of units rented.

b. How many units can be rented if the rent is \$680.00?

c. If there are 28 units rented, what was the rent?

d. What does the slope tell you about the relationship between rent and the number of units rented?

70. A tour company bus has a capacity of 52 people; the company has found that if they charge \$320 for their tour they can fill their bus to capacity. If they charge \$380 for their tour they have only 40 passengers.

a. Assuming the relationship between the number of passengers and the price charged is linear, find the linear function $P(x)$, where x is the price charged and P is the number of passengers.

b. How many passengers will buy tour tickets if the price of the tour is \$410?

c. If 48 passengers buy tickets what was the tour price?

d. What does the slope tell you about the relationship between the price and the number of passengers?

71. \$5000 was invested in an account in 1990, at simple interest, which is a linear function. There is \$5200 in the account in 1995.

a. Create a function $A(t)$ that gives the amount of money in the account for any given year. A represents the amount of money in the account and t the number of years, where $t = 0$ in the year 1995.

b. How much money will be in the account in the year 2010?

c. When will there be \$6200 in the account?

d. What does the slope tell you about the relationship between the amount of money invested and how long the money is left in the account?

72. A warehouse is worth \$2,000,000 on the company books in the year 2000. A company depreciates the value of the warehouse by the same amount each year. In 2010 the warehouse will be worth \$900,000 on the company books.
    a. Find the linear function $V(t)$ that gives the value of the warehouse on the company books for any given year. V represents the value of the warehouse and t the number of years, where $t = 0$ in the year 2000.
    b. How much will the warehouse be worth on the company books in 2015?
    c. What year will the warehouse be worth \$0 on the company books?
    d. What does the slope tell you about the relationship between the worth of the warehouse on the books and the number of years?

73. A car company charges \$150 a week and \$0.15 per mile for renting a mid–size car.
    a. Find the linear function $C(m)$ that gives the cost of renting the car for a week in terms of the number of miles driven. C is the cost for the week and m is the number of miles driven.
    b. How much will it cost to drive 520 miles for the week?
    c. A salesman is given an allowance of \$210 a week for a rental car. How many miles can the salesman drive and not have to pay any money out of his on pocket?
    d. What does the slope tell you about the relationship between the number of miles driven in a week and cost of renting the car?

74. A car company charges \$200 a week and \$0.10 per mile for every mile above 200 for renting a mid–size car.
    a. Find the linear function $A(m)$ that gives the cost of renting the car for a week in terms of the number of miles driven. A is the cost for the week and m is the number of miles driven.
    b. How much will it cost to drive 520 miles for the week?
    c. Comparing the rental with problem 73, how many miles must be driven for problem 74 to be the best deal?
    d. What does the slope tell you about the relationship between the number of miles driven in a week and the cost of renting the car?

75.

Infant Mortality Rate per 1,000 live births:

1993 (t = 0)	8.4
1994 (t = 1)	8.0
1995 (t = 2)	7.6

(Source World Almanac 2000 page 248)

a. Find the linear function $M(t)$ that gives the infant mortality rate as a function of time. M is the mortality rate and t is time in years.

b. Predict the infant mortality rate per 1,000 live births in the year 2002.

c. Predict the year the infant mortality rate will reach 5.4 per 1,000 live births.

76.

Average cost of a lb. of lead in dollars:

1996 ($t = 0$)	\$0.49
1997 ($t = 1$)	\$0.47
1998 ($t = 2$)	\$0.45

(Source Statistical Abstract of the U.S. 2000 page 698)

a. Find the linear function $L(t)$ that gives the cost of a lb. of lead as a function of time. L is the cost per a lb. of lead and t is time.

b. Predict the cost of a lb. of lead in the year 2001.

c. Predict the year the cost of a lb. of lead will be \$0.32.

77.

Number of students per computer in U.S. public schools:

1990 ($t = 0$)	22
1992 ($t = 2$)	18
1994 ($t = 4$)	14

(Source World Almanac 2000 page 248)

a. Find the linear function $C(t)$ that gives the number of student per computer in U.S. public school as a function of time. C is the number of students per computer and t is time.

b. Predict the number of students per computer in the year 2000.

c. Predict the year when the number of students per computer will be 4.

78.

Sales in millions of dollars for monochrome televisions:

1995 ($t = 0$)	34
1997 ($t = 2$)	27
1999 ($t = 4$)	20

(Source Statistical Abstract of the U.S. 2000 page 750)

a. Find the linear function $M(t)$ that gives the sales in millions of dollars for monochrome televisions as a function of time. M is the sales in millions of dollars of monochrome televisions and t is time in years.
b. Predict the sales in millions of dollars of monochrome televisions in 2005.
c. Predict the year when \$15,000,000 of monochrome televisions will be sold.

79.

Social Security Trust Funds

Old–Age and Survivors Insurance Trust Fund

1996 ($t = 0$)	356,000
1997 ($t = 1$)	386,000
1998 ($t = 2$)	416,000

(Source World Almanac 2000 page 766)

a. Find the linear function $S(t)$ that gives the amount of money in millions of dollars in the Old–Age and Survivors Insurance Trust Fund as a function of time. S is the amount of money in millions of dollars in the Old–Age and Survivors Insurance Trust Fund and t is time in years.
b. Predict the amount of money in millions of dollars in the Old–Age and Survivors Trust Fund in the year 2004.
c. Predict the year when there will be 500,000 millions in Trust Fund.

80.

Energy production in North America in quadrillion Btu:

1993 ($t = 0$)	351.2
1995 ($t = 2$)	365.2
1997 ($t = 4$)	379.2

(Source Statistical Abstract of the U.S. 2000 page 843)

a. Find the linear function $E(t)$ that gives the amount of energy in quadrillion Btu in North America as a function of time. E is the amount of energy produced in quadrillion Btu and t is the time in years.
b. Predict the amount of energy in quadrillion Btu that will be produced in 2002.
c. Predict the year when 400 quadrillion Btu will be produced.

Section 5.2 Systems of Two-Variable Equations

Let's move on to discuss systems of equations. Once again, you have probably seen this topic before but we will spend a little time reviewing it.

A **system of equations** is a group of equations that give us information about the same problem.

Many real-life problems can be described by a group, or system, of equations. To create solutions to these problems, we need to be able to solve the system of equations, that is, find the point or points that make all the equations true.

The **graph** of an equation is a picture of all the solutions of the equation.

An important thing to remember about equations is that they have a corresponding graph. One way to solve a system of equations is to plot all of the equations on one graph and look for any points (solutions) they have in common. In other words, we want to find where all the graphs intersect.

The graphing calculator is a wonderful tool for just such a job as long as there are only two variables.

Example 1: Solving a System of Equations Graphically

Find the solution to the system of linear equations by using a graphing calculator.

$$\begin{cases} 19x - 2y = 34 \\ 15x + 22y = -38 \end{cases}$$

Solution:

<table>
<tr><td>To find the solution on a graphing calculator we must first solve each equation for y so that we can type the equations into the calculator.</td><td>$-2y = 34 - 19x$
$y = -17 + \frac{19}{2}x$</td><td>$22y = -15x - 38$
$y = \frac{-15}{22}x - \frac{38}{22}$</td></tr>
<tr><td>Graph both equations.</td><td colspan="2">

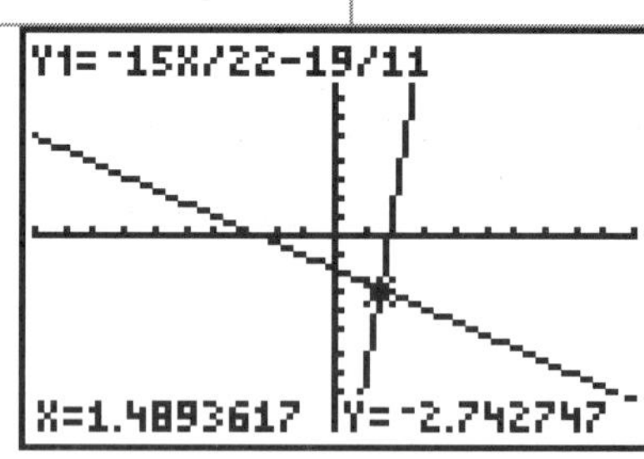

</td></tr>
<tr><td>By tracing we can see that the answer is close to (1.5, -2.7). Now use the [2nd] [TRACE] keys and then the intersect command to find our answer. (When you use</td><td>

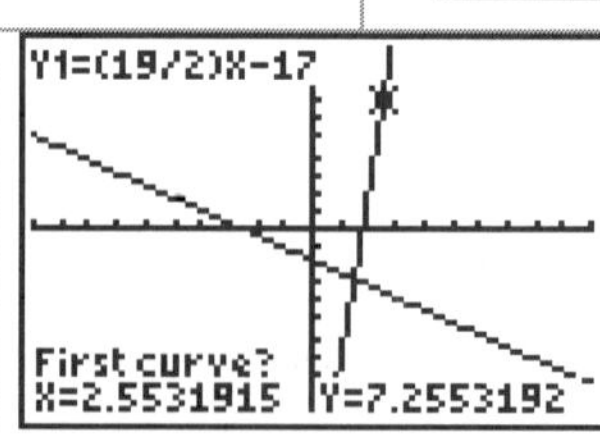

</td><td>

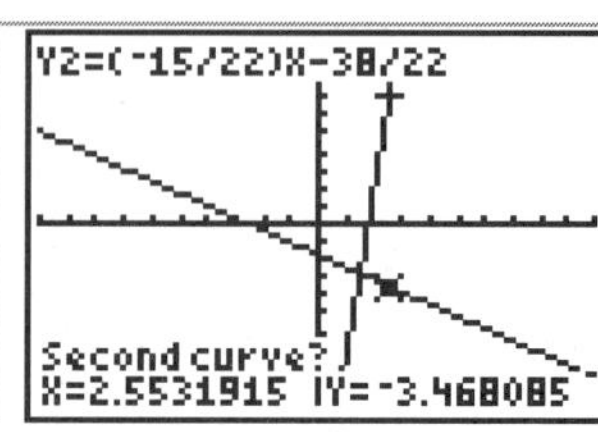

</td></tr>
</table>

the intersect command you must push ENTER to select the 1st graph and push ENTER to select the 2nd graph. Then move the cursor close to the intersection and push ENTER again, like you did with the root/zero command.)	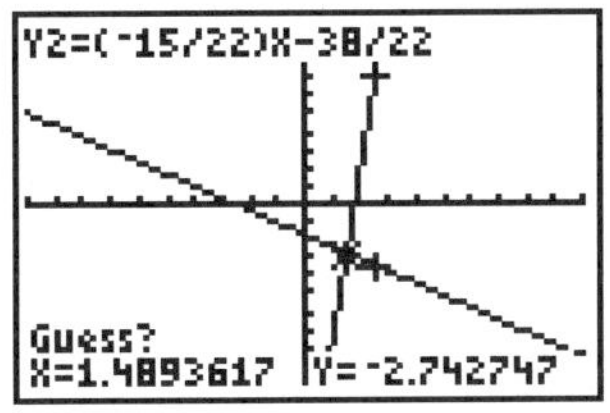 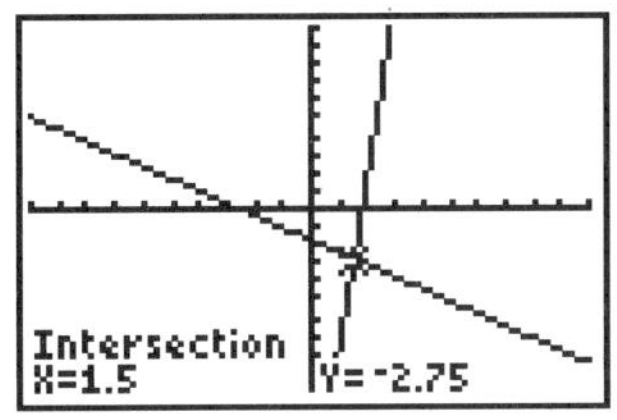

Now we can use the Frac command on our calculator to get the fractional solution: Go to the main screen. Get the x-value (X,T, θ,n key) on the screen "Frac" it. Push the ALPHA key followed by the 1 key (this gives you Y). "Frac" it. Notice these lines intersect at only *one* point. We call a system of equations with this kind of solution **consistent** and **independent.**	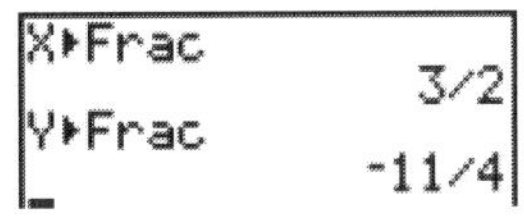 The point of intersection is (3/2, –11/4).

Question 1: What would the solution be if the lines representing a system of equations didn't intersect? What would have to be true about the two lines for them to not intersect? (By the way, we would call this system **inconsistent**.)

A lot of times, especially with real-world data, the point(s) of intersection won't turn out to be rational answers, so we can only approximate them with our calculators. Let's look at two ways in which we can find the solution to a system of equations exactly.

Example 2: Solving a System of Equations Algebraically

Find the solution to this system of linear equations.

$$\begin{cases} 2x+3y=12 \\ x-\ y=1 \end{cases}$$

Solution:

Another method used to solve systems is called the **substitution method**. To use this method you must solve one of the equations for one of the variables then substitute into the other equation.	$x-y=1$ could be written $x=y+1$ Now replace the x in $2x + 3y = 12$ with $(\underline{y+1})$ $2(\underline{y+1}) + 3y = 12$

Solve the linear equation for *y*.	$2(\underline{y+1}) + 3y = 12$ $2y + 2 + 3y = 12$ $5y + 2 = 12$ $5y = 10$ $y = 2$
Now plug 2 in for *y* in either of the two original equations to find the value of *x*.	$x - (2) = 1$ $x = 3$
The solution is the point	(3, 2) [where the lines intersect]

Example 3: A Dependent System of Equations

Find the solution to this system of linear equations.

$$\begin{cases} 2x + 3y = 6 \\ -4x - 6y = -12 \end{cases}$$

Solution:

To use **substitution** you must solve one of the equations for one of the variables then substitute into the other equation.	$2x + 3y = 6 \Rightarrow 2x = -3y + 6 \Rightarrow x = \dfrac{-3y+6}{2}$ Now replace the *x* in $-4x - 6y = -12$ with $\left(\dfrac{\mathbf{-3y+6}}{\mathbf{2}}\right)$ $-4\left(\dfrac{\mathbf{-3y+6}}{\mathbf{2}}\right) - 6y = -12$
Solve the linear equation for *y*.	$-4\left(\dfrac{-3y+6}{2}\right) - 6y = -12$ $6y - 12 - 6y = -12$ $-12 = -12$
$-12 = -12$ always (regardless of *x* and *y* in this case), so this means that *any point on the graph of these equations is the solution.* The equations must be the same equation. In this example, if you multiply the first equation by –2 it's the same as the second one. Here is the graph of both equations. They are identical.	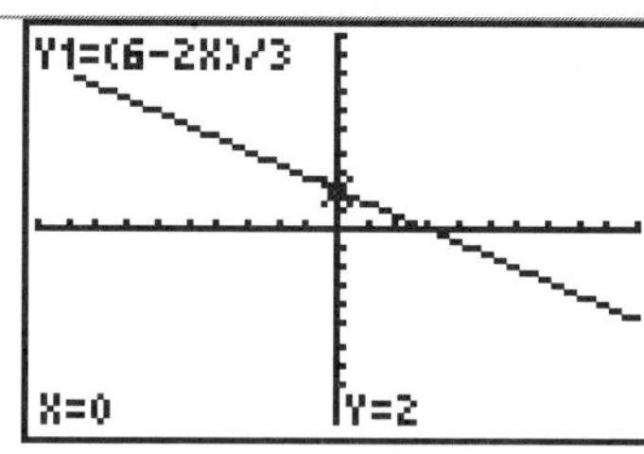 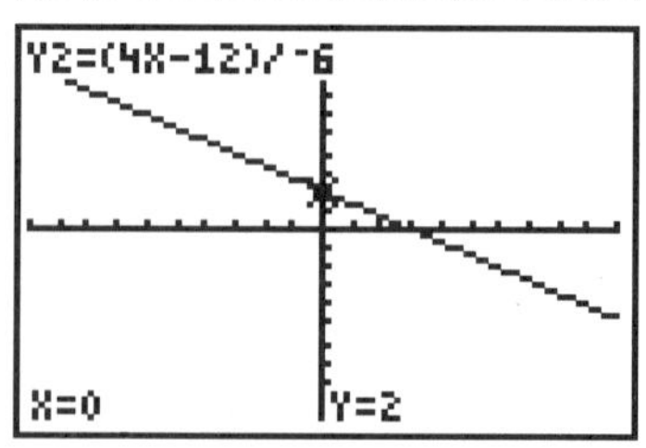 These are the *same line.*

In mathematics, this type of system is called **consistent** and **dependent**.	

Another algebraic method used to find a solution to a system of equations is called the **addition method.** With the addition method, you want to start with each equation written in standard form. The idea is to get the coefficients of one of the variables to be the same magnitude but have the opposite sign, so that when we add the equations together one of the variables will vanish.

We can add one equation to another, because we are really just adding equal amounts to both sides of the equation. For example:

If we know that the following two equations are true statements,	$2+5=7$ $+\quad \underline{4-1=3}$
Then it should seem reasonable that this must also be true.	$(2+5)+(4-1)=7+3$
So if we have this,	$2x+5y=7$ $+\quad \underline{4x-1y=3}$
Then this must be true.	$(2x+5y)+(4x-1y)=7+3$
Or, more specifically, we hope that we can get something like this to happen.	$7x-3y=14$ $+\quad \underline{2x+3y=-5}$
When the two equations are added together it gives us this, which only has one variable now.	$9x \qquad = 9$

Answer Q1: The answer would be "No solution" since they wouldn't have any points in common. The lines would have to be parallel.

This method works well with linear systems of equations. Let's do an example.

Example 4: Solving a System of Equations Using the Addition Method

Find the solution to this system of linear equations.

$$\begin{cases} 3x+5y=-7 \\ 2x-y=5 \end{cases}$$

Solution:

To solve a system by the **addition method** we need to get one of the variables in the two equations to have the same magnitude but opposite sign. Here we can do that by multiplying the second equation by 5	$3x+5y=-7$ $5(2x-y=5)$ $3x+5y=-7$ $\underline{10x-5y=25}$

When we add these two equations together we now get	$13x = 18$
Which, when solved for x, gives	$x = \frac{18}{13}$
Now substitute x into one of the equations to find the y-value.	$2(\frac{18}{13}) - y = 5 \Rightarrow \frac{36}{13} + 5 = y \Rightarrow \frac{101}{13} = y$

So our solution is the point $\left(\frac{18}{13}, \frac{101}{13}\right)$.

Question 2: Find the solution to the system of linear equations below by using the addition method.

$$\begin{cases} -x + 3y = -11 \\ 5x - 15y = 5 \end{cases}$$

When you get a false statement (like $0 = -50$) it means that what you are trying to do is impossible. Here is a graph of the two linear functions in the last question.

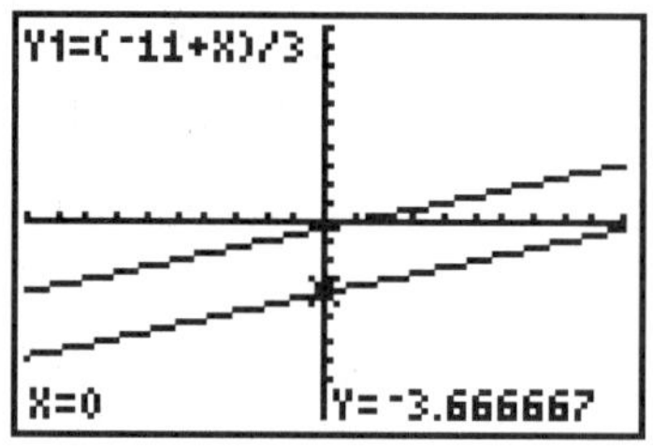

Notice that the *algebraic* outcome of a false statement means *graphically* that the two lines are parallel, which means that the system is inconsistent (it has no solution).

Example 5: Solving a System of Equations Using the Addition Method

Find the solution to the system $\begin{cases} 2x - 5y = 1 \\ 3x - 2y = 7 \end{cases}$.

Solution:

We can multiply *both* equations by numbers in order to eliminate one of the variables.

Multiply the first equation by 3 and the second equation by -2 to make both coefficients of x equal ± 6	$6x - 15y = 3$ $\underline{-6x + 4y = -14}$
When we add these together we get	$-11y = -11$ $y = 1$
Substituting back into the first equation we find	$2x - 5(1) = 1$ $2x - 5 = 1$ $2x = 6$

	$x = 3$
Our solution is the point of intersection at	(3, 1)

Example 6: A Dependent System of Equations

Find the solution to the system $\begin{cases} 2x - 2y = \dfrac{14}{5} \\ -5x + 5y = -7 \end{cases}$.

Solution:

Multiply the first equation by 5 and the second equation by 2 to make both coefficients on y equal $\pm$ 10. It doesn't matter which variable you try to eliminate.	$10x - 10y = 14$ $\underline{-10x + 10y = -14}$
When we add these together we get	$0 = 0$
As in example 3, this means we have two lines that are the same. So, every point on the line is a solution. We like to write the solutions in point form (as an ordered pair) if we have them, and so we will do that here by solving for one variable and using that to make the points.	$-5x + 5y = -7$ $5y = 5x - 7$ (add $5x$) $y = x - \frac{7}{5}$ (divide by 5)
We have an infinite number of solutions since there are an infinite number of points on a line.	$(x, x - \frac{7}{5})$ (x can be any real number)

We now have a general form of all the points that make this dependent system of equations true. If you pick any value for x and plug it into the form $(x, x - \frac{7}{5})$ you will get a point that satisfies the system.

Question 3: Find the expression for the points of solutions to example 3.

Discussion 1: Setting Up and Solving a System of Equations

Let's say that you get a job working for a nut company, and they have all different kinds of nuts. Let's also say that your boss has asked you to come up with a new can of mixed nuts and you've decided to use cashews and macadamias. If cashews cost \$9.50 a pound and macadamias cost \$16.00 a pound and you've been told to keep the price around \$12, then let's see how many pounds of each type of nut would yield a \$12 price on a one-pound can.

There are two equations we need to set up. One has to deal with <u>how much we have</u> (amount equation) → and the other has to do with <u>how much it will cost</u> (cost equation). → Let c be how many pounds of cashews we will use and let m be how many pounds of macadamias we will use.	$c + m = 1$ (want one-pound can) $\underline{9.5(c)} + \underline{16(m)} = \underline{12(1)}$ ↑ ↑ ↑ cost per pound times amount
Now we have two equations and two variables so let's use the addition method to solve our problem. We will multiply the first equation by –9.5 then add.	$-9.5(c + m = 1)$ $9.5c + 16m = 12$ $-9.5c - 9.5m = -9.5$ $\underline{9.5c + 16m = 12}$ $6.5m = 2.5$ $m = 0.3846 = 5/13$
By substituting m into the equation $c + m = 1$, we determine that c (cashews) = 8/13. So, we need to mix 5/13 pounds macadamias with 8/13 pounds cashews to create a new mixed nuts product for the market place with a cost of \$12 per pound.	$c = 1 - 5/13 = 8/13$ The point of intersection of the two lines is (5/13, 8/13).

Answer Q2:
$-5x + 15y = -55$
$\underline{5x - 15y = 5}$
$0 = -50$
Parallel lines

Example 7: System of Equations – Cell Phones

You are looking for a cell phone company and you have narrowed your choices down to two. Company A will give you 300 minutes each month for a fee of \$40 a month and then charge you \$0.25 for each additional minute beyond the initial 300. Company B will give you 400 minutes a month for a fee of \$40 a month and then charge you \$0.40 a minute for each additional minute over the initial 400. How many minutes a month would you need to use in order to make the cost of both phone companies equal if you use more than the initial minutes in each plan? (In economics terms this is called the breakeven point, the point at which one choice is no different than the other.)

Solution:

There are two equations we need to set up. One has to do with the phone bill from company A if we use more than 300 minutes and the other has to do with the phone bill from company B if we use more than 400 minutes. Let m be how many minutes we use and $B(m)$ the total bill (B). Notice that 400 minutes with company A would cost \$40 + \$0.25(100) = \$65 and with company B would cost \$40. But as time goes by, company A will be cheaper because the cost per minute beyond 300 at \$0.25 is less than the \$0.40 cost per minute beyond 400 of company B.

We get the following two equations which are only correct for $m > 400$. Clearly company B is best for anything between 300 and 400 minutes.	Company A: $B(m) = 40 + 0.25\ (m - 300)$ Company B: $B(m) = 40 + 0.40\ (m - 400)$
Use substitution. Clear the parentheses. Collect like terms. Add 120 and subtract $0.25m$. Divide by 0.15.	$40 + 0.25\ (m - 300) = 40 + 0.40\ (m - 400)$ $40 + 0.25m - 75 = 40 + 0.4m - 160$ $0.25m - 35 = 0.4m - 120$ $85 = 0.15m$ $567 = m$ (rounded up)
So, the breakeven point is	(567, \$106.67) [$B(567) = \$106.67$]

Answer Q3: From example 3 we know that, $x = \dfrac{-3y+6}{2}$ so our solutions are $\left(\dfrac{-3y+6}{2}, y\right)$. Or, if we use $y = \dfrac{-2x+6}{3}$, then our solutions are $\left(x, \dfrac{-2x+6}{3}\right)$.

Question 4: Using you graphing calculator graph the two equations in example 8. Does your graph confirm our solution?

We have discussed how to solve systems of linear equations. Here are the important facts to remember:

- The **graph** of an equation is a picture of all the solutions of the equation.
- We can solve systems by graphing, substituting or by adding equations together.
- When we have a system of two lines only three things can happen: one point of intersection, no points of intersection (parallel lines), intersect at all the points on the lines (same lines).

EXPONENTIAL AND LOGARITHMIC SYSTEMS

Since we have finished reviewing how to solve systems of equations we can now talk about how to find exponential and logarithmic equations to model real-world phenomenon. The basic exponential function is $f(t) = ae^{kt}$ (section 4.7), and the basic logarithmic function is $f(t) = a\ln(t) + b$. These formulas have two unknown constants. In the exponential function they are a and k. In the logarithmic function they are a and b. In order to find the unknown constants you must have as many points as you have unknowns. If you noticed, back in section 5.1 we needed two points to find a line. That was because a line has two unknown constants namely m and b in the formula $y = mx + b$.

Now you might be thinking, "we did this back in chapter 4," and you would be right. But, life was a little different back there. We didn't need systems of equations to find exponential models because one point in each problem was the starting point, which we knew was the constant a in the formula. So, we really only had one constant we didn't know and thus one point was all we needed to complete the model. Here we want to talk about the more general case where we might not know the starting point, or we don't want any of the points we do know to be our starting point.

Discussion 2: Solving Systems of Exponential and Logarithmic Equations

Let's see how we go about finding exponential and logarithmic functions that contain the points (3, 100) and (5, 150).

To find the functions we substitute the two points into the general models and then solve the resulting system for the two constants.

Exponential Function

Points (3, 100) and (5, 150) substituted into formula $E(t) = ae^{kt}$

$$\begin{cases} 100 = ae^{3k} \\ 150 = ae^{5k} \end{cases}$$

Now solve this system for a and k.

In chapter 4 we learned how to solve equations of this type by taking the natural log of both sides. So, let's try that here and then use the substitution method we learned in this section.

$\ln(100) = \ln(ae^{3k})$

$\ln(150) = \ln(ae^{5k})$

Using log properties

$\log MN = \log M + \log N$

$\ln(e^x) = x$

we get the following:

$\ln(100) = \ln(a) + \ln(e^{3k}) = \ln(a) + 3k$

$\ln(150) = \ln(a) + \ln(e^{5k}) = \ln(a) + 5k$

Now solve the first equation for $\ln(a)$:

$\ln(a) = \ln(100) - 3k$

Substitute into the second equation:

$\ln(150) = [\ \ln(100) - 3k\ ] + 5k$

Simplify and solve for k:

$\ln(150) = \ln(100) + 2k$

$\ln(150) - \ln(100) = 2k$

$$\frac{\ln(150) - \ln(100)}{2} = k$$

$k \approx 0.2027$

Now substitute back to find a.

Logarithmic Function

Points (3, 100) and (5, 150) substituted into formula $L(t) = a\ln(t) + b$

$$\begin{cases} 100 = a\ln(3) + b \\ 150 = a\ln(5) + b \end{cases}$$

Now solve this system for a and b.

It looks like the addition method will work well here. These two equations are *linear*, not exponential like the problem on the left.

Multiply the second equation by –1 and add equations:

$100 = a\ln(3) + b$

$\underline{-150 = -a\ln(5) - b}$ $\quad$ [$-1 \bullet 2^{\text{nd}}$ equation]

$-50 = a\ln(3) - a\ln(5)$

Simplify and solve for a:

$-50 = a(\ln(3) - \ln(5))$ $\quad$ [factored a]

$$\frac{-50}{\ln(3) - \ln(5)} = a$$

$a \approx 97.881$

Now substitute back to find b.

$100 = ae^{3(0.2027)}$

$100 \approx 1.8369a$

$a \approx 54.44$

Our function then is

$E(t) = 54.44\, e^{0.2027t}$

We used E(t) since this is exponential (E for exponential).

Here is the graph of the two points and the exponential function we found

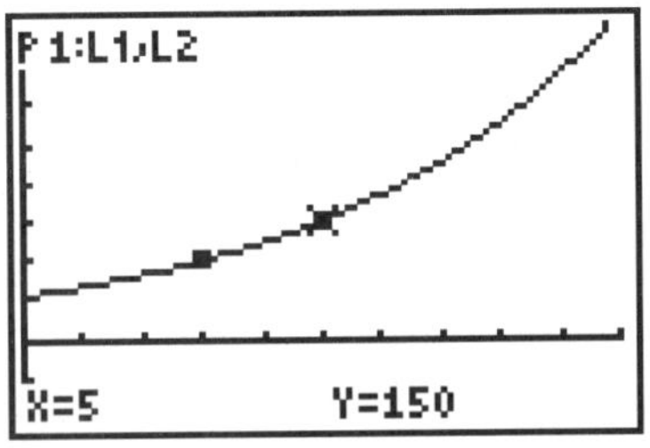

$100 = 97.881 \ln(3) + b$

$100 - 97.881 \ln(3) = b$

$b \approx -7.533$

Our function then is

$L(t) = 97.881 \ln(t) - 7.533$

We used $L(t)$ since this is logarithmic (L for logarithmic).

Here is the graph of the two points and the logarithmic function we found

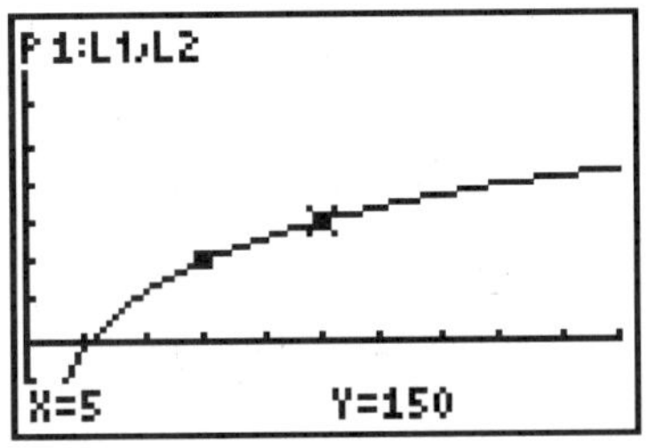

Answer Q4:

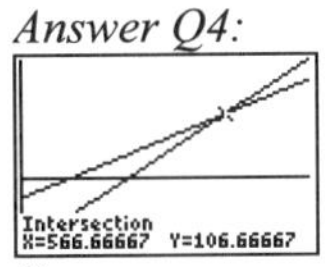

Yes.

You have just seen that finding a logarithmic model is easier than an exponential model. This is true because solving for the coefficients in a logarithmic model amounts to solving a linear system.

Also make note of the fact that though these models both fit the two points given (where $t = 3$ and $t = 5$), they differ greatly in their beginning and ending behavior. If we are going to model real-world phenomenon, we need to find models that fit many points over a long period of time. We will discuss this more in the coming sections of this chapter. For now let's look at a real-world example that has just a few points.

Discussion 3: Modeling Salaries

Here is a table of average salary offers to people graduating college with a bachelor's degree in general business. We will let $t = 0$ represent 1995.

Average Salary Offered to Bachelor's Graduates in General Business			
Year (t)	1997 ($t = 2$)	1998 ($t = 3$)	1999 ($t = 4$)
Salary $S(t)$	\$29300	\$31500	\$33300

Source: 2000 Statistical Abstract of the United States

Now we only need two points to find a function that might fit this table of values, so let's use the first two points. We will use 1995 as are starting point. We will follow the same steps as in the last example.

Exponential Function	Logarithmic Function
Plug in points (2, 29300), (3, 31500) $\begin{cases} 29300 = ae^{2k} \\ 31500 = ae^{3k} \end{cases}$	Plug in points (2, 29300), (3, 31500) $\begin{cases} 29300 = a\ln(2) + b \\ 31500 = a\ln(3) + b \end{cases}$
$\ln(29300) = \ln(a) + \ln(e^{2k})$ $\ln(31500) = \ln(a) + \ln(e^{3k})$	Using the addition method we get
$\ln(29300) = \ln(a) + 2k$ $\ln(31500) = \ln(a) + 3k$ Solve one equation for $\ln(a)$ and substitute: $\ln(a) = \ln(29300) - 2k$	$29300 = a\ln(2) + b$ $\underline{-31500 = -a\ln(3) - b}$ $-2200 = a\ln(2) - a\ln(3)$ Now solve for a:
$\ln(31500) = \underline{[\ \ln(29300) - 2k\ ]} + 3k$	$-2200 = a[\ln(2) - \ln(3)]$ (factored)
Solve for k: $\ln(31500) = \ln(29300) + k$ $k \approx 0.0724$	So divide by the [ ] and $a = \dfrac{-2200}{\ln(2) - \ln(3)} = 5425.8676 \approx 5426$
So, $(29300) = ae^{0.0724\,(2)}$ $a \approx \dfrac{(29300)}{e^{0.0724(2)}} \approx 25350$	Now, $29300 = 5426\ln(2) + b$ $b \approx 29300 - 5426\ln(2) \approx 25539$
Our function then is as follows:	Our function then is as follows:
$S(t) = 25350\, e^{0.0724\, t}$	$S(t) = 5426\ln(t) + 25539$
Here is the graph of this function 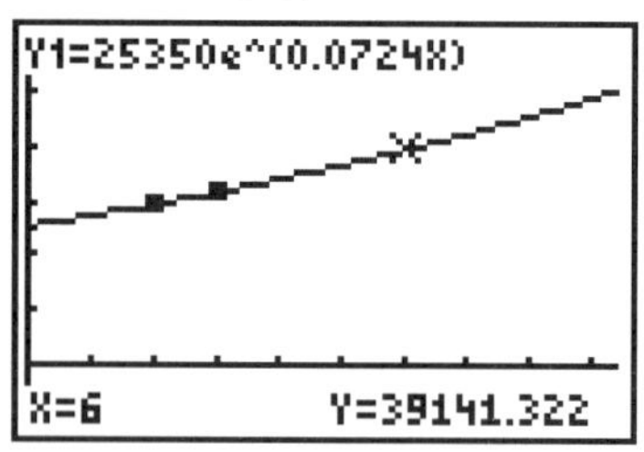 	Here is the graph of this function 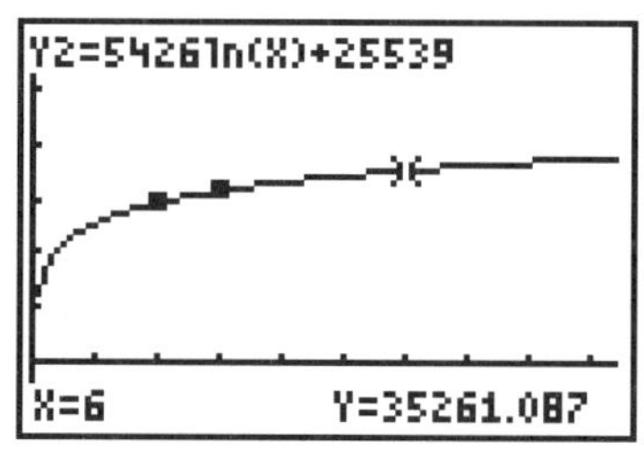

Question 5: What does the indicated point mean in both graphs above?

If I'm a business graduate I hope that the first model is the one that best represents the real world!

Question 6: If the general form of a quadratic function is $y = ax^2 + bx + c$ then how many points would you need to know in order to be able to find the constants a, b, c?

Section Summary

- The **graph** of an equation is a picture of all the solutions of the equation.
- A solution to a system of equations is a point(s) where all the graphs of the equations intersect.
- With linear systems of two variables and two equations you can have either: no solution (parallel), one solution (intersect one point), or an infinite number of solutions (same lines).
- We need two points in order to find a logarithmic ($a\ln(x) + b$) or exponential (ae^{kt}) function that can be used to model a real-world phenomena.
- Once you have two points you substitute them into the function and solve for the constants.
- We need to know as many points as the general formula we are trying to use as a model has constants.

SECTION 5.2 HOMEWORK

Answer Q5: That in the year 2001 (1995 + 6) the average salary offer given to general business graduates with a Bachelor's degree will be \$39,141 in the first function and \$35,261 in the second. Note: According to the National Association of Colleges and Employers the average salary offer to business majors in 2001 was around \$39,500.

(1–6) Find the x and y values that are common solutions for each of the following pairs of linear equations. Use the graphing calculator and the intersection method shown in the reading.

1. $3x + 2y = 2$
 $6x + y = -8$
2. $5x - y = 5$
 $10x + y = 4$
3. $2x - 5y = -2$
 $4x + 3y = 9$
4. $3x + 5y = 1$
 $4x - 3y = -18$
5. $\frac{1}{2}x + \frac{2}{3}y = 8$
 $\frac{3}{4}x - \frac{5}{6}y = 1$
6. $\frac{1}{6}x - \frac{3}{4}y = 4$
 $\frac{5}{12}x + \frac{5}{8}y = -10$

(7–24) Find the x and y values that are the common solutions for each of the following pairs of linear equations. Use any appropriate method, addition or substitution.

7. $x - 2y = 5$
 $3x + 2y = 7$
8. $3x - 5y = 19$
 $4x + 5y = 2$
9. $3x + 5y = -2$
 $7x + 10y = 2$
10. $2x - 7y = 8$
 $6x + 2y = -22$
11. $x = 3y - 20$
 $2x + 3y = 5$
12. $y = 25 - 2x$
 $3x - 2y = 6$
13. $2x + 3y = 15$
 $3x - 2y = -9$
14. $5x + 4y = -23$
 $3x + 7y = 11$
15. $6x = 5y - 3$
 $3y = 4x + 5$
16. $4y = 5 - 3x$
 $7x = 10y - 7$
17. $\frac{3}{4}x - \frac{2}{3}y = \frac{5}{6}$
 $\frac{2}{5}x - \frac{3}{5}y = \frac{7}{15}$
18. $\frac{4}{9}x - \frac{2}{3}y = \frac{7}{18}$
 $\frac{5}{6}x - \frac{7}{18}y = \frac{5}{9}$
19. $3x + 5y = 10$
 $9x = 30 - 15y$
20. $2x - 3y = 15$
 $\frac{4}{5}x - \frac{6}{5}y = 6$
21. $2x - 7y = 12$
 $14y = 4x + 18$
22. $2x - \frac{16}{3}y = 3$
 $3x - 8y = 7$
23. $x = \frac{3}{4}y + 2$
 $4x - 3y = 8$
24. $y = \frac{2}{3}x + 5$
 $2x - 3y = 2$

Answer Q6: Three. You need to know as many points as you have constants in your formula. We will talk about this some in the chapter on matrices.

(25–28) Find the exponential function $E(t) = ae^{kt}$ and logarithm function $L(t) = a\ln(t) + b$ that fits the given pair of points. (Approximate the values of each answer to three decimal places and check the answer with a graph of each function and a stat plot of the two points.)

25. (5, 3) and (10,15)
26. (2, 100) and (6, 250)
27. (5, 80) and (8, 30)
28. (3, 135) and (7, 50)

29. A bacteria colony has 300 bacteria after 10 days and 754 bacteria after 14 days.

    a. If the bacteria colony is growing exponentially, find a function $B(t)$ that will predict the number of bacteria, B, in the colony after t days.

    b. If the bacteria colony is growing as a logarithmically, find a function $B(t)$ that will predict the number of bacteria, B, in the colony after t days.

30. Twenty years after saplings were planted a forest had 570 trees and thirty years after the saplings were planed the forest had 785 trees.

    a. If the number of trees is growing exponentially, find a function $P(t)$ that will predict the number of trees, P, in the forest after t years.

    b. If the number of trees is growing logarithmically, find a function $P(t)$ that will predict the number of trees, P, in the forest after t years.

31. An area was accidentally sprayed with a very toxic chemical. Ten days after the spraying the population of rabbits in the area had dropped to 1,000 and fifteen days after the spraying the rabbit population in the area had dropped to 800.

    a. If the drop in population is exponential, find a function $R(t)$ that will predict the rabbit population, R, in the area t days after spraying.

    b. If the drop in population is logarithmic, find a function $R(t)$ that will predict the rabbit population, R, in the area t days after spraying.

32. The population of a city was 35,000 people two years after a major manufacturing plant closed in the city. Five years after the plan closing there was a population of 32,500 in the city.

    a. If the drop in population is exponential, find a function $P(t)$ that will predict the population, P, when the plant has been closed for t years.

    b. If the drop in population is logarithmic, find a function $P(t)$ that will predict the population, P, when the plant has been closed for t years.

(33–36) Use the first pair of data points to construct the exponential and logarithmic functions.

33. National Income from wages and salaries

(Billions of dollars)

Year	Amount in billions of dollars
1993 ($t = 3$)	3085.2
1995 ($t = 5$)	3424.7
1999 ($t = 9$)	4472.3

(Statistical Abstract of the U.S. 2000)

    a. Find an exponential function $I(t)$ that will predict the income from wages and salaries in billions of dollars, I, after t years.

    b. Find a logarithmic function $I(t)$ that will predict the income from wages and salaries in billions of dollars, I, after t years.

c. Use your graphing calculator to graph the exponential function, logarithmic function and a stat plot of the given data on the same graph. By inspecting these graphs do you think the exponential function or logarithmic function best fits the data?

34. Maximum amount of social security each month for retirees

Year	Monthly social security payments
1990 ($t = 10$)	\$975.00
1994 ($t = 14$)	\$1197.50
1998 ($t = 18$)	\$1342.80

(World Almanac 2000)

a. Find an exponential function $S(t)$ that will predict the monthly social security payments, S, to retirees after t years.

b. Find a logarithmic function $S(t)$ that will predict the monthly social security payments, S, to retirees after t years.

c. Use your graphing calculator to graph the exponential function, logarithmic function and a stat plot of the given data on the same graph. By inspecting these graphs do you think the exponential function or logarithmic function best fits the data?

35. The number of U.S. Farms

(In 1000's)

Year	Number of U.S. Farms in 1000's
1993 ($t = 3$)	2,083,430
1995 ($t = 5$)	2,071,520
1997 ($t = 7$)	2,057,910

(www. fedstats)

a. Find an exponential function $F(t)$ that will predict the number of U.S. farms in 1000's, F, after t years.

b. Find a logarithmic function $F(t)$ that will predict the number of U.S. farms in 1000's, F, after t years.

c. Use your graphing calculator to graph the exponential function, logarithmic function and a stat plot of the given data on the same graph. By inspecting these graphs do you think the exponential function or logarithmic function best fits the data?

36. Number of Mayflies Alive after t days

Day	Mayflies still alive
3	1197098
10	1105164
20	968202

a. Find an exponential function $F(t)$ that will predict the number of mayflies, F, that are still alive after t days.

b. Find a logarithmic function $F(t)$ that will predict the number of mayflies, F, that are still alive after t days.

c. Use your graphing calculator to graph the exponential function, logarithmic function and a stat plot of the given data on the same graph. By inspecting these graphs do you think the exponential function or logarithmic function best fits the data?

Section 5.3 Modeling with the Graphing Calculator

In this section we are going to expand what we have already learned with regard to how we find a function to model real-world phenomenon. In previous sections of the book we showed you how to find the models by hand. In each of those sections we were careful about what data we chose to use for examples and exercises since we wanted to keep the problems at a manageable level. But now, we'd like to talk about how to use technology so that we can handle data more typically found in the real world.

Specifically, we are going to show you how to use your graphing calculators to arrive at a good choice for the function you can use as your model. Before we can start creating models to help "finding solutions to the world's problems" we need to learn how to use some new features on our calculators. As we work through the following example we'll discuss how to use your calculators to find the "best model" for the real-world data.

Discussion 1: Using Your Graphing Calculator to Model Data

Below is a table of values for the total number of civil airports operating in the U.S.A.

Year ($t = 0$ in 1960)	1970 ($t = 10$)	1980 ($t = 20$)	1985 ($t = 25$)	1990 ($t = 30$)	1995 ($t = 35$)	1996 ($t = 36$)	1997 ($t = 37$)
Airports in Operation	11260	15160	16320	17490	18220	18290	18350

Source: 2000 Statistical Abstract of the United States

First, notice that it would be difficult to compute common differences or common ratios because we don't have equal time intervals. Also, remember that in previous sections we only required two data points to create a model; here we've got seven data points. So, we'll turn to our graphing calculator for help.

What we are beginning to look at in this section is called regression analysis. There is much more to this topic than we can cover in this section. If we had the time we would learn about finding what's called "the curve of best fit." One technique used to find the best curve is something called Least Squares. There are other techniques discussed in statistic course(s) but at this level of Mathematics we are going to let the calculator evaluate possible curves and we'll decide which ones might be the best. All we need to know is how to use our calculator and, even more importantly, how to use what it gives us. The first thing we need to do in this example is enter the data into the calculator as follows:

TI-83	TI-86
From the home screen push the STAT key and this is what you'll see: 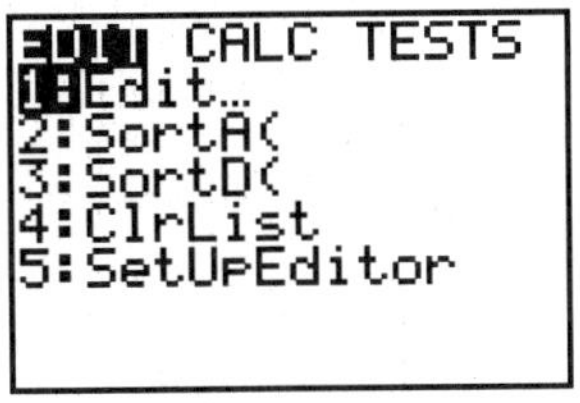 	From the home screen push 2nd and + to get the STAT menu; this is what you'll see: 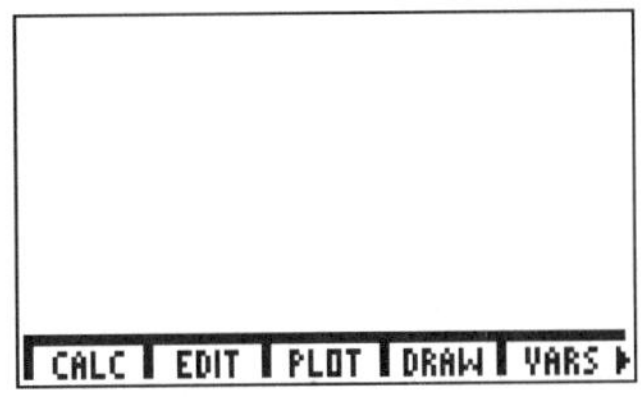
Next, push the 1 key or enter key 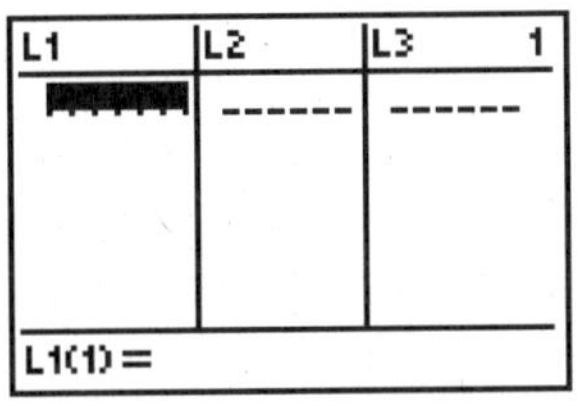 	Next, push the F2 key (EDIT) 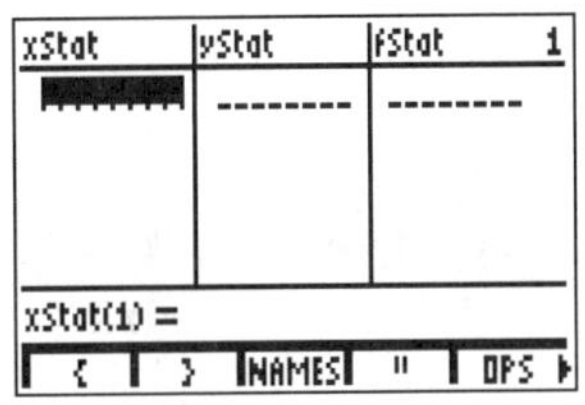
If there are any values already entered on this screen, arrow over and up to the L1, L2, etc. boxes and push CLEAR followed by ENTER to erase the old values. Enter the input values in column L1; hit ENTER after each value. Arrow right to enter the output values in column L2. 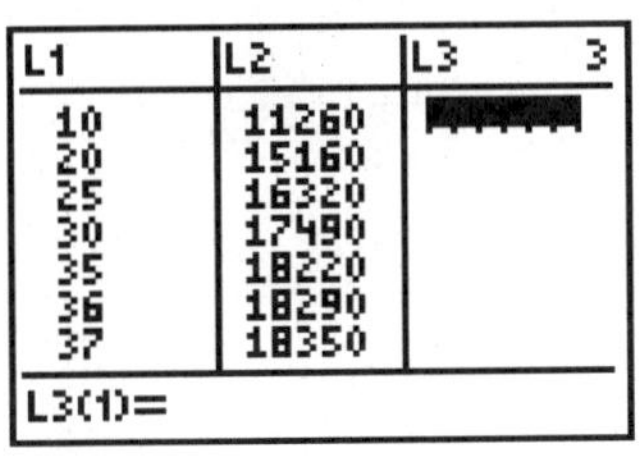 	If there are any values already entered on this screen, arrow over and up to the xStat, yStat, etc. boxes and push CLEAR followed by ENTER to erase the old values. Enter the input values in the xStat column; hit ENTER after each value. Arrow right to enter the output values in the yStat column. 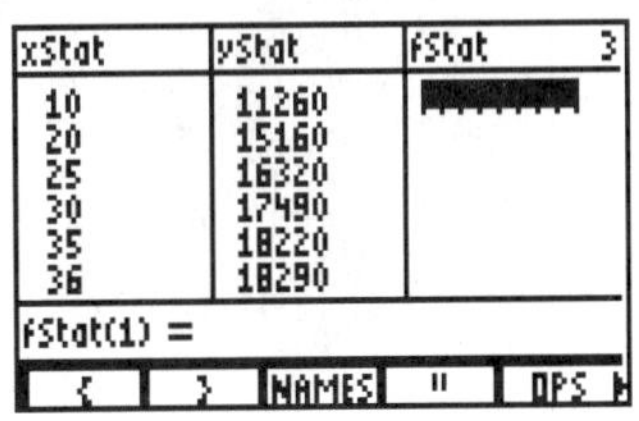

Once you have all the data entered in your calculator make sure that:

1. You have no functions entered in Y= (go to the Y= screen)
2. You turn on the STAT PLOT feature of the calculator by using the up arrow to move the cursor to Plot1 and then push ENTER. It is "on" if it is highlighted.

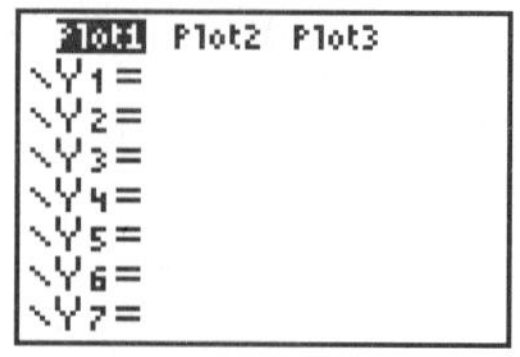

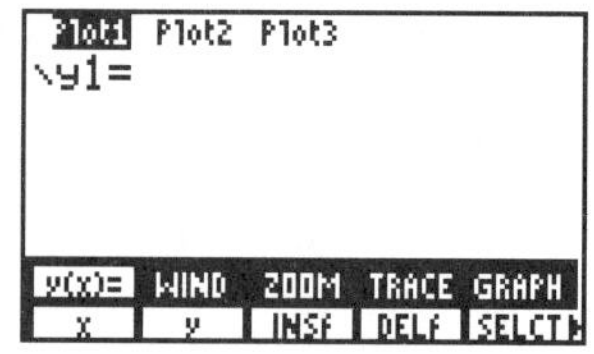

Now use ZoomStat (TI-83) or ZDATA (TI-86) under ZOOM and you'll see a scatter plot (points plotted on a graph).

TI-83	TI-86
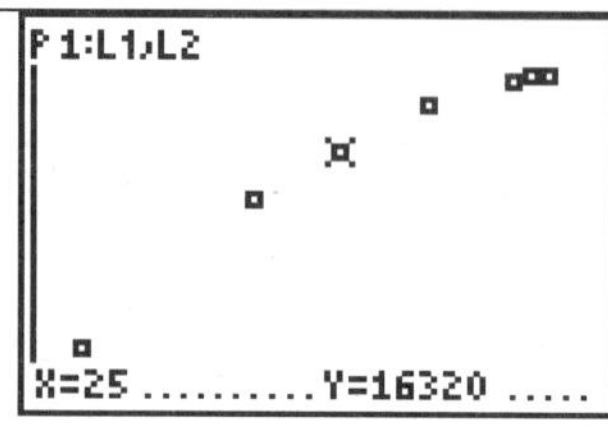	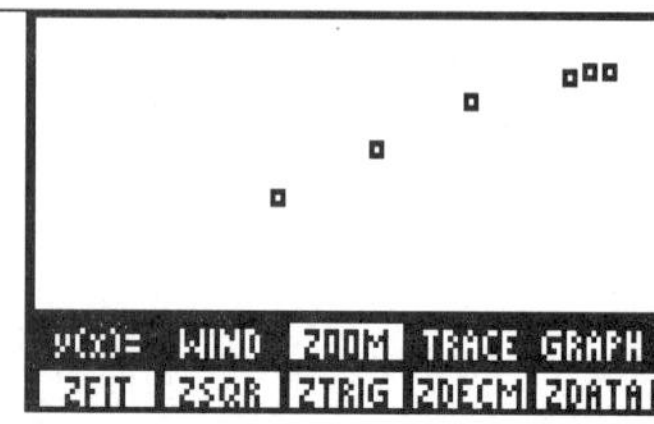

Question 1: Which, of the functions that we have studied so far in this book, would you think might be the best curve to fit this data (linear, quadratic, natural log, exponential)?

Now we use the calculator to find the best linear, exponential, logarithmic, etc. model. Then, given the different answers we get, we'll decide which we think might really be the best to use.

TI-83	TI-86
Push the STAT key, then arrow to the right to highlight the word CALC, and push 4 to get LinReg(ax + b) 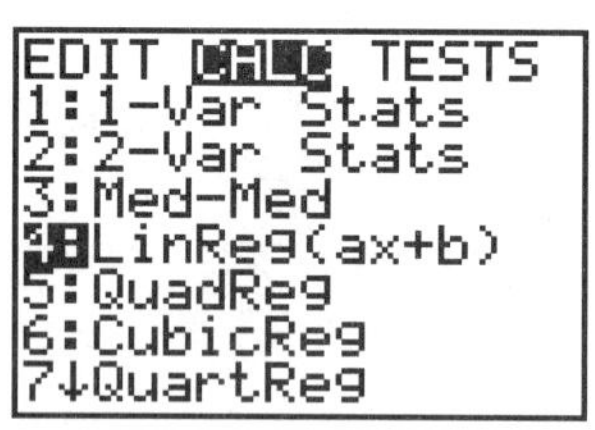	Push 2nd and + keys to get the STAT menu, then push F1 for CALC, then F3 to get LinR 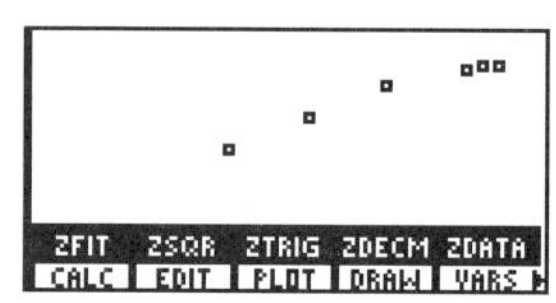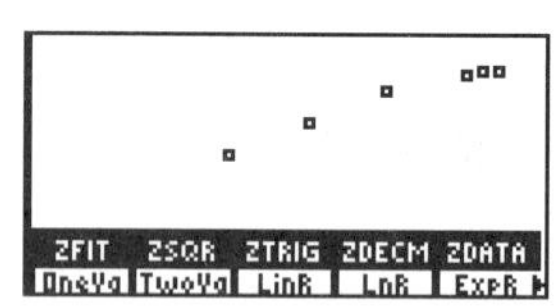
Now push ENTER to tell the calculator to find the "best" line through the points	Push CUSTOM, then xStat followed by , then yStat, then ENTER. If you have not entered these two options in your custom menu then you will need to go to the LIST

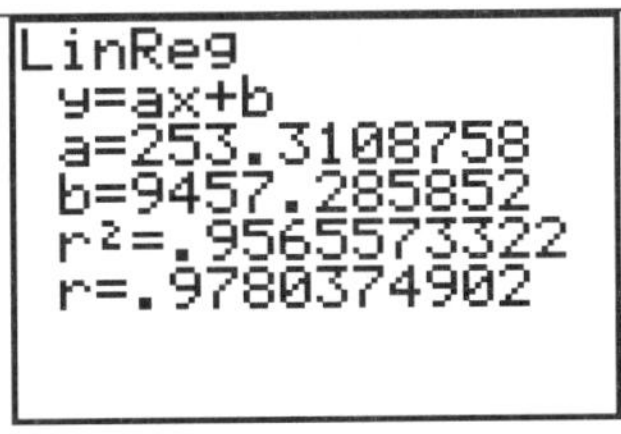 This gives us a the equation of a line that should fit the data: $A(t) = 253t + 9457$	menu (2nd –), then F3 for NAMES, then F2, then ,, then F3. 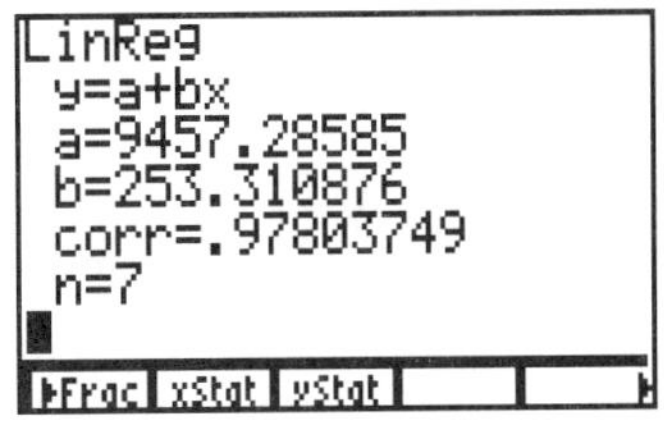 This gives us the equation of a line that should fit the data: $A(t) = 253t + 9457$
Let's graph this linear model and compare it to the scatter plot of our data points. Push Y= then VARS followed by 5. Arrow to the right two spaces to EQ and push 1.  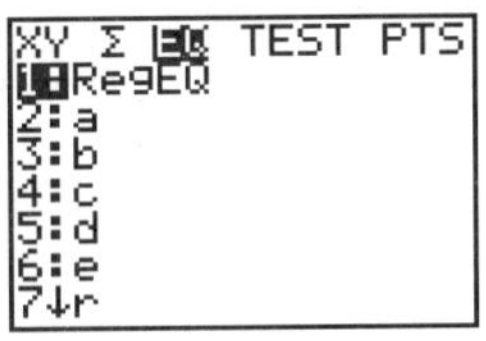 Now graph. 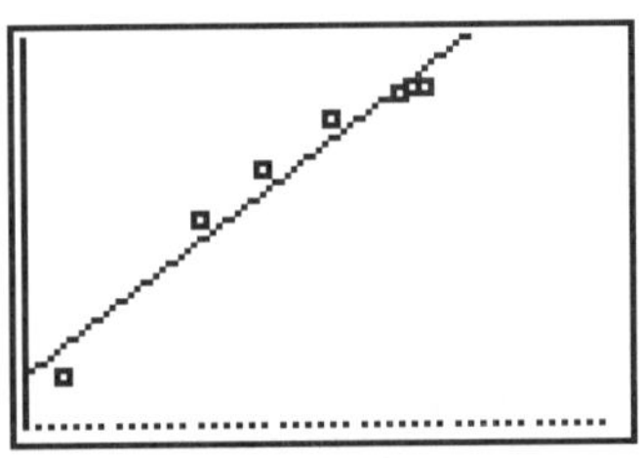 $A(t) = 253t + 9457$	Graph this linear model and compare it to the scatter plot of our data points. Go to Y=, then push 2nd and + followed by F5, then push MORE *twice* followed by F2. 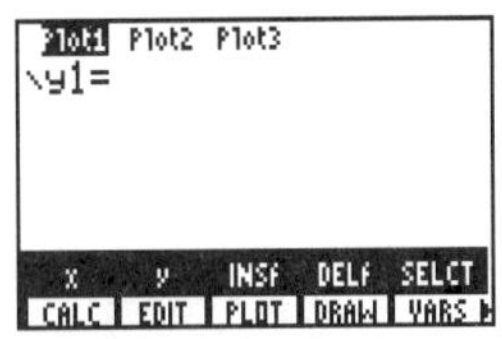 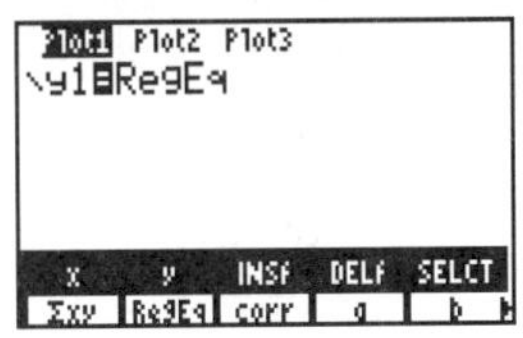 Now graph. 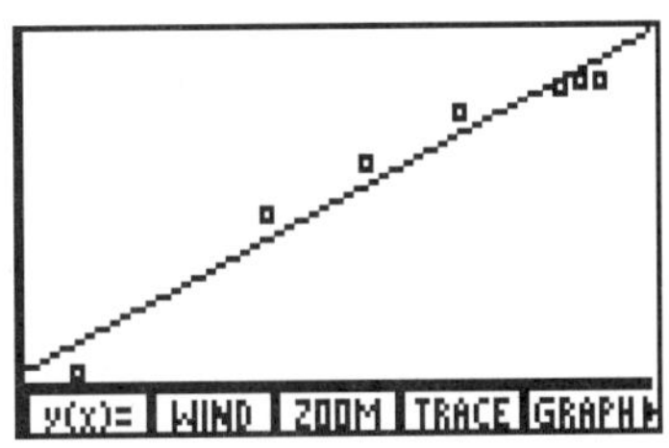 $A(t) = 253t + 9457$
This line doesn't look like it's the best choice possible. But of course we didn't think the best answer would be linear. Here are some of the other possibilities.	This line doesn't look like it's the best choice possible. But of course we didn't think the best answer would be linear. Here are some of the other possibilities.
To find the quadratic possibility follow the same steps above but find this screen on your calculator	To find the quadratic possibility follow the same steps above but find screens like these on your calculator

Answer Q1: Does not seem to be straight enough for linear, It looks like it is not exponential (not growing faster and faster) maybe logarithmic or polynomial (quadratic to be specific).

<table>
<tr>
<td>

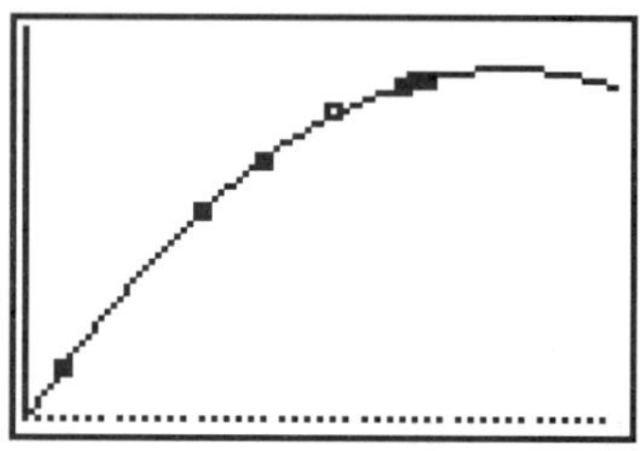

$A(t) = -6.769\,t^2 + 581t + 6148$
</td>
<td>
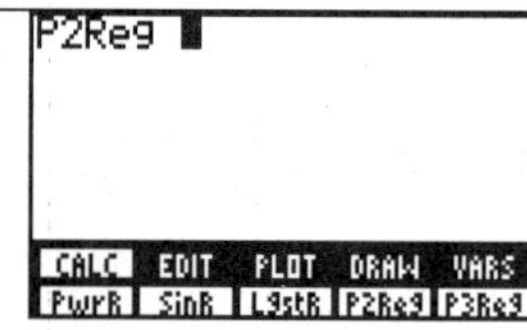

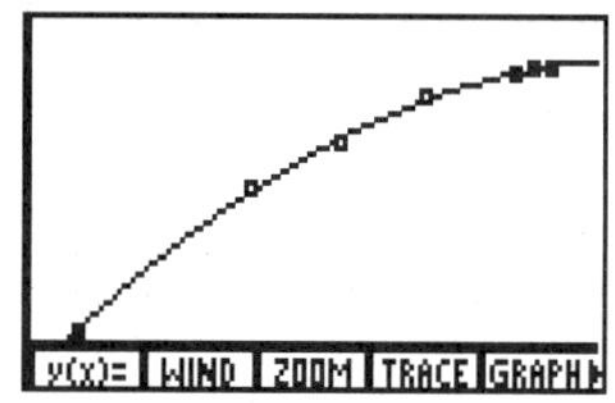

$A(t) = -6.769\,t^2 + 581t + 6148$
</td>
</tr>
<tr>
<td>
To find the natural log (ln) possibility follow the same steps above but find this screen on your calculator

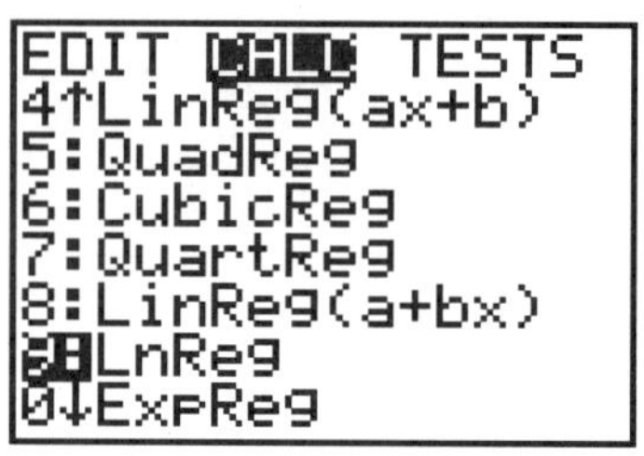

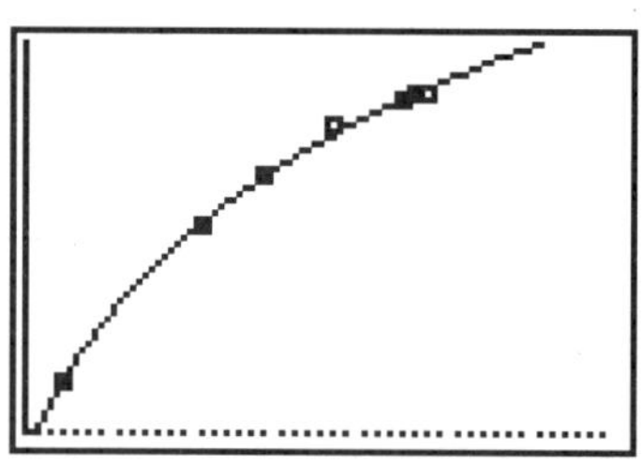

$A(t) = -1336 + 5489.7 \ln t$
</td>
<td>
To find the natural log (ln) possibility follow the same steps above but find screens like these on your calculator

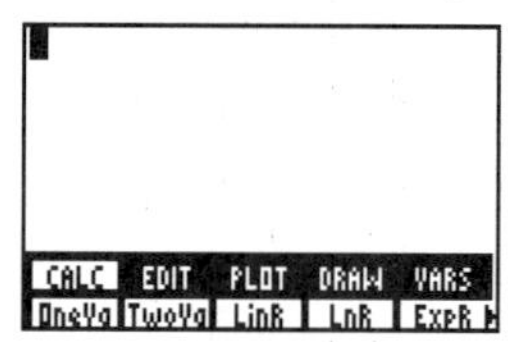

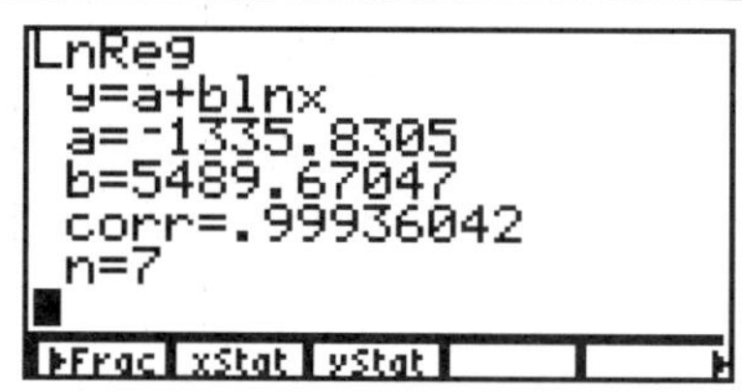

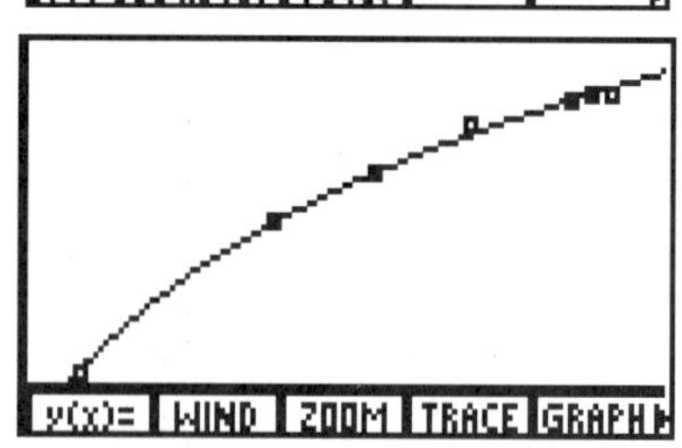

$A(t) = -1336 + 5489.7 \ln t$
</td>
</tr>
<tr>
<td>To find the exponential possibility follow the same steps above but find this screen on</td>
<td>To find the exponential possibility follow the same steps above but find screens like these</td>
</tr>
</table>

your calculator

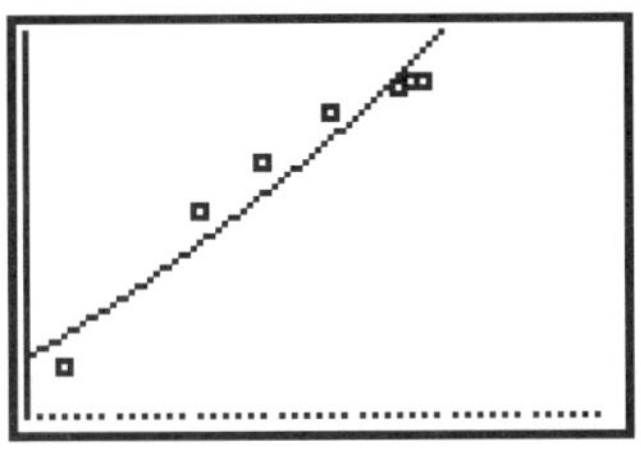

$A(t) = 10145(1.0172)^t$

on your calculator

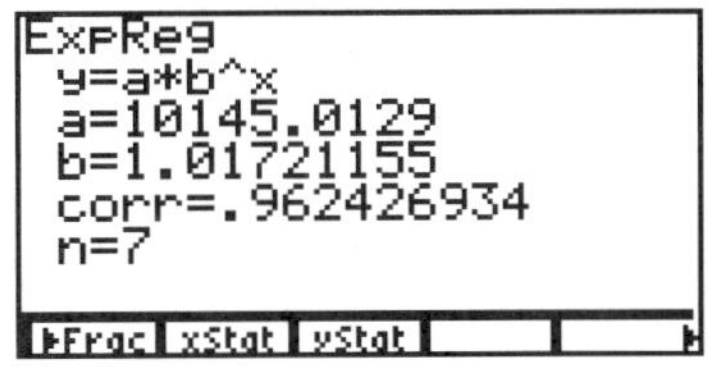

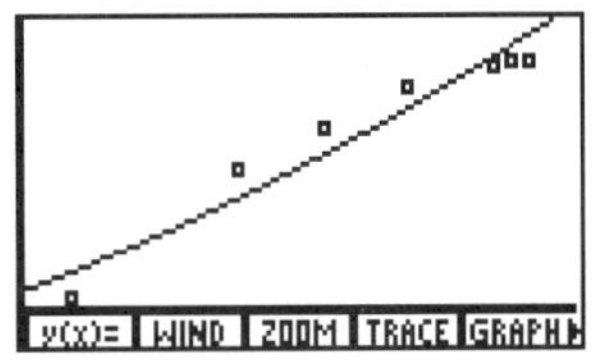

$A(t) = 10145(1.0172)^t$

Question 2: Which of the four graphs, linear, quadratic, natural log, or exponential do you think might be the best function to use as a model for how many civil airports are operating in the U.S.A.?

There are other functions we could look at but we'll leave them for another time. Our biggest concern right now is that you learn how to use your calculator and make good logical choices.

Example 1: Modeling Stock Market Behavior

Find a linear, quadratic, natural log and exponential model for the data below about the Standard and Poor's 500 composite indices from the stock market.

Solution:

Year ($t = 0$ in 1990)	1994 ($t = 4$)	1995 ($t = 5$)	1996 ($t = 6$)	1997 ($t = 7$)	1998 ($t = 8$)	1999 ($t = 9$)
S & P 500	461	615	741	970	1229	1469

Source: 2000 Statistical Abstract of the United States

TI-83	TI-86
First, from the home screen push the [STAT] key.	First, from the home screen push the [2nd] and [+] keys.
Next, push [1] or [ENTER].	Next, push [F2].
If there are any values already entered on	If there are any values already entered on

<table>
<tr>
<td>this screen. simply arrow over and up to the L1, L2, etc. boxes and push the clear key followed by the enter key to erase the old values. Now you are ready to enter the data. You enter the data by first inputting all the input values, each one followed by the ENTER key, then arrow right and input all the output values.
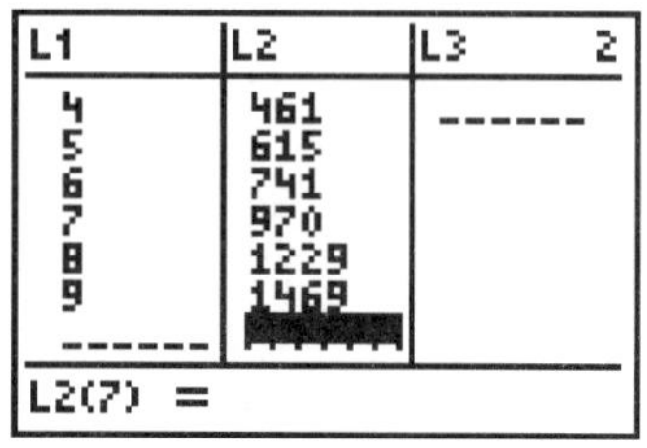
</td>
<td>this screen, simply arrow over and up to the xStat, yStat, etc. boxes and push the clear key followed by the enter key to erase the old values. Now you are ready to enter the data. You enter the data by first inputting all the input values each one followed by the ENTER key then arrow right and input all the output values.
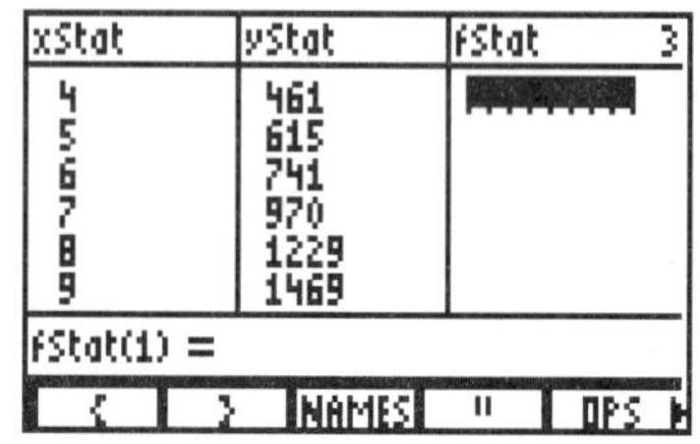
</td>
</tr>
</table>

Now do ZoomStat or ZDATA under ZOOM and you'll see a scatter plot (points plotted on a graph).

<table>
<tr>
<td>TI-83
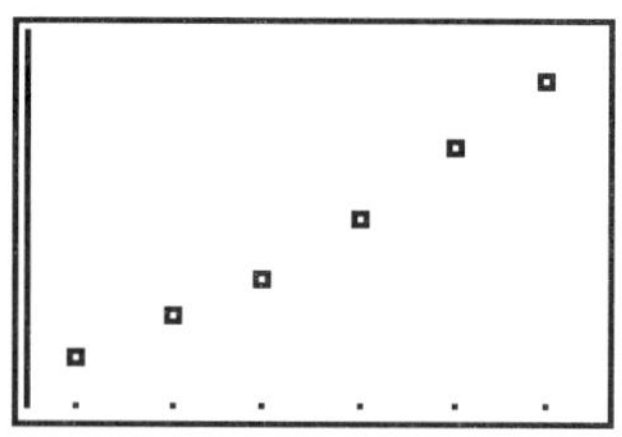</td>
<td>TI-86
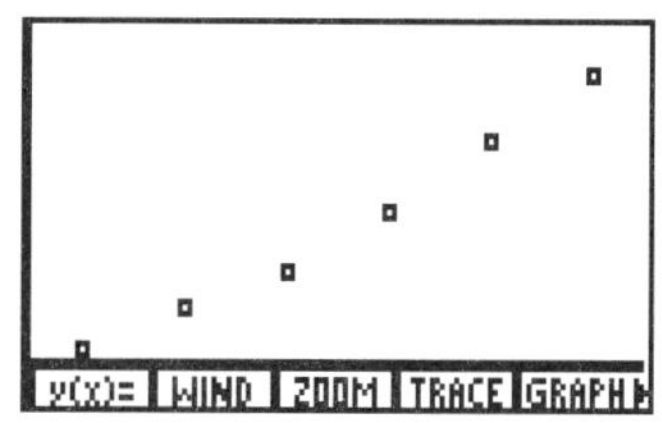
</td>
</tr>
<tr>
<td>Now let's find a linear model
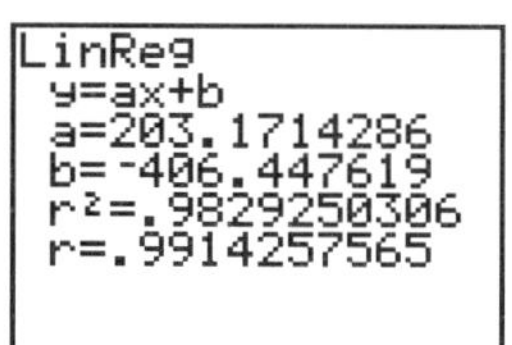

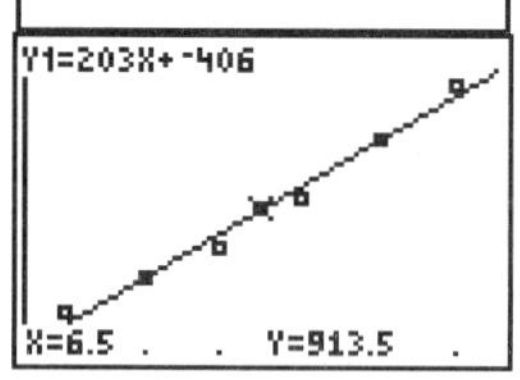
</td>
<td>Now let's find a linear model

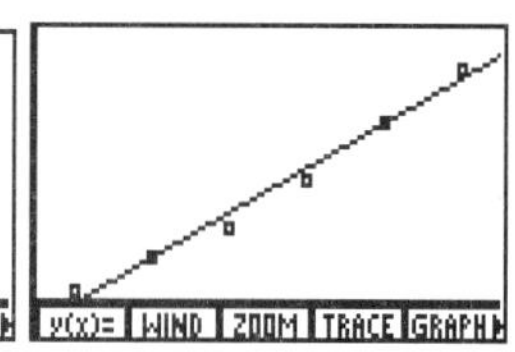
</td>
</tr>
<tr>
<td>Now let's find a quadratic model
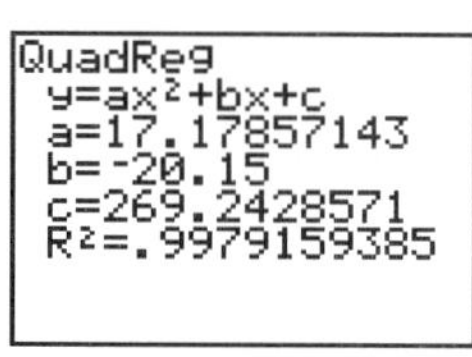

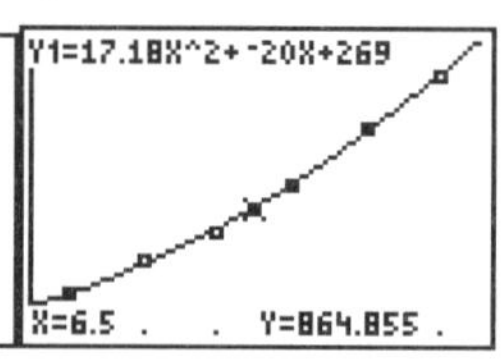
</td>
<td>Now let's find a quadratic model

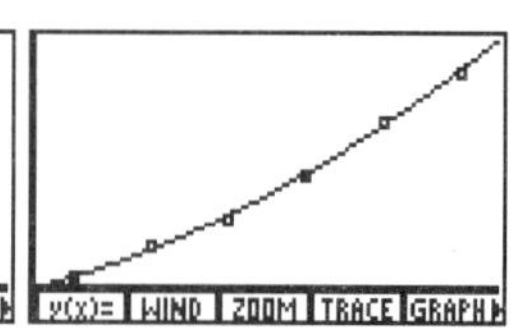
</td>
</tr>
<tr>
<td>Now let's find a natural log model</td>
<td>Now let's find a natural log model</td>
</tr>
</table>

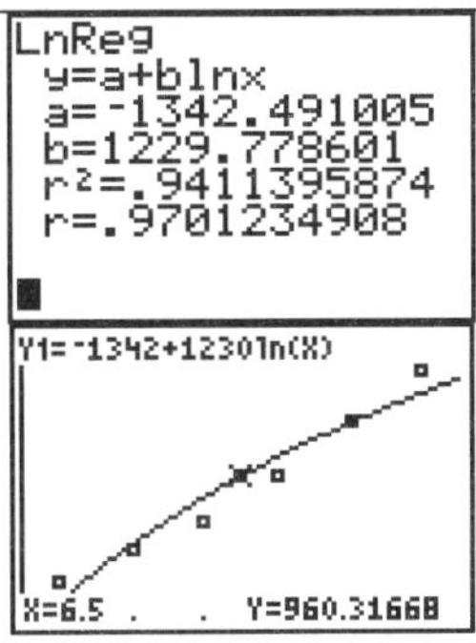	 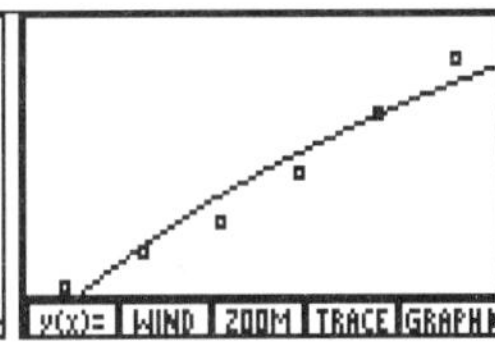
Now let's find an exponential model	Now let's find an exponential model
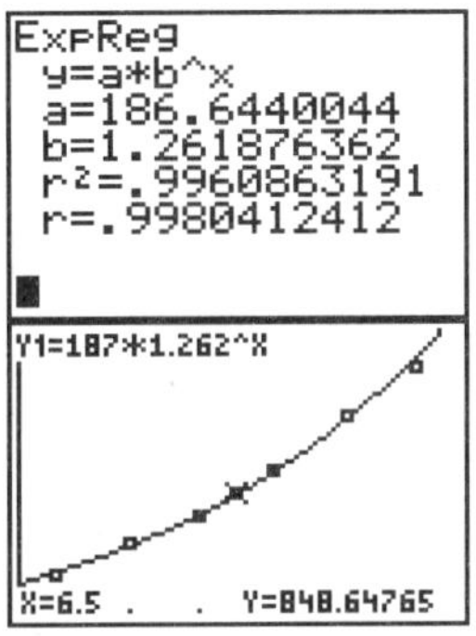	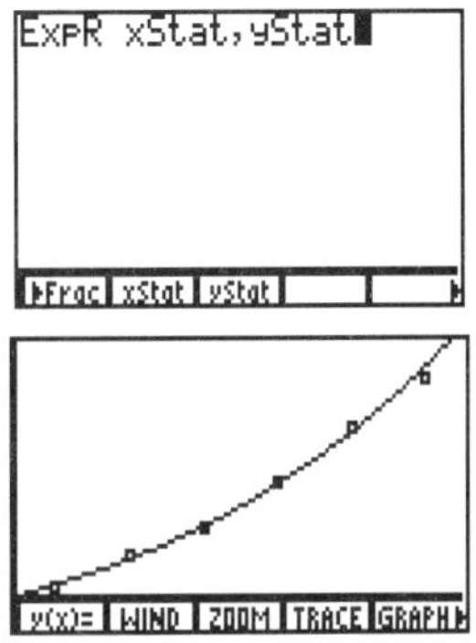

You might notice that when we put the functions into Y1 we rounded off a little just for simplicity.

Answer Q2: Natural log model, since it passes close to all of our points and continues to grow. Quadratic looks good too but it begins to decrease after a while which seems unlikely to happen.

Question 3: So which function looks like the best model to you?

Section Summary

The table below contains all the keystrokes necessary to get data points and an equation both graphed on the graphing calculator, starting from the home screen.

TI-83	TI-86
1. Push [STAT] key (go to STAT). 2. Push [ENTER] (EDIT). 3. Arrow over and up to clear each column, push [CLEAR] then [ENTER]. 4. Arrow over to first spot in the first column and enter the input value of each data point followed by the	1. Push [2nd] and [+] keys (go to STAT). 2. Push [F2] (EDIT). 3. Arrow over and up to clear each column, push [CLEAR] then [ENTER]. 4. Arrow over to first spot in the first column and enter the input value of each data point followed by

ENTER key. Enter the output values in the second column. 5. Push the Y= key and arrow up to turn on the Plot1 (push ENTER to highlight). 6. Push the ZOOM key, then arrow down to ZoomStat, then push ENTER (now you can see the points graphed). 7. Push 2nd and MODE keys to go back to the home screen (QUIT). 8. Push STAT 9. Arrow over to CALC. 10. Arrow down to the function you want to create and push ENTER 11. Push ENTER to get the function. 12. Go to Y= and, push VARS 13. Push 5 (Statistics) and then arrow over to EQ and push ENTER (now you can round off to create a function that will be easier to work with). 14. Push GRAPH to see both the points and the function graphed.	the ENTER key. Enter the output values in the second column. 5. Push GRAPH then Y= and arrow up to turn on the Plot1 (push ENTER to highlight). 6. Push GRAPH F3 (ZOOM) MORE, then F5 (ZDATA) (now you can see the points graphed). 7. Push 2nd and EXIT keys to go back to the home screen. 8. Push 2nd and + keys (go to STAT). 9. Push F1 (CALC). 10. Find the function you want to create and push the appropriate F key. 11. Push the CUSTOM key, then xStat followed by the , key, then yStat, then ENTER to calculate the function. 12. Go to Y= and, push 2nd and + keys (go to STAT). 13. Push F5 (VARS) followed by MORE *twice*, then F2 (RegEq). 14. Now, push the GRAPH key then F5 (GRAPH). Now you can see both the points and the equation used to model the data.

SECTION 5.3 HOMEWORK

(1–18) For each of the data sets:

a. Enter the data into your graphing calculators and create a scatter plot.
b. Find the best fit linear function $f(t)$, then graph it along with the scatter plot.
c. Find the best fit quadratic function $f(t)$, then graph it along with the scatter plot.
d. Find the best fit logarithmic function $f(t)$, then graph it along with the scatter plot.
e. Find the best fit exponential function $f(t)$, then graph it along with the scatter plot.
f. Choose which function or functions seem to best fit the data.

1.

Percent of U.S. population without health care

Year	1987 $t = 7$	1990 $t = 10$	1993 $t = 13$	1996 $t = 16$	1999 $t = 19$
Percent without healthcare	14.4%	15.7%	17.2%	17.6%	17.4%

www.fedstats.gov

Answer Q3: It looks like the quadratic or the exponential might be best for this data set.

2.

Percent of the population between the ages of 45–64 in the U.S. with health care coverage

Year	1984 $t = 4$	1995 $t = 15$	1996 $t = 16$	1997 $t = 17$
Percent with health care	83.3%	80.4%	79.5%	79.1%

World Almanac 2000

3.

Life expectancy in years at birth

Year	1950 $t = 1$	1960 $t = 11$	1970 $t = 21$	1980 $t = 31$	1985 $t = 36$	1990 $t = 41$	1995 $t = 46$	1998 $t = 49$
Life expectancy	68.2	69.7	70.8	73.7	74.7	75.4	75.8	76.7

www.fedstats.gov

4.

Receipts of private employers firms in billions

Year	1990 $t = 1$	1991 $t = 2$	1992 $t = 3$	1993 $t = 4$	1994 $t = 5$	1995 $t = 6$	1996 $t = 7$	1997 $t = 8$
Receipts	9.451	12.961	13.605	14.099	14.840	15.751	16.655	18.243

Statistical Abstract of the U.S. 2000

5.

Death rates for immune deficiency virus (HIV) per 100,000 population

Year	1987 $t = 7$	1990 $t = 10$	1992 $t = 12$	1993 $t = 13$	1994 $t = 14$	1995 $t = 15$
Death rates per 100,000	5.6	10.2	13.2	14.5	16.2	16.3

www.fedstats.gov

6.

Total crimes in 1000's in the U.S.

Year	1993 $t = 3$	1994 $t = 4$	1995 $t = 5$	1996 $t = 6$	1997 $t = 7$	1998 $t = 8$
Crimes in 1000's	14,145	13,990	13,863	13,494	13,195	12,476

Statistical Abstract of the U.S. 2000

7.

U.S. army enlisted personnel in 1,000's on active duty

Year	1985 $t = 1$	1990 $t = 6$	1994 $t = 10$	1995 $t = 11$	1996 $t = 12$	1997 $t = 13$	1998 $t = 14$	1999 $t = 15$
Number of personnel	666.56	639.71	465.69	438.80	408.51	403.07	402.00	388.21

World Almanac 2000

8.

Monthly pay of a General in the U.S. Army according to years of service

Years of Service	4	8	12	16	20	26
Monthly Pay	\$8,114.40	\$8,425.80	\$8,892.60	\$9,528.20	\$10,167.00	\$10,800.00

World Almanac 2000

9.

Federal expenditures in billions

Year	1960 $t = 1$	1965 $t = 6$	1970 $t = 11$	1975 $t = 16$	1980 $t = 21$	1985 $t = 26$	1990 $t = 31$	1995 $t = 36$
Expenditures in billions	92	118	196	332	591	946	1,253	1,516

Statistical Abstract of the U.S. 2000

10.

U.S. National Debt in billions

Year	1965 $t = 5$	1970 $t = 10$	1975 $t = 15$	1980 $t = 20$	1985 $t = 25$	1990 $t = 30$	1995 $t = 35$
U.S. debt in billions	261	283	395	712	1,507	2,412	3,605

Statistical Abstract of the U.S. 2000

11.

Percent of high school seniors using marijuana

Year	1992 $t = 2$	1993 $t = 3$	1994 $t = 4$	1995 $t = 5$	1996 $t = 6$	1997 $t = 7$
Percent of high school students using marijuana	32.6	35.3	38.2	41.7	44.9	49.6

World Almanac 2000

12.

Percent of high school seniors that have used alcohol heavily

Year	1993 $t = 3$	1994 $t = 4$	1995 $t = 5$	1996 $t = 6$	1997 $t = 7$	1998 $t = 8$
Percent of high school students using alcohol	27.5	28.2	29.8	30.2	31.3	31.5

World Almanac 2000

13.

Air carriers in the U.S. (carrying passengers or freight)

Year	1960 $t = 1$	1970 $t = 11$	1975 $t = 16$	1980 $t = 21$	1985 $t = 26$	1990 $t = 31$	1995 $t = 36$	1998 $t = 39$
Number of air carriers	2,135	2,679	3,495	3,808	4,678	6,083	7,411	8,111

www.fedstats.com

14.

Freight cars in the U.S. in thousands

Year	1960 $t = 10$	1965 $t = 15$	1970 $t = 20$	1975 $t = 25$	1980 $t = 30$	1985 $t = 35$	1990 $t = 40$	1995 $t = 45$
Freight cars in 1000's	1,658	1,478	1,423	1,359	1,168	867	659	583

www.fedstats.com

15.

Amount of ordinary life insurance in force by U.S. life insurance companies in billions of dollars

Year	1991 $t = 1$	1992 $t = 2$	1993 $t = 3$	1994 $t = 4$	1995 $t = 5$	1996 $t = 6$	1997 $t = 7$
Amount of ordinary life insurance ($ in billions)	4.057	4.240	4.456	4.443	4.604	5.067	5.279

World Almanac 2000

16.

Hospital care expenses in the U.S. in billions

Year	1980 $t = 10$	1990 $t = 20$	1995 $t = 25$	1996 $t = 26$	1997 $t = 27$	1998 $t = 28$
Hospital care expenses in billions	102	254	344	357	367	380

www.fedstats.com

17.

Payroll of private employers in billions

Year	1990 $t = 10$	1991 $t = 11$	1992 $t = 12$	1993 $t = 13$	1994 $t = 14$	1995 $t = 15$	1996 $t = 16$	1997 $t = 17$
Payroll in billions	2.103	2.145	2.272	2.363	2.488	2.666	2.849	3.048

Statistical Abstract of the U.S. 2000

18.

Federal and state prison population in thousands

Year	1990 $t = 10$	1991 $t = 11$	1992 $t = 12$	1993 $t = 13$	1994 $t = 14$	1995 $t = 15$	1996 $t = 16$	1997 $t = 17$
Prison population in thousands	740	790	846	932	1,017	1,085	1,138	1,195

Statistical Abstract of the U.S. 2000

19.

Percent of the U.S. population enrolled in a Health Maintenance Organizations (HMO)

Year	1980 $t = 10$	1990 $t = 20$	1991 $t = 21$	1992 $t = 22$	1993 $t = 23$
Percent of the population in HMOs	9.1	33.0	34.0	36.1	38.4

World Almanac 2000

a. Use the calculator to find the best fit quadratic function for the above information.

b. Use the calculator to find the best fit exponential function for the above information.

c. Graph the two functions above and the stat plot on the same graph; which graph looks like it works best?

d. Predict the percent of the population with both the quadratic and exponential for 1994.

e. Predict the percent of the population with both the quadratic and exponential for 1995.

f. The actual percent population in 1994 was 45.1 and in 1995 it was 50.9; which one of the functions actually predicted the percent the most accurately?

20.

Number of soldiers with chest size according to survey done by U.S. Army

Chest size in inches	Number of soldiers with chest size
33	3
34	18
35	81
36	185
37	420
38	749
39	1,073
40	1,079
41	934
42	658
43	370
44	92
45	50
46	21
47	4
48	1

www.lib.stats.cmu.edu

a. Use the graphing calculator to find the best fit line function, quadratic function, logarithm function and exponential function.

b. Graph each of the above functions and the stat plot on the same graph. Do any of the functions seem to fit the stat plot?

c. Graph $y = 1100e^{\frac{-(x-39.7)^2}{8.9}}$ and the stat plot of the above data on the same curve; does this look like a good fit? This equation is called a Gaussian model and used in probability and statistics to represent populations.

(21–22) The following is data derived from a scientific experiment with radioactive isotopes:

a. Use the calculator to put the data in a table and find the best fit exponential function for the data.

b. Change the exponential function to an equivalent exponential function of base *e*, $R(t)=ae^{kt}$. (Round k to 6 decimal places when writing the function with base e.)

c. Find the half–life of the isotope.

d. Use the table from homework section 4.7 problems 5–12 to determine which radioactive isotope was used for the experiment.

21.

Day	1	2	3	4	5	6
Grams of isotope	9.527	9.076	8.647	8.238	7.848	7.477

22.

Year	1	2	3	4	5	6	10
Grams	9.7716	9.5485	9.3305	9.1174	8.9092	8.7058	7.9374

(23–24) In each problem, the data was taken from the statement for an investment that is compounded continuously:

a. Use the calculator to put the data in a table and find the best fit exponential function for the data.

b. Change the exponential function to an equivalent exponential function of base e, $A(t)=e^{kt}$. Round k to 3 decimal places.

c. What is the interest rate?

23.

Year	2	5	10	15	20
Value of the investment	\$586.76	\$745.91	\$1112.77	\$1,660.06	\$2,476.52

24.

Year	2	5	10	15	20
Value of the investment	\$1,221.40	\$1,648.72	\$2,718.28	\$4,481.69	\$7,389.06

Section 5.4 Additional Models

In this section we will continue our discussion and do several examples of finding candidates for appropriate models we could use to make predictions about real-world phenomenon.

Example 1: Modeling Marital Status

Find the model of best fit for the following data set. Give a written explanation defending your choice. (Data about the number of unmarried couples in the U.S.A. in millions.)

Year ($t = 0$ in 1975)	1980 ($t = 5$)	1985 ($t = 10$)	1990 ($t = 15$)	1995 ($t = 20$)	1999 ($t = 24$)
Unmarried couples (in millions)	1.59	1.98	2.86	3.67	4.49

Source: 2000 Statistical Abstract of the United States

Solution:

We will first calculate all of the possible function models that are on the graphing calculator. Then we will decide which one(s) might best fit this phenomenon.

TI-83

First, let's enter the data into the calculator and see its graph.

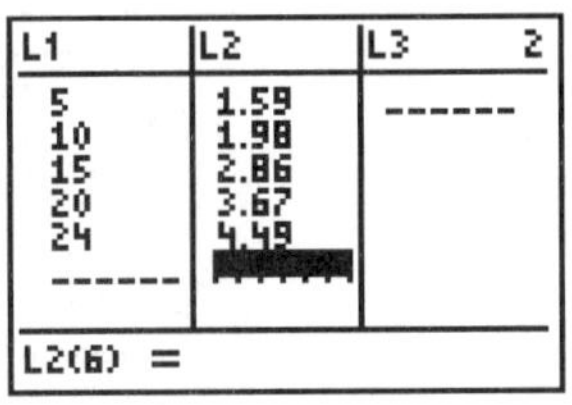

ZoomStat

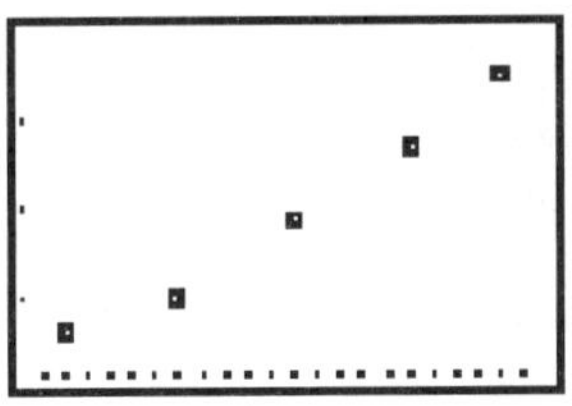

Let's get a little bigger view by changing the window to the following:

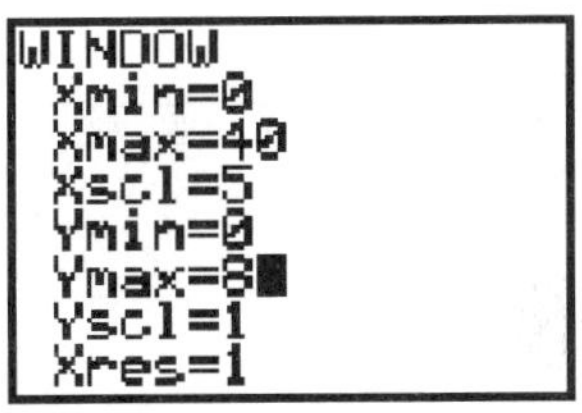

TI-86

First, let's enter the data into the calculator and see its graph.

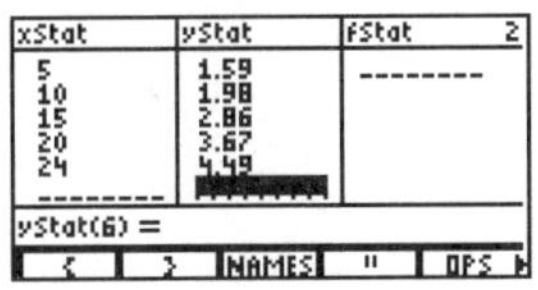

Zdata

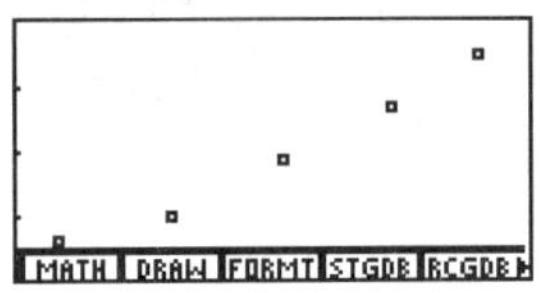

Let's get a little bigger view by changing the window to the following:

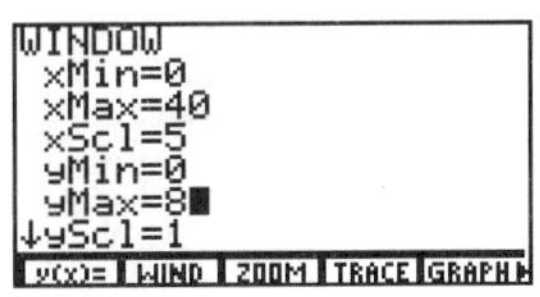

<table>
<tr>
<td>Now we go to the stat menu and calculate different function models.
Linear Function Model
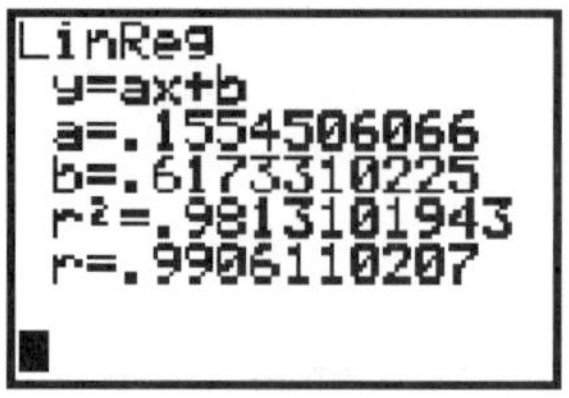
Graph this with our data points.
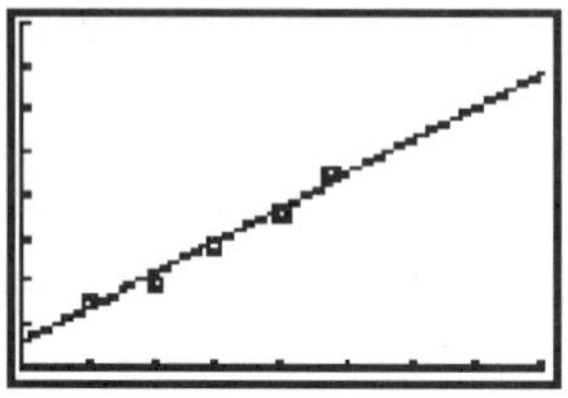</td>
<td>Now we go to the stat menu and calculate different function models.
Linear Function Model
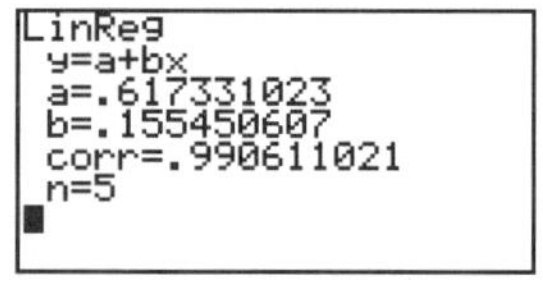
Graph this with our data points.
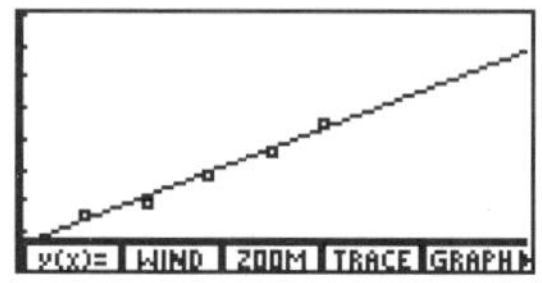</td>
</tr>
<tr>
<td>Quadratic Function Model
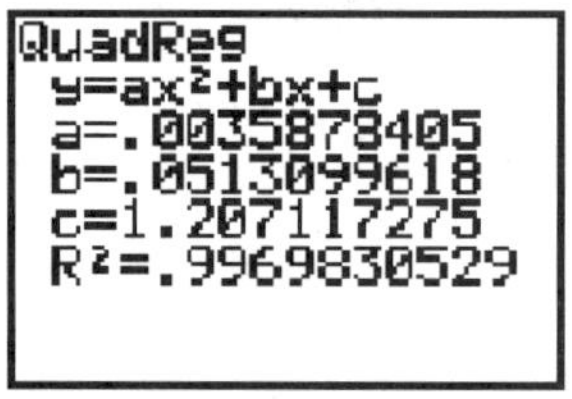
Graph
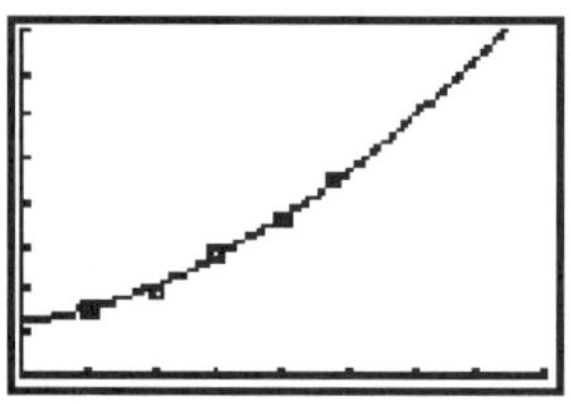</td>
<td>Quadratic Function Model (P2Reg)

Graph
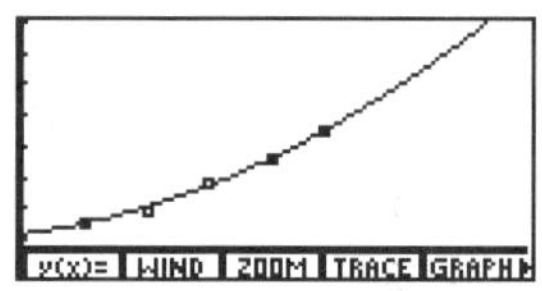</td>
</tr>
<tr>
<td>Cubic Function Model
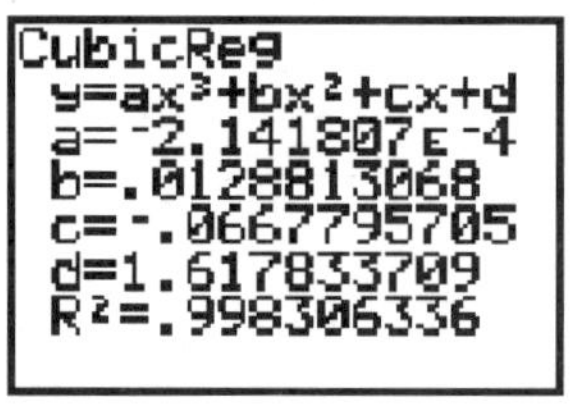
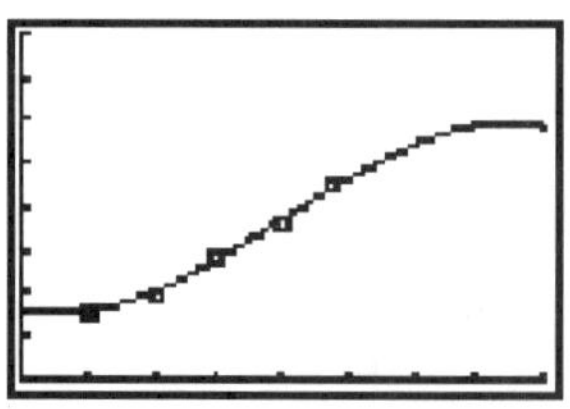</td>
<td>Cubic Function Model (P3Reg)
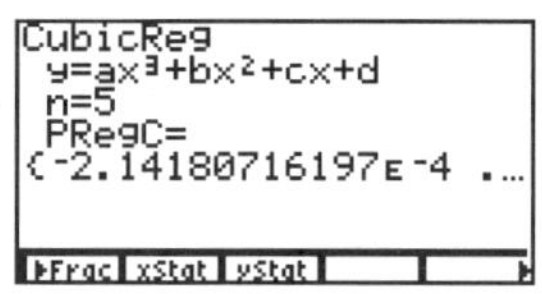
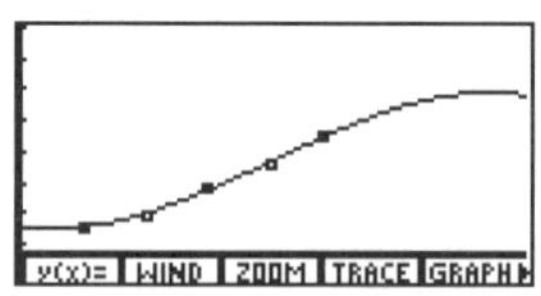</td>
</tr>
</table>

<table>
<tr><td>

Quartic Function Model

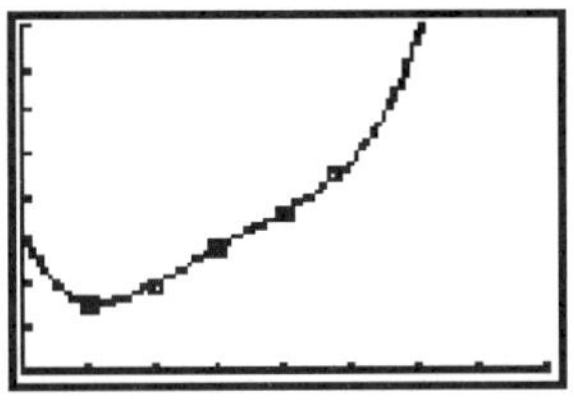

</td><td>

Quartic Function Model (P4Reg)

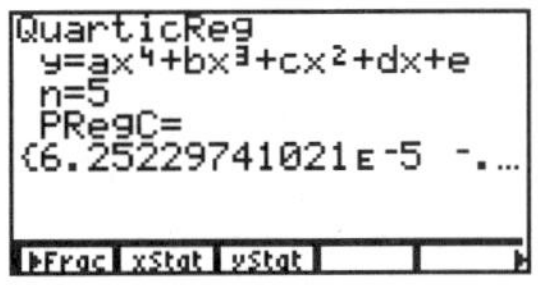

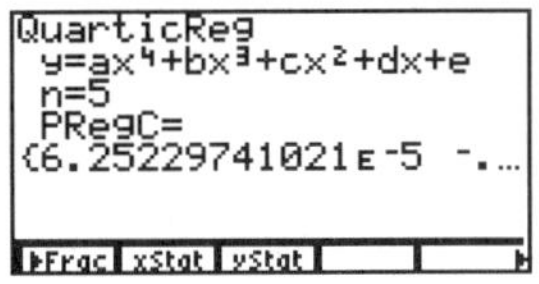

</td></tr>
<tr><td>

Natural Log Function Model

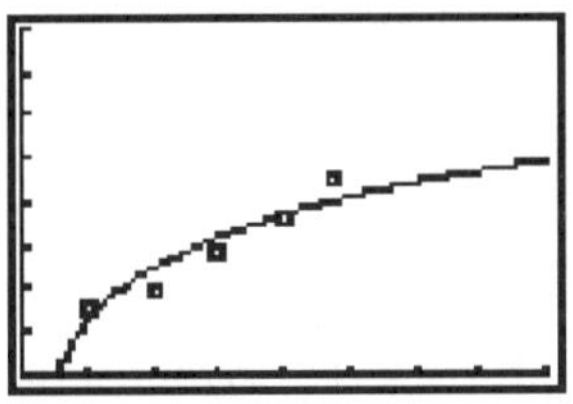

</td><td>

Natural Log Function Model

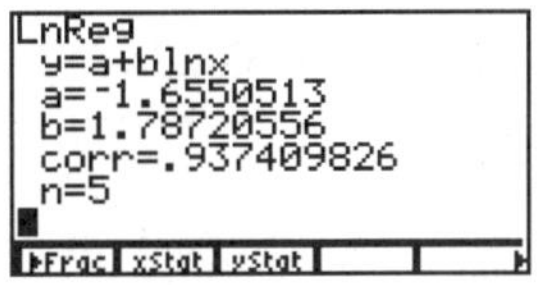

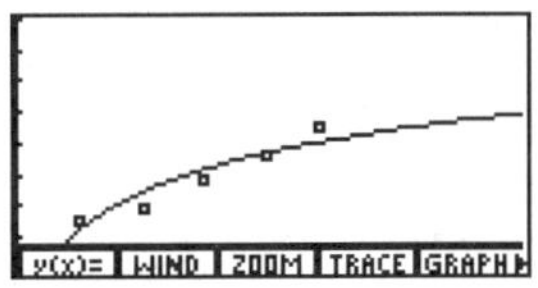

</td></tr>
<tr><td>

Exponential Function Model

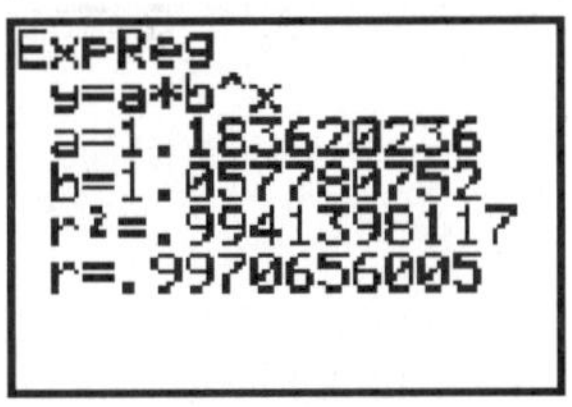

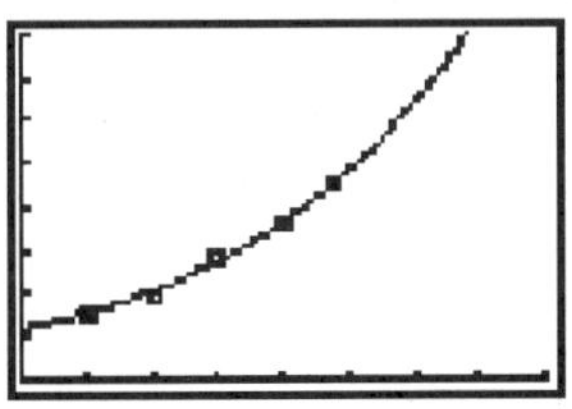

</td><td>

Exponential Function Model

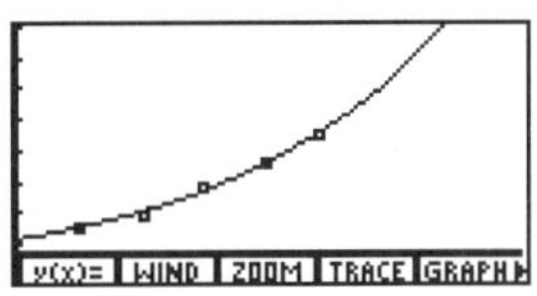

</td></tr>
</table>

Power Function Model

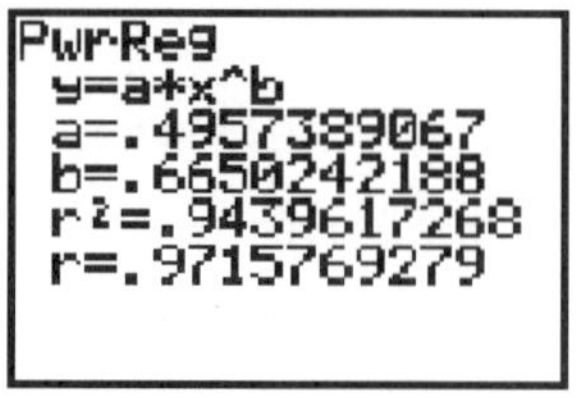

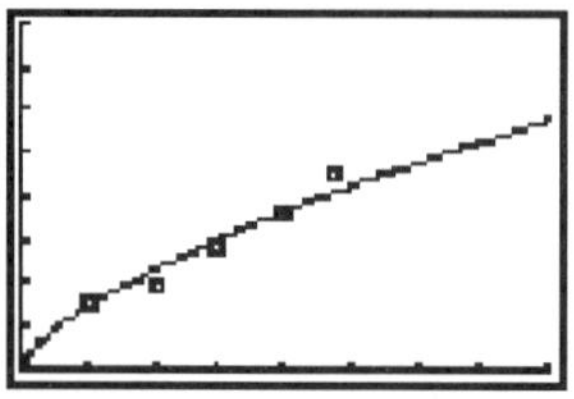

Power Function Model

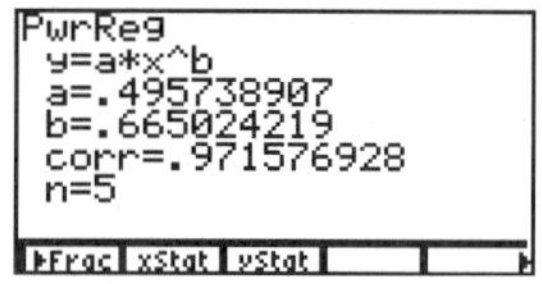

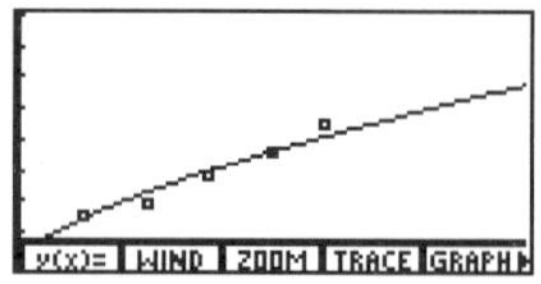

Let's start by throwing out those models that are obviously not going to be good ones. The poor models appear to be the natural log function and the power function. Our points seem to be growing faster and faster (bending upwards or possibly straighten out, linear) but not bending downwards (like the power and natural log models above which grow fast then slow down).

The quartic function fits the data exactly since we have five data points and a quartic function has five unknown coefficients ($ax^4 + bx^3 + cx^2 + dx + e$). But as you look at the graph of the quartic model in relation to the data, it seems unlikely that the model would fit the real-world phenomenon, before 1980 and after 1999.

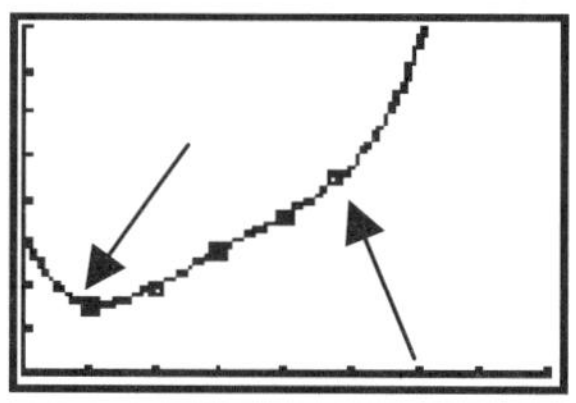

THE r AND r^2 VALUES

One tool that can help us determine if a model fits the data is the r^2-value for the model.

If your calculator isn't displaying the r or r^2 values, then you must turn them on as follows (shown for the TI-83):

For the TI-83 you must push the [2nd] key followed by the [0] key, then push the [x^{-1}] key (D) then arrow down to DiagnosticOn and push [ENTER].

The r-value is what is called the **correlation coefficient** and r^2 is related to that value. If you look back at some of the models we found using the graphing calculator for this example, you will see the

r and r^2 values. The values give you an indication of how close the data points are to the curve (model function). If the r-value is 1 (or r^2 is 1), then the points fall exactly on the curve. The farther the r-values and r^2-values are from 1, the more spread out the data points are in relation to the curve.

Of course this doesn't tell the whole story. We can see in our example with respect to the quartic function that even though its r^2-value is 1, it is very unlikely that this is a good model for this data set when looking backwards or forwards in time.

The other models in this example that have r^2-values close to 1 are:

- The linear function ($r^2 = 0.9813$)
- The quadratic function ($r^2 = 0.99698$)
- The cubic function ($r^2 = 0.9983$)
- The exponential function ($r^2 = 0.994$)

Out of these four, the r^2-value of the linear model is the farthest from 1. So we might want to concentrate on the other three. The cubic function has an r^2-value closest to 1 but not by much, and the graph appears to have a similar problem as the quartic had. There doesn't appear to be any reason to believe that the values are going to be leveling off and then turning back down. Remember a cubic often has two humps in it. In fact if we increase the size of our window we see this:

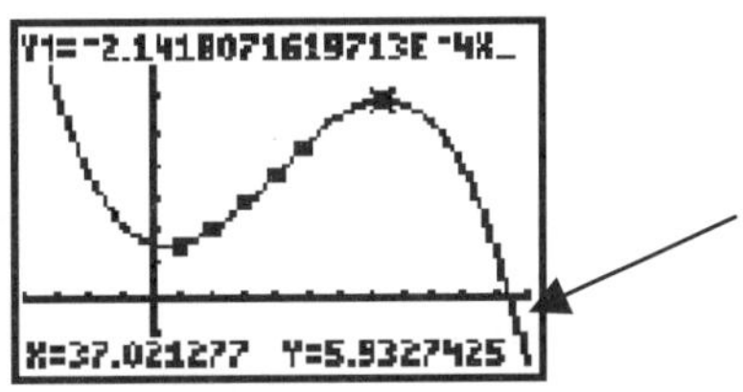

If we follow this model down the road a few years we see that eventually there will be a negative number of unmarried couples (not possible). This brings us to the quadratic and exponential function models. They look quite similar and have very similar r^2-values.

<table>
<tr><th>Quadratic</th><th>Exponential</th></tr>
<tr><td>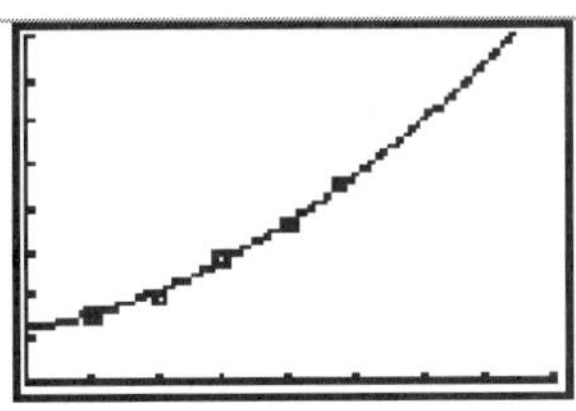
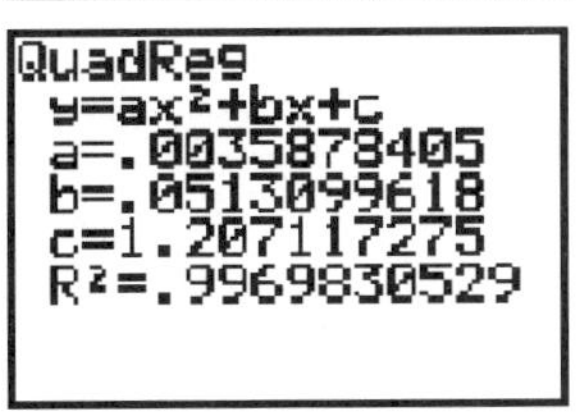</td><td>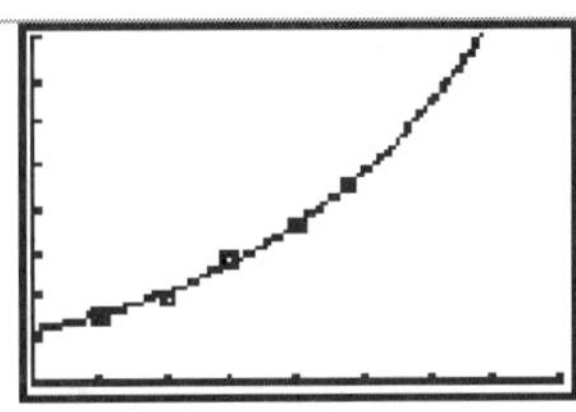
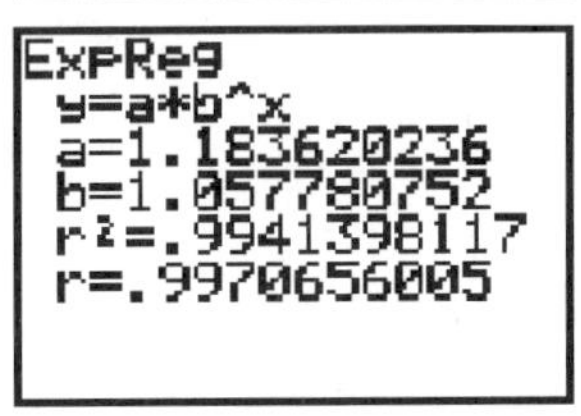</td></tr>
</table>

We do know from our study of the two different functions that the exponential function will really grow much more rapidly than the quadratic. Taking that into consideration along with the fact that the r^2-value for the quadratic is closer to 1, as well as the nature of what we are finding a model for (unmarried couples), we might decide to choose the quadratic model for this data set. Given all of these factors, we decide that the curve of "best fit" is $\boldsymbol{U(t) = 0.003588t^2 + 0.05131t + 1.207}$.

Question 1: What would be the equation of the curve of best fit if we decided to go with the exponential function?

Question 2: Calculate how many couples will be unmarried in the year 2010, according to both the quadratic and exponential models.

This kind of modeling has applications across many disciplines outside of mathematics. For example, as a social behavioral scientist you could use these models to predict trends and future behavior. By incorporating other factors, you might develop theories about why this is happening, how it will affect our society, and whether there is anything that can or should be done about it.

One of the other models that is in your calculator but we haven't talked about yet is the logistics function model. The logistics function is one that often shows up in population problems where there is a limit to how much something can grow. For this example the logistics function is:

<table>
<tr><th>TI-83</th><th>TI-86</th></tr>
<tr><td>
Logistic Function Model

Because of the nature of this function we need a different window size.

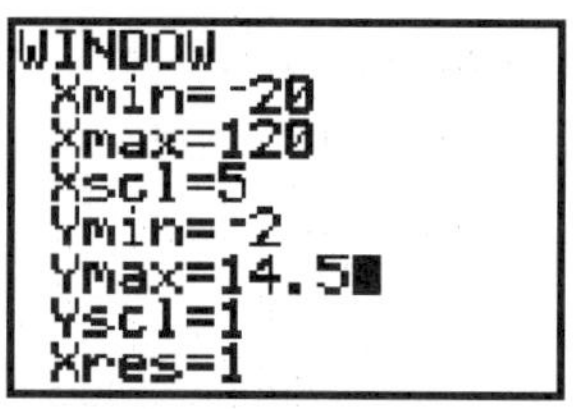

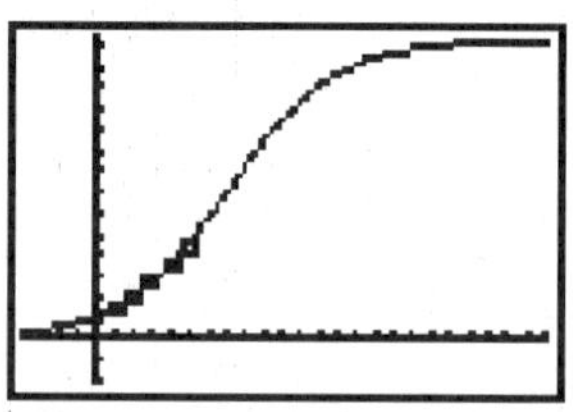
</td><td>
Logistic Function Model (LgstR)

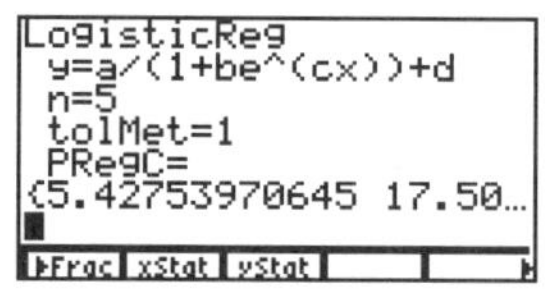

Because of the nature of this function we need a different window size.

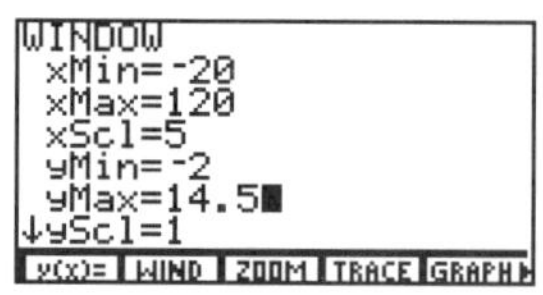

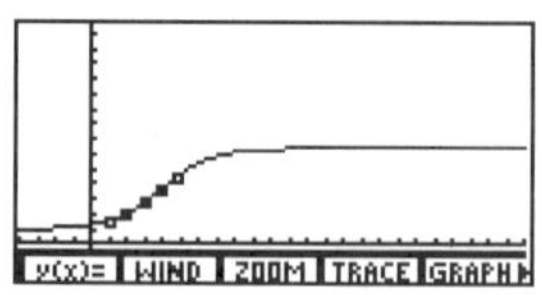

As you can see the TI-86 uses a different type of logistic formula as compared to the TI-83.
</td></tr>
</table>

It would be hard to claim that this is the "best" model without some other information that would give us an indication that there is a limit to the growth of unmarried couples and what that limit might be.

Example 2: Modeling the Dow Jones Industrial Average

Find the best fit model for the following data (Dow Jones Industrial Average from 1929-2002). Give an explanation as to why you believe you have found the "best."

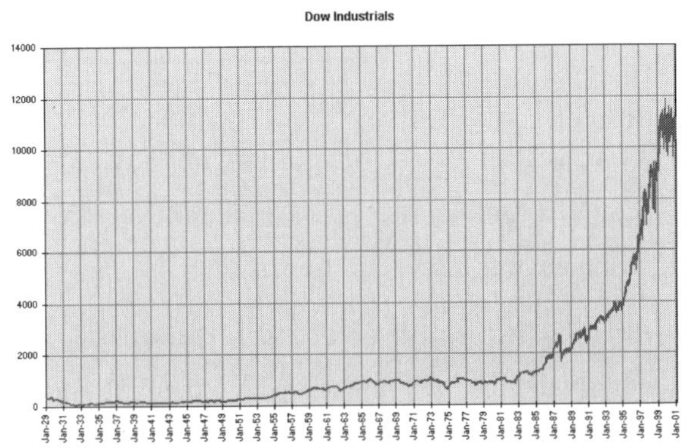

Year since 1900	32	42	50	62	74	80	90	102
DJIA low for each 10-year period	50	100	200	550	600	750	2200	7300

Source: http://www.investorsgrapevine.org/Dow.html

Solution:

First, type in the data and then graph. Second, investigate only those functions that might fit the data. From the historical graph and the data point graph we might want to look at exponential, or polynomials (quadratic or larger) only.	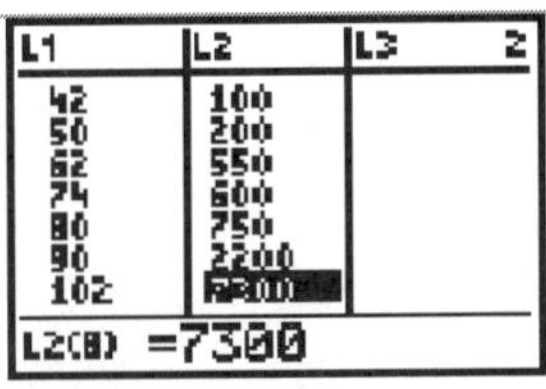	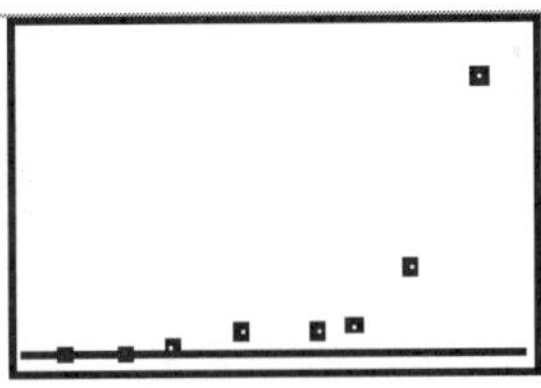
Exponential Fits the data fairly well.		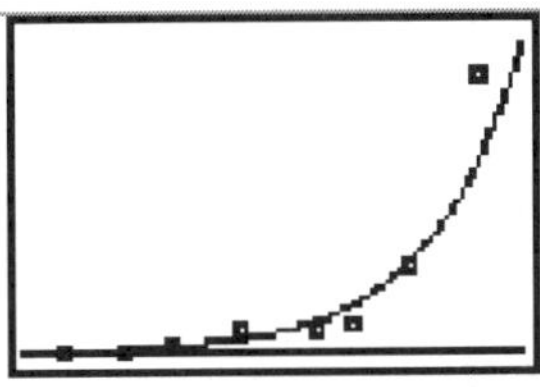
Quadratic This doesn't really follow the data points very well.	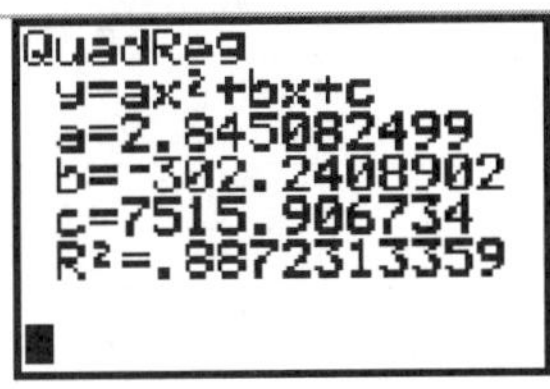	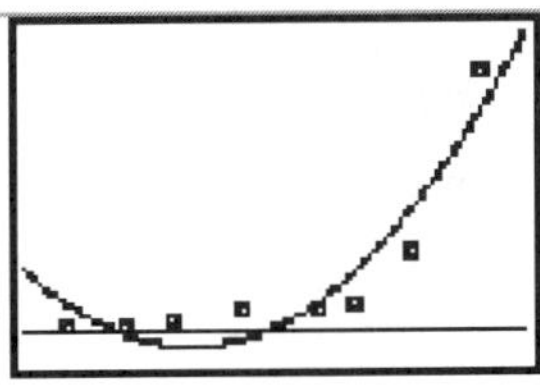

Answer Q1: In the form $y = a(b)^t$ we get $U(t) = 1.184(1.0578)^t$ or in the form $y = a(e)^{kt}$, we get $U(t) = 1.184(e)^{0.0562t}$ (remember $b=e^k$).

Model	Regression	Graph
<u>Cubic</u> Closer than the quadratic but the beginning is off. It might be good at predicting the future though.	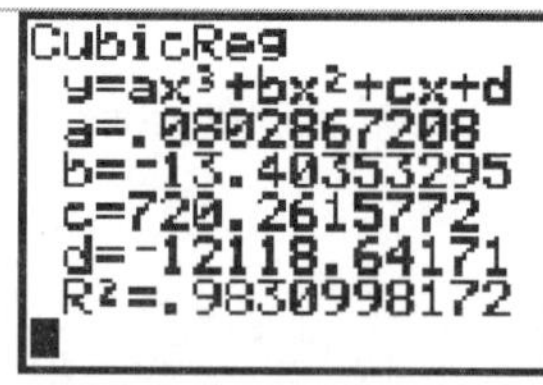	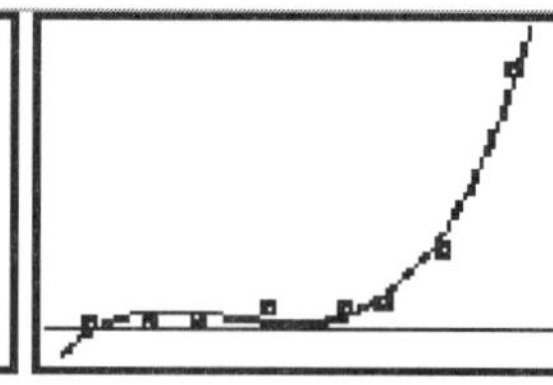
<u>Quartic</u> This has the best r^2 value. It wouldn't work for the past but should be good for the future.		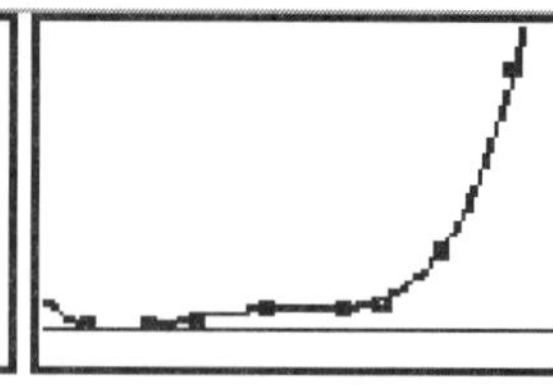

To determine which model is really best would require more investigation and more knowledge about modeling than we've covered. But, knowing that limitation, let's look at both the exponential model and the quartic model and answer a few questions.

Question 3: If we were to take the exponential model as the function of best fit, what does the base tell us about the approximate rate of return in the stock market over the last 70 years?

One advantage of using the exponential model is that it easily tells us an approximate average rate of return on investments made in the stock market. Of course, this value is only valid if the model is an accurate one that fits the data well. We might also want to investigate the highs over the past seven decades as well. The two together, high and low models, might give us a good indication of the range of return to expect when investing in the stock market. But once again, this is only if we believe that the exponential model is a good model to use when modeling this data set.

Answer Q2: The quadratic model yields $U(35) = 0.003588(35)^2 + 0.05131(35) + 1.207 = 7.4$ million. The exponential model yields $U(35) = 1.184(1.0578)^{35} = 8.46$ million.

Question 4: Using the quartic model $\text{DJIA}(t) = 0.00203t^4 - 0.4606589t^3 + 38.0204t^2 - 1331.305t + 16698$, find the DJIA in the year 2010. Also, how does this differ from the exponential model?

If we were to look at the year 2060, the quartic model yields a value of 220,533 and the exponential yields a value of 227,130, a higher number. We might caution that looking too far into the future with any model is likely to be way off target. We can't always predict major events that would affect the phenomenon being modeled.

Question 5: Find the exponential model that fits the highs for the last seven decades.

Year since 1900	37	46	59	66	73	89	99
DJIA high for each 10-year period	200	220	680	990	1070	2750	11500

It looks like no matter how we view the stock market, the rate of return based on an exponential model is around 6.4% to 6.8%. Of course our values might be substantially different if we used just the last three decades.

Exponential: last three decades lows in 80, 90, and 102.

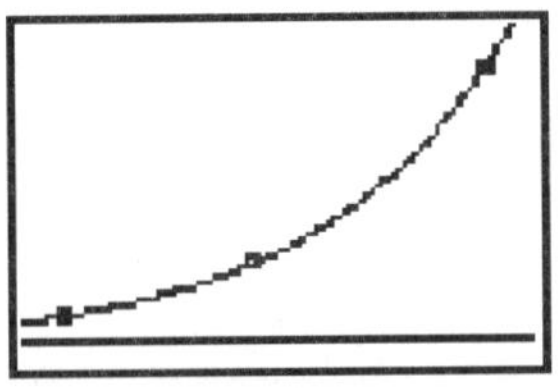

A base of 1.109 means a 10.9% return.

Exponential: last three decades highs in 73, 89, and 99.

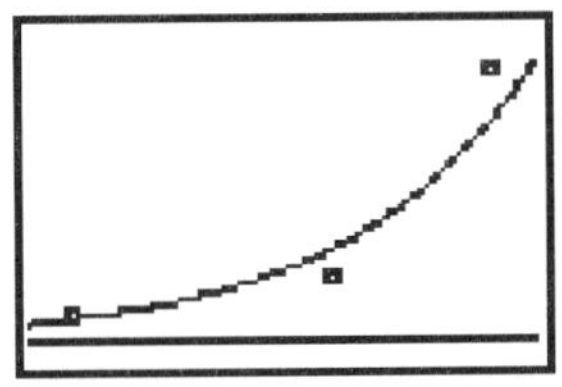

A base of 1.092 means a 9.2% return.

Something you should notice right away is that the average rate of return has been higher in the recent years than over the past 70 years. Also, the exponential model fits the data very well with the lows over the past 22 years whereas the quartic model was better over the past 70 years.

Let's look at one more example using a function called **sine**. The general shape of this function looks like this:

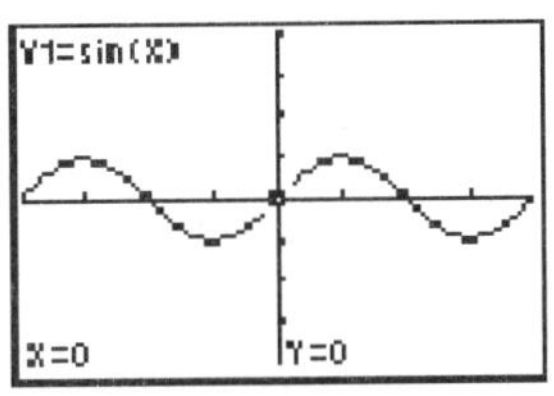

This function is usually studied in a trigonometry course, but we'd like you to see how it can be a good model for some kinds of phenomena in life.

Example 3: Modeling Monthly Temperature

Model the following data on average monthly temperatures in Portland, Oregon over the past 30 years by using the sine function:

Month	Jan	Feb	Mar	Apr	May	Jun	Jul	Aug	Sept	Oct	Nov	Dec
Temp	39.6	43.6	47.3	51	57.1	63.5	68.2	68.6	63.3	54.5	46.1	40.2

Solution:

Answer Q3: Since the formula is DJIA(t) = 6.47(1.0676)^t, using the lows over the past seven decades, the rate of return would be the base minus 1, which is 6.76% (1.0676 – 1 = .0676).

We type in the data using 1-12 in column L1 for the months.		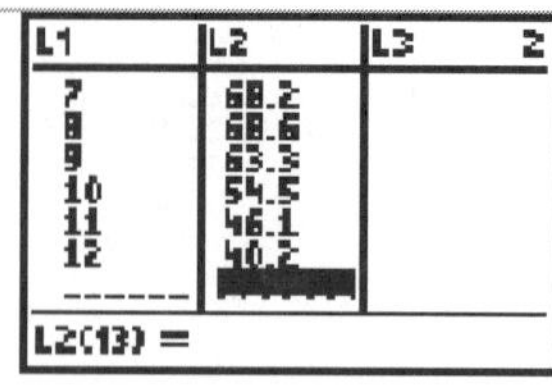
Now, graph the data.		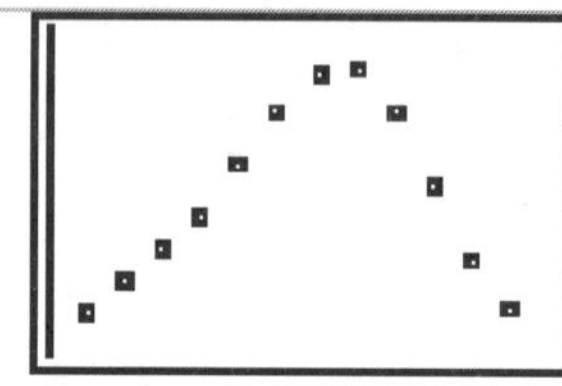
The sine curve is one that repeats itself just like the temperatures do year in and year out.		
Let's graph the points and the model and see what we have.		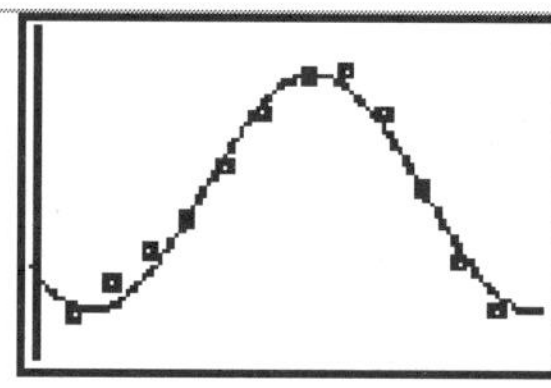

Answer Q4: DJIA(110) = 14376. The exponential model yielded 8626, 5750 points lower.

Any data that repeats itself over time is a good candidate for using the sine function as its model.

There is much to study in mathematics and if this idea of modeling interests you, you should plan on taking some modeling classes in the future to learn more about how to model real-world data.

Section Summary

Answer Q5: Exponential. Notice that the base is 1.064! That is roughly a 6.4% return.

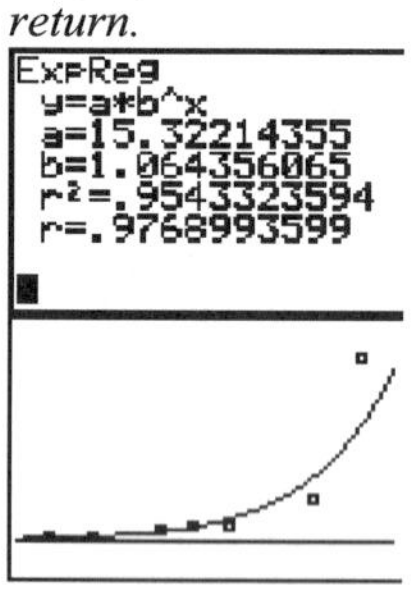

- Use the r^2 values along with the graphs as compared to the data points to help you decide upon the "curve of best fit."
- Also consider the graph's endpoints when choosing the best model.
- Keep in mind that other factors influence the outcome of the phenomena we model.

SECTION 5.4 HOMEWORK

1. The data in the following table is linear.

x	2	5	8	11	14
y	11	20	29	38	47

   a. Check to make sure the data is linear by using the common difference.
   b. Graph the set of data to see if it looks linear.
   c. Find the equation that generates the data with algebra. (Hint: use any two data points.)
   d. Find the equation that generates the data with the calculator and regression.
   e. Graph the equation and the scatter plot on the same graph to check accuracy.

2. The data in the following table is quadratic.

x	2	5	8	11	14
y	9	66	177	342	561

   a. Check to make sure the data is quadratic by using the common difference.
   b. Graph the set of data to see if it looks quadratic.
   c. Find the equation that generates the data with the calculator and regression. (In 6.5 we will learn how to algebraically solve a quadratic given three points and it will give the same answer as the regression formula.)
   d. Graph the equation and the scatter plot on the same graph to check accuracy.

3. The data in the following table is exponential.

x	2	5	8	11	14
y	12	96	768	6,144	49,152

   a. Check to make sure the data is exponential by using the common ratio.
   b. Graph the set of data to see if it looks exponential.
   c. Find the equation that generates the data with algebra. (Hint: use any two data points.) (We learned how to solve $y = ae^{kx}$ in section 5.2; to check against the regression formula in your calculator let $b = e^k$ to get the formula $y = ab^x$)
   d. Find the equation that generates the data with the calculator and regression.
   e. Graph the equation and the scatter plot on the same graph to check accuracy.

4. The data in the following table is logarithmic.

x	2	5	8	11	14
y	4.386294361	6.218875825	7.158883083	7.795790546	8.278114659

   a. Graph the set of data to see if it looks logarithmic.
   b. Find the equation that generates the data with algebra. (Hint: use any two data points.)
   c. Find the equation that generates the data with the calculator and regression.
   d. Graph the equation and the scatter plot on the same graph to check accuracy.

5. The following table gives the amount of money in a savings account after a number of years.

Years	5	10	15	20	25
Amount of money	\$6,750	\$8,500	\$10,250	\$12,000	\$13,750

a. Find an appropriate regression formula that will give the amount of money m in the account after t years.

b. How much money will be in the account after 35 years?

6. The following table gives the amount of money in a savings account after a number of years.

Years	10	20	30	40	50
Amount of money	\$1,830	\$2,660	\$3,490	\$4,320	\$5,150

a. Find an appropriate regression formula that will give the amount of money m in the account after t years.

b. How much money will be in the account after 25 years?

7. The following table gives the amount of radioactive material R in a sample after t days.

Days	5	10	15	20	25
Amount of material	5.48 grams	3.01 grams	1.65 grams	0.91 grams	0.50 grams

a. Find an appropriate regression formula that will give the amount of radioactive material in the sample after t days.

b. How much radioactive material will be in the sample after 32 days?

8. The following table gives the amount of radioactive material R in a sample after t years.

Years	100	200	300	400	500
Amount of material	12.13 grams	7.36 grams	4.46 grams	2.71 grams	1.64 grams

a. Find an appropriate regression formula that will give the amount of radioactive substance in the sample after t years.

b. How much radioactive material will be in the sample after 720 days?

9. The following table gives the amount of money m in an investment after t years with no added deposits.

Years	5	10	15	20	25
Amount of money	\$7,346.64	\$10,794.62	\$15,860.85	\$23,304.79	\$34,242.38

a. Find an appropriate regression formula that will give the amount of money in the investment after t years.

b. How much money will be in the investment after 35 years?

c. If you can determine the interest rate, what is it?

10. The following table gives the amount of money m in an investment after t years with no added deposits.

Years	5	10	15	20	25
Amount of money	\$1,417.62	\$2,009.66	\$2,848.95	\$4,038.74	\$5,725.42

a. Find an appropriate regression formula that will give the amount of money in the investment after t years.

b. How much money will be in the investment after 35 years?

c. If you can determine the interest rate, what is it?

(11–30) For each of the sets of data use the regression formula ability of your calculator to find the best fit curve and check the last data point to see how accurate your model is.

11. The following data is the U.S. population in thousands for years 1995 to 1999. (Let $x = 0$ for the year 1995)

Year	1995	1996	1997	1998	1999
Population of the U.S.	261,452	263,943	266,531	269,027	271,491

Source: U.S. Census Bureau

12. The following data is the U.S. population in thousands for the years 1950 to 1960. (Let $x = 0$ for the year 1950)

Year	1950	1960	1970	1980	1990
Population of the U.S.	151,326	179,323	203,302	226,542	248,718

Source: U.S. Census Bureau

13. The following data is the amount of dollars in billions spent on alcohol in the U.S. (Let $x = 0$ for the year 1994)

Year	1994	1995	1996	1997	1998
Dollars spent of alcohol	77.6	80.0	83.9	88.1	92.4

Source: U.S. Dept. of Agriculture, Economic Research Service

14. The following data is the amount in dollars spent in millions for all college and universities in the U.S. (Let $x = 0$ for the year 1995.)

~~Source: U.S. Center for Education Statistics~~

Year	1995	1996	1997	1998	1999
College and university expenditure	220,461	225,086	231,424	238,763	246,700

Source: U.S. Center for Education Statistics

15. Average tuition and fees for public 2–year schools in the U.S.(Let $x = 0$ for the year 1990)

Year	1991	1993	1995	1997	1999
Tuition and fees	$937	$1,125	$1,239	$1,314	$1,338

Source: National Center for Education Statistics U.S. Dept. of Education

16. Average tuition and fees for private 4–year schools in the U.S.
(Let $x = 0$ for the year 1993)

Year	1994	1995	1996	1997	1998
Tuition and fees	$11,481	$12,243	$12,881	$13,344	$13,973

Source: National Center for Education Statistics U.S. Dept. of Education

17. Total taxes paid as a percentage of income in thousands for Albuquerque New Mexico.

Income	25	50	75	100	150
Taxes percent of income	6.2	7.3	8.4	8.9	9.5

Source: Statistical Abstract of the U.S.

18. Lottery sales in thousands for the U.S. (Let $x = 0$ for the year 1999)

Year	1995	1996	1997	1998	1999
Lottery sales	31,931	34,031	35,486	35,588	35,966

Source: Statistical Abstract of the U.S.

19. Teen births in the U.S. (let $x = 0$ for the year 1994)

Year	1994	1995	1996	1997	1998
Number of teen births	505,488	499,873	494,272	489,211	484,975

Source: U.S. Center for Health Statistics

20. Government receipts in millions. (Let $x = 0$ for the year 1994.)

Year	1995	1996	1997	1998	1999
Receipts	1,351,495	1,452,763	1,578,955	1,721,421	1,827,302

Source: Financial Management Services U.S. Dept. of the Treasury

21. Persons living alone in thousands. (Let $x = 0$ for the year 1980.)

Year	1980	1985	1990	1995	1999
Number of people	18,296	20,602	22,999	24,732	26,606

Source: U.S. Census Bureau

22. Total school expenditures in millions for private colleges and universities in the U.S. (Let $x = 0$ for the year 1995)

Year	1995	1996	1997	1998	1999
Expenditures	80,877	83,706	86,460	89,523	92,600

Source: U.S. National Center for Education Statistics

23. Total taxes paid by gross family income in thousands for Albuquerque New Mexico.

Income	25	50	75	100	150
Taxes Paid	$1,592	$3,637	$6,303	$8,886	$14,261

Source: Statistical Abstract of the U.S.

24. Average tuition and fees for 2-year private colleges in U.S. (Let $x = 0$ for the year 1993.)

Year	1994	1995	1996	1997	1998
Tuition and fees	$6,914	$7,094	$7,236	$7,464	$7,854

Source: National Center for Education Statistics U.S. Dept. of Education

25. Net contributions in millions for social security funds. (Let $x = 0$ for the year 1995.)

Year	1996	1997	1998	1999	2000
Net contributions	321,557	349,946	371,207	396,352	418,219

Source: The World Almanac Book of Facts

26. Distributions in millions for social security funds. (Let $x = 0$ for the year 1995.)

Year	1996	1997	1998	1999	2000
Net distributions	302,861	316,257	326,462	334,383	347,868

Source: The World Almanac Book of Facts

27. Personal consumption in billions by U.S. citizens. (Let $x = 0$ for the year 1995.)

Year	1995	1996	1997	1998	1999
Total personal consumption	4,969.0	5,237.5	5,529.3	5,856.0	6,246.5

Source: Bureau of Economic Analysis U.S. Dept. of Commerce

28. Billions spent on food and tobacco in the U.S. (Let $x = 0$ for the year 1995.)

Year	1995	1996	1997	1998	1999
Total spent on food and tobacco	802.5	834.1	862.0	906.9	964.7

Source: Bureau of Economic Analysis U.S. Dept. of Commerce

29. The number of cigarettes smoked per 100 population versus the deaths per 100,000 from lung cancer for various states in the U.S.

Cigarettes smoked per 100 population	18.24	23.75	28.60	33.60	40.46
Death per 100,000 from lung cancer	15.98	19.50	22.07	24.55	27.27

Source: Journal of the National Cancer Institute

30. The number of cigarettes smoked per 100 population versus the deaths per 100,000 from bladder cancer for various states in the U.S.

Cigarettes smoked per 100 population	18.24	23.75	28.60	33.60	40.46
Death per 100,000 from bladder cancer	2.99	3.95	4.46	4.78	5.40

Source: Journal of the National Cancer Institute

Chapter 5 Review

Topic	Section	
Basics about lines	5.1	• $m = \dfrac{y_2 - y_1}{x_2 - x_1}$. The slope formula (need **two points**). • $y = mx + b$. The slope-intercept form of a line (need the **slope** and ***y*-intercept**). • $y - y_1 = m(x - x_1)$. The point-slope form of a line (need a **point** and a **slope**). • $Ax + By = C$. The standard form of a line (needed for systems of equations). • $y = D$. The form of a horizontal line ($m = 0$). • $x = D$. The form of a vertical line ($m =$ undefined). • $m_1 = m_2$. The lines are parallel. • $m_1 = \dfrac{-1}{m_2}$. The lines are perpendicular. • Also remember that slope is the most important aspect of a line, so look for it first. Then, when you are trying to find an equation, the name of the formula will tell you the information you need in order to use it.
Find the LN or EXP models	5.2	• We need two points in order to find a log ($\boldsymbol{a\ln(x) + b}$) or exponential ($\boldsymbol{ae^{kt}}$) function that could be used to model a real-world phenomena. • Once you have two points then you substitute them into the function and solve for the constants. • We need to know as many points as the model we are using has constants.
Calculator key strokes	5.3	See end of section 5.3
r and r^2	5.4	• Use the r and r^2 values along with the graphs as compared to the data points to help you decide upon the "curve of best fit."

CHAPTER 5 REVIEW HOMEWORK

Section 5.1

(1–4) Find the slope of the line through the given points.

1. (3, –2) and (7, 3)
2. (–5, 1) and (4, –3)
3. $(5, \frac{-1}{2})$ and $(-3, \frac{-1}{2})$
4. $(\frac{2}{3}, -3)$ and $(\frac{2}{3}, 5)$

(5–8) For each of the following equations find the slope and *y*–intercept.

5. $y = \frac{3}{5}x + 7$
6. $2x + 3y = 5$
7. $x = \frac{3}{4}$
8. $y = -3$

(9–14) Write the slope–intercept equation of each line that fits the following information.

9. Slope is $\frac{3}{5}$ and the *y*–intercept is (0, 5).
10. Slope is $\frac{-3}{5}$ and the line passes through (–3, –2).
11. Slope is undefined and the *x*–intercept is $(\frac{-3}{8}, 0)$.
12. Slope is zero and passes through (5, –2).
13. Passes through (3, –5) and (7, 2).
14. Passes through (–2, 3) and (–7, 5).

(15–16) Find the equations of the lines through the given point that are parallel and perpendicular to the given line.

15. $(-2, -1)$; $y = \frac{-3}{5}x + 2$
16. $(-3, 5)$; $2x - 3y = 7$

17. A market survey finds that for \$100 each 60 baseball gloves are sold. Another market survey in a similar market finds for \$120 each 50 baseball gloves of the same type are sold. Assuming the relationship between the number of gloves sold, $g(x)$, and the price per glove, x, is linear answer the following:
    a. Give the linear equation $g(x)$.
    b. How many gloves can be sold if the price per glove is \$90?
    c. What is the price of a baseball glove if you sell 52 gloves?
    d. What does the slope tell you about the relationship between price and gloves sold?
18. A manufacturing company has a piece of equipment that cost \$1,200,000 when it was first purchased. The company is allowed to depreciate the value of equipment an equal amount over a 10-year period.

a. Find the linear equation $E(t)$ that gives the value of the equipment on the company book after time t in years. (The value of the equipment is the original value minus depreciation.)

b. How much is the equipment worth after 6 years?

c. When will the equipment be worth $840,000?

19. U.S. Personnel on Active Duty (in hundreds of thousands)

2000 ($t = 0$)	4.71
2001 ($t = 1$)	4.78
2002 ($t = 2$)	4.85

Source: Dept. of the Army, U.S. Dept. of Defense

a. Find the linear function $M(t)$ that gives the number of U.S. Army personnel as a function of time. M is the number of U.S. Army personnel on active duty and t is the time in years.

b. Predict how many U.S. Army personnel will be on active duty in the year 2005.

c. Predict the year the U.S. Army personnel on active duty will number 541,000.

20. Percent of Population Covered by Private or Government Health Insurance

1994 ($t = 0$)	84.8
1995 ($t = 1$)	84.6
1996 ($t = 2$)	84.4

Source: U.S. Census Bureau

a. Find the linear function $H(t)$ that gives the percent of the population covered by private or government health insurance as a function of time. H is the percent of the population covered and t is the time in years.

b. Predict the percentage of the population covered in 1998.

c. Predict the year 83.6% of the population will be covered.

Section 5.2

(21–28) Find the x and y values that are the common solutions for each of the following pairs of equations.

21. $2x - y = 8$
$3x + y = 22$

22. $3x + 2y = 6$
$5x - y = 10$

23. $3x + 5y = 8$
$2x - 7y = 15$

24. $2x = 5y - 9$
$3y = 2x + 7$

25. $\frac{3}{5}x + \frac{2}{3}y = \frac{7}{15}$
$\frac{5}{4}x - \frac{7}{8}y = 8$

26. $x = \frac{2}{3}y - 4$
$3x + 7y = 9$

27. $2x = 4y + 7$
$x - 2y = 9$

28. $5y = 15 - 15x$
$3x + y = 3$

29. Find the exponential function $E(t) = ae^{kt}$ and logarithmic function $L(t) = a\ln(t) + b$ that contain the points (3, 20) and (9, 80). (Approximate each value to three decimal places and the check the answer with a graph of each equation and a stat plot of the points.)

30. A beetle infestation has killed several trees in a section of the national forest. After 10 months of infestation 5,000,000 trees are still alive in the section; after 20 months of infestation 4,000,000 trees are still alive in the section.

    a. Find an exponential function $B(t) = ae^{kt}$ that fits the data.

    b. Find a logarithmic function $B(t) = a\ln(t) + b$ that fits the data.

31. Railroad Passenger and Freight Outlays (In billions)

Year	1985 ($t = 5$)	1990 ($t = 10$)	1992 ($t = 12$)
Dollars	492.2	630.4	644.6

Source: Statistical Abstract of the U.S.

   a. Using the first two data points, find an exponential function $F(t) = ae^{kt}$ that computes passenger and freight outlays in billions. F represents the dollars in billions and t represents time.

   b. Using the first two data points, find the logarithmic function $F(t) = a\ln(t) + b$ that computes passenger and freight outlays in billions.

   c. Use your graphing calculator to graph the exponential function, the logarithmic function, and the stat plot of the data points. By inspecting these graphs do you think the exponential function or logarithmic function best fits the data?

Sections 5.3 and 5.4

(32–35) For each of the sets of data, use the regression formula ability of your calculator to find the best fit curve for each.

32. The following data is the number of male employees for state and local government (in thousands; $x = 0$ for the year 1983):

Year	1983	1985	1987	1989	1991
Salary	2,674	2,789	2,918	3,030	3,140

Source: U.S. Equal Opportunity Commission

33. The following data is the number of commercial country–western stations in existence for each year ($x = 0$ for the year 1995):

Year	1995	1996	1997	1998	1999
Number of Stations	2,613	2,525	2,491	2,368	2,306

Source: The World Almanac

34. The following data represents the total national income in billions ($x = 0$ for the year 1995):

Year	1995	1998	1999	2000
Total Income	5,876.7	7,041.4	7,468.7	7,984.4

Source: Bureau of Economic Analysis, U.S. Dept. of Commerce

35. The following data represents the percentage of the total workers in U.S. that are farm workers ($x = 0$ for the year 1900):

Year	1920	1940	1960	1980	1994
Percentage	27.0	17.4	6.1	2.7	2.5

Source: U.S. Dept. of Agriculture, Economic Research Service

CHAPTER 5 EXAM

(1–7) Write the slope-intercept equation of each line that fits the following information.

1. Slope is $\frac{-3}{4}$, y-intercept is (0, 8).

2. Slope is 5, passes through (–2, –5).

3. Slope is 0, passes through (3, –4).

4. Passes through (–1, 3) and (–5, –2).

5. Passes through (–3, 5) and (–3, –2).

6. Parallel to the line $3x - 2y = 6$ and passes through (2, 7).

7. Perpendicular to $2x + 5y = 8$ and passes through (–2, –4).

8. A real-estate developer owns an apartment complex with 100 apartments. If he rents the apartments for $600 all the apartments will be rented. If the rent is $650 per apartment the developer can rent 85 apartments.

   a. Assuming the relationship between rent and the number of apartments rented is linear, find a linear function $A(x)$, where x is the rental price and A the number of apartments rented.

   b. How many apartments will be rented when the rent is $750?

   c. If he wants to rent 64 apartments, what rent should he charge?

9. Average Salary of Classroom Teachers (In Thousands)

Year	Average Salary
1994 ($t = 0$)	35.7
1995 ($t = 1$)	36.7
1996 ($t = 2$)	37.7

Source: U.S. Equal Employment Opportunity Commission
Elementary-Secondary Staff Information

   a. Find the linear function $S(t)$ that gives the average classroom teacher's salary S in thousands of dollars for year t.

   b. Predict the average salary for the year 2000.

   c. Predict the year the average salary will be $47,700.

(10–15) Find the x and y values that are common solutions for each of the following pairs of linear equations.

10. $x - 3y = 8$
    $3x + 3y = 24$

11. $2x + 3y = 12$
    $3x - 2y = 15$

12. $\frac{2}{3}x + \frac{1}{2}y = \frac{5}{6}$
    $\frac{3}{5}x - \frac{1}{10}y = 3$

13. $\frac{1}{2}x - \frac{3}{4}y = 2$
    $6x = 9y + 24$

14. $0.3x + 0.12y = 0.5$

$0.2x - 0.24y = 0.12$

15. $8x + 12y = 36$

$3y = 9 - 2x$

16. For the pair of points (3, 20) and (9, 60) find:

a. The exponential function $E(t) = ae^{kt}$ that fits the given pair of points.

b. The logarithm function $F(t) = a\ln(t) + b$ that fits the given pair of points.

17. Two years after the walking catfish was introduced in the Florida Everglades there were 10,000 in the Florida Everglades; after five years there were 40,000.

a. If the number of catfish is growing exponentially, find algebraically the exponential function that will predict how many catfish there will be in the future.

b. If the number of catfish is growing logarithmically, find algebraically the logarithmic function that will predict how many catfish there will be in the future.

18. Use the first pair of points to construct the exponential and logarithmic functions.

Personal Income in the U.S. (in billions of dollars)

Year	1980 ($t = 10$)	1990 ($t = 20$)	2000 ($t = 30$)
Personal Income	4,191.3	5,705.4	7,770.0

Source: U.S. Census Bureau

a. Find an exponential function $P(t)$ that will predict personal income for year t.

b. Find a logarithmic function $P(t)$ that will predict personal income for year t.

c. Use the graphing calculator to graph the exponential function, the logarithmic function, and stat plot of the given data on the same graph. By inspecting these graphs do you think the exponential function or logarithmic function best fits the data?

(19–20) Use the regression mode of your calculator to find the best fit curve for each set of data.

19. Percent of occupied households that are owned ($x = 0$ for the year 1993)

Year	1994	1995	1996	1997	1998
Percent	64.0	64.7	65.4	65.7	66.3

Source: U.S. Census Bureau

20. Total Farm Real Estate Debt (in millions of dollars; $x = 0$ for the year 1994)

Year	1995	1996	1997	1998	1999
Debt	79,287	81,657	85,359	89,615	94,226

Source: Economic Research Service, U.S. Dept. of Agriculture

(21–22) For each of the tables of data:

a. Use common difference or common ratio to find what equation best fits the data.

b. Use algebra to find the equation.

c. Use your calculator and the regression mode to find the equation.

21.

x	2	4	6	8	10
y	0.008	0.032	0.128	0.512	2.048

22.

x	2	4	6	8	10
y	4.634	8.92	13.206	17.492	21.778

Chapter 6 Matrices

In chapter 5, section 2 we discussed how to solve a system of equations with two equations by hand, but there is a better way to solve these systems. We have now reached the point where we can talk about **Matrices**: the better way to solve a system of equations.

Section 6.1 Matrix Operations

There are a few things about matrices that you need to learn before we can talk about how to use them to solve systems of equations. First, let's give you the definition of a matrix.

> A **matrix** is a rectangular array of values made up of rows and columns.

Here is the general form of a matrix, called A.

$$A_{mn} = \begin{bmatrix} a_{11} & a_{12} & a_{13} & \cdot\cdot & a_{1n} \\ a_{21} & a_{22} & a_{23} & \cdot\cdot & a_{2n} \\ a_{31} & a_{32} & a_{33} & \cdot\cdot & a_{3n} \\ \cdot & \cdot & \cdot & \cdot\cdot & \cdot \\ \cdot & \cdot & \cdot & \cdot\cdot & \cdot \\ a_{m1} & a_{m2} & a_{m3} & \cdot\cdot & a_{mn} \end{bmatrix}$$

Typically, we use capital letters from the first part of the alphabet to name matrices. On the inside of the matrix, we label the position of each **element** (a number) in the matrix. For example, the first position in the matrix is labeled a_{11} (read "*a* one one") which represents the value in the first row, first column position. Likewise, a_{23} (read "*a* two three") represents the value in the second row, third column position. We always state the number of the row first, then the number of the column.

Also, in the general form, notice that the rectangle has m rows and n columns, so we call this matrix an ***m* x *n*** (read "m by n") matrix. This is the **order** of the matrix or what we will call the **dimension** (dim) of the matrix and thus the matrix is labeled A_{mn}. We will use the term dimension since that is the term used in our graphing calculators.

Question 1: What would a_{42} represent in the matrix?

Example 1: Dimension of a Matrix

What is the dimension of the matrix $E = \begin{bmatrix} 3 & -7 & 0 \\ 3 & 5 & -3 \end{bmatrix}$?

Solution:

The dimension of matrix E is 2 x 3 because it has two rows of numbers and three columns.

Question 2: What are the dimensions of the following matrices?

a) $B=\begin{bmatrix}-1 & 2\\ 0 & -5\\ 7 & 1\end{bmatrix}$ b) $C=\begin{bmatrix}1 & -3 & 5 & 2 & -11\\ -2 & 4 & 7 & 0 & 10\end{bmatrix}$ c) $D=\begin{bmatrix}1 & 2 & 3\\ 4 & 5 & 6\\ 7 & 8 & 9\end{bmatrix}$

In question 2, $c_{14}=2$, since in the C matrix the number 2 is in the first row fourth column position. Likewise, $d_{32}=8$ since 8 is in the third row, second column position of the D matrix.

Question 3: What are the values of the following from the matrices in question 2?

1) $b_{21}=$ 2) $c_{25}=$ 3) $d_{13}=$

Here are the steps necessary to enter matrix B into a graphing calculator.

<table>
<tr><th>TI-83</th><th>TI-86</th></tr>
<tr><td>Push the MATRX key then arrow right to EDIT
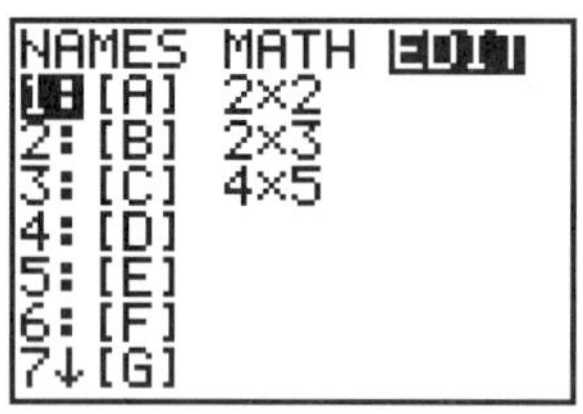
</td><td>Push 2nd 7 to get MATRX, then F2 for EDIT
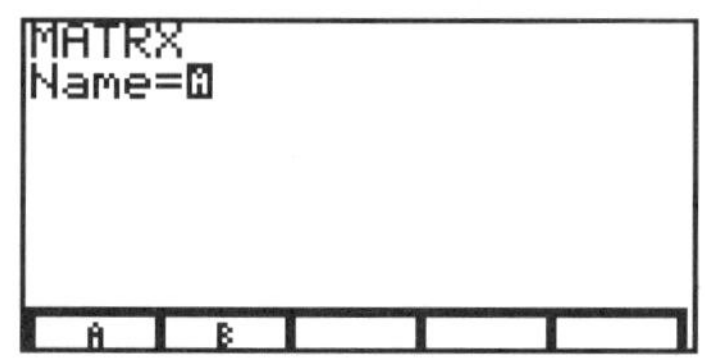
</td></tr>
<tr><td>Choose which matrix name you want to use (push 2 to select B as the name); then enter the dimension of the matrix.
(3 ENTER then 2 ENTER).
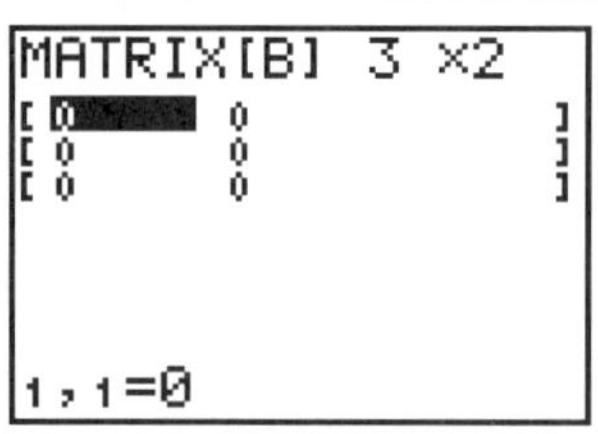
</td><td>Now type in a name for your matrix (push F2 to select B as the name, or we could just type the letter b by pushing the SIN key on the calculator); then push ENTER and enter the dimension of the matrix
(3 ENTER then 2 ENTER).
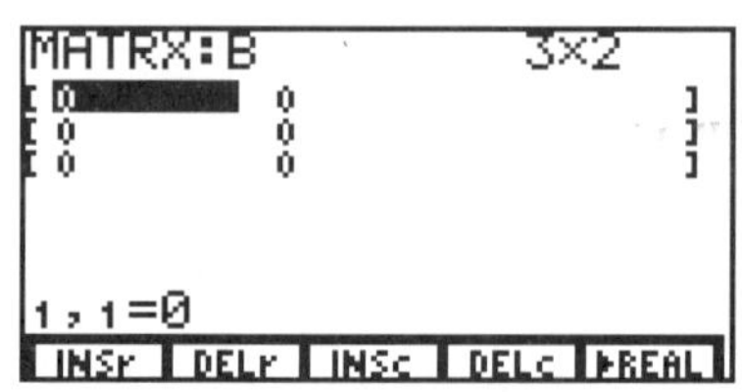
</td></tr>
<tr><td>Now, enter the numbers from matrix B one at a time; go from left to right, and from the</td><td>Now, enter the numbers from matrix B one at a time; go from left to right, and from the first</td></tr>
</table>

<table>
<tr>
<td>
first row down to the last.

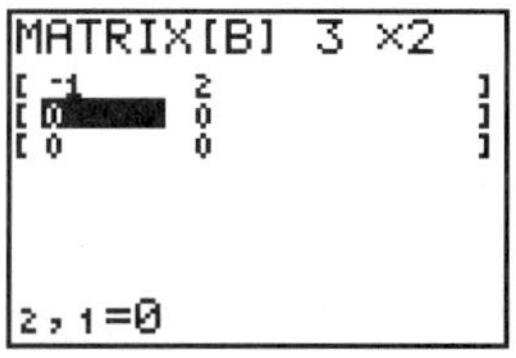

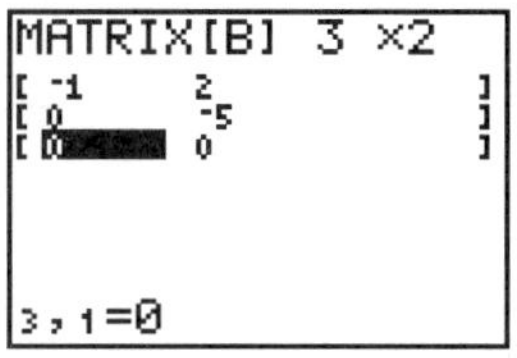

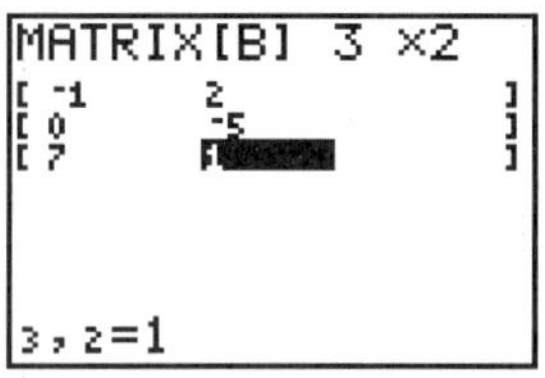

</td>
<td>
row down to the last.

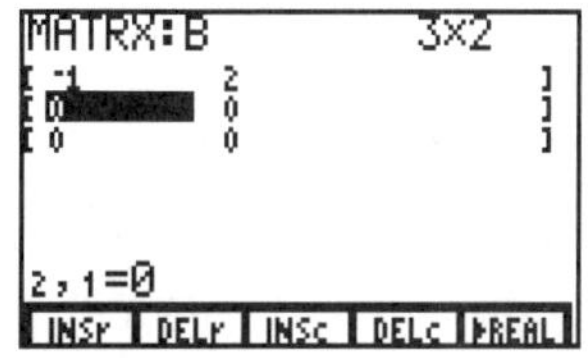

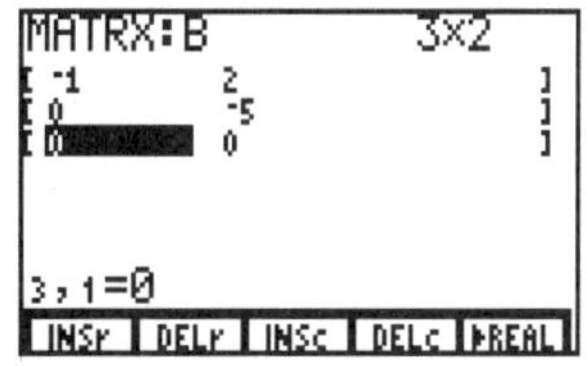

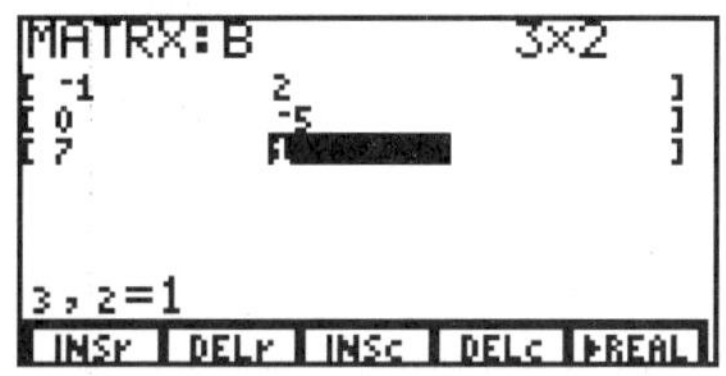

</td>
</tr>
<tr>
<td>
Our matrix is now entered into our calculators. If we want to see the matrix, then we must do the following:

- Go to the home screen.
- Push MATRX
- Choose the name of the matrix.
- Push ENTER

You should now see this on your screen:

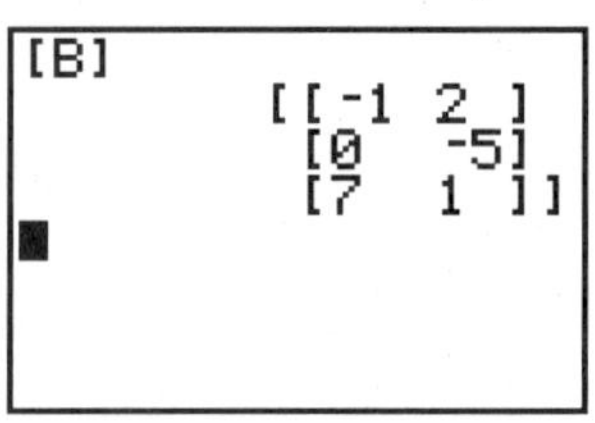

</td>
<td>
Our matrix is now entered into our calculators. If we want to see the matrix, then we must do the following:

- Go to the home screen.
- Push 2nd 7 to get MATRX.
- Push names (F1) and choose your matrix.
- Push ENTER

You should now see this on your screen:

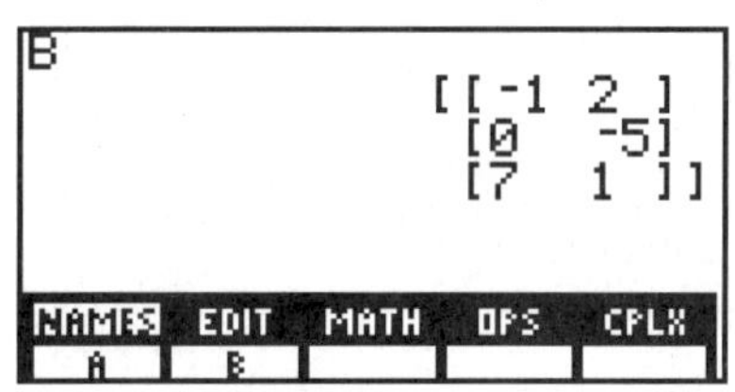

</td>
</tr>
</table>

Answer Q1: The value in the fourth row second column position.

Let's now take a look at how the basic operations (+, –, •) work with matrices.

Discussion 1: Matrix Addition

Add matrix $A = \begin{bmatrix} -2 & 6 \\ 3 & 0 \end{bmatrix}$ to matrix $B = \begin{bmatrix} 7 & -4 \\ -9 & 5 \end{bmatrix}$.

The addition of matrices works the way you might expect it to. You simply add the numbers that are in the exact same positions.

$$A+B=\begin{bmatrix}-2 & 6\\ 3 & 0\end{bmatrix}+\begin{bmatrix}7 & -4\\ -9 & 5\end{bmatrix}=\begin{bmatrix}(-2+7) & (6+(-4))\\ (3+(-9)) & (0+5)\end{bmatrix}=\begin{bmatrix}5 & 2\\ -6 & 5\end{bmatrix}$$

We can check ourselves on the graphing calculators.

Enter the two matrices [A] and [B] by the directions above and then put [A] + [B] on the home screen.

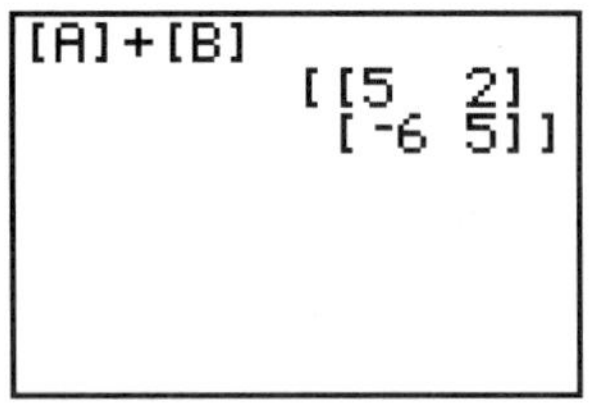

Notice that with the TI-83 you can't just type in the letters A and B you need to go to MATRIX then NAMES.

Enter the two matrices [A] and [B] and then put A + B on the home screen.

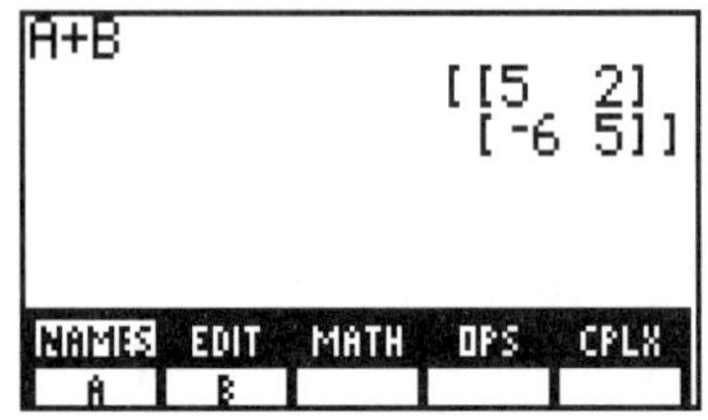

Answer Q2:
a) 3 x 2 since it has three rows of numbers going across and two columns of numbers going down.
b) 2 x 5 since it has two rows and 5 columns
c) 3 x 3.

Discussion 2: Matrix Addition That Doesn't Work

Add matrix $C=\begin{bmatrix}0 & -3 & -5\\ 4 & 11 & -1\end{bmatrix}$ to matrix $D=\begin{bmatrix}7 & 8\\ -3 & 4\\ 2 & 0\end{bmatrix}$.

$$C+D=\begin{bmatrix}0 & -3 & -5\\ 4 & 11 & -1\end{bmatrix}+\begin{bmatrix}7 & 8\\ -3 & 4\\ 2 & 0\end{bmatrix}=\begin{bmatrix}0+7 & -3+8 & -5+?\\ & & \end{bmatrix}$$

Answer Q3:
1) $b_{21}=0$
2) $c_{25}=10$
3) $d_{13}=3$

Notice, as we begin to add these together, that we find ourselves in a dilemma. There isn't a number in the D matrix at the first row, third column position because the D matrix *doesn't have a third column.* We can't add matrices that don't have *exactly the same dimension.* Here is what it looks like on your calculators.

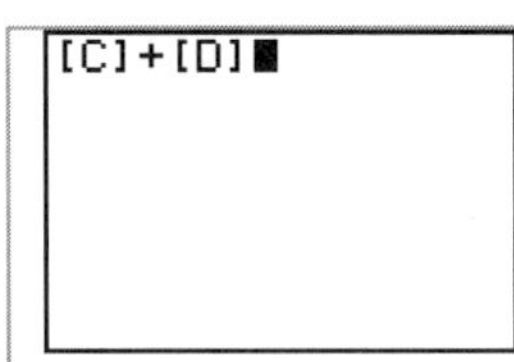

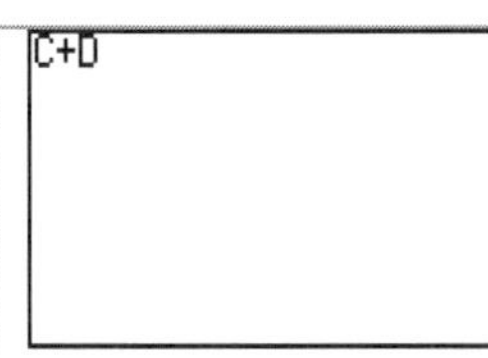

Question 4: Add matrix $A=\begin{bmatrix}23 & -11 & 0\\ -15 & 12 & 3\end{bmatrix}$ to matrix $B=\begin{bmatrix}-14 & 5 & 9\\ 18 & 4 & -5\end{bmatrix}$.

Subtraction works the same as addition; first, check that the two matrices have the same dimension, and then subtract the numbers that are in the same positions.

Example 2: Matrix Subtraction

Perform the indicated matrix operation. $\begin{bmatrix} -2 & 6 \\ 3 & 0 \end{bmatrix} - \begin{bmatrix} 7 & -4 \\ -9 & 5 \end{bmatrix}$

Solution:

$\begin{bmatrix} -2 & 6 \\ 3 & 0 \end{bmatrix} - \begin{bmatrix} 7 & -4 \\ -9 & 5 \end{bmatrix} = \begin{bmatrix} ((-2)-7) & (6-(-4)) \\ (3-(-9)) & (0-5) \end{bmatrix} = \begin{bmatrix} -9 & 10 \\ 12 & -5 \end{bmatrix}$. Here it is on your calculators.

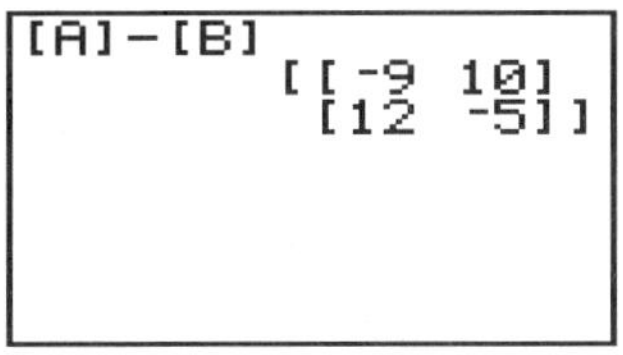

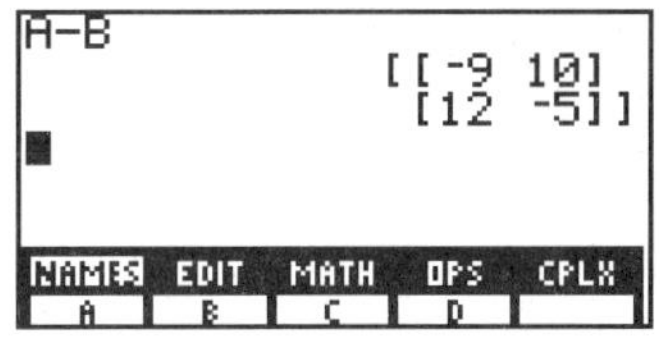

Question 5: Perform the indicated matrix operation. $\begin{bmatrix} 7 & 8 \\ -3 & 4 \\ 2 & 0 \end{bmatrix} - \begin{bmatrix} -14 & 5 & 9 \\ 18 & 4 & -5 \end{bmatrix}$

To reiterate, subtraction has the same requirements as addition. Whether we are adding or subtracting matrices, we need the matrices to have exactly the same dimensions. You should also remember from the properties of real numbers that $(5 - 7)$, a subtraction problem, is just like $(5 + (-7))$ which is an addition problem.

Since we have brought up the topic of properties of real numbers, let's make a few more observations. What number when added to 5 gives us an answer of 5? Obviously the answer is 0. Likewise, 5 added to what equals 5? Again, the answer is 0. The property works both ways $(0 + 5 = 5, 5 + 0 = 5$, so $0 + 5 = 5 + 0)$.

Question 6: What is the special name we give to the number 0 when we are talking about the properties of real numbers with respect to addition?

So given this fact what do you think is the additive identity for matrices? For example, if we have the matrix $\begin{bmatrix} -2 & 5 \\ 3 & -1 \end{bmatrix}$ then what added to it will equal $\begin{bmatrix} -2 & 5 \\ 3 & -1 \end{bmatrix}$?

Much like the additive identity for real numbers, since we add the same positions in order to add matrices and matrices are made up of real numbers, we need to add a 0 *to each position in the matrix.* Since matrices must be the same dimension to be added, the **identity matrix** must be the same dimension as the matrix it is being added to.

In this example we have: $\begin{bmatrix} -2 & 5 \\ 3 & -1 \end{bmatrix} + \begin{bmatrix} 0 & 0 \\ 0 & 0 \end{bmatrix} = \begin{bmatrix} 0 & 0 \\ 0 & 0 \end{bmatrix} + \begin{bmatrix} -2 & 5 \\ 3 & -1 \end{bmatrix} = \begin{bmatrix} -2 & 5 \\ 3 & -1 \end{bmatrix}$.

The matrix $\begin{bmatrix} 0 & 0 \\ 0 & 0 \end{bmatrix}$ is the additive identity for the matrix $\begin{bmatrix} -2 & 5 \\ 3 & -1 \end{bmatrix}$.

Question 7: What is the identity matrix for the matrix $\begin{bmatrix} -1 & 3 & 0 \\ -11 & 9 & -5 \end{bmatrix}$, and is there only one additive identity for all matrices?

The additive identity will be different depending on the dimension of the matrix. Here is the first indication that the properties of matrices will *not all be the same* as they are with real numbers.

Question 8: Write down what you think might be the answer to the following operation.

$$5\begin{bmatrix} -1 & 0 & 2 \\ 5 & -3 & 1 \\ 7 & 4 & -6 \end{bmatrix} \quad \text{(This is a number, 5, times a matrix.)}$$

Answer Q4:
$A + B =$
$$\begin{bmatrix} 23 & -11 & 0 \\ -15 & 12 & 3 \end{bmatrix} + \begin{bmatrix} -14 & 5 & 9 \\ 18 & 4 & -5 \end{bmatrix} = \begin{bmatrix} 9 & -6 & 9 \\ 3 & 16 & -2 \end{bmatrix}.$$

If we use our calculators to answer question 8, it would look like the following:

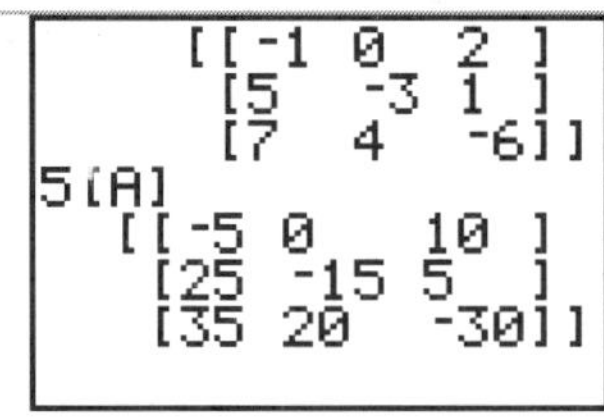

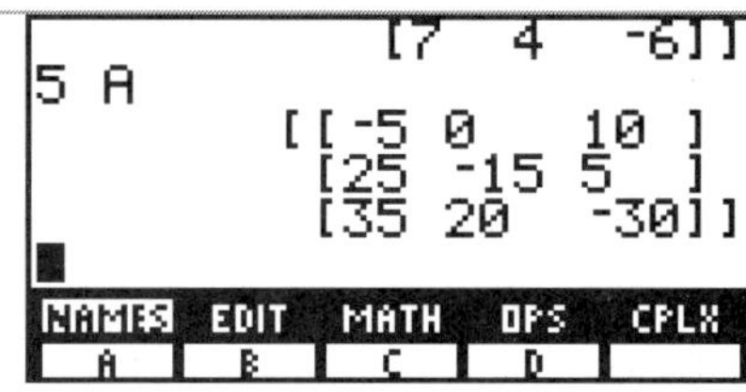

Once again, we hope that this seems intuitive to you. To multiply a whole matrix by a number we just multiply each element in the matrix by that number. This is called **scalar multiplication**. Here is another example for you.

Example 3: Scalar Multiplication

Perform the indicated matrix operation. $\quad -3\begin{bmatrix} 1 & -2 \\ 3 & 5 \\ -1 & 0 \\ -7 & 4 \end{bmatrix}$

Solution:

Multiply each element of the matrix by (–3). $-3\begin{bmatrix} 1 & -2 \\ 3 & 5 \\ -1 & 0 \\ -7 & 4 \end{bmatrix} = \begin{bmatrix} -3\cdot 1 & -3\cdot(-2) \\ -3\cdot 3 & -3\cdot 5 \\ -3\cdot(-1) & -3\cdot 0 \\ -3\cdot(-7) & -3\cdot 4 \end{bmatrix} = \begin{bmatrix} -3 & 6 \\ -9 & -15 \\ 3 & 0 \\ 21 & -12 \end{bmatrix}.$

Now, of course, any combination of these operations is possible. Here is an example.

Example 4: Matrix Operations

Simplify the matrix expression $(3A - 2B)$, given $A = \begin{bmatrix} 2 & -1 \\ 0 & 5 \\ 3 & 8 \end{bmatrix}$ and $B = \begin{bmatrix} 1 & 3 \\ 2 & -4 \\ -1 & 0 \end{bmatrix}$.

Solution:

Write the expression to be simplified.	$3\begin{bmatrix} 2 & -1 \\ 0 & 5 \\ 3 & 8 \end{bmatrix} - 2\begin{bmatrix} 1 & 3 \\ 2 & -4 \\ -1 & 0 \end{bmatrix}$
Perform the scalar multiplication. It will make life easier if we view the second matrix as being multiplied by –2.	$\begin{bmatrix} 6 & -3 \\ 0 & 15 \\ 9 & 24 \end{bmatrix} + \begin{bmatrix} -2 & -6 \\ -4 & 8 \\ 2 & 0 \end{bmatrix}$
Now, add the two matrices.	$\begin{bmatrix} 4 & -9 \\ -4 & 23 \\ 11 & 24 \end{bmatrix}$
On the calculator we would see this:	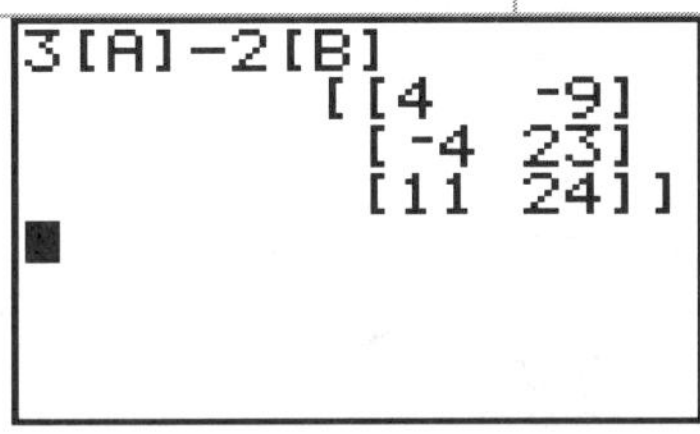 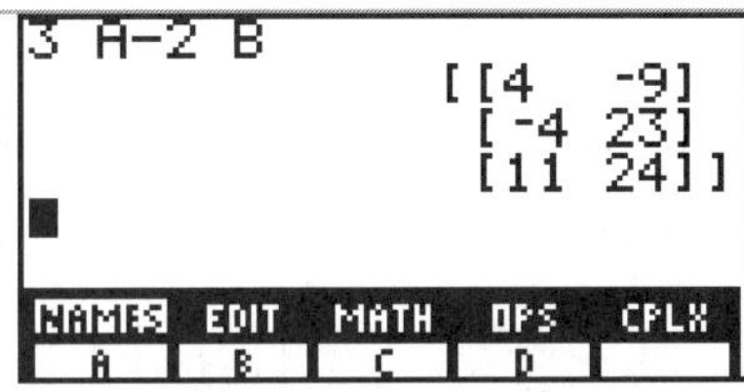

Answer Q5: We can't perform the subtraction because these two matrices don't have the same dimension. One matrix is a 3 x 2 and the other is a 2 x 3.

So far we have learned how to add and subtract matrices, multiply by a scalar, or a combination of these operations. Now we'll turn to multiplication of one matrix with another.

Up until now, each of the operations has worked like we might have expected them to. That is not necessarily the case here.

Answer Q6: 0 is called the additive identity. That is, 0 added to any number or any number added to 0 gives that same number back as an answer.

Discussion 3: Matrix Multiplication

Let's look at finding the product of the two matrices A and B given below.

$$A = \begin{bmatrix} 2 & -1 & 3 \\ 0 & 4 & -5 \end{bmatrix},\ B = \begin{bmatrix} -3 & 1 \\ 7 & 6 \\ 0 & -4 \end{bmatrix}$$

When we multiply two matrices, we multiply the elements in the *rows of the first matrix* by the elements of the *columns in the second matrix*. We are in a sense distributing the rows of the first matrix with the columns of the second. Here is how it works.

Step 1: Multiply the elements from the first row of matrix A by the elements in the first column of matrix B, in order (first with first, second with second, etc.). Add the three products together; that sum goes into the first position in the solution matrix.
$\begin{bmatrix} \mathbf{2} & \mathbf{-1} & \mathbf{3} \\ 0 & 4 & -5 \end{bmatrix}\begin{bmatrix} \mathbf{-3} & 1 \\ \mathbf{7} & 6 \\ \mathbf{0} & -4 \end{bmatrix} = \begin{bmatrix} [(2)(-3)+(-1)(7)+(3)(0)] & \_ \\ \_ & \_ \end{bmatrix} = \begin{bmatrix} -13 & \_ \\ \_ & \_ \end{bmatrix}$

Answer Q7: The additive identity for $\begin{bmatrix} -1 & 3 & 0 \\ -11 & 9 & -5 \end{bmatrix}$ *is* $\begin{bmatrix} 0 & 0 & 0 \\ 0 & 0 & 0 \end{bmatrix}$.

No, there isn't only one additive identity! We have just seen two different ones $\left(\begin{bmatrix} 0 & 0 \\ 0 & 0 \end{bmatrix}, \begin{bmatrix} 0 & 0 & 0 \\ 0 & 0 & 0 \end{bmatrix}\right)$

Step 2: Now we apply the same procedure to the first row of matrix A and the second column of matrix B. The sum goes into row one, column two of the solution matrix.

$$\begin{bmatrix} \mathbf{2} & \mathbf{-1} & \mathbf{3} \\ 0 & 4 & -5 \end{bmatrix} \begin{bmatrix} -3 & \mathbf{1} \\ 7 & \mathbf{6} \\ 0 & \mathbf{-4} \end{bmatrix} = \begin{bmatrix} -13 & [(2)(1)+(-1)(6)+(3)(-4)] \\ \_ & \_ \end{bmatrix} = \begin{bmatrix} -13 & -16 \\ \_ & \_ \end{bmatrix}$$

Step 3: Continue with the second row of matrix A and the first column of matrix B. The sum goes into row two, column one of the solution matrix.

$$\begin{bmatrix} 2 & -1 & 3 \\ \mathbf{0} & \mathbf{4} & \mathbf{-5} \end{bmatrix} \begin{bmatrix} \mathbf{-3} & 1 \\ \mathbf{7} & 6 \\ \mathbf{0} & -4 \end{bmatrix} = \begin{bmatrix} -13 & -16 \\ [(0)(-3)+(4)(7)+(-5)(0)] & \_ \end{bmatrix} = \begin{bmatrix} -13 & -16 \\ 28 & \_ \end{bmatrix}$$

Step 4: Last, we multiply the second row of matrix A by the second column of matrix B and add the products. The sum goes into row two, column two of the solution matrix.

$$\begin{bmatrix} 2 & -1 & 3 \\ \mathbf{0} & \mathbf{4} & \mathbf{-5} \end{bmatrix} \begin{bmatrix} -3 & \mathbf{1} \\ 7 & \mathbf{6} \\ 0 & \mathbf{-4} \end{bmatrix} = \begin{bmatrix} -13 & -16 \\ 28 & [(0)(1)+(4)(6)+(-5)(-4)] \end{bmatrix} = \begin{bmatrix} -13 & -16 \\ 28 & 44 \end{bmatrix}$$

So, after four steps, we have finally arrived at the answer.

$$\begin{bmatrix} 2 & -1 & 3 \\ 0 & 4 & -5 \end{bmatrix}_{2x3} \begin{bmatrix} -3 & 1 \\ 7 & 6 \\ 0 & -4 \end{bmatrix}_{3x2} = \begin{bmatrix} -13 & -16 \\ 28 & 44 \end{bmatrix}_{2x2}$$

Answer Q8:

$$5\begin{bmatrix} -1 & 0 & 2 \\ 5 & -3 & 1 \\ 7 & 4 & -6 \end{bmatrix} = \begin{bmatrix} 5(-1) & 5 \cdot 0 & 5 \cdot 2 \\ 5 \cdot 5 & 5(-3) & 5 \cdot 1 \\ 5 \cdot 7 & 5 \cdot 4 & 5(-6) \end{bmatrix} = \begin{bmatrix} -5 & 0 & 10 \\ 25 & -15 & 5 \\ 35 & 20 & -30 \end{bmatrix}$$

On the calculator we would see this

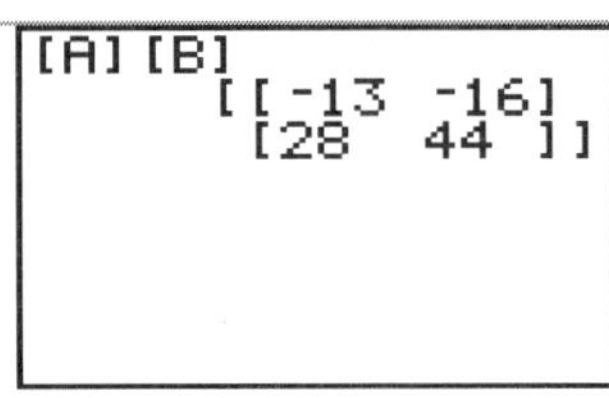

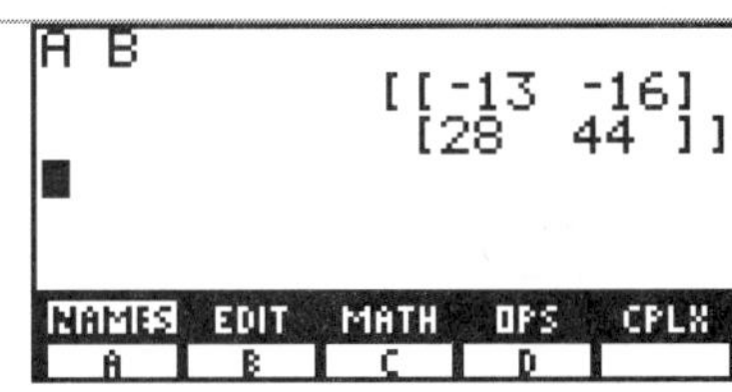

Question 9: Find the product of the two matrices A and B. $A = \begin{bmatrix} 1 & -2 \\ 3 & -5 \end{bmatrix}$, $B = \begin{bmatrix} 0 & 4 \\ -1 & 2 \end{bmatrix}$.

Example 5: Matrix Multiplication

Let's take the two matrices from question 9 and multiply them the other way around.

Solution:

$$BA = \begin{bmatrix} 0 & 4 \\ -1 & 2 \end{bmatrix}_{2x2} \begin{bmatrix} 1 & -2 \\ 3 & -5 \end{bmatrix}_{2x2} = \begin{bmatrix} [(0)(1)+(4)(3)] & [(0)(-2)+(4)(-5)] \\ [(-1)(1)+(2)(3)] & [(-1)(-2)+(2)(-5)] \end{bmatrix} = \begin{bmatrix} 12 & -20 \\ 5 & -8 \end{bmatrix}_{2x2}$$

Question 10: Does $AB = BA$? If not, did you think that they should have been equal?

You should notice from question 10 that matrices don't necessarily work like real numbers do. The commutative property of multiplication of real numbers states $ab = ba$ when a and b are real numbers. We have just seen an example showing that the commutative property doesn't usually hold true for matrices. Let's revisit discussion 3 and multiply the matrices the other way around.

Example 6: Matrix Multiplication

Find the product $\begin{bmatrix} -3 & 1 \\ 7 & 6 \\ 0 & -4 \end{bmatrix}\begin{bmatrix} 2 & -1 & 3 \\ 0 & 4 & -5 \end{bmatrix}$.

Solution:

$$\begin{bmatrix} -3 & 1 \\ 7 & 6 \\ 0 & -4 \end{bmatrix}_{3x2}\begin{bmatrix} 2 & -1 & 3 \\ 0 & 4 & -5 \end{bmatrix}_{2x3}$$

$$= \begin{bmatrix} [(-3)(2)+(1)(0)] & [(-3)(-1)+(1)(4)] & [(-3)(3)+(1)(-5)] \\ [(7)(2)+(6)(0)] & [(7)(-1)+(6)(4)] & [(7)(3)+(6)(-5)] \\ [(0)(2)+(-4)(0)] & [(0)(-1)+(-4)(4)] & [(0)(3)+(-4)(-5)] \end{bmatrix}$$

$$= \begin{bmatrix} -6 & 7 & -14 \\ 14 & 17 & -9 \\ 0 & -16 & 20 \end{bmatrix}_{3x3}.$$

Notice that this answer isn't equal, or even the same size (dimension), as the answer in discussion 3.

In discussion 3:	In example 6:
$\begin{bmatrix} 2 & -1 & 3 \\ 0 & 4 & -5 \end{bmatrix}_{2x3}\begin{bmatrix} -3 & 1 \\ 7 & 6 \\ 0 & -4 \end{bmatrix}_{3x2} = \begin{bmatrix} -13 & -16 \\ 28 & 44 \end{bmatrix}_{2x2}$	$\begin{bmatrix} -3 & 1 \\ 7 & 6 \\ 0 & -4 \end{bmatrix}_{3x2}\begin{bmatrix} 2 & -1 & 3 \\ 0 & 4 & -5 \end{bmatrix}_{2x3} = \begin{bmatrix} -6 & 7 & -14 \\ 14 & 17 & -9 \\ 0 & -16 & 20 \end{bmatrix}_{3x3}$

Now, this is bad enough, but an even more drastic difference can occur between the two multiplications, AB and BA.

Example 7: Matrix Multiplication

Find the products of AB and BA, given $A = \begin{bmatrix} 1 & -2 \\ 3 & -5 \end{bmatrix}$ and $B = \begin{bmatrix} 2 & -1 & 3 \\ 0 & 4 & -5 \end{bmatrix}$.

Solutions:

$$AB = \begin{bmatrix} 1 & -2 \\ 3 & -5 \end{bmatrix}_{2x2}\begin{bmatrix} 2 & -1 & 3 \\ 0 & 4 & -5 \end{bmatrix}_{2x3}$$

$$= \begin{bmatrix} [(1)(2)]+(-2)(0)] & [(1)(-1)]+(-2)(4)] & [(1)(3)]+(-2)(-5)] \\ [(3)(2)]+(-5)(0)] & [(3)(-1)]+(-5)(4)] & [(3)(3)]+(-5)(-5)] \end{bmatrix}$$

$$= \begin{bmatrix} 2 & -9 & 13 \\ 6 & -23 & 34 \end{bmatrix}_{2x3}.$$

$$BA = \begin{bmatrix} 2 & -1 & 3 \\ 0 & 4 & -5 \end{bmatrix}_{2x3} \begin{bmatrix} 1 & -2 \\ 3 & -5 \end{bmatrix}_{2x2}$$

$$= \begin{bmatrix} [(2)(1)+(-1)(3)+(3)(\textbf{?}) & \_ \\ \_ & \_ \end{bmatrix}_{?x?}.$$

We can't multiply B times A. The dimensions don't match up properly. Just like with addition, we can't multiply just any two matrices together. So this is an example where not only does $AB \neq BA$ but you can't even perform one of the multiplications. Here is what this looks like on your calculator.

Answer Q9: $AB = \begin{bmatrix} 1 & -2 \\ 3 & -5 \end{bmatrix}_{2x2} \begin{bmatrix} 0 & 4 \\ -1 & 2 \end{bmatrix}_{2x2} = \begin{bmatrix} 2 & 0 \\ 5 & 2 \end{bmatrix}_{2x2}$

You might be wondering, "How can I tell if it is possible to multiply two matrices together, and what will the final answer look like?" Let's look at some of the examples we have done so far, along with two new ones and see if we can discern a pattern.

$\begin{bmatrix} 2 & -1 & 3 \\ 0 & 4 & -5 \end{bmatrix} \begin{bmatrix} -3 & 1 \\ 7 & 6 \\ 0 & -4 \end{bmatrix} = \begin{bmatrix} -13 & -16 \\ 28 & 44 \end{bmatrix}$ 2 x 3 3 x 2 2 x 2	$\begin{bmatrix} -3 & 1 \\ 7 & 6 \\ 0 & -4 \end{bmatrix} \begin{bmatrix} 2 & -1 & 3 \\ 0 & 4 & -5 \end{bmatrix} = \begin{bmatrix} -6 & 7 & -14 \\ 14 & 17 & -9 \\ 0 & -16 & 20 \end{bmatrix}$ 3 x 2 2 x 3 3 x 3
$\begin{bmatrix} 2 & -1 & 6 \end{bmatrix} \begin{bmatrix} 0 & 1 \\ -3 & 4 \\ 6 & -7 \end{bmatrix} = \begin{bmatrix} 39 & -44 \end{bmatrix}$ 1 x 3 3 x 2 1 x 2	$\begin{bmatrix} 3 \\ -2 \end{bmatrix} \begin{bmatrix} -4 & 7 \end{bmatrix} = \begin{bmatrix} -12 & 21 \\ 8 & -14 \end{bmatrix}$ 2 x 1 1 x 2 2 x 2
$\begin{bmatrix} 1 & -2 \\ 3 & -5 \end{bmatrix} \begin{bmatrix} 2 & -1 & 3 \\ 0 & 4 & -5 \end{bmatrix} = \begin{bmatrix} 2 & -9 & 13 \\ 6 & -23 & 34 \end{bmatrix}$ 2 x 2 2 x 3 2 x 3	$\begin{bmatrix} 2 & -1 & 3 \\ 0 & 4 & -5 \end{bmatrix} \begin{bmatrix} 1 & -2 \\ 3 & -5 \end{bmatrix}$ = Can't Do 2 x 3 2 x 2 ?

Answer Q10: No! Well, $5 \bullet 7 = 7 \bullet 5$ so it seems like matrices should have worked the same way.

Hopefully, you have taken some time to try to figure out the two patterns. If you see and remember the patterns, it will give you a way to quickly decide if you can multiply two matrices together and a way to check your answer by seeing if you have the correct dimensions. Here are the patterns you should have spotted:

$\begin{bmatrix} 1 & -2 \\ 3 & -5 \end{bmatrix} \begin{bmatrix} 2 & -1 & 3 \\ 0 & 4 & -5 \end{bmatrix} = \begin{bmatrix} 2 & -9 & 13 \\ 6 & -23 & 34 \end{bmatrix}$ 2 x 2 2 x 3 2 x 3 These are equal so we can multiply the two matrices Tells the dim of the answer	$\begin{bmatrix} 2 & -1 & 3 \\ 0 & 4 & -5 \end{bmatrix} \begin{bmatrix} 1 & -2 \\ 3 & -5 \end{bmatrix}$ = Can't Do 2 x 3 2 x 2 ? These are **not** equal so we **can't** multiply the two matrices

The number of columns in the first matrix must equal the number of rows in the second matrix. The dimension of your answer will be the number of rows in the first matrix followed by the number of columns in the second. In matrix notation form we have this,

$$A_{m\,x\,n} \bullet B_{n\,x\,r} = C_{m\,x\,r}.$$

Look back at the other examples and see if this worked for every one of them.

Question 11: Find the product of the following two matrices (AB) and state the dim of the answer.

$$A = \begin{bmatrix} 3 & -1 \\ -5 & 2 \end{bmatrix}, B = \begin{bmatrix} x \\ y \end{bmatrix}$$

Finally, here are two practical applications of matrix addition and multiplication.

Example 8: A Manufacturing Application

A company manufactures and sells three different types of window shades: economical, standard and deluxe. There are three different plants manufacturing each of the types; one factory is in Phoenix, one is in Sacramento, and the other is in Dallas. Table A gives the amount manufactured by each site for the first half of the year and table B gives the amount manufactured by each site for the second half of the year.

A =

First half of year	Economical	Standard	Deluxe
Phoenix	10,053	8,935	6,859
Sacramento	9,253	7,855	4,893
Dallas	12,358	9,354	8,933

B =

Second half of year	Economical	Standard	Deluxe
Phoenix	12,385	10,835	8,332
Sacramento	10,832	8,354	7,055
Dallas	15,383	14,854	12,356

a) What are the two matrices we could make from this information?

b) What would the sum of the two matrices tell us in this example?

Solution:

a) We just create two matrices using the number in the two tables

$$A = \begin{bmatrix} 10{,}053 & 8{,}935 & 6{,}859 \\ 9{,}253 & 7{,}855 & 4{,}893 \\ 12{,}358 & 9{,}354 & 8{,}933 \end{bmatrix} \quad B = \begin{bmatrix} 12{,}385 & 10{,}835 & 8{,}332 \\ 10{,}832 & 8{,}354 & 7{,}055 \\ 15{,}383 & 14{,}854 & 12{,}356 \end{bmatrix}$$

b) When we add the two matrices together we get the number of shades of each type made at each city for the whole year.

$$\begin{bmatrix} 10{,}053 & 8{,}935 & 6{,}859 \\ 9{,}253 & 7{,}855 & 4{,}893 \\ 12{,}358 & 9{,}354 & 8{,}933 \end{bmatrix} + \begin{bmatrix} 12{,}385 & 10{,}835 & 8{,}332 \\ 10{,}832 & 8{,}354 & 7{,}055 \\ 15{,}383 & 14{,}854 & 12{,}356 \end{bmatrix} = \begin{bmatrix} 22438 & 19770 & 15191 \\ 20085 & 16209 & 11948 \\ 27741 & 24208 & 21289 \end{bmatrix}$$

Example 9: Car Sales

A new car company sells three different models of cars in the U.S.; the Royale, the Tempter and the Creast. It has a dealership in Las Vegas, Los Angeles and San Diego. The company is using a new idea for selling the cars. They will sale the cars fully loaded at one constant price for each model. Table A represents how many of each model was sold at each dealership and table B represents the price of each model.

A =

Cities	Royale	Tempter	Creast
Las Vegas	1,450	1,283	1,038
Los Angeles	2,384	3,222	3,835
San Diego	2,223	1,983	3,543

B =

Car type	Selling Price
Royale	$23,899
Tempter	$30,599
Creast	$38,299

a) What are the two matrices we could make from this information?

b) What would the two matrices times each other tell us in this example?

Solution:

To answer part a) the two matrices would be:

$$A = \begin{bmatrix} 1,450 & 1,283 & 1,038 \\ 2,384 & 3,222 & 3,835 \\ 2,223 & 1,983 & 3,543 \end{bmatrix} \quad B = \begin{bmatrix} 23,899 \\ 30,599 \\ 38,299 \end{bmatrix}$$

Notice that if we take A times B, that we would be multiplying rows of A by the column of B.

$$\begin{bmatrix} 1,450 & 1,283 & 1,038 \\ 2,384 & 3,222 & 3,835 \\ 2,223 & 1,983 & 3,543 \end{bmatrix} \begin{bmatrix} 23,899 \\ 30,599 \\ 38,299 \end{bmatrix} = \begin{bmatrix} 113,666,429 \\ 302,441,859 \\ 249,498,651 \end{bmatrix}$$

(3 cities **by** 3 types of cars) • (3 types of cars **by** 1 cost per car) = (3 cities **by** 1 cost for cars)

See how the dimensions match up (3 x 3 times 3 x 1 = 3 x 1). Also, the units on each row or column helps you to figure out what the product will mean [(city x type) (type x cost) = (city x cost)]. We have number of cars at a city times cost per car, so we end up with total cost of all cars at each particular city.

Section Summary

- Matrices are an array of numbers or stated another way just a rectangle full of numbers.
- In order to add or subtract matrices, the matrices must be of the same dimension and you perform the operations by simply adding or subtracting the numbers in the same positions.
- Scalar multiplication is just distributing the number throughout the matrix.
- Finally, with multiplication, you multiply rows times columns to perform the operation ($A_{m\,x\,n} \bullet B_{n\,x\,r} = C_{m\,x\,r}$).

SECTION 6.1 HOMEWORK

$A = \begin{bmatrix} 1 & -2 \\ 3 & 4 \end{bmatrix}$ $B = \begin{bmatrix} 3 & -1 \\ 2 & 3 \end{bmatrix}$ $C = \begin{bmatrix} -2 & 3 \\ 1 & 2 \\ 4 & 1 \end{bmatrix}$ $D = \begin{bmatrix} 3 & -2 \\ 4 & 5 \\ 2 & 3 \end{bmatrix}$

$E = \begin{bmatrix} 1 & -1 & 2 \\ 3 & 2 & 4 \\ 2 & 1 & 3 \end{bmatrix}$ $F = \begin{bmatrix} -3 & 5 & -4 \\ 0 & 3 & 5 \\ -4 & 2 & -1 \end{bmatrix}$ $G = \begin{bmatrix} 1 & -2 & 3 \\ 2 & 3 & 5 \end{bmatrix}$ $H = \begin{bmatrix} 3 & 1 & -5 \\ 0 & -8 & -1 \end{bmatrix}$

$X = \begin{bmatrix} x & 2 \\ 3 & 2x \end{bmatrix}$ $Y = \begin{bmatrix} x \\ y \\ z \end{bmatrix}$

(1–62) Using the matrices above find each of the following:

1. Dimension of A
2. Dimension of B
3. Dimension of F
4. Dimension of H
5. $a_{21} =$
6. $c_{32} =$
7. $e_{23} =$
8. $h_{21} =$
9. $d_{23} =$
10. $f_{43} =$
11. $A + B$
12. $B + A$
13. $C + D$
14. $D + C$
15. $E + F$
16. $F + E$
17. $G + H$
18. $H + G$
19. $A + C$
20. $E + G$
21. $A - B$
22. $B - A$
23. $G - H$
24. $H - G$
25. $3A$
26. $5E$
27. $3A + B$
28. $5E - F$
29. $5C - 2D$
30. $4H + 2G$
31. AB
32. BA
33. EF
34. FE
35. CH
36. HC
37. AC
38. CA
39. GE
40. EG
41. AX
42. XG
43. EY
44. GY
45. $(AB)G$
46. $A(BG)$
47. $E(FC)$
48. $(EF)C$
49. B^2
50. E^2
51. A^3
52. F^3
53. D^2
54. G^2
55. a. Give the additive identity matrix, I, for A.
    b. $A + I =$
    c. $I + A =$
56. a. Give the additive identity matrix, I, for C.
    b. $C + I =$
    c. $I + C =$
57. a. $-A =$ b. $A + (-A)$ c. $(-A) + A =$
    d. What is the relationship between A and –A?
58. a. $-C =$ b. $C + (-C) =$ c. $(-C) + C =$
    d. What is the relationship between C and $-C$?
59. Problems 11–18 exhibit what property of matrix addition?
60. Problems 45–48 exhibit what property of matrix multiplication?

Answer Q11:

$\begin{bmatrix} 3 & -1 \\ -5 & 2 \end{bmatrix}_{2x2} \begin{bmatrix} x \\ y \end{bmatrix}_{2x1} = \begin{bmatrix} (3x-y) \\ (-5x+2y) \end{bmatrix}_{2x1}$

. The answer is a 2 x 1 because we had a (2 x 2) (2 x 1)

61. Problems 31–40 show what property of matrix multiplication doesn't always work?

62. Problems 21–24 show what property of matrix subtraction doesn't always work?

63. $A = \begin{bmatrix} 2x-3 & 5 \\ 4 & 3y+2 \end{bmatrix} \quad B = \begin{bmatrix} 5 & 5 \\ 4 & 11 \end{bmatrix}$

What values of x and y would make $A = B$?

64. $A = \begin{bmatrix} 3x+1 & 2y-3 \\ 2x-5 & 3y+1 \end{bmatrix} \quad B = \begin{bmatrix} 7 & 7 \\ -1 & 16 \end{bmatrix}$

What values of x and y would make $A = B$?

65. A large state college on the East Coast has two campuses, a north campus and a south campus. Each has degrees in accounting, law and mechanical engineering. Table A shows information about the number of students trying for a degree in each discipline at the north campus and table B shows information about the number of students trying for a degree in each discipline on the south campus.

A =

	Accounting	Law	Mechanical engineering
Sophomore	3583	1283	1583
Junior	3258	1099	1325
Senior	2933	933	1079

B =

	Accounting	Law	Mechanical engineering
Sophomore	2038	1283	1583
Junior	1923	1400	635
Senior	1834	1283	529

a. Create a Matrix A and B from the above table of information.

b. Find $A + B$.

c. How many sophomores were in law school?

d. How many juniors were in mechanical engineering?

e. How many seniors were in accounting?

66. A company producing natural vitamins has two production companies, one on the West Coast and another on the East Coast. They produce three different bottles of Vitamin C; one bottle contains 50 tablets of Vitamin C, another 100 tablets of Vitamin C and the third 250 tablets of Vitamin C. Table A represents how many bottles of each was produced over a 4 week period in the West Coast plant and table B represent how many bottles of each was produced over a 4 week period in the East Coast plant.

A =

	Bottles with 50 tablets	Bottles with 100 tablets	Bottles with 250 tablets
Wk 1	5038	4833	3529
Wk 2	5263	4628	2933
Wk 3	4859	5029	2758
Wk 4	4623	5381	4033

B =

	Bottles with 50 tablets	Bottles with 100 tablets	Bottles with 250 tablets
Wk 1	6345	5835	4003
Wk 2	6583	6035	4323
Wk 3	6834	6530	4832
Wk 4	6583	7234	5234

a. Create a matrix A and B from the tables.

b. Find $A + B$.

c. How many 50 tablet bottles was produced in week three?

d. How many 100 tablet bottles was produced in week four?

e. How many 250 tablet bottles was produced in week two?

67. A computer store chain sells four different models of Pentium 4 computers out of three different stores. Table A represents the number of computers sold at each store and table P represents the selling price for each computer.

A =

	Model I	Model II	Model III	Model IV
Store A	53	48	35	29
Store B	61	53	49	32
Store C	49	45	39	43

P =

	Selling price
Model I	\$893
Model II	\$957
Model III	\$1238
Model IV	\$1393

a. Create a matrix A for the number of computers sold and P for the price of each model of computer.

b. Find AP.

c. How much revenue did store A receive?

d. How much revenue did store B receive?

e. How much revenue did store C receive?

68. Table C represents the store's cost for each type of computer in problem 67.

C =

	Cost
Model I	\$630
Model II	\$720
Model III	\$933
Model IV	\$1022

a. Create a matrix C for the cost to the stores for each model of computer.

b. Find AC.

c. How much did store A pay for the computers the store sold?

d. How much did store B pay for the computers the store sold?

e. How much did store C pay for the computers the store sold?

f. Find $AP–AC$ and interpret the entries.

69. A company makes tapes, discs and records in locations in Los Angeles and Atlanta. Each tape, disc and record must be manufactured and tested for quality. Table A gives the number of hours it takes to produce each item and table B gives the cost (in dollars per hour) in each city for each operation.

A =

	Manufacturing time	Testing time
Record	0.6 hours	0.4 hours
Tape	0.5 hours	0.2 hours
Disk	0.3 hours	0.3 hours

B =

	Los Angeles	Atlanta
Manufacturing labor cost	\$11	\$9
Testing labor cost	\$14	\$11

a. Give a matrix A for production time and matrix B for cost of production.

b. Find AB.

c. What does each entry of AB tell you?

70. A company sells five models of televisions through four different outlet stores. Table A represents the number of televisions sold by model at each outlet store and table B represents the cost of production selling price of each model.

A =

	Model I	Model II	Model III	Model IV	Model V
Outlet store 1	4	8	3	5	2
Outlet store 2	7	5	6	2	4
Outlet store 3	5	10	8	3	1
Outlet store 4	9	7	6	7	7

B =

	Cost	Selling price
Model I	\$135	\$228
Model II	\$158	\$283
Model III	\$183	\$318
Model IV	\$257	\$409
Model V	\$383	\$527

a. Give a matrix A for number of televisions sold and matrix B for cost of production and selling price.

b. Find AB.

c. What does each entry of AB tell you?

(71-74) Use the following tables to answer the questions:

Percent of voters that registered by age group to vote in the presidential elections 1980–1996

	1980	1988	1992	1996
18–20 years old	44.7	44.9	48.5	45.6
21–24 years old	52.7	50.6	55.3	51.2
25–34 years old	62.0	57.8	60.6	56.9
35–44 years old	70.6	69.3	69.02	66.5
45–64 years old	75.8	75.5	75.3	73.5
64 years old and over	74.6	78.4	78.0	77.0

Percent of voters that actually voted by age group
in the presidential elections 1980–1996

	1980	1988	1992	1996
18–20 years old	35.7	33.2	28.5	31.2
21–24 years old	28.3	45.7	33.4	24.2
25–34 years old	54.6	48.0	53.2	43.1
35–44 years old	64.4	61.3	63.6	54.9
45–64 years old	69.3	67.9	70.0	69.3
64 years old and over	65.1	68.8	70.1	70.2

Number of possible voters by age
in millions

	18–20 Years old	21–24 Years old	25–34 Years old	35–44 Years old	45–64 Years old	64 Years old and over
1980	12.3	15.9	35.7	25.6	43.6	24.1
1988	10.7	14.8	42.7	35.2	45.9	28.8
1992	9.7	14.6	41.6	39.7	49.1	30.8
1996	10.8	13.9	40.1	43.3	53.7	31.9

(Statistical Abstract of the U.S. 2000)

A column matrix is a matrix consisting of one column and n rows, a row matrix is a matrix consisting of one row and m columns.

e.g. Row matrix $\begin{bmatrix} 1 \\ -2 \\ 3 \\ 4 \end{bmatrix}$ Column matrix $\begin{bmatrix} 2 & 4 & -5 \end{bmatrix}$

71. a. Create a row matrix A using the percentage of 18–20 year olds that registered to vote and a column matrix B of the total population for 18–20 year olds for each year of the table.
    b. Find AB and interpret that value..
    c. Create a row matrix C using the percent of the 18–20 year olds that voted for each year of the table.
    d. Find CB and interpret that value.
    e. Find $AB - CB$ and interpret that value.
72. a. Create a row matrix A using the percent of the 45–64 year olds that registered to vote and a column matrix B of the total population of 45–64 year olds for each year of the table.
    b. Find AB and interpret that value.
    c. Create a row matrix C using the percent of the 45–64 year olds that voted for each year of the table.
    d. Find CB and interpret that value.
    e. Find $AB - CB$ and interpret that value.
73. a. Create a column matrix A using the year 1980 and the percent registered to vote for each age group and a row matrix B using the year 1980 and the total population of each age group.
    b. Find BA and interpret the value.
    c. Create a column matrix C using the year 1980 and the percent that voted for each age group.
    d. Find BC and interpret the value.
    e. Find $BA - BC$ and interpret the value.
74. a. Create a column matrix A using the year 1996 and the percent registered to vote for each age group and a row matrix B using the year 1996 and the total population of each age group.

b. Find BA and interpret the value.
c. Create a column matrix C using the year 1996 and the percent that voted for each age group.
d. Find BC and interpret the value.
e. Find $BA - BC$ and interpret the value.

Section 6.2 Matrix Identities and Inverses

We have learned how to add, subtract and multiply matrices. Now, let's discuss a few more attributes of matrices. First, you might have noticed that in the last section we didn't talk about division. Division isn't defined for matrices. This isn't necessarily a problem though. For example, if we wanted to solve the linear equation $5x = 15$, we might divide both sides of the equation by 5 to get $x = 3$. But there is another way to solve this equation that doesn't require division. The other way is to multiply both sides of the equation by the multiplicative inverse of 5, which is $\frac{1}{5}$. This way leads to the same result: $x = 3$.

Question 1: Why is $\frac{1}{5}$ the multiplicative inverse of 5?

You can always multiply by a multiplicative inverse instead of using division. That is what we will need to do with matrices since division is not defined. Let's take a few moments here to tie up any loose ends you might have with respect to inverses.

ADDITIVE INVERSE AND IDENTITY

Example 1: Additive Inverse of a Number

What is the additive inverse of 13?

Solution:

The additive inverse of 13 is the number that when added to 13 results in an answer which is the additive identity (0). So, for 13 the additive inverse would be -13 since $13 + (-13) = (-13) + 13 = 0$.

Question 2: What would be the additive inverse ($-A$) for the matrix $A = \begin{bmatrix} 2 & -5 & 0 \\ 1 & 3 & -7 \end{bmatrix}$? (Hint: remember that the matrix $\begin{bmatrix} 0 & 0 & 0 \\ 0 & 0 & 0 \end{bmatrix}$ is the additive identity for this matrix (section 6.1).)

Let's talk about the multiplicative identity now.

Question 3: What is the multiplicative identity for real numbers?

MULTIPLICATIVE IDENTITY

We know that 1 times any other real number is that real number, but what is the multiplicative identity for matrices? Your first thought might be that there will be more than one identity, since in section 6.1 we saw that there was more than one additive identity. If this was your first thought, you are correct. There is more than one multiplicative identity for matrices in general. Now, the question is, what are they? Let's try some possibilities.

Discussion 1: Multiplicative Identity for Matrices

The matrix of all zeros worked for the additive identity for matrices so maybe a matrix made up of all ones might be the multiplicative identity. For our original matrix let's use the matrix $\begin{bmatrix} 8 & -3 \\ -5 & 2 \end{bmatrix}$.

$$\begin{bmatrix} 8 & -3 \\ -5 & 2 \end{bmatrix} \begin{bmatrix} 1 & 1 \\ 1 & 1 \end{bmatrix} = \begin{bmatrix} ((8)1+(-3)1) & ((8)1+(-3)1) \\ ((-5)1+(2)1) & ((-5)1+(2)1) \end{bmatrix} = \begin{bmatrix} 5 & 5 \\ -3 & -3 \end{bmatrix}$$

Well, it doesn't look like that idea is working. This might have been your first thought, but its not correct. It's a good first guess though. Maybe the real multiplicative identity is close to this.

Remember, multiplication with matrices wasn't as intuitive as addition and scalar multiplication was. What we need is an identity matrix such that when it is multiplied by the original matrix, it will only multiply the first number of the original matrix by 1 when we take the first row times the first column, and it will only multiply the second number of the original matrix by 1 when we multiply the first row by the second column and so on. It looks like we might need 0's in some places. Let's give it another try.

$$\begin{bmatrix} 8 & -3 \\ -5 & 2 \end{bmatrix} \begin{bmatrix} 1 & 0 \\ 0 & 1 \end{bmatrix} = \begin{bmatrix} ((8)1+(-3)0) & ((8)0+(-3)1) \\ ((-5)1+(2)0) & ((-5)0+(2)1) \end{bmatrix} = \begin{bmatrix} 8 & -3 \\ -5 & 2 \end{bmatrix}$$

That seems to work! We must remember, though, that for something to be an identity it needs to work no matter which way the two are multiplied and for matrices, that can be a problem since, in general, matrix multiplication isn't commutative. Let's try the same problem again but this time reverse the order of multiplication.

$$\begin{bmatrix} 1 & 0 \\ 0 & 1 \end{bmatrix} \begin{bmatrix} 8 & -3 \\ -5 & 2 \end{bmatrix} = \begin{bmatrix} ((1)8+(0)(-3)) & ((0)8+(0)(-3)) \\ ((1)(-5)+(0)2) & ((0)(-5)+(1)2) \end{bmatrix} = \begin{bmatrix} 8 & -3 \\ -5 & 2 \end{bmatrix}.$$

It is clear now, that $\begin{bmatrix} 1 & 0 \\ 0 & 1 \end{bmatrix}$ is the identity for this matrix. Let's try another example.

Example 2: Multiplicative Identity of a Matrix

What is the multiplicative identity for the matrix $\begin{bmatrix} -2 & 5 & 0 \\ -1 & -3 & 7 \end{bmatrix}$?

Solution:

This matrix is a 2 x 3 so we need to multiply by a 3 x ?. It turns out that multiplicative identities are always square matrices. That is, they always have the same number of rows and columns. So, we will need to try using a 3 x 3 here.

$$\begin{bmatrix} -2 & 5 & 0 \\ -1 & -3 & 7 \end{bmatrix} \begin{bmatrix} 1 & 0 & 0 \\ 0 & 1 & 0 \\ 0 & 0 & 1 \end{bmatrix}$$

$$= \begin{bmatrix} ((-2)1+(5)0+(0)0) & ((-2)0+(5)1+(0)0) & ((-2)0+(5)0+(0)1) \\ ((-1)1+(-3)0+(7)0) & ((-1)0+(-3)1+(7)0) & ((-1)0+(-3)0+(7)1) \end{bmatrix} = \begin{bmatrix} -2 & 5 & 0 \\ -1 & -3 & 7 \end{bmatrix}$$

Let's multiply them the other way around this time.

$\begin{bmatrix} 1 & 0 & 0 \\ 0 & 1 & 0 \\ 0 & 0 & 1 \end{bmatrix} \begin{bmatrix} -2 & 5 & 0 \\ -1 & -3 & 7 \end{bmatrix} = ?$ This can't be multiplied because the dimensions don't match.

As you can see, this identity matrix (3 x 3) doesn't work on both sides, but maybe there is another identity matrix that exists that will work on the left side. Let's try $\begin{bmatrix} 1 & 0 \\ 0 & 1 \end{bmatrix} \begin{bmatrix} -2 & 5 & 0 \\ -1 & -3 & 7 \end{bmatrix}$.

$$\begin{bmatrix} 1 & 0 \\ 0 & 1 \end{bmatrix} \begin{bmatrix} -2 & 5 & 0 \\ -1 & -3 & 7 \end{bmatrix} = \begin{bmatrix} ((-2)1+(-1)0) & ((5)1+(-3)0) & ((0)0+(7)0) \\ ((-2)0+(-1)1) & ((5)0+(-3)1) & ((0)0+(7)1) \end{bmatrix} = \begin{bmatrix} -2 & 5 & 0 \\ -1 & -3 & 7 \end{bmatrix}$$

There are two matrices, that when multiplied by the original matrix, yields the original matrix as the answer. We might call these left $\begin{bmatrix} 1 & 0 \\ 0 & 1 \end{bmatrix}$ and right $\begin{bmatrix} 1 & 0 & 0 \\ 0 & 1 & 0 \\ 0 & 0 & 1 \end{bmatrix}$ identities. But, for something to be truly an identity for something else, you need to have **only one** possibility. So, as you can see, non-square matrices won't have an identity and thus we will not be able to find inverses for them (in general, anything times its inverse must equal an identity).

So, only square matrices have identities, and **multiplicative identities** for matrices must be square matrices of the appropriate dimension with 1's down the diagonal and 0's elsewhere in the matrix. Also we notice that there are many multiplicative identities, just like there are many additive identities (1 x 1, 2 x 2, 3 x 3, …).

MULTIPLICATIVE INVERSES FOR MATRICES

Now that we know what the multiplicative identities are for matrices, we need to figure out what the multiplicative inverses are for them. Remember that a multiplicative inverse for a real number is that number, which, when multiplied by the original number, yields the identity, 1 for the answer. The same concept is true for matrices. The multiplicative inverse will be that matrix, which, when multiplied by the original matrix, will yield the multiplicative identity as the answer.

Answer Q1: Whenever you can multiply two things together, AB and BA, and the result is the same, (in this case 1), then the two things multiplied together are called multiplicative inverses

Answer Q2: The additive inverse is –A $= \begin{bmatrix} -2 & 5 & 0 \\ -1 & -3 & 7 \end{bmatrix}$ since $\begin{bmatrix} 2 & -5 & 0 \\ 1 & 3 & -7 \end{bmatrix} + \begin{bmatrix} -2 & 5 & 0 \\ -1 & -3 & 7 \end{bmatrix} = \begin{bmatrix} -2 & 5 & 0 \\ -1 & -3 & 7 \end{bmatrix} + \begin{bmatrix} 2 & -5 & 0 \\ 1 & 3 & -7 \end{bmatrix} = \begin{bmatrix} 0 & 0 & 0 \\ 0 & 0 & 0 \end{bmatrix}$. *Since matrices are made up of real numbers, the additive inverse matrix is just the matrix made up of opposite signed numbers.*

Answer Q3: The multiplicative identity for real numbers is the number 1. Any number times one is itself (x(1)=(1)x=x).

Discussion 2: Multiplicative Inverse of a Matrix

Let's find the multiplicative inverse (A^{-1}) for the matrix $A = \begin{bmatrix} 8 & -3 \\ -5 & 2 \end{bmatrix}$.

We need to find a matrix, which, when multiplied by matrix A, will equal $\begin{bmatrix} 1 & 0 \\ 0 & 1 \end{bmatrix}$.

Question 4: What will the dimension of the inverse matrix have to be?

The answer to question 4 tells us what the dimension of the inverse must be, so we get the following:

$$\begin{bmatrix} 8 & -3 \\ -5 & 2 \end{bmatrix} \begin{bmatrix} x & y \\ z & w \end{bmatrix} = \begin{bmatrix} (8(x)+(-3)(z)) & (8(y)+(-3)(w)) \\ ((-5)(x)+2(z)) & ((-5)(y)+2(w)) \end{bmatrix} = \begin{bmatrix} 1 & 0 \\ 0 & 1 \end{bmatrix}$$

If we look at just the first position (a_{11}) in both of the last two matrices, we see that $8x - 3z = 1$. We also see in the a_{21} position that $-5x + 2z = 0$. Between these two equations we can figure out what x and z must equal. We'll use the elimination method to solve.

Multiply the first equation by 2 and the second by 3 to eliminate the z's.	$\begin{cases} 16x - 6z = 2 \\ -15x + 6z = 0 \end{cases}$
Added together we get	$x \quad = 2$
Plug the answer for x back into one of the equation and we find that z is	$8(2) - 3z = 1$ $16 - 3z = 1$ $-3z = -15$ $z = 5$

If we do the same for the other two variables, we will find that we get $y = 3$ and $w = 8$. So the inverse matrix is $\begin{bmatrix} 2 & 3 \\ 5 & 8 \end{bmatrix}$. Let's verify that on **both sides**.

$$\begin{bmatrix} 8 & -3 \\ -5 & 2 \end{bmatrix} \begin{bmatrix} 2 & 3 \\ 5 & 8 \end{bmatrix} = \begin{bmatrix} (8(2)+(-3)(5)) & (8(3)+(-3)(8)) \\ ((-5)(2)+2(5)) & ((-5)(3)+2(8)) \end{bmatrix} = \begin{bmatrix} 1 & 0 \\ 0 & 1 \end{bmatrix}$$

$$\begin{bmatrix} 2 & 3 \\ 5 & 8 \end{bmatrix} \begin{bmatrix} 8 & -3 \\ -5 & 2 \end{bmatrix} = \begin{bmatrix} (2(8)+(3)(-5)) & (2(-3)+(3)(2)) \\ ((5)(8)+8(-5)) & ((5)(-3)+8(2)) \end{bmatrix} = \begin{bmatrix} 1 & 0 \\ 0 & 1 \end{bmatrix}.$$

Therefore, $A^{-1} = \begin{bmatrix} 2 & 3 \\ 5 & 8 \end{bmatrix}$.

There just has to be better ways to find an inverse, and there are. One method for finding an inverse uses something called the **reduced row echelon form of a matrix**, but that requires an understanding of topics to be covered in a later section (6.4) so we can't use that method yet. There is another method, but it requires going much farther than we'd like into a branch of mathematics called Linear Algebra. The method we will choose to show you here uses the graphing calculator. Let's redo that last problem on the calculator.

Discussion 3: Finding the Inverse of a Matrix

Let's find the multiplicative inverse of $\begin{bmatrix} 8 & -3 \\ -5 & 2 \end{bmatrix}$ using our calculators.

	TI-83	TI-86
Input the matrix into your calculator and check to make sure that all the numbers are correct.	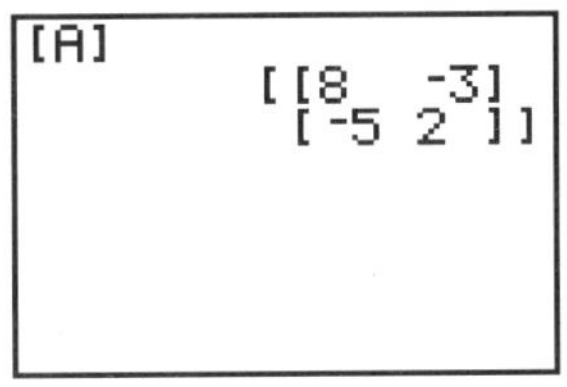	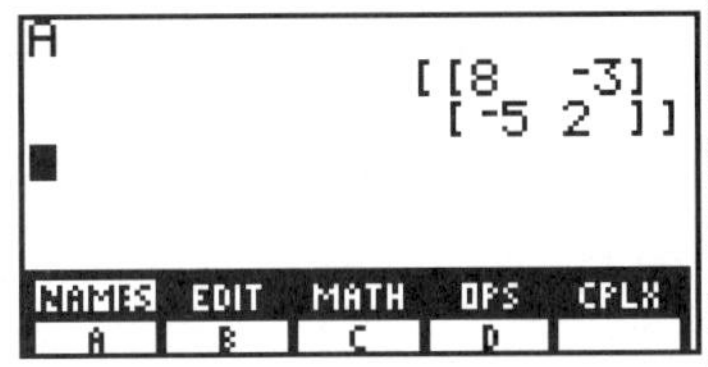
Tell the calculator to find the inverse by getting the *A* matrix on the home screen. Then push the x^{-1} key on the left hand side of your calculator, and then push ENTER.	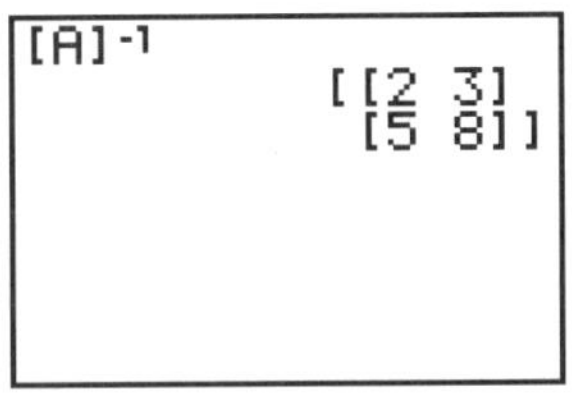	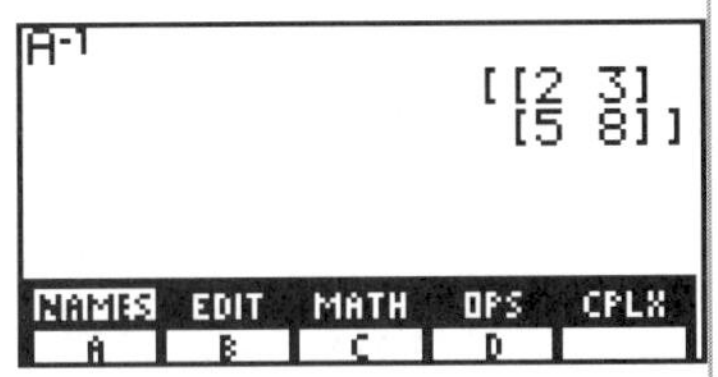
Now your answer for the inverse is on your screen.		

Example 3: Finding the Inverse of a Matrix

Find the inverse of $\begin{bmatrix} -2 & 5 & 1 \\ 0 & 3 & 4 \\ -1 & 2 & 6 \end{bmatrix}$.

Solution:

	TI-83	TI-86
Input the matrix into your calculator and check to make sure that all the numbers are correct.	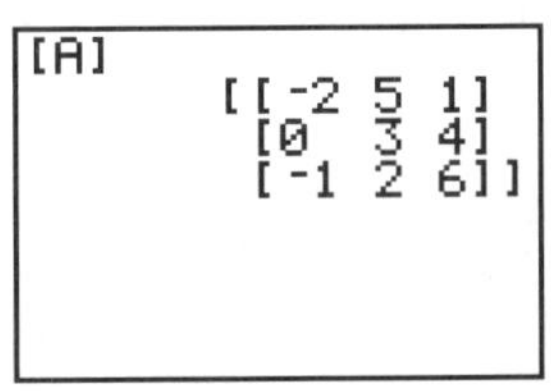	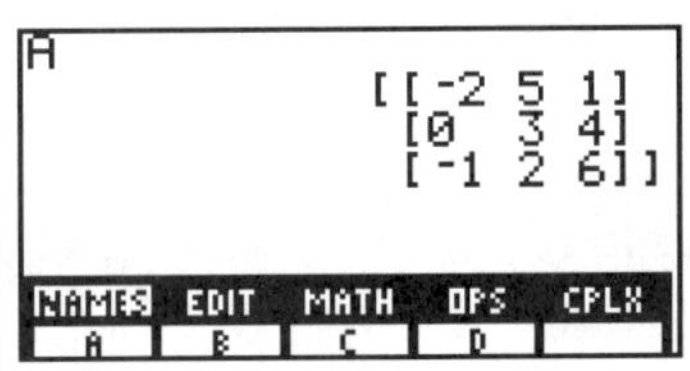
Tell the calculator to find the inverse by getting the *A* matrix on the home screen. Then push the x^{-1} key on the left hand side of your calculator. Finally, push ENTER.	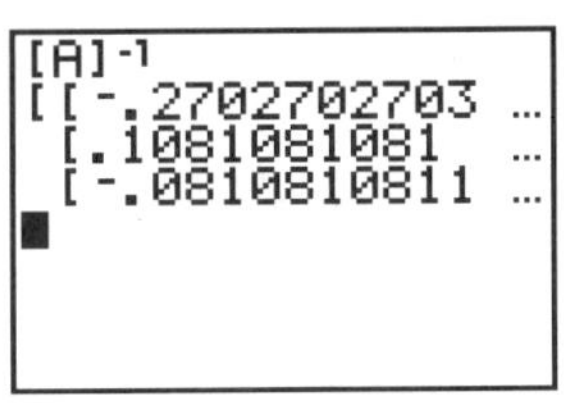	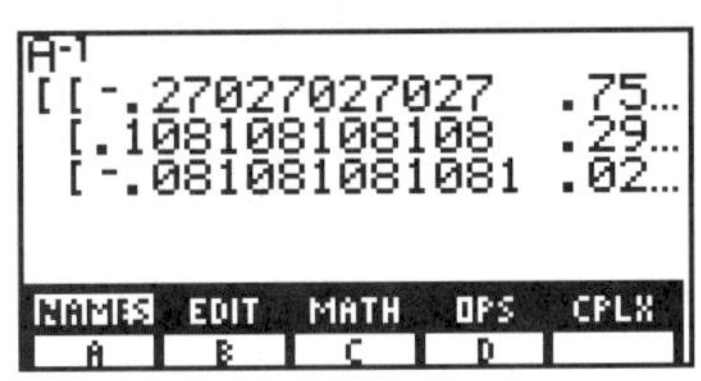
As you can see the matrix doesn't fit on the screen. To see the rest of the matrix use the right arrow to move to the right.		

Also an answer with all those decimals is not desirable, so let's FRAC it.	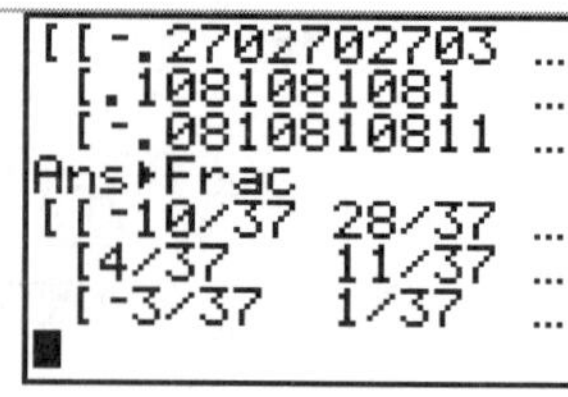	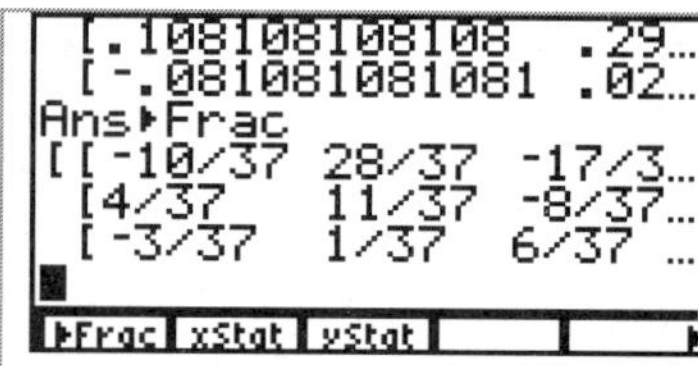
Now our final answer looks like this:		$A^{-1}=\begin{bmatrix} \frac{-10}{37} & \frac{28}{37} & \frac{-17}{37} \\ \frac{4}{37} & \frac{11}{37} & \frac{-8}{37} \\ \frac{-3}{37} & \frac{1}{37} & \frac{6}{37} \end{bmatrix}$

Answer Q4: 2 x 2 since we have a (2 x 2) times (? x ?) = (2 x 2).

Let's check our answer on the calculator just to be sure it really is the inverse.

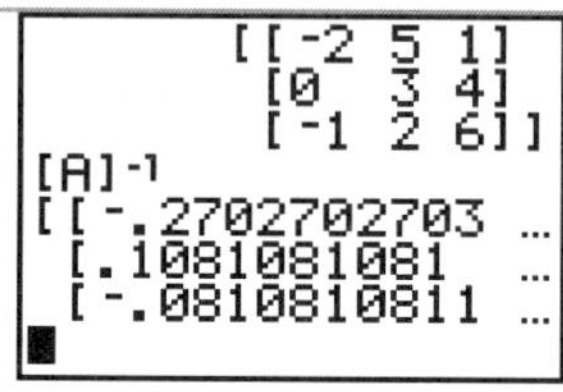

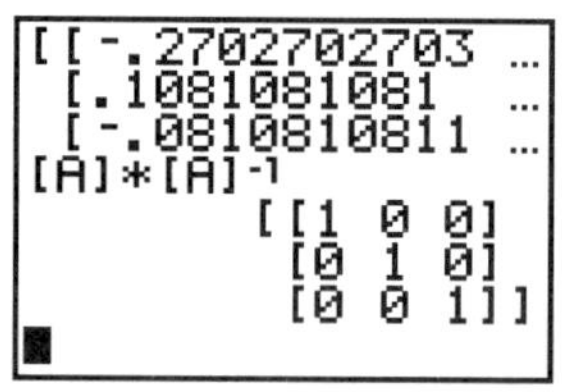

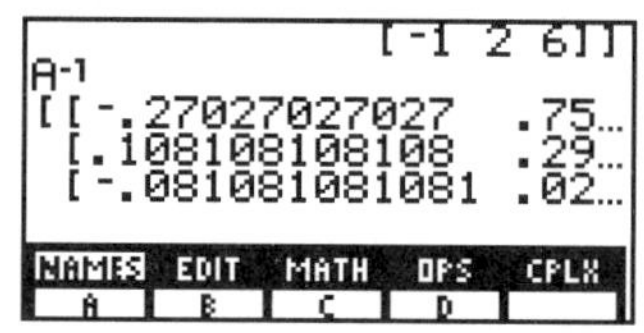

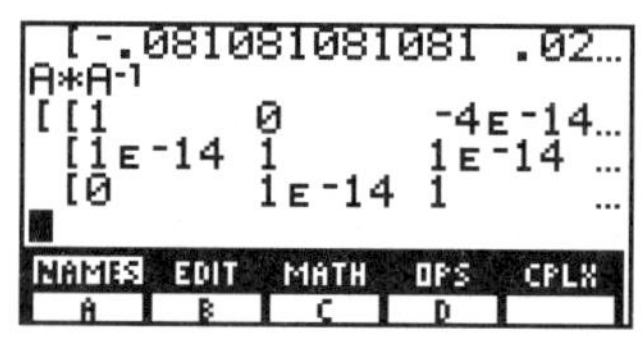

In the TI-86 the E ⁻14 = 1 x 10^{-14} = 0.00000000000001, which means basically that the answer for that position in the matrix is 0. Also remember to check it on both sides.

Question 5: Find the inverse of $\begin{bmatrix} 5 & 7 & -2 \\ 1 & -3 & 9 \end{bmatrix}$.

As you tried to do question 5, you should have run into a problem. Your calculator should have given you an error message like one of these:

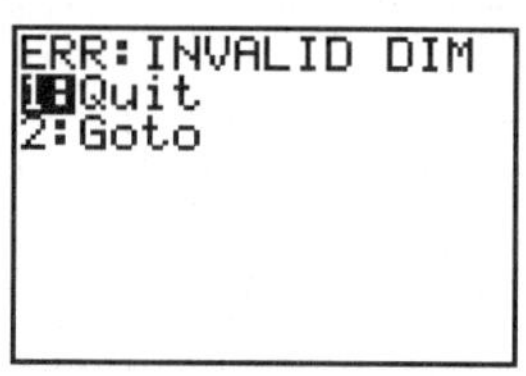

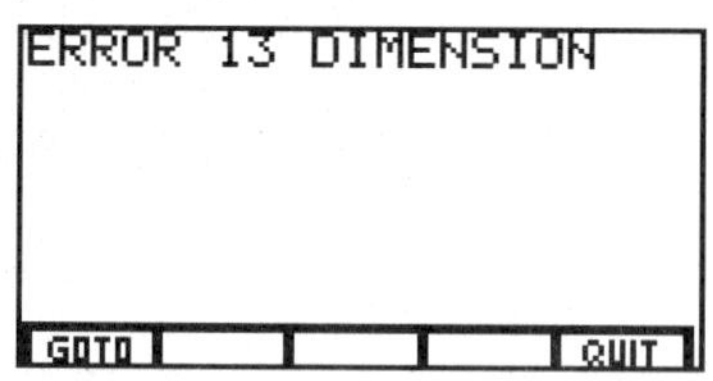

Notice the error message. It tells us that the problem had to do with the dimension. Only square matrices can have inverses but even some of these don't have an inverse.

Example 4: Finding the Inverse of a Matrix

Find the inverse of $\begin{bmatrix} 1 & 2 & 3 \\ 4 & 5 & -1 \\ 1 & -1 & -10 \end{bmatrix}$.

Solution:

Let's enter this into our calculators and see what we get.

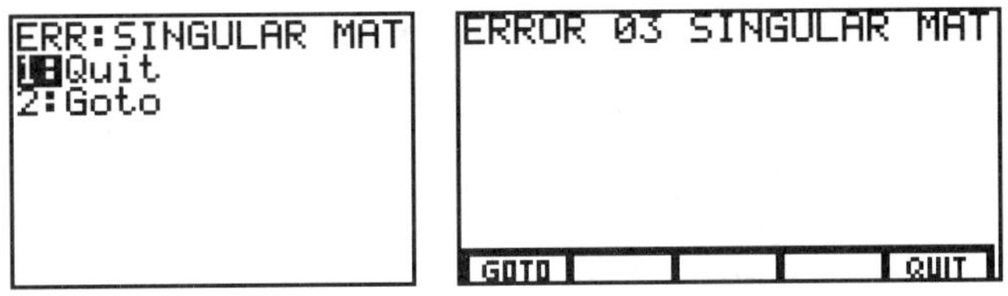

This time the error message tells us that this is something called a **singular matrix**. For now, it's enough to know that this means that this matrix doesn't have an inverse. You might think it strange that not every matrix has an inverse but not all real numbers do either. For example, the number 0 doesn't have a multiplicative inverse. There doesn't exist any number that when multiplied by 0 gives 1 for the answer.

DETERMINANTS

Let's take a look at one more topic called **determinants**.

The **determinant** of a square matrix is a number that is associated with that matrix and tells us if the matrix has an inverse.

Here is how we find the determinant of a 2 x 2 matrix by hand.

Discussion 4: Finding the Determinant of a Matrix by Hand

Let's find the determinant of the matrix $\begin{bmatrix} 8 & -3 \\ -5 & 2 \end{bmatrix}$.

To find the determinant (det) of a 2 x 2 matrix you simply cross multiply the numbers and subtract. (Diagonal down and to the right minus the diagonal up and to the right.)	$\det \begin{bmatrix} 8 & -3 \\ -5 & 2 \end{bmatrix} = 8(2) - (-5)(-3) = 1$

So, the determinant of this matrix is 1. In a general form it looks like this:

$$\det A = \det \begin{bmatrix} a_1 & b_1 \\ a_2 & b_2 \end{bmatrix} = a_1 b_2 - a_2 b_1$$

To find the determinant of a larger dimensioned square matrix is a bit more involved. We will show you one example here and then go to the calculator.

Discussion 5: Finding the Determinant of a 3 x 3 Matrix by Hand

The determinant of the matrix: $\begin{bmatrix} -2 & 5 & 1 \\ 0 & 3 & 4 \\ -1 & 2 & 6 \end{bmatrix}$ (this is the same matrix we used in example 6).

To find the determinant of a 3 x 3 matrix or larger, we must do something called **expansion about a row or column** (this procedure works for finding a determinant of a 2 x 2 as well, but thinking in terms of cross multiplying is easier). In this example we are going to show you how to expand about the first row in order to find the determinant.

Answer Q5: There doesn't exist an inverse matrix for this matrix since it's not a square matrix.

First, you must find what is called a **minor** and then a **cofactor** for the a_{11} position. You find a minor by eliminating the row and column that a_{11} is in and what remains is the minor. You then find the determinant of the minor.	$\begin{bmatrix} -2 & 5 & 1 \\ 0 & 3 & 4 \\ -1 & 2 & 6 \end{bmatrix} \rightarrow \begin{vmatrix} 3 & 4 \\ 2 & 6 \end{vmatrix} = 18 - 8 = 10$ minor
The cofactor is the minor times either 1 or −1. Each location in a matrix has a sign associated with it. Here is the pattern $\begin{bmatrix} \textcircled{+} & - & + & \cdots \\ - & + & - & \cdots \\ + & - & + & \cdots \\ \cdots & \cdots & \cdots & \cdots \end{bmatrix}$	So, this cofactor is $+1(10) = 10$
Now do this for each element in the first row. Here is the minor and cofactor for the a_{12} position.	$\begin{bmatrix} -2 & 5 & 1 \\ 0 & 3 & 4 \\ -1 & 2 & 6 \end{bmatrix} \rightarrow \begin{vmatrix} 0 & 4 \\ -1 & 6 \end{vmatrix} = 0 - (-4) = 4$ minor So the cofactor is $-1(4) = -4$
The minor and cofactor for the a_{13} position are	$\begin{bmatrix} -2 & 5 & 1 \\ 0 & 3 & 4 \\ -1 & 2 & 6 \end{bmatrix} \rightarrow \begin{vmatrix} 0 & 3 \\ -1 & 2 \end{vmatrix} = 0 - (-3) = 3$ So the cofactor is $+1(3) = 3$
Now the determinant equals each element times its corresponding cofactor added together.	
$(\mathbf{-2})(1)(10) + (\mathbf{5})(-1)(4) + (\mathbf{1})(1)(3) =$	
$-20 + -20 + 3 = -37$	

Enough of this! Let's go to the calculator now.

Discussion 6: Finding the Determinant with a Calculator

Let's find the determinant of the matrix in discussion 5 by using the calculator.

<table>
<tr>
<td>First, input your matrix; then return to the home screen; then go to MATRIX, then MATH, then det.</td>
<td>
</td>
<td>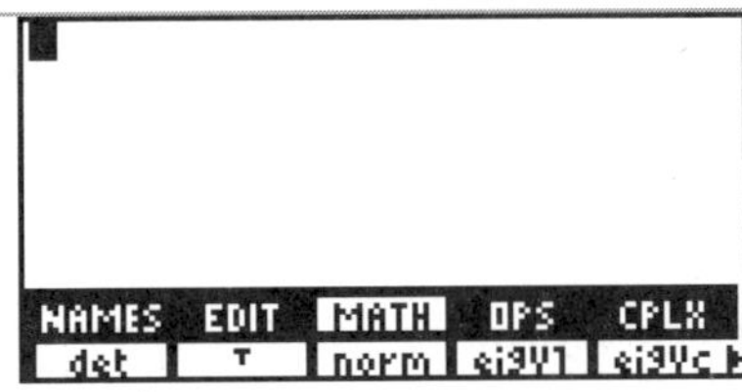
</td>
</tr>
<tr>
<td>Push ENTER (F1 if the TI-86); get the name of your matrix, and then push ENTER again.</td>
<td>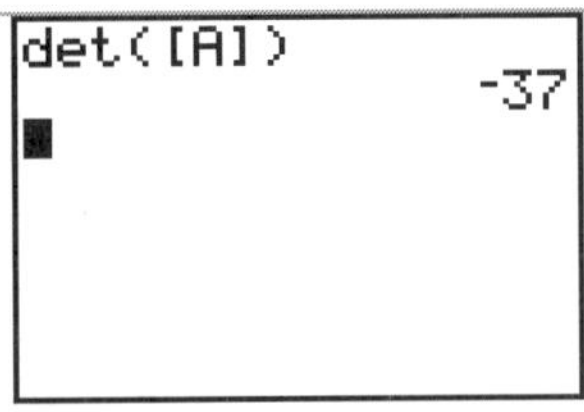
</td>
<td>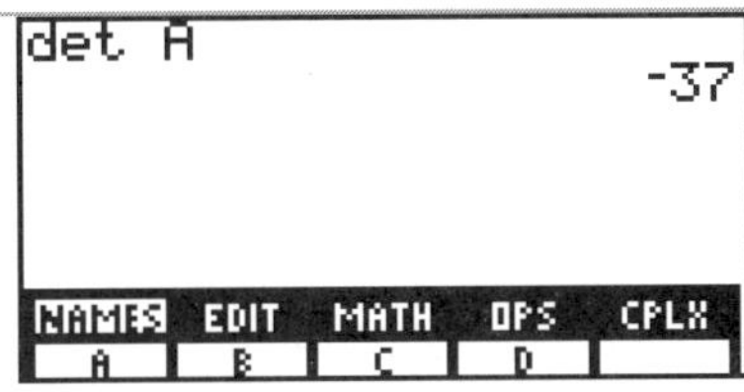
</td>
</tr>
</table>

As you can see, the calculator returns to us the same answer we arrived at by hand. You might also notice that the answer we got (–37) is in the denominator of the elements of the inverse of the matrix in example 6. Here is the formula for finding an inverse taught in a Linear Algebra course:

$$A^{-1} = \frac{1}{\det(A)}(\text{adj }(A))$$

Notice how the determinant ends up in the denominator in this formula. That is why we see the (37) in the denominators in example 6. But we will leave the explanation of this way of finding an inverse for another math course.

Question 6: Find the determinant of $\begin{bmatrix} 1 & 2 & 3 \\ 4 & 5 & -1 \\ 1 & -1 & -10 \end{bmatrix}$, the matrix from example 7.

There is a theorem that states

A matrix has a multiplicative inverse if and only if the determinant of the matrix isn't 0.

So, only square matrices can have an inverse and only if their determinant isn't equal to 0.

Section Summary

- The additive identities for matrices are matrices made up entirely of 0's.
- The additive inverses for matrices are matrices made up of opposite signed elements.
- The multiplicative identities for square matrices are matrices of the same dimension with 1's down the diagonal and 0's elsewhere.
- Only square matrices have a determinant.
- Only square matrices can have a multiplicative inverse.
- Only square matrices whose determinants *aren't* 0 have inverses.
- To find the inverse of a square matrix (with a nonzero determinant), type it into your calculator and use the $\boxed{x^{-1}}$ key.
- To find the determinant of a matrix, type it into your calculator and use the **det** function of the calculator.

SECTION 6.2 HOMEWORK

(1–16) Use the matrices given below to answer the questions for 1–16:

$A=\begin{bmatrix}1 & 0 & 2\\ 0 & -3 & 2\end{bmatrix}$ $B=\begin{bmatrix}0 & -3\\ 2 & -4\\ 1 & 5\end{bmatrix}$ $C=\begin{bmatrix}1 & -3\\ 3 & 4\end{bmatrix}$

$D=\begin{bmatrix}3 & -5\\ 2 & 4\end{bmatrix}$ $E=\begin{bmatrix}-3 & 2 & 5\\ 1 & 3 & -2\\ 4 & 1 & 2\end{bmatrix}$ $F=\begin{bmatrix}2 & -3 & 1\\ 5 & -2 & 3\\ 3 & 2 & -1\end{bmatrix}$

1. Find the additive inverse of A.
2. Find the additive inverse of B.
3. Find the additive inverse of C.
4. Find the additive inverse of D.
5. Find the additive inverse of E.
6. Find the additive inverse of F.
7. Find the additive identity of A.
8. Find the additive identity of B.
9. Find the additive identity of C.
10. Find the additive identity of D.
11. Find the additive identity of E.
12. Find the additive identity of F.
13. What matrix is the answer for the addition of a 2 x 3 matrix and its additive inverse?
14. What matrix is the answer for the addition of a 3 x 3 matrix and its additive inverse?
15. What is the additive identity matrix for a 4 x 4 matrix?
16. What is the additive identity matrix for a 3 x 5 matrix?

Answer Q6: det=0. Notice from example 6 that this square matrix didn't have a multiplicative inverse and the calculator stated that it is a singular matrix.

(17–42) Use the matrices given below to answer the questions for 17–42:

(If asked to find the inverse or determinant of a matrix, use your calculator)

$A=\begin{bmatrix}-2 & 3 & 1\\ 2 & 1 & 3\end{bmatrix}$ $B=\begin{bmatrix}3 & -2\\ 1 & 3\\ 1 & 2\end{bmatrix}$ $C=\begin{bmatrix}3 & -2\\ 4 & 3\end{bmatrix}$ $D=\begin{bmatrix}-2 & 1\\ 3 & 5\end{bmatrix}$

$E=\begin{bmatrix}1 & -2 & 3\\ 1 & 2 & 3\\ -2 & 3 & 1\end{bmatrix}$ $F=\begin{bmatrix}2 & -3 & 4\\ 2 & 1 & -3\\ 3 & -2 & 1\end{bmatrix}$ $G=\begin{bmatrix}3 & 6\\ 5 & 10\end{bmatrix}$ $H=\begin{bmatrix}1 & -2 & 3\\ 3 & 1 & 2\\ -2 & 4 & -6\end{bmatrix}$

17. Find the multiplicative inverse of C.
18. Find the multiplicative inverse of D.
19. Find the multiplicative inverse of E.
20. Find the multiplicative inverse of F.
21. Find the multiplicative inverse of A.
22. Find the multiplicative inverse of B.
23. Find the multiplicative inverse of G?
24. Find the multiplicative inverse of H?
25. What is the multiplicative identity of C?
26. What is the multiplicative identity of D?
27. What is the multiplicative identity of E?
28. What is the multiplicative identity of F?
29. What is the multiplicative identity of A?
30. What is the multiplicative identity of B?
31. Find the determinant of C.
32. Find the determinant of D.
33. Find the determinant of E.
34. Find the determinant of F.
35. Find the determinant of A.
36. Find the determinant of B.
37. Find the determinant of G.
38. Find the determinant of H.
39. What happened when you tried to find the multiplicative inverse of A with your calculator?
40. What happened when you tried to find the multiplicative inverse of B with your calculator?

41. What happened when you tried to find the multiplicative inverse of G with your calculator and what was the determinant value of G?

42. What happened when you tried to find the multiplicative inverse of H with your calculator and what was the determinant value of H?

43. What matrix is the answer for the multiplication of a 3 x 3 matrix and its multiplicative inverse?

44. What matrix is the answer for the multiplication of a 4 x 4 matrix and its multiplicative inverse?

45. What is the multiplicative identity matrix for a 4 x 4 matrix?

46. What is the multiplicative identity matrix for a 6 x 6 matrix?

47. What is the multiplicative identity for a 3 x 4 matrix?

48. What is the multiplicative identity for a 6 x 4 matrix?

49. $A = \begin{bmatrix} 4 & 5 \\ 2 & 3 \end{bmatrix}$

    a. Find the multiplicative inverse of the matrix A by hand and check your answer with your calculator.

    b. Find the determinant of matrix A by hand then check your answer with your calculator.

50. $B = \begin{bmatrix} 2 & -3 \\ 1 & 3 \end{bmatrix}$

    a. Find the multiplicative inverse of the matrix B by hand and check your answer with your calculator.

    b. Find the determinant of matrix B by hand and check your answer with your calculator.

Section 6.3 Systems of Equations

It is now time to talk about how to solve a system of equations by using matrices. There are two methods that we will discuss in this section. One of the two methods we will call the **inverse matrix method**. With this method we will use the inverse of a matrix in order to find the solution to the system. First, though, we need to talk about something called a **matrix equation**.

A **matrix equation** is an equation of the form $AX = B$, where A, X and B are all matrices.

A is called the **coefficient matrix**. It is the matrix made up of all the coefficients of the variables in the system of equations. X is called the **variable matrix** and it is made up of all the variables in the system. And B is called the **constant matrix,** made up of the constants on the right side of the equations.

Discussion 1: System of Equations in Matrix Form

Writing the matrix equation that corresponds to the system of equations $\begin{cases} 19x - 2y = 34 \\ 15x + 22y = -38 \end{cases}$.

First, write the coefficient matrix (A). Remember this is a matrix made up of all the coefficients on the variables.	A $\begin{bmatrix} 19 & -2 \\ 15 & 22 \end{bmatrix}$
Now, write the variable matrix next to the A matrix. We now have AX.	$A \bullet X$ $\begin{bmatrix} 19 & -2 \\ 15 & 22 \end{bmatrix} \begin{bmatrix} x \\ y \end{bmatrix}$
Finally, set what we have (AX) equal to the constant matrix (B).	$A \bullet X = B$ $\begin{bmatrix} 19 & -2 \\ 15 & 22 \end{bmatrix} \begin{bmatrix} x \\ y \end{bmatrix} = \begin{bmatrix} 34 \\ -38 \end{bmatrix}$

Question 1: From example 1, find the answer to ($A \bullet X$) and its resulting dimension.

Notice that matrix B in Discussion 1 is also a 2 x 1 matrix. So, when we set AX equal to B $\left(\begin{bmatrix} 19x - 2y \\ 15x + 22y \end{bmatrix} = \begin{bmatrix} 34 \\ -38 \end{bmatrix} \right)$ in the matrix equation it means, since they are of the same dimension, that $19x - 2y = 34$ and $15x + 22y = -38$. These are exactly the equations in the original system of equations. Because of the way we define matrix multiplication,

$$\begin{cases} 19x - 2y = 34 \\ 15x + 22y = -38 \end{cases} \text{ is equivalent to } \begin{bmatrix} 19 & -2 \\ 15 & 22 \end{bmatrix} \begin{bmatrix} x \\ y \end{bmatrix} = \begin{bmatrix} 34 \\ -38 \end{bmatrix}.$$

Question 2: Write the matrix equation for this system $\begin{cases} x-4y+z=13 \\ 3x-10y+2z=34 \\ 5x-2y+7z=31 \end{cases}$.

Example 1: Converting from a Matrix to a System of Equations

Given this matrix equation $\begin{bmatrix} 0 & -3 & 2 \\ 2 & -1 & 0 \\ 3 & -2 & 3 \end{bmatrix}\begin{bmatrix} x \\ y \\ z \end{bmatrix}=\begin{bmatrix} 3 \\ -1 \\ 4 \end{bmatrix}$ write its corresponding system of equations.

Solution:

$\begin{bmatrix} 0 & -3 & 2 \\ 2 & -1 & 0 \\ 3 & -2 & 3 \end{bmatrix}$ is the coefficient matrix, so each number is the coefficient of the variables in the system. So we have	$0x - 3y + 2z$ $2x - 1y + 0z$ $3x - 2y + 3z$
$\begin{bmatrix} 3 \\ -1 \\ 4 \end{bmatrix}$ is the constant matrix. These are the values that the equations in the system are equal to, so the system of equations is	$\begin{cases} 0x-3y+2z=3 \\ 2x-y+0z=-1 \\ 3x-2y+3z=4 \end{cases}$

Now we can show you how the inverse matrix comes into play. Remember from previous sections that we discussed how you can solve the equation $5x = 15$ by multiplying both sides of the equation by $\frac{1}{5}$, the inverse of 5. We will do the same thing for matrix equations. If we have $AX = B$, then we will multiply both sides of the equation by A^{-1} to get X by itself. In a sense, we treat this just like we would a linear equation.

But, one thing we need to remember is that matrix multiplication isn't commutative. With matrices we have what is called **left** or **right** hand multiplication. So, if we multiply one side of the matrix equation on the left, then we must multiply the other side of the equation on the left also. So, to get X by itself, in the matrix equation, we multiply both sides of the equation on the left by A^{-1}.

Matrix equation	$AX = B$
Multiply both sides on the left by A^{-1}	$A^{-1}AX = A^{-1}B$
The result is	$IX = A^{-1}B$ (I is the identity: $A^{-1}A = I$)
Now we see	$X = A^{-1}B$ ($IX = X$: definition of identity)

THE INVERSE METHOD

We now have another way to find the solution to a system of equations, which involves taking the inverse of the coefficient matrix times the constant matrix ($X = A^{-1}B$).

Discussion 2: Solving a System of Equations Using the Inverse Method

Solving the system of equation from question 2, $\begin{cases} x - 4y + z = 13 \\ 3x - 10y + 2z = 34 \\ 5x - 2y + 7z = 31 \end{cases}$.

First, we need to write this in the matrix equation form.	$\begin{bmatrix} 1 & -4 & 1 \\ 3 & -10 & 2 \\ 5 & -2 & 7 \end{bmatrix} \begin{bmatrix} x \\ y \\ z \end{bmatrix} = \begin{bmatrix} 13 \\ 34 \\ 31 \end{bmatrix}$ $(A \quad \bullet \quad X = B)$
Now, input matrices A and B into our calculator.	[[1 -4 1] [3 -10 2] [5 -2 7]] [B] [[13] [34] [31]]
Tell the calculator to find $A^{-1}B$.	[[13] [34] [31]] [A]-1[B] [[4] [-2] [1]]
We know that $\begin{bmatrix} x \\ y \\ z \end{bmatrix} = X = A^{-1}B = \begin{bmatrix} 4 \\ -2 \\ 1 \end{bmatrix}$ from our calculator, so we know that $x = 4$, $y = -2$, $z = 1$. Our solution to this system is the point $(4, -2, 1)$.	

Answer Q1:

$\begin{bmatrix} 19 & -2 \\ 15 & 22 \end{bmatrix} \begin{bmatrix} x \\ y \end{bmatrix} = \begin{bmatrix} 19x + (-2)y \\ 15x + 22y \end{bmatrix}$.

This is a 2 x 1 matrix.

This is what we call the inverse method for solving a system of equations. Let's do another.

Example 2: Solving a System of Equations by Finding the Inverse Matrix

Solve the system $\begin{cases} 2x + 3y = 6 \\ -4x - 6y = -12 \end{cases}$ by using the inverse method.

Solution:

First, we need to write this in the matrix equation form.	$\begin{bmatrix} 2 & 3 \\ -4 & -6 \end{bmatrix} \begin{bmatrix} x \\ y \end{bmatrix} = \begin{bmatrix} 6 \\ -12 \end{bmatrix}$ $(A \quad \bullet \quad X = B)$

Answer Q2:

$$\begin{bmatrix} 1 & -4 & 1 \\ 3 & -10 & 2 \\ 5 & -2 & 7 \end{bmatrix} \begin{bmatrix} x \\ y \\ z \end{bmatrix} = \begin{bmatrix} 13 \\ 34 \\ 31 \end{bmatrix}.$$

Now, input matrices A and B into our calculator.	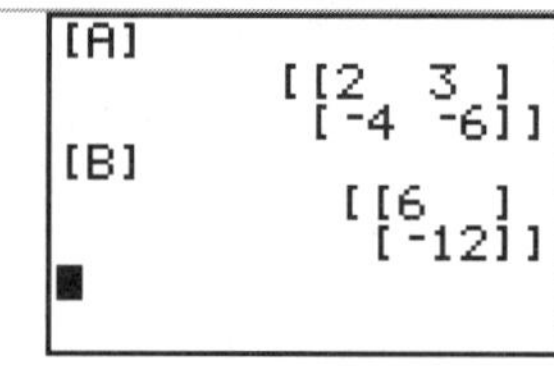
Tell the calculator to find $A^{-1}B$.	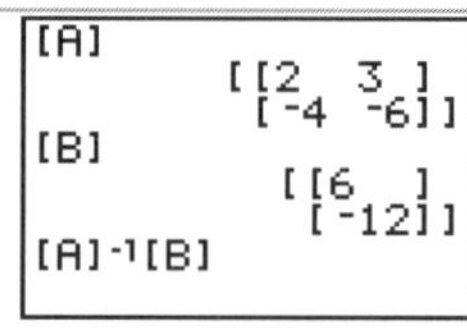 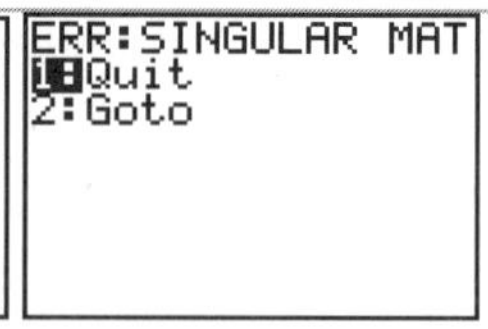

The calculator is telling us we have a singular matrix. If you remember, this means that the determinant of the A matrix is 0 and thus A doesn't have an inverse. It looks as if the inverse method isn't one we can use on this system, since it requires us to be able to get the inverse of the A matrix.

We can still do this last problem by using either the elimination or substitution methods learned in 5.2. We find from this example that though the inverse method is fairly easy with the help of the calculator, it won't solve every system of equations. In the next section we will talk about a way to use matrices that will be able to help us solve this sort of problem. But for now, let's investigate the other method we wish to discuss in this section.

CRAMER'S RULE

The next method we want to talk about is called **Cramer's Rule, which** uses determinants to solve systems of equations. This method comes from using the elimination method on a general system of equations and then noticing a connection to determinants. Let us show you what we mean.

We'll begin with a general system of equations with two linear variables.	$\begin{cases} a_1x + b_1y = c_1 \\ a_2x + b_2y = c_2 \end{cases}$
To use the elimination method we need one of the variables to have the same coefficients but opposite signs (we've chosen to eliminate y).	$\begin{matrix} b_2 \bullet \\ -b_1 \bullet \end{matrix} \begin{cases} a_1x + b_1y = c_1 \\ a_2x + b_2y = c_2 \end{cases} = \begin{cases} a_1b_2x + b_1b_2y = c_1b_2 \\ -a_2b_1x - b_1b_2y = -c_2b_1 \end{cases}$
Now, we add the two equations to eliminate the y's and get	$a_1b_2x - a_2b_1x = c_1b_2 - c_2b_1$
To solve, let's factor out x and then divide by the parentheses.	$x(a_1b_2 - a_2b_1) = c_1b_2 - c_2b_1$ $x = \dfrac{c_1b_2 - c_2b_1}{a_1b_2 - a_2b_1}$

Just like with the quadratic formula, where we began with the general form of a quadratic and then solved for x, we have found a formula for finding the x-coordinate solution of the system of equations. The form of the numerator and denominator of the formula should look familiar to you (the formula for finding the determinant of a 2 x 2 matrix is in section 6.2).	
First, for the denominator, if we view $a_1b_2 - a_2b_1$ as a determinant, it must have come from the coefficient matrix A (the coefficient matrix of the system of equations).	$\begin{cases} a_1x + b_1y = c_1 \\ a_2x + b_2y = c_2 \end{cases}$ $A = \begin{bmatrix} a_1 & b_1 \\ a_2 & b_2 \end{bmatrix}$ $\det A = a_1b_2 - a_2b_1$
Likewise, for the numerator, if we view $c_1b_2 - c_2b_1$ as a determinant, it must have come from the matrix we will call A_x (the coefficient matrix of the system of equations with the x coefficients replaced by the constant matrix).	$\begin{cases} a_1x + b_1y = c_1 \\ a_2x + b_2y = c_2 \end{cases}$ $A_x = \begin{bmatrix} c_1 & b_1 \\ c_2 & b_2 \end{bmatrix}$ $\det A_x = c_1b_2 - c_2b_1$
Therefore, x equals the division of the two determinants. Finding y is similar. You make a matrix called A_y (the coefficient matrix of the system of equations with the y coefficients replaced by the constant matrix), find its determinant, and then divide that by det A.	$x = \dfrac{\det\begin{bmatrix} c_1 & b_1 \\ c_2 & b_2 \end{bmatrix}}{\det\begin{bmatrix} a_1 & b_1 \\ a_2 & b_2 \end{bmatrix}}$ $y = \dfrac{\det\begin{bmatrix} a_1 & c_1 \\ a_2 & c_2 \end{bmatrix}}{\det\begin{bmatrix} a_1 & b_1 \\ a_2 & b_2 \end{bmatrix}}$

So, we have an easy way to solve a system of equations by finding the determinants of matrices. One word of caution-

Question 3: What would happen if the determinant in the denominator were equal to 0?

We will be "stuck" once again if the determinant is 0 (in this case the determinant in the denominator). We'll have a situation where the coefficient matrix of the system of equations is a singular matrix. We saw this earlier in the section. When this happened we were unable to solve the system by the inverse method. We will find a way to handle this problem in the next section. For now, let's do an example.

Discussion 3: Solving a System of Equations Using Cramer's Rule

Solve the system $\begin{cases} 3x - 2y = 16 \\ x - 2y = 4 \end{cases}$ by Cramer's Rule.

Write down all the different matrices for which we need a determinant.	$A=\begin{bmatrix}3 & -2\\ 1 & -2\end{bmatrix}$ $A_x=\begin{bmatrix}16 & -2\\ 4 & -2\end{bmatrix}$ $A_y=\begin{bmatrix}3 & 16\\ 1 & 4\end{bmatrix}$
Find the determinants.	$\det A = 3(-2)-1(-2) = -6+2 = -4$ $\det A_x = 16(-2)-4(-2) = -32+8 = -24$ $\det A_y = 3(4)-1(16) = 12-16 = -4$
So, x will be	$x = \dfrac{\det A_x}{\det A} = \dfrac{-24}{-4} = 6$
And y will be	$y = \dfrac{\det A_y}{\det A} = \dfrac{-4}{-4} = 1$
The final solution for this system is	(6, 1)

Question 4: Try using this new method to solve example 4 ($\begin{cases} 2x+3y=6 \\ -4x-6y=-12 \end{cases}$).

Example 3: Solving a System of Equations Using Cramer's Rule

Solve the system $\begin{cases} x-2y+3z=-1 \\ 3x+y-2z=3 \\ 5x-2y+3z=-2 \end{cases}$ using Cramer's Rule.

Solution:

Write down the coefficient matrix (A). (We start with just the A matrix since its determinant might be 0 and then we wouldn't be able to use Cramer's Rule.)	$A=\begin{bmatrix}1 & -2 & 3\\ 3 & 1 & -2\\ 5 & -2 & 3\end{bmatrix}$	
Find det A. Since the determinant is not 0, we know that we can proceed with Cramer's Rule.	$\det A = 4$	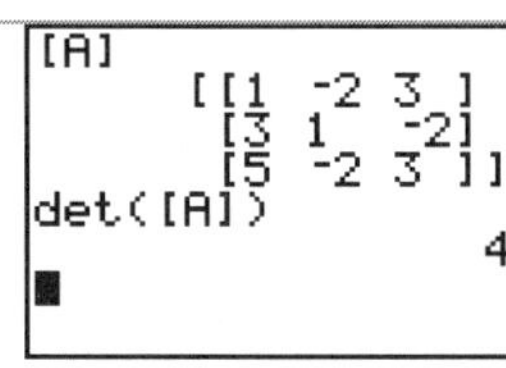
Now, find det A_x, det A_y, and det A_z.	$A_x=\begin{bmatrix}-1 & -2 & 3\\ 3 & 1 & -2\\ -2 & -2 & 3\end{bmatrix}$ so $\det A_x = -1$	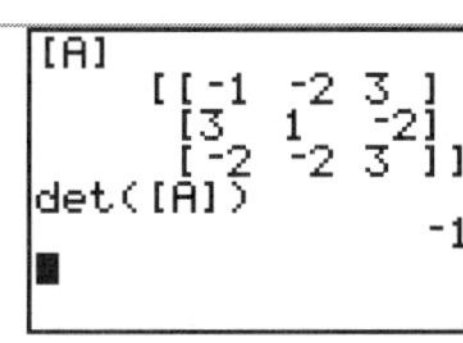
	$A_y=\begin{bmatrix}1 & -1 & 3\\ 3 & 3 & -2\\ 5 & -2 & 3\end{bmatrix}$ so $\det A_x = -39$	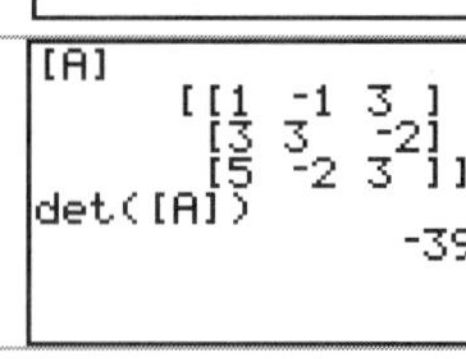

	$A_z = \begin{bmatrix} 1 & -2 & -1 \\ 3 & 1 & 3 \\ 5 & -2 & -2 \end{bmatrix}$ so $\det A_x = -27$	[A] [[1 -2 -1] [3 1 3] [5 -2 -2]] det([A]) -27
$x = \dfrac{\det A_x}{\det A}, y = \dfrac{\det A_y}{\det A}, z = \dfrac{\det A_z}{\det A}$	$x = \dfrac{-1}{4}, y = \dfrac{-39}{4}, z = \dfrac{-27}{4}$	

The solution to our system of equations is the point $\left(\dfrac{-1}{4}, \dfrac{-39}{4}, \dfrac{-27}{4}\right)$.

Example 4: Solving a System of Equations Using Cramer's Rule

Solve the system $\begin{cases} -2x+2y-4z=9 \\ 2x-3y+6z=3 \\ 3x+4y-8z=7 \end{cases}$ using Cramer's Rule.

Solution:

Write down the coefficient matrix (A). (Again, we start with just the A matrix since its determinant might be 0 and then we wouldn't need to do any other more work.)	$A = \begin{bmatrix} -2 & 2 & -4 \\ 2 & -3 & 6 \\ 3 & 4 & -8 \end{bmatrix}$
Find the determinant.	$\det A = 0$ [A] [[-2 2 -4] [2 -3 6] [3 4 -8]] det([A])
This method will not help us in finding the solution to this system.	

Answer Q3: We would be dividing by 0, which is undefined. This means that we can't use this method to solve a system with a determinant equal to 0, similar to the inverse method.

In each of the two methods we have talked about, the inverse method and Cramer's Rule, if the determinant of the coefficient matrix *isn't* 0, then the system will yield just one point as the solution. If, on the other hand, the determinant is 0 then the system has either no solution or an infinite number of solutions.

GRAPHICAL REPRESENTATIONS OF SOLUTIONS IN THREE VARIABLES

Let's take a moment to show you what three equations in three linear variables looks like graphically. Here are the graphs of the three equations in example 7. Notice that all three of the graphed planes do not intersect at a common point.

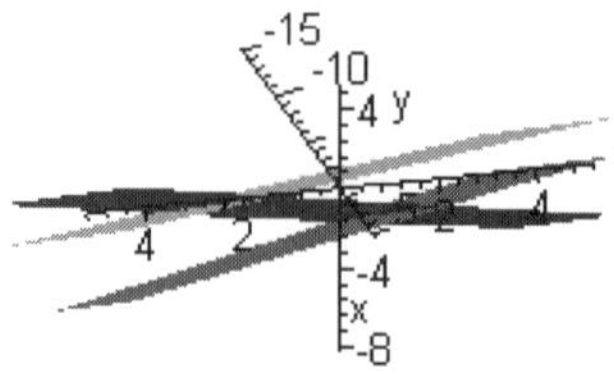

As you can see when we have three linear variables in an equation, the graph is a plane in space. So, three equations in three linear variables create three planes in space. There is a pattern here that is good to see.

Answer Q4:
$A = \begin{bmatrix} 2 & 3 \\ -4 & -6 \end{bmatrix}$
so the det A = (−12) − (−12) = 0. Hence, we can't use this method to solve this system ($x = \dfrac{\det A_x}{\det A} = \dfrac{\#}{0} =$ *undefined*).

Example of a one-variable linear equation: $x = 5$ (a *point* which is zero-dimensional, on a number *line* which is one-dimensional)	
Example of a two-variable linear equation: $x + y = 1$ (a *line*, which is one-dimensional, on a *plane*, which is two-dimensional)	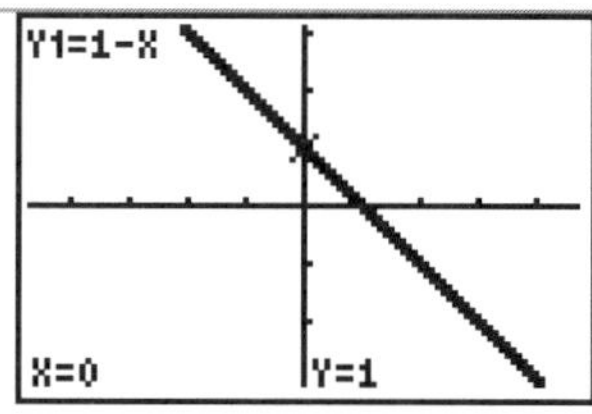
Example of a three-variable linear equation: $x + y + z = 1$ (a *plane*, which is two-dimensional, in a *cube*, which is three-dimensional)	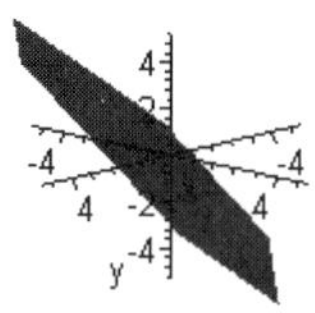

Notice that the dimensions of the space that your graph is in equals the number of linear variables you have in the equation. Also notice that the graph of the linear equation will always be one dimension less than the number of linear variables in the equation. This means that if you are solving a system of linear equations, you can have anywhere from zero-dimensional answers to the dimensions of the graphs of your system of linear equations, plus the possibility of no solution. In general we see the following:

	Example	Possible Solutions
One equation, one linear variable:	$ax + b = c$	*One* solution
Two equations, two linear variables:	$\begin{cases} a_1x + b_1y = c_1 \\ a_2x + b_2y = c_2 \end{cases}$	*No* solution *One* solution A *line* as the solution

Three equations and three linear variables:	$\begin{cases} a_1x+b_1y+c_1z=d_1 \\ a_2x+b_2y+c_2z=d_2 \\ a_3x+b_3y+c_3z=d_3 \end{cases}$	*No* solution *One* solution A *line* as the solution A *plane* as the solution

Remember from section 5.2 that we saw how "no solution" was represented graphically as two parallel lines. "One solution" was where two lines intersected, and "a line as the solution" was where the two lines were exactly the same line. With three variables we get the following:

No solution - where at least one of the planes is parallel to another.	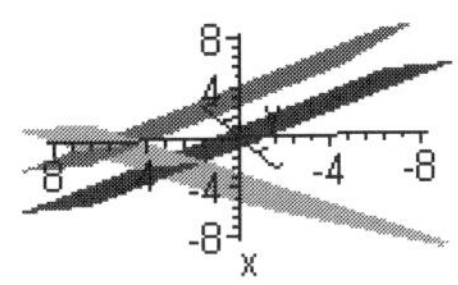
One solution - where they all intersect at one point.	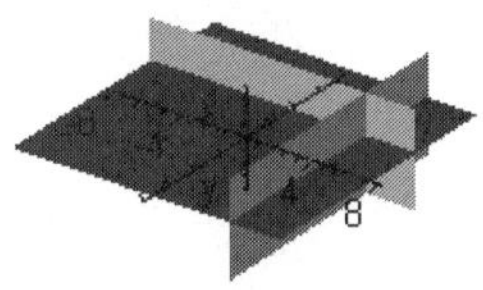
Line solution - where they all intersect at a line.	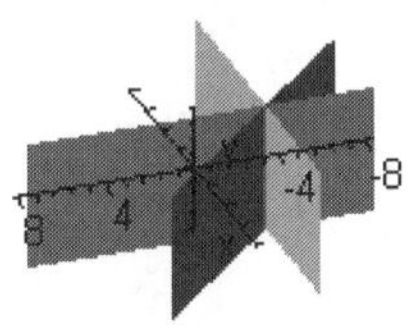
Plane solution - where they all are the same plane.	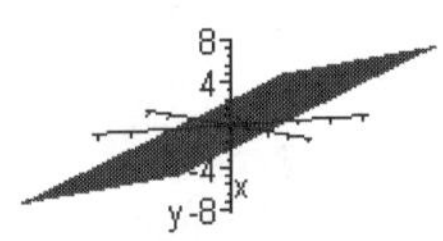

Section Summary

- A system of linear equations can be written as a matrix equation and $A^{-1}B = X$ is the solution, if the inverse matrix exists.
- A system of linear equations can be solved using determinants if the determinant of the coefficient matrix (det A) is not equal to 0 ($x = \dfrac{\det A_x}{\det A}$, $y = \dfrac{\det A_y}{\det A}$, $z = \dfrac{\det A_z}{\det A}$).

- If the determinant of the coefficient matrix (det A) is equal to 0, then the solution to the system of equations can't be a single point but must be either "no solution" or have an infinite number of solutions.

SECTION 6.3 HOMEWORK

(1–8) For each system of equations give:

a. Coefficient matrix A

b. Variable matrix X

c. Constant matrix B

d. Write a matrix equation corresponding to the system of equations

1. $3x + 4y = 8$
   $8x - 9y = 15$

2. $5x - 7y = 9$
   $-11x + 9y = -3$

3. $3x = 2y - 15$
   $12 - 3y = 4x$

4. $7y = 3x + 19$
   $8x - 15 = 5y$

5. $2x - 3y + 2z = 8$
   $5x + 7y - 8z = -12$
   $3x - 5y - 9z = 23$

6. $5x + 2y - 8z = 4$
   $-2x - 4y + 3z = 9$
   $x - 2y - z = 21$

7. $x = 3y - 9$
   $2z = 3x - y + 10$
   $5y - 9 = z$

8. $2y - 3z = 23 - 5x$
   $3x - 8 = z$
   $9y = 11 - x$

(9–16) Translate each matrix equation to a corresponding system of equations:

9. $\begin{bmatrix} 3 & 1 \\ 2 & -4 \end{bmatrix}\begin{bmatrix} x \\ y \end{bmatrix} = \begin{bmatrix} 3 \\ 9 \end{bmatrix}$

10. $\begin{bmatrix} -2 & 3 \\ 1 & -2 \end{bmatrix}\begin{bmatrix} x \\ y \end{bmatrix} = \begin{bmatrix} 21 \\ 15 \end{bmatrix}$

11. $\begin{bmatrix} 1.3 & 2.8 \\ 3.4 & -1.7 \end{bmatrix}\begin{bmatrix} r \\ s \end{bmatrix} = \begin{bmatrix} 3.54 \\ 2.78 \end{bmatrix}$

12. $\begin{bmatrix} \frac{2}{3} & \frac{-3}{5} \\ \frac{1}{6} & \frac{4}{15} \end{bmatrix}\begin{bmatrix} a \\ b \end{bmatrix} = \begin{bmatrix} 5 \\ -3 \end{bmatrix}$

13. $\begin{bmatrix} -2 & 3 & -4 \\ 1 & 1 & -2 \\ 3 & -2 & 1 \end{bmatrix}\begin{bmatrix} x \\ y \\ z \end{bmatrix} = \begin{bmatrix} -23 \\ 35 \\ 18 \end{bmatrix}$

14. $\begin{bmatrix} 1 & -2 & 7 \\ 3 & 4 & -8 \\ -3 & 1 & -1 \end{bmatrix}\begin{bmatrix} r \\ s \\ t \end{bmatrix} = \begin{bmatrix} -150 \\ 33 \\ 128 \end{bmatrix}$

15. $\begin{bmatrix} 3 & -2 & 1 & 2 \\ 1 & -3 & 0 & 5 \\ 0 & 3 & -2 & 1 \\ 1 & 0 & 0 & -4 \end{bmatrix}\begin{bmatrix} x \\ y \\ z \\ w \end{bmatrix} = \begin{bmatrix} 58 \\ -29 \\ 120 \\ -83 \end{bmatrix}$

16. $\begin{bmatrix} -1 & 0 & 0 & 3 \\ 0 & 2 & -5 & 0 \\ 1 & 3 & 0 & -4 \\ 0 & 0 & 3 & 2 \end{bmatrix}\begin{bmatrix} r \\ s \\ t \\ w \end{bmatrix} = \begin{bmatrix} 48 \\ -34 \\ 120 \\ 65 \end{bmatrix}$

(17–36) Use the inverse method to solve each of the following systems of equations:

17. $2x - 3y = 12$
    $5x + 4y = 7$

18. $-3x + 5y = 35$
    $7x - 9y = -71$

19. $\frac{3}{4}x + \frac{5}{3}y = \frac{-17}{12}$
    $\frac{-2}{3}x + \frac{7}{4}y = \frac{-503}{432}$

20. $0.35x - 0.03y = -1.174$
    $0.27x + 0.19y = -0.522$

21. $2x - 3y + 5z = 3$
    $4x + 9y - 10z = 4$

22. $7x - 10y - 12z = 2$
    $-14x + 5y - 8z = 14$

$-8x + 6y - 15z = -10$

$21x - 15y + 16z = -27$

23. $\frac{3}{5}x + \frac{2}{5}y - \frac{1}{8}z = 8$

$\frac{2}{3}x - \frac{5}{2}y + \frac{3}{4}z = 82$

$\frac{-5}{2}x + \frac{3}{4}y + \frac{3}{2}z = -66$

24. $-1.08x + 2.35y - 0.34z = 13.78$

$2.35x + 1.84y + 0.24z = -3.67$

$0.05x - 0.24y - 2.3z = -8.11$

25. $2x + z = -1$

$3y - 2z = 11$

$-x + 3y - z = 16$

26. $2y + 3z = -3$

$x - y + 5z = 10$

$3x + 5y = -9$

27. $3x + 2y - 4z + 3w = -22$

$2x - 5y + 3z - w = 21$

$-3x + 7y - 8z + 2w = -43$

$x - 2y + 3z + w = 18$

28. $x - y + z - w = -4$

$2x + 3y - z + 3w = 16$

$-3x + y + 2z - 2w = -15$

$5x - 4y - 3z + w = 17$

29. $x + z = 7$

$2y + 3w = -13$

$2x + 3z - 5w = 33$

$3x + 2y - 4z = -11$

30. $2x + 5w = -19$

$3y - 2z = -7$

$3x + 3z - 4w = 21$

$2y - z + w = -6$

31. $x + 2y = 3z - 19$

$2x = y - z + 12$

$y = 3z - x - 20$

32. $2x = 3y - 4z + 16$

$2y = 3x + z + 6$

$y = 3z - 20$

33. $5x + 2y - 4z = 33$

$2x + y - z = 14$

$x - y - 5z = 2$

34. $7x - 3y + 41z = -6$

$-2x + y - 12z = 3$

$x + 2y + z = 22$

35. $3x + 2y + 4z = -3$

$x + y + z = -2$

$x - y + 3z = 4$

36. $5x + 2y - 4z = 25$

$2x + y - z = 11$

$-x + y + 5z = 2$

(37–56) Use the systems of equations from problems (17–36) and solve each system using Cramer's Rule. (You can use this as a check of your answers for 17–36.)

37. Use problem 17
38. Use problem 18
39. Use problem 19
40. Use problem 20
41. Use problem 21
42. Use problem 22
43. Use problem 23
44. Use problem 24
45. Use problem 25
46. Use problem 26
47. Use problem 27
48. Use problem 28
49. Use problem 29
50. Use problem 30
51. Use problem 31
52. Use problem 32

53. Use problem 33

54. Use problem 34

55. Use problem 35

56. Use problem 36

57. A number other than 0 divided by 0 has no solution; however, 0 divided by 0 has an infinite number of solutions. For problems 53, 54, 55 and 56 you found there was not a unique solution because the determinant of the coefficient matrix was 0, and this meant there was either an infinite number of solutions or no solutions. Find all the determinants for these four problems. Using the arithmetic idea given and Cramer's Rule, state when you think a system of equations such as these four will have no solution or an infinite number of solutions.

58. The following system of equations has an infinite number of solutions. See if your idea from problem 57 works for this system of equations.

$7x + 2y + 13z = 10$

$3x + y + 6z = 4$

$5x + y + 8z = 8$

59. The following system of equations has no solution. See if your idea from problem 57 works for this system of equations.

$5x + 2y - 16z = 29$

$2x + y - 7z = 12$

$x + 2y - 8z = 10$

Section 6.4 Applications of Matrices

We have learned what a matrix is and how to add, subtract, scalar multiply, multiply, find inverses and solve equations with them. Now, let's look at another way in which we can use matrices to help us solve problems.

Back in section 5.2 we talked about solving systems of equations by hand. At the time we didn't look at more than two equations with two variables, but we did mention that we would solve bigger systems later in the book. In section 6.3 we talked about how to solve systems of equations that had more than two equations and two variables. But we couldn't solve every system we ran into. Well, we have finally arrived at a place where we can solve every system with the help of the calculator.

Matrices give us a good way to input the system of equations into our calculators and use technology to do mundane calculations. Assuming that the variables throughout the system of equations are the same (i.e.. x's everywhere, not x's and x^2's, and y's everywhere, not y's and y^2's, for instance), we can rewrite the system into *one* matrix using the variable coefficients and the constants on the right side of the equations.

Discussion 1: Creating an Augmented Matrix

Write the given system $\begin{cases} x+2y-2z=-2 \\ 5x+9y-4z=-3 \\ 3x+4y-5z=-3 \end{cases}$ as a matrix.

One way to write this system as a matrix is to rewrite the system in what's called the **augmented** form (this isn't the matrix equation form mentioned in 6.3). The augmented matrix for this system is

$$\left[\begin{array}{ccc|c} 1 & 2 & -2 & -2 \\ 5 & 9 & -4 & -3 \\ 3 & 4 & -5 & -3 \end{array}\right].$$

All that is needed to change a system of equations into an augmented matrix is to write *all* of the coefficients and constants of the system in a matrix. Or more specifically, as mentioned in section 6.3, $[A|B]$ where A was the coefficient matrix and B was the constant matrix.

Now, in the augmented matrix, each row represents the original equations. Where the equal signs were, we now draw a line in the matrix. It is very important to have each equation written in *standard form* before you convert the system of equations into the augmented matrix form. The standard form for linear equations of several variables is $Ax + By + Cz + \ldots =$ some constant. The order of the variables really isn't that important; what is important is that *each equation in the system is written in the same order*. So, in summary, to write a system of equations as an augmented matrix you follow these steps:

1. Make sure every equation is written with the variables in the same order (standard form).
2. Write the coefficients from the equations in a matrix.

Question 1: Write this system $\begin{cases} -3y+2z=3 \\ 2x-y=-1 \\ 3x-2y+3z=4 \end{cases}$ as an augmented matrix.

Question 2: Given the augmented matrix $\left[\begin{array}{cc|c} -2 & 5 & 23 \\ 7 & 1 & 10 \end{array}\right]$, what is a possible system of equations that could have created this augmented matrix?

Augmented matrices are to systems of equations what synthetic division is to long division. By using just the coefficients, we can greatly simplify the amount of work necessary when trying to perform an operation. In this case, we are trying to solve the system.

Let's work on solving a system of equations from chapter 5 by using an augmented matrix. We call this new method for solving systems of equations the **Gaussian Elimination Method**, or in short, the "Gaussian method."

Discussion 2: Solving a System of Equations Using the Gaussian Elimination Method

Solving this system of equations $\begin{cases} x-y=1 \\ 2x+3y=12 \end{cases}$ by the Gaussian Elimination method.

This is the system of equations from example 2 from section 5.2. We will show you how solving this problem by using an augmented matrix is similar to solving this by the addition method.

Addition Method	Gaussian Elimination
$\begin{cases} x-y=1 \\ 2x+3y=12 \end{cases}$ The system is:	$\left[\begin{array}{cc|c} 1 & -1 & 1 \\ 2 & 3 & 12 \end{array}\right]$ (The augmented matrix.)
One way that you might use the addition method on this system is to multiply the first equation by –2 and add to the second equation to eliminate the *x*.	With the augmented matrix we do the same as with the addition method, but we don't show the variables. Also we will replace the second equation with the answer we get from adding the two equations together.
$-2x+2y=-2$ $\underline{2x+3y=12}$ $5y=10$	$\begin{array}{ccc} -2 & 2 & -2 \\ 2 & 3 & 12 \\ \hline & 5 & 10 \end{array}$ So our new matrix will look like this: $\left[\begin{array}{cc|c} 1 & -1 & 1 \\ 0 & 5 & 10 \end{array}\right]$

	We replaced the second row with the new equation found by taking –2 times the first row and adding it to the second row.
Now, in the addition method we solve for the y. $y = 2$ (divided both sides by 5)	Remember, each row of the augmented matrix represents an equation so the second row represents $5y = 10$ $y = 2$
In the addition method we now substitute our answer for y back into one of the equations to find x.	We now back substitute (plug our answer for y into the first row of the matrix, which has two variables in it).
$x - 2 = 1$ so, $x = 3$ The answer to the system is (3, 2).	$x - y = 1$ so $x - 2 = 1$ $x = 3$ The answer to the system is (3, 2).

The Gaussian method is just the addition method (5.2) without the variables, which means we can do it on our calculators, which will most likely make our lives easier.

The goal of the Gaussian Method for solving a system of equations is like that of the addition method: get down to one equation with one variable. Since each row in the augmented matrix represents an equation, our goal is to work the augmented matrix down to a point where it looks like this (numbers everywhere except in the lower left corner of the matrix):

$\left[\begin{array}{ccc\|c} \# & \# & \# & \# \\ 0 & \# & \# & \# \\ 0 & 0 & \# & \# \end{array}\right]$	Top row has all three variables in it (#x + #y + #z = #)
	Middle row has only two variables in it (0x + #y + #z = #)
	Bottom row has only one variable in it (0x + 0y + #z = #)

Let's do a longer example this time.

Example 1: Solving a System of Equations Using the Gaussian Elimination Method

Solve the system $\begin{cases} x - 4y + z = 13 \\ 3x - 10y + 2z = 34 \\ 5x - 2y + 7z = 31 \end{cases}$ by the Gaussian Elimination method.

Solution:

We must make sure that the equations are written in the same order (standard form); then we write the system as an augmented matrix.	$\left[\begin{array}{ccc\|c} 1 & -4 & 1 & 13 \\ 3 & -10 & 2 & 34 \\ 5 & -2 & 7 & 31 \end{array}\right]$
We like to eliminate the x's first, so let's multiply the first row by –3 and add that to the second row.	–3 times Row 1: (–3 12 –3 –39) + Row 2: (3 –10 2 34) 0 2 –1 –5
Now, put our answer where the second row used to be.	$\left[\begin{array}{ccc\|c} 1 & -4 & 1 & 13 \\ 0 & 2 & -1 & -5 \\ 5 & -2 & 7 & 31 \end{array}\right]$
We now eliminate the x in the third row by multiplying row one by –5 and adding it to row three	–5 times Row 1: (–5 20 –5 –65) + Row 3: (5 –2 7 31) 0 18 2 –34
Now, put our answer where the third row used to be.	$\left[\begin{array}{ccc\|c} 1 & -4 & 1 & 13 \\ 0 & 2 & -1 & -5 \\ 0 & 18 & 2 & -34 \end{array}\right]$
We need one equation in one variable if we are to solve this system. So, now we need to eliminate the y's in the bottom equation. We can eliminate the y in the third row by multiplying the second row by –9 and adding it to the third.	–9 times Row 2: (0 –18 9 45) + Row 3: (0 18 2 –34) 0 0 11 11 Notice that by using the second row instead of the first, we preserve the 0's in the first column.
Now, put our answer where the third row use to be. We have reached our goal mentioned earlier.	$\left[\begin{array}{ccc\|c} 1 & -4 & 1 & 13 \\ 0 & 2 & -1 & -5 \\ 0 & 0 & 11 & 11 \end{array}\right]$
The last row says	$11z = 11$ so, $z = 1$
Now, back-substitute the answer for z into the second row to find y.	$2y - z = -5$ so, $2y - (1) = -5$ $2y - 1 = -5$ $2y = -4$ $y = -2$
Now, back-substitute the answers for z and y into the first row to find x.	$x - 4y + z = 13$ so, $x - 4(-2) + (1) = 13$ $x + 8 + 1 = 13$ $x = 4$
Our solution is	(4, –2, 1)

Answer Q1:

$$\left[\begin{array}{ccc|c} 0 & -3 & 2 & 3 \\ 2 & -1 & 0 & -1 \\ 3 & -2 & 3 & 4 \end{array}\right]$$

Notice the 0's; we put them in anywhere we are missing a variable.

AnswerQ2:

$$\begin{cases} -2x+5y=23 \\ 7x+y=10 \end{cases}.$$

As you can see we treat each row as an equation and work our way down to the point where we have one equation with one variable.

Now, since each row is representing an equation, we can do several different operations to each row. Here is a list of what mathematicians call **elementary row operations:**

1. You may interchange any two rows (each row in the augmented matrix represents an equation and so it doesn't matter which equation (row) you write first, second, etc.).
2. You may multiply all elements of a row by any nonzero number. (You can multiply any equation by a nonzero number and still have an equivalent equation [row = equation]).
3. You may multiply all the elements of one row and add them to the elements of another row and then replace that second row. (This is just the elimination method where you multiply one equation by a number then add to another equation in order to get a new equation (row).)

Question 3: Given the augmented matrix, with the Gaussian method already applied to it, find the solution to the system that generated this reduced augmented matrix. $\left[\begin{array}{ccc|c} 1 & -2 & 3 & -10 \\ 0 & 4 & 5 & 22 \\ 0 & 0 & -3 & -6 \end{array}\right]$

Let's look at how we can do this exact process on our calculators. We will redo example 3 to demonstrate how this is done.

Discussion 3: The Gaussian Elimination Method and Your Calculator

Solving the system $\begin{cases} x - 4y + z = 13 \\ 3x - 10y + 2z = 34 \\ 5x - 2y + 7z = 31 \end{cases}$ by the Gaussian Elimination method using your calculator.

This is the same system of equations as was used in example 1.

TI-83	TI-86
Input the augmented matrix 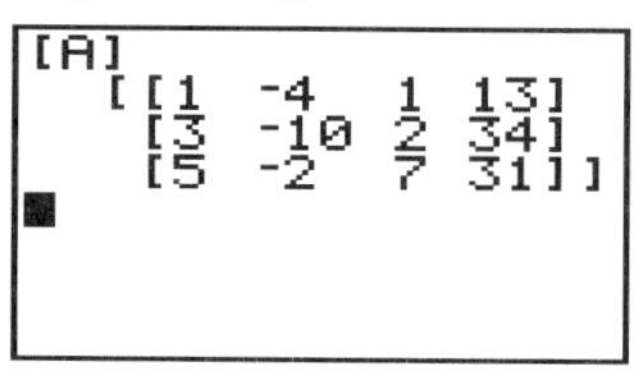	Input the augmented matrix 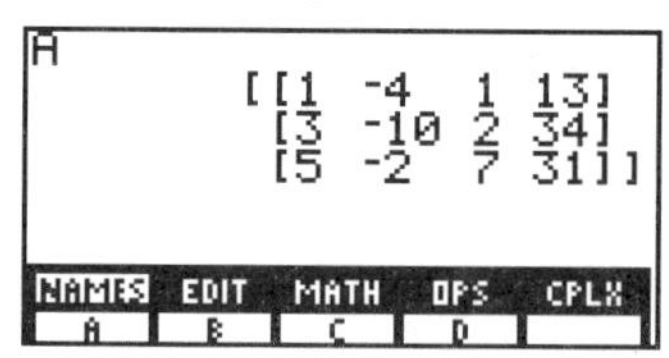

<table>
<tr>
<td>Multiply Row 1 by –3 and add to Row 2. You go to MATRIX; then arrow right to MATH; then choose F *row+(and type in what you see on the screen below.

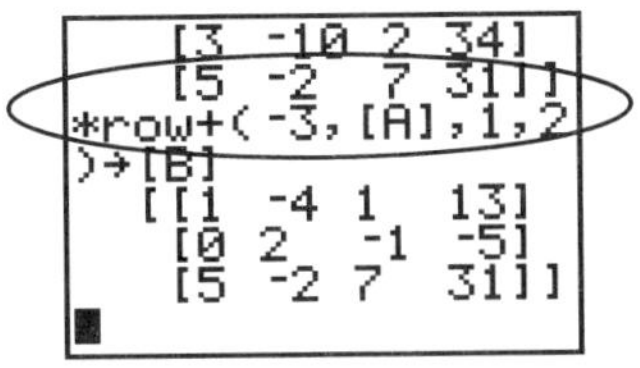
</td>
<td>Multiply Row 1 by –3 and add to Row 2. You go to MATRIX; then F4 for OPS, then MORE; then choose F5 mRAdd and type in what you see on the screen.
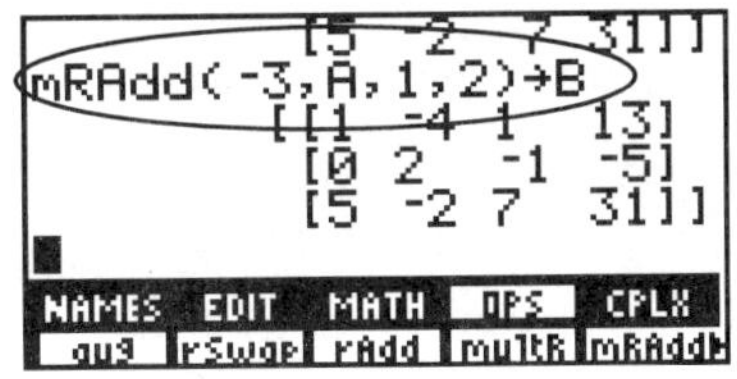
</td>
</tr>
<tr>
<td>The way to tell the calculator what to do is as follows:
*row+(multiplier, matrix name, row being multiplied, row being added to) then the arrow (STO key) means store answer in, and then the matrix name where you want to store your answer.</td>
<td>The way to tell the calculator what to do is as follows:
mRAdd(multiplier, matrix name, row being multiplied, row being added to) then the arrow (STO key) means store answer in, and then the matrix name where you want to store your answer.</td>
</tr>
<tr>
<td>Now multiply Row 1 by –5, and add to Row 3.
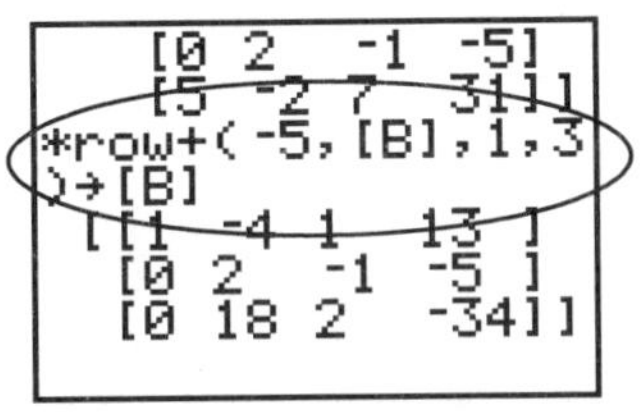
</td>
<td>Now multiply Row 1 by –5, and add to Row 3.
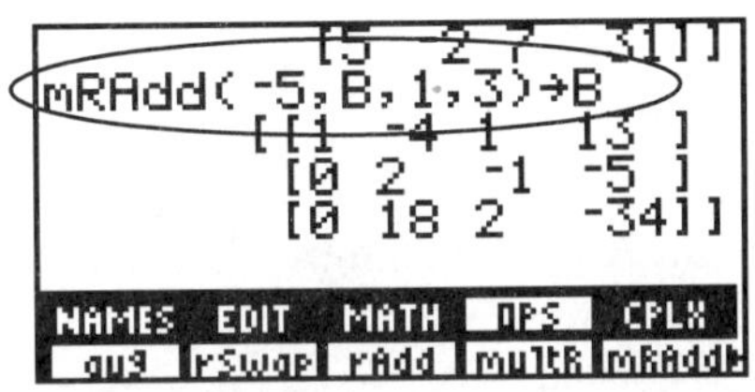
</td>
</tr>
<tr>
<td>Now multiply Row 2 by –9, and add to Row 3.
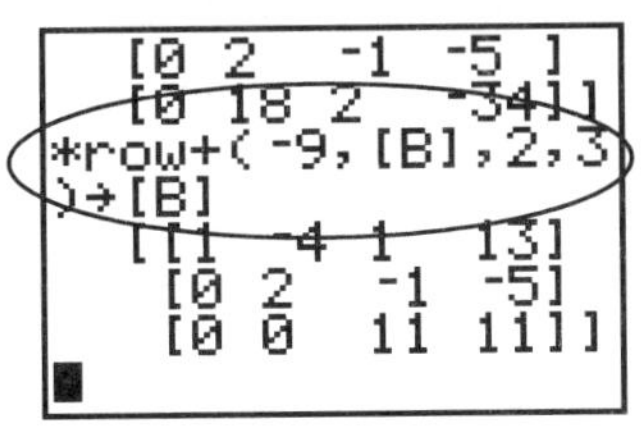
</td>
<td>Now multiply Row 2 by –9, and add to Row 3.
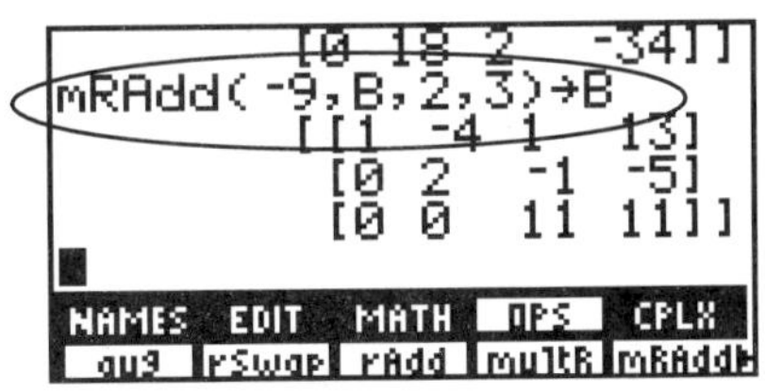
</td>
</tr>
</table>

<table>
<tr><td>Notice that the last matrix on the calculator screen is the same one we came up with in example 3 where we did it by hand.

$$\left[\begin{array}{ccc|c} 1 & -4 & 1 & 13 \\ 0 & 2 & -1 & -5 \\ 0 & 0 & 11 & 11 \end{array}\right]$$</td></tr>
<tr><td>We can now do the back-substitution by hand and arrive at the correct answer we found in example 3. The only difference is that the calculator took care of all the adding and multiplying flawlessly.</td></tr>
</table>

Answer Q3: The last row tells us that $-3z = -6$, so $z = 2$. So the second row now tells us that $4y + 5z = 22$ but $z = 2$ so, $4y + 5(2) = 22$ which means $y = 3$. Now the first row says that $x - 2y + 3z = -10$, but $y = 3$, $z = 2$ so, $x - 2(3) + 3(2) = -10$ or $x = -10$. The solution is $(-10, 3, 2)$.

We have seen other ways to use the calculators to solve systems that are easier than this, but as we have also seen, they didn't always lead to a resolution of the problem. The Gaussian method will always lead to a resolution.

Question 4: Back in section 5.2, what were the different things that could happen when you were trying to solve a system of linear equations in two variables?

When you used the elimination method, how did you know which type of solution you had? Well, you knew that you had one point as a solution if you were able to get values for the variables. You knew that you had no solution if you got a statement that wasn't true, like $0 = -4$, when you were eliminating variables. Lastly, you knew that you had every point on the line as the solution if when trying to eliminate the variables the result was a statement that was always true, like $5 = 5$. Those same things and more can happen here, when we solve systems of linear equations with two or more variables. Let's do another example.

Example 2: An Inconsistent System

Solve the system $\begin{cases} x - 2y = -4 \\ 2x - 4y = -10 \end{cases}$ by the Gaussian Elimination method.

Solution:

<table>
<tr><td>Do by hand or input the augmented matrix into the calculator

$$\left[\begin{array}{cc|c} 1 & -2 & -4 \\ 2 & -4 & -10 \end{array}\right]$$</td><td>TI-83

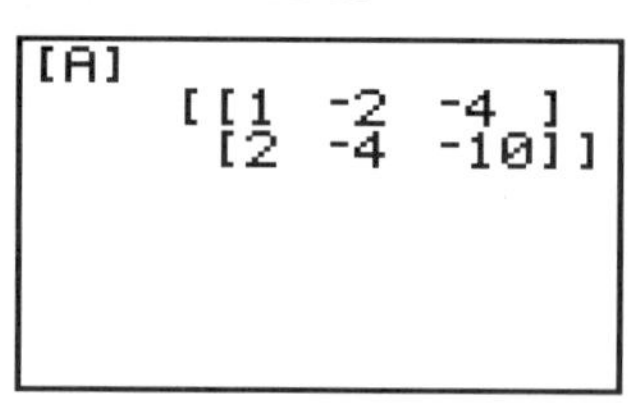
</td><td>TI-86

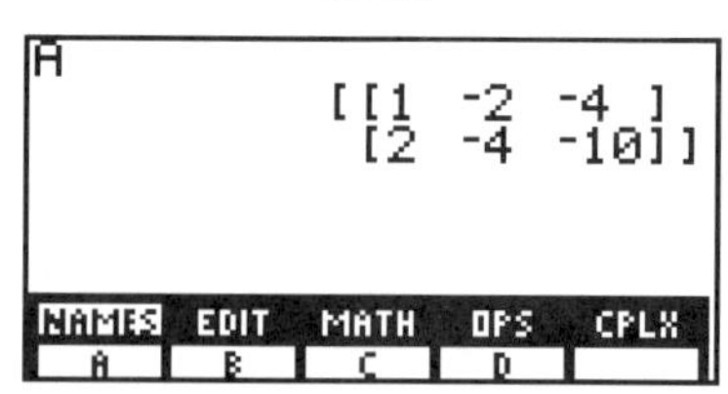
</td></tr>
<tr><td>Multiply Row 1 by −2 and add to Row 2.</td><td></td><td></td></tr>
</table>

-2 • Row 1: (-2 4 8) + Row 2: <u>(2 -4 -10)</u> 0 0 −2	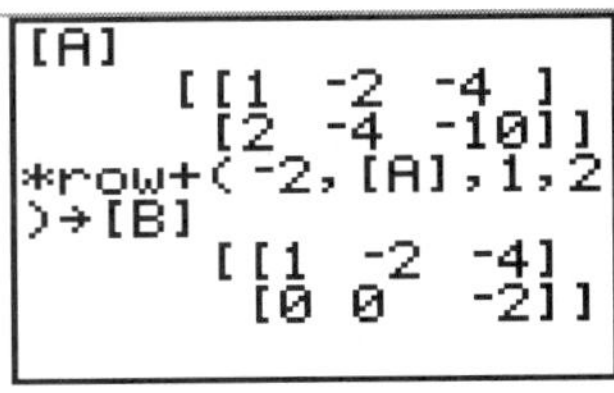	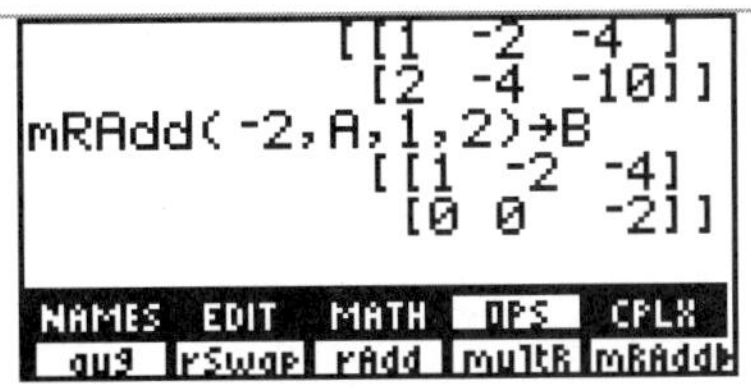
The final matrix is $\left[\begin{array}{cc\|c} 1 & -2 & -4 \\ 0 & 0 & -2 \end{array}\right]$	The last row says that $0x + 0y = -2$ which means we have $0 = -2$ a false statement.	This system doesn't have a solution. The two lines must be parallel

So, as mentioned in section 5.2, question 1, this system, since it has no solution, is called an inconsistent system.

Example 3: A System with an Infinite number of Solutions

Solve this system by the Gaussian Elimination method $\begin{cases} 2x+3y = 6 \\ -4x-6y = -12 \end{cases}$.

Solution:

	TI 83	TI 86
Do by hand or input the augmented matrix into the calculator $\left[\begin{array}{cc\|c} 2 & 3 & 6 \\ -4 & -6 & -12 \end{array}\right]$	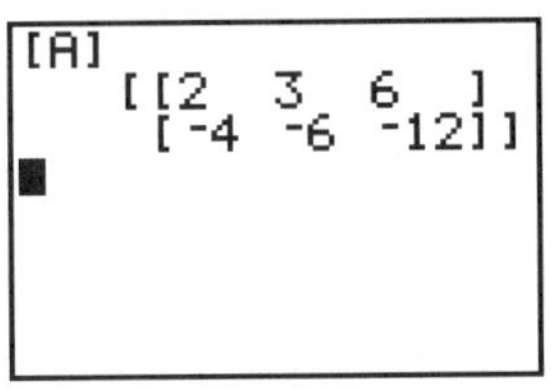	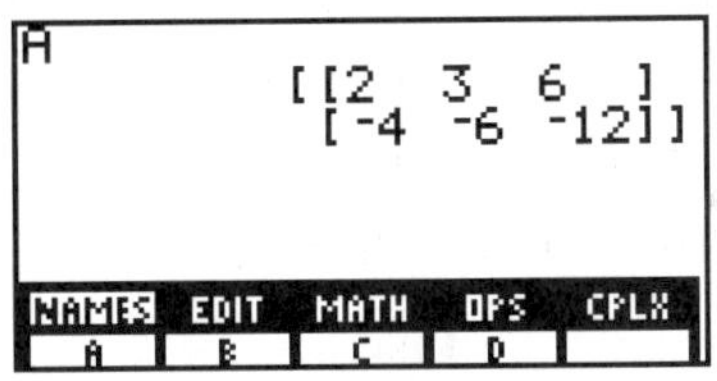
Multiply Row 1 by 2 and add to Row 2. 2• Row 1: (4 6 12) + Row 2: <u>(-4 -6 -12)</u> 0 0 0	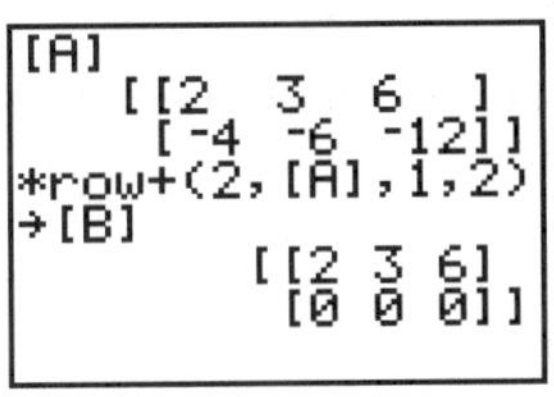	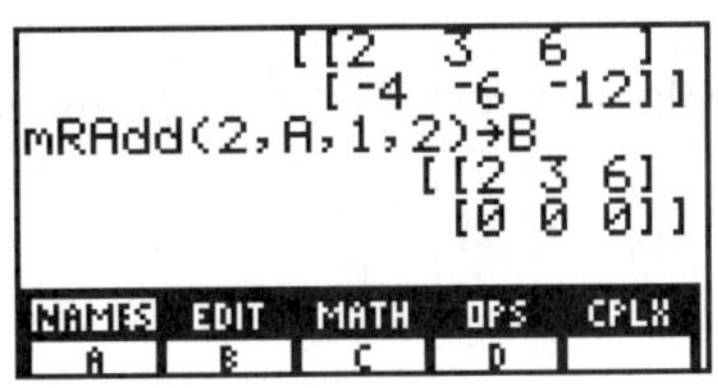

Answer Q4: You could get <u>one point</u> as your solution (lines intersected) or <u>no points</u> for the solution (the lines were parallel) or <u>every point on the line</u> as a solution (the same line).

The final matrix is $\left[\begin{array}{cc|c} 2 & 3 & 6 \\ 0 & 0 & 0 \end{array}\right]$ The last row says that $0x + 0y = 0$	which means we have $0 = 0$, a true statement. This system has an infinite number of solutions. The two lines must be the same.	We would take the next row up from the bottom and solve it for one of the variables. $2x + 3y = 6$ $3y = 6 - 2x$ $y = \frac{6-2x}{3}$ Our solution, then, in point form is $(x, \frac{6-2x}{3})$.

In section 6.3 we were unable to solve systems of the types in examples 5 and 6 (where the determinant = 0). This is one reason the Gaussian method is needed. It can solve problems regardless of the type of solutions they have. Besides this, there are many uses for row operations on matrices in different areas of mathematics, and so it is important for us to know about them. But when it comes to solving a system of equations, our calculators have some buttons built into them to make our job easier. Here is an example of how our calculators can make this job of finding a solution to a system of equations by the Gaussian method easier.

Example 4: Gaussian Elimination and Calculator Programs

Solve the system $\begin{cases} x - 4y + z = 13 \\ 3x - 10y + 2z = 34 \\ 5x - 2y + 7z = 31 \end{cases}$ by the Gaussian elimination method with the help of your calculator. This is the same system as in examples 3 and 4.

Solution:

TI-83	TI-86
Input the augmented matrix. 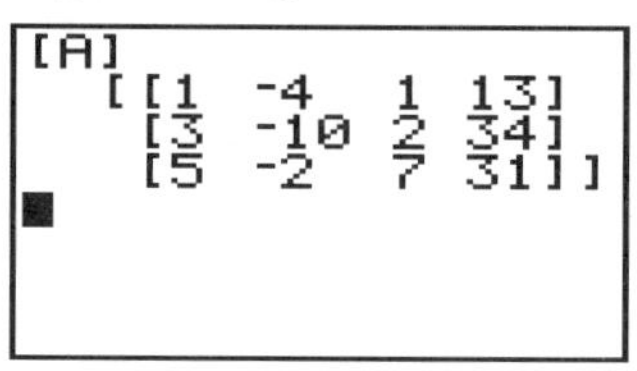	Input the augmented matrix. 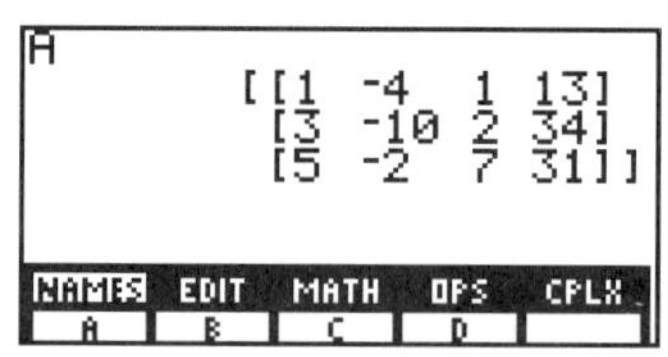

<table>
<tr>
<td>Go to MATRIX; then arrow over to MATH; then arrow down and choose A ref(

</td>
<td>Go to MATRIX, then F4 for OPS, then choose F4 ref
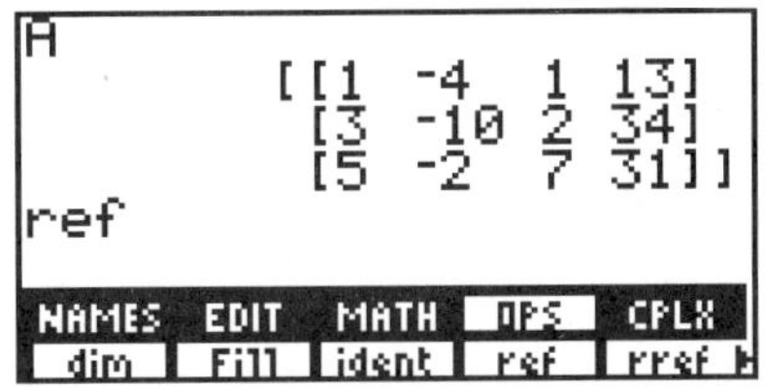
</td>
</tr>
<tr>
<td>Now, the correct format is ref([A]).
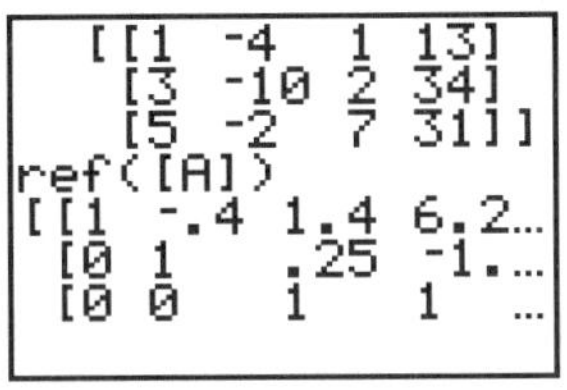
</td>
<td>Now, the correct format is ref A.
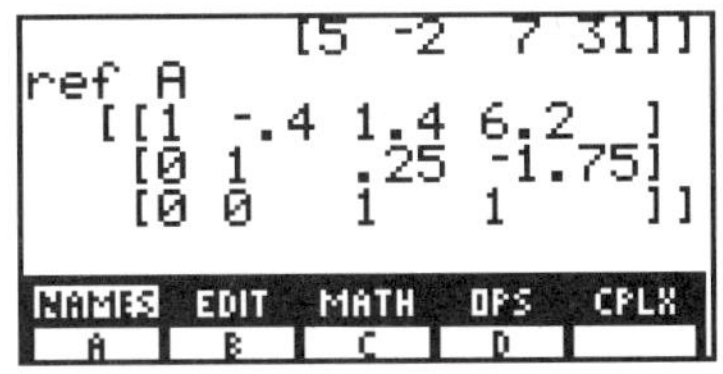
</td>
</tr>
<tr>
<td>Of course we could FRAC this and get
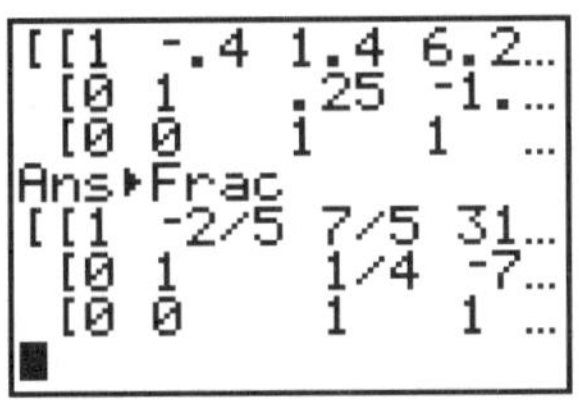

The whole matrix doesn't fit on the screen so you must arrow across to see the rest of it.</td>
<td>Of course we could FRAC this and get
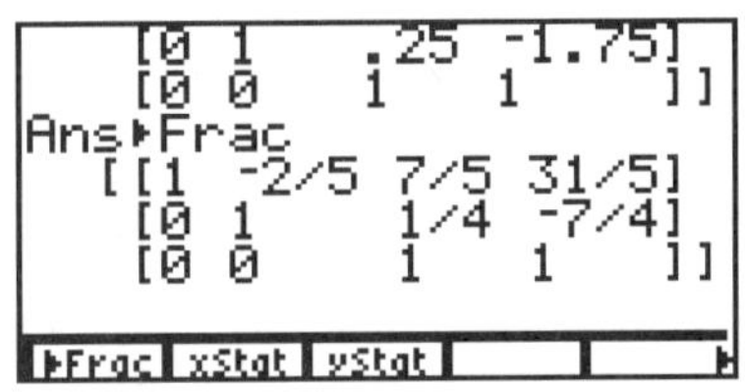

If the whole matrix doesn't fit on the screen, you must arrow across to see the rest of it.</td>
</tr>
<tr>
<td colspan="2">The final row reduced matrix is
$$\left[\begin{array}{ccc|c} 1 & \frac{-2}{5} & \frac{7}{5} & \frac{31}{5} \\ 0 & 1 & \frac{1}{4} & \frac{-7}{4} \\ 0 & 0 & 1 & 1 \end{array}\right]$$</td>
</tr>
<tr>
<td colspan="2">Comparing this to examples 3 and 4 we see that the calculator uses the row operations in such a way that each row is made to have a first coefficient equal to 1.</td>
</tr>
<tr>
<td colspan="2">From here you can back-substitute like we did in example 3 to arrive at the correct solution.
(4, –2, 1)</td>
</tr>
</table>

This example shows you how calculators can be used to do the same procedures we did by hand. But the calculator can go even farther. It can even help us to skip the back-substituting step. Here is the same example, but this time we will use the **rref(** button.

<table>
<tr><th>TI-83</th><th>TI-86</th></tr>
<tr><td>Input the augmented matrix.
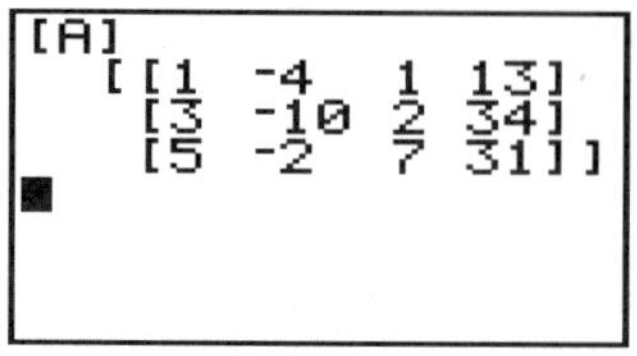</td><td>Input the augmented matrix.
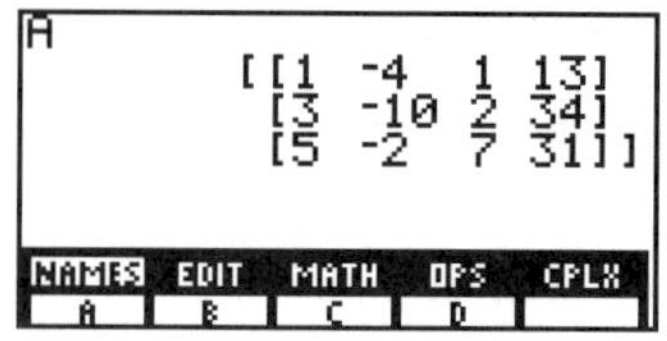</td></tr>
<tr><td>Go to MATRIX, then arrow over to MATH, then arrow down and choose A rref(
</td><td>Go to MATRIX, then F4 for OPS, then choose F5 rref
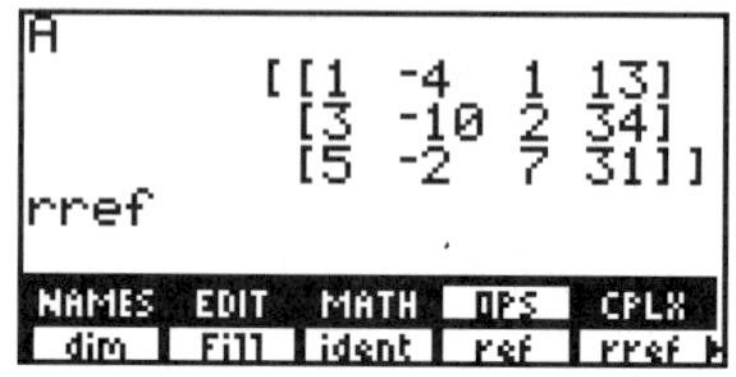</td></tr>
<tr><td>Now, the correct format is rref([A])
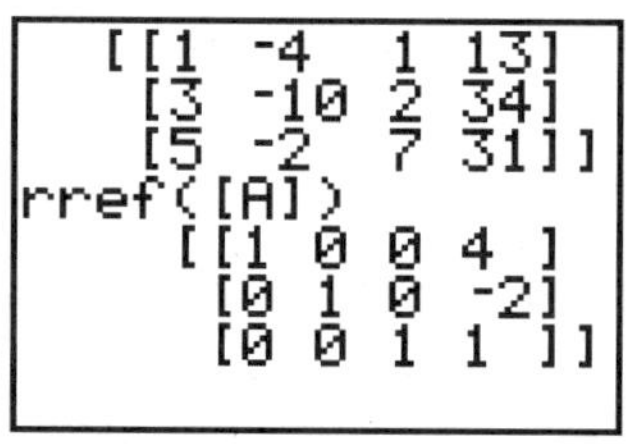</td><td>Now, the correct format is rref A
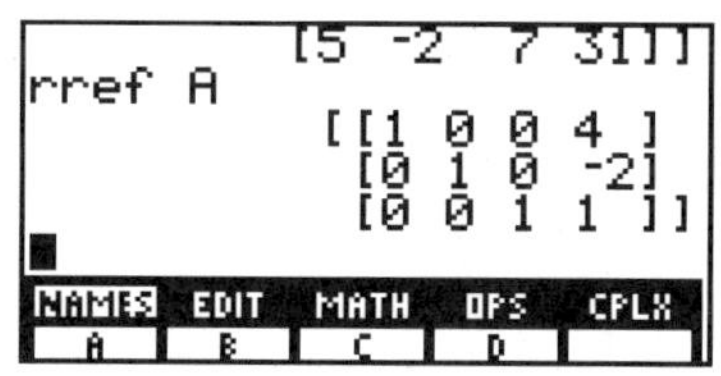</td></tr>
<tr><td colspan="2">Now, look at each row and notice how there is only one variable in each. The answer is right there staring us in the face. (This is called the Reduced Row Echelon Form - rref.)
$1x+0y+0z=4$
$0x+1y+0z=-2$
$0x+0y+1z=1$
Our solution must be (4, –2, 1).</td></tr>
</table>

Example 5: The Gaussian Method and Your Calculator

Solve the following three problems using the Gaussian method and your calculator.

a.) $\begin{cases} x-2y+3z=5 \\ 3x-3y+z=9 \\ 5x+y-3z=3 \end{cases}$ b.) $\begin{cases} x+y-2z=3 \\ 3x-2y+4z=1 \\ 2x-3y+6z=8 \end{cases}$ c.) $\begin{cases} x \quad +2z=5 \\ -3x+y+2z=4 \\ -13x+3y-2z=-8 \end{cases}$

Solutions:

<table>
<tr><td colspan="3">We will type in the augmented matrix for each problem and then use rref(</td></tr>
<tr><td>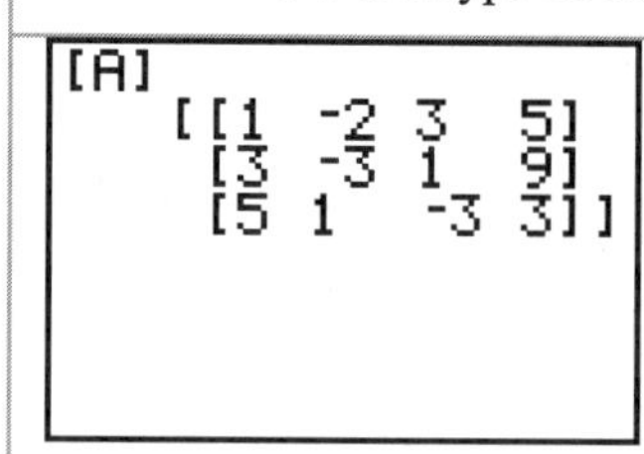</td><td>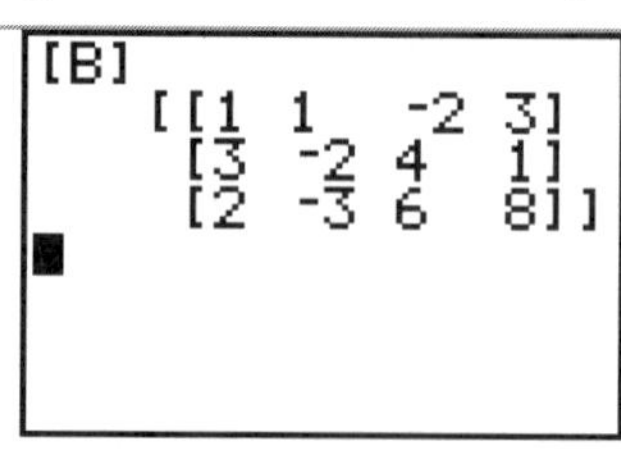</td><td>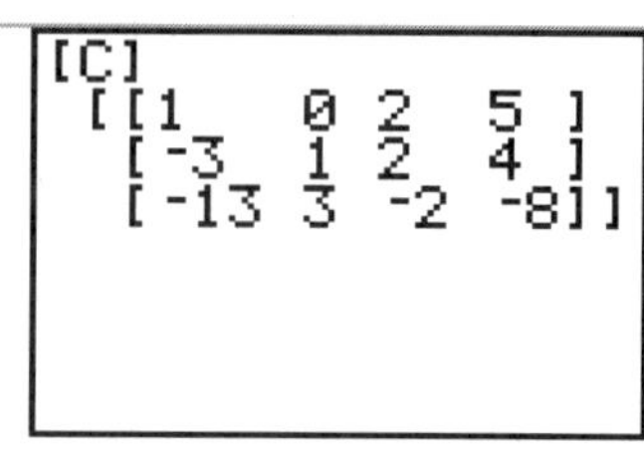</td></tr>
</table>

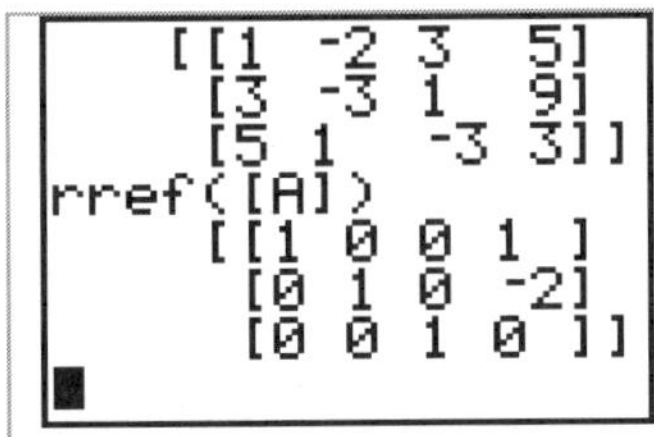	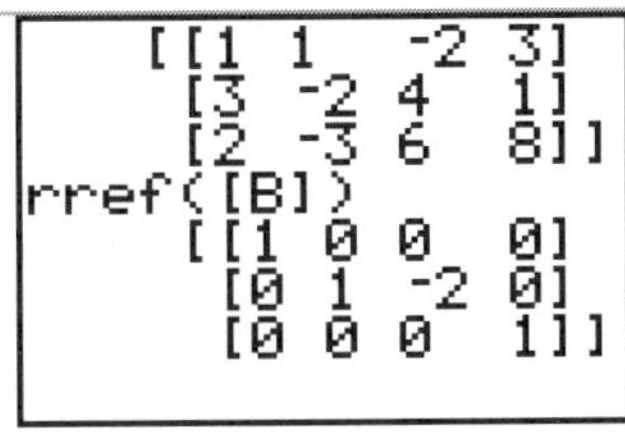	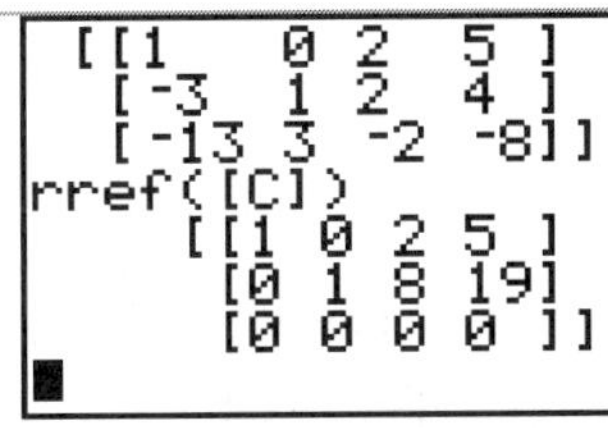
It looks like $z = 0$, $y = -2$ and $x = 1$, so the solution is (1, −2, 0).	Since the last row in the matrix is 0 0 0 1, which means that we have the equation 0 = 1, a false statement, the answer is that this system has no solution.	Since the last row in the matrix is 0 0 0 0, which means that we have the equation 0 = 0, a true statement, the answer is that this system has many solutions.

When possible, we would like to write our solutions in point form (as an ordered pair or triple or etc.). In example 8(c) we found that the system has many solutions just like in example 6, since the last row was 0 = 0. Once again, to write the solution in point form, we need to go to the next row up and solve for one of the variables.

Here is the matrix for example 8(c) in rref form.	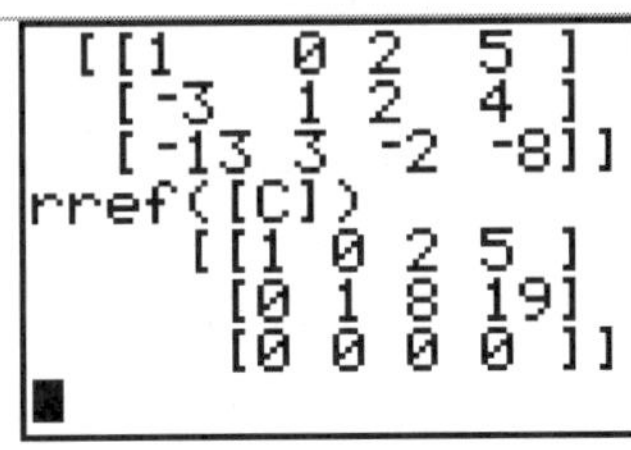
The second row from the bottom is	[0 1 8 19] which means $0x + y + 8z = 19$
Solve for y since it's the easiest to solve for	$y = 19 - 8z$
Now, go to the first row ([1 0 2 5]) and put (9 – 8z) in for y and solve for x.	$x + 0y + 2z = 5$ $x + 0(19 - 8z) + 2z = 5$ $x + 2z = 5$ $x = 5 - 2z$
Our solution in point form describes the line of intersection between the three planes in terms of the variable z. Any value you plug in for the variable z will yield a point of intersection.	(5 – 2z, 19 – 8z, z)

Example 6: Systems Where the Number of Variables and Equations Don't Match

Solve the following two systems of equations by using the rref key on your calculator.

a.) $\begin{cases} 2x + y - 3z = 4 \\ 3x + 2y + z = 2 \end{cases}$ b.) $\begin{cases} 3x - 5y = 2 \\ -2x + y = 10 \\ 7x + 4y = 8 \end{cases}$

Solution:

Notice that we haven't seen any problems like this yet. We have one system where there are more variables than equations (a) and one where there are more equations than variables (b). But we can still try to find their solution by using their augmented matrices.

(a) Input and solve using the calculator and the rref key.	(b)Input and solve using the calculator and the rref key.
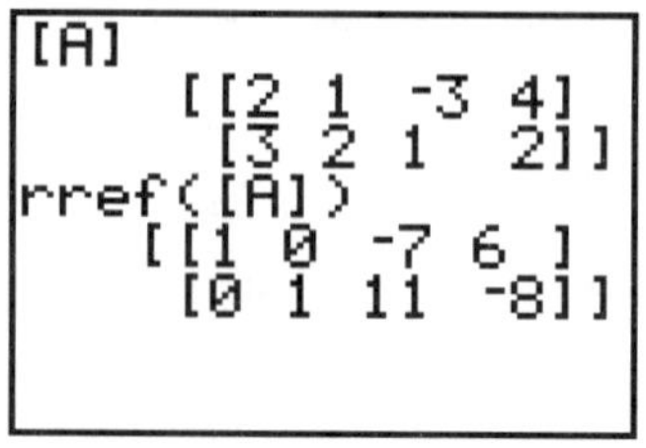	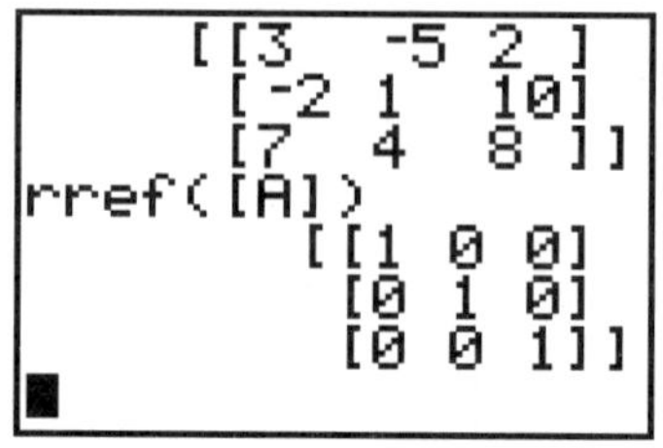
So, we now know that $0x + y + 11z = -8$ (bottom row) and $x + 0y - 7z = 6$ (top row). Since we didn't have enough equations we are not getting a single point solution. But we know the following: $y = -11z - 8$ $x = 7z + 6$ Our solution then is $(7z + 6, -11z - 8, z)$, a line of intersection between the two planes.	Now, in this problem we get the following three equations from the first, second and third rows $x + 0y = 0$ $0x + y = 0$ $0x + 0y = 1$ The last row tells us that we have a "no solution" answer for this problem $(0 = 1)$.

Section Summary

- The augmented matrix is constructed by simply arranging all of the coefficients and constants in a single matrix.
- You can perform row operations on augmented matrices in order to solve system of equations. The most used row operation on the TI-83 graphing calculator is *row+(**multiplier, matrix name, row being multiplied, row being added to**) and mRAdd(**multiplier, matrix name, row being multiplied, row being added to**) on the TI-86.
- Other frequently used operations are ref and rref.
- The Gaussian Elimination method can be quite lengthy if done by *hand*. It's a little better on the calculator, doing each row operation, because the calculator doesn't make simple arithmetic errors. It is important to learn about the row operations because they are necessary in certain areas in mathematics like Linear Programming, a finite math course topic for most business majors.

SECTION 6.4 HOMEWORK

(1–10) Given the following systems of equations, write an augmented matrix for each.

1. $\begin{cases} 2x-5y=23 \\ 3x+7y=-13 \end{cases}$

2. $\begin{cases} \frac{2}{3}x+\frac{5}{4}y=\frac{1}{3} \\ \frac{3}{8}x-\frac{9}{5}y=\frac{4}{7} \end{cases}$

3. $\begin{cases} 3x-5y+7z=9 \\ 2x+3y-8z=-3 \\ -3x+2y+z=7 \end{cases}$

4. $\begin{cases} 0.34x-0.23y-0.38z=580 \\ 0.3x+3.24y+0.83z=0.234 \\ -x-2y+5z=689 \end{cases}$

5. $\begin{cases} x-3z=7 \\ 2x+y=9 \\ -3y+2z=-9 \end{cases}$

6. $\begin{cases} x+3z=-8 \\ y=6 \\ -2x+7y=15 \end{cases}$

7. $\begin{cases} x+2y=3-5z \\ y-3z=5+2x \\ 3z+2x-5y=8 \end{cases}$

8. $\begin{cases} x=y+z+9 \\ y=2z \\ y+z=3x \end{cases}$

9. $\begin{cases} x+2y-3z+w=9 \\ 2x-y+z-3w=15 \\ 3x+2y-5z+2w=18 \\ 2x-y+3z+w=-5 \end{cases}$

10. $\begin{cases} 3x-z=8 \\ 3y+5w=39 \\ x-3z+2w=7 \\ y+z=3w-\frac{1}{2}x \end{cases}$

(11–18) Given the following augmented matrix, write the system of equations it represents for the indicated variables.

11. $\left[\begin{array}{cc|c} 1 & -2 & 3 \\ 3 & 4 & 4 \end{array}\right]$ (x, y)

12. $\left[\begin{array}{cc|c} 2 & \frac{1}{3} & 3 \\ \frac{3}{4} & -3 & -2 \end{array}\right]$ (x, y)

13. $\left[\begin{array}{ccc|c} 1 & -3 & 2 & -2 \\ 1 & 4 & -5 & -3 \\ 2 & -1 & 3 & 4 \end{array}\right]$ (x, y, z)

14. $\left[\begin{array}{ccc|c} 3 & -1 & 3 & -2 \\ \frac{3}{4} & \frac{5}{6} & \frac{1}{2} & 7 \\ \frac{-2}{3} & 3 & \frac{3}{4} & -3 \end{array}\right]$ (r, s, t)

15. $\left[\begin{array}{ccc|c} 1 & 0 & 3 & -2 \\ 0 & 3 & -2 & 4 \\ -1 & 3 & 2 & -3 \end{array}\right]$ (x, y, z)

16. $\left[\begin{array}{ccc|c} 2 & -1 & 0 & 3 \\ 3 & 0 & 1 & -2 \\ 0 & 3 & 4 & 5 \end{array}\right]$ (a, b, c)

17. $\left[\begin{array}{cccc|c} 1 & -1 & 2 & 0 & 3 \\ 3 & 1 & 0 & -2 & 5 \\ -2 & 1 & 3 & 4 & -3 \\ 0 & 1 & 0 & 3 & -3 \end{array}\right]$ (x, y, z, w)

18. $\left[\begin{array}{cccc|c} 2 & 0 & 3 & -1 & 5 \\ 0 & 2 & 0 & 5 & 3 \\ 1 & -3 & 5 & 0 & 4 \\ 2 & 3 & -3 & 1 & -5 \end{array}\right]$ (r, s, t, v)

(19–24) Given the following augmented matrix, row reduce by hand and solve for the indicated variables.

19. $\left[\begin{array}{cc|c} 3 & 2 & 4 \\ 2 & -3 & -19 \end{array}\right]$

Variables x and y

20. $\left[\begin{array}{cc|c} 5 & -4 & 1 \\ 3 & 2 & 1 \end{array}\right]$

Variables x and y

21. $\left[\begin{array}{ccc|c} 1 & -2 & 3 & 26 \\ 2 & -3 & 5 & 43 \\ 3 & 2 & -1 & -4 \end{array}\right]$

Variables r, s, and t

22. $\left[\begin{array}{ccc|c} 1 & -1 & 2 & 4 \\ 2 & -3 & 2 & 7 \\ -3 & 4 & 2 & -10 \end{array}\right]$

Variables a, b, and c

23. $\left[\begin{array}{ccc|c} 1 & -2 & 3 & -15 \\ 0 & 3 & 2 & -2 \\ 0 & 4 & -3 & 20 \end{array}\right]$

Variables x, y, and z

24. $\left[\begin{array}{ccc|c} 1 & 3 & -2 & 12 \\ 0 & 4 & -3 & 18 \\ 0 & 2 & 5 & -4 \end{array}\right]$

Variables x, y and z

(25–36) Given the following row reduced matrices solve for the indicated variables.

25. $\left[\begin{array}{ccc|c} 1 & -2 & 3 & -15 \\ 0 & 1 & 2 & -3 \\ 0 & 0 & 1 & 5 \end{array}\right]$

Variables x, y, and z

26. $\left[\begin{array}{ccc|c} 1 & 3 & -1 & 23 \\ 0 & 1 & -3 & -11 \\ 0 & 0 & 1 & 7 \end{array}\right]$

Variables x, y and z

27. $\left[\begin{array}{ccc|c} 1 & -2 & 3 & 2 \\ 0 & 1 & -2 & -3 \\ 0 & 0 & 0 & 0 \end{array}\right]$

Variables x, y and z

28. $\left[\begin{array}{ccc|c} 1 & 3 & 2 & -3 \\ 0 & 1 & -3 & -11 \\ 0 & 0 & 0 & 0 \end{array}\right]$

Variables x, y and z

29. $\left[\begin{array}{cccc|c} 1 & -2 & 3 & 1 & 9 \\ 0 & 1 & -2 & 3 & -13 \\ 0 & 0 & 1 & -2 & 7 \\ 0 & 0 & 0 & 1 & -2 \end{array}\right]$

Variables r, s, t and v

30. $\left[\begin{array}{cccc|c} 1 & 3 & -2 & -2 & -10 \\ 0 & 1 & 3 & -2 & -19 \\ 0 & 0 & 1 & 3 & 6 \\ 0 & 0 & 0 & 1 & 5 \end{array}\right]$

Variables a, b, c and d

31. $\left[\begin{array}{ccc|c} 1 & -3 & 4 & 5 \\ 0 & 1 & -3 & 2 \\ 0 & 0 & 0 & 1 \end{array}\right]$

Variables A, B and C

32. $\left[\begin{array}{ccc|c} 1 & 3 & 4 & -10 \\ 0 & 1 & 4 & -33 \\ 0 & 0 & 0 & 1 \end{array}\right]$

Variables x, y and z

33. $\left[\begin{array}{ccc|c} 1 & \frac{3}{4} & \frac{5}{8} & \frac{7}{16} \\ 0 & 1 & \frac{1}{4} & \frac{5}{16} \\ 0 & 0 & 1 & \frac{3}{8} \end{array}\right]$

Variables x, y and z

34. $\left[\begin{array}{ccc|c} 1 & \frac{3}{2} & \frac{7}{4} & \frac{3}{4} \\ 0 & 1 & \frac{3}{8} & \frac{1}{4} \\ 0 & 0 & 1 & \frac{1}{2} \end{array}\right]$

Variables x, y and z

35. $\left[\begin{array}{ccc|c} 1 & 0 & 0 & 3 \\ 0 & 1 & 2 & 5 \\ 0 & 0 & 0 & 0 \end{array}\right]$

Variables r, s and t

36. $\left[\begin{array}{ccc|c} 1 & -1 & 1 & 3 \\ 0 & 1 & 0 & -3 \\ 0 & 0 & 0 & 0 \end{array}\right]$

Variables A, B and C

(37– 76) Given the following systems of equations, solve each with the graphing calculator. (Use either the rref or the row operations functions of the calculator.)

37. $2x - 3y = 8$

$3x + 2y = 9$

38. $3x + 5y = -23$

$2x - 3y = 37$

39. $\frac{2}{3}x+5y=\frac{3}{8}$

$\frac{3}{4}x-\frac{3}{16}y=\frac{1}{2}$

40. $\frac{3}{5}x-\frac{7}{10}y=\frac{11}{20}$

$\frac{3}{20}x+\frac{4}{5}y=\frac{7}{10}$

41. $3x-2y=10$

$\frac{-3}{2}x+y=5$

42. $5x+10y=21$

$\frac{5}{6}x+\frac{5}{3}y=\frac{7}{2}$

43. $\frac{1}{3}x=\frac{2}{3}y+2$

$5x-10y=30$

44. $\frac{1}{2}x=\frac{4}{3}-\frac{5}{6}y$

$3x+5y=8$

45. $0.3A+0.2B=0.5$

$1.5A-0.9B=3.2$

46. $0.23r-0.35s=0.83$

$0.35r-0.9s=0.75$

47. $2x-3y+5z=3$

$3x+y-3z=9$

$-3x+7y+z=-29$

48. $3x+y-z=2$

$-2x+3y+2z=-1$

$x-2y+3z=13$

49. $m+3n-2p=-1$

$2m+5n-3p=-1$

$-3m+2n-5p=-8$

50. $-R-6S+3T=1$

$R-2S+T=3$

$3R+2S-T=5$

51. $2x-y+z=5$

$x+3y-2z=20$

$-3x-2y+z=-20$

52. $x-y+z=2$

$2x+y-z=1$

$4x+5y-5z=2$

53. $2a+3b=5$

$3a-5c=-1$

$3b+2c=7$

54. $3y-2z=-16$

$3x+5z=19$

$x-5y=23$

55. $x-2y=-3z-24$

$2x+5z=3y-39$

$3y-2z=5x+45$

56. $3x+5z=3y+1$

$5y-2z=8-2x$

$7x-4y=3z+55$

57. $0.25x-0.32y+0.28z=-0.34$

$0.32x+0.24y-0.15z=1.33$

$0.18x+0.12y-0.52z=0.26$

58. $1.34x-2.35y+1.5z=-2.84$

$2.12x+3.14y-3.2z=6.97$

$0.74x-2.8y+5.2z=3.01$

59. $\frac{1}{2}x-\frac{3}{4}y+\frac{5}{8}z=\frac{43}{8}$

$\frac{3}{5}x-\frac{2}{3}y-\frac{4}{9}z=\frac{12}{5}$

$\frac{2}{3}x+\frac{1}{9}y+\frac{2}{5}z=\frac{164}{45}$

60. $\frac{9}{10}x-\frac{3}{5}y+\frac{2}{15}z=\frac{23}{120}$

$\frac{2}{3}x+\frac{7}{18}y-\frac{5}{24}z=\frac{1537}{1728}$

$\frac{3}{4}x+\frac{5}{12}y+\frac{7}{20}z=\frac{179}{288}$

61. $-2x+3y+z=5$

62. $x+2y-z=3$

$3x - 2y - 3z = -20$

$x + y - 2z = -15$

$2x - 5y + 3z = 4$

$3x - 3y + 2z = 7$

63. $x - 2y + 3z - 5w = 7$

$3x + y - 2z + 3w = 6$

$2x + 3y + 2z + 4w = 1$

$-x + 5y - 4z - w = 34$

64. $2x + 3y - 5z + w = 23$

$x - y + z - 3w = -5$

$-3x - 5y + 4z + 2w = -42$

$3x - 2y + 3z - w = -31$

65. $-5x + 3y - 16z = 0$

$-2x + y - 6z = -1$

$-x + 2y - 6z = 7$

66. $7x + 2y - 8z = 27$

$3x + y - 3z = 11$

$x + y + z = 1$

67. $-2a - b + 11c = -13$

$3a + b - 14c = 15$

$a - 2b + 7c = -15$

68. $3a - b + 23c = 18$

$-2a + b - 18c = -7$

$a + 2b - 11c = 42$

69. $5x - 2y - 4z = 31$

$-2x + y + 2z = -13$

$x + y + 2z = 2$

70. $4x + y - 8z = 33$

$3x + y - 6z = 24$

$2x - 2y - 4z = 24$

71. $-2r - s - 2t = -31$

$3r + s + 2t = 18$

$4r + 2s + 4t = 27$

72. $4m + n - 36p = 63$

$3m + \mathrm{n} - 27\mathrm{p} = 48$

$m + n - 9p = 19$

73. $5x + 2y + 7z = 19$

$2x + y + 3z = 8$

74. $5x - 2y + 3z = -20$

$-2x + y - z = 9$

75. $5x + 2y + 5z = 19$

$2x + y + 2z = 8$

76. $\frac{13}{12}x + \frac{1}{6}y + \frac{17}{12}z = \frac{-4}{3}$

$\frac{1}{2}x + y + \frac{5}{2}z = 4$

Section 6.5 Matrices in Real Life

In this final section of chapter 6, we'll apply what we've learned about matrices to a variety of "real life" related problems. Systems of equations can be used to solve everyday problems when they involve several different variables or choices. Let's start with a simple example that you have most likely seen in a math text sometime in your past.

Example 1: A Classic Distance, Rate, Time Problem

Find the solution to this classic distance problem by using matrices.

> A boat on a river can travel 25 miles in three hours going with the current, but that same boat can only travel 10 miles in two hours going against the current. What would be the rate of speed of the boat on still water and what is the rate of speed of the current of the river in miles per hour?

Solution:

First notice that there are two things we need to find: the speed of the boat, b, and the speed of the current of the river, c. Also, there are two events taking place: going upstream and going downstream. This suggests that we will need two variables in two equations, a system of equations.

The two equations in this case are $\begin{cases} 3b+3c=25 \\ 2b-2c=10 \end{cases}$.

Remember, rate times time equals distance and going with the river adds to your speed [$(b+c)$], so adds to distance traveled [$3(b+c)$], while going against the river subtracts from your speed [$b-c$)] so it subtracts from distance traveled [$2(b-c)$]. Now we can translate this system of equations into an augmented matrix and then **rref** it on our calculators to arrive at a solution. (Of course you could use the inverse matrix method or Cramer's Rule or several other methods to go about solving this system.)

Gaussian Elimination Method	Inverse Matrix Method
Change to its augmented matrix form. $\left[\begin{array}{cc\|c} 3 & 3 & 25 \\ 2 & -2 & 10 \end{array}\right]$	Input the coefficient matrix and the constant matrix. [A] [[3 3] [2 -2]] [B] [[25] [10]]
rref the matrix.	Now take $A^{-1}B$.

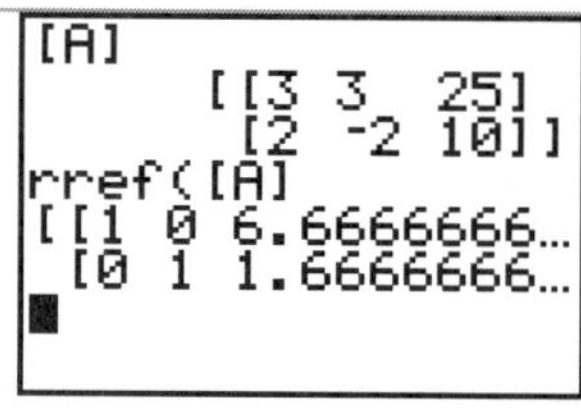	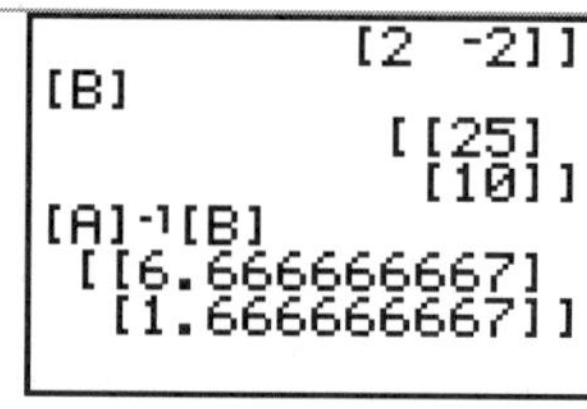
The answer might be easier to see if we **FRAC** this matrix. 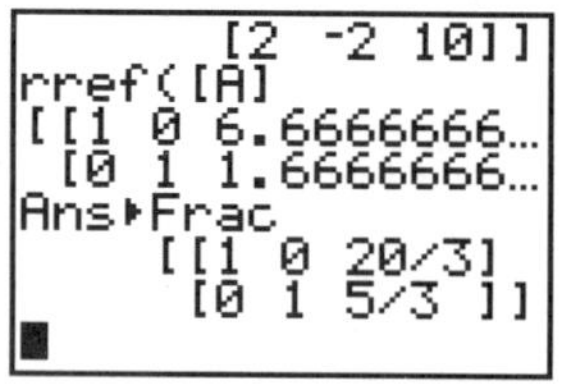	The answer might be easier to see if we **FRAC** this matrix. 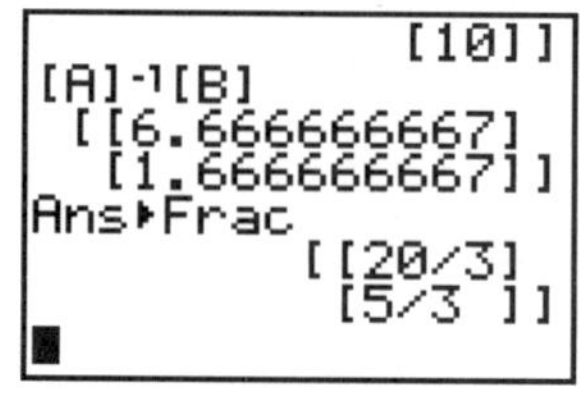

From this we can see that the speed of the boat is $\frac{20}{3}$ mph and the speed of the current is $\frac{5}{3}$ mph.

This was a simple example that you probably did in a beginning algebra class. Let's look at a more complicated and possibly more realistic problem.

Discussion 1: Chemical Content of Soil

The following table shows the active chemical content of three different soil additives (Fertilizers).

ADDITIIVES	AMMONIUM NITRATE (grams)	PHOSPHORUS (grams)	IRON (grams)
1	30	10	10
2	40	15	10
3	50	5	5

A soil chemist wants to prepare two chemical samples. The chemist wants the first sample to contain 380 grams of ammonium nitrate, 95 grams of phosphorus, and 85 grams of iron. The chemist wants the second sample to contain 380 grams of ammonium nitrate, 110 grams of phosphorus, and 90 grams of iron. Let's look at how we will determine how many grams of each additive (containing three active chemicals) are needed to create the samples?

Let's first concentrate on the creation of sample 1. Notice that we have three different additives (variables) that can be used to create this new additive and that each additive contains three parts (equations). We will let x, y, z represent the amounts of additives #1, #2, #3, respectively, that will be used to create sample 1.

So, the amount of sample 1 will equal x • (Additive #1) plus y • (Additive #2) plus z • (Additive #3). Having x amount of additive #1 means you will get x amount of each of its parts, likewise with additives #2 and #3. The following chart shows the make-up of sample 1.

The composition of sample 1			
	Ammonium Nitrate	Phosphorus	Iron
Additive #1	30x	10x	10x
Additive #2	40y	15y	10y
Additive #3	50z	5z	5z
Sample 1 needs	380	95	85

Now, the chemist knows that sample 1 needs to contain 380 grams of ammonium nitrate, 95 grams of phosphorus, and 85 grams of iron. So, we finally get to our equations.

The composition of sample 1:		
Ammonium Nitrate $30x + 40y + 50z = 380$	Phosphorus $10x + 15y + 5z = 95$	Iron $10x + 10y + 5z = 85$

We can now write our system of equations, translate them into an augmented matrix, and find the solution to this problem.

The system of equations for sample 1:	$\begin{cases} 30x+40y+50z=380 \\ 10x+15y+5z= \;\; 95 \\ 10x+10y+5z= \;\; 85 \end{cases}$
The augmented matrix:	$\left[\begin{array}{ccc\|c} 30 & 40 & 50 & 380 \\ 10 & 15 & 5 & 95 \\ 10 & 10 & 5 & 85 \end{array}\right]$

rref the matrix.	`[[30 40 50 380]`
	` [10 15 5  95 ]`
	` [10 10 5  85 ]]`
	`rref([A]`
	`    [[1 0 0 5]`
	`     [0 1 0 2]`
	`     [0 0 1 3]]`

The answer is:	Mix 5 parts Additive #1 with 2 parts Additive #2 and 3 parts Additive #3.

Question 1: Try to find the solution for making sample 2 in example 2.

Example 2: Chemical Content of Soil, Part II

What if the chemist from discussion 1 wanted to make a third sample composed of 380 grams of ammonium nitrate, 100 grams of phosphorus, and 110 grams of iron. Could the chemist do it? What would the proper mixture need to be to create it?

Solution:

First, let's write out the composition for this third sample.

The composition of sample 3			
	Ammonium Nitrate	Phosphorus	Iron
Additive #1	$30x$	$10x$	$10x$
Additive #2	$40y$	$15y$	$10y$
Additive #3	$50z$	$5z$	$5z$
Sample 3 needs	380	100	110

Second, our equations would turn out to be the following:

The composition of sample 3		
Ammonium Nitrate $30x + 40y + 50z = 380$	Phosphorus $10x + 15y + 5z = 100$	Iron $10x + 10y + 5z = 110$

Now, let's try solving with the help of our graphing calculator.

The system of equations for sample 3:	$\begin{cases} 30x + 40y + 50z = 380 \\ 10x + 15y + 5z = 100 \\ 10x + 10y + 5z = 110 \end{cases}$
The augmented matrix:	$\left[\begin{array}{ccc\|c} 30 & 40 & 50 & 380 \\ 10 & 15 & 5 & 100 \\ 10 & 10 & 5 & 110 \end{array}\right]$
rref the matrix.	[[30 40 50 380] [10 15 5 100] [10 10 5 110]] rref([A] [[1 0 0 12] [0 1 0 -2] [0 0 1 2]]

Question 2: What do the numbers in the final matrix tell us?

Notice that the calculator works very well at doing the mundane arithmetic calculations, but it can't think for us. It's a great tool to use but we need to understand the problem and the basics of mathematics in order to interpret what the calculator gives us. Let's do some more examples.

Example 3: Investing in Stock

An individual invests \$10,000 in three stocks that pay dividends of 6 %, 8 % and 10%. The amount invested at 10 % is twice the amount invested at 6 %, and the return on the three investments combined is \$860. What is the amount invested in each stock?

Solution:

Once again, here is a fairly standard problem found in math texts. Notice that we have three things we want to find (variables), and we are told three things (equations) about them ($10,000; one investment is twice another; and the total interest is $860). It looks like we need a system of equations again. Let *x, y, z* be the amounts invested in the 10%, 8%, and 6% earning stocks respectively.

First, we see a total investment of $10,000.	$x + y + z = 10000$
Next, we see that the amount invested at 10% is twice that invested at 6 %.	$x = 2z$
Lastly, we see that the total dividend (percents times amounts) earned is $860.	$0.1x + 0.08y + 0.06z = 860$
Now, write the system of equations in standard form.	$\begin{cases} x + y + z = 10000 \\ x + 0 - 2z = 0 \\ 0.1x + 0.08y + 0.06z = 860 \end{cases}$
Now, use the calculator to solve. Augmented matrix →	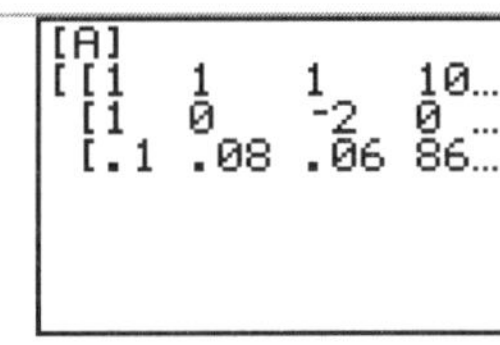
We can see from the matrix that we must have had $6,000 invested at 10 %, $1,000 invested at 8 %, and $3,000 invested at 6%.	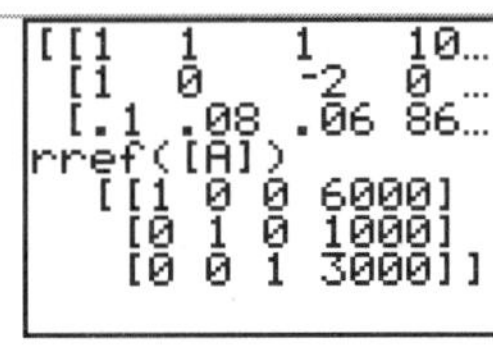

Answer Q1:
The system is

$$\begin{cases} 30x+40y+50z=380 \\ 10x+15y+5z=110 \\ 10x+10y+5z=90 \end{cases}$$

the answer is 4 parts Additive #1 with 4 parts Additive #2 and 2 parts Additive #3.

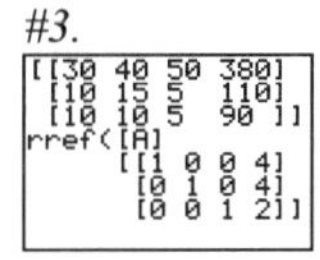

There are many areas where we can use systems and matrices to solve problems. Look at our next example.

Example 4: Finding a Polynomial Through Three Points

Find a quadratic polynomial that passes through these three points: (2, –15), (–1, 6), (–3, 0).

Solution:

In chapter 5 we saw that we needed two points to find a line since we have two unknowns in the most basic form of a line ($y = mx + b$). In chapter 3 we discussed how the standard form for a quadratic is $y = ax^2 + bx + c$. So, we see that we have three unknowns (a, b, c) in the standard form of a quadratic. Since we have three points, we have enough to find the equation that will go through all three of these points. Here is how we find the equation.

We plug each point into the standard form of the quadratic and get three different equations	$(2, -15) \rightarrow -15 = a(2)^2 + b(2) + c$ $(-1, 6) \rightarrow 6 = a(-1)^2 + b(-1) + c$

involving the three unknown coefficients.	$(-3, 0) \rightarrow \quad 0 = a(-3)^2 + b(-3) + c$
This gives us three equations in three unknowns.	$\begin{cases} 4a + 2b + c = -15 \\ 1a - 1b + c = \;\;6 \\ 9a - 3b + c = \;\;0 \end{cases}$
Now we can use one of the methods discussed in this chapter to solve this system.	[[4 2 1 -15] [1 -1 1 6] [9 -3 1 0]] rref([A] [[1 0 0 -2] [0 1 0 -5] [0 0 1 3]]
We see our answers for the coefficients *a*, *b*, *c*	$y = ax^2 + bx + c$ $\mathbf{y = -2x^2 - 5x + 3}$

Question 3: Find a quadratic polynomial that passes through the points (2, 0), (–4, 6), (–1, 3).

$(2, 0) \rightarrow 0 = a(2)^2 + b(2) + c$ $(-4, 6) \rightarrow 6 = a(-4)^2 + b(-4) + c$ $(-1, 3) \rightarrow 3 = a(-1)^2 + b(-1) + c$	$\begin{cases} 4a + 2b + c = 0 \\ 16a - 4b + c = 6 \\ 1a - 1b + c = 3 \end{cases}$	[[4 2 1 0] [16 -4 1 6] [1 -1 1 3]] rref([A] [[1 0 0 0] [0 1 0 -1] [0 0 1 2]]

Answer Q2: This matrix says that for this sample to be made we would need –2 parts of Additive #2, which means we can't make this sample.

The calculator does the arithmetic; we do the thinking! We found the solution even though we were mistaken about it being a quadratic. We will do one more example.

Example 5: Fast Food and Diet

Let's assume that we are on a diet that allows us to eat at a fast food restaurant on certain days. At this restaurant, we may choose between a hamburger, a chicken sandwich, and a fish sandwich. We need to fulfill 10% of our USRDA (U.S. recommended daily allowance) for vitamin A, and 10% for vitamin C, and 15% for iron, as we eat this meal. The following table gives us the nutritional information on each of the food items.

	Vitamin A (%)	Vitamin C (%)	Iron (%)
Chicken	8	15	8
Fish	2	0	10
Hamburger	15	6	20
Total % Needed	**10**	**10**	**15**

Find out how much of each item we will need to eat to fulfill our recommended amounts of these nutritional items.

Solution:

As in example 3, we can now create three equations and then solve the system using any of the methods we have learned.

<table>
<tr><td>We create the system from the information in the problem (x = amount of chicken, y = amount of fish, and z = amount of hamburger).</td><td colspan="2">$$\begin{cases} 8x+2y+15z=10 & \text{vitamin A} \\ 15x+0y+6z=10 & \text{vitamin C} \\ 8x+10y+20z=15 & \text{Iron} \end{cases}$$</td></tr>
<tr><td>We input the augmented matrix into our calculator and find the solution to our dietary question.</td><td>[A]
[[8 2 15 10]
[15 0 6 10]
[8 10 20 15]]
rref([A])</td><td>[[8 2 15 10]
[15 0 6 10]
[8 10 20 15]]
rref([A])
[[1 0 0 .537124...
[0 1 0 .422590...
[0 0 1 .323854...</td></tr>
<tr><td colspan="3">Rounding off, we see that we need to eat roughly 0.54 of the chicken sandwich, 0.42 of the fish sandwich and 0.32 of the hamburger. Bon Appétit!</td></tr>
</table>

We could go on forever showing you examples, from those that are fairly "real" to those that are very contrived. So, we will stop here and let you try your hand at them in the homework section.

Section Summary

- Identifying *what* the unknowns are (variables, things you want to find) and *how many* pieces of information you know about them is vital to starting word problems.
- The first part gives you the *number of variables* you will need for the problem and the second will give you the *number of equations* you have to work with.

SECTION 6.5 HOMEWORK

For each of the following problems:

a) Name all of the unknowns with a variable,

b) Write a system of equations that will solve the problem, and

c) Give the solution for the problem.

Answer Q3: Notice, in the final matrix we have c = 2, b = -1 and ***a = 0****! This really means we don't get a quadratic but a linear equation that goes through these points ($y = -x + 2$).*

1. An airplane travels 1200 miles in 2 hours going with the wind and 1120 miles in 2 hours going against the wind. What is the speed of the airplane and the speed of the wind?
2. A boat going against the current travels 24 miles in 3 hours and 60 miles in 5 hours going with the current. What is the speed of the boat and the speed of the current?
3. A steer needs 3,150 pounds of sorghum silage and 535 pounds of grain sorghum to go from 500 pounds to 750 pounds in 100 days. A heifer needs 2,825 pounds of sorghum silage and 485 pounds of grain sorghum to go from 500 to 675 pounds in 100 days. If, in the 100 days, the farmer uses 456,250 pounds of sorghum silage and 77,750 pounds of grain silage, how many steers and how many heifers are there?
4. From problem 3 above, if the total starting weight of all the cattle was 50,000 pounds and after the 100 days the total weight of all the cattle is 72,000 pounds, how many steers and how many heifers are there?
5. A chemist mixes an 8% sulfuric acid mixture with a 5% sulfuric acid mixture to form 30 liters of a 7% sulfuric acid mixture. How much of the 8% and 5% sulfuric acid mixture should be used?
6. An alloy containing 40% silver is melted with an alloy containing 20% silver to form 130 ounces of an alloy containing 35% silver. How much of the 40% and 20% silver alloy must be mixed together?
7. An alloy containing 20% gold is melted with an alloy containing 12% gold to form 30 ounces of an alloy containing 15% gold. How much of the 20% and 10% gold alloy must be mixed together?
8. A chemist mixes an 8% hydrochloric acid mixture with a 3% hydrochloric mixture to form 10 liters of a 5% hydrochloric acid mixture. How much of the 8% and 3% hydrochloric acid mixture should be used?
9. An inheritance of $200,000 is invested in three different simple interest investments at 5%, 7% and 8.5%. If the interest after one year is $12,800 and $2,000 more is invested at 5% than 7%, how much was invested in each investment?
10. A lottery winner of $3,800,000 invested the winnings in three different simple interest investments of 8%, 7.5% and 6.2%. If the interest after one year is $268,200 and the amount invested at 6.2% is twice the amount invested at 8%, how much was invested in each investment?
11. An IRA of $100,000 is reinvested in three accounts: a bond earning 6.4% per year, an IRA earning 5.8% per year, and a mutual fund at 7.3% per year. The amount invested in the mutual

fund is three times the amount invested in the IRA and the amount invested in the bond is twice as much as the amount invested in the mutual fund. How much interest will be earned after one year?

12. Three investments are made. The investments are for \$10,000, \$30,000 and \$40,000. The interest return after one year of simple interest is \$5900; the interest rate of the \$10,000 investment is 2% more that the interest rate of the \$40,000 investment and the interest rate of the \$30,000 investment is 1% less than the interest rate of the \$40,000 investment. What is the interest rate on each investment?
13. In the last election 191,200 people voted in a city that has 360,000 possible voters. If 48% of the Republicans voted in the election, 52% of the Democrats voted in the election, 65% of the Independents voted in the election and there are twice as many possible Republican voters as possible Independent voters, how many possible voters are Republican?
14. In the last national election 3,012,000 people voted in a state that has 5,800,000 possible voters. If 52.4% of the Republicans voted in the election, 49.5% of the Democrats voted in the election, and 64.5% of the Independents voted in the election and the total number of the Democrats that voted was 200,000 more that the total number of the Republicans and Independents that voted, how many Democrats voted?
15. A two–hour zoo tour costs \$7.50 for children ages 6–15, \$12.50 for adults and is free for children below 6 years of age. The receipts for all of the two–hour tours on Monday were \$1560 and 206 people took the tour. The number of children below 6 and the number of adults equaled the number of children between the age of 6 and 15. How many of each group took the tour?
16. A pro–basketball team has three season ticket prices: \$3,066 for the lower level seating, \$2,016 for mid–level, and \$1,050 for the upper level. A total of 11,000 season tickets are available for sale and if all the season tickets are sold, the revenue is \$23,436,000. The total of the lower level tickets and upper level tickets is 1,000 more than the total of the mid-level tickets. How many of each type of season ticket is available?
17. Airline tickets from Phoenix, Arizona, to Kansas City, Missouri have three different roundtrip prices: first class tickets sell for \$349.50, coach sells for \$248.50 and internet specials sell for \$195.50. The plane has 264 total seats available; the revenue if the plane is full is \$66,636 and coach has three times as many seats as first class. How many of each ticket is available?
18. The Amtrak train from Los Angeles to New York has several different ways to travel and several different prices. A deluxe bedroom costs \$1,466.00 for a one–way ticket, a standard bedroom costs \$1,291.00 for a one–way ticket and reserved coach costs \$606.00 for a one–way ticket. The total number of tickets available is 300. The number of reserved coaches available is equal to the total of the deluxe bedroom and standard bedrooms added together. If all of the above tickets are sold, the revenue would be \$295,925.00. How many of each type of accommodation is available?

19. 5,000 acres of farmland were planted with wheat, corn and soybeans; wheat yielded 54 bushels per acre, corn yielded 107 bushels per acre and soybeans yielded 35 bushels per acre with a total yield for all the grains of 374,000 bushels. Wheat sold for \$2.70 per bushel, corn sold for \$2.08 per bushel and soybeans sold for \$3.90 per bushel; the total revenue for all the grains sold was \$906,950.00. How many acres of each grain were planted?
20. Peanuts, cashews and almonds are mixed to form 350 pounds of mixed nuts that will be sold for \$6.10 a pound. Peanuts sell for \$3.50 per pound, cashews sell for \$9.90 per pound and almonds sell for \$7.50 per pound. There are twice as many peanuts as cashews in the mix. How many pounds of each type of nut are there in this mix?
21. The general quadratic equation is $y = ax^2 + bx + c$. Find the specific quadratic equation with the following three solutions: (2, –9), (–3, –49), and (1, –5). (Hint: each point gives you an x and y value and the unknowns you are looking for are a, b, and c.)
22. The general third degree equation is $y = ax^3 + bx^2 + cx + d$. Find the specific third degree equation with the following four solutions: (–2, –24), (1, 0), (2, 4), and (3, 16).
23. Using the general quadratic equation from 21, find the specific quadratic equation with the following three solutions: (–3, 5), (–1, 15), and (2, 30). What type of equation do you end up with? (Check with the common differences idea from chapter 3.)
24. Using the general third degree equation from 22, find the specific third degree equation with the following four solutions: (–3, 0), (1, 0), (3, 12), and (–4, 5). What type of equation do you end up with? (Check with the common differences idea from chapter 3.)
25. The following table gives the amount of time in hours it takes to manufacture sofas, chairs and love seats for one day.

	Finishing time	Covering time	Assembly time
One sofa	0.3 hours	0.5 hours	0.4 hours
One chair	0.2 hours	0.3 hours	0.2 hours
One love seat	0.2 hours	0.4 hours	0.2 hours
Totals	16 hours	27 hours	18 hours

How many of each were built in that day?

26. The following table gives the amount of time in hours it takes to manufacture dining room tables, dining room chairs, and buffets for a one-week period.

	Lathe time	Finishing time	Assembly time
One table	0.6 hours	0.4 hours	0.3 hours
One chair	0.4 hours	0.2 hours	0.1 hours
One buffet	0.8 hours	0.5 hours	0.4 hours
Totals	340 hours	185 hours	110 hours

How many tables, chairs and buffets were built that week?

27. The following table gives the cost and time in hours it takes to manufacture stuffed bears for one week.

	Cost	Stuffing time	Sewing time
One small bear	$1.00	0.1 hours	0.15 hours
One medium bear	$1.50	0.2 hours	0.3 hours
One large bear	$2.25	0.25 hours	0.4 hours
Totals	$1100	125 hours	195 hours

How many of each type of bear were manufactured that week?

28. The following table gives the manufacturing time in hours, cost and selling price of three different types of shirts manufactured for one month.

	Manufacturing time in hours	Cost	Selling price
One economy shirt	0.1 hours	$6.50	$12.95
One basic shirt	0.15 hours	$9.50	$15.95
One Fancy shirt	0.2 hours	$15.50	$32.95
Totals	6100 hours	$422,000.00	$812,000.00

How many of each type of shirt were produced that month?

29. The following table gives the manufacturing time in hours, cost and selling price of three different CD players for one month.

	Manufacturing time in hours	Cost	Selling price
Model 250	0.5 hours	$50.00	$85.95
Model 350	0.6 hours	$82.00	$120.95
Model 450	0.8 hours	$129.00	$195.95
Totals	2,640 hours	$358,200.00	$545,085.00

How many of each model were produced that month?

30. The following table gives the cost per unit, selling price per unit and labor cost per unit for three different laundry products for a three-month period.

	Cost of manufacturing	Selling price	Labor cost
One bottle of liquid softener	$2.83	$5.95	$0.95
One bottle of liquid detergent	$5.74	$12.95	$1.03
One bottle of liquid bleach	$1.57	$3.95	$0.83
Totals	$189,290.00	$425,350.00	$50,150.00

How many of each product were produced for that three months?

31. The following table represents the ideal nutritional diet.

	Children ages 2 to 6 years, women, some older adults	Older children, teen girls, active women, most men	Teen boys, active men
Bread, Cereal, Rice and pasta Group	6 servings	9 servings	11servings
Vegetable Group	3 servings	4 servings	5 servings
Fruit Group	2 servings	3 servings	4 servings
Milk, Yogurt, and Cheese Preferably fat free or low fat (Milk Group)	2 servings	2 servings	2 servings
Meat, Poultry, Fish, Dry Beans, Eggs and Nuts (Meat and Beans Group)	5 ounces	6 ounces	7 ounces
Calories	1,600 calories	2,200 calories	3,000 calories

An ideal meal was served for a group that had 186 servings of the Bread Group, 85 servings of the Vegetable Group and 64 servings of the Fruit Group. How many of each type of person in the chart above were served?

32. Use the table from 31 above.

    An ideal meal was served for a group that had 48 servings from the Milk Group, 148 ounces from the Mean and Beans Group and 56,800 calories. How many of each type of person in the chart above were served?

33. A hardware store sells four types of 6-foot ladders. Each extra heavy duty ladder sells for \$119.00, each heavy duty ladder sells for \$108.00, each standard ladder sells for \$74.00 and each economy ladder sells for \$60.00. Last year the hardware store sold 360 ladders and the revenue for these ladders was \$28,920.00. There were twice as many standard ladders sold as extra heavy duty ladders and twice as many economy ladders sold as heavy duty ladders. How many of each type of ladder were sold?

34. A concert has four different ticket prices. The floor tickets sell for \$90.00 each, the lower rows sell for \$75.00 each, the middle rows sell for \$65.00 and the upper rows sell for \$50.00 each. The total number of seats available is 20,900 and if the concert is a sell–out the total revenue is \$1,468,500.00. There are twice as many \$65.00 seats than there are \$50.00 seats, and the sum of the \$50.00 seats and \$75.00 seats is 3,000 more than the \$65.00 seats. How many of each type of seat are there?

35. The following table gives the cost of production and selling price of four different models of 35" television sets.

	T–750	T–700	T–650	T–600
Cost of production per television set	\$385.90	\$325.70	\$295.80	\$263.40
Selling price of each television set	\$502.85	\$433.50	\$402.50	\$352.80

A total of 3500 television sets were sold in a three-month period. The total cost of production was \$1,113,105.00 and the total revenue was \$1,483, 690.00 for the three month period. The total number of T–750 and T–700 television sets produced for the three month period was equal to the total number of T–650 and T–600 television sets produced during that same three month period. How many of each television set were produced and sold during this three month period?

36. The following table gives the cost and selling price for each of four different models of basketball shoe that was sold by a shoe store.

	Rim Rattler	Dunken Master	All World	Sky High
Cost per pair of shoes	\$68.50	\$59.25	\$75.25	\$83.45
Selling price per pair of shoes	\$92.70	\$89.35	\$106.50	\$115.90

For the year the store sold 970 pairs of these basketball shoes; their total cost for the basketball shoes was \$68,145.00 and their total revenue for the basketball shoes was \$96,590.00. There were 30 more pairs of Dunken Masters and Sky High basketball shoes sold then there were pairs of Rim Rattler and All World basketball shoes sold. How many of each model of basketball shoes were sold?

Chapter 6 Review

Topic	Section	
Matrices and matrix operations	6.1	• Matrices are an array of numbers or, stated another way, just a rectangle full of numbers. • In order to add or subtract matrices, the matrices must be of the same dimension and you perform the operations by simply adding or subtracting the numbers in the same positions. • Scalar multiplication is just distributing the number throughout the matrix. • With multiplication, you multiply rows times columns to perform the operation ($A_{m \times n} \bullet B_{n \times r} = C_{m \times r}$).
Matrix properties	6.2	• The additive identities for matrices are matrices made up of 0's. • The additive inverses for matrices are matrices made up of opposite signed elements. • The multiplicative identities for square matrices are matrices of the same dimension with 1's down the diagonal and 0's elsewhere. • Only square matrices have a determinant. • Only square matrices can have a multiplicative inverse. • Only square matrices with determinants equal to 0 have inverses. • To find the inverse of a square matrix (with a determinant not equal to 0, type it into your calculator and use the $\boxed{x^{-1}}$ key. • To find the determinant of a matrix, type it into your calculator and use the **det** function of the calculator.
Cramer's Rule and Inverse Matrix methods	6.3	• A system of linear equations can be written as a matrix equation and then $A^{-1}B = X$ is the solution, if the inverse matrix exists. • A system of linear equations can be solved using determinants if the determinant of the coefficient matrix (det A) is not equal to 0 ($x = \frac{\det A_x}{\det A}$, $y = \frac{\det A_y}{\det A}$, $z = \frac{\det A_z}{\det A}$). • If the determinant of the coefficient matrix (det A) is equal to 0, then the solution to the system of equations can't be a single point but must be either "no solution" or have an infinite number of solutions.

Augmented matrix and ways to use it	6.4	• The augmented matrix is constructed by simply arranging all of the coefficients and constants in a single matrix. • You can perform row operations on augmented matrices in order to solve system of equations. The most used row operation on the TI-83 graphing calculator is *row+(**multiplier, matrix name, row being multiplied, row being added to**) and mRAdd(**multiplier, matrix name, row being multiplied, row being added to**) on the TI-86. • Other frequently used operations are ref and rref. • The Gaussian Elimination method can be quite lengthy if done by *hand.* It's a little better on the calculator, doing each row operation, because the calculator doesn't make simple arithmetic errors. It is important to learn about the row operations because they are necessary in certain areas in mathematics like Linear Programming, a finite math course topic for most business majors.
Solving "real-life" problems	6.5	• Identifying *what* the unknowns (variables, things you want to find) are and identifying *how many* pieces of information you know about them is vital to starting word problems. • The first part (unknowns) gives you the *number of variables* you will need for the problem and the second part (information about the variables) will give you the *number of equations* you have to work with.

CHAPTER 6 REVIEW HOMEWORK

Section 6.1

$$A=\begin{bmatrix}1 & -2 & 3\\ 2 & -3 & 2\\ 1 & -1 & 1\end{bmatrix} \quad B=\begin{bmatrix}0 & 3 & -1\\ 2 & 3 & 1\\ 4 & 2 & 1\end{bmatrix} \quad C=\begin{bmatrix}1 & -2\\ 3 & 2\\ 1 & 4\end{bmatrix} \quad D=\begin{bmatrix}2 & -3\\ 3 & 4\\ -1 & 5\end{bmatrix}$$

(1–15) Using the matrices above, find each of the following:

1. Dimension of A
2. Dimension of C
3. $b_{23}=$
4. $d_{33}=$
5. $A+B$
6. $B+D$
7. $5D$
8. $3A+2B$
9. $2C-3D$
10. B^2
11. C^2
12. AB
13. BC
14. CB
15. Find the values for x and y that make $A=B$, $A=\begin{bmatrix}2x+3 & 3y-5\\ 5 & -2\end{bmatrix}$ $B=\begin{bmatrix}9 & -14\\ 5 & -2\end{bmatrix}$.
16. A furniture manufacture has two different plants: one in Sacramento, California and the other in Dallas, Texas. They manufacture easy chairs, sofas and love seats. Table A represents the number of easy chairs, sofas and love seats produced during the months of May, June and July in the Sacramento plant. Table B represents the number of easy chairs, sofas and love seats produced during these same months in the Dallas plant.

A =

	Easy chairs	Sofas	Love Seats
May	533	489	481
June	628	595	574
July	493	432	359

B =

	Easy Chairs	Sofas	Love Seats
May	653	602	586
June	789	757	709
July	587	519	495

    a. Create matrices A and B from the above table of information.
    b. Find $A+B$.
    c. How many love seats were produced in June?
    d. How many sofas were produced in July?
    e. How many easy chairs were produced in May?

17. An electronics company sells three models of CD players out of three different stores. Table A represents the number of each model of CDs sold at each store. Table P represents the selling price of each CD at each store.

A =

	Model 350–CD	Model 300–CD	Model 250–CD
Store A	323	503	438
Store B	259	439	357
Store C	301	485	399

P =

	Selling price
Model 350–CD	\$213.00
Model 300–CD	\$185.00
Model 250–CD	\$163.00

a. Create a matrix A for the number of CDs sold at each store.

b. Create a matrix P for the selling price of each model of CD.

c. Find AP.

d. How much revenue did store B receive?

e. How much revenue did store C receive?

f. How much revenue did store A receive?

Section 6.2

$$A=\begin{bmatrix}1 & -2\\ 3 & 1\\ 2 & 5\end{bmatrix} \qquad B=\begin{bmatrix}3 & -2\\ 5 & -4\end{bmatrix} \qquad C=\begin{bmatrix}1 & -2 & 3\\ -2 & 1 & 0\\ 3 & -1 & 2\end{bmatrix}$$

(18–27) Using the matrices above, find each of the following:

18. Find the additive inverse of A.
19. Find the additive inverse of B.
20. Find the additive identity of A.
21. Find the additive identity of C.
22. Find the multiplicative inverse of B.
23. Find the multiplicative inverse of C.
24. Find the multiplicative inverse of A.
25. Find the determinant of B.
26. Find the determinant of A.
27. Find the determinant of C.
28. Find the multiplicative inverse of B by hand and check with your answer from 22.

Section 6.3

(29–30) For each system of equations give:

a. Coefficient matrix A

b. Variable matrix X

c. Constant matrix B

d. Write a matrix equation corresponding to the systems of equations

29. $3x+2y-4z=8$
    $2x-y+3z=9$
    $x+y-z=-12$

30. $2x-z=3$
    $x+2y=5-3z$
    $3y=2z+8$

(31–32) Translate each matrix equation to a corresponding system of equations:

31. $\begin{bmatrix}-2 & 3\\ 1 & 5\end{bmatrix}\begin{bmatrix}x\\ y\end{bmatrix}=\begin{bmatrix}-3\\ 5\end{bmatrix}$

32. $\begin{bmatrix}1 & -3 & 1\\ 3 & -1 & 2\\ -2 & 0 & 3\end{bmatrix}\begin{bmatrix}a\\ b\\ c\end{bmatrix}=\begin{bmatrix}2\\ -1\\ 5\end{bmatrix}$

(33–39) Solve each system of equations by both the inverse method and Cramer's rule. (Compare the two answers.)

33. $3x - 2y + 2z = -10$

$5x + y - 3z = -10$

$-2x + 2y + z = 11$

34. $\frac{3}{2}x - \frac{4}{3}y - 2z = \frac{23}{6}$

$\frac{1}{3}x + \frac{3}{2}y + \frac{5}{6}z = \frac{-5}{2}$

$3x - \frac{5}{6}y + \frac{2}{3}z = \frac{43}{3}$

35. $0.38x - 2.7y + 0.5z = -13.74$
$-0.5x + 0.74y - 3.2z = 9.1$
$3.1x + 0.7y - 0.28z = 10.26$

36. $3x + y = 7$
$2y = -7 - 3z$
$5z = 3x - 14$

37. $2x + y - w + z = -4$
$3x - 2y + w - 2z = 13$
$2x + 3y - 2w + 3z = -14$
$x - y + 3w - 2z = 12$

38. $x + 3y + 9z = -7$
$-2x - 5y - 16z = 11$
$x + y + 5z = -1$

39. $x + 2y + 2z = 13$
$2x + 5y + 6z = 31$
$-3x - 8y - 10z = -48$

Section 6.4

(40–43) Given the following systems of equations, write an augmented matrix for each:

40. $3x + 5y = -9$
$2x - 3y = 8$

41. $2x - 3y + z = 7$
$3x + 2y - 3z = 12$
$x + 3y - 3z = 12$

42. $x + y - z + 2w = 9$
$2x + 3y + 2z - w = 15$
$3x - y - 2z - 3w = -12$
$-2x + y - 3z + w = 3$

43. $x + 2y = 5 - z$
$2x = 3z + 8$
$5y - 7 = 4z$

(44–47) Given the augmented matrices, write the system of equations each one represents for the indicated variables:

44. $\left[\begin{array}{cc|c} 3 & 2 & -5 \\ -3 & 4 & 7 \end{array}\right]$ Variables x, y

45. $\left[\begin{array}{ccc|c} 1 & -3 & 5 & 3 \\ -2 & 2 & -4 & -7 \\ 3 & 1 & -2 & 10 \end{array}\right]$ Variables x, y, and z

46. $\left[\begin{array}{cccc|c} 2 & -3 & 1 & 3 & 10 \\ -1 & 2 & -3 & 1 & -4 \\ 3 & 5 & -2 & 4 & 15 \\ 1 & -1 & 4 & -2 & 9 \end{array}\right]$ Variables x, y, z, and w

47. $\left[\begin{array}{ccc|c} 1 & 0 & 3 & 2 \\ 0 & 1 & 3 & -5 \\ 2 & 0 & 3 & 7 \end{array}\right]$ Variables r, s, and t

(48–49) Row reduce the matrix by hand and solve for the indicated variables:

48. $\left[\begin{array}{cc|c} 2 & -3 & 4 \\ 3 & -1 & -1 \end{array}\right]$ Variables x, y

49. $\left[\begin{array}{ccc|c} 1 & -3 & 2 & 1 \\ 0 & 2 & 4 & -10 \\ 0 & -2 & 3 & -4 \end{array}\right]$ Variables r, s and t

(50–55) Given the following row reduced matrices, solve for the indicated variables:

50. $\left[\begin{array}{cc|c} 1 & 0 & -2 \\ 0 & 1 & 3 \end{array}\right]$ Variables a, b

51. $\left[\begin{array}{ccc|c} 1 & 0 & 0 & 3 \\ 0 & 1 & 0 & -5 \\ 0 & 0 & 1 & 4 \end{array}\right]$ Variables x, y, and z

52. $\left[\begin{array}{ccc|c} 1 & 0 & 1 & 2 \\ 0 & 1 & -2 & 3 \\ 0 & 0 & 1 & 4 \end{array}\right]$ Variables *r*, *s* and *t*

53. $\left[\begin{array}{ccc|c} 1 & -2 & 3 & 1 \\ 0 & 1 & 3 & 5 \\ 0 & 0 & 1 & -2 \end{array}\right]$ Variables x, y, and z

54. $\left[\begin{array}{ccc|c} 1 & 0 & 2 & 3 \\ 0 & 1 & -3 & 4 \\ 0 & 0 & 0 & 0 \end{array}\right]$ Variables *a*, *b*, and *c*

55. $\left[\begin{array}{ccc|c} 1 & 0 & -3 & 2 \\ 0 & 1 & 2 & -3 \\ 0 & 0 & 0 & 1 \end{array}\right]$ Variables *m*, *n*, and *p*

(56–65) Given the following systems of equations, solve each with the graphing calculator:

(Use either rref or the row operation functions of the calculator.)

56. $2x - 5y = -31$

$7x + 9y = 24$

57. $0.35x - 3.82y = 79.9$

$2.3x + 0.95y = 4$

58. $2x + 7y - 3z = -57$

$x - 8y + z = 49$

$3x + 5y - z = -30$

59. $\frac{3}{4}x + \frac{2}{5}y - \frac{1}{10}z = \frac{3}{5}$

$2x - 5y + z = -29$

$\frac{2}{3}x - \frac{3}{4}y + \frac{5}{6}z = \frac{-17}{2}$

60. $2x - 3y + z - 2w = 18$

$x + 2y - 3z + 2w = -17$

$-3x + y + 5z - w = 12$

$3x - y + 2z + 3w = 1$

61. $a = 2b + 5$

$5b = 11 - 7c$

$3c = 5a + 4$

62. $-3x - 2y - 2z = -1$

$2x + y + 2z = -1$

$9x + 3y + 12z = -11$

63. $3x - 2y - 12z = 19$

$-x + y + 5z = -7$

$6x - 3y - 21z = 36$

64. Solve Problem 38. What is the difference between this solution and what happened when you did 38 the other two ways?

65. Solve Problem 39. What is the difference between this solution and what happened when you did 39 the other two ways?

Section 6.5

66. A total of \$250,000.00 is invested in into two simple interest accounts, one at 8.5% and the other at 6.9%. The interest for one year for both accounts was \$19,410.00. How much was invested in each account?

67. A high school play charges \$5.00 a ticket for adults and \$2.50 a ticket for high school students with I.D. The total revenue for one night of the play was \$4,275.00 and twice as many adults as students attended the play. How many students and adults attended the play?

68. An electronics store sells two different models of receivers, Model 550R and Model 400R. The following table gives the store's cost and revenue for each receiver and the total of all the store's costs and revenues for the sale of all units in June.

	Cost per unit	Selling price per unit
Model 550R	$323.89	$582.99
Model 400R	$223.59	$355.99
Total	$11,789.64	$19,780.54

How many of each model were sold in June?

69. A chemist mixes three solutions of boric acid: a 12% solution, a 15% solution and a 34% solution to obtain 40 liters of a 19% boric acid solution. There is twice as much 15% solution as 34% solution. How many liters of each solution should the chemist use?

70. The Museum of Natural History in Denver, Colorado, has three admission prices: $6.00 for adults, $4.00 for children 4–12 years old and $4.00 for senior citizens over 65. One Sunday afternoon the Museum's receipts were $5,670 for 1,233 admissions and there was 236 more children (4-12) than senior citizens. How many seniors purchased admission that Sunday?

71. A company that makes clothes makes three different types of blue jeans. The following table gives the cutting time in hours, sewing time in hours and production cost for each pair of blue jeans. The table also gives the total time spent cutting, total time spent sewing and total cost of production for one day in December.

	Cutting time per pair	Sewing time per pair	Production cost per pair
Westerner	0.5 hours	0.3 hours	$15.89
Boot Leg	0.4 hours	0.25 hours	$13.23
Straight Leg	0.35 hours	0.2 hours	$10.17
Totals for one day	101 hours	60.5 hours	$3,161.10

How many of each type of blue jean were produced for that one day in December?

CHAPTER 6 EXAM

(1–20) Use the following matrices to answer the questions below:

$A=\begin{bmatrix}-2 & 3\\ -1 & 4\end{bmatrix}$ $B=\begin{bmatrix}3 & -1\\ 2 & -2\end{bmatrix}$ $C=\begin{bmatrix}1 & -2 & 3\\ 0 & 2 & -1\\ 2 & -3 & 4\end{bmatrix}$

$D=\begin{bmatrix}2 & 3 & -1\\ 4 & -3 & 2\\ -2 & 5 & 0\end{bmatrix}$ $E=\begin{bmatrix}-2 & -3\\ 1 & 4\\ -5 & 3\end{bmatrix}$ $F=\begin{bmatrix}5 & -4\\ 7 & -9\\ 2 & -8\end{bmatrix}$

1. $A+B=$
2. $E+F=$
3. $C+F=$
4. $3B=$
5. $-4C=$
6. $3C-4D=$
7. $AB=$
8. $CD=$
9. $EF=$
10. $CE=$
11. $c_{23}=$
12. $f_{33}=$
13. Additive identity of $E=$
14. Multiplicative identity of $C=$
15. Additive inverse of $E=$
16. Multiplicative inverse of $D=$
17. Determinant of $A=$
18. Determinant of $C=$
19. Determinant of $F=$
20. Multiplicative inverse of $E=$

(21–23) Using the following system of equations answer the questions below:

$$\begin{aligned}3x-2y+z&=16\\ 5x+3y-3z&=-11\\ -2x-4y+5z&=28\end{aligned}$$

21. a. Give the coefficient matrix, variable matrix and constant matrix.
    b. Write a matrix equation corresponding to the system of equations.
    b. Find the solution of the system of equations by using the inverse method.
22. a. Give the determinant matrices you would use to solve by Cramer's Rule.
    b. Find the solution of the system of equations by using Cramer's Rule.
    c. Find the solution of the system of equations by using the inverse method.
23. a. Give the augmented matrix.
    b. Find the solution of the system of equations by using rref or row reduction using your calculator.

(24–28) Solve the following system of equations by any method:

24. $$\begin{aligned}0.3x-2.3y+0.4z&=6.1\\ 0.12x+.05y-0.25z&=-.73\\ 2.3x+3.5y+1.4z&=\end{aligned}$$

25. $$\begin{aligned}5x+2y+7z&=16\\ 2x+y+3z&=7\\ -0.5x+2y+3z&=8\end{aligned}$$

26. $$\begin{aligned}\frac{1}{2}x-\frac{2}{3}y+\frac{5}{6}z&=\frac{-5}{6}\\ \frac{3}{5}x+\frac{3}{10}y+\frac{1}{5}z&=\frac{18}{5}\\ \frac{2}{3}x+\frac{5}{6}y-\frac{7}{9}z&=8\end{aligned}$$

27. $$\begin{aligned}7x+2y+9z&=17\\ 3x+y+4z&=7\\ 4x+2y+6z&=11\end{aligned}$$

28. $x + y - 2z + w = 2$
$2x - y + 3z - 3w = 5$
$3x + 2y + 2z - 5w = 14$
$-3x + 3y - 5z - w = 10$

29. The state fair has two admission prices: \$5.00 for general admission and \$13.00 for an admission that includes unlimited rides. A total of 10,258 tickets were sold with a total revenue of \$117,346.00. How many of each type of ticket were sold?

30. Flying with the wind, a plane flew 800 miles in 3 hours. Flying against the wind, it took the plane 5 hours to travel the 800 miles. Find the rate of the plane in calm air and the rate of the wind.

31. A certain fast food restaurant serves special combination meals. A combination meal consisting of a deluxe hamburger, small fries and small drink costs \$2.99. If you upgrade this meal with large fries and large drink, it costs \$.49 extra and if you upgrade the original meal with extra large fries and extra large drink, it costs \$.89 extra. During one Monday the restaurant sold 833 of these meals and made a total revenue of \$2,745.65. The restaurant sold 174 more of the regular combination meal than it did of the \$.49 upgrade. How many of each type of combination meal did the restaurant sell on that Monday?

32. The following table gives the manufacturing time in hours, cost and selling price of three different models of cell phones for one month.

	Manufacturing time in hours	Cost	Selling price
Model C–100	0.5 hours	\$23.95	\$45.80
Model C–80	0.45 hours	\$20.50	\$40.45
Model C–60	0.4 hours	\$16.75	\$32.65
Totals for the month	899.6 hours	\$40,503.60	\$78,754.75

How many of each model of all phones were manufactured for the month?

Chapter 7 Conics

We have spent the majority of this book looking at functions. We've studied polynomial functions, exponential functions, and logarithmic functions. We have also looked at how we solve all kinds of problems involving these three basic families of functions. But we are now going to turn our attention to a group of mathematical formulas called **conic sections**.

This group contains one function (a quadratic polynomial) and three quadratic relations (non-functions). Conics have applications in the "real" world, so we study them even though most of them aren't functions. Graphically, conic sections are the curves you get when slicing your typical sugar ice cream cone in different ways.

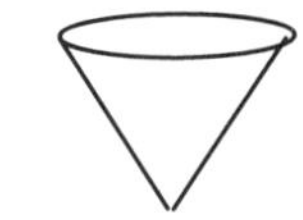

If you put two cones tip to tip, then slice through them at different angles, you create the graphs of the four different types of conics.

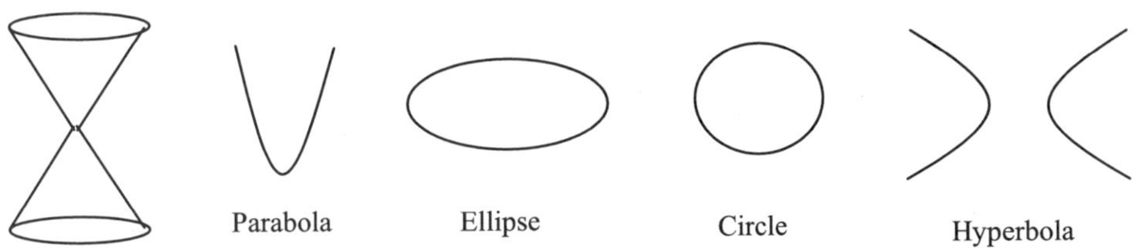

Section 7.1 Parabolas

Before we go any further with our discussion about conics, we need to take some time to look at the distance formula (between two points). All of the definitions of the different conic sections involve the idea of distance. The distance formula for the distance between two points comes from the **Pythagorean Theorem**:

> The square of the hypotenuse of a right triangle is equal to the sum of the squares of the other two legs: $c^2 = a^2 + b^2$

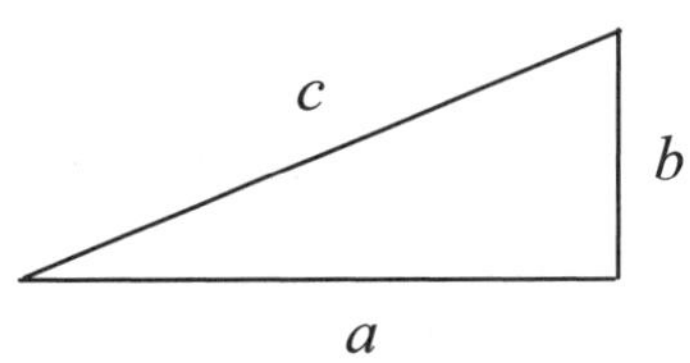

The distance between two points is related to this because you can form a right triangle with the two points and use the Pythagorean Theorem to determine the distance between the two points in the xy-plane.

Point 1: $P_1\ (x_1, y_1)$ Point 2: $P_2\ (x_2, y_2)$

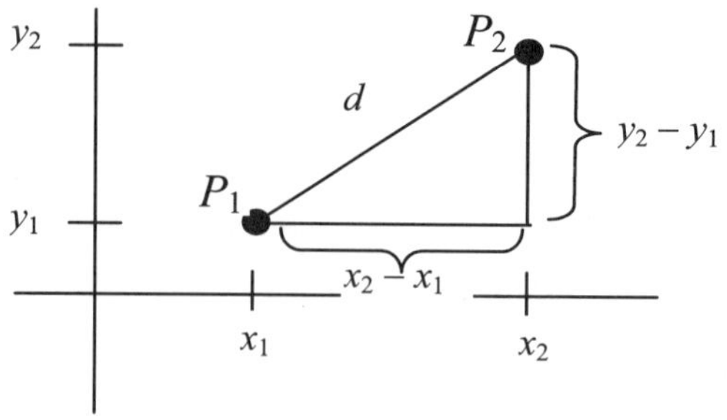

So using the Pythagorean theorem ($c^2 = a^2 + b^2$), we get $d^2 = (x_2 - x_1)^2 + (y_2 - y_1)^2$. Now by taking the square root of each side, we come up with the formula for finding the distance between two points.

> **Distance Formula:** The distance between two points $P_1(x_1, y_1)$ and $P_2(x_2, y_2)$ in the xy-plane is
> $$d = \sqrt{(x_2 - x_1)^2 + (y_2 - y_1)^2}$$

Example 1: Distance Between Two Points

Find the distance between $P_1(-3, 5)$ and $P_2(2, -1)$.

Solution:

Plug our numbers into the formula.	$d = \sqrt{(2-(-3))^2 + ((-1)-5)^2}$
Now simplify the expression.	$d = \sqrt{5^2 + (-6)^2}$ $= \sqrt{25+36}$ $= \sqrt{61}$

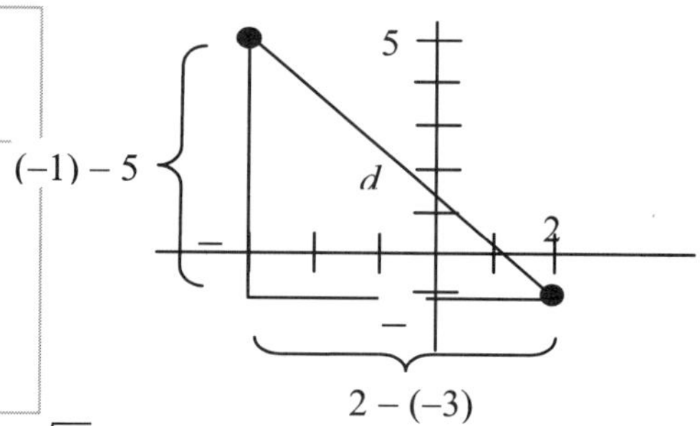

The distance between these two points on the xy-plane is $\sqrt{61}$.

PARABOLAS

Let's begin with the first of the four conic sections, the parabola. We have already seen the parabola (a quadratic polynomial) in chapters 1, 2, 3, and 5. Now, we will focus on its definition and properties as a useful conic section. Here is the definition of a parabola.

> **Parabola**: A parabola is a set of points that are equidistant from a fixed point (called a focus) and a line (called a directrix), all of which are in the same plane.

We will now use this definition to derive the formula for a parabola where the vertex is at the origin; the focus is at the point $(0, p)$, and the directrix is the line $y = -p$.

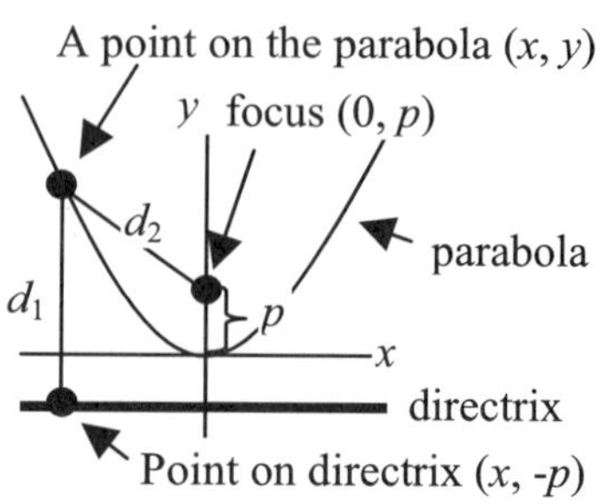

The point $(0, p)$ is the fixed point and the line $(y = -p)$ is the directrix. The p represents the distance from the vertex to the focus. According to the definition, the distance d_1 should equal the distance d_2. So we will use the distance formula for each distance, set them equal to each other, and then solve the equation for y.

The two distances are	$d_1 = \sqrt{(x-x)^2 + (y-(-p))^2}$ $= \sqrt{0 + (y+p)^2}$ $d_2 = \sqrt{(x-0)^2 + (y-p)^2}$ $= \sqrt{x^2 + (y-p)^2}$
Set the distances equal to each other. Square both sides to eliminate the $\sqrt{\ }$. Clear parentheses. Subtract y^2, p^2 and add $2py$ to both sides. Solving for y now gives us a formula for any parabola of this type.	$d_1 = d_2$ $\sqrt{0 + (y+p)^2} = \sqrt{x^2 + (y-p)^2}$ $(y+p)^2 = x^2 + (y-p)^2$ $y^2 + 2py + p^2 = x^2 + y^2 - 2py + p^2$ $4py = x^2$ $y = \frac{1}{4p}x^2$

For simplicity sake, let's replace $\frac{1}{4p}$ with the letter a. So a parabola of this type will be said to be in **standard position** and have the form $y = ax^2$ (where $a = \frac{1}{4p}$).

We might mention here that every parabola has what is called an axis of symmetry.

> The **axis of symmetry** is the imaginary line that can be drawn on a graph, where the part of the curve on one side of the line is the mirror image of the part of the curve on the other side of the line.

Another view of the concept of axis of symmetry is that if we fold the graph paper along this imaginary line, one part of the graph will fall on top of the other part. For a parabola in standard position, the axis of symmetry will be the line $x = 0$ (the y-axis). It divides the parabola into two equal parts.

There is a practical reason why we care about where the focus and the axis of symmetry are with respect to the parabola. A physical property of the parabola is that anything coming into a parabola, parallel to the axis of symmetry, will be reflected off of the parabola back towards the focus. We call this the reflective property of the parabola. This means that if light is coming into a parabola (or parabolic shape), parallel to the axis of symmetry, it will all be gathered at that one point, the focus. A good example of this happens in telescopes. Here is a cross section of one type of telescope.

In this type of telescope, a camera is positioned at the focus in order to take advantage of this property.

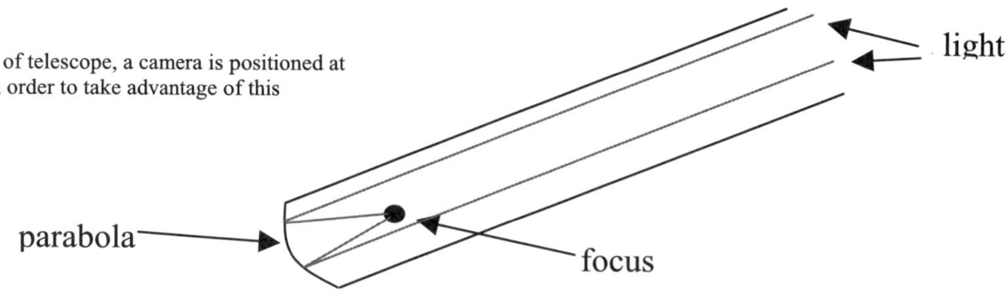

The other way to use this reflective property is to put a light source at the focus so that when the light rays hit the parabola, they will all be reflected out in a beam of light (e.g. a spot light). As you can see, it is critical to know exactly where the focus is when working with practical applications. We'll work more with these applications in the exercises. Let's graph a few parabolas in standard position.

Example 2: Parabolas in Standard Position

Sketch the following parabolas along with the focus and directrix.

a.) $y = 2x^2$ b.) $y = \frac{1}{12}x^2$ c.) $y = \frac{-1}{8}x^2$ d.) $y = -x^2$

Solutions:

$y = ax^2$ is the form of a parabola in standard position and $a = \frac{1}{4p}$.

a.) $y = 2x^2$	b.) $y = \frac{1}{12}x^2$	c.) $y = \frac{-1}{8}x^2$	d.) $y = -x^2$
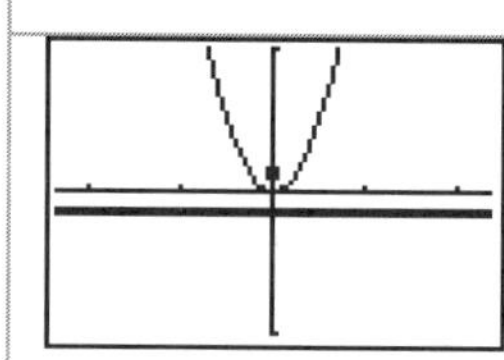	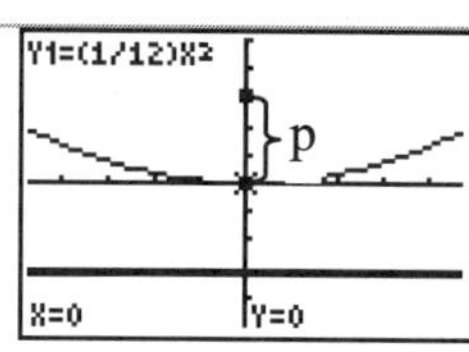	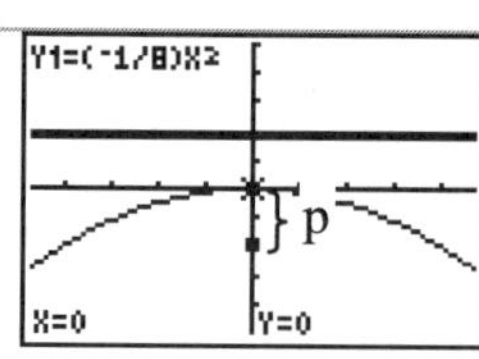	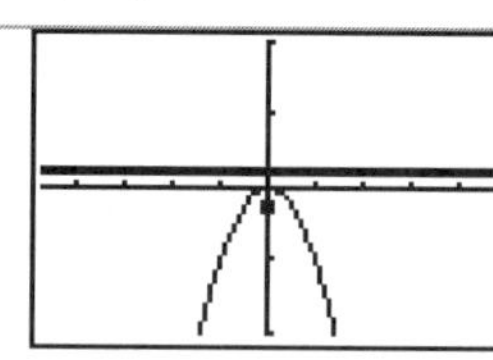
$2 = a$, therefore $2 = \frac{1}{4p}$, and so $8p = 1$ and thus $p = \frac{1}{8}$ (distance to	$\frac{1}{12} = a$; therefore $\frac{1}{12} = \frac{1}{4p}$, and so $4p = 12$ and thus $p = 3$	$\frac{-1}{8} = a$; therefore $\frac{-1}{8} = \frac{1}{4p}$, and so $4p = -8$ and thus $p = -2$	$-1 = a$; therefore $-1 = \frac{1}{4p}$, and so $4p = -1$ and thus $p = \frac{-1}{4}$

focus from vertex).	(distance to focus).	(distance to focus).	(distance to focus).
focus (0, 1/8),	focus (0, 3),	focus (0, -2),	focus (0, -1/4),
directrix y = -1/8	directrix y = -3	directrix y = 2	directrix y = 1/4

Question 1: What is the axis of symmetry for all of the parabolas in example 2?

There are many things to notice from these four examples.

- First, if a is positive, go up from the vertex to get to the focus and down to get to the directrix (parabola goes up).
- Second, on the other hand, if a is negative you do the opposite: go down from the vertex to the focus and up to the directrix (parabola goes down).
- Third, if the magnitude of a is greater than 1, then the focus is close to the vertex and the parabola is narrow.
- Fourth, on the other hand, if the magnitude of a is less than 1, then the focus is farther from the vertex and the parabola is wider.

Question 2: Sketch a picture of the parabola $y = \frac{1}{4}x^2$ along with its focus and directrix. Also state the domain and range for this function.

Now, parabolas don't only go up or down. They can go sideways as well. If we put the vertex at the origin, but this time we put the focus at $(p, 0)$ (on the x-axis) and the directrix at $x = -p$, then our parabola looks like the following:

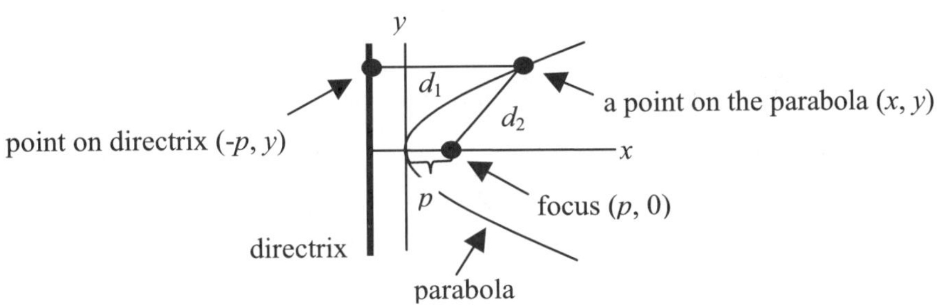

With everything just turned 90 degrees from the standard position parabola, the formula for this parabola will be the same as before, but x and y change places ($x = ay^2$, where $a = \frac{1}{4p}$). By the way, this is the inverse relation to the parabola $y = ax^2$.

Example 3: Sideways Parabolas

Graph the following parabolas along with the focus and directrix.

a.) $x = \frac{1}{2}y^2$ b.) $x = -4y^2$

First, before we do example 3, try answering the following question.

Question 3: Are these two relations (*a* and *b* in example 3) such that *y* is a function of *x*? (Note: from chapter 1, a function is just a special relation where for each input there is only one output.)

Solution:

So, here are the graphs of (a) and (b) for example 3:

a.) $x = ½y^2$	b.) $x = -4y^2$
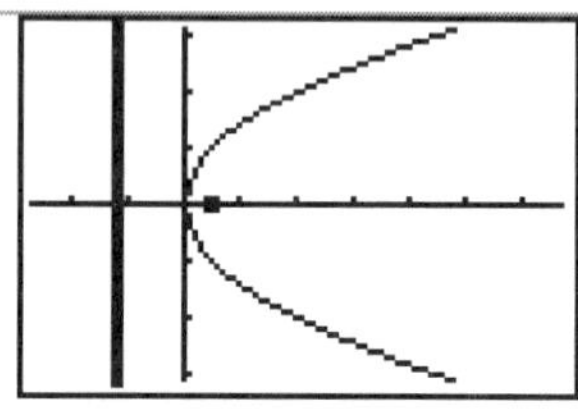	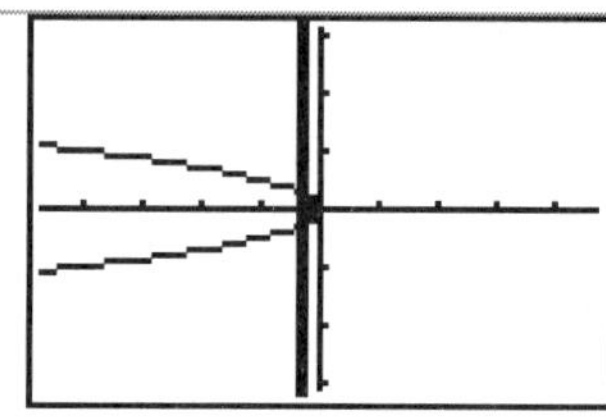
$x = ay^2$ is the form of a parabola so, $1/2 = a$ in this example. But $a = \frac{1}{4p}$; therefore $\frac{1}{2} = \frac{1}{4p}$ $4p = 2$ $p = \frac{1}{2}$ (distance to focus from vertex).	Here $-4 = a$; therefore $-4 = \frac{1}{4p}$ $-16p = 1$ $p = \frac{-1}{16}$ (distance to focus). Very hard to see on a graph.

Of course, all parabolas don't have their vertex at the origin. So we need to talk about how parabolas get translated, reflected and stretched.

There are several ways in which a parabola in standard position can be translated (shifted) or altered. It can be moved up or down from the origin (shifted vertically). It can be moved left or right from the origin (shifted horizontally). It can be flipped upside down or rotated (reflected). It can also be made to be narrower or wider (stretched or compressed).

Everything that we are about to learn can be applied to any function and most relations fairly easily. We choose here to emphasize how the parabola in the standard position ($y = ax^2$) is affected in each of these different ways. We will give you some examples later, on how to apply this learning to all functions.

Let's take these different ways of translating or altering the parabola one at a time.

VERTICAL TRANSLATIONS

Notice that the parabola ($y = x^2$) is a function. Let's rewrite this parabola in function notation form ($f(x) = x^2$). What would happen to $f(x)$ if we were to add some number k to the x^2 term? We would in essence be adding that number to $f(x)$ since $f(x) = x^2$. But, remember that $f(x)$ just refers to the output variable (y), which graphically represents the height of the function. Thus, for example, if we add a positive number to the function, we will be adding a positive amount to the height of the function. We shift the graph upwards. Here are our rules for vertical translations in function notation form.

> **Vertical Translation:** If k is a positive number and you have a function $y = f(x)$, then
>
> 1.) The graph of the function $h(x) = f(x) + k$ is translated (shifted) up k units from the $f(x)$ graph.
>
> 2.) The graph of the function $h(x) = f(x) - k$ is translated (shifted) down k units from the $f(x)$ graph.

Answer Q1: The line x = 0

Discussion 1: Vertical Translations of x^2

We will graph the following parabolas using the idea of vertically translating the function $f(x) = x^2$.

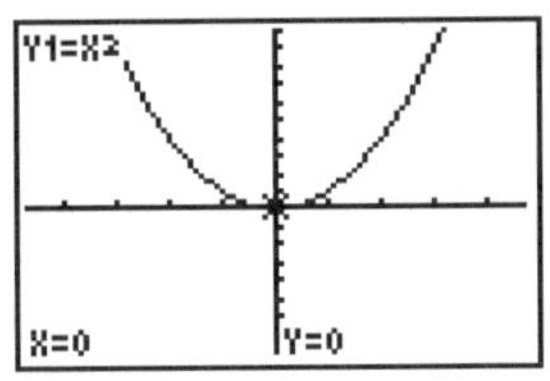

a.) $h(x) = x^2 - 3$

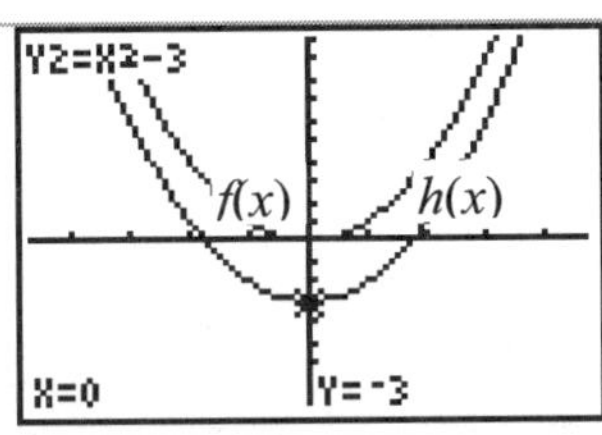

X	Y1	Y2
-3	9	6
-2	4	1
-1	1	-2
0	0	-3
1	1	-2
2	4	1
3	9	6

Y2=-3

b.) $k(x) = x^2 + 5$

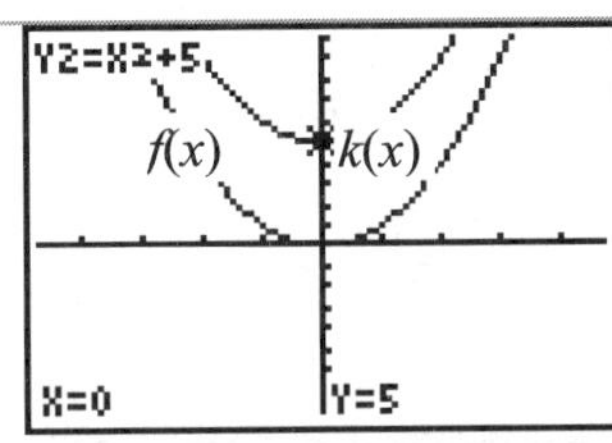

X	Y1	Y2
-3	9	14
-2	4	9
-1	1	6
0	0	5
1	1	6
2	4	9
3	9	14

Y2=5

Answer Q2:

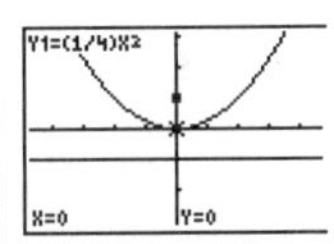

$a = \frac{1}{4}$ so, $\frac{1}{4p} = \frac{1}{4}$ *thus* $4p = 4$ *therefore* $p = 1$*. Domain is* $(-\infty, \infty)$ *and the range is* $[0, \infty)$*.*

Notice how in the table of part (a) the third column of values, $h(x)$, are all 3 less than the values in the second column. This happens because $h(x)$ subtracted 3 from all of the y-values of $f(x)$. The graph of $h(x)$, as you can see, is shifted down three units from the graph of $f(x)$.

Likewise, in the table of part (b), the third column of values, $k(x)$, are all 5 more than the values in the second column because $k(x)$ added 5 to all of the y-values of $f(x)$. The graph of $k(x)$ is shifted up five units from the graph of $f(x)$. We use k when referring to a change in y-values.

HORIZONTAL TRANSLATIONS

Given the parabola $f(x) = x^2$, what would happen if we add some number h to the x variable in the x^2 term? Since the x variable (input variable) has to do with how far left or right we go on the graph, it only seems right that this change will cause the graph to move left or right. But it is a little less intuitive than our discussion on vertical translations.

Answer Q3: These are not functions with respect to the variable x. If we try solving for "y", we would find that we get $y = \pm\sqrt{2x}$ in part a, and $y = \pm\sqrt{\frac{-x}{4}}$ in part b. The $\pm$ symbol tells us that we are getting two possibilities for the output variable. This isn't allowed with a function.

Question 4: Graph the following three parabolas on your graphing calculator and see if you can identify how adding to the x variable can change the position of the graph.

a.) $f(x) = x^2$ b.) $g(x) = (x+2)^2$ c.) $h(x) = (x-4)^2$

As you graphed these three functions on your graphing calculator, you should have noticed that adding to the x variable actually moved you in the opposite direction of what you might have thought would happen. Here are our rules for horizontal translations in function notation form.

Horizontal Translation: If h is a positive number and you have a function $y = f(x)$, then

1.) The graph of the function $h(x) = f(x + h)$ is translated (shifted) left h units from the $f(x)$ graph.

2.) The graph of the function $h(x) = f(x - h)$ is translated (shifted) right h units from the $f(x)$ graph.

We use h when referring to a change in x-values.

MULTIPLICATION BY A NEGATIVE

Given the parabola $f(x) = x^2$, what would happen if we were to multiply the term x^2 by a negative or just the independent variable x by a negative?

Discussion 2: Flipping and Rotating

Graph the following parabolas and explain how their graphs differ from the graph of the basic parabola $f(x) = x^2$.

a.) $h(x) = -x^2$	b.) $k(x) = (-x)^2$

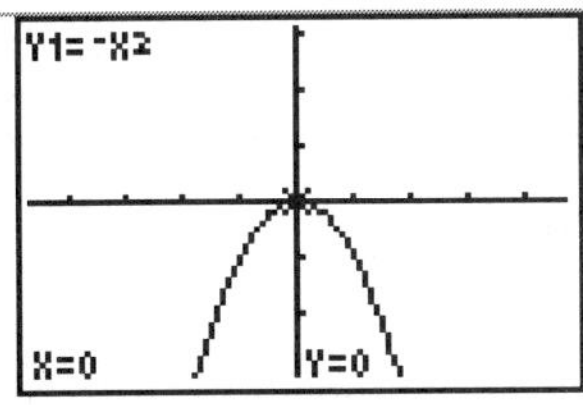	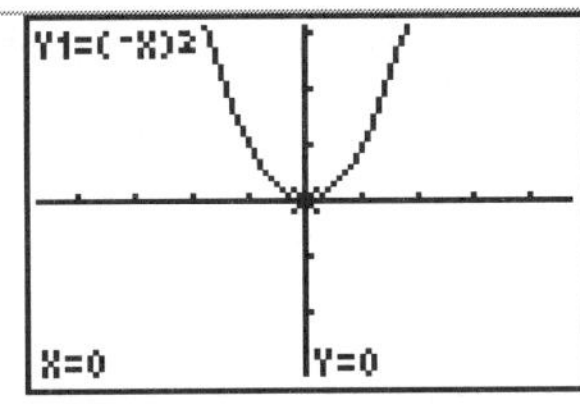
It looks as though multiplying the function by a negative flipped it upside down. This shouldn't be a surprise given the fact that we have just multiplied all of the y-values by a negative (the negative y-axis is just the flip of the positive y-axis).	It looks as though this didn't change the graph any at all. But multiplying all of the x-values by a negative really rotated the graph around the y-axis. Since this graph is symmetric to the y-axis, the graph only *appears* unchanged.

Let's do another example similar to discussion 2(b).

Discussion 3: Horizontal and Rotation Translations

Graph $y = (-x + 3)^2$ and see if you can tell that the graph rotates around the y-axis.

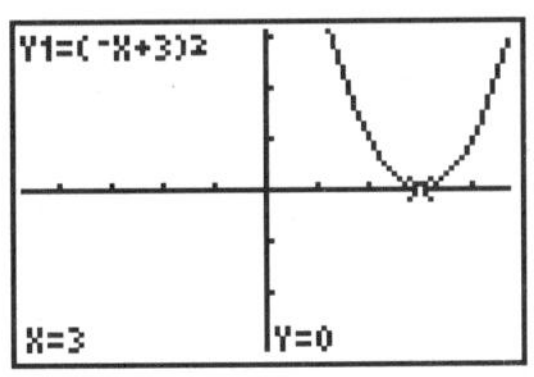

Remember that adding 3 should move the graph to the **left,** but since we multiplied the x variable by a negative, it actually rotated the graph around the y-axis; thus it looks like the graph has been shifted to the right.

> **Rotating:** If you have a function $y = f(x)$, then
>
> 1.) The graph of the function $h(x) = -f(x)$ is flipped about the x-axis, compared to the $f(x)$ graph (reflected about the x-axis).
>
> 2.) The graph of the function $h(x) = f(-x)$ is rotated about the y-axis, compared to the $f(x)$ graph (reflected about the y-axis).

COMPRESS AND STRETCH

Given the parabola $f(x) = x^2$, what would happen if we multiply the x^2 term by some number a? We would in essence be multiplying $f(x)$, since $f(x) = x^2$. But $f(x)$ just refers to the output variable (y), which graphically represents the height of the function. Thus, if we multiply by a number, we will be multiplying all of the output values by that number. This will have the affect of stretching or compressing the graph.

Example 4: Compressing and Stretching

Graph the following parabolas and explain how their graphs differ from the graph of the basic parabola $f(x) = x^2$.

Solution:

Answer Q4:

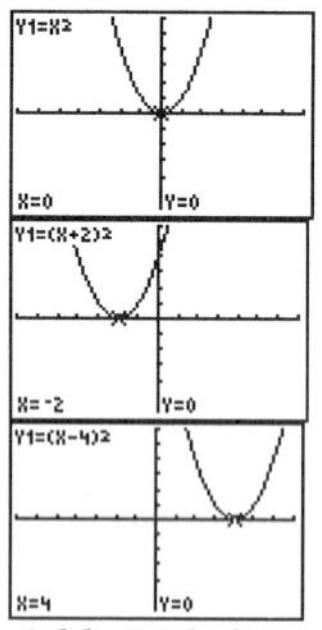

Adding shifts the graph to the left (opposite of what you would think) and subtracting shifts the graph to the right.

a.) $h(x) = 3x^2$	b.) $k(x) = \frac{1}{4}x^2$
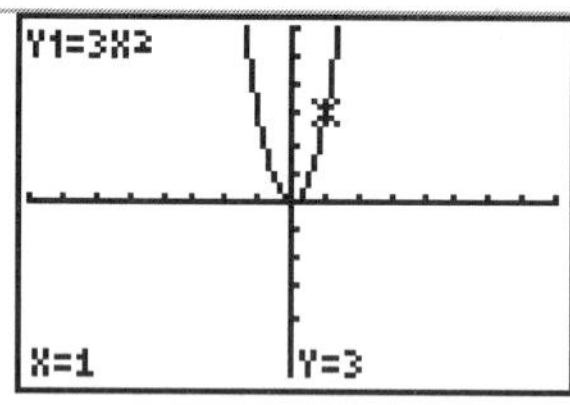	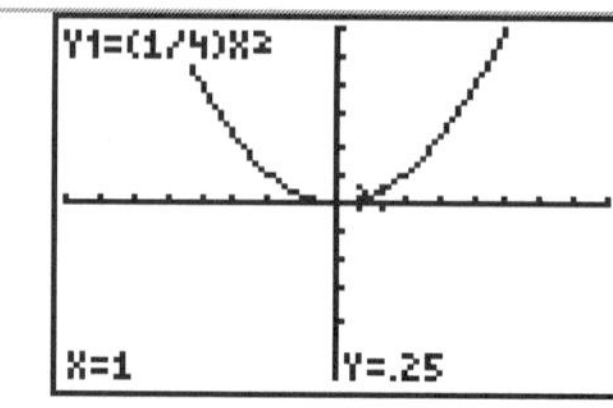
It looks as though multiplying the function by a number with magnitude greater than 1 makes the graph become narrower. This shouldn't be a surprise given the fact that we have just multiplied all of the y-values by that number (the number makes all the previous heights higher), so it appears to have stretched upwards.	It looks as though multiplying the function by a number with magnitude less than 1 makes the graph become wider. This shouldn't be a surprise given the fact that we have just multiplied all of the y-values by that number (the number makes all the previous heights lower), so it appears to have been compressed downwards.

We might make a statement as follows:

> **Stretching and compressing:** If you have a function $y = f(x)$, then the graph of the function $h(x) = af(x)$ is
>
> 1.) Stretched if $|a|$ is larger than one (the graph is narrower than x^2).
>
> 2.) Compressed if $|a|$ is smaller than one (the graph is wider than x^2).

When it comes to altering the graph of a function there are four things that might take place:

1. Adding or subtracting a number from a function ($f(x) \pm k$) will *shift* the original graph either up or down.
2. Adding or subtracting a number from the independent variable $f(x \pm h)$ will *shift* the original graph either left or right.
3. Multiplying the function by a negative will *flip* the original graph up or down and multiplying the independent variable by a negative will *rotate* the original graph around the y-axis.
4. Multiplying the function by a number of magnitude greater than 1 will stretch the original graph (make it narrower) and multiplying the function by a number of magnitude less than 1 will compress the original graph (make it wider).

Question 5: Sketch a graph of the following functions without your graphing calculator. Use the graph of $f(x) = x^2$ as your guide; then check your results with the graphing calculator.

a) $g(x) = (x + 2)^2 + 3$ b) $h(x) = -(x - 5)^2$

As you have seen, we can combine these different effects together in the same function.

Example 5: Combinations of Translations

Graph the following functions without your graphing calculator. Use the graph of $f(x) = x^2$ as your guide; then check your results with the graphing calculator.

a.) $g(x) = 2(x + 3)^2 - 1$ b.) $h(x) = -\frac{1}{3}(x - 2)^2 + 5$

Solution:

When sketching a graph using the idea of translations, if we go from left to right applying each alteration as it comes, we will usually have the best chance of sketching the correct graph. For example, in part (a) we see a multiplier of 2 in front of the squared term, so that will stretch the graph (make it more narrow). Then 3 is added to the independent variable; that will shift the graph to the left, and then subtracting 1 will shift the graph down. When we graph this function we will lightly sketch it as narrower than x^2, then lightly sketch it three to the left, and then sketch the final picture one down from there.

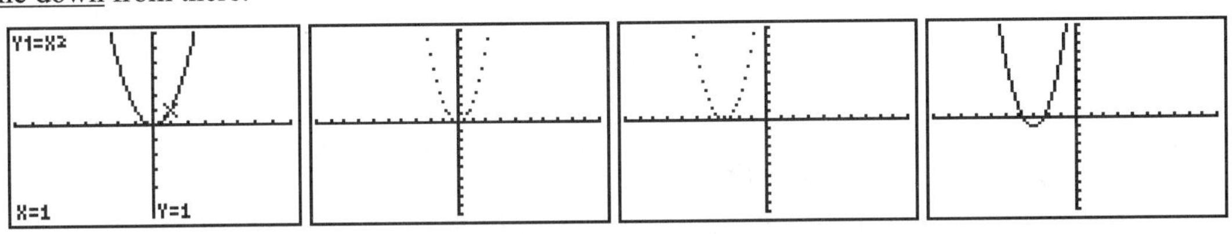

Notice that the vertex of the parabola is now at (–3, –1), the distances we moved left and down.

In part (b), we see a multiplier of $-\frac{1}{3}$ in front of the squared term, so that will flip the graph (turn it upside down) and compress the graph (make it wider). Then 2 is subtracted from the independent variable; that will shift the graph to the right, and then adding 5 will shift the graph up. When we graph this function we will lightly sketch it upside down and wider than x^2, then lightly sketch it 2 units to the right, and then sketch the final picture 5 units up from there.

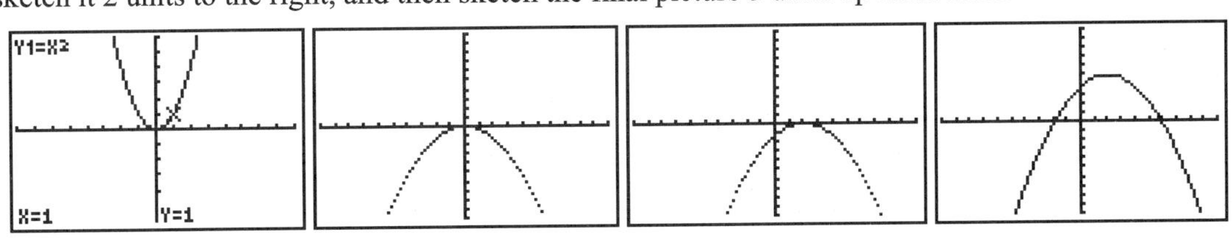

Notice that the vertex of this parabola is now at (2, 5), the distances we moved right and up. So, with all the possible alterations discussed, we can now state the most general form of a parabola function. For the non-function parabola just switch the x and y.

The Standard Form of the Parabola Function: $y = a(x - h)^2 + k$, where (h, k) is the vertex (the result of the parabola in standard position being translated) and $a = \dfrac{1}{4p}$ gives us a way to find the focus; p is the distance form the vertex to the focus along the axis of symmetry.

Example 6: Sketching and Labeling a Parabola

Sketch a graph of the following parabola. Label the vertex and focus. Check the sketch with the graphing calculator. Also state the domain, range and axis of symmetry.

$$y = \frac{-1}{20}(x + 5) + 3$$

Solution:

The vertex will be at (–5, 3) since this parabola is translated 5 to the left and up 3 from the standard position. Also, since $a = \dfrac{-1}{20}$, we know that $p = -5$ ($\dfrac{1}{4p} = \dfrac{-1}{20} \Rightarrow 4p = -20 \Rightarrow p = -5$).

vertex (-5, 3)

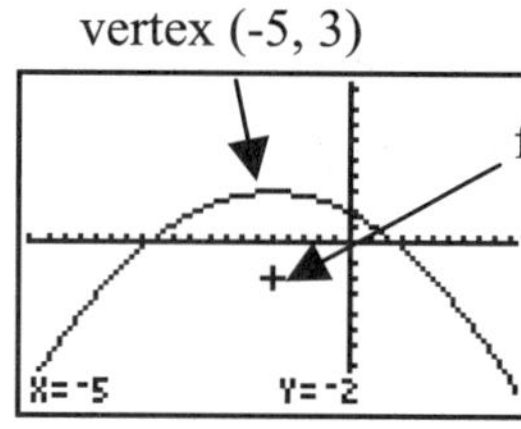

focus (-5, -2): Down 5 from vertex

The parabola turns down since a was negative. The domain is all real numbers since we don't have division by zero or negative numbers in a radical. The range will be $[3, -\infty)$ since (–5,3) is the highest point on the graph (look back at section 1.2 if you have forgotten how to find domain and range). The axis of symmetry is $x = -5$.

Question 6: Sketch a graph of the following parabola. Label the vertex and focus. Check your sketch with the graphing calculator. Also state the domain, range and axis of symmetry.

$$y = \frac{1}{8}(x - 2) + 1$$

Of course if the parabola goes sideways instead of up or down, then the general form is

The Standard Form of a non-function (sideways) Parabola: $x = a(y - k)^2 + h$, where (h, k)

is the vertex and, $a = \dfrac{1}{4p}$ gives us a way to find the focus. (Notice, h is always associated with x and k is always associated with y.)

Answer Q5:
(a) Adding 2 to x will shift the graph to the left and the +3 will shift it up three units.
(b) The negative in front of the function will flip the graph and subtracting 5 from x will shift it to the right five units.

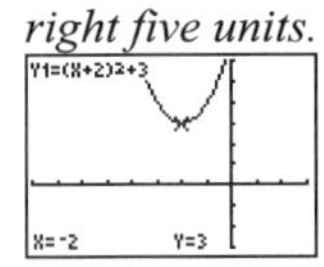

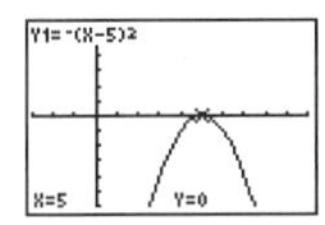

Example 7: Sketching a Sideways Parabola

Sketch a graph of the following parabola. Label the vertex and focus. Check the sketch with the graphing calculator. Also state the domain, range, and the axis of symmetry.

$$x = (y - 3)^2 + 2$$

Solution:

The vertex is (2, 3): $a = 1$ so $p = ¼$; so the focus is at (2.25, 3).

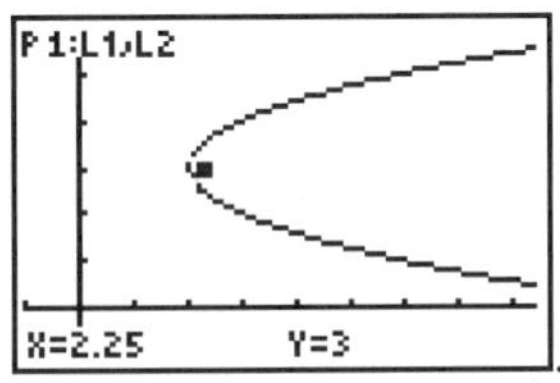

For this function the domain is $[2, \infty)$. The farthest to the left you can go on the graph is 2. The range is (∞,∞) and the axis of symmetry is $y = 3$.

Let's now direct our attention to shifting non-parabolic functions.

Discussion 4: Translations of Functions

Given the graph of the function $f(x)$,

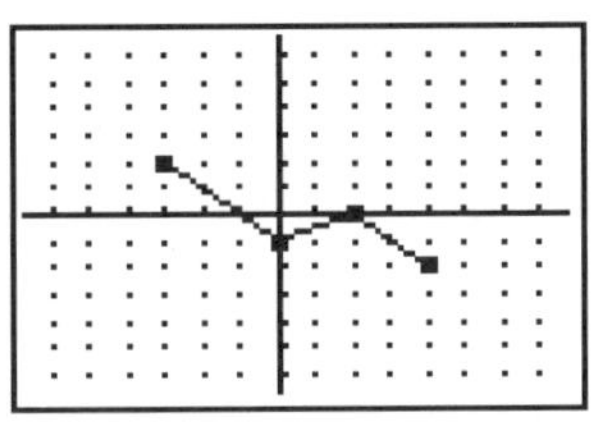

sketch the graphs of the following functions.

a) $f(x + 1) + 2$ b) $-f(x - 3)$ c) $-f(x) - 4$

Solutions:

a) $f(x + 1) + 2$	b) $-f(x - 3)$	c) $-f(x) - 4$
According to what we have learned about translations, this should translate the original $f(x)$ 1 unit to the left, and 2 units up. All we need to do is move each of the four major turning points and then	According to what we have learned about translations, this should flip the original $f(x)$ upside down, and then three units to the right. All we need to do is move each of the four major turning points	According to what we have learned about translations, this should flip the original $f(x)$ upside down, and then 4 units down. All we need to do is move each of the four major turning points and then

connect the dots.	the four major turning points and then connect the dots.	major turning points and then connect the dots.
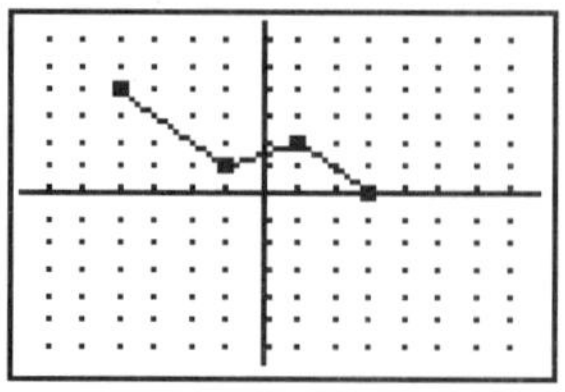	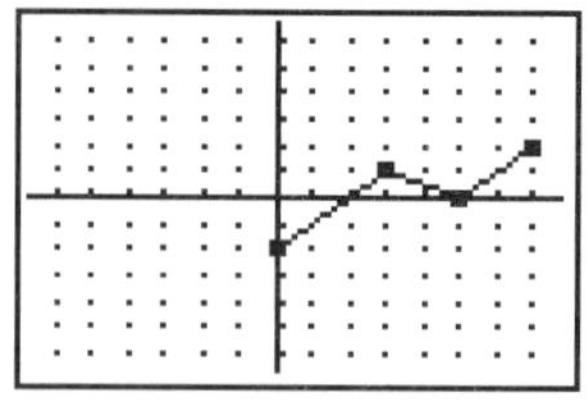	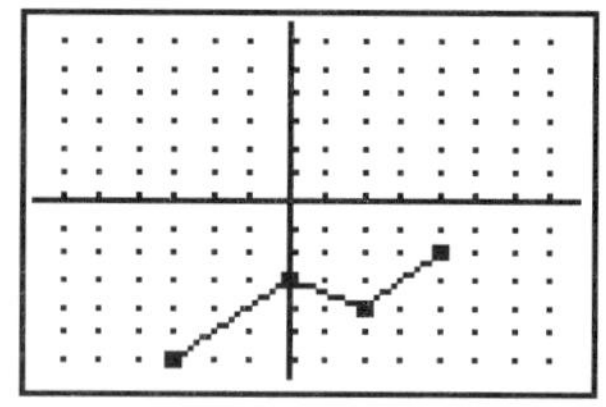

Question 7: Given the function $g(x)$ in the table below, fill in the table values that would result from the given translations. a) $g(x) - 3$. What would the new y-values be? b) $g(x + 2)$. What would the x-values need to be in order to yield the same y-values as $g(x)$?

X	g(x)	
-6	9	
-4	4	
-2	1	
0	0	
2	1	
4	4	
6	9	
Y1=9		

X	g(x) – 3	
-6		
-4		
-2		
0		
2		
4		
6		
X= -6		

X	Y1 g(x + 2)
-8	
-6	
-4	
-2	
0	
2	
4	
X= -8	

Answer Q6:

Since $a = \frac{1}{8}$, we know that $p = 2$

$\frac{1}{4p} = \frac{1}{8} \Rightarrow$

$4p = 8 \Rightarrow$

$p = 2$. Also with $(x - 2)$ we see a shift to the right 2 units and the plus one will shift the graph up 1.

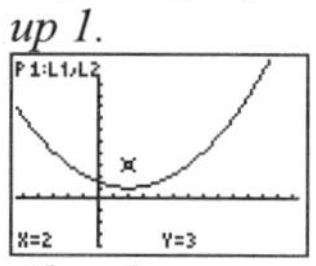

Also the domain is $(-\infty, \infty)$ and the range is $[1, \infty)$. Axis of symmetry is $x = 2$.

Finding the numerical effects of translations may be difficult, but it's also quite revealing. It shows us numerically, why, when we subtract from a function, the graph moves down. Likewise, it shows us numerically, why, when we add to the input variable, the graph actually moves to the left. When we change the inputs, the outputs change also, so we need to move to different inputs than before in order to see the same outputs as before.

Section Summary:

Here is a list of ideas expressed in this section:

- Distance formula: $d = \sqrt{(x_2 - x_1)^2 + (y_2 - y_1)^2}$, needed to define the parabola.
- Standard positions for the parabola: $y = ax^2$ or $x = ay^2$ (where $a = \frac{1}{4p}$); the vertex is at the origin (0, 0).
- The constant p represents the distance from the vertex to the focus of a parabola.
- Standard form of the function parabola: $y = a(x - h)^2 + k$, where (h, k) is the vertex, $a = \frac{1}{4p}$, the focus is $(h, k + p)$, the axis of symmetry is $x = h$ and the directrix is $y = k - p$.

- Standard form of the non-function parabola: $x = a(y - k)^2 + h$, where (h, k) is the vertex, $a = \frac{1}{4p}$, the focus is $(h + p, k)$, the axis of symmetry is $y = k$ and the directrix is $x = h - p$.
- Vertical translation: $f(x) + k$, means the graph shifts up or down k units.
- Horizontal translation: $f(x + h)$, means the graph shifts left or right h units.
- Flipping the graph: $-f(x)$, means the graph flips upside down.
- Rotating the graph: $f(-x)$, means the graph rotates around the y-axis.
- Stretching or compressing: $af(x)$, means the graph is wider if $|a|$ is larger than 1 and narrower if $|a|$ is less than 1.
- In physical applications, the reflective property of a parabola means that when anything comes into the parabola parallel to the axis of symmetry, it will be reflected back towards the focus and vice versa.

SECTION 7.1 HOMEWORK

(1–6) Find the distance between the two given points.

1. $(-3, 7)$ and $(1, 4)$
2. $(2, -5)$ and $(-10, 0)$
3. $(3, -1)$ and $(7, -5)$
4. $(-5, 10)$ and $(3, -1)$
5. $\left(\frac{-1}{2}, \frac{1}{3}\right)$ and $\left(\frac{3}{2}, \frac{5}{6}\right)$
6. $(a-3, b+2)$ and $(a+5, b-4)$

(7–28) Without using the graphing calculator, sketch the graphs of the following parabolas (check your graph with the graphing calculator):

7. $y = 3x^2$
8. $y = -2x^2$
9. $y = \frac{-1}{4}x^2$
10. $y = \frac{1}{2}x^2$
11. $y = (x+3)^2$
12. $y = (x-2)^2$

Answer Q7:

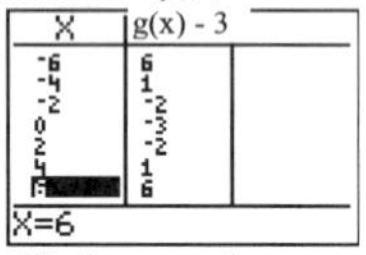

All the y-values are 3 less.

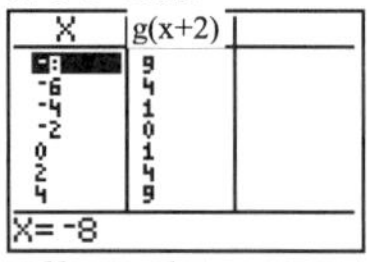

All x-values must decrease by 2 in order to produce the same outputs as before.

13. $y = x^2 - 3$
14. $y = x^2 + 2$
15. $y = (x-2)^2 + 5$
16. $y = (x+3)^2 - 2$
17. $y = 2(x+3)^2 + 5$
18. $y = -3(x-5)^2 - 1$
19. $x = 2y^2$
20. $x = -3y^2$
21. $x = (y-4)^2$
22. $x = (y+6)^2$
23. $x = y^2 + 1$
24. $x = y^2 - 7$
25. $x = (y-4)^2 + 2$
26. $x = (y+1)^2 - 2$
27. $x = -2(y+5)^2 + 3$
28. $x = 4(y-2)^2 - 7$

(29–40) For each of the following functions, explain verbally the translation each represents with respect to $y = f(x)$:

29. $y = f(x-2)$
30. $y = f(x+3)$
31. $y = f(x) + 5$
32. $y = f(x) - 21$
33. $y = 3f(x)$
34. $y = \frac{-1}{2}f(x)$
35. $y = f(-x)$
36. $y = -2f(-x)$
37. $y = f(x+5) - 9$
38. $y = f(x-15) + 23$
39. $y = \frac{-1}{3}f(x-12) - 8$
40. $y = 5f(x+9) + 1$

(41 – 50) For each of the following parabolas determine the:

a. Vertex
b. Axis of symmetry
c. Focus
d. Directrix
e. Maximum or minimum value of y
f. Effect of the translation with respect to $y = x^2$; explain in words what happened.
g. Domain and range of the function

(Check answers to all parts with the graphing calculator.)

41. $y = -2x^2$
42. $y = \frac{2}{3}x^2$
43. $y = (x-3)^2$

44. $y = (x + 9)^2$

45. $y = x^2 + 12$

46. $y = x^2 - 23$

47. $y = (x - 15)^2 + 7$

48. $y = -(x + 21)^2 - 19$

49. $y = \frac{-1}{8}(x + 11)^2 + 32$

50. $y = \frac{1}{4}(x - 8)^2 - 18$

(51 – 60) For each of the following parabolas determine the:

a. Vertex

b. Axis of symmetry

c. Focus

d. Directrix

e. Effect of the translation with respect to $x = y^2$; explain in words what happened.

f. Domain and range of the relation

(Check with the graphing calculator)

51. $x = 2y^2$

52. $x = \frac{-1}{2}y^2$

53. $x = (y + 15)^2$

54. $x = (y - 5)^2$

55. $x = y^2 - 23$

56. $x = y^2 + 18$

57. $x = (y - 7)^2 + 12$

58. $x = (y + 15)^2 - 11$

59. $x = \frac{-1}{16}(y + 6)^2 + 1$

60. $x = \frac{1}{8}(y - 17)^2 - 8$

(61–68)

$f(x) =$

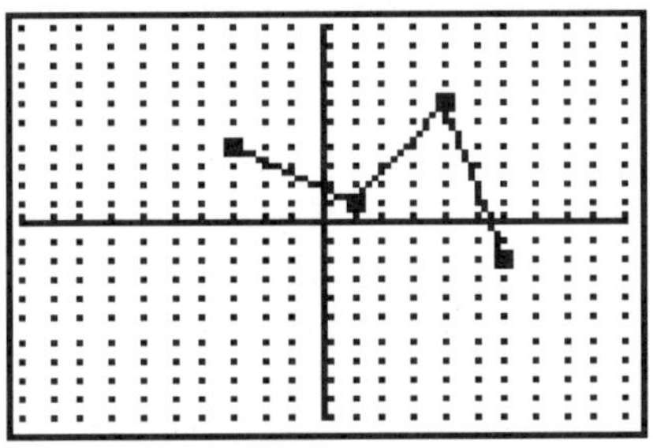

Using the above graph of $f(x)$ sketch the graph of the following:

61. $f(x + 2)$

62. $f(x - 3)$

63. $f(x) - 4$

64. $f(x) + 3$

65. $f(x - 5) + 2$

66. $f(x + 3) - 2$

67. $-f(x + 4)$

68. $-f(x) - 3$

(69 – 70)

x	$g(x)$
−5	5
−3	7
−1	4
0	−2
1	−9
2	−15
3	−25

Using the table above create a table for the following functions:

69. $g(x) - 5$

70. $g(x) + 3$

71. Given the function $y = \frac{1}{8}x^2$:

a.) Find the focus.

b.) Find the directrix.

c.) The points (0,0), (4,2) and (8,8) are on the graph of the function. For each point find the distance to the focus and the distance to the directrix and show the distances are equal.

72. Given the function $y = \frac{-1}{4}x^2$:

a.) Find the focus.

b.) Find the directrix.

c.) The points (0,0), (2,–1) and (4,–4) are on the graph of the function. For each point find the distance to the focus and the distance to the directrix and show the distances are equal.

Section 7.2 Ellipses and Circles

Let's continue now with our next two conic sections: the ellipse and the circle. These two are very similar; in fact, a circle is just a special ellipse. Let's begin with the definition of an ellipse.

> **Ellipse**: An ellipse is a set of points in the xy-plane, where the sum of the distances between two fixed points (foci) in the xy-plane is a constant sum ($d_1 + d_2 = d$ (constant sum)).

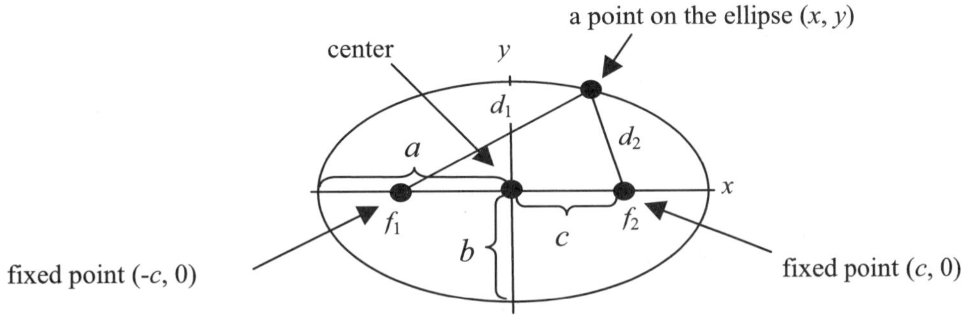

In the figure, a represents the distance from the center (point of symmetry) to the edge horizontally, b represents the distance from the center (point of symmetry) to the edge vertically (note: a is assumed to be larger than b) and c represents the distance from the center (point of symmetry) to the fixed points (foci).

> **Point of Symmetry (center):** A point in space will be a point of symmetry (or center) if for any point on the graph (P_1) you draw a line from it through the center and beyond to a point opposite on the graph (P_2) and the two points are always equal distance from the center. **($d_1 = d_2$)**
>
> 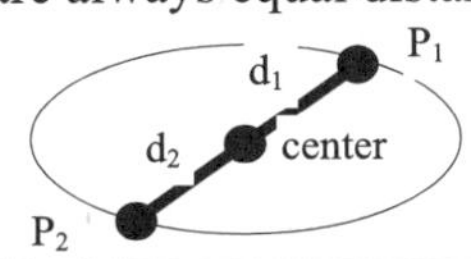
>

According to the definition of an ellipse, the distance d_1 from the first fixed point to a point on the ellipse, plus the distance d_2 from the second fixed point to the same point on the ellipse should equal a constant sum d, for any point you choose on the ellipse. For simplicity, we will choose to center the ellipse at the origin and use the point on the ellipse right above the center and assume that a is larger than b. We will label the graph as follows:

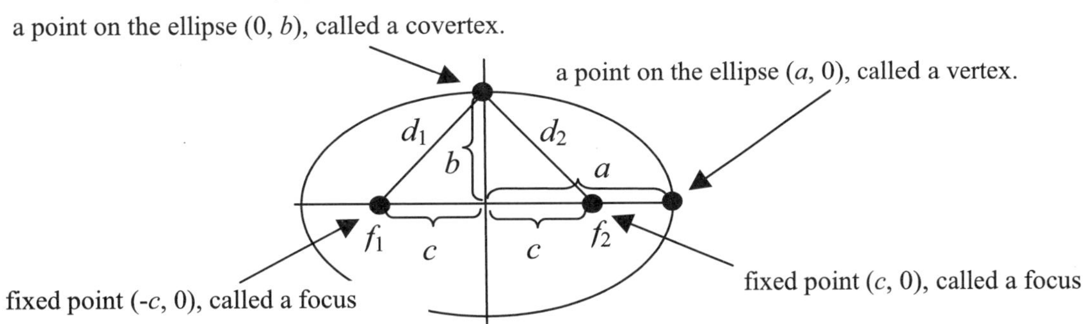

Definitions:
Covertices – Points on the ellipse that are the shortest distance from the center.
Vertices – Points on the ellipse that are the farthest distance from the center.
Major Axis – The line segment connecting the vertices (contains foci).
Minor Axis – The line segment connecting the covertices.

To find the formula for an ellipse we:

1. Add together the two distances from the point on the ellipse to the two fixed points
2. Set the two distances equal to a constant we will call d
3. Then, simplify the equation.

For simplicity's sake let's assume that a is bigger than b, then the constant sum d turns out to be equal to $2a$. This is true

1. Because if we look at the point $(a, 0)$ on the ellipse, the distance to f_1 is $(a + c)$ and the distance to f_2 is $(a - c)$, so the sum of the two distances is $(a + c + a - c = 2a = d)$.
2. And from the graph above we can also see that the point $(0, b)$ is on the ellipse and that d_1 equals d_2, so each of these distances is equal to a, since $d_1 + d_2 = d = 2a$ ($2d_1 = 2a$ or $2d_2 = 2a$).

So, the formula for an ellipse in standard position comes from the following:

Add the two distances from any point (x, y) on the ellipse to the two fixed points and set equal to $2a$.	$d_1 = a = \sqrt{(x-(-c))^2 + (y-0)^2}$ $= \sqrt{(x+c)^2 + y^2}$ $d_2 = a = \sqrt{(x-c)^2 + (y-0)^2}$ $= \sqrt{(x-c)^2 + y^2}$ $\sqrt{(x+c)^2 + y^2} + \sqrt{(x-c)^2 + y^2} = 2a$
Isolate one of the square roots.	$\sqrt{(x+c)^2 + y^2} = 2a - \sqrt{(x-c)^2 + y^2}$
Square the equation and simplify.	$(x + c)^2 + y^2 = 4a^2 - 4a\sqrt{(x-c)^2 + y^2} + (x - c)^2 + y^2$ $\cancel{x^2} + 2cx + \cancel{c^2} + \cancel{y^2} = 4a^2 - 4a\sqrt{(x-c)^2 + y^2} + \cancel{x^2} - 2cx + \cancel{c^2} + \cancel{y^2}$
Isolate the square root, divide by 4, then square and simplify.	$4a\sqrt{(x-c)^2 + y^2} = 4a^2 - 4cx$ $a\sqrt{(x-c)^2 + y^2} = a^2 - cx$ $a^2((x-c)^2 + y^2) = a^4 - 2ca^2x + c^2x^2$ $a^2(x^2 - 2cx + c^2 + y^2) = a^4 - 2ca^2x + c^2x^2$ $a^2x^2 - \cancel{2ca^2x} + a^2c^2 + a^2y^2 = a^4 - \cancel{2ca^2x} + c^2x^2$

	$a^2x^2 - c^2x^2 + a^2y^2 = a^4 - a^2c^2$
To put the formula into a better form, we need to use the fact that $b^2 = a^2 - c^2$ (assuming a is bigger than b). We know this because of the Pythagorean Theorem and the triangle formed by the two fixed points and our point $(0, b)$.	
Factor and replace our algebra with $b^2 = a^2 - c^2$.	$(a^2 - c^2)x^2 + a^2y^2 = a^2(a^2 - c^2)$ $b^2x^2 + a^2y^2 = a^2b^2$
Now, divide both sides by a^2b^2.	$\frac{x^2}{a^2} + \frac{y^2}{b^2} = 1$

Whew, we made it to the formula for an ellipse in standard position (if a isn't bigger than b everything ends up similarly)!

> **Ellipse in Standard Position**: $\frac{x^2}{a^2} + \frac{y^2}{b^2} = 1$, where the center is (0,0), a is the distance from the center to the left and right edges of the ellipse; b is the distance from the center to the top and bottom of the ellipse, and c is the distance from the center to the foci along the major axis $(c^2 = |a^2 - b^2|)$.

You may have noticed that we slipped in some absolute value symbols in that last definition. They show up here because the ellipse could be longer in the vertical direction than the horizontal. In that case, b would be bigger than a, and c^2 would end up negative if it were not for the absolute value sign.

Discussion 1: Sketching Ellipses

Let's sketch the graphs of the following ellipses by hand and label the centers and foci.

a.) $\frac{x^2}{25} + \frac{y^2}{9} = 1$ (horizontal ellipse) b.) $\frac{x^2}{16} + \frac{y^2}{25} = 1$ (vertical ellipse)

a) Since 25 is under x^2, $a^2 = 25$, so $a = 5$. Likewise, 9 is under y^2, so $b^2 = 9$, which means that $b = 3$. Also, $c^2 = \left|a^2 - b^2\right|$, which means that $c^2 = \|25 - 9\| = 16$, so $c = 4$.	b) Since 16 is under x^2, $a^2 = 16$, so $a = 4$. Likewise, 25 is under y^2, so $b^2 = 25$, which means that $b = 5$. Also, $c^2 = \left|a^2 - b^2\right|$, which means that $c^2 = \|16 - 25\| = \|-9\| = 9$, so $c = 3$.

We are just about ready to graph these two ellipses, and remember the foci points are always on the major axis. For example in part (a), since a is bigger than b, the major axis is the part of the x-axis (horizontal line) that connects the vertices (–5,0) and (5,0). In part (b), since b is bigger than a, the major axis is the part of the y-axis (vertical line) that connects the vertices (0, 5) and (0, –5). We now graph our two examples.

a.) $\frac{x^2}{25}+\frac{y^2}{9}=1$	b.) $\frac{x^2}{16}+\frac{y^2}{25}=1$
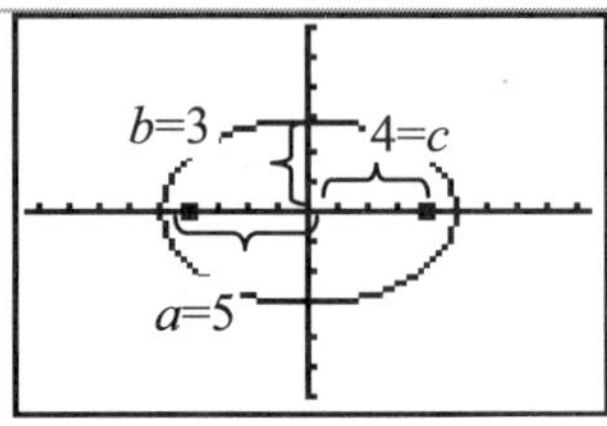	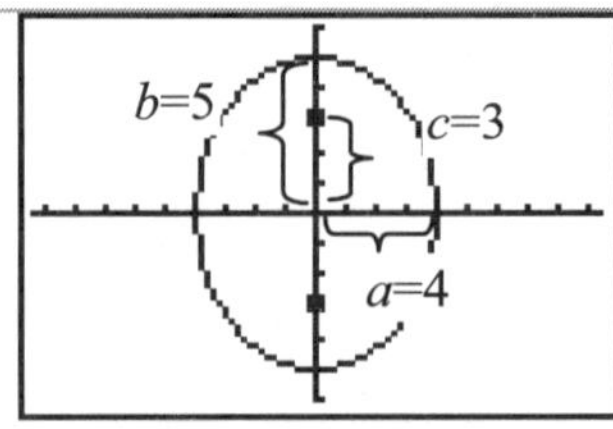
Center –> (0, 0) Foci –> (–4,0) and (4,0) Vertices –> (–5,0) and (5,0) Covertices –> (–3,0) and (3,0)	Center –> (0, 0) Foci –> (0,–3) and (0,3) Vertices –> (0,–5) and (0,5) Covertices –> (0,–4) and (0,4)

In order to stay consistent, we will always label the number under the x^2 term as a^2. Likewise, we will always label the number under the y^2 term as b^2. Given what was discussed in the last section, try sketching a graph of the next ellipse. Think about translating the center.

Question 1: Sketch the vertical ellipse $\frac{(x-2)^2}{25}+\frac{(y+3)^2}{169}=1$, and label the center and foci.

First, notice that we see x being changed from the standard position by 2, $(x-2)$. This will mean a shift to the right, two units. Second, the y term $(y+3)$ has a look very similar to the x term, and thus will react similarly. The graph will shift 3 units **down,** since as with the x term, the shift acts "opposite" to the sign present. Let's do another example.

Example 1: Sketching a Horizontal Ellipse

Sketch a graph of this horizontal ellipse $\frac{(x+1)^2}{49}+\frac{(y-5)^2}{36}=1$, and label the center and foci.

Solution:

First, $a^2=49$ so $a=7$. Second, $b^2=36$ so $b=6$. Third, $c^2=|49-36|=13$, so $c=\sqrt{13}\approx 3.6$.

Finally, the center is (–1, 5). We can now sketch a graph:

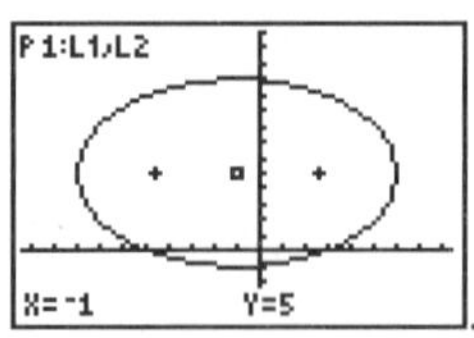

The foci turn out to be (2.6, 5) and (–4.6, 5) because the major axis is horizontal this time.

> **The Standard Form of an Ellipse:** $\frac{(x-h)^2}{a^2}+\frac{(y-k)^2}{b^2}=1$, where (h, k) is the center of the ellipse, a is the distance from the center to the left and right edges of the ellipse, b is the distance from the center to the top and bottom of the ellipse, and c is the distance from the center to the foci ($c^2 = \left|a^2-b^2\right|$) along the major axis. If $a^2 > b^2$, then you have a horizontal ellipse and if $b^2 > a^2$, then you have a vertical ellipse.

If we wanted to graph the last example on our graphing calculators, we would need to solve for y. Remember, the calculator only likes to graph functions. This clearly isn't a function. From the graph we can see that it doesn't pass the vertical line test that we discussed back in section 1.1. When we solve this for y, we will get two different answers, both of which we will graph in order to get the entire picture of the ellipse.

Get the y term alone.	$\frac{(y-5)^2}{36}=1-\frac{(x+1)^2}{49}$
Multiply by 36, then take the square root.	$(y-5)=\pm\sqrt{36\left(1-\frac{(x+1)^2}{49}\right)}$
Now, add 5.	$y=\pm\sqrt{36\left(1-\frac{(x+1)^2}{49}\right)}+5$
We have two functions (the two halves of the ellipse) to input into our graphing calculators.	$y=\sqrt{36\left(1-\frac{(x+1)^2}{49}\right)}+5$ $y=-\sqrt{36\left(1-\frac{(x+1)^2}{49}\right)}+5$

Since we are just concerned with seeing a graph on the calculator, there isn't any great need to simplify these two expressions any more than absolutely necessary. Here is the graph.

Zoom Standard	Zoom Decimal (zoom out by a factor of 4)	Y= screen
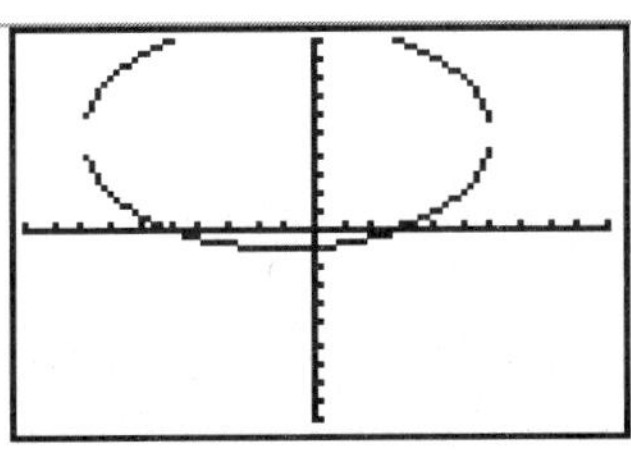	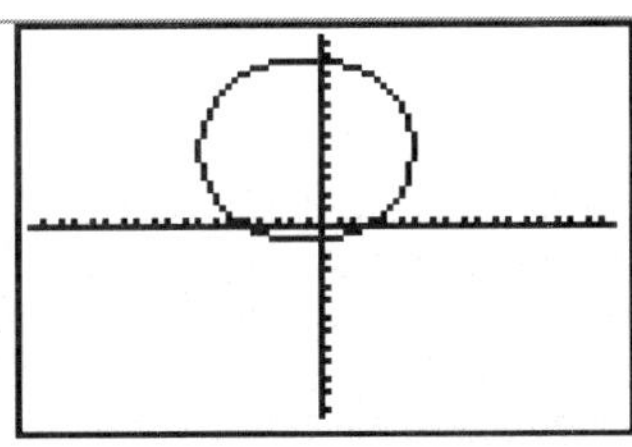	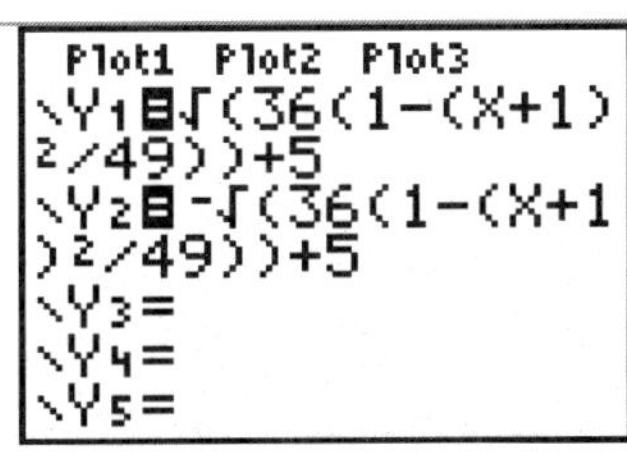

Notice how one graph is complete and the other isn't. Remember, the calculator has a pixel problem and often doesn't graph correctly with zoom standard.

Answer Q1:

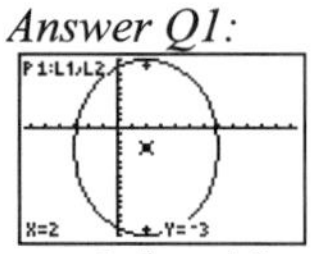

a = 5, b = 13, c = $\sqrt{169-25}=\sqrt{144}$ *= 12. The major axis is vertical since b is bigger than a. The center is (2, –3); the foci are (2, 9), (2, –15); the vertices are (2, 10),(2, –16) and the covertices are (7, –3),(–3, –3).*

Example 2: Sketching a Vertical Ellipse

Sketch a graph of this vertical ellipse $\frac{(y-2)^2}{9}+(x-4)^2=1$, and label the center, foci, vertices and covertices.

Question 2: What are *a, b, c* and the center for this ellipse?

Solution:

This problem is easy to get incorrect if you don't notice that the *y* term is the first term this time. Look at the equation again, if you didn't notice that fact while you were answering question 2. Here is the graph of the vertical ellipse.

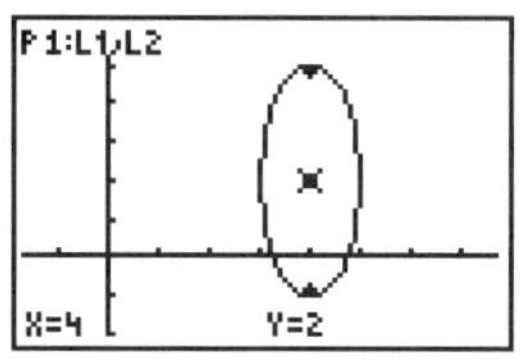

The center is (4, 2); the foci are (4, 4.83) and (4, –0.83); the vertices are (4, 5) and (4, –1), and the covertices are (3, 2) and (5, 2).

Example 3: Sketching a Special Ellipse

Sketch a graph of this ellipse $\frac{(x+1)^2}{16}+\frac{(y-3)^2}{16}=1$, and label the center and foci.

Solution:

Once again, we must look to see what *a, b, c* and the center are, and then graph. In this case, $a^2 = 16$ so $a = 4$: $b^2 = 16$. Thus $b = 4$, $c = \sqrt{16-16} = 0$, and the center is (–1, 3). Now in this example, *c* has turned out to be equal to 0.

Question 3: What does it mean if *c* is equal to 0?

Here is a graph of the ellipse.

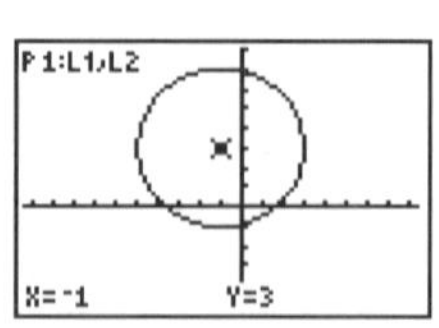

Since it is equal distances in all directions from the center of the ellipse to the edges ($a = b = 4$), this must be the special ellipse we call a **circle**. If you look back at examples 2 and 3, you will notice that the closer a^2 and b^2 are in value, the closer the foci are to the center. The closer the foci are to the center, the more rounded the ellipse. If the foci both end up at the center then the distances from the foci to the edge will always be the same. In this example, that distance is **4**. Here is the definition of a circle.

A circle is: The set of all points in the *xy*-plane that are equidistant from a fixed point in the *xy*-plane. Namely, the center.

Let's manipulate the last example.

Here is the formula of the ellipse:	$\frac{(x+1)^2}{16}+\frac{(y-3)^2}{16}=1$
Multiply both sides of the equation by 16.	$(x+1)^2+(y-3)^2=16$
From this form we can see the center (point of symmetry) and the radius (distance from center to edge of circle).	Center is $(-1, 3)$ Radius is 4.

This brings us to the standard form of a circle.

The Standard Form of a Circle: $(x-h)^2+(y-k)^2=r^2$, where (h, k) is the center of the circle, and r is the radius of the circle.

Question 4: Sketch a graph of the circle $(x-2)^2+(y+7)^2=36$, and label the center and radius.

REFLECTIVE PROPERTY OF AN ELLIPSE

Finally, let's discuss the reflective properties of an ellipse. You should remember from the last section that the reflective property of a parabola means that anything that travels *towards* a parabola parallel to the axis of symmetry will reflect off of the parabola back towards its focus. With an ellipse the property means that anything *emitted from* one focus is reflected off of the ellipse back towards the other focus.

Here is a practical application of this property.

In the medical field a treatment for kidney stones called lithotripsy makes use of the reflective property of an ellipse. The way the treatment works is this: the patient sits in a liquid-filled elliptical bathtub, positioned with the kidney stone at one of the foci points; at the other foci point is located a high energy sound wave device. Then the device emits a high energy sound wave which is completely reflected back towards the kidney stone and breaks it up into pieces without hurting

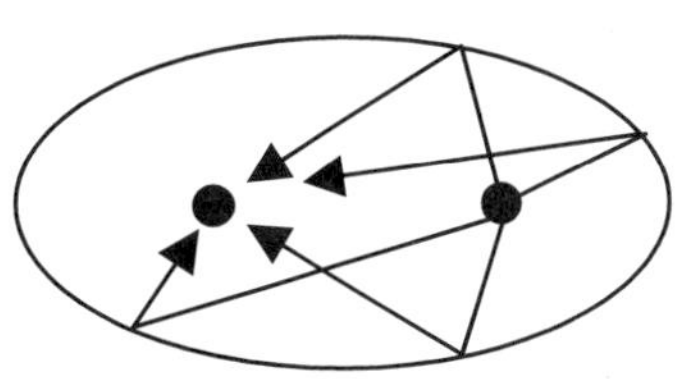

the rest of the patient. This gives the patient a way to avoid surgery.

Another example of ellipses in use is the light in a dentist's chair.

Section Summary:

Answer Q2:
$a^2 = 1$ so
$a = 1$: $b^2 = 9$
so $b = 3$
,$c = \sqrt{9-1} =$
$\sqrt{8} \approx 2.83$,
and the center is (4,2).

Answer Q3:
If c is equal to zero, that just means that the distance from the center to the foci is zero. This would mean that the foci are both at the center of the ellipse.

- A circle is a special ellipse where a and b are equal and thus the foci are located at the center.
- The standard formula for a circle is $(x-h)^2+(y-k)^2=r^2$, where (h, k) is the center and r is the radius.
- The standard formula for an ellipse is $\frac{(x-h)^2}{a^2}+\frac{(y-k)^2}{b^2}=1$, where ($h$, k) is the center; a is the distance from the center to the left and right edges; b is the distance from the center to the top and bottom; c is the distance from the center to the foci ($c^2 = \left|a^2-b^2\right|$) along the major axis. If $a^2 > b^2$, then you have a horizontal ellipse. However, if $b^2 > a^2$, then you have a vertical ellipse.
- The reflective property of an ellipse means that anything that is emitted from one focal point is reflected towards the other focal point.

SECTION 7.2 HOMEWORK

(1–8) For each of the ellipses in standard position write the standard equation and find the following:

a) Horizontal or vertical ellipse

b) Center

c) Foci

d) Vertices

e) Covertices

f) Domain and Range

Then sketch the graph.

Answer Q4: The center is (2,–7) and the radius is 6.

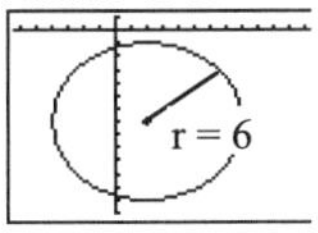

1. $\frac{x^2}{9}+\frac{y^2}{25}=1$ 2. $\frac{x^2}{169}+\frac{y^2}{25}=1$ 3. $\frac{x^2}{40}+\frac{y^2}{16}=1$

4. $\frac{x^2}{16}+\frac{y^2}{40}=1$ 5. $\frac{x^2}{4}=1-\frac{y^2}{9}$ 6. $\frac{y^2}{25}=1-\frac{x^2}{36}$

7. $x^2+4y^2=16$ 8. $3x^2+2y^2=24$

(9–16) For each of the ellipses in translated position write the standard equation and find the following:

a) Horizontal or vertical ellipse

b) The translation from standard position

c) Center

d) Foci

e) Vertices

f) Covertices

g) Major axis

h) Minor axis

i) Domain and Range

Then sketch the graph

9. $\frac{(x-3)^2}{100}+\frac{(y-2)^2}{36}=1$ 10. $\frac{(x-5)^2}{576}+\frac{(y-4)^2}{676}=1$ 11. $\frac{(x+2)^2}{8}+\frac{(y+3)^2}{12}=1$

12. $\frac{(x+5)^2}{25}+\frac{(y+1)^2}{10}=1$ 13. $\frac{(x-4)^2}{9}+\frac{(y+5)^2}{4}=1$ 14. $\frac{(x-2)^2}{9}+\frac{(y+7)^2}{16}=1$

15. $\frac{(x+6)^2}{6}=1-\frac{(y-3)^2}{15}$ 16. $\frac{(y-8)^2}{10}=1-\frac{(x+1)^2}{14}$ 17. $4(x-2)^2+5(y-5)^2=20$

18. $5(x+6)^2+3(y-2)^2=60$

19. Find the standard equation of an ellipse with a center of (0, 0); vertices of (5, 0) and (–5, 0); and foci points of (3, 0) and (–3, 0).

20. Find the standard equation of an ellipse with a center of (0, 0); vertices of (0, 13) and (0,–13); and foci points of (0, 12) and (0, –12).
21. Find the standard equation of an ellipse with a center of (0, 0); a constant sum of 8; and foci points of (0, 3) and (0, –3).
22. Find the standard equation of an ellipse with a center of (0, 0); a constant sum of 12 and foci points of (4, 0) and (–4, 0).
23. Find the standard equation of an ellipse with a center of (0, 0); vertices of (10, 0) and (–10,0); and covertices of (0, 8) and (0, –8).
24. Find the standard equation of an ellipse with a center of (0, 0); vertices of (0, 26) and (0,–26); and covertices of (24, 0) and (–24, 0).
25. Find the standard equation of an ellipse with a center of (–3, 2); covertices of (1, 2) and (–7,2); foci points of (–3, 5) and (–3, –1).
26. Find the standard equation of an ellipse with a center of (3, –5); vertices of (13, –5) and (–7,–5); and foci points of (11, –5) and (–5, –5).
27. Find the standard equation of an ellipse with a center of (2, 4); constant sum of 10; and covertices of (6, 4) and (–2, 4).
28. Find the standard equation of an ellipse with a center of (–4, –6); constant sum of 12; and foci points of (–4, –3) and (–4, –9).
29. Find the standard equation of an ellipse with a center of (3, 0); vertices of (3, 8) and (3, –8); and covertices of (9, 0) and (–3, 0).
30. Find the standard equation of an ellipse with a center of (0, –3); vertices of (10, –3) and (–10, –3); and covertices of (0, –2) and (0, –4).
31. The ellipse $\frac{x^2}{4}+\frac{y^2}{9}=1$ is translated to the right 3 units and up 5 units. What is the equation in standard form for the new ellipse?
32. The ellipse $\frac{x^2}{25}+\frac{y^2}{16}=1$ is translated to the left 4 units and down 2 units. What is the equation in standard form for the new ellipse?

(33–38) For each of the following circles in standard position, write the standard equation and find the:

a) Center
b) Radius
c) Domain and range.

33. $x^2+y^2=16$
34. $x^2+y^2=8$
35. $x^2=25-y^2$
36. $x^2+y^2-36=0$
37. $3x^2+3y^2=81$
38. $\frac{x^2}{4}+\frac{y^2}{4}=9$

(39–46) For each of the following circles in translated position, write the standard equation and find the:

a) Center

b) Radius

c) Domain and range

Explain the translation from standard position.

39. $(x-3)^2 + y^2 = 12$ 40. $x^2 + (y+2)^2 = 24$ 41. $(x+4)^2 + (y-2)^2 = 30$

42. $(x-5)^2 + (y+1)^2 = 28$ 43. $3(x-2)^2 + 3(y-3)^2 = 27$ 44. $\frac{(x+1)^2}{5} + \frac{(y+4)^2}{5} = 125$

45. $(x+3)^2 = 15 - (y+3)^2$ 46. $(y-5)^2 = 18 - (x-4)^2$

47. Write the standard equation of a circle with center of (3, –2) and radius equal to 8.

48. Write the standard equation of a circle with center of (–5, 4) and radius equal to $\sqrt{15}$.

49. Write the standard equation of a circle with center of (–3, 4) and a point on the circle of (1,1). (Hint: use the distance formula to find the radius.)

50. Write the standard equation of a circle with center of (3, 5) and a point on the circle of (–2,8).

51. Write the standard equation of a circle with (–3, 5) and (1, –1) being endpoints of a diameter of the circle. (Hint: use the distance formula and the midpoint formula.)

52. Write the standard equation of a circle with (2, –3) and (5, 4) being endpoints of a diameter of the circle.

53. A lithotripter is based on the ellipse $\frac{x^2}{45} + \frac{y}{36} = 1$. How many units from the center of the elliptical tub must the kidney stone and the source for the beam, used to destroy the kidney stone, be placed for this lithotripter to destroy the kidney stone?

54. A lithotripter is based on the ellipse $\frac{x^2}{25} + \frac{y^2}{9} = 1$. How many units from the side of the elliptical tub must the kidney stone and the source for the beam, used to destroy the kidney stone, be placed for this lithotripter to destroy the kidney stone?

55. A whispering gallery is based on the reflective property of an ellipse. A room is constructed with an elliptical ceiling so that a person standing on one focus of the ellipse can hear anything said by a person standing on the other focus. The height of the whispering gallery at the center is 20 feet and the length of the room is 104 feet. What is the equation of the ellipse and how far from the sides of each wall should two people stand to carry on a conversation with each other?

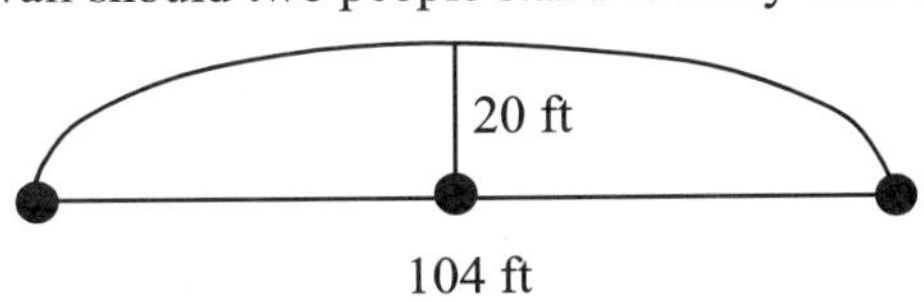

56. A person in a whispering gallery is standing on one focus 5 feet from the wall. A friend is standing on the other focus 120 feet away. What is the length of the room and how high is the ceiling at the center?

(57–58) The orbits of the planets around the sun are elliptical with the sun located at a focus of the ellipse. The ***aphelion*** of a planet is its greatest distance from the sun and the ***perihelion*** of a planet is its shortest distance from the Sun. The ***mean distance*** is half the distance of the major axis.

57. Earth has a perihelion of 91.5 million miles and a mean distance of 93 million miles.
    a. What is the equation in standard form of Earth's orbit around the sun?
    b. What is the aphelion of Earth's orbit?

58. Jupiter has a aphelion of 507 million miles and a mean distance of 483.8 million miles.
    a. What is the equation in standard form of Jupiter's orbit around the sun?
    b. What is the perihelion of Jupiter's orbit?

59. A one-way road passes under an overpass that is in the form of half an ellipse. The opening is 18 feet high at the center and 24 feet wide.
    a. What is the equation in standard form of the ellipse?
    b. What is the tallest, 12 feet wide truck that can pass through the overpass? (Hint: Use the equation above and assume the truck will drive through centered on the overpass.)

60. If the same road (from Problem 59) has an elliptical shaped overpass where the opening is 16 feet high at the center and 20 feet wide,
    a. What is the equation in standard form of the ellipse?
    b. What is the tallest, 10 feet wide truck, that can pass through the overpass?

(61–64) Using the graph, give the equation of each ellipse.

61.

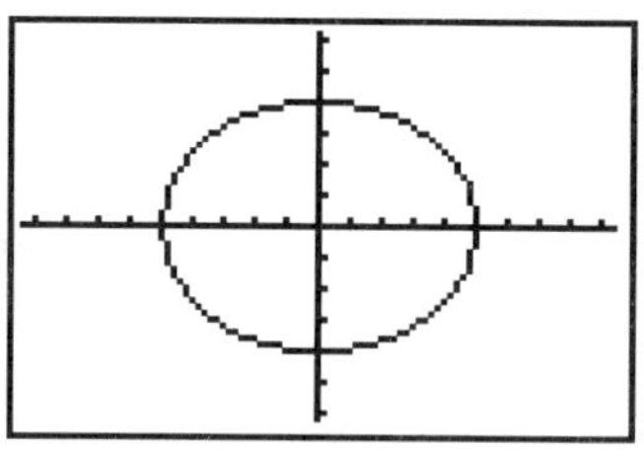

62.

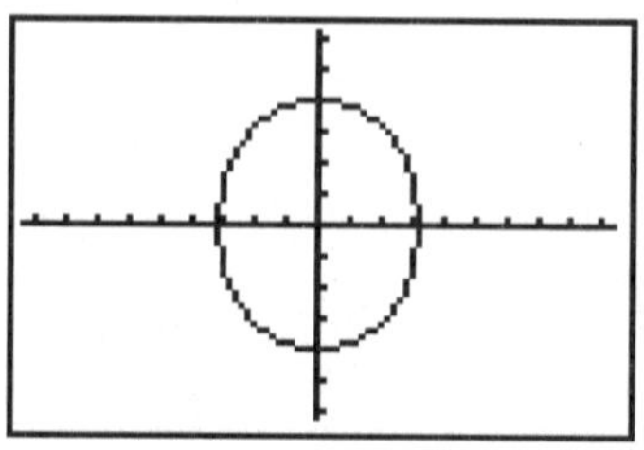

63.

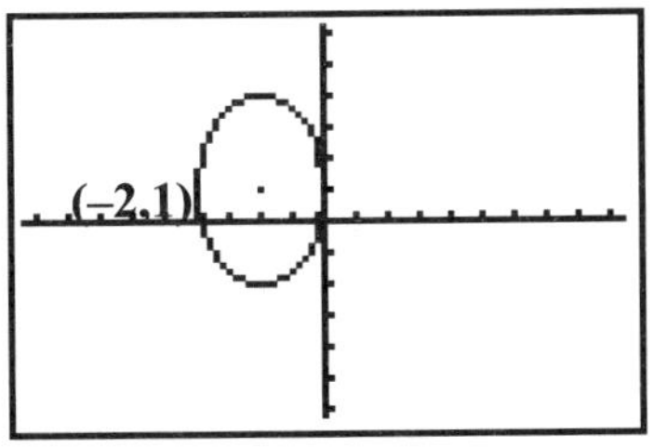

64.

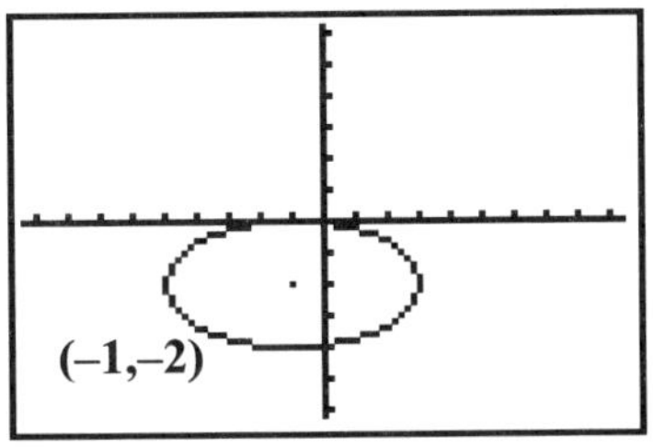

65. You are playing pool on an elliptical pool table with one pocket located at one of the foci points of the ellipse. If the ball you are trying to put into the pocket is located on the other foci point, where would you need to hit the ball off the side in order to make the shot?

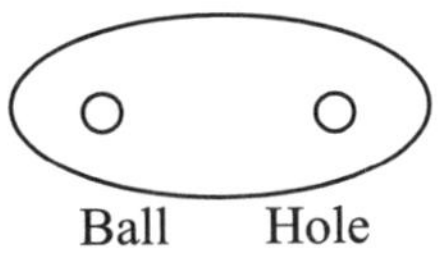

Section 7.3 Hyperbolas

In this section we are going to look at our last conic section, the hyperbola. Like the ellipse and the circle, the hyperbola is not a function. Let's begin our discussion with the definition of the hyperbola.

Hyperbola: A hyperbola is a set of points in the xy-plane, where the difference of the distances between two fixed points (foci) in the xy-plane is a constant difference ($d_1 - d_2 = d$).

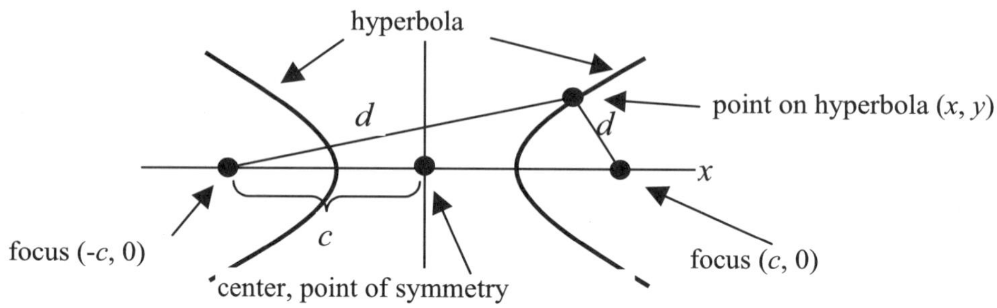

The definition of a hyperbola is very similar to the definition of an ellipse. The only difference between the definitions is that there exists a constant difference with the hyperbola instead of a constant sum. Although there is a slight change between the ellipse and the hyperbola, the proof of where the formula for a hyperbola comes from is very similar. Instead of going through all of the algebra again, we will only state what the formula is for the hyperbola in standard position. But first, we need to define a few terms.

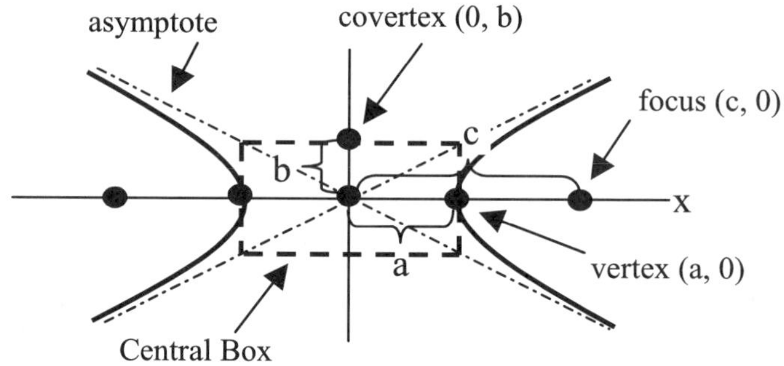

Definitions:

Transverse Axis – The line segment through the center connecting the vertices.
Conjugate Axis – The line segment through the center perpendicular to the transverse axis that connects the covertices.
Vertices – Points on the hyperbola where the hyperbola and the transverse axis intersect.
Covertices – Points that create the conjugate axis.
Central Box – The rectangle, centered around the center of the hyperbola, that is $2b$ units in length vertically and $2a$ units in length horizontally.
Asymptote – The lines, $y = \pm\frac{b}{a}x$ with the hyperbola in standard position, that the graph of the hyperbola gets closer and closer to, as the magnitude of x gets large.

With the change of constant difference instead of constant sum (ellipse), we see a sign change in a couple of places in the formula for a hyperbola compared to the formula for an ellipse. We see that we will now end up subtracting the two squared variables instead of adding them, and we see that we add a^2 and b^2 to get c^2 instead of subtracting, as we do with ellipses. Here are the formulas for hyperbolas.

> **Hyperbolas in Standard Position**: $\frac{x^2}{a^2} - \frac{y^2}{b^2} = 1$ (horizontal hyperbola) or $\frac{y^2}{b^2} - \frac{x^2}{a^2} = 1$ (vertical hyperbola), where the center is (0,0), a is the distance from the center to the left and right edges of the central box, b is the distance from the center to the top and bottom of the central box, and c is the distance from the center to the foci ($c^2 = a^2 + b^2$) along the transverse axis.

We have two different formulas for a hyperbola because, depending upon which squared term is being subtracted, there is a great difference in the graphs. If the x^2 term is subtracting the y^2 term, then the hyperbola will go left and right as in the figures above and have two x-intercepts. But if the y^2 term is subtracting the x^2 term, then the hyperbola will go up and down and have two y-intercepts. Here are two examples.

Example 1: Graphing Hyperbolas on a Graphing Calculator

Graph the following two hyperbolas using your graphing calculator.

a.) $\frac{x^2}{9} - \frac{y^2}{16} = 1$ b.) $\frac{y^2}{16} - \frac{x^2}{9} = 1$

Solution:

To graph these on a graphing calculator we will need to solve them both for y first.

a) $-\frac{y^2}{16} = -\frac{x^2}{9} + 1$	b) $\frac{y^2}{16} = \frac{x^2}{9} + 1$
$y^2 = \frac{16x^2}{9} - 16$ $y = \pm\sqrt{\frac{16x^2}{9} - 16}$	$y^2 = \frac{16x^2}{9} + 16$ $y = \pm\sqrt{\frac{16x^2}{9} + 16}$
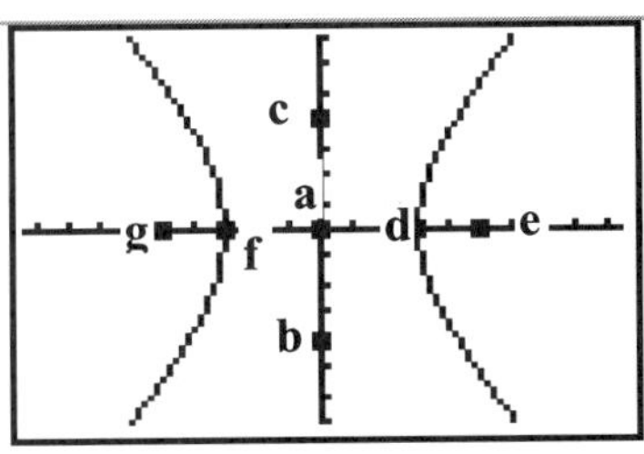	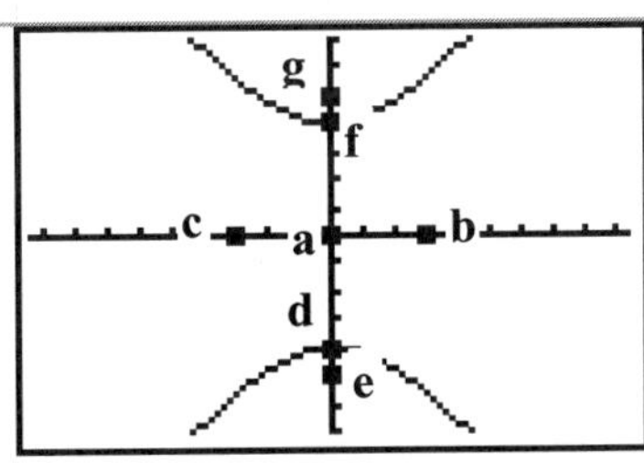

As you can see, if the x term comes first, then the graph of the hyperbola goes left and right (a). If, on the other hand, the y term comes first, then the graph of the hyperbola goes up and down (b). Also on the two graphs we have marked the vertices, covertices, center and the foci.

Question 1: Identify which points (a…g), are the vertices, covertices, center and foci in both graphs of example 1.

In order to sketch hyperbolas by hand we need to follow these steps:

1. Identify what the center is, and find a, b, c for the hyperbola.
2. Plot the center and create the central box.
3. Draw the asymptotes.
4. Draw the hyperbola and label the vertices.
5. Mark and label the foci.

Example 2: Sketching a Hyperbola by Hand

Sketch the hyperbolas from example 1 by hand and label all of the important parts.

Solution:

a.) $\frac{x^2}{9} - \frac{y^2}{16} = 1$	b.) $\frac{y^2}{16} - \frac{x^2}{9} = 1$
The center is (0, 0) since this is in standard position. $a = 3$, $b = 4$ and $c = \sqrt{9+16} = \sqrt{25} = 5$.	The center is (0, 0) since this is in standard position. $a = 3$, $b = 4$ and $c = \sqrt{9+16} = \sqrt{25} = 5$.
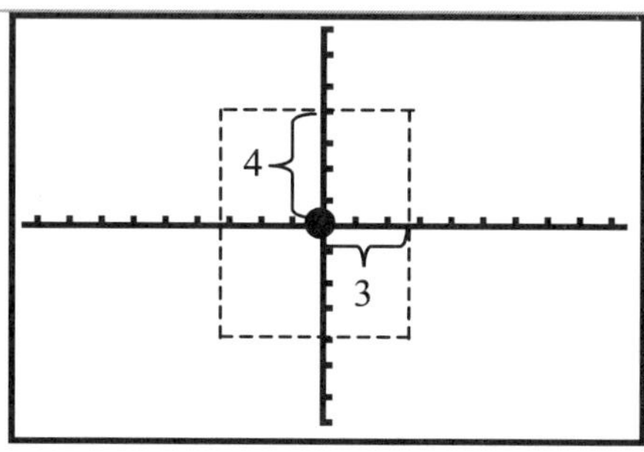	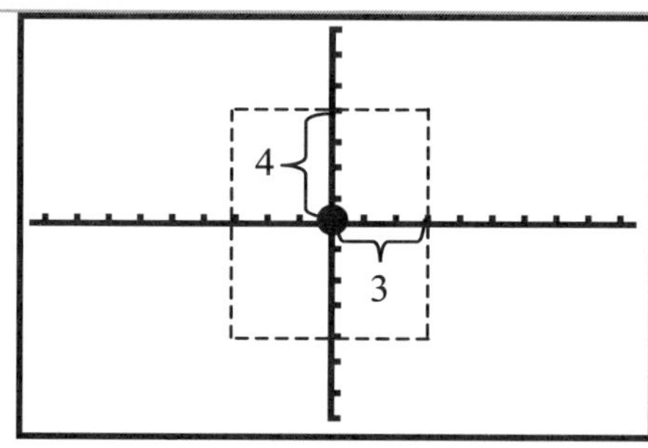
The central box is created by using the a and b values to make a box centered on (0, 0) that is $2a$ x $2b$ in size.	The central box is created by using the a and b values to make a box centered on (0, 0) that is $2a$ x $2b$ in size.

<table>
<tr>
<td>We draw the asymptotes by drawing two lines through the opposite corners of the central box.
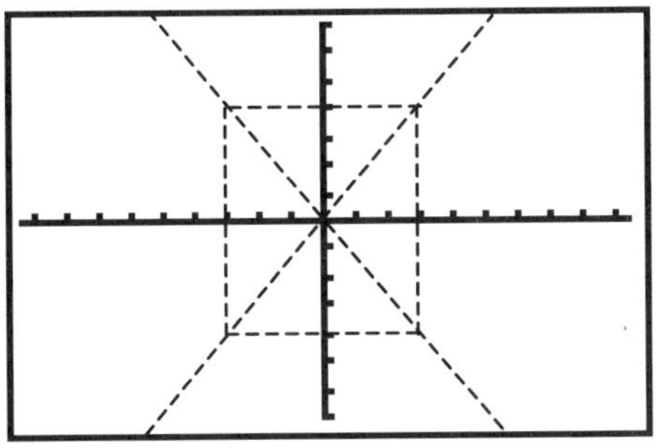</td>
<td>We draw the asymptotes by drawing two lines through the opposite corners of the central box.
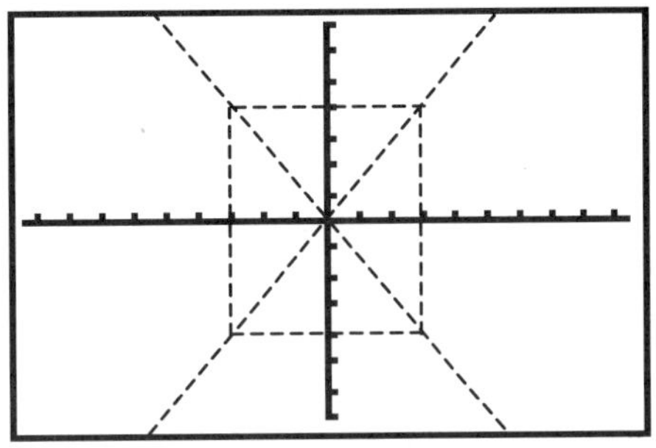</td>
</tr>
<tr>
<td>Since x^2 is first (positive), we know that the hyperbola goes left and right. So we sketch in the graph using the x-intercepts of the central box as the vertices and the asymptotes as guides. Lastly, label the foci.
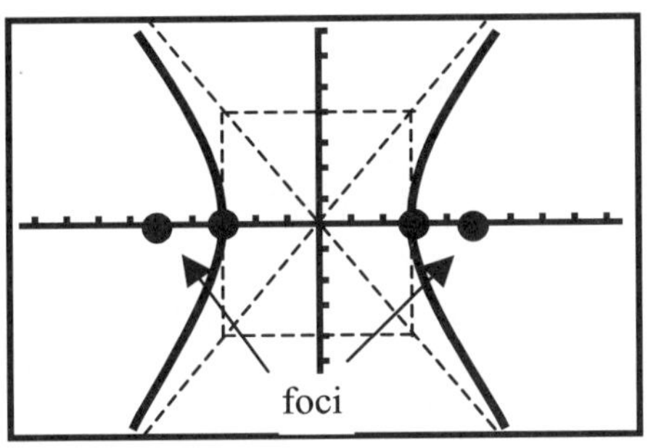
</td>
<td>Since y^2 is first (positive), we know that the hyperbola goes up and down. So we sketch in the graph using the y-intercepts of the central box as the vertices and the asymptotes as guides. Lastly, label the foci.
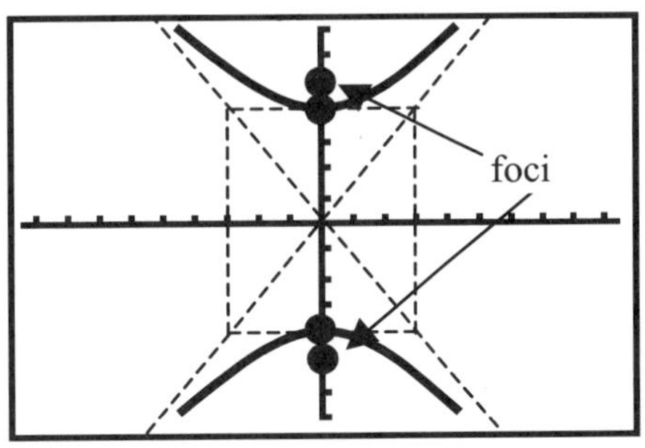
</td>
</tr>
<tr>
<td>vertices (–3, 0), (3, 0), foci (–5, 0), (5, 0)</td>
<td>vertices (0, –4), (0, 4), foci (0, –5), (0, 5)</td>
</tr>
</table>

Notice that the asymptotes go through the opposite corners of the central box and through the center. This makes it very easy to find the formula for these lines. In the last example both central boxes were the same, so they have the same asymptotes. Since the hyperbolas were in standard position (i.e. centered at the origin), the slope of one asymptote will be $\frac{4}{3}$, since the upper right corner of the box is (3, 4) and you must rise 4 and run 3 from the origin to get to that corner ($m = \frac{rise}{run}$). The other asymptote has slope $-\frac{4}{3}$, since the upper left corner of the box is (–3, 4) and you must rise 4 and run –3 from the origin to get to that corner. Also since the y-intercept is (0, 0), we can now use the slope-intercept form of a line to get the equations. The equations of the two asymptotic lines are $y_1 = \frac{4}{3}x$ and $y_2 = -\frac{4}{3}x$.

Answer Q1: a – center, b & c – covertices, d & f – vertices, and e & g – foci.

Example 3: Sketching a Vertical Hyperbola

Graph the following hyperbola by hand and label the vertices, foci, and asymptotes. $\frac{y^2}{25} - \frac{x^2}{4} = 1$

Solution:

<table>
<tr><td colspan="2">First, notice that the center is (0, 0), and that $b = 5$, $a = 2$, and $c = \sqrt{25+4} = \sqrt{29} \approx 5.385$.</td></tr>
<tr><td>We now make the central box and add in the asymptotes.</td><td>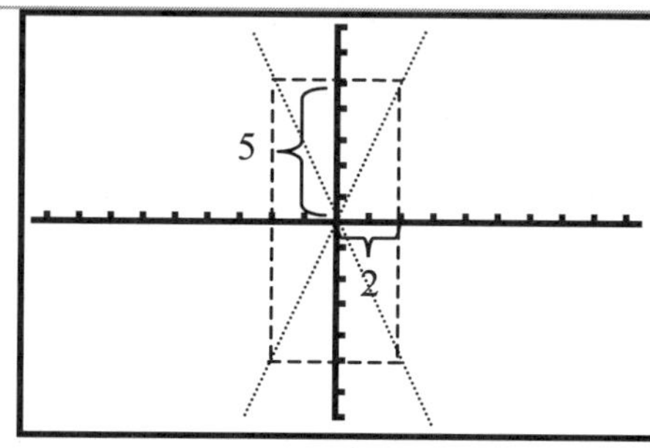
</td></tr>
<tr><td>Since the y^2 term is subtracting the x^2 term the hyperbola will be going up and down, so the vertices are (0, –5), (0, 5) and the foci are (0, –5.39), (0, 5.39). The asymptotes have the formulas $y = \frac{5}{2}x$ and $y = -\frac{5}{2}x$.</td><td>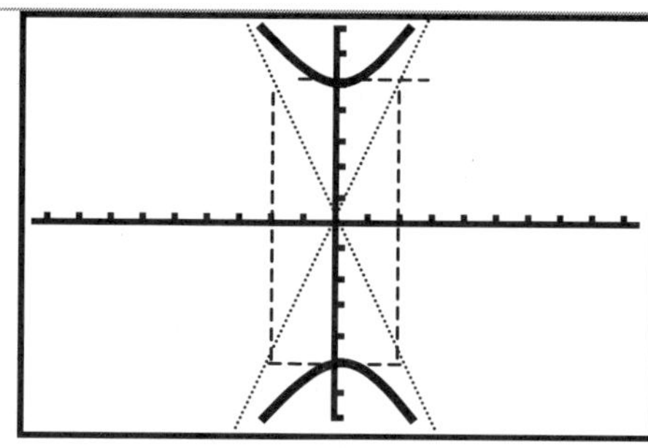</td></tr>
</table>

Question 2: Given what we have done in 7.1, try sketching a graph of this hyperbola.

$$\frac{(x+2)^2}{36} - \frac{(y-3)^2}{64} = 1$$

After this question we can now state the standard form of a hyperbola.

The Standard Forms of a Hyperbola are: $\frac{(x-h)^2}{a^2} - \frac{(y-k)^2}{b^2} = 1$ (horizontal hyperbola) and $\frac{(y-k)^2}{b^2} - \frac{(x-h)^2}{a^2} = 1$ (vertical hyperbola), where the center is (h, k), a is the distance from the center to the left and right edges of the central box, b is the distance from the center to the top and bottom of the central box, and c is the distance from the center to the foci ($c^2 = a^2 + b^2$) along the transverse axis.

Example 4: Sketching a Hyperbola

Graph the following hyperbola by hand and label the vertices, foci, and asymptotes. Also state the domain and range of this relation. $\frac{(x-5)^2}{3} - y^2 = 1$

Solution:

Let's rewrite this equation so that it will be easier to see what the values are for *a*, *b*, *c* and the center.	$\frac{(x-5)^2}{3} - \frac{(y-0)^2}{1} = 1$
From this form, it is easy to see that $a = \sqrt{3}$, $b = 1$, $c = \sqrt{3+1} = \sqrt{4} = 2$, and the center is (5, 0). With this information we can now make the central box and add in the asymptotes.	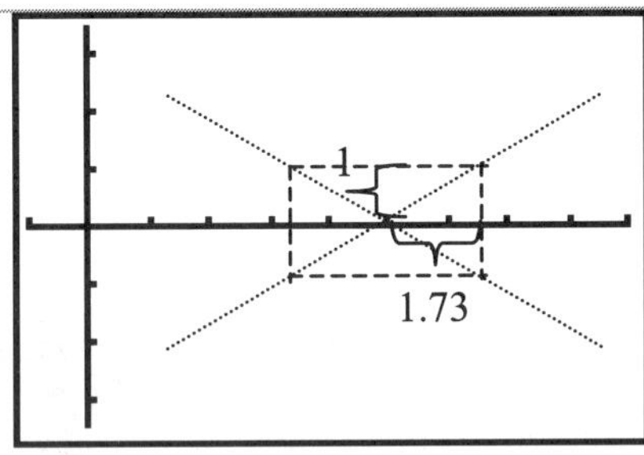
Since the x^2 term is subtracting the y^2 term the hyperbola will be going left and right, so the vertices are $(5-\sqrt{3}, 0)$, $(5+\sqrt{3}, 0)$ and the foci are (3, 0), (7, 0). The asymptotes have the formulas $y = \frac{1}{\sqrt{3}}(x-5)$ and $y = -\frac{1}{\sqrt{3}}(x-5)$, since the center is translated 5 units to the right from the hyperbola in standard position.	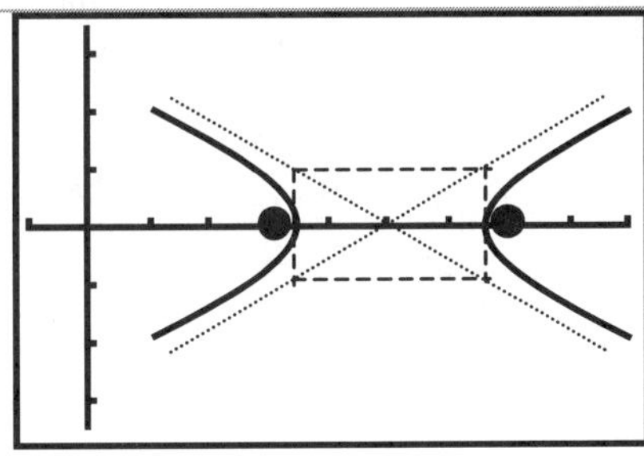

REFLECTIVE PROPERTY OF A HYPERBOLA

Let's talk about the reflective property of a hyperbola: anything coming in towards one of the foci points will be reflected off of one branch of the hyperbola back towards the other foci point. Here is an illustration of the property and one way in which it is used in telescopes.

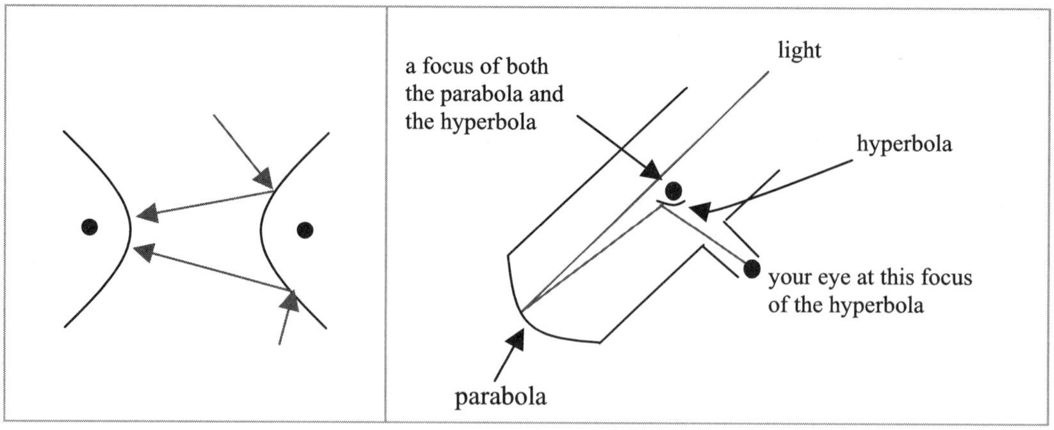

Answer Q2: Since we have (x+2) and (y−3), the center will be (−2, 3) because this hyperbola has been translated from the standard position. Also, since the x term is positive, a = 6, b = 8, and c = $\sqrt{100}$ *= 10, the graph is as follows:*

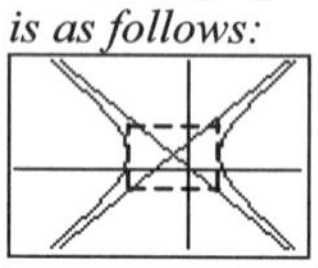

Section Summary:

- The standard form of a hyperbola has two possible forms: $\frac{(x-h)^2}{a^2}-\frac{(y-k)^2}{b^2}=1$ (horizontal hyperbola) or $\frac{(y-k)^2}{b^2}-\frac{(x-h)^2}{a^2}=1$ (vertical hyperbola).
- $c^2 = a^2 + b^2$, where c is the distance from the center to the foci.
- The center is (h, k).
- a and b define the edges of the central box.
- The reflective property of a hyperbola means that anything coming in towards one focal point is reflected back towards the other.

SECTION 7.3 HOMEWORK

(1–8) For each of the hyperbolas in standard position write the standard equation and find the following:

a.) Horizontal or vertical hyperbola

b.) Center

c.) Foci

d.) Vertices

e.) Covertices

f.) Asymptotes

g.) Domain and range

Then sketch a graph of the hyperbola.

1. $\frac{x^2}{16}-\frac{y^2}{9}=1$

2. $\frac{y^2}{25}-\frac{x^2}{144}=1$

3. $\frac{y^2}{16}-\frac{x^2}{16}=1$

4. $\frac{x^2}{25}-\frac{y^2}{25}=1$

5. $\frac{x^2}{8}=\frac{y^2}{12}+1$

6. $\frac{y^2}{18}=1+\frac{x^2}{24}$

7. $3x^2-5y^2=45$

8. $4x^2-4y^2=100$

(9–18) For each of the hyperbolas in translated position write the standard equation and find the following:

a.) Horizontal or vertical hyperbola

b.) The translation from standard position

c.) Center

d.) Foci

e.) Vertices

f.) Covertices

g.) Asymptote

h.) Domain and range

Then sketch the graph.

9. $\frac{(x-3)^2}{144}-\frac{(y+3)^2}{25}=1$

10. $\frac{(y+5)^2}{16}-\frac{(x+4)^2}{9}=1$

11. $\frac{(y-5)^2}{9}-\frac{(x-3)^2}{25}=1$

12. $\frac{(x+5)^2}{169}-\frac{(y+3)^2}{25}=1$

13. $\frac{(x+1)^2}{16}-\frac{(y-2)^2}{16}=1$

14. $\frac{y^2}{25}-\frac{(x-3)^2}{25}=1$

15. $\frac{(y-6)^2}{8}=\frac{x^2}{12}+1$

16. $\frac{(x-2)^2}{10}=1+\frac{(y-7)^2}{4}$

17. $3(x+2)^2-4(y-1)^2=36$

18. $5(y+3)^2-5(x-3)^2=45$

19. Find the standard equation of a hyperbola with center of (0, 0); vertices of (3, 0) and (–3, 0); and foci points of (5, 0) and (–5, 0).

20. Find the standard equation of a hyperbola with center of (0, 0); vertices of (0, 4) and (0, –4); and foci points of (0, 5) and (0, –5).

21. Find the standard equation of a hyperbola with center of (0, 0); vertices of (0, 8) and (0, –8); and covertices of (4, 0) and (0, –4).

22. Find the standard equation of a hyperbola with center of (0, 0); vertices of (6, 0) and (–6, 0); and covertices of (0, 8) and (0, –8).

23. Find the standard equation of a hyperbola with center (0, 0); constant difference of 10; and foci points of (13, 0) and (–13, 0).

24. Find the standard equation of a hyperbola with center (0, 0); constant difference of 12; and foci points of (0, 10) and (0, –10).

25. Find the standard equation of a hyperbola with center (–2, 3); vertices of (–2, 8) and (–2,–2); and covertices of (2, 3) and (–6, 3).

26. Find the standard equation of a hyperbola with center (3, –2); vertices of (8, –2) and (–2,–2); and covertices of (3, 3) and (3, –7).

27. Find the standard equation of a hyperbola with center (2, 5); foci points of (12, 5) and (–8,5); and vertices of (8, 5) and (–4, 5).

28. Find the standard equation of a hyperbola with center (–3, –1); foci points of (–3, 7) and (–3, –9); and vertices of (–3, 3) and (–3, –5).

29. Find the standard equation of a hyperbola with center (0, –2); constant difference of 10; and covertices of (0, 2) and (0, –6).

30. Find the standard equation of a hyperbola with center (3, 0); constant difference of 16; and covertices of (13, 0) and (–7, 0).

31. The hyperbola $\frac{y^2}{9}-\frac{x^2}{16}=1$ is translated to the left 5 and down 4. What is the equation in standard form of the resulting hyperbola?

32. The hyperbola $\frac{x^2}{25}-\frac{y^2}{144}=1$ is translated to the right 2 and up 5. What is the equation in standard form of the resulting hyperbola?

(33–48) Write each of the equations in standard form and answer the following questions:

a.) Does the equation represent a hyperbola, an ellipse, a circle or a parabola?

b.) Is the conic in standard position or translated position?

c.) If the conic is an ellipse or hyperbola, is it a vertical or horizontal type?

33. $(x-2)^2+(y-1)^2=25$
34. $y=3(x-2)^2+5$
35. $3(x-1)^2-4(y+3)^2=24$
36. $5(x+3)^2+4y^2=40$
37. $x=-2(y+5)^2-8$
38. $3(y-2)^2=18+(x-1)^2$
39. $8(y+7)^2=48-6(x-9)^2$
40. $3(x-4)^2=48-3(y+1)^2$
41. $3x^2+4y^2=12$
42. $y=3x^2$
43. $x^2+y^2=36$
44. $8x^2-9y^2=72$

45. $x = -2y^2$

46. $5x^2 + 3y^2 = 60$

47. $3y^2 = 5x^2 + 30$

48. $3x^2 = 27 - 3y^2$

(49–52) Using the graphs, write an equation for each hyperbola.

49.

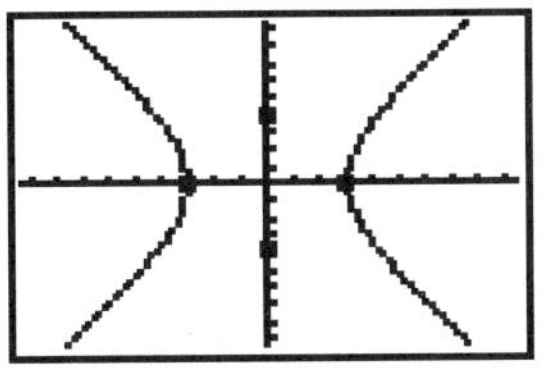

50.

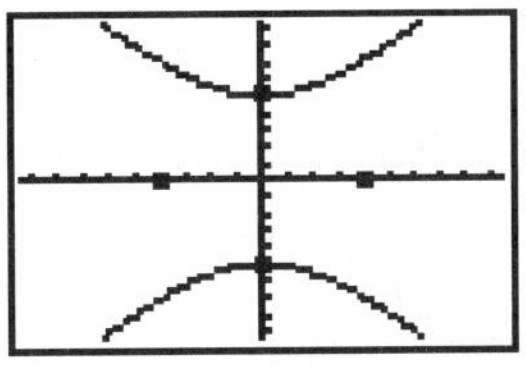

51.

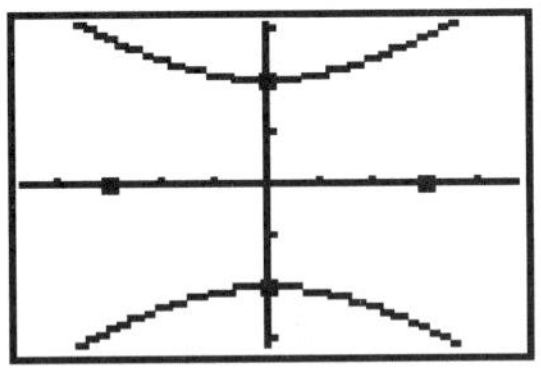

52.

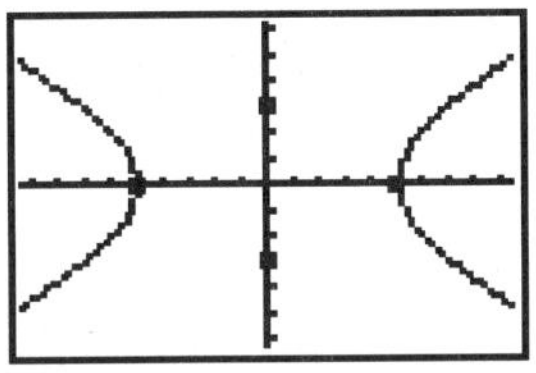

Section 7.4 Identifying Conics

In this section we are going to be looking at all four of the conics we have learned about so far in this chapter. We will be learning how we can distinguish each from the others, and how we can take the general forms of the conics and change them into their standard forms. But before we begin our discussion, let's review the four conics by answering the following question.

Question 1: Given the following four conics in standard form, draw a graph of each one.

Parabola	Ellipse	Circle	Hyperbola
a) $y=\frac{1}{8}(x-3)^2-2$	b) $\frac{(x+1)^2}{36}+\frac{(y-5)^2}{49}=1$	c) $x^2+(y+4)^2=16$	d) $\frac{(x-2)^2}{4}-\frac{(y+1)^2}{9}=1$

The general form of *any* conic, from which each of the previous four come, is defined as follows:

General Form of a Conic: Any second-degree equation of the form

$$Ax^2+Bxy+Cy^2+Dx+Ey+F=0$$

is an equation for a conic section in general form.

This general form can be used to express more than just conics. Here are a few examples.

If A, B, and C are equal to 0, then you have	$Dx+Ey+F=0$, the standard form of a line
If B, C, and D are equal to 0, then you have	$Ax^2+Ey+F=0$, a parabola $\left(y=\frac{-A}{E}x^2-\frac{F}{E}\right)$
If B, D, and E are equal to 0, then you have	$Ax^2+Cy^2+F=0$, an ellipse, hyperbola or a circle in standard position (centered at the origin)

As you can see, depending on which of the coefficients are 0, you get completely different types of functions and relations.

Question 2: Which coefficients need to be 0 in order to get a vertical line?

We saw above that if B, D and E are 0, then we get $Ax^2+Cy^2+F=0$. This is a non-function conic in standard position. For example,

If $A=9$ and $C=25$ and $F=-225$, then we would have	$9x^2+25y^2-225=0$, which simplifies to $\frac{x^2}{25}+\frac{y^2}{9}=1$, an ellipse

Here is its graph.	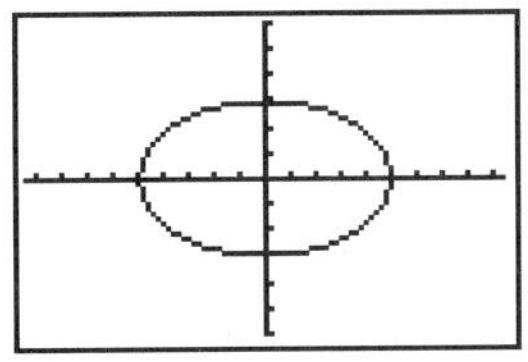

Notice, that this ellipse is in what we call standard position, that is, centered at the origin. If, on the other hand, we make only $B = 0$, then we have $Ax^2 + Cy^2 + Dx + Ey + F = 0$. This turns out to be a translated conic. For example,

If we adjust our last example and let $D = 5$, $E = 7$ and $F = 36$, then we get	$9x^2 + 25y^2 - 36x + 150y + 36 = 0$
When we complete the square, to put it into standard form, we get	$\frac{(x-2)^2}{25} + \frac{(y+3)^2}{9} = 1$
This ellipse has a center at (2, –3). It has been translated 2 units to the right and 3down.	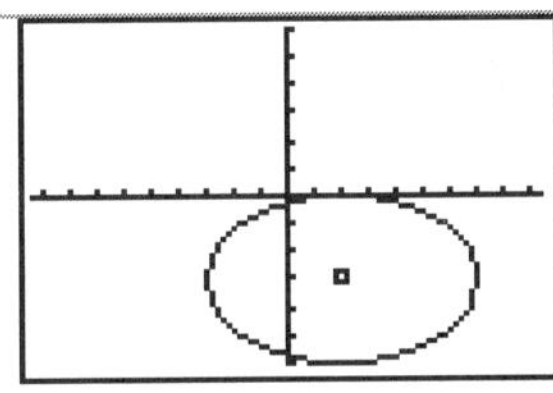

In this book we are not going to take the time to look at equations that have a Bxy term. If there is such a term in a second-degree equation, then you get a conic that is rotated. For example, if we have the equation $x^2 + 2xy + 2y^2 - x - 3y - 10 = 0$, then we get this for its graph.

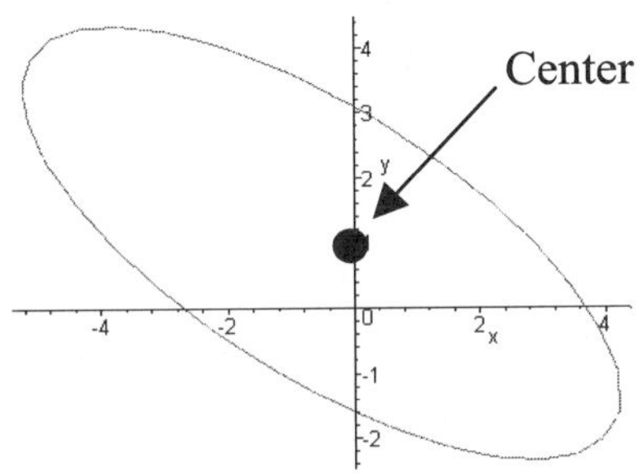

As you can see, this is just an ellipse that has been translated, left and up, and rotated. The x and y terms translate and the xy term rotates.

It is now time to quickly review how to **complete the square**, which was discussed in section 2.2. In this current chapter you should have noticed that in every standard form of a conic there was at least one term of the form $(x \pm h)^2$ or $(y \pm k)^2$, binomial squares, which will produce a perfect square trinomial $(x^2 - 2hx + h^2)$ or $(y^2 - 2ky + k^2)$.

If you want something in the form of a binomial square, you first need to find a perfect square trinomial and then factor it into the binomial square that you are looking for. Usually, the equation you begin with doesn't have a perfect square trinomial in it and that's when you need to complete the square. That is, you perform the proper steps so that you end up with a perfect square trinomial, which in turn is then factored into a binomial square. Let's do one example to refresh your memory.

Example 1: Completing the Square

Complete the square on $-3x^2 + 12x$ in order to change it into a binomial square.

Solution:

In this problem we are looking to add some amount to what we already have in order to create a perfect square trinomial, which then can be factored into a binomial square. We need to factor out the coefficient of the squared term first, before we can begin to complete the square. Remember, we need $x^2 - 2hx + h^2$ so that it will factor into $(x - h)^2$. As you can see, this will not allow for any coefficient other than a one to be in front of the squared term.

Answer Q1:

a)

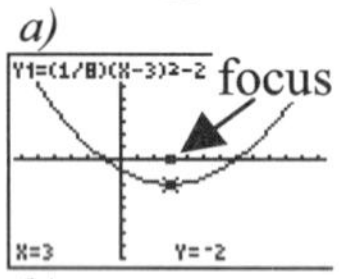

b)

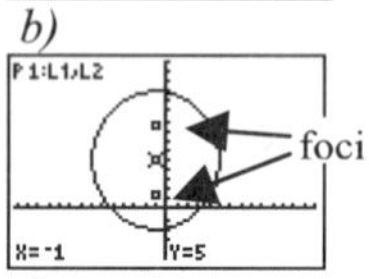

c)

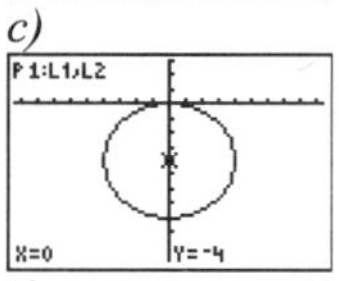

d)

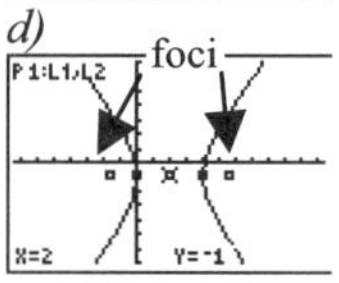

What we need to get	What we have
$x^2 - 2hx + h^2$	$-3x^2 + 12x$
First, factor out a factor of –3. This gives us a coefficient of 1; we can now complete the square on the part inside the parentheses.	$-3(x^2 - 4x)$
$-2h$ So h must equal 2, and h^2 must equal 4	needs to be equal to -4 So, we need to add 4 inside the parentheses.
By adding 4 inside the parentheses, what we had becomes	$-3(x^2 - 2x + \mathbf{4}) = -3\,(x - 2)^2$

We have now completed the square.

Remember, in order to complete the square you will need to do the following steps:

- Concentrate on just the squared term and the linear term.
- Factor out the coefficient of the squared term in the expression, if there is one.
- Take half of the coefficient on the linear term and square it and add that amount to your expression.
- Now factor the perfect square trinomial, which will always be the variable plus the half amount from the previous step.

Answer Q2: A, B, C, and E would need to be zero leaving $Dx + F = 0$.

Question 3: Complete the square on the expression $5x^2 + 15x$.

Now, we will use this process to take conics in the general form and change them into the standard form.

Discussion 1: Transforming a Parabola Into Its Standard Form

Let's put the parabola, $-x^2 + 6x + y - 10 = 0$, into its standard form.

What we need to get $y = a(x-h)^2 + k$	**What we have** $-x^2 + 6x + y - 10 = 0$
We need to get y by itself (add x^2 and 10 and subtract $6x$ on both sides of the equation).	$y = x^2 - 6x + 10$
Complete the square on the x variable. Now, take half of the 6, square it, and add that inside the parentheses.	$y = (\mathbf{x^2 - 6x} \quad\quad) + 10$ $y = (x^2 - 6x + \mathbf{9}) + 10$

This completes the square but it changes the equation. We need to make some kind of adjustment to counter the fact that we just added 9 to the right side of the equation. There are several ways to do this.

Question 4: Write down one way that you can think of to counter what we did so that the equation will remain equivalent.

The typical response to our question is to add 9 to the other side of the equation. Another way to make the adjustment is to subtract 9 from the same side that you added 9. This is, in our opinion, a good way to go about making this adjustment when you are dealing with conics. Let's continue with our example.

When we add 9 inside the parentheses we will also subtract 9 in the same place. This equation has not changed, since in essence we have added 0 to the right side.	$y = (x^2 - 6x + \mathbf{9 - 9}) + 10$
We factor the perfect square trinomial and move the –9 to the outside of the parentheses.	$y = (\mathbf{x^2 - 6x + 9} \underline{-9}) + 10 = \mathbf{(x-3)^2} \underline{-9} + 10$
Add the like terms and we're done.	$y = (x-3)^2 + 1$

Example 3: Transforming a Hyperbola in to its Standard Form

Put the hyperbola, $4x^2 - 25y^2 + 16x + 50y - 109 = 0$, into its standard form.

Solution:

What we need to get $\dfrac{(x-h)^2}{a^2}-\dfrac{(y-k)^2}{b^2}=1$	**What we have** $4x^2 - 25y^2 + 16x + 50y - 109 = 0$
Our first step is to complete the square on both variables.	$(4x^2 + 16x) + (-25y^2 + 50y) - 109 = 0$
We begin by factoring out the coefficients in front of the squared terms.	$4(x^2 + 4x) - 25(y^2 - 2y) - 109 = 0$
We then complete the square by: • halving the coefficients in front of the linear terms • squaring that result • adding and subtracting the squared amounts to each parenthesis.	$4(x^2 + \underline{4}x + \mathbf{4 - 4}) - 25(y^2 - \underline{2}y + \mathbf{1 - 1}) - 109 = 0$
Factor the perfect square trinomials.	$4[(x+2)^2 - 4] - 25[(y-1)^2 - 1] - 109 = 0$
Here is where you need to be careful. The 4 and –25 in front of the brackets will change the –4 and –1 inside the brackets because of the distributive law.	$4[(x+2)^2 - 4] - 25[(y-1)^2 - 1] - 109 = 0$ $4(x+2)^2 \mathbf{- 16} - 25(y-1)^2 \mathbf{+ 25} - 109 = 0$ $4(x+2)^2 - 25(y-1)^2 \mathbf{+ 25} - 109 \mathbf{- 16} = 0$
We now move all of the numeric values to the other side of the equation.	$4(x+2)^2 - 25(y-1)^2 = 100$
We almost have what we want. Now, just divide both sides by 100 and simplify and we will have our hyperbola in standard form.	$\dfrac{(x+2)^2}{25}-\dfrac{(y-1)^2}{4}=1$

Answer Q3: Factor out 5 to get $5(x^2 + 3x)$. Now, complete the square in the parentheses by taking half of 3 and squaring it and adding the result $5(x^2+3x+\mathbf{9/4}) = 5(x + 3/2)^2$.

Question 5: Put this ellipse $x^2 + 9y^2 + 6x - 36y + 36 = 0$ into its standard form.

Another thing we need to discuss is how to identify what kind of conic you might be looking at when it is in the general form. Let's look at question 1 again.

Discussion 2: Observations about General Forms

Let's take the following four conics in standard form, and write them in the standard form.

Parabola	Ellipse	Circle	Hyperbola
a) $y=\frac{1}{8}(x-3)^2-2$	b) $\frac{(x+1)^2}{36}+\frac{(y-5)^2}{49}=1$	c) $x^2+(y+4)^2=16$	d) $\frac{(x-2)^2}{4}-\frac{(y+1)^2}{9}=1$

Here is what the four conics look like in the general form.

Parabola	Ellipse	Circle	Hyperbola
$x^2-6x-8y-7=0$	$49x^2+36y^2+98x-360y-815=0$	$x^2+y^2+8y=0$	$9x^2-4y^2-36x-8y-4=0$

Question 6: What is the main difference between the parabola and all the rest of the conics with regards to the general forms?

When you're looking at a conic written in its general form, ask yourself the following questions:

- Does it have just one squared term? If so, it is a parabola.
- Does it have a positive squared term and a negative squared term? If so, it is a hyperbola.
- Are the coefficients of the squared terms the same? If they are, then you have a circle.

If you have determined that you don't have a parabola (one squared term) and you don't have a hyperbola (different signs on the squared terms) and that you don't have a circle (same coefficients on squared terms), then you are looking at an ellipse.

We must say here that there are cases where you don't get any of these. You might have an equation that isn't true for any points in the plane like ($x^2+y^2=-4$). You might also have what are called degenerate cases. Here is an example of such a case: $3x^2-\frac{4}{3}y^2-24x-8y=0$. If we change this into its standard form, we get $\frac{(x-4)^2}{12}-\frac{(y+3)^2}{27}=0$. This turns out to be two intersecting lines.

Answer Q4: If you added 9 to the right side then add 9 to the left side.

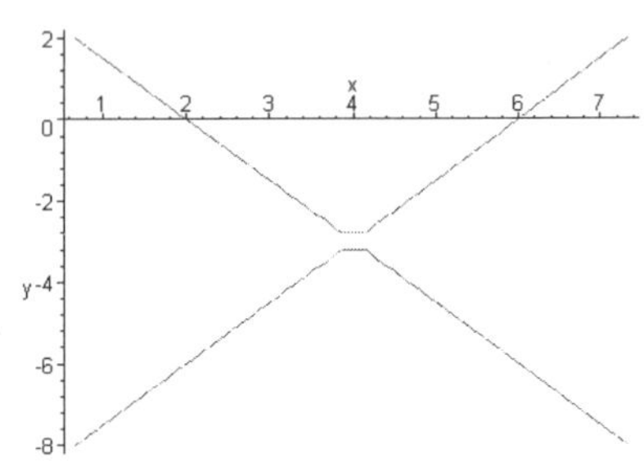

We are not going to focus on these special cases in this book. But their existence is worth noting.

Question 7: Identify the following conics.

a) $2x^2+2y^2-6x=10$ b) $y^2-9y+6x-25=9x^2$ c) $x+7=y^2-y$

We have one last topic to discuss here: using your calculator to graph conics in general form. If you are given an equation for a conic section in general form, then there is an easy way to get the graphing calculator to graph it for you.

Example 4: Using the Graphing Calculator with General Forms

Use the graphing calculator to graph $7x^2 + 2y^2 - 3x + 5y - 10 = 0$.

Solution:

Answer Q5: Complete the square on both variables by adding and subtracting 9 with the x's and 4 with the y's $(x^2+6x+9-9)+9(y^2-4y+4-4)+36=0$. This now becomes $(x+3)^2+9(y-2)^2=9+36-36$ which equals $(x+3)^2+9(y-2)^2=9$, which, in standard form is $\frac{(x+3)^2}{9}+\frac{(y-2)^2}{1}=1$

First, we rewrite this equation in a form that will allow us to use the quadratic formula (i.e. in descending order with respect to the y variable).	$\mathbf{2}y^2 + \underline{5}y + 7x^2 - 3x - 10 = 0$
Now, in the quadratic formula, $a = \mathbf{2}$ (the number in front of the squared term), $b = \underline{5}$ (the number in front of the linear term), and $c = 7x^2 - 3x - 10$ (all the rest of the terms in the equation that don't have a y in them).	$y = \dfrac{-5 \pm \sqrt{25 - 4(2)(7x^2 - 3x - 10)}}{2(2)}$
Now, you can type the following two equations into your calculator.	$Y1 = \dfrac{-5 + \sqrt{25 - 8(7x^2 - 3x - 10)}}{4}$ $Y2 = \dfrac{-5 - \sqrt{25 - 8(7x^2 - 3x - 10)}}{4}$
Here is what the graph looks like on a zoom decimal screen.	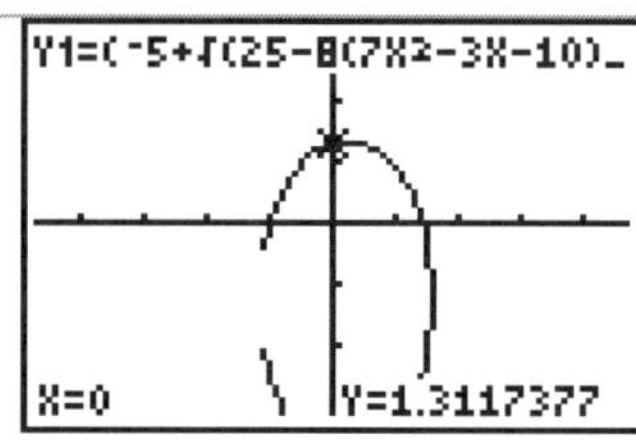

Section Summary

- The general form of a conic is $Ax^2 + Bxy + Cy^2 + Dx + Ey + F = 0$.
- For our purposes we will use this $Ax^2 + Cy^2 + Dx + Ey + F = 0$ as the general form. We will not look at any equations that contain a Bxy term in this text.
- To complete the square you must do the following:
    1. Concentrate on just the squared term and the linear term.
    2. Factor out the coefficient of the squared term in the expression, if there is one.

3. Take half of the coefficient of the linear term, square it, and add that amount to your expression.
4. Now factor the perfect square trinomial, which will always be the variable plus the amount found in step 3.

- You can identify the conics in the **general form** ($Ax^2 + Cy^2 + Dx + Ey + F = 0$) in the following ways:
    1. Parabolas have only one squared variable.
    2. Hyperbolas have squared variables with opposite signs.
    3. Circles have positive squared variables with the exact same coefficients.
    4. Ellipses have positive squared variables with different coefficients.

Answer Q6: The parabola only has one squared term.

Answer Q7: a) Is a circle because the coefficients are the same. b) Is a hyperbola because when you rewrite it in the general form you have $y^2 - 9x^2$ (opposite signs). c) Is a parabola because only one of the variables is squared.

SECTION 7.4 HOMEWORK

(1–24) For each of the following general equations state what type of conic each represents.

1. $4x^2 + 9y^2 - 24x - 36y + 36 = 0$
2. $25x^2 + 16y^2 + 50x + 64y - 311 = 0$
3. $4x^2 - 9y^2 - 16x - 54y - 101 = 0$
4. $-4x^2 + 9y^2 + 8x - 36y - 4 = 0$
5. $2x^2 - 12x - y + 23 = 0$
6. $3y^2 - x + 6y - 4 = 0$
7. $x^2 + y^2 - 4x - 6y - 3 = 0$
8. $x^2 + y^2 + 10x + 6y + 16 = 0$
9. $-x^2 + 2y^2 + 8x - 24y + 52 = 0$
10. $x^2 - 2y^2 + 10x - 16y - 11 = 0$
11. $2x^2 + y^2 - 20x + 6y + 49 = 0$
12. $x^2 + 2y^2 + 8x - 8y - 6 = 0$
13. $5y^2 + x - 30y + 37 = 0$
14. $3x^2 + 24x + y + 55 = 0$
15. $3x^2 + 3y^2 + 24x - 30y + 119 = 0$
16. $4x^2 + 4y^2 - 24x + 32y + 91 = 0$
17. $48x^2 - 120x - 16y + 81 = 0$
18. $32y^2 - 8x + 48y + 13 = 0$
19. $8x^2 - 12y^2 + 24x + 60y - 105 = 0$
20. $-32x^2 + 48y^2 + 80x + 72y - 215 = 0$
21. $8x^2 + 16y^2 + 24x + 24y - 5 = 0$
22. $24x^2 + 12y^2 - 32x - 20y - 53 = 0$
23. $27x^2 + 27y^2 + 36x + 18y + 11 = 0$
24. $500x^2 + 500y^2 - 400x - 300y + 89 = 0$

(25– 48) For each of the general equations (1–24) rewrite in standard form and sketch a graph.

25. (1) 26. (2) 27. (3) 28. (4)
29. (5) 30. (6) 31. (7) 32. (8)
33. (9) 34. (10) 35. (11) 36. (12)
37. (13) 38. (14) 39. (15) 40. (16)
41. (17) 42. (18) 43. (19) 44. (20)
45. (21) 46. (22) 47. (23) 48. (24)

(49–52) **Using calculus, the area of an ellipse is πab and the circumference of the ellipse is approximately $2\pi\sqrt{\frac{a^2+b^2}{2}}$, where a^2 and b^2 are the denominators of an ellipse written in standard form.** For each of the ellipses below find the area and approximate circumference.

49. $16x^2 + 9y^2 - 144 = 0$
50. $4x^2 + 25y^2 - 100 = 0$
51. $9x^2 + 16y^2 - 54x + 64y + 1 = 0$
52. $36x^2 + 4y^2 + 216x - 8y + 184 = 0$

(53–58) For each of the following parabolas find the maximum or minimum value by using the standard form and the vertex, check your answer with the graphing calculator and maximum or minimum under **calc**.

53. $x^2 - 24x + y - 354 = 0$
54. $x^2 - 100x - y + 300 = 0$
55. $x^2 - 50x - y + 2500 = 0$
56. $x^2 - 20x + y - 500 = 0$
57. $3x^2 - 18x + y - 10 = 0$
58. $4x^2 - 32x - y + 300 = 0$
59. A company's revenue is determined by the equation $R = 400x - x^2$, where x is the number of items produced and R is the revenue. Rewrite the quadratic in standard form; find the number

of items to produce for maximum revenue and determine the maximum revenue. (Use the maximizing capability of your graphing calculator to check your answer.)

60. A company's profit is determined by the equation $P = -x^2 + 60x - 100$, where x is the number of items produced and P is the profit. Rewrite the quadratic in standard form; find the number of items to produce for maximum profit and determine the maximum profit. (Use the maximizing capability of your graphing calculator to check you answer.)

61. A company's profit is determined by the equation $P = 230 + 20x - 0.5x^2$, where x is the number of items produced and P is the profit. Rewrite the quadratic in standard form; find the number of items to produce for maximum profit and determine the maximum profit. (Use the maximizing capability of your graphing calculator to check your answer.)

62. A piece of sheet metal with dimensions of 15 inches by 18 inches is used to form a gutter by bending up the side that is 15 inches long.
    a. Give an equation for the volume of that gutter; let x equal the width of the bent up piece and V equal the volume.
    b. Rewrite the quadratic equation in part (a) in standard form and find the value of x that maximizes volume.
    c. What is the maximum volume?

63. A toy rocket is shot up vertically from ground level with an initial velocity of 240 feet per second. The height of the rocket in feet and the time it has traveled in seconds is given by the equation $s = -16t^2 + 240t$, where s represents the height and t the time. Rewrite the quadratic equation in standard form; find at what time the toy rocket reaches the maximum height and determine the maximum height. (Use the maximizing capability of your graphing calculator to check your answer.)

64. A ball is thrown vertically upward with an initial velocity of 48 feet per second and an initial height of 8 feet above the ground. The height of the ball in feet and the time it has traveled in seconds is given by the equation $s = -16t^2 + 48t + 8$, where s represents the height and t the time. Rewrite the quadratic equation in standard form; find at what time the toy rocket reaches the maximum height and determine the maximum height. (Use the maximizing capability of your graphing calculator to check your answer.)

Section 7.5 Nonlinear Systems

In section 5.2 we first discussed how to solve systems of linear equations. Later on, in chapter 6, we learned how to solve linear systems by using matrices. Now, we are going to learn how to solve systems of equations that involve some non-linear equations. One method that will be used is the substitution method, which we discussed in section 5.2. Let's begin with an example.

Discussion 1: Solving a Non Linear System

Solving the system of equations $\begin{cases} x^2 - y = -2 \\ x - y = -4 \end{cases}$.

You should remember that with the substitution method we solve one of the equations for one of the variables and then substitute into the other equation.	$\begin{cases} x^2 - y = -2 \\ x + 4 = y \end{cases}$ We added 4 and y to both sides
We substitute for y in the first equation.	$x^2 - (x + 4) = -2$
Now, solve for x.	$x^2 - x - 4 = -2$ $x^2 - x - 2 = 0$ $(x - 2)(x + 1) = 0$ $x = 2$ or $x = -1$
We have found that there are two solutions for x. We simply substitute each one into an equation and we will find the two points where these two graphs must be intersecting.	Take $x = 2$, and substitute into $y = x + 4$: $y = \mathbf{2} + 4 = 6$ Take $x = -1$, and substitute into $y = x + 4$: $y = \mathbf{-1} + 4 = 3$
The two points of intersection are	$(-1, 3)$ and $(2, 6)$
Here are the graphs.	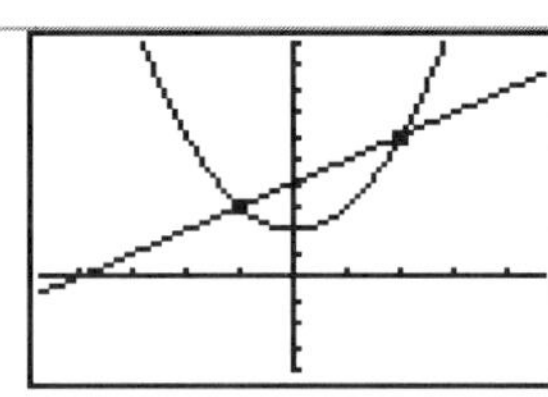

In this particular case we could have used the addition method to solve this system, since we could have eliminated the y's by multiplying one equation by -1 and then adding the two equations.

With linear systems of two variables (studied in section 5.2) there were only three possible outcomes: no solution (parallel lines), one solution (a point of intersection), or an infinite number of solutions (they are the same line). When working with non-linear systems, however, we can have all kinds of different outcomes. As we have just seen, we found two solutions in discussion 1. With non-linear equations the graphs can bend and weave in all kinds of directions, so they can intersect in many different ways. In this last example, we saw one possible outcome that can happen between a line ($x - y = -4$) and a parabola ($x^2 - y = -2$): two solutions.

Question 1: How many different kinds of answers could result, when you are looking at a two variable systems of equations where one equation is a line and the other equation is a parabola?

Example 1: Solving for Two Parabolas

Solve the system of equations $\begin{cases} x^2 + y = 0 \\ x^2 - 4x = y \end{cases}$.

Solution:

Since the second equation is already solved for y, we will substitute for y in the first equation.	$x^2 + x^2 - 4x = 0$
Now, solve for x.	$2x^2 - 4x = 0$ $2x(x - 2) = 0$ $x = 0$ or $x = 2$
We have found that there are two solutions for x. We simply substitute each one into an equation and we will find the two points where these two graphs must be intersecting.	Take $x = 2$, and substitute into $y = x^2 - 4x$: $y = \mathbf{2}^2 - 4(\mathbf{2}) = -4$ Take $x = 0$, and substitute into $y = x^2 - 4x$: $y = \mathbf{0}^2 - 4(\mathbf{0}) = 0$
The two points of intersection are	(0, 0) and (2, –4)
Here are the graphs. We can see how the two parabolas intersect at two points.	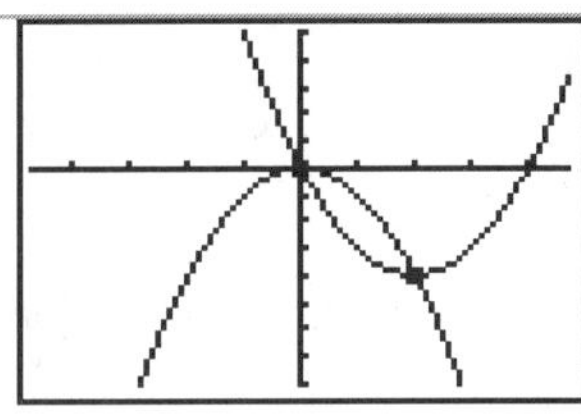

Question 2: Given the system of equations $\begin{cases} x^2 + y^2 - 169 = 0 \\ x^2 - 8y - 104 = 0 \end{cases}$, what names do we give to each equation (line, parabola, ellipse, etc.) and what are the possible outcomes (no solution, one solution, etc.) if we were to try to solve this system?

Example 2: Solving Two Conics: a Circle and a Parabola

Solve the system in question 2 two ways (substitution, addition).

Solution:

Substitution method: Solve for y in the second equation.	$x^2 - 104 = 8y \Rightarrow (x^2 - 104)/8 = y$
We will substitute for y in the first equation.	$x^2 + \left(\frac{x^2-104}{8}\right)^2 - 169 = 0$
Now, solve for x.	$x^2 + \left(\frac{x^4 - 208x^2 + 10816}{64}\right) - 169 = 0$ $64x^2 + x^4 - 208x^2 + 10816 - 10816 = 0$ $x^4 - 144x^2 = 0$ $x^2(x^2 - 144) = 0$ $x^2(x - 12)(x + 12) = 0$ $x = 0$ or $x = 12$ or $x = -12$
We have found that there are three solutions for x. We simply substitute each one into an equation to find the three points where these two graphs must be intersecting.	Plug $x = 0$ into $y = (x^2 - 104)/8$. $y = (\mathbf{0}^2 - 104)/8 = -13$ Plug $x = 12$ in to the equation. $y = (\mathbf{12}^2 - 104)/8 = 5$ Plug $x = -12$ into the equation. $y = ((\mathbf{-12})^2 - 104)/8 = 5$
The three points of intersection are	$(-12, 5)$, $(0, -13)$ and $(12, 5)$
Here are the graphs.	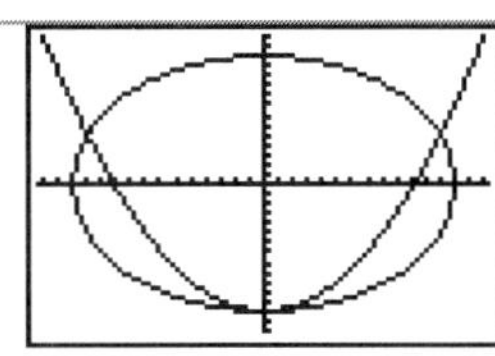

Addition method: We want to add equations such that we eliminate one of the variables. (Multiply the second equation by – 1.)	$x^2 + y^2 \quad - 169 = 0$ $\underline{-x^2 \quad + 8y + 104 = 0}$ $y^2 + 8y - 65 = 0$
Solve for y.	$y^2 + 8y - 65 = 0$ $(y + 13)(y - 5)$ $y = -13$ or $y = 5$
We have found that there are two solutions for y. We simply substitute each one into an equation to find the points where these two graphs must be intersecting.	Plug $y = -13$ into $x^2 + y^2 - 169 = 0$ $x^2 + (-13)^2 - 169 = 0$ $x^2 + 169 - 169 = 0$ $x^2 = 0$

Take the square root of both sides.	$x = 0$ Plug $y = 5$into $x^2 + y^2 - 169 = 0$ $x^2 + (5)^2 - 169 = 0$ $x^2 + 25 - 169 = 0$ $x^2 = 144$ $x = \pm\sqrt{144} = \pm 12$
The three points of intersection are	$(0, -13)$, $(-12, 5)$, and $(12, 5)$

Answer Q1: You could have two solutions (they intersect in two points),

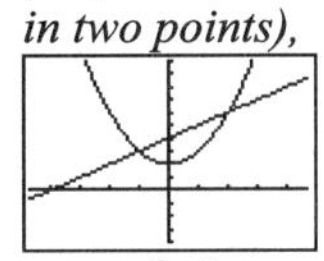

one solution (they intersect at one point),

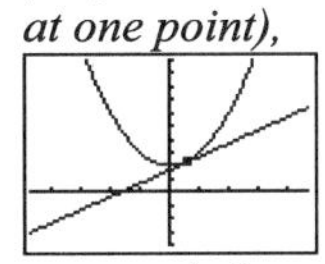

or no solution (they don't intersect anywhere).

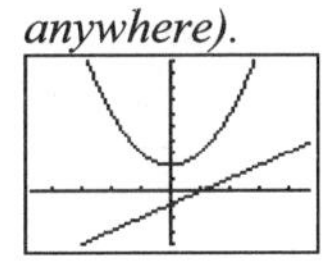

As you can see, the addition method might very well be the better method to use when you can. Let's do another example to show how the addition method might be used to solve non-linear systems.

Example 3: Solving an Ellipse and a Hyperbola

Solve the system of equations $\begin{cases} 2x^2 + y^2 = 9 \\ x^2 - y^2 = 3 \end{cases}$ by the addition method.

Solution:

Since the y^2 terms are set up perfectly for the addition method, we only need to add the two equations together.	$2x^2 + y^2 = 9$ $\underline{x^2 - y^2 = 3}$ $3x^2 \quad\quad = 12$
Solve for x by isolating the x^2 term and then take the square root.	$x^2 = 4$ $x = \pm 2$
We have found two values for x, so we substitute those into an equation to find the y-values.	Let $x = 2$, use $x^2 - y^2 = 3$ $2^2 - y^2 = 3 \Rightarrow 4 - y^2 = 3 \Rightarrow 1 = y^2 \Rightarrow$ $y = \pm 1$ Let $x = -2$, use $x^2 - y^2 = 3$ $(-2)^2 - y^2 = 3 \Rightarrow 4 - y^2 = 3 \Rightarrow 1 = y^2 \Rightarrow$ $y = \pm 1$
We have four solutions and they are	$(2, 1), (2, -1), (-2, 1), (-2, -1)$
Here is a graph of the system. As you can see, this system is made up of an ellipse and a hyperbola and they intersect in four places.	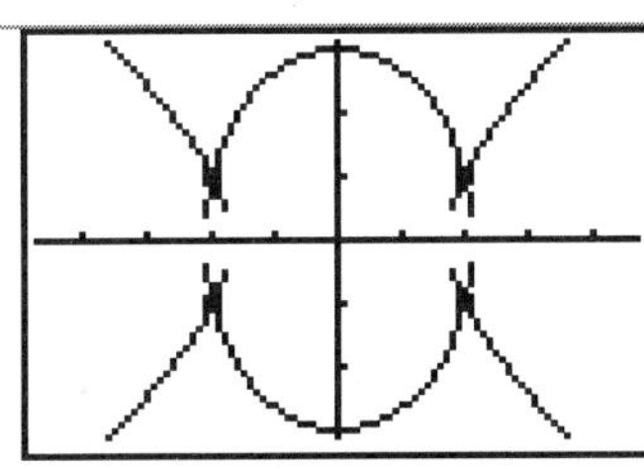

Answer Q2: The first equation is a circle and the second is a parabola. The possible outcomes when trying to solve this system are no solution,

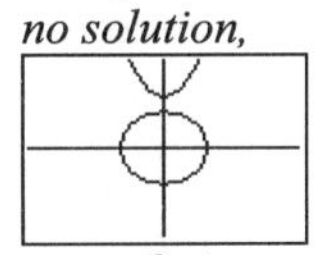

one solution,

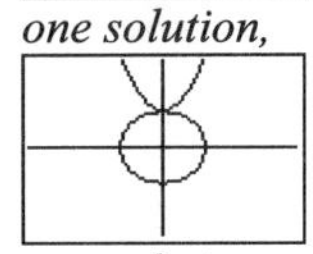

two solutions,

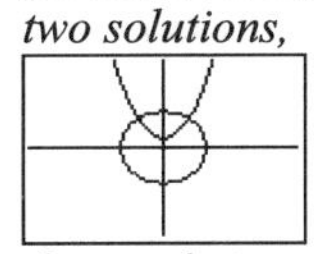

three solutions

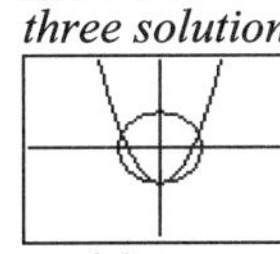

and four solutions.

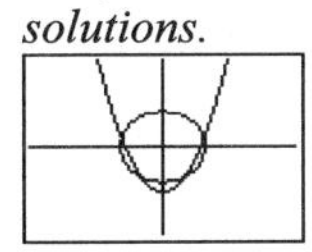

Let's do one example where there is no solution.

Example 4: A No Solution System

Solve this system of equations $\begin{cases} y = 2x - 1 \\ x^2 - 2y^2 = 4 \end{cases}$.

Solution:

Since the first equation is already solved for y, we will substitute for y in the second. equation.	$x^2 - 2(2x - 1)^2 = 4$
Now, solve for x.	$x^2 - 2(4x^2 - 4x + 1) = 4$ $x^2 - 8x^2 + 8x - 2 = 4$ $-7x^2 + 8x - 2 = 4$ $0 = 7x^2 - 8x + 6$ $x = \dfrac{8 \pm \sqrt{64 - 4(7)(6)}}{14}$ $x = \dfrac{8 \pm \sqrt{-104}}{14} = \dfrac{8 \pm 2i\sqrt{26}}{14}$
We have found that there are two solutions for x, but they are both complex numbers. These values for x are in the complex plane, not the real plane in which we are working. Hence, there are no solutions to this system of equations.	$x = \dfrac{4 + i\sqrt{26}}{7}$ and $x = \dfrac{4 - i\sqrt{26}}{7}$
Here are the graphs. We can see why there isn't a solution by the fact that the two graphs don't intersect anywhere.	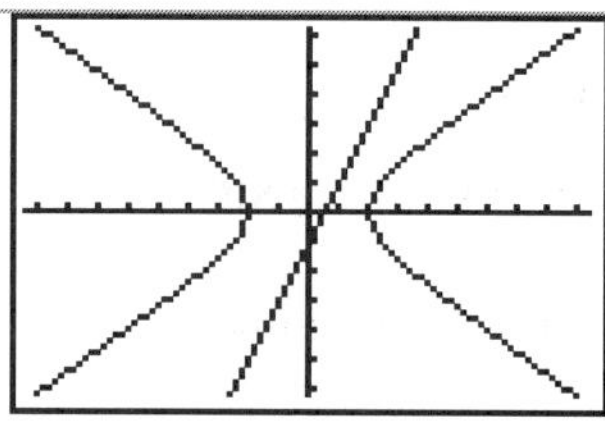

Non-linear systems are just systems of equations where at least one of the equations is not linear. Of course, if a system has a conic in it, then the system will have to be a non-linear system. But there are other non-linear equations.

Question 3: Can you name another non-linear equation that we have looked at so far in this book?

Example 5: Solving a Logarithmic System

Solve this system of equations $\begin{cases} y = \ln x + 1 \\ y = \ln(x - e + 1) + 2 \end{cases}$.

Solution:

We will substitute for y in the first equation.	$\ln(x-e+1)+2=\ln(x)+1$
Now, solve for x. Combine logs. → Convert to exponent form. → Linear, so get x's on same side. → Factor. Get x alone.	$1=\ln(x)-\ln(x-e+1)$ $1=\ln\dfrac{x}{x-e+1}$ $e=\dfrac{x}{x-e+1}$ $ex-e^2+e=x$ $ex-x=e^2-e$ $x(e-1)=e(e-1)$ $x=\dfrac{e(e-1)}{e-1}=e$
Substitute $x=e$ to find y.	$y=\ln(e)+1=1+1=2$
The point of intersection is $(e, 2)$.	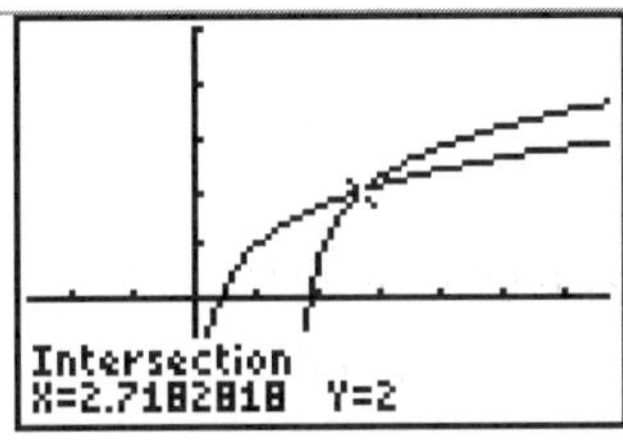

Most likely, when you have systems of non-linear equations and the two equations aren't of the same type, then you will not be able to solve them by hand. Here is an example of what we are talking about.

Example 6: Solving a System with the Calculator

Solve this system of equations $\begin{cases} y=e^{-x+1}+1 \\ y=\ln x+2 \end{cases}$.

Solution:

If we try solving this by hand, we find that there isn't any way for us to solve this type of problem. This doesn't fit anything we learned in chapter 4.	$\ln x+2=e^{-x+1}+1$ $\ln x+1=e^{-x+1}$
We will now use the calculator to approximate the point(s) of intersection. We use the intersect command at 2nd TRACE (CALC).	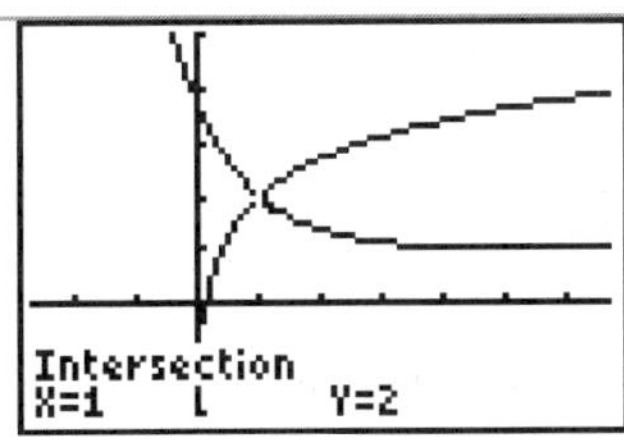
From the calculator we see that the solution is	(1, 2)

Question 4: Solve this system $\begin{cases} y = |3x| \\ y = 2^x \end{cases}$ using the graphing calculator.

Answer Q3: You may have thought of exponential and logarithmic equations or possibly polynomial equations (of degree higher than 1).

In example 6, we had an exponential and a logarithmic equation, and in question 4, we had an absolute value and an exponential equation. In both cases, the functions used were from different families of functions. If you mix functions from different families, usually you will not be able to solve the system by hand. So far in this book, we have learned about three basic families of functions, and they are polynomials, exponentials and logarithmic functions.

Section Summary:

We learned the following items:

- You can use both the substitution method and the addition method to solve non-linear systems of equations. (Note: there are times when the addition method won't work.)
- Typically, if you have a linear equation mixed with a non-linear equation, you will need to use the substitution method to solve the system.
- If you have a mixture of different types of equations (different families– polynomials, logarithmic, exponential, to name a few) you most likely can only solve the system by using the graphing calculator or a computer to approximate the solutions.

SECTION 7.5 HOMEWORK

(1–42) For each of the following system of equations state what type of graph each equation represents and find the common **real number** solutions. (If there are no real solutions, state "no real solutions.")

1. $3x + y = 5$
   $x^2 + y^2 = 5$

2. $x - 3y = 7$
   $x^2 + y^2 = 5$

3. $x + 2y = 1$
   $3x^2 + 2y^2 = 35$

4. $2x - y = 1$
   $2x^2 + 3y^2 = 83$

5. $4x - y = 5$
   $x^2 - 3y^2 = -23$

6. $x + 3y = 5$
   $2x^2 - y^2 = 23$

7. $x - 2y = 3$
   $4x^2 - y^2 = -4$

8. $2x + y = 8$
   $x^2 + 4y^2 = 12$

9. $3x - y = 10$
   $x^2 - y = -4$

10. $x + 2y = 6$
    $3x^2 + 3y^2 = 4$

11. $x - y = 5$
    $x^2 + 2x - y = 11$

12. $2x + y = -5$
    $y^2 + 3x - 2y = -10$

13. $2x + 3y = 10$
    $x^2 - 3x + y = 0$

14. $2x - y = 2$
    $x^2 + 4x - y = -4$

15. $x - 2y = 1$
    $y^2 - 3x + 2y = 5$

16. $y - 2x = 3$
    $x^2 + y^2 = 4$

17. $x^2 + y^2 = 5$
    $2x^2 - y^2 = 7$

18. $2x^2 + 2y^2 = 26$
    $x^2 - 3y^2 = -3$

19. $x^2 + y^2 = 5$
    $x^2 + 2y^2 = 9$

20. $x^2 + 2y^2 = 22$
    $2x^2 - y^2 = -1$

21. $3x^2 - 2y^2 = -6$
    $x^2 + 2y^2 = 22$

22. $3x^2 + 3y^2 = 39$
    $4x^2 + y^2 = 25$

23. $2x^2 + 3y^2 = 21$
    $3x^2 + 2y^2 = 29$

24. $5x^2 - 3y^2 = -3$
    $2x^2 - 5y^2 = -62$

25. $x^2 - 3y = 13$
    $3x^2 + 3y^2 = 123$

26. $y^2 + 2x = 5$
    $2x^2 + 2y^2 = 26$

27. $2x^2 + y^2 = 3$
    $3x^2 - y^2 = -5$

28. $3x^2 + 5y^2 = 15$
    $2x^2 + y^2 = 24$

29. $x^2 - y^2 = 24$
    $3x^2 - 2y^2 = 6$

30. $x^2 + 3y^2 = 6$
    $x^2 + y = 8$

31. $x^2 - 3y^2 = 2$
    $2x^2 + y^2 = 18$

32. $x^2 + 3y^2 = 18$
    $3x^2 - y^2 = 4$

33. $2x + 4y = 14$

34. $x - 3y = 3$

$x^2 + y^2 - 2x + 3y = 13$

$2x^2 + 2y^2 - 3x + 4y = 27$

35. $x + 2y = 10$
$3x^2 + 4y^2 - 6x - 2y = 56$

36. $3x - y = 13$
$3x^2 + 2y^2 - 9x + 3y = 20$

37. $x - 5y = 7$
$2x^2 - 3y^2 + 3x - y = -1$

38. $4x + y = 1$
$3x^2 - 2y^2 - 13x + 2y = -34$

39. $6x + 3y = 15$
$2x^2 - y^2 + 2x - 3y = 4$

40. $2x + 4y = 6$
$3x^2 + 2y^2 - 3x - y = 4$

Answer Q4:

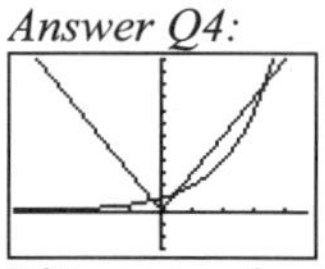

This system has three approximate solutions: (–0.2754…,0.8262…)(0.4578…,1.3734…), (3.313…, 9.939…).

41. $2x - y = 2$
$x^2 - 2y^2 - 8x + y = -4$

42. $3x + y = 4$
$x^2 + 2y^2 + 4x + 4y = -2$

(43–44) Solve each system of equations for x:

43. $y = \ln(x) - 2$
$y = \ln(2x + e - 1) - 3$

44. $y = \ln(x + 2) + 4$
$y = \ln(3x - 3e + 1) + 3$

(45–58) For each of the following system of equations find the approximate answers for x and y with your graphing calculator, using the intersection feature on your calculator.

45. $y = \ln(x + 3) - 2$
$2x + y = 3$

46. $y = \ln(x - 2) + 1$
$x + 2y = 8$

47. $y = \ln(x + 2) + 1$
$x - 2y + 2 = 0$

48. $y = \ln(2x - 1) + 3$
$x - 3y + 6 = 0$

49. $y = \ln(2x + 1) + 3$
$3y = 4x + 18$

50. $y = \ln(3 - 2x) + 1$
$2x - 3y = 9$

51. $y = \ln(3x - 1)$
$y = e^{0.5x} - 4$

52. $y = \ln(2x - 3)$
$y = e^{0.3x} - 5$

53. $y = \ln(x + 3) + 2$
$y = e^{-0.3x} + 3$

54. $y = \ln(x - 2) + 1$
$y = e^{-0.7x} + 2$

55. $y = \ln(2x + 3)$
$y = e^{0.3x} - 3$

56. $y = \ln(x + 5) - 3$
$y = e^{0.2x} - 4$

57. $y = \ln(x + 2) - 5$
$y = e^{0.7x} + 3$

58. $y = \ln(x - 3) - 1$
$y = e^{0.4x} + 2$

Chapter 7 Review

Topic	Section	
Distance formula	7.1	Distance formula, $d=\sqrt{(x_2-x_1)^2+(y_2-y_1)^2}$, needed to define the parabola.
Standard form of a parabola function	7.1	Standard form of the parabola function, $y = a(x-h)^2 + k$, where (h, k) is the vertex, $a=\dfrac{1}{4p}$, and the axis of symmetry is $x = h$. And p is the distance from the vertex to the focus $(h, k+p)$; directrix $y = k - p$.
Standard form of the non-function parabola	7.1	$a(y-k)^2 + h$, where (h, k) is the vertex, $a=\dfrac{1}{4p}$, and the axis of symmetry is $y = k$, $(h+p, k)$, directrix $x = h - p$.
Translations	7.1	• Vertical translation: $f(x) + k$, means the graph shifts up or down k units. • Horizontal translation: $f(x + h)$, means the graph shifts left or right h units. • Flipping the graph: $-f(x)$, means the graph flips upside down. • Rotating the graph: $f(-x)$, means the graph rotates around the y-axis. • Stretching or compressing: $af(x)$, means the graph is wider if $\lvert a\rvert$ is greater than 1 and narrower if $\lvert a\rvert$ is less than 1.
Reflective property parabolas	7.1	The reflective property of a parabola is that when anything comes into the parabola parallel to the axis of symmetry, it will be reflected back towards the focus and vice versa.
Ellipse	7.2	• The standard formula for an ellipse is $\dfrac{(x-h)^2}{a^2}+\dfrac{(y-k)^2}{b^2}=1$, where (h, k) is the center; a is the distance from the center to the left and right edges; b is the distance from the center to the top and bottom; c is the distance from the center to the foci ($c^2 = \lvert a^2-b^2\rvert$) along the major axis, and if $a^2 > b^2$, then you have a horizontal ellipse. However, if $b^2 > a^2$, then you have a vertical ellipse. • The reflective property of an ellipse means that anything that is emitted from one focal point is reflected towards the other focal point.
Circle	7.2	• A circle is a special ellipse where a and b are equal and thus the foci are located at the center. • The standard formula for a circle is $(x-h)^2+(y-k)^2=r^2$, where (h, k) is the center and r is the radius.

<table>
<tr><td>Hyperbola</td><td>7.3</td><td>

- The standard form of a hyperbola has two possible forms: $\frac{(x-h)^2}{a^2}-\frac{(y-k)^2}{b^2}=1$ (horizontal hyperbola) or $\frac{(y-k)^2}{b^2}-\frac{(x-h)^2}{a^2}=1$ (vertical hyperbola).
- $c^2 = a^2 + b^2$, where c is the distance from the center to the foci.
- The center is (h, k).
- a and b define the edges of the central box.
- The reflective property of a hyperbola means that anything coming in towards one focal point is reflected back towards the other.

</td></tr>
<tr><td>General form of conics</td><td>7.4</td><td>

- The general form of a conic is $Ax^2 + Bxy + Cy^2 + Dx + Ey + F = 0$. For our purposes though, we will use $Ax^2 + Cy^2 + Dx + Ey + F = 0$ as the general form. We will not look at any equations that contain a Bxy term in this text.
- You can identify the conics in the **general form** ($Ax^2 + Cy^2 + Dx + Ey + F = 0$) in the following ways:
    1. Parabolas have only one squared variable.
    2. Hyperbolas have squared variables with opposite signs.
    3. Circles have positive squared variables with the exact same coefficients.
    4. Ellipses have positive squared variables with different coefficients.

</td></tr>
<tr><td>Completing the square</td><td>7.4</td><td>

To complete the square you must do the following:

1. Concentrate on just the squared term and the linear term.
2. Factor out the coefficient on the squared term in the expression, if there is one.
3. Take half of the coefficient of the linear term, square it, and add that amount to your expression.
4. Now factor the perfect square trinomial, which will always be the variable plus the number found in step 3.

</td></tr>
<tr><td>Solving non-linear systems</td><td>7.5</td><td>

- You can use both the substitution method and the addition method to solve non-linear systems of equations (Note: there are times when the addition method won't work).
- Typically, if you have a linear equation mixed with a non-linear equation, you will need to use the substitution method to solve the system.
- If you have a mixture of different types of equations (different families, polynomials, logarithmic, exponential, to name a few) you most likely can only solve the system by using the graphing calculator or a computer and approximating the solutions.

</td></tr>
</table>

CHAPTER 7 REVIEW HOMEWORK

Section 7.1

(1–2) Find the distance between the two given points.

1. (3, 2) and (–9, 7)
2. (5, –2) and (3, –8)

(3–10) Sketch the graph of the following parabolas.

3. $y = 2x^2$
4. $y = -3x^2$
5. $x = \frac{1}{4}y^2$
6. $x = -\frac{1}{3}y^2$
7. $y = -3(x-1)^2 + 3$
8. $y = 2(x+3)^2 - 5$
9. $x = -(y+3)^2 + 4$
10. $x = 2(y-1)^2 - 3$

(11–14) For each of the following functions explain the translation each represents with respect to the graph of $f(x)$.

11. $y = f(x+2) - 3$
12. $y = f(x-3) + 2$
13. $y = 2f(x-1) - 2$
14. $y = -3f(x+1) + 2$

(15–17) For each of the following parabolas give the:

a.) Vertex
b.) Axis of symmetry;
c.) Focus
d.) Directrix
e.) Maximum or minimum value of y
f.) Written explanation of the translation with respect to the graph of $y = x^2$
g.) Domain and range.

15. $y = (1/3)x^2$
16. $y = (x-1)^2 + 3$
17. $y = -3(x+1)^2 + 4$

(18 – 20) For each of the following parabolas give the:

a.) Vertex
b.) Axis of symmetry
c.) Focus
d.) Directrix
e.) Written explanation of the translation with respect to the graph of $x = y^2$
f.) Domain and range.

18. $x = -2y^2$
19. $x = (y+3)^2 - 4$
20. $x = -2(y-1)^2 - 3$

(21–22) Use the given table to create a table for the following functions:

x	–2	1	4	7	10
$g(x)$	–10	–5	0	10	15

21. $g(x) + 2$
22. $g(x) - 3$

Section 7.2

(23–30) For each of the following ellipses write the standard equation and give the:

a.) Horizontal or vertical ellipse

b.) Center

c.) Foci

d.) Vertices

e.) Covertices

f.) Domain and range

g.) Written explanation of the translation from standard position if one exists

Then sketch a graph.

23. $\frac{x^2}{9}+\frac{y^2}{25}=1$ 24. $\frac{x^2}{169}+\frac{y^2}{25}=1$ 25. $\frac{(x-2)^2}{25}+\frac{(y-2)^2}{9}=1$

26. $\frac{(x+2)^2}{25}+\frac{(y+1)^2}{169}=1$ 27. $3x^2+8y^2=48$ 28. $5x^2+2y^2=50$

29. $4(x+3)^2+2(y-2)^2=36$ 30. $3(x-1)^2+9(y+3)^2=27$

31. Find the standard equation of an ellipse with center (0,0); constant sum of 10; and foci points (3,0) and (–3,0).

32. Find the standard equation of an ellipse with center (0,0); vertices (13,0) and (–13,0); and foci points (5,0) and (–5,0).

33. Find the standard equation of an ellipse with vertices (2,2) and (2,–8) and covertices of (5,–3) and (–1,–3).

34. Find the standard equation of an ellipse with center (3,2); vertices (16,2) and (–10,2); and foci points of (8,2) and (–2,2).

35. The ellipse $\frac{x^2}{9}+\frac{y^2}{16}=1$ is translated left 2 units and up 3 units. What is the equation in standard form of the new ellipse?

36. Use the graph to write the equation of the ellipse:

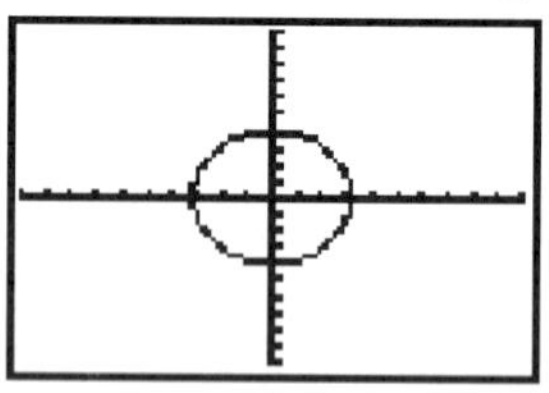

(37–40) For each of the following circles write the standard equation and give the:

a.) Center

b.) Radius

c.) Domain and range

d.) Written explanation of the translation from standard position, if one exists.

37. $5x^2 + 5y^2 = 125$

38. $\frac{x^2}{3} + \frac{y^2}{3} = 12$

39. $3(x-2)^2 + 3(y+3)^2 = 27$

40. $\frac{(x+3)^2}{2} + \frac{y^2}{2} = 32$

41. Find the standard equation of a circle with center (–2,–1) and radius of 15.

42. Find the standard equation of a circle with center (3,2) and point on the circle (–1,5).

43. A lithotripter is based on the ellipse $\frac{x^2}{64} + \frac{y^2}{28} = 1$. How many units from the center of the elliptical tub must the kidney stone and the source of the beam, used to destroy the kidney stone, be placed for this lithotripter to destroy the kidney stone?

44. A one-way road passes under and overpass that is in the form of half an ellipse. The opening is 20 feet high at the center and 22 feet wide.

    a.) What is the equation in standard form of the ellipse?

    b.) What is the tallest 12 feet wide truck that can pass through the overpass?

Section 7.3

(45–49) For each of the following hyperbolas write the standard equation and give the:

a.) Horizontal or vertical hyperbolas

b.) Center

c.) Foci

d.) Vertices

e.) Covertices

f.) Asymptotes

g.) Domain and range

h.) Written explanation of the translation from standard position, if one exists

Then sketch the graph.

45. $4x^2 - 9y^2 = 36$

46. $8y^2 - 12x^2 = 24$

47. $9(x-2)^2 - 16(y+3)^2 = 144$

48. $5(y+3)^2 - 15(x+2)^2 = 150$

49. Find the standard equation of a hyperbola with vertices (4,0) and (–4,0); and Covertices (0,3) and (0,–3).

50. Find the standard equation of a hyperbola with vertices (0,5) and (0,–5); and foci (0,13) and (0,–13).

51. Find the standard equation of a hyperbola with center (–2,3); covertices (1,3) and (–5,3); and constant difference of 6.

52. Find the standard equation of a hyperbola with vertices (3,4) and (3,–6); and covertices (8,–1) and (–2,–1).

53. The hyperbola $\frac{x^2}{25} - \frac{y^2}{36} = 1$ is translated left 5 units and up 2 units. What is the standard equation of the new hyperbola?

54. The hyperbola $\frac{y^2}{9} - \frac{x^2}{9} = 1$ is translated right 3 units and down 4 units. What is the standard equation of the new hyperbola?

(55–60) Write each of the following equations in standard form and answer these questions.

a.) Does the equation represent a hyperbola, ellipse, circle or parabola?

b.) Is the conic in standard position or translated position?

c.) If the conic is an ellipse or hyperbola is it a vertical or horizontal type?

55. $5(x-1)^2 - 9(y+2)^2 = 45$ 56. $12x^2 + 3y^2 = 72$ 57. $5x^2 + 5y^2 = 50$

58. $x = -3(y-1)^2 + 3$ 59. $4(x+3)^2 + 8(y-1)^2 = 64$ 60. $\frac{(x-1)^2}{3} + \frac{(y+1)^2}{3} = 27$

(61–62) Use the graphs to write the equation for each hyperbola.

61.

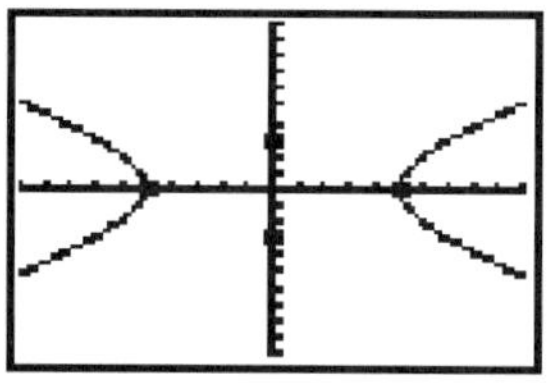

62.

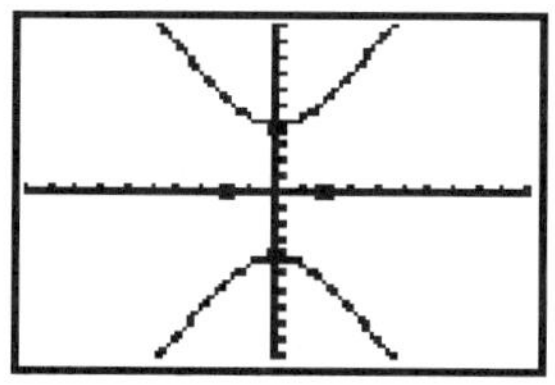

Section 7.4

(63–70) For each of the following general equations:

a.) State the type of conic the equation represents.

b.) Rewrite each general equation in standard form.

63. $9x^2 + 4y^2 - 54x - 16y + 61 = 0$ 64. $16x^2 - 9y^2 + 64x - 54y + 31 = 0$

65. $2x^2 - 12x - y + 21 = 0$ 66. $3x^2 + 3y^2 - 12x + 6y - 12 = 0$

67. $-16x^2 + 25y^2 - 64x - 100y - 364 = 0$ 68. $25x^2 + 16y^2 - 250x + 64y + 289 = 0$

69. $4x^2 + 4y^2 + 24x + 16y + 16 = 0$ 70. $3y^2 - x + 12y + 7 = 0$

(71–72) For each of the following parabolas rewrite in standard form and find the maximum or minimum value of y.

71. $y = x^2 - 20x + 130$ 72. $y = -2x^2 + 36x + 38$

73. A company's profit is determined by the equation $P = 60x - 0.3x^2$, where P is the profit and x is the number of items sold. Rewrite the equation in standard form and find the number of items to produce to maximize profit. What is the maximum profit?

74. An object is shot into the air from 6 feet above ground. The height in feet of the object after t seconds is given by the equation $s = -16t^2 + 384t + 6$. Rewrite the quadratic equation in standard form and find the time when the object reaches its maximum height. What is the maximum height?

Section 7.5

(75–88) For each system of equations find the common **real number** solution(s). (If there are no real number solutions, state "no real solutions.")

75. $x + 2y = 3$
$2x^2 + y^2 = 6$

76. $2x - y = 7$
$y = x^2 + 2x - 16$

77. $3x + y = 11$
$x^2 - 2y^2 = 1$

78. $x + 3y = 1$
$x^2 + y^2 = 5$

79. $2x^2 - y^2 = 9$
$3x^2 - 2y^2 = 9$

80. $x^2 + 2y^2 = 6$
$2x^2 + y^2 = 9$

81. $2x^2 - 3y^2 = 6$
$3x^2 + 2y^2 = 35$

82. $3x^2 + 3y^2 = 39$
$4x^2 + 3y^2 = 48$

83. $x - y = 3$
$2x^2 + y^2 - 3x + y = 41$

84. $x - 2y = 7$
$x^2 - 2y^2 + 2x - y = 9$

85. $x - 2y = 10$
$2x^2 + 3y^2 = 12$

86. $2x + y = 1$
$3x^2 - 2y^2 = 24$

87. $3x^2 + 2y^2 = 12$
$x^2 + y^2 = 9$

88. $2x^2 + y^2 = 16$
$8x^2 + 3y^2 = 24$

(89–90) Solve each system of equations for x.

89. $y = \ln(x + 1) - 1$
$y = \ln(3x + e - 2) - 2$

90. $y = \ln(x - 3) + 2$
$y = \ln(3x + e - 2)$

(91–95) For each system of equations, using the graphing calculator, find the approximate common solution(s) for x and y.

91. $y = \ln(x + 2) - 3$
$y = 4x - 1$

92. $y = 3\ln(x - 1) - 4$
$y = 2x - 5$

93. $y = 3\ln(3x + 2) + 3$
$y = e^{-0.2x} + 1$

94. $y = \ln(2x + 1) + 1$
$y = e^{0.3x} - 1$

95. $y = 2x + 4$
$y = 3\ln(2x - 3) - 2$

96. $y = 2\ln(2x + 1) - 3$
$y = e^{2x} + 1$

CHAPTER 7 EXAM

1. Explain the translations of $y = -2f(x+3) + 5$ with respect to $y = f(x)$.

(2–3) For the following parabolas:

a.) What is the vertex?
b.) What is the axis of symmetry?
c.) What is the focus?
d.) What is the directrix?
e.) What is the maximum or minimum value of y?
f.) Explain the translation with respect to $y = x^2$.

2. $y = -2(x-50)^2 + 30000$
3. $y = 3(x-20)^2 + 1500$

(4–5) For the following ellipses:

a.) Is the ellipse vertical or horizontal?
b.) What is the center?
c.) What are the foci points?
d.) What are the vertices?
e.) What are the covertices?
f.) Explain the translation from standard position.

4. $\dfrac{(x-2)^2}{169} + \dfrac{(y+3)^2}{144} = 1$
5. $\dfrac{(x+1)^2}{16} + \dfrac{(y-2)^2}{25} = 1$

(6–7) For the following hyperbolas:

a.) Is the hyperbola vertical or horizontal?
b.) What is the center?
c.) What are the foci points?
d.) What are the vertices?
e.) What are the covertices?
f.) What are the asymptotes?
g.) Explain the translation from standard position.

6. $\dfrac{(y-3)^2}{16} - \dfrac{(x-5)^2}{9} = 1$
7. $\dfrac{(x+2)^2}{25} - \dfrac{(y+3)^2}{144} = 1$

8. For the circle $5(x-2)^2 + 5(y+3)^2 = 80$:

a.) What is the center?
b.) What is the radius?
c.) Explain the translation from standard position.

9. Find the standard equation of an ellipse with vertices (8,–2) and (–2,–2) and covertices (3,2) and (3,–6).

10. Find the standard equation of a hyperbola with center (–3,4); vertices (–3,8) and (–3,0); and and foci points (–3,9) and (–3,–1).

11. Find the standard equation of a circle with center (–2,–3) and a points on the circle is (3,1).

(12–17) For each general equation:

a.) State the type of conic the general equation represents.
b.) Rewrite each general equation in standard form.

12. $x = 3y^2 - 18y + 5$
13. $4x^2 + 9y^2 - 16x + 54y + 61 = 0$
14. $4x^2 - 9y^2 + 16x + 54y - 29 = 0$
15. $3x^2 + 3y^2 + 30x + 12y + 51 = 0$
16. $y = -2x^2 + 16x - 24$

(17–19) For each system of equations find the common real solution(s).

17. $x - 2y = 4$

 $3x^2 + 2y^2 = 30$

18. $2x^2 - y^2 = 14$

 $3x^2 + 2y^2 = 35$

19. $3x^2 + 2y^2 = 6$

 $6x^2 + y^2 = 24$

20. Solve for x:

 $y = \ln(x + e - 2) - 3$

 $y = \ln(1 - x) - 2$

21. $R = 900x - 0.1x^2$ is an equation that finds the revenue of a business when x items are produced. How many items should be produced to maximize revenue?

22. A race track is in the shape of an ellipse 120 feet long and 60 feet wide. What is the width of the race track 30 feet from the side? In other words, find d.

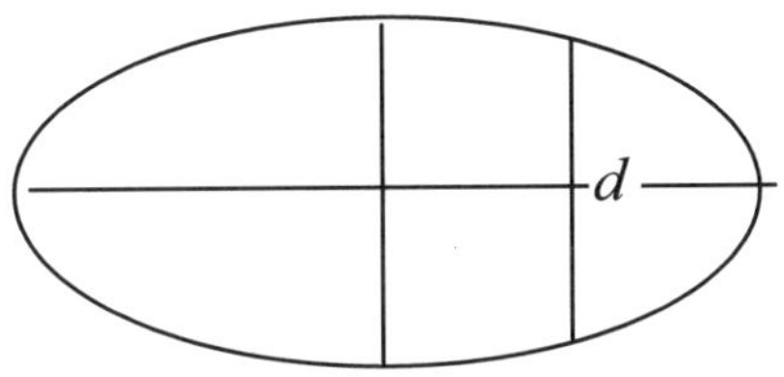

Chapter 8 Sequences, Series, and Probability

We are going to cover several different topics in this chapter. The topics are: sequences, series, counting and probability.

Section 8.1 Sequences

In this section we are going to discuss **sequences**.

> **Sequence** – An infinite sequence is a function whose domain is all positive integers and which is expressed in subscript notation (a_n) instead of function notation $f(n)$. The function values a_1, a_2, a_3... are called the terms of the sequence. If there are only a finite number of terms in a sequence, then we call that a finite sequence.

To state it simply, a sequence is just an ordered list of numbers. Let's look at an example of a simple sequence.

Discussion 1: Finding Terms of a Sequence

How to find the first five terms of the sequence $a_n = 2n + 3$.

To find the first term, evaluate a_n when n is 1 (the first positive integer).	$a_1 = 2(1) + 3 = 5$
To find the second term, evaluate a_n when n is 2 (the second positive integer).	$a_2 = 2(2) + 3 = 7$
Now evaluate a_n when n is 3.	$a_3 = 2(3) + 3 = 9$
Now evaluate a_n when n is 4.	$a_4 = 2(4) + 3 = 11$
Now evaluate a_n when n is 5.	$a_5 = 2(5) + 3 = 13$

Notice how the subscript on the a corresponds to the position of the term in the sequence. That is, a_1 is the first term, a_2 is the second term, a_3 is the third term and so on. Also notice that the subscript is the domain value that was plugged into the function that gave us the value of the term.

Question 1: Find the first five terms of the sequence $a_n = 12 - 3n$.

A sequence does not always have to start with the domain value of 1, but for simplicity's sake we will assume this unless otherwise stated. For example, in question 1 we could have written the formula as $a_n = 9 - 3n$ with the domain of (0, 1, 2, …) and arrived at the exact same sequence (9, 6, 3, 0, −3, …).

Since we are looking at functions that have only integer domain values, we can now add a new kind of function to the ones we already know. This new function is called **factorial**.

> **Factorial** – The symbol that means factorial is "!" and if n is a positive integer, then it is defined by $\boldsymbol{n!} = n(n-1)(n-2)(n-3)\ \ldots\ 1$. We also define **0!** to equal 1.

Example 1: Factorials

Find the first six factorials.

Solution:

First is 0!	$0! = 1$
Second is 1!	$1! = 1$
Third is 2!	$2! = 2 \bullet 1 = 2$
Fourth is 3!	$3! = 3 \bullet 2 \bullet 1 = 6$
Fifth is 4!	$4! = 4 \bullet 3 \bullet 2 \bullet 1 = 24$
Sixth is 5!	$5! = 5 \bullet 4 \bullet 3 \bullet 2 \bullet 1 = 120$

Question 2: Find the first five terms of the sequence $a_n = \dfrac{n!}{4}$.

Example 2: Finding Terms of a Sequence

Find the first five terms of the sequence $b_n = \dfrac{2^n}{n^2+1}$.

Notice that this time we used the notation b_n instead of a_n. Usually, when we are working with sequences, we use the first part of the alphabet as our function names (a_n, b_n, c_n, d_n).

Solution:

Let n be 1 and find our first term.	$b_1 = \dfrac{2^{(1)}}{(1)^2+1} = \dfrac{2}{2} = 1$
Now let n be 2 and find the second term.	$b_2 = \dfrac{2^{(2)}}{(2)^2+1} = \dfrac{4}{5}$
Now let n be 3 and find the third term.	$b_3 = \dfrac{2^{(3)}}{(3)^2+1} = \dfrac{8}{10} = \dfrac{4}{5}$
Now let n be 4 and find the fourth term.	$b_4 = \dfrac{2^{(4)}}{(4)^2+1} = \dfrac{16}{17}$
Now let n be 5 and find the fifth term.	$b_5 = \dfrac{2^{(5)}}{(5)^2+1} = \dfrac{32}{26} = \dfrac{16}{13}$

Let's take a moment now and see how we could use the calculator to help us with sequences. Of course, there is more than one way to view sequences on your calculators. First, let's go to the home screen and then bring up the sequence operation. The format for simple sequences like the ones we have seen so far is

seq(expression, variable, domain start value, domain ending value).

This format is the same for both the TI-83 and TI-86.

	TI-83	TI-86
On the home screen we want to see seq($2x + 3$, x, 1, 5). This is the sequence in example 1. $2x + 3$ is the formula, x is the variable, 1 is the first domain value, and 5 is the last domain value.	To get the sequence operation on the TI-83 • Push 2nd STAT (to go to the LIST menu). • Arrow over to OPS, and then push 5 for seq(• Now type in $2x + 3$ followed by a comma, x followed by a comma, 1 followed by a comma, 5 followed by a right parenthesis. 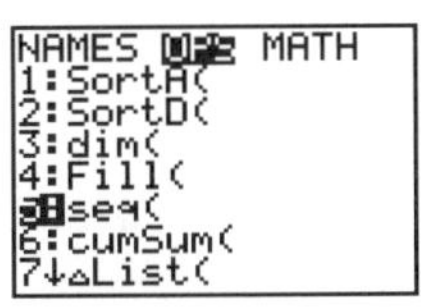 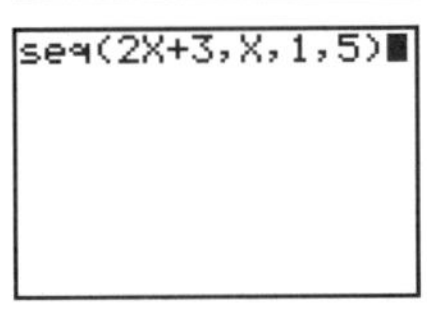 • Now push ENTER to see the first five terms of the sequence. 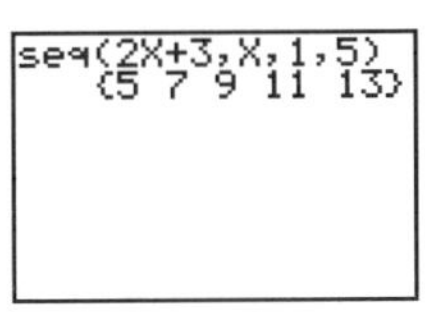 	To get the sequence operation on the TI-86 • Push 2nd, − (to go to the LIST menu), F5 for OPS, MORE, and finally F3 for seq(. • Now type in $2x + 3$ followed by a comma, x followed by a comma, 1 followed by a comma 5 followed by a right parenthesis. • Push ENTER to see the first five terms of the sequence. 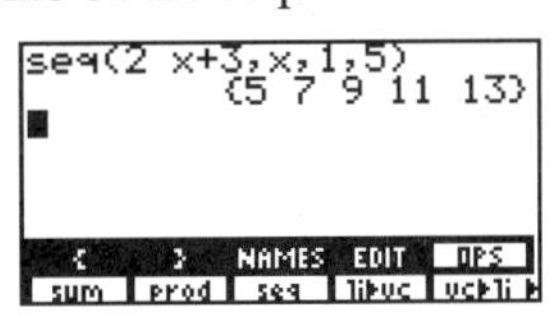

Answer Q1:
a_1=12−3(1)=**9**,
a_2=1−3(2)= **6**,
a_3 =12−3(3 =**3**,
a_4=12−3(4)=**0**,
a_5=12−3(5)=**-3**

Answer Q2:

$a_1 = \frac{1!}{4} = \frac{1}{4}$,

$a_2 = \frac{2!}{4} = \frac{2}{4} = \frac{1}{2}$,

$a_3 = \frac{3!}{4} = \frac{6}{4} = \frac{3}{2}$,

$a_4 = \frac{4!}{4} = \frac{24}{4} = 6$,

$a_5 = \frac{5!}{4} = \frac{120}{4} = 30$

<table>
<tr>
<td>Another way to work with a sequence is to first type it into the Y= window and then go to the home screen. At that point we can use the seq(format to get a list of the first few terms of the sequence.

With the function typed into the Y= screen, we can also look at a graph of the function (TI-83) if we wish and/or look at a table of values of the function (TI-83 and TI-86).</td>
<td>Go to the Y= screen and type in $2x + 3$.

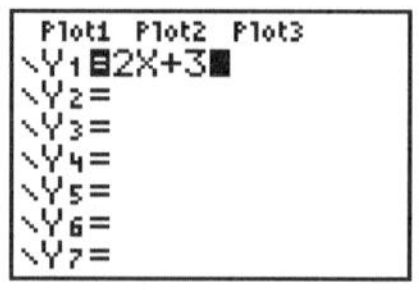

Now, go to the home screen and bring up the seq(operation symbol from the list menu.

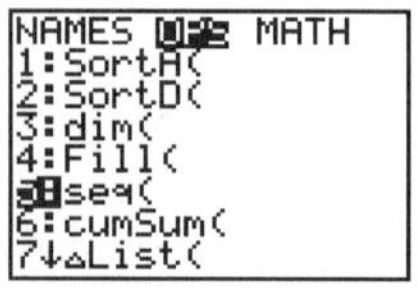

Now, push VARS, arrow over to Y-VARS, push ENTER, then choose the $Y_\#$ that you typed the function into on the graph screen.

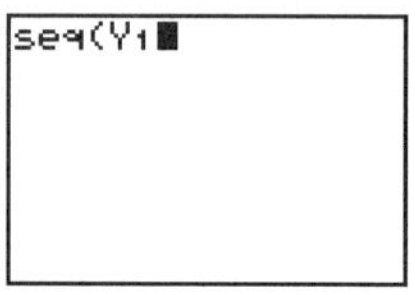

We now complete the process by typing in **,x,1,5)**.

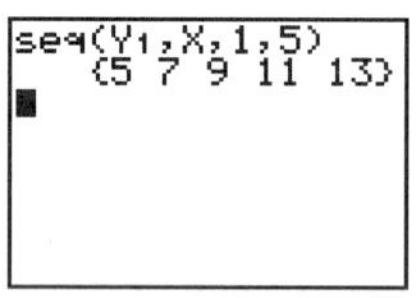

</td>
<td>Go to the Y= screen and type in $2x + 3$.

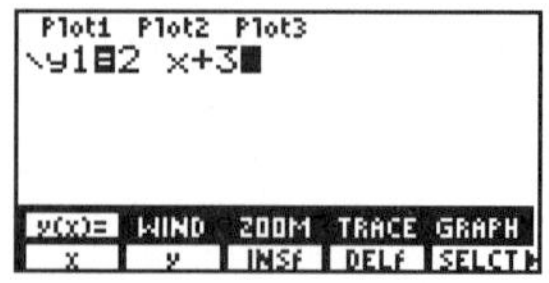

Now, go to the home screen and bring up the seq(operation symbol from the list menu.

Push 2nd, CUSTOM (for CATLG-VARS), MORE, then F4 for EQU (equations).

Arrow down to Y_1 (where the equation is typed in), and push ENTER.

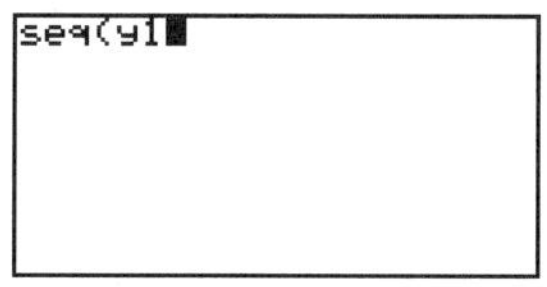

We now complete the process by typing in **,x,1,5)**.

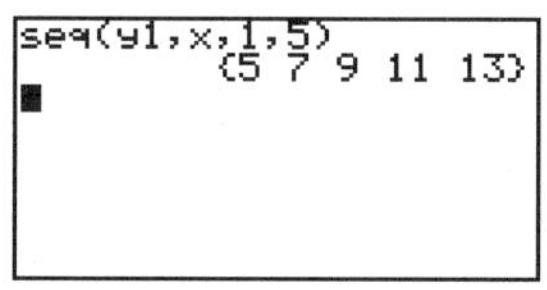

</td>
</tr>
<tr>
<td>Let's show you how to see a table of values for this sequence.</td>
<td>First, you need to set the parameters on your table.
-Push 2nd then WINDOW (TBLSET).
-Type 1 for TblStart (what input</td>
<td>First, you need to set the parameters on your table.
-Push TABLE, then F2 (TBLST).
-Type 1 for TblStart (what input value to use as a starting point for the table)</td>
</tr>
</table>

A graph isn't the best method to use here because the calculator will use all real numbers as the domain, instead of just the positive integers.

value to use as a starting point for the table) and [1] for ΔTbl (how far it is between each domain value).

-Push [2nd] followed by [GRAPH] in order to see the table.

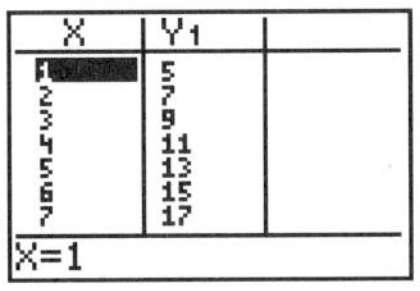

By using the up and down arrows, you can now move up and down the table to see other values.

and [1] for ΔTbl (how far it is between each domain value).

-Push [F1] in order to see the table.

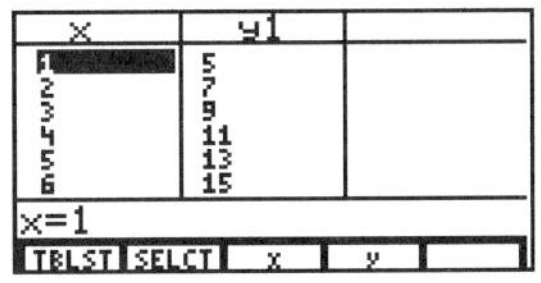

Question 3: Find the first 7 terms of the sequence in example 3 by using your calculator.

On the last question, if you want to see the entire answer just arrow to the right

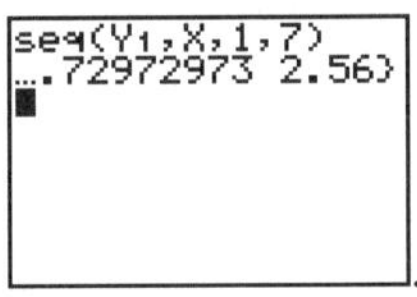

. To put the answer in fraction form just FRAC it

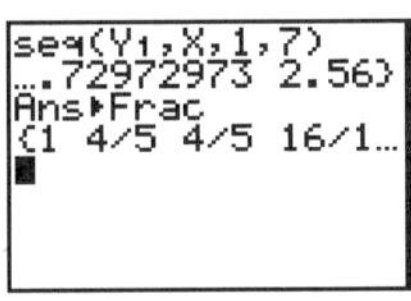

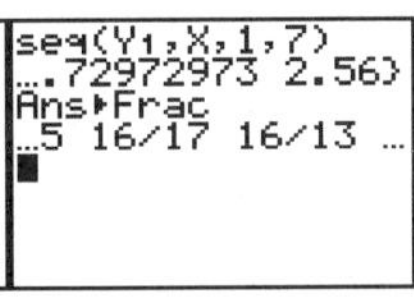

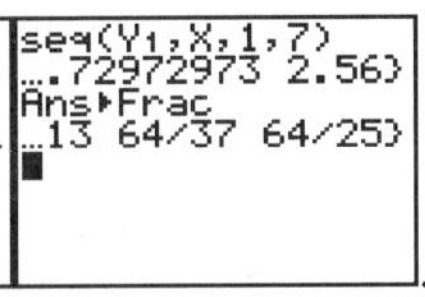

.

Example 3: Table of Values on the Graphing Calculator

Create a table of values for question 2.

Solution:

First type in the sequence as Y_1; then set up the table and view the values.

Hint: On the TI-83 the factorial symbol is found by pushing [MATH] arrowing over to PRB (probability menu), then pushing [4] (!). On the TI-86 you need to push [2nd], then [x] (times key) (MATH menu), [F2] for PROB, and then [F1] for !.

$$a_n = \frac{n!}{4}$$

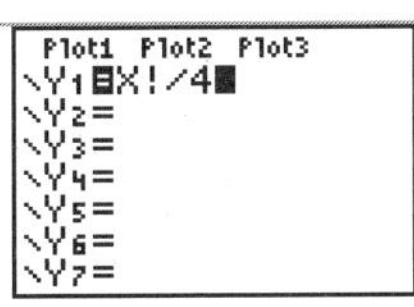

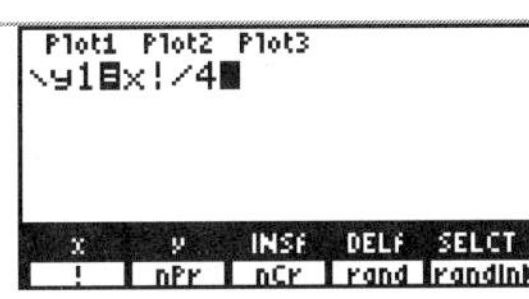

	 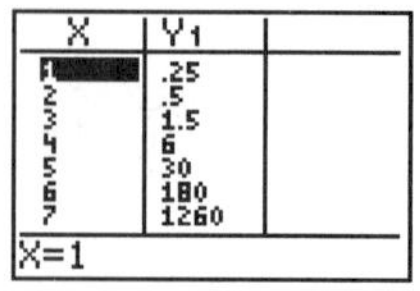	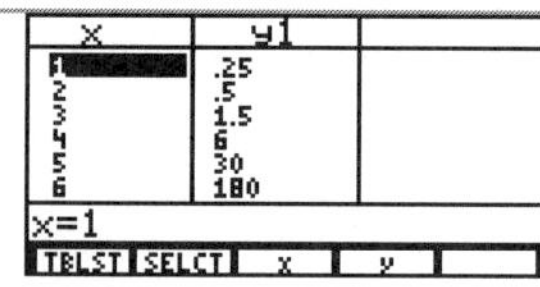

There are some special types of sequences that we will be investigating. The first type is called a **recursive sequence**.

> **Recursive Sequence** – A sequence for which the formula defines the nth term of the sequence as a function of the previous terms.

So far through this section the terms of our sequences have been defined based solely upon the domain value used, but with recursive sequences they are based upon the previous terms in the sequence. So the outputs become the inputs for the next terms. Here is an example.

Example 4: Finding Terms of a Recursive Sequence

Find the first five terms of the sequence given that $a_1 = 3$ and $a_n = a_{n-1} + 4$.

Solution:

The first term is given to us.	$a_1 = 3$
To find the second term, evaluate a_n when n is a 2 (the second positive integer). Remember $a_1 = 3$.	$a_2 = a_{2-1} + 4 = \underline{a_1} + 4 = \underline{3} + 4 = 7$ Notice that this is just the first term plus 4.
Now evaluate a_n when n is 3. Remember we found that $a_2 = 7$.	$a_3 = a_{3-1} + 4 = a_2 + 4 = 7 + 4 = 11$ Notice that this is just the second term plus 4.
Now evaluate a_n when n is 4. Remember we found that $a_3 = 11$.	$a_4 = a_{4-1} + 4 = a_3 + 4 = 11 + 4 = 15$ Notice that this is just the third term plus 4.
Now evaluate a_n when n is 5. Remember we found that $a_4 = 15$.	$a_5 = a_{5-1} + 4 = a_4 + 4 = 15 + 4 = 19$ Notice that this is just the fourth term plus 4.

Notice how each answer involved the previous one in its calculation. Here is another example.

Example 5: Recursive Sequence

Find the first six terms of the sequence given $a_1 = -3$, $a_2 = -1$ and $a_n = 5a_{n-1} - 2a_{n-2}$.

Solution:

The first two terms are given to us.	$a_1 = -3,\ a_2 = -1$
Now evaluate a_n when n is 3. Remember, we know $a_1 = -3,\ a_2 = -1$.	$a_3 = 5a_{3-1} - 2a_{3-2} = 5a_2 - 2a_1$ $= 5(-1) - 2(-3) = -5 + 6 = 1$
Now evaluate a_n when n is 4. Remember we found that $a_3 = 1$.	$a_4 = 5a_{4-1} - 2a_{4-2} = 5a_3 - 2a_2$ $= 5(1) - 2(-1) = 5 + 2 = 7$
Now evaluate a_n when n is 5. Remember we found that $a_4 = 7$.	$a_5 = 5a_{5-1} - 2a_{5-2} = 5a_4 - 2a_3$ $= 5(7) - 2(1) = 35 - 2 = 33$
Now evaluate a_n when n is 6. Remember we found that $a_5 = 33$.	$a_6 = 5a_{6-1} - 2a_{6-2} = 5a_5 - 2a_4$ $= 5(33) - 2(7) = 165 - 14 = 151$

As you can see, a recursion formula can involve more than one of the previous terms.

Question 4: Find the first eight terms of the recursive sequence called the Fibonacci sequence which is defined as follows: $a_1 = 1,\ a_2 = 1,\ a_n = a_{n-1} + a_{n-2}$.

Another special type of sequence that needs to be discussed is called an **arithmetic sequence**.

Answer Q3:

> **Arithmetic Sequence** – A sequence where each term is an equal difference apart from the next term (called the common difference).

An informal way to define this is that an arithmetic sequence is one where we add the same amount each time to get to the next term in the sequence. Formally, mathematicians use the term "common difference" because we must find the difference between consecutive terms in order to determine the amount being added to each term to get to the next term in the sequence.

Yes, these are exactly the same common differences that we discussed in chapter 3. As you might remember, when we found that common differences between data points were the same, we knew that we had a linear function. Here, when we see a sequence that has the same common differences between the terms of the sequence, we will know that it is an arithmetic sequence. The implication here, which is true, is that every arithmetic sequence is linear.

For every arithmetic sequence, there exists a linear function whose range values would be the same as the sequence terms, if you used only the positive integers as the domain.

Example 6: Arithmetic Sequence

Given the following sequence $\{-3, 4, 11, 18, \ldots\}$, is it arithmetic or not?

Solution:

Let's begin by looking at the common differences.

−3	
4	$4-(-3)=7$
11	$11-4=7$
18	$18-11=7$

The common differences are the same (7), so this must be arithmetic.

Now, since we know that this is arithmetic, we know that it is also linear; therefore, we can use previously learned knowledge to help us find the formula for this sequence.

First, we might remember that the common difference is the slope of the linear function when the difference in the domain values is 1. Second, given a point and the slope, you can find the formula for a linear function.	$d=7$ implies that $m=7$. We know that when $n=1$, $a_n=-3$. This is the point (1,-3). $y-(-3)=7(x-1)$ point-slope formula $y+3=7(x-1)$ $y=7(x-1)-3$
Since we know that the linear function is $y=7(x-1)-3$, we know that the formula for the sequence is	$a_n=7(n-1)-3$

Hence, all you need in order to find a formula for an arithmetic sequence is the common difference, d, and a term of the sequence (preferably the first term, a_1).

Given this fact, the general formula for a_n is defined as follows:

> **Arithmetic general formula:** $a_n = d(n-1) + a_1$, where d is the common difference, n the independent variable with domain (1, 2, 3,…) and a_1 the first term in the sequence.

Question 5: Given the arithmetic sequence $\{8, 5, 2, -1,\ldots\}$, find the formula for a_n.

There is one more type of sequence we want to introduce here and that is the **geometric sequence**.

> **Geometric Sequence**: A sequence where each term is an equal ratio (or multiple) of the next term (called the common ratio or the common multiplier).

An informal way to define this is that a geometric sequence is one where you multiply by the same amount each time to get the next term in the sequence. Formally, mathematicians use the term common ratio because we must find the ratio between consecutive terms in order to determine the amount each term is being multiplied by to get the next term in the sequence.

Yes, these are exactly the same common ratios that we discussed in chapter 4. As you might remember, when we found that common ratios between data points were the same, we knew that we had an exponential function. Here, when we see a sequence that has the same common ratios between the terms of the sequence, we will know that it is a geometric sequence. The implication here, which is true, is that every geometric sequence is exponential. For every geometric sequence there exists an exponential function whose range values would be the same as the sequence terms, if you used only the positive integers as the domain.

Example 7: Geometric Sequence

Given the following sequence {3, 6, 12, 24, ...}, is it geometric or not?

Solution:

Let's begin by looking at the common ratios.		
	3	
	6	$\frac{6}{3} = \mathbf{2}$
	12	$\frac{12}{6} = \mathbf{2}$
	24	$\frac{24}{12} = \mathbf{2}$
The common ratios are the same (2) so this must be geometric.		

Answer Q4: $a_1 = 1$, $a_2 = 1$, $a_3 = a_2 + a_1 = 1 + 1 = 2$, $a_4 = a_3 + a_2 = 2 + 1 = 3$, $a_5 = 3 + 2 = 5$, $a_6 = 5 + 3 = 8$, $a_7 = 8 + 5 = 13$, *and* $a_8 = 13 + 8 = 21$.

Now, since we know that this is geometric we know that it is also exponential; therefore, we can use previously learned knowledge to help us find the formula for this sequence.

First, we might remember, that the common ratio is the base of the exponential function.	We know that $r = 2$, and that we have the point (1, 3) from $a_1 = 3$. Thus we have, $3 = a(2)^1$ (plugged in point (1,3))
Second, given that the general form for an exponential function is, $f(t) = a(r)^{t/n}$ *or* $y = a(r)^{t/n}$ where, a = start amount at $t = 0$, r = common ratio, n = time interval, we can just plug in the r and use a point to solve for a. Also $n = 1$ when the difference in the domain values is 1.	$\frac{3}{2} = a$ (divided by 2) $y = \frac{3}{2}(2)^x$ (put a into formula) $y = 3\left(\frac{1}{2}\right)2^x$ (break up $\frac{3}{2}$ into $3*\left(\frac{1}{2}\right)$) $y = 3(2^{-1})2^x$ (write $\left(\frac{1}{2}\right)$ as 2^{-1}) $y = 3(2)^{x-1}$ ($2^{-1} \bullet 2^x = 2^{x-1}$ exp. rule)
[Does the following belong anywhere?--a = start amount at t=0 (exponent = 0)]	Now this is set up correctly so that we start with an exponent of zero (n begins at one).
Since we know that an exponent function for these points is $y = 3(2)^{x-1}$, we know that the formula for the sequence is	$a_n = 3(2)^{n-1}$

All you need in order to find the formula for a geometric sequence is the <u>common ratio, r,</u> and a term of the sequence (preferably the <u>first term, a_1</u>).

Given this fact, the general formula for a_n is defined as follows:

> **Geometric General Formula** – $a_n = a_1(r)^{n-1}$, where r is the common ratio, n the independent variable with domain (1, 2, 3,…) and a_1 the first term in the sequence.

Question 6: Given the geometric sequence {45, 15, 5, …}, find the formula for a_n.

Example 8: Finding Formulas for Sequences

Find the common ratio or common difference and then the general formula for the following sequences.

a) {−2, 6, −18, 54, …} b) {13, 9, 5, 1, …}

Solution:

Answer Q5: $d = -3$, $a_1 = 8$ so $a_n = -3(n-1)+8$.

(a) There isn't a common difference, but there is a common ratio (−3). This must be a geometric sequence.

Sequence	d	r
−2		
6	(6−(−2))= 8	$\frac{6}{-2} = -3$
−18	(−18−6)= −24	$\frac{-18}{6} = -3$
54	(54−(−18))= 72	$\frac{54}{-18} = -3$

We know that $r = -3$ and that the first term is −2, so the formula is $a_n = -2(-3)^{n-1}$

(b) There isn't a common ratio, but there is a common difference (−4). This must be an arithmetic sequence.

Sequence	d	r
13		
9	(9 − 13) = −4	$\frac{9}{13}$
5	(5 − 9) = −4	$\frac{5}{9}$
1	(1 − 5) = −4	$\frac{1}{5}$

We know that $d = -4$ and that the first term is 13, so the formula is $a_n = -4(n-1) + 13$

Not every sequence is arithmetic or geometric. But, some sequences are simple enough that even though they aren't arithmetic or geometric, you can still discern a pattern then write a formula for the a_n term.

Example 9: Finding the nth term of a Sequence

Find the a_n term for the following sequences:

a) $\{1, 4, 9, 16, \ldots\}$ b) $\{1, \frac{1}{2}, \frac{1}{6}, \frac{1}{24}, \frac{1}{120}, \ldots\}$ c) $\{-1, \frac{1}{2}, \frac{-1}{4}, \frac{1}{8}, \frac{-1}{16}, \ldots\}$

Solution:

(a) We notice that the first term is 1^2, then the second term is 2^2, then the third term is 3^2 and so on. Thus the formula must be	$a_n = n^2$
(b) We have a fraction, so let's take the numerators and denominators separately $(1 = \frac{1}{1})$.	Numerators 1, 1, 1, 1, 1 Denominators 1, 2, 6, 24, 120
The numerator's pattern is easy to spot. It is just 1 all the time. The denominator's pattern is a little harder to see unless you notice that the numbers are the first couple of factorials.	Thus the formula is $a_n = \frac{1}{n!}$
(c) We once again have fractions, so we will break this one down like the last example.	Numerators −1, 1, −1, 1 Denominators 2, 4, 8, 16
The denominator appears to be the powers of the number 2. Since the first number must be 1 then the sequence in the denominator must be $2^0, 2^1, 2^2, 2^3, 2^4$, so the formula for it is	2^{n-1} By having $(n - 1)$ in the exponent position, we make the first term begin with (1-1) which gives us the zero exponent necessary to get a 1 in the first denominator.
As for the numerator, we see that, as in the last example, they all have an absolute value of 1 but they alternate signs. To create alternating sign we could try using the formula $(-1)^n$, which creates alternating signs $((-1)^1 = -1, (-1)^2 = 1, (-1)^3 = -1$, etc.). Thus the formula for the numerator is	$(-1)^n$
The formula for the whole sequence is	$a^n = \frac{(-1)^n}{2^{n-1}}$

Question 7: Find the a_n term for the sequence $\{\frac{1}{3}, \frac{8}{5}, \frac{27}{7}, \frac{64}{9}, \ldots\}$

Sometimes you aren't given the information in a straightforward manner and must do some calculating in order to get the formula.

Example 10: Finding the nth Term

Find the formula for a_n given the following information.

a) $a_3 = 5$, $a_8 = 12.5$, and is arithmetic b) $a_7 = 768$, $a_5 = 192$, and is geometric

Solution:

Answer Q6: $\frac{15}{45} = \frac{5}{15} = \frac{1}{3}$ *$= r$ and $a_1 = 45$, so the formula is* $a_n = 45\left(\frac{1}{3}\right)^{n-1}$.

(a) We could approach this two ways.

First, we could just plug what we know into the formula for arithmetic sequence ($a_n = d(n-1) + a_1$) and then solve the two equations for d and a_1.

Method 1) When $n = 3$ we have

$$5 = d(3-1) + a_1$$

When $n = 8$ we have

$$12.5 = d(8-1) + a_1$$

We now have a system of equations to solve.

$$\begin{cases} d(2) + a_1 = 5 \\ d(7) + a_1 = 12.5 \end{cases}$$

We could use any method we want; let's use the rref key on the calculator.

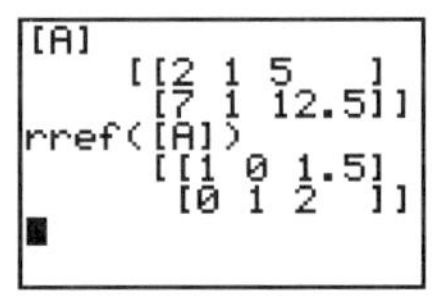

Thus, $d = 1.5$ and $a_1 = 2$. The formula for this sequence, then, is

$a_n = 1.5(n-1) + 2$ or simplified

$a_n = 1.5n + 0.5$

Second, we could try figuring out how many times the common difference was added and then figure out from that what d and a_1 must be.

Method 2) From a_3 to a_8, d was added 5 times $(8 - 3 = 5)$.

$$a_3 \xrightarrow{+d} a_4 \xrightarrow{+d} a_5 \xrightarrow{+d} a_6 \xrightarrow{+d} a_7 \xrightarrow{+d} a_8$$

Now, $12.5 - 5 = 7.5$ (the total amount added to get from a_3 to a_8) and if we divide that by 5 (the number of times d was added) we get $\frac{7.5}{5} = 1.5$, which is d. To find a_1 we simply plug one of the values we know into the formula and solve for a_1.

$$a_3 = d(n-1) + a_1$$

$$5 = 1.5(3-1) + a_1$$

$$5 = 1.5(2) + a_1$$

	$5 = 3 + a_1$ $2 = a_1$ The formula is $a_n = 1.5(n - 1) + 2$.
(b) We can use either of the two methods we used in part (a). Let's solve the system of equations and use the substitution method here.	When $n = 7$ we have $768 = a_1(r)^{7-1}$ When $n = 5$ we have $192 = a_1(r)^{5-1}$ We now have a system of equations to solve. $\begin{cases} a_1(r)^6 = 768 \\ a_1(r)^4 = 192 \end{cases}$ We will solve the second equation for a_1 $\left(a_1 = \dfrac{192}{r^4}\right)$ and then substitute into the first. $\left(\dfrac{192}{r^4}\right)(r)^6 = 768$ $192r^2 = 768$ $r^2 = 4$ $r = \pm 2$ Substituting back we find $a_1 = \dfrac{192}{2^4} = 12$ or $a_1 = \dfrac{192}{(-2)^4} = 12$ Thus, $r = 2$ or -2 and $a_1 = 12$ for both. The formulas for the two possible sequences are $a_n = 12(2)^{n-1}$, or $a_n = 12(-2)^{n-1}$

Answer Q7: The numerator is cubing and the denominator is arithmetic adding 2 each time, so $a_n = \dfrac{n^3}{2(n-1)+3}$ *or preferably simplified as* $a_n = \dfrac{n^3}{2n+1}$.

Section Summary:

- A sequence is a function where the domain usually is all positive integers and is expressed in subscript notation (a_n).
- ! means factorial; $\boldsymbol{n!} = n(n-1)(n-2)(n-3) \ldots 1$, if n is a whole number.
- You can put sequences in your graphing calculators and see them either on the home screen or in a table.
- A recursive sequence is a sequence for which the formula defines the nth term of the sequence as a function of the previous term or terms.
- An arithmetic sequence means adding the same amount each time to get to the next term in the sequence. The general formula is $a_n = d(n-1) + a_1$, where you only need to find d and a_1.

- A geometric sequence means multiplying the same amount each time to get to the next term in the sequence. The general formula is $a_n = a_1(r)^{n-1}$, where you only need to find r and a_1.
- When you are trying to find the formula for a sequence, remember to look for patterns. You might try looking to see if the sequence is arithmetic or geometric first and if it is neither, then look for other types of patterns.

SECTION 8.1 HOMEWORK

(1–16) For each of the following sequences give the first five terms

1. $a_n = 5n + 2$
2. $b_n = 3n - 8$
3. $c_n = 2 \cdot 3^n$
4. $a_n = 3\left(\frac{1}{3}\right)^n$
5. $b_1 = 3;\ n > 1;\ b_n = 2b_{n-1} + 5$
6. $a_1 = -3;\ n > 1;\ a_n = 5a_{n-1} - 8$
7. $c_1 = 5;\ c_2 = -3;\ n > 2;\ c_n = 2c_{n-2} - 3c_{n-1}$
8. $a_1 = 4;\ a_2 = 5;\ n > 2;\ a_n = a_{n-1} \bullet a_{n-2} + 7$
9. $b_n = 3^n - n!$
10. $c_n = 2^n - n!$
11. $a_1 = 2;\ n > 1;\ a_n = \dfrac{5}{a_{n-1} + 3}$
12. $b_1 = 5;\ n > 1;\ b_n = \dfrac{4}{b_{n-1} - 3}$
13. $a_n = \dfrac{n!}{n^2}$
14. $c_n = \dfrac{n^2 + 3}{n! + 3}$
15. $a_n = (-2)^n$
16. $b_n = (-3)^n + 5$

(17–28) Assuming the pattern does not change for each of the following what would be the next term of the sequence?

17. 5, 9, 13,17,...
18. 9, 2, –5, –12,...
19. 3, 6, 12, 24,...
20. $-2, \frac{-2}{3}, \frac{-2}{9}, \frac{-2}{27}, \ldots$
21. $-5, \frac{5}{2}, \frac{-5}{4}, \frac{5}{8}, \ldots$
22. 3, –15, 75, –375,...
23. 3, 5, 8, 13, 21,...
24. 9, 4, 5, –1, 6, –7,...
25. $1, \frac{1}{2}, \frac{1}{3}, \frac{1}{4}, \ldots$
26. $-2, 1, \frac{-1}{2}, \frac{1}{4}, \frac{-1}{8}, \ldots$
27. 1, 4, 9, 16,...
28. 1, 8, 27, 64, 125,...

(29–38) Answer the following questions for each problem:

a. Find the general equation of the arithmetic sequence.

b. Find the tenth term of the arithmetic sequence.

29. $a_1 = 8;\ d = 12$
30. $a_1 = 6;\ d = -3$
31. $a_1 = -5;\ a_7 = 43$
32. $a_1 = 6;\ a_{12} = -71$
33. $a_1 = 7;\ a_5 = -3$
34. $a_1 = -39;\ a_7 = -5$
35. $a_3 = -20;\ a_7 = -44$
36. $a_4 = 40;\ a_9 = 95$
37. $a_3 = 15;\ a_8 = 39$
38. $a_5 = -15;\ a_{12} = -53$

(39–46) Answer the following questions for each problem:

a. Find the general equation of the geometric sequence.

b. Find the sixth term of the geometric sequence.

39. $a_1 = 5;\ r = 3$
40. $a_1 = 7;\ r = -2$
41. $a_1 = 8;\ r = \frac{-1}{4}$
42. $a_1 = -10;\ r = \frac{1}{2}$
43. $a_1 = 4;\ a_4 = 500$
44. $a_1 = -5;\ a_8 = -10{,}935$

45. $a_2 = 28;\ a_4 = 448;\ r > 0$

46. $a_3 = 36;\ a_7 = 2{,}916;\ r < 0$

(47–58) For each of the following sequences:

a. Is the sequence arithmetic, geometric or neither?

b. If the sequence is arithmetic or geometric, give the general equation and a_8.

c. If not arithmetic or geometric, give the next term of the sequence.

47. $5, -2, -9, -16, \ldots$

48. $5, 28, 51, 74, \ldots$

49. $3, 7, 12, 18, \ldots$

50. $2, 6, 24, 120, \ldots$

51. $2, 10, 50, 250, \ldots$

52. $4, -16, 64, -256, \ldots$

53. $3, -1, \frac{1}{3}, \frac{-1}{9}, \ldots$

54. $4, \frac{4}{5}, \frac{4}{25}, \frac{4}{125}, \ldots$

55. $1, 3, 5, 11, 21, \ldots$

56. $1, 3, 1, -5, -7, \ldots$

57. $\frac{3}{5}, \frac{-1}{15}, \frac{-11}{15}, \frac{-7}{5}, \ldots$

58. $\frac{4}{7}, \frac{41}{35}, \frac{62}{35}, \frac{83}{35}, \ldots$

59. $f(x) = 3x + 5$ is a linear function written in slope–intercept form. Assume the domain of this linear function is natural numbers, then give the equivalent general equation of the arithmetic sequence defined by the linear function.

60. $a_n = 3 - 5(n - 1)$ is an arithmetic sequence. Give the slope–intercept form of an equivalent linear function, assuming the domain is natural numbers.

61. $f(x) = 3(2)^x$ is an exponential function. Assume the domain of this exponential function is natural numbers, then give the equivalent general equation of the geometric sequence defined by the exponential function.

62. $a_n = 15(3)^{n-1}$ is a geometric sequence. Give the equation of an equivalent exponential function, assuming the domain is natural numbers.

63. You have taken a job with starting pay of \$250.00 a week, with a raise each week of \$5.00.

a. If $a_1 = 250$, does this information define an arithmetic or geometric sequence?

b. Give the general equation of the sequence defined.

c. How much will the weekly pay be in the 50th week?

64. You have taken a job that pays 1 cent for the first hour and the wages double each hour after the first hour. (So, for instance, you make 2 cents for the 2nd hour, 4 cents for the 3rd hour and so on.)

a. If $a_1 = \$0.01$, does this information define an arithmetic or geometric sequence?

b. Give the general equation of the sequence defined.

c. How much will you be paid for the 30th hour?

65. \$5,000 is invested in an account at 8% interest compounded yearly.

a. Give a function that tells how much money is in the account at the end of each year. (Hint: check chapter 5 to find the compound interest formula.)

b. What is the value of the account after 9 years?

c. An equivalent geometric sequence is $a_n = 5000(1.08)^{n-1}$. What is the value of a_1 and what does this value represent?

d. Using the geometric sequence what is the value of a_{10} and what does that value represent?

e. In general what does the value of a_n represent?

66. $5,000 is invested in an account at 8% yearly simple interest.

a. Give a function that tells how much money in the account at the end of each year.

b. What is the value of the account after 9 years?

c. An equivalent arithmetic sequence is $a_n = 5000 + 400(n - 1)$. What is the value of a_1 and what does that value represent?

d. Using the arithmetic sequence, what is the value of a_{10} and what does that value represent?

e. In general what does the value of a_n represent?

Section 8.2 Series

In the last section we learned what a sequence is. Primarily we learned about two specific types of sequences, arithmetic and geometric. Next, we are going to discuss a topic called **series**.

Series – A series is a sum of the terms of a sequence.

For example, if we have the sequence {2, 5, 8, 11, 14, …} and we add together the first five terms of this sequence, then we would have the series $S_5 = 2 + 5 + 8 + 11 + 14$. The symbol S_5 represents the series that uses the first five terms of a sequence. If you were asked to find S_3, then you would be adding the first three terms of a series. For this example you'd have $S_3 = 2 + 5 + 8$. This brings us to another definition.

***n*^th^ Partial Sum** – An nth partial sum is a series where you add the first n terms of a sequence.

S_5, and S_3 that we have just discussed are both called partial sums. Specifically, these are called the 5th partial sum and the 3rd partial sum respectively.

Here is another term.

Infinite Series – An infinite series is a sum of all the terms of an infinite sequence (S_∞).

Now it is pretty hard to add up an infinite number of numbers. We can do this though, but we go about it by using a special sequence called the sequence of partial sums.

Sequence of Partial Sums – A sequence of partial sums is the sequence made up of all of the possible partial sums for a particular sequence.

Let's use the sequence {2, 5, 8, 11, 14, …} again to find its first five partial sums.

$S_1 = 2$
$S_2 = 2 + 5 = 7$
$S_3 = 2 + 5 + 8 = 15$
$S_4 = 2 + 5 + 8 + 11 = 26$
$S_5 = 2 + 5 + 8 + 11 + 14 = 40$

Notice that this creates another sequence {2, 7, 15, 26, 40, …}, called the sequence of partial sums. If you can figure out what is happening to this sequence as you work your way towards infinity, you can calculate the value for the infinite series (S_∞). If you are starting to feel a little overwhelmed, don't feel bad. Calculus students struggle with these concepts, too. Let's take a moment to recap the definitions informally:

- A <u>sequence</u> is a <u>list</u> of numbers.
- A <u>series</u> is a <u>sum</u> of the numbers from a sequence.
- An <u>n^{th} partial sum</u> is a <u>series</u> where you have added the first n terms of a sequence.
- The <u>list</u> of partial sums is itself another <u>sequence</u>.

Basically just remember this: if you are listing numbers it is a sequence, and if you are summing numbers, then it is a series.

There is a special notation mathematicians use to express sums. It is called **summation notation**, and is defined as follows.

$\sum_{i=1}^{n} a_i = a_1 + a_2 + a_3 + \ldots + a_n$, where a_n is the formula for the sequence, i is called the index of the summation, n is the upper limit on the summation and 1 is the lower limit.

The notation $\sum_{i=1}^{n} a_i$ is read "The summation as i goes from 1 to n of a_n." The lower limit tells you from which term in the sequence (a_n) to start summing and the upper limit tells you at which term in the sequence to stop summing.

Discussion 1: Summation Notation

Let's look at how we find values from summation notation with the following examples.

a) $\sum_{i=3}^{6} i^2$ b) $\sum_{i=1}^{8} 3i-1$ c) $\sum_{i=0}^{4} i!$

(a) The lower index is 3, so we start with i = 3 and continue to add terms of the sequence up through i = 6, the upper limit. $\sum_{i=3}^{6} a_i = a_3 + a_4 + a_5 + a_6$	$a_3 = 3^2 = 9$ $a_4 = 4^2 = 16$ $a_5 = 5^2 = 25$ $a_6 = 6^2 = 36$ $\sum_{i=3}^{6} i^2 = 9 + 16 + 25 + 36 = 86$
(b) The lower index is 1, so we start with i = 1 and continue to add terms up through i = 8. $\sum_{i=1}^{8} a_i = a_1 + a_2 + a_3 + a_4 + a_5 + a_6 + a_7 + a_8$ $= S_8$, the 8th partial sum.	$a_1 = 3(1)-1=2$; $a_2 = 3(2)-1=5$ $a_3 = 3(3)-1=8$; $a_4 = 3(4)-1=11$ $a_5 = 3(5)-1=14$; $a_6 = 3(6)-1=17$ $a_7 = 3(7)-1=20$; $a_8 = 3(8)-1=23$ $\sum_{i=1}^{8} 3i-1$ $= 2 + 5 + 8 + 11 + 14 + 17 + 20 + 23$ $= 100 = S_8$

(c) The lower index is 0, so we start with $i =$ 0 and go through $i = 4$. The index can start at any integer as long as all of the integers from there on are in the domain of the function we are working with. As in this example, we wouldn't be able to have an index that starts with a negative number.	$a_0 = 0! = 1$ $a_1 = 1! = 1$ $a_2 = 2! = 2$ $a_3 = 3! = 6$ $a_4 = 4! = 24$ $\sum_{i=0}^{4} i! = 1 + 1 + 2 + 6 + 24 = 34$

Question 1: Find the value of $\sum_{i=1}^{8} 2i - 1$.

There are special summations that show up in later math courses that have useful formulas for computing their sums. Here are just three of them,

1. $\sum_{i=1}^{n} i = 1 + 2 + 3 + 4 + 5 + ... + n = \frac{n(n+1)}{2}$

2. $\sum_{i=1}^{n} i^2 = 1 + 4 + 9 + 16 + 25 + ... + n^2 = \frac{n(n+1)(2n+1)}{6}$

3. $\sum_{i=1}^{n} i^3 = 1 + 8 + 27 + 64 + 125 + ... + n^3 = \frac{n^2(n+1)^2}{4}$

For any value of $n > 1$ you can simply plug into these formulas the value of n and determine their sum.

Example 1: Sums of Integers and Squares of Integers

Find the following sums.

a) $\sum_{i=1}^{37} i$ b) $\sum_{i=1}^{18} i^2$

Solution:

(a) The upper limit is 37, so we plug $n = 37$ into formula 1 and compute the sum.	$\sum_{i=1}^{37} i = 1 + 2 + 3 + \ldots + 37 = \frac{37(37+1)}{2}$ $= \frac{37(38)}{2} = 37(19) = 703$
(b) The upper limit is 18, so we plug $n = 18$ into formula 2 and compute the sum.	$\sum_{i=1}^{18} i^2 = 1 + 4 + 9 + \ldots + 18^2$ $= \frac{18(18+1)(2(18)+1)}{6}$

	$= \dfrac{18(19)(37)}{6} = 2109$

Notice that these three formulas are able to give us the value of a partial sum (S_n) for any value of n. This means that these formulas are themselves the functions that generate a sequence, specifically, the sequence of partial sums for $a_n = n$, $a_n = n^2$, and $a_n = n^3$ respectively.

> Question 2: Find the first five terms and the 20th term of the sequence of partial sums for the sequence $a_n = n$. (Hint: $S_n = \sum_{i=1}^{n} a_n = \sum_{i=1}^{n} i = \dfrac{n(n+1)}{2}$)

These formulas are proven by something called **mathematical induction**, which is in the appendix of this book. We will leave that discussion for another day.

Next, just like with sequences, we have two special series we need to discuss, arithmetic and geometric. An arithmetic series is one where we are adding the terms of an arithmetic sequence, and a geometric series is one where we are adding the terms of a geometric sequence. Let's concentrate on arithmetic series first.

Discussion 2: Finding the nth Partial Sum (S_n)

Let's look at how we find a formula for the nth partial sum of the arithmetic sequence $a_n = 3n - 1$. First we look at a specific example, like finding S_9, and then we write down what might be the general formula for S_n.

$a_1 = 3(1) - 1 = 2$ $a_2 = 3(2) - 1 = 5$ $a_3 = 3(3) - 1 = 8$ $a_4 = 3(4) - 1 = 11$ . . . $a_9 = 3(9) - 1 = 26$	$S_9 = 2 + 5 + 8 + 11 + 14 + 17 + 20 + 23 + 26$ Since the partial sum is made up of an arithmetic sequence, each pair of consecutive terms differs by the same amount. Also, we notice that if we were to add the first and last terms of the partial sum together, we get the same answer as when we add the second and next to last terms and so on. $S_9 = 2 + 5 + 8 + 11 + 14 + 17 + 20 + 23 + 26$ 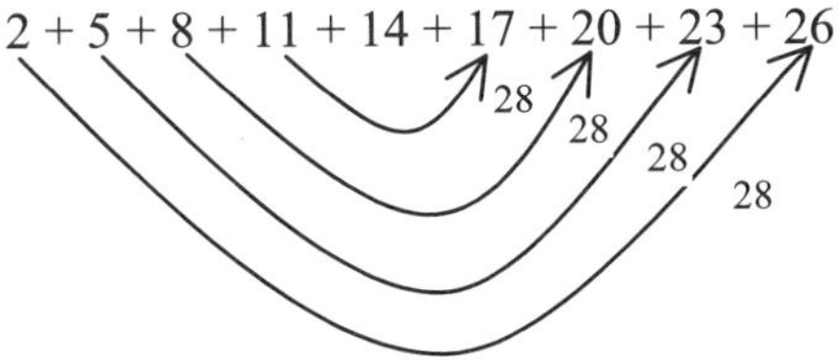 Therefore, S_9 is just 28 + 28 + 28 + 28 + 14 (the middle number that was left over), which is 126.
In the computing of S_9 the 28's came from adding two numbers together. Thus, the average	

of the two numbers is 14 (28/2). That just happens to be the middle number that didn't have a partner to go with it. That means that the nine terms together average 14 apiece. So the total would be the average (14) times the number of terms (9) for a total of 126 ($14 \cdot 9 = 126$). Since each pair of numbers we added was the same (28), all we really needed to do was add the first and last terms together, average their sum, and then multiply by the number of terms. There's our formula $S_n = \left(\frac{a_1 + a_n}{2}\right)n$. Let's test this on S_8.

Answer Q1:
$a_1 = 1$, $a_2 = 3$, $a_3 = 5$, ... $a_8 = 15$, thus
$\sum_{i=1}^{8} 2i - 1 =$
1+3+5+7+9+11+13+15=64.

$S_8 = 2 + 5 + 8 + 11 + 14 + 17 + 20 + 23$

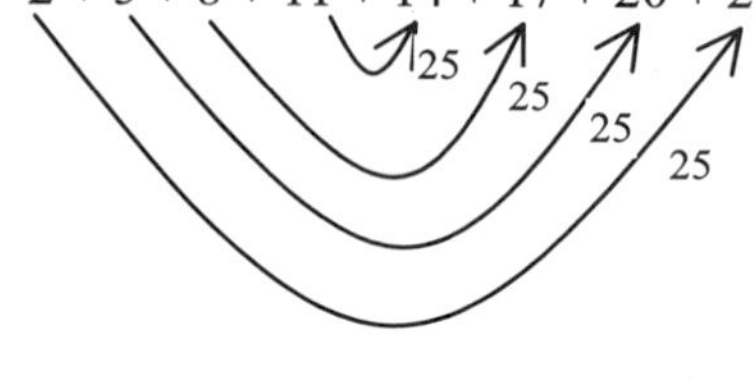

Therefore, S_8 is just 25 + 25 + 25 + 25, which is 100.

Using our new formula we have $S_8 = \left(\frac{2+23}{2}\right)8 = \left(\frac{25}{2}\right)8 = 12.5 \cdot 8 = 100.$

Therefore, the formula for S_n for this particual sequence is the first term, plus the formula for a_n, divided by 2, and multiplied by n.

$$S_n = \left(\frac{a_1 + a_n}{2}\right)n = \left(\frac{2+(3n-1)}{2}\right)n = \left(\frac{3n+1}{2}\right)n = 1.5n^2 + 0.5n$$

The formula for the nth partial sum of any **arithmetic series** is $S_n = \left(\frac{a_1 + a_n}{2}\right)n$.

We are not going to go through the proof of this formula, but we hope that discussion 2 convinces you that it will always work for an arithmetic series.

Example 2: Finding S_n

Given $a_n = 73 - 5n$, find the nth partial sum for this arithmetic sequence.

Solution:

We can see that the formula for a_n is linear, so this really is an arithmetic sequence. We need the first and last terms and the number of terms to find a partial sum for this sequence. If we are asked to find the nth sequence. If we are asked to find the nth partial sum, then a_n is the last term and n is the number of terms. All we need to do is find a_1 and we'll basically be done.	$a_1 = 73 - 5(1) = 68$ $S_n = \left(\frac{a_1 + a_n}{2}\right)n = \left(\frac{68+(73-5n)}{2}\right)n$ $= \left(\frac{141-5n}{2}\right)n = 70.5n - 2.5n^2$

find a_1 and we'll basically be done.	
We now have the general formula for the nth partial sum of the sequence $a_n = 73 - 5n$.	$S_n = 70.5n - 2.5n^2$

Question 3: Find the nth partial sum of the sequence $a_n = 7n - 38$ and then the value of S_{12}.

Answer Q2:
$S_1 = 1$,
$S_2 = \frac{2(3)}{2} = 3$
$S_3 = \frac{3(4)}{2} = 6$
$S_4 = \frac{4(5)}{2} = 10$
$S_5 = \frac{5(6)}{2} = 15$
$S_{20} = \frac{20(21)}{2} = 210$
The sequence of partial sums is $\{1, 3, 6, 10, 15, \ldots, 210, \ldots \frac{n(n+1)}{2}\}$.

Example 3: Finding Total Energy Used from 1979 Through 2003

Given the following data from discussion 1 of section 5.1, approximate the total amount of energy used from 1979 through 2003. (Quad. Btu stands for quadrillion Btu's a measurement of energy.)

Year	Energy Consumption by Residential and Commercial users in (quad. Btu)
1979 $t = 0$	26.6
1989 $t = 10$	30.4
1999 $t = 20$	34.2

Source: 2000 Statistical Abstract of the United States

Solution:

In discussion 1 of section 5.1, we found that the formula for this data is linear and that the formula is $E(t) = 0.38t + 26.6$, which means that this is an arithmetic sequence. In sequence notation we write this as $a_n = 0.38n + 26.6$ with $n = 0$ as our first input value. The question posed to us is a series question since it is asking us to find a sum (total amount used over last 25 years (2003 – 1979 = 24 + 1 since we include both end points of the interval).

The first term is 26.6 and the 25th term is the value of a_{24} (started with $n = 0$, not 1) in the year 2003.	$a_0 = 26.6$ $a_{24} = 0.38(24) + 26.6 = 35.72$
The total is going to be the value for S_{25}, the sum of the first 25 terms.	$S_{25} = \left(\frac{26.6 + 35.72}{2}\right)25 = 779$ quad Btu

The approximate amount of energy used by residential and commercial users in quad. Btu's from 1979 through 2003 is 779 quad Btu's. Let's turn our attention to geometric series now.

The proof of this formula is pretty straightforward, so we will work through it here.

The nth partial sums of a geometric sequence $a_n = a_1(r)^{n-1}$ and the series rS_n.	$S_n = a_1 + a_2 + a_3 + \ldots + a_{n-1} + a_n$ $= a_1 + a_1r + a_1r^2 + \ldots + a_1r^{n-3} + a_1r^{n-2} + a_1r^{n-1}$ $rS_n = a_1r + a_1r^2 + a_1r^3 + \ldots + a_1r^{n-2} + a_1r^{n-1} + a_1r^n$

Now, subtract rS_n from S_n.	$S_n = a_1 + a_1r + a_1r^2 + a_1r^3 + \ldots + a_1r^{n-2} + a_1r^{n-1}$ $\underline{-rS_n = \quad a_1r + a_1r^2 + a_1r^3 + \ldots + a_1r^{n-2} + a_1r^{n-1} + a_1r^n}$
Factor and solve for S_n.	$S_n - rS_n = a_1 - a_1r^n$ $S_n(1 - r) = a_1(1 - r^n)$ $S_n = \dfrac{a_1(1-r^n)}{1-r}$
The formula for the nth partial sum of a **geometric series** is $S_n = \dfrac{a_1(1-r^n)}{1-r}$, $r \neq 1$.	

Note: Since the formula for the nth partial sum of a geometric series has $(1 - r)$ in the denominator, and denominators aren't allowed to equal 0, r can't equal 1. The sum of a geometric series, if r is 1, is trivial since that means you are adding the same amount each time. The sum in this special case would just be $S_n = a_1n$.

Example 4: Finding S_n

Find a formula for arriving at the nth partial sum of the geometric sequence $a_n = -2(0.5)^{n-1}$, then find S_7.

We need to find a_1 and plug it, along with r, into our formula.	$a_1 = -2(0.5)^{1-1} = -2$ $S_n = \dfrac{-2(1-0.5^n)}{1-0.5} = \dfrac{-2(1-0.5^n)}{0.5}$ $= -4(1 - 0.5^n)$
To find S_7 let $n = 7$.	$S_7 = -4(1 - 0.5^7) = -4(1 - 0.0078125)$ $= -3.96875$ or $\dfrac{-127}{32}$ (Frac key)

Question 4: Find the nth partial sum of the sequence $a_n = 50000(1.02)^n$ and then the value of S_{18}.

This last question could very well be a problem about the total pay a teacher might earn given that he or she would receive a 2% increase in salary each of the next 18 years. He or she would earn a cumulative pay of roughly $1,092,000 over the 18-year period.

Discussion 3: Infinite Sum

Given the sequence $a_n = 25(0.9)^{n-1}$, let's look at how we would find the sum of *all* the terms.

We have a geometric sequence and we have been asked to find the infinite sum of all the terms (n approaching ∞). Our formula for a geometric n^{th} partial sum is $S_n = \dfrac{a_1(1-r^n)}{1-r}$, which, for this example, would imply we have $S_n = \dfrac{25(1-0.9^n)}{1-0.9}$ as n approaches ∞.

What is going to happen to 0.9^n as n approaches ∞? Well, if we think about it, we are taking a number smaller than 1 times itself, endlessly. That would make the products get smaller and smaller. The result should end up close to 0. In mathematics we say that as n approaches ∞, 0.9^n approaches 0. Therefore, since 0.9^n becomes 0 for this example, the formula simply becomes $S_\infty = \dfrac{25(1-0)}{1-0.9} = \dfrac{25}{0.1} = 250$. The symbol S_∞ signifies n approaching ∞.

Answer Q3: $a_1 = -31$, $S_n = \left(\dfrac{-31+(7n-38)}{2}\right)n = \left(\dfrac{7n-69}{2}\right)n = 3.5n^2 - 34.5n$, $S_{12} = 3.5(12)^2 - 34.5(12) = 90$.

Question 5: What happens to 1.2^n as n approaches ∞?

> The formula for the sum of an infinite **geometric series** is $S = \dfrac{a_1}{1-r}$, $|r| < 1$.

Notice that we need the restriction on r ($|r| < 1$) because if the absolute value of r is larger than 1 the answer is ∞.

Example 5: Finding the Infinite Sum (S_∞)

Find the infinite sum for the sequence $a_n = 3(-0.5)^{n-1}$.

Solution:

We need to find a_1 and plug it, along with r, into our formula. Note: $\lvert -0.5 \rvert = 0.5 < 1$.	$a_1 = 3(-0.5)^{1-1} = 3$ $S_n = \dfrac{3}{1-(-0.5)} = 2$

Therefore, the sum of the terms of the sequence $\{3, -1.5, 0.75, -0.375, 0.1875, \ldots 3(-0.5)^{n-1}, \ldots\}$ is 2. In other words, the sequence of partial sums for this geometric sequence approach the value of 2 as n approaches ∞.

Section Summary

- A <u>series</u> is a <u>sum</u> of the terms of a sequence.
- An infinite series is a sum of all the terms of an infinite sequence (S_∞).

- An nth partial sum is a series where we add the first n terms of a sequence.
- A <u>sequence</u> of partial sums is the sequence made up of all of the possible partial sums for a particular sequence (<u>list</u> of possible sums).
- Summation notation means, $\sum_{i=1}^{n} a_i = a_1 + a_2 + a_3 + \ldots + a_n$
- The formula for the nth partial sum of an **arithmetic series** is $S_n = \left(\frac{a_1 + a_n}{2}\right)n$.
- The formula for the nth partial sum of a **geometric series** is $S_n = \frac{a_1(1-r^n)}{1-r}$ $r \neq 1$.
- The formula for the sum of an **infinite geometric series** is $S = \frac{a_1}{1-r}$, $|r| < 1$.

Answer Q4:
$a_1 = 51000$, $S_n = \frac{51000(1-1.02^n)}{1-1.02}$, $S_{18} = \frac{51000(1-1.02^{18})}{1-1.02} = 1{,}092{,}027.93$

SECTION 8.2 HOMEWORK

(1–6) Find the indicated sum.

1. $\sum_{i=1}^{5}(2i+3)$
2. $\sum_{i=1}^{6}3i$
3. $\sum_{i=1}^{4}(i^2+3)$
4. $\sum_{i=1}^{5}3i^2$
5. $\sum_{i=1}^{4}\left(\frac{1}{2}\right)^{1+i}$
6. $\sum_{i=1}^{6}\left(\frac{1}{i+1}\right)$

(7–12) Find the indicated partial sum for each problem.

7. 1, 3, 6, 10, 15, 21, … $S_5=$
8. –2, –3, –5, –8, –12, –17, … $S_5=$
9. 1, 2, 3, 5, 8, … $S_8=$
10. –2, 3, 5, 2, –3, –5, … $S_9=$
11. $\frac{-1}{2}, \frac{-1}{3}, \frac{-1}{4}, \frac{-1}{5}, \ldots$ $S_8=$
12. $\frac{3}{4}, \frac{4}{5}, \frac{5}{6}, \frac{6}{7}, \ldots$ $S_7=$

(13–26) Determine if the following sequences are arithmetic or geometric; then use the arithmetic or geometric series formulas to find the indicated partial sum.

13. 5, 9, 13, 17, … $S_{20}=$
14. 3, –2, –7, –12, … $S_{15}=$
15. 1, 3, 9, 27, … $S_8=$
16. 2, –4, 8, –16, 32, … $S_{10}=$
17. $\frac{3}{5}, \frac{1}{5}, \frac{-1}{5}, \frac{-3}{5}, \ldots$ $S_{12}=$
18. $\frac{9}{10}, \frac{4}{5}, \frac{7}{10}, \frac{3}{5}, \ldots$ $S_{10}=$
19. $a_n = 3 - 7(n-1)$ $S_{16}=$
20. $a_n = -15 + 3(n-1)$ $S_{19}=$
21. $a_n = 2(3)^{n-1}$ $S_8=$
22. $a_n = 3(-2)^{n-1}$ $S_{12}=$
23. $a_n = 5 + \frac{2}{3}(n-1)$ $S_{10}=$
24. $a_n = -7 - \frac{3}{4}(n-1)$ $S_{10}=$
25. $a_n = 2\left(\frac{-3}{2}\right)^{n-1}$ $S_9=$

Answer Q5: 1.2 is a number larger than one, so as you multiply it by itself many times, the products would be getting larger and larger. Therefore, as n approaches ∞, 1.2^n approaches ∞.

26. $a_n = -3\left(\frac{2}{5}\right)^{n-1}$ $S_6=$

27. $a_n = 5n + 9$ $S_{30}=$

28. $a_n = 8 - 7n$ $S_{24}=$

(Hint: For 29 and 30, think of finding the value of two different types of partial sums and adding the answers.)

29. $a_n = 5(2)^{n-1} + 3(n-1)$ $S_8=$

30. $a_n = 2(-3)^{n-1} - 5(n-1)$ $S_9=$

(31–34) For each of the following geometric sequences, find the infinite series value.

31. $3, 1, \frac{1}{3}, \frac{1}{9}, \frac{1}{27}, \ldots$ $S_\infty=$

32. $5, -1, \frac{1}{5}, -\frac{1}{25}, \frac{1}{125}, \ldots$ $S_\infty=$

33. $a_n = 4\left(-\frac{2}{5}\right)^{n-1}$ $S_\infty=$

34. $a_n = 8\left(\frac{3}{4}\right)^{n-1}$ $S_\infty=$

35. $a_n = 6\left(\frac{5}{4}\right)^{n-1}$ $S_\infty=$

36. $a_n = 10\left(-\frac{7}{3}\right)^{n-1}$ $S_\infty=$

37. Looking at a pyramid, you notice the base has 100 blocks facing you and each successive level has one less block facing you. If there are 100 layers of blocks to the top of the pyramid, how many total blocks are facing you?

38. An amphitheater has 40 seats on the bottom row and each successive row has two additional seats. If there are 40 rows in all, how many seats are in the amphitheater?

39. You are paid one cent the first hour you work and your wages are doubled every hour after the first hour. How much money will you have at the end of the 20th hour?

40. You have been promised by a wealthy uncle that he will give you $4,096 on July 1, and he will give you one-half of what he gave you the previous day for 10 days. How much money will he have given you after the 10 days?

41. A company offers you $35,000 the first year with a raise of $1,500 each year over the next 10 years. How much total salary do you receive for the 10-year period?

42. You have been offered two different jobs. The first job will pay you $36,000 the first year with a raise of $1,200 a year for the next 8 years and the second job will pay you $35,000 the first

year with a raise of $1,500 a year for the next 8 years. Which job should you take in order to have made the most money after 8 years?

43. You have been offered a job that will pay you $38,000 the first year with an increase of 4% each year for 8 years. How much salary do you receive for the 8-year period?

44. You have been offered two different jobs. The first job will pay you $34,000 the first year with an increase of 5% every year for five years and the second job will pay you $36,000 the first year with an increase of 3% every year for five years. Which job should you take to accumulate the most money after 5 years and which job will have the largest annual salary after 5 years?

Section 8.3 Counting Theory

We have finished discussing sequences and series. In this section we will take a look at counting problems. It is sometimes important to be able to figure out how many different possibilities exist. For example, how many different automobile license plates can a state produce, or how many different telephone numbers can you have in one area code? To answer these questions we need to develop a way to determine the maximum number of possibilities. There are a few formulas and ideas we need to discuss with regards to this topic. Let's start with what is called the **counting principle**.

> **Counting Principle** – The number of different ways that a progression of choices can happen is a product of the number of possibilities available at each choice.

Example 1: License Plates

If a state uses three letters followed by three numbers in its license plates, how many different license plates are possible?

Solution:

There are six *events*, one after the other. The license plate begins with a letter, then another letter, then one more letter, then a number, followed by another number, and then the last number.	The state license plates look like this: $\underline{?}\ \underline{?}\ \underline{?}\quad \underline{?}\ \underline{?}\ \underline{?}$
We apply the counting principle to this problem by taking the number of choices at each position on the license plate, times each other. Thus, the total number of possible license plates for this state without any restrictions on the combinations of letters and numbers would be	There are 26 letters in the alphabet and 10 digits used in numbers. $\underline{26} \cdot \underline{26} \cdot \underline{26} \cdot \underline{10} \cdot \underline{10} \cdot \underline{10} = 17{,}576{,}000$

If the state limits some of the choices because they don't want the three letters to form swear words or words like "DIP" or "FAT", then the total possibilities would be different.

> Question 1: How many different license plates could a state have if it wanted three letters followed by three numbers, but couldn't use any vowels (a, e, i, o, u, y)?

Example 2: Multiply Choice Tests

How many different ways could you answer a 10-question multiply choice exam with four choices for each question?

Solution:

There is a progression of 10 events (questions) each one having 4 choices, so our calculation would be	$\underline{4}\bullet\underline{4}\bullet\underline{4}\bullet\underline{4}\bullet\underline{4}\bullet\underline{4}\bullet\underline{4}\bullet\underline{4}\bullet\underline{4}\bullet\underline{4}=4^{10}=$ 1,048,576

Of course, there is only one exact way to get 100% on the test. Guessing is probably not the best option given that there are 1,048,576 different ways to answer the test.

Let's look at a simple illustration to show why multiplication is involved in the counting principle. Suppose that you wanted to order a pizza. You are going to choose one type of crust, one meat, and one vegetable. Here is a tree diagram to illustrate your choices given that there are two types of crust, three types of meats and two types of vegetables.

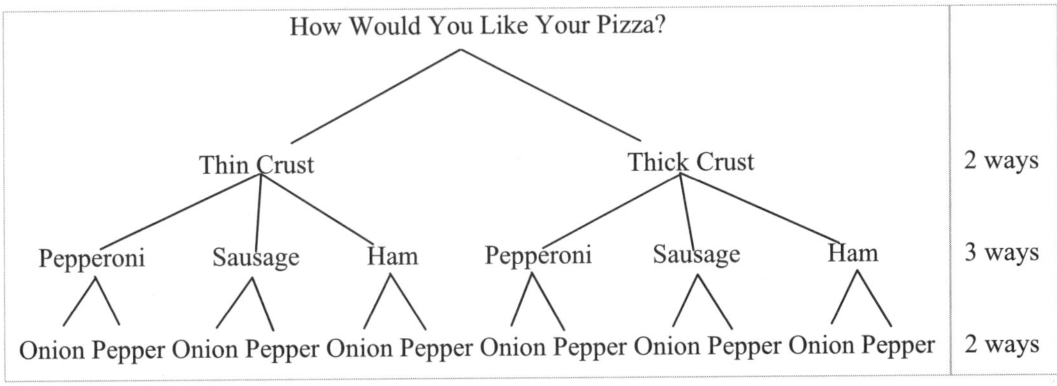

As you follow the lines down to the final choices, you see that there are 12 different pizzas possible. If you take the number of choices at each step along the way to reaching your final decision, you have $2\bullet 3\bullet 2=12$, the same number as when we followed the lines in the diagram.

Of course, the counting principle doesn't fit every kind of counting problem. When you want to determine how many different choices are possible in a given situation, there are two key questions that need to be considered. These key questions are

1. Can you repeat the choices?
2. Does the order of the choices matter?

Let's see how the answers to these two questions affect whether we can use the counting principle.

Discussion 1: Repeat or Not

How does repeating or not repeating affect the number of ways to arrange the numbers {1, 2, 3}?

If you can not repeat the numbers 1, 2, and 3, then the number of arrangements of the three	

values are $3 \cdot 2 \cdot 1 = 6$ (you have three choices for the first number, then only two numbers left for the second choice and then only one number for the last choice)

1 2 3	2 1 3	3 1 2
1 3 2	2 3 1	3 2 1

If you *can* repeat the numbers, then you have three choices at each spot ($3 \cdot 3 \cdot 3 = 27$). Here are the first few options.

1 1 1	1 1 2	1 1 3
1 2 1	1 2 2	1 2 3
1 3 1	1 3 2	1 3 3

and so on starting with a 2 and then a 3.

Notice that in either case we were able to use the counting principle to find the total number of possibilities. Therefore, regardless of the answer to the question of whether something can repeat or not, we can see that the counting principle will apply.

Question 2: How many telephone numbers are possible if you use 6 digits but you can't repeat any digits within the same telephone number (some places in Europe have six-digit telephone numbers).

Answer Q1:

*If the vowels are (a, e, i, o, u, y), then the total would be 20*20*20*10*10*10 = 8,000,000.*

Notice once again that with the counting principle it doesn't really matter whether you are allowed to have repeated choices or not, the principle works either way. However within the counting principle there is a subgroup called **permutations**.

Permutations – An arrangement of choices in which the choices may not repeat and the order of the choices is important (**no repeat, order matters**).

If we look back at question 2, we see that the answer was $10 \cdot 9 \cdot 8 \cdot 7 \cdot 6 \cdot 5 = 151{,}200$, which looks very similar to factorials ($n!$) from section 8.1. If we had $10 \cdot 9 \cdot 8 \cdot 7 \cdot 6 \cdot 5 \cdot 4 \cdot 3 \cdot 2 \cdot 1$, that would be exactly 10!. The difference between these two products is that 10! has $4 \cdot 3 \cdot 2 \cdot 1$ as part of the product, whereas the product in question 2 ends at 5. Therefore it appears that the answer to the telephone problem in Europe could be $\frac{10!}{4!} = \frac{10 \cdot 9 \cdot 8 \cdot 7 \cdot 6 \cdot 5 \cdot 4 \cdot 3 \cdot 2 \cdot 1}{4 \cdot 3 \cdot 2 \cdot 1} = 10 \cdot 9 \cdot 8 \cdot 7 \cdot 6 \cdot 5$.

$\frac{10!}{4!}$ could also be written as $\frac{10!}{(10-6)!}$. This leads us to a formula used to determine permutations.

Permutations formula – The number of permutations (different arrangements of choices) of n choices taken r at a time is

$$_nP_r = \frac{n!}{(n-r)!}$$

Example 3: Club Leadership

If you are in a math club and the club wants to elect people to be the president, secretary and treasurer, how many different election outcomes can happen if there are 10 members in the club?

Solution:

Here the order matters. For example if Sarah, Jose, and John are members of the club and Sarah is elected as the president, Jose as the treasurer and John as the secretary, that is different than Jose as the president John as the treasurer and Sarah as the secretary. The answer to this question is easy. We take 10 people (n) 3 at a time (r).	${}_{10}P_3 = \dfrac{10!}{(10-3)!}$ $= \dfrac{10\bullet 9\bullet 8\bullet 7\bullet 6\bullet 5\bullet 4\bullet 3\bullet 2\bullet 1}{7\bullet 6\bullet 5\bullet 4\bullet 3\bullet 2\bullet 1}$ $= 10\bullet 9\bullet 8$ $= 720$

The big picture here is that anywhere you can use permutations you can use the counting principle, but the nice thing about permutations is that they have a formula that goes with them. Since there is a nice formula we could use our calculators to evaluate the answer to the last problem. Here is how we use the calculators to evaluate ${}_{10}P_3$.

	TI-83	TI-86
First, you type in the number for n in the formula, then find the ${}_nP_r$ command in the calculator (go to the MATH menu and then to PRB), and finally type in the number for r and press ENTER.	 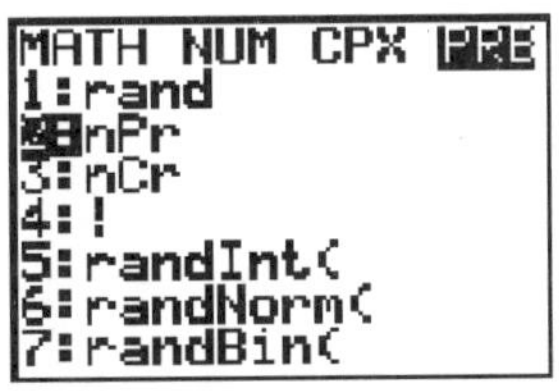 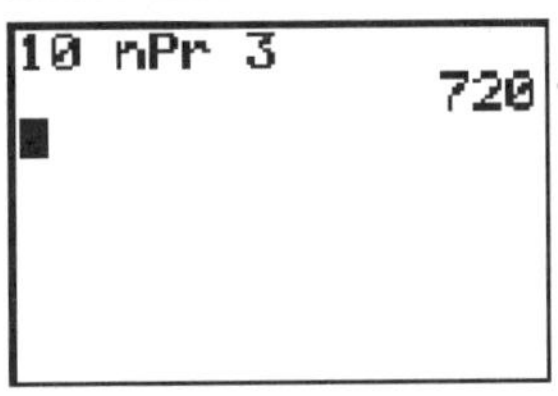	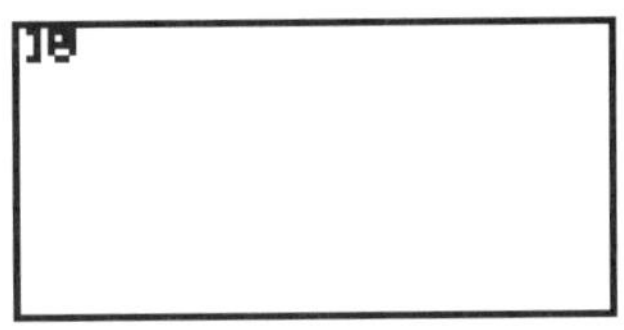 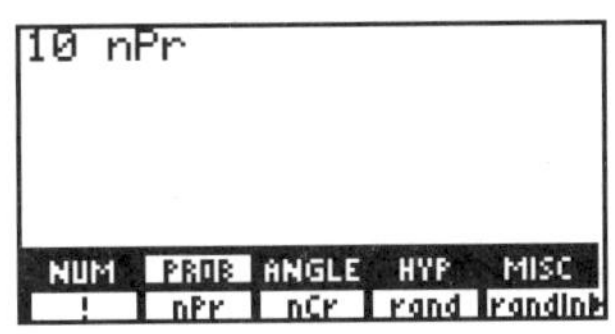 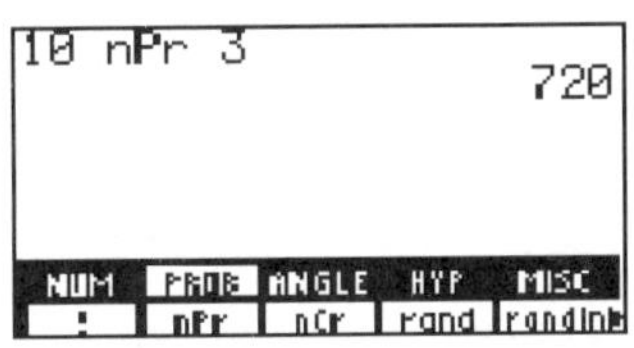

Let's now take a look at the second key question, which is about whether the order matters or not.

Discussion 2: Does the Order Matter?

In this math club of yours, you want to pick a three-person committee to plan your next club activity. How many different committees can be chosen if there are 10 people in the club?

Your first thought might be that it is $10\bullet 9\bullet 8 = 720$. Of course this is a good guess because you can't repeat people (they haven't cloned people yet), but it's not correct. In all of the examples that we have used to demonstrate the counting principle, the order of the choices has been important. For example, in a phone number like 473-0981, if you input the numbers in a different order, 743-0981, you would reach a different person. When calling someone, the order in which you input the digits into the phone is very important. If a policeman was looking for a criminal who is known to be driving a car with the license number TGH-762, you'd be in trouble if the order didn't matter and your license number is THG-762. The reason that the counting principle doesn't work in example 5 is that the order in which you select people for a committee doesn't matter. Let's illustrate this problem.

*Answer Q2: 10*9*8*7*6*5 = 151,200. There are 10 digits to choose from and 6 places to put them. But after you have chosen one number for the first slot, there are only 9 digits left to work with, since you can't repeat. Then, for the third slot, there would only be 8 left, and so on.*

Let's work with the club members John, Sarah and Jose again. If the order mattered, choosing Sarah, then Jose, then John would make the committee different from choosing John, then Sarah, then Jose. But of course, having Jose, John and Sarah on the same committee, no matter who was chosen first, second or third, leaves the same three people on the committee. Therefore, if the order did matter, you would expect there to be 6 choices, but since the order doesn't matter, there is really only one choice with these three people.	John, Sarah, Jose John, Jose, Sarah Jose, John, Sarah Jose, Sarah, John Sarah, Jose, John Sarah, John, Jose $3 \cdot 2 \cdot 1 = 6$ All six different combinations of these three people still constitute just one committee, not six different committees.

We can't finish this example yet. We don't have all of the tools that we need. It doesn't matter whether something repeats or not with the counting principle, but the order matters, so we can't use the counting principle here (**repeat or not, order matters**). Thus, this brings us to what is called **combinations**.

> **Combinations** – An arrangement of choices that do not repeat, in which the order of the choices is not important (**no repeat, order doesn't matter**).

If we once again look back at example 5, you see that John, Jose, and Sarah can be arranged into six different orders. But notice that $6 = 3! = 3 \cdot 2 \cdot 1$; so for every six choices where order could matter (like in elections), it boils down to just one choice with committees. We can find the answers to combinations by simply dividing out the different ordered possibilities that would exist in permutations. We get the following formula for combinations.

> **Combinations formula** – The number of combinations (different arrangements of choices) of n choices taken r at a time is
>
> $${}_nC_r = \frac{n!}{(n-r)!r!}$$

We now come back to the question posed in example 5. How many different committees of three can be created out of a group of 10 people?

With committees the order doesn't matter and you can't repeat. Therefore, we take 10 people (n) 3 at a time (r) and plug those numbers into our formula for combinations.	$_{10}C_3 = \frac{10!}{(10-3)!3!}$ $= \frac{10 \bullet 9 \bullet 8 \bullet 7 \bullet 6 \bullet 5 \bullet 4 \bullet 3 \bullet 2 \bullet 1}{7 \bullet 6 \bullet 5 \bullet 4 \bullet 3 \bullet 2 \bullet 1 \bullet 3 \bullet 2 \bullet 1}$ $= \frac{10 \bullet \cancel{9} \bullet \cancel{8}}{\cancel{3} \bullet \cancel{2} \bullet 1}$ $= 10 \bullet 3 \bullet 4$ $= 120$

Notice that we have 120 arrangements for committees but 720 (six times the amount) of arrangements for elected officers. The same steps used to find $_{10}P_3$ on the graphing calculator can be used to find $_{10}C_3$.

If the numbers in a counting problem get very large, then we need to find a better way to simplify factorials.

Example 4: Simplifying Factorials

Simplify the following factorials.

a) $\frac{42!}{(42-6)!}$ b) $\frac{52!}{(52-5)!5!}$

Solution:

(a) Instead of writing out every number in 42!, we will write out the numbers in 42! in the numerator until we reach the factorial that is in the denominator.	$\frac{42!}{(42-6)!} = \frac{42 \bullet 41 \bullet 40 \bullet 39 \bullet 38 \bullet 37 \bullet \cancel{36}!}{\cancel{36}!}$ $= 42 \bullet 41 \bullet 40 \bullet 39 \bullet 38 \bullet 37$ $= 3{,}776{,}965{,}920$ "This happens to be the number of lottery possibilities given that you have 42 numbers and that the order matters."
(b) We will do the same as in part (a).	$\frac{52!}{(52-5)!5!} = \frac{52 \bullet 51 \bullet \cancel{50} \bullet \cancel{49} \bullet \cancel{48} \bullet \cancel{47}!}{\cancel{47}! \bullet \cancel{5} \bullet \cancel{4} \bullet \cancel{3} \bullet \cancel{2} \bullet 1}$ $= 52 \bullet 51 \bullet 10 \bullet 49 \bullet 2$ $= 2{,}598{,}960$ "This happens to be the number of 5 card hands possible when playing poker with a 52 card deck. Here the order doesn't matter."

Now, there are situations where we need to use a combination of the aforementioned counting techniques (counting principle, permutations, combinations).

Example 5: Forming a Committee

How many ways can we form a committee that needs 3 men and 4 women on it, if we have 8 men and 7 women from which to choose?

Solution:

First of all, we are forming a committee, so the order doesn't matter. Second of all, we basically need to just figure out how many ways to get 3 men and 4 women and multiply those two answers together.	Since the order doesn't matter, we will assume that the 3 men are the first three people selected for the committee. $\underbrace{\underline{\text{man}}\ \underline{\text{man}}\ \underline{\text{man}}}_{{}_8C_3} \bullet \underbrace{\underline{\text{woman}}\ \underline{\text{woman}}\ \underline{\text{woman}}\ \underline{\text{woman}}}_{{}_7C_4}$ $\frac{8!}{(8-3)!3!} \bullet \frac{7!}{(7-4)!4!}$ $56 \bullet 35$ 1960
Notice that we treated this as a two choice counting principle problem where each of the two choices was a combination problem.	

Question 3: In some lotteries the first 5 numbers of the six numbers drawn can be in any order, but the last number must be a certain number. For this kind of a lottery find the total number of different outcomes, given that we can use any number from 1- 40 for the first 5 numbers (no repeat) and then any number from 1- 40 for the last number (which may be the same as one of the previous 5)?

There are many different counting problems, but let's look at one more type.

Discussion 3: Distinguishable Permutations

How many distinct ways are there to arrange the letters in the word Mississippi?

Here we have a unique challenge. The order does matter; you can have some repeating, but repeating can also cause you to repeat like arrangements. For example, you could move an "i" or two around and still spell Mississippi the same way. Here we have a formula given to the right that divides out equal arrangements because we have letters duplicated.	$\frac{n!}{r_1!r_2!r_3!\ldots}$ n is the total number of items r_n is the number of times each item is duplicated.

In the word Mississippi we have 11 letters: one M, four i's, four s's, and two p's.	# of arrangements $= \dfrac{11!}{1!4!4!2!}$ $= 34{,}650$
Let's use the calculator on this one. On a TI-83, you type in the calculation just like it looks and you get the factorial symbol under MATH under PROB, then choice 4. 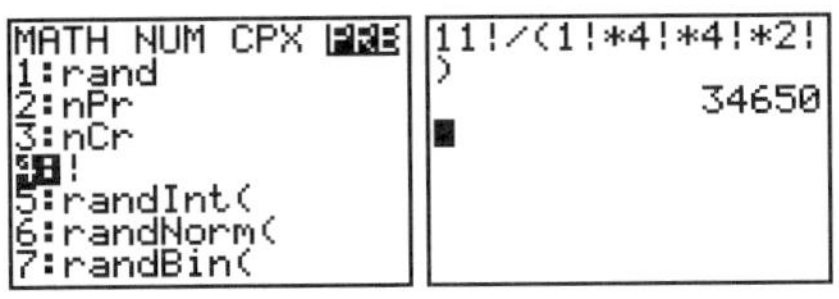	Let's use the calculator on this one. On a TI-86, you type in the calculation just like it looks and you get the factorial symbol under MATH under PROB, then choice F1. 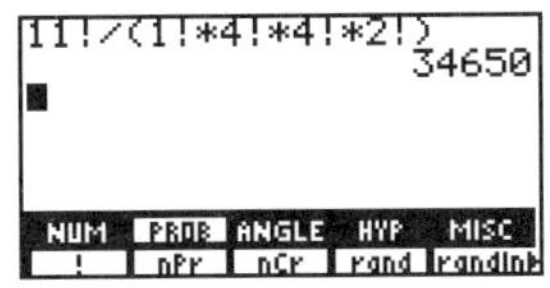

Section Summary:

When you wish to count the number of possibilities that exist in a given situation, you will need to remember the following.

- There are two key questions when faced with a counting problem: 1) Can there be repeating or not? 2) Does the order matter or not?
- Counting Principle – the number of different ways that a progression of choices can happen is a product of the number of possibilities you have at each choice in the progression of choices (**repeat or not, order matters**).
- Permutations – An arrangement of choices in which the choices may not repeat and the order of the choices is important (**no repeat, order matters**). $P(n, r) = \dfrac{n!}{(n-r)!}$.
- Combinations – An arrangement of choices that do not repeat, in which the order of the choices is not important (**no repeat, order doesn't matter**). $C(n, r) = \dfrac{n!}{(n-r)!r!}$.
- In more complicated problems you may need to use these methods of counting together.

SECTION 8.3 HOMEWORK

(1–12) Use your graphing calculator to find the value of each of the following:

Answer Q3:

${}_{40}C_5 \bullet \underline{40} =$ 658,008 • 40 = 26,320,320.

1.	${}_{10}P_4$	2.	${}_{10}P_6$	3.	${}_{5}P_1$
4.	${}_{5}P_4$	5.	${}_{20}P_3$	6.	${}_{20}P_{17}$
7.	${}_{10}C_4$	8.	${}_{10}C_6$	9.	${}_{5}C_1$
10.	${}_{5}C_4$	11.	${}_{20}C_3$	12.	${}_{20}C_{17}$

(13–24) Each of the following is an example of a counting principle problem. Solve each of the following problems and show what you multiplied to get the answer.

13. A state needs 12,000,000 license plates. If the state decides to use 2 letters of the alphabet followed by 4 numbers, how many different license plates can it make? Is this enough for the need?

14. From problem 13, if the state adds one letter of the alphabet at the front, how many different license plates can it make? Is this enough for the need?

15. A phone company needs 10,000,000 phones for its area. They will use seven numbers to make up their phone numbers. If they are not allowed to use 0, 1 or 5 as the first number, how many phone numbers can be made up? Is this enough for the need?

16. Phone numbers consist of an area code of three numbers followed by seven numbers. If the area code is not allowed to start with 0 or 1 and the seven numbers are not allowed to use 0, 1, or 5 as the first number of the seven, how many different phone numbers can be made up?

17. You want to arrange 7 different mathematics books on a shelf. How many different ways can you arrange those books?

18. A worldwide conference consists of delegates from 12 different countries. If one delegate from each country were to be seated on the same side of a long conference table, how many different seating arrangements would there be?

19. A roller coaster ride allows only one person per seat. If the roller coaster is long enough to have 20 different seats, how many different ways can 20 people be seated on the roller coaster assuming the order matters?

20. How many different arrangements are there of the word "**Help**"?

21. How many different five digit numbers can be made from the numbers 1, 2, 3, 4, 5?
    a. If repeats are allowed?
    b. If repeats are not allowed?

22. You want to make up a four symbol code words using the symbols #, @, *, %:
    a. How many code words can you make up if the symbols can be repeated?
    b. How many code words can you make up if the symbols cannot be repeated?

23. A restaurant offers a breakfast menu with 3 choices of juices, 4 choices of eggs, and 2 choices of drinks. You must choose one juice, one type of egg, and one type of drink. How many different breakfasts can you choose?

24. A seafood restaurant has a special for $20.95 where you can choose one seafood dish from a choice of 10, one salad from a choice of 4, one potato dish from a choice of 5, one vegetable from a choice of 4, one drink from a choice of 8, and one dessert from a choice of 5. How many different dinners can be chosen?

(25–36) Each of the following is an example of a permutation problem. Solve each of the following permutation problems and show what permutation you used solve the problem.

25. If 4 letters are used at a time, how many permutations of the word "famous" can you make?

26. Six books out of ten different books are to be arranged on a shelf; order matters. How many different ways can the books be arranged on a shelf?

27. For a lottery, 4 numbers are drawn from 48 numbers, one at a time. If the order the numbers is drawn in counts, how many possible outcomes are there?

28. A baseball team consisting of 12 players must choose 9 out of the 12 for a batting order, how many possible batting orders can be made from these 12 players?

29. An instructor is constructing a mathematics exam by choosing 10 questions out of 20 questions. If the order of the questions matters, how many different exams can the instructor create?

30. A club, which has 20 members, is going to choose a president, vice–president, secretary, and a treasurer from the 20 members. How many different ways can the 4 officers be chosen?

31. A survey asks the participants to choose their 5 favorite ice cream flavors from 10 different flavors and rank them 1 thru 5. How many different possible rankings are there?

32. A disc jockey must choose 15 songs out of 30 to play during a one–hour segment. If the order of the songs makes a difference, how many different ways can the disc jockey play the songs?

33. A roller coaster ride allows only one person per seat. If the roller coaster has room only for 20 people and has 50 people waiting, how many different possibilities are there for filling the roller coaster for a ride? (It makes a difference in the enjoyment of the ride depending upon which seat the person is sitting in.)

34. There are 40 people trying to get into a basketball game that has only 10 reserved seats left. If 10 people at random are chosen to get the reserved seats, how many different ways could the seats be assigned?

(35–44) Each of the following is an example of a combination problem. Solve each of the following and show what combination use used to solve the problem.

35. How many different ways can 4 letters be chosen from the word "famous", if order doesn't matter?

36. Six books out of 10 different books are to be placed on shelf and order doesn't matter. How many different ways can the books be placed on the shelf?

37. For a lottery, 4 numbers are drawn from 48 numbers. If the order of the numbers makes no difference, how many possible outcomes are there?

38. An instructor is constructing a mathematics exam by choosing 10 questions from 20. If the order the questions appear on the exam makes no difference, how many different exams can the instructor create?
39. A club with 20 members is choosing a committee of 4. How many different committees can be chosen?
40. There are 40 people trying to get into a basketball game that has only 10 general admission tickets left. If 10 people at random are chosen to get the general admission tickets, how many different ways can the tickets be assigned?
41. Stud poker is a card game in which each person is dealt 5 cards. If a deck of playing cards used in stud poker consists of 52 cards, how many different stud poker hands can be dealt?
42. Bridge is a card game in which each person is dealt 13 cards. If a deck of playing cards used in bridge consists of 52 different cards, how many different bridge hands can be dealt?
43. At a state political convention there are 100 delegates. If 6 delegates are to be chosen to send to the national convention, how many different possible choices are there?
44. A car manufacturer has a special price for a package of 4 options. If the consumer can make his choice of 4 options from 10 different options, how many different packages are possible?

(45–74) Use an appropriate method to solve each of the following problems.

45. A pro–sport team does random drug testing. If the team has 28 players and 4 are to be tested, how many possibilities are there?
46. Four mathematics books are to be chosen from 10 different mathematics books. How many different choices are possible?
47. A true/false test has 15 different questions. How many different possible outcomes are there?
48. How many different ways are there to fill in an answer sheet if a quiz consists of 10 multiple–choice questions with 4 possible answers to each question?
49. A truck driver starts in Los Angeles and must make deliveries in order to Phoenix, Dallas, New Orleans, and Atlanta. There is a choice of 2 routes from Los Angles to Phoenix, 3 routes from Phoenix to Dallas, 4 routes from Dallas to New Orleans, and 3 routes from New Orleans to Atlanta. The truck driver must drive this schedule every month and wants to vary his route to avoid boredom. How many months will it take to drive every possible route?
50. At each intersection a driver of a car can make either a left–hand turn, a right–hand turn, or go straight ahead.
    a. How many different routes are possible if 2 different intersections are encountered?
    b. How many different routes are possible if 3 different intersections are encountered?
    c. How many different routes are possible if 10 different intersections are encountered?
    d. In general how many different routes are possible if n different intersections are encountered?

51. A man has 5 suits, 10 ties, 15 shirts, and 8 pairs of shoes. How many different outfits does he have?
52. A sport shirt can be purchased in a choice of sizes: small, medium, large, and extra large. It is also available in colors white, blue, green, gray, and brown. Suppose a store wishes to keep in stock exactly one shirt of this style in each combination of size and color. How many shirts must the store keep in stock?
53. A tennis tournament awards money to first, second, third, and fourth place finishers. If there are 16 entries in the tournament, how many different ways can the money be awarded?
54. A team in a basketball tournament is playing for first, second, or third place. If the tournament starts with 32 teams, how many possible outcomes are there for first, second, and third place?
55. A coast guard uses flags on a flagpole to signal ships concerning matters such as weather conditions. There are 6 flags available and 3 at a time are to be used to signal ships.
    a. If the order counts, how many different signals can be sent?
    b. If the order doesn't count, how many different signals can be sent?
56. A lottery uses 40 balls numbered 1 through 40. Five balls are drawn one at a time without replacement.
    a. How many possible outcomes are there for the lottery if order counts?
    b. How many possible outcomes are there for the lottery of order doesn't count?
57. At a certain army base there are 4 officers, 10 sergeants, and 40 enlisted personnel. How many ways can a 4-person committee be formed with the following conditions:
    a. The committee consists of 4 officers?
    b. The committee consists of 4 sergeants?
    c. The committee consists of 4 enlisted personnel?
    d. The committee consists of 1 officer, 2 sergeants, and 1 enlisted personnel?
    e. The committee has at least one officer as a member?
58. A committee of 6 is formed from 9 women and 10 men. How many different committees can be formed with the following restrictions:
    a. Any 6 can be chosen?
    b. All 6 must be women?
    c. All 6 must be men?
    d. 3 women and 3 men must be chosen?
    e. 4 women and 2 men must be chosen?
    f. 2 women and 4 men must be chosen?
    g. At least 3 women must be chosen?
59. An agricultural expert has designed an experiment to test the effects of 2 types of insecticide and 4 different types of fertilize on six different crops. Each plot uses 1 insecticide, 1 fertilizer, and 1 crop. How many plots are needed?

60. A veterinarian wants to test 5 different dosages of a vaccine on 3 types of animals using 3 different age groups. How many different tests will the veterinarian need to perform to check all possibilities if the test consists of 1 dosage of vaccine with 1 animal from 1 age group?

61. A pizza place has a special price of $5.95 for a pizza with exactly three toppings. Assuming cheese is automatic (not one of the choices of toppings) and you will chose 3 other toppings from 9 possible toppings, how many different pizzas could be chosen?

62. A librarian is allowed to order 10 new books for the library and has narrowed the choice to 15 different books. How many different orders could be made?

63. A bank has a created an 8 digit identification number for each of its customers. How many different customers can have a distinct identification number if
   a. Digits are allowed to repeat?
   b. Digits are not allowed to repeat?

64. A manufacturer of television sets places an identification tag on the back of each television set it manufacturers. The identification tag consists of 3 letter followed by 4 numbers. How many television sets can be produced with different identification tags if:
   a. The letters can repeat and the digits cannot repeat?
   b. The letters and digits can both repeat?
   c. There are no repetitions of digits or letters?
   d. The letters cannot repeat and the digits can repeat?

65. There are 10 racecar drivers:
   a. How many ways can these racecar drivers be assigned to 4 different cars?
   b. How many ways can these racecar drivers be assigned to 6 different cars?
   c. How many ways can these racecar drivers be assigned to 8 different cars?

66. There are 20 people applying for jobs:
   a. How many ways can these applicants be assigned to 4 different job positions?
   b. How many ways can these applicants be assigned to 5 different job positions?
   c. How many ways can these applicants be assigned to 7 different job positions?

67. A candy shop has 5 different types of canned nuts available, 8 different small boxes of chocolates available, and 4 different types of sacked hard candy available. If the candy shop wants to make up different gift packages using the nuts, chocolates, and sacked hard candy, how many different gift packages could be made if:
   a. The gift package has 1 type of canned nut, 2 boxes of different chocolates, and 1 type of sacked hard candy?
   b. The gift package has 2 different types of canned nuts, 1 box of chocolates, and 2 different types of sacked hard candy?

68. A small midwestern town consists of 20 families living on the west side of the highway and 15 families living on the east side of the highway. A company wants to do a survey. How many different choices are there for this survey if:
    a. The company wants to choose 7 families from the town?
    b. The company wants to choose 3 families from the west side and 4 families from the east side?
    c. The company wants to choose 6 families from the west side and 1 family from the east side?
69. A pizza has 7 meats and 5 vegetables available for toppings. Assume you are not allowed to have double toppings of any meat or vegetable and no extra cheese is allowed.
    a. How many different 5 topping pizzas are possible?
    b. How many different 3 meat and 2 vegetable topping pizzas are possible?
    c. How many different 2 meat and 3 vegetable topping pizzas are possible?
    d. How many different pizzas with 5 meats only toppings are possible?
    e. How many different pizzas with 3 vegetables only toppings are possible?
70. Using the information from problem (41) about stud poker:
    a. How many hands are possible with 2 kings and 3 queens?
    b. How many hands are possible with 5 diamonds?
    c. How many hands are possible with 5 of the same suit?
    d. How many hands are possible with 4 aces?
    e. How many hands are possible with 4 of the same kind?
71. How many distinct permutations are there for the letters of the word "**success**"?
72. How many distinct permutations are there for the letters of the word "**mathematics**"?
73. How many distinct permutations are there for the letters of the word "**bookkeeping**"**?**
74. How many distinct permutations are there for the letters of the word "**Mississippi**"?
75. Four balls numbered 1 through 4 are in a hat. Two balls are drawn from the hat one at a time.
    a. If order doesn't count, how many possible outcomes are there?
    b. List all the possible outcomes.
76. Using problem 75:
    a. If order does count, how many possible outcomes are there?
    b. List all the possible outcomes.

Section 8.4 Probability

We spent the last section talking about counting; now we will use that to solve probability questions. To study probability we need to define a few terms first.

> **Outcome** – One of the possible results in an experiment.
>
> **Sample space** – The set of all possible outcomes in a particular experiment.
>
> **Event** – All similar outcomes. A subset of the sample space. (These outcomes are called successes, that is, getting what we where looking for.)

For example, if we toss a coin in the air there are two possible outcomes (heads, tails). Getting a head or a tail is an outcome; the two possibilities together make up the whole sample space. To determine the probability of getting a head (event) when tossing a coin into the air is the number of ways of getting a head (1) divided by the total number of possibilities (2, the whole sample space).

> **Probability** – The study of random events where $n(E)$ is the number of outcomes (successes) in an event and $n(S)$ is the total number of outcomes that are possible. Thus we have the formula:
>
> $$P(E) = \frac{n(E)}{n(S)} \qquad \frac{\textit{number of successes}}{\textit{number of total outcomes}}$$

Example 1: Roll the Die

What is the probability that you will get a two (an event) when you roll a six-sided die?

Solution:

There are 6 total possible outcomes (1, 2, 3, 4, 5, 6), so $n(S) = 6$. There is only one way to get a 2, so $n(E) = 1$. The probability of getting a 2 is	$P(\text{getting a 2}) = \frac{n(E)}{n(S)}$ $= \frac{1}{6} = 0.125$

Question 1: If you have a standard 52-card deck that has 13 spades, 13 hearts, 13 diamonds, and 13 clubs, what is the probability that if you draw one card from the deck it will be a heart?

Example 2: The Price Is Right

On the Price Is Right game show on television, there is a game where there are 8 disks in a bag. Five of the disks have numbers on them that correspond to the price of a car. Three of the disks have X's on them called "strikes". If you fail to guess the price of the car before you have drawn the third X out of the bag, you lose. What is the probability that your first disk drawn from the bag isn't an X?

Solution:

There are 8 total possible outcomes (five numbers, three X's), so $n(S) = 8$. There are five ways not to get an X, so $n(E) = 5$ (successful outcomes). The probability of not getting an X is	$P(\text{not getting an X}) = \dfrac{n(E)}{n(S)}$ $= \dfrac{5}{8}$ $= 0.6$

What would be the probability of getting a 7 when you roll a six-sided die? Well, obviously the answer is 0! Mathematically speaking we see that we have 0 ways to get a 7 out of 6 total possibilities. Thus $P(\text{getting a 7}) = \dfrac{n(E)}{n(S)} = \dfrac{0}{6} = 0$. We call this an **impossible event**. We could also say that there is a 0% chance of this event happening.

Can there be **certain events**? Sure! What is the probability of getting a number between 1 and 6 when we roll a six-sided die? The answer is $\text{P(getting a 1-6)} = \dfrac{n(E)}{n(S)} = \dfrac{6}{6} = 1$. Here we could say that there is a 100% chance that this event will happen. Notice that the probability of an event for any sample space must fall between 0 and 1 since $0 \leq n(E) \leq n(S)$ and we can write these in decimal form and or put them into percent form if we wish.

Example 3: Craps

When rolling two six-sided dice, what is the number of possible outcomes, $n(S)$? Also, what is the probability of rolling an 8 with two six-sided dice?

Solution:

First, we need to do a counting problem to figure out $n(S)$.	Order isn't a concern here since we have two separate situations (die 1 and die 2). There isn't anything to order. As far as repeating or not, the second die can have the same value as the first. Hence the counting principle should work for this problem. ___ • ___ die 1 die 2
There are 6 possible numbers on each die.	$n(S) = 6 \cdot 6 = 36$
Now we need to figure out how many ways there are to get the two dice to add up to 8.	2 6 3 5 4 4

	5 3 6 2
n(getting an 8) is	5
So the P(getting an 8) is	$P(E) = \frac{n(E)}{n(S)} = \frac{5}{36} = .1388888...$

Notice that P(of getting an 8) is equal to a number between 0 and 1. We could use percents to describe this situation and say that there is a $13.\overline{8}\,\%$ chance of rolling an 8 with two six-sided dice. Also here is a good place to bring up the term "independent events." The outcome of the first die roll has no bearing on the outcome of the second die. Thus we call these independent events.

Independent events – Events where the outcomes of one event do not affect the outcomes of another event.

Question 2: What is the possibility of rolling a 13 on two 8-sided dice?

Knowing some simple probabilities like these can come in handy when making decisions about strategies in games like Risk, or Yahtzee. In fact, if you don't want to lose a fair amount of money you need to know something about this topic before you make a trip to Las Vegas!

Let's continue with the next example.

Example 4: Playing Cards

Answer Q1: $n(E) = 13$ (13 hearts in the deck, similar outcomes, successes) and $n(S) = 52$ (52 cards in the deck, total outcomes), so the answer would be P(getting a heart) = $\frac{13}{52} = \frac{1}{4}$.

Find the probabilities in the following two questions [picture of cards].

a) What is the probability of drawing either a heart or a club from a standard 52-card deck?	b) What is the probability of drawing either a spade or a king from a standard 52-card deck?

Solution:

(a) We can find the probability of getting a heart or a club by adding the probabilities of each separately.	P(getting a heart) + P(getting a club) $\frac{13}{52} + \frac{13}{52} = \frac{1}{4} + \frac{1}{4} = \frac{2}{4} = .5$ P(getting a heart or a club) = .5
(b) We have a little overlap since one of the kings is a spade. So we need to adjust our answer when we are done.	P(getting a spade) + P(getting a king) $\frac{13}{52} + \frac{4}{52} = \frac{17}{52}$
We have counted the king of spades twice since it is both a king and a spade. So we	$\frac{17}{52} - \frac{1}{52} = \frac{16}{52} = \frac{4}{13} = .30769...$

need to subtract the overlap.	

In part (a) we would say that these two events are mutually exclusive. In part (b) we would say that these are not mutually exclusive.

Mutually Exclusive Events – Events where none of the outcomes of either event are in common.

If two events are mutually exclusive, then we can use the following formula.

Given two events E and F that are mutually exclusive, the probability of E or F is given by $P(E \text{ or } F) = P(E) + P(F)$ which may also be written as $P(E \cup F) = P(E) + P(F)$.
(Note: "or" and union ($\cup$) are synonymous.)

If two events are not mutually exclusive, then we can use this formula.

Given two events E and F that are not mutually exclusive, the probability of E or F is given by $P(E \text{ or } F) = P(E) + P(F) - P(E \text{ and } F)$ which may also be written as
$P(E \cup F) = P(E) + P(F) - P(E \cap F)$.
(Note: "and" and intersection ($\cap$) are synonymous.)

Notice that just as we did in our previous example, in this formula we subtract what was in common ($P(E \text{ and } F)$).

Question 3: In a standard deck of cards a jack, queen, and king are called face cards. Find the probability of drawing either a face card or a diamond from the deck.

We've discussed how to find the probability of a single event like rolling a 5, or drawing a king or a diamond, or getting an 8 when rolling two six-sided dice. Let's look at how we find the probability of several events in a row.

Example 5: Tossing Coins

Find the probability of tossing three heads in a row with a coin.

Solution:

Your chance of tossing a head the first time is .5 and that is the same for the second and third tries.

Each toss of the coin is independent of the others; to find the probability of tossing three heads in a row, we multiply the probabilities of tossing a head each time. (Similar to the	$P(\text{three heads}) = P(H) \cdot P(H) \cdot P(H)$ $= \frac{1}{2} \cdot \frac{1}{2} \cdot \frac{1}{2} = \frac{1}{8}$

counting principle where we select three people to be a president, treasure and secretary. We use multiplication.)	$= .125$
Notice how multiplication yields the same answer as when we list all the possible outcomes and use the formula for probability to get the answer.	HHH, HHT, HTH, HTT THH, THT, TTH, TTT $P(\text{Three heads}) = \frac{n(E)}{n(S)} = \frac{1}{8} = .125$

You can do this whenever each event is independent of the others.

*Answer Q2: 8*8 = 64 which is n(S). The ways to get 13 are (5,8) (6,7) (7,6) (8,5), so n(E) = 4. Thus P(getting a 13) =* $\frac{4}{64} = \frac{1}{16} = 0.0625$ *or we could say a 6.25% chance.*

Question 4: Find the probability of drawing two cards from a standard deck, where you place the first card drawn back into the deck before you draw the second card, and where the first one is a spade the second one is a diamond.

Question 5: What happens in question 4 if you don't place the first card back into the deck?

Discussion 1: Balls in a Jar

If you have a jar containing 7 white balls, 3 red balls, and 4 blue balls and you randomly draw out three balls, what is the probability that at least one of them is blue?

To go after this answer directly would be difficult since there are many combinations of three draws that would yield at least one blue. It is sometimes better to look at the opposite of what you want. In this example we want at least one blue (although we could have more than one). The opposite of this would be drawing NO blue balls. This is easier to calculate.

This situation is a combinations problem. We don't care about the order in which the balls are drawn, and no single ball, once drawn, can be drawn again; thus, there is no repeating the same ball. No repeat and order doesn't matter is a combination problem. We have a total of 14 balls of which 10 aren't blue and we are choosing 3. Therefore we need to find ${}_{10}C_3$ (10 things taken 3 at a time) for $n(E)$, and ${}_{14}C_3$ (14 things taken 3 at a time) for $n(S)$.	$P(\text{not Blue}) = \frac{n(E)}{n(S)} = \frac{n(\textit{not blue})}{n(\textit{all possibilities})}$ $= \frac{{}_{10}C_3}{{}_{14}C_3}$ $= \frac{\frac{10!}{3!(10-3)!}}{\frac{14!}{3!(14-3)!}} = \frac{120}{364}$ $\approx .32967$

We haven't quite answered our question yet. If the question was "What is the possibility of not drawing a blue?" then we'd have the answer (.32967). One way in which we could answer our question is to notice that if 120 combinations don't have a blue in them, out of the total of 364 combinations, then the other 244 must have at least one blue in them. Thus the *P*(at least one blue) = $\frac{244}{364} \approx .67033$. Now, either a combination has a blue in it or not; thus, these two combinations together have to cover all the possibilities that exist. Hence, the sum of the two probabilities must equal one, which they do (.32967 + .67033 = 1). This fact shows that we could have answered our question by simply taking 1 – .32967 and arrived at the answer.

The probability of *E*, knowing the probability of <u>not *E*</u> (E'), is $$P(E) = 1 - P(E')$$

Let's do another example of this type using one of the most famous problems of all time.

Example 6: The Birthday Problem!

What is the probability that at least two people in a group of 30 people have birthdays on the same day?

Solution:

This situation is like the previous one. There are so many combinations we'd go crazy trying to figure them all out. Instead, we will look at the probability of no one having birthdays on the same day $P(E')$. Then, we will use the formula $P(E) = 1 - P(E')$	*P*(same B-days) = 1 – *P*(no same B-days) To find $P(E')$ we will look at each person independently of the others and multiply the probabilities together.
person 1: *P*(a B-day some time in the year) = $\frac{365}{365}$ person 2: *P*(B-day not the same as person 1) = $\frac{364}{365}$ person 3: *P*(B-day not the same as person 1 and 2) = $\frac{363}{365}$ person 4: *P*(B-day not the same as person 1, 2 and 3) = $\frac{362}{365}$ etc…	
We have independent events so we use the formula that says we multiply these together and we get	$\overset{1}{\frac{365}{365}} \bullet \overset{2}{\frac{364}{365}} \bullet \overset{3}{\frac{363}{365}} \bullet \ldots \bullet \overset{30}{\frac{336}{365}} \approx .2937$
That is the *P*(no birthdays the same), so we	*P*(same B-day) = 1 – .2937

Answer Q3: These are not mutually exclusive because there are three face cards that are diamonds; thus, our answer is P(getting either a face card or a diamond) = P(face card) + P(diamond) – P(a face card diamond) = $\frac{12}{52} + \frac{13}{52} - \frac{3}{52} = \frac{22}{52} = \frac{11}{26} =$ *.42307…*

now find our answer.	≈ .7063 or roughly 70%

Getting the answer to *P*(not same B-days) from the calculator is a bit complicated. Here is how you can get the calculator to evaluate this for you.

*Answer Q4: P(spade then a diamond) = P(spade)*P(diamond) =* $\frac{13}{52}*\frac{13}{52}=$ $\frac{1}{4}*\frac{1}{4}=\frac{1}{16}=$ *.0625.*

First, create a sequence of the numerators in the factions to be multiplied (30 total numbers).

$$\frac{365}{365}\bullet\frac{364}{365}\bullet\frac{363}{365}\bullet\cdots\bullet\frac{336}{365}$$

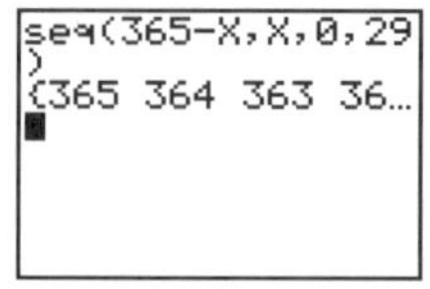

Now go to List, then arrow over to MATH, then choose 6 **prod(** which means product. With the TI-86 you go to LIST, push F5 for OPS, then push MORE, then F2.

Now, multiply all the numbers in the sequence by pushing 2nd (-) to get ANS. Now push ENTER to get the answer for the product of all of the numerators.

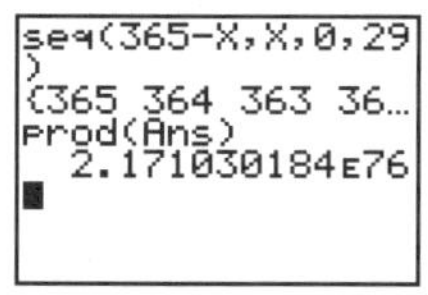

*Answer Q5: P(spade then a diamond) = P(spade)*P(diamond) =* $\frac{13}{52}*\frac{13}{51}=$ $\frac{1}{4}*\frac{13}{51}=\frac{13}{204}$ *= .0637.*

Now do the same for the denominator and then divide the two answers.

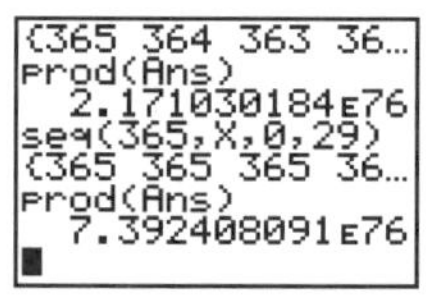

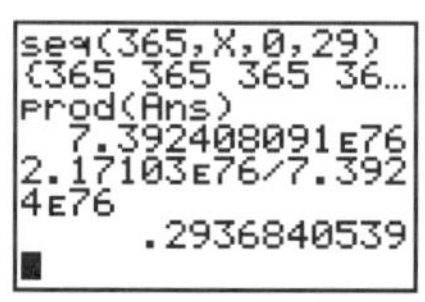

Question 6: If a company knows that, on average, seven out of 700 products produced will be defective [NOTE: *P*(defective) = $\frac{7}{700}$ and *P*(not defective) = $\frac{693}{700}$], then what is the probability that at least one out of 20 products checked will be defective (similar to B-day problem)?

Section Summary:

Here is the vocabulary and formulas discussed in this section.

- An **outcome** is one of the possible results in an experiment.
- A **sample space** is the set of all possible outcomes in a particular experiment.

- An **event** is all similar outcomes, and it is a subset of the sample space. (These outcomes are called successes; that is, you are getting what you were looking for.)
- $P(E) = \dfrac{n(E)}{n(S)}$ $\quad \dfrac{\textit{number of successes}}{\textit{number of total outcomes}}$
- **Independent events** are events where the outcomes of one event do not affect the outcomes of the other event.
- **Mutually exclusive events** are events where none of the outcomes of either event are in common.
- $P(E \text{ or } F) = P(E) + P(F)$ if E and F are mutually exclusive.
- $P(E \text{ or } F) = P(E) + P(F) - P(E \text{ and } F)$ if E and F are not mutually exclusive.
- Lastly, remember that sometimes it is easier to find $P(E')$ than it is to find $P(E)$, in which case $P(E) = 1 - P(E')$.

SECTION 8.4 HOMEWORK

Answer Q6:
P(at least one defective) = 1 − P(none defective) = $1-\frac{693}{700}*\frac{693}{700}*\ldots*\frac{693}{700}$ *(20 times)*
$=1-\left(\frac{693}{700}\right)^{20}$
= 1 − .8179 = .1821

(1–4) A jar contains 3 red, 5 black and 9 green balls and one ball is drawn from the jar.

1. What is the probability the ball is red?
2. What is the probability the ball is black?
3. What is the probability the ball is black or green?
4. What is the probability the ball is not red?

(5– 8) A dice is thrown once.

5. What is the probability the face will be a six?
6. What is the probability the face will be a two?
7. What is the probability the face will be an even number?
8. What is the probability the face value will be greater than two?

(9–18) One card is drawn from a deck of cards.

9. What is the probability the card will be a four?
10. What is the probability the card will be red?
11. What is the probability the card will be a face card? (Ace, king, queen or jack)
12. What is the probability the card will not be a face card?
13. What is the probability the card will be a diamond?
14. What is the probability the card will be a red three?
15. What is the probability the card will be a four or a black card?
16. What is the probability the card will be an ace or a four?
17. What is the probability the card will be face card or a red card?
18. What is the probability the card will be a face card or a numbered card greater than six?

(19–26) Two dice are tossed.

19. Set up a table and list all the possible outcomes.

(**need a table set up for problem**)

20. What is the total number of outcomes possible?
21. What is the probability the faces add up to seven?
22. What is the probability the faces add up to twelve?
23. What is the probability the faces add up to a number greater than four?
24. What is the probability the faces add up to a number less than nine?
25. What is the probability the faces don't add up to six?
26. What is the probability the faces add up to five or nine?

(27–32) Four coins are tossed.

27. List all the possible outcomes.
28. How many different outcomes are possible?
29. What is the probability the toss has exactly two heads?
30. What is the probability the toss has exactly one tail?

31. What is the probability the toss has at least one tail?

32. What is the probability the toss has three heads or less?

(33–34) A lottery is designed so that six numbers out of forty-six are chosen.

33. A person buys one ticket. What is the probability that this person will win?

34. A person buys ten tickets with different numbers on each, what is the probability that this person will win?

(35–36) A lottery consists of forty–six white balls and forty–six red balls. Five white and one red ball are chosen at random.

35. A person buys one ticket. What is the probability that this person will win?

36. How many different ticket does the person need to draw to assume winning?

(37–40) Two cards are chosen from a deck of 52 cards.

37. What is the probability the two cards are black?

38. What is the probability the two cards are face cards?

39. What is the probability the two cards are both hearts?

40. What is the probability the cards are both aces?

(41–46) A poker hand consists of five cards from a deck of 52 cards.

41. What is the probability the poker hand has 3 aces?

42. What is the probability the poker hand had 3 of a kind?

43. What is the probability the poker hand has five hearts?

44. What is the probability the poker hand has all five cards of the same suit, called a flush?

45. What is the probability the poker hand has 3 aces and 2 kings?

46. What is the probability the poker hand has 3 of one kind and two of another kind, called a full house?

(47–50) A committee of 6 people is to be chosen at random from 18 people consisting of 7 Republicans, 8 Democrats and 3 Independents.

47. What is the probability that exactly 3 Democrats are on the committee?

48. What is the probability the committee consists of 3 Republicans, 2 Democrats and one Independent?

49. What is the probability there are no Independents on the committee?

50. What is the probability the committee has at least one Republican on the committee?

51. What is the probability that at least two people in a group of 20 have birthdays on the same day?

52. A company knows that on average 5 out of 600 products produced will be defective. What is the probability that at least one out of 30 products produced will be defective?

53. The following table represents the educational attainment of the population 18 to 19 years old in thousands.

None	1st–4th grade	5th–6th grade	7th–8th grade	9th grade	10th grade	11th grade	High school	Some college no degree	Associate degree	Total
14	17	60	100	223	521	2,629	2,294	2,201	25	8,084

(Source U.S. Census Bureau)

a. What is the probability that if you choose a person between the ages of 18 to 19, he or she will have attained exactly a high school education?

b. What is the probability that if you choose a person between the ages of 18 to 19 he or she will have attained some college with no degree?

c. What is the probability that if you choose a person between the ages of 18 to 19 he or she will have attained at least a high school education?

d. What is the probability that if you choose a person between the ages of 18 to 19 he or she will have attained below a high school education?

54. The following table represents the educational attainment of the male population 18 to 19 years old in thousands.

None	1st–4th grade	5th–6th grade	7th–8th grade	9th grade	10th grade	11th grade	High school	Some college no degree	Associate degree	Total
10	7	8	40	74	258	1,732	1,541	1,585	10	5,265

(Source U.S. Census Bureau)

a. What is the probability that if you choose a male between the ages of 18 to 19, he will have attained exactly a high school education?

b. What is the probability that if you choose a male between the ages of 18 to 19, he will have attained some college with no degree?

c. What is the probability that if you choose a male between the ages of 18 to 19, he will have attained at least a high school education?

d. What is the probability that if you choose a male between the ages of 18 to 19, he will have attained below a high school education?

55. The following table represents the population of the U.S for the year 2000.

Total Population	273,643,274
Male	133,551,361
Female	140,091,913
Under 9 years old	39,380,598
10 to 19 years old	39,461,504
20 to 34 years old	56,062,053
35 to 44 years old	44,342,850
45 to 54 years old	37,337,894
55 to 64 years old	23,945,041
65 to 74 years old	18,108,981
75 to 84 years old	11,708,844
85 years and over	3,295,509

(U.S. Census Bureau)

a. What is the probability that if a person is chosen at random, that person will be a female?

b. What is the probability that if a person is chosen at random, that person will be between 45 to 54 years old?

c. What is the probability that if a person is chosen at random, that person will be between 10 to 19 years old or between 65 to 74 years old?

d. What is the probability that if a person is chosen at random, that person will be older than 9 years?

56. The following table represents the population of Arizona for the year 2000.

Total Population	5,130,632
Male	2,461,057
Female	2,569,575
Under 9 years old	772,255
10 to 19 years old	745,933
20 to 34 years old	1,105,525
35 to 44 years old	768,804
45 to 54 years old	627,904
55 to 64 years old	442,372
65 to 74 years old	363,841
75 to 84 years old	235,473
85 years and over	68,525

(U.S. Census Bureau)

a. What is the probability that if a person is chosen at random, that person will be male?

b. What is probability that if a person is chosen at random, that person will be between 20 to 34 years old?

c. What is the probability that if a person is chosen at random, that person will be between 55 to 64 years old or between 35 to 44 years old?

d. What is the probability that if a person is chosen at random, that person will be less than 75 years old?

(57–64) The odds of an event happening are given by the ratio $\dfrac{\textit{Number of successes of the event}}{\textit{Number of failures of the event}}$.

What are the odds of getting a head on a single flip of a coin?

The number of successes is one head and the number of failures is one tail, so the odds are $\dfrac{1}{1}$.

This would normally be written 1:1 which is read "1 to 1."

57. What are the odds of getting a six on the roll of a single die?

58. One card is drawn from a deck of 52 cards. What are the odds the card is a king?

59. Two dice are rolled. What are the odds the sum of the faces is seven?

60. Three cards are drawn from a deck of 52 cards. What are the odds the three cards are tens?

61. The odds of picking a defective part off an assembly line are 4:96. What is the probability of getting a defective part when you choose one part off the assembly line?

62. The probability of getting a defective part off an assembly line when you choose one part is $\frac{3}{140}$. What are the odds of picking a defective part of an assembly line?

63. A lottery consists of 48 numbers and 5 numbers are drawn at random. What are the odds of wining the lottery if you purchase one ticket?

64. A lottery consists of 48 white balls numbered 1–48 and 48 red balls numbered 1–48, 4 white balls and 1 red ball are drawn at random. What are the odds of wining the lottery if you purchase one ticket?

65. Two balls are drawn from a jar that contains 5 red, 6 blue and 8 white balls.
   a. What is the probability the first ball is red and the second ball is blue?
   b. What is the probability the first ball is white and the second ball is white?
   c. If the first draw is put back in the jar and the second draw is made, what is the probability the first ball is red and the second ball is blue?
   d. If the first draw is put back in the jar and the second draw is made, what is the probability the first ball is white and the second ball is white?

66. A hat contains 15 slips of paper with a one written on the paper, 12 slips of paper with a two written on the paper, 10 slips of paper with a three written on the paper, and 8 slips of paper with a four written on the paper. Two slips of paper are drawn from the hat.
   a. What is the probability the first slip has a number one and the second has a number three?
   b. What is the probability the first slip has a number two and the second has a number four?
   c. If the first draw is put back in the hat and the second draw is made, what is the probability the first slip has a number one and the second has a number three?
   d. If the first draw is put back in the hat and the second draw is made, what is the probability the first slip has a number two and the second has a number four?

Chapter 8 Review

Topic	Section	
Sequences	8.1	• A sequence is a function where the domain usually is all positive integers and is expressed in subscript notation (a_n). • ! means factorial and if n is a positive integer then it is defined by **$n!$** $= n(n-1)(n-2)(n-3)\ldots 1$. • You can put sequences in your graphing calculators and see them either on the home screen or in a table. • A recursive sequence is a sequence for which the formula defines the nth term of the sequence as a function of the previous term or terms. • An arithmetic sequence means adding the same amount each time to get to the next term in the sequence. The general formula is $a_n = d(n-1) + a_1$. • A geometric sequence means multiplying the same amount each time to get to the next term in the sequence. The general formula is $a_n = a_1(r)^{n-1}$. • When you are trying to find the formula for a sequence, remember to look for patterns. You might try looking to see if the sequence is arithmetic or geometric first and if it is neither, then look for other types of patterns.
Series	8.2	• A series is a sum of the terms of a sequence. • An infinite series is a sum of all the terms of an infinite sequence (S_∞). • An nth partial sum is a series where we add the first n terms of a sequence. • A sequence of partial sums is the sequence made up of all of the possible partial sums for a particular sequence (list of possible sums). • Summation notation means, $\sum_{i=1}^{n} a_i = a_1 + a_2 + a_3 + \ldots + a_n$ • The formula for the nth partial sum of an **arithmetic series** is $S_n = \left(\frac{a_1 + a_n}{2}\right)n$. • The formula for the nth partial sum of a **geometric series** is $S_n = \frac{a_1(1-r^n)}{1-r}$, $r \neq 1$. • The sum of an **infinite geometric series** is $S = \frac{a_1}{1-r}$, $\lvert r\rvert < 1$.

Counting	8.3	• There are two key questions when faced with a counting problem: 1) Can there be repeating or not? 2) Does the order matter or not? • Counting Principle – the number of different ways that a progression of choices can happen is a product of the number of possibilities you have at each choice in the progression of choices (**repeat or not, order matters**). • Permutations – An arrangement of choices in which the choices may not repeat and the order of the choices is important (**no repeat, order matters**). $P(n, r) = \dfrac{n!}{(n-r)!}$. • Combinations – An arrangement of choices that do not repeat, in which the order of the choices is not important (**no repeat, order doesn't matter**). $C(n, r) = \dfrac{n!}{(n-r)!r!}$. • In more complicated problems you may need to use these methods of counting together.
Probability	8.4	• An **outcome** is one of the possible results in an experiment. • A **sample space** is the set of all possible outcomes in a particular experiment. • An **event** is all similar outcomes, and it is a subset of the sample space. (These outcomes are called successes; that is, you are getting what you were looking for.) • $P(E) = \dfrac{n(E)}{n(S)} \qquad \dfrac{\textit{number of successes}}{\textit{number of total outcomes}}$ • **Independent events** are events where the outcomes of one event do not affect the outcomes of the other event. • **Mutually exclusive events** are events where none of the outcomes of either event are in common. • $P(E \text{ or } F) = P(E) + P(F)$ if E and F are mutually exclusive. • $P(E \text{ or } F) = P(E) + P(F) - P(E \text{ and } F)$ if E and F are <u>not</u> mutually exclusive. • Lastly, remember that sometimes it is easier to find $P(E')$ than it is to find $P(E)$, in which case $P(E) = 1 - P(E')$.

CHAPTER 8 REVIEW HOMEWORK

Section 8.1

(1–6) For each of the following sequences give the first five terms.

1. $a_n = 3n + 7$
2. $b_n = 5n - 9$
3. $c_n = 3(2)^n$
4. $a_n = 2\left(-\frac{1}{2}\right)^n$
5. $b_n = \frac{n^3}{n!}$
6. $c_1 = 3,\ c_2 = 1,\ c_n = 3c_{n-2} + 2c_{n-1}$

(7–12) Assuming the pattern does not change for each of the following, what would be the next term of the sequence?

7. $5, -1, -7, -13, \ldots$
8. $2, 9, 16, 23, \ldots$
9. $-2, 4, -8, 16, \ldots$
10. $8, 2, \frac{1}{2}, \frac{1}{8}, \ldots$
11. $\frac{2}{3}, \frac{3}{4}, \frac{4}{5}, \frac{5}{6}, \ldots$
12. $5, 4, 9, 13, 22, \ldots$

(13–16) Using the given information about an arithmetic sequence:

a. Give the general equation of the arithmetic sequence.

b. Give the 10th term of the arithmetic sequence.

13. $a_1 = -3;\ \ d = 5$
14. $a_1 = 5;\ \ d = -4$
15. $a_3 = 1;\ \ a_8 = -14$
16. $a_4 = 26;\ \ a_9 = 61$

(17–20) Using the given information about an geometric sequence:

a. Give the general equation of the geometric sequence.

b. Give the 6th term of the geometric sequence.

17. $a_1 = -3;\ \ r = 2$
18. $a_1 = 5;\ \ r = -2$
19. $a_1 = 6;\ \ r = \frac{1}{3}$
20. $a_2 = 15;\ \ a_4 = 375;\ \ r > 0$

(21–26) For each of the following sequences:

a. Is the sequence arithmetic, geometric or neither?

b. If the sequence is arithmetic or geometric, give the general equation and find a_8.

c. If not arithmetic or geometric, give the next term of the sequence.

21. $7, 14, 28, 56, \ldots$
22. $5, -2, -9, -16, \ldots$
23. $5, 3, 2, 1, 1, \ldots$
24. $3, -1, \frac{1}{3}, \frac{-1}{9}, \ldots$
25. $2, 3, 7, 13, 27, \ldots$
26. $-8, -1, 6, 13, \ldots$

27. $f(x) = -2x + 3$ is a linear function in slope–intercept form. Assume the domain of the linear function is natural numbers; then give the equivalent general equation of the arithmetic sequence defined by the linear function.

28. $f(x) = 2(3)^x$ is an exponential function. Assume the domain of the exponential function is natural numbers; then give the equivalent general equation of the geometric sequence defined by the exponential function.

29. You have a job offer with a starting salary of \$40,000 a year, with a promised raise of \$1,200 each year.
    a. Give the general equation of the sequence defined with $a_1 = \$40{,}000$.
    b. What is your salary the sixth year of your employment?

30. You have a job offer with a starting salary of \$40,000 a year, with a promised raise of 2.5% each year.
    a. Give the general equation of the sequence defined with $a_1 = \$40{,}000$. (Hint: Use the compound interest formula.)
    b. What is your salary the sixth year of your employment?

Section 8.2

(31–34) Find the indicated sum.

31. $\sum_{i=1}^{6}(3i+2)$

32. $\sum_{i=1}^{5}(-2i)$

33. $\sum_{i=1}^{4}\left(2i^2+3i-5\right)$

34. $\sum_{i=3}^{6}\left(i^2+1\right)$

(35–36) Find the indicated partial sum for each problem.

35. 2, 5, 9, 14, 20, … $S_4 =$

36. –5, –8, –13, –21, –34, –55, … $S_5 =$

(37–42) Determine if the following sequences are arithmetic or geometric; then use the arithmetic or geometric series formulas to find the indicated sum.

37. –3, 8, 19, 30, … $S_{12} =$

38. –2, –6, –18, –54, … $S_8 =$

39. $a_n = 5 - 4(n-1)$ $S_{11} =$

40. $1, \frac{1}{2}, 0, \frac{-1}{2}, -1, \ldots$ $S_8 =$

41. $a_n = 2(-3)^{n-1}$ $S_{10} =$

42. $2, 1, \frac{1}{2}, \frac{1}{4}, \frac{1}{8}, \ldots$ $S_8 =$

(43–44) For each of the following geometric sequences, find the infinite series values.

43. $4, -1, \frac{1}{4}, \frac{-1}{16}, \ldots$ $S_\infty =$

44. $a_n = 16\left(\frac{-3}{4}\right)^{n-1}$ $S_\infty =$

45. A mall has a mosaic pattern is the shape of a pyramid built into the floor. The base of the pyramid has 20 squares and each successive layer has one less square than the previous layer.
    a. This ends up with one square on top. How many layers are there?
    b. How many tiles are used in this pattern?

46. A company is willing to give you a starting salary of \$30,000.
    a. If the company will raise your salary \$400 a year for 10 years, how much salary will you receive after 10 years?
    b. If the company will raise your salary 3% a year for 10 years, how much salary will you receive after 10 years? (Hint: Use the compound interest formula)
    c. Which is the best deal for salary?

Section 8.3

(47–50) Use your graphing calculator to find the value of each of the following:

47. ${}_{12}P_3$

48. ${}_{15}P_{10}$

49. ${}_{12}C_3$

50. ${}_{15}C_{10}$

(51–58) Use an appropriate method to solve each of the following problems.

51. A local bank is assigning PIN numbers for their ATM. cards. The PIN numbers consist of 7 digits. How many different PIN numbers can be assigned?

52. A certain club has 50 members. The club must chose 5 members for a committee to make plans for a banquet. How many different committees could be chosen?

53. The club from problem (52) is electing a president, a vice–president, a secretary, a treasurer, and a sergeant of arms. How many different choices are there?

54. The membership of a young Republicans club consists of 30 men and 25 women. How many different ways can a committee of 5 be chosen if:

    a. All are men
    b. All are women
    c. 2 are men and 3 women
    d. 3 are men and 2 women
    e. At least one member is a woman

55. How many different permutations are there for the word fraction?

56. How many different permutations are there for the word committee?

57. Ten men and ten women are seated at a long table. A woman must occupy the first seat and the rest of the seating must alternate men and women. How many different seating arrangements are there?

58. Suppose you have 6 different math books and 9 different computer books to place on a shelf with 8 slots. If the first 2 slots are filled with math books and the next 4 slots with computer books, how many different ways can the books be arranged on the shelf?

Section 8.4

59. A cookie jar contains 10 chocolate chip cookies, 12 pecan cookies, and 9 oatmeal cookies.

    a. If one cookie is drawn from the jar, what is the probability the cookie is an oatmeal cookie?
    b. If one cookie is drawn from the jar, what is the probability the cookie is a chocolate chip cookie or a pecan cookie?
    c. If two cookies are drawn from the jar, what is the probability the first cookie is an oatmeal cookie and the second cookie is a pecan cookie?
    d. One cookie is drawn from a jar then replaced and a second cookie is drawn from the jar. What is the probability the first cookie is an oatmeal cookie and the second cookie is a pecan cookie?

60. One card is drawn from a deck of 52 cards.

    a. What is the probability the card is a spade?

    b. What is the probability the card is not a spade?

    c. What is the probability the card is a three or king?

    d. What is the probability the card is a five or a red card?

61. A board of directors consists of 6 men and 8 women.

    a. What is the probability a committee of 4 will be all men?

    b. What is the probability a committee of 4 will be 1 man and 3 women?

    c. What is the probability a committee of 4 will have at least 3 women?

62. A company knows that on average 3 out of 200 products will be defective. What is the probability at least one out of 50 products will be defective?

63. The following is a table representing household population.

Total Household Population	273,643,274
Householder	106,905,819
Spouse	54,216,796
Child	81,964,816
Other relatives	16,853,625
Nonrelatives	13,702,218

(U.S. Census Bureau)

    a. What is the probability that if one person is chosen randomly, that person is a householder?

    b. What is the probability that if one person is chosen randomly, that person is a spouse or child?

    c. What is the probability that if one person is chosen randomly, that person is not a child?

64. What are the odds that if a person chooses one card from a deck of 52 cards, that card will be a diamond?

65. What are the odds with a toss of a coin that the coin will be a head?

66. If the odds of choosing a defective part off an assembly line are 2:148, what is the probability of picking a defective part off an assembly line?

67. If the probability of dying by the age of 78 is $\frac{5}{7}$, what are the odds of dying by the age of 78?

CHAPTER 8 EXAM

(1–4) For each of the following sequences, give the first five terms:

1. $a_n = 3-4n$

2. $a_n = 5(-3)^n$

3. $a_n = \dfrac{n!}{n^2}$

4. $a_1 = -2,\ a_2 = 3,\ a_n = 2a_{n-2} - 3a_{n-1}$

5. An arithmetic sequence has a first term of 3 and a common difference of 5.
   a. Give the general equation for the arithmetic sequence.
   b. Find the 10th term of the sequence.

6. A geometric sequence has a first term of –2 and a common ratio of $\dfrac{1}{2}$.
   a. Give the general equation for the geometric sequence.
   b. Find the 6th term of the sequence.

(7–9) For each of the following sequences:
   a. Is the sequence arithmetic, geometric, or neither?
   b. If the sequence is arithmetic or geometric, give the general equation and find a_7.
   c. If it is not arithmetic or geometric, give the next term of the sequence.

7. 3, –6, 12, –24,…

8. 2, –3, 1, –5, –3, …

9. 4, –1, –6, –11, …

10. Find the indicated sum. for $\sum_{i=1}^{4}(2i^2 - 5i)$.

(11–13) Determine if the following sequences are arithmetic, geometric or neither; then find the indicated sum. (If arithmetic or geometric use the series formulas.)

11. 3, 6, 12, 24, … $S_8 =$

12. $a_n = -3 - 4(n-1)$ $S_{10} =$

13. 2, –5, –7, –2, 5,… $S_6 =$

14. Find the infinite series value of $a_n = 12\left(\dfrac{-2}{3}\right)^{n-1}$.

15. Use your calculator to find the value of $_{10}P_5$.

16. Use your calculator to find the value of $_{10}C_5$.

17. If social security numbers are made from 9 digits, how many different social security numbers can be assigned?

18. A club that has 40 members must choose 5 of those members to represent the club at a national meeting. How many different ways can the 5 members be chosen?

19. The rotary club has 20 members. How many ways can the club choose a president, vice–president, secretary and treasurer?

20. How many different permutations are there for the word calculus?

21. The student senate of a local college consists of 8 men and 7 women and 5 of the senators are to represent the college at a national meeting.
    a. How many different ways can the 5 senators be chosen if all are women?
    b. How many different ways can the 5 senators be chosen if 3 are men and 2 are women?
    c. How many different ways can the 5 senators be chosen if at least one is a man?
22. A bowl is filled with 6 lemon truffles, 5 blueberry truffles and 8 chocolate truffles.
    a. If one truffle is drawn from the bowl, what is the probability the truffle is a lemon truffle?
    b. If one truffle is drawn from the bowl, what is the probability the truffle is a chocolate truffle or a lemon truffle?
    c. If two truffles are drawn from the bowl, what is the probability the first truffle is a lemon truffle and the second truffle is a blueberry truffle?
    d. If one truffle is drawn from the bowl and put back in the bowl; then a second truffle is drawn from the bowl, what is the probability that the first truffle is a chocolate truffle and the second truffle is a lemon truffle?
23. Use the information from problem 20.
    a. What is the probability all 5 senators chosen are women?
    b. What is the probability the choice of senators is 3 men and 2 women?
24. A company knows that on average 2 out of 150 items produced on an assembly line will be defective. What is the probability at least one out of 50 will be defective?
25. The following is a table representing housing occupation.

Total occupied housing units	104,733,569
Owner occupied	69,345,619
Renter occupied	35,387,950

(U.S. Census Bureau)

    a. A person that lives in an occupied housing unit is chosen at random. What is the probability the person lives in an owner occupied unit?
    b. A person that lives in an occupied housing unit is chosen at random. What is the probability the person lives in either an owner occupied or renter occupied unit?
26. A lottery consists of 48 different numbers and 5 numbers are drawn, order doesn't count. What are the odds of winning the lottery if one ticket is purchased?

Chapter 0 Odd Homework Answers

SECTION 0.1

1) 81	3) $\frac{1}{243}$	5) 125
7) 8	9) $x^{\frac{2}{3}}$	11) $x^{\frac{2}{3}}y^{\frac{7}{3}}$
13) x^8y^4	15) $15a^5b^6$	17) $\frac{b}{a^8}$
19) $\frac{y^2}{x^8z^3}$	21) $\frac{-15y^4}{x^3z^4}$	23) $a^{\frac{7}{8}}b^{\frac{5}{6}}$
25) $x^{\frac{3}{8}}y^{\frac{1}{4}}z^{\frac{1}{4}}$	27) $4x^{11}y^{12}$	29) $\frac{8x^4y}{9}$
31) $x^{10}y^6$	33) $x^{\frac{1}{2}}y^{\frac{3}{10}}$	35) $x^{\frac{27}{40}}y^{\frac{8}{5}}$
37) $\frac{a^2x^2}{b^3y}$	39) $\frac{2ab^6}{7xy^2}$	41) $\frac{a^7x}{b^2y^7}$
43) $\frac{a}{by^2}$	45) $\frac{3a^4b^5}{4x^3y^4}$	47) 3)84 x 10^{-5}
49) 5)948 x 10^{-3}	51) 3)838 x 10^{9}	53) 5)83 x 10^{12}
55) 0)000385	57) 0)0000000098	59) 217,000
61) 8,345,600,000	63) $2\sqrt{7}$	65) $25\sqrt{3}$
67) $3\sqrt[3]{2}$	69) $2xy^2\sqrt{2x}$	71) $xy^4z^4\sqrt[3]{x^2y}$
73) $x^2y^3z^3\sqrt[4]{xy^3}$	75) $\frac{\sqrt{6}}{3}$	77) $\frac{\sqrt{10}}{4}$
79) $\frac{\sqrt{6}}{2}$	81) $\frac{4\sqrt{3}}{3}$	83) $\frac{\sqrt{2}}{2}$
85) $4-2\sqrt{3}$	87) $-1+\sqrt{5}$	89) $\frac{10+2\sqrt{2}+5\sqrt{3}+\sqrt{6}}{23}$
91) $9\sqrt{5}-3\sqrt{3}$	93) $-11\sqrt{2}$	95) $9\sqrt{2}-19\sqrt{3}$
97) $42\sqrt{2}+2\sqrt{5}-19\sqrt{3}$	99) $15\sqrt{2}-42$	101) $50-30\sqrt{2}$
103) $-8+11\sqrt{3}$	105) $-6-14\sqrt{6}$	107) 13
109) -38	111) $9\sqrt{3}+36\sqrt{2}+6\sqrt{5}+8\sqrt{30}$	113) $\frac{5}{2\sqrt{3}}$
115) $\frac{3}{6-\sqrt{3}}$		

Answers to Odd-Numbered Homework Problems

SECTION 0.2

1) 3	3) 5	5) 6
7) $7x^3-8x^2+5x+9$	9) $2x^3y^2-5x^2y+3xy^3$	11) $7x^3-16x-4$
13) $8x^3-6x^2-5x+11$	15) $3a^3b+10a^2b^2-3ab^3$	17) $-3x^2+5x-2$
19) $9y^3-8y^2-3y+22$	21) $-4x^3y-17x^2y^2+18xy^3$	23) $24x^3-15x$
25) $54a^5b^3+30a^3b^4$	27) $15x^4-24x^3-27x^2$	29) $-10a^3b^4+35a^2b^3+40ab^2$
31) $15x^2+31x+14$	33) $35x^2-11x-6$	35) $15x^2+31xy+14y^2$
37) $5a^2+42ab-27b^2$	39) $25x^2-9$	41) $25a^2-9b^2$
43) $10x^3-x^2-29x+12$	45) $35a^3-6a^2b+2ab^2+4b^3$	47) $27x^3-8$
49) $8x^3+y^3$	51) $9x^2+30x+25$	53) $4y^2-12y+9$
55) $4x^2+12xy+9y^2$	57) $9a^2-30ab+25b^2$	59) $21x^4-26x^3-39x^2+46x-16$
61) $5xy$	63) $\frac{-7x^2}{y^2}$	65) $2x^2-3x$
67) $\frac{2}{3}a^2-\frac{1}{2}a+\frac{4}{3}$	69) $4a-3b$	73) $5x^2-x-3$
71) $2x^2-3x+2$	75) $4x^2-3x-2$	77) $3x^2-5x+17-\frac{32}{x+2}$
79) $3x^2+13x+18+\frac{39}{x-2}$	81) $3x^2+5x-4+\frac{3}{2x+5}$	83) $2x^3+4x^2+3x+6+\frac{56}{x-2}$
85) $3x^3-6x^2+12x+28-\frac{14}{x+2}$		

SECTION 0.3

1) $5a(b-2)$	3) $8a(3x+5y)$	5) $5ab^2(2a+5b)$
7) $7a^2b^2(5a^3-4b^2)$	9) $3a(5x-2y+3)$	11) $5xy(2x-3y-7)$
13) $6x^2y^2(5x-2y-4c)$	15) $5a^{-3}b^{-4}(b^2+4a^2)$	17) $5x^{\frac{-3}{2}}\left(2-3x^2\right)$
19) $(x-4)(x-6)$	21) $(x-9)(x+6)$	23) $(x+4y)(x+5y)$
25) $(x-12a)(x+4a)$	27) $5(x+9)(x-4)$	29) $3x(a+4b)(a+9b)$
31) Prime	33) $(ax+8y)(ax+15y)$	35) $(x^2+2)(x^2+7)$
37) $(a^2+24b)(a^2+b)$	39) $(2x+3)(5x+2)$	41) $(3x-4)(5x-3)$
43) $(5x-y)(x-3y)$	45) $(4x-5y)(3x+2y)$	47) $3(3x+5)(x+1)$
49) $4a(5x+1)(x-2)$	51) $(3a^2+5b)(2a^2+b)$	53) $(3ab-2xy)(4ab+3xy)$
55) $(x+4)^2$	57) $(a-6)^2$	59) $(2x-3)^2$
61) $(3x+4y)^2$	63) $3(x+9)^2$	65) $(x-9y)^2$
67) $(2x+7y)^2$	69) $(x^2-3y)^2$	71) $(x-6)(x+6)$

SECTION 0.3

73) (2a–5)(2a+5)	75) Prime	77) 5(a–2)(a+2)
79) (5x–6y)(5x+6y)	81) $(y^2-8)(y^2+8)$	83) $(x^2+4)(x-2)(x+2)$
85) (x–2)(x+2)(x–3)(x+3)	87) $(y-2)(y^2+2y+4)$	89) $(a+5)(a^2-5a+25)$
91) $(2x-3)(4x^2+6x+9)$	93) $(4x+1)(16x^2-4x+1)$	95) $5(2x-1)(4x^2+2x+1)$
97) $2a(x+4)(x^2-4x+16)$	99) (a+x)(b+3)	101) $(x^2+2)(a+5)$
103)(a–5)(x–3)	105) (x+4)(7–a)	107) (x–3)(x+3)(a+3)
109)(x+3–y)(x+3+y)	111) (y+5–3a)(y+5+3a)	113) (2x+5–2a)(2x+5+2a)

SECTION 0.4

1) $\frac{-x}{2y^2}$	3) $\frac{3x^2}{2}$	5) $\frac{5}{3}$
7) $\frac{5}{4}$	9) 2x + 4	11) –2
13) $\frac{5}{x+2}$	15) $\frac{-(y+3)}{5}$	17) $\frac{a+5}{a+1}$
19) $\frac{x+5}{x+3}$	21) $\frac{x+3}{x-5}$	23) $\frac{x^2+3x+9}{x+5}$
25) $\frac{x^2+9x+18}{x^2+2x-15}$	27) $\frac{x^6}{2y}$	29) $\frac{5b^5}{6}$
31) $\frac{(x-2)(x-3)}{4(x+2)(x-5)}$	33) $\frac{x+2}{x+4}$	35) $\frac{x-5}{x-3}$
37) $\frac{(y+3)^2}{(y-3)(y-2)}$	39) 1	41) y^2-3y+9
43) $\frac{2(x-2)(x+3)}{(x+2)^2}$	45) $\frac{a+5}{4(a+2)}$	47) $\frac{x+5}{x+3}$
49) $\frac{(x-5)(x+5)}{(x+7)(x-7)}$	51) $\frac{x-5}{x+4}$	53) 1
55) 3	57) x + 3	59) $\frac{12x+29}{(x+2)(x+3)}$
61) $\frac{x^2+17x}{(x-1)(x+5)}$	63) $\frac{-3}{x-3}$	65) $\frac{x}{(x-3)(x+3)(x-2)}$
67) $\frac{7x+6}{5(x-3)(x+3)}$	69) $\frac{2x^2+16x}{(x-2)(x+2)(x+1)}$	71) $\frac{x^2-5x+3}{(x-5)(x+5)}$
73) $\frac{a^2-2a+12}{(a-3)^2(a-2)}$	75) $\frac{-3x-10}{(x-5)(x+5)}$	77) $\frac{10x-25}{(x-3)(x+2)(x-4)}$

SECTION 0.4

79) $\frac{14}{19}$	81) $\frac{2x+3}{5x+4}$	83) $\frac{3x-4}{2x-5}$
85) $\frac{3ab+b}{5ab+a}$	87) $\frac{2a^2b+3b^2}{5ab^2-3a^2}$	89) $\frac{2x-11}{5x-12}$
91) $\frac{3(x-2)}{x+3}$		

SECTION 0.5

1) x=13	3) x= −5	5) x= −2
7) $x=\frac{5}{2}$	9) No solution	11) All real numbers
13) $x=\frac{19}{26}$	15) x= −1	17) No solution
19) All real numbers	21) x=63	23) $x=\frac{7}{2}$
25) $x=\frac{18}{5}$	27) $x=\frac{27}{4}$	29) No solution
31) All real numbers	33) $x=\frac{17}{6}$	35) $x=\frac{-12}{47}$
37) $x=\frac{-89}{27}$	39) x= −8 or x= 14	41) No solution
43) x=4 or x= −3	45) x=2 or $x=\frac{-8}{7}$	47) x= −1 or x= −5
49) No solution	51) x= −1 or $x=\frac{5}{3}$	53) $x=\frac{7}{3}$
55) x=3 or x= −18	57) x=7 or $x=\frac{-19}{3}$	59) $x=\frac{43}{8}$ or $x=\frac{-37}{8}$
61) No solution	63) x= 0 or $x=\frac{-20}{3}$	

SECTION 0.6

1) a) 1	b) **Step 1:** divided 10 by 5; **Step 2:** added 2 to the answer from Step 1; **Step 3:** subtracted 3 from the answer from Step 2)
3) a) −0.6	b) **Step 1:** added 2 and 10; **Step 2:** divided the answer from Step 1 by 5; **Step 3:** subtracted 3 from the answer from step 2)
5) a) 6	b) **Step 1:** added 2 and 10; **Step 2:** subtracted 3 from 5; **Step 3:** divided the answer from Step 1 by the answer from Step 2)

SECTION 0.6

	a)	b)
7)	a) 14	b) **Step 1:** squared 3; **Step 2:** added 5 to the answer from Step 1)
9)	a) 2187	b) **Step 1:** added and 2 and 5; **Step 2:** raised 3 to the power of the answer from Step 1)
11)	a) 32	b) **Step 1:** raised 4 to the 3rd power; **Step 2:** divided the answer from Step 1 by 2)
13)	a) 8	b) **Step 1:** divided 3 by 2; **Step 2**: raised 4 to the power of the answer from Step 1)
15)	a) 32	b) **Step 1:** multiplied 3 by 6; **Step 2:** added 5 to the answer from Step 1; **Step 3:** added 9 to the answer from Step 2)
17)	a) 57	b) **Step 1:** added 5 and 3; **Step 2:** multiplied 6 by the answer from Step 1; **Step 3:** added 9 to the answer from Step 2)
19)	a) 120	b) **Step 1:** added 5 and 3; **Step 2:** added 6 and 9; **Step 3:** multiplied the answer from Step 1 times the answer from Step 2)
21)	a) 5467.86	b) **Step 1:** divided 0)08 by 12; **Step 2:** added one to Step 1; **Step 3:** multiplied 12 times 30; **Step 4:** raised the answer from Step 2 to the power of the answer from Step 3; **Step 5:** multiplied 500 times the answer from Step 4)
23)	a) 405.53	b) **Step 1:** divided 0)08 by 12 in the numerator; **Step 2:** multiplied step 1 by 20000 in the numerator; **Step 3:** divided 0)08 by 12 in the denominator; **Step 4:** added 1 to step 3; **Step 5:** subtracted the answer from Step 4 from 1; **Step 6:** multiplied –12 times 5; **Step 7:** raised the answer from Step 5 to the power of the answer from Step 6; **Step 8:** divided the answer from Step 2 by the answer from Step 7)

25) a) 5^2 b) 25	27) a) 2^-5 b) 0.03125	29) a) (3+5)/(6+4) b) 0.8
31) a) 9^(3/2) b) 27	33) a) (3*2+8)/7 b) 2	35) a) (5*3^2–7)/(4*2+11) b) 2
37) a) (5/8) + (3/4) b) 1.375	39) a) $3\sqrt[x]{13824}$ or (13824)^(1/3) b) 24	
41) a) 5000(1+.09/12)^(12*15)	b) Approx. 19190.22	
43) a) 24000(.085/12)/(1–(1+.085/12)^(-12*5))	b) Approx. 492.40	
45) $\frac{-32}{7}$	47) $\frac{36}{11}$	49) $\frac{95}{72}$

SECTION 0.7

1) 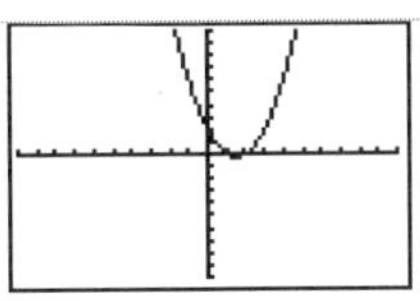 a) 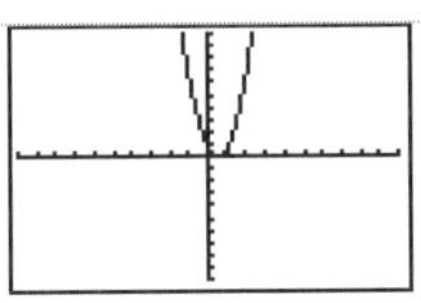 b) 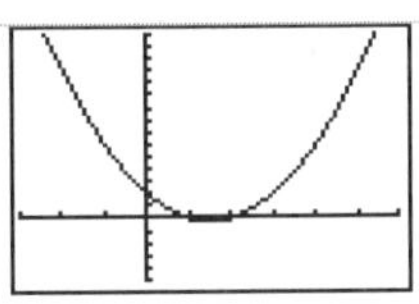 c) Window *b*

3) 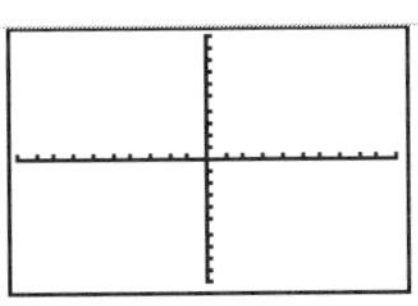 a) 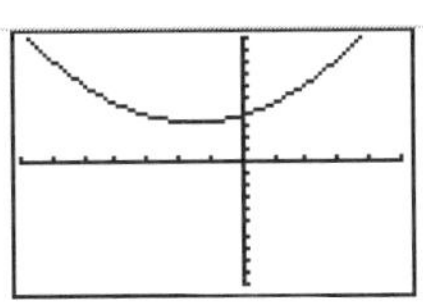 b) 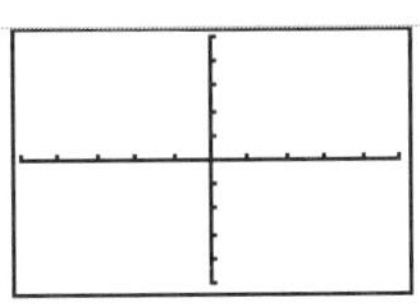 c) Window *a*

5) 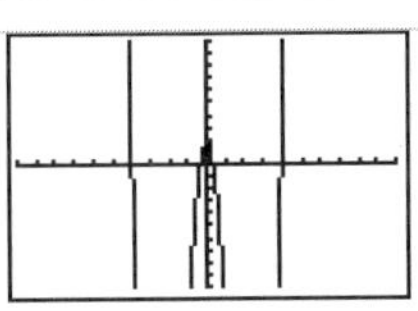 a) 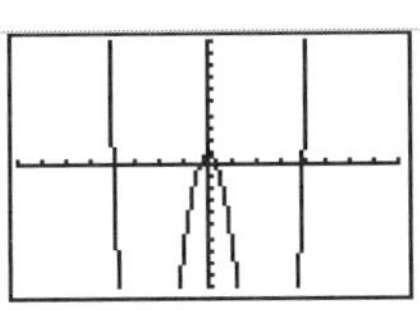 b) 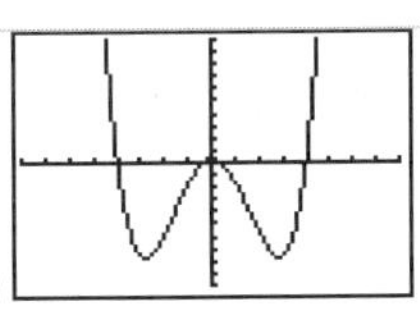 c) Window *b*

SECTION 0.7

7) 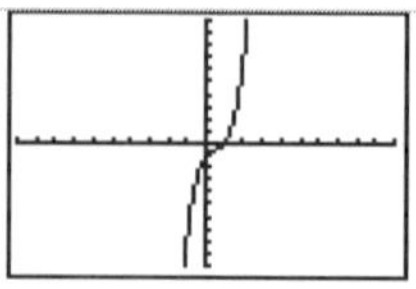 a) 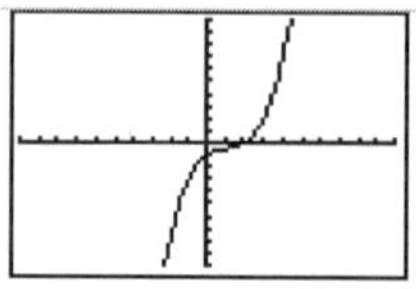 b) 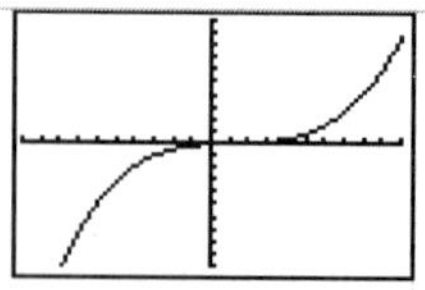 c) Window *b*

9) 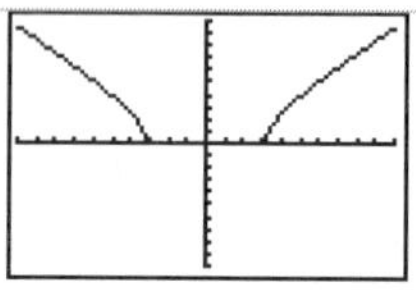 a) 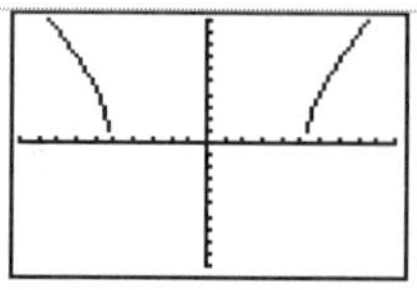 b) 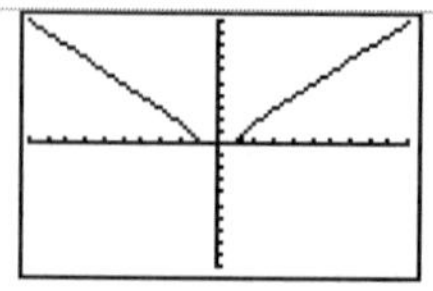 c) Standard window and window *b* and *c*

11) 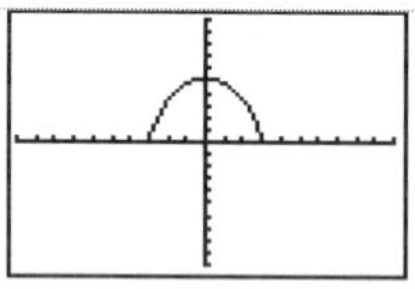 a) 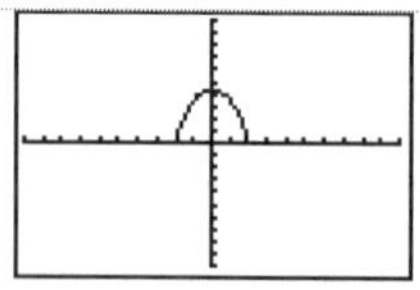 b) 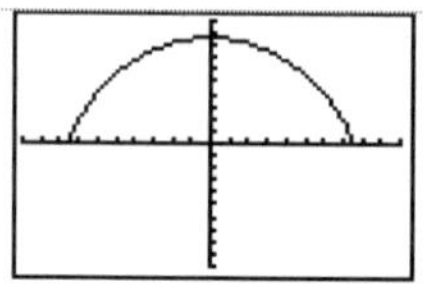 c) Window *b*

13) 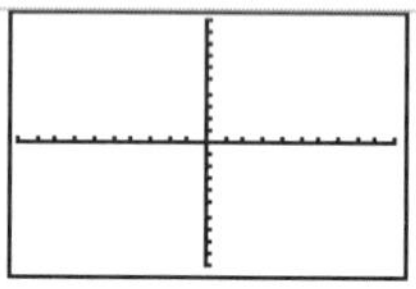 a) 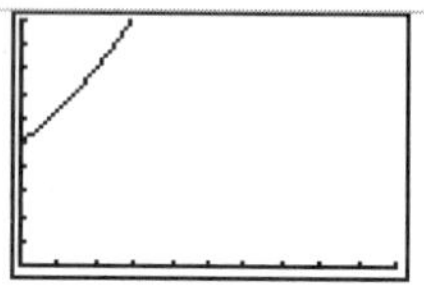 b) 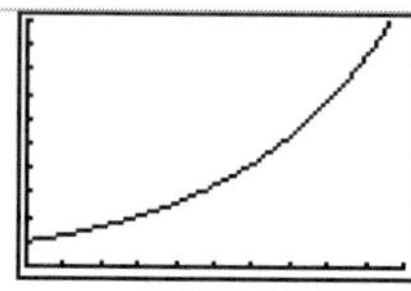 c) Window *b*

For 15–30 the following are suggested windows:

15) –5≤x≤5 Scale 0.5 –15≤y≤15 Scale 1.5	17) –5≤x≤5 Scale 0.5 –62≤y≤62 Scale 6
19) –6≤x≤6 Scale 0.6 –40≤y≤40 Scale 4	21) –6≤x≤6 Scale 0.6 –45≤y≤45 Scale 4
23) –10≤x≤10 Scale 1 –10≤y≤30 Scale 2	25) –13≤x≤13 Scale 1 –13≤y≤13 Scale 1
27) 0≤x≤30 Scale of 3 0≤y≤5,200 Scale of 500	29) –2≤x≤10 Scale of 1 0≤y≤3,000 Scale of 300

31) y = 603	33) y = 653	35) y = –40,790

37) y is approximately equal to 23,304.79

39) a) Gives error message **Invalid**

b) x = 15 was not in the window as x values are from –10 to 10 only.

SECTION 0.8

1) x = 1.5 or –1.25	3) x = 0.4 or –3 or 5.25	5) x = –3 or –0.25 or 1 or 5
7) x = –5	9) x ≈ –0.3028 or 3.3028	11) x ≈ –1.5414 or 4.5414
13) x= –3 or x≈–1.3028 or 2.3028	15) Minimum = –13	17) Maximum = 29
19) Minimum ≈ –5.0357	21) Maximum = 47 Minimum = –78	23) Maximum = 22 Minimum = –234

25) Maximum = 30 Minimum = –418 or –2.0625	27) Maximum ≈ 24.5849 Minimum ≈–131.5849

29) 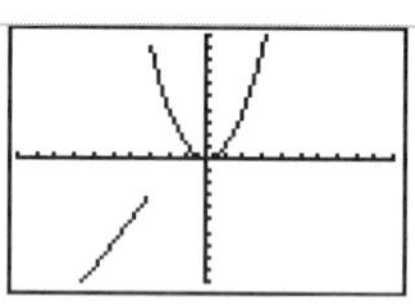 31) 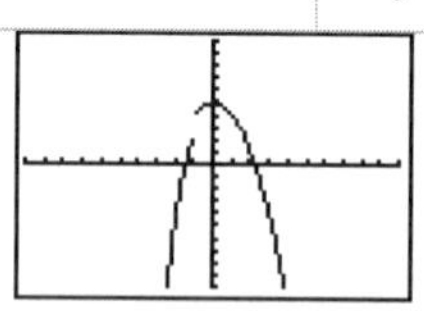 33) 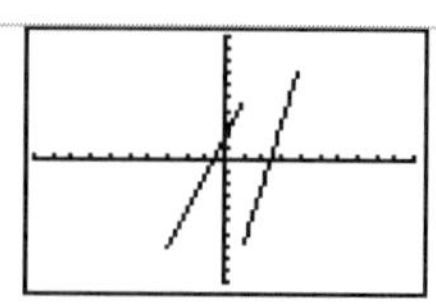

35) 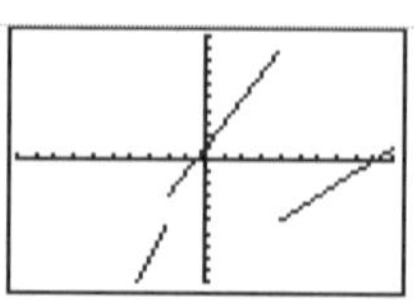

CHAPTER 0 REVIEW

1) 729	3) 27	5) a^9b^{13}
7) $\dfrac{6y^2}{x^7}$	9) $\dfrac{27y^{12}}{8x^6}$	11) $\dfrac{15a^4x^2y^2}{16b^8}$
13) $\dfrac{2b^4x^2}{3ay^3}$	15) 8)32 x 10^{-6}	17) 0)0000000293
19) $4\sqrt{5}$	21) $3\sqrt[3]{5}$	23) $x^3y^2z^2\sqrt[4]{x^3y}$
25) $2\sqrt{3}$	27) $10-5\sqrt{3}$	29) $12\sqrt{7}-6\sqrt{5}$
31) $42+12\sqrt{7}$	33) 61	35) $13x^3-19x^2-6$
37) $3y^3+16y-22$	39) $15a^3-40a^2+10a$	41) $15a^2-22a+8$
43) $10x^3-11x^2-9$	45) $9x^2-25$	47) $25b^2+30b+9$
49) $\dfrac{3a}{2b^3}$	51) $3x^2+x-5+\dfrac{12}{x+2}$	53) $9a(x-3)$
55) $(x-9)(x+7)$	57) $(a+15b)(a+8b)$	59) $(3x+5)(x+3)$
61) $(2y-5b)(4y+3b)$	63) $(x^2+7)(x^2+5)$	65) $(4a-7b)^2$
67) $(y-11)(y+11)$	69) Prime	71) $(2x+3y)(4x^2-6xy+9y^2)$
73) $(y-3)(b+3)$	75) $(x-3)(x+3)(a-5)$	77) $2b(2y+3)(3y-5)$
79) $(x-3)(x+3)(x-2)(x+2)$	81) $\dfrac{7y^3}{6x^3}$	83) $\dfrac{-3}{2}$
85) $\dfrac{y-4}{y-8}$	87) $\dfrac{5a^3}{12b^2}$	89) $\dfrac{x+1}{5(x-5)}$
91) $\dfrac{3x+1}{5x-2}$	93) $\dfrac{x+5}{x-3}$	95) 5
97) $\dfrac{3x^2-x}{(x+3)(x-2)}$	99) $\dfrac{11a+15}{2(a-3)(a+3)}$	101) $\dfrac{-52}{(x+3)(x-3)(x+5)}$
103) $\dfrac{3b+5}{7b-8}$	105) $\dfrac{2x^2+x}{3x^2-1}$	107) $\dfrac{y^2(y+5)}{y-5}$
109) $x=3$	111) $x=-3$	113) No solution
115) No solution	117) $x=\dfrac{-19}{3}$	119) $x=\dfrac{5}{3}$ or -3
121) No solutions	123) $x=-2$ or $\dfrac{4}{5}$	125) $x=\dfrac{1}{2}$ or $\dfrac{-3}{2}$
127) 78,125	129) 2	131) 25
133) -7	135) Approximately 98,018.82	137) $\dfrac{236}{345}$

Answers to Odd-Numbered Homework Problems

CHAPTER 0 REVIEW

139) $\frac{-31}{11}$

141) $\frac{16}{9}$

143) 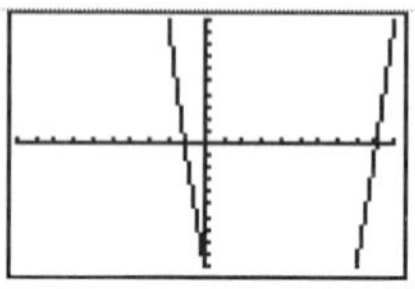 a) 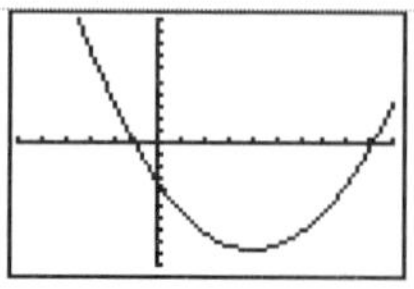 b) 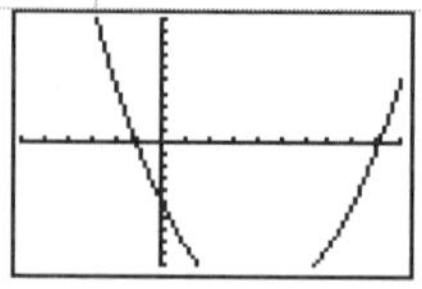 c) Window *a*

145) 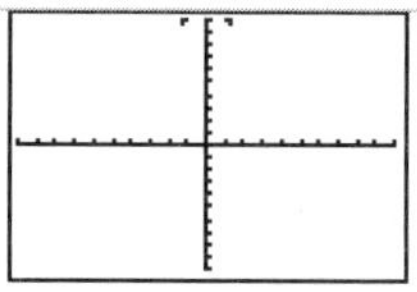 a) 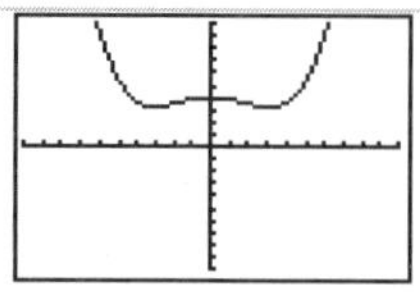 b) 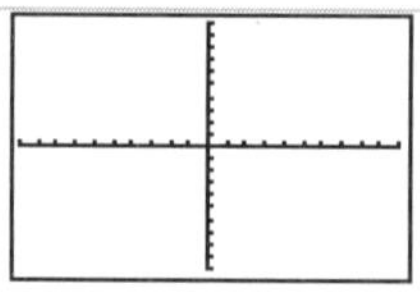 c) Window *a*

147) $-2 \le x \le 12$ Scale 1 $-6 \le y \le 22$ Scale 2	149) $-4 \le x \le 4$ Scale 0.4 $-18 \le y \le 18$ Scale 2

151) y=211	153) y=11,000	155) x= –4 or 1.6
157) x≈–0.3723 or 5.3723	159) Minimum= –5	161) Maximum=179 Minimum= –77

163) 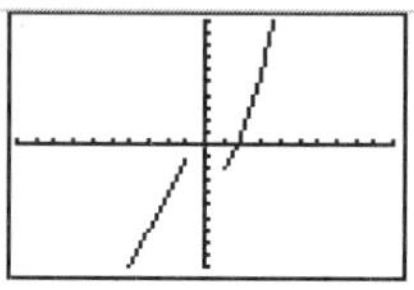

165) 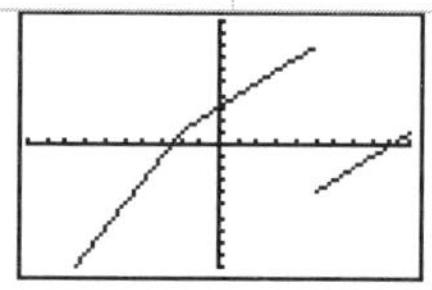

CHAPTER 0 EXAM

1) $\frac{1}{64}$	3) $\frac{-20x^7}{yz^4}$	5) $x^{\frac{17}{12}} y^{\frac{9}{8}}$
7) 2)857 x 10^{-5}	9) $5\sqrt[3]{7}$	11) $\frac{\sqrt{35}}{7}$
13) $\frac{17-8\sqrt{2}}{23}$	15) $34\sqrt{2}-30\sqrt{3}$	17) $28+14\sqrt{2}$
19) $-2x^2+13xy-15y^2$	21) $15x^2-11xy-14y^2$	23) $49a^2-42ab+9b^2$
25) $3x^2+x+3+\frac{12}{x-3}$	27) (y–12)(y+5)	29) (3x+2)(x+8)
31) (5x–7)(5x+7)	33) $(y-4)(y^2+4y+16)$	35) $\frac{x+3}{2x+3}$
37) $\frac{6a^2x^3}{5b^4y}$	39) 1	41) $\frac{2x-21}{(x-2)(x+2)(x-3)}$

43) $\frac{3b+2a^2b}{5a^2-3a^2b}$	45) $x = 60$	47) $x = -2$ or $\frac{2}{3}$	49) $x = \frac{11}{2}$ or $\frac{3}{2}$

51) $\frac{125}{8}$	53) 216
55) $-2 \le x \le 7$ Scale 1 $-15 \le y \le 15$ Scale 1	57) y=684
59) x= –3 or 0.75	61) Maximum=14

63) 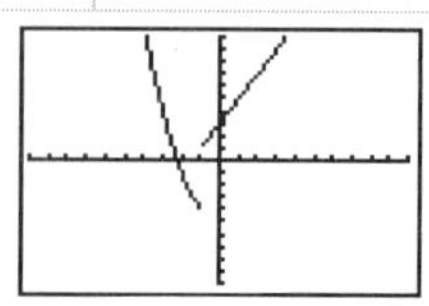

Chapter 1 Odd Homework Answers

SECTION 1.1

1) a. between \$800 and \$900 b. between \$1300 and \$1400 c. yes; For every input of time there is only one output for money.

3) a. 1,817 b. 804 c. 904 d. yes; For every input of a year there is only one output for the number immigrants.

5) a. 120 to 130 b. 400 to 410 c. yes; For each input of year there is only one output for the number of state prisoners. d. 1984 e. 1990 f. yes; For each input of number of state prisoners there is only one output for the year.

7) a. \$3,664.04 b. \$16,319.33 c. yes; For each input of year there is only one output of money. d. 25 e. 10 f. yes; For each input of money there is only one output of year.

9) a. 67,432 b. 60,267 c. yes; For each input of year there is only one output for student population. d. 1995 e. 1965 f. yes; For each input of student population there is only one output for year. g. Yes, if the number of students sometime in the future is the same as the number of previous years.

11) a. 23 b. 23 c. yes; For each input of year there is only one output for retirements. d. 1972 e. 1992 f. no; For the input of 23 retirements there are two outputs in years, 1968 and 1988.

13) a. \$200 b. \$400 c. yes; For each input of year there is only one output of interest earned

15) a. 288π cubic inches b. $\frac{256}{3}\pi$ cubic feet c. yes; For each input of radius there is only one output for volume.

17) a. t b. I(t) c. 1,817 d. 804 e. {1989, 1990, 1991, 1992, 1993, 1994, 1995, 1996} f. {1,090, 1,536, 1,817, 904, 804, 720}

19) a. t b. P(t) c. 57,591 d. 64,803 e. {1965, 1970, 1975, 1980, 1981, 1982, 1983, 1984, 1985, 1986, 1987, 1988, 1989, 1990, 1991, 1992, 1993, 1994, 1995, 1996} f. {54,394, 59,838, 61,004, 58,305, 57,916, 57,591, 57,432, 57,150, 57,226, 57,709, 58,254, 58,485, 59,436, 60,267, 61,605, 62,686, 63,241, 63,986, 64,803, 65,674}

21) a. t b. I(t) c. \$1,000 d. \$360

23) a. r b. V(r) c. 972π cubic inches d. 36π cubic feet

25) a. yes b.For every purchase price there would only be one sales tax c. no d. For one sales tax there would be more than one purchase price.

27) 12	29) 37	31) 11
33) 3	35) 4	37) $\frac{2}{9}$
39) $\frac{-5}{2}$	41) No Answer	43) $3a^2 - 2a + 4$
45) $3a^2+6ah+3h^2-2a-2h+4$	47) $\frac{b+1}{b+8}$	

49) a. The independent value was the value that was substituted for x in each problem.
b. The dependent value was the answer that resulted from that substitution in a. above.

51) function	53) not a function	55. not a function
57) function	59) not a function	61. function
63) not a function	65) not a function	67. 36
69) 21	71) 2	73. 3
75) 9–3a–3b	77) 7	

SECTION 1.2

1) a. {(1,3); (2,8); (3,16); (4,32); (5,63)} b. {1,2,3,4,5} c. {3,8,16,32,63} d. yes; For every domain there is only one range.

3) a. {(0,0); (4,2); (4,–2); (9,3); (9,–3); (16,4); (16,–4)} b. {0,4,9,16} c. {0,2,–2,3,–3,4,–4} d. no; For domain values 4, 9, 16 there is more than one range.

5) a. {(1960,49.7); (1964,61.1); (1968,42.7); (1972, 37.5); (1976,50.1); (1980,41.0); (1984,40.2); (1988, 45.6); (1992,43.0); (1996,49.2)} b. {1960, 1964, 1968, 1972, 1976, 1980, 1984, 1988, 1992, 1996} c. {37.5, 40.2, 41.0, 42.7, 43.0, 45.6, 49.2, 49.7, 50.1, 61.1)} d. yes; For every domain there is only one range.

7) a. {(1,1); (1,–1); (16,2); (16,–2); (64,4); (64,–4); (125,5); (125,–5); (216,6); (216,–6); (343,7); (343,–7)} b. {1, 16, 64, 125, 216, 343} c. {1,–1,2,–2,4,–4,5,–5,6,–6,7,–7} d. no; Each of the domain elements has more than one range element paired with them.

9) $(-\infty,\infty)$	11) $(-\infty,3)$ or $(3,\infty)$	12) $(-\infty,-5)$ or $(-5,\infty)$
13) $(-\infty,\infty)$	15) $(-\infty,-3)$ or $(-3,3)$ or $(3,\infty)$	17) $[2,\infty)$
19) $(-\infty,\infty)$	21) $(-\infty,2]$	23) $(-\infty,3)$ or $(3,\infty)$
25) $(-\infty,\infty)$	27) $(-\infty,\infty)$	29) $(5,\infty)$
31) $[0,\infty)$	33) $[3,\infty)$	35) $[0,\infty)$
37) $[2,\infty)$	39) $(-\infty,3]$	41) $[4,\infty)$
43) Domain: $(-\infty,\infty)$ Range: $[2,\infty)$	45) Domain: $(-\infty,\infty)$ Range: $[2,\infty)$	47) Domain: $(-\infty,\infty)$ Range: $[0,\infty)$
49) Domain: $(-\infty,\infty)$ Range: $[3,\infty)$	51) Domain: $(-\infty,\infty)$ Range: $(-\infty,\infty)$	53) Domain: $[4,\infty)$ Range: $[2,\infty)$
55) Domain: $(-\infty,-3]$ or $[3,\infty)$ Range: $[0,\infty)$	57) Domain: $(-\infty,-4)$ or $(-4,\infty)$ Range: $(-\infty,0)$ or $(0,\infty)$	
59) Domain: $(-\infty,\infty)$ Range: $(-\infty,\infty)$	61) Domain: $(-\infty,\infty)$ Range: $(-\infty,\infty)$	

63) a. 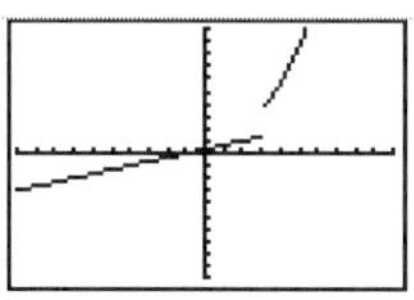 b. $(-\infty,\infty)$ c. $(-\infty,4)$ or $[10,\infty)$

65) a. 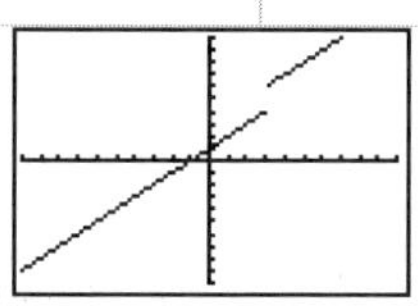 b. $(-\infty,\infty)$ c. $(-\infty,4)$ or $[6,\infty)$

67) a. 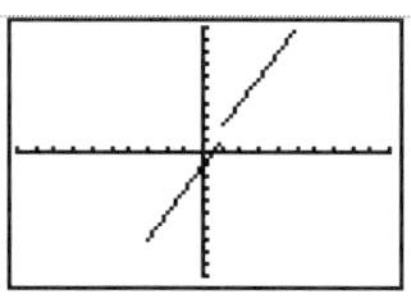 b. $[-3,5]$ c. $[-7,1)$, $[2,10]$

69) a. 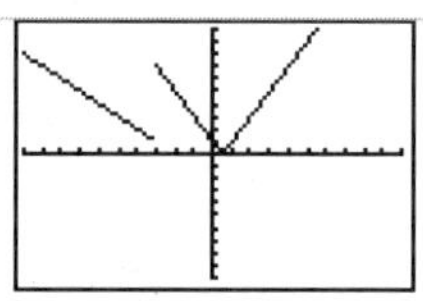 b. $(-\infty,\infty)$ c. $[0,\infty)$

71) a. 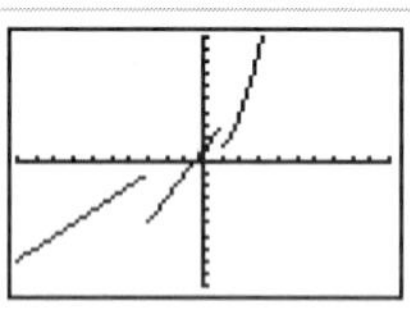 b. $(-\infty,\infty)$ c. $(-\infty,\infty)$

73) a. $[0,5]$ b. $[0,74.07]$ 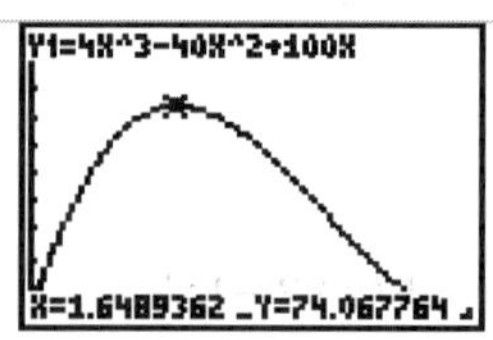

SECTION 1.3

1) a. Jan. to July b. July to Jan. c. 52.2 d. 81.4

3) a. 9:30 to 11:30 or 2:30 to 3:30 b. 11:30 to 2:30 or 3:30 to 4:30 c. $79\frac{1}{8}$ and $77\frac{3}{8}$ d. $77\frac{1}{8}$ and $77\frac{1}{4}$

SECTION 1.3

5) a. none b. –2.25 c. (–0.5,∞) d. (–∞,–0.5)	7) a. none b. –6.125 c. (–1.25, ∞) d. (–∞,–1.25)
9) a. 8 b. none c. (–∞,4) d. (4, ∞)	11) a. 6.125 b. none c. (–∞,–1.25) d. (–1.25, ∞)
13) a. 9.1 b. –3.4 c. (–∞,–3) or (2, ∞) d. (–3,2)	15) a. 11 b. –16 c. (–∞,–1) or (2, ∞) d. (–1,2)
17) a. 4 b. –2.25 c. (–1.58,0) or (1.58, ∞) d. (–∞,–1.58) or (0,1.58)	19) a. none b. 1.145 c. (0.607, ∞) d. (–∞,0.607)
21) a. 1.105 b. none c. (–∞,0.734) d. (0.734, ∞)	23) a. none b. 3 c. (2.5, ∞) d. (–∞,2.5)
25) a. 5 b. none c. (–∞,–1.5) d. (–1.5, ∞)	27) a. none b. none c. (–∞,∞) d. never
29) a. none b. none c. (–∞,∞) d. never	31) a. none b. none c. (0, ∞) d. never
33) a. 7.5 seconds b. 900 ft. c. (0,7.5) d. (7.5,15) e. [0,15] f. [0,900]	
35) a. (0 7.5) b. 3.75 inches c. 506.25 inches^3 d. (0,3.75) e. (3.75, 7.5) f. (0, 506.25]	
37) a. $2,000 b. $43,000 c. [0,2000) d. (2000,4932.58] e. [0,4932.58] f. [0,43000]	
39) a. 10 b. $300 c. (10, ∞) d. (0,10)	
41) a. 10 b. 172 c. (0,10) d. (10,20]	

SECTION 1.4

1) a. $8x-1$; (–∞,∞) b. $-2x + 5$; (–∞,∞) c. $15x^2+x-6$; (–∞,∞) d. $\frac{3x+2}{5x-3}$; $(-\infty,\frac{3}{5})$ or $(\frac{3}{5},\infty)$ e. $15x-7$; (–∞,∞) f. $15x+7$; (–∞,∞)	3) a. $4x^2+4x-2$; (–∞,∞) b. $2x^2-4x-2$; (–∞,∞) c. $3x^4+12x^3-2x^2-8x$; (–∞,∞) d. $\frac{3x^2-2}{x^2+4x}$; (–∞,–4) or (–4,0) or (0, ∞) e. $3x^4+24x^3+48x^2-2$: (–∞,∞) f. $9x^4-4$; (–∞,∞)
5) a. x^2+x-14; (–∞,∞) b. $-x^2+x+18$; (–∞,∞) c. $x^3+2x^2-16x-32$; (–∞,∞) d. $\frac{x+2}{x^2-16}$; (–∞,–4) or (–4,4) or (4, ∞) e. x^2-14; (–∞,∞) f. $x^2+4x-12$; (–∞,∞)	7) a. $\frac{10x+4}{3}$; (–∞,∞) b. $\frac{8x+8}{3}$; (–∞,∞) c. $\frac{3x^2-4x-4}{3}$; (–∞,∞) d. $\frac{9x+6}{x-2}$; (–∞,2) or (2, ∞) e. x; (–∞,∞) f. x; (–∞,∞)
9) a. $3x-1+\sqrt{x+2}$; [–2, ∞) b. $3x-1-\sqrt{x+2}$; [–2, ∞) c. $(3x-1)(\sqrt{x+2})$; [–2, ∞) d. $\frac{(3x-1)\sqrt{x+2}}{x+2}$; (–2, ∞) e. $3\sqrt{x+2}-1$; [–2, ∞) f. $\sqrt{3x+1}$; $[\frac{-1}{3},\infty)$	11) a. $\frac{3\sqrt{3x-1}+x^2+1}{3}$; $[\frac{1}{3},\infty)$ b. $\frac{3\sqrt{3x-1}-x^2-1}{3}$; $[\frac{1}{3},\infty)$ c. $\frac{(\sqrt{3x-1})(x^2+1)}{3}$; $[\frac{1}{3},\infty)$ d. $\frac{3\sqrt{3x-1}}{x^2+1}$; $[\frac{1}{3},\infty)$ e. $\lvert x\rvert$; (–∞,∞) f. x; $[\frac{1}{3},\infty)$

13) 10	15) 5	17) –13
19) 3	21) 9	23) 8
25) $\frac{3}{16}$	27) $\frac{1}{34}$	29) 9

SECTION 1.4

31) a^2+4a-2	33) $\dfrac{8c-6}{2c+1}$	35) $(a^4+2a^2+1)(\lvert a\rvert)$
37) −1	39) $\dfrac{2}{3}$	41) −17
43) −3	45) −7	47) −13
49) −4	51) not possible	53) $\dfrac{1}{2}$
55) undefined	57) −8	59) not possible
61) 10	63) 64	65) {−3,−2,−1,0,1,2,3,4}
67) {−2,0,2,4}	69) 5	71) −7
73) 18	75) $\dfrac{-3}{2}$	77) 0
79) 5	81) 2	83) −10
85) −15	87) $\dfrac{-5}{3}$	89) 2
91) 0	93) 21	95) −16
97) −7	99) −40	101) 8
103) −1	105) −11	107) 600
109) $h(x) = -x^2 + 3x$	111) $h(x) = -x^2 + 2x + 2$	

SECTION 1.5

1) $D = \dfrac{C}{\pi}$	3) $r = \sqrt{\dfrac{A}{\pi}}$	5) $R = \dfrac{100}{C}$
7) $s = \sqrt[3]{V}$	9) $x(y) = \dfrac{y+5}{3}$	11) $t(w) = \dfrac{w^2-6w+7}{3}$
13) $s(m) = \dfrac{8m+5}{6}$	15) $x(y) = \dfrac{3y+1}{2y-3}$	17) $w(r) = \dfrac{-2r-7}{r}$

19) $t(s) = 6s + 7$	21) $y(x) = 18x^2-12x+5$	23) $y(r) = \dfrac{9-3x^2}{x^2}$	25) $m(w) = \lvert w\rvert$

27) a. $R(x) = 80x-x^2$ b. $R(p) = 80p-p^2$	29) a. $P(x)= -.03x^2+70x-300$
31) a. $V(r)=3\pi r^3$ b. $V(h)=\dfrac{1}{9}\pi h^3$	33) a. $V(W)= 12W^2 + 18W$ b. $V(L)= 3L^2 - 9L$
35) a. $A(W)= 5000W-W^2$ b. $A(L)= 5000L-L^2$	37) a. $F(W)=\dfrac{80000}{W}+2W$ b. $F(L)=2L+\dfrac{80000}{L}$
39) a. $A(a)=\dfrac{1}{2}a\sqrt{16-a^2}$ b. $A(b)=\dfrac{1}{2}b\sqrt{16-b^2}$	41) a. $V(r)=\dfrac{2}{3}\pi r^3$ b. $V(h)=\dfrac{1}{12}\pi h^3$

SECTION 1.5

43) a. $A(L)=\frac{10800}{L}+12L+1044$ b. $A(W)=\frac{10800}{W}+12W+1044$

45) 3

47) 6x+3h–2

49) $6x^2+6xh+2h^2-6x-3h$

51) 192 ft/sec

53) 280 dollars/unit

55) 168 ft^3/sec

SECTION 1.6

1. a. Multiply both sides of the equation by 3. b. Subtract 1 from both sides of the equation. c. Add 2x to both sides of the equation. d. Divide both sides of the equation by x–5. Result: $\frac{5x-1}{x-5}=\frac{2x-10}{x-5}$

3) a. b. yes c. yes

x	$f^{-1}(x)$
5	–3
3	–2
1	–1
–1	0
–3	1
–5	2
–7	3

5) a. b. no c. no

x	$f^{-1}(x)$
9	–3
4	–2
1	–1
0	0
1	1
4	2
9	3

7) a. b. yes c. yes

x	$f^{-1}(x)$
7	49
6	36
5	25
4	16
3	9
2	4
1	1

9) a. b. no c. no

x	$f^{-1}(x)$
76	–3
11	–2
–4	–1
–5	0
–4	1
11	2
76	3

11) one–to–one

13) not one–to–one

15) not one–to–one

17) one–to–one

19) not one–to–one

21) one–to–one

23) a. appears to be inverses b. f(g(x)) = x and g(f(x)) = x, are inverses

25) a. appears to be inverses b. f(g(x)) = x and g(f(x)) = x, are inverses

27) a. appears to be inverses b. f(g(x)) = x and g(f(x)) = x; are inverses

29) a. do not appear to be inverses b. $f(g(x))=\frac{4x+9}{15}$, are not inverses

31) a. do not appear to be inverses b. f(g(x)) = 8x–28, are not inverses

33) a. appear to be inverses b. f(g(x)) = x and g(f(x)) = x, are inverses

35) $f^{-1}(x)=\frac{x+5}{3}$

37) $f^{-1}(x)=\frac{5x+3}{2}$

39) $f^{-1}(x)=\frac{x^2+1}{2}$

41) $f^{-1}(x)=\pm\sqrt{\frac{x-1}{3}}$, not a function

43) $f^{-1}(x)=\frac{x^3-3}{2}$

45) $f^{-1}(x)=\sqrt[3]{\frac{x+1}{2}}$

47) $f^{-1}(x)=\frac{3x+1}{1-x}$

49) $f^{-1}(x)=\frac{2x+3}{x-2}$

SECTION 1.6

51) a. $t(A)=\dfrac{A-5000}{400}$ b. Allows you to find easily how long it takes the account to be worth a certain amount of money.

53) a. $t(V)=\sqrt[3]{\dfrac{V}{36\pi}}$ b. Allows you to find easily how long it takes for the balloon to have a certain volume.

55) Domain: [1,9], Range: [–3,5]	57) (3, –2)

CHAPTER 1 REVIEW

1) a. 35,886 b. 26,071 c. yes d. w e. A(w)	3) a. \$32.01 b. yes c. p d. A(p)	
5) a. 22 b. 1900, 1930 c. no	7) a. \$5.50 b. \$6.90 c. yes	9) –4
11) a^2	13) yes	15) no
17) a. {(1960, 52), (1970, 54), (1980, 54), (1985, 60), (1990, 58), (1995, 58), (1998, 59)} b. {1960, 1970, 1980, 1985, 1990, 1995, 1998} c. {52, 54, 58, 59, 60} d. yes		
19) $(-\infty, 3)$ or $(3, \infty)$	21) $[3, 4)$ or $(4, \infty)$	23) $[0, \infty)$
25) $[-4, \infty)$	27) D: $(-\infty, \infty)$ R: $(-\infty, 0)$ or $[1, \infty)$	29) D: $[5, \infty)$ R: $[0, \infty)$
31) a. 1975 to 1980 b. 1980 to 1998 c. 16.2 d. 14.0 and 14.4		
33) a. 7.5 sec. b. 980 ft. c. (0, 7.5) d. (7.5, 15.3] e. [0, 15.3] f. [0, 980]		
35) a. $(1.5, \infty)$ b. $(-\infty, 1.5)$ c. none d. 9.75		
37) a. $(-\infty, -1.155)$ or $(1.155, \infty)$ b. $(-1.155, 1.155)$ c. 3.079 d. –3.079		
39) a. $(-2.55, 0)$ or $(2.55, \infty)$ b. $(-\infty, -2.55)$ or $(0, 2.55)$ c. –16 d. –26.25		
41) a. $(2.5, \infty)$ b. $(-\infty, 2.5)$ c. none d. 7	43) a. [0, 10) b. (10, 30] c. 10 minutes	
45) 28	47) undefined	49) 29
51) $\dfrac{5x-3}{2}$	53) $-a^2+2a-7$	55) $\lvert x\rvert$
57) $[4, \infty)$	59) $x(y)=\dfrac{y+1}{3}$	61) $t(r)=r^2-10r+22$
63) $r(s)=6s-11$	65) $\lvert t\rvert$	67) $4x-5+2h$
69) $l=\dfrac{3000}{w}$	71) $R=100x-x^2$	73) 897.00 dollars/widgets sold
75) no	77) yes	79) Not one–to–one function
81) yes	83) no, $f(g(x))=x$ but $g(f(x))=\lvert x\rvert$	85) $f^{-1}(x)=\dfrac{3x+1}{x-1}$, yes
87) a. $x(C)=\dfrac{C-3000}{500}$ b. 18 items		

CHAPTER 1 TEST

1) a. approx) 48% b. yes

3) $\sqrt{5}$

5) y is not expressed as a function of x.

7) $[-2,\infty)$

9) Domain: $(-\infty,\infty)$ Range: $[-1,\infty)$

11) a. 3.75 seconds b. 225 feet c. [0, 3.75) d. (3.75, 7.5] e. [0, 7.5] f. [0, 225]

13) increasing $(0, \infty)$; decreasing $(-\infty, 0)$; relative minimum = 65

15) increasing $(0, \infty)$; decreasing $(-\infty, 0)$; relative minimum = –16

17) $2x^3 - 7x^2 + 8x - 3$

19) –4

21) $-a^2 - 2a - 1$

23) $t(s) = \dfrac{3s+3}{s-2}$

25) $A(t) = 9\pi t^2$

27)

x	$f^{-1}(x)$
–9	–2
–4	–1
–1	0
0	1
7	2

29) $f^{-1}(x) = \dfrac{3x+1}{2x-1}$

31) $t(A) = \dfrac{A-10000}{1000}$

Chapter 2 Odd Homework Answers

SECTION 2.1

1) 0 + 5i	3) $0+2\sqrt{10}\ i$	5) 5+7i	7) $4+\sqrt{3}\ i$
9) 8+0i	11) When b is 0)	13) 13+5i	15) –5+3i
17) 8–2i	19) 15–10i	21) –9+18i	23) –10+0i
25) –24+0i	27) 25+19i	29) 0+13i	31) 14–8i
33) 13+0i	35) 100+0i	37) 1	39) i
41) –1	43) –i		

45) a. 3+2i b. 13+0i

47) a. 2–3i b. 13+0i

49) a. 0–3i b. 9+0i

51) $\dfrac{15}{29} - \dfrac{6}{29}i$

53) 0+i

55) $\dfrac{-21}{65} + \dfrac{12}{65}i$

57) $\dfrac{3}{5} - \dfrac{2}{5}i$

59) $\dfrac{30}{13} + \dfrac{7}{13}i$

61) $\dfrac{19}{27} + \dfrac{7\sqrt{2}}{27}i$

63) V = 46+20i

65) R = 5+3i

67) a. 3 iterations b. 5 iterations

SECTION 2.2

1) –1, 7	3) –3, –5/3	5) 2/3
7) $2\sqrt{2},\ -2\sqrt{2}$	9) –4, 4	11) i, –i
13) 2/3, 8/3	15) $\dfrac{-1+\sqrt{5}}{3}, \dfrac{-1-\sqrt{5}}{3}$	17) 1/2 + 3/2 i, 1/2 – 3/2 i

SECTION 2.2

19) $\frac{3+\sqrt{29}}{2}, \frac{3-\sqrt{29}}{2}$	21) $\frac{-5+\sqrt{33}}{4}, \frac{-5+\sqrt{33}}{4}$	23) 3 + i, 3 − i
25) 2 + 1/2 i, 2 − 1/2 i	27) $\frac{11}{6}+\frac{\sqrt{23}}{6}i, \frac{11}{6}-\frac{\sqrt{23}}{6}i$	29) −1, 8
31) $\frac{1}{3}+\frac{\sqrt{14}}{3}i, \frac{1}{3}-\frac{\sqrt{14}}{3}i$	33) $\frac{3+\sqrt{19}}{5}, \frac{3-\sqrt{19}}{5}$	35) −1/3, 1
37) $\frac{-5+4\sqrt{10}}{9}, \frac{-5-4\sqrt{10}}{9}$	39) $\frac{-9}{20}+\frac{\sqrt{119}}{20}i, \frac{-9}{20}-\frac{\sqrt{119}}{20}i$	41) $\frac{-7+\sqrt{57}}{2}, \frac{-7-\sqrt{57}}{2}$
43) $\frac{3}{4}+\frac{\sqrt{31}}{4}i, \frac{3}{4}-\frac{\sqrt{31}}{4}$	45) $\frac{7+\sqrt{145}}{12}, \frac{7-\sqrt{145}}{12}$	47) $-\sqrt{2}, \frac{-\sqrt{2}}{2}$
49) $\frac{-\sqrt{5}+\sqrt{29}}{4}, \frac{-\sqrt{5}-\sqrt{29}}{4}$	51) $-i, \frac{-2}{3}i$	53) −)548, 2)00
55) −1)172, −)164	57) $x^2 + 8x + \underline{16}$	59) $x^2 - 5x + \frac{25}{4}$
61) $2(x^2 + 6x + \underline{9})$	63) $-(x^2 - 16x + \underline{64})$	65) $2(x^2 - \frac{9}{2}x + \frac{81}{16})$

67) 200 feet by 600 feet

69) 20 feet by 26 feet

71) a. 10” by 10” by 6” b. 15” by 15” by 6”

73) One car is 30 feet from the intersection and the other 40 feet from the intersection

75) a. 2 seconds or 6 seconds b. 3 seconds or 5 seconds c. 4 seconds d. time= 4 seconds; height= 262 feet e. approximately 8.05 seconds

77) a. 30 widgets or 170 widgets b. 40 widgets or 160 widgets c. 100 widgets d. 100 widgets; Maximum profit= $9,900

79) a. 14” by 14” piece of cardboard b. $2\frac{1}{3}$ ”

SECTION 2.3

1) 13	3) $\frac{13}{2}$	5) No Solution	7) −3, −1
9) 3	11) $\frac{-1-\sqrt{5}}{2}, \frac{-1+\sqrt{5}}{2}$	13) 3	15) No Solution
17) 3	19) 1, −3	21) 2	23) 4
25) 3, −1	27) 7	29) $9+2\sqrt{17}$	31) $\frac{-1+\sqrt{17}}{2}$
33) 1	35) 4	37) 64, −64	39) 256
41) 64, −64	43) 1	45) $-11, \frac{31}{3}$	47) $\frac{1}{32}$

SECTION 2.3

49) $\frac{1}{125}, \frac{-1}{125}$	51) $\frac{17}{48}$	53) 0, 1	55) 0, 32
57) 0, 243, –243	59) $4(\sqrt[3]{4})$	61) $16\sqrt{2}, -16\sqrt{2}$	63) 16, 0
65) (1, 2) and (3, 6)	67) 225 chairs	69) Either 4 miles or $\frac{7}{8}$ miles down the shoreline	
71) 16 seconds			

SECTION 2.4

1) $\frac{11}{9}$	3) $\frac{-57}{25}$	5) 11	7) –1, 4
9) –2, 3	11) No Solution	13) 2	15) No Solution
17) 2	19) No Solution	21) $\frac{-1}{5}, 2$	23) $\frac{-11}{2}, 2$
25) –8	27) –4, –1	29) –10	31) No Solution
33) $\frac{-17 \pm \sqrt{57}}{4}$	35) $\frac{-19 \pm \sqrt{505}}{6}$	37) $-5, \frac{-1}{3}$	39) $\frac{-9 \pm \sqrt{321}}{10}$
41) $\pm 2, \pm 1$	43) $\pm 2i, \pm \sqrt{7}$	45) ±2i, ±4i	47) $\pm 2, \pm \sqrt{6}$
49) $\pm 2, \pm i\sqrt{5}$	51) $\pm 1, \pm \frac{5}{2}$	53) $\pm 1, \pm \frac{2}{5}i$	55) $\pm i, \pm \frac{2}{9}i$
57) $\pm \frac{5\sqrt{3}}{3}, \pm \frac{\sqrt{2}}{2}$	59) 16, 81	61) $16, \frac{625}{81}$	63) 9
65) ±8, ±27	67) 16	69) $\frac{1}{5}, \frac{1}{3}$	71) $\frac{-1}{4}, \frac{-1}{6}$
73) $\pm \frac{1}{3}, \pm \frac{1}{2}$	75) $\frac{1}{27}, \frac{-1}{512}$	77) $\frac{1}{25}$	79) –8
81) $\pm \sqrt{\frac{3+\sqrt{17}}{2}}, \pm \sqrt{\frac{\sqrt{17}-3}{2}}\, i$		83) $\pm \sqrt{\frac{1+\sqrt{7}}{2}}, \pm \sqrt{\frac{\sqrt{7}-1}{2}}\, i$	
85) 1, 2	87) –4, –1	89) $-2, \sqrt[3]{5}$	91) 24 hours
93) 15 hours or 30 hours		95) 100 items or 8,100 items	

SECTION 2.5

1) (–∞, –1)	3) (45/2, ∞)	5) [43/15, ∞)	7) [25, ∞)
9) {20/3, –10/3}	11) {8/3, –2}	13) No Solution	15) 3
17) {8, –2/3}	19) (–∞, –5) or (19/3, ∞)	21) (–11/2, 5/2)	23) (–∞, –10] or [7, ∞)
25) [–2, 5]	27) No Solution	29) (–∞, ∞)	31) (–∞, ∞)
33) (–∞, –4] or [7, ∞)	35) (–∞, –9) or (4, ∞)	37) [–5, 5]	39) (–9, 3)

SECTION 2.5

41) $(-\infty, \infty)$	43) No Solution	45) $(-\infty, -2-\sqrt{2}\,]$ or $(-2+\sqrt{2}, \infty)$
47) $\left(\frac{-3-\sqrt{29}}{2}, \frac{-3+\sqrt{29}}{2}\right)$	49) $(-\infty, -2)$ or $[3, \infty)$	51) $(-1, 5)$
53) $(-\infty, -3)$ or $(-2, 3)$ or $(5, \infty)$	55) $(-6, -4]$ or $[-3, 3)$	
57) $\left(-\infty, \frac{-3-\sqrt{5}}{2}\right]$ or $\left(-\sqrt{3}, \frac{-3+\sqrt{5}}{2}\right]$ or $\left(\sqrt{3}, \infty\right)$	59) $(-4, 4)$	
61) $(3, 8]$	63) $(-\infty, 5/2)$ or $(3, \infty)$	65) $(-\infty,-3)$ or $[-1, 2)$
67) $(-\infty, -2]$ or $(-1, 1/2]$ or $(2, \infty)$	69) $(-4, -2]$ or $(0, 1]$ or $(2, 3]$	71) $[-2, 1]$ or $[5, \infty)$
73) $[-2, -5/3]$ or $[1, 3/2]$	75) From 5 years to 7 years	77) From 3 seconds to 17 seconds
79) From 200 go-carts to 400 go-carts		

CHAPTER 2 REVIEW

1) $0 + 7i$	3) $7-5i\sqrt{2}$	5) $19 + 8i$
7) $15 + 9i$	9) 145	11) $1 - 41i$
13) $\frac{16}{29}+\frac{11}{29}i$	15) i 16) −i	17) −2, 5
19) $\frac{\sqrt{30}}{3}, \frac{-\sqrt{30}}{3}$	21) $\frac{3+\sqrt{7}}{5}, \frac{3-\sqrt{7}}{5}$	23) $\frac{-9+\sqrt{21}}{10}, \frac{-9-\sqrt{21}}{10}$
25) $\frac{-1}{3}+\frac{i\sqrt{11}}{3}, \frac{-1}{3}-\frac{i\sqrt{11}}{3}$	27) $\frac{-7\sqrt{5}+\sqrt{145}}{10}, \frac{-7\sqrt{5}-\sqrt{145}}{10}$	29) −0.425, 1.142
31) 16 inches by 36 inches	33) 9 km or 12 km	
35) a. $R(x) = -10x^2 + 1050x$ b. Rent 16 homes @ \$890 c. 20 at \$850, Revenue is \$17,000		
37) $\frac{97}{7}$	39) 1	41) −3, −2
43) 1, 5	45) $22-4\sqrt{31}$	47) No Solution
49) 243	51) $\frac{17}{3}$	53) $\frac{-1}{64}, \frac{1}{64}$
55) $\frac{5}{9}$	57) 100 shoes	
59) No Solution; Domain: All real numbers except −3 and 3		
61) $\frac{-13+\sqrt{73}}{6}, \frac{-13-\sqrt{73}}{6}$; Domain: All real numbers except −3 and −2		
63) $3+\sqrt{5}, 3-\sqrt{5}$; Domain: All real numbers except 2 and 3		
65) $\frac{-8+\sqrt{10}}{3}, \frac{-8-\sqrt{10}}{3}$; Domain: All real numbers except −3 and −2		

CHAPTER 2 REVIEW

67) $\frac{-1}{3}$; Domain: All real numbers except −2, 1, 3

69) −2, 2, −5, 5

71) $-2i, 2i, -i\sqrt{5}, i\sqrt{5}$

73) $\frac{-3}{2}, \frac{3}{2}, \frac{-i\sqrt{3}}{2}, \frac{i\sqrt{3}}{2}$

75) 8, 64

77) $\frac{-1}{3}, \frac{-1}{7}$

79) Approx. 46.5 hours

81) $\left[\frac{9}{8}, \infty\right)$

83) $(-\infty, -2)$ or $(5, \infty)$

85) No Solution

87) $[-3, 5]$

89) No Solution

91) $(-\infty, -5)$ or $(0, 8)$

93) $\left(-\infty, \frac{-5}{2}\right)$ or $\left[\frac{1}{3}, \infty\right)$

95) $\left(\frac{-9}{2}, -2\right)$

97) $(-\infty, -3]$ or $(-1, 0)$ or $[1, 2)$ or $[3, \infty)$

99) $\left(-2-\sqrt{3}, -\sqrt{3}\right]$ or $\left(-2+\sqrt{3}, \sqrt{3}\right]$

CHAPTER 2 TEST

1) a. $10 - 7i$ b. $-5 + 11i$ c. $27 - 11i$ d. $\frac{-3}{13} + \frac{11}{13}i$ e. $-3 - 3i$

3) −366, 1.052

5) $\frac{1+\sqrt{5}}{2}, \frac{1-\sqrt{5}}{2}$

7) 4

9) $\frac{40}{27}, \frac{41}{27}$

11) $\frac{-12+\sqrt{19}}{5}, \frac{-12-\sqrt{19}}{5}$

13) $3, -3, -i\sqrt{6}, i\sqrt{6}$

15) $\frac{-1}{4}, \frac{1}{5}$

17) $(-4, -1)$

19) $(-\infty, -4)$ or $(9, \infty)$

21) No Solution

23) $(-\infty, 3)$ or $(8, \infty)$

25) a. 800 or 1200 umbrellas

27) 7.12 hours for one machine and 9.12 for the other machine

Chapter 3 Odd Homework Answers

SECTION 3.1

1) LINEAR

1	9	
2	−4	−13
3	−17	−13
4	−30	−13
5	−43	−13

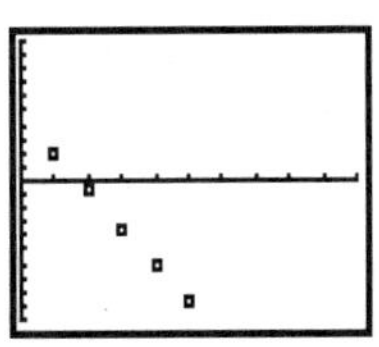

3) QUADRATIC

1	−2		
2	7	9	
3	22	15	6
4	43	21	6
5	70	27	6

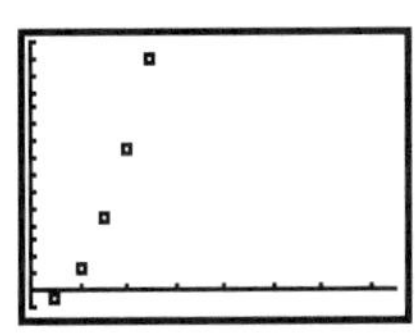

SECTION 3.1

5) CONSTANT

3	−7	
4	−7	0
5	−7	0
6	−7	0
7	−7	0

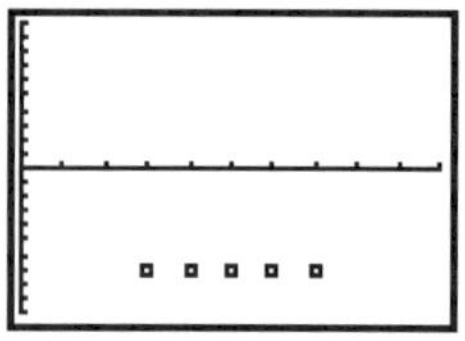

7) LINEAR

5	17	
0	2	−15
−5	−13	−15
−10	−28	−15
−15	−43	−15

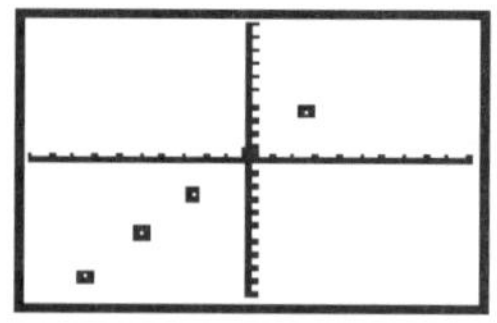

9) NONE

0	1		
1	3	2	
2	9	6	4
3	27	18	12
4	81	54	36

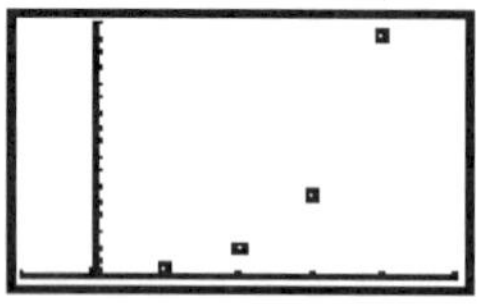

11) QUADRATIC

0	−3		
2	1	4	
4	13	12	8
6	33	20	8
8	61	28	8

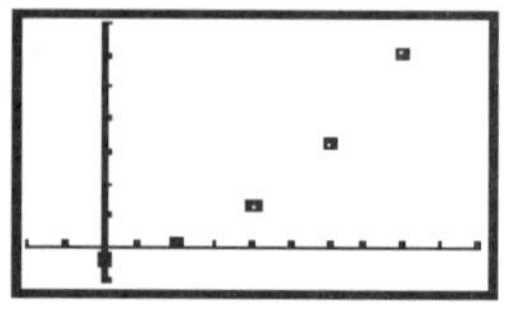

13)

0	−2		
1	1	3	
2	4	3	
3	7	3	
4	10	3	
5	13	3	

LINEAR

15)

0	3		
1	1	−2	
2	−5	−6	−4
3	−15	−10	−4
4	−29	−14	−4
5	−47	−18	−4

QUADRATIC

17)

0	−8		
1	−8	0	
2	−8	0	
3	−8	0	
4	−8	0	
5	−8	0	

CONSTANT

19)

0	1		
1	−1	−2	
2	3	4	6
3	13	10	6
4	29	16	6
5	51	22	6

QUADRATIC

21)

0	−4		
1	$-3\frac{1}{3}$	$\frac{2}{3}$	
2	$-2\frac{2}{3}$	$\frac{2}{3}$	
3	−2	$\frac{2}{3}$	
4	$-1\frac{1}{3}$	$\frac{2}{3}$	
5	$-\frac{2}{3}$	$\frac{2}{3}$	

LINEAR

23) LINEAR

0	15		
1	13	−2	
2	11	−2	
3	9	−2	

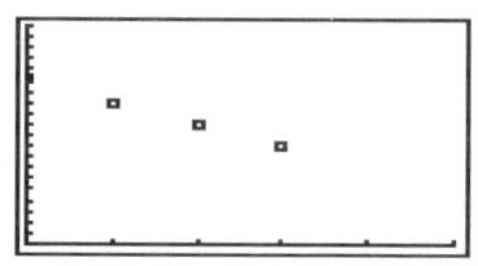

25) QUADRATIC

0	0		
2	448	448	
4	768	320	-128
6	960	192	-128
8	1024	64	-128

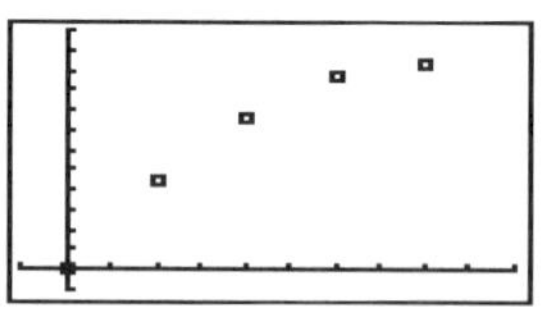

27) LINEAR

10	7,00		
20	12,000	5,000	
30	17,000	5,000	
40	22,000	5,000	
50	27,000	5,000	

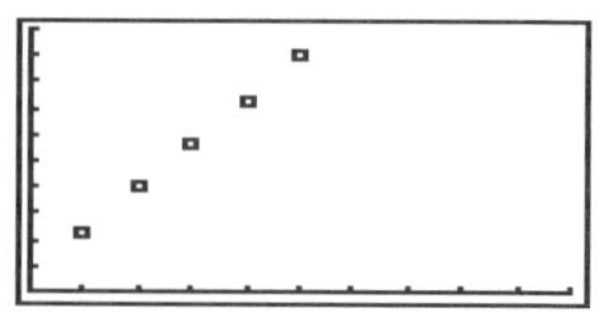

29) QUADRATIC

1	480		
2	768	288	
3	864	96	−192
4	768	−96	−192
5	480	−288	−192

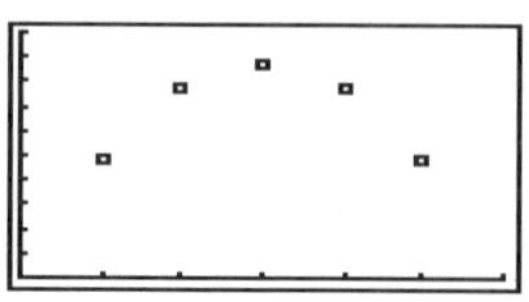

31) CONSTANT

1	1,240)28		
2	1,240)28	0	
3	1,240)28	0	
4	1,240)28	0	
5	1,240)23	0	

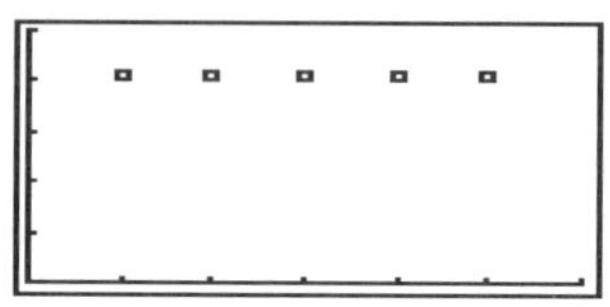

SECTION 3.1

33) NONE

10	100,000		
30	85,441	−14,559	
50	69,804	−15,637	−1,078
70	38,569	−31,235	−15,598
90	847	−37,722	−6,487

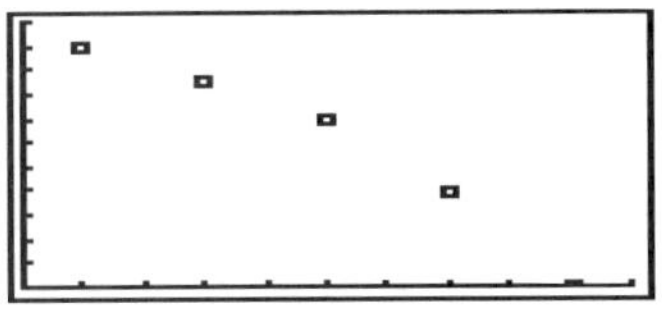

SECTION 3.2

1) a. Can be a polynomial b. Even

3) a. Can be a polynomial b. Odd

5) a. Not a polynomial

7) a. Can be a polynomial b. Odd

9) Not a polynomial

11) a. Can be a polynomial b. Even

13) a. Can be a polynomial b. Even

17) a. 1 b. 2 c. at the top d. at the top

e. 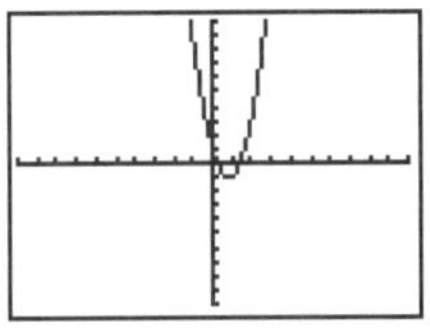 f. inc.(.833,∞) dec. (−∞,.833) min. = -1.083

19) a. 0, 2 b. 3 c. at the bottom d. at the top

e. 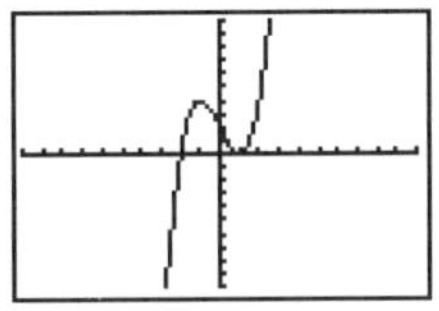 f. inc. (−∞,−1) or (1,∞) dec. (−1,1) min = 1 max = 3.37

21) a. 0, 2 b. 3 c. at the top d. at the bottom

e. 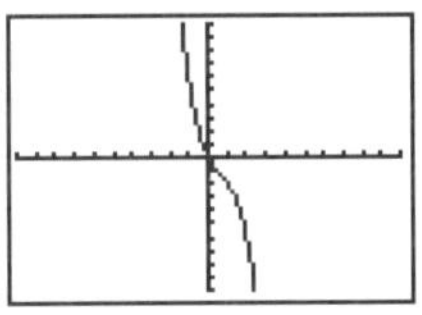 f. dec. (−∞, ∞) No max. or min

23) a. 1–3 b. 4 c. at the bottom d. at the bottom

e. 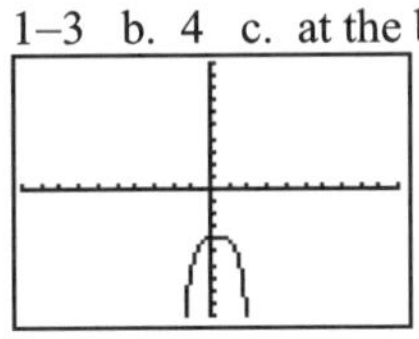 f. inc (−∞, .75) dec (.75, ∞) max = −3.89

25) a. 1–3 b. 4 c. at the top d. at the top

e. 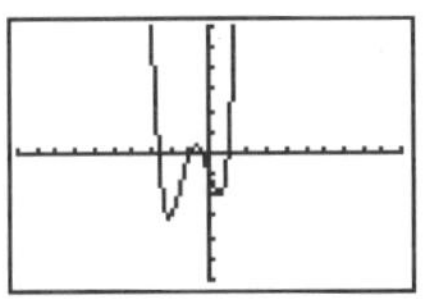 f. inc. (−2.076, −.694) or (.520, ∞) min = −2.065 dec. (−∞, -2.076) or (−.694, .520) max = 0.311

27) a. 0–4 b. 5 c. at the bottom d. at the top

e. 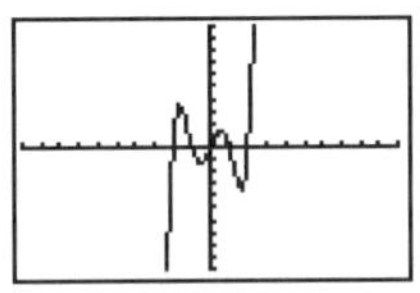 f. inc. (−∞, −1.644) or (−.544, .544) or (1.644, ∞) dec. (−1.644, −.544) or (.544, 1.644) min. = −1.142 and −3.631 max. = 3.631 and 1.142

29) a. 0–4 b. 5 c. at the bottom d. at the top

e. 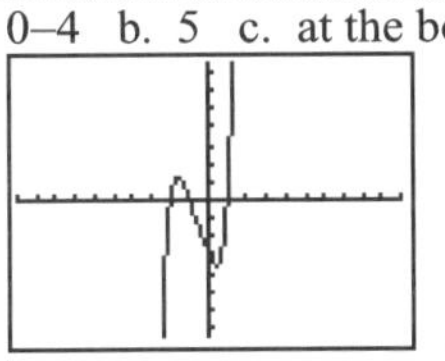 f. inc. (−∞, −1.612) and (.390, ∞) dec. (−1.612, .390) max = 2.473 min = −7.178

31) 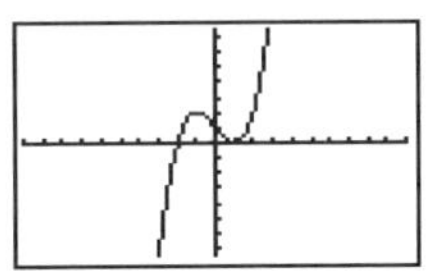

SECTION 3.2

33) 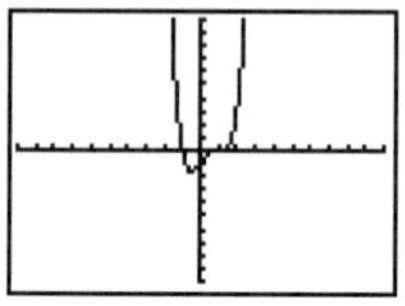	35) 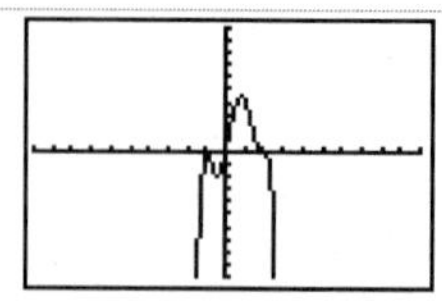	37) 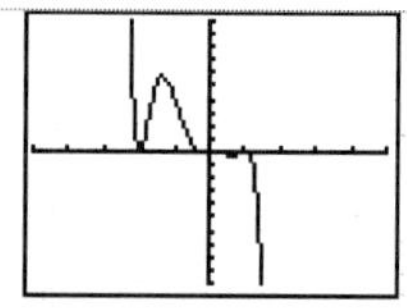
39) $f(x) = -(x-2)^2(x+3)$ 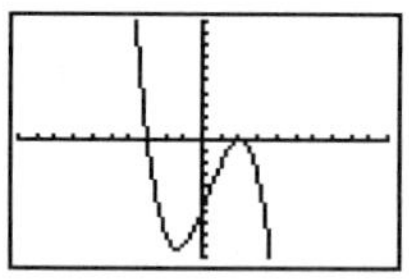	41) $f(x) = (x-3)^3(x-1)^2$ 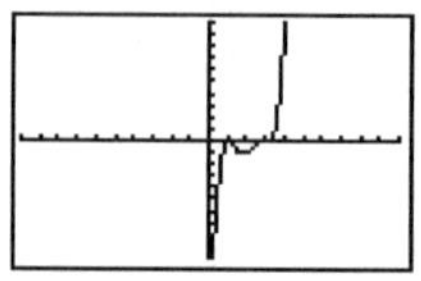	43) $f(x) = -x(x+3)^3(x-2)$ 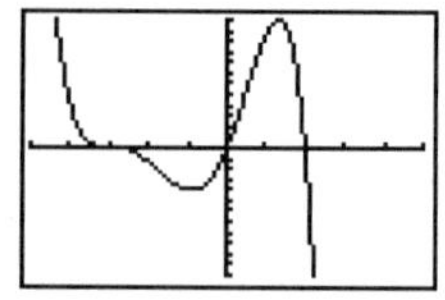
45. odd	47. even	49. neither
51. even	53. neither	55. odd

SECTION 3.3

1) $x^2 - 4x + 11 - \dfrac{24}{x+2}$	3) $x^3 + 2x + 1 + \dfrac{2}{x-3}$
5) $2x^2 - 3x + 12 - \dfrac{52}{x+5}$	7) $x^3 + 3x^2 + 6x + 18 + \dfrac{57}{x-3}$
9) $x^3 + x^2 + x + 1 - \dfrac{18}{x-1}$	11) $x^7 - x^3 + 4x^2 - 16x + 64 - \dfrac{251}{x+4}$
13) $3x^3 - 3x^2 - 9x - 12 - \dfrac{8}{x - \frac{4}{3}}$	15) $4x^2 - 3x + \dfrac{9}{2} - \dfrac{\frac{7}{4}}{x + \frac{3}{2}}$

17) -7	19) -62	21) 29
23) 9	25) -3	27) 11
29) $(x+5)(3x^2 - 5x + 3)$	31) $(x-3)(x^3 - 3x^2 + x - 2)$	33) Not a factor
35) $(x-5)(x^4 - x^3 + 2x^2 - x + 3)$	37) $(x+5)(x^3 - 5x^2 - 9x + 45)$	39) Not a factor
41) $(x - \frac{1}{2})(6x^2 - 10x + 6)$	43) $(x - \frac{3}{2})(2x^3 - 4x + 6)$	45) $(x-2i)[x^3 + (-1+2i)x^2 + (-2-2i)x - 4i]$

47) $(x-1+2i)[x^3 - 2ix^2 + (1-2i)x + (-2-4i)]$

49) 6

51) When n is odd $(-1)^n = -1$ and $x^n + 1 = 0$. Therefore by the remainder theorem $x + 1$ is a factor.

53) Because you are adding and all the exponents are even, the answer you get when you plug in any value of c will always be positive and therefore it would be impossible to have a remainder of 0.

SECTION 3.4

1) $\pm 1, \pm 2, \pm 5, \pm 10$	3) $\pm 1, \pm 3, \pm 5, \pm 15, \pm\frac{1}{5}, \pm\frac{3}{5}$

SECTION 3.4

5) $\pm1, \pm2, \pm4, \pm8, \pm16, \pm\frac{1}{6}, \pm\frac{1}{3}, \pm\frac{2}{3}, \pm\frac{4}{3}, \pm\frac{8}{3}, \pm\frac{1}{2}, \pm\frac{16}{3}$

7) 1, 3, –2; $(x-1)(x-3)(x+2)$

9) $5, -2\pm\sqrt{2}$; $(x-5)(x-(-2+\sqrt{2}))(x-(-2-\sqrt{2}))$

11) $-4, -1\pm 2i$; $(x+4)(x-(-1+2i))(x-(-1-2i))$

13) –3, –1, 3, 5; $(x+3)(x+1)(x-3)(x-5)$

15) $-3, -1, -2\pm\sqrt{3}$; $(x+3)(x+1)(x-(-2+\sqrt{3}))(x-(-2-\sqrt{3}))$

17) –2, 3, $-1\pm2i$; $(x+2)(x-3)(x-(-1+2i))(x-(-1-2i))$

19) $\frac{-1}{2}, \frac{1}{2}, \frac{2}{3}$; $(x+\frac{1}{2})(x-\frac{1}{2})(x-\frac{2}{3})$

21) $\frac{-2}{3}, \frac{3\pm\sqrt{17}}{2}$; $(x+\frac{2}{3})(x-\frac{3+\sqrt{17}}{2})(x-\frac{3-\sqrt{17}}{2})$

23) $\frac{1}{4}, \frac{-3\pm i\sqrt{23}}{4}$; $(x-\frac{1}{4})(x-\frac{-3+i\sqrt{23}}{4})(x-\frac{-3-i\sqrt{23}}{4})$

25) –2, 3; $(x+2)(x-3)^2$

27) –1, 2; $(x+1)^2(x-2)$

29) –3, 2; $(x+3)^2(x-2)^2$

31) $-3, \frac{3\pm\sqrt{5}}{2}$; $(x+3)^2(x-\frac{3+\sqrt{5}}{2})(x-\frac{3-\sqrt{5}}{2})$

33) 2, $-2\pm4i$; $(x-2)^2(x-(-2+4i))(x-(-2-4i))$

35) $3, \frac{-3\pm\sqrt{13}}{2}$; $(x-3)^3(x-\frac{-3+\sqrt{13}}{2})(x-\frac{-3-\sqrt{13}}{2})$

37) $\frac{-3}{4}, \frac{1}{3}$; $(x+\frac{3}{4})^2(x-\frac{1}{3})^2$

39) $\frac{-5}{3}, \frac{-1}{2}, \frac{3}{4}$; $(x+\frac{5}{3})(x+\frac{1}{2})(x-\frac{3}{4})$

41) x^3-6x^2-x+30

43) $x^4-2x^3-13x^2+14x+24$

45) x^3-3x^2+4

47) $x^4-8x^3-128x-256$

49) x^4-8x^2+16

51) $x^3-8x^2+16x-8$

53) $x^3-2x^2-14x+40$

55) $x^4-14x^3+82x^2-246x+377$

57) x^3-4x^2-x+22

61) –3.951

63) –1.569, 2.290

65) 4.498

67) –5.656, 0.684, 2.627

69) No solution

71) a. None b. All answers are complex c. $2\pm i$, $1\pm2i$

73) $(-\infty, -3)$ or $(2, 4)$

75) $(-\infty, -3]$ or $[-1, 1]$ or $[5, \infty)$

77) $(-\infty, -3)$

79) $(-\infty, \infty)$

81) 40 air conditioners

83) a. $R(x) = 1800x$ b. 80 or 97 air conditioners c. 89 air conditioners

85) 3 feet x 5 feet x 6 feet

SECTION 3.5

1) a. $y = 0$ b. $x = 4$ c. none d. $(-\infty, 4)$ or $(4, \infty)$

3) a. $y = -3/5$ b. $x = -6/5$ c. (0,0) d. $(-\infty, -6/5)$ or $(-6/5, \infty)$

5) a. $y = 3$ b. $x = 0$ c. (2/9, 0) d. $(-\infty, 0)$ or $(0, \infty)$

7) a. $y = 1$ b. $x = -2$ c. (3, 0) d. $(-\infty, -2)$ or $(-2, \infty)$

9) a. $y = 0$ b. $x = 2$ and $x = -2$ c. $(\frac{3}{2}, 0)$ d. $(-\infty, -2)$ or $(-2, 2)$ or $(2, \infty)$

11) a. $y = 2$ b. $x = -1$ and $x = 5/2$ c. (–2, 0) and (2, 0) d. $(-\infty, -1)$ or $(-1, 5/2)$ or $(5/2, \infty)$

13) a. $y = 0$ b. $x = -2$ and $x = 5$ c. No x–intercept d. $(-\infty, -2)$ or $(-2, 5)$ or $(5, \infty)$

15) a. $y = 2/5$ b. No vertical asymptotes c. $(\frac{-\sqrt{2}}{2}, 0)$ and $(\frac{\sqrt{2}}{2}, 0)$ d. $(-\infty, \infty)$

17) a. $y = 3$ b. $x = -4$ c. $(\sqrt[3]{9}, 0)$ d. $(-\infty, -4)$ or $(-4, \infty)$

SECTION 3.5

19) $f(x) = \dfrac{2x+6}{x-2}$

21) $f(x) = \dfrac{x+2}{x^2+2x-3}$

23) $f(x) = \dfrac{2(x+2)(x-3)}{(x+3)(x-1)}$

25) $f(x) = \dfrac{3(x-2)^2}{x^2+1}$ **(denominator can be any 2nd degree polynomial with no real number roots and a leading coefficient of 1)**

27) $f(x) = \dfrac{1}{(x+2)(x-2)}$ **(Numerator can be any real number)**

29) $f(x) = \dfrac{x-2}{x+1}$

31) $f(x) = \dfrac{-2(x+1)}{x+3}$

33) $f(x) = \dfrac{(x+1)(x-2)}{(x-3)(x+2)}$

35) $y = x - 1$

37) $y = 3x - 11$

39) $y = x^2 + 2x + 5$

41) (–2, –4)

43) (–5, –1/6)

45) a. y = 0 b. x = 5/2 c. none d. none e.

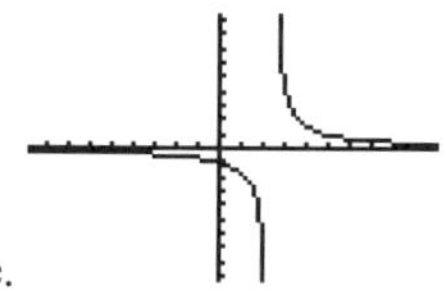

47) a. y = 2/3 b. x = 2/3 c. none d. none e.

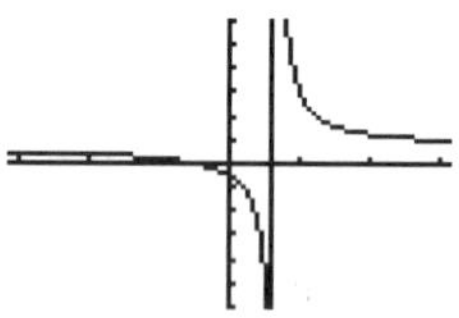

49) a. y = 2/3 b. x = –1 and x = 2/3 c. none d. none e.

51) a. none b. none c. y = 1/3 x + 2/3 d. (2, 4/3) e.

53) a. none b. x = –3 c. y = –2x + 8 d. none e.

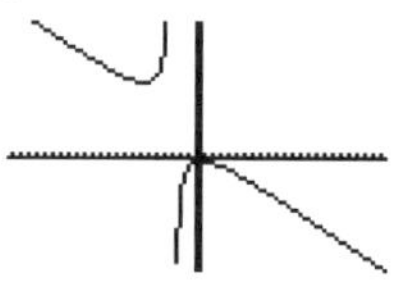

55) a. y = 1 b. x = 1 c. none d. (3, 3) e.

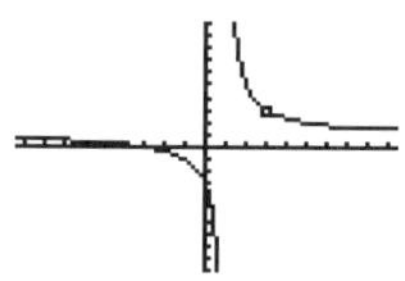

57) a. y = 1 b. x = –1 and x = 2 c. none d. (–2, 15/4) e.

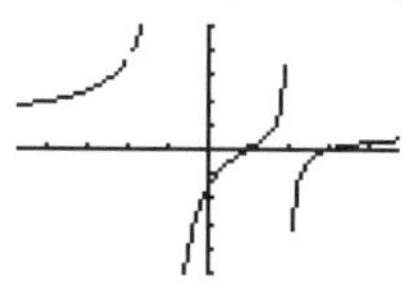

59) a. none b. x = 2 c. y = x + 4 d. (–3, –3/5) e.

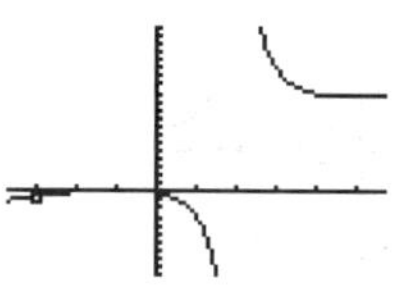

61) a. y = 0 b. x = –2 and x = 2 c. none d. (1, –1) e.

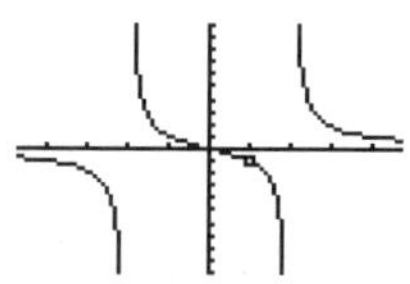

63) If the denominator of the rational polynomial function is not equal to zero for any real number value.

65) The degree of the numerator must be greater than the degree of the denominator.

67) The numerator and denominator have the same degree and the ratio of the leading coefficients is 3.

69) If the numerator and denominator have a common factor.

71) The numerator factors to a perfect square, the value that makes the numerator zero has a multiplicity of two.

CHAPTER 3 REVIEW

1) Linear

3) Constant

5) a. Linear

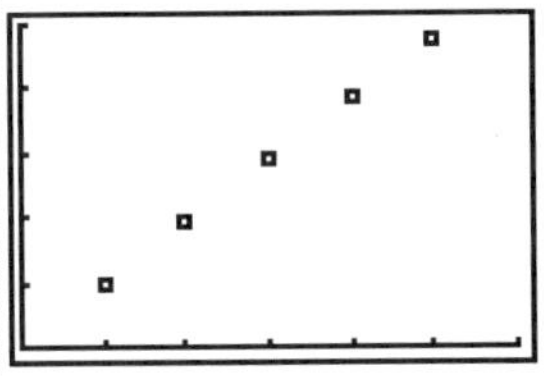

b. Not linear is quadratic

10	495		
20	980	485	
30	1455	475	−10
40	1920	465	−10
50	2375	455	−10

7) **Consider Monday as day 1, Tuesday as day 2, Wednesday as day 3, Thursday** as **day 4, and Friday as day 5.**

a. Constant

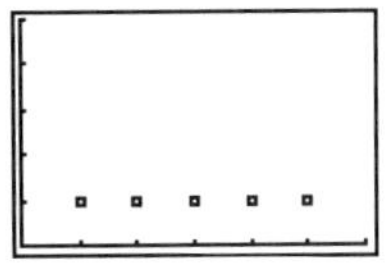

b. Constant

1	1	
2	1	0
3	1	0
4	1	0
5	1	0

9) Yes, even

11) No

13) No

15) a. 2, 0 b. 3 c. on top d. at bottom

17) a. 4, 0 b. 5 c. at bottom d. on top

19) 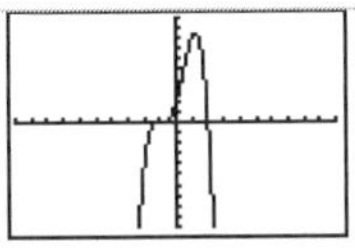

21) $f(x) = (x-3)(x+2)^2$

23. $f(x) = -x^3(x-2)^2(x+2)$

25) even

27) $2x^2 - x - \dfrac{3}{x-2}$

29. $x^3 - 2x^2 + 4x - 8 + \dfrac{8}{x+2}$

31) 9

33) −8

35. $(x-5)(x^2+2x-3)$

37) Not a factor

39) $\pm 1, \pm 2, \pm 4, \pm\frac{1}{3}, \pm\frac{2}{3}, \pm\frac{4}{3}$

41) $\pm 1, \pm 3, \pm 5, \pm 15,$

$\pm\frac{1}{2}, \pm\frac{3}{2}, \pm\frac{5}{2}, \pm\frac{15}{2}, \pm\frac{1}{3}, \pm\frac{5}{3}, \pm\frac{1}{6}, \pm\frac{5}{6}$

43)) 2, 3, −3; $(x-2)(x-3)(x+3)$

45) 2, $2 \pm \sqrt{3}$; $(x-2)(x-2+\sqrt{3})(x-2-\sqrt{3})$

47) −3, 1; $(x+3)(x-1)^2$

49) 1, $\frac{-2}{3}$; $(x-1)^3\left(x+\frac{2}{3}\right)$

51) 2, 2± i; $(x-2)(x-2+i)(x-2-i)$

53) 2±3i, 1±2i; $(x-2-3i)(x-2+3i)(x-1-2i)(x-1+2i)$

55) $x^4+2x^3-13x^2-14x+24$

57) $x^3-8x^2+22x-20$

59) 3.196

61) a. 100 b. 75 tool kits c. $2,926.25

63) a. x=6, x= −1 b. y=0 c. none d. none e. $\left(\frac{1}{3}, 0\right)$ f. all reals except −1 and 6

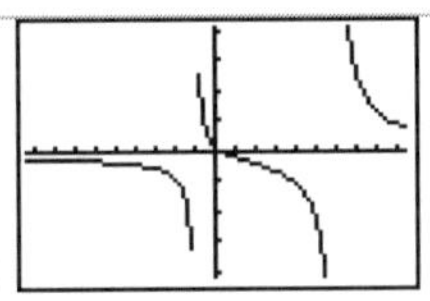

65) a. x=2 b. y=1 c. none d. (3, 6) e. (−3,0) f. all reals except 3 and −3

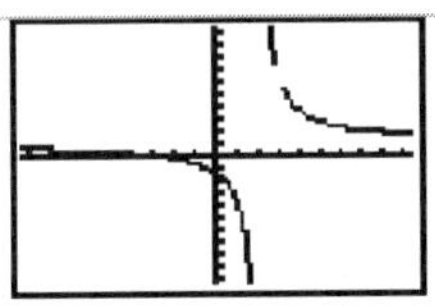

CHAPTER 3 REVIEW

67) $f(x) = \dfrac{-(x-2)}{x+3}$	69) $f(x) = \dfrac{x^3}{(x+2)(x-2)}$ numerator any power greater than two
71) a. x=–1 b. y=1 c. (3,0)	73) a. x=3, x=–3 b. y=–2 c. (2,0), (–2,0)

CHAPTER 3 EXAM

1) a. Linear	b. Quadratic	c. Constant	d. Neither
3) a. two	b. three	c. enters at the top	d. exits at the top

5) $-x^2(x+3)^2(x-2)$

7) a. f(3) = 57 b. f(–3) = 158

3	3	–5	12	-15
		9	12	72
	3	4	24	57

–3	2	0	0	9	23
		–6	18	–54	135
	2	–6	18	–45	158

9) a. $3, \dfrac{-5+\sqrt{17}}{2}, \dfrac{-5-\sqrt{17}}{2}$ Factor $(x-3)(x^2+5x+2)$ b. $\dfrac{2}{3}, 1+2i, 1-2i$ Factor $(3x-2)(x^2-2x+5)$

c. –2, 1 Factor $(x+2)^2(x-1)^3$

11. A. a. $y = 2$ b. $x = -3$ c. none d. none e. $\left(\dfrac{1}{2}, 0\right)$ f. All real nos. except –3

B. a. $y = 0$ b. $x = -2$ c. none d. $\left(2, \dfrac{1}{4}\right)$ e. none f. All real nos. except –2 and 2

C. a. none b. $x = -2$ c. $y = x - 5$ d. none e. (4, 0) and (–1, 0) f. All real nos. except –2

13. $\dfrac{2x^2-2x-4}{x^2+2x-3}$

Chapter 4 Odd Homework Answers

SECTION 4.1

1) Exponential function	3) Exponential function	5) Linear function
7) Neither	9) Exponential function	11) Linear function
13) Linear function	15) Exponential function	17) Linear function

19) Exponential function

21) a. 1.08,1.08, 1.08 b. 1.08 c. $A(t) = 5000(1.08)^t$ d. \$23,304.79 e. Growth

23) a. 1.1, 1.1, 1.1 b. 1.1 c. $F(t) = 100{,}000(1.1)^t$ d. 161,051 fish e. Growth

25) a. 0.96, 0.96, 0.96 b. 0.96 c. $P(t) = 1{,}000{,}000(0.96)^t$ d. 721,390 people e. Decay

27) a. 0.8, 0.8, 0.8 b. 0.8 c. $S(t) = 150(0.8)^t$ d. 16,106 whales e. Decay

29) a. $f(x) = 2^x$ b. Growth	31) a. $f(x) = (0.5)^x$ b. Decay	33. a. $f(x) = 2(3)^x$ b. Growth

SECTION 4.2

1)

x	–1	0	1	2	3
f(x)	1/2	1	2	4	8

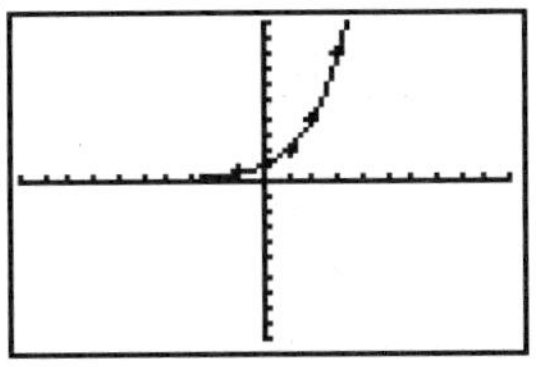

3)

x	–1	0	1	2	3
f(x)	4	1	1/4	1/16	1/64

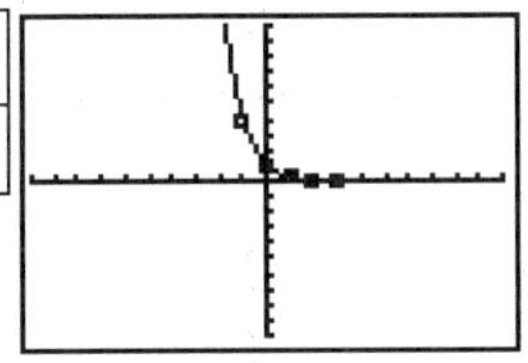

5)

x	–1	0	1	2	3
f(x)	4	1	1/4	1/16	1/64

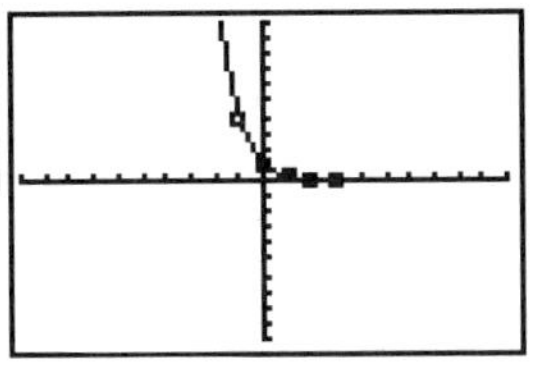

7)

x	–1	0	1	2	3
f(x)	7/2	4	5	7	11

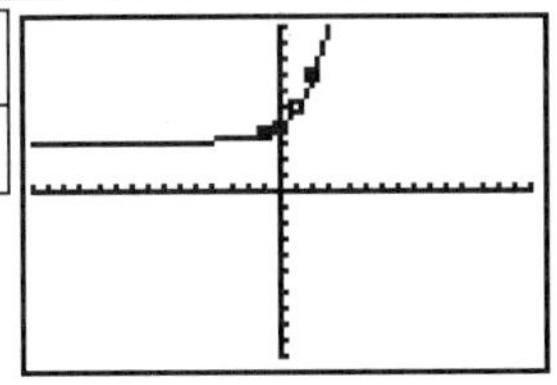

9)

x	–1	0	1	2	3
f(x)	2	0	–2/3	–8/9	–26/27

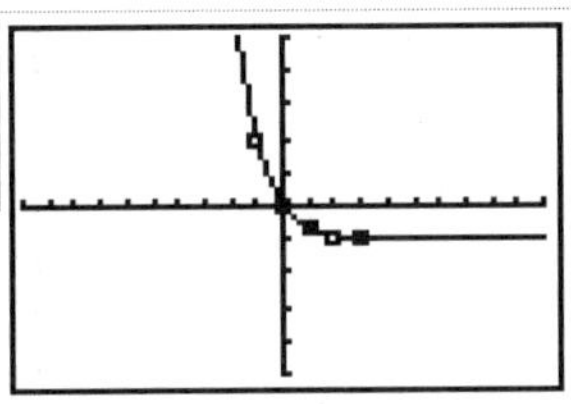

11)

x	–1	0	1	2	3
f(x)	3/2	3	6	12	24

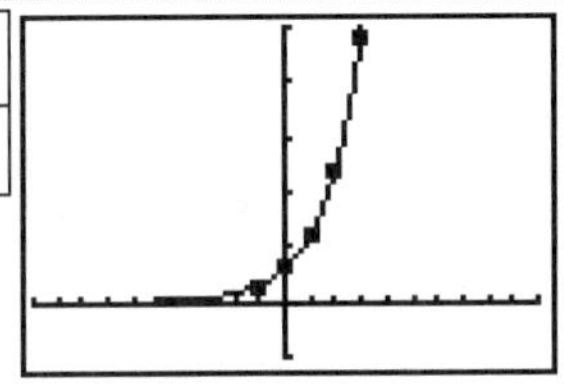

13)

x	–1	0	1	2	3
f(x)	1	2	4	8	16

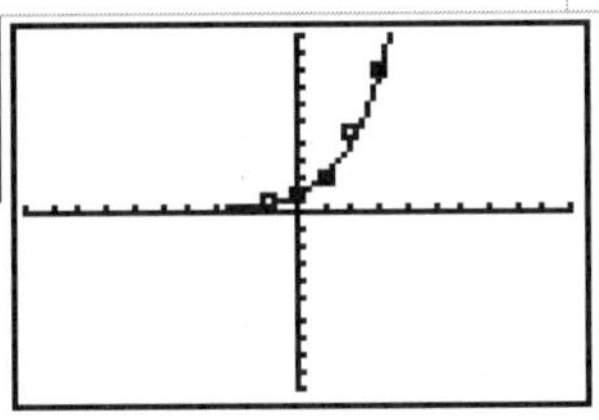

15) a. Decay b. All real numbers c. $(0, \infty)$ d. (0, 20)

17) a. Growth b. All real numbers c. $(0, \infty)$ d. (0, 50)

19) a. Growth b. All real numbers c. $(0, \infty)$ d. (0, 5000)

21) a. Decay b. All real numbers c. $(0, \infty)$ d. (0, 1)

23) a. Decay b. All real numbers c. $(5, \infty)$ d. (0, 6)

25) a. Growth b. All real numbers c. $(-3, \infty)$ d. (0, –2)

27) a. Growth b. All real numbers c. $(10, \infty)$ d. (0, 20)

29) 4

31) 6

33) –2

35) –4

37) 0

39) 1

41) No solution

43) a. –1 b. 0 c. 1 d. 2 e. 3

45) a. –1 b. 0 c. 1 d. 2 e. 3

47) a. –1 b. 0 c. 1 d. 2 e. 3

49) a. –1 b. 0 c. 1 d. 2 e. 3

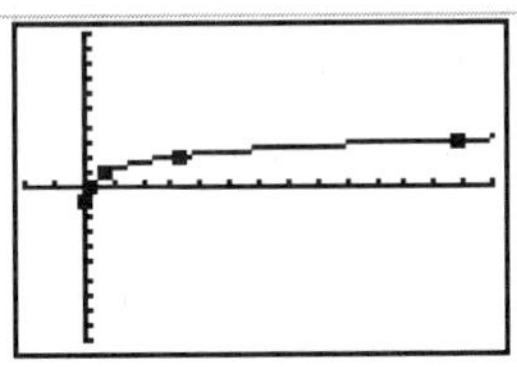

51)

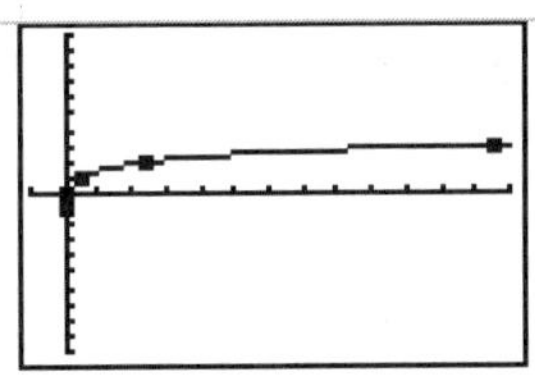

53)

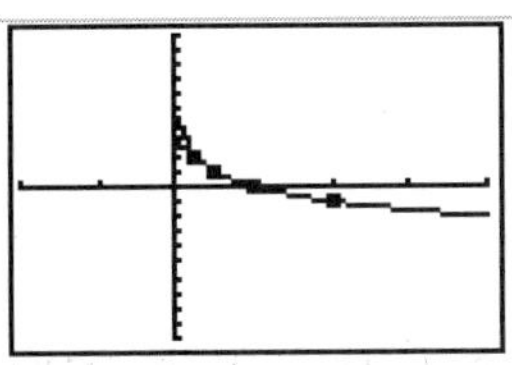

55)

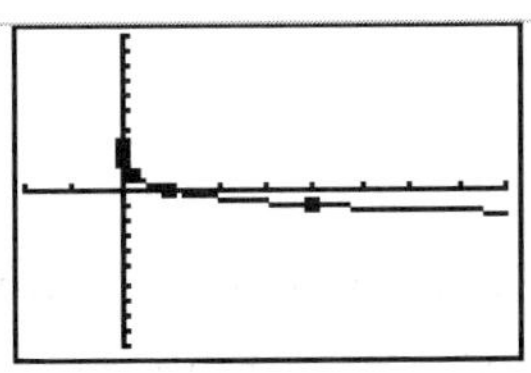

57)

59) 2.26

61) 6.64

63) 3.07

SECTION 4.2

65. $y = 10(3)^x$ has y–intercept at (0,10) and $y = 3^x$ has a y–intercept at (0,1).

67. $y = 8(1/2)^x$ has a y–intercept at (0,8) and $y = (1/2)^x$ has a y–intercept at (0,1).

69. a. $y = 2^x + 5$ has a y–intercept at (0,6) and $y = 2^x$ has a y–intercept at (0,1).
    b. $y = 2^x + 5$ has a range of $(5, \infty)$ any $y = 2^x$ has a range of $(0, \infty)$.

71. a. $y = (1/2)^x + 3$ has a y–intercept at (0, 4) and $y = (1/2)^x$ has a y–intercept at (0,1).
    b. $y = (1/2)^x + 3$ has a range of $(3, \infty)$ and $y = (1/2)^x$ has a range of $(0, \infty)$.

73. a. $y = 3(2)^x + 2$ has a y–intercept at (0,5) and $y = 2^x$ has a y–intercept at (0,1).
    b. $y = 3(2)^x + 2$ has a range of $(2, \infty)$ and $y = 2^x$ has a range of $(0, \infty)$.

SECTION 4.3

1) $3^y = x$

3) $5^x = 2$

5) $2^a = x + y$

7) $\log_3 m = x$

9) $\log_a 5 = 3x - 1$

11) $\log_a (x+y) = 5$

13) $\log_3 3 + \log_3 x$

15) $\log_2 x - \log_2 3$

17) $\log x + \log y - \log 5 - \log z$

19) $2\log_5 x + \log_5 y$

21) $3\ln a - 2\ln x - \ln y$

23) $\log_4 3 + \log_4 y + (1/3)\log_4 x - \log_4 a - 2\log_4 b$

25) $(2/3)\ln x + (1/3)\ln y - (1/3)\ln a$

27) $\log_2 3x$

29) $\log_5 (r/4)$

31) $\log(x(2x-1))$

33) $\log \dfrac{x}{3x+2}$

35) $\ln \dfrac{3(2x-1)}{x+3}$

37) $\log_7 \dfrac{3x^2}{y^3}$

39) $\log \dfrac{x^2(2x-1)}{y^3}$

41) $\ln \dfrac{x^2 x^3}{x^5} = \ln 1 = 0$

43) a. 1.5563 b. 1.5563 c. 1.5563 d. 1.5563
Property $\log_a bc = \log_a b + \log_a c$

45) a. 1.6094 b. 1.6094 c. 1.6094 d. 1.6094
Property: $\log_a (b/c) = \log_a b - \log_a c$

47) a. 2.0969 b. 2.0969 Property: $\log_a b^n = n\log_a b$

49) a. –3.2189 b. –3.2189 Property: $\log_a b^n = n\log_a b$

51) $\ln 15 \approx 2.7081$ $\ln 3 + \ln 5 \approx 2.7081$

53) $\log 3 \approx 0.4771$ $\log a - \log b$, where $(a/b) = 3$ i.e. $\log 24 - \log 8 \approx 0.4771$

55) $2\log 5 + 2\log 8 - 2\log 10 \approx 1.2041$, $\log \dfrac{5^2 8^2}{10^2} \approx 1.2041$

57) 2.5335 check $5^{2.5335} \approx 58.9984$

59) 2.6049 check $7^{2.6049} \approx 158.9995$

61) 4.5236 check $(1/2)^{4.5236} \approx 0.0435$ **($(1/23) \approx 0.0435$)**

63) –2.2619 check $3^{-2.2619} \approx 0.0833$ **($(1/12) \approx 0.0833$)**

65) No answer

67) No answer

69) Domain: $(0, \infty)$ Range: $(-\infty, \infty)$

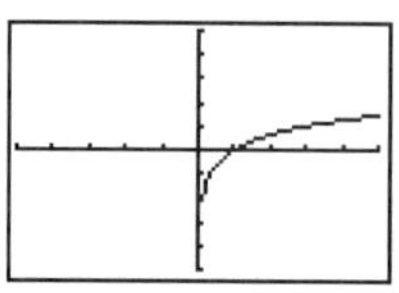

71) Domain: $(0, \infty)$ Range: $(-\infty, \infty)$

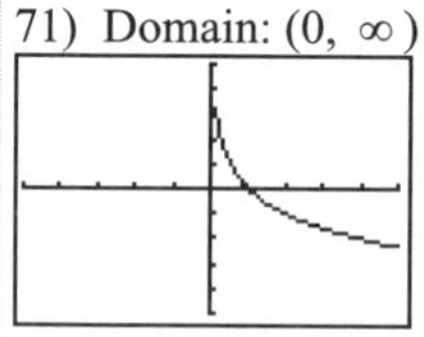

73. Domain: $(0, \infty)$ Range: $(-\infty, \infty)$

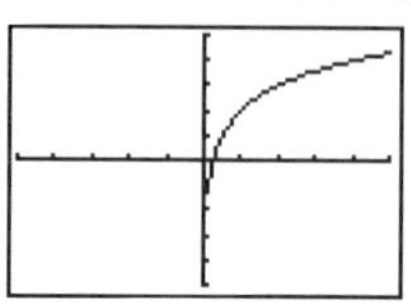

75. Domain: $(0, \infty)$ Range: $(-\infty, \infty)$

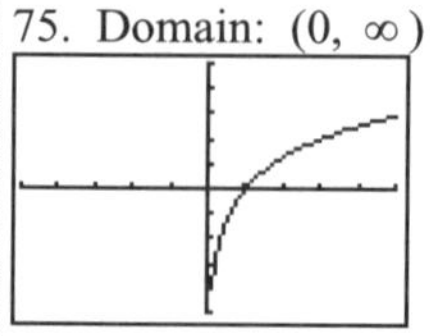

SECTION 4.3

77) $(-\infty, 0)$	79) $(-(1/3), \infty)$	81) $(-\infty, -4)$ or $(4, \infty)$

SECTION 4.4

1) $x = 1.5$	3) $x = -5$	5) $x = 3.5$
7) $x = \frac{4}{5}$	9) $x = \frac{\ln 23}{3\ln 5} \approx 1.682$	11) $x = \frac{\ln 19 - 2(\ln 3)}{\ln 3} \approx 0.680$
13) $x = \frac{\ln 51 + \ln 2}{2\ln 2} \approx 3.336$	15) $x = \frac{\ln 15 - 2}{3} \approx 0.236$	17) $x = \frac{\ln 9 - \ln 5}{2\ln 5} \approx 0.183$
19) $x = \frac{3 + \ln 3}{2} \approx 2.049$	21) $x = \frac{\ln 2}{\ln(1 + 0.09)} \approx 8.043$	23) $x = \frac{\ln 3}{0.08} \approx 13.733$
25) $x = \frac{\ln 3}{12\ln\left(1 + \frac{0.08}{12}\right)} \approx 13.778$	27) $x = 5$	29) $x = 9$
31) No solution	33) $x = 2$	35) $x = \frac{-1 + \sqrt{23}}{2}$
37) $x = \frac{-3 + \sqrt{39}}{6}$	39) No solution	41) No solution
43) $x = 10$	45) $x = 14$	47) $x = -2$
49) $x = 13/5$	51) $x = 11/5$	53) $x = 4$
57) No solution	59) No solution	

61) a. 7 b. 3.55 c. 3.98×10^{-7} d. 6.3×10^{-4}

63) a. Approx. 56.2 minutes b. Approx. 15.8 hours c. Approx. 90.5^0

65) a. Approx. 338 snails b. Approx. 15.4 months | 67) a. Approx. $49.530.32 b. Approx. 20.1 years

69) a. Approx. 6.3 grams b. Approx. 34.8 hours

SECTION 4.5

1)

Initial Investment	Compounded	Annual Percentage Rate	Doubling time	Amount after 10 years	Amount after 20 years	Amount after 40 years
$10,000	**Yearly**	**8%**	9 years	$21,589.25	$46,609.57	$217,245.22
$30,000	**Monthly**	**9.5%**	7.3years	$77,281.66	$199,081.84	$1,321,119.33
$11,172.92	**Quarterly**	**10%**	7 years	**$30,000.00**	$80,551.92	$580,744.04
$11,168.57	**Weekly**	**7.5%**	9.25years	$23,613.09	**$50,000.00**	$223,842.39
$12,000.00	**Yearly**	8.5%	**8.5 years**	$27,131.80	$61.344.55	$313,596.19
$6,000.00	**Monthly**	12%	**5.8 years**	$19,802.32	$65,355.32	$711,886.35
$10,000.00	**Yearly**	10.5%	6.9 years	**$27,200.00**	$73,662.35	$542,614.16
$5,000.00	**Monthly**	11%	6.3 years	**$15,000.00**	$44,675.08	$399,172.50

3) $25,113.19	5) $192,106.56	7) $3,109.43

9) a. Approx. 57.8 years b. Approx. 46.2 years c. Approx. 57.6 years d. Approx. 46 years

SECTION 4.5

11) a. Approx. 9 years b. Approx. 8 years c. Approx. 7.3 years d. Approx. 6.1 years

13) a. Approx. 15.7 years b. Approx. 12.2 years c. Approx. 10 years

15) \$50,664.11	17) Approx. 6.3 years	19) Approx. 6.7 years
21) Approx. \$3,966.87	23) Approx. 11.75%	25) \$1,121,46

SECTION 4.6

1) \$104,520.91	3) \$898,255.09	5) a. \$118,825.28 b. \$91,825.28

7) a. Approx. \$104.74 b. \$25,137.60 of own money and \$32,862.40 interest

9) a. Approx. \$57.70, b. \$18,002.40 of own money and \$5,997.60 interest

11) Approx. \$78.64	13) Approx. \$197.64
15) Approx. 25 years	17) Approx. 9%

19) a. Monthly payments of approx. \$1,100.65 amount of total interest paid is \$246,234
 b. Monthly payments of approx. \$1,048.82 amount of total interest paid is \$227,575.20
 c. Monthly payments of approx. \$1,348.24 amount of total interest paid is \$92,683.20
 d. Monthly payments of approx. \$1,306.66 amount of total interest paid is \$85,198.80

21) Best deal is 8% with \$3,000 back as a down payment

23) a. You can buy a home for \$147,186.17 b. You can buy a home for \$158,316.85

25) a. You can buy furniture worth \$4,682.68
 b. You can buy furniture worth \$5,891.89

27) Approx. 7.9% interest rate

29) Approx. 48 months

31) a. Approximately 6.4 years or 77 months
 b. Approximately \$16,940 total payments
 c. Approximately \$6,940 interest paid

33) a. $A(R) = R\left[\dfrac{\left(1+\dfrac{0.08}{12}\right)^{12\bullet 30}-1}{\left(\dfrac{0.08}{12}\right)}\right]$ b. Looks Linear

35) a. $A(n) = 25\left[\dfrac{\left(1+.\dfrac{0.079}{n}\right)^{20n}-1}{\left(\dfrac{0.079}{n}\right)}\right]$ b. Looks Linear

37) $R(t) = \dfrac{20{,}000\left(\dfrac{0.08}{12}\right)}{1-\left(1+\dfrac{0.08}{12}\right)^{-12t}}$ Looks Linear

a. Approximately \$488.26 b. Approximately \$405.53 c. Approximately \$242.66

SECTION 4.7

1) a. $R = 10e^{-0.05776t}$ b. Approx. 1.4866 grams

3) a. $R = 15e^{-0.000428t}$ b. Approx. 10.2047 grams

5) a. $R = 10e^{-0.0231t}$ b. Approx. 30 years c. Approx. 3.1506 grams d. Cesium–137

7) a. $R = 50e^{-0.165t}$ b. Approx. 4.2 days c. Approx. 0.3542 grams d. Thallium–206

SECTION 4.7

9) Approx. 2,820,903,237 years old	11) Approx. 7,575 years old
13) a. Approx. 28 years b. Approx. 280 years	15) a. Approx. 6 hours b. Approx. 60 hours
17) a. $A = 20e^{0.32189t}$ b. Growth c. 312,523 ants	19) a. $P = 12500e^{0.03646t}$ b. Growth c. 37,320 people d. 19 years
21) a. $P = 5000^{-0.07438t}$ b. Decay c. Approx. 30 days	23) a. Approx. 4.5 minutes b. Approx. 13 minutes

25) a. Approx. 112.6° b. Approx. 128 minutes c. Approx. 156 minutes

27) a. $y = 3,321,000(1.006153)^t$ b. $y = 3,321,000e^{0.006135t}$
c. $y = 3,321,000(1.006153)^8 \approx 3,488,037$ votes $y = 3,321,000\, e^{(0.006135)(8)} \approx 3,488,061$ votes

29) Number 27 is an increasing function and number 28 is a decreasing function.

31) a. $k = -0.05776$ b. Approx. 50 grams c. $50e^{-0.05776t}$ d. Approx. 12 days e. Barium–131

33) a. $k = -0.1318$ b. Approx. 30 grams c. $30e^{-0.1318t}$ d. Approx. 5.26 years e. Cobalt–60

SECTION 4.8

1) Approx. 20.6 decibels	3) Approx. 89.5 decibels	5) $1,000,000\ I_0$
7) Approx. $1.088930093 \text{x} 10^{14} I_0$	9) 4	11) 6
13) 6.5	15) 4.3	17) $10,000\ I_0$
19) $5011.87\ I_0$	21) $19.952.62\ I_0$	23) $7,943,282.35\ I_0$
25) $pH \approx 3.8$	27) $pH \approx 7.8$	29) $H^+ = 0.0000001$

31) $H^+ = 0.00000000001$	33) a. Approx. 79.8% b. Approx. 74 inches tall
35) a. Approx. 14 years old b. Approx. 78.8 inches tall	37) Approx. 8.7 years

39) Approx. 7.4 years	41) Approx. 5%	43) Approx. 0.7219
45) 1	47) Approx. 33 or 300	

CHAPTER 4 REVIEW

1) Linear	3) Exponential

5) a. $21,600/20,000 = 1.08$ $23,328/21,600 = 1.08$ $25,194.15/23,328 = 1.08$
b. 1.08 c. $A(t) = 20,000(1.08)^t$ d. $93,219.14 e. Growth

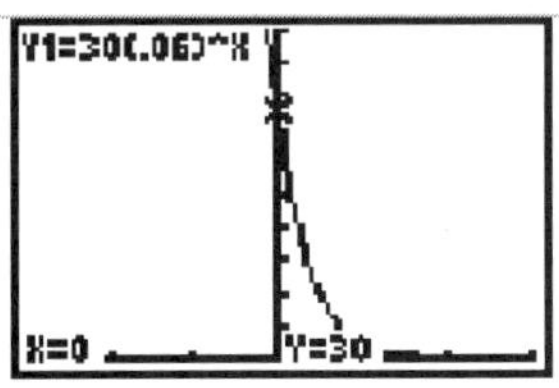

7) a. Decay b. All real numbers c. $(0, \infty)$ d. (0, 30)

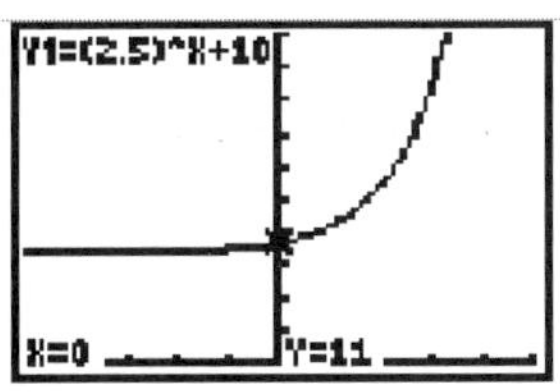

9) a. Growth b. All real numbers c. $(10, \infty)$ d. (0,11)

CHAPTER 4 REVIEW

11) –2	13) 1	15) 0
17) –3	19) Approx. 1.95	21) Approx. 3.10
23) $a^x = 5$	24) $3^b = y+3$	25) $\log_5 y = x$
27) $\log_3 3 + 3\log_3 x + 2\log_3 y$	29) $\log_5 3 + 4\log_5 x - 3\log_5 a$	31) $\log_3 ab^5$
33) $\log\frac{3x^5}{b^5\sqrt[3]{a}}$	35) 2log3 + (1/2)log4 – 3log5 ≈ –0.8416	37) Approx. 2.2757
39) Approx. –3.6309	41) (0, ∞)	43) (3/2, ∞)
45) x = 2	47) Approx. 0.475	49) $x = \frac{\ln 54 - 3\ln 6}{2\ln 6} \approx -0.387$
51) $x = \frac{\ln 24 + \ln 3}{5\ln 3} \approx 0.779$	53) x = 5	55) x = 14/3
57) x = 5	59) x = 11/5	
61) a. Approx. 0.1564 grams b. 10 grams c. Approx. 11.6 days		
63. Approx. $317,312.47	65. Approx. 7.3 years	67. Approx. $18,925.20
69. Approx. $1,268.31	71. Approx. $567,170.14	
73) a. $R(t) = 5e^{-0.08664t}$ b. Approx. 9 days	75) a. $P = 4{,}500e^{0.00656t}$ b. Growth c. Approx. the year 2012	
77) a. Approx. 51.76 decibels b. 10,000,000,000 I_o		

CHAPTER 4 EXAM

1) a. 125,829/123,000 = 1.023 128,723/125,829 = 1.023 b. $P = 123{,}000e^{0.004548t}$ c. Approx. 130,491 people d. The year 2010		
3) –5	5) log3 + 3logx – 2loga – (1/3)logb	7) (–2/3, ∞)
9) $x = \frac{\ln 71 + 2\ln 5}{3\ln 5} \approx 1.5495$	11) x = 14	13) x = 28/3
15) Approx. $29,772.78	17) Approx. $244.84	
19) a. Approx. 13.6079 grams b. Approx. 2,141.5 years	21) a. pH ≈ 3.3 b. pH ≈ 5.3 **(leave out the oranges)**	

Chapter 5 Odd Homework Answers

SECTION 5.1

1) m = –1	3) m = -4	5) Undefined
7) m = –1	9) m = 8/5	11) m = –13/6
13) m = 0	15) m = 2/3; y–intercept is (0,7)	17) m = –8/3; y–intercept is (0,7)
19) m = 0; y–intercept is (0,–4)	21) m = 3/2; y–intercept is (0,-4)	23) m = –5/2; y–intercept is (0,4)
25) m = –9/8; y–intercept is (0,9/2)	27) y = (–2/3)x – 4	29) y = 4x + 3/8
31) y = –5	33) y = 3x + 17	35) y = (–3/5)x – 46/5

SECTION 5.1

37) x = –10	39) y = –4x – 11	41) y = (9/5)x + 22/5
43) y = –6x – 4	45) y = –5.4	47) Parallel
49) Neither	51) Perpendicular	

53) Parallel: y = (2/3)x + 19/3, Perpendicular:(–3/2)x + 2

55) Parallel: y = –0.7x–9.1, Perpendicular: y = (10/7)x – 19/7

57) Parallel: f(x) = 3x, Perpendicular: f(x) = (–1/3)x + 10

59) Parallel: y = (–3/2)x–1, Perpendicular: y = (2/3)x–16/3

61) Parallel: y = 2x–1, Perpendicular: y = (–1/2)x–6

63) Parallel: y = (3/10)x + 17/2, Perpendicular: y = (–10/3)x–29/3

65) Parallel: x = 3, Perpendicular: y = –4

67) a. P(x)= –2x + 220 b. 30 pairs of shoes c. $86 d. Two fewer pairs of shoes are sold if the price is raised $1.

69) a. A(x)=(–1/8)x + 120 b. 35 units c. $736 d. One less units is rented when the rent is raised $8.

71) a. A(t)=40t + 5200 b. $5,800 in the account c. The year 2020 d. Each year the money in the account increases $40.

73) a. C(m)=0.15m + 150 b. $228 for 520 miles in one week c. 400 miles d. Each mile driven costs $0.15.

75) a. m(t)=–0.4t + 8.4 b. 4.8 per1,000 c. Approximately the year 2000

77) a. C(t)=–2t + 22 b. 2 students per computer c. 1999

79) a. S(t)=30,000t + 356,000 b. 596,000 millions of dollars c. Approximately the year 2000

SECTION 5.2

1) x = –2, y = 4	3) x = 1.5, y = 1	5) x = 8, y = 6
7) x = 3, y = –1	9) x = 6, y = –4	11) x = –5, y = 5
13) x = 3/13, y = 63/13	15) x = –8, y = –9	17) x = 34/33, y = –1/11
19) Same line	21) Parallel lines no solution	23) Same line

25) $E(t) = 0.6e^{0.32189t}$ $L(t) = 17.312\ln(t) - 24.862$

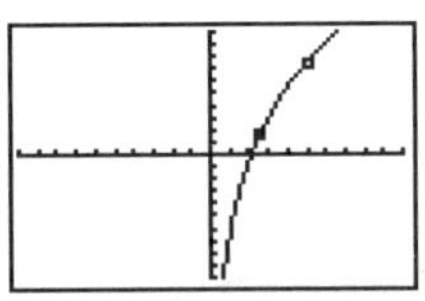

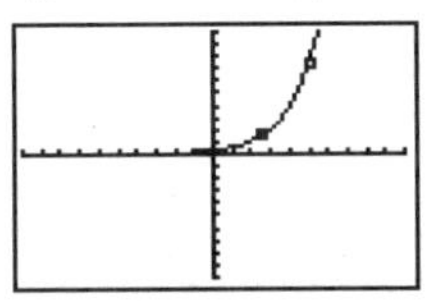

27) $E(t) = 410.15e^{-0.3269t}$ $L(t) = -106.382\ln(t) + 251.215$

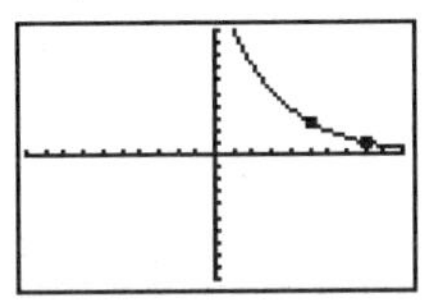

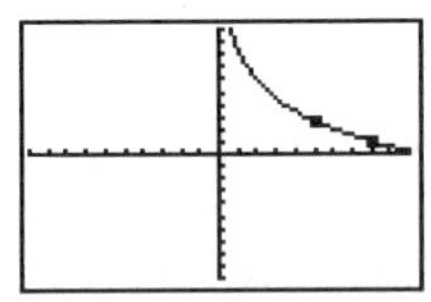

29) a. $B(t) = 30e^{0.2304t}$ b. $B(t) = 1{,}349.294\ln(t) - 2{,}806.86$

31) a. $R(t) = 1562e^{-0.0446t}$ b. $R(t) = -493.26\ln(t) + 2135.77$

33) a. $I(t) = 2638e^{0.0522t}$ b. $I(t) = 664.62\ln(t) + 2355.05$

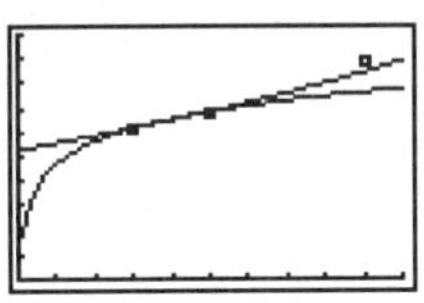

Exponential function best

35) a. $F(t) = 2{,}101{,}424e^{-0.0028665t}$

b. $F(t) = -23{,}315.2\ln(t) + 2{,}109{,}044.36$

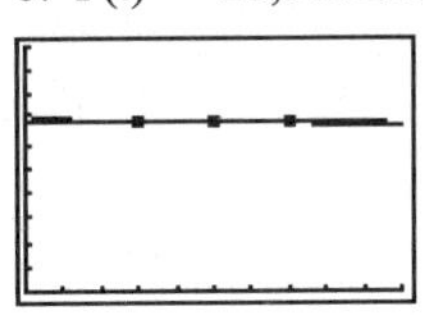

Exponential & Logarithmic good

SECTION 5.3

1) a. y = 0.00263x + .13036 b. $y = -3.254x^2 + 0.011x + 0.1812$ c. y = 0.0813 + 0.0333ln(x) d. $y = (0.1326)(1.01656)^x$

e. linear, quadratic, and exponential

SECTION 5.3

3) a. $y = 0.1822x + 67.75$ b. $y = 0.0002822x^2 + 0.1678x + 67.8569$ c. $y = 66.8589 + 2.098\ln(x)$ d. $y = (67.8358)(1.0025)^x$ e. linear, quadratic and exponential

5) a. $y = -3.9276 + 1.4023x$ b. $y = -0.04905x^2 + 2.4766x - 9.4216$ c. $y = -23.195 + 14.6919\ln(x)$ d. $y = (2.3782)(1.146)^x$ e. logarithmic and quadratic

7) a. $y = -23.082x + 713.1602$ b. $y = -0.1456x^2 - 20.7519x + 707.3348$ c. $y = 706.7761 - 109.0435\ln(x)$ d. $y = (736.532)(0.9564)^x$ e. exponential

9) a. $y = 43.219x - 169.0524$ b. $y = 1.0919x^2\ 2.8186x + 61.3395$ c. $y = -262.1561 + 352.0914\ln(x)$ d. $y = (80.693)(1.0914)^x$ e. quadratic

11) a. $y = 3.3514x + 25.3019$ b. $y = 0.2x^2 + 1.5514x + 28.7686$ c. $y = 21.8387 + 13.0517\ln(x)$ d. $y = (27.508)(1.0866)^x$ e. linear, quadratic and exponential

13) a. $y = 163.0207x + 1111.6568$ b. $y = 3.8014x^2 - 5.7535x + 2164.212$ c. $y = 996.4759+1373.4747\ln(x)$ d. $(1905.4835)(1.03725)^x$ e. quadratic and exponential

15) a. $y = 0.1953x + 3.8111$ b. $y = 0.02057x^2 + 0.03071x + 4.058$ c. $y = 3.8977 + 0.5703\ln(x)$ d. $y = (3.8657)(1.043)^x$ e. linear and exponential

17) a. $y = 0.1362x + 0.6529$ b. $y = 0.0118x^2 - 0.1823x + 2.741$ c. $y = -2.00975 + 1.7733\ln(x)$ d. $y = (1.1902)(1.0556)^x$ e. quadratic

19) a. $y = -0.0376x^2 + 3.4715x - 21.8458$ b. $y = (2.9645)(1.1218)^x$ c. quadratic function d. quadratic ≈ 39.8 exponential ≈ 46.8 e. quadratic ≈ 41.4 exponential ≈ 52.5 f. exponential

21) a. $y = 10(0.9527)^t$ b. $R(t) = 10e^{-0.048455t}$ c. Approximately 14.3 days d. Phosphorus–32

23) a. $y = 500(1.083)^t$ b. $A(t) = 500e^{0.08t}$ c. 8%

SECTION 5.4

1) a.

x	y	cd
2	11	
5	20	9
8	29	9
11	38	9
14	47	9

b. 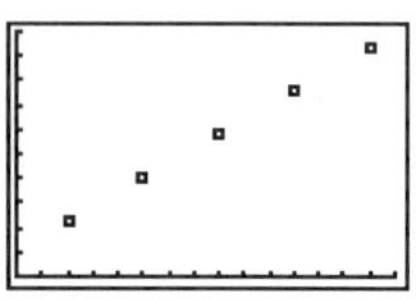

c. $y = 3x + 5$

d. $y = 3x + 5$

e. 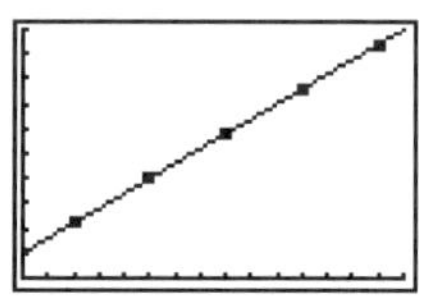

3) a.

x	y	cd
2	12	
5	96	8
8	768	8
11	6,144	8
14	49,152	8

b. 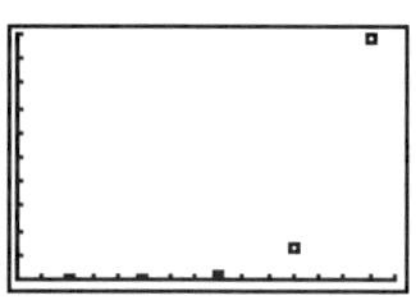

c. $y = 3(2)^x$

d. $y = 3(2)^x$

e. 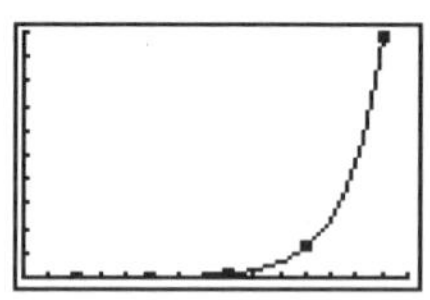

5) a. $y = 350t + 5000$ b. $17,250

7) a. $y = 9.961215655(0.8871908943)^t$ b. Approximately 0.22 grams

9) a. $y = 5,000(1.08)^t$ b. Approximately $73,926.72 c. 8%

11) $y = 2,516.2x + 261,456.4$

13) $y = 0.2928571429x^2 + 2.598571429x + 77.44571429$

15) $y = 932.3885107 + 188.4429424\ \ln(x)$

17) $y = 9.795424857/(1 + 1.067453932e^{-0.0238215966x})$

19) $y = 235.5714286x^2 - 6,111.085714x + 505,572.5429$

21) $y = 432.3700173x + 18,409.77383$

23) $y = 0.0966397644x^2 + 85.35755523x - 690.0824742$

SECTION 5.4

25) y = 23,973x + 299,537.2

27) $y = 19.92142857x^2 + 237.6642857x + 4{,}972.802857$

29) y = –25.38938181 + 14.20221005 ln(x)

CHAPTER 5 REVIEW

1) $\frac{5}{4}$

3) Zero

5) Slope = $\frac{3}{5}$; y–intercept (0, 7)

7) Undefined slope; No y–intercept

9) $y = \frac{3}{5}x + 5$

11) $x = \frac{-3}{8}$

13) $y = \frac{7}{4}x - \frac{41}{4}$

15) Parallel: $y = \frac{-3}{5}x - \frac{11}{5}$ Perpendicular: $y = \frac{5}{3}x + \frac{7}{3}$

17) a. $g(x) = \frac{-1}{2}x + 110$ b. 65 c. $116 d. For every increase of $2 per glove one less glove will be sold.

19) a. M(t) = 0.07t + 4.71

21) x = 6; y = 4

24) x = –2; y = 1

25) $x = \frac{689}{163}$; $y = \frac{-506}{163}$

27) No solution Parallel lines

29) $E(t) = 10.001e^{0.231t}$ L(t) = –40 + 54.614 ln (t)

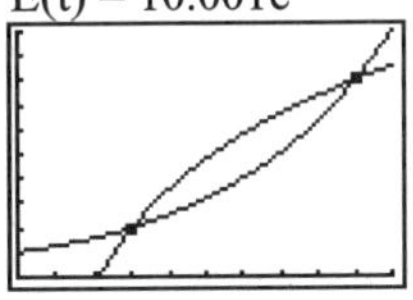

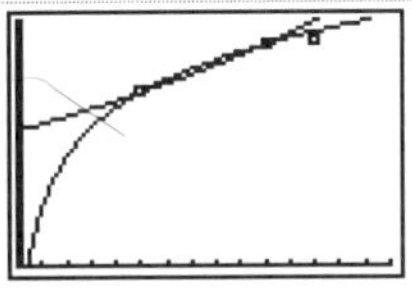

31) a. $F(t) = 384.3\,e^{0.0494938801t}$ b. F(t) = 199.4 ln(t) + 171.3 c. Logarithmic

33) y = –77.1x + 2691.9

35) y = 78.45566727 – 17.08148562 ln(x)

CHAPTER 5 EXAM

1) $y = \frac{-3}{4}x + 8$

3) y = –4

5) x = –3

7) $y = \frac{5}{2}x + 1$

9) a) S(t) = 5+35.7 b) $41,700 average salary c) in the year 2006

11) $x = \frac{69}{13}$, $y = \frac{6}{13}$

13) Same line

15) No Solution "Parallel"

17) a. $F(t) = 3968.5e^{0.462098t}$ b. F(t) = 32740.72ln5−12694.12

19) Linear seems best: y = 0.56x+63.54

CHAPTER 5 EXAM

21) a.

x	y	cd	cr
2	0.008		
4	0.032	0.032−0.008=0.024	0.008÷0.008=4
6	0.128	0.128−0.032=0.096	0.128÷0.032=4
8	0.512	0.512−0.128=0.394	0.512÷0.128=4
10	2.048	2.048−0.512=1.536	2.048÷0.512=4

Common ratios are equal so the best fit would be an exponential function

b. $y = 0.002(2)^x$ c. $y = 0.002(2)^x$

Chapter 6 Odd Homework Answers

SECTION 6.1

1) 2 x 2

3) 3 x 3

5) 3

7) 4

9) Doesn't exist

11) $\begin{bmatrix} 4 & -3 \\ 5 & 7 \end{bmatrix}$

13) $\begin{bmatrix} 1 & 1 \\ 5 & 7 \\ 6 & 4 \end{bmatrix}$

15) $\begin{bmatrix} -2 & 4 & -2 \\ 3 & 5 & 9 \\ -2 & 3 & 2 \end{bmatrix}$

17) $\begin{bmatrix} 4 & -1 & -2 \\ 2 & -5 & 4 \end{bmatrix}$

19) Not possible

21) $\begin{bmatrix} -2 & -1 \\ 1 & 1 \end{bmatrix}$

23) $\begin{bmatrix} -2 & -3 & 8 \\ 2 & 11 & 6 \end{bmatrix}$

25) $\begin{bmatrix} 3 & -6 \\ 9 & 12 \end{bmatrix}$

27) $\begin{bmatrix} 6 & -7 \\ 11 & 15 \end{bmatrix}$

29) $\begin{bmatrix} -16 & 19 \\ -3 & 0 \\ 16 & -1 \end{bmatrix}$

31) $\begin{bmatrix} -1 & -7 \\ 17 & 9 \end{bmatrix}$

33) $\begin{bmatrix} -11 & 6 & -11 \\ -25 & 29 & -6 \\ -18 & 19 & -6 \end{bmatrix}$

35) $\begin{bmatrix} -6 & -26 & 7 \\ 3 & -15 & -7 \\ 12 & -4 & -21 \end{bmatrix}$

37) Not possible

39) $\begin{bmatrix} 1 & -2 & 3 \\ 21 & 9 & 31 \end{bmatrix}$

41) $\begin{bmatrix} x-6 & 2-4x \\ 3x+12 & 8x+6 \end{bmatrix}$

43) $\begin{bmatrix} x-y+2z \\ 3x+2y+4z \\ 2x+y+3z \end{bmatrix}$

45) $\begin{bmatrix} -15 & -19 & -38 \\ 35 & -7 & 96 \end{bmatrix}$

47) $\begin{bmatrix} -16 & -32 \\ 55 & -23 \\ 31 & -22 \end{bmatrix}$

49) $\begin{bmatrix} 7 & -6 \\ 12 & 7 \end{bmatrix}$

51) $\begin{bmatrix} -35 & -30 \\ 45 & 10 \end{bmatrix}$

53) Not possible

55) a. $\begin{bmatrix} 0 & 0 \\ 0 & 0 \end{bmatrix}$ b. $\begin{bmatrix} 1 & -2 \\ 3 & 4 \end{bmatrix} = A$ c. $\begin{bmatrix} 1 & -2 \\ 3 & 4 \end{bmatrix} = A$

57) a. $\begin{bmatrix} -1 & 2 \\ -3 & -4 \end{bmatrix}$ b. $\begin{bmatrix} 0 & 0 \\ 0 & 0 \end{bmatrix}$ addition identity c. $\begin{bmatrix} 0 & 0 \\ 0 & 0 \end{bmatrix}$ addition identity d. Matrix A and –A are additive inverses.

59) Commutative property

61) Commutative property

63) x = 4; y = 3

SECTION 6.1

65) a. $A=\begin{bmatrix} 3,583 & 1,283 & 1,583 \\ 3,258 & 1,099 & 1,325 \\ 2.933 & 933 & 1,079 \end{bmatrix}$ $B=\begin{bmatrix} 2,038 & 1,283 & 1,583 \\ 1.923 & 1.400 & 635 \\ 1,834 & 1,283 & 529 \end{bmatrix}$ b. $\begin{bmatrix} 5,621 & 2,566 & 3,166 \\ 5,181 & 2,499 & 1,960 \\ 4,767 & 2,216 & 1,608 \end{bmatrix}$ c. 2,566 d. 1,960 e. 4,767

67) a. $A=\begin{bmatrix} 53 & 48 & 35 & 29 \\ 61 & 53 & 49 & 32 \\ 49 & 45 & 39 & 43 \end{bmatrix}$ $P=\begin{bmatrix} 893 \\ 957 \\ 1,238 \\ 1,393 \end{bmatrix}$ b. $AP=\begin{bmatrix} 176,992 \\ 210,432 \\ 195,003 \end{bmatrix}$ c. \$176,992 d. \$210,432 e. \$195,003

69) a. $A=\begin{bmatrix} 0.6 & 0.4 \\ 0.5 & 0.2 \\ 0.3 & 0.3 \end{bmatrix}$ $B=\begin{bmatrix} 11 & 9 \\ 14 & 11 \end{bmatrix}$ b. $AB=\begin{bmatrix} 12.2 & 9.8 \\ 8.3 & 6.7 \\ 7.5 & 6 \end{bmatrix}$ c. Los Angeles costs are: \$12.20 per record; \$8.30 per tape; and \$7.50 per disk, Atlanta costs are \$9.80 per record; \$6.70 per tape; and \$6.00 per disk

71) a. $A=\begin{bmatrix} 44.7 & 44.9 & 48.5 & 45.6 \end{bmatrix}$ $B=\begin{bmatrix} 12.3 \\ 10.7 \\ 9.7 \\ 10.8 \end{bmatrix}$ b. $AB=\begin{bmatrix} 1993.17 \end{bmatrix}$ In the age group (18–20), 19,931,700 registered to vote in the presidential elections 1980–1996. c. $C=\begin{bmatrix} 35.7 & 33.2 & 28.5 & 31.2 \end{bmatrix}$ d. $CB=\begin{bmatrix} 1407.76 \end{bmatrix}$ In the age group (18–20), 14,077,600 actually voted in the presidential elections 1980–1996. e. $AB-CB=\begin{bmatrix} 585.41 \end{bmatrix}$ In the age group (18–20), 5,854,100 that registered to vote did not vote in the presidential elections 1980–1996.

73) a. $A=\begin{bmatrix} 44.7 \\ 52.7 \\ 62.0 \\ 70.6 \\ 75.8 \\ 74.6 \end{bmatrix}$ $B=\begin{bmatrix} 12.3 & 15.9 & 35.7 & 25.6 & 43.6 & 24.1 \end{bmatrix}$ b. $BA=\begin{bmatrix} 10511.24 \end{bmatrix}$ In the 1980 election, 105,112,400 voters registered to vote. c. $C=\begin{bmatrix} 35.7 \\ 28.3 \\ 54.6 \\ 64.4 \\ 69.3 \\ 65.1 \end{bmatrix}$ d. $BC=\begin{bmatrix} 9,077.33 \end{bmatrix}$ In the 1980 election 90,773,300 actually voted.

e. $BA-BC=\begin{bmatrix} 1,433.91 \end{bmatrix}$ In the 1980 election 14,339,100 that registered to vote didn't vote.

SECTION 6.2

1) $\begin{bmatrix} -1 & 0 & -2 \\ 0 & 3 & -2 \end{bmatrix}$	3) $\begin{bmatrix} -1 & 3 \\ -3 & -4 \end{bmatrix}$	5) $\begin{bmatrix} 3 & -2 & -5 \\ -1 & -3 & 2 \\ -4 & -1 & -2 \end{bmatrix}$

SECTION 6.2

7) $\begin{bmatrix} 0 & 0 & 0 \\ 0 & 0 & 0 \end{bmatrix}$	9) $\begin{bmatrix} 0 & 0 \\ 0 & 0 \end{bmatrix}$	11) $\begin{bmatrix} 0 & 0 & 0 \\ 0 & 0 & 0 \\ 0 & 0 & 0 \end{bmatrix}$
13) $\begin{bmatrix} 0 & 0 & 0 \\ 0 & 0 & 0 \end{bmatrix}$	15) $\begin{bmatrix} 0 & 0 & 0 & 0 \\ 0 & 0 & 0 & 0 \\ 0 & 0 & 0 & 0 \\ 0 & 0 & 0 & 0 \end{bmatrix}$	17) $\begin{bmatrix} \frac{3}{17} & \frac{2}{17} \\ \frac{-4}{17} & \frac{3}{17} \end{bmatrix}$
19) $\begin{bmatrix} \frac{-1}{4} & \frac{11}{28} & \frac{-3}{7} \\ \frac{-1}{4} & \frac{1}{4} & 0 \\ \frac{1}{4} & \frac{1}{28} & \frac{1}{7} \end{bmatrix}$	21) Doesn't exist	23) Doesn't exist
25) $\begin{bmatrix} 1 & 0 \\ 0 & 1 \end{bmatrix}$	27) $\begin{bmatrix} 1 & 0 & 0 \\ 0 & 1 & 0 \\ 0 & 0 & 1 \end{bmatrix}$	29) Doesn't exist
31) 17	33) 28	35) Doesn't exist
37) 0	39) Error invalid dimension (A isn't a square matrices)	41) Error singular matrix (determinant value was zero)
43) $\begin{bmatrix} 1 & 0 & 0 \\ 0 & 1 & 0 \\ 0 & 0 & 1 \end{bmatrix}$	45) $\begin{bmatrix} 1 & 0 & 0 & 0 \\ 0 & 1 & 0 & 0 \\ 0 & 0 & 1 & 0 \\ 0 & 0 & 0 & 1 \end{bmatrix}$	47) Doesn't exist
49) a. $\begin{bmatrix} \frac{3}{2} & \frac{-5}{2} \\ -1 & 2 \end{bmatrix}$ b. 2		

SECTION 6.3

1) a. $\begin{bmatrix} 3 & 4 \\ 8 & -9 \end{bmatrix}$ b. $\begin{bmatrix} x \\ y \end{bmatrix}$ c. $\begin{bmatrix} 8 \\ 15 \end{bmatrix}$ d. $\begin{bmatrix} 3 & 4 \\ 8 & -9 \end{bmatrix}\begin{bmatrix} x \\ y \end{bmatrix}=\begin{bmatrix} 8 \\ 15 \end{bmatrix}$	3) a. $\begin{bmatrix} 3 & -2 \\ -4 & -3 \end{bmatrix}$ b. $\begin{bmatrix} x \\ y \end{bmatrix}$ c. $\begin{bmatrix} -15 \\ -12 \end{bmatrix}$ d. $\begin{bmatrix} 3 & -2 \\ -4 & -3 \end{bmatrix}\begin{bmatrix} x \\ y \end{bmatrix}=\begin{bmatrix} -15 \\ -12 \end{bmatrix}$
5) a. $\begin{bmatrix} 2 & -3 & 2 \\ 5 & 7 & -8 \\ 3 & -5 & -9 \end{bmatrix}$ b. $\begin{bmatrix} x \\ y \\ z \end{bmatrix}$ c. $\begin{bmatrix} 8 \\ -12 \\ 23 \end{bmatrix}$ d. $\begin{bmatrix} 2 & -3 & 2 \\ 5 & 7 & -8 \\ 3 & -5 & -9 \end{bmatrix}\begin{bmatrix} x \\ y \\ z \end{bmatrix}=\begin{bmatrix} 8 \\ -12 \\ 23 \end{bmatrix}$	
7) a. $\begin{bmatrix} 1 & -3 & 0 \\ -3 & 1 & 2 \\ 0 & 5 & -1 \end{bmatrix}$ b. $\begin{bmatrix} x \\ y \\ z \end{bmatrix}$ c. $\begin{bmatrix} -9 \\ 10 \\ 9 \end{bmatrix}$ d. $\begin{bmatrix} 1 & -3 & 0 \\ -3 & 1 & 2 \\ 0 & 5 & -1 \end{bmatrix}\begin{bmatrix} x \\ y \\ z \end{bmatrix}=\begin{bmatrix} -9 \\ 10 \\ 9 \end{bmatrix}$	

SECTION 6.3

9) $\begin{cases} 3x + y = 3 \\ 2x - 4y = 9 \end{cases}$

11) $\begin{cases} 1.3r + 2.8s = 3.54 \\ 3.4r - 1.7s = 2.78 \end{cases}$

13) $\begin{cases} -2x + 3y - 4z = -23 \\ x + y - 2z = 35 \\ 3x - 2y + z = 18 \end{cases}$

15) $\begin{cases} 3x - 2y + z + 2w = 58 \\ x - 3y + 5w = -29 \\ 3y - 2z + w = 120 \\ x - 4w = -83 \end{cases}$

17) (3,–2)

19) (–2.9,–3/4)

21) (3/2, –2/3,–2/5)

23) (30,–20, 16)

25) (–2, 17/3, 3)

27) (–2, –3, 4, 2)

29) (3, –2, 4, –3)

31) (18/7, 1, 55/7)

33) Can't find a solution by the inverse method

35) Can't find a solution by the inverse method

37) (3, –2)

39) (–2/9, –3/4)

41) (3/2, –2/3, –2/5)

43) (30, –20, 16)

47) (–2, –3, 4, 2)

49) (3, –2, 4, –3)

51) (18/7, 1, 55/7)

53) Can't find a solution with Cramer's Rule

55) Can't find a solution with Cramer's Rule

57) Problems 53 and 54 don't have all determinants equal to zero therefore no solutions. Problems 55 and 56 have all determinants equal to zero therefore infinite solutions.

59) Some determinants are not equal to zero and there is no solutions

SECTION 6.4

1) $\left[\begin{array}{cc|c} 2 & -5 & 23 \\ 3 & 7 & -13 \end{array}\right]$

3) $\left[\begin{array}{ccc|c} 3 & -5 & 7 & 9 \\ 2 & 3 & -8 & -3 \\ -3 & 2 & 1 & 7 \end{array}\right]$

5) $\left[\begin{array}{ccc|c} 1 & 0 & -3 & 7 \\ 2 & 1 & 0 & 9 \\ 0 & -3 & 2 & -9 \end{array}\right]$

7) $\left[\begin{array}{ccc|c} 1 & 2 & 5 & 3 \\ -2 & 1 & -3 & 5 \\ 2 & -5 & 3 & 8 \end{array}\right]$

9) $\left[\begin{array}{cccc|c} 1 & 2 & -3 & 1 & 9 \\ 2 & -1 & 1 & -3 & 15 \\ 3 & 2 & -5 & 2 & 18 \\ 2 & -1 & 3 & 1 & -5 \end{array}\right]$

11) $\begin{cases} x - 2y = 3 \\ 3x + 4y = 4 \end{cases}$

13) $\begin{cases} x - 3y + 2z = -2 \\ x + 4y - 5z = -3 \\ 2x - y + 3z = 4 \end{cases}$

15) $\begin{cases} x + 3z = -2 \\ 3y - 2z = 4 \\ -x + 3y + 2z = -3 \end{cases}$

17) $\begin{cases} x - y + 2z = 3 \\ 3x + y - 2w = 5 \\ -2x + y + 3z + 4w = -3 \\ y + 3w = -3 \end{cases}$

19. (–2, 5)

21) (3, –4, 5)

23) (1, 2, –4)

25) (–56, –13, 5)

27) (z–4, 2z–3, z)

29) (0, –1, 3, –2)

31) No solution

33) (5/128, 7/32, 3/8)

35) (3, 5 – 2t, t)

37) (43/13, –6/13)

39) (329/496, –5/372)

41) No solution

43) (2y + 6, y)

45) (109/57 , –7/19)

47) (2, –3, –2)

49) (2 – p, p – 1, p)

51) No solution

53) (–2, 3, –1)

55) (–4, 7, –2)

57) (2, 3.5, 1)

59) (4, –2, 3)

SECTION 6.4

61) (7/5 z – 10, 3/5 z – 5, z)	63) (3, 5, –2, –4)	65) (3 – 2z, 2z + 5, z)
67) No solution	69) (5, –2z – 3, z)	71) No solution
73) (3 – z, 2 – z, z)	75) (3 – z, 2, z)	

SECTION 6.5

1) a. A = speed of the airplane; W = speed of the wind
   b. 2(A + W) = 1200; 2(A – W) = 1120
   c. The speed of the airplane is 580 mph and the speed of the wind is 20 mph.

3) a. S = the number of steers; H = the number of heifers
   b. 3,150S + 2,825H = 456,250; 535S + 485H = 77,750
   c. There is 100 steers and 50 heifers.

5) a. x = the liters of 8% acid solution; y = the liters of 5% acid solution
   b. x + y = 30; 0.08x + 0.05y = 0.07(30)
   c. There is 20 liters of 8% acid solution and 10 liters of 5% acid solution

7) a. x = the number of ounces of 20% gold alloy; y = the number of ounces of 12% gold alloy.
   b. x + y = 30; 0.20x + 0.12y = 0.15(30)
   c. There is 11.25 ounces of the20% gold alloy and 18.75 ounces of the 12% gold alloy.

9) a. x = the amount invested at 5% simple interest; y = the amount invested at 7% simple interest; z = the amount invested at 8.5% simple interest
   b. x + y + z = 200,000; 0.05x + 0.07y + 0.085z = 12,800; x = y + 2000
   c. The amount invested at 5% is $84,600, the amount invested at 7% is $82,600 and the amount invested at 8.25% is $32,800.

11) a. x = the amount invested in the 6.4% bond; y = the amount invested in the 5.8% IRA; z = the amount invested in the 7.3% mutual fund.
   b. x + y + z = 100,000; z = 3y; x = 2z
   c. The amount of interest earned after one year is $6,610.

13) a. R = the number of possible republican voters; D = the number of possible democratic voters; I = the number of possible independent voters
   b. R + D + I = 360,000; 0.48R + 0.52D + 0.65I = 191,200; R = 2I
   c. The number of possible republican voters is 160,000.

15) a. a = the number of adults; b = the number of children (6–15); c = the number of children below 5
   b. a + b + c = 206; 12.50a + 7.50b = 1,560; b = a + c
   c. There was 63 adults, 103 children (6–15) and 40 children under 5 that took the tour on Monday.

17) a. F = the total number of 1st class tickets available; C = the number of coach tickets available; I = the number of internet specials available
   b. F + C + I = 264; 349.50F + 248.50C + 195.50I = 66,636; C = 3F
   c. There was 48 total 1st class tickets available, 144 total coach tickets available and 72 total internet tickets available.

19) a. W = the total number of acres of wheat planted; C = the total number of acres of corn planted; S = the total number of acres of soybeans planted
   b. 54W + 107C + 35S = 374,000; W + C + S = 5,000; 2.70(54W) + 2.08(107C) + 3.90(35S) = 906,950.00
   c. There was 1,000 total acres of wheat planted, 2,500 total acres of corn planted and 1,500 total acres of soybeans planted.

SECTION 6.5	
21) a. a = the coefficient of x^2; b = the coefficient of x; c = the constant b. $4a + 2b + c = -9$; $9a - 3b + c = -49$; $a + b + c = -5$ c. $y = -3x^2 + 5x - 7$	23) a. a = the coefficient of x^2; b = the coefficient of x; c = the constant b. $9a - 3b + c = 5$; $a - b + c = 15$; $4a + 2b + c = 30$ c. $y = 5x + 20$, this is a linear equation.
25) a. s = the number of sofas manufactured in one day; c = the number of chairs manufactured in one day; L = the number of love seats manufactured in one day b. $0.3s + 0.2c + 0.2L = 16$; $0.5s + 0.3c + 0.4L = 27$; $0.4s + 0.2c + 0.2L = 18$ c. There was 20 sofas , 30 chairs and 20 loveseat built in one day.	27) a. s = the number of small bears manufactured in one week; m = the number of medium bears manufactured in one week; L = the number of large bears manufactured in one week b. $1.00s + 1.50m + 2.25L = 1{,}100$; $0.1s + 0.2m + 0.25L = 125$; $0.15s + 0.3m + 0.4L = 195$ c. There was 200 small bears, 150 medium bears and 300 large bears manufactured in one week.
29) a. x = the number of Model 250 manufactured in one month; y = the number of Model 350 manufactured in one month; z = the number of Model 450 manufactured in one month b. $0.5x + 0.6y + 0.8z = 2{,}640$; $50.00x + 82.00y + 129.00z = 358{,}200$; $85.95x + 120.95y + 195.95z = 545{,}085$ c. There was 1,000 Model 250, 2,500 Model 350 and 800 Model 450 produced in one month	31) a. c = the total number of children (2–6) served; w = the total number of older children, teen girls and active women served; m = the total number of teen boys and active men served b. $6c + 9w + 11m = 186$; $3c + 4w + 5m = 85$; $2c + 3w + 4m = 64$ c. There was 5 in the children (2-6) category served, 10 in the older children category served and 6 in the teen boys category served.
33) a. x = the number of extra heavy duty ladders sold; y = the number of heavy duty ladders sold; w = the number of standard ladders sold; z = the number of economy ladders sold b. $x + y + w + z = 360$; $119.00x + 108.00y + 74.00w + 60.00z = 28{,}920$; $w = 2x$; $z = 2y$ c. There was 40 extra heavy duty ladders , 80 of the heavy duty ladders, 80 of the standard ladders and 160 of the economy ladders sold.	35) a. x = the number of T–750 television sets produced and sold; y = the number of T–700 television sets produced and sold; w = the number of T–650 television sets produced and sold; z = the number of T–600 television sets produced and sold b. $x + y + w + z = 3{,}500$; $385.90x + 325.70y + 295.80w + 263.40z = 1{,}113{,}105$; $502.85x + 433.50y + 402.50w + 352.80z = 1{,}483{,}690$; $x + y = w + z$ c. There was 800 T–750, 950 T–700, 1,050 T–650 and 700 T–600 television sets produced and sold in the three month period.

CHAPTER 6 REVIEW

1) 3 x3	3) 1	5) $\begin{bmatrix} 1 & 1 & 2 \\ 4 & 0 & 3 \\ 5 & 1 & 2 \end{bmatrix}$
7) $\begin{bmatrix} 10 & -15 \\ 15 & 20 \\ -5 & 25 \end{bmatrix}$	9) $\begin{bmatrix} -4 & 5 \\ -3 & -8 \\ 5 & -7 \end{bmatrix}$	11) Not possible
13) $\begin{bmatrix} 8 & 2 \\ 12 & 6 \\ 11 & 0 \end{bmatrix}$	15) (3, –3)	
17) a. $A = \begin{bmatrix} 323 & 503 & 438 \\ 259 & 439 & 357 \\ 301 & 485 & 399 \end{bmatrix}$ b. $P = \begin{bmatrix} 213.00 \\ 185.00 \\ 163.00 \end{bmatrix}$ c. $\begin{bmatrix} 233{,}248 \\ 194{,}573 \\ 218{,}875 \end{bmatrix}$ d. \$194,573.00 e. \$218,857.00 f. \$233,248		
19) $\begin{bmatrix} -3 & 2 \\ -5 & 4 \end{bmatrix}$	21) $\begin{bmatrix} 0 & 0 & 0 \\ 0 & 0 & 0 \\ 0 & 0 & 0 \end{bmatrix}$	23) $\begin{bmatrix} \frac{-2}{9} & \frac{-1}{9} & \frac{1}{3} \\ \frac{-4}{9} & \frac{7}{9} & \frac{2}{3} \\ \frac{1}{9} & \frac{5}{9} & \frac{1}{3} \end{bmatrix}$
25) –2	27) –9	
29) a. $\begin{bmatrix} 3 & 2 & -4 \\ 2 & -1 & 3 \\ 1 & 1 & -1 \end{bmatrix}$ b. $\begin{bmatrix} x \\ y \\ z \end{bmatrix}$ c. $\begin{bmatrix} 8 \\ 9 \\ 12 \end{bmatrix}$ d. $\begin{bmatrix} 3 & 2 & -4 \\ 2 & -1 & 3 \\ 1 & 1 & -1 \end{bmatrix}\begin{bmatrix} x \\ y \\ z \end{bmatrix} = \begin{bmatrix} 8 \\ 9 \\ 12 \end{bmatrix}$		
31) $\begin{cases} -2x + 3y = -3 \\ x + 5y = 5 \end{cases}$	33) (–2, 3, 1)	35) (2, 5, –2)
37) (1, –3, 2, –1)	39) Can't solve with these methods	41) $\begin{bmatrix} 2 & -3 & 1 & 7 \\ 3 & 2 & -3 & 12 \\ 1 & 3 & -3 & 12 \end{bmatrix}$
43) $\begin{bmatrix} 1 & 2 & 1 & 5 \\ 2 & 0 & -3 & 8 \\ 0 & 5 & -4 & 7 \end{bmatrix}$	45) $\begin{cases} x - 3y + 5z = 3 \\ -2x + 2y - 4z = -7 \\ 3x + y - 2z = 10 \end{cases}$	47) $\begin{cases} r + 3t = 2 \\ s + 3t = -5 \\ 2r + 3t = 7 \end{cases}$
49) (2, –1 –2)	51) (3, –5, 4)	53) (29, 11, –2)
55) No solution	57) (10, –20)	59) (–2, 4, –5)
61) (1, –2, 3)	63) (2z+5, –3z–2, z)	
65) No solution: With this solution method you can see you have no solutions.		
67) There was 342 students and 684 adults attending the play.		
69) There was 10 liters of the 12% solution, 20 liters of the 15% solution and 10 liters of the 34% solution in the mixture.		
71) The company produced 60 Westerner jeans, 90 Boot Leg jeans and100 Straight Leg jeans that one day in December.		

CHAPTER 6 EXAM

1) $\begin{bmatrix} 1 & 2 \\ 1 & 2 \end{bmatrix}$	3) Not possible	5) $\begin{bmatrix} -4 & 8 & -12 \\ 0 & -8 & 4 \\ -8 & 12 & -16 \end{bmatrix}$
7) $\begin{bmatrix} 0 & -4 \\ 5 & -7 \end{bmatrix}$	9) Not possible	11) -1

13) $\begin{bmatrix} 0 & 0 \\ 0 & 0 \\ 0 & 0 \end{bmatrix}$	15) $\begin{bmatrix} 2 & 3 \\ -1 & -4 \\ 5 & -3 \end{bmatrix}$	17) -5	19) Doesn't exist

21) a. Coefficient matrix $\begin{bmatrix} 3 & -2 & 1 \\ 5 & 3 & -3 \\ -2 & -4 & 5 \end{bmatrix}$ Variable matrix $\begin{bmatrix} x \\ y \\ z \end{bmatrix}$ Constant matrix $\begin{bmatrix} 16 \\ -11 \\ 28 \end{bmatrix}$ b. $\begin{bmatrix} 3 & -2 & 1 \\ 5 & 3 & -3 \\ -2 & -4 & 5 \end{bmatrix}\begin{bmatrix} x \\ y \\ z \end{bmatrix} = \begin{bmatrix} 16 \\ -11 \\ 28 \end{bmatrix}$

c. $(2, -3, 4)$

23) a. $\left[\begin{array}{ccc|c} 3 & -2 & 1 & 16 \\ 5 & 3 & -3 & -11 \\ -2 & -4 & 5 & 28 \end{array}\right]$ b. $(2, -3, 4)$	25) (2–z, 3-z, z)
27) No solution	29) There was 2,001 general admission and 8,257 unlimited ride tickets sold.

31) The restaurant sold 431 of the \$2.99 meal, 257 of the \$0.49 upgrade and 145 of the \$0.89 upgrade on that Monday.

Chapter 7 Odd Homework Answers

SECTION 7.1

1) 5	3) $4\sqrt{2}$	5) $\dfrac{\sqrt{17}}{2}$
7) 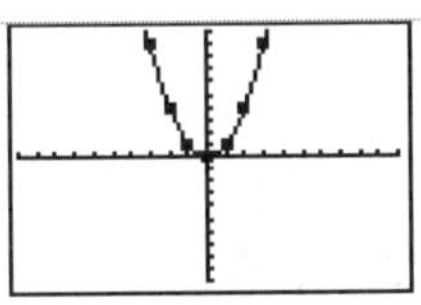	9) 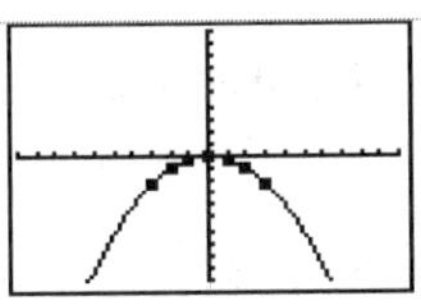	11) 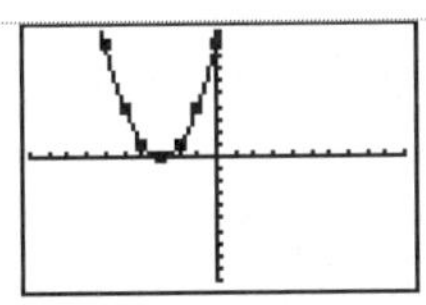
13) 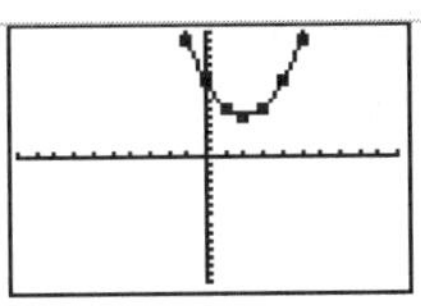	15) 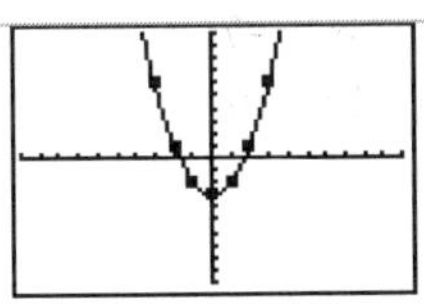	17) 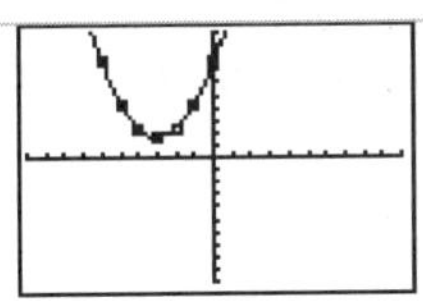
19) 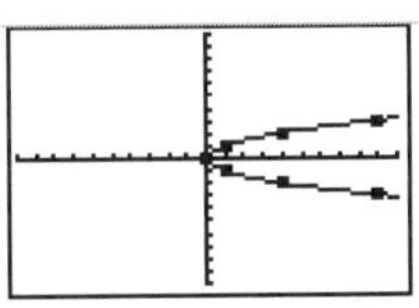	21) 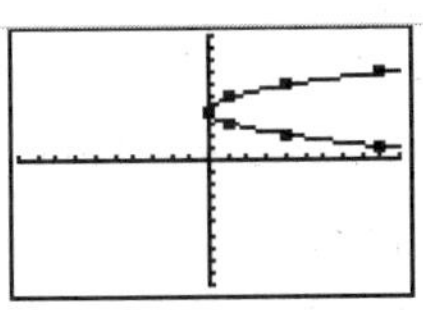	23) 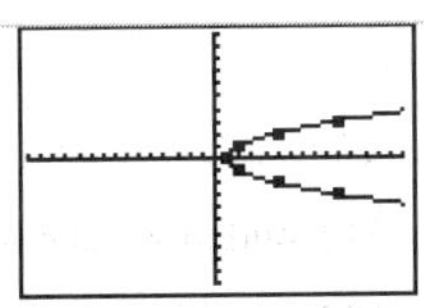

SECTION 7.1

25) 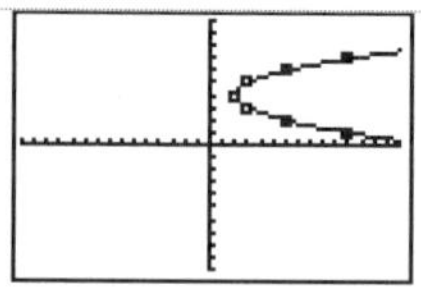

27) 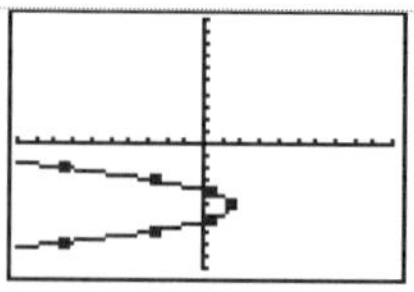

29) shifted right 2 units

31) shifted up 5 units

33) stretched

35) reflected about the y–axis

37) shifted left 5 units and shifted down 9 units

39) stretched, reflected about the x–axis, shifted right 12 units and shifted down 8 units

41) a. (0,0) b. $x = 0$ c. $(0, \frac{-1}{8})$ d. $y = \frac{1}{8}$
e. Maximum = 0 f. stretched and reflected about the x–axis g. Domain: All real numbers Range: $[0, \infty)$

43) a. (3,0) b. $x = 3$ c. $(3, \frac{1}{4})$ d. $y = \frac{-1}{4}$
e. Minimum = 0 f. shifted right 3 units
g. Domain: All real numbers Range: $[0, \infty)$

45) a. (0,12) b. $x = 0$ c. $(0, 12\frac{1}{4})$ d. $y = 11\frac{3}{4}$
e. Minimum = 12 f. shifted up 12 units
g. Domain: All real numbers Range: $[12, \infty)$

47) a. (15,7) b. $x = 15$ c. $(15, 7\frac{1}{4})$ d. $y = 6\frac{3}{4}$
e. Minimum = 7 f. shifted right 15 units and up 7 units
g. Domain: All real numbers Range: $[7, \infty)$

49) a. (–11,32) b. $x = -11$ c. (–11,30) d. $y = 34$
e. Maximum = 32 f. reflected about the x–axis, stretched, shifted left 11 units and shifted up 32 units
g. Domain: All real numbers Range: $(-\infty, 32]$

51) a. (0,0) b. $y = 0$ c. $(\frac{1}{8}, 0)$ d. $x = \frac{-1}{8}$
e. stretched f. Domain: $[0, \infty)$
Range: All real numbers

53) a. (0,–15) b. $y = -15$ c. $(\frac{1}{4}, -15)$ d. $x = \frac{-1}{4}$
e. shifted down 15 units f. Domain: $[0, \infty)$
Range: All real numbers

55) a. (–23,0) b. $y = 0$ c. $(-22\frac{3}{4}, 0)$ d. $x = -23\frac{1}{4}$
e. shifted left 23 units f. Domain: $[-23, \infty)$
Range: All real numbers

57) a. (12,7) b. $y = 7$ c. $(12\frac{1}{4}, 7)$ d. $x = 11\frac{3}{4}$
e. shifted right 12 units and shifted up 7 units
f. Domain: [12,0) Range: All real numbers

59) a. (1,–6) b. $y = -6$ c. (–3,–6) d. $x = 5$
e. reflected about the y–axis, stretched, shifted right 1 and shifted down 6 f. Domain: $(-\infty, 1]$
Range: All real numbers

61) 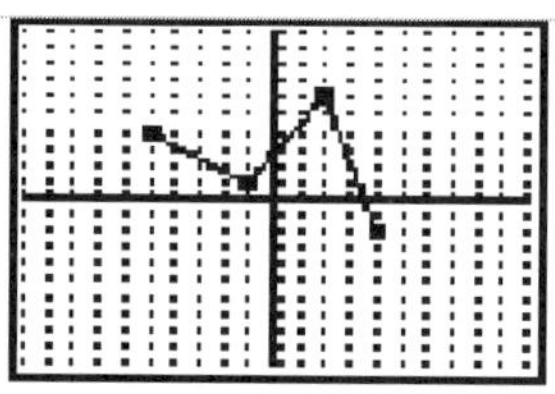

63) 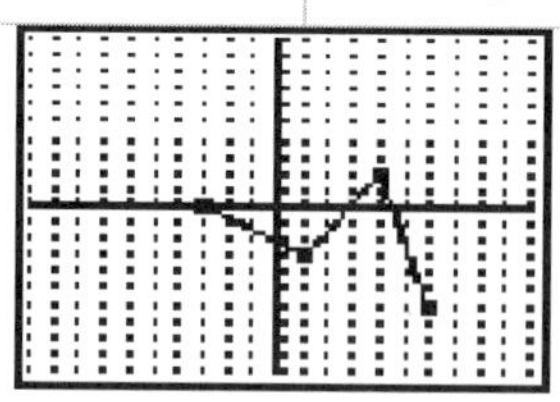

65) 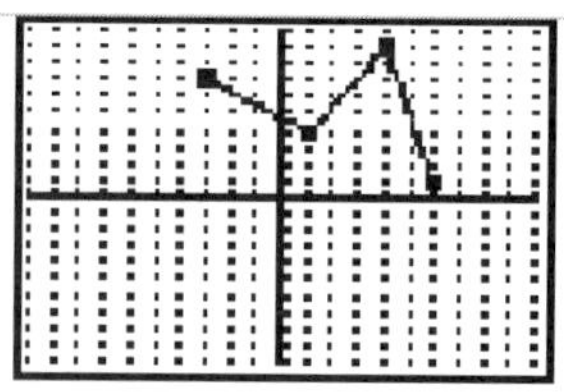

67) 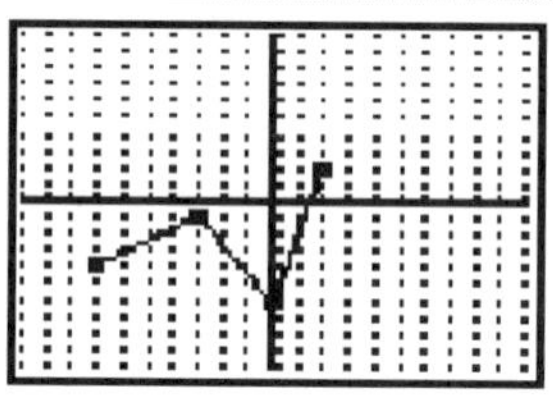

69)

x	g(x)	g(x)–5
–5	5	0
–3	7	2
–1	4	–1
0	–2	–7
1	–9	–14
2	–15	–20
3	–25	–30

SECTION 7.1

71) a. focus (0,2) b. directrix $y=-2$ c. the distance from (0,0) to (0,2) is 2, and the distance from $y=-2$ to (0,0) is 2, the distance from (4,2) to (0,2) is 4, and the distance from $y=-2$ to (4,2) is 4, the distance from (8,8) to (0,2) is 10, and the distance from $y=-2$ to (8,8) is 10

SECTION 7.2

1) A. Vertical B. (0,0) C. $(0,\pm 4)$ D. $(0,\pm 5)$ E. $(\pm 3,0)$ F. Domain: $[-3,3]$ Range: $[-5,5]$

G. 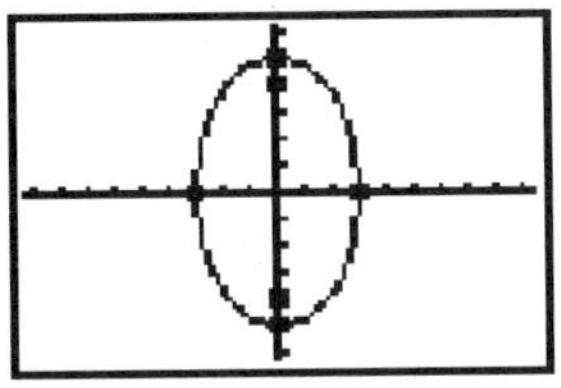

3) A. Horizontal B. (0,0) C. $(\pm 2\sqrt{6},0)$ D. $(\pm 2\sqrt{10},0)$ E. $(0,\pm 4)$ F. Domain: $[-2\sqrt{10},2\sqrt{10}]$ Range: $[-4,4]$

G. 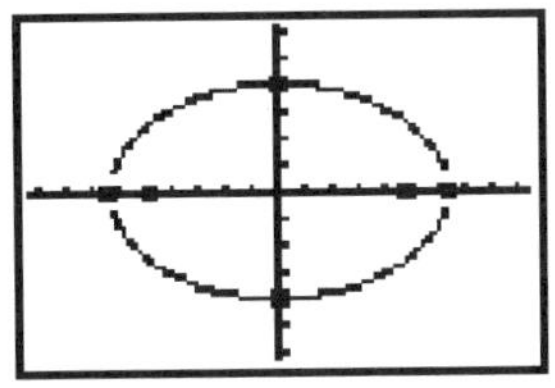

5) A. Vertical B. (0,0) C. $(0,\pm\sqrt{5})$ D. $(0,\pm 3)$ E. $(\pm 2,0)$ F. Domain: $[-2,2]$ Range: $[-3,3]$

G. 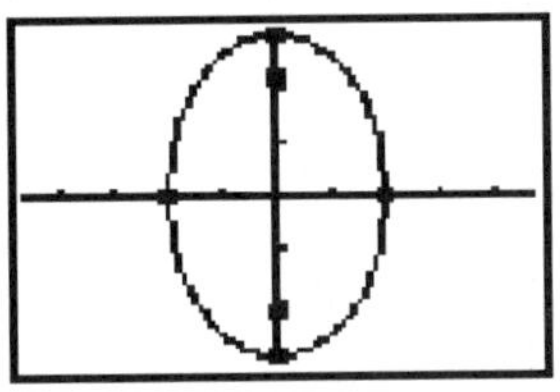

7) A. Horizontal B. (0,0) C. $(\pm 2\sqrt{3},0)$ D. $(\pm 4,0)$ E. $(0,\pm 2)$ F. Domain: $[-4,4]$ Range: $[-2,2]$

G. 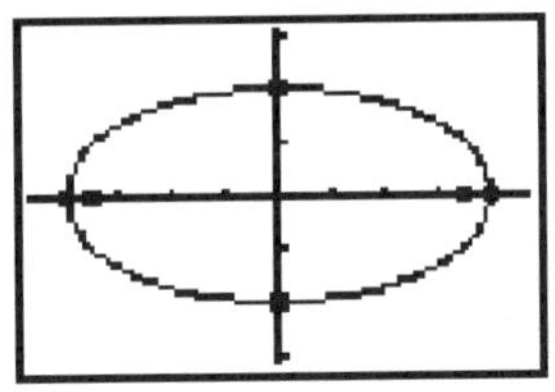

9) A. Horizontal B. Shifted right 3 units and up 2 units C. (3,2) D. $(3\pm 8,2)$ E. $(3\pm 10,2)$ F. $(3,2\pm 6)$ G. $y=2$ H. $x=3$ I. Domain: $[-7,13]$ Range: $[-4,8]$

J. 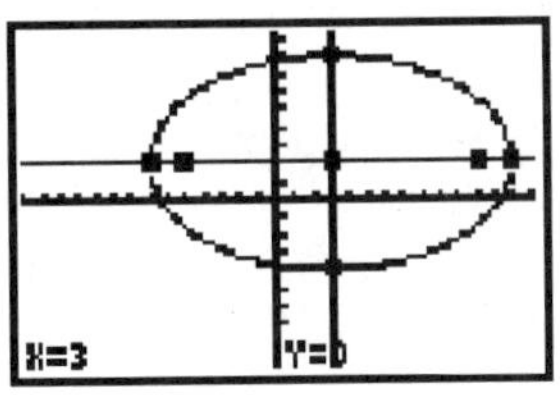

11) A. Vertical B. Shifted left 2 units and down 3 units C. $(-2,-3)$ D. $(-2,-3\pm 2)$ E. $(-2,-3\pm 2\sqrt{3})$ F. $(-2\pm 2\sqrt{2})$ G. $x=-2$ H. $y=-3$ I. Domain: $[-2-2\sqrt{2},-2+2\sqrt{2}]$ Range: $[-3-3\sqrt{2},-3+3\sqrt{2}]$

J. 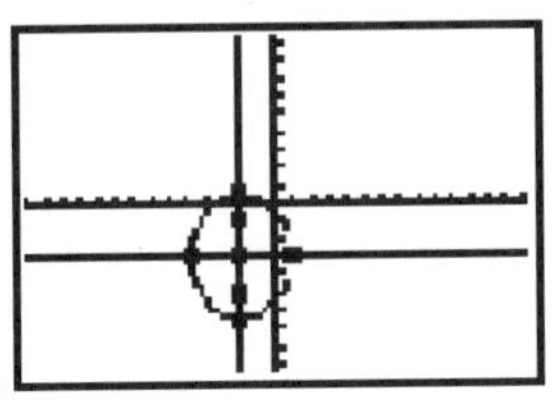

13) A. Horizontal B. Shifted right 4 units and down 5 units C. $(4,-5)$ D. $(4\pm\sqrt{5},-5)$ E. $(4\pm 3,-5)$ F. $(4,-5\pm 2)$ G. $y=-5$ H. $x=4$ I. Domain: $[1,7]$ Range: $[-7,-3]$ J. 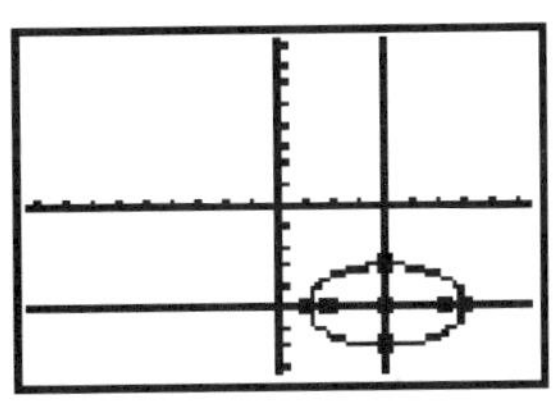

SECTION 7.2

15) A. Vertical B. Shifted left 6 units and up 3 units
C. (–6,3) D. (–6,3±3) E. $(-6,3\pm\sqrt{15})$ F. $(-6\pm\sqrt{6},3)$
G. x = –6 H. y = 3 I. Domain: $[-6-\sqrt{6},-6+\sqrt{6}]$
Range: $[3-\sqrt{15},3+\sqrt{15}]$
J. 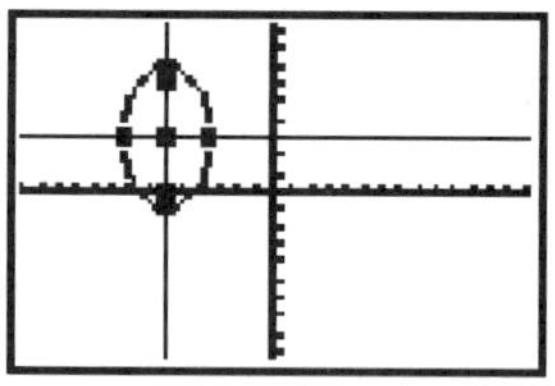

17) A. Horizontal B. Shifted right 2 units and up 5 units
C. (2,5) D. (2±1,5) E. $(2\pm\sqrt{5},5)$ F. (2,5±2)
G. y = 5 H. x = 2 I. Domain: $[2-\sqrt{5},2+\sqrt{5}]$
Range: [3, 7] J. 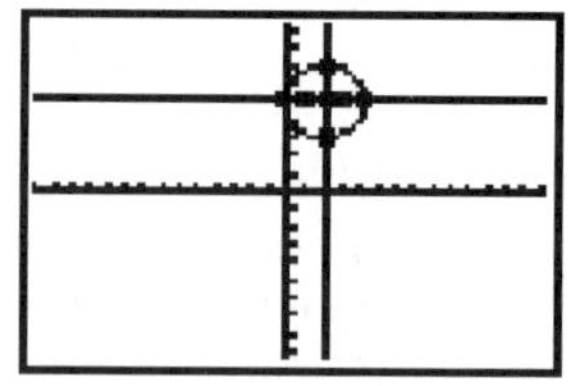

19) $\frac{x^2}{25}+\frac{y^2}{16}=1$

21) $\frac{x^2}{7}+\frac{y^2}{16}=1$

23) $\frac{x^2}{100}+\frac{y^2}{64}=1$

25) $\frac{(x+3)^2}{16}+\frac{(y-2)^2}{25}=1$

27) $\frac{(x-2)^2}{16}+\frac{(y-4)^2}{25}=1$

29) $\frac{(x-3)^2}{36}+\frac{y^2}{64}=1$

31) $\frac{(x-3)^2}{4}+\frac{(y-5)^2}{9}=1$

33) A. (0,0) B. 4 C. Domain: [–4,4] Range: [–4,4]

35) A. (0,0) B. 5 C. Domain: [–5,5] Range: [–5,5]

37) A. (0,0) B. $3\sqrt{3}$ C. Domain: $[-3\sqrt{3},3\sqrt{3}]$ Range: $[-3\sqrt{3},3\sqrt{3}]$

39) A. (3,0) B. $2\sqrt{3}$ C. Domain: $[3-2\sqrt{3},3+2\sqrt{3}]$ Range: $[-2\sqrt{3},2\sqrt{3}]$ D. Shifted right 3 units

41) A. (–4,2) B. $\sqrt{30}$ C. Domain: $[-4-\sqrt{30},-4+\sqrt{30}]$ Range: $[2-\sqrt{30},2+\sqrt{30}]$ D. Shifted left 4 units and up 2 units

43) A. (2,3) B. 3 C. Domain: [–1,5] Range: [0,6] D. Shifted right 2 units and up 3 units

45) A. (–3,–3) B. $\sqrt{15}$ C. Domain: $[-3-\sqrt{15},-3+\sqrt{15}]$ Range: $[-3-\sqrt{15},-3+\sqrt{15}]$ D. Shifted left 3 units and down 3 units

47) $(x-3)^2+(y+2)^2=64$

49) $(x+3)^2+(y-4)^2=25$

51) $(x+1)^2+(y-2)^2=13$

53) 3 units

55) $\frac{x^2}{2{,}704}+\frac{y^2}{400}=1$ The people should stand 4 feet from the side of the room.

57) A. $\frac{x^2}{8{,}649}+\frac{y^2}{8{,}646.75}=1$ B. The aphelion of the Earth's orbit is 94.5 million miles.

59) A. $\frac{x^2}{144}+\frac{y^2}{324}=1$ B. Approximately 15.6 feet high

61) $\frac{x^2}{25}+\frac{y^2}{16}=1$

63) $\frac{(x+2)^2}{4}+\frac{(y-1)^2}{9}=1$

65) By the reflective property of the ellipse you can bank the ball anywhere.

SECTION 7.3

1) A. Horizontal B. (0,0) C. (±5,0) D. (±4,0) E. (0, ±3) F. $y=\pm\frac{3}{4}x$ G. Domain: (–∞,-4] or [4,∞) Range: All real numbers H.

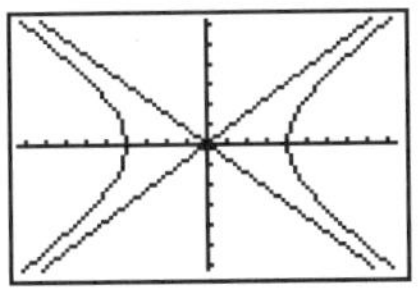

3) A. Vertical B. (0,0) C. $(0, \pm4\sqrt{2})$ D. (0, ±4) E. (±4,0) F. $y=\pm x$ G. Domain: All real numbers Range: (–∞,–4] or [4,∞) H.

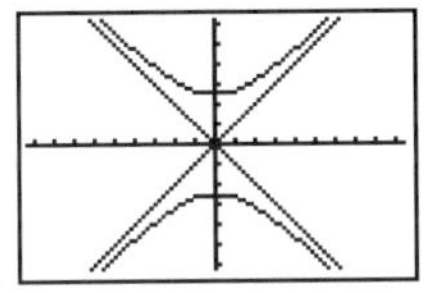

5) A. Horizontal B. (0,0) C. $(\pm2\sqrt{5},0)$ D. $(\pm2\sqrt{2},0)$ E. $(0,\pm2\sqrt{3})$ F. $y=\pm\frac{\sqrt{6}}{2}x$ G. Domain: $(-\infty,-2\sqrt{2}]$ or $[2\sqrt{2},\infty)$ Range: All real numbers H.

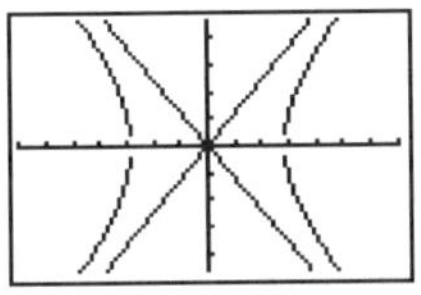

7) A. Horizontal B. (0,0) C. $(\pm2\sqrt{6},0)$ D. (0, ±3) E. $(\pm\sqrt{15},0)$ F. $y=\pm\frac{3}{\sqrt{15}}x$ G. Domain: $(-\infty,-\sqrt{15}]$ or $[\sqrt{15},\infty)$ Range: All real numbers H.

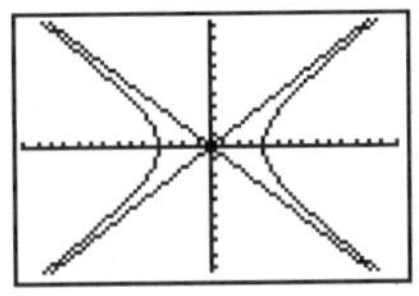

9) A. Horizontal B. Shifted right 3 and down 3 units C. (3,–3) D. (3±13,–3) E. (3±12,–3) F. (3,–3±5) G. $y+3=\pm\frac{5}{12}(x-3)$ H. Domain: (–∞,–9] or [15,∞) Range: All real numbers I.

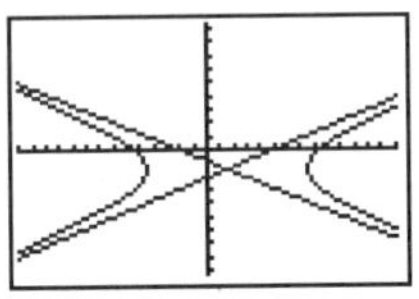

11) A. Vertical B. Shifted right 3 and up 5 units C. (3,5) D. $(3,5\pm\sqrt{34})$ E. (3,5±3) F. (3±5,5) G. $y-5=\pm\frac{3}{5}(x-3)$ H. Domain: All real numbers Range: (–∞,2] or [8,∞) I.

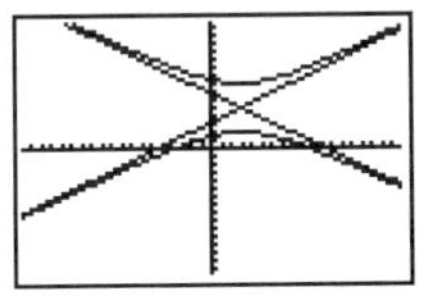

13) A. Horizontal B. Shifted left 1 and up 2 units C. (–1,2) D. $(-1\pm4\sqrt{2},2)$ E. (–1±4,2) F. (–1,2±4) G. $y-2=\pm(x+1)$ H. Domain: (–∞,–5] or [3,∞) Range: All real numbers I.

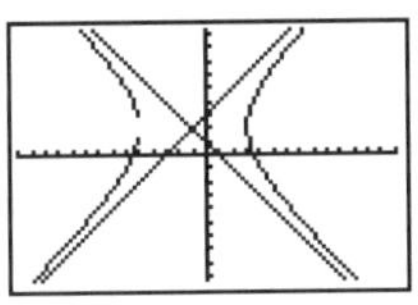

15) A. Vertical B. Shifted up 6 units C. (0,6) D. $(0,6\pm2\sqrt{5})$ E. $(0,6\pm2\sqrt{2})$ F. $(0\pm2\sqrt{3},6)$ G. $y-6=\pm\frac{\sqrt{6}}{3}x$ H. Domain: All real numbers Range: $(-\infty,6-2\sqrt{2}]$ or $[6+2\sqrt{2},\infty)$ I.

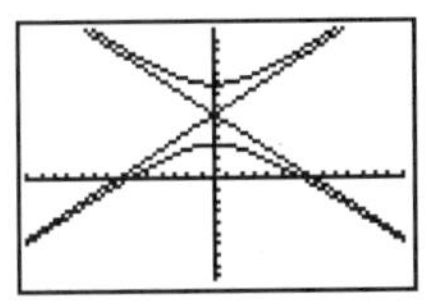

17) A. Horizontal B. Shifted left 2 and up 1 unit C. (–2,1) D. $(-2\pm\sqrt{21},1)$ E. $(-2\pm2\sqrt{3},1)$ F. (–2,1±3) G. $y-1=\pm\frac{\sqrt{3}}{2}(x+2)$ H. Domain: $(-\infty,-2-2\sqrt{3}]$ or $[-2+2\sqrt{3},\infty)$ Range: All real numbers I.

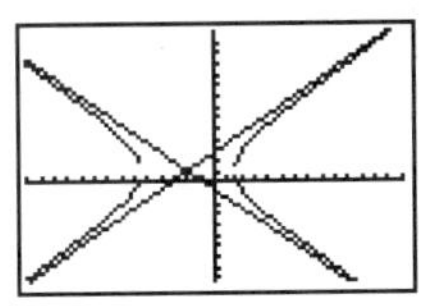

19) $\frac{x^2}{9}-\frac{y^2}{16}=1$

21) $\frac{y^2}{64}-\frac{x^2}{16}=1$

23) $\frac{x^2}{25}-\frac{y^2}{144}=1$

SECTION 7.3

25) $\frac{(y-3)^2}{25}-\frac{(x+2)^2}{16}=1$

27) $\frac{(x-2)^2}{36}-\frac{(y-5)^2}{64}=1$

29) $\frac{x^2}{25}-\frac{(y+2)^2}{16}=1$

31) $\frac{(y+4)^2}{9}-\frac{(x+5)^2}{16}=1$

33) A. Circle B. Translated position

35) A. Hyperbola B. Translated position C. Horizontal

37) A. Parabola B. Translated position

39) A. Ellipse B. Translated position C. Horizontal

41) A. Ellipse B. Standard position C. Horizontal

43) A. Circle B. Standard position

45) A. Parabola B. Standard position

47) A. Hyperbola B. Standard position C. Vertical

49) $\frac{x^2}{9}-\frac{y^2}{16}=1$

51) $\frac{y^2}{4}-\frac{x^2}{9}=1$

SECTION 7.4

1) Ellipse

3) Hyperbola

5) Parabola

7) Circle

9) Hyperbola

11) Ellipse

13) Parabola

15) Circle

17) Parabola

19) Hyperbola

21) Ellipse

23) Circle

25) $\frac{(x-3)^2}{9}+\frac{(y-2)^2}{4}=1$

27) $\frac{(x-2)^2}{9}-\frac{(y+3)^2}{4}=1$

29) $y = 2(x-3)^2 + 5$

31) $(x-2)^2 + (y-3)^2 = 16$

33) $\frac{(y-6)^2}{2}-\frac{(x-4)^2}{4}=1$

35) $\frac{(x-5)^2}{5}+\frac{(y+3)^2}{10}=1$

37) $x = -5(y-3)^2 + 8$

39) $(x+4)^2 + (y-5)^2 = \frac{4}{3}$

41) $y = 3(x-\frac{5}{4})^2 + \frac{3}{8}$

43) $\frac{(x+\frac{3}{2})^2}{6}-\frac{(y-\frac{5}{2})^2}{4}=1$

45) $\frac{(x+\frac{3}{2})^2}{4}+\frac{(y+\frac{3}{4})^2}{2}=1$

47) $(x+\frac{2}{3})^2 + (y+\frac{1}{3})^2 = \frac{4}{27}$

49) 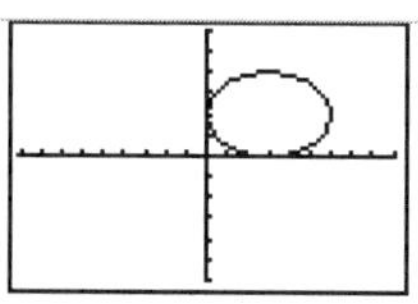

51) 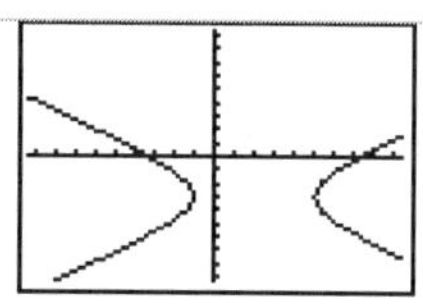

53) 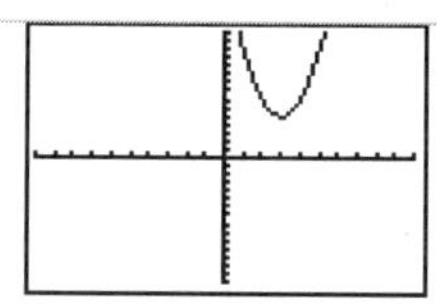

55) 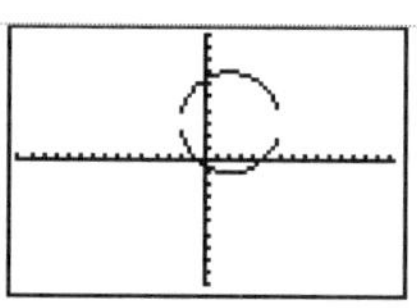

57) 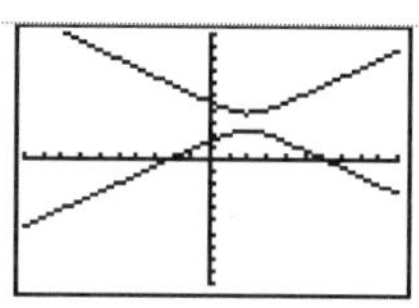

59) 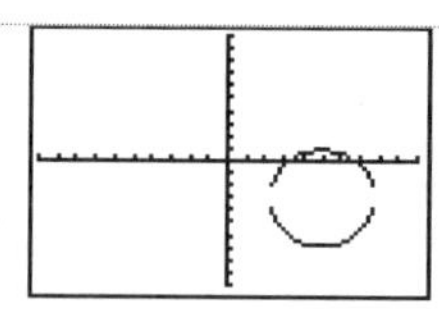

61) Area = 12π Circumference ≈22.2144

63) Area = 12π Circumference≈22.2144

65) $y = -(x-12)^2 + 498$ Maximum = 498

67) $y = (x-25)^2 + 1875$ Minimum = 1875

69) $y = -3(x-3)^2 + 37$ Minimum = 236

71) $R = -(x-200)^2 + 40{,}000$ 200 items sold Maximum revenue $40,000

73) $P = -0.5(x-20)^2 + 480$ 20 items sold
Maximum profit $430

75) $s = -16(t-15/2)^2 + 900$ time = 7.5 seconds
Maximum height = 900 feet

SECTION 7.5

1) (2,–1) and (1,2) **(Line and Circle)**

3) (–3,2) and (23/7, –8/7) **(Line and Ellipse)**

5) (2,3) and (26/47, –131/47) **(Line and Hyperbola)**

7) No real solutions **(Line and Hyperbola)**

9) No real solutions **(Line and Parabola)**

11) (2,–3) and (–3,–8) **(Line and Parabola)**

13) (2,2) and (5/3, 20/9) **(Line and Parabola)**

15) $(5+4\sqrt{3}, 2+2\sqrt{3})$ and $(5-4\sqrt{3}, 2-2\sqrt{3})$ **(Line and Parabola)**

17) (2,1); (2,–1); (–2,1); (–2,–1) **(Circle and Hyperbola)**

19) (1,2); (1,–2); (–1,2); (–1,–2) **(Circle and Ellipse)**

21) (2,3); (2,–3); (–2,3); (–2,–3) **(Hyperbola and Ellipse)**

23) (3,1); (3,–1); (–3,1); (–3, –1) **(Ellipse and Ellipse)**

25) (5,4) and (–5,4) **(Hyperbola and Circle)**

27) No real numbers **(Ellipse and Hyperbola)**

29) No real numbers **(Hyperbola and Hyperbola)**

31) $(2\sqrt{2}, \sqrt{2})$; $(2\sqrt{2}, -\sqrt{2})$; $(-2\sqrt{2}, \sqrt{2})$; $(-2\sqrt{2}, -\sqrt{2})$ **(Hyperbola and Ellipse)**

33) (3,2) and (13/5, 11/5) **(Line and Circle)**

35) (2,4) and (17/4, 23/8) **(Line and Ellipse)**

37) (–3,–2) and (29/47, –60/47) **(Line and Hyperbola)**

39) $(7+3\sqrt{3}, -9-6\sqrt{3})$ and $(7-3\sqrt{3}, -9+6\sqrt{3})$ **(Line and Hyperbola)**

41) No real solutions **(Line and Hyperbola)**

43) $x = \dfrac{1-e}{2-e}$

45) $x \approx 1.7237$ $y \approx -0.4474$

47) $x \approx -1.5361$ $y \approx 0.2320$ and $x \approx 3.3569$ $y \approx 2.6783$

49) No solution

51) $x \approx 3.6835$ $y \approx 2.3076$

53) $x \approx 1.8371$ $y \approx 3.5763$

55) $x \approx -1.4525$ $y \approx -2.3532$ and $x \approx 5.7896$ $y \approx 2.6796$

57) No solution

CHAPTER 7 REVIEW

1) 13

3) 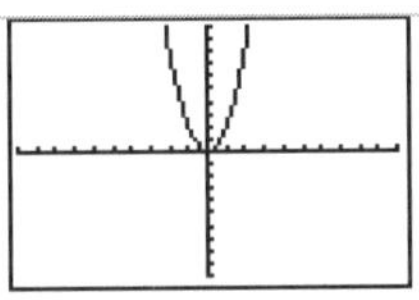

5) 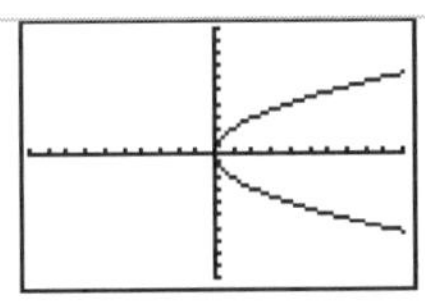

7) 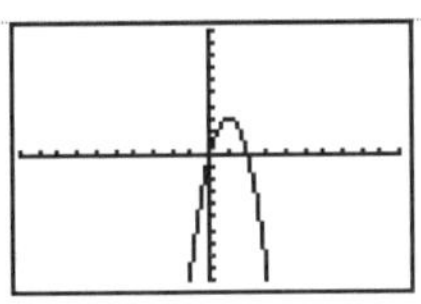

9) 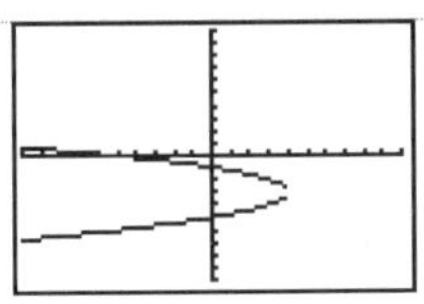

11) Shifted left 2; down 3

13) Stretched; shifted right 1; down 2

15) a.) (0,0) b.) y–axis c.) (0, ¾) d.) $y = \dfrac{-3}{4}x$
e.) Maximum 0 f.) Stretched
g.) Domain: $(-\infty, \infty)$ Range: $[0, \infty)$

17) a.) (–1, 4) b.) x = –1 c.) $(-1, \dfrac{47}{12})$ d.) $y = \dfrac{49}{12}$
e.) Maximum 4 f.) Reflected; stretched; shifted down 1 and up 4. g.) Domain: $(-\infty,\infty)$ Range: $(-\infty, 4]$

19) a.) (–4, –3) b.) y = –3 c.) $(\dfrac{-15}{4}, -3)$ d.) $x = \dfrac{-17}{4}$
e.) Shifted left 4 and down 3
f.) Domain: $[-4,\infty)$ Range: $(-\infty, \infty)$

21)

x	–2	1	4	7	10
g(x)+2	–8	–3	2	12	17

CHAPTER 7 REVIEW

23) A.) Vertical B.) (0,0) C.) (0, ±4) D.) (0, ±5) E.) (±3, 0) F.) Domain: [–3, 3] G.) Range: [-5,5] H.) No translation I.)

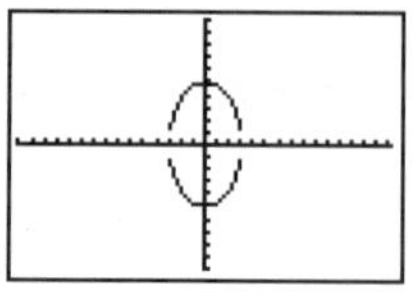

25) A.) Horizontal B.) (2,2) C.) (6,2) and (–2,2) D.) (7,2) and (–3,2) E.) (2,5) and (2,–1) F.) Domain: [–3,7] G.) Range: [–1,5] H.) Shifted right 2 and up 2 I.)

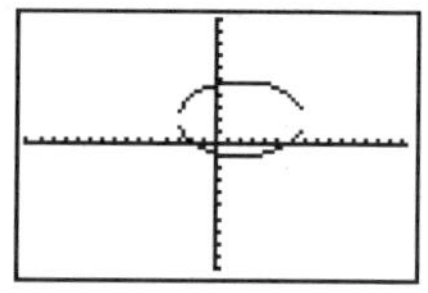

27) A.) Horizontal B.) (0,0) C.) ($\pm\sqrt{10}$,0) D.) (±4,0) E.) (0, $\pm\sqrt{6}$) F.) Domain: [–4,4] G.) Range: [$-\sqrt{6},\sqrt{6}$] H.) No translation I.)

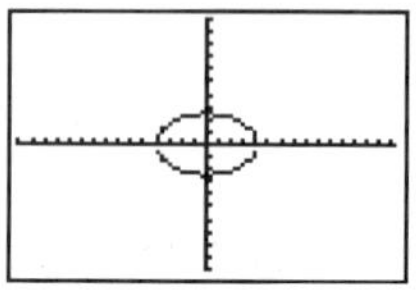

29) A.) Vertical B.) (–3,2) C.) (–3,5) and (–3,–1) D.) (–3, 2+3$\sqrt{2}$) and (–3, 2–3$\sqrt{2}$) E.) (–6,2) and (0,2) F.) Domain: [–6,0] G.) Range: [2–3$\sqrt{2}$,2+3$\sqrt{2}$] H.) Shifted left 3 and up 2 I.)

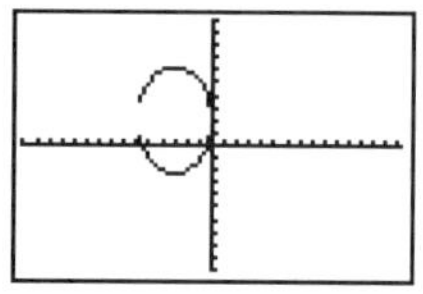

31) $\frac{x^2}{25}+\frac{y^2}{16}=1$

33) $\frac{(x-2)^2}{9}+\frac{(y+3)^2}{25}=1$

35) $\frac{(x+2)^2}{9}+\frac{(y-3)^2}{16}=1$

37) $x^2 + y^2 = 25$ A.) (0,0) B.) 5 C.) Domain: [–5,5] Range: [–5,5] D.) No translation

39) $(x-2)^2+(y+3)^2=9$ A.) (2,–3) B.) 3 C.) Domain: [–1,5] Range: [–6,0] D.) Shifted right 2 and down 3

41) $(x+2)^2+(y+1)^2 = 225$

43) 6 units

45) A.) Horizontal B.) (0,0) C.) ($\pm\sqrt{13}$,0) D.) (±3,0) E.) (0, ±2) F.) $y=\pm\frac{2}{3}x$ G.) Domain:(–∞,–3] or [3,∞) Range: (–∞,∞) H.) No translation I.)

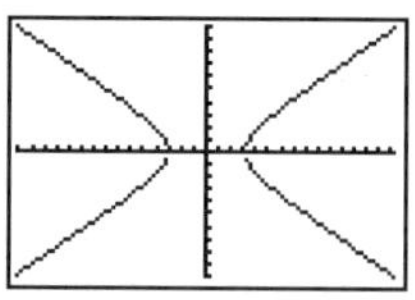

47) A.) Horizontal B.) (2,–3) C.) (7,–3) and (–3,–3) D.) (6,–3) and (–2,–3) E.) (2,0) and (2,–6) F.) $y+3=\pm\frac{3}{4}(x-2)$ G.) Domain: (–∞,–2] or [6,∞) Range: (–∞,∞) H.) Shifted right 2 and down 3 I.)

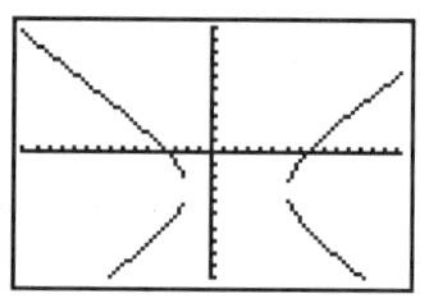

49) $\frac{x^2}{16}-\frac{y^2}{9}=1$

51) $\frac{(y-3)^2}{9}-\frac{(x+2)^2}{9}=1$

53) $\frac{(x+5)^2}{25}-\frac{(y-2)^2}{36}=1$

55) A.) Hyperbola B.) Translated C.) Horizontal

57) A.) Circle B.) Standard C.) Doesn't apply

59) A.) Ellipse B.) Translated C.) Horizontal

61) $\frac{x^2}{25}-\frac{y^2}{9}=1$

63) A.) Ellipse B.) $\frac{(x-3)^2}{4}+\frac{(y-2)^2}{9}=1$

65) A.) Parabola B.) $y = 2(x-3)^2 + 3$

CHAPTER 7 REVIEW

67) A.) Hyperbola B.) $\frac{(y-2)^2}{16}-\frac{(x+2)^2}{25}=1$	69) A.) Circle B.) $(x+3)^2+(y+2)^2=9$
71) $y=(x-10)^2+30$ Minimum of 30	73) $P=-0.3(x-100)^2+3000$ Maximum profit is $3,000.00

75) (–1, 2) and $(\frac{5}{3},\frac{2}{3})$	77) (3,2) and $(\frac{81}{17},\frac{-56}{17})$	79) (±3, ±3)
81) (±3, ±2)	83) (5,2), $(\frac{-7}{3},\frac{-16}{3})$	85) No real solutions
87) No real solutions	89) $x=\frac{-2}{e-3}$	91) $x\approx-0.3793$ and $y\approx-2.5171$

93) $x\approx-0.4207$ and $y\approx2.0878$	95) No real solutions

CHAPTER 7 EXAM

1) Stretched; reflected; shifted left 3 and up 5	3) A.) (20,1500) B.) $x=20$ C.) $(20, 1499\frac{11}{12})$ D.) $y=1500\frac{1}{12}$ E.) Minimum is 1,500 F.) Stretched; shifted right 20 units and up 1,500 units
5) A.) Vertical B.) (–1,2) C.) (–1,5) and (–1,–1) D.) (–1,7) and (–1,–3) E.) (3,2) and (–5,2) F.) $y+3=\pm\frac{12}{5}(x+2)$ G.) Shifted left 1 unit and up 2 units	7) A.) Horizontal B.) (–2,–3) C.) (11,–3) and (–15,–3) D.) (3,–3) and (–7,–3) E.) (–2,9) and (–2,–15) F.) $y+3=\pm\frac{12}{5}(x+2)$ G.) Shifted left 2 units and down 3 units
9) $\frac{(x-3)^2}{25}+\frac{(y+2)^2}{16}=1$	11) $(x+2)^2+(y+3)^2=41$
13) A.) Ellipse B.) $\frac{(x-2)^2}{9}+\frac{(y+3)^2}{4}=1$	15) A.) Circle B.) $(x+5)^2+(y+2)^2=12$

17) (–2,–3), $\left(\frac{22}{7},\frac{-3}{7}\right)$	19) No real number solutions	21) 4,500 items

Chapter 8 Odd Homework Answers

SECTION 8.1

1) 7,12,17,22,27	3) 6,18,54,162,486	5) 3,11,27,59,123
7) 5,–3,19,–63,227	9) 2,7,21,57,123	11) $2,1,\frac{5}{4},\frac{20}{17},\frac{85}{71}$

SECTION 8.1

13) $1, \frac{1}{2}, \frac{2}{3}, \frac{3}{2}, \frac{24}{5}$	15) –2,4,–8,16,–32	17) 21
19) 48	21) $\frac{-5}{16}$	23) 34
25) $\frac{1}{5}$	27) 25	29) a. $a_n=8+12(n-1)$ b. 116
31) a. $a_n=-5+8(n-1)$ b. 67	33) a. $a_n=7-\frac{5}{2}(n-1)$ b. $\frac{-31}{2}$	35) a. $a_n=-8-6(n-1)$ b. –62
37) a. $\frac{27}{5}+\frac{24}{5}(n-1)$ b. $\frac{243}{5}$	39) a. $a_n=5(3)^{n-1}$ b. 1215	41) a. $a_n=8\left(\frac{-1}{4}\right)^{n-1}$ b. $\frac{-1}{128}$
43) a. $a_n=4(5)^{n-1}$ b. 12500	45) a. $a_n=7(4)^{n-1}$ b. 7168	49) a. Neither c. 25

47) a. Arithmetic b. $a_n=5-7(n-1)$; $a_8=-44$	51) a. Geometric b. $a_n=2(5)^{n-1}$; $a_8=156{,}250$
53) a. Geometric b. $a_n=3\left(\frac{-1}{3}\right)^{n-1}$; $a_8=\frac{-1}{729}$	55) a. Neither c. 43
57) a. Arithmetic b. $a_n=\frac{3}{5}-\frac{2}{3}(n-1)$; $a_8=\frac{-79}{15}$	59) $a_n=8+3(n-1)$
61) $a_n=6(2)^{n-1}$	63) a. Arithmetic b. $a_n=250+5(n-1)$ c. $a_{50}=\$495.00$

65) a. $A=5000(1+0.08)^t$ b. $A\approx\$9{,}995.02$ c. $a_1=5000$; a_1 represents the original value of the account d. $a_{10}\approx 9{,}995.02$; a_{10} represents the value of the account after 9 years. e. a_n represents the value of the account after n–1 years.

SECTION 8.2

1) 45	3) 42	5) $\frac{15}{32}$
7) 35	9) 87	11) $\frac{-4{,}609}{2{,}520}$
13) 860	15) 3,280	17) $\frac{-96}{5}$
19) –792	21) 6,560	23) 80
25) $\frac{4{,}039}{128}$	27) 2,595	29) 1,359
31) $\frac{9}{2}$	33) $\frac{20}{7}$	35) No numerical answer

37) 5,050 total blocks facing you.	39) You will be paid a total of \$10,485.75.
41) Your total salary for the 10 years will be \$417,500.	43) You would receive a total salary of \$350,140.60 for the eight years.

SECTION 8.3

1) 5,040	3) 5	5) 6,840
7) 210	9) 5	11) 1,140
13) 26 x 26 x 10 x 10 x 10x10 = 6,760,000 Not enough	15) 7 x 10 x 10 x 10 x 10 x 10 = 7,000,000 Not enough	
17) 7 x 6 x 5 x 4 x 3 x 2 x 1 = 5,040	19) 20x19x18x17x16x15x14x13x12x11x10x9x8x7x6x5x4x3x2x1 = $2.432902008 \text{x} 10^{18}$	
21) a. 5 x 5 x 5 x 5 x 5 = 3,125 b . 5 x 4 x 3 x 2 x 1 = 120	23) 3 x 4 x 2 = 24	
25) ${}_6P_4 = 360$	27) ${}_{48}P_4 = 4,669,920$	29) ${}_{20}P_{10} = 6.704425728 \text{x} 10^{11}$
31) ${}_{10}P_5 = 30,240$	33) ${}_{50}P_{20} = 1.146607551 \text{x} 10^{32}$	35) ${}_6C_4 = 15$
37) ${}_{48}C_4 = 194,580$	39) ${}_{20}C_4 = 4,845$	41) ${}_{52}C_5 = 2,598,960$
43) ${}_{100}C_6 = 1,192,052,400$	45) 20,475	49) 72 months
51) 6,000	53) 43,680	55) a. 120 b. 20
57) a. 1 b. 210 c. 91,390 d. 7,200 e. 85,951	59. 48	
61) 84	63) a. 100,000,000 b. 1,814,400	65) a. 5,040 b. 151,200 c. 1,814,400
67) a. 560 b. 480	69) a. 792 b. 350 c. 210 d. 21 e. 10	71. 420
73) 4,989,600	75) a. ${}_4C_2 = 6$ b. {12, 13, 14, 23, 24, 34}	

SECTION 8.4

1) $\frac{3}{17}$	3) $\frac{14}{17}$	5) $\frac{1}{6}$	7. $\frac{1}{2}$
9) $\frac{1}{13}$	11) $\frac{4}{13}$	13) $\frac{1}{4}$	15. $\frac{7}{13}$
17) $\frac{17}{26}$	19) (see table below)		21) $\frac{1}{6}$
23) $\frac{5}{6}$	25) $\frac{31}{36}$	27) TTTT, TTTH, TTHT, THTT, HTTT, TTHH, THTH, THHT, HHTT, HTHT, HTTH, HHHT, HHTH, HTHH, THHH, HHHH	

19)

1,1	1,2	1,3	1,4	1,5	1,6
2,1	2,2	2,3	2,4	2,5	2,6
3,1	3,2	3,3	3,4	3,5	3,6
4,1	4,2	4,3	4,4	4,5	4,6
5,1	5,2	5,3	5,4	5,5	5,6
6,1	6,2	6,3	6,4	6,5	6,6

29) $\frac{3}{8}$	31) $\frac{15}{16}$	33) $\frac{1}{9,366,819} \approx 0.0000001676$
35) $\frac{1}{63,054,084} \approx 0.00000001586$	37) $\frac{25}{102} \approx .245$	39) $\frac{1}{17} \approx 0.0588$
41) $\frac{91}{54,145} \approx 0.0017$	43) $\frac{429}{866,320} \approx 0.0004952$	45) $\frac{1}{108,290} \approx 0.000009234$
47) $\frac{80}{221} \approx 0.36199$	49) $\frac{55}{204} \approx 0.2696$	51) ≈ 0.4114

SECTION 8.3

53) a. $\frac{1,147}{4,042} \approx 0.28377$ b. $\frac{2,201}{8,084} \approx 0.27227$ c. $\frac{1,130}{2,021} \approx 0.5591$ d. $\frac{891}{2,021} \approx 0.4409$

55) a. $\frac{140,091,913}{273,643,274} \approx 0.51195$ b. $\frac{18,668,947}{136,821,637} \approx 0.13645$ c. $\frac{57,570,485}{273,643,274} \approx 0.21039$ d. $\frac{117,131,338}{136,821,637} \approx 0.85609$

57) 1:5	59) 1:5	61) $\frac{1}{25}$	63) 1:1,712,303

65) a. $\frac{5}{57} \approx 0.0877$ b. $\frac{28}{171} \approx 0.1637$ c. $\frac{30}{361} \approx 0.0831$ d. $\frac{64}{361} \approx 0.1773$

CHAPTER 8 REVIEW

1) 10, 13, 16, 19, 22	3) 6, 12, 24, 48, 96	5) 1, 4, $\frac{9}{2}, \frac{8}{3}, \frac{25}{24}$
7) −19	9) −32	11) $\frac{6}{7}$
13) $a_n = -3 + 5(n-1)$; $a_{10} = 42$	15) $a_n = 7-3(n-1)$; $a_{10} = -20$	17) $a_n = -3(2)^{n-1}$; $a_6 = -96$
19) $a_n = 6(\frac{1}{3})^{n-1}$; $a_6 = \frac{2}{81}$	21) a. Geometric b. $a_n = 7(2)^{n-1}$; $a_8 = 896$	23) a. Neither c. 0
25) a. Neither c. 53	27) $a_n = 1-2(n-1)$	29) a. $a_n = 40,000 + 1,200(n-1)$ b. \$46,000
31) 75	33) 70	35) 30
37) Arithmetic; 690	39) Arithmetic; −165	41) Geometric; −29,524
43) $\frac{16}{5}$	45) a. 20 b. 210	47. 1,320
49. 220	51. 10,000,000	53. 254,251,200
55. 40,320	57. $\approx 1.316818944 \times 10^{13}$	59) a. $\frac{9}{31}$ b. $\frac{22}{31}$ c. $\frac{18}{155}$ d. $\frac{108}{961}$
61) a. $\frac{15}{1001}$ b. $\frac{48}{143}$ c. $\frac{58}{143}$	65) 1:1	67) 5:2

63) a. $\frac{106,905,819}{273,643,274} = \approx 39\%$ b. $\frac{136,181,612}{273,643,274} \approx 49.8\%$ c. $\frac{191,678,458}{273,643,274} \approx 70\%$

CHAPTER 8 EXAM

1) −1, −5, −9, −13, −17	3) 1, $\frac{1}{2}, \frac{2}{3}, \frac{3}{2}, \frac{24}{5}$	5) a. $a_n = 3 + 5(n-1)$ b. 48
7) a. Geometric b. $a_n = 3(-2)^{n-1}$; $a_7 = 192$	9) a. Arithmetic b. $a_n = 4 - 5(n-1)$; $a_7 = -26$	
11) $S_8 = 765$	13) $S_6 = 0$	15) 30,240

CHAPTER 8 EXAM		
17) 1,000,000,000	19) 116,280	21) a. 21 b. 1,176 c. 2,982
23) a. $\frac{21}{3003} \approx 0.007$ b. $\frac{1176}{3003} \approx 0.3916$	25) a. $\frac{69,345,619}{104,733,569} \approx 0.6621$ b. 1	